PRACTICAL ELECTROPHYSIOLOGY

PRACTICAL ELECTROPHYSIOLOGY

EDITORS

Jasbir S. Sra, MD, FACC, FAHA, FHRS
Aurora Sinai/Aurora St. Luke's Medical Centers,
University of Wisconsin School of Medicine and
Public Health, Milwaukee, Wisconsin

Masood Akhtar, MD, FACC, MACP, FAHA, FHRS
Aurora Sinai/Aurora St. Luke's Medical Centers,
University of Wisconsin School of Medicine and
Public Health, Milwaukee, Wisconsin

ASSOCIATE EDITORS

Andrea Natale, MD, FACC, FHRS, FESC
Texas Cardiac Arrhythmia Institute,
St. David's Medical Center, Austin, Texas

David J. Wilber, MD, FAHA, FACC
Center for Heart & Vascular Medicine,
Loyola University Health System,
Maywood, Illinois

cardiotext
PUBLISHING
Minneapolis, Minnesota

Cardiotext Publishing, LLC
3405 W. 44th Street
Minneapolis, Minnesota 55410
USA

www.cardiotextpublishing.com

Any updates to this book may be found at: http://cardiotextpublishing.com/practical-electrophysiology

Comments, inquiries, and requests for bulk sales can be directed to the publisher at: info@cardiotextpublishing.com.

Library of Congress Control Number: 2014955006

ISBN: 978-1-935395-17-1

eISBN: 978-1-935395-37-9 (pdf)

DEDICATION

To my wife Lori and my children, whose understanding made this work possible. This book is dedicated to all those interested in advancing the field of electrophysiology.

Jasbir S. Sra, Milwaukee

CONTENTS

CONTRIBUTORS

Editors

Jasbir S. Sra, MD, FACC, FAHA, FHRS
Clinical Adjunct Professor of Medicine, Department of Electrophysiology, Aurora Cardiovascular Services, Aurora Sinai/Aurora St. Luke's Medical Centers, University of Wisconsin School of Medicine and Public Health, Milwaukee, Wisconsin

Masood Akhtar, MD, FACC, MACP, FAHA, FHRS
Clinical Adjunct Professor of Medicine, Department of Electrophysiology, Aurora Cardiovascular Services, Aurora Sinai/Aurora St. Luke's Medical Centers, University of Wisconsin School of Medicine and Public Health, Milwaukee, Wisconsin

Associate Editors

Andrea Natale, MD, FACC, FHRS, FESC
Executive Medical Director, Texas Cardiac Arrhythmia Institute, St. David's Medical Center, Austin, Texas; Consulting Professor, Division of Cardiology, Stanford University, Palo Alto, California; Clinical Professor of Medicine, Case Western Reserve University, Cleveland, Ohio; Director, Interventional Electrophysiology, Scripps Clinic, San Diego, California; Senior Clinical Director, EP Services, California Pacific Medical Center, San Francisco, California

David J. Wilber, MD, FAHA, FACC
George M. Eisenberg Professor of Cardiovascular Sciences, Director of Cardiology and Clinical Electrophysiology, Center for Heart & Vascular Medicine, Loyola University Health System, Maywood, Illinois

Contributors

William T. Abraham, MD, FACP, FACC, FAHA, FESC
Professor of Medicine, Physiology, and Cell Biology, Chair of Excellence in Cardiovascular Medicine; Director, Division of Cardiovascular Medicine, Deputy Director, Davis Heart and Lung Research Institute, The Ohio State University, Columbus, Ohio

Dominic J. Abrams, BSc, MBBS, MD, MRCP
Inherited Arrhythmia Programme, Consultant Cardiologist, Barts & The London NHS Trust, Honorary Consultant Cardiologist, Great Ormond Street Hospital for Children, London, UK

Wayne O. Adkisson, MD, FACC
Assistant Professor, Cardiovascular Division, University of Minnesota Medical School, Minneapolis, Minnesota

Khawaja Afzal Ammar, MD
Cardiology, Aurora Cardiovascular Services, Aurora Sinai/Aurora St. Luke's Medical Centers, Milwaukee, Wisconsin

Charles Antzelevitch, PhD, FACC, FAHA, FHRS
Gordon K. Moe Scholar, Professor of Pharmacology, Masonic Medical Research Laboratory, Utica, New York

Evgeny P. Anyukhovsky, PhD
Senior Research Scientist, Department of Physiology and Biophysics, Stony Brook University, Stony Brook, New York

Samuel J. Asirvatham, MD, FACC, FHRS
Consultant, Division of Cardiovascular Diseases and Internal Medicine, Division of Pediatric Cardiology; Professor of Medicine and Pediatrics, Mayo Clinic College of Medicine; Research Translation Cardiovascular Liaison; Program Director, EP Fellowship Program, Mayo Clinic, Rochester, Minnesota

Nitish Badhwar, MD, MBBS, FACC, FHRS
Director, Cardiac Electrophysiology Training Program, Associate Chief, Cardiac Electrophysiology, University of California, San Francisco, San Francisco, California

Tanvir Bajwa, MD, FACC, FSCAI
Clinical Adjunct Professor Cardiology, Aurora Cardiovascular Services, Aurora Sinai/Aurora St. Luke's Medical Centers, University of Wisconsin School of Medicine and Public Health, Milwaukee, Wisconsin

Rupa Bala, MD
Assistant Professor of Medicine, Division of Electrophysiology, Hospital of the University of Pennsylvania, Philadelphia, Pennsylvania

Ramprakash Balasundaram, MD
CARE Hospitals and CARE Foundation, Hyderabad, India

Meir Bar-Tal, MSc
Biosense Webster LTD, Yokneam, Israel

Mohamed Bassiouny, MD
Department of Cardiovascular Medicine, Section of Electrophysiology and Cardiac Pacing, Cleveland Clinic Foundation, Cleveland, Ohio

Karen J. Beckman, MD, FACC, FHRS
Professor of Medicine, Cardiovascular Section, University of Oklahoma Health Sciences Center, Oklahoma City, Oklahoma

Barry F. Belanger, PhD
Retired from GE Healthcare, Clinical Research Manager and X-Ray Imaging Physicist in Interventional X-Ray Division, Waukesha, Wisconsin

David G. Benditt, MD, FACC, FHRS, FRCPC, FESC
Cardiac Arrhythmia / Syncope Center, Cardiovascular Division, University of Minnesota Medical School, Minneapolis, Minnesota

Atul Bhatia, MD, FACC
Clinical Adjunct Associate Professor, Department of Electrophysiology, Aurora Cardiovascular Services, Aurora Sinai/Aurora St. Luke's Medical Centers, University of Wisconsin School of Medicine and Public Health, Milwaukee, Wisconsin

Zalmen Blanck, MD, FACC, FHRS
Clinical Adjunct Associate Professor of Medicine, Electrophysiology Department, Aurora Cardiovascular Services, Aurora Sinai/Aurora St. Luke's Medical Centers, University of Wisconsin School of Medicine and Public Health, Milwaukee, Wisconsin

Hugh Calkins, MD
Professor of Medicine, Nicholas J. Fortuin Professor of Cardiology, Johns Hopkins Medical Institutions, Johns Hopkins, Baltimore, Maryland

Indrajit Choudhuri, MD
Clinical Adjunct Assistant Professor of Medicine, Electrophysiology Department, Aurora Cardiovascular Services, Aurora Sinai/Aurora St. Luke's Medical Centers, University of Wisconsin School of Medicine and Public Health, Milwaukee, Wisconsin

Aman Chugh, MD, FACC
Associate Professor of Medicine, Cardiac Electrophysiology Service, Division of Cardiovascular Medicine, University of Michigan Cardiovascular Center, Ann Arbor, Michigan

Ryan Cooley, MD
Clinical Adjunct Assistant Professor, Electrophysiology, Aurora Cardiovascular Services, Aurora Sinai/Aurora St. Luke's Medical Centers, University of Wisconsin School of Medicine and Public Health, Milwaukee, Wisconsin

Andrew E. Darby, MD
Assistant Professor of Cardiovascular Medicine, Cardiovascular Division, Department of Medicine, University of Virginia School of Medicine, Charlottesville, Virginia

Mithilesh K. Das, MD, MRCP, FACC
Associate Professor of Clinical Medicine, Krannert Institute of Cardiology, Indiana University School of Medicine, Indianapolis, Indiana

Nicolas Derval, MD
Department of Rhythmologie, Hôpital Cardiologique du Haut-Lévêque and the Université Bordeaux II, Bordeaux, France

Kinjal Desai
Rutgers, Robert Wood Johnson Medical School (Class of 2016), New Brunswick, New Jersey

Anwer Dhala, MD, FACC, FHRS
Clinical Professor of Pediatrics, Electrophysiology, Aurora Cardiovascular Services, Aurora Sinai/Aurora St. Luke's Medical Centers, Children's Hospital of Wisconsin, Milwaukee, Wisconsin

Luigi Di Biase, MD, PhD, FACC, FHRS
Cardiologist, Electrophysiologist, Senior Researcher, Texas Cardiac Arrhythmia Institute at St. David's Medical Center, Austin, Texas; Adjunct Associate Professor, Department of Biomedical Engineering, University of Texas, Austin, Texas; Clinical Assistant Professor, Department of Cardiology, University of Foggia, Foggia, Italy

John P. DiMarco, MD, PhD
Professor of Cardiovascular Medicine, Cardiovascular Division, Department of Medicine, University of Virginia School of Medicine, Charlottesville, Virginia

Mohamed Djelmami-Hani, MD
Clinical Adjunct Assistant Professor, Electrophysiology, Aurora Cardiovascular Services, Aurora Sinai/Aurora St. Luke's Medical Centers, University of Wisconsin School of Medicine and Public Health, Milwaukee, Wisconsin

Adrian Dusa, MD
Assistant Professor of Medicine, University of Oklahoma Health Sciences Center, Oklahoma City, Oklahoma

Kenneth A. Ellenbogen, MD, FACC
Director of Clinical Cardiac Electrophysiology and Pacing, Chair of the Division of Cardiology at the Pauley Heart Center, Kontos Endowed Professor at Virginia Commonwealth University's Medical College of Virginia, Richmond, Virginia

Sitaramesh Emani, MD
Division of Cardiovascular Medicine (Department of Medicine), The Ohio State University Wexner Medical Center, Columbus, Ohio

Peter C. Frommelt, MD
Pediatric Cardiologist, Children's Hospital of Wisconsin; Professor, Pediatrics, Division of Pediatric Cardiology, Medical College of Wisconsin, Milwaukee, Wisconsin

Maggie Fung, MEng
GE Healthcare, MR, Waukesha, Wisconsin

G. Joseph Gallinghouse, MD
Cardiac Electrophysiologist, Texas Cardiac Arrhythmia Institute, St. David's Medical Center, Austin, Texas

Carol J. Gilbert, RN, MSN, MBA, FHRS
Aurora Cardiovascular Services, Aurora St. Luke's Medical Center, Milwaukee, Wisconsin

Ilan Goldenberg, MD
Professor of Cardiology; Director, Department of Cardiology and Israeli Association for Cardiovascular Trials, Leviev Heart Center, Sheba Medical Center, Tel Aviv University, Sheba Medical Center, University of Rochester Medical Center, Tel Hashomer, Israel

Babak Haddadian, MD
Department of Cardiology, Gundersen Lutheran Health System, LaCrosse, Wisconsin

Michel Haïssaguerre, MD
Hôpital Cardiologique du Haut-Lévêque and the Université Victor Segalen Bordeaux II, Bordeaux, France

Mélèze Hocini, MD
Hôpital Cardiologique du Haut-Lévêque and the Université Victor Segalen Bordeaux II, Bordeaux, France

José F. Huizar, MD, FACC, FHRS, FAHA
Associate Professor of Medicine, Virginia Commonwealth University/Medical College of Virginia; Director, Arryhythmia and Device Clinic, McGuire VA Medical Center, Richmond, Virginia

Ross J. Hunter, BSc, MBBS, MRCP, PhD
Department of Cardiology and Electrophysiology, St. Bartholomew's Hospital, Barts Health NHS Trust and QMUL, London, UK

Raymond Ideker, MD, PhD
Professor Emeritus of Medicine, Department of Medicine, Division of Cardiovascular Disease, University of Alabama Birmingham, Birmingham, Alabama

Warren M. Jackman, MD, FACC, FHRS
Professor of Medicine, George Lynn Cross Research Professor, University of Oklahoma Health Sciences Center, Oklahoma City, Oklahoma

Amir S. Jadidi, MD
Department of Cardiac Arrhythmias, Hopital Cardiologique du Haut-Leveque, CHU de Bordeaux, Pessac - Bordeaux, France; Department of Cardiac Arrhythmias, University Heart Centre Freiburg - Bad Krozingen, Bad Krozingen, Germany

Pierre Jaïs, MD
Department of Rhythmologie, Hôpital Cardiologique du Haut-Lévêque and the Université Bordeaux II, Bordeaux, France

José Jalife, MD, FHRS, FAHA
Cyrus and Jane Farrehi Professor of Cardiovascular Research; Professor of Internal Medicine/Cardiovascular Medicine; Professor of Molecular and Integrative Physiology, University of Michigan, Ann Arbor, Michigan

M. Fuad Jan, MBBS, MD
Aurora Cardiovascular Services, Aurora Sinai/Aurora St. Luke's Medical Centers, University of Wisconsin School of Medicine and Public Health, Milwaukee, Wisconsin

Mohamed Kanj, MD
Associate Director of Electrophysiology Laboratories at the Robert and Suzanne Tomsich Department of Cardiovascular Medicine, Cleveland Clinic, Cleveland, Ohio

Karoly Kaszala, MD, PhD, FACC, FHRS
Associate Professor of Medicine, Director, Cardiac Electrophysiology Laboratory, Hunter Holmes McGuire VA Medical Center, Medical College of Virginia / VCU School of Medicine, Richmond, Virginia

Bijoy K. Khandheria, MD, FACP, FESC, FASE, FAHA, FACC
Adjunct Clinical Professor of Medicine; Director, Echocardiography Services; Director, Echocardiography Center of Innovation and Research, Aurora Sinai/Aurora St. Luke's Medical Center/University of Wisconsin School of Medicine and Public Health, Milwaukee, Wisconsin

Sébastien Knecht, MD, PhD
Professor, Electrophysiology, CHU Brugmann, Brussels, Belgium

Jayanthi N. Koneru, MBBS
Assistant Professor of Medicine, VCU School of Medicine, Medical College of Virginia Hospitals, Richmond, Virginia

David C. Kress, MD, FACS, FACC
Chief, Department of Cardiothoracic Surgery, Aurora Sinai/Aurora St. Luke's Medical Center, Milwaukee, Wisconsin

David Krum, MS
Electrophysiology, Aurora Cardiovascular Services, Aurora Sinai/Aurora St. Luke's Medical Centers, Milwaukee, Wisconsin

Nirusha Lachman, PhD
Associate Professor of Anatomy, Mayo Clinic, Rochester, Minnesota

Ralph Lazzara, MD, FACC, FHRS
Department of Medicine, Section of Cardiovascular Disease, Heart Rhythm Institute, University of Oklahoma Health Sciences Center, Oklahoma City, Oklahoma

Deborah Lockwood, BM, BCh, MA, FACC, FHRS
Department of Medicine, Cardiovascular Division, Oklahoma University Medical Center Heart Rhythm Institute, Oklahoma University Health Sciences Center, Oklahoma City, Oklahoma

Paula Gonçalves Macedo, MD
Fellow at the Cardiovascular Research Department, Coordinator of the Atrial Fibrillation Center, Mayo Clinic, Rochester, Minnesota, Hospital de Base do Distrito Federal Brasilia, DF, Brazil

Rajiv Mahajan, MD, DM
Centre for Heart Rhythm Disorders (CHRD), South Australian Health and Medical Research Institute (SAHMRI), University of Adelaide and Royal Adelaide Hospital Adelaide, Australia

Rohit Malhotra, MD
Assistant Professor, Department of Internal Medicine, Cardiology Division, Heart Rhythm Center, University of Virginia Health Sciences Center, Charlottesville, Virginia

Saurabh Malhotra, MD, MPH
Heart and Vascular Institute,University of Pittsburgh Medical Center, Pittsburgh, Pennsylvania

Francis E. Marchlinski, MD, FACC, FHRS
Director, Cardiac Electrophysiology, Department of Medicine, Division of Cardiovascular Medicine, University of Pennsylvania Health System; Professor of Medicine, University of Pennsylvania School of Medicine, Philadelphia, Pennsylvania

George M. McDaniel, MD, MS
Assistant Professor of Pediatrics and Internal Medicine; Director, Pediatric Electrophysiology, University of Virginia, Charlottesville, Virginia

Shinsuke Miyazaki, MD
Hôpital Cardiologique du Haut-Lévêque and the Université Victor Segalen Bordeaux II, Bordeaux, France

Fred Morady, MD
McKay Professor of Cardiovascular Disease, Professor of Medicine, University of Michigan Health System, Ann Arbor, Michigan

Arthur J. Moss, MD, FACC
Department of Medicine, Division of Cardiology, University of Rochester Medical Center, Rochester, New York

Hiroshi Nakagawa, MD, PhD
Professor of Medicine; Director, Clinical Catheter Ablation Program; Director, Translational Electrophysiology, Heart Rhythm Institute, University of Oklahoma Health Sciences Center, Oklahoma City, Oklahoma

Calambur Narasimhan, MD, DM
Division of Electrophysiology, Department of Cardiology, CARE Hospitals and CARE Foundation, Hyderabad, India

Saman Nazarian, MD, PhD
Assistant Professor of Medicine & Epidemiology; Associate Director of Internal Medicine, Johns Hopkins Clinical Research Network; Director, Ventricular Arrhythmia Ablation Service, Johns Hopkins Hospital, Baltimore, Maryland

Imran Niazi, MD, FACC, FACP, FHRS
Clinical Adjunct Assistant Professor, Electrophysiology, Aurora Cardiovascular Services, Aurora Sinai/Aurora St. Luke's Medical Centers, University of Wisconsin School of Medicine and Public Health, Milwaukee, Wisconsin

Hakan Oral, MD, FACC
Huetwell Professor of Cardiovascular Medicine, Professor of Internal Medicine Director, Cardiac Arrhythmia Service, University of Michigan, Ann Arbor, Michigan

Douglas L. Packer, MD, FHRS, FAHA, FACC
Professor of Medicine; Director, Heart Rhythm Services, Cardiovascular Division, Department of Medicine, Mayo Clinic/St. Mary's Campus, Rochester, Minnesota

Patrizio Pascale, MD
Electrophysiology Unit, Cardiology Service, Centre Hospitalier Universitaire Vaudois and University of Lausanne, Lausanne, Switzerland

Chinmay P. Patel, MD
Pinnacle Health Cardiovascular Institute, Harrisburg, Pennsylvania

Timothy E. Paterick, MD, JD, MBA
Cardiology, Aurora Cardiovascular Services, Aurora Sinai/Aurora St. Luke's Medical Centers, Milwaukee, Wisconsin

Michala Pedersen, MD, DPhil
Hôpital-Haut-Lévêque, Service Rhythmologie, Hôpital-Haut-Lévêque and the Université Victor Segalen, Bordeaux II, Bordeaux-Pessac, France

Sunny S. Po, MD, PhD
Professor of Medicine; Director, Clinical Electrophysiology; Associate Director, Heart Rhythm Institute, University of Oklahoma Health Sciences Center, Oklahoma City, Oklahoma

Brian D. Powell, MD, FACC, FHRS
Cardiovascular Division, Sanger Heart & Vascular Institute, Carolinas HealthCare System, Charlotte, North Carolina

Hygriv Rao, MD, DM
Senior Consultant Cardiologist and Electrophysiologist, KIMS Hospitals, Hyderabad, India

Lena Rivard, MD
Department of Rhythmologie, Hôpital Cardiologique du Haut-Lévêque and the Université Bordeaux II, Bordeaux, France

Dan M. Roden, MD
Assistant Vice Chancellor for Personalized Medicine, Professor of Medicine, William Stokes Chair in Experimental Therapeutics, Professor of Pharmacology, Vanderbilt University School of Medicine, Nashville, Tennessee

Manoj Kumar Rohit, MD, DM
Additional Professor, Department of Cardiology, PGIMER, Chandigarh, India

Michael R. Rosen, MD, FHRS
Pfeiffer Professor of Pharmacology, Professor of Pediatrics, Director of the Center for Molecular Therapeutics at Columbia University, Adjunct Professor of Physiology and Biophysics at Stony Brook University, New York, New York

Laurent Roten, MD
Department of Cardiology, Inselspital, Bern University Hospital, and University of Bern, Bern, Switzerland

Frederic Sacher, MD
Department of Rhythmologie, Hôpital Cardiologique du Haut-Lévêque and the Université Bordeaux II, Bordeaux, France

Sanjeev Saksena, MD, FHRS, FACC, FESC, FAHA, FRSM
Clinical Professor of Medicine, Rutgers, Robert Wood Johnson Medical School; Medical Director, Electrophysiology Research Foundation, New Brunswick and Warren, New Jersey

Pasquale Santangeli, MD
Texas Cardiac Arrhythmia Institute, St. David's Medical Center, Austin, Texas

Takeshi Sasaki, MD, PhD
Division of Cardiology, Johns Hopkins University, Baltimore, Maryland

Melvin M. Scheinman, MD
Professor of Medicine, Walter H. Shorenstein Endowed Chair of Cardiology, Cardiac Electrophysiology, University of California, San Francisco, California

Benjamin J. Scherlag, PhD, FHRS
Professor of Medicine, Division of Cardiology, Heart Rhythm Institute, University of Oklahoma Health Sciences Center, Oklahoma City, Oklahoma

Daniel Scherr, MD, PD
Assistant Professor of Medicine, Division of Cardiology, Department of Medicine, Medical University of Graz, Graz, Austria

Richard J. Schilling, MD
Department of Cardiology and Electrophysiology, St. Bartholomew's Hospital Barts Health NHS Trust and QMUL, London, UK

Nandini Sehar, MBBS
Sri Venkateshwaraa Medical College Hospital and Research Centre, Puducherry, India

Ashok J. Shah, MD
Hôpital Cardiologique du Haut-Lévêque and the Université Victor Segalen Bordeaux II, Bordeaux, France

Dipen Shah, MD, FHRS
Department of Internal Medicine, Cardiology Division, University Hospital of Geneva, Geneva, Switzerland

Shlomo Shpun, DSc
Director of Medical Science, Biosense Wesbter Inc., Diamond Bar, CA; Currently - Consultant for Electrophysiology and Medical Devices, West Berlin, New Jersey

Serge Sicouri, MD
Research Scientist, Masonic Medical Research Laboratory, Utica, New York

Eugene A. Sosunov, PhD
Department of Quality Initiatives, The Mount Sinai Hospital, New York, New York

Stavros Stavrakis, MD, PhD
Assistant Professor of Medicine, Heart Rhythm Institute, University of Oklahoma Health Sciences Center, Oklahoma City, Oklahoma

William G. Stevenson, MD
Director, Cardiac Electrophysiology Program, Brigham and Women's Hospital; Professor of Medicine, Harvard Medical School, Boston, Massachusetts

A. Jamil Tajik, MD, FACC, FAHA
Clinical Adjunct Professor, Cardiology, Aurora Cardiovascular Services, Aurora Sinai/Aurora St. Luke's Medical Centers, University of Wisconsin School of Medicine and Public Health, Milwaukee, Wisconsin

Kewal K. Talwar, MD, DM, DSc (Hon Causa), FAMS, FNA, FACC
Chairman, Department of Cardiology, Max Healthcare Institute Ltd., Former Director, Professor & Head Department of Cardiology, PGIMER, Chandigarh, Max Healthcare Institute Ltd., New Delhi, India

Khaldoun G. Tarakji, MD, MPH, FHRS
Section of Cardiac Pacing and Electrophysiology, Robert and Suzanne Tomsich Department of Cardiovascular Medicine, Sydell and Arnold Miller Family Heart and Vascular Institute, Cleveland Clinic, Cleveland, Ohio

Patrick Tchou, MD
Section of Cardiac Electrophysiology and Pacing, Department of Cardiovascular Medicine, Heart and Vascular Institute, Cleveland Clinic, Cleveland, Ohio

Usha B. Tedrow, MD, MSc
Assistant Professor of Medicine, Harvard Medical School; Director, Clinical Cardiac Electrophysiology Program, Cardiovascular Division, Department of Medicine, Brigham and Women's Hospital, Boston, Massachusetts

Matt M. Umland, RDCS, FASE
Aurora Health Care, Aurora Cardiovascular Services, Milwaukee, Wisconsin

Régis Vaillant, PhD
Cardiovascular Engineering, DGS, GE Healthcare, Buc, France

Rajesh Venkataraman, MD, MPH
Assistant Professor of Medicine, Complex Arrhythmia Center, The University of Texas Health Science Center Houston, Memorial Hermann-Texas Medical Center, Heart and Vascular Institute, Houston, Texas

Hein J. J. Wellens, MD, PhD, FACC, FAHA, FESC
Professor of Cardiology, Cardiovascular Research Center, University of Maastricht, The Netherlands

Bruce L. Wilkoff, MD
Institute Medical Information Officer, Sydell and Arnold Miller Family Heart and Vascular Institute; Director, Cardiac Pacing & Tachyarrhythmia Devices, Robert and Suzanne Tomsich Department of Cardiovascular Medicine; Professor of Medicine, Cleveland Clinic Lerner College of Medicine of Case Western Reserve University, Cleveland, Ohio

Stephen B. Wilton, MD, MSc, FRCPC
Cardiac Electrophysiologist, Libin Cardiovascular Institute of Alberta, University of Calgary, Calgary, Alberta, Canada

Andrew L. Wit, PhD, FACC
Department of Pharmacology, College of Physicians and Surgeons, Columbia University New York City, New York

Seth J. Worley, MD
Heart Center – Lancaster General Hospital and the Lancaster Heart and Stroke Foundation, Lancaster, Pennsylvania

PREFACE

A plethora of significant new research and findings makes it difficult to keep up with the ever-changing field of electrophysiology. Despite these constant advances, there are fundamental aspects of the science that need to be understood by students of electrophysiology. This book was created to educate, and uses cases and questions to keep the reader engaged. Chapter and case topics were chosen so that the information presented is useful for years to come. My associate editors and I are hopeful that this book will prove a useful tool for those interested in the field of electrophysiology. We also are very grateful to all the contributing authors for spending their time and effort to help create this handy but comprehensive and interesting work.

Jasbir S. Sra, Milwaukee

ABBREVIATIONS

2D	two-dimensional
3D	three-dimensional
ADHF	acute decompensated heart failure
ADP	antidiuretic hormone
AEGM	atrial electrogram
AF	atrial fibrillation
AFCL	atrial fibrillatory cycle length
AFl	atrial flutter
AFP	atriofascicular pathways
AMI	acute myocardial infarction
ANP	atrial natriuretic peptide
AP	accessory pathway
APC	atrial premature contraction
ARVC/D	arrhythmogenic right ventricular cardiomyopathy/dysplasia
AS	atrial sensed
ASD	atrial septal defect
AT	atrial tachycardia
ATP	antitachycardia pacing
AUC	area under the curve
AV	atrioventricular
AVJ	atrioventricular junction
AVN	atrioventricular node
AVNRT	atrioventricular nodal reentrant tachycardia
BBB	bundle branch block
BBR	bundle branch reentry
BMI	body mass index
BNP	brain natriuretic peptide
CABG	coronary artery bypass grafting
CAD	coronary artery disease
CFAE	complex fractionated atrial electrogram
CHF	congestive heart failure
CI	confidence interval
CL	cycle length
CMR	cardiac magnetic resonance
CNP	C-type natriuretic peptide
CNS	central nervous system
CO	cardiac output
CPR	cardiopulmonary resuscitation
CPVA	circumferential pulmonary vein ablation
CPVT	catecholaminergic polymorphic ventricular tachycardia
CRP	C-reactive protein
CRT	cardiac resynchronization therapy
CS	coronary sinus
CT	computed tomograpy
CTI	cavotricuspid isthmus
CVD	cardiovascular disease
CVP	central venous pressure
DAD	delayed afterdepolarization
EAD	early afterdepolarization
ECG	electrocardiogram
ED	emergency department
EDPVR	end-diastolic pressure-volume relationship
EF	ejection fraction
EGM	electrogram
EMI	electromagnetic interference
EP	electrophysiology; electrophysiological
ERP	effective refractory period
ESA	electrically silent area
FDA	Food and Drug Administration
FFRW	far-field R wave
FM	fibrillar muscle
HB	His bundle
HBd	His-bundle distal
HBm	His-bundle mid
HBp	His-bundle proximal
HCM	hypertrophic cardiomyopathy
HR	hazard ratio
HRA	high right atrium
IART	intra-atrial reentrant tachycardia
ICD	implantable cardioverter-defibrillator

IV intravenous
IVC inferior vena cava

J joules
JVD jugular venous distention

LAD left anterior descending artery
LAO left anterior oblique
LBBB left bundle branch block
LCP lower common pathway
LCX left circumflex
LGE-CMR late gadolinium-enhancement cardiac magnetic resonance
LIE leftward inferior extension
LIPV left inferior pulmonary vein
LQTS long QT syndrome
LSPV left superior pulmonary vein
LV left ventricle; left ventricular
LVAD left ventricular assist device
LVEF left ventricular ejection fraction
LVOT left ventricular outflow tract

mA milliampere
MA mitral annulus
MAP mean arterial pressure
MCV middle cardiac vein
MI myocardial infarction
MRI magnetic resonance imaging
MSNA muscle sympathetic nerve activity

NC no capture
NICMP nonischemic dilated cardiomyopathy
NO nitric oxide
NYHA New York Heart Association

OR odds ratio
ORT orthodromic reentry tachycardia

PA pulmonary artery
PAC premature atrial contraction
PCI percutaneous coronary intervention
PCL pacing cycle length
PCWP pulmonary capillary wedge pressure
PI preexcitation index
PJRT permanent junctional reciprocating tachycardia
PMT pacemaker-mediated tachycardia
PPI postpacing interval
PRR plasma refill rate
PTCA percutaneous coronary angioplasty
PV pulmonary vein
PVAB postventricular atrial blanking period
PVARP postventricular atrial refractory period
PVC premature ventricular contraction
PVI pulmonary vein isolation

RAA right atrial appendage
RAO right anterior oblique
RAP right arterial pressure
RBB right bundle branch
RBBB right bundle branch block
RCA right coronary artery
RFA radiofrequency ablation
RIE rightward inferior extension
RIPV right inferior pulmonary vein
RSPV right superior pulmonary vein
RV right ventricle; right ventricular
RVA right ventricular apex
RVG radionuclide ventriculography
RVOT right ventricular outflow tract
RVR repetitive ventricular response

SNS sympathetic nervous system
SPECT single photon emission computed tomography
SQTS short QT syndrome
SR sinus rhythm
SVC superior vena cava
SVT supraventricular tachycardia

TARP total atrial refractory period
TCL tachycardia cycle length
TdP torsades de pointes
TDR transmural dispersion of repolarization
TEE transesophageal echocardiogram
TENS transcutaneous electrical nerve stimulation
ToT tendon of Todaro
TVA tricuspid valve annulus
TVI time velocity integral

V ventricular
VA ventriculoatrial
VEGM ventricular electrogram
VERP ventricular effective refractory period
VF ventricular fibrillation
VPC ventricular premature contraction
VRR ventricular rate regulation
VS ventricular sensed
VSD ventricular septal defect
VT ventricular tachycardia

WPW Wolff-Parkinson-White

CHAPTER 1

Cellular Basis for Cardiac Electrophysiology

Evgeny P. Anyukhovsky, PhD,
Eugene A. Sosunov, PhD,
and Michael R. Rosen, MD

INTRODUCTION

Cardiac muscle is an excitable tissue in which electrical activity reflects the organized gating of ion channels that activate and control cardiac impulse initiation, conduction, and contraction. Activation normally starts in the pacemaker cells in the sinoatrial node (SAN), and propagation of the electrical impulse through the atria, the atrioventricular conduction system, and the ventricular myocardium coordinates excitation and contraction. Each myocyte is excitable and can excite its neighboring cells. Cardiac cellular electrophysiology (EP) investigates the unique electrical properties of myocytes. Studying these electrical properties is important because: it provides knowledge that is the basis for clinical cardiac EP; and electric diseases of the heart (arrhythmias) are a major health problem, most of which incorporate alterations at the cellular level. Consequently, the objective of this chapter is to briefly review morphology of the action potential in different regions of the heart, the molecular structure and biophysical properties of multiple membrane ion channels and pumps in myocytes, the ionic basis of the resting and action potential, and the propagation of the action potential.

MORPHOLOGY OF THE ACTION POTENTIAL IN DIFFERENT REGIONS OF THE HEART

During the cardiac cycle, individual cells generate action potentials that represent the change in transmembrane voltage over time. The action potential of one cell provides excitatory stimuli for its neighbors and initiates contraction by transmitting the extracellular signal to the cell interior. Thus, the cardiac action potential underlies the normal spread of electrical activity and resultant contraction of the heart muscle.

The action potential has 5 distinct phases (numbered 0–4) corresponding to clearly recognizable landmarks in its contour (Figure 1.1A). Phase 0 is the rapid depolarization (or action potential upstroke) that quickly brings the membrane potential from a negative value during diastole to a positive potential. Following phase 0

Practical Electrophysiology © 2015 Jasbir S. Sra, Masood Akhtar, Andrea Natale, David J. Wilber
Cardiotext Publishing, SBN: 978-1-935395-17-1

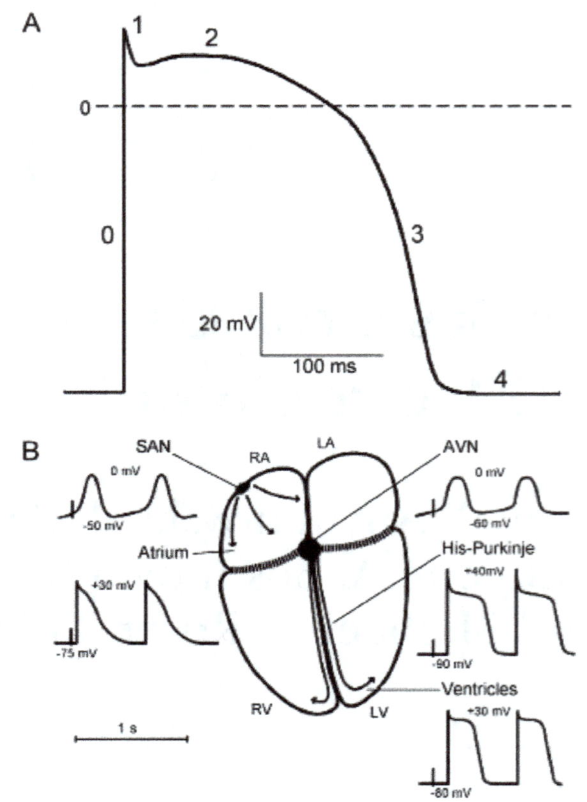

Figure 1.1. **A:** Schematic diagram of the cardiac action potential. The numbers refer to the action potential phases. **B:** Schematic diagram of action potential properties in different regions of the heart. RA and LA, right and left atrium; RV and LV, right and left ventricle; SAN, sinoatrial node; AVN, atrioventricular node. Maximum diastolic potential and overshoot potential are shown for each region. (Modified from Schram et al.[49])

(at the end of the action potential upstroke), a brief partial repolarization (phase 1) sometimes occurs, which is interrupted when the cell reaches the plateau (phase 2). Phase 1 can result in the appearance of a notch between phases 0 and 2. During the plateau, repolarization progresses slowly, and the plateau phase is eventually ended by the final rapid repolarization (phase 3). Phase 4 is the period between termination of repolarization and the onset of the subsequent action potential. Membrane potential during phase 4 (diastole) can be a stable resting potential (atrial and ventricular myocytes) or a slow diastolic depolarization (nodal structures and His-Purkinje cells).

Large differences exist in the shapes of the action potentials among different regions of the heart. Differences are seen in diastolic potential, in upstroke and amplitude, in the shape of the plateau, and in the action potential duration throughout the heart (see Figure 1.1B). Based on a difference in origin of the inward current responsible for the action potential upstroke, cardiac cells can be subdivided in 2 groups: slow- and fast-response cells. The action potential upstroke of the slow-response cells is due to Ca^{2+} current and the depolarization of the fast-response cells is caused by Na^{+} current.

In pacemaker *SAN cells,* phase 4 is not constant, and spontaneous diastolic depolarization brings the membrane potential from the most negative value achieved at the end of the preceding repolarization (maximum diastolic potential) to the threshold of activation, eliciting a new response. The maximum diastolic depolarization is in the range between –50 and –65 mV and the threshold for activation is about –35 mV. Phase 0, carried by inward Ca^{2+} current, has a low upstroke velocity (1–10 V/s), therefore SAN cells are slow-response cells. The action potential duration is from 100 to 200 ms. The intrinsic frequency of discharge in SAN pacemaker cells is faster than that of any other pacemaker in the heart.

Atrial cells are not spontaneously active and have a stable resting potential of about –75 to –80 mV. Phase 0 depolarization, carried by Na^{+} current, has a maximum upstroke velocity (V_{max}) of 100 to 200 V/s and an overshoot (the potential at the peak of the upstroke) of about 30 mV. After phase 1, a short plateau may occur, but often there is no clear separation between a plateau and final repolarization. The action potential duration is from 100 to 200 ms and is shorter in left atrium than in right atrium.

The primary function of the atrioventricular node (AVN) is to govern the ventricular response to supraventricular activation. Conduction through the AVN is slow and the cells have long refractory periods preventing dangerously rapid ventricular rates in response to supraventricular tachyarrhythmias. Action potential morphology in the AVN varies widely, depending on the nodal region. In the center of the AVN, action potential characteristics are similar to those of the SAN (see Figure 1.1B). The AVN cells have a low diastolic potential of –60 to –70 mV, and the potential is more negative close to His bundle. The AVN demonstrates a slow diastolic depolarization, but the rate of pacemaker activity is slower than in the SAN. The ionic currents contributing to AVN automaticity are similar to those in SAN. Spontaneous activity in AVN cells is suppressed by the higher SAN rate and by atrial electrotonic influences. However, the AVN can act as a subsidiary pacemaker: if the SAN fails, AVN cells can discharge spontaneously and excite the entire heart.

The *His-Purkinje system* forms the fast conducting system of the heart. The His bundle runs from the AVN through the connective tissue separating atria and ventricles. It then divides into left and right bundle branches, which continue into a network of Purkinje fibers electrically contacting the myocardium. Thus, the AVN-His-Purkinje system is the only electrical pathway connecting atria and ventricles. In contrast to nodal cells, Purkinje cells are highly polarized; their action potential upstroke velocities mediated by Na^+ current are very fast (up to 800 V/s), resulting in a high propagation velocity (2–5 m/s). The action potential shows an overshoot of about 40 mV, which is followed by a fast repolarization (phase 1). A short notch can precede the plateau. The action potential duration is highly frequency-dependable and can vary from 200 to 500 ms. An important characteristic of His-Purkinje cells is their ability to generate spontaneous activity. The frequency of these spontaneous discharges is significantly slower than in nodal cells.

The resting membrane potential in *ventricular myocytes* ranges from –80 to –85 mV; the maximum upstroke velocity is 100 to 200 V/s, and there is an overshoot of about 30 mV. The plateau is relatively positive (10–20 mV) and phase 3 repolarization is rapid. Similar to atrial muscle, there is no phase-4 depolarization or automaticity. Regional (apex–base and transmural) ventricular heterogeneity of the action potential is well-established. Action potential duration is shorter in cells at the apex than at the base of the ventricle. Compared with endocardium, epicardial action potentials have a smaller overshoot, a more prominent phase 1 followed by a notch, and a shorter duration, whereas resting potential and V_{max} are not different. The action potential propagates in ventricular myocardium at a velocity up to 0.5 m/s.

Summarizing the preceding paragraphs, they demonstrate that the magnitude and time course of action potentials vary significantly from one cardiac region to another. In this way electrical activity in each region of the heart is adapted to its specialized function. The regional differences in morphology of the action potential reflect differences in the expression and/or the properties of the ion channels. Therefore, to provide insight into ionic mechanisms governing regionally specialized cardiac electrical activity, the properties of the channels most important for normal electrical activity in the heart must be understood.

ION CHANNELS

The myocyte membrane is a phospholipid bilayer, a very thin nonconductive medium that separates 2 conductive media: the intra- and extracellular environments. Generation of the myocyte action potential results from the sequential activation and inactivation of different ion channels that conduct either inward (depolarizing) or outward (repolarizing) currents. Ion channels are protein structures that span the membrane; effectively, they are water-filled pores through which ions can move from one side of the membrane to the other according to their electrochemical gradients. The electrochemical gradient for each ion is established by the transmembrane potential, which is in the range of –60 to –95 mV (depending on the region of the heart), and the extracellular and intracellular concentrations of an ion. For the ions that determine the major depolarizing and repolarizing currents in myocytes, the extra- and intracellular concentrations are, respectively: sodium (Na^+) 140 and 10 mmol/L (ratio 14:1), potassium (K^+) 5 and 150 mmol/L (ratio 1:30), calcium (Ca^{2+}) 2 and 10^{-4} mmol/L (ratio $2{\times}10^4$:1). Although intracellular Ca^{2+} content is about 2 mmol/L, most of this is bound or sequestered in intracellular organelles—mitochondria and sarcoplasmic reticulum (SR).

Two major functional properties characterize ion channels: their selectivity and gating. Selectivity refers to the property of a channel to discriminate between ions of similar size and charge, permitting passage of one type to the exclusion of all others. Another important feature of ion channel function is gating—the process by which channels transition between open and closed states. Depending on their gating mechanism, all ion channels can be divided into 3 major groups: voltage-gated, ligand-gated, and gap junction channels. Voltage-gated channels have a sensor within their protein structure to monitor the electric field; opening or closing of the pore is influenced by transmembrane potential. Channel opening of ligand-gated ion channels is controlled by an appropriate ligand present in intracellular or extracellular spaces. Gap junction channels are responsible for electrochemical and metabolic intercellular coupling. They provide a low-resistance pathway between adjacent cells facilitating the spread of excitation.

Sodium Channels

Voltage-gated Na^+ channels carry the inward current (I_{Na}) that initiates the action potential and are responsible for propagation of excitation in atria, ventricles, and the His-Purkinje system. They are also present at low density in the SAN and AVN cells, but seem to play little or no physiological role in those tissues because the channels are inactivated at the low diastolic potential (about –60 mV). However, I_{Na} can contribute to the total inward current in neonatal SAN pacemaker cells.[1] Sodium channels are protein complexes consisting of a large α subunit (which contains the ion-conducting pore and the voltage sensor to gate the channel) and one or two auxiliary β subunits[2] (Figure 1.2A). The human cardiac α subunit (Nav1.5) is encoded by the gene SCN5A, which is located on chromosome 19.[3] Interestingly, recent data demonstrate an interaction between KChIP2, a K^+ channel interacting protein that is an accessory subunit for the Kv4 family of channels, and SCN5A. Specifically, KChIP2 appears to affect expression of Na^+ channel α and β1 subunits.[4]

Sodium channels have 3 functional states: closed, open, and inactivated. The channels are normally closed at the resting membrane potential

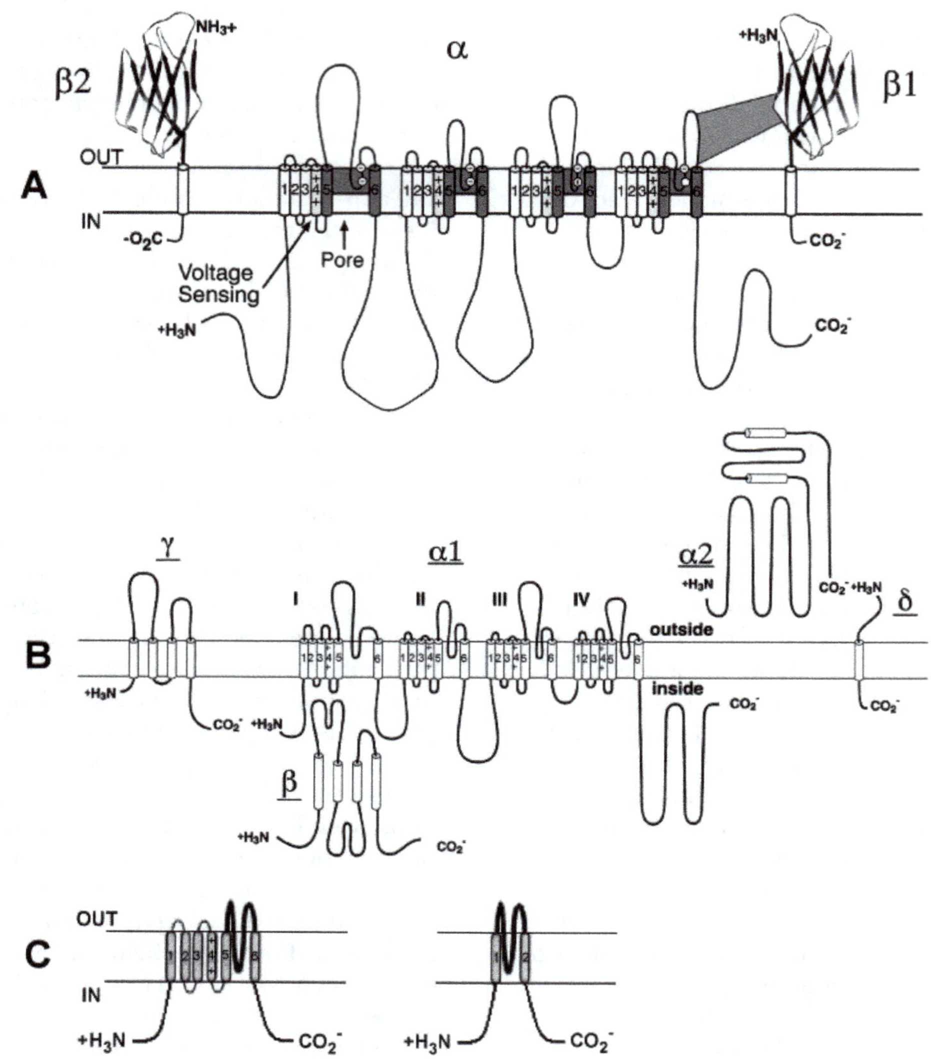

Figure 1.2. Molecular structure of Na^+, Ca^{2+} and K^+ channels. A: Voltage-gated Na^+ channel folding model depicting the α, β1, and β2 subunits. The α subunit consists of 4 homologous domains each with 6 transmembrane segments. The ion-conducting pore is formed by transmembrane segments 5 and 6 (S5 and S6) of each homologous domain. The S4 segment contains positive charges every third amino acid that sense transmembrane potential. Modified from Catterall et al.[50] B: Subunit structure of Ca^{2+} (Ca_v1) channels. The primary structure of the pore-forming α_1 subunit is similar to α subunit of Na^+ channel and composed of 4 homologous domains, each of which consists of 6 transmembrane segments. The α_2, δ, and γ subunits contain transmembrane domains, whereas the β subunit is entirely intracellular. Modified from Catterall et al.[51] C: The transmembrane folding of α subunits of voltage-gated (left) and 2 transmembrane-domain (right) K^+ channels. The α subunit of voltage-gated K^+ channels is homologous to the domains of the Na^+ and Ca^{2+} channels and the channels are formed as tetramers. Inward-rectifying K^+ channel proteins consist of 2 transmembrane segments with a pore loop in between.

(approximately –85 mV) and they open upon depolarization positive to threshold (from –70 to –60 mV) (Figure 1.3A), allowing Na^+ ions to go down their electrochemical gradient and further depolarize the cell. The inward current (I_{Na}) increases with depolarization due to the voltage dependence of channel activation. This induces a regenerative increase in Na^+ conductance and causes the membrane potential to depolarize at a rate of about 200 V/s in ventricular muscle (~500 V/s in Purkinje fibers). Maximum peak I_{Na} is seen at membrane depolarization to between –30 and –20 mV (see Figure 1.3B). On depolarization to more positive potentials, the amplitude of I_{Na} decreases due to the reduction in driving force (see Figure 1.3B). The sodium channel opening is immediately followed by inactivation that is also voltage-dependent and initiated by depolarization. Inactivation shortens I_{Na} and results in its termination (see Figure 1.3A). An inactivated channel can open again only after it has recovered from inactivation. This requires a return to negative membrane potentials and is fastest at the resting potential.

Inactivation and reactivation of I_{Na} are relatively rapid processes (2–10 ms duration) and their voltage dependency explains the time course and the voltage dependence of sodium current. Steady-state activation and inactivation have been described by a Boltzman distribution (see Figure 1.3C), which showed that in a limited range of potentials, activation and inactivation curves overlap. This crossover produces a small, sustained sodium current, the so-called window current.[5] Also a small fraction of sodium channels inactivates slowly, generating a small current over a broad range of potentials.[6,7] The range of window current includes the voltages of the action potential plateau, and the window and slowly inactivating currents contribute a small amount of inward current that sustains the action potential plateau.

Single sodium channel conductance is about 20 to 25 pS,[8] the whole-cell peak current at physiological extracellular Na^+ concentration and temperature is on the order of 70 to 140 nA, and current densities at peak currents are 0.5 to 1 nA/nF.[9]

Calcium Channels

Calcium influx through Ca^{2+} channels plays an important role in maintenance of the action potential plateau in atrial, ventricular, and His-Purkinje cells and also triggers SR Ca^{2+} release that facilitates contraction. Two main types of Ca^{2+} channels, T- and L-type, are expressed in myocytes. Both are voltage-activated. However, they have different electrophysiologic and pharmacological characteristics, and are differently expressed in the heart. The names T-type and L-type reflect the different kinetic behaviors of these channels (transient and long-lasting currents, respectively).[10]

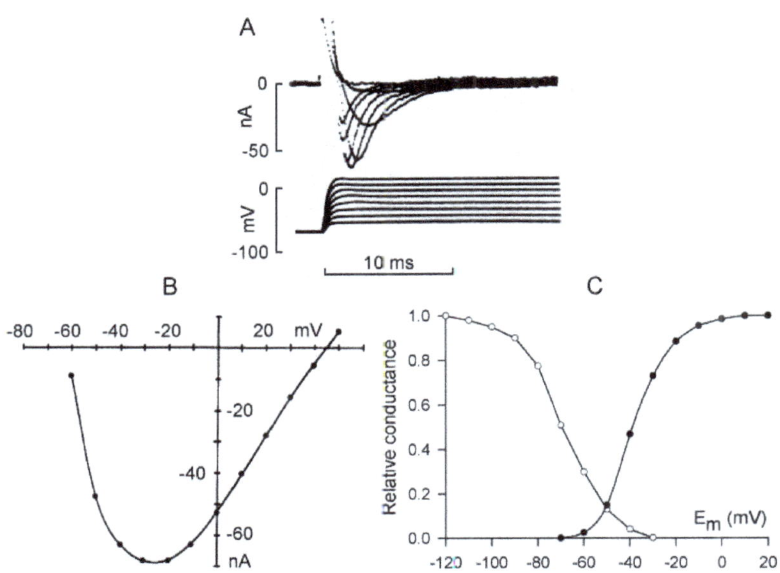

Figure 1.3. Sodium current. A: Time course of sodium current in rabbit Purkinje fibers elicited by depolarization to different membrane potentials. Modified from Colatsky.[52] B: Current–voltage relationship of peak I_{Na}. C: Steady-state activation (solid symbols) and inactivation (open symbols) curves of sodium current in rat ventricular cells. Modified from Brown et al.[9]

The T-type calcium current (I_{CaT}) is activated relatively rapidly and by weak depolarization. In contrast, the L-type current (I_{CaL}) is activated relatively slowly and by strong depolarization. L-type channels are present in all regions of the heart, whereas T-type channels are nearly absent in atrial and ventricular cells and are expressed in Purkinje and SAN cells but with a lesser density than L-type.[10] The L-type channels are characterized by a larger single channel conductance in comparison to T-type and are the dominant Ca^{2+} channels in the heart. Molecular structure of the L-type calcium channel is shown in Figure 1.2B. The $Ca_v1.2$ nomenclature (vs α_{1C}) is more formally accepted now. The human $Ca_v1.2$ subunit that is sufficient to express channel activity is encoded by the CACLN1A1 gene on chromosome 12p13.3. It resembles the sodium channel with 4 homologous domains and a region between 5 and 6 segments of each domain. A highly charged S4 segment acts as a voltage sensor for activation. The β subunit and α2 subunit (which is linked to a small intramembrane δ subunit) are involved in modulating kinetics and in affecting expression. In addition, KChIP2, the accessory subunit for Kv4 family of channels, functionally modulates the cardiac $Ca_v1.2$.[11]

Compared to I_{Na}, relatively large depolarizations are needed to activate I_{CaL}, and it has relatively slower activation properties. The threshold for I_{CaL} activation is in the voltage range of –35 to –25 mV; half-maximum activation is attained at about –15 mV; activation develops over several ms and is maximum by 0 to 10 mV (Figure 1.4A and B).[12] Inactivation and reactivation of I_{CaL} are slow as well (30–300 ms). Steady-state activation and inactivation curves are shown in Figure 1.4C. The overlap of these 2 curves indicates a window current resulting in a small steady-state I_{CaL} (≤1% of maximal) entering the cell.[13] The single channel conductance at physiological Ca^{2+} concentrations is 5 to 7 pS[14] and peak I_{CaL} density is 10 to 30 pA/pF.[15]

Inactivation of I_{CaL} is not only voltage-dependent but intracellular Ca^{2+} concentration-dependent.[16] When intracellular Ca^{2+} is buffered at its resting level or Ca^{2+} ions are replaced by Ba^{2+}, the current through L-type channels decays very slowly. In contrast, if intracellular Ca^{2+} rises to micromolar levels, I_{CaL} decays substantially. The mechanism of calcium-dependent I_{CaL} inactivation is as follows: when the local Ca^{2+} concentration at the inner mouth of the channel (see Figure 1.2B) increases, Ca^{2+} binds to the high-affinity calcium binding sites on the C-terminal, which leads to a conformational change and subsequent closure of the channel.[17] Calcium-dependent inactivation of I_{CaL} provides a negative feedback to limit further Ca^{2+}entry and

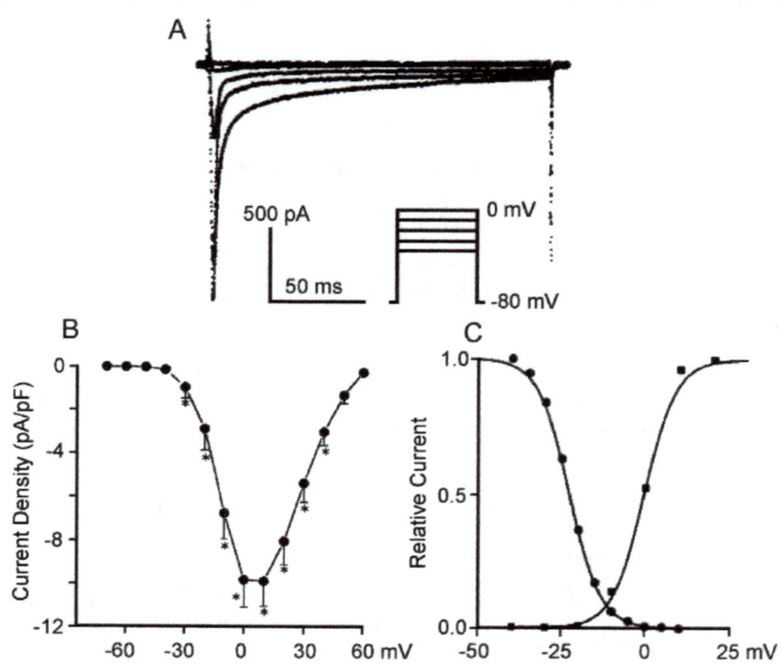

Figure 1.4. **A:** Time course of I_{CaL} in human right ventricular myocytes at different membrane potentials. **B:** Current–voltage relations of I_{CaL} expressed in terms of current density from holding potential of –80 mV. Modified from Li et al.[53] **C:** Steady-state activation (squares) and inactivation (circles) curves for I_{CaL} in canine Purkinje cells. Modified from Hirano et al.[13]

prevent Ca^{2+} overload of the cell. The quantity and time course of Ca^{2+} entry through I_{CaL} are important for the action potential and excitation-contraction coupling. Typically, Ca^{2+} channels are located in membrane sections close to the ryanodine receptors, the sarcoplasmic release channel for Ca^{2+}. Ca^{2+} diffuses slowly, therefore the concentration seen by the channel may be much higher than the general cytoplasmic Ca^{2+}.[18]

In addition to the 2 important roles of I_{CaL} described above (determining the plateau of the action potential and coupling excitation and contraction), it is responsible for the upstroke of the action potential in SAN and AVN. These structures have relatively few Na^+ channels, and these are inactivated at the relatively positive diastolic potentials. With regard to I_{CaT}, it is prominent in embryonic and neonatal atrial and ventricular cells. I_{CaT} declines with maturation but remains important in Purkinje and SAN cells. The channel has been proposed to play a role in pacemaking.

Potassium Channels

The cardiac action potential is controlled by the complex interaction between inward (depolarizing) currents and outward (repolarizing) currents. Potassium channels are key contributors to repolarization. There is a large variety of K^+ channels in cardiac cells. The major cardiac potassium channels can be divided into 2 major functional groups: voltage-gated (I_{to}, I_{Kur}, I_K (I_{Kr} + I_{Ks}), and I_{K1}) and ligand-activated ($I_{K,Ach}$ and $I_{K,ATP}$). In addition, many K^+ currents exhibit rectification. The current is ohmic (unrectified) when there is a linear relationship between the membrane potential and current (in other words, membrane resistance is constant over a range of membrane potentials). If a channel's ability to conduct current varies with the membrane potential, the resultant dependence of membrane resistance on voltage is called rectification. When depolarization facilitates the channels opening outward, rectification occurs. In contrast, the channels exhibit inward rectification if they tend to close with depolarization. Potassium currents in the heart can exhibit both inward and outward rectification as described below.

Voltage-gated potassium channels: The molecular composition of voltage-gated K^+ channels is highly homologous with that of Na^+ and Ca^{2+} channels. The difference is that whereas the α subunit of Na^+ and Ca^{2+} channels has 4 domains, each consisting of 6 transmembrane segments, the α subunit of K^+ channels has only one domain with 6 segments (see Figure 1.2C, left panel). There is a typical, positively charged S4 (the voltage sensor) and a P region between S5 and S6, conferring K^+ selectivity.[19] Voltage-gated K^+ channels are formed as tetramers. The channels are a major determinant of the voltage time course of the action potential repolarization.

Transient outward current (I_{to}): Most cardiac cells have a K^+ channel, generating a transient outward current, I_{to}, that initiates repolarization after the I_{Na}-induced spike (phase 1 repolarization of the action potential). The major pore-forming channel proteins expressed in the mammalian heart having biophysical properties consistent with I_{to} are Kv1.4 (KCNA4) and Kv4.2/4.3 (KCND2/3). Kv4.2 and Kv4.3 interact with a small protein MiRP1 (Min-K-related peptide 1) encoded by KCNE2. The Kv4 family of channels has accessory subunits called Kv channel-interacting proteins (KChIPs).

I_{to} is rapidly activated and is also inactivated during depolarization steps (Figure 1.5A). Because recovery from inactivation is rather slow, I_{to} amplitude is smaller at fast rates, attenuating the shortening of the action potential at elevated frequencies. The current is differently expressed in the different parts of the heart. I_{to} density is higher in right ventricular myocardium versus left ventricular myocardium,[20] and is higher in subepicardial and midmyocardial cells than in subendocardial myocytes. The differences in expression of I_{to} across the ventricular wall may result from different expression of the KChIP2 subunit rather than the Kv4.3 α subunit.[21] Single channel conductance is in the order of 10 pS.

The ultrarapid delayed rectifier current (I_{Kur}): I_{Kur} is one of the primary K^+ currents involved in atrial repolarization. In the human atrium, this current corresponds to the Kv1.5 protein encoded by KCNA5 on 12p13.[22] The current is very rapidly activated and shows no or very slow inactivation (see Figure 1.5B). It has been considered a noninactivating component of I_{to} (see Figure 1.5A). Recovery from inactivation is slow, therefore I_{Kur} is markedly reduced at high rates. The current–voltage relationships reveal outward rectification of the current.[23] Single channel conductance at normal extracellular potassium concentration is 14 to 20 pS.[24]

The delayed rectifier K^+ current (I_K): The cardiac delayed rectifier K^+ current initially was identified on the basis of its relatively slow onset in comparison to other K^+ currents. I_K initiates repolarization and is the primary determinant of late phase 2 and phase 3 repolarization. I_K has 2 components; these are rapidly (I_{Kr}) and slowly (I_{Ks}) activated currents that can be distinguished based on the biophysical and pharmacological properties of the channel and its molecular structure. Which of these 2 currents plays a greater role in terminating the plateau depends on

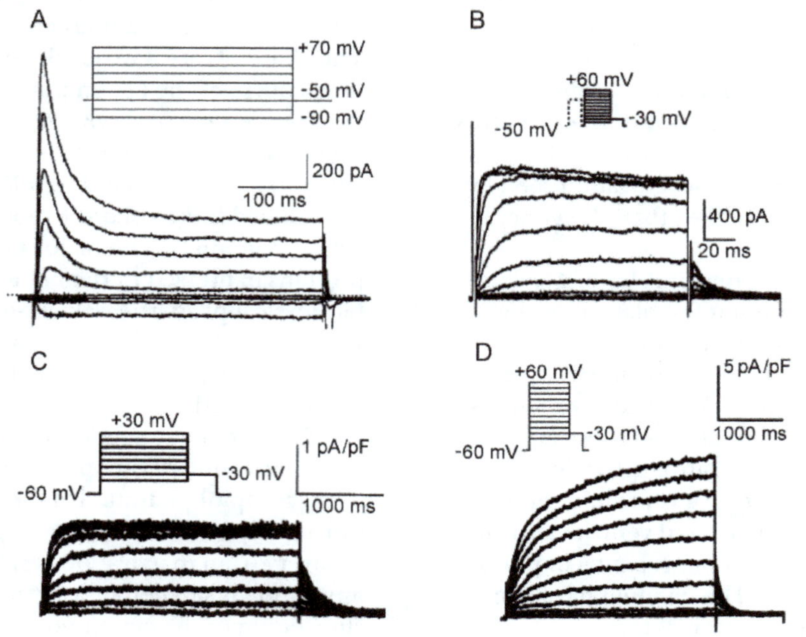

Figure 1.5. Representative K+ channels in cardiac cells. A: I_{to} in human atrial cells. Modified from Van Wagoner et al.[54] B: I_{Kur} in canine atrial cells. Modified from Yue et al.[24] C and D: I_{Kr} and I_{Ks}, respectively, in canine atrial myocytes. Modified from Li et al.[55]

experimental conditions. Under voltage-clamp conditions with long depolarizing steps, the value of I_{Ks} may be several folds greater than I_{Kr}. However, during an action potential, due to differences in voltage dependence of activation and kinetics, the currents can be of comparable magnitude.

The rapid delayed K+ current (I_{Kr})*:* I_{Kr} contributes significantly to the repolarization of the action potential. In human cardiac myocytes, the HERG gene located on chromosome 7q35-36 determines I_{Kr} expression. Kinetics and current density result from co-assembling of HERG with the β subunit minK-related peptide 1 (MiRP1), a small protein with only one membrane segment coded by KCNE2 on chromosome 21q22.12.[25] The channel shows rapid activation upon depolarization (it is half-activated at –30 mV) but also very rapid inactivation (see Figure 1.5C). At depolarization to positive membrane potentials, inactivation proceeds at a faster rate than channel activation.[26] As a result, I_{Kr} displays marked inward rectification.[27] Consequently, the current has a relatively low conductance during action potential plateau, but during repolarization the channels reenter the open state. Because deactivation from the open state is slow, the large current occurs upon repolarization, terminating the action potential plateau. The single channel conductance under physiological conditions is about 2 pS.

The slowly activated delayed K+ current (I_{Ks})*:* Cardiac I_{Ks} is generated by co-assembly of KvLQT1 (KCNQ1) and minK. The I_{Ks} current activates slowly over a broad range of depolarizing potentials (half-maximum activation is about +20 mV) and shows no inactivation (see Figure 1.5D). I_{Ks} demonstrates no rectification—the current–voltage relation approaches linearity.[28] In contrast to I_{Kr}, I_{Ks} increases gradually during the plateau phase of the action potential but, similar to I_{Kr}, remains activated during the repolarization phase as the result of slow deactivation. At rapid heart rates, the shortened diastolic interval and slow inactivation result in subpopulations of both channels remaining in the activated state, thereby contributing to a faster rate of repolarization. Single channel conductance has been estimated at 3 to 5.4 pS and does not depend on extracelluar K^+.

The inward rectifier K+ current (I_{K1})*:* Kir2.1 protein encoded by the gene KCNJ2 on chromosome 17q23.1-q24.2 underlies cardiac I_{K1}. The α subunit of the channel is distinct from Kv channels and consists of only 2 transmembrane segments with a pore loop sequence in between (this is analogous to the S5, S6, and P loop region of Kv channels) (see Figure 1.2C, right panel). The mechanism for voltage gating is not due to an S4 voltage sensor, since the channels do not have an S4 structure. There are 2 mechanisms for this process, block by intracellular Mg^{2+} and block by polyamines[29] at a divalent binding site on the channel. Normally, these blocking particles are prevented from binding by the electrical field. Depolarization facilitates block and repolarization relieves it. Initially I_{K1} was considered as a time-independent current. Then, it was found that block and unblock by Mg^{2+} and polyamines was time-dependent. I_{K1} is large at negative membrane potentials and small in depolarized state. The current–voltage relation for I_{K1} has a region of negative slope (Figure 1.6A), and the strong rectification properties of I_{K1} keep the cell at a stable, negative resting potential. They also prevent these channels from counteracting the membrane depolarization induced by Na^+ influx through voltage-gated sodium channels. I_{K1} also plays a major role in the late phase of repolarization of the action potential. I_{K1} density is highest in Purkinje fibers and ventricular myocardium, less in atria, and relatively scanty in nodal cells.[30] Single channel conductance ranges from 20 to 45 pS.

Ligand-activated K^+ channels: The major cardiac ligand-activated K^+ channels are the acetylcholine-activated (I_{KACh}) and the adenosine triphosphate (ATP)-sensitive (I_{KATP}) potassium channels. Similar to I_{K1}, these K^+ channels belong to the Kir channel family.[31] I_{KACh} and I_{KATP} show inward rectification and, together with I_{K1}, play a critical role in stabilizing the resting membrane potential and modulating repolarization of the action potential.

Acetylcholine-activated K^+ current (I_{KACh}): Cardiac I_{KACh} channels consist of 2 subunits Kir3.1 (GIRK1) coded by the KCNJ3 gene on chromosome 2q24.1, and Kir3.4 (GIRK4) coded by the KCNJ5 gene on chromosome 11q24. The channels are directly coupled to M2 muscarinic receptors and the coupling mechanism involves a pertussis toxin-sensitive guanosine triphosphate–binding protein (G protein). I_{KACh} is characterized by inward rectification, which is weaker than that of I_{K1} (see Figure 1.6B). The channels are expressed in atrial, SAN, AVN, and Purkinje cells and much less in ventricular cells. In atrial cells, where I_{K1} is relatively weak,[30] I_{KACh} activated by background parasympathetic tone can contribute substantially to the atrial resting potential and to the action potential duration. In pacemaker structures, I_{KACh} activation results in membrane hyperpolarization and slowing of pacemaker rate. A baseline I_{KACh} can be observed in the absence of agonists (background openings). I_{KACh} unitary conductance is similar to that recorded for I_{K1} and is approximately 30 to 40 pS.

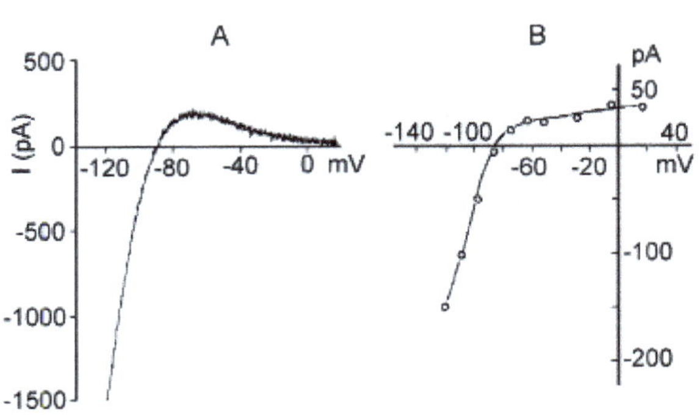

Figure 1.6. Two inward rectifier K^+ currents.
A: Current–voltage relation of I_{K1} in canine Purkinje cells. Modified from Sosunov et al.[56] B: Current–voltage relation of ACh-induced current in human atrial cells. Modified from Heidbuechel et al.[57]

ATP-sensitive K^+ current (I_{KATP}): The I_{KATP} channels provide a link between cellular metabolism and excitability. They are expressed uniformly throughout the heart at much higher density than other sarcolemmal K^+ channels. The molecular correlate of I_{KATP} is a composite of Kir6.2 (encoded by the gene KCNJ11 on chromosome 11p15.1) and the ATP-binding cassette proteins SUR2A that encode the sulfonylurea receptor. I_{KATP} channels are blocked rapidly and reversibly by cytoplasmic ATP. In myocardial cells, intracellular ATP concentration is in the millimolar range, whereas the half-maximum inhibitory concentration is in the micromolar range. Thus, under physiological conditions, the channels are closed and do not contribute to the action potential shape. During metabolic stress, as occurs during ischemia or hypoxia, intracellular ATP concentration decreases, favoring the opening of the channels. Activation of I_{KATP} leads to a shift of the membrane potential closer to the equilibrium potential for K^+ ions, shortening of the action potential duration, reduction of contractile force, and decrease of K^+ efflux.[32] At physiological extracellular K^+ concentrations, the single channel conductance is relatively high (about 25 pS). Thus, the opening of these channels is a critical factor in action potential shortening during hypoxia/ischemia.

Hyperpolarization-activated Pacemaker Current (I_f)

The channels generating I_f are named hyperpolarization-activated cyclic nucleotide-gated (HCN) channels because they are voltage-sensitive and can be activated by cAMP. I_f is an important contributor to pacemaking.[33] The molecular basis of I_f was identified by cloning the HCN channel gene family. Four different genes (HCN1 to HCN4) have been identified. HCN1, HCN2, and HCN4 are expressed in the heart more than HCN3. In SAN, the dominant HCN transcript is HCN4. The HCN4 gene is located on chromosome 15q24-q25, and

HCN2 on chromosome 19p13.3. Similar to the Kv family (see Figure 1.2C, left panel), the HCN channel protein is composed of 6 transmembrane segments, including a positively charged S4 domain (voltage sensor). The cytoplasmic C terminus contains a cAMP-binding domain mediating the modulatory effect of cAMP. The ion-conducting pore region is located between transmembrane domains 5 and 6. The channel subunits assemble to form tetramers. MiRP1 (in addition to regulating I_{Kr}) can act as a β subunit: coexpression of MiRP1 with HCN1 or HCN2 increases current density and accelerates activation kinetics.

Unlike all other voltage-gated depolarizing cardiac currents, I_f is activated not on depolarization of the membrane but on membrane hyperpolarization; hence the extent and rate of its activation are larger at more negative potentials (Figure 1.7A). The channel is permeable to K^+ and Na^+ ions, and its reversal potential is approximately –20 mV. Membrane potentials at which the channel opens are close to the K^+ equilibrium potential and far distant from the Na^+ equilibrium potential. Therefore, the driving force on K^+ is low and the predominant current through this channel is a depolarizing inward Na^+ current. In SAN cells, the steady-state activation curve is in the range of –45 mV to –100 mV (see Figure 1.7B). At the maximum diastolic potential of SAN cells (–60 to –70 mV), I_f channels carry an inward current that contributes to phase 4 depolarization and rhythmic activity. β-adrenergic agonists (which increase the intracellular cAMP level) shift the activation curve to more depolarized voltages and facilitate the channel opening. In addition, activation kinetics of the channel are accelerated. Lowering of cAMP by muscarinic agonists inhibits opening of the channel (see Figure 1.7B). The unitary conductance of the I_f channel was estimated as 1 pS, which is rather small for a nonselective channel. The I_f current has been described primarily in SAN, AVN, and Purkinje cells.

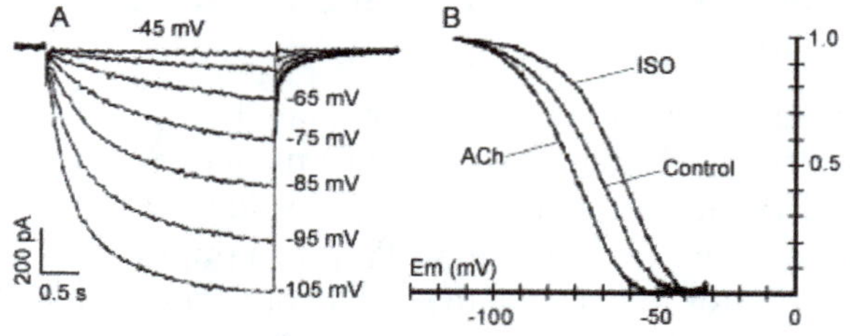

Figure 1.7. A: I_f current in rabbit SAN at different hyperpolarizing voltage steps. Modified from Shi et al.[58] B: Steady-state activation curves for I_f in rabbit SAN during control and exposure to isoproterenol (Iso) and acetylcholine (ACh). Modified from Accili et al.[59]

The channel is also present in ventricular and atrial myocardium. However, because the activation range of I_f in these cells is more negative than the resting potential, it does not play a role in physiological conditions.

Gap Junction Channels

Gap junctions are responsible for the syncytial function of cardiac tissue, as they provide the necessary low-resistance pathways for the spread of excitation from one heart cell to its neighbor. Unlike other channels, gap junction channels are located in the intercalated disk—a membrane separating the interiors of 2 adjacent myocardial cells. Each gap junction channel consists of 2 hemichannels or connexons; each connexon is composed of 6 proteins called connexins (Figure 1.8A). Each connexin traverses the membrane 4 times (domains M1–M4), with both the C terminus and the N terminus projecting into the cytoplasmic space (see Figure 1.8B). Two extracellular loops, E_1 and E_2, provide continuity between domains M1 and M2 and between M3 and M4, whereas the region between M2 and M3 (cytoplasmic loop) is located in the intracellular space. The C terminus together with the cytoplasmic loop forms the so-called proton gate. In the formation of gap junctions, hemichannels act as ligands for each other. Cardiac myocytes express 3 different connexin isoforms Cx40, Cx43, and Cx45. Cx43 is the most widely distributed. Human Cx43 is coded by the gene GJA1 located on chromosome 6q21-6q23.2, Cx40 by GJA5 on chromosome 1q21.1, and Cx45 by GJA7 on chromosome 17.

Gap junction channels have chemical- and voltage-dependent gating mechanisms. All cardiac gap junctions can be closed by protons; however, the pH sensitivity varies with type of connexin. Gap junctions containing Cx45 are most sensitive and are closed by acidification to pH 6.7, whereas Cx43 and Cx40 are less susceptible to acidification-induced uncoupling: these channels are closed by intracellular pH ≤6.4.[34] Increasing intracellular Ca^{2+} also decreases junctional conductance. The effects of H^+ and Ca^{2+} are synergistic. This is important because simultaneous increase in intracellular concentrations of these 2 ions occurs during myocardial ischemia. Closure of the gap junctions uncouples healthy cells from damaged cells and can limit the size of the ischemic region.

Figure 1.8C and D illustrate the voltage-dependent behavior of gap junctions made of Cx43, the most abundant connexin in the heart. Figure 1.8C demonstrates that increasing transjunctional potential (V_j) gives rise to both

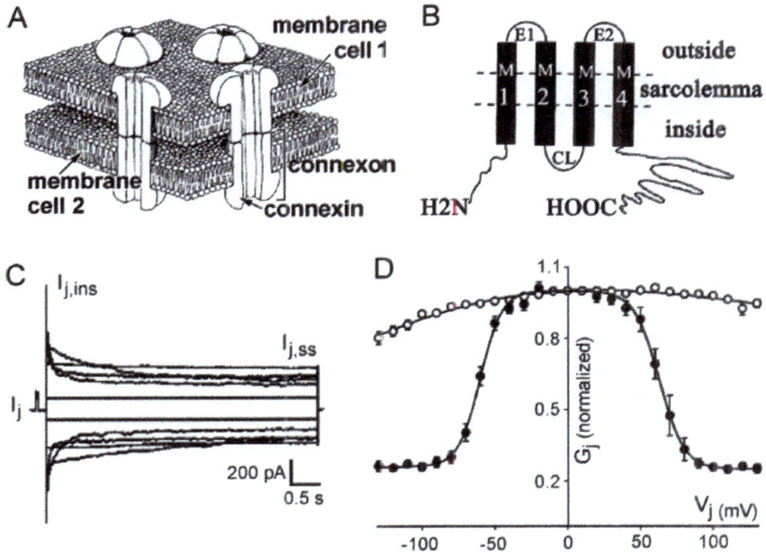

Figure 1.8. **A:** Schematic representation of part of a gap junction with individual gap junction channels, each consisting of 2 connexons representing oligomerizations of hexagonally arranged connexins. **B:** Predicted molecular structure of a connexin. From hydropathy plots 4 membrane spanning regions (M1–M4), 2 extracellular loops (E1 and E2) and one cytoplasmic loop (CL) are predicted. N and C termini are indicated. Modified from van der Velden et al.[60] **C** and **D:** Dependence of the gap junction conductance G_j on junctional potential V_j for homotypic Cx43-Cx43 channels. C: Family of gap junction current signals I_j recorded at different transjuctional potentials. $I_{j,ins}$, instantaneous current; $I_{j,ss}$, steady-state current. D: Normalized instantaneous (open symbols) and steady-state (solid symbols) conductances as functions of transjuctional potential. Modified from Desplantez et al.[61]

instantaneous ($I_{j,inst}$) and steady-state ($I_{j,ss}$) components of transjunctional current. Figure 1.8D shows that application of transjunctional potential causes fast closing of the channels to a non-zero conductance state. In homotypic channels the current–voltage relation is symmetric. Voltage gating accelerates cells uncoupling during ischemia and, together with the effects of intracellular acidification and Ca^{2+}, increasingly isolates injured myocardium from healthy tissue.

Permeation studies indicate that the gap junction channels are aqueous pores with large diameters (~16 A) that do not discriminate between anions and cations. They exhibit several conductance states. Unitary conductance for Cx43 channels ranges from 20 to 100 pS. In addition to conducting electrical current, gap junctions facilitate intercellular exchange of molecular information and are responsible for metabolic coupling. A gap junction consists of an array from tens to thousands of closely packed channels that permit intercellular passage of small molecules up to 1.2 kDa in molecular weight.

MEMBRANE PUMPS AND EXCHANGERS

Ion movements across the membrane during each excitation change their transmembrane gradients. However, for ion channels to perform their task repetitively, maintaining normal gradients is crucial. Furthermore, keeping the cell out of equilibrium with its milieu and sustaining the ionic environment within cells is important for optimal functioning of the numerous secondary processes. Active ion transport is responsible for this maintenance of the ionic balance of the myocyte. Two major contributors to ionic control of Ca^{2+}, Na^+, and K^+ (the Na^+/K^+-ATPase system and the Na^+-Ca^{2+} exchange system) are briefly discussed below.

Na^+/K^+ Pump

The Mg^{2+}-activated Na^+/K^+-ATPase constitutes the sodium pump of the cardiac sarcolemma—the enzyme responsible for the maintenance of sodium- and potassium-concentration gradients. It also provides energy for numerous secondary active transport processes. The Na^+/K^+-ATPase protein is a member of the P-type ATPase superfamily. It is a heterodimer incorporating α and β subunits (Figure 1.9A). Up to 4 distinct isoforms of the α and β subunits have been identified. Both subunits must be present for functional activity. The α subunit is a large (~110 kD) protein having 10 transmembrane-spanning segments: it carries the ATP-binding site and a phosphorylation site at the intracellular portion, and the digitalis binding site

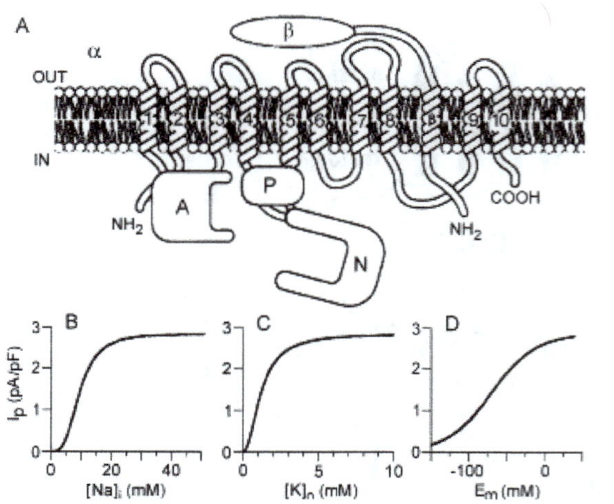

Figure 1.9. A: Structure of the Na^+/K^+-ATPase. The α subunit has 10 transmembrane segments and cytoplasmic N and C termini. Three distinct structural domains, A (involved in dephosphorylation step), P (phosphorilation site), and N (ATP binding site), form the intracellular part of the protein. Ouabain binding site is located in the extracellular loop between the 5th and 6th segments. The β subunit has a single transmembrane segment, a short intracellular NH_2 terminal, and a large extracellular domain. It is located close to the 7th and 8th transmembrane segments of the α subunit. Modified from Horisberger.[62] Sensitivity of the Na^+/K^+ pump to intracellular Na^+ concentration (B), extracellular K^+ concentration (C), and membrane potential (D). Modified from Stimers.[63]

facing extracellularly.[35] The β subunit is a type II membrane protein, having a single transmembrane segment with the major portion of its mass located extracellularly.

The Na^+/K^+ pump transports Na^+ out from the cell and K^+ into the cell. Because these directions are opposite to the electrochemical gradients of these ions, pump function requires energy. The active transport of Na^+ and K^+ ions is coupled: during optimal operation for each ATP molecule hydrolyzed, 3 Na^+ ions and 2 K^+ ions are transported. Thus, under physiological conditions, the operation of this transporter is electrogenic, producing an outwardly directed current (I_{NaK}). Thus, activation of the pump hyperpolarizes the resting membrane potential and influences the repolarization phase of the action potential. Figures 1.9B and C show the sensitivity of the Na^+/K^+ pump current to intracellular Na^+ ($[Na^+]_i$) and extracellular K^+ ($[K^+]_o$) concentrations, respectively. Whereas at physiological $[K^+]_o$ (~5 mM) the pump is almost maximally activated, normal $[Na^+]_i$ (~10 mM) activates the pump current only partially. Hence, the pump is mainly regulated by intracellular Na^+ concentration. This explains the sensitivity of the pump to pacing rate. During activity at a normal rate, each action potential results in a transient increase in $[Na^+]_i$, which is restored by the activity of the Na^+/K^+ pump. An increase in the rate of activation leads to an overload of $[Na^+]_i$. This activates the pump, as more Na^+ ions must be extruded from the cell to restore the normal $[Na^+]_i$. Enhanced outward pump current will hyperpolarize and accelerate repolarization of the action potential, resulting in shortening of its duration. Therefore, the Na^+/K^+ pump is an essential contributor to the dependence of the action potential duration on stimulation frequency.

In pacemaking structures, the pump-induced hyperpolarization and enhanced outward current counteract the inward currents that induce spontaneous diastolic depolarization. This is the mechanism of overdrive suppression—the inhibition of spontaneous rate by stimulation at higher frequencies. Overdrive suppression plays an important role in the maintenance of sinus rhythm by inhibiting the activity of subsidiary pacemakers. Because the operation of Na^+/K^+ pump is electrogenic, its activity depends on the membrane potential: depolarization of the membrane decreases the amount of energy required to move positive charge out of the cell and facilitates the pump (see Figure 1.9D).

Na^+-Ca^{2+} Exchanger

The Na^+-Ca^{2+} exchanger is present in all cardiac plasma membranes. The primary role of the Na^+-Ca^{2+} exchanger in cardiac cells is to extrude Ca^{2+} that enters the cell during each action potential.[36] The human cardiac Na^+-Ca^{2+} exchanger is encoded by the gene NCX1 on chromosome 2p21-p23. The NCX1 protein consists of 938 amino acids and is approximately 110 kD in size. It has 9 transmembrane segments, with a large cytoplasmic loop containing many of the regulatory sites (Figure 1.10A).

To pump Ca^{2+} against its electrochemical gradient, the Na^+-Ca^{2+} exchanger uses the energy of the Na^+ concentration gradient across the cell membrane established by the Na^+/K^+ pump. The Na^+-Ca^{2+} exchanger is an ion counter-transporter with a generally accepted stoichiometry of 3 Na^+:1 Ca^{2+}, which means that the exchanger is electrogenic and generates current. As a consequence, the operation of the exchanger both influences and is influenced by the transmembrane potential. Under normal conditions the reversal potential for exchanger current (I_{NCX}) is between –10 and –50 mV (see Figure 1.10B). Negative to the reversal potential, the exchanger moves Na^+ in and Ca^{2+} out of the

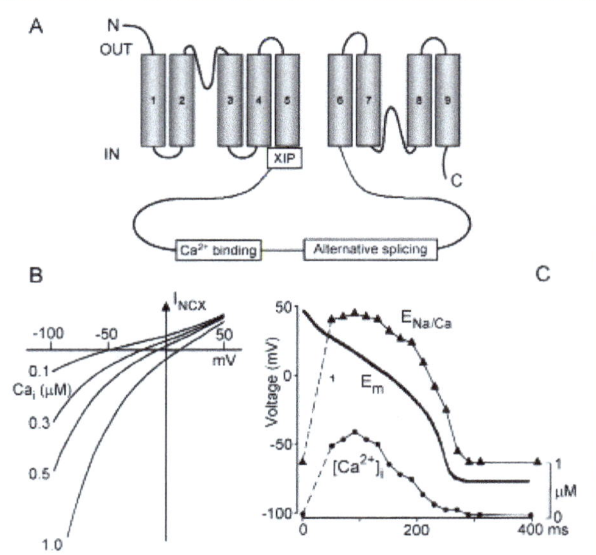

Figure 1.10. **A:** Topological model of the cardiac Na^+-Ca^{2+} exchanger. The NCX protein contains 9 transmembrane segment helices. There is a large cytoplasmic loop between the 5th and 6th segments, and the N and C termini are located on the extracellular and intracellular sides, respectively. This loop contains 520 amino acids and represents more than half the size of the total protein. It also contains many of the regulatory sites, including a Ca^{2+} binding site, a phosphorylation site, and a Na^+-dependent inactivation site, which is also the site of action of an inhibitory peptide (XIP). Modified from Ren et al.[64] **B:** Current–voltage relation of the Na^+-Ca^{2+} exchange current at different $[Ca^{2+}]_i$ and 10 mM $[Na^+]_i$ (computer simulation). Modified from Kimure.[65] **C:** The time course of the equilibrium potential of the Na^+-Ca^{2+} exchanger (E_{NaCa}) and intracellular Ca^{2+} concentration ($[Ca^{2+}]_i$) during the action potential (E_m) in guinea pig ventricular cells. Modified from Egan et al.[66]

cell (forward mode) and generates inward current. Positive to the reversal potential, Na^+ is extruded and Ca^{2+} moved into the cell (reverse mode) and outward current is generated.

There are 3 phases to exchanger activity during myocardial excitation. At rest, there is a very small inward current, indicating that the exchanger is pumping Ca^{2+}out to balance any inward Ca^{2+} leak. During the initial part of the action potential, when the voltage becomes positive (exceeds the reversal potential) and $[Ca^{2+}]_i$ is still low, the carrier transiently works in the reverse mode. During this time it contributes to Ca^{2+} entry to activate contraction. As $[Ca^{2+}]_i$ rises (due to release from SR), the reversal potential becomes positive to the plateau potential, and the exchanger reverses direction again, resulting in Ca^{2+} extrusion (see Figure 1.10C). During this phase, the exchanger contributes to the inward current and maintains the late plateau phase of the action potential.

IONIC BASIS OF RESTING AND ACTION POTENTIALS

The generation of the action potential reflects the sequential activation and inactivation of ion channels that conduct depolarizing and repolarizing currents. Figures 1.11 and 1.12 give a schematic overview of the time course of the most important ionic currents during the cardiac cycle in different heart regions. The different phases of the action potentials in relation to the contributions of these ionic currents in different cardiac cell types in normal conditions are discussed below.

Resting Membrane Potential

Resting membrane potential in cardiomyocytes is determined primarily by the concentrations of ions inside and outside the cell and the permeability of the cell membrane to those ions. Among the ions that have different concentrations across the cell

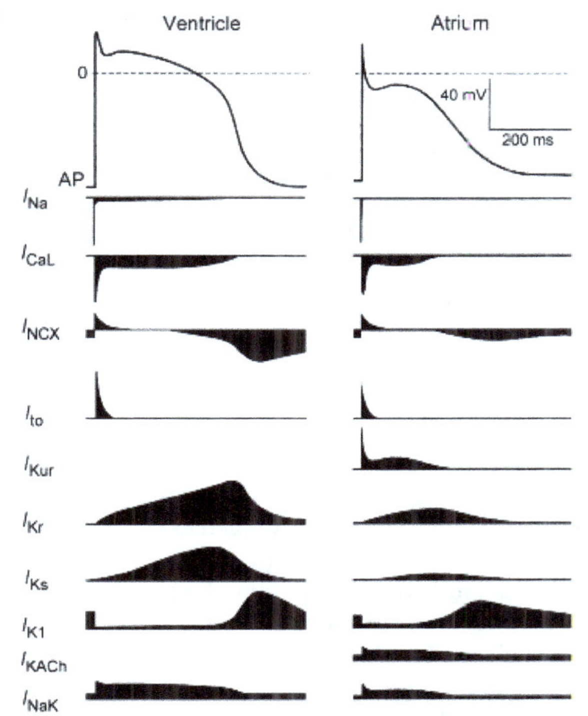

Figure 1.11. Schematic of the typical action potential (AP) configuration in ventricular (left) and atrial (right) myocytes (the 5 phases are labeled) and of the time courses of the most important currents during the cardiac cycle. The amplitudes of the currents are not to scale (the fast sodium current, eg, is orders of magnitude larger than plateau currents). Current traces have been drawn by hand based on experimental results and model simulations. Changes in trace thickness are meant to illustrate approximately the phases of the cycle in which the current is important. Modified from Reference 67

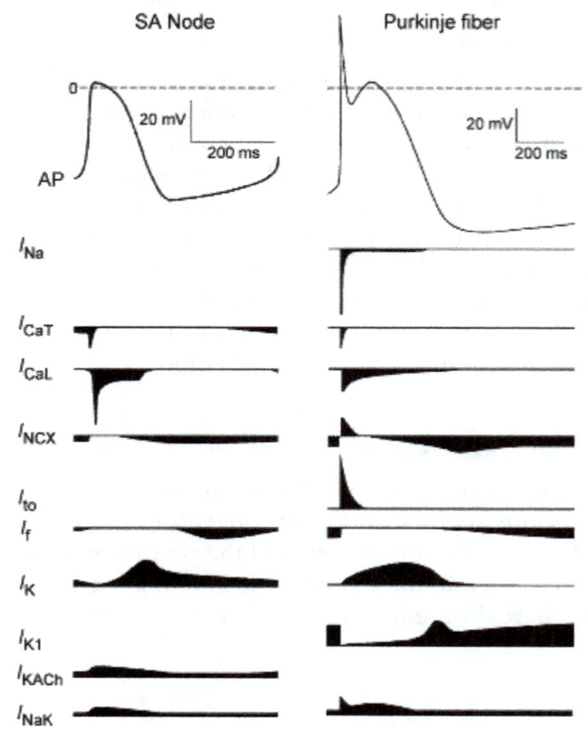

Figure 1.12. Schematic presentation of the typical action potential (AP) configuration of spontaneously beating sinoatrial (SA) node (left) and Purkinje (right) cells (the 5 phases are labeled) and of the time course of the most important currents during the cardiac cycle. Like in the previous figure, the currents are not to scale and a large size of a current indicates the phases of the cycle in which the current is important. The delayed rectifier K^+ current (I_K) is the sum of I_{Kr} and I_{Ks}. Modified from Reference 67.

membrane are Na^+, K^+, Cl^-, and Ca^{2+}. There are also negatively charged proteins within the cell itself. Generally, the potential at which steady-state occurs (the net charge transfer across the membrane is zero) can be found using the Goldman–Hodgkin–Katz potential equation that includes the concentration gradients and permeabilities for all ions. Because in the resting state the membrane is most permeable to K^+ and relatively impermeable to other ions, the resting potential is dominated by the K^+ equilibrium potential. In this case, the Goldman–Hodgkin–Kats equation reduces to the Nernst equation giving the equilibrium potential for one particular ion:

$$E_K=(RT/zF)\ \ln\ ([K]_i/[K]_o),$$

where E_K is the equilibrium potential for K^+, R is the universal gas constant, T is the absolute temperature, F is the Faraday, and z is the charge number of the ion (+1 for monovalent cations). At 37 °C the Nernst equation can be written as:

$$E_K= -61\ \log\ ([K]_i/[K]_o)$$

for $[K^+]_i$ of 140 and $[K^+]_o$ of 4 mmol/L, the equilibrium potential E_K= –94 mV (ie, the potential of the cytoplasm is more negative than that of the extracellular space). However, the actual value of resting potential for ventricular myocytes (–85 to –90 mV) is slightly less negative. This is because the cells show a small (in comparison to K^+) but finite permeability for other ions (eg, Na^+, whose equilibrium potential is about +50 mV). Therefore, the final resting potential of a cell is established by the balance among the equilibrium potentials for the ions to which a cell is permeable and the conductivities for those ions. Since the resting potential is more positive than the potassium equilibrium potential, there is a net electrochemical driving force on K^+ in an outward direction resulting in a slow leak of K^+ from the cell. Also, there is a prominent electrochemical force driving sodium into the cell and, although the permeability of the membrane to Na^+ is very low, a small amount of Na^+ leaks into the cell at rest. Thus, K^+ and Na^+ chemical gradients would be lost over time; however, the Na^+/K^+ pump returns K^+ into the cell and Na^+ out of the cell, thereby maintaining the chemical gradients of the ions.

The high potassium permeability of the resting cell membrane is mainly due to I_{K1} channels. Because the channels are differentially expressed throughout the heart, the resting membrane potential (or maximum diastolic potential for pacemaker cells) varies in different regions of the heart from –50 to –90 mV (see Figure 1.2B).

Ionic Basis of Action Potentials in Different Heart Regions

Ventricular action potential: The ventricular myocyte is frequently employed to facilitate teaching the generation of the cardiac action potential (see Figure 1.11, left panels). The action potentials of ventricular myocytes (as well as atrial and specialized conducting fibers) are dependent largely on fast inward Na^+ current. Attainment of potentials positive to threshold leads to opening of some Na^+ channels, allowing sodium ions to move along their electrochemical gradient. Thus, at threshold, the membrane rapidly switches from being mostly permeable to K^+ to being largely permeable to Na^+. This inward movement of Na^+ overcomes the outward K^+ current causing further depolarization and opening of additional Na^+ channels, leading to a regenerative response (phase 0). Since the kinetics of Na^+ entry are fast, there is a rapid and large increase in I_{Na}; the current reaches a peak at a potential about –20 mV within < 1 ms. Then I_{Na} declines (due to a decrease of the driving force and to inactivation of the channel) but it still continues to depolarize the cell. The membrane potential becomes positive and reaches a peak of about +30 mV (the action potential overshoot). During phase 0 the cell is mostly permeable to Na^+, therefore, the upstroke velocity of phase 0 (dV/dt_{max} or V_{max}) is frequently used as an indirect measure of I_{Na}. Because of I_{Na} inactivation after phase 0, a second action potential can be elicited only after recovery of Na^+ channels from inactivation, which occurs when the membrane repolarizes to potentials negative of –60 to –70 mV. The time during which the membrane remains refractory to activation regardless of the magnitude of the excitatory input is known as the absolute refractory period. As the cell repolarizes, it once again becomes excitable. However, there is an interval during which the amount of current needed to induce the active response is much larger than in control and, due to partial I_{Na} availability, action potentials with slower upstrokes may be elicited (the relative refractory period). Under physiological conditions, recovery of I_{Na} is fast and the end of the relative refractory period precedes the completion of repolarization. The widely used parameter describing refractoriness of cardiac tissue is the effective refractory period, which is measured from the onset of normally timed depolarization to the onset of the earliest premature depolarization *that can propagate* in adjacent tissues.

Phase 0 of ventricular action potentials is followed by a transient rapid repolarization (phase 1) reflecting I_{Na} inactivation and activation of the transient voltage-gated K^+ current (I_{to}). This transient repolarization or "notch" is prominent in subepicardial myocytes and minimally present in subendocardial myocytes reflecting differences in the expression of I_{to} channels. The transient nature of phase 1 results from fast inactivation of I_{to}. Depolarization-induced activation of time-independent I_{NCX} (that briefly works in the reverse mode) and I_{NaK} can contribute to the slope of phase 1 as well.

Initial repolarization is then followed by the action potential plateau (phase 2), which reflects a relative balance between outward and inward currents. Due to inward rectification, the conductance for I_{K1} drops dramatically resulting in the high membrane resistance during this phase of the action potential. This implies that even a small perturbation in an individual current component during the plateau may lead to relatively large changes in membrane potential and hence in the rate of repolarization and the action potential duration. The major inward current is I_{CaL}, which activates on depolarization more slowly than I_{Na} and does not contribute to phase 0. In ventricular subepicardial cells, where I_{to} is prominent, simultaneous activation of I_{CaL} and inactivation of I_{to} can result in a brief phase of secondary depolarization, which gives the action potential a spike and dome appearance. Window Na^+ current and the slowly inactivating component of I_{Na} may also contribute to plateau inward currents. The major outward currents are the delayed rectifier currents I_{Kr} and I_{Ks}, which gradually increase during the plateau phase. I_{Kr} rises faster but its inward rectification limits the maximum size of the current. I_{Ks} activates with slower kinetics, but, due to little rectification, its amplitude at the end of plateau can be comparable to I_{Kr}. Depolarization activates I_{NaK}, which contributes to outward plateau currents. During a large part of the plateau, I_{NCX} is relatively small, since its reversal potential is not very different from the plateau potential.

Final repolarization (phase 3), which follows the plateau, results from inactivation of I_{CaL} and activation of the delayed K^+ currents. Superimposition of I_{CaL} inactivation on time-dependent activation of I_{Kr} and I_{Ks} eventually leads to a predominance of outward (repolarizing) currents over inward (depolarizing) currents and initiates repolarization. The slow repolarization during phase 2 may move membrane potential out of the I_{CaL} and I_{Na} window ranges and remove the window inward currents components.

I_{K1} plays an important role in the final phase of repolarization. Due to its strong inward rectification, the channel is closed during the action potential plateau. However, as repolarization proceeds, the membrane voltage becomes more negative and moves into a range in which I_{K1} conductance becomes significant and increases steeply with further repolarization. This behavior of I_{K1} explains a relatively fast final repolarization.

The diastolic potential (phase 4) in ventricular cells is determined by a large K^+ conductance due to I_{K1}. Therefore, the resting potential is close to the K^+ reversal potential and remains constant throughout the diastolic interval. Stable resting potential reflects a balance between small inward and outward currents. Inward currents result from a small background inflow of Na^+ and Ca^{2+} ions and the exchanger current I_{NCX} that pumps Ca^{2+} out to balance an inward Ca^{2+} leak. The electrogenic Na^+/K^+ pump provides a small outward current and balances inward currents. Ventricular myocytes remain at this constant level of potential until a new excitatory stimulus brings the membrane potential to the threshold of I_{Na} activation, thus eliciting a new active response.

Atrial action potential: The time courses of the currents that determine the morphology of the ventricular action potential are present in atrial cells as well (see Figure 1.11, right panels). The most notable differences in the action potentials are the less-negative resting potential, the lower plateau potential, and shorter action potential duration in atrial than ventricular myocardium. All these differences can be explained by the lower expression of I_{K1} in atrial cells and presence of 2 additional currents that are absent in ventricular myocytes: I_{Kur} and I_{KACh}. The smaller size of I_{K1} results in ~5 to 10 mV less-negative resting potential and explains the more gradual final repolarization of atrial cells. In atrial cells, ACh-induced K^+ conductance can be of the same order as the basal K^+ conductance and, therefore, I_{KACh} channels contribute substantially to the normal resting potential. The lower plateau potential results from activation of I_{Kur} and I_{KACh}, which is active during phase 3 due to a weak inward rectification. Operation of these 2 currents also leads to a shorter action potential in atrial than ventricular myocytes.

Action potential in SAN cells: mechanisms of pacemaking: The SAN, located in the right atrial roof between the venae cavae, is specialized for physiological pacemaker function and its pacemaker cells generate spontaneous action potentials under physiological conditions. The cell type gradually changes from typical nodal cells centrally to atrial cells toward the periphery. The longest action potential durations are in the central node, preventing invasion by ectopic impulses and preserving the dominance of SAN. Electrical coupling to the atrium is designed to allow the SAN to excite the large mass of atrial muscle while insulating it from hyperpolarizing atrial influences.[37,38]

The shape of the SAN action potential differs markedly from that in ventricular and atrial myocytes (see Figures 1.1B and 1.12, left panels). In SAN cells of the primary pacemaker region, I_{Na} channels are either absent or minimally expressed[39] (and those present are completely inactivated by the low value of the diastolic potentials). Here, the depolarizing current is largely carried by Ca^{2+} (I_{CaT} and I_{CaL}). Due to smaller magnitude and slower kinetics of Ca^{2+} currents, as compared to I_{Na}, the maximal upstroke velocities of Ca^{2+}-dependent action potentials are quite slow compared to those of ventricular myocytes (2–5 V/s in the central node vs 200–250 in ventricular myocytes). Because I_{to} is practically absent in SAN cells,[40] the action potential plateau starts from the peak of phase 0, without a transient rapid repolarization (phase 1). The shape of phase 2 depends on activation of I_{Kr} and I_{Ks} (I_K=I_{Kr}+I_{Ks} in Figure 1.12). Depolarization-induced activation of I_{NaK} and I_{KACh} can also contribute to repolarizing currents during the plateau. Final repolarization (phase 3) results from inactivation of I_{CaL} and activation of I_K.

A critical difference in the population of ionic currents in the SAN and those in the ventricle is that the background K^+ current (I_{K1}) is almost absent in central SAN cells.[40–42] The important consequences of this are: a more positive diastolic potential, the absence of I_{K1} stabilizing effect on membrane potential during diastole, and a high membrane resistance. The latter means that even small net inward current can generate depolarization. As a result, membrane potential during diastole is not constant and displays a progressive slow depolarization from the negative value achieved at the end of an action potential (maximum diastolic potential) to the threshold level required to initiate a new action potential.

The net current is outward during final repolarization and becomes zero at the maximum diastolic potential. Therefore, the diastolic depolarization must be initiated by either a time-dependent increase of inward currents or a time-dependent decrease of outward currents (or a combination of both). The major time-dependent inward currents activated in the diastolic range of potential are: the hyperpolarization-activated current I_f and 2 Ca^{2+} currents I_{CaT} and I_{CaL}.

The cationic current, I_f, is a key contributor to pacemaker function activating in voltage range –40 to –65 mV in SAN cells. These voltages are within the diastolic range and, therefore, I_f becomes available at the end of phase 3 repolarization and can contribute to the initiation of diastolic depolarization. The low-threshold I_{CaT}, which activates at about –50 mV, can significantly contribute inward current to mid-late diastole due to its high channel density. I_{CaL}, which activates near –30 mV and provides the inward current for the upstroke of the SAN action potential, can also contribute to the final phase of diastolic depolarization. The delayed rectifier I_K that is responsible for repolarization can also contribute to diastolic depolarization as follows: because it activates following action potential initiation and deactivates during repolarization, its deactivation will shift a balance between outward and inward currents in favor of inward currents, facilitating depolarization.

The mechanism of spontaneous diastolic depolarization described above constitutes a voltage (or membrane) clock mechanism. Recent studies, however, show that cardiac impulse initiation is more complex and involves a so-called Ca^{2+} clock.[43] In spontaneously firing SAN cells, SR is responsible not only for action potential-induced global cytosolic Ca^{2+} transients but also for the occurrence of localized Ca^{2+} releases beneath the cell surface membrane during late diastolic depolarization. These rhythmic local Ca^{2+} releases activate the Na^+-Ca^{2+} exchanger current (I_{NCX}) and initiate action potentials. The Ca^{2+} clock and membrane clock dynamically interact and coordinately regulate sinus rate.

Thus, not a single but multiple mechanisms determine the pacemaker function of the SAN. In addition, the SAN is a heterogeneous structure and due to regional differences in ion channels, the relative contribution of each pacemaker mechanism to automaticity may differ in different regions of the node.[44] The marked redundancy in pacemaker mechanisms and their heterogeneity make the SAN a very robust and reliable pacemaker; if a particular mechanism is inhibited by diverse conditions, other mechanisms can substitute it and/or a leading pacemaker site can shift to a region within the SAN where contribution of this mechanism is relatively small.

Action potentials and automaticity in His-Purkinje system: Purkinje cells are found in the bundle of His and Purkinje system and are specialized for rapid conduction. Their maximum diastolic potential (–85 to –90 mV) is more negative, V_{max} (400–800 V/s) mediated by I_{Na} is greater, and action potential plateau is lower than in ventricular myocytes (see Figure 1.12, right panels). Both Ca^{2+} currents are present in Purkinje cells with I_{CaL} being smaller than in ventricular cells (consistent with their more negative plateau). I_{to} is well-expressed, leading to a prominent phase 1 notch and a resultant spike and dome configuration. I_K in Purkinje cells resembles ventricular and atrial I_K; but a smaller I_{K1} has been reported in Purkinje cells in comparison to ventricular muscle.[45] Purkinje cells show prominent diastolic depolarization and are able to generate spontaneous action potentials. In contrast to the SAN, I_{CaT}, which is quite substantial in Purkinje cells, does not contribute to diastolic depolarization because its activation threshold is more positive than the diastolic range of potentials. The major pacemaker current of Purkinje cells is I_f, which is large enough to counteract the small I_{K1}. The frequency of Purkinje cells automaticity is significantly slower than in the SAN and AVN. These cells can pace the ventricles at low rates when activation of supraventricular origin fails.

CONDUCTION OF THE CARDIAC IMPULSE

Cardiac tissue consists of millions of cells and each cell is connected to its neighbors through low-resistance intercellular channels—gap junctions that electrically couple the cells and make up the cardiac syncytium. Current entering the cytoplasm of a given cell not only spreads along this cell and discharges its capacitor but also moves in all directions and depolarizes neighboring cells. In other words, if a local regenerative reaction (action potential) is initiated, the resulting voltage difference between the excited and adjacent resting tissue drives local circuit currents that excite the resting tissue thereby supporting spread of excitation throughout the whole heart. Two main factors govern the propagation of excitation: the passive electrical properties of myocardial tissue and the properties of the ion channels.

The passive electrical properties determine the behavior of the tissue during flow of subthreshold current—so-called electrotonic interaction. The subthreshold level of current means that the membrane resistance can be taken as constant. In other words, an electrotonic potential is a response whose characteristics are determined by the passive tissue properties, whereas the active generator properties of the membrane play no role. In this case, the current is governed by Ohm's law and the spread of the electrical signal in a cell or through multicellular fiber bundles resembles the spread in an electric cable. Therefore, electrotonus or passive spread of current can be analyzed with the help of linear cable theory.[46] The cell can be represented as a cylinder of electrically conducting cytoplasm separated from the conducting external medium by the membrane. The equivalent electrical circuit of 4 adjoining cells is shown in Figure 1.13A. The equivalent circuit of each cell is represented by a parallel circuit of membrane capacitance, c_m, and membrane resistance, r_m. Each cell is connected to its neighboring cells by the intercellular resistance, r_i, equivalent to the lump sum of the resistance of the gap junction and the cytoplasmic resistance. The resistance of the extracellular space (which is provided by the relatively low resistance of extracellular solution bathing the cable) is shown as r_o. If a subthreshold current pulse is applied to such a system, the current will distribute along the cable according to the values of the resistors described above and produce changes in potential across the intracellular resistance. However, because the

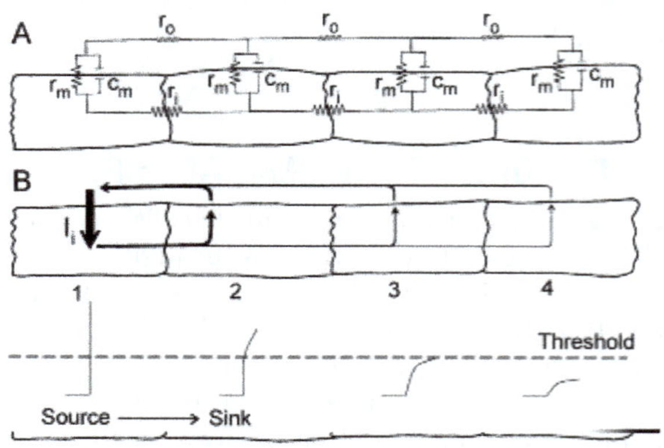

Figure 1.13. A: Equivalent electrical circuit of four-myocyte array: r_m, membrane resistance; c_m, membrane capacitance; r_o, extracellular resistance; r_i, intracellular resistance.
B: Distribution of currents during propagation of an action potential from depolarized cell (Source) to resting cells (Sink).

membrane is an imperfect insulator, some of the charge will leak out of the cell through r_m and cause a gradual decrease of the axial current flow, resulting in a decline of the potential. At steady state, the cable equation describes the distribution of voltage along the cable:

$$V_x = V_o e^{-x/\lambda},$$

where V_x is the potential at point x, V_o is the potential at the point of current injection, x is the distance from the point of current injection, and $\lambda = (r_m/(r_i+r_o))^{1/2}$ (where r values are normalized for unit length). The equation shows that the elicited electrotonic potential decays exponentially from a point of current injection. The space constant of the voltage decay, λ, is a useful parameter to characterize electrotonic interaction. At a given value of membrane resistance, λ is a measure of both intracellular and extracellular resistances in series and is affected by cell-to-cell coupling. It represents the distance over which the amplitude of an electrotonic potential is reduced to 1/e (~37%) of its peak value at the injection point. In cardiac tissue, λ varies from about 0.2 to 2 mm, indicating that the space constant of electrotonic interaction exceeds the length of a single cell. The cable equation describes the distribution of voltage along a continuous, passive, unidirectional fiber of infinite length. Therefore, application of the cable equations to cardiac tissue is a simplification of the real situation. The decay of the electrotonic potential is more complex in linear cables with sealed or open ends, and in two- and three-dimensional (3D) structures. However, the one-dimensional linear cable equation is a very useful model and, from its solution, a number of important conclusions can be made. It helps in understanding electrical propagation, well-describes the propagation in multicellular systems that are organized in bundles (such as Purkinje fibers), and can give a reasonable approximation in 3D structures (such as myocardial wall).

To analyze propagation of the action potential in cardiac muscle, the fixed membrane resistor should be replaced with a variable (nonlinear) resistor without which no action potentials can be generated. The value of this resistor is voltage- and time-dependent to allow the flow of inward ionic current through specific ion channels (I_i). In atrial myocardium, ventricular myocardium, and the specialized conduction system, this current is carried by Na^+ ions and in SAN and AVN cells by Ca^{2+} ions. In addition to I_i, the axial current is generated. Figure 1.13B illustrates schematically the action potential propagation in a uniform array of cardiac cells. As cell 1 depolarizes due to the flow of I_i, a voltage difference develops between this cell and its immediate neighbors, which generates a current spreading to the adjacent cells and causing depolarization of these cells by discharging their membrane capacitance. Thus, the action potential becomes a source of current, and the resting potential of cells downstream is the sink. The depolarizing current generated by the action potential propagates electrotonically and, in agreement with the cable theory, its amplitude decays exponentially with distance. In addition, cell capacitance introduces time dependence in response to depolarizing current, and it takes some time for an electrotonic potential to develop. The potential rises relatively rapidly close to the point of the action potential initiation, but more slowly at more distant sites (see Figure 1.13B, bottom). For cells located far away from the source, the electrotonic potential is subthreshold; however, for cell 2 the depolarization is sufficient to bring the membrane to its threshold potential and cell 2 then generates an action potential. Consequently, cell 2 becomes a source of current for cells located further downstream and the process repeats itself, propagating the action potential along the fiber bundle. Thus, the conduction of the action potential results from a combination of a passive spread of electrical charge along the cable and the active generator properties of the membrane (threshold phenomenon).

It is important to note that in the case of the fiber bundle, in contrast to a single myocyte where all inward current discharges that cell's capacitor, only a fraction of this current depolarizes the cell: a portion of the current flows along the cable and depolarizes the cells downstream. Nevertheless, the changes in the maximum upstroke velocity (V_{max}) of the action potential can still be taken as an indirect measure of changes in inward current flow. The theory of impulse propagation in a continuous conducting medium states that changes in ionic depolarizing current flow and changes in r_o and r_i have independent effects on conduction velocity, θ.[47] Formally, θ is proportional to the square root of V_{max} and, independently, inversely proportional to the square root of ($r_o + r_i$).[48]

The continuous electrical cable is a model that relates the resistive tissue properties and ionic channel function to propagation of excitation. In this model, cytoplasmic and gap junctional resistances are lumped together in a single intracellular resistive space forming an electrical continuum. The application of this theory is limited to electrically well-coupled cells, a small amount of connective tissue, and to propagation along the main axis of cardiac strands. In the presence of a large amount of connective tissue or electrical cell-to-cell uncoupling or during transverse conduction, the discontinuous nature of cardiac tissue becomes more prominent and limits the application of the continuous cable model.

CONCLUSIONS

The preceding has been an overview of a complex and detailed literature representing more than a century of discovery. In preparing this chapter, we often have resorted to simplification and generalization. Nonetheless, the examples provided and the principals stated do reflect the structure and function of the components determining the action potentials of cardiac cells. The composite activities of these cells provide the signals recorded via electrodes on a macroscale from cardiac tissues, and are routinely read during both noninvasive and invasive recordings of cardiac electrical activity.

REFERENCES

1. Baruscotti M, DiFrancesco D, Robinson RB. Na^+ current contribution to the diastolic depolarization in newborn rabbit SA node cells. *Am J Physiol Heart Circ Physiol.* 2000;279:H2303–H2309.
2. Catterall WA. From ionic currents to molecular mechanisms: the structure and function of voltage-gated sodium channels. *Neuron.* 2000;26:13–25.
3. Balser JR. Structure and function of the cardiac sodium channels. *Cardiovasc Res.* 1999;42:327–328.
4. Deschenes I, Armoundas AA, Jones SP, Tomaselli GF. Post-transcriptional gene silencing of KChIP2 and Nav[beta]1 in neonatal rat cardiac myocytes reveals a functional association between Na and Itc currents. *J Mol Cell Cardiol.* 2008;45:336–346.
5. Attwell D, Cohen I, Eisner D, Ohba M, Ojeda C. The steady state TTX-sensitive ("window") sodium current in cardiac Purkinje fibres. *Pfluegers Archiv Eur J Physiol.* 1979;379:137–142.
6. Richmond JE, Featherstone DE, Hartmann HA, Ruben PC. Slow inactivation in human cardiac sodium channels. *Biophys J.* 1998;74:2945–2952.
7. Carmeliet E. Slow inactivation of the sodium current in rabbit cardiac Purkinje fibres. *Pfluegers Archiv Eur J Physiol.* 1987;408:18–26.
8. Sheets MF, Scanley BE, Hanck DA, Makielski JC, Fozzard HA. Open sodium channel properties of single canine cardiac Purkinje cells. *Biophys J.* 1987;52:13–22.
9. Brown AM, Lee KS, Powell T. Sodium current in single rat heart muscle cells. *J Physiol.* 1981;318:479–500.
10. Bean BP. Two kinds of calcium channels in canine atrial cells. Differences in kinetics, selectivity, and pharmacology. *J Gen Physiol.* 1985;86:1–30.
11. Thomsen MB, Wang C, Ozgen N, et al. Accessory subunit KChIP2 modulates the cardiac L-type calcium current. *Circ Res.* 2009;104:1382–1389.
12. McDonald TF, Pelzer S, Trautwein W, Pelzer DJ. Regulation and modulation of calcium channels in cardiac, skeletal, and smooth muscle cells. *Physiol Rev.* 1994;74:365–507.
13. Hirano Y, Moscucci A, January CT. Direct measurement of L-type Ca^{2+} window current in heart cells. *Circ Res.* 1992;70:445–455.
14. Guia A, Stern MD, Lakatta EG, Josephson IR. Ion concentration-dependence of rat cardiac unitary L-Type calcium channel conductance. *Biophys J.* 2001;80:2742–2750.
15. Pelzer D, Pelzer S, McDonald T. Properties and regulation of calcium channels in muscle cells. *Rev Physiol Bioch P.* 1990;114:107–207.
16. Kass RS, Sanguinetti MC. Inactivation of calcium channel current in the calf cardiac Purkinje fiber. Evidence for voltage- and calcium-mediated mechanisms. *J Gen Physiol.* 1984;84:705–726.
17. Kim J, Ghosh S, Nunziato DA, Pitt GS. Identification of the components controlling inactivation of voltage-gated Ca^{2+} channels. *Neuron.* 2004;41:745–754.
18. Lipp P, Mechmann S, Pott L. Effects of calcium release from sarcoplasmic reticulum on membrane currents in guinea pig atrial cardioballs. *Pfluegers Archiv Eur J Physiol.* 1987;410:121–131.
19. Biggin PC, Roosild T, Choe S. Potassium channel structure: domain by domain. *Curr Opin Struc Biol.* 2000;10:456–461.
20. Volders PGA, Sipido KR, Carmeliet E, et al. Repolarizing K^+ Currents I_{to1} and I_{Ks} are larger in right than left canine ventricular midmyocardium. *Circulation.* 1999;99:206–210.
21. Rosati B, Pan Z, Lypen S, et al. Regulation of KChIP2 potassium channel beta subunit gene expression underlies the gradient of transient

outward current in canine and human ventricle. *J Physiol.* 2001;533:119–125.
22. Fedida D, Wible B, Wang Z, et al. Identity of a novel delayed rectifier current from human heart with a cloned K^+ channel current. *Circ Res.* 1993;73:210–216.
23. Wang Z, Fermini B, Nattel S. Sustained depolarization-induced outward current in human atrial myocytes. Evidence for a novel delayed rectifier K+ current similar to Kv1.5 cloned channel currents. *Circ Res.* 1993;73:1061–1076.
24. Yue L, Feng J, Li GR, Nattel S. Characterization of an ultrarapid delayed rectifier potassium channel involved in canine atrial repolarization. *J Physiol.* 1996;496:647–662.
25. Mazhari R, Greenstein JL, Winslow RL, Marban E, Nuss HB. Molecular interactions between two long-QT syndrome gene products, HERG and KCNE2, rationalized by in vitro and in silico analysis. *Circ Res.* 2001;89:33–38.
26. Tseng GN. I_{Kr}: The hERG channel. *J Mol Cell Cardiol.* 2001;33:835–849.
27. Smith PL, Baukrowitz T, Yellen G. The inward rectification mechanism of the HERG cardiac potassium channel. *Nature.* 1996;379:833–836.
28. Li GR, Feng J, Yue L, Carrier M, Nattel S. Evidence for two components of delayed rectifier K^+ current in human ventricular myocytes. *Circ Res.* 1996;78:689–696.
29. Lopatin AN, Nichols CG. Inward rectifiers in the heart: an update on I_{K1}. *J Mol Cell Cardiol.* 2001;33:625–638.
30. Hume JR, Uehara A. Ionic basis of the different action potential configurations of single guinea-pig atrial and ventricular myocytes. *J Physiol.* 1985;368:525–544.
31. Nerbonne JM, Kass RS. Molecular physiology of cardiac repolarization. *Physiol Rev.* 2005;85:1205–1253.
32. Isomoto S, Kurachi Y. Function, regulation, pharmacology, and molecular structure of ATP-sensitive K^+ channels in the cardiovascular system. *J Cardiovasc Electr.* 1997;8:1431–1446.
33. DiFrancesco D. Serious workings of the funny current. *Progr Biophy Mol Biol.* 2001;90:13–25.
34. Stergiopoulos K, Alvarado JL, Mastroianni M, et al. Hetero-domain interactions as a mechanism for the regulation of connexin channels. *Circ Res.* 1999;84:1144–1155.
35. Rice WJ, Young HS, Martin DW, Sachs JR, Stokes DL. Structure of Na^+, K^+-ATPase at 11-A resolution: comparison with Ca^{2+}-ATPase in E1 and E2 states. *Biophys J.* 2001;80:2187–2197.
36. Shigekawa M, Iwamoto T. Cardiac Na^+-Ca^{2+} exchange: molecular and pharmacological aspects. *Circ Res.* 2001;88:864–876.
37. Bleeker WK, Mackaay AJ, Masson-Pevet M, Bouman LN, Becker AE. Functional and morphological organization of the rabbit sinus node. *Circ Res.* 1980;46:11–22.
38. Coppen SR, Kodama I, Boyett MR, et al. Connexin45, a major connexin of the rabbit sinoatrial node, is co-expressed with Connexin43 in a restricted zone at the nodalûcrista terminalis border. *J Histochem Cytochem.* 1999;47:907–918.
39. Dobrzynski H, Boyett MR, Anderson RH. New insights into pacemaker activity: Promoting understanding of sick sinus syndrome. *Circulation.* 2007;115:1921–1932.
40. Guo J, Mitsuiye T, Noma A. The sustained inward current in sino-atrial node cells of guinea-pig heart. *Pfluegers Archiv Euro J Physiol.* 1997;433:390–396.
41. Shinagawa Y, Satoh H, Noma A. The sustained inward current and inward rectifier K^+ current in pacemaker cells dissociated from rat sinoatrial node. *J Physiol.* 2000;523:593–605.
42. Dobrzynski H, Boyett MR, Anderson RH. New insights into pacemaker activity: Promoting understanding of sick sinus syndrome. *Circulation.* 2007;115:1921–1932.
43. Maltsev VA, Lakatta EG. Dynamic interactions of an intracellular Ca^{2+} clock and membrane ion channel clock underlie robust initiation and regulation of cardiac pacemaker function. *Cardiovasc Res.* 2008;77:274–284.
44. Boyett MR, Honjo H, Kodama I. The sinoatrial node, a heterogeneous pacemaker structure. *Cardiovasc Res.* 2000;47:658–687.
45. Verkerk AO, Veldkamp MW, Abbate F, et al. Two types of action potential configuration in single cardiac Purkinje cells of sheep. *Am J Physiol - Heart CPhysiol.* 1999;277:H1299–H1310.
46. Fozzard HA. Conduction of the action potential. In: Berne RM, Geiger SR, Sperelakis N, eds. *Handbook of Physiology. Sect.2. The cardiovascular system.* Bethesda, MD: American Physiological Society; 1979:335–356.
47. Kleber AG, Janse MJ, Fast VG. Normal and abnormal conduction in the heart. In: Page E, Fozzard HA, Solaro J, eds. *Handbook of Physiology: The Cardiovascular System.* New York, NY: Oxford University Press; 2002:455–530.
48. Walton MK, Fozzard HA. The conducted action potential. Models and comparison to experiments. *Biophys J.* 1983;44:9–26.
49. Schram G, Pourrier M, Melnyk P, Nattel S. Differential distribution of cardiac ion channel expression as a basis for regional specialization in electrical function. *Circ Res.* 2002;90:939–950.
50. Catterall WA, Goldin AL, Waxman SG. International union of pharmacology. XlVII. Nomenclature and structure-function relationships of voltage-gated sodium channels. *Pharmacol Rev* 2005;57:397–409.
51. Catterall WA, Perez-Reyes E, Snutch TP, Striessnig J. International union of pharmacology. XlVIII. Nomenclature and structure-function relationships of voltage-gated calcium channels. *Pharmacol Rev* 2005;57:411–425.
52. Colatsky TJ. Voltage clamp measurements of sodium channel properties in rabbit cardiac Purkinje fibres. *J Physiol* 1980;305:215–234.
53. Li GR, Yang B, Feng J, et al. Transmembrane I_{Ca} contributes to rate-dependent changes of action potentials in human ventricular myocytes. *Am J Physiol Heart Circ Physiol.* 1999;276:H98–H106.

54. Van Wagoner DR, Nerbonne JM. Molecular basis of electrical remodeling in atrial fibrillation. *J Mol Cell Cardiol.* 2000;32:1101–1117.
55. Li D, Zhang L, Kneller J, Nattel S. Potential ionic mechanism for repolarization differences between canine right and left atrium. *Circ Res.* 2001;88:1168–1175.
56. Sosunov EA, Obreztchikova MN, Anyukhovsky EP, et al. Mechanisms of beta-adrenergic potentiation of ventricular arrhythmias in dogs with inherited arrhythmic sudden death. *Cardiovasc Res.* 2004;61:715–723.
57. Heidbuechel H, Vereecke J, Carmeliet E. The electrophysiological effects of acetylcholine in single human atrial cells. *J Mol Cell Cardiol.* 1987;19:1207–1219.
58. Shi W, Wymore R, Yu H, et al. Distribution and prevalence of hyperpolarization-activated cation channel (HCN) mRNA expression in cardiac tissues. *Circ Res.* 1999;85:e1–e6.
59. Accili EA, Robinson RB, DiFrancesco D. Properties and modulation of If in newborn versus adult cardiac SA node. *Am J Physiol Heart Circ Physiol.* 1997;272:H1549–H1552.
60. van der Velden HMW, Jongsma HJ. Cardiac gap junctions and connexins: Their role in atrial fibrillation and potential as therapeutic targets. *Cardiovasc Res.* 2002;54:270–279.
61. Desplantez T, Halliday D, Dupont E, Weingart R. Cardiac connexins Cx43 and Cx45: formation of diverse gap junction channels with diverse electrical properties. *Pfluegers Archiv Eur J Physiol.* 2004;448:363–375.
62. Horisberger J-D. Recent insights into the structure and mechanism of the sodium pump. *Physiology.* 2004;19:377–387.
63. Stimers JR. Cardiac Na^+/K^+ pump. In: Sperelakis N Kurachi Y, Terzic A, Cohen MV, eds. *Heart Physiology and Pathophysiology.* San Diego: Academic Press; 2001:407–416.
64. Ren X, Nicoll DA, Philipson KD. Helix packing of the cardiac Na^+-Ca^{2+} exchanger. *J Biol Chem* 2006;281:22808–22814.
65. Kimure J. Cardiac Na^+-Ca^{2+} exchanger: Pathophysiology and pharmacology. In: Sperelakis N Kurachi Y, Terzic A, Cohen MV, eds. *Heart Physiology and Pathophysiology.* San Diego: Academic Press; 2001:417–425.
66. Egan TM, Noble D, Noble SJ, et al. Sodium-calcium exchange during the action potential in guinea-pig ventricular cells. *J Physiol* 1989;411:639–661.
67. The Sicilian gambit. A new approach to the classification of antiarrhythmic drugs based on their actions on arrhythmogenic mechanisms. Task Force of the Working Group on Arrhythmias of the European Society of Cardiology. *Circulation.* 1991;84 1831–1851.

CHAPTER 2

Regional Anatomy for the Invasive Electrophysiologist

Paula Gonçalves Macedo, MD,
Nirusha Lachman, PhD,
Nandini Sehar, MBBS, and
Samuel J. Asirvatham, MD

INTRODUCTION

The extraordinary advances over the last 2 decades with the invasive management of cardiac arrhythmia can be traced to the years of understanding the anatomy and electrophysiology (EP) of arrhythmogenic substrates.

The contemporary invasive electrophysiologist needs to be just as conversant with the knowledge of regional anatomy and arrhythmia correlation. This is necessary not only to understand the rationale for existing and future ablation approaches, but also important in maximizing the safety by avoiding inadvertent damage to surrounding structures.

In this chapter, we review the major clinically relevant arrhythmia (atrial fibrillation [AF], atrial flutter [AFl], and ventricular tachycardia [VT]) and sequentially discuss the major anatomic principles that guide present day ablation for these disorders. For each of the principle arrhythmia categories, a discussion of the regional anatomy for the sites that house the primary ablation targets followed by a brief discussion of the correlative value in detailed understanding of these structures to the invasive electrophysiologist follows.

ATRIAL FIBRILLATION

While the circuit of the typical AFl is anatomically well recognized, the substrate and triggers of AF are not fully understood. Circumferential ablation around the pulmonary veins (PVs) is an established management option for patients with symptomatic AF. However, other thoracic veins and the retroatrial structures are also involved in the pathogenesis and sometimes should be targeted to achieve a successful ablation. The importance of knowing the regional anatomy of the PVs is to accurately diagnose the origin of the electrograms (EGMs), which enables the electrophysiologist to perform the ablation with higher precision and safety. The other thoracic veins potentially involved in the AF mechanism are the superior vena cava (SVC) and the

Practical Electrophysiology © 2015 Jasbir S. Sra, Masood Akhtar, Andrea Natale, David J. Wilber
Cardiotext Publishing, ISBN: 978-1-935395-17-1

vein of Marshall (VOM). Both structures have complex anatomic relationships, and improving the knowledge of these boundaries will allow accurate mapping and completeness of ablation. In addition, the left atrium (LA) and thoracic veins are related to important and vital structures that should be recognized in order to avoid serious complications. In turn, the relevant anatomy of the PVs, SVC, VOM, and the LA are detailed.

The Pulmonary Veins

Gross anatomy and relations: In most human hearts, there are 4 veins draining into the LA.[1] A relatively frequent anatomic variation is the presence of an accessory middle right PV or a common trunk for the left PVs (Figure 2.1).[2] In normal hearts, the mean distance between the left PVs is 8.4 mm and between the right PVs is 6.5 mm. The orifices of the right PVs are adjacent to the interatrial septum, while the left-sided PVs are just posterior to the LAA.[1] The VOM lies in between the LAA ostium and the left PVs, an area also rich in autonomic nerve fibers.[3]

The right superior pulmonary vein (RSPV) courses immediately posterior to the junction of the SVC with the right atrium (RA). The azygous vein and a branch of the right pulmonary artery (PA) are located superior and posterior to the right upper PV. The RIPV is the only area where a branch of the right PA courses anteriorly to the vein, which separates it from the myocardial structures.[4]

Importance of the PV ostium and myocardial sleeves: For AF ablation, the PV ostium has a crucial role. The heterogeneous histology in and around the vein ostium is mainly responsible for AF arrhythmogenesis, and why it is the principal target for ablation. The anatomical recognition of the PV ostium is also important to avoid radiofrequency (RF) applications deep within the veins, which can lead to PV stenosis. However, with all the imaging tools available, it is still difficult to determine their exact location. Even in macroscopic histological examinations, there are no landmarks demarcating the transition from vein to atrium. Microscopically, syncytial atrial myocardium extends into the veins in a way to form a sphincter-like structure with the function of preventing venous regurgitation.[5] Close to the ostium, the myocardium is thicker and ablation in this site leads to thinning of the myocardial layers and aneurismal dilation. Deep in the vein, however, the myocardium is very thin or absent and RF applications cause endothelial proliferation and luminal narrowing. In turn, the electrophysiologist should target the myocardium proximal to the ostium to avoid PV stenosis.[4] The use of preprocedural computed tomography (CT) scans or magnetic resonance imaging can help in determining the approximate location where the PVs join to the LA. During ablation, an intracardiac ultrasound, rotational angiography, retrograde pulmonary venography, or electroanatomic mapping systems can also be used.[6]

Myocardial sleeves are extensions of atrial muscle fibers into the PVs. The longest sleeves are usually seen in the superior PVs (10 mm on average comparing to 5 mm in the inferior PVs).[1,5] Although the fibers are most circularly oriented in the transmural extension, there are also fascicles of longitudinal and oblique fibers that interdigitate between the circular bundles. This change in orientation can produce nonuniform anisotropy, which can then function as a substrate for micro-reentry or automatic foci.[7]

Recently, node-like cells were described in the PVs of patients with AF. They were called P cells (pale cells) and had characteristics of automaticity.[8] Also, Cajal cells and PAS-positive cells were encountered in the PVs, especially in AF patients.[9,10] Although it looks like it is a promising mechanistic explanation, there is no sufficient data to affirm that these cells trigger AF.

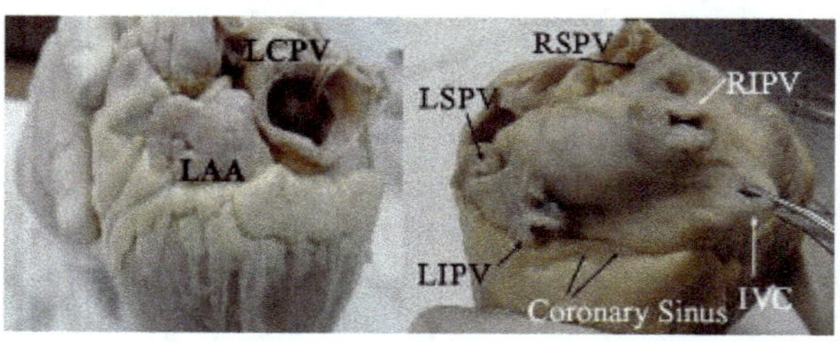

Figure 2.1. **Left panel:** Left lateral view of an explanted heart showing a large left common pulmonary vein (LCPV). The posterior wall of the left atrial appendage (LAA) is in close relation with the anterior wall of this vein, which can generate contaminant signals when recording from the left-sided PV. **Right panel:** Posterior view of an explanted heart. The RSPV is related to the posterior wall of the RA and the left veins are close to the coronary sinus (CS) and LAA. As the right inferior pulmonary vein (RIPV) has no close anatomical relations, it is rare to record far-field potentials in this location. (Reproduced from Macedo et al[4] with permission from John Wiley and Sons.)

The Superior Vena Cava

The presence of myocardial sleeves in the SVC has been related to non-PV triggers of AF. In addition, automatic atrial tachycardias (ATs) and reentrant circuits (upper loop reentry) were also described in this site.[11–13] Knowledge of the anatomic relationships allows the ablationist to better interpret the EGMs, avoid important surrounding structures, and perform a successful and safe procedure.

Gross anatomy and relations: The SVC is formed by the confluence of the right and left brachiocephalic veins, which lie just below the junction of the right first rib to the sternum.[14] Its origin landmark in imaging studies is the right tracheobronchial angle, which is always caudal 0.1 to 3.8 cm.[15] From the origin to the junction with the RA, the SVC measures 6 to 8 cm and courses with a slight right convexity.[4] The azygous vein enters into the posteromedial region of the SVC at variable distances from the SVC/RA junction. The lower part of the SVC is involved by the pericardium.

Anteriorly to the SVC are the anterior margins of the right lung and pleural, and posteriorly is also the right lung, the vagus nerve, and more inferiorly the RSPV. Laterally, there is the right phrenic nerve, which courses posteroinferiorly to the vein in the direction of the RA, in between the RSPV and the SVC.[16] On the left lies the origin of the brachiocephalic artery and the ascending aorta (AO). Among the SVC, the AO, and the trachea, there is a reflection of pericardium called the aortocaval reflex. The floor of this reflex is formed by the right PA.[14]

In contrast to the inferior vena cava (IVC) wall, the SVC is endowed with muscular extensions from the RA, similar to what occurs within the CS and the PV.[4,17] The muscle fibers are more circularly oriented at the ostium, which can be responsible for a delayed conduction into the vein. This region also has a rich parasympathetic plexus (see below).

The sinus node is an epicardial structure located posterolaterally at the junction of the SVC to the RA in the region of the sulcus terminalis. The nodal cells can be fanned out, occasionally reaching the adjacent veins.[14]

Anatomy-electrophysiology correlations: The SVC-related mechanism of AF triggering and sustaining is not completely understood. It was already demonstrated that the muscular sleeves in this vein can trigger cardiac activation and be the source of AF and other atrial arrhythmias.[11–13,18] The SVC musculature may also be responsible for sustaining AF, as rapid firing activity has been described.[19] Both phase IV depolarization-initiated automatic activity and abnormal automaticity have been recorded inside this vein, which could yield AF.[20] Occasionally, the muscular sleeves in the SVC may connect not only to the RA but also to the RSPV, making up a potential connection to the LA.

Because of the draining azygous vein, the adjacent RSPV, and the RA itself, the EGM in this region can display contaminant signals. However, pacing from the RA and then from the SVC can differentiate the PV signal and allow safe ablation in the SVC/RA junction or within the RA.[4]

Phrenic nerve injury is possible not only with RSPV ablation but also while targeting SVC isolation. The nerve may be spared with continuous pacing from the distal mapping/ablation electrode at 5 V output while performing cryoablation.[4,21]

Furthermore, extensive ablation at the RA close to the SVC junction may damage parts of the sinus node. If venous isolation requires many RF applications at the RA, close monitoring of the sinus rhythm should be done watching for any decrease in automaticity.[4]

In the aortocaval reflection described above, there is an important structure of the autonomic innervation of the heart: the aortocaval ganglion. As it comprises the autonomic tissue of the surrounding AO, right PVs, and PAs, ablation at this site can alter atrial repolarization and conduction times. Thus, ablating these ganglia can modulate atrial arrhythmias beyond SVC musculature isolation.[22,23]

The Vein of Marshall

The VOM is a remnant of the left SVC, and it is located between the ostium of the LAA and the left-sided PVs. It is of great importance to the electrophysiologist because it can be the source of AF triggers or a bridge from the PVs to the LA after conventional PV isolation. In addition, it can serve as a vantage point to access and ablate the retroatrial ganglionated plexuses in this location.[4]

The atrial myocardium drains into the CS through various veins, VOM being the most consistent. It runs obliquely in the posterior wall of the LA to join the great cardiac vein and the main posterior lateral vein to form the CS (Figure 2.2).[24] Similar to the SVC, it is possible to encounter nodal tissue and autonomic innervation in the VOM. Atrial muscle fibers are found in the CS and at the ostium of the draining atrial (VOM) and ventricular (V) veins.[4]

Anatomy-electrophysiology correlation: The VOM can be a trigger site for initiating AF. When re-initiation of AF occurs after PV and possibly SVC isolation, origin in the VOM should be considered. Ablation approaches include energy delivery within the vein, isolation at the ostium of this vein at its junction with the CS, and endocardial transmural ablation, or epicardial ablation via subxiphoid access can be done.[4,25,26] The VOM may possibly also act as a bridge electrically connecting the CS, LA, and epicardial surface of the PV.

The VOM can also be involved in a macro-reentrant LA flutter circuit or inter-AFl circuit. It can be suspect when there is still electrical

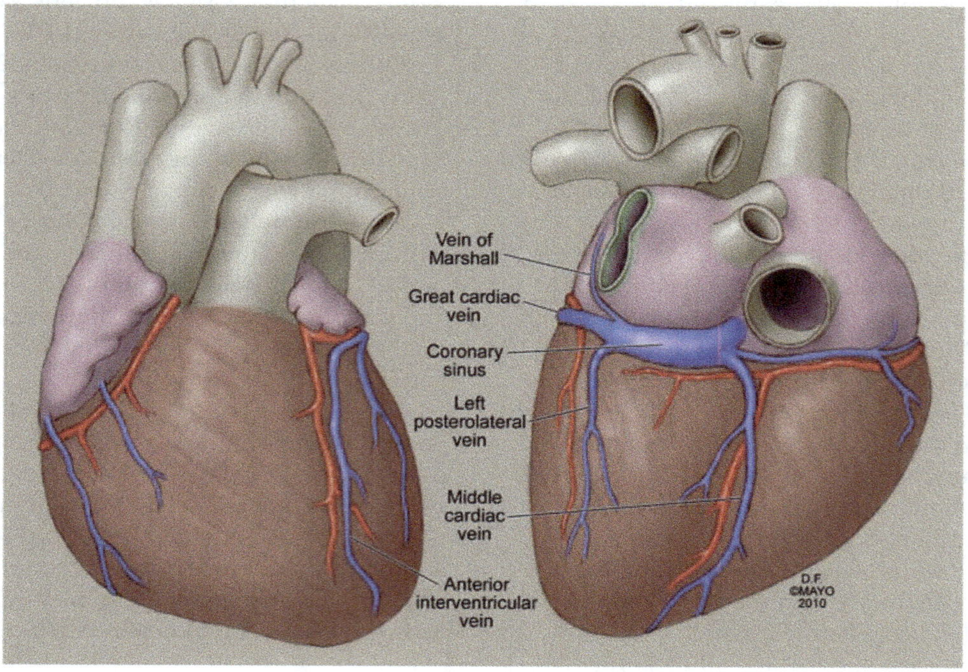

Figure 2.2. Coronary venous system. Note the course of the VOM obliquely in the posterior wall of the LA. It joins the great cardiac vein and the left posterolateral vein to form the CS. (Reproduced from Lachman et al[95] with permission from John Wiley and Sons.)

conduction after a surgical maze procedure or after linear ablation between the left inferior PV and the mitral annulus. In such cases, electrical conduction may occur from the lateral wall of the LA to the CS via the VOM and then back to the medial LA wall or to the RA completing the circuit. Ablation of either the VOM or the CS can terminate these macro-reentrant circuits.[4]

Ablation of the VOM can be performed either directly via vein cannulation or through the atrial endocardium.[27] As the endocardial ridge separating the LAA and the left PVs is a recognizable internal mark of the VOM, RF applications at this point can successfully ablate this vein without the need of cannulating the vein.[4]

Occasionally, it is necessary to ablate the retroatrial ganglia close to the VOM. Here, this vein can allow easy access to the epicardial ganglia, which may be targeted for ablation or modulation.

The Left Atrium

Knowledge of the LA regional anatomy is essential for avoiding collateral damage, as this is the most posterior chamber of the heart and is in close proximity to important structures.

The LA is posterior and superior to the RA. Posteriorly, the LA is related to the tracheal bifurcation and to the esophagus, from which it is separated only by the fibrous pericardium.[28] Anteriorly, this chamber accommodates the aortic root and posteroinferiorly it is connected to the CS. The LA can be divided in 4 regions: the septum, the vestibule, the appendage, and the venous component.[1] The septum is recognizable by the flap valve of the foramen ovale. The vestibule is the region around the mitral annulus. The LAA is small and tubular, and is the only component of the LA where pectinate muscles can be found. The venous component is the largest part of the LA and receives the PVs.

Anatomy-electrophysiology correlation: The atrial myocardium thickness varies considerably from region to region in the LA. Importantly, the atrial roof may be so thin that it can be transilluminated. Therefore, catheter manipulation must be done carefully in this area in order to avoid cardiac perforation. In addition, whenever ablating in the atrial roof, RF application should only be performed if a contact near-field EGM is displayed and should be avoided where EGMs are absent or no longer present.[28]

In contrast, the thickness of atrial myocardium can make ablation challenging in different regions, such as the mitral isthmus, the Bachmann's bundle, and the endocardial ridge between the LAA ostium and the left-sided PVs.[4]

The mitral isthmus consists of the atrial myocardium separating the mitral valve and the left PVs, and it can participate in atypical AFl circuits.[7] Linear ablation across this isthmus in the level of the left inferior PV is the typical approach. This ablation is best done after ostial isolation of the PV that serves as an anchoring point.

The primary site of interatrial conduction is the Bachmann's bundle, which is located in the anterior portion of the LA roof. It is composed of traversing fibers from the right atrial appendage (RAA) to the LAA.[29] Ablation is this site is currently used to terminate inter-AFls occurring after surgical maze procedure. In such circuits, the Bachmann's bundle connects both atria, and the return limb is through the CS muscular sleeves.[28]

Important Anatomic Relations to Consider in Order to Avoid the Collateral Damage

With the increasing number of AF ablations, collateral damage has also become more frequent. Since the LA and the thoracic veins are adjacent to important cardiac and noncardiac structures, knowing the regional anatomy is essential for avoiding inadvertent injury of surrounding tissue. The occurrence and methods of preventing PV stenosis has been described above; therefore, in this section the esophageal trauma and coronary artery (CA) damage will be detailed.

Esophageal injury: The esophagus lies immediately behind the posterior wall of the LA, separating only the oblique sinus and a variable amount of connective tissue. The oblique sinus is a pericardial reflection and has been used to monitor and prevent esophageal trauma during ablation.[28]

The occurrence of atrial-esophageal fistula is a rare but catastrophic collateral damage of ablation in the posterior wall of the LA.[30,31] Since this region has a thin and sometimes absent myocardial layer, careful RF application should be attempted. Also to avoid this complication, monitoring the luminal esophageal temperature is used. However, thermal injury may occur even before temperature rise on the luminal probe.[32,33] Therefore, also direct visualization of the esophagus through intracardiac ultrasound is used to allow early detection of any echo density formation or effusion. Nevertheless, so far there is no endpoint for temperature increase, duration of energy delivery, or power setting as a recommendation for different atrial regions.

In addition, the placement of a balloon in the oblique sinus has been attempted to move the esophagus away from the LA posterior wall.[33] Also, mechanical deflection of the esophagus using an echocardiography probe has been tried in few patients with successful results.[34]

Coronary artery damage: CA occlusion can occur not only during AF ablation but also while ablating epicardial accessory pathways (APs) and outflow tract VTs.[35,36] Nonetheless, this section will address the anatomic relations important to CA collateral damage when ablation of AF is attempted.

While targeting the mitral isthmus, a potential injury to the circumflex CA may occur (Figure 2.3). This artery is typically located more V than the CS in this region.[24,37] Therefore, if higher energy delivery is necessary, either from the endocardium or the epicardium, one should avoid the V portion of the mitral annulus.

Following isolation of the PVs, automatic ATs arising from either the aortic cusp and/or the mitral annulus can occur.[38,39] In this region, the distal left main CA (LMCA) and proximal circumflex CA are within 5 to 10 mm from the aortic-mitral continuity. Judicious energy delivery should be done just sufficient to terminate tachycardia and arrhythmogenic atrial substrate.

Occasionally, the myocardium of the CS is targeted for ablation. This includes ablation of the VOM or of an automatic foci in the CS itself or the interruption of a macro-reentrant inter-AFl (either connections with the LA or with the RA).[4] A posterolateral branch of the RCA can be injured while ablation of the proximal CS is attempted. Otherwise, the main circumflex CA can be damaged

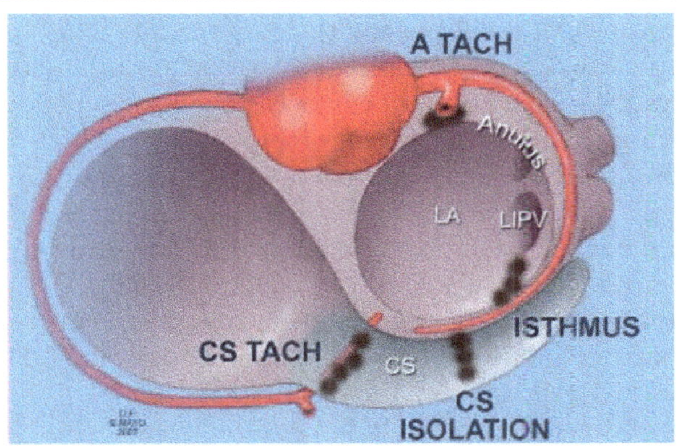

Figure 2.3. Potential sites for coronary damage during AF ablation. Damage to the circumflex artery may occur while attempting ablation between the LIPV and the mitral valve or while targeting the CS near the VOM. Ablation in the proximal CS may damage a distal branch of the right coronary artery (RCA). Following PV isolation, one may encounter a tachycardia focus arising from the septal mitral annulus. Ablation in this site may lesion the proximal circumflex and left anterior descending arteries (LAD). (Reproduced from Macedo et al[28] with permission from John Wiley and Sons.)

when ablating deeper in the CS, especially if closer to the V portion of the annulus.[28]

ATRIAL FLUTTER

The AFl was identified as an intra-atrial circuit around the vena cava in 1921 by Sir Thomas Lewis et al.[40] Forty years ago, Puech et al discriminated the typical and rare types of AFl and described that the typical AFl was the result of an activation proceeding cranially along the interatrial septum and caudally along the free wall of the RA in a large reentrant circuit (RA).[41] In 1990, an area of slow conduction in the posteroinferior region of the RA was associated with the presence of AFl.[42] The so-called cavotricuspid isthmus (CTI) was later characterized as the atrial myocardium between the hinge point of the tricuspid valve (TCV) and the IVC and was recognized as the critical isthmus for typical AFl.[43,44] The RF delivery across the CTI with the creation of a bidirectional conduction block is currently the endpoint criteria for typical AFl ablation and has been correlated with high rates of success in terminating the arrhythmia and preventing recurrences.[44,45] In this section, the normal CTI and its boundaries, as well as anatomic variations, will be described and correlated with the difficulties possibly encountered during AFl ablation.

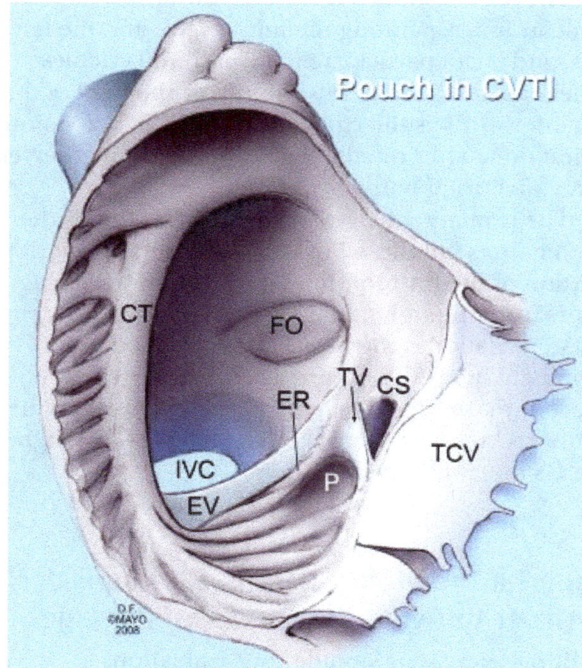

Figure 2.4. Illustration showing that the CTI is not a flat structure. Encroaching pectinate muscles, a sub-Eustachian pouch, and the Eustachian valve (EV) provide "hills" and "valleys" to the CTI surface. Here a deep pouch is displayed, which usually is associated with a prominent Thebesian valve (TV). CT, crista terminalis; CS, coronary sinus; ER, Eustachian ridge; FO, fossa ovalis; IVC, inferior vena cava; TCV, tricuspid valve. (Reproduced from Asirvatham SJ.[46])

Normal Cavotricuspid Isthmus

The CTI is an isthmus bounded anteriorly by the tricuspid annulus (TA) and posteroinferiorly by the transition of the RA to the IVC (Figure 2.4). The myocardial fibers in this region are usually oriented parallel to the TCV, but can occasionally be septally arranged, in which case they merge with the CS.[46] The CTI is broader and thicker in its lateral portion and is thinner in the center; however, the thickness in the inferolateral portion is highly influenced by the presence and size of traversing pectinate muscles.[47,48] In an anatomic study of 602 human hearts, it was found that the mean length of the CTI was 17.7 mm and that the pectinate muscles extended into the CTI in 70% of the hearts.[49] In such cases, they tend to be more laterally located. In turn, the atrial myocardium close to the TA is usually smooth, and this part of the CTI is called vestibular portion.

Otherwise, the topography of the CTI is hardly smooth or flat. The ER is an elevation of variable heights which divides the isthmus into 2 portions: one from the TA to the ER (sub-Eustachian isthmus) and the other between the crest of the ER descending to the IVC.[46]

In the sub-Eustachian isthmus, the CTI is in close proximity to the CS, and in 62% of the hearts a TV is found partially or completely bordering the CS.[49] When a prominent TV is encountered, a sub-Eustachian pouch is frequently present. Importantly to the electrophysiologist, a prominent TV can make CS catheterization difficult, and the visualization of a pouch at angiography can give a clue to the presence of a TV.[49]

On the posterior interatrial septum, the CS separates the CTI from the compact atrioventricular (AV) node, and in fact, the septal part of the vestibular CTI lies in close proximity to the slow pathway input to the AV node.[46] External to the CTI near the annulus are the small cardiac veins and the distal branches of the RCA, which in turn are relatively close to the septal portion of the CTI.[48] In a CT-based study, Saremi et al found that the mean distances from the RCA and from the nodal artery to the RA wall were 3.2 mm and 2.1 mm, respectively.[50]

The portion between the crest of the ER and the IVC is relatively unruffled, and no important anatomic variations have been described. Nevertheless, the ER itself is occasionally pronounced and has been related to resistant gaps during AFl ablation.[51]

Considering those anatomic landmarks, the CTI ablation line should be performed in the mid portion (6 o'clock as viewed in the left anterior oblique projection). As it is a region where the myocardium tends to be thinner, the AV node is spared compared to a more septal ablation, and the isthmus is not so long and thick as it is in the inferolateral region.[46]

Causes for Difficulty with Atrial Flutter Ablation

RF ablation of the CTI with demonstrated bidirectional conductional block is successful in more than 90% of the cases.[49,52] When it is not straightforward, the use of large-tipped or irrigated catheters and increasing maximum power delivery is usually sufficient to completely ablate the isthmus.[52,53] Rarely, resistant "gaps" are seen and recurrent AFl may occur. The potential causes are poor catheter contact, inadequate power delivery, thick myocardium, and discontinuous ablation.[46] The anatomic variations that contribute to these difficulties are a deep sub-Eustachian pouch, large pectinate muscles trespassing upon the CTI, and a prominent ER.

Sub-Eustachian Pouch: In autopsied human hearts, a distinct sub-Eustachian pouch was found in 16% of the hearts.[49] They were always bound by the EV, the TV, and the TA, with an average depth of 6.1 mm. In a study using CT, a pouch was seen in 41% of the patients.[50]

When prominent, the pouch can be difficult to ablate. As a result of poor blood flow, power delivery is suboptimal, resulting in quick temperature and impedance rise and potential coagulum formation.[46] The presence of a sub-Eustachian pouch has been related to longer ablation times and more RF applications.[54–56] It can easily be visualized in angiographic studies, intracardiac ultrasound, CT scan, or in three-dimensional (3D) electroanatomic mappings, which enable the electrophysiologist to perform maneuvers to avoid the deep pouch, which can also be detected during an ablation if there is unusual catheter movement at the tip with catheter instability.

In the autopsy study described above, a sub-Eustachian pouch was never found in the lateral portion of the CTI.[49] Therefore, in the presence of a prominent pouch, the CTI ablation line should be performed laterally in order to avoid power delivery problems and potential perforation within the pouch.[46]

Pectinate muscles encroaching onto the isthmus: The extension of pectinate muscles from the *crista terminalis* into the lateral portion of the CTI has been described in 2005, but only recently their frequency and real length were detailed.[49,57] It was found that in 70% of the hearts the pectinates encroached onto the CTI, comprising a mean of 23% of the CTI total length.[49] In 7% of the cases, they even crossed into the CS, which was correlated with an absent or vestigial TV. Furthermore, the height of the pectinate muscles was always more prominent in the lateral portion compared to the height in the septal CTI.

During ablation, the presence of encroaching pectinate muscles can be suspected when high-amplitude atrial EGMs surrounded by areas where impedance rises and minimal power delivery (secondary to poor blood circulation) occur. In such cases, ablation may be challenging for several reasons: (1) the thickness of the pectinates may impede the creation of a transmural lesion; (2) in between muscles, inadequate power delivery with impedance increase and coagulum formation may occur because of poor blood flow; (3) catheter instability on a "hilly" topography (Figure 2.5).[46]

In this situation, one should try to approach the isthmus ablation more medially, close to the septum, where the presence of pectinate muscles is less likely.

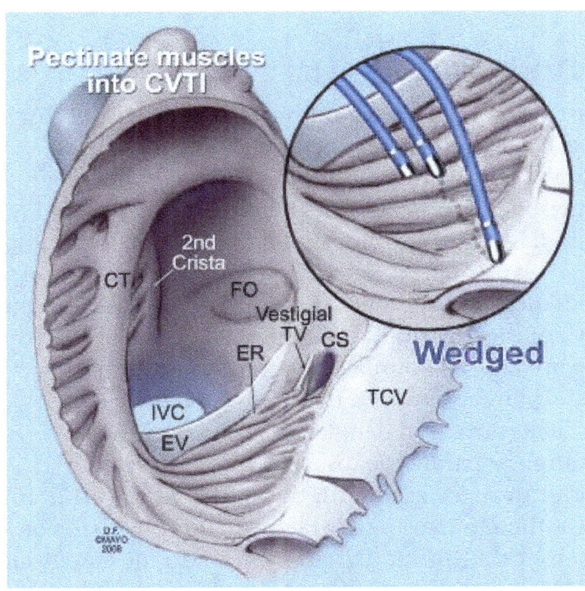

Figure 2.5. Illustration of encroaching pectinate muscles in the CTI and consequent ablation challenges. Thickness of the pectinate muscle may impede transmural ablation while the catheter may be wedged between 2 pectinate muscles, causing rapid impedance rise and local tissue heating. Abbreviations as in Figure 2.4. (Reproduced from Asirvatham SJ[45] with permission from John Wiley and Sons.)

In addition, the use of irrigated catheters can better tailor that ruffled surface, allowing enhanced power delivery on the thick muscles and increasing the maximum energy that can be delivered in the "valleys."[46]

Prominent Eustachian ridge: Various studies have described a prominent ER in 24% to 30% of human hearts.[50,51,57] This finding was related to gaps in the ablation line requiring greater RF applications to achieve a complete bidirectional block on the CTI.[51] However, the ER itself is largely made of fibrous tissue and not much ablation is needed directly on this structure.[46] The main cause of difficulty, in fact, is related to catheter manipulation and contact.

During a usual AFl ablation, a clockwise torque will bring the ablation catheter to the septum. In patients with a prominent ER, the catheter will turn in the ridge crest and move to the lateral annulus instead. Thus, the ER will function as the sole contact point. To an experienced electrophysiologist, this is the clue for a prominent ER, which can be confirmed by intracardiac ultrasound imaging.

Another challenge is to maintain contact on the upslope of the ridge, between the ER crest and the TA. In order to solve both the inability to torque the catheter on the septum and the poor contact on the upslope of the ridge, the ablationist should use an appropriate guiding sheath.[46]

AUTONOMIC GANGLIA

Cardiac ganglia are known to exist throughout the epicardial fat. Primary locations for the cardiac ganglia may be divided on the basis of either their functional significance or their anatomic location. Ganglia are typically located in 3 main pericardiac regions: retroatrial (ie, along the posterior portion of the left and right atria), annular V (ie, around the area of the junction between atria and ventricles in the area of the annular apparatus), and peri-great vessel ganglia (around the great vessels).

The retroatrial ganglia are also comprised of multiple separate plexuses distributed over the posterior surfaces of both the right and left atria. In turn, these plexuses may be divided on the basis of whether they primarily supply input to the sinoatrial or AV nodes, with the sinoatrial ganglia mainly consisting of those inferior and posterior to the sinoatrial node, extending from the right PVs to the sulcus terminalis, and the AV ganglia mainly consisting of those in and near the interatrial groove and at the coronary sulcus junction (Figure 2.6).

The mixed autonomic cardiac plexus receives its major contributions from the vagus nerve and

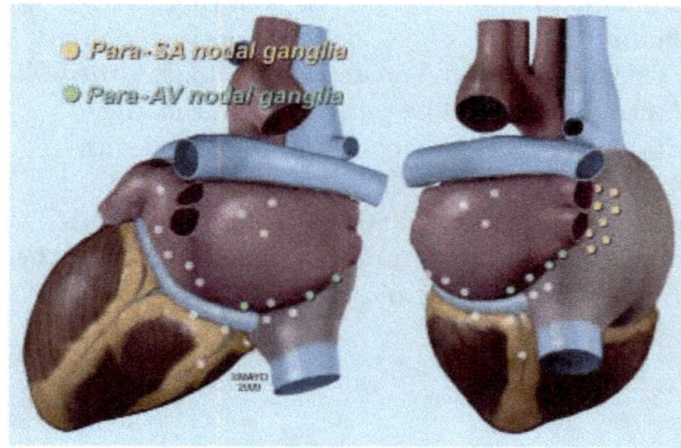

Figure 2.6. Illustration of sites of the retroatrial ganglia. The para-SA nodal ganglia extend from the right PVs to the sulcus terminalis. The para-AV nodal ganglia consist of plexuses in and around the interatrial groove and at the coronary sulcus. (Reproduced from Lachman et al[95] with permission from John Wiley and Sons.)

superior cervical ganglion. Anatomically, the cardiac plexus is described in terms of its location as deep and superficial. The deep plexus contains contributions from the cervical and upper thoracic sympathetic ganglia and vagal recurrent laryngeal nerves while the superficial plexus contains the cardiac branch of the left superior cervical ganglion, the lower of the 2 cervical cardiac branches of the left vagus and a small cardiac ganglion to the right of the ligamentum arteriosum. Ganglion cells are distributed to the heart along branches of the plexus.

Detailed descriptions of the intrinsic cardiac ganglia have not been extensively published and are often limited to early stages of heart development or present discrepancies.[58] In a study conducted by Singh et al,[58] cardiac ganglia were identified primarily in the epicardial layer within a bed of adipose tissue, more infrequently within the myocardium but never endo- or sub-endocardially. Typically, cardiac ganglia have been reported to be closely associated with cardiac fat pads. However, smaller collections are known to be distributed across the atrial and V regions, independent of fat pads.[58]

Atrial ganglia may be located over an extensive area of myocardium. Although there is great variability in the location of smaller collections, the largest aggregates of atrial ganglia appear to be adjacent to the nodal cardiac tissue:

- *Para SA nodal ganglia*—extend from right PVs to sulcus terminalis (and sometimes as far as the atrial-auricular junction).
- *Para AV nodal ganglia*—epicardial surface surrounding the node near interatrial groove and coronary sulcus junction.

Regions devoid of ganglia include the free wall of the RA, lateral to sulcus terminalis, anterior and superior surfaces of PVs, and the atrial appendages.

These ganglia appear to lie around the base of the AO and PAs, close to their V junction. The venae cavae have moderate aggregations of ganglia on their posterior surfaces with the more definite collections around the atrial junction of the SVC.

One of the difficulties in studying the significance of these ganglia in electrophysiologic processes has been the complex interactions between them. This has been suggested in numerous studies which have shown that there is extensive unidirectional and bidirectional feedback between ganglia.[59,60] Furthermore, the input to any specific cardiovascular structure may depend on which group of ganglia is considered. For example, those located in the vicinity of the SVC/atrial junction, when transected, may interrupt autonomic innervation to the sinoatrial node but not to the AV node, which is more affected by ganglia located in the vicinity of the IVC and LA.[61] Given the extensive anatomic understanding of the location of these ganglia and their relationship to the pericardial recesses, it would be reasonable to assume that ganglia identification and ablation could be done on a purely anatomic basis. However, similar to the complex fractionated EGMs used endocardially to differentiate sites where ganglia are located, it would also be reasonable to assume that their differing electrophysiologic properties may be even more apparent when mapped directly from within the pericardial space.

Electrophysiology Correlation

The role of the autonomic tone in arrhythmia diseases is being more and more studied, as it appears to be a promising target for treatment. Considerable information has been gathered regarding autonomics in AF, V arrhythmias, and channelopathies.

In order to understand the relationship of AF and the autonomic cardiac innervation, it is important to highlight the regional anatomy of the PVs. In this context, PV ostia comprise a high density of nerves compared to the rest of the LA. Thus, the success of circumferential PV ablation could also be attributed to disruption of the vein-neural interface.[62–64] Actually, some authors advocate that ablation of these ganglia may treat AF, either as a single treatment or coupled with PV isolation.[65,66] Although this approach appears to reduce AF inducibility, it is still unclear how these ganglia can be accurately identified. The surrogate markers used most often for the presence of ganglionated plexus are the complex fractionated EGMs and a parasympathetic response to high-frequency electrical stimulation.[63,67,68] The former was also correlated to proximity to the PV; however, neither reproducibility of these EGMs nor their relationship to autonomic activity has been clarified. Regarding the induction of a vagal response, both sympathetic and parasympathetic responses would be expected in these mixed ganglia; nevertheless, the latter has typically a faster reaction. In summary, the modulation of the PV autonomic innervation probably has an important role in treating AF, although the best method to perform it has not been revealed yet.

The interaction of autonomic activity and V arrhythmias is well-studied. An unopposed sympathetic activation may predispose to VT and sudden cardiac death. The high incidence of sudden death during sleep has been related to surges of sympathetic activity during rapid eye movement sleep and apneic events.[69,70] Also the presence of CA disease, myocardial infarction, and diabetes was associated with decreased parasympathetic activity at sleep, which could play a role in the high prevalence of nocturnal sudden cardiac death in this population.[71] Another theory to explain the occurrence of V arrhythmias in patient with myocardial infarction is the regional denervation and nerve sprouting in the affected myocardium. These could lead to a sympathetic unbalance between the healthy and ischemic tissues, increasing the susceptibility to premature V contraction-initiated arrhythmias.[72,73] Although sympathetic hyperactivity has been consistently related to arrhythmogenesis, therapeutic approaches as the cervical sympathectomy are controversial.

In contrast, left cardiac sympathetic denervation has been successfully used to reduce sudden death in patients with high risk long QT syndrome. Although there is still a high rate of cardiac events in the long term, this therapy appears to be useful in patients with recurrent syncope despite beta-blockage in the setting of electrical storms with implantable cardiac defibrillators.[74,75]

ENDOCAVITARY STRUCTURES

The term “endocavitary structure” (ECS) is used to describe cardiac structures typically within the V chambers and includes: (1) papillary muscles; (2) moderator band; and (3) false tendons. Although these structures are part of the normal anatomical makeup, myocardial tissue extending into the V cavity results in ECS becoming electrically active, resulting in V arrhythmias (Figure 2.7).[7]

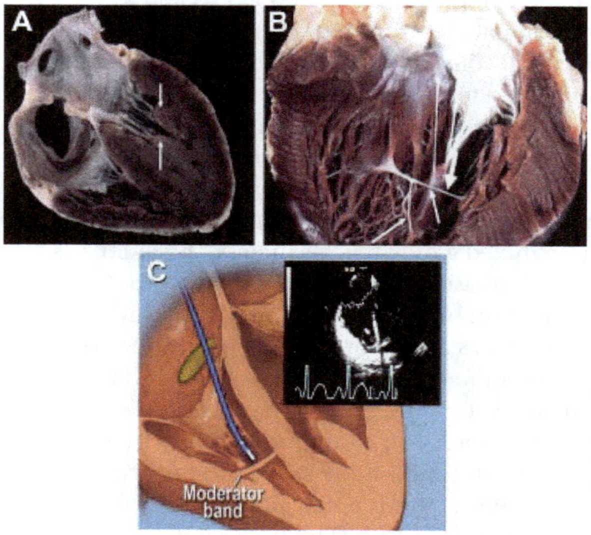

Figure 2.7. ECSs. **A:** Autopsied heart showing the papillary muscles as shown by arrows. **B:** False tendons are found traversing the left ventricular (LV) cavity and between the septum and the papillary muscle. **C:** Illustration with corresponding intracardiac ultrasound image of direct catheter mapping on the moderator band.

False tendons in particular have received more attention as common anatomic variants of the normal LV and their potential to be misinterpreted as pathologic entities on an echocardiogram. Although the embryonic basis for the presence of false tendons appears to be unknown, pathological evidence of shared fibromuscular histology with interspersed conduction elements may suggest a common embryologic pathway for false tendon formation.[76] While the incidence of LV false tendons ranges between 40% and 55% they appear to be equally distributed by gender and age.[77,78] They are said to predominate between the posteromedial papillary muscle and the V septum and are less present between the free wall and the septum and rarely between the 2 aspects of the free wall.[77] In addition, anomalous bands have been observed in the LV and both atria.[79] Atrial bands appear to be associated with a patent foramen ovalis, atrial malformations, and a prominent Chiari network within the RA.[79]

Electrophysiology Correlation

In a retrospective study, ECS were found to affect VT ablation in 24% of patients.[80] In one-third of this sample, the structures were actively involved in the arrhythmia, either through reentrancy or abnormal automaticity mechanism. In the rest of patients, ECS were associated with difficulty in catheter manipulation or in pace or scar mapping interpretation.

The role of ECS in idiopathic VT has been recognized for more than 10 years, but only recently more detailed epidemiology and ablation outcomes have been reported.[80–82] While false tendons and papillary muscles are more commonly associated with VT, only one study demonstrated the involvement of the moderator band.[80] However, there is evidence that the bundle branches may extend to false tendons (left bundle) and to the moderator band (right bundle), in which cases those arrhythmias could be in fact variants of fascicular tachycardias.[76,83]

In contrast to false tendon and moderator band-related VT, papillary muscle-related VT is significantly more encountered in patients with ischemic cardiomyopathy.[80,84,85]

Ablation in the ECS, especially in the papillary muscles, can be challenging due to the risk of damaging the valve apparatus. Therefore, just enough energy delivery to eliminate the arrhythmogenic substrate is advocated. In addition, direct visualization of the ECS, such as with intracardiac ultrasound imaging, should be used to help in avoiding the disruption of calcification and monitor for new valve regurgitation.

OUTFLOW TRACTS

Understanding the regional anatomy of the great artery ventricular outflow tract (VOFT) junction is the key to maximize ablation success in decreasing the risk of damage to surrounding structures. Developmentally, the roots and proximal portion of the AO and PA arise from the truncus arteriosus. Septum formation in the truncus arteriosus and its subsequent differentiation begins during the fifth week of development.[86] With division of the AO and PA, surrounding mesenchyme fuses with endocardial cushion, remodels, and results in the formation of semilunar valves. The outflow tract septum continues to differentiate until its division into aortic and pulmonary streams.[87]

The aortic valve has important anatomic relationships with each of the other cardiac valves and the conduction system (Figure 2.8A).[88] The left coronary cusp (LCC) of the aortic valve is essentially continuous with the anterior leaflet of the mitral valve forming the aortic-mitral continuity.[88] This region is generally thought to be devoid of myocardium. While the proximity of the CA to the aortic valve cusps was generally appreciated, the relationship with the right ventricular outflow tract (RVOT) and pulmonic valve which courses from

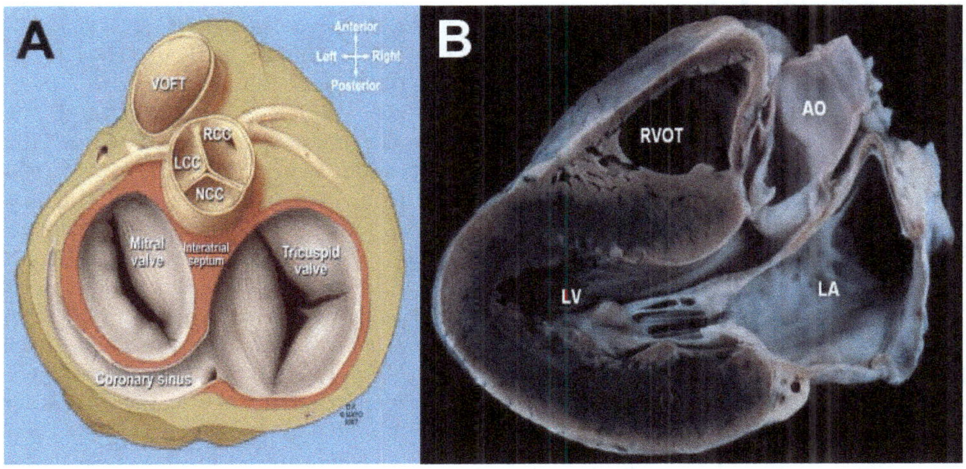

Figure 2.8. Anatomic relations of the outflow tracts. A: Cross-sectional illustration through the base of the heart showing the regional anatomy and relations of the aortic cusps (see text for details). B: Anatomic dissection through the VOFT. Note that the RVOT is anterior to the left ventricular outflow tract (LVOT) and the aortic valve is relatively caudal to the pulmonic valve. VOFT, ventricular outflow tract; LCC, left coronary cusp; RCC, right coronary cusp; NCC, noncoronary cusp; LV, left ventricle; AO, aorta; LA, left atrium. (Reproduced from Asirvatham SJ[46] with permission from John Wiley and Sons.)

right to left anterior to the aortic root is less clear. The LCC portion adjacent to the right coronary cusp (RCC) is immediately posterior to the infundibular myocardium of the posterior RVOT. The noncoronary cusp (NCC) is immediately adjacent to the atrial myocardium of the interatrial septum, whereas the RCC was immediately adjacent in all cases to the infundibular portion of the RVOT. The NCC occasionally harbors V myocardial extensions (especially near the junction with the RCC) and was simultaneously adjacent to atrial myocardium (interatrial septum).[39]

The commissure between the RCC and NCC lies directly in apposition to the junction of the septal and anterior leaflets with the TCV, the membranous portion of the intraventricular septum where the penetrating bundle of His is located. Although the membranous septum is generally devoid of myocardium, myocardial sleeves have been reported in a minority of cases.

The junction of the RVOT and PV lies to the left and cephalad to the aortic valve (see Figure 2.8B). The RVOT (infundibulum) is anterior to the LVOT, and since the aortic valve is relatively caudal to the pulmonic valve, the RCC lies immediately posterior to the infundibulum from the pulmonic valve. Unlike the aortic valve, there appears to be no contiguous atrial myocardium posteriorly, but a consistent relationship between the LAA and the left pulmonary cusp and the RAA and the right pulmonary cusp exists.

The closest portion of the VOFT myocardium to the LMCA is the posterior RV tract infundibulum at the level of the left cusp of the pulmonary valve. This may be explained through the crossing of the RVOT anterior to the aortic root, since both the pulmonary valve and the ostia of the coronary arteries are cephalad to the aortic valve. The RCA is most closely related to the posterior portion of the right cusp of the pulmonary valve but is characteristically separated by epicardial fat and portions of the overlying RAA. The left anterior descending artery (LAD) courses anterior and leftward of the margin of the RVOT. The anterior intraventricular vein crosses anterior to the LAD and lies to the right of the LAD with no immediate relation to any of the pulmonary valve cusps.

It is important to appreciate that myocardium and myocardial sleeves may be present above and between the cusps of the semilunar valve and into the valve tissue itself. In the aortic valve, although these sleeves are most prominent above the RCC, they do occur, albeit less commonly, in the LCC and NCC. Knowledge of the detailed regional anatomy of the great arteries and semilunar valve will help ablationists target the appropriate tissue for ablation, maximize efficacy, and decrease complications.

Electrophysiology Correlation

Based on the anatomy described above, it is easy to understand why the outflow tracts and the great arteries can house the substrate for VTs, ATs, and APs.[89–91] The fact that V myocardial sleeves extend into the great arteries explains why frequent premature V complexes and sustained VTs either from RVOT or LVOT origin can be ablated above the pulmonic valve or in the RCC or LCC.[92]

However, the anatomy-EP correlation for the substrate of atrial arrhythmias and APs is not as straightforward. The intimate relation of the NCC with the interatrial septum and atria is undoubtedly implicated. Even ATs arising from the septal TA or from the anteroseptal annulus are all mapped very close to the location of the NCC.[93]

The intricate anatomy of the outflow tracts can generate difficulty during pace mapping. For example, high-output pacing from the RCC may reproduce the ECG or clinical arrhythmia even when the origin is in the posterior RVOT or subvalvar anterior LVOT. Also, pacing from the proximal anterior intra-ventricular vein, the LCC, and the aortic-mitral continuity can all elicit a similar tracing.[93]

Knowing the regional relationships is also important to prevent collateral damage to surrounding structures. The LMCA arises cephalad to the LCC. As the pulmonic valve is also cephalad and leftward of the aortic valve, the posterior wall of the PA is in close relation to the LMCA. Therefore, whenever ablation above the pulmonic valve posteriorly or in the posterior RVOT is attempted, intracardiac ultrasound imaging or coronary angiography should be performed. Perforation of the valve cusps can also result from catheter manipulation or inadvertent energy delivery. Another main concern is to avoid damaging the conduction system close to the aortic annulus. The penetrating bundle of His is at the membranous septum formed in part by the commissure between the RCC and NCC of the aortic valve. Cryoablation can add to safety in the aforementioned situations.[94]

REFERENCES

1. Ho SY, Sanchez-Quintana D, Cabrera JA, Anderson RH. Anatomy of the left atrium: Implications for radiofrequency ablation of atrial fibrillation. *J Cardiovasc Electrophysiol.* 1999;10(11):1525–1533.
2. Hof I, Chilukuri K, Arbab-Zadeh A, et al. Does left atrial volume and pulmonary venous anatomy predict the outcome of catheter ablation of atrial fibrillation? *J Cardiovasc Electrophysiol.* 2009;20(9):1005–1010.
3. Cabrera JA, Ho SY, Climent V, Sanchez-Quintana D. The architecture of the left lateral atrial wall: A particular anatomic region with implications for ablation of atrial fibrillation. *Eur Heart J.* 2008;29(3):356–362.
4. Macedo PG, Kapa S, Mears JA, Fratianni A, Asirvatham SJ. Correlative anatomy for the electrophysiologist: Ablation for atrial fibrillation. Part I: Pulmonary vein ostia, superior vena cava, vein of Marshall. *J Cardiovasc Electrophysiol.* 2010;21(6):721–730.
5. Ho SY, Cabrera JA, Tran VH, et al. Architecture of the pulmonary veins: Relevance to radiofrequency ablation. *Heart.* 2001;86(3):265–270.
6. Asirvatham SJ. Methods to identify the pulmonary vein ostium. *Heart Rhythm.* 2005;2(10):1090–1093.
7. Ho SY, Sanchez-Quintana D. The importance of atrial structure and fibers. *Clin Anat.* 2009;22(1):52–63.
8. Perez-Lugones A, McMahon JT, Ratliff NB, et al. Evidence of specialized conduction cells in human pulmonary veins of patients with atrial fibrillation. *J Cardiovasc Electrophysiol.* 2003;14(8):803–809.
9. Morel E, Meyronet D, Thivolet-Bejuy F, Chevalier P. Identification and distribution of interstitial Cajal cells in human pulmonary veins. *Heart Rhythm.* 2008;5(7):1063–1067.
10. Nguyen BL, Fishbein MC, Chen LS, Chen PS, Masroor S. Histopathological substrate for chronic atrial fibrillation in humans. *Heart Rhythm.* 2009;6(4):454–460.
11. Li J, Wang L. Catheter ablation of atrial fibrillation originating from superior vena cava. *Arch Med Res.* 2006;37(3):415–418.
12. Okamoto K, Ito H, Kuma F, et al. Reentrant atrial tachycardia originating from the superior vena cava. *J Interv Card Electrophysiol.* 2003;8(3):203–208.
13. Tsai CF, Tai CT, Hsieh MH, et al. Initiation of atrial fibrillation by ectopic beats originating from the superior vena cava: Electrophysiological characteristics and results of radiofrequency ablation. *Circulation.* 2000;102(1):67–74.
14. Asirvatham S. Anatomy of the vena cava: An electrophysiological perspective. In: Chen S, Haissaguerre M, and Zipes D, eds. *Thoracic Vein Arrhythmias: Mechanisms and Treatment.* Malden, MA: Blackwell/Futura; 2004:54–65.
15. Aslamy Z, Dewald CL, Heffner JE. MRI of central venous anatomy: Implications for central venous catheter insertion. *Chest.* 1998;114(3):820–826.
16. Mears JL, Lachman N, Christensen K, Asirvatham S. The phrenic nerve and atrial fibrillation ablation procedures. *J Atr Fibrillation.* 2009;7:430–477.
17. Volkmer M, Antz M, Hebe J, Kuck KH. Focal atrial tachycardia originating from the musculature of the coronary sinus. *J Cardiovasc Electrophysiol.* 2002;13(1):68–71.
18. Tai CT, Huang JL, Lin YK, et al. Noncontact three-dimensional mapping and ablation of upper loop re-entry originating in the right atrium. *J Am Coll Cardiol.* 2002;40(4):746–753.
19. Chang KC, Lin YC, Chen JY, Chou HT, Hung JS. Electrophysiological characteristics and radiofrequency ablation of focal atrial tachycardia originating from the superior vena cava. *Jpn Circ J.* 2001;65(12):1034–1040.
20. Chen YJ, Chen YC, Yeh HI, Lin CI, Chen SA. Electrophysiology and arrhythmogenic activity of single cardiomyocytes from canine superior vena cava. *Circulation.* 2002;105(22):2679–2685.

21. Dib C, Kapa S, Powell BD, Packer DL, Asirvatham SJ. Successful use of "cryo-mapping" to avoid phrenic nerve damage during ostial superior vena caval ablation despite nerve proximity. *J Interv Card Electrophysiol.* 2008;22(1):23–30.
22. Asirvatham SJ, Kapa S. Sleep apnea and atrial fibrillation: The autonomic link. *J Am Coll Cardiol.* 2009;54(22):2084–2086.
23. Ghias M, Scherlag BJ, Lu Z, et al. The role of ganglionated plexi in apnea-related atrial fibrillation. *J Am Coll Cardiol.* 2009;54(22):2075–2083.
24. Asirvatham S. Anatomy of the coronary sinus. In: Yu CM, Hayes DL, and Aurrichio A, eds. *Cardiac Resynchronization Therapy.* Oxford: Blackwell/Futura; 2006:211–238.
25. Doshi RN, Wu TJ, Yashima M, et al. Relation between ligament of Marshall and adrenergic atrial tachyarrhythmia. *Circulation.* 1999;100(8):876–883.
26. Hwang C, Wu TJ, Doshi RN, Peter CT, Chen PS. Vein of Marshall cannulation for the analysis of electrical activity in patients with focal atrial fibrillation. *Circulation.* 2000;101(13):1503–1505.
27. Hwang C, Fishbein MC, Chen PS. How and when to ablate the ligament of Marshall. *Heart Rhythm.* 2006;3(12):1505–1507.
28. Macedo PG, Kapa S, Mears JA, Fratianni A, Asirvatham SJ. Correlative anatomy for the electrophysiologist: Ablation for atrial fibrillation. Part II: Regional anatomy of the atria and relevance to damage of adjacent structures during AF ablation. *J Cardiovasc Electrophysiol.* 2010;21(7):829–936.
29. Lemery R, Guiraudon G, Veinot JP. Anatomic description of Bachmann's bundle and its relation to the atrial septum. *Am J Cardiol.* 2003;91(12):1482–1485, A8.
30. Pappone C, Oral H, Santinelli V, et al. Atrio-esophageal fistula as a complication of percutaneous transcatheter ablation of atrial fibrillation. *Circulation.* 2004;109(22):2724–2726.
31. Scanavacca MI, D'Avila A, Parga J, Sosa E. Left atrial-esophageal fistula following radiofrequency catheter ablation of atrial fibrillation. *J Cardiovasc Electrophysiol.* 2004;15(8):960–962.
32. Bunch TJ, Day JD. Novel ablative approach for atrial fibrillation to decrease risk of esophageal injury. *Heart Rhythm.* 2008;5(4):624–627.
33. Buch E, Nakahara S, Shivkumar K. Intra-pericardial balloon retraction of the left atrium: A novel method to prevent esophageal injury during catheter ablation. *Heart Rhythm.* 2008;5(10):1473–1475.
34. Herweg B, Johnson N, Postler G, et al. Mechanical esophageal deflection during ablation of atrial fibrillation. *Pacing Clin Electrophysiol.* 2006;29(9):957–961.
35. Duong T, Hui P, Mailhot J. Acute right coronary artery occlusion in an adult patient after radiofrequency catheter ablation of a posteroseptal accessory pathway. *J Invasive Cardiol.* 2004;16(11):657–659.
36. Schneider HE, Kriebel T, Gravenhorst VD, Paul T. Incidence of coronary artery injury immediately after catheter ablation for supraventricular tachycardias in infants and children. *Heart Rhythm.* 2009;6(4):461–467.
37. Habib A, Lachman N, Christensen KN, Asirvatham SJ. The anatomy of the coronary sinus venous system for the cardiac electrophysiologist. *Europace.* 2009;11(Suppl. 5):v15–v21.
38. Yamada T, Huizar JF, McElderry HT, Kay GN. Atrial tachycardia originating from the noncoronary aortic cusp and musculature connection with the atria: Relevance for catheter ablation. *Heart Rhythm.* 2006;3(12):1494–1496.
39. Gami AS, Venkatachalam KL, Friedman PA, Asirvatham SJ. Successful ablation of atrial tachycardia in the right coronary cusp of the aortic valve in a patient with atrial fibrillation: What is the substrate? *J Cardiovasc Electrophysiol.* 2008;19(9):982–986.
40. Rytand DA. The circus movement (entrapped circuit wave) hypothesis and atrial flutter. *Ann Intern Med.* 1966;65(1):125–159.
41. Puech P, Latour H, Grolleau R. [Flutter and his limits]. *Arch Mal Coeur Vaiss.* 1970;63(1):116–144.
42. Olshansky B, Okumura K, Hess PG, Waldo AL. Demonstration of an area of slow conduction in human atrial flutter. *J Am Coll Cardiol.* 1990;16(7):1639–1648.
43. Schwartzman D, Callans DJ, Gottlieb CD, et al. Conduction block in the inferior vena caval-tricuspid valve isthmus: Association with outcome of radiofrequency ablation of type I atrial flutter. *J Am Coll Cardiol.* 1996;28(6):1519–1531.
44. Schumacher B, Pfeiffer D, Tebbenjohanns J, et al. Acute and long-term effects of consecutive radiofrequency applications on conduction properties of the subeustachian isthmus in type I atrial flutter. *J Cardiovasc Electrophysiol.* 1998;9(2):152–163.
45. Poty H, Saoudi N, Abdel Aziz A, Nair M, Letac B. Radiofrequency catheter ablation of type 1 atrial flutter. Prediction of late success by electrophysiological criteria. *Circulation.* 1995;92(6):1389–1392.
46. Asirvatham SJ. Correlative anatomy and electrophysiology for the interventional electrophysiologist: Right atrial flutter. *J Cardiovasc Electrophysiol.* 2009;20(1):113–122.
47. Cabrera JA, Sanchez-Quintana D, Ho SY, Medina A, Anderson RH. The architecture of the atrial musculature between the orifice of the inferior caval vein and the tricuspid valve: The anatomy of the isthmus. *J Cardiovasc Electrophysiol.* 1998;9(11):1186–1195.
48. Sanchez-Quintana D, Ho SY, Cabrera JA, Farre J, Anderson RH. Topographic anatomy of the inferior pyramidal space: Relevance to radiofrequency catheter ablation. *J Cardiovasc Electrophysiol.* 2001;12(2):210–217.
49. Gami AS, Edwards WD, Lachman N, et al. Electrophysiological anatomy of typical atrial flutter: The posterior boundary and causes for difficulty with ablation. *J Cardiovasc Electrophysiol.* 2010;21(2):144–149.
50. Saremi F, Pourzand L, Krishnan S, et al. Right atrial cavotricuspid isthmus: Anatomic characterization with

multi-detector row CT. *Radiology.* 2008;247(3):658–668.

51. Scaglione M, Caponi D, Di Donna P, et al. Typical atrial flutter ablation outcome: Correlation with isthmus anatomy using intracardiac echo 3D reconstruction. *Europace.* 2004;6(5):407–417.
52. Jais P, Haissaguerre M, Shah DC, et al. Successful irrigated-tip catheter ablation of atrial flutter resistant to conventional radiofrequency ablation. *Circulation.* 1998;98(9):835–838.
53. Feld GK. Radiofrequency ablation of atrial flutter using large-tip electrode catheters. *J Cardiovasc Electrophysiol.* 2004;15(10 Suppl.):S18–S23.
54. Chang SL, Tai CT, Lin YJ, et al. The electroanatomic characteristics of the cavotricuspid isthmus: Implications for the catheter ablation of atrial flutter. *J Cardiovasc Electrophysiol.* 2007;18(1):18–22.
55. Da Costa A, Faure E, Thevenin J, et al. Effect of isthmus anatomy and ablation catheter on radiofrequency catheter ablation of the cavotricuspid isthmus. *Circulation.* 2004;110(9):1030–1035.
56. Heidbuchel H, Willems R, van Rensburg H, et al. Right atrial angiographic evaluation of the posterior isthmus: Relevance for ablation of typical atrial flutter. *Circulation.* 2000;101(18):2178–2184.
57. Cabrera JA, Sanchez-Quintana D, Farre J, Rubio JM, Ho SY. The inferior right atrial isthmus: Further architectural insights for current and coming ablation technologies. *J Cardiovasc Electrophysiol.* 2005;16(4):402–408.
58. Singh S, Johnson PI, Lee RE, et al. Topography of cardiac ganglia in the adult human heart. *J Thorac Cardiovasc Surg.* 1996;112(4):943–953.
59. Mizeres NJ. The anatomy of the autonomic nervous system in the dog. *Am J Anat.* 1955;96(2):285–318.
60. Norris JE, Randall WC. Responses of the canine myocardium to stimulation of thoracic cardiac nerves. *Am J Physiol.* 1977;232(5):H485–H494.
61. Yanowitz F, Preston JB, Abildskov JA. Functional distribution of right and left stellate innervation to the ventricles. Production of neurogenic electrocardiographic changes by unilateral alteration of sympathetic tone. *Circ Res.* 1966;18(4):416–428.
62. Chevalier P, Tabib A, Meyronnet D, et al. Quantitative study of nerves of the human left atrium. *Heart Rhythm.* 2005;2(5):518–522.
63. Po SS, Nakagawa H, Jackman WM. Localization of left atrial ganglionated plexi in patients with atrial fibrillation. *J Cardiovasc Electrophysiol.* 2009;20(10):1186–1189.
64. Vaitkevicius R, Saburkina I, Rysevaite K, et al. Nerve supply of the human pulmonary veins: An anatomical study. *Heart Rhythm.* 2009;6(2):221–228.
65. Pokushalov E. The role of autonomic denervation during catheter ablation of atrial fibrillation. *Curr Opin Cardiol.* 2008;23(1):55–59.
66. Kapa S, Asirvatham SJ. Atrial fibrillation: Focal or reentrant or both? A new autonomic lens to examine an old riddle. *Circ Arrhythm Electrophysiol.* 2009;2(4):345–348.
67. Nademanee K, McKenzie J, Kosar E, et al. A new approach for catheter ablation of atrial fibrillation: Mapping of the electrophysiologic substrate. *J Am Coll Cardiol.* 2004;43(11):2044–2053.
68. Katritsis D, Giazitzoglou E, Sougiannis D, Voridis E, Po SS. Complex fractionated atrial electrograms at anatomic sites of ganglionated plexi in atrial fibrillation. *Europace.* 2009;11(3):308–315.
69. Lavery CE, Mittleman MA, Cohen MC, Muller JE, Verrier RL. Nonuniform nighttime distribution of acute cardiac events: A possible effect of sleep states. *Circulation.* 1997;96(10):3321–3327.
70. Gami AS, Howard DE, Olson EJ, Somers VK. Day-night pattern of sudden death in obstructive sleep apnea. *N Engl J Med.* 2005;352(12):1206–1214.
71. Verrier RL, Josephson ME. Impact of sleep on arrhythmogenesis. *Circ Arrhythm Electrophysiol.* 2009;2(4):450–459.
72. Stanton MS, Tuli MM, Radtke NL, et al. Regional sympathetic denervation after myocardial infarction in humans detected noninvasively using I-123-metaiodobenzylguanidine. *J Am Coll Cardiol.* 1989;14(6):1519–1526.
73. Paul M, Schafers M, Kies P, et al. Impact of sympathetic innervation on recurrent life-threatening arrhythmias in the follow-up of patients with idiopathic ventricular fibrillation. *Eur J Nucl Med Mol Imaging.* 2006;33(8):866–870.
74. Goldenberg I, Moss AJ. Long QT syndrome. *J Am Coll Cardiol.* 2008;51(24):2291–2300.
75. Schwartz PJ, Priori SG, Cerrone M, et al. Left cardiac sympathetic denervation in the management of high-risk patients affected by the long-QT syndrome. *Circulation.* 2004;109(15):1826–1833.
76. Jongbloed MR, Schalij MJ, Poelmann RE, et al. Embryonic conduction tissue: A spatial correlation with adult arrhythmogenic areas. *J Cardiovasc Electrophysiol.* 2004;15(3):349–355.
77. Luetmer PH, Edwards WD, Seward JB, Tajik AJ. Incidence and distribution of left ventricular false tendons: An autopsy study of 483 normal human hearts. *J Am Coll Cardiol.* 1986;8(1):179–183.
78. Tamborini G, Pepi M, Celeste F, et al. Incidence and characteristics of left ventricular false tendons and trabeculations in the normal and pathologic heart by second harmonic echocardiography. *J Am Soc Echocardiogr.* 2004;17(4):367–374.
79. Ho SY. Anatomic insights for catheter ablation of ventricular tachycardia. *Heart Rhythm.* 2009;6(8 Suppl.):S77–S80.
80. Abouezzeddine O, Suleiman M, Buescher T, et al. Relevance of endocavitary structures in ablation procedures for ventricular tachycardia. *J Cardiovasc Electrophysiol.* 2010;21(3):245–254.
81. Thakur RK, Klein GJ, Sivaram CA, et al. Anatomic substrate for idiopathic left ventricular tachycardia. *Circulation.* 1996;93(3):497–501.
82. Doppalapudi H, Yamada T, McElderry HT, et al. Ventricular tachycardia originating from the posterior papillary muscle in the left ventricle: A distinct clinical syndrome. *Circ Arrhythm Electrophysiol.* 2008;1(1):23–29.

83. Nakagawa H, Beckman KJ, McClelland JH, et al. Radiofrequency catheter ablation of idiopathic left ventricular tachycardia guided by a Purkinje potential. *Circulation.* 1993;88(6):2607–2617.
84. Bogun F, Desjardins B, Crawford T, et al. Post-infarction ventricular arrhythmias originating in papillary muscles. *J Am Coll Cardiol.* 2008;51(18):1794–1802.
85. Liu XK, Barrett R, Packer DL, Asirvatham SJ. Successful management of recurrent ventricular tachycardia by electrical isolation of anterolateral papillary muscle. *Heart Rhythm.* 2008;5(3):479–482.
86. Sadler TW. *Langman's Medical Embryology.* 8th ed. Baltimore, MD: Lippincott Williams and Wilkins; 2000.
87. McBride RE, Moore GW, Hutchins GM. Development of the outflow tract and closure of the interventricular septum in the normal human heart. *Am J Anat.* 1981;160(3):309–331.
88. Anderson RH, Razavi R, Taylor AM. Cardiac anatomy revisited. *J Anat.* 2004;205(3):159–177.
89. Suleiman M, Asirvatham SJ. Ablation above the semilunar valves: When, why, and how? Part II. *Heart Rhythm.* 2008;5(11):1625–1630.
90. Kanagaratnam L, Tomassoni G, Schweikert R, et al. Ventricular tachycardias arising from the aortic sinus of valsalva: An under-recognized variant of left outflow tract ventricular tachycardia. *J Am Coll Cardiol.* 2001;37(5):1408–1414.
91. Timmermans C, Rodriguez LM, Crijns HJ, Moorman AF, Wellens HJ. Idiopathic left bundle-branch block-shaped ventricular tachycardia may originate above the pulmonary valve. *Circulation.* 2003;108(16):1960–1967.
92. Suleiman M, Asirvatham SJ. Ablation above the semilunar valves: When, why, and how? Part I. *Heart Rhythm.* 2008;5(10):1485–1492.
93. Tabatabaei N, Asirvatham SJ. Supravalvular arrhythmia: Identifying and ablating the substrate. *Circ Arrhythm Electrophysiol.* 2009;2(3):316–326.
94. d'Avila A, Thiagalingam A, Holmvang G, et al. What is the most appropriate energy source for aortic cusp ablation? A comparison of standard RF, cooled-tip RF and cryothermal ablation. *J Interv Card Electrophysiol.* 2006;16(1):31–38.
95. Lachman N, Syed FF, Habib A, et al. Correlative anatomy for the electrophysiologist, part II: cardiac ganglia, phrenic nerve, coronary venous system. *J Cardiovasc Electrophysiol.* 2011;22:104–110.

CHAPTER 3

Physiological Responses During Electrophysiologic Evaluation

Masood Akhtar, MD and Jasbir S. Sra, MD

INTRODUCTION

The growth of electrophysiology (EP) as a unique medical discipline can be traced to the fact that arrhythmia-related symptoms are quite common, but elusive, and the arrhythmia is seldom present at the time of patient encounter. Early on, it necessitated the discovery of electrical stimulation of the cardiac tissue to replicate the underlying arrhythmia to make EP evaluation of practical value.[1,2] Once that ability was achieved, the concurrent intracardiac electrical recording allowed a detailed examination of the mechanism, origin, and direction of propagation to arrive at a more rational diagnosis.[3–10] To have an accurate interpretation of observed signals, however, one must also distinguish normal from the abnormal. While this may sound intuitive and somewhat obvious, it actually has been learned over many years, after numerous studies in both the normal and the cardiac disease population.[11–22] Without such an understanding, errors can be made relatively easily. If the study population has a prevalence of cardiac pathology, in the elderly, for example, a physiological response may be labeled abnormal, which could lead to therapeutic intervention. Atrial (A) pacing-induced atrioventricular (AV) nodal Wenckebach phenomena (WP) or prolongation of His-Purkinje conduction time following atrial premature extrastimulus may be interpreted as abnormal, until one realizes that these responses are universal and seen in all population groups.[11–22]

The purpose of this chapter is to present a range of intervals, timing of events, and phenomena that occur during the conduct of EP and are considered normal. It should be pointed out that "normal" is a working definition for events seen in any population with no arrhythmic consequences and hence no need for therapeutic intervention. This does not mean, however, that numerical values at the fringes of so-called normal will necessarily have an untoward outcome.

Normal Rhythm and Sinoatrial (SA) Conduction

The SA node, situated in the right upper part of the right atrium, is the normal pacemaker due to its fastest rate of automaticity in the human heart, hence it dominates the rhythm. The resting rate ranges from 60 to 100 bpm during sinus rhythm (SR) with a characteristic P-wave morphology, upright in leads II, III, and avF (axis of about 60° with a range of –30 to +90) and biphasic in the V_1, that is, positive deflection in the first half and negative in the second. Atrial activity arising anywhere else in the atria is termed as atrial. During routine EP evaluation, at least one pair of electrodes from an electrode catheter is dedicated to the recording of the right atrial electrical activity. In general, the intracardiac signals are used to time the event and sequence of activation with the use of multiple recordings. Since the intracardiac electrical signals are commonly filtered at 30 to 40 Hz high-pass frequency, they are not useful for morphology recognition. Concomitant surface ECG recording (leads I, II, and V_1 or some other leads) are necessary to appreciate P-QRST variations (Figure 3.1). Continuous and concurrent timelines are also desirable to accommodate any changes in the speed of the recording apparatus (paper, tape, disk, etc.).

At the usual magnification the sinus impulse origin is not recognizable, even by electrode catheters in close vicinity to the SA node. However, at higher magnification, filtering and appropriate clipping of other signals' SA node activity preceding the atrial signal can be identified (Figure 3.2).[23,24] With direct measurement, the range of the SA interval (SA conduction time) is 46 to 116 ms.[23,24] Without the benefit of SA recording, the SA interval traditionally has been measured indirectly. This is done by way of A pacing as close to the SA node as possible, interrupting the pacing and then measuring the interval between the last paced atrial signal to the onset of the return cycle (RC) (on high right atrial [HRA] recording).[25–27] The sinus cycle length (CL) is deducted from this interval, and the difference is divided by 2 assuming the A-pace-to-sinus equals the sinus-to-atrial signal.

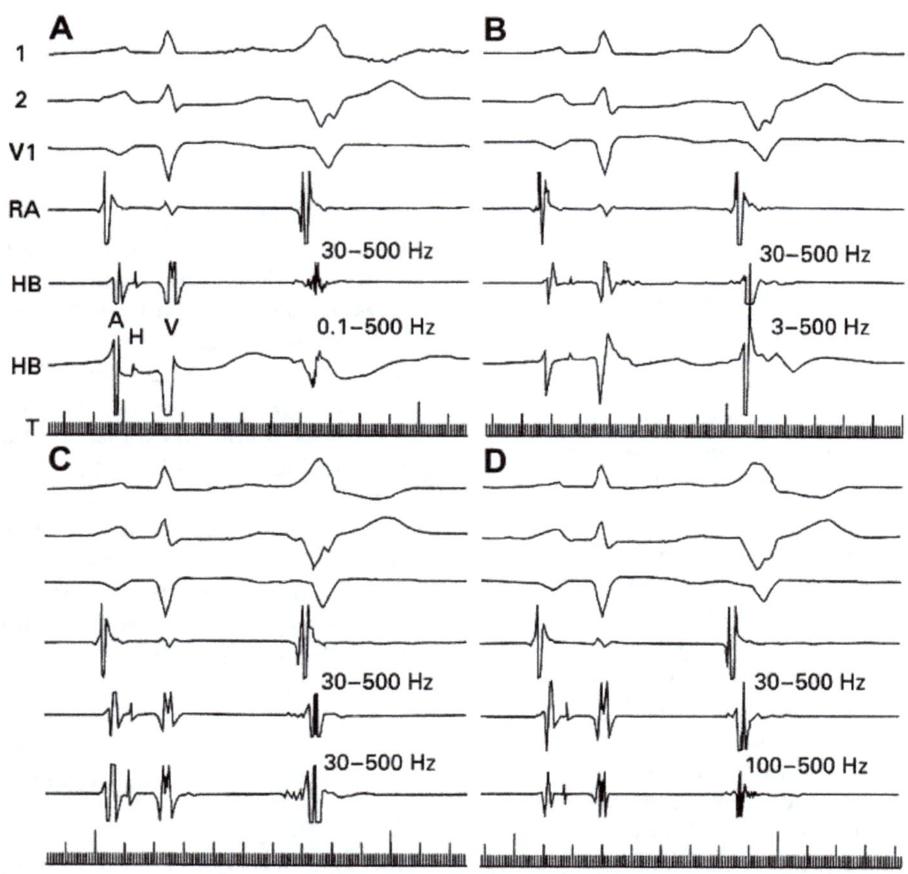

Figure 3.1. Effect of following on intracardiac signals. (**A** and **B**) From top to bottom in each panel are surface ECG leads I, II, and V_1, intracardiac right AEGM (RA) and the 2 recordings from the AV junction in the region of the HB. Time lines are depicted at the bottom continuously running at 10 × 100 ms. The ECG signals are essentially unfiltered. Alteration of high-pass filter on surface ECG has significant effect and should not be done. The 2 HB tracings are shown for comparison. The top HB signals are recorded at 30 to 500 Hz in all panels. The high-pass filter in the bottom tracing is changed, and one can see the effect. The low-frequency signals, such as T wave, are completely eliminated in **Panels C** and **D**. Similar labeling format is used in subsequent tracings unless stated otherwise. (Reproduced from Akhtar with permission from Wolters Kluwer Publishing.[111])

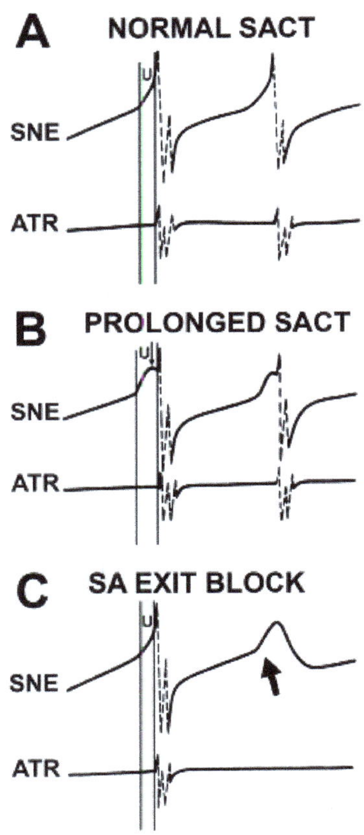

Figure 3.2. Sinoatrial conduction time (SACT). Direct recording from sinus node (SNE) starting slow upward deflection (U) and abruptly interrupted by rapid ATR. The 2 vertical lines enclose the SACT, which is normal in **Panel A** and somewhat prolonged in **Panel B**. In **Panel C**, the first sinus beat shows a normal SACT, but the second sinus firing (arrow) is not followed by ATR (SA exit block). (Reproduced from Reiffel et al with permission from Wolters Kluwer Publishing and the American Heart Association.[23])

The SA conduction time, measured indirectly with the A pacing method, has been reported to range from 40 to 156 ms.[28]

A more commonly used EP method for determining SA nodal function is the use of sinus node recovery time (SNRT) as an absolute measurement or corrected for the dominant sinus CL, that is, corrected SNRT or CSNRT.[28,29] HRA pacing is done for 30 to 60 seconds with the rate starting at a CL just shorter than sinus CL down to 400 ms, and the interval between the last paced atrial complex to the RC is measured in the same electrogram (EGM) as the pacing site. Normal SNRT calculated in this manner ranges from 1400 to 1600 ms. When sinus CL is deducted from this interval, CSNRT is achieved, which measures <680 ms in normal individuals in a nonmedicated state (Figure 3.3).

Sensitivity of SNRT or CSNRT in patients suspected of sinus node dysfunction (SND) varies from 35% to 96%.[28] The reason for this rather unacceptable scatter is worth examining. While it cannot be stated with certainty, it can be speculated that in the early years of clinical EP the distinction between sinus arrest or SND from vagal versus atrial abnormalities was not clearly made.[30–33] For instance, bradycardia and asystole from sinus node (SN) pauses due to neurocardiogenic mechanisms (Figure 3.4) and true SA disease were lumped together at times and tested for abnormal SNRT or CSNRT. It is not surprising that in such a heterogeneous population, the predictive accuracy of SNRT or CSNRT will be poor. While this is generally understood now, no systematic studies have been done to improve the value of testing in a

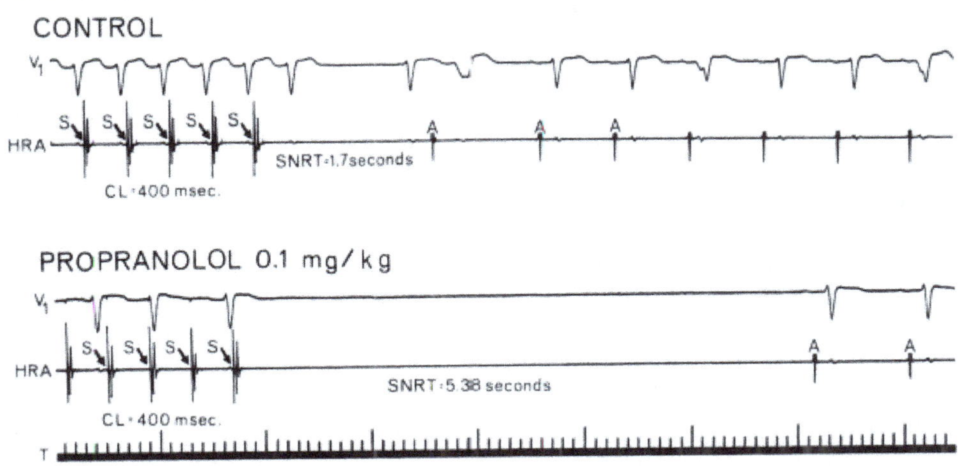

Figure 3.3. SNRT and CSNRT. At termination of atrial overdrive pacing at 400 ms (top panel), the SNRT is 1.7 seconds and, when stable, sinus CL of 700 ms (average of last 4 sinus CL) is deducted, that is, 1 second of CSNRT is clearly abnormal. At similar atrial PCL, postpropranolol SNRT is markedly prolonged (bottom panel).

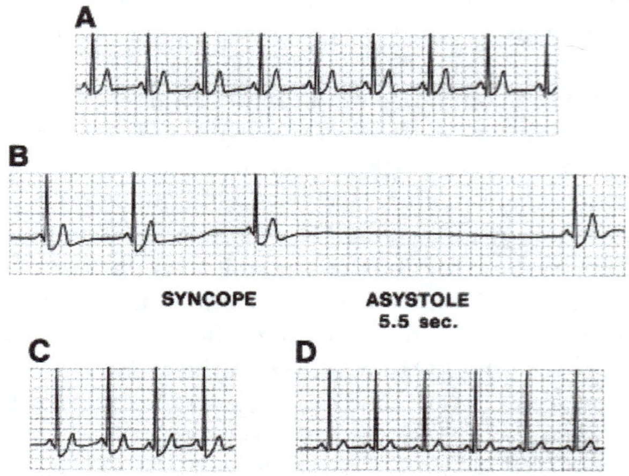

Figure 3.4. Surface ECG during an episode of syncope. Surface ECG lead II tracings prior to (**A**), during (**B**), and after vasovagal syncope (**C** and **D**) are shown. Drop in blood pressure tracing typically occurs before the slowing of sinus CL (and many times before AV prolongation as well). SNRTs are usually normal in this group of patients.

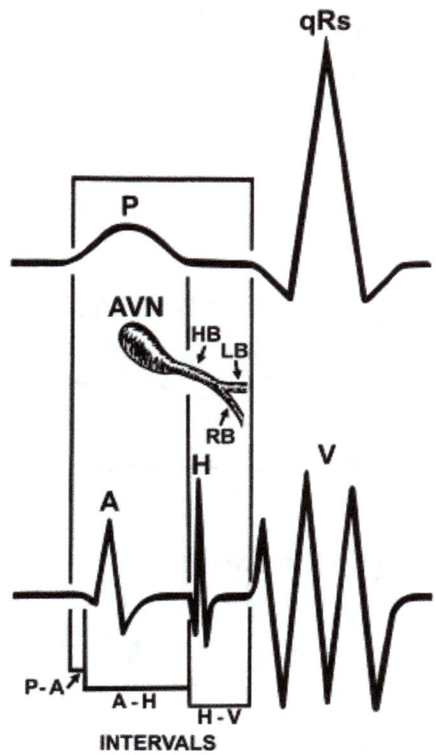

Figure 3.5. Schema of HB EGM recordings. Schematic representation of surface ECG (top), underlying anatomic structures (AVN, atrioventricular node; LB, left bundle; RB, right bundle), and HB EGM tracings are shown. The traditional divisions of the main components of the PR interval are depicted with self-explanatory labels.

better defined population with apparent SND. If patients with true SND were studied, at least one nagging question regarding SND might be answered, that is, is the so-called intermittent sinus arrest (due to cessation of SN firing) purely a vagal phenomenon, or does there exist a significant population of patients in whom sinus arrest represents a true SN disease? It may turn out that true SND is primarily a SA (or atrial) disease, in which case the predictive accuracy of SNRT/CSNRT/SA conduction time would be better than previously reported.

AV Conduction

The PR interval: When the intracardiac tracings are employed along with surface ECG leads, the PR interval can be divided into various subcomponents (Figure 3.5). As a rough estimate of intra-atrial (mostly right atrial) conduction, one can use the onset of surface ECG P waves to the beginning of atrial electrograms (AEGMs) on the His bundle (HB) tracing. This is termed as PA interval. If there is an interest in measuring the total intra-atrial conduction time, a lead placed in the distal coronary sinus (CS) can be similarly employed. Such conduction times are seldom used in routine clinical settings unless a specific aspect of intra-atrial conduction is of clinical relevance. The time from the onset of atrial signal or HB EGM to the first detectable HB potential (AH interval) is used as an approximation of AV nodal conduction time (normal AH interval range is 60–140 ms). The His-Purkinje conduction time (HV interval) is from the onset of HB potential to the first detectable ventricular activity on the surface ECG or intracardiac EGM, whichever comes first (Figure 3.6). The usual quoted normal range for HV interval is 35 to 55 ms. The duration of HB deflection is 10 to 20 ms. A wider or split His potential or a prolonged HV with a narrow QRS is considered as an evidence of intra-HB pathology.

Atrial Pacing

Electrical stimulation of the various cardiac structures is an integral part of EP evaluation, the purpose being to replicate the arrhythmia in the laboratory. The mere recording of SR or spontaneous frequent or incessant arrhythmias would have limited usefulness and hence minimum appeal to clinicians. Typically these maneuvers are

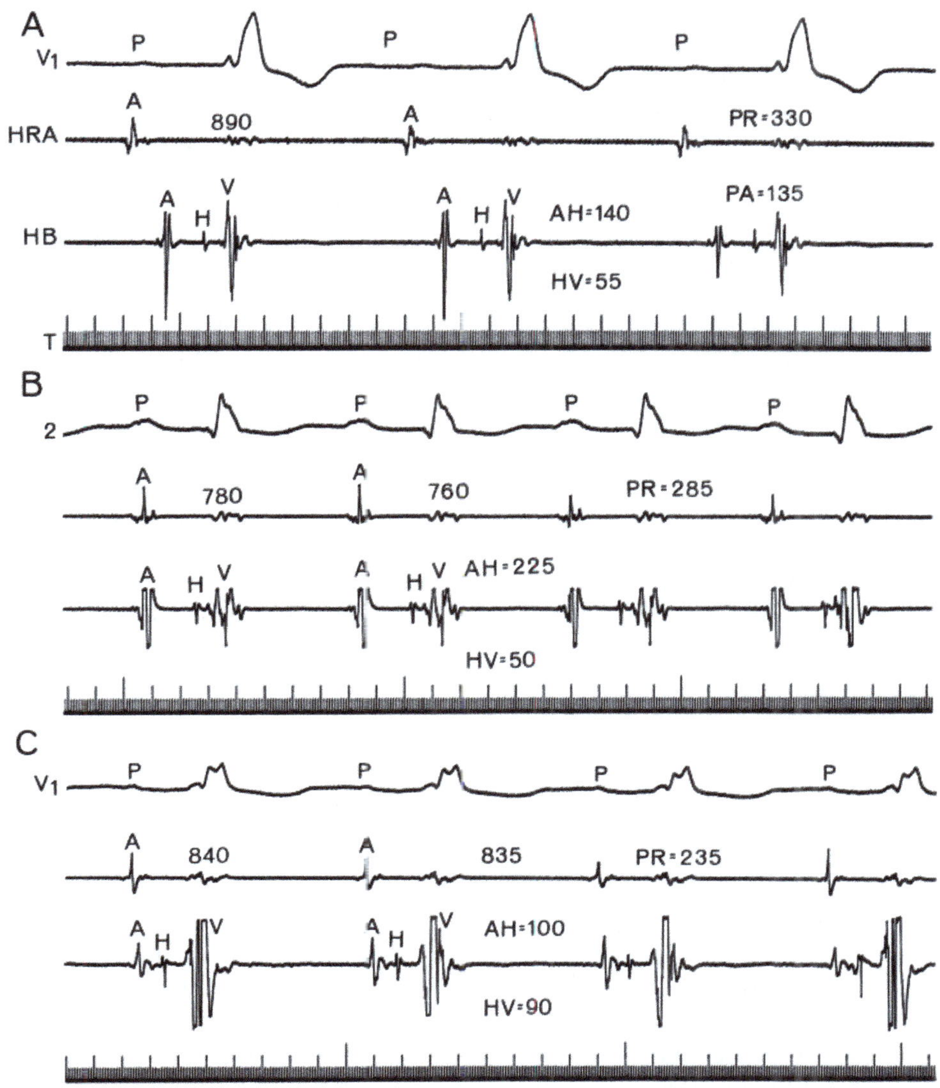

Figure 3.6. The PR interval. Three examples of long PR interval are displayed, and the main components contributing to PR prolongation, that is, the PA (A), AH (B), and HV intervals (C), are labeled. In all examples, the QRS shows a RBBB pattern.

done to reproduce the clinically documented or suspected arrhythmia and, therefore, some routine pacing protocols are followed by most cardiac electrophysiologists. It is preferred that these protocols are to be carried out in a nonsedated postabsorptive state. However, if the purpose is solely to reproduce an arrhythmia that occurred under a medicated state or if the medication has to be continued for other necessary indications, then studies may be carried out with the medications on board. Use of sedation to varying degrees is currently a routine practice in many laboratories. The topic of cardiac pacing or stimulation will be covered under incremental atrial pacing, atrial extrastimulation, incremental ventricular (V) pacing, and ventricular extrastimulation.

Incremental Atrial Pacing

Response of the AV node: The term "incremental pacing" implies that overdrive A pacing is initiated at a CL just shorter than the sinus CL and that the pacing rate is gradually increased (ie, CL is decreased) until the desired arrhythmia is induced or AV nodal block is encountered. The latter makes it difficult to stress the AV conduction further or, for that matter, increase ventricular rates. Although the intention is to achieve constant CL pacing, in reality, from the moment stimulation is first turned on, the first few CL behave like a series of premature complexes. Depending upon the preceding CL, the initial atrial coupling (the last sinus to first atrial-stimulated complex), the duration of pacing CL, and its train duration, a variety of responses may be observed. For example at near sinus cycle or long CL (700–800 ms), the AH interval and AV nodal conduction and its refractoriness are very similar to sinus CL. This creates a situation where the His-Purkinje system (HPS) can be penetrated with less conduction delay encountered in the AV node. With progressively shorter atrial CL of pacing, AV nodal conduction time and refractoriness increase. At various CL, generally in the range of 300 to 400 ms, AV nodal WP is encountered. Further shortening of the CL leads to 2:1 or 3:1 AV nodal block. It is generally believed that AV nodal WP at paced CL (PCL) of ≥600 ms is unusual, as a physiological block in an awake individual and a 1:1 AV response at a CL of ≤300 ms is a reflection of enhanced AV nodal conduction. These figures are only useful as reference points. From a clinical decision-making perspective, however, worsening of AV conduction or development of AV block with sympathomimetic or anticholinergic agents or exercise will be considered abnormal. In our experience, subjects with 1:1 AV nodal response at CL of ≤300 ms tend to have a more rapid ventricular response during atrial arrhythmias and are more resistant to AV nodal conduction-blocking agents. After the initiation of A pacing, it takes several CL before the AH stabilizes at a constant AH value. This in turn depends upon the CL of pacing, AV nodal conduction, and refractory period (RP) properties, the baseline AH interval, autonomic tone, on-board drugs, etc. This so-called AV nodal accommodation occurs over a number of CL, and has been systematically explored (Figure 3.7).[15] With long atrial coupling at the onset of pacing train, the first AH may be shorter than the subsequent AH interval (see Figure 3.7A and B) until a stable AH interval is achieved (crescendo pattern). If the atrial coupling is short, the subsequent AH will stabilize to a shorter AH interval and a crescendo pattern is observed (see Figure 3.7D and E). Almost all patients will display a pattern of immediate accommodation, that is, all AH intervals will measure the same at certain coupling intervals (see Figure 3.7C).[15] Conduction of supraventicular impulses occur over Kent or atriofascicular pathways (AFP). In other words, an atrial coupling interval exists for all patients where AV nodal accommodation is instantaneous. To study various EP responses more systematically during incremental pacing, we developed a mode of pacing that avoided the variation of sinus CL by overdrive pacing (see Figure 3.7). Similarly, the atrial coupling and the subsequent constant CL were programed appropriately to deal with the randomness of the initial pause.

The effect of cardioactive drugs, and reproducibility of various EP parameters regarding atria, AV node, HPS, and ventricular myocardium (VM) over several hours, has been studied and published previously.[21,34–40] Briefly, it can be stated that AV nodal conduction and RP show prolongation with agents such as digitalis, beta-blockers, and calcium channel blockers as well as other drugs with similar properties. This would include amiodarone, sotalol, and propafenone. Additionally, flecainide is also noted to depress AV nodal conduction. Atrial muscle, HPS, and VM show depressive responses to Class Ia, Class Ic, and Class III antiarrhythmic agents.

On the other hand, a shortening of AV nodal conduction time and refractoriness is noted with sympathomimetic and anticholinergic agents.[37] Abbreviation of refractoriness at the atrial and ventricular muscle level is observed with catecholamines. Furthermore, prolongation of sinus CL, PR interval, and brief periods of type 1 second-degree AV nodal block may be noted during sleep, even in the normal population.

Occasionally an abrupt increase in AH interval may be noted during incremental A pacing. This has often been referred to as evidence of dual AV nodal physiology, that is, AV nodal conduction switches from a fast to a slow pathway.[41] This may or may not be associated with initiation of a single atrial reciprocal response, nonsustained or sustained AV nodal reentry. The initiation of AV nodal delay and the concomitant onset of tachycardia *per se* is not proof of AV nodal reentry; they can occur at the start of AV reentry. Another AV node phenomenon that occurs with pacing, or even spontaneously, is 1:2 HB activation (ie, $P_1 > V_1 > V_2$) over a fast and then slow pathway in response to a single P wave. When it occurs spontaneously, it results in a supraventricular tachycardia (SVT) for which the rate of QRS complexes (R-R interval) is twice that of the P wave (Figure 3.8). Since the PR over the slow as compared to the fast pathway is not necessarily twice as long, alternation or some irregularity of the R-R interval may be noted. This is a rare form of narrow QRS SVT in which there are less P waves than QRS complexes but there is still AV association, that is, one P wave is connected to the 2 QRS complexes.

Response of the HPS: In contrast to the AV node, the HPS conduction time (HV interval) remains fairly constant during incremental A pacing. When the rate is gradually increased, even with a 1:1 AV response at a CL of ≤300 ms, bundle branch block (BBB) should not be encountered as long as there is

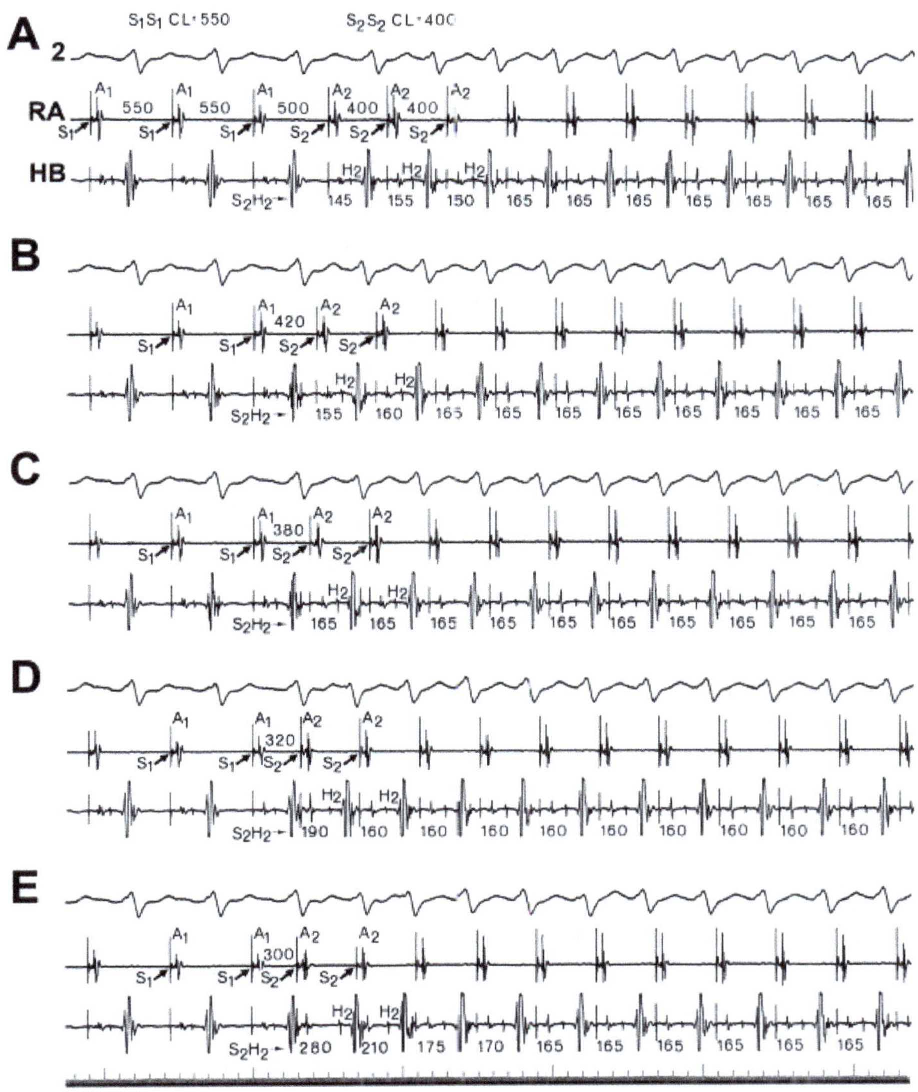

Figure 3.7. AV nodal accommodation. A controlled atrial pacing model at an initial constant CL pacing train of 550 ms (S_1-A_1) is depicted, but only the last 2 CL (A_1-A_1) are shown (A–E). Following the initial train is another fixed train of 400-ms CL (S_2-A_2) (Panels A–E). The coupling interval between the 2 trains of atrial pacing is progressively abbreviated from 500 ms to 300 ms. The AV nodal conduction (S_2-H_2) that follows in each train is labeled. The crescendo (Panels A and B), instant (Panel C), and decrescendo (Panels D and E) patterns can be observed. (Reproduced from Lehmann et al with permission from Elsevier Publishing.[15])

no HH CL variation. Development of BBB (right or left or hemiblock) during incremental A pacing can occur in the following circumstances:

1. Rate-related. The RP of the HPS is abnormally prolonged and, at a given acceleration of heart rate, the refractoriness of the HPS will be reached and aberrant conduction will be noted (tachycardia or phase 3-dependent). The same response in the same patient will be noted whenever that critical rate is reached. Once established, the aberrant conduction may persist even if the atrial CL is increased because of retrograde invasion of the bundle branch (BB) via the contralateral ventricular activation. However, it quickly dissipates with further prolongation of the atrial CL. The reverse of this may be true when aberrant conduction occurs at critically slow heart rates (bradycardia or phase 4-dependent) (Figure 3.9).[42] The rate-dependent aberrant conduction often precedes fixed BBB by a variable period in most, if not all, patients.
2. During overdrive constant CL pacing, HPS conduction may be normal initially, but

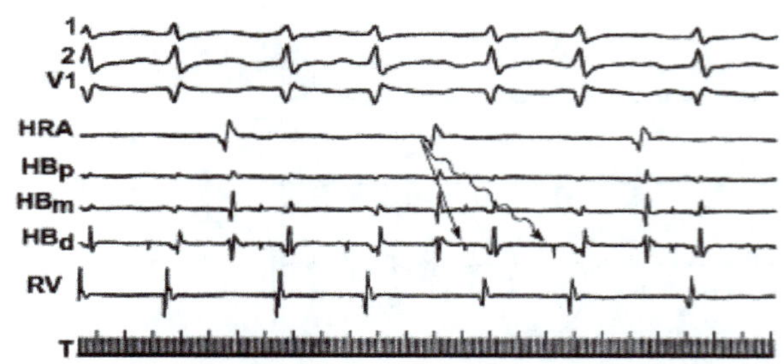

Figure 3.8. 1:2 AV nodal response and resultant SVT. This figure demonstrates SR (HRA lead) and a sustained episode of narrow QRS tachycardia with alternating CL of 320 × 410 ms and sinus CL of 730 ms. The lines drawn show the alternate conduction over the fast (AH: 120 ms) and slow (AH: 410 ms) pathways. The combined alternate HH (and R-R) interval exactly equals the sinus CL and is a clue to the association between the atrial activity and HB response. p, proximal; m, middle; d, distal; HRA, high right atrial EGM; RV, right ventricular EGM. (Reproduced from Lehmann et al with permission from Elsevier Publishing.[15])

subsequently develop abnormality even though the CL is not changed. The underlying mechanism for this manifestation of aberrant conduction is poorly understood at this time, and vaguely referred to as fatigue phenomenon.[43]

3. Functional or physiological aberrant conduction. As a practical matter, HPS block at any constant CL pacing is abnormal when the rate increment is gradual. In patients with enhanced AV nodal conduction with 1:1 AV response at CL <300 ms, the HPS's ability to conduct normally is not well understood. Such rates are seldom encountered at the HPS level due to intervening AV nodal refractoriness that would result in AV nodal WP. This will set up a short–long–short sequence and consequent aberrant conduction or block in the HPS, but at that point HPS is no longer responding to constant CL pacing (Figure 3.10). The criterion for functional block is met when the HPS is shown to be capable of conducting normally and also results in HPS block at the same CL. The reason for aberrancy versus no aberrancy is encroachment of the first paced or spontaneous complex upon the refractoriness of HPS in relation to the previous CL.

Abrupt onset of rapid A pacing or onset of SVT during normal SR may set up a stable 2:1 AV block in the HPS.[22,44–46] Often such a block is preceded by 3:2 HPS Wenckebach cycle (Figures 3.10–3.12), and usually the 2:1 AV block returns to 1:1 conduction via an HPS Wenckebach cycle or aberrant complex (see Figure 3.11).[44–46]

While abrupt onset of HPS block is normal at short CL, once 1:1 AV conduction with a constant AH interval is established, return of AV block in the HPS is abnormal. Clinical significance of block in the HPS (intra- or infra-Hisian block) during stable, constant atrial CL is similar to AV block during exercise; the latter would generally be in the HPS. When there is any doubt regarding the site of block, EP evaluation may be carried out. If prolonged HV interval, BBB, etc., is noted but no AV block within the HPS is seen, the latter can be stressed with overdrive A pacing or intravenous (IV) procainamide (10 mg/kg/min IV infusion) (Figures 3.13 and 3.14).[47–50] If AV nodal refractoriness prevents 1:1 access to HPS, 0.05 mg of IV atropine can be helpful.[21] A larger dose is not recommended because of concomitant shortening of the sinus CL. In some cases, with markedly prolonged HV interval, that is, >100 ms with unexplained syncope and no other etiology such as ventricular tachycardia (VT), etc., pacemaker insertion may be indicated.

Physiological block in the HPS, once initiated, can be terminated by sudden shortening of the HPS CL via spontaneous or induced ventricular premature complexes. Most of the time, however, it resolves by downstream migration of the site of block, leading to progressive shortening of the RP at the point the impulse reached during prior propagation (Figure 3.15).[14]

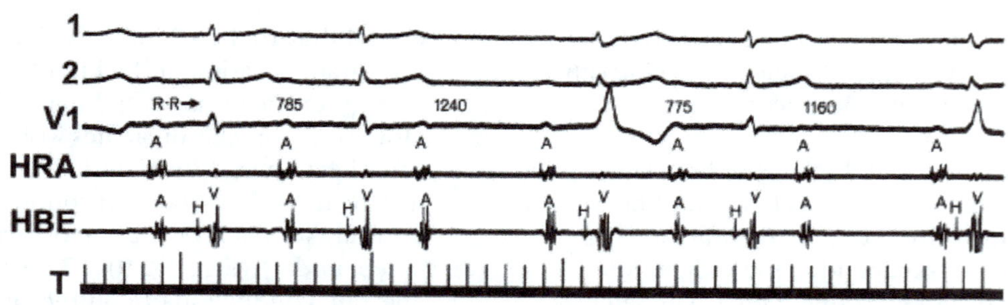

Figure 3.9. Bradycardia-dependent BBB. This tracing shows first- and second-degree (WP) AV nodal block. The long CL after the block shows a RBBB morphology not seen with any other cycle (bradycardia, pause-dependent, or phase-4 block).

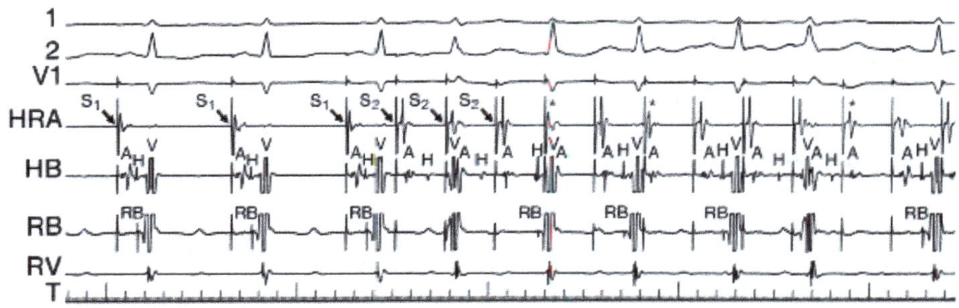

Figure 3.10. Variable degree of block in the HPS with constant CL pacing. Two sets of constant CL, S_1-S_1, and S_2-S_2 (ie, 600 ms followed by 260 ms), are displayed. There is a concomitant AVN WP (progressive AH prolongation where fifth paced atrial beat blocks in the AVN) (*) and HPS WP (the second short CL-paced conducts with long HV and RBB block and the third CL blocks below the HB recording site [no QRS]). All of the HPS blocks are set up by critical changes in the HH cycle, in turn initiated by AVN WP, or simply a long–short CL sequence such as the onset of pacing. To avoid this at the onset of incremental pacing, the CL and S_1-S_2 coupling intervals have to be chosen carefully. Of interest also is that there is no RB deflection following the H with either BB-His or RBB block, suggesting that the site of delay and/or block is proximal to the RBB recording location. This will be discussed in greater detail further in the text. All atrial paced complexes blocking in the AVN are depicted by asterisks. (Reproduced from Akhtar et al with permission from Elsevier Publishing.[51]).

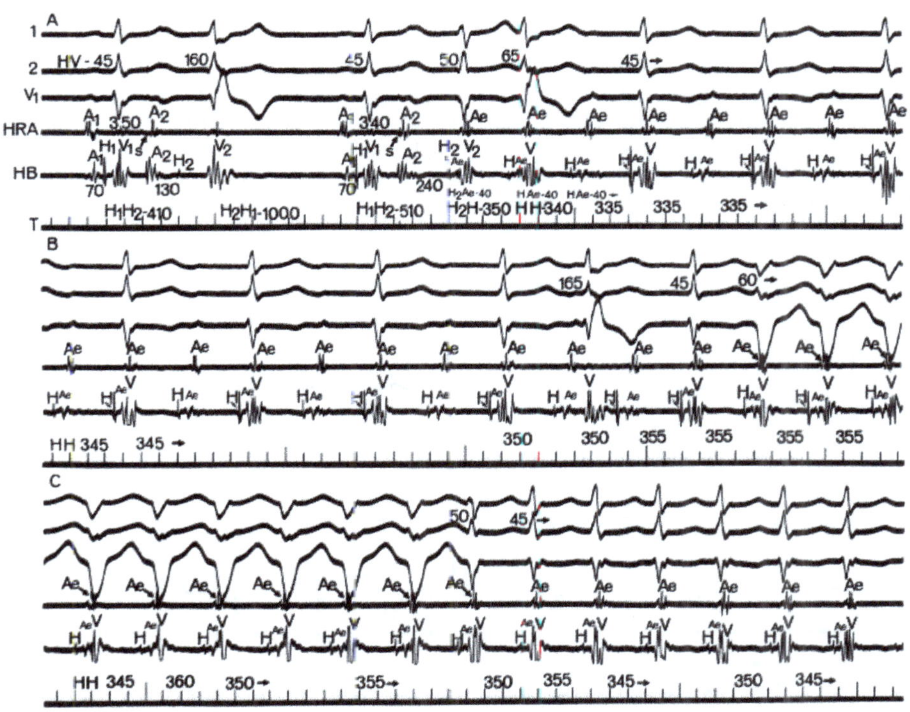

Figure 3.11. Variable degree of block in the HPS during atrioventricular nodal reentry tachycardia (AVNRT). **Panel A** shows atrial bigeminy. The first A_2 conduction is over the fast AVN pathway (A_2-H_2 130 ms) and H_2-V_2 of 160 ms followed by RBBB. The HV intervals are labeled above ECG lead II. The atrial coupling of 350 ms and 340 ms are depicted in Panel A. The second A_2 blocks in the fast pathway and conducts over the slow pathway (A_2-H_2 240 ms) and starts a slow–fast AVNRT. The values of AH, H-A_e (ie, H-atrial echo), and HH intervals are labeled. The rightward-directed arrow at the bottom of the panel depicts stable measurement beyond that point and is only relabeled when the measurements change. Please note that at the onset of AVNRT, there is 3:2 HPS block (normal → longer HV + RBBB) → block below the H deflection and then stable 2:1 block below the HB potential. The transition of 2:1 to 1:1 conduction (B) with a LBBB is seen at the end of **Panel B**. A cycle of 3:2 WP in the HPS (very similar to what is seen at the onset of AVNRT in Panel A) precedes 1:1 conduction. Also, worth noting are 2 other aspects: a fusion QRS, almost always sandwiched between the LBBB and normal QRS (middle of **Panel C**); and a relatively narrow retrograde P wave (better appreciated during 2:1 block below the His) due to inferior atrial septal breakthroughs of atrial impulse from the fast pathway of the AVN and hence resulting in simultaneous activation of the 2 atria. Neither of these findings are uncommon; they are typical of slow–fast AVNRT at the time of tachycardia onset, whether induced or spontaneous. (Reproduced from Akhtar with permission from Wolters Kluwer Publishing.[44])

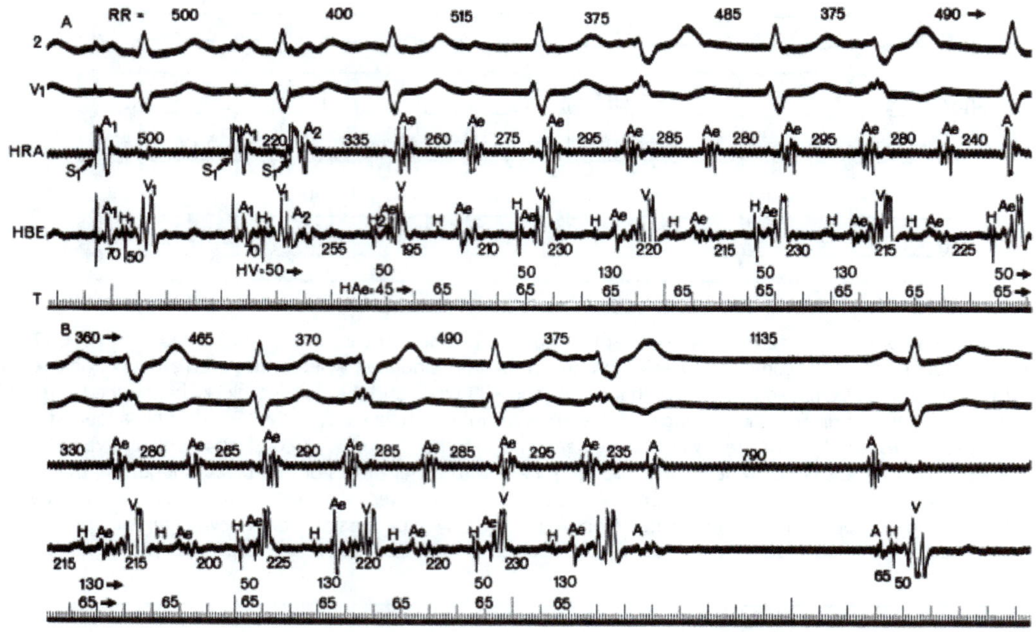

Figure 3.12. SVT and HPS WP. At a basic atrial CL of 500 ms, SVT induction with A_2 is depicted in **Panel A**. A_2 conducts with a longer A_2-H_2 and starts slow–fast AVNRT. The first echo beat blocks in the HPS (no QRS) but AVNRT continues. Subsequent to 2:1, a 3:2 block HPS (normal HV 50 ms → prolonged HV of 130 ms → BB-His) is initiated. The WP is going on in both the RBB and left bundle branch (LBB) when one examines the QRS morphology, that is, RBB block, left axis, and progressive prolongation in the HV interval and subsequent block. Sustained 3:2 WP in the HPS is interrupted by an early echo or an early (premature) atrial complex (the last A before sinus in **Panel B**). Panels A and B are continuous; the RR, AA, AH, HV, and H-RB cycles as measured in milliseconds are labeled.

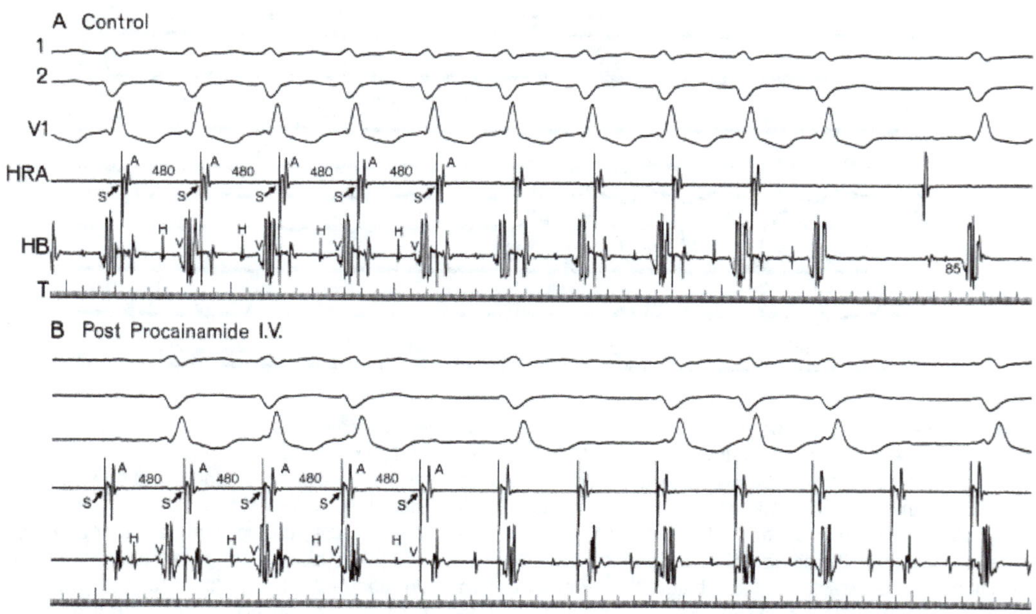

Figure 3.13. AV block within HPS post IV procainamide (PA interval). In someone suspected to have intermittent AV block (bifascicular block and long HV of 85 ms during SR), no AV block is induced during an atrial pacing CL of 480 ms (**A**). However, following procainamide infusion, the same paced atrial CL induces further HV prolongation 85 ms → 130 ms and infra-His AV block (no QRS following the fourth, sixth, and tenth paced atrial complexes) (**B**), an infrequent finding in asymptomatic patients with BBB.

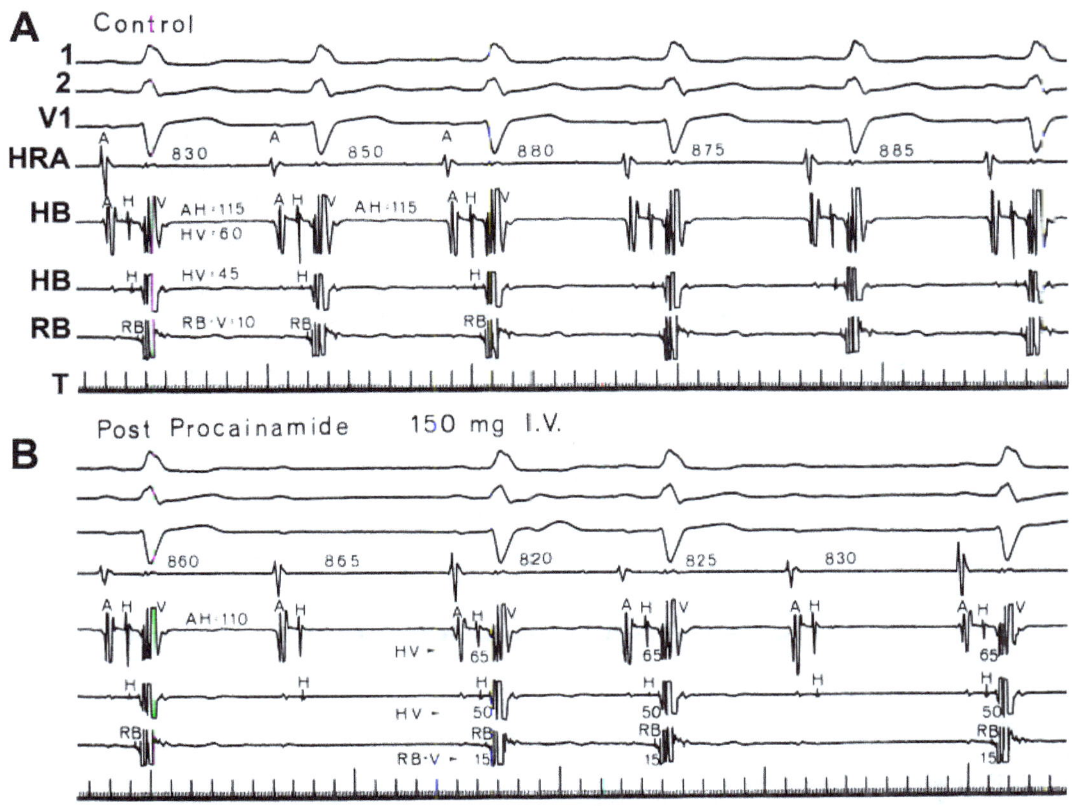

Figure 3.14. Spontaneous AV block within HPS post IV procainamide. Tracings in **Panel A** are from a patient with recurrent syncope and no inducible arrhythmia. Minimal HV prolongation develops AV block within the HPS during SR with a small amount of IV procainamide (**B**). Note the site of block along the RB axis is quite proximal (ie, H but no RB). Furthermore, it is of interest that almost all of the HV prolongation can be accounted for by prolongation of RB-V intervals during conducted beats. (Reproduced from Akhtar with permission from McGraw Hill Publishing.[50])

A word of caution is warranted in patients with preexisting BBB regarding placement of the catheters. Introduction and placement of the electrode or luminal catheters poses a risk of traumatic contralateral BBB, which may lead to development of complete AV block in the HPS requiring urgent V pacing (Figure 3.16).[19] This dramatic scenario can be avoided by being prepared for V pacing in patients with preexisting BBB when there is a possibility of trauma to the ipsilateral BB before HB recording is attempted. Particular attention should also be paid to the HV interval prolongation and axis shift during EP studies to avoid misinterpretation of the data, particularly in patients with suspected AV block-related symptoms such as syncope, presyncope, etc. On occasion, VT may also result from catheter trauma.

Premature (Extrastimulation) Techniques

The idea here is to test the integrity of the conduction system or to replicate an arrhythmia in response to premature stimulus or a series of extrastimuli. During a routine test, a train of constant CL is chosen, at the end of which a single extrastimulus is introduced. At the beginning of the process, the premature stimulus is delivered in late diastole and the coupling interval between the last CL (S_1, A_1, V_1, etc.) and the extrastimulus (S_2, A_2, V_2) is progressively shortened until one of the following is observed:

1. The desired arrhythmia is induced (Figures 3.17–3.19). If a single premature stimulus is not successful in triggering the arrhythmia in question up to the point of atrial effective refractory period (ERP), then additional extrastimuli (S_3, S_4, A_3, A_4, V_3, V_4, and so on) are introduced.[50–51] Sometimes other sites of pacing, such as the left atrium via the CS, are tested. Additionally, several atrial CLs and at least one short-to-long CL change (eg, 400 to >600 ms) followed by an extrastimulus could be scanned.

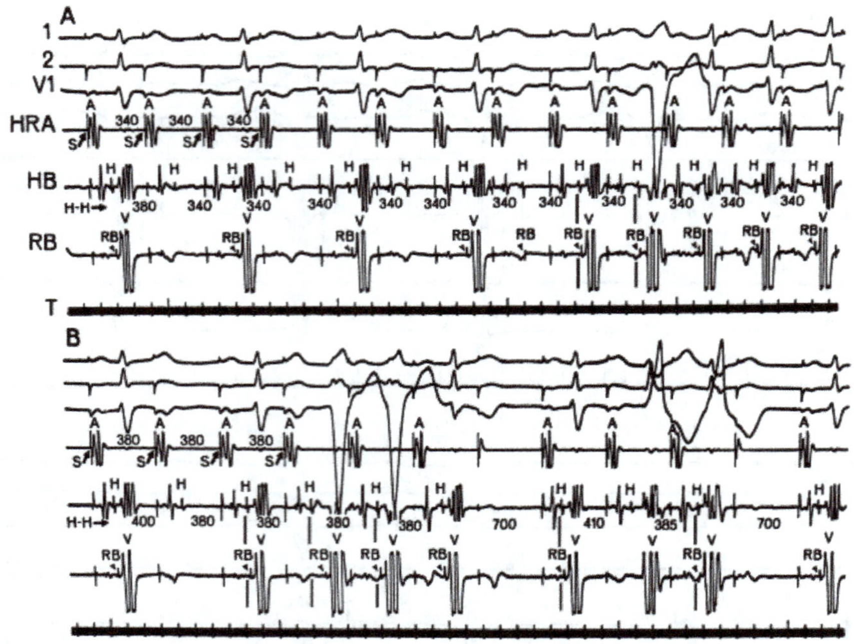

Figure 3.15. Downstream migration of HPS block. During constant CL pacing, a stable 2:1 block below HB is seen in **Panel A**. The first AH interval increase with the second atrial paced beat is due to shortening of CL from 380 ms to 340 ms (crescendo AH pattern). Block during the initial cycle is between the H and RB recording site along the RBB (ie, H but no RB). However, after 3 blocked cycles, the site of block shifts to a more distal location (the RB deflection reappears) but still no QRS is seen. During the next cycle 1:1 conduction resumes, and the first conducted cycle shows a LBBB pattern instead of AV block in the HPS. It is noteworthy that the H-RB interval prior to LBBB and the previous or subsequent normal complex is the same. Typically, the first aberrant complex shows an H-RB interval delay as seen in LBBB in **Panel B**. This is due to the fact that site of delay or block has already migrated to a more distal location. A missed atrial capture (Panel B) sets up another cycle of long–short sequence and a RBBB pattern. Again the first RBBB complex shows an increase in H-RB interval such that RB deflection merges into the QRS or the impulse blocks proximal to the RB recording site. RB deflection returns to its usual location with the second complex slowing a RBBB pattern. (Reproduced from Akhtar et al with permission from Wolters Kluwer Publishing and the American Heart Association.[14])

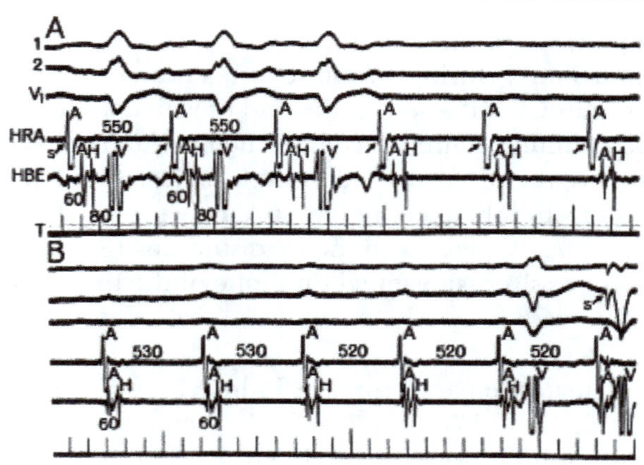

Figure 3.16. Traumatic induction of HPS block. During atrial pacing (**A**), a patient with LBBB and prolonged HV interval suddenly develops unexpected AV block. At this point, atrial pacing is suspended and ventricular pacing instituted (**B**), but spontaneous ventricular escape occurs after the eighth blocked atrial impulse in the HPS. Catheter trauma was considered and the catheter was repositioned with no further recurrence of infra-His block. (Reproduced from Akhtar with permission from Wolters Kluwer Publishing.[111])

2. AV nodal block is encountered (see Figure 3.19C). This could happen during the basic drive (A_1A_1) if the CL is too short or the AV nodal conduction is prolonged to begin with. However, a CL just short of sinus (down to 400 ms) should work in most patients with 1:1 AV conduction during baseline rhythm.
3. Atrial fibrillation (AF) is induced. This can be prevented by gentle placement of catheters and avoiding delivery of atrial extrastimuli during the atrial relative refractory period (RRP). Patients with a history of spontaneous AF are particularly vulnerable. Inadvertent AF induction may occur in any patient when multiple extrastimuli S_2, S_3, S_4, etc., are delivered at a very short coupling intervals. At these couplings, clear evidence of stimulus-to-response latency (S_2-A_2, S_3-A_3, etc.) is usually observed and can be avoided. If continuation of atrial stimulation is desirable, 3 approaches are often taken: (1) waiting for a variable period for spontaneous termination;

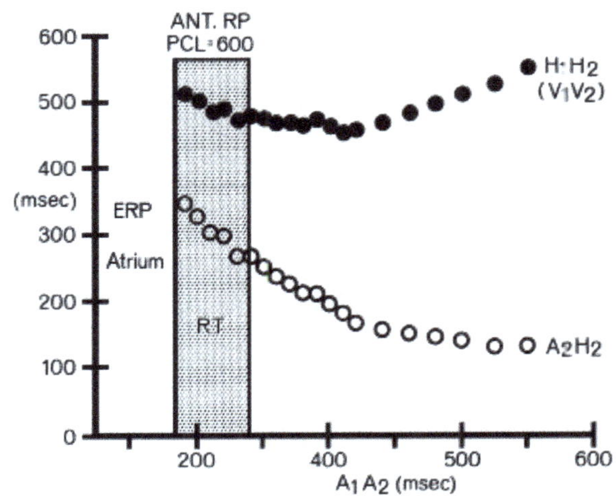

Figure 3.17. Antegrade refractory period (ANT RP) curves in a patient with slow–fast AVN RT. During scanning of PCL of 600 ms, A_2-H_2 and H_1-H_2 intervals (V_1-V_2 parallel since no H_2-V_2 changes are noted) are plotted against A_1-A_2 intervals in milliseconds. A progressive prolongation of A_2-H_2 occurs and the shortest H_1-H_2 interval, that is, the FRP–AVN interval (at 420 ms), is observed. The RT is initiated (shaded area) until the ERP of the atrium is reached. There is no clear separation or a switch from fast-to-slow pathway conduction.

(2) administering IV medication such as procainamide; and (3) direct current cardioversion (DCCV). The latter, of course, will require sedation if the patient is not already sedated.

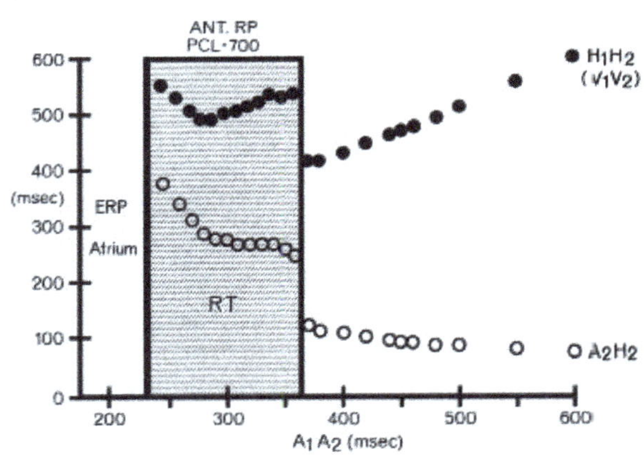

Figure 3.18. Typical dual AV nodal physiology. This figure is similar to Figure 3.17, but there is a sudden jump in A_2-H_2 and, consequently, H_1-H_2 curves, which coincides with the onset of slow–fast AV nodal RT. One can see 2 sets of FRPs (ie, first time it coincides with ERP of fast pathway, then there is a dip seen on the slow pathway H_1-H_2 curve). The basic PCL here is 700 ms. ANT. RP, antegrade refractory period. (Reproduced from Akhtar et al with permission from Elsevier Publishing [51])

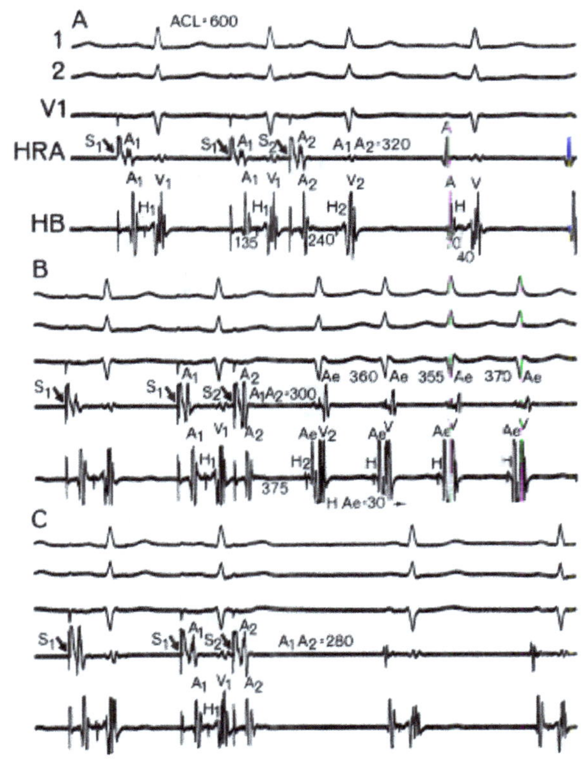

Figure 3.19. SVT induction in patients with dual AVN physiology. BCL of atrial pacing (ACL) is 600 ms and A_2 is delivered at an A_1-A_2 of 320 ms. (A). The corresponding A_2-H_2 measures 240 ms. There is no atrial reciprocal response and yet there is Pseudoterminal R wave in V1. At a 20-ms decrease in atrial coupling (B), the A_2-H_2 jumps to 375 ms. ERP fast pathway conducts via slow pathway and returns via the fast pathway (H-A_e 30 ms). A_e on HB EGM precedes local VEGM, mimicking shortening of HV interval. A sustained AVNRT is initiated. **Panel C** shows that A_2 blocks proximal to the HB recording site (ERP–AVN). AVNRT is no longer initiated. (Modified from Akhtar with permission from Wolters Kluwer Publishing.[44])

Routine atrial/anterograde RP determinations and working definitions are illustrated in Figure 3.20.[50–52] At any given basic cycle length (BCL), the ERP of the HPS is defined as the longest H_1-H_2 interval at which the impulse blocks in the HPS (ie, below the HB recording site) (see Figure 3.20A). Similarly, ERP of the AV node is defined as the longest A_1-A_2 interval that blocks in the AV node (ie, proximal to the HB) (see Figure 3.20B). Atrial ERP is the longest S_1-S_2 at which S_2 (premature stimulus artifact) is not followed by an atrial response (ATR) (see Figure 3.20C). S_1, A_1, H_1, and V_1 are the labels used for basic drive cycle. In the same way, S_2, A_2, H_2, and V_2 designations are employed for first extrastimulus and corresponding response. When more than one extrastimulus is utilized, the terms S_3, S_4, etc., are used. RRP is the

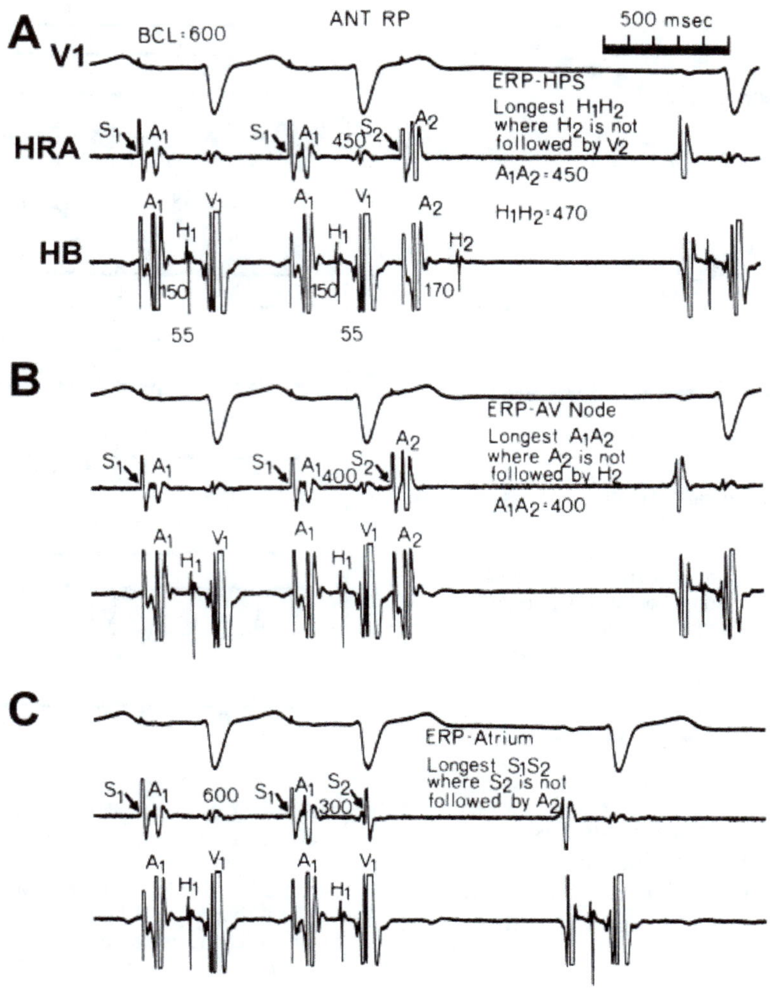

Figure 3.20. Demonstration of successive ERP during ANT RP determination. BCL and various coupling intervals (A_1-A_2) as well as the definitions of various EP parameters are labeled. **Panel A** is HPS block (H_2 but no V_2), **Panel B** is AV node block (A_2 but no H_2), and **Panel C** is atrial ERP (ie, S_2 but no A_2 response). The longest coupling interval at which these findings are expressed is the ERP. (Reproduced from Akhtar with permission from Wolters Kluwer Publishing.[111])

longest coupling interval at which the response shows delay as compared to the baseline value. These values vary at different scanned CLs. Aside from revealing some general idea about the parameters in a given individual in a given state, the values can also be compared among individuals to define the range or study the effect of cardioactive drugs. When a specific goal is in mind, the pacing protocol can be modified to fit the need.

During propagation of an atrial impulse, several tissue layers are encountered. When atrial recordings are available from various sites, such as high, low, lateral, septal, right atrium and/or left atrium, intra-atrial conduction may be delayed, and if a critical amount of delay/block intra-atrial reentry occurs, evaluation of sustained atrial reentrant tachycardia (RT), SN reentry, and various responses of the SN is possible as noted below.

Response of SN to atrial extrastimulus: Traditionally, when a premature atrial stimulus (S_2, A_2) is delivered during SR (A_1) (Figure 3.21A), it responds in 4 different ways, depending upon the timing of the stimulus (see Figure 3.21B–E).[28,53,54]

1. Complete compensation. Extrastimuli during the late-diastolic interval collide with oncoming sinus impulses, such that the last sinus → premature → return sinus cycle (RC) equals 2 sinus cycles, that is, $A_1 - A_2 + A_2 - RC = 2\ A_1\ A_1$ (see Figure 3.21B).
2. With further reduction in the coupling interval, the atrial impulse penetrates the SA node and

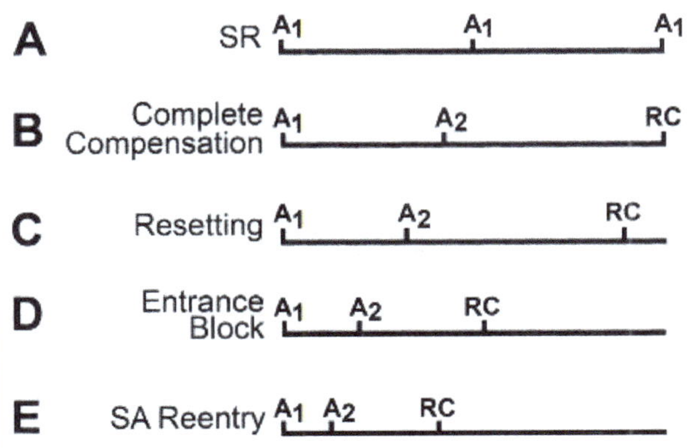

Figure 3.21. Schematic responses of SA node to extrastimulation. **Panel A** reflects SR. **Panels B–E** show various deliveries of atrial extrastimuli (A), and the intervals between A_1 (sinus) and A_2 and RC are measured. See text for explanation.

resets the sinus so that $A_1 A_2 + A_2 - RC < 2 A_1 A_1$ (see Figure 3.21C).

3. SA entrance block. If the $A_1 - A_2$ shortens further, the A_2 may not be able to enter the sinus and thus would not disturb the impulse generation. Consequently $A_1–A_2 + A_2 - RC = 1 A_1–A_1$ (see Figure 3.21D).
4. SA reentry.[54,55] At closer $A_1 A_2$'s, the atrial impulse may enter some portion of the SN and exit from the SA node to re-excite the atrium. This is called SN or SA reentry in which $A_1 A_2 + A_2 - RC < 1 A_1 A_1$ (see Figure 3.21E). Patients with clinical SN reentry may have a wider range of atrial coupling intervals where nonsustained or sustained SN reentry may be initiated such that $A_1 A_2 + A_2 - RC$ may exceed one sinus cycle ($A_1 A_1$) (Figure 3.22).[45]

While the SN reentry displays the same P-wave morphology (see Figure 3.18) as garden variety sinus tachycardia, the latter differs in 3 key areas:

1. Garden variety sinus tachycardia is secondary to increased adrenergic drive and/or parasympathetic withdrawal and hence is secondary to other triggers, such as exercise, fever, congestive heart failure (CHF), medications, etc., and does not present as a clinical problem *per se.*
2. It is gradual in onset and termination as the primary cause starts and improves. Hence, it feels like appropriate rate acceleration to meet the body needs and is seldom sensed by the individual as inappropriate. SN reentry is triggered by extrastimuli (see Figure 3.22). It is abrupt in onset and termination and hence perceived by the patient as abnormal and is frequently the reason for a patient to seek medical advice. Inappropriate sinus tachycardia (different clinical entity) has a somewhat excessive rate for the needs of the body and is also felt by patients as an unusually fast heart rate requiring medical attention.
3. Unlike the usual form of sinus tachycardia, SN reentry is not triggered by the adrenergic drive and therefore, concomitant AV node conduction is prolonged in the form of a longer PR interval (AV nodal delay) or occasionally as type 1, second-degree AV nodal block (see Figure 3.22). This is akin to A pacing where type 1, second-degree AV nodal block is the rule; the atrial rate increases as the CL of the A pacing is shortened. This type of AV nodal block is not

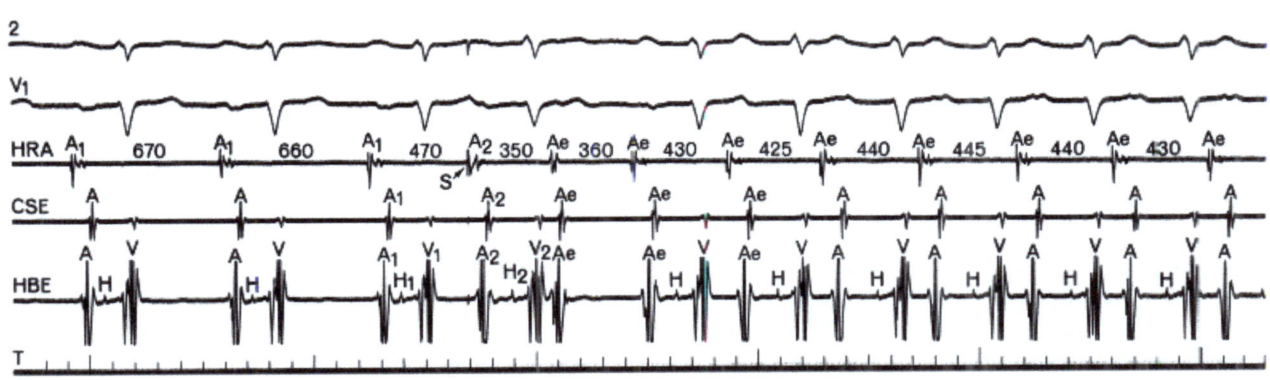

Figure 3.22. SA reentry. During SR (A_1-A_1), a relatively late single atrial premature beat (A_2) is delivered in the HRA region. SA RT is initiated. Note the sequence of atrial activation is similar to SR. During the tachycardia, which represents a sudden acceleration of atrial rate, a first- or second-degree AVN block (longer AH or no H) is observed. CSE, coronary sinus electrogram. (Reproduced from Akhtar et al with permission from Elsevier.[45])

much different than seen during the atrial rate acceleration that occurs in clinical atrial tachycardia (AT), atrial flutter (AFl), and AF. Since increased adrenergic drive is the mechanism of rate acceleration in the usual variety of sinus tachycardia and most cases of inappropriate sinus tachycardia, there is concomitant enhancement of AV nodal conduction such that the PR interval (AV nodal conduction) does not seem to change.

Response of the AV node to atrial extrastimuli: Generally, a minimal A_2-H_2 prolongation is noted with extrastimulation (A_2) delivered in late diastole compared to the baseline A_1-H_1 interval (RRP-AV node) (around 500 ms in both) (see Figures 3.17 and 3.18). As the A_1-A_2 decreases, further A_2-H_2 increases are seen. At some point during the scanning, the shortest H_1-H_2 interval is reached in response to progressive shortening in the atrial coupling interval. This so-called functional refractory period (FRP) of the AV node (see Figures 3.17 and 3.18) bears a somewhat parallel relationship to the shortest H_1-H_2 interval achieved during rapid input, such as AF. The V_1-V_2 (R_1-R_2) interval curve parallels the H_1-H_2 curve, assuming the corresponding H_2-V_2 interval remains unchanged from the baseline. Further abbreviation in A_1-A_2 will result in one or more of the following:

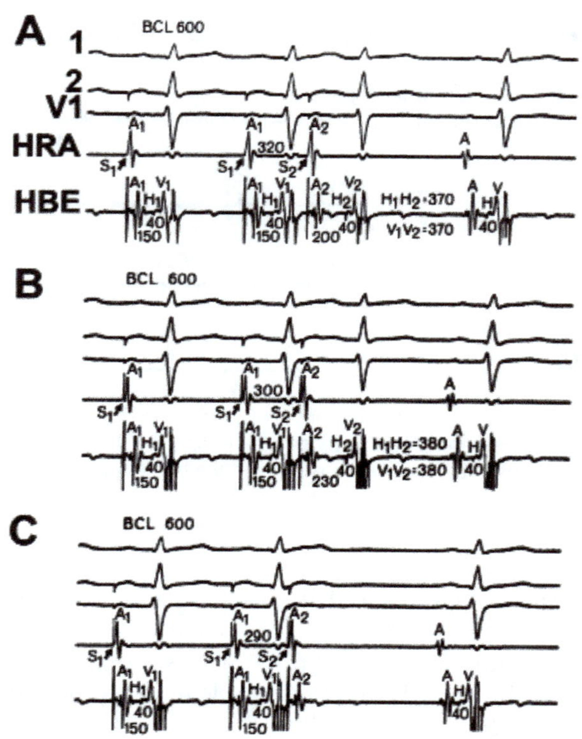

Figure 3.23. AVN response to A_2. BCL of 600 ms is scanned with progressive shorter A_1-A_2 (only 320 to 290 ms are shown). A_2-H_2 prolongation is noted in **Panel A** and **Panel B** (200 and 230 ms, respectively) compared to the baseline A_1-H_1 (150 ms), and A_2 fails to reach the HB (ie, blocks in the AVN-ERP-AVN, **Panel C**). ERP, effective refractory period.

1. A sudden jump in A_2-H_2 (and hence H_1-H_2) interval may occur suggesting functional duality of the AV nodal pathways, meaning a switch of conduction from a fast- to a slow-conducting pathway (see Figure 3.18). This jump in A_2-H_2 may (see Figure 3.18) or may not be associated with initiation of a SVT. A 50-ms increase in A_2-H_2 with a 10-ms decrease in atrial coupling interval is generally considered as a sufficient manifestation of dual AV nodal physiology.
2. ERP of the AV node is encountered and continued AV nodal block is noted with further reduction in the A_1-A_2 interval (Figure 3.23).
3. RRP of the atrium is encountered—although the S_1-S_2 is being shortened, A_1-A_2 cannot be further abbreviated due to increasing latency between the stimulus and ATR (Figure 3.24).[56]
4. On rare occasions A_2 may conduct at shorter coupling intervals, while it is blocked at longer coupling intervals. This is akin to the gap phenomenon in which a proximal AV nodal delay will allow a later arrival at the site of initial block in the AV node.[56] This type of horizontal dissociation has not been systematically studied. A similar phenomenon referred to as 1,2,3,4 phenomenon has been studied in canine models and both horizontal and longitudinal dissociation of AV nodal pathways have been offered in explanation.[57,58] Another mechanism for such an observation may be simply a change in the autonomic tone, which is at play at all times. The AV node is richly innervated and easily influenced by autonomic tone.
5. With dual AV nodal physiology, a double response may occur exclusively via the AV node, conducting over the fast and then the slow AV nodal pathway (Figure 3.25). Both of the QRS complexes will be preceded by an HB potential as well as an H_2 and H_3 with HV interval equal to or greater than the baseline HV interval. The V_2 and V_3 morphology will be more of a function of relative timing of H_1, H_2, or H_3 intervals and other associated prior or concurrent electrophysiologic events.

Response of HPS to atrial extrastimuli: At the usual changes in heart rate, the normal HPS does not show much variation. However, HPS is very quick to respond to abrupt CL changes. The AV node is intervening tissue encountered between the atria

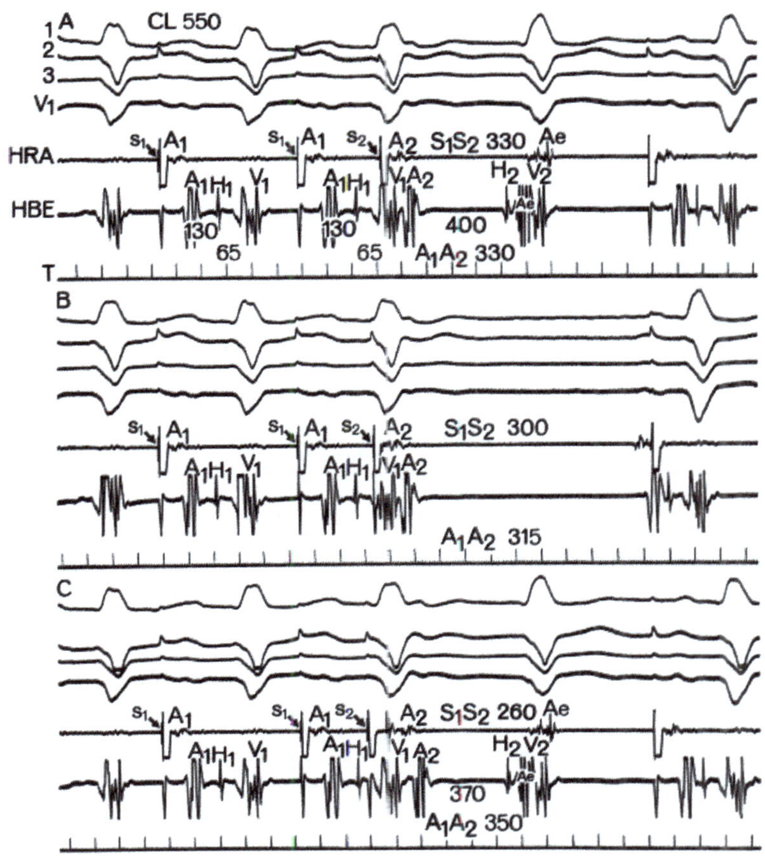

Figure 3.24. AV nodal ERP and atrial latency. The illustration depicts progressively shorter S_1-S_2 of 330, 300, and 260 ms in **Panels A** through **C**, respectively. Basic atrial-PCL is 550 ms, and the QRS complex shows a LBBB pattern. In Panel A, S_1-S_2 = A_1-A_2 and A_2 (330 ms) conducts with A_2-H_2 of 400 ms and is followed by QRS and an AV nodal reentrant A_e beat (low–high atrial sequence). The shorter S_1-S_2 (Panel B) shows a stimulus-to-atrial-response latency such that the resulting A_1-A_2 is 315 ms. A_2 nonetheless blocks in the AV node (ie, no H_2 response). However, at further shortening of S_1-S_2 (Panel C), there is a marked latency (90 ms), and the A_1-A_2 now measures 350 ms. AV conduction resumes and so does the AV nodal reciprocal (A_e) beat.

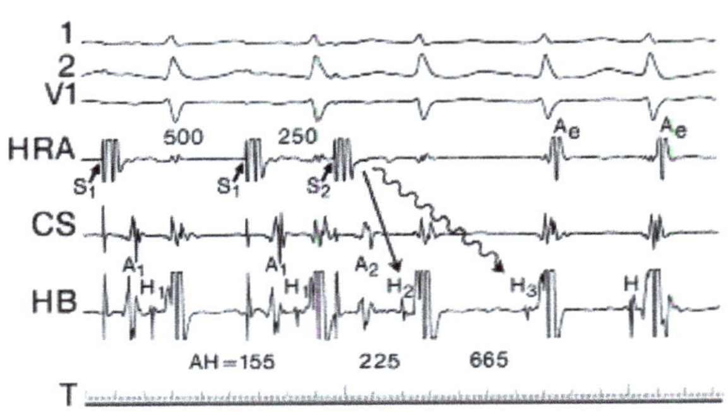

Figure 3.25. 1:2 response in the AV node. At a drive CL of 500 ms, the premature atrial complex (S_2, A_2) at A_1-A_2 of 250 ms conducts to H_2 via the fast pathway and concurrently to H_3 via the slow AV nodal pathway; in essence, 2 QRS complexes result from one extrastimulus. Slow–fast AV nodal RT with the usual AV relationship begins after H_3. (Reproduced from Jazayeri et al with permission from Elsevier Publishing.[52])

and the HPS during atrial extrastimulation and its conduction and RP properties control what reaches the HPS. At an atrial CL of ≥500 ms, particularly at a CL nearing sinus, some HPS electrophysiologic properties can usually be studied (Figure 3.26).[59] With normal intraventricular AV nodal conduction and RP, aberrant conduction can be demonstrated in roughly 30% of cases. This is particularly true during atrial bigeminy (Figure 3.27), in which the longest HH cycles are followed by the shortest HH cycle achieved without using a specifically designed pacing protocol.[60,61]

It has been previously stated that, at relatively long CLs, right bundle branch block (RBBB) tends to have a longer RP than left bundle branch block (LBBB).[62] While this may be the case in some patients, in many others, both LBBB and RBBB can be observed at the same CL (Figure 3.28). As a

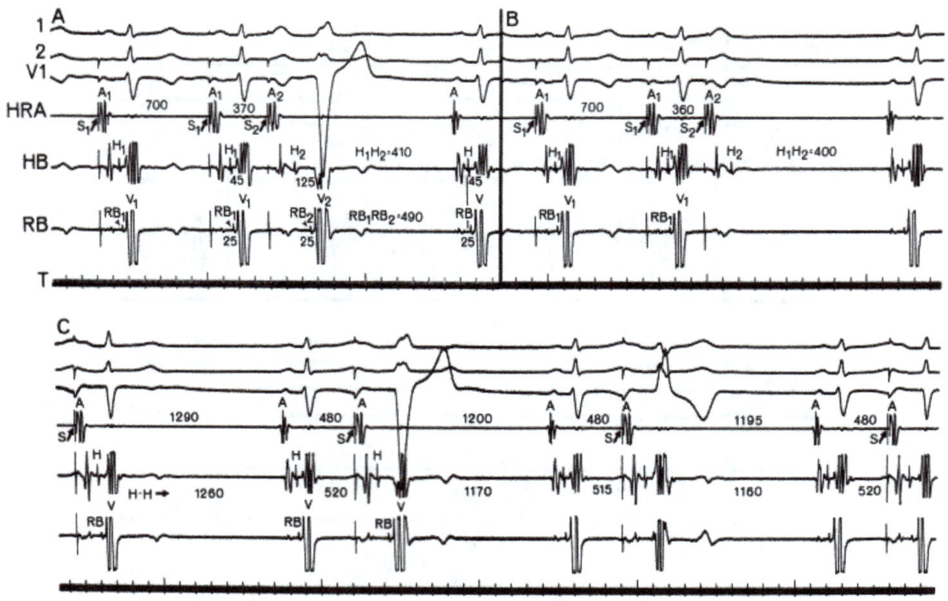

Figure 3.26. HPS response to A_2. In **Panels A** and **B**, the BCL is 700 ms. A_2, delivered at 370-ms coupling (Panel A), conducts with minimum A_2-H_2 prolongation but marked prolongation in H_2-V_2 and all of this additional delay is between the H_2-RB_2 recording site. Prolonged HV and LBBB demonstrate that there is even a greater delay along the His-LB axis. At 10 ms, shorter atrial coupling of the A_2 blocks distal to the HB recording site. Note there is H_2 but no RB_2 nor QRS. During the atrial bigeminal rhythm, both LBBB and RBBB are seen with alternate A_2 complex coupled to a sinus escape CL of similar duration, not an uncommon phenomenon. Again, the conduction delay is between the H_2-RB_2 recording sites. During RBBB, the RB recording either merges into V_2 or H_2 blocks between H_2-RB_2. The alternation between RBBB and LBBB is explained in Figure 3.27 because all of the CLs are controlled. See text for details (Reproduced from Akhtar et al with permission from Wolters Kluwer Publishing and the American Heart Association.[14])

rule, however, while RBBB may occur without H_2-V_2 prolongation in response to A_2, it is rare to see a lack of H_2-V_2 prolongation preceding QRS showing a LBBB pattern, regardless of the BCL (see Figure 3.26).[61] A block below the HB recording site can be achieved in about 20% of patients with normal AV conduction at baseline. RBBB and LBBB, as well as block below the HB, can be demonstrated even more frequently with atrial bigeminy (see Figure 3.27).[61,62] During the conduction of anterograde RP studies, both RBBB and LBBB and/or bilateral block may be encountered on occasion and normal conduction may resume at shorter coupling intervals. This is called gap phenomenon and is one explanation why a shorter A_1-A_2 conducts normally, whereas longer A_1-A_2 encounters AV block.[63] In type-I gap, the initial site of delay or block is in the HPS and the site of proximal delay is in the AV node, allowing distal recovery and resumption of conduction through the HPS (Figure 3.29).[63,64]

In type-II gap, the site of initial block as well as the proximal region of slower conduction permitting distal recovery (and hence return of AV conduction) is also in the HPS (Figures 3.30 and 3.31). Both of the foregoing are alternative explanations for what may be interpreted as supernormal conduction on the surface ECG.[10,63,64]

When clinicians encounter BBB, it is often difficult to distinguish it from a complex originating in the ventricle. This is a particularly important issue during AF and sustained wide QRS tachycardias. Gouaux and Ashman described the sequences of long–short as more diagnostic for aberrant conduction, and this can be used in conjunction with characteristic QRS morphology.[17,65–69] QRS configuration in various forms of broad QRS complex tachycardia is separately presented elsewhere in this book and therefore, will not be discussed here. However, several related points should be mentioned:

1. Aberrant conduction occurs as a result of conduction delay or block in the HPS, which precedes the QRS and the delay could be marked (Figures 3.31–3.33). Therefore, depending on the HH interval preceding the QRS, aberrant conduction may actually have a

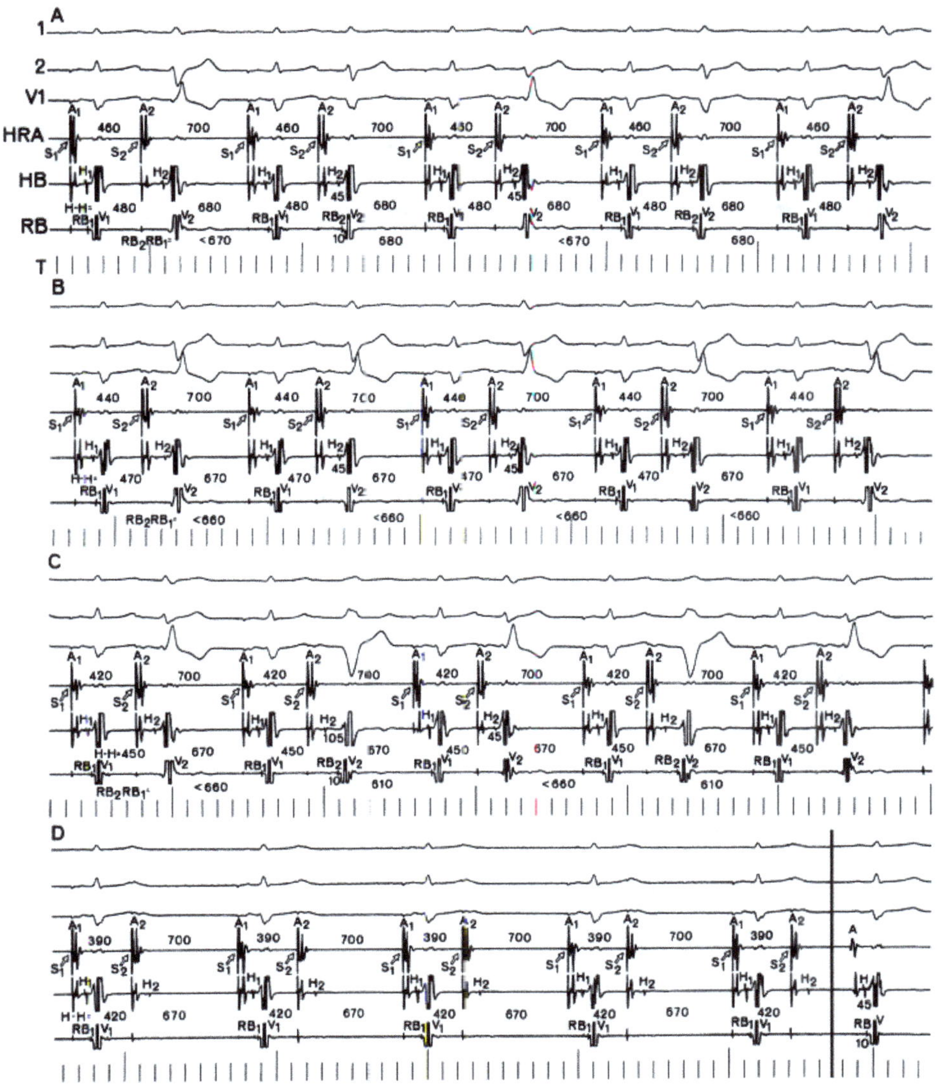

Figure 3.27. Response of HPS during atrial bigeminy. A controlled atrial bigeminy where the long CL is 700 ms (ie, A_2-A_1) and the atrial coupling (A_1-A_2) is 460 ms in **Panel A**, 440 ms in **Panel B**, 420 ms in **Panel C**, and 390 ms in **Panel D**. The last complex in Panel D is a reference sinus beat with H-RB interval of 10 ms. All of the relevant values are labeled under the corresponding EGM tracing. Since H deflection can be recognized in all panels, the values are labeled as measured, but RB_2 potentials are not identified preceding QRS complexes showing RBBB (Panels A–C) or BB-His (Panel D). The arrows indicate that the measurement is made from V_2 EGMs preceding RBBB configuration QRS because RB_2 potential is not recognizable, which shortens the RB_2-RB_1 CL and hence the next A_2 conducts normally. The site of conduction delay and/or block is proximal to the RB recording site (Panel C: preceding LBBB; Panel D: block below the His recording site). The mechanism of alternation is similar to the one between the RBBB and the normal A_2 conduction seen in Panel A. The most likely reason for abbreviation of the refractoriness of the BB is that the block is caused by dual excitation from retrograde preexcitation and forward migration from the previous site of local block. This would result in relative prolongation of refractoriness of contralateral BB at similar A_2-A_1 and A_1-A_2 CL. (Reproduced from Denker et al with permission from the American Society for Clinical Investigation.[61])

short–long sequence or a long–short sequence. So the gold standard for distinction between impulse of ventricular versus aberrant conduction is HH interval preceding VV interval, not the latter preceding the former. When the former is not available, some doubt will remain about the accuracy of the diagnosis (see Figures 3.32 and 3.33). Most examples of Ashman phenomenon are based on the analysis of only long–short CL sequence. If the CL before

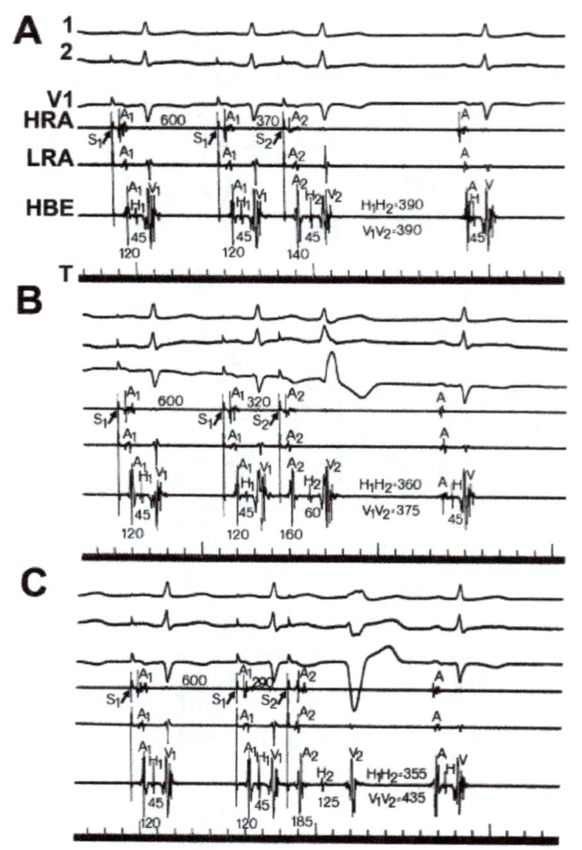

Figure 3.28. **A:** HPS response to A_2. At the same BCL of 600 ms, both RBBB and LBBB aberrancy is noted (**Panels B** and **C**, respectively). In both instances, however, the H_2-V_2 prolongation is noted prior to the V_2, suggesting that RRP/ERP of the contralateral bundle or HB has already been encountered. (Reproduced from Denker et al with permission from Elsevier Publishing.[67])

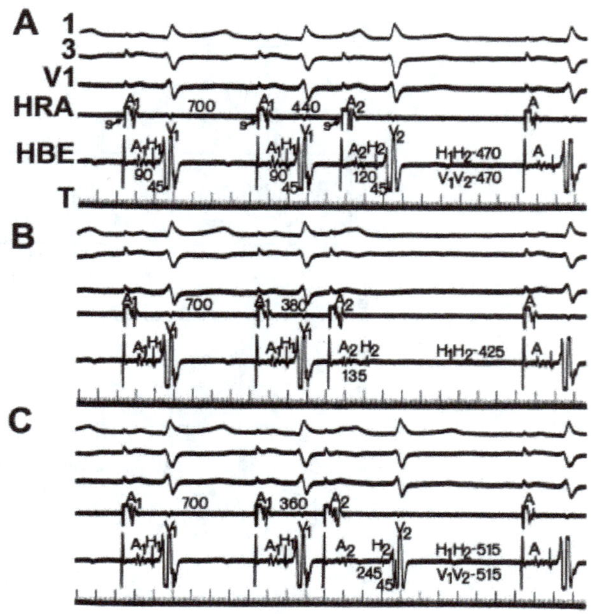

Figure 3.29. Type-I gap phenomenon in the HPS. A_2 conducts to the ventricle in **Panel A**, and there is BB-His in **Panel B** at a shorter H_1-H_2 of 425 ms. At an even shorter coupling interval of 360 ms (**C**) compared to Panel B, A_2 resumes conduction because the resultant increase in A_2-H_2 now produces an H_1-H_2 interval of 515 ms, which is outside the ERP and RRP of the HPS. (Reproduced from Akhtar et al with permission from Wolters Kluwer Publishing and the American Heart Association.[63])

the long CL is further dissected, many aspects of aberrancy not generally realized emerge.[68] For example, short–long–short sequence produces more frequent aberrant conduction than long–long–short or long–short–short sequence (Figure 3.34).

2. Aberrancy also occurs at long–short CL sequence, but it requires shorter H_1-H_2 coupling when compared to constant CL and long–short–short CL (Figures 3.35 and 3.36).[68,69]
3. Different fascicles may respond differently to these abrupt CL changes.[70] The mechanism of this almost universal electrophysiologic behavior was not readily obvious based on simple CL analysis. However, another explanation relating to the role of diastolic interval duration of the HPS rather than the action potential duration (APD) seemed a more likely mechanism and is schematically depicted in Figure 3.37.[70]

During atrial extrastimulation in most patients showing aberrant conduction with normal intraventricular conduction (IVC) at baseline, the location of initial conduction delay or block is proximal, ie, within the HB or proximal BB. This can be demonstrated with the use of additional recording along the His-BB axis (Figure 3.38).[14] Occasionally it may be distal to the right-bundle recording site, at least during RBBB aberrancy (Figure 3.39).[14] As a rule, the site of block shifts to a more distal location with second, third or more extrastimuli, or even constant CL (Figure 3.40).[14] Many years ago, Narula showed normalization of QRS in patients with LBBB and left anterior fascicular (LAF) block by distal HB pacing.[71] This suggested that the site of block is quite proximal, seen even in these pathological examples.[71,72] A small lesion within the HB, along with anisotropic conduction, could create this clinical scenario of asymmetric conduction downstream.

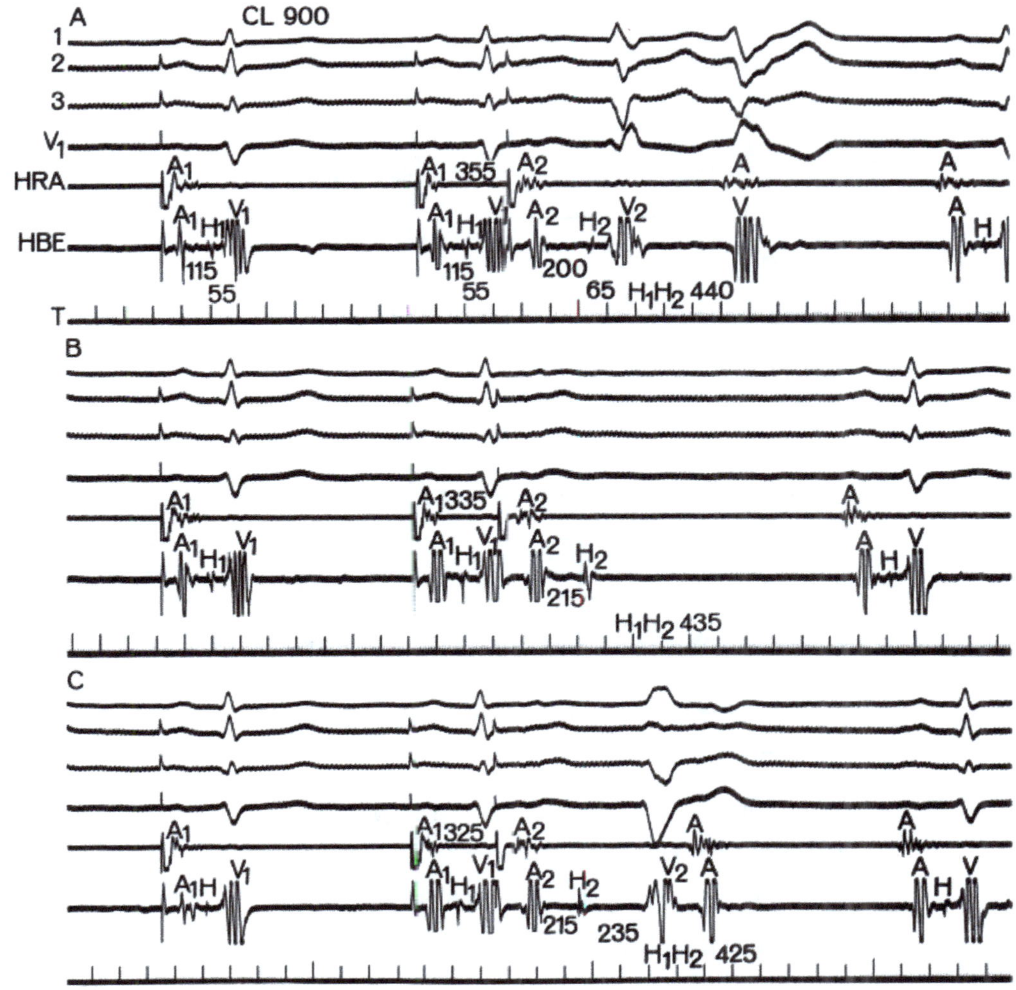

Figure 3.30. Type-II gap phenomenon in the HPS. Like Type 1 gap phenomenon, the initial site of block is the HPS (**B**). However, unlike Type-I gap, the site of proximal delay allowing recovery of the initial site of block in HPS is also in the HPS. So in **Panel C**, with A_2 at a shorter A_1-A_2 of 325 ms and an H_1-H_2 of 425 ms, the H_2 impulse manages to reach the ventricle. Due to the marked prolongation of H_2-V_2 (235 ms), the postblock V_1-V_2 (605 ms, Panel C) is much longer than preblock (450 ms, A), representing marked conduction delay along the RBB. (Modified from Akhtar et al with permission from Wolters Kluwer Publishing and the American Heart Association.[63])

Forward migration of block in the HPS is probably the most common reason for spontaneous resolution of functional HPS block in the setting of a constant CL (see Figure 3.15). However, with abrupt change in CL (during aberrancy), other mechanisms may be involved in resolution of the block. It is generally believed that once a BBB occurs, its continuation is maintained by a retrograde invasion from the contralateral ventricle. This is referred to as linking phenomenon, which in turn is sustained either by collision of the 2 impulses or by interference created by the refractoriness at the level of penetration by the previous impulse (Figure 3.41).[73] The link can be broken by spontaneous or induced premature complexes. The premature impulse will collide or preexcite the HPS retrogradely and shorten its RP for the oncoming supraventricular impulse, which is then likely to conduct with a normal or narrower QRS complex (Figure 3.42). Another mechanism of spontaneous disappearance of HPS block is the slowing of the CL for any reason; this is most commonly seen in clinical settings with the use of medications or vagal stimulation. Spontaneous or induced premature atrial or ventricular complexes also abolish aberrancy by prolonging the post-ectopic RC, which may then exceed the RRP-ERP of the HPS. Similarly, continuous pacing or SVT with persistent aberrancy may change to

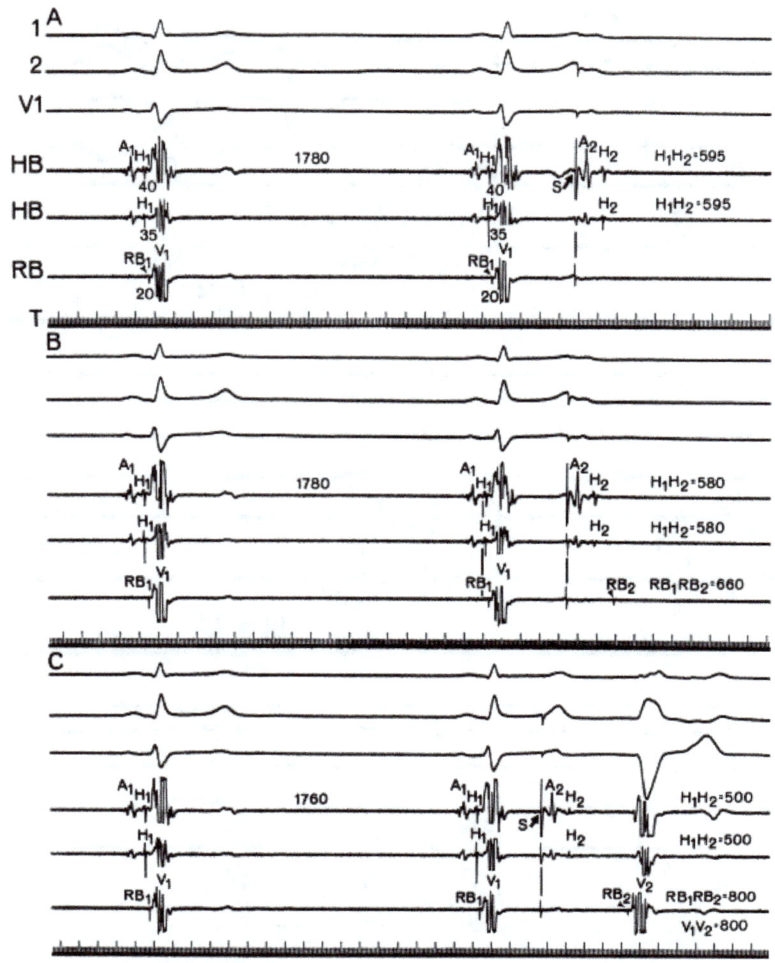

Figure 3.31. Proximal delay and serial distal recovery sites in HPS. Baseline normal HPS conduction, that is, HV 40 and 35 and RB-V of 20 ms during SR are depicted. In **Panel A**, A_2 impulse blocks distal to the His but proximal to the RB recording site at H_1-H_2 of 595 ms. **Panel B** shows that the location of the H_2 impulse block migrates distally beyond the RB recording site (H_2-RB_2 = 80 ms vs baseline of 15–20 ms) and activates RB_2, but still blocks more distally (no QRS) even though the H_1-H_2 is shorter (580 ms). **Panel C:** Marked prolongation of the H_2-RB_2 interval is illustrated at an H_1-H_2 of 500 ms, which restores conduction to the ventricle via the RBBB. All of the conduction delays noted are proximal to the RB recording site. The resulting QRS has a LBBB pattern and exemplifies that, during a functional aberrancy, there can be a marked discrepancy between the underlying HPS conduction and the surface R-R intervals, even in the setting of baseline normal HPS. (Modified from Akhtar et al with permission from Wolters Kluwer Publishing and the American Heart Association.[14])

normal conduction of an atrial complex blocking in the AV node or HPS (Figure 3.43).

While persistent aberrancy can be abolished with extrastimuli, the published examples are mostly generated in the EP laboratory. Seldom is a clinical spontaneous persistent aberrancy like this documented hours after onset.[66] Forward migration of block by progressive deeper penetration of the successive impulses is the likely reason for spontaneous disappearance of HPS block as stated earlier because it can be shown that retrograde penetration from contralateral ventricle (ie, linking) is not necessary to maintain or abolish the block (see Figures 3.15 and 3.43). Transient functional HPS blocks are common but rarely persist for a long time and other mechanisms such as premature complexes are not critical to abolish the block (see Figures 3.15 and 3.43).

Response of accessory pathways (AP): This chapter will not be complete without some mention of AP and how they respond to A pacing and extrastimulation. Common AV connection (the so-called Kent bundle, which may be overt or concealed) is an example.[74,75] For our purposes, a clarification of the term concealed AP is in order. While most definitions imply that concealed pathway

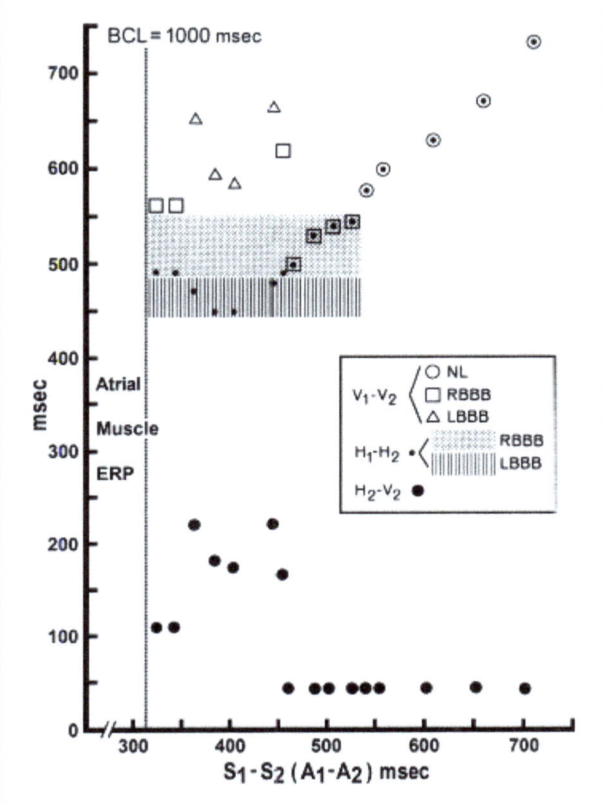

Figure 3.32. Surface ECG versus HH correlation. During scanning of a basic atrial cycle length (BCL) of 1000 ms, H_1-H_2, H_2-V_2, and V_1-V_2 are illustrated at various A_1-A_2 intervals in milliseconds. The type of resulting QRS morphology, that is, normal (circle), RBBB (square), and LBBB (triangle), are also displayed. Note that during normal (NL) QRS and initial onset of RBBB (up to A_1-A_2 of 460 ms), there is no H_2-V_2 prolongation and therefore H_1-H_2 = V_1-V_2. Significant H_2-V_2 prolongation occurs at shorter A_1-A_2 and corresponding H_1-H_2 both with RBBB and LBBB (above the shaded area). This is particularly the case at A_1-A_2 of 440, 430, and 350 ms. The resultant H_1-H_2 are associated with marked H_2-V_2 prolongation, such that the V_1-V_2 measure 620, 660, and 630 ms, respectively. When compared with the shortest NL V_1-V_2, R-R interval recorded in this patient at this BCL was 580 ms. From a surface ECG, this will give one the impression that the wide QRS complexes with V_1-V_2 ≥ 620 ms are of ventricular origin. This point is made in schematic form in the next figure. ERP, effective refractory period. (Reproduced from Denker et al with permission from Elsevier Publishing.[67])

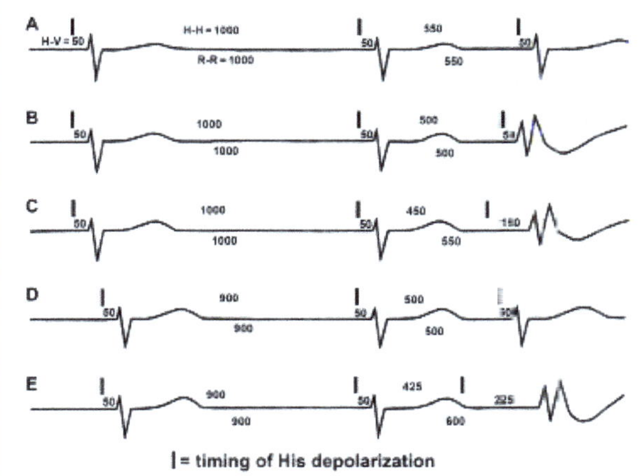

Figure 3.33. Schematic representation of HH and R-R interval relationship. The duration of HH and R-R intervals for long and short CL are labeled. **Panels A** to **C** show a long CL of 1000 ms, while the CL is 900 in **Panels D** and **E**. The vertical solid marker is the projected timing of the His potential. Aberrantly conducted QRS complexes (notched) are either preceded by the same HV as baseline (Panel B) or prolonged HV (Panels C and E). The whole depiction in Panel B fits the classical description for aberrant conduction, while wide QRS in Panels C and E would be interpreted as ventricular in origin when compared to Panels A and D. (Reproduced from Denker et al with permission from Elsevier Publishing.[67])

means only retrograde conduction with no preexcitation manifested on surface ECG, this definition is deficient, because while AP may not be detectable on the surface ECG, anterograde conduction over the AP may still be intact. At least 4 scenarios can be described where this is true:

1. The manifest preexcitation is a result of relative velocity of AV conduction via the normal pathway (NP) versus the AP. In AFP, the preexcitation is generally concealed due to slow conduction and only manifests when AV conduction over the NP is slower (Figure 3.44) than that of the AFP.[76,77] A combination of block in NP and slow conduction over AFP usually starts the preexcited tachycardia (see Figure 3.44B).
2. Occasionally, a more rapidly conducting pathway (Kent bundle) may not show ventricular preexcitation when located farthest from the SN or pacing site (Figure 3.45), such as left free wall, particularly when the NP conduction is enhanced.
3. Intra-atrial block may change the direction of impulse propagation and, because of its diversion, only conduction via the NP reaches the ventricle.[74] In all of the examples mentioned above, the overt preexcitation can be demonstrated by pacing near the atrial insertion of the AP. This would indicate that true concealed AP is a diagnosis confidently made only with intracardiac studies, unless the patient has already demonstrated lack of AP conduction with a spontaneous bout of atrial premature complexes, tachycardia, flutter, or fibrillation.
4. When there are 2 or more AP, one or more may remain concealed (Figure 3.46). For example, right or septal pathways may conceal

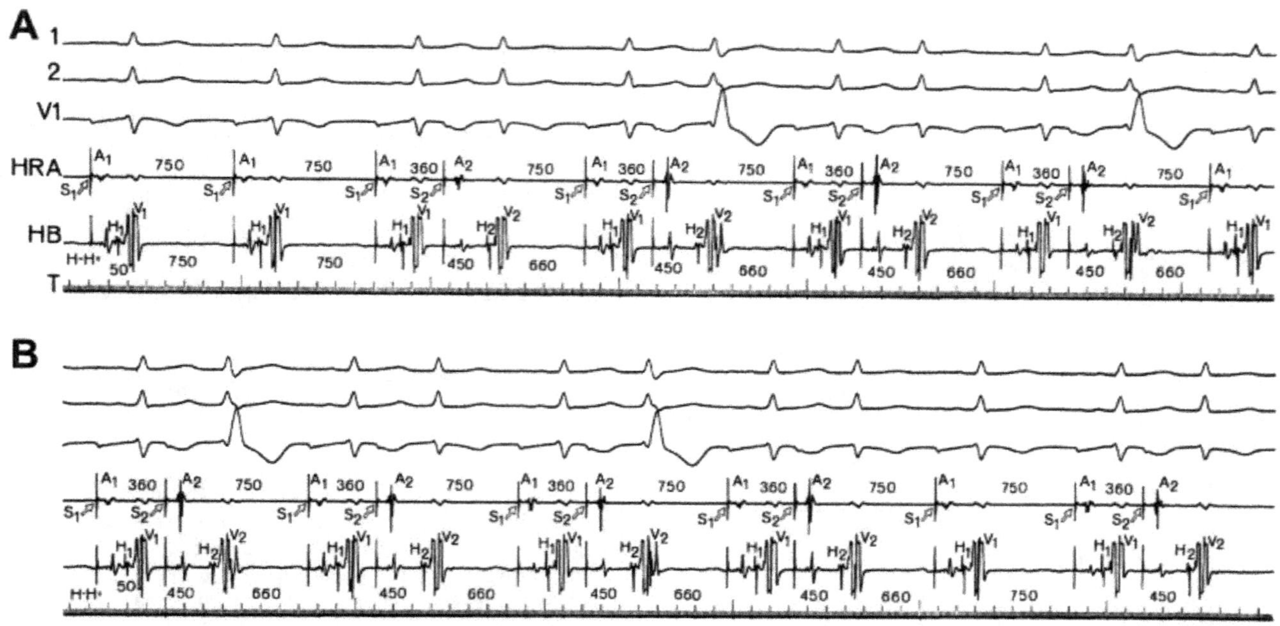

Figure 3.34. **A and B:** HPS response to constant ventricular stimulation: short-to-long CL. The tracings start and end with 2 constant atrial CL of 750 ms (A_1-A_1). An atrial or premature complex (A_2) with an atrial coupling interval of 360 ms is delivered, which alternates with the longer CL of 750 ms. This long–short bigeminal pattern is repeated (the 2 panels are not continuous). The first H_1-H_2 of 450 ms is not followed by aberrancy (long–long–short). The next H_1-H_2 is the same but is preceded by a much shorter H_2-H_1 of 660 ms, due to A_2-H_2 prolongation in response to prior A_2. Despite that, the resulting QRS shows a RBBB (short–long–short). This shortens the RB_2-RB_1 CL significantly and also causes a long–short sequence, consequently the next A_2 conducts without aberrancy. This process continues until the end at which time 2 constant A_1-A_1 CL end the process without aberrant conduction. (Reproduced from Denker et al with permission from the American Society for Clinical Investigation.[61])

a left free wall AP because of the proximity of the former to the SN and AV node to the septum, despite similar AP properties (see Figure 3.46A and B).[78]

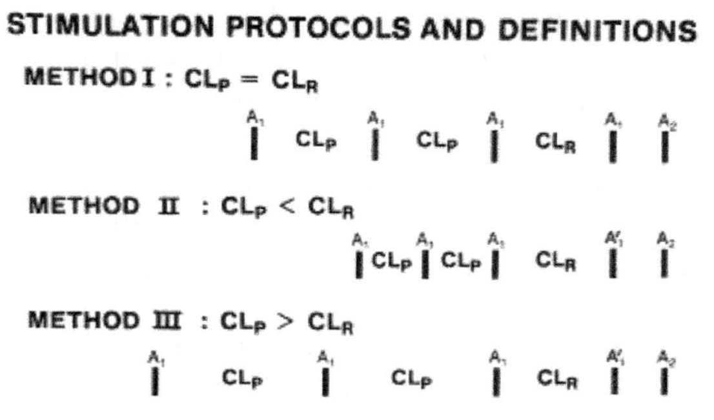

Figure 3.35. Protocols for constant S-L and L-S sequences. Schema shows a series of constant cycle lengths preceding (CL_P) the last so-called reference cycle length (CL_R). The A_2 is delivered, and the HPS's response is noted. In Method I, CL_P and CL_R are the same, whereas in Method II, a series of short CL_P are followed by a long CL_R that equals the CL_R in all panels. The reverse is true when CL_P is longer than CL_R (Method III). (Reproduced from Denker et al with permission from Wolters Kluwer Publishing and the American Heart Association.[17])

Incremental A pacing produces a progressive increase in AH interval, leading to a shorter HV and greater preexcitation. Depending upon conduction delay or block in the AV node, various degrees of fusion to fully preexcited complexes can be noted. During maximum preexcitation, it is not always clear whether the atrial impulse is simply delayed or blocked in the AV node. When the RP of the AP is longer than the NP, a block in the AP causes conduction over the NP with initiation of orthodromic tachycardia. The reverse, ie, antidromic SVT, is infrequent due to the differences in the conduction and RP properties of AP versus the NP in the 2 directions. This has been quite extensively studied, and is said to be the general experience among clinical electrophysiologists and has been illustrated in a clinical model of ventricular preexcitation (Figures 3.47–3.49), which shows a low likelihood of antidromic reentry quite similar to clinical observation.[79] In patients with ventricular preexcitation and AV nodal reentry (coexistence is about 15%), the latter may be initiated during A pacing with a narrow or preexcited QRS. Preexcited tachycardia can also occur through other mechanisms such as 2 AP (see Figure 3.46), one used

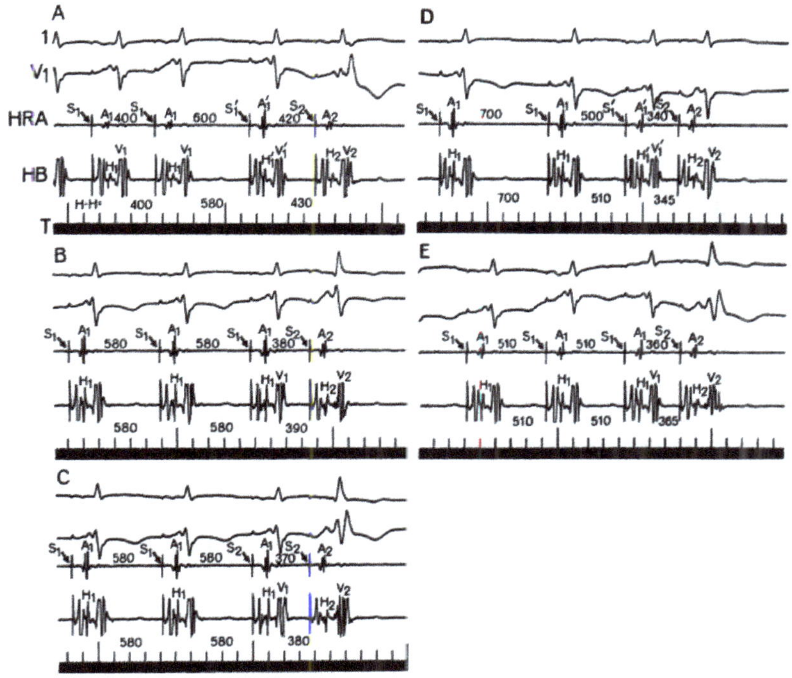

Figure 3.36. Graphic illustration of HPS response with sudden changes in CL. The tracing shows HPS quickly adjusts and, in fact, overshoots the refractoriness in the direction of CL change. A 400 to 600 ms change in CL (**A**) is compared to a constant CL of 580 ms (**B** and **C**). The 580-ms interval is chosen for the constant CL because there is a 20-ms shortening of AH with a 400 to 600-ms atrial CL change. Note at least a 40-ms increase in the RRP/ERP of RBB. The reverse is true (**D** and **E**) when the HPS refractoriness shortens with a long-to-short change in CL. These examples convey a convincing message that the CL, prior to the long–short CL sequence, should be examined in cases where the mechanism of wide QRS, that is, aberrant versus ventricular origin, is uncertain, particularly in the setting of AF. (Reproduced from Denker et al with permission from Wolters Kluwer Publishing and the American Heart Association.[17])

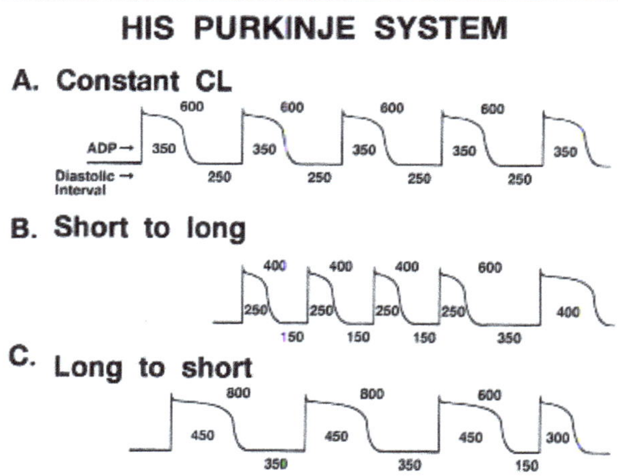

Figure 3.37. Proposed mechanism for HPS response with sudden change in CL. Schema depicts HPS-APD and diastolic interval in milliseconds. The CL_P is 600 in **Panel A**, 400 in **Panel B**, and 800 in **Panel C**, while the CL_R is 600 in all 3 panels. The APD of the last complex reflects changes in refractoriness before the premature atrial beat (A_2, not shown). The duration of the diastolic interval and its effect on the duration of the next APD is the reason why a short-to-long change in CL prolongs RRP–ERP–HPS and a long-to-short change does the opposite (see Figure 3.36). (Reproduced from Akhtar et al with permission from Elsevier Publishing.[68])

anterogradely and the other retrogradely, flutter AT, etc. This makes for an interesting differential diagnosis and will be detailed elsewhere in this book under a separate title dealing with preexcited tachycardia.

Rapid A pacing should be avoided to prevent inadvertent induction of AF unless it is intentional and desirable for clinical reasons.

Response of Common AV Connection (Kent Bundle) to Atrial Extrastimulation

1. With progressive shortening of A_1-A_2, AV nodal delay makes the preexcitation more obvious. During this time, the H_2-V_2 interval gets progressively shorter to a point where H_2 is obscured by the local V_2 EGM and may occasionally emerge on the other side of the V_2 EGM, or A_2 simply blocks in the NP. At a negative HV value of around 50 ms transseptal conduction time, ie, now VH interval (His deflection after the QRS), fully preexcited QRS complex is expected. At full preexcitation, the A_1-A_2 is, in fact, more

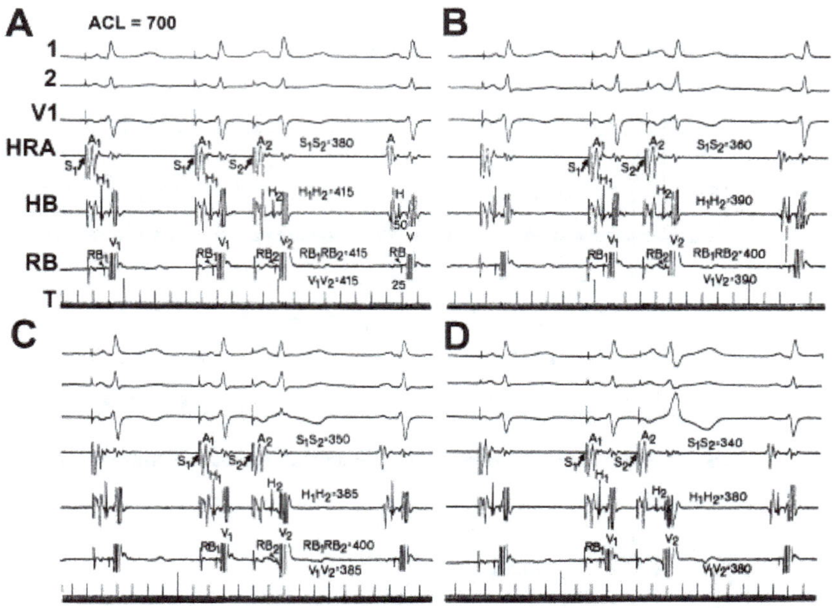

Figure 3.38. Proximal site of block in HPS. ACL is 700 ms in all panels. The H-RB interval measures 25 ms during sinus (**A**). In **Panels B** to **D**, A_2 is delivered at progressively shorter A_1-A_2 intervals, and a gradually increasing conduction delay in the RBB on the surface ECG is more obvious in lead V_1. Concurrent with surface ECG changes, there is also an H_2-RB_2 interval increase (Panels B and C), and RBB potential can no longer be recognized in Panel D. It either merges into the V_2 EGM, or there is simply block between the H and RB recording sites. See Figures 3.26, 3.27, and 3.31 for proximal and distal H-RB delay/block and resumption of RB conduction with marked delay. (Reproduced from Akhtar et al with permission from Wolters Kluwer Publishing and the American Heart Association.[14])

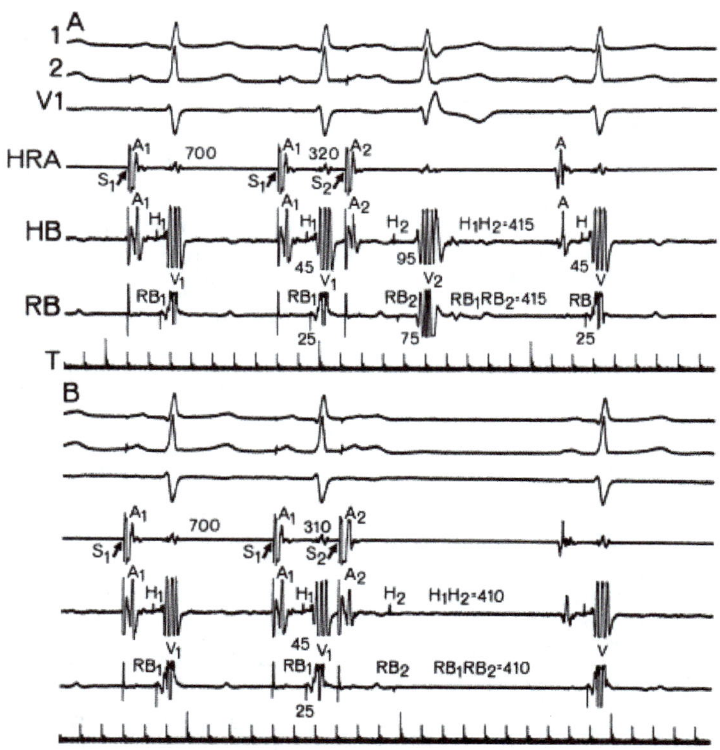

Figure 3.39. Distal block in the HPS. In this case, the A_2 shows a conduction delay (**Panel A**) and block distal to the RB recording site (**Panel B**), an observation less common for the first aberrant complex. See also Figure 3.31. (Reproduced from Akhtar et al with permission from Wolters Kluwer Publishing and the American Heart Association.[14])

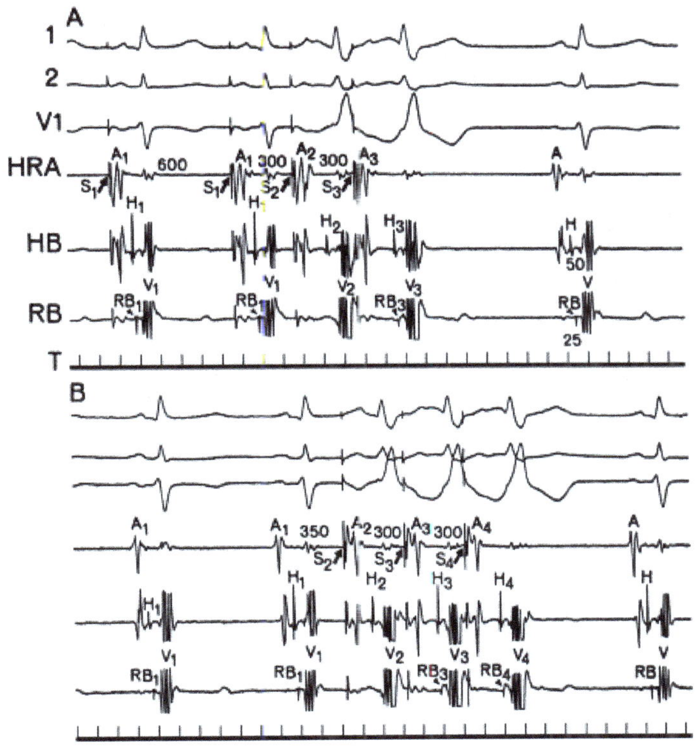

Figure 3.40. Distal migration in site of HPS block with successive impulses. During a paced atrial CL of 600 ms (A) and SR (B), 2 and 3 extrastimuli, respectively, are delivered. Note that in both panels, A_2 blocks proximal to the RB recording site (ie, RB_2 is not identifiable). While A_3 and A_4 block distally, RB_2 deflection returns, suggesting more distal migration of the block site along the H-RB axis. The mechanism of this migration is either local shortening of the RP at or beyond the site of block by anterograde impulse (nonpropagated) or retrograde penetration of the right bundle to the area of block, or both. (Reproduced from Akhtar et al with permission from Wolters Kluwer Publishing and the American Heart Association.[14])

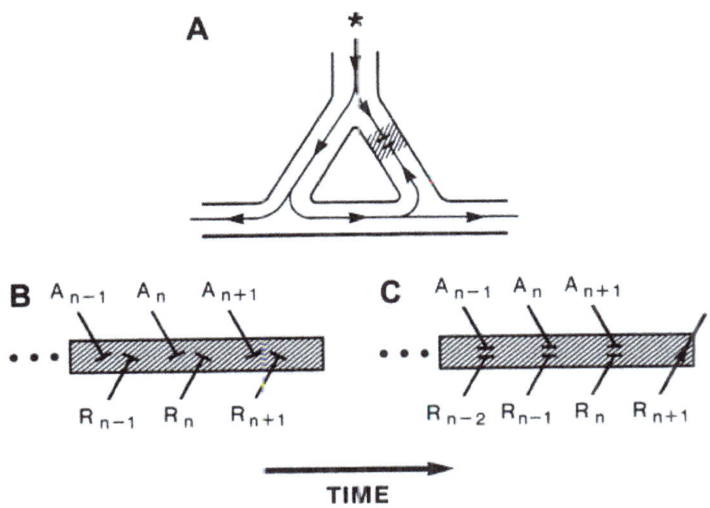

Figure 3.41. Linking phenomenon. **Panel A** shows a classical schema of reentry in which there is block of anterograde impulse (hatched area) followed by penetration by retrograde impulse. This block can be sustained by a link of the 2 impulses; the anterograde (A) and retrograde (R). At least 2 mechanisms can be offered for perpetuation of this block. **Panel B** depicts one mechanism (linking by interference) in which an impulse blocks but, in its wake, leaves the area refractory for the next impulse traveling from another direction. Linking by collision is shown in **Panel C**. There is no RP barrier, but the 2 opposing impulses collide by virtue of arriving at the same time. The site of linking by collision is not predetermined, unlike linking by refractoriness (interference), in which the sites of previous blocked impulses determine the site of linking. (Reproduced from Lehmann et al with permission from Wolters Kluwer Publishing and the American Heart Association.[73])

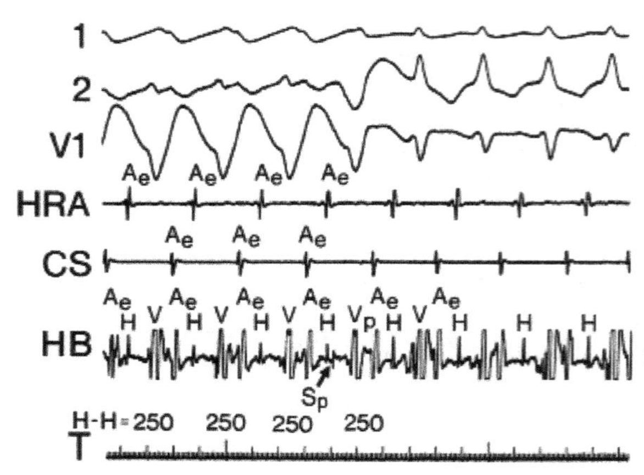

Figure 3.42. Abolition of aberrant conduction with a single ventricular extrastimulus. Orthodromic SVT with LBBB aberrancy is shown. First, 3 complexes are aberrantly conducted coincident with introduction of the complex ventricular extrastimulus (S_p, V_p) soon after the anterograde His is activated, which retrogradely preexcites the distal right bundle and then the LB and shortens the RP of both BBs. Subsequent tachycardia shows narrow QRS, despite the fact that there is no change in HH interval. (Reproduced from Jazayeri et al with permission from John Wiley and Sons Publishing.[59])

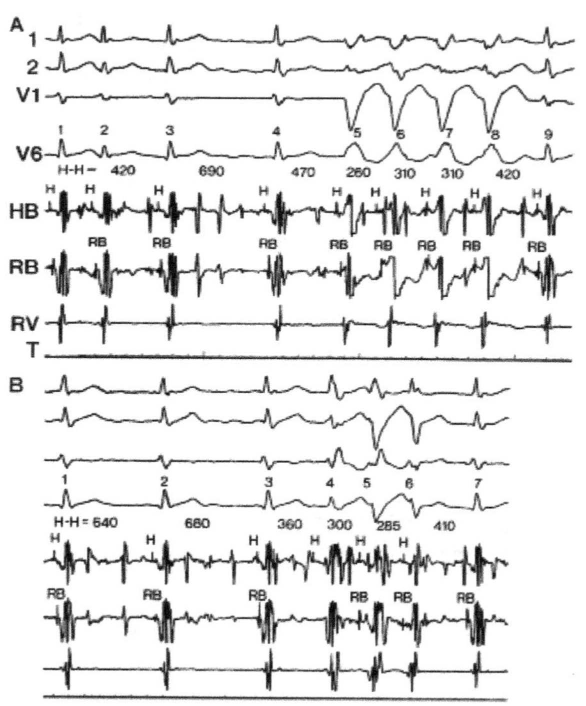

Figure 3.43. Spontaneous abolition of aberrancy. In both panels, the underlying rhythm is AF. Short runs of left- (**A**) and right-bundle block patterns (**B**) are followed by a pause long enough that the duration is outside the RPs of both branches and QRS normalizes. (Reproduced from Jazayeri et al with permission from John Wiley and Sons Publishing.[59])

like V pacing (V_1-V_2) at the site of ventricular insertion of the AP, and many of the phenomena outlined later under ventricular extrastimulation apply here (see Figures 3.47 and 3.48). When the ERP of the AP is longer than the NP, A_2 blocks in the AP and conducts over the NP to initiate the orthodromic tachycardia (Figure 3.50). If the ERP of the AP is shorter than the NP, orthodromic tachycardia is less likely to be initiated with A_2, but antidromic tachycardia can be initiated. Other methods may be necessary to initiate orthodromic reentry: (a) pacing near the site of atrial insertion of the AP; in fact, it is advisable that both the right and left atrial sites be stimulated as a rather simple way to expose multiple AP as well as determine if there is truly a concealed AP; (b) multiple extrastimuli A_3, A_4, etc., might be successful in initiating the orthodromic SVT; (c) ventricular extrastimuli are more reliable in starting orthodromic SVT when anterograde AP-ERP is shorter than NP-ERP; (d) if it is critical to start orthodromic SVT with A pacing, IV procainamide may prolong AP-ERP sufficiently to allow SVT induction.

2. The rapidity of ventricular response in AF in patients with Wolff-Parkinson-White (WPW) syndrome is one of the most feared arrhythmias. It is important to define this risk in symptomatic patients and those individuals at high risk, such as athletes and professionals in public transportation (pilots, bus drivers, etc.). Wellens et al pointed out many years ago that when the AP-ERP is ≤270 ms, these patients are prone to develop rapid ventricular rate during AF via the AP. This identifies a high-risk group.[80] Since adrenergic stimulation drives the AP-ERP even lower, it is not clear if AP-ERP of >270 ms with adrenergic influence may identify a larger group at high risk. Many investigators believe AF should be induced in symptomatic patients to assess true risk, and this should be dealt with while the patient is still in the laboratory. The use of exercise stress testing is the noninvasive option to assess the integrity of AP, but it mostly works to identify the group at lower risk, showing a block in the AP during exercise. More often than not, exercise enhances AV conduction via the NP and may lead to lesser AP contribution to ventricular activation, giving an erroneous impression of AP block. Similarly, it has been reported that patients who show AP block after administration of IV medications such as procainamide may be less prone to adverse consequences during AF.[81]
3. With overt preexcitation, at times, the AV nodal conduction delay may be sufficient to produce a

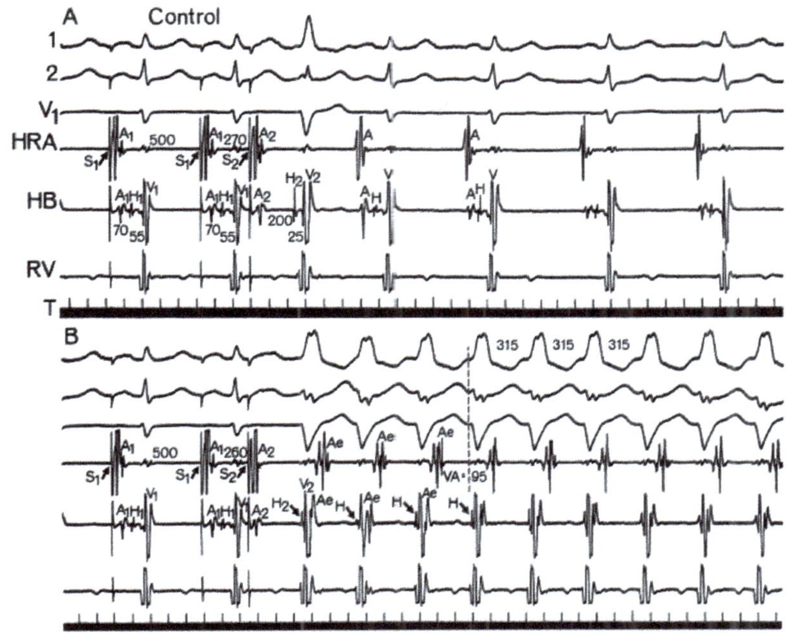

Figure 3.44. Concealed conduction in AFP. **Panel A** shows SR, normal PR, AH, and HV intervals and no evidence of AP on the surface ECG. At a 500-ms atrial pacing drive and an A_1-A_2 of 270 ms, there is a marked PR prolongation (from A_1-V_1 of 125 ms to A_2-V_2 of 225 ms). Note the decrease in HV interval to 25 ms and the concurrent appearance of preexcitation on the surface ECG with an incomplete LBBB appearance. A 10-ms shortening in A_1-A_2 (**B**) results in full preexcitation and the start of a preexcited (antidromic) tachycardia with a classic LBBB and left anterior descending (LAD) artery morphology typical of SVT in patients with AFP. (Reproduced from Akhtar with permission from John Wiley and Sons Publishing.[110])

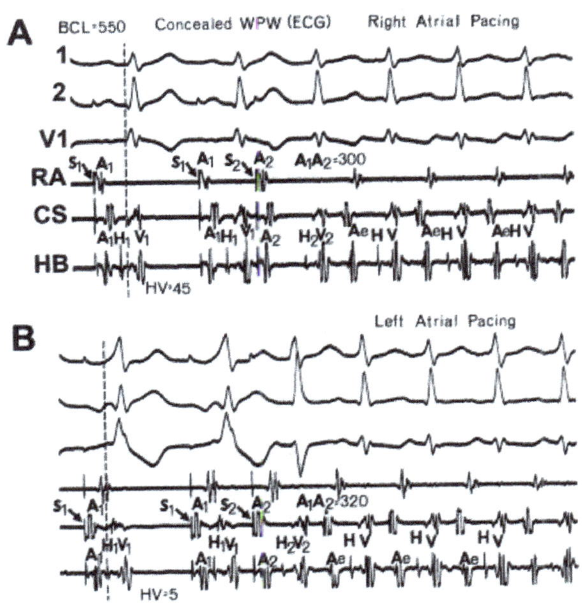

Figure 3.45. Concealed conduction in AP on surface ECG. **Panel A** shows no ventricular preexcitation during right-atrial pacing and induction of orthodromic tachycardia (retrograde atrial activation via a left-sided AP; CSE is the earliest). When paced close to atrial AP insertion (**B**, CS pacing), preexcitation is quite obvious. BCL, basic cycle length.

1:2 response, that is, first, QRS over the AP and second, QRS via the AV node-HPS (Figure 3.51).[82] Invariably, the second QRS is preceded by H_2 deflection, with H_2-V_3 intervals equal to or longer than the baseline and a V_3 complex that is wide and often mimics V_2. It is important to recognize that, in patients with manifest preexcitation, A_1-A_2 intervals also produce corresponding V_1-V_2 stimulation of equal duration due to direct input into the VM via the accessory connection. Hence, V_3 may not necessarily be secondary to A_2 with anterograde H_2, but instead may be a BB reentrant complex (see Figure 3.51) via retrograde H_2. To clearly distinguish the 2, right bundle branch (RBB) recordings are particularly helpful.

4. HPS plays a significant role, both in orthodromic and antidromic SVT in WPW syndrome. During the performance of atrial RP, for example, occurrence of BBB may facilitate orthodromic tachycardia initiation when the AP is ipsilateral to the BBB by providing additional but critical delay (Figure 3.52). This also plays a role in diagnosis, where a change in ventriculoatrial (VA) time is noted when the BBB or fascicular block occurs or is abolished during both ortho- and antidromic

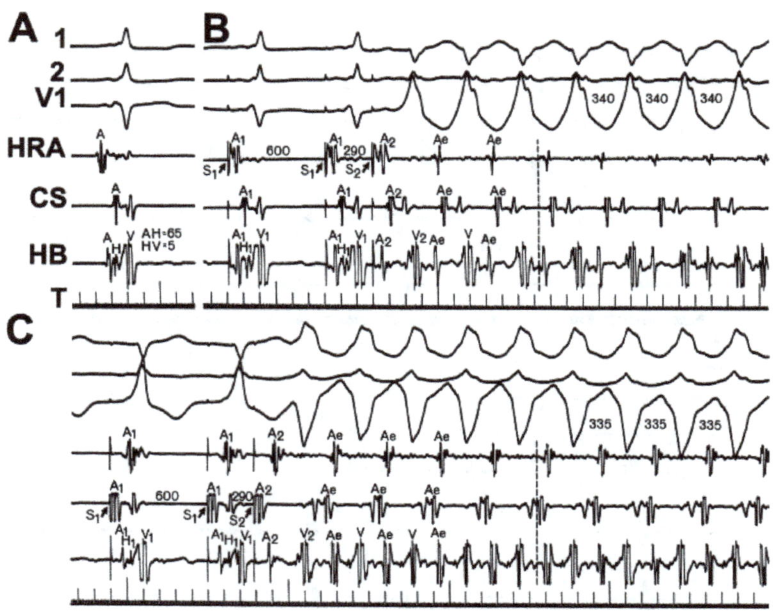

Figure 3.46. Concealed conduction along one pathway in the presence of 2 AP on surface ECG. In **Panel A** (sinus) and **Panel B** (right-atrial pacing) there is evidence of anteroseptal preexcitation but no evidence of left free wall (LFW) AP. A_2 delivered at 290 ms blocks in the anteroseptal AP, conducts via the LFW AP and starts a preexcited tachycardia. **Panel C**, on the other hand, shows CS pacing with no evidence of anteroseptal preexcitation until A_2 blocks in the LFW AP, and the preexcited tachycardia, using the anteroseptal AP, starts. Retrograde atrial activation sequences are shown by the perpendiculars. (Reproduced from Akhtar with permission from John Wiley and Sons Publishing.[110])

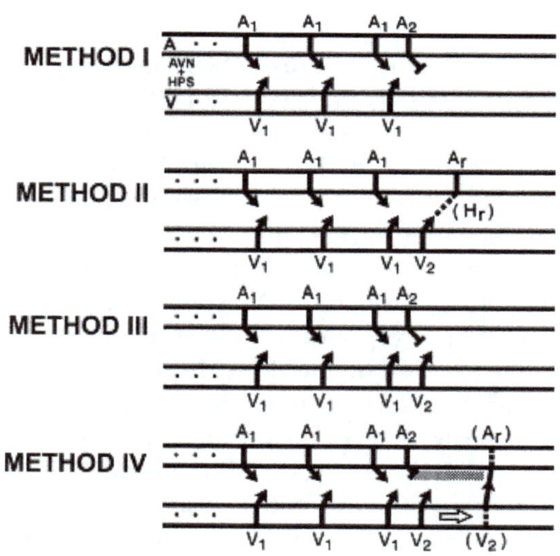

Figure 3.47. Model of antidromic reentry with A_2. The pacing methodology is depicted, where A_1-V_1 sequential pacing during the basic drive, with a predetermined AV interval, as PR (in all panels) is scanned. A_1-A_2 and V_1-V_2 of equal duration with A_2 or V_2 alone is seen in Method I or II, respectively; Method III is the complete preexcitation model, that is, A_1-V_1 = A_2-V_2. Method IV illustrates how long a V_1-V_2 interval conducting to the atria must be before it is possible to get a single ATR of antidromic reentry following A_2 block in the AVN. (Reproduced from Lehmann et al with permission from Wolters Kluwer Publishing and the American Heart Association.[79])

SVT (Figures 3.50–3.54).[69,83] The CL of SVT on the surface ECG/intracardiogram generally shows shortening of R-R interval with narrow QRS versus ipsilateral BBB. On occasion, the converse may occur, in which case disappearance of ipsilateral BBB is associated with a switch of AV nodal conduction from fast to slow pathway (Figure 3.55) and R-R interval may show prolongation of CL. In such instances, the R-R interval change on the surface ECG may be misleading. Regardless, VA time at transition remains the most reliable reproducible change for which a ≥25 ms increase in VA with ipsilateral BBB is indicative of ipsilateral free wall AP. Concurrent changes on the surface ECG are noted less frequently in anteroseptal pathways with RBBB and posteroseptal pathways with LBBB.

AFP: Mahaim?

Not long ago, this entity did not exist in the literature. There were extensive reviews of Mahaim connection, traditionally divided into nodoventricular and fasciculoventricular Mahaim. A systematic diagnosis of the existence of atriofascicular fibers was made in the EP laboratory

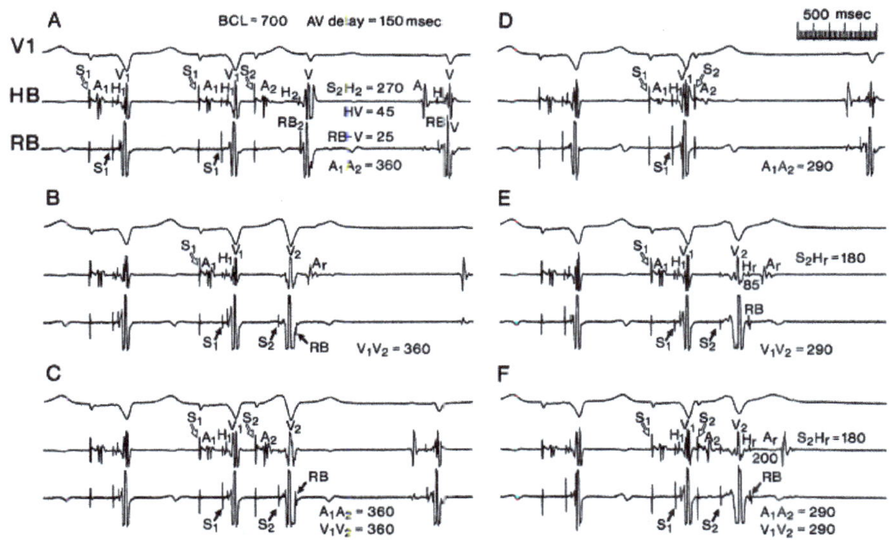

Figure 3.48. Graphic illustration of preexcitation model with AV delay of 150 ms. BCL S_1-S_1 (A_1-A_1 and V_1-V_1 is 700 ms and A_1-V_1 interval is 150 ms in all panels). It can be noted that in this case, A_1-V_1 is longer than A_1-H_1, so the A and V collide in the HPS. Response to A_2 alone at A_1-A_1 of 360 ms is seen in **Panel A**. All the pertinent values are labeled. In **Panel B**, V_2 alone with V_1-V_1 of 360 ms produces recognizably retrograde RB_2 activation and retrograde A_r can be identified. A_2-V_2 of 150 ms, introduced with the same V_2 coupling of a 360-ms impulse blocks in the AV node, that is, no A_r (**C**). At A_1-A_2 of 290 ms, A_2 alone blocks in the AV node (**D**) and V_2 alone at V_1-V_2 of 290 ms conducts via LB branch (H-RB shorter) as compared to 20 ms in sinus (Panel A). When both A_2 and V_2 are delivered with A_2-V_2 of 150 ms, A_2 blocks in the AV node. V_2 conducts with a longer retrograde H-A_r delay (200 ms), revealing concealed conduction of A_2 in the AV node (**E**). If antidromic tachycardia was to follow, it would be as a result of V_1-V_2, although one would assume that A_2 initiates the antidromic reentry (**F**). The model helps to explain the individual role of AV nodal-HPS in both directions; it is not possible to illustrate this in an actual patient with overt preexcitation. (Reproduced from Lehmann et al with permission from Wolters Kluwer Publishing and the American Heart Association.[79])

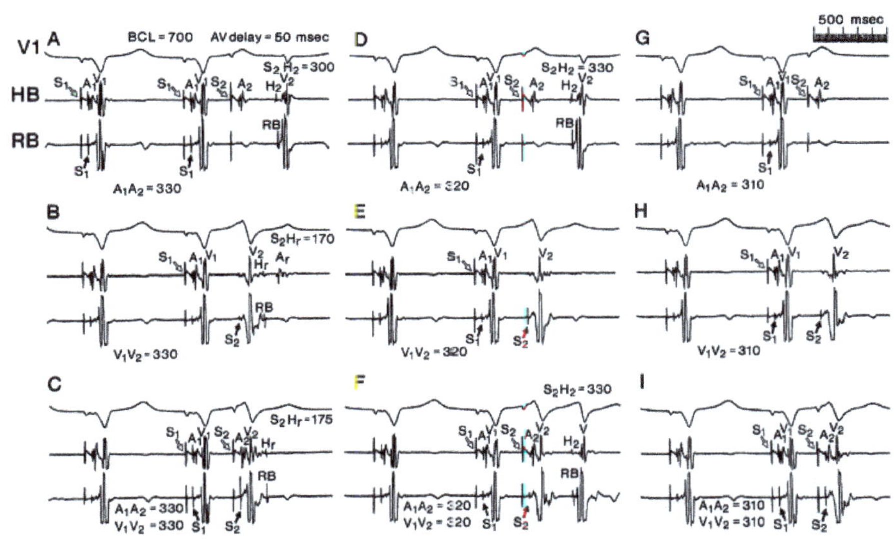

Figure 3.49. Initiation of antidromic reentry with A_2 in simulated WPW with short PR (50 ms). Basic drive, format, and patient are the same as in Figure 3.48. The AV interval is 50 ms; this time, A_1-V_1 collides in the AV node. A_1-A_2, V_1-V_2, and A_2-V_2 coupling intervals are 330 ms (**A–C**), 320 ms (**D–F**), and 310 ms (**G–I**). Response to A_2 alone (Panels A, D, and G), V_2 alone (Panels B, E, and H) and both A_2 and V_2 (complete preexcitation model, Panels C, F, and I) is seen. S_2-H_2 is 300 ms in Panel A, 330 in Panel D, and A_2 blocks in the AV node (no H_2) in Panel G. V_2 produces retrograde A (A_r) via H_2. V_2 blocks retrogradely in the HPS (no H_r or RB_r) (Panel E); because of that, S_2-H_2 measures the same as in Panel D (no concealment in the AV node). A_2 alone blocks in the AV node (Panel G), V_2 alone blocks in the HPS (Panel H), and with A_2 V_2 together (Panel I), no ATR follows or is expected, but the sites of block for A_2 and V_2 are different. (Reproduced from Lehmann et al with permission from Wolters Kluwer Publishing and the American Heart Association.[79])

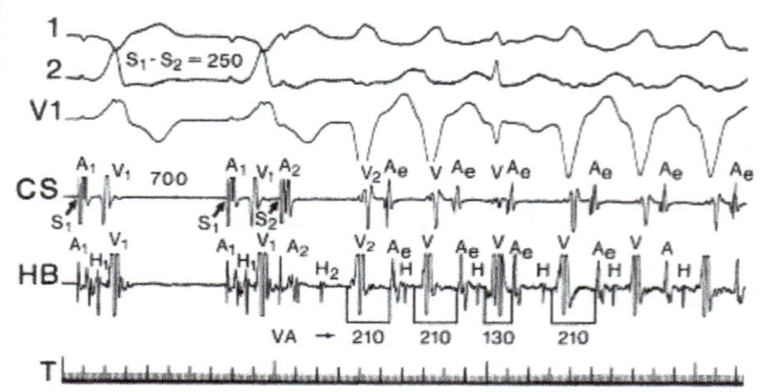

Figure 3.50. Orthodromic reentry in response to A_2. A_2 delivered at A_1-A_2 of 250 ms blocks in the AP and conducts via the NP with AV nodal and HPS delays. The resulting QRS shows LBBB and LAD artery at the onset of orthodromic AV reentry, and the left free wall location of the AP is clear from the short A_1-V_1 during CS pacing and subsequent QRS morphology (VA interval prolongation of 80 ms with LBBB vs normal QRS) as well as early CS activation. (Reproduced from Jazayeri et al with permission from the American Society for Clinical Investigation.[83])

for the first time in 1988 (Figures 3.56 and 3.57).[76] Two independent cases by Klein et al[77] were presented simultaneously, diagnosing this connection in an operating room setting. Since then, most of the published occurrences have been about AFP, indicative of the fact that much of the previously published work on nodoventricular Mahaim may have actually been on AFP.

Most AFP are connections between the right atrium and distal right bundle or adjacent right ventricular (RV) myocardium with the following characteristics:

1. During SR, the AFP is concealed because of slow conduction, unlike the NP (see Figure 3.44). Occasionally preexcitation may be visible on the surface ECG. Among maneuvers such as A pacing and vagal stimulation, administration of IV drugs will bring out the preexcitation by slowing AV nodal conduction with characteristic long AV interval and concomitant shortening of the HV interval (see Figure 3.44). A distinct AFP potential can be recorded along its course. Once a complete block occurs in the AV node, the conduction proceeds along the AFP, retrogradely engaging the NP, and an antidromic SVT is initiated (see Figures 3.44 and 3.57). Since the connection is in the RV most of the time, the QRS shows LBBB and normal or left-axis configuration. If the connection is with the right bundle, the QRS morphology is typical of LBBB due to impulse spread via the HPS, but direct ventricular insertion may create slurring of the initial part of the QRS.

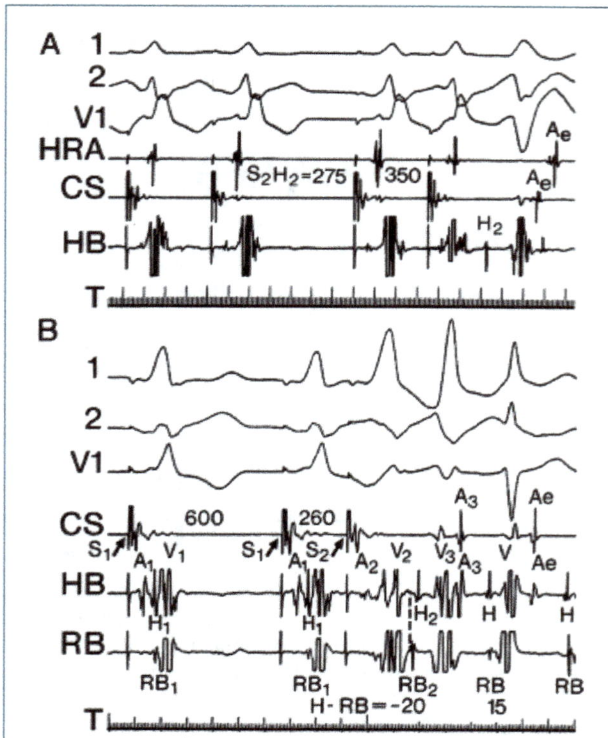

Figure 3.51. Double response to A_2 in ventricular preexcitation. **Panels A** and **B** demonstrate 2 QRS complexes in response to A_2. In Panel A, the A_1-A_2 at 350 ms is delivered at a CL of a 400 to 700-ms change, and at A_1-A_2 of 260 ms (Panel B), a CL of 600 ms follows. In both panels, there is a left-sided ventricular preexcitation. The mechanism for V_2 in each case is full preexcitation. The subsequent QRS (complex following V_2 in Panel A, not labeled, and V_3 in Panel B) have different mechanisms. Most likely the second QRS in Panel A is due to conduction of A_2 over the NP with a long HV and LBBB and left axis (Panel A), due to delayed recovery following V_2, block retrogradely in the HPS. The availability of the RB recording clarifies the mechanism of the V_3 (Panel B), which is morphologically similar to V_2 (ie, RBBB and LAD artery), and is preceded by an RB_2 and H_2 sequence, appropriate for the QRS morphology, due to BB reentry. In both examples, V_3 blocks retrogradely in the HPS to initiate orthodromic SVT. If the site of V_3 block was in the AV node, orthodromic SVT would not occur due to retrograde concealed conduction in the AV node, as is usually the case in WPW syndromes. (Modified from Jazayeri et al with permission from John Wiley and Sons Publishing.[59])

2. The retrograde limb is usually the NP. Generally patients with reentrant antidromic SVT have rapid retrograde AV nodal conduction and AV nodal reentry, which may result in AFP involvement as a bystander. Since the retrograde impulse may engage either the right or left bundle (LB), variation in conduction time may be due to HPS conduction times in the retrograde direction leading to variation in CL (see Figure 3.57).

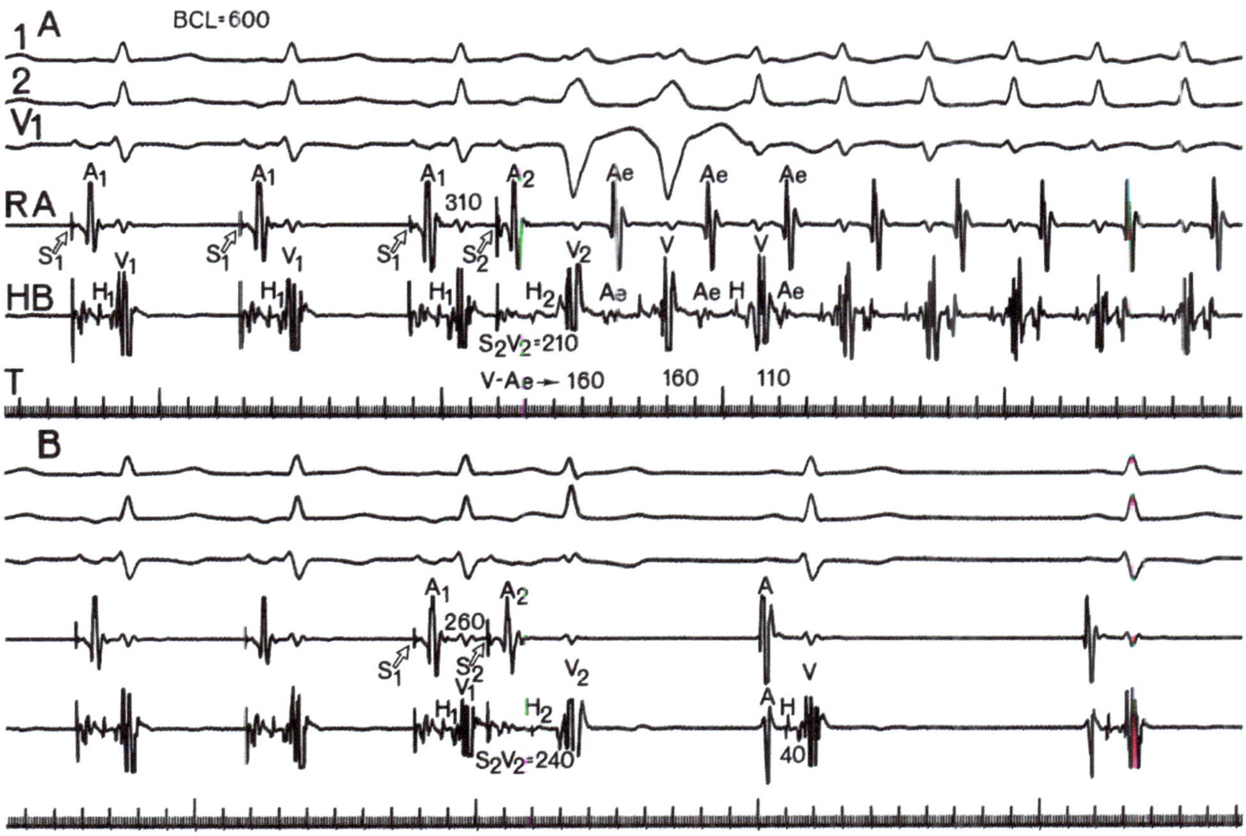

Figure 3.52. Role of LBBB in orthodromic SVT initiation. Note the onset of orthodromic SVT with A_2 in **Panel A** and not in **Panel B** despite shorter A_1-A_2 and longer A_2-V_2, the reason being that there is a concealed left-sided AP, and LBBB during V_2 allows additional recovery time for left free wall AP to initiate SVT. Also, note that VA decreases when the QRS normalizes by 50 ms and the CL of tachycardia shortens. (Reproduced from Akhtar et al with Permission from Brill Publishing.[69])

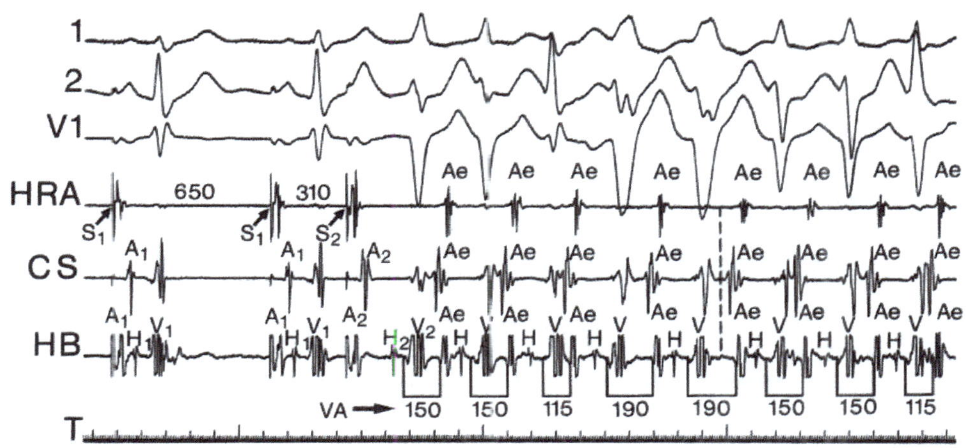

Figure 3.53. Left anterior fascicular block (LAFB) and VA in orthodromic SVT. Initiation of orthodromic tachycardia in response to A_2 is shown. Note the VA prolongation with LB and left atrial block (190 ms), LAFB (150 ms), and normal or RBBB and right axis A_2 (115 ms). This information is sufficient to localize AP to the left free wall location even if the CS (left atrial) was not available. (Reproduced from Jazayeri et al with permission from the American Society for Clinical Investigation.[83])

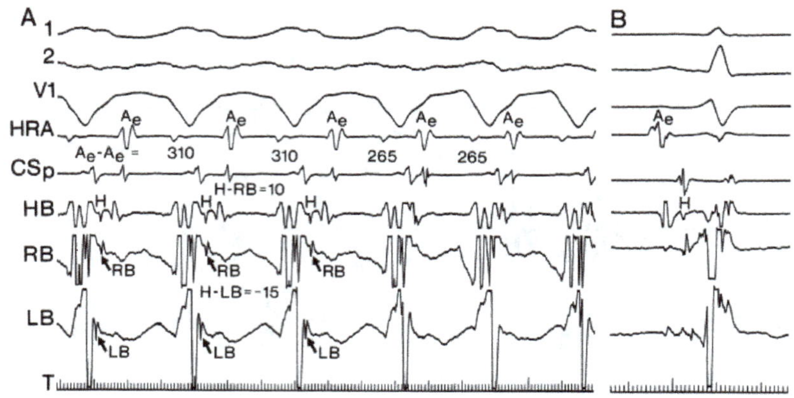

Figure 3.54. HPS block and antidromic tachycardia. Ongoing antidromic tachycardia using a right-sided AP is depicted in **Panel A**. The CL of tachycardia is initially 310 ms when there is associated retrograde RBBB. The sequence of retrograde BB activation is LB > RB > H > atrium. Abolition of retrograde RBBB shortens the VA time considerably with the CL decreasing by 45 ms (310–265), and the onset of atrial activation on the HB trace merges toward the local VEGM but is still separately recognizable, typical of the Kent bundle-type of AP at any location. **Panel B** shows a reference sinus beat with an H-RB-LB recording.

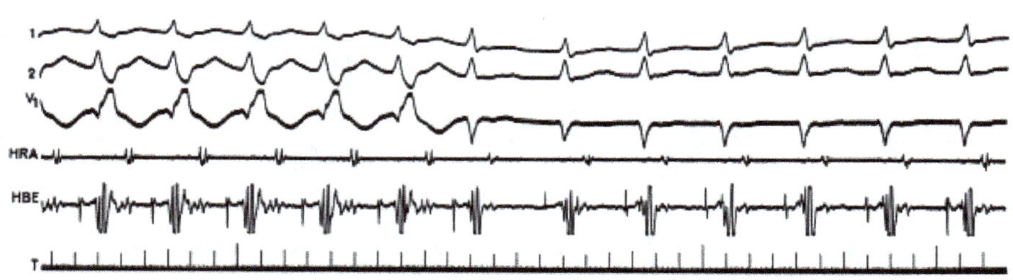

Figure 3.55. Abolition of RBBB and CL of orthodromic tachycardia (OT). During OT with RBBB, the VA is longer in this patient with right-sided AV pathway. Abolition of RBBB shortens VA, and hence the AA CL, which engages a slow AV nodal pathway, and the OT CL increases, contrary to what is usually observed in these cases. (Reproduced from Akhtar et al with permission from Elsevier Publishing.).[8]

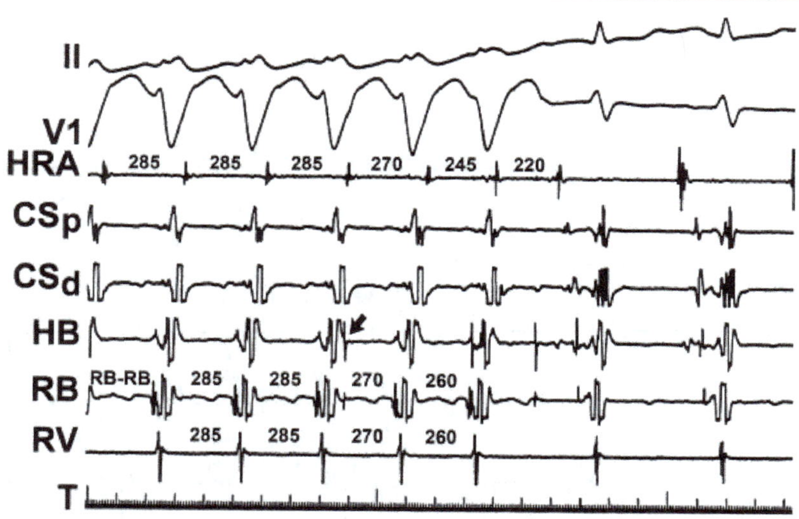

Figure 3.56. AFP. An atriofascicular nature of AP, with no anterograde involvement of the AV node, is demonstrated in this tracing. During the tachycardia, at a CL of 285 ms and a LBBB pattern, the HR-B sequence (see during sinus complex) is reversed and the RB deflection precedes the QRS (ie, atrial-RB connection) by 10 ms. During right-lateral atrial pacing, the QRS is advanced by 15 ms (following the arrow), while all of the displayed AEGMs are already activated retrogradely (all AA measure the same.) The advancement of QRS (V-V CL) indicates that atrial pacing is engaging a slow-conducting pathway directly connecting the right atrium to the right bundle (AFP). (Reproduced from Tchou et al with permission from Wolters Kluwer Publishing and the American Heart Association.[76])

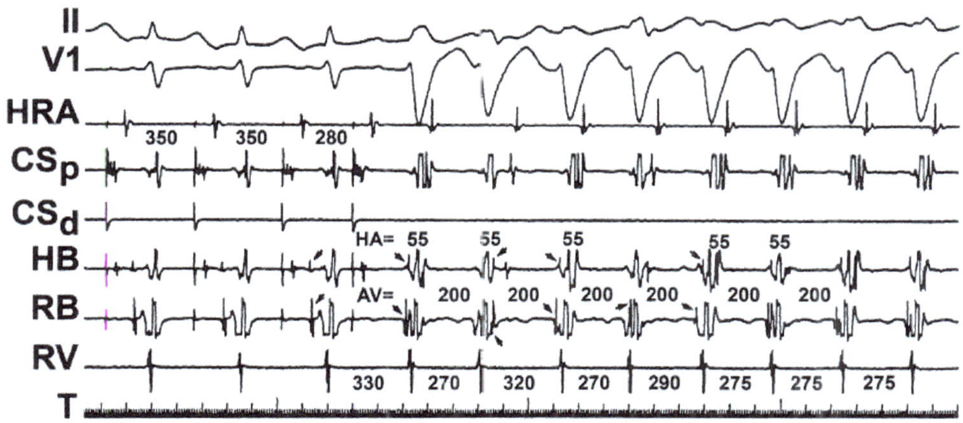

Figure 3.57. Antidromic SVT CL and HPS conduction delay in patients with AFP. Antidromic SVT induction using AFP is seen. The stimulation site is the proximal coronary sinus (CS_p) and atrial coupling is 280 ms. Note that the AV and HA intervals remain the same, but the VH variation affects the CL of tachycardia at the onset due to retrograde right bundle block above the AP insertion site, and more proximal to the RB recording location. The ventricular activation is still via AP > distal RB > RV. It is also noteworthy that, at the onset, right AEGMs on the HB trace are difficult to recognize once the retrograde RBBB resolves, which would be unlikely in the Kent bundle-type of AV-AP (see also Figure 3.54). (Reproduced from Tchou et al with permission from Wolters Kluwer Publishing and the American Heart Association.[76])

The shortest CL is noted when the retrograde conduction comes via the fully recovered right bundle. This phenomenon can also occur during antidromic SVT using Kent bundle with a right- or left-sided AV connection (see Figure 3.54). The simplest way to distinguish the 2 is to examine the separation of the AEGM from the ventricular electrogram (VEGM) on the HB tracing. When antidromic SVT with AFP conducts retrogradely via the RBB with no delay, the beginning of the AEGM merges into the local ventricular EGM. In similar situations, with the Kent bundle for example, the impulse must travel through the RV myocardium before the NP to reach the atrium. So as a general rule, the onset of the AEGM on the HB tracing is almost always separated from the local ventricular EGM (compare HB tracing of Figures 3.54 and 3.56). Interestingly, this separation of the septal AEGM from the ventricular electrogram EGM is an important clue to the possible participation of the Kent bundle, regardless of whether it is ortho- or antidromic SVT or preexcited tachycardia using 2 AP (see HB in Figures 3.45, 3.46, and 3.53–3.55).

3. The typical clinical presentation of patients with AFP is that of palpitation, normal ECG at baseline and the documented tachycardias with a LBBB pattern. The differential diagnosis of these preexcited tachycardias will be discussed in the case presentations section of this book.

Other AP

Nodoventricular Mahaim is seldom seen as a clinical tachycardia since the recognition of AFP. Fasciculoventricular Mahaim does exist, and is an ECG curiosity. It has no associated clinical tachycardia. The presence of normal PR, short HV, and ventricular preexcitation may warrant analysis in symptomatic patients to rule out other possibilities. It can cause diagnostic confusion but is otherwise of little clinical relevance. The so-called James fiber (partial or complete bypass of the AV node) producing a short PR but normal QRS also has not panned out as an arrhythmia syndrome. Another entity, that is, Lown-Ganong-Levine syndrome, has proven unimportant for any arrhythmic association. The syndrome is a combination of short PR, narrow QRS, and a history of tachycardia. The length of the PR has a bell-shaped curve among the patients with SVT. Most patients with AV nodal reentry have a normal PR and a small portion of these cases have a short or long PR with no causal relationship to the SVT. Slow- and rapid-conducting AP in a retrograde direction will be discussed in the ventricular extrastimulation section of this chapter.

Incremental Ventricular Pacing (General Comments)

Routine EP evaluation employs electrical stimulation from the RV apex. However, temporary pacing from the RV outflow tract and septum has the same validity for the general purpose of EP study. As was outlined earlier under incremental A pacing, abrupt onset of V pacing generally leads to a series of premature complexes with different consequences compared to the intent of incremental V pacing. To avoid the randomness of initial coupling, we developed pacing protocols for systematic and reproducible evaluation of electrophysiologic events (Figure 3.58).[84]

Across the population, different responses are encountered. With straight V pacing, some patients show no evidence of VA conduction. Even in these patients, the paced impulse does traverse the HPS and then the block occurs in the AV node (see Figure 3.58). This is not due to anatomic reasons and is a function of how the impulse approaches the AV node retrogradely via the HB or due to prolonged recovery after concealed penetration from dissociated sinus impulses (Figure 3.59).[85] Therefore, many of these individuals can be shown to have intact VA conduction through the AV node. This can be illustrated with manipulation of the AV conduction by simultaneous or AV sequential pacing, causing intranodal collision of impulses and shortening of the RP of the AV node, and/or with IV drugs such as isoproterenol or atropine (see Figure 3.59), or simply by prevention of concealed AV nodal penetration (Figure 3.60). A higher percentage of patients do have intact VA conduction via the AV node at baseline. AV nodal WP can be demonstrated with progressive decrease in CL. The PCL at which VA WP occurs is usually the same or longer than the corresponding AV block with A pacing. This group constitutes the majority of the cases with normal baseline AV conduction.[12]

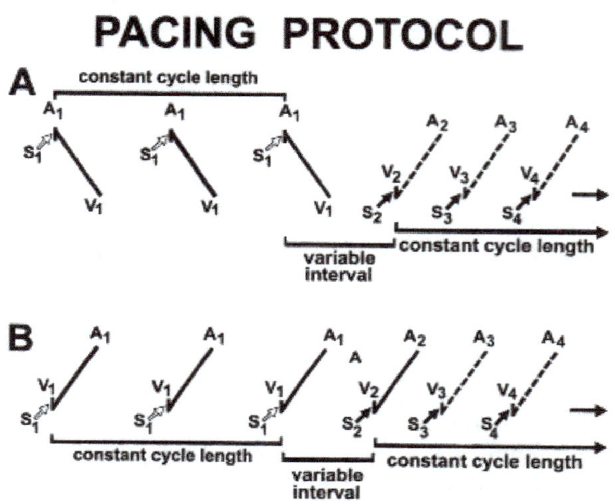

Figure 3.58. Pacing protocol of constant CL pacing. Since the so-called incremental pacing generally functions initially as a series of premature complexes with unintended and variable consequences, to avoid some of these issues, a pacing protocol is created (**Panel A**). The atria and/or ventricles (S_1, A_1, V_1) or both are paced for a series of constant CL starting at just shorter than sinus CL, then a subsequent train of constant CL is introduced after a variable coupling interval between the 2 trains. Even though the complexes are numbered as S_2, S_3, A_2, A_3, V_2, V_3, etc., the coupling of the first constant CL train is varied and limited at the long end by spontaneous escape rhythm and at the short end by ventricular refractoriness. The pacing protocol can be further modified by simultaneous (**Panel B**) AV or VA sequential pacing with different levels of collision in the AV node and HPS depending upon the parameter to be evaluated. This and other pacing protocols we have used, besides being constant, reliable, reproducible, and not random (as opposed to spontaneous sinus or other rhythms, such as escape, etc.), are also very helpful in understanding various aspects of the human AV and VA conduction system. To further refine the pacing protocol, only one CL (S_2, S_3) of the second train is initially introduced. Subsequent CL in the second train were added once the response of the previous CL was understood. (Reproduced from Akhtar M et al with permission from Elsevier Publishing.[84])

A smaller number of individuals have better VA than AV conduction, as defined by the ventricular-PCL at which VA WP occurs. The degree of prolongation in VA conduction from beat-to-beat is minimal. The aforementioned group does not include patients with AV nodal reentry or AP.[12]

Narula compared antegrade versus retrograde conduction times in several patients and concluded that, in many patients, VA intervals from the site of pacing to A recording were shorter than from the site of A pacing to V recording, from the catheter used for V pacing. He used the expression "retrograde preexcitation" to explain his results.[11] Our comparison[84] of AV and VA conduction was based on the CL of pacing at which AV or VA WP occurred. Therefore, one should be careful in interpreting the Narula study with our results because they are not directly comparable. We had the specific intention to study the HPS and ventricular extrastimulation as well to ascertain whether VA (and HA) conduction flowed via the HPS and AV node or some form of AP.

Incremental Ventricular Pacing

Similar to A pacing, constant CL V pacing is started at the CL just short of sinus. Even though the VA conduction is intact, it may take several cycles before the VA occurs via the paced complex. As the CL of pacing is decreased, several patterns emerge.

1. VA capture is intermittent but can be 1:1 if a longer-PCL is attempted.
2. VA interval will progressively increase until VA block is noted.[12] The cycle will repeat itself with further pacing. In some cases the VA change is minimal before VA block occurs.

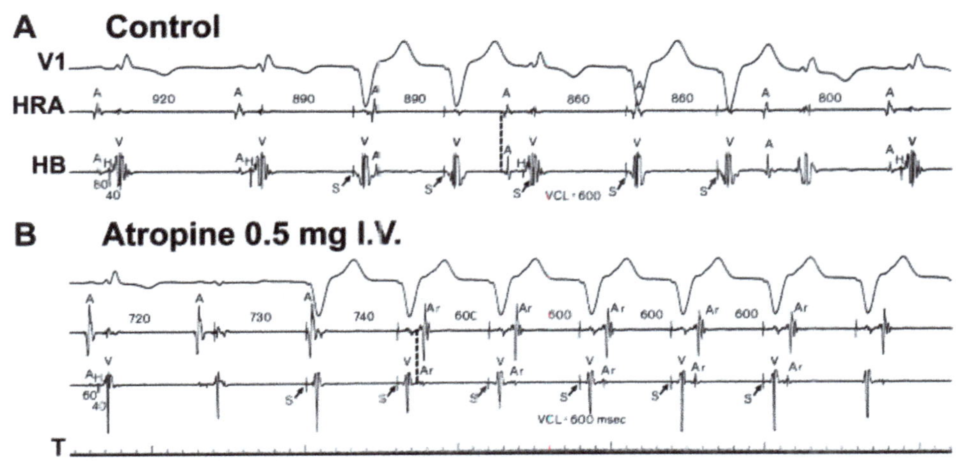

Figure 3.59. Ventricular (V) pacing and VA conduction. **Panel A** shows a dissociated sinus CL during V pacing (VCL) at 600 ms. The site of block is the AV node, since the dissociated sinus impulses are either blocked or delayed in the AV node, that is, followed by prolonged AH or no H as compared to sinus AH of 80 ms (concealed conduction in the AV node). With a small amount of IV atropine (0.5 mg), 1:1 VA response is noted at the same ventricular-PCL. A high–low and low–high (Ar) sequence of atrial activation is depicted with the perpendicular in Panels A and **B**, respectively.

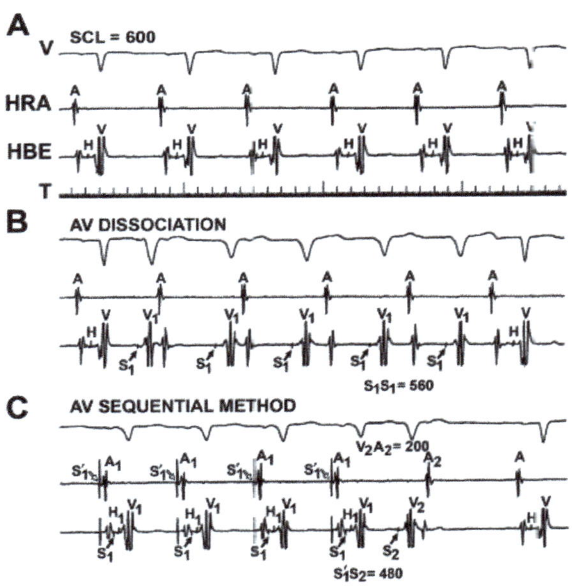

Figure 3.60. Concealed AV nodal conduction and no VA conduction. SR (SCL, 600 ms) (**A**) and V pacing at a CL (S_1-S_1) of 560 ms (**B**) are shown. AH prolongation (compared to sinus) and block of dissociated sinus indicates block of paced beats in the AV node (concealed conduction). This cycle of A and V block (linking due to AV nodal refractoriness) can be broken with control of sinus by AV sequential drive (AV = 100) (**C**). Now the site of collision is in the HPS (H_1 is anterograde). This allows more time for the AV node to recover and subsequent ventricular pacing, that is, V_1-V_2 (480 ms), now demonstrates the ability of the AV node to conduct (V_2-A_2 = 200 ms). (Reproduced from Mahmud et al with permission from Elsevier Publishing.[85])

3. VA WP occurs and pacing CL is interrupted by a ventricular echo beat (Figure 3.61).[12,86]
4. The VA WP is somewhat different than its antegrade counterpart. Unlike an atrial reciprocal beat, in which the ventricular echo or reciprocal beat occurs, it is preceded twice by HB activation (ie, once in the retrograde and then in the antegrade direction) before the QRS is inscribed. Thus, it is quite likely that the emerging AV nodal impulse will collide with the next paced ventricular impulse. The ventricular echo beat may not reach the ventricle in the process and a VA block will be noted (see Figure 3.61). In many cases, this collision of impulses will occur outside the AV node. This has led to the hypothesis that reentry may be the mechanism of VA WP.[86] The fact that block is outside the node explains AV nodal recovery because the shortest VA interval is noted after the block. Extending this observation further, A pacing induces WP and the subsequent block is followed by shortening of the AH (PR) interval. The block occurs outside the periphery of the AV node where the AV nodal reentrant impulse collides with the next atrial paced impulse. In other words, is antegrade AV nodal WP caused by concealed AV nodal reentry? The natural question arises as to whether paced WP cycles fundamentally differ from clinical spontaneous WP. The latter occurs in the setting of an abnormal AV node at normal sinus rates, whereas paced WP is seen in the normal AV

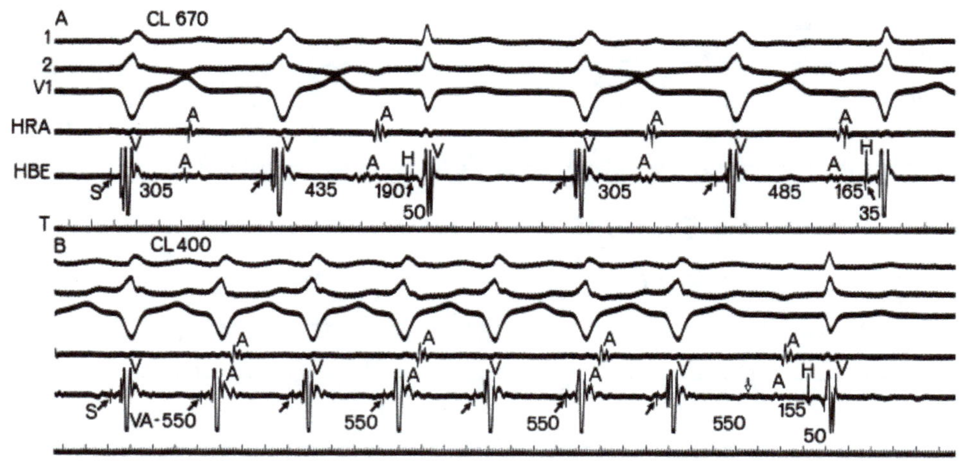

Figure 3.61. Retrograde WP in AV node. At a CL of 670 ms (A), retrograde conduction (low–high atrial activation sequence) and ventricular echo beats are seen (every third supraventricular complex is preceded by His deflection). Stimulus artifact is depicted by the arrow near the H deflection. In fact, the last QRS complex is a fusion beat. The ventricular pacing rate is accelerated. **Panel B** shows a retrograde 2:1 block and a very long VA conduction time and results in a ventricular echo beat when the ventricular stimulus is omitted (white arrow in Panel B). This shows that the AV nodal reentrant impulse was on its way to the HB and was met by ventricular paced impulse during the other all-blocked complexes (second, fourth, and sixth V pace). Expressed differently, during ventricular-paced AV nodal WP, the blocked impulse may be colliding with reentrant impulses and, depending upon the rate and timing, the site of VA block can be outside the AV node. See text for discussion.

node at the accelerated atrial rate. Many of the theories that have been put forth to explain the mechanism of spontaneous WP are based on cellular physiology and are beyond the scope of this chapter. Nonetheless, these varied possible scenarios further add to the mystique of WP for which a uniform mechanism is still debated.

5. Further shortening of V pacing CL leads to 2:1, 3:1 VA nodes (see Figure 3.61).
6. A variable pacing CL, particularly at short CL, SVT, or VT, may be initiated (Figures 3.62 and 3.63). VT and some forms of SVT in WPW are more reliably initiated with V rather than A pacing. This is particularly so in patients with Kent bundle-type AP with short ERP.
7. As in A pacing, the start of V pacing is a series of premature beats (Figures 3.62–3.64) unless the V pacing stimulus is synchronized to the previous sinus complex with a long coupling interval. If the coupling interval is short HPS, conduction delay or block will ensue. For this reason we developed several pacing models to synchronize the V-PCL with a previous sinus or paced interval. The subsequent constant CL was added one CL at a time so that the response of the previous CL was known before the next CL was introduced; using this pacing protocol, at least 4 different patterns can be identified. Three of these are outlined later under bullet #12 (patterns a, b, and c).
8. V pacing in some patients causes significant hypotension leading to adrenergic stimulation, which causes sinus rate acceleration or tachycardia. Often this leads to isorhythmic AV dissociation and limits the study of certain phenomenology.
9. During V pacing as a rule, the HB potential is not clearly recognizable and is obscured by the local ventricular EGM during the basic drive (V_1-V_1) (Figure 3.65) such that the 2 components of VA (ie, VH and HA) are indirectly deduced. If it is critical to record the HB deflection during V pacing, a close bipolar (1 mm) electrode pair or orthogonal lead design is preferable. The latter lead design has been used for recording AP potential and can also be helpful in this situation.
10. Sometimes the HB potential is identifiable in individuals with normal ipsilateral IVC (Figure 3.66). During RV pacing, the potential will precede the local ventricular EGM. Although no ideal method exists to measure VH interval during basic drive, the stimulus to the onset of His deflection of basic drive generally exceeds the HV interval during sinus, as measured from the end of HB potential. Furthermore, for comparison purposes, the HV interval should be measured to the electrode used for V pacing; but this has not been done systematically. It seems fair to say that, for all

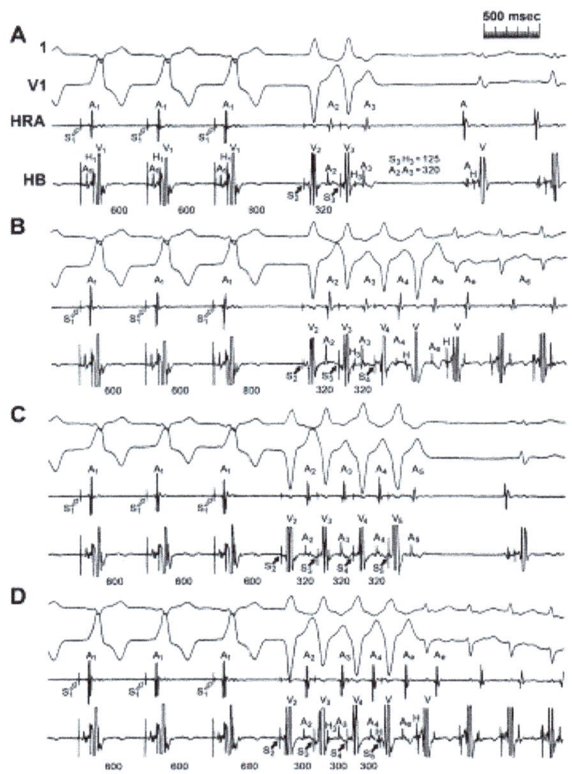

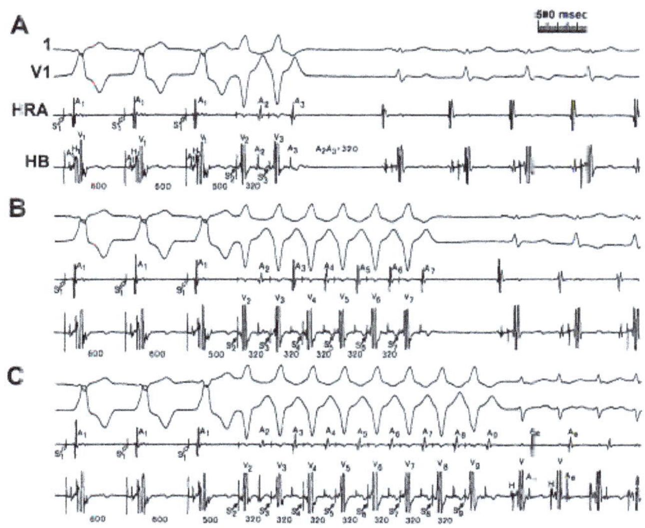

Figures 3.62 and 3.63. Orthodromic tachycardia (OT) onset with constant CL pacing. Both figures are taken from the same patient, and in all panels the initial drive is CS pacing at a CL of 600 ms in a patient with a left free wall accessory pathway (LFW-AP). After a pause of 800 ms in **Panel A**, a constant-CL V-V train (V_2-V_3, etc.) is initiated at 320 ms. Panel A shows response to the first cycle. Retrograde H_3 emerges and there is no change in A_2-A_3, indicating that A_3 was not related to H_3 and no SVT is initiated. With introduction of another CL of 320 ms, V_4 blocks in the HPS (**B**) and conducts to A_4 via the AP, which then conducts antegradely without delay in the AV node to start an orthodromic tachycardia. The lack of S_4 penetration into the AV node can be deduced from the relatively short AH, which subsequently increases. When the pause (S_1-S_2) coupling is reduced to 600 ms (**C** and **D**), and the V-V CL of the next train is reduced to 300 ms, it sets the stage for the same phenomenon to occur, and SVT is started. In Figure 3.63 (same patient), S_1-S_1 CL is the same. The S_1-S_2 coupling is reduced to 500 ms, and the second train is 320 ms. Here the HPS accommodates early (Panel B), and SVT onset needs an AV nodal block, which takes several cycles, such that the SVT starts after 8 ventricular-paced beats (Panel C). Six paced complexes were insufficient (Panel B) to produce sufficiently distal block in the AV node. Note evidence of AV nodal concealed conduction, as with the first AH of the tachycardia, which is longer and shortens subsequently. Some points can be made from the figures: (1) Random onset of incremental pacing is in essence a series of premature complexes; (2) When orthodromic SVT starts soon after ventricular pacing, it is invariably due to block in the HPS; and (3) Linking by interference in the AV node can be appreciated in Panel B of Figure 3.63 and forward migration through zone of block in the AV node (see Figure 3.63C). (Reproduced from Akhtar et al with permission from Elsevier Publishing.[84])

practical purposes, a ventricular-stimulated VH impulse will probably not reach the AV node in less time than its antegrade counterpart, that is, the HV interval.

11. With impaired HPS conduction in the antegrade direction (prolonged HV, RBBB, LBBB), one can anticipate that the paced impulse at basic CL will encounter some delay. Ordinarily the RV-paced impulse will travel via the right bundle to reach the His/AV node, unless there is also a retrograde RBBB in which case the RV impulse will travel across the septum and the HB potential will be identified at the tail end of the V_1 EGM (Figure 3.67). Assuming the left-bundle conduction is normal, the VH interval will be longer but stable. Since the impulses originating in the ventricle, whether spontaneous or paced, reach the HPS before the AV node, integrity of the HPS has a direct influence on the route of propagation to the AV node and to the atria (Figure 3.68). Understanding this aspect of VA conduction is important for accurate interpretation of certain tachycardias, conduction intervals, or changes in the CL.
12. When HPS delays are encountered at the onset of V pacing (Figure 3.69), several scenarios can evolve:[16,84] (a) HPS may quickly accommodate to the subsequent constant CL changes by

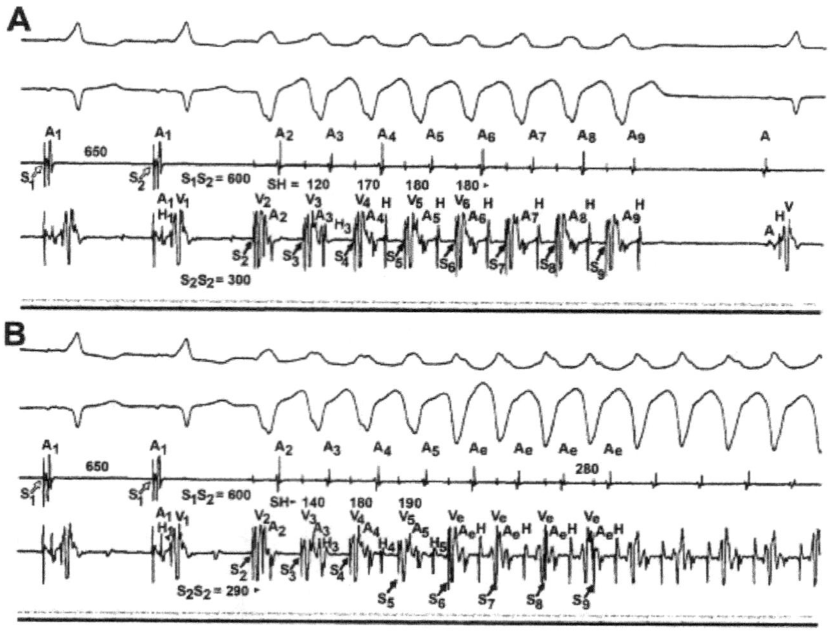

Figure 3.64. Orthodromic tachycardia with onset of incremental pacing. In a patient with right-sided pathway after an atrial PCL of 650 ms, a constant train of ventricular pacing of 300 ms (S_2-S_3 CL) is introduced at an S_1-S_2 coupling of 600 ms (**A**). Retrograde HPS conduction delays (V_2-H_2) and subsequent linking in the HPS (by interference) are observed after V_4 blocks in the HPS and the HB. H_4 (not labeled) is activated anterogradely. From this point onward, the V impulse blocks below the His (BB-H) and activates the atria via the AP, which in turn blocks the BB-H anterogradely. A linking phenomenon develops indefinitely with sequence V-A-H. Shortening of the V-V train (**B**) to 290 ms allows for forward migration in the HPS and the linking phenomenon ends, which initiates orthodromic tachycardia with a LBBB. (Reproduced from Lehmann et al with permission from Wolters Kluwer Publishing and the American Heart Association.[73])

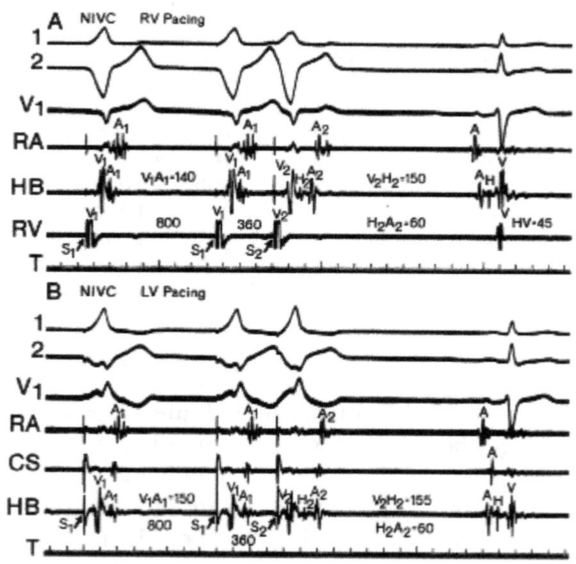

Figure 3.65. VA conduction in a patient with normal IVC. **Panels A** and **B** show RV or LV apical pacing, respectively, at the same CL of 800 ms in the same patient. The retrograde H_1 EGM is obscured by the local V_1 EGM. At a comparable V_1-V_2 of 360 ms, the retrograde H_2 emerges from the local V_2 EGM and has a fairly comparable retrograde HPS conduction delay. The retrograde H_2-A_2 remains the same and is identical in both situations.

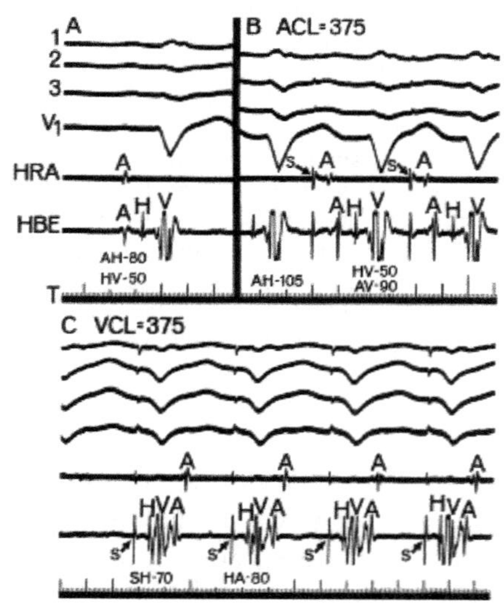

Figure 3.66. AV and VA conduction during constant CL pacing. A 1:1 AV and VA conduction is depicted. In all panels, the pertinent values are listed. SR shows LBBB and A and V pacing is at the same CL of 375 ms. Regardless of how the measurements are made (that is, from the end or beginning of the deflection) the HV is shorter than VH. When the retrograde H is activated via the intact RBB, retrograde H deflection may precede the local VEGM (as in this case). (Reproduced from Akhtar et al with permission from Wolters Kluwer Publishing and the American Heart Association.[12])

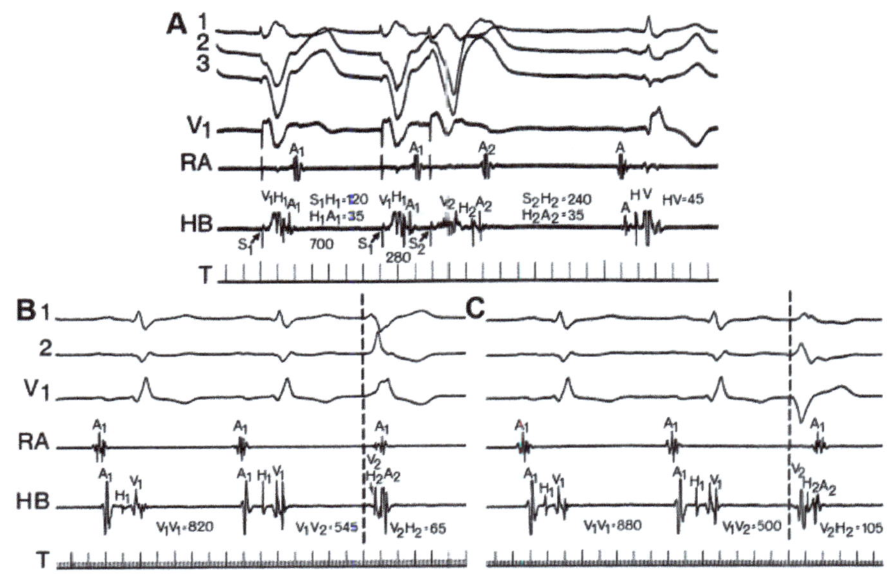

Figure 3.67. Retrograde HPS conduction (VH) in patients with RBBB. The last complex in **Panel A** shows RV apical pacing in a patient with preexisting RBBB. The retrograde H_1 follows the local V_1 due to retrograde RBBB, and the impulse reaches the H_1 via the LBB. H_2 separation from V_2 can be appreciated with premature beat. In another patient (B) with preexisting RBBB, a spontaneous V_2 occurs, arising from the LV. The retrograde H_2 precedes V_2. The following A is a fusion complex. In the same patient, when a spontaneous ventricular complex arises from the RV (C), the retrograde H_2 follows V_2 (retrograde RBBB).

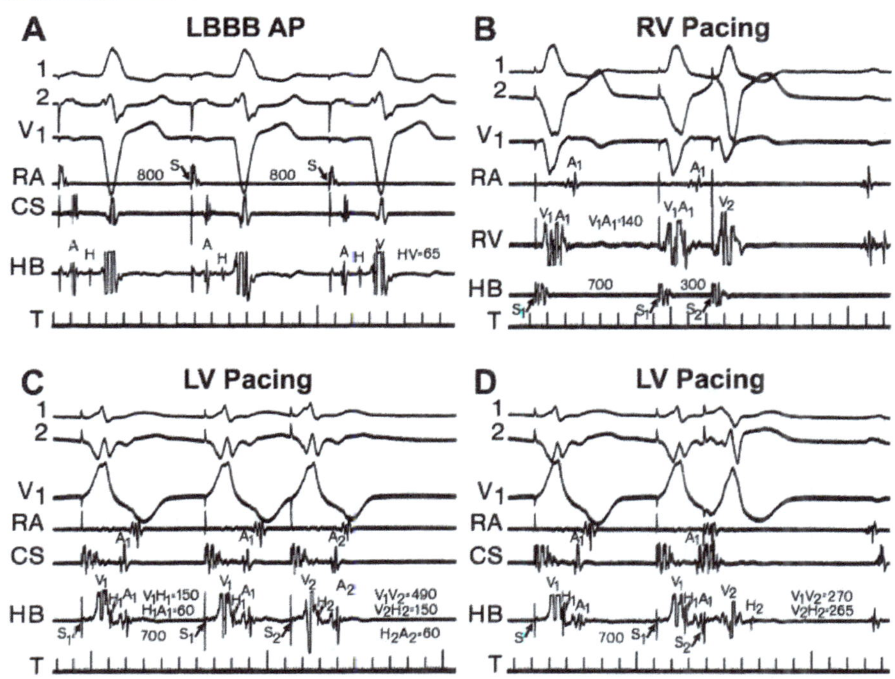

Figure 3.68. RV and LV pacing in patients with preexisting LBBB during sinus and atrial pacing AP. AP conducts with a LBBB pattern (A) with an HV of 65 ms. H_1 is not visible with retrograde RV pacing (B) and probably conducts via the RBB and, with V_2, it blocks retrogradely in the RB, crosses the septum and blocks retrogradely in the LBB system. In patients with normal inferior vena cava at baseline, the retrograde H_2 would have emerged from the V_2 EGM at V_1-V_2 of <50% of the V_1-V_1 CL due to HPS delay and/or retrograde ipsilateral BBB with transseptal conduction, such that the H_2 is activated via the contralateral BBB (see Figure 3.65A and B). When the LV is paced (C), the retrograde His deflection is seen after V_1, similar to that seen in RV pacing in patients with retrograde RBB block (see Figure 3.67). At a V_1-V_2 of 270 ms, H_2 moves out of the V_2 EGM and blocks in the AV node (H_2, no A_2), suggesting intra-RBB conduction delay in a retrograde direction.

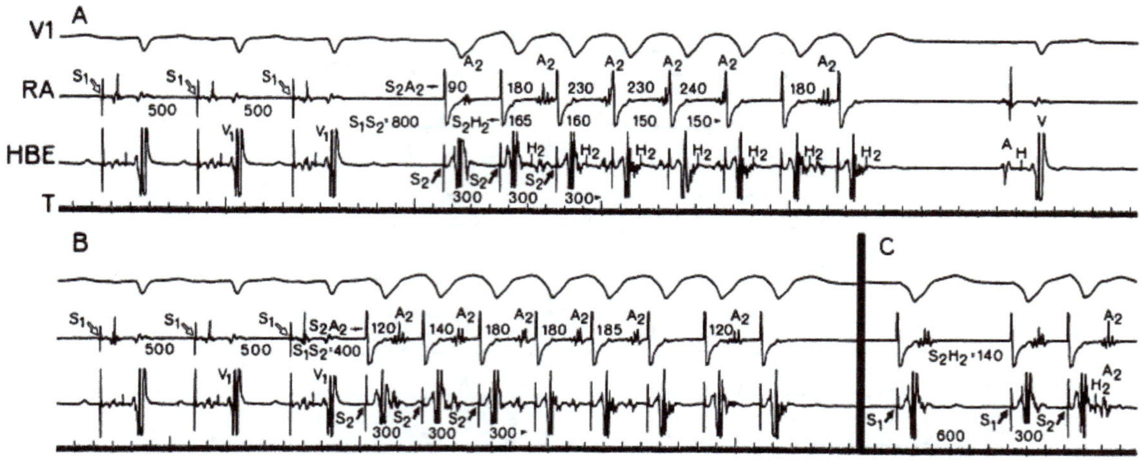

Figure 3.69. Interplay between HPS and AV node during V pacing. An atrial paced train of 500 ms of constant CL and the V pacing train of 300 ms (**A** and **B**, respectively) are shown. Top panel depicts the AV coupling interval of 800 ms with initial HPS delay and subsequent shortening of VH to 150 ms with a sudden block. The block is above the H_2 (AV node) and there is HA prolongation before the block. A 2:1 HA block is seen in the last 2 cycles. In the bottom panel, the AV coupling of 400 ms leads to instant HPS accommodation (H_2 obscured by V_2 EGM) and then 2:1 block in the AV node. **Panel C** is a reference CL showing that when V_1-V_2 >50% of V_1-V_2 retrograde, H_2 is expected to emerge from the V_2 EGM. (Reproduced from Lehmann et al with permission from Wolters Kluwer Publishing and the American Heart Association.[16])

progressive upstream migration of impulse as occurs in an antegrade direction; (b) 2:1, 3:2 WP type of block may occur with either accommodation (see Figure 3.69) or persistence for variable duration; (c) linking may develop and be sustained for a period (see Figures 3.63–3.69); (d) VT or SVT may be initiated. If the CL of tachycardia is faster, it will override the paced cycles (see Figure 3.63C). On the other hand, entrainment may follow when the PCL is shorter than the tachycardia. Also, tachycardia may be initiated, terminated or reinitiated, and each phenomenon may be difficult to decipher unless the pacing is periodically interrupted at the appropriate time.

13. A bidirectional block may be noted at certain CL but the level of block may be different. For example, in individuals with third-degree AV block in the HPS, HPS conduction may be intact, even though there is no VA conduction during V pacing. The site of block may be the AV node, as evidenced by concealed conduction in the AV node of the dissociated sinus or atrial impulses. It may also happen in either the RBBB or LBBB, where conduction may be intact in one direction. Whether some of these HPS blocks can be reversed by medications such as isoproterenol or with electrophysiologic maneuvers has not been systematically studied.
14. Some patients have occasional fast VA conduction to the atria via the AV node. This is thought to be some form of bypass track. VA block can invariably be produced with the use of agents known to have a depressant effect on the AV node, such as verapamil, beta-blockers, and even by vagal maneuvers (Figure 3.70), leaving little doubt that the AV node in the antegrade and retrograde direction can behave both in similar and dissimilar ways. The rapidity of conduction velocity or VA block in the AV node is more of a physiological property than an anatomic issue. In all of these instances, it can be illustrated that HPS activation is necessary for VA conduction to occur.

Responses to ventricular extrastimulation:[12,13,87] Regardless of the arrhythmias documented previously, a complete EP evaluation includes introduction of ventricular extrastimulation (S_2, V_2), scanning the diastolic interval for the point of the ERP of VM, that is, the longest S_1-S_2 interval in which no V_2 response is noted. Late in diastole, the V_2 impulse does not encounter much delay in the HPS, hence it reaches the AV node almost at the same interval as the V_1-V_2, that is, the retrograde H_1-H_2 = V_1-V_2 (Figure 3.71). At this state, AV nodal response may be observed at the interval measured by S_1-S_2 (V_1-V_2). However, at variable V_1-V_2 intervals, HPS delays begin and closer input into the AV node is impeded by the tissue (HPS) encountered initially by V_2 (see Figure 3.71). Therefore, it makes logical sense to address the HPS response to the V_2 impulse first.

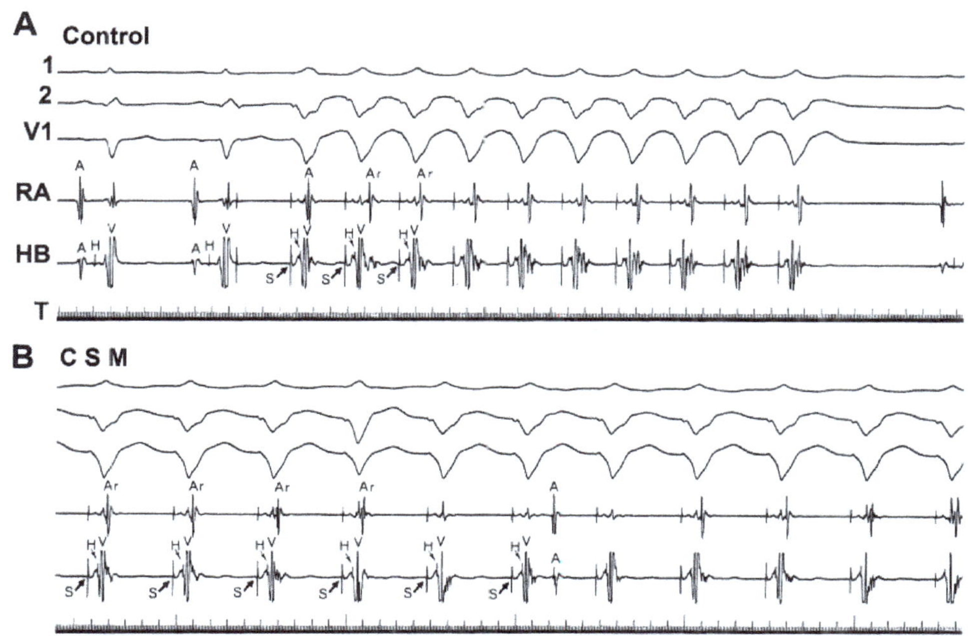

Figure 3.70. Carotid sinus massage (CSM) and VA conduction. VA conduction is intact at a PCL of 330 ms (A) and at 500 ms (B). Note the short VA as measured on H-RA (125 ms). The onset of A on HB is not clearly defined. In the lower panel, VA is even shorter (110 ms) due to the longer-PCL; the retrograde H is identifiable and VH is 65 ms, with all paced beats, while the HV is 45 ms. If it could be measured, the HA on the HB trace would be <110 ms, so that calculated HA is no more than 50 ms. With CSM, abrupt block of HA occurs with sinus escape, and there is no change in VH conduction (Panel B). This block in HA conduction with CSM is fairly typical of AVN response. (Ar, retrograde atrial capture.) (Reproduced from Akhtar with permission from Wolters Kluwer Publishing.[44])

Responses of HPS to ventricular extrastimulation:[13,87] After a series of basic drive cycles (S_1-S_1, V_1-V_1), an extrastimulus (S_2, V_2) is delivered. When the sinus or atrial impulses are dissociated from the V-paced cycle: (a) the V pacing cycle is interrupted; (b) there is a variable degree of concealed conduction in the AV node; (c) anterograde and ventricular HPS and ventricular capture are unpredictable; (d) ventricular fusion occurs with consequent divergent impulse propagation; and (e) artificial prolongation of ventricular ERP occurs as a result of ventricular fusion. To avoid all these problems, one can simultaneously pace the atrium and ventricle, realizing the 2 impulses will collide somewhere in the NP, but at least there will be a constant, reliable, stable, and reproducible basic CL before the premature (V_2) complex is introduced.

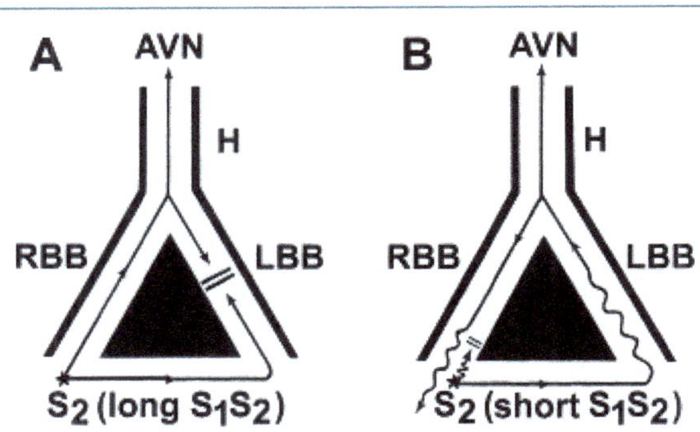

Figure 3.71. Route of retrograde impulse propagation during RV pacing. Schema shows that, with RV pacing, the impulse travels along the RB during S_1-S_1 and long S_1-S_2 (>65% of the BCL S_1-S_1) (A) and shorter S_1-S_2 (B), the main differences being the lack of conduction delay in Panel A and retrograde delay and block in the RBB system in Panel B, transseptal conduction and then retrograde delay along the LBB and subsequent H_2 activation, anterograde RBB penetration and possible V_3 from the BB reentry. Depending upon the HB and RB recordings at their respective locations, the His (H) and RB can have a different relationship to each other.

During the basic drive and the long V_1-V_2 intervals (in patients with normal HPS) during pacing of the right ventricle, the V_1 impulse will reach the HB via the RBB and concurrently will cross the septum and then penetrate the LBB. While this impulse will reach the AV node via the RBB, it will begin to descend along the LBB antegradely (see Figure 3.71), and the 2 wavefronts will collide in some portion of the LBB. The level of collision will vary among individuals. With S_1 and S_2 delivered late in the diastolic phase, the retrograde H_1 and H_2 are usually not identifiable and are obscured by the local V_2 EGM (Figure 3.72) during V_1 or V_2 until the H_2

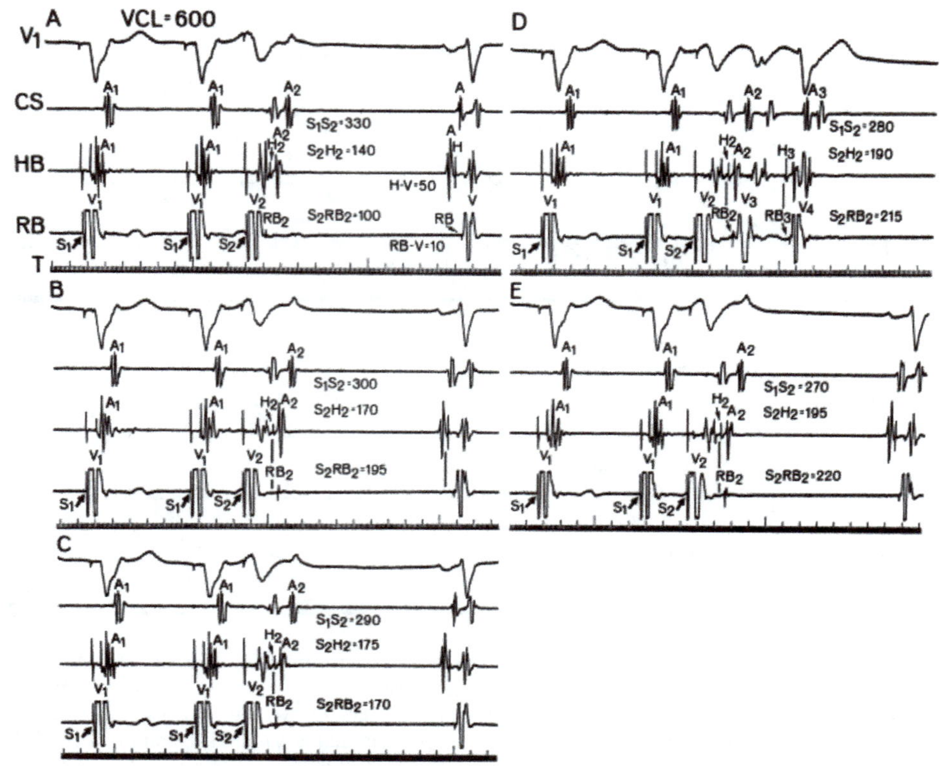

Figure 3.72. HPS propagation following short V_1-V_2. **Panel A** depicts retrograde RB_2, and H_2 emerges from the V_2 EGM. The sequence is RB_2-H_2 with a 40-ms interval, approximately the same as the H-RB interval during reference sinus complex, suggesting H_2 activation via the RBB. All pertinent intervals are labeled. At a slightly shorter S_1-S_2 of 300 ms (**B**), there is a reversal of H-RB sequence and now the H_2-RB_2 interval measures 25 ms, as would be expected since both the HB and RB are being activated via the LBB. In **Panel C**, the timing suggests the RB was activated via RB and the H_2 via LB (retrograde gap phenomenon in the RBB). Further delays in HPS, retrograde block in RB, retrograde H_2 activation via LB and anterograde penetration of RB leads to V_3 and then V_4 due to BB reentry (**D**). At longer VH delays (**E**) and no further change in H-RB sequence, the impulse blocks below the RB site, a typical site of conduction delay and/or block (ie, distal) during V pacing (retrograde concealed conducted in the HPS).

emerges from the V_2 EGM. That does not occur until the V_1-V_2 interval approaches approximately 65% of the basic CL and is quite similar to antegrade HPS behaviors. This delay in VH interval could be along the right or LB. It is generally assumed that the emergence of H_2 from the V_2 EGM is indicative of retrograde block in the right bundle, and H_2 emergence means the impulse propagation to the HB is via the LBB. However, this cannot be certain unless recordings from the HB and one of the BBs are available. When HB recordings are available, the retrograde H_2 activation may recur via the RBB or LBB from different sites of pacing, but this is not consistent behavior (see Figure 3.72). In patients with retrograde RBBB or LBBB, HB potential emergence is expected during basic CL as well as V_2 during pacing of the ipsilateral ventricle (see Figures 3.66–3.68).

With progressive shortening of the V_1-V_2, further V_2-H_2 delays are encountered (Figure 3.73). The retrograde H_2-A_2 interval (AV nodal conduction) depends upon the retrograde H_1-H_2 interval but since H_1 is not always visible and V_1-H_1 is constant, one can use V_1-H_2 as a surrogate for retrograde H_1-H_2. The retrograde behavior of the AV node will be dealt with later. At a certain level of V_2-H_2 delay, other nonpaced complexes may appear (Figure 3.74). The most common nonstimulated V_3 after the paced V_2 is called macroreentry within the HPS or BB reentry (see Figures 3.72–3.74). It has the following mechanisms and characteristics.[16,87]

1. When pacing the RV, the V_2 blocks retrogradely in the RBB, and conducts with retrograde delay along the LBB to reach the HB and then antegradely along the right bundle to produce V_3 with an H_2-V_3 interval usually longer or equal to the HV interval of sinus complexes (Figure 3.75).

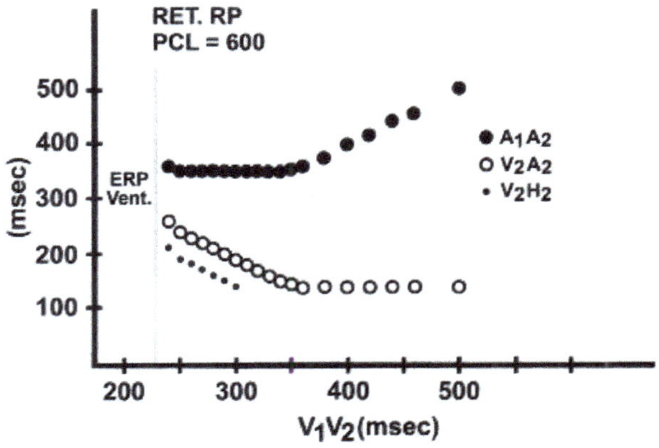

Figure 3.73. Scanning of PCL of 600 ms with V_2. V_1-V_2 from 500 ms to the ERP of the ventricle is shown. Corresponding V_2-A_2, V_2-H_2, and A_1-A_2 are depicted in milliseconds. Note progressive increase in V_2-A_2 at a V_1-V_2 of 350 ms. After the H_2 emerges (at 300 ms), H_2-A_2 is constant and the A_1-A_2 curve is quite flat. The shorter V_1-V_2 impulse cannot penetrate the AV node earlier due to progressive V_2-H_2 delay, which almost equals V_1-V_2 decrease. In patients with short H_2-A_2 intervals as in this case, only shortening of the basic length or shorter retrograde H_1-H_2 will allow earlier AV nodal penetration and H_2-A_2 increase, as was seen in Figure 3.69A. (Reproduced from Akhtar et al with permission from Elsevier Publishing.[51])

2. These critical V_2-H_2 delays, which are necessary for V_3 to occur, vary among patients.
3. If the block occurs below the HB, that is, V_2 no H_2, the V_3 phenomenon does not occur (Figure 3.76).[16,87]
4. Resumption of V_2-H_2 conduction while V_2 is blocked at longer V_1-V_2 intervals, the so-called retrograde gap phenomenon in the HPS (see Figure 3.76), is associated with return of V_3 if it occurs before the block at longer V_1-V_2 intervals.
5. There is an inverse relationship between V_1-V_2 and V_2-V_3 intervals; similarly there is an inverse relationship between V_2-H_2 and H_2-V_3 intervals (see Figure 3.76).
6. When V_1-V_2 > V_2-V_3, V_3 configuration and axis are similar to V_2 but when the V_1-V_2 < V_2-V_3, V_3 begins to have more of a typical LBBB and shift in axis toward normal because longer V_2-H_2 delays allow a greater degree of recovery in the right bundle and VM.
7. In 5% to 10% of the cases, the V_2 impulse from the RV blocks in the LBB retrogradely and yet reaches the HB H_2 via right-bundle delay. This often results in V_3 with a RBBB pattern and RB-H-V sequence appropriate for the morphology of the QRS (Figure 3.77).
8. V_3 from bundle branch reentry (BBR) is not affected by events in the AV node, atria or AP. For example, it will occur in approximately 50% of individuals with normal IVC regardless of whether they have AV nodal reentry, AV nodal block, AF (Figure 3.78), concealed or manifest WPW, etc.[16]
9. The V_3 usually blocks retrogradely in the HPS (no H_3) and therefore is self-terminating. If V_3

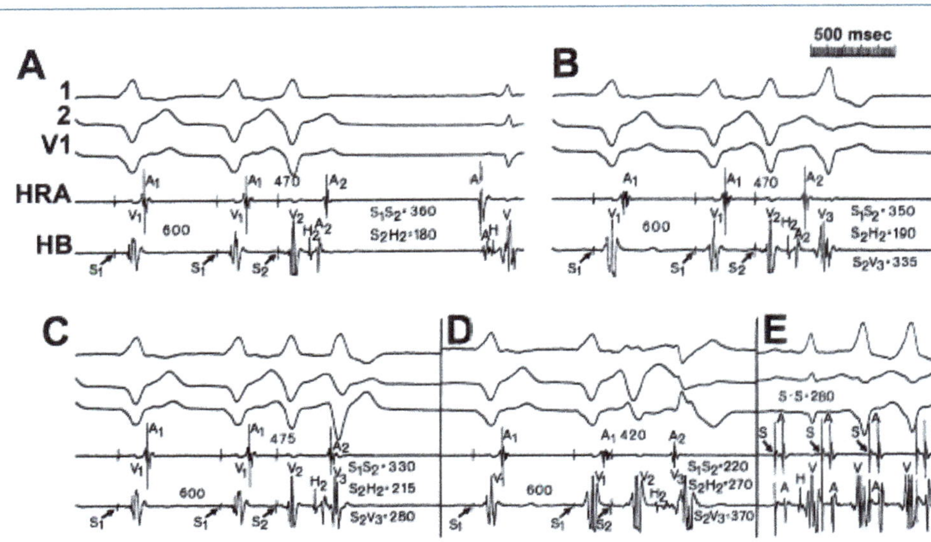

Figure 3.74. Nonpaced wide QRS complexes following V_2. All 5 panels are taken from the same patient. At a basic drive of 600 ms (V_1-V_1), S_1-S_2 are progressively shortened (**A–D**). There is subtle preexcitation during SR (Panel A). However, there is no VA conduction via the AP (unidirectional block). In Panel B, V_2 conducts via the H_2 to the atria. An antidromic conduction over V_3 via a right-sided AP follows A_2. At longer V_2-H_2, V_3 occurs due to BBR (Panel C) following V_2, reaching the H_2 via the LBB and then anterogradely over the RBB. The reversal of route (RB to HB to anterior fascicle) produces BBR with a RBBB and left posterior hemiblock pattern. **Panel E** shows overt preexcitation during right pacing, as seen in Panel B. (Reproduced from Akhtar et al with permission from Elsevier Publishing.[51])

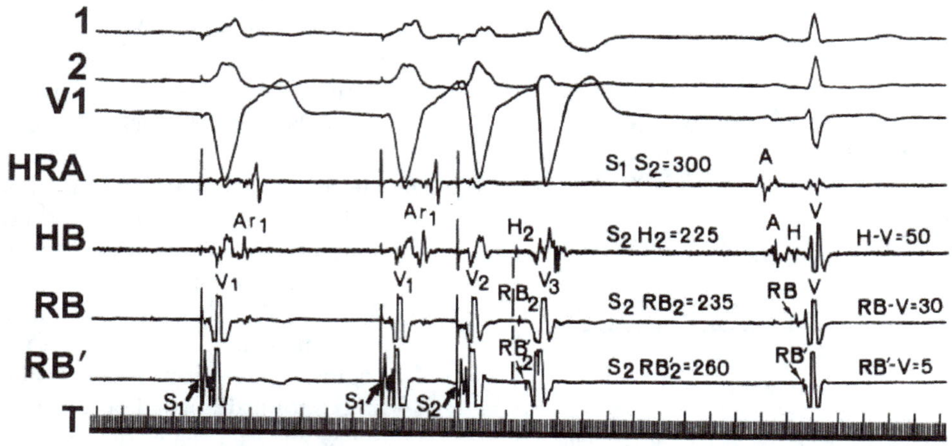

Figure 3.75. H-RB sequence during BBR with a LBBB pattern. V_2 is followed by V_3 due to BBR. Note the sequence of H_2, then RB_2. RB_2 is the same as sinus. The H_2-V_3 is 60 ms (not labeled). The H_2-RB_2 interval is only 10 ms, as opposed to sinus which is 20 ms. This is due to the fact that the proximal RB and H are both activated via the LBB by the diverging impulse and not sequential, as in sinus.

does reach the HB (H_3), V_4 will generally follow, which in turn will block in the HPS (see Figure 3.72D).

10. The state of HPS conduction has a marked influence on BBR. Complete RBBB or LBBB in either direction will not allow the impulse to propagate and will abolish BBR. This, of course, depends upon the site of V pacing and route of propagation. In fact, this information is used to ablate a BB to produce complete bidirectional block (Figures 3.79 and 3.80).
11. Small doses of procainamide-like drugs can create a scenario where V_3, V_4, etc., may occur, leading to a nonsustained BBR.[88]

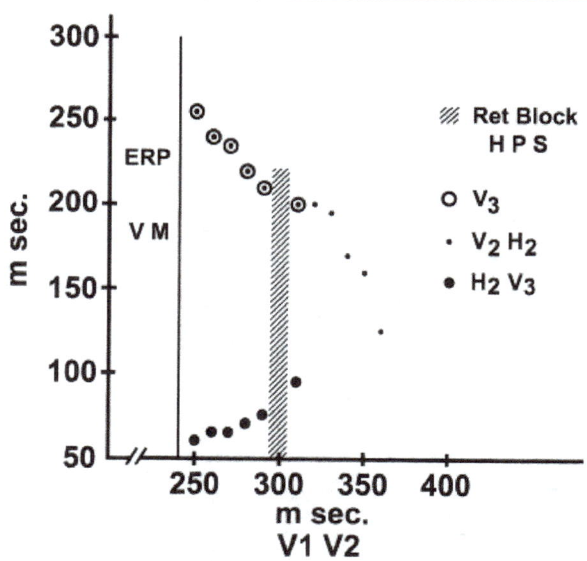

Figure 3.76. V_1-V_2, V_2-H_2, and V_2-V_3 relationship during BBR. Scan of a ventricular-PCL of 700 ms is shown with decreasing V_1-V_2 interval. Note inverse relationship between V_1-V_2 and V_2-H_2. After the BBR (V_3) starts, there is also an inverse relationship between V_2-H_2 and H_2-V_3; also remarkable is the fact that there is a retrograde gap phenomenon in the HPS and, when V_2 blocks in the HPS (ie, no H_2), V_2 does not occur, but then V_2-H_2 conduction and V_3 resumes at shorter V_1-V_2.

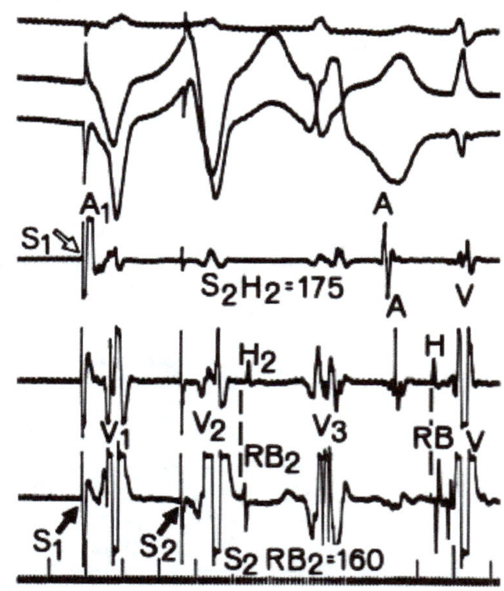

Figure 3.77. BBR with a RBBB pattern. Only the last S_1 is shown. V_2 blocks in the LBB, activates H_2 via RB (RB_2 precedes H_2) (RB_2-H_2 = H-RB during sinus), and produces V_3 with a RBBB pattern and left axis. The H_2-V_3 measures 110 ms > HV of sinus (150 vs 40 ms), indicating concealed conduction in the LBB but less so in the left posterior fascicle. (Modified from Akhtar et al with permission from Wolters Kluwer Publishing and the American Heart Association.[87])

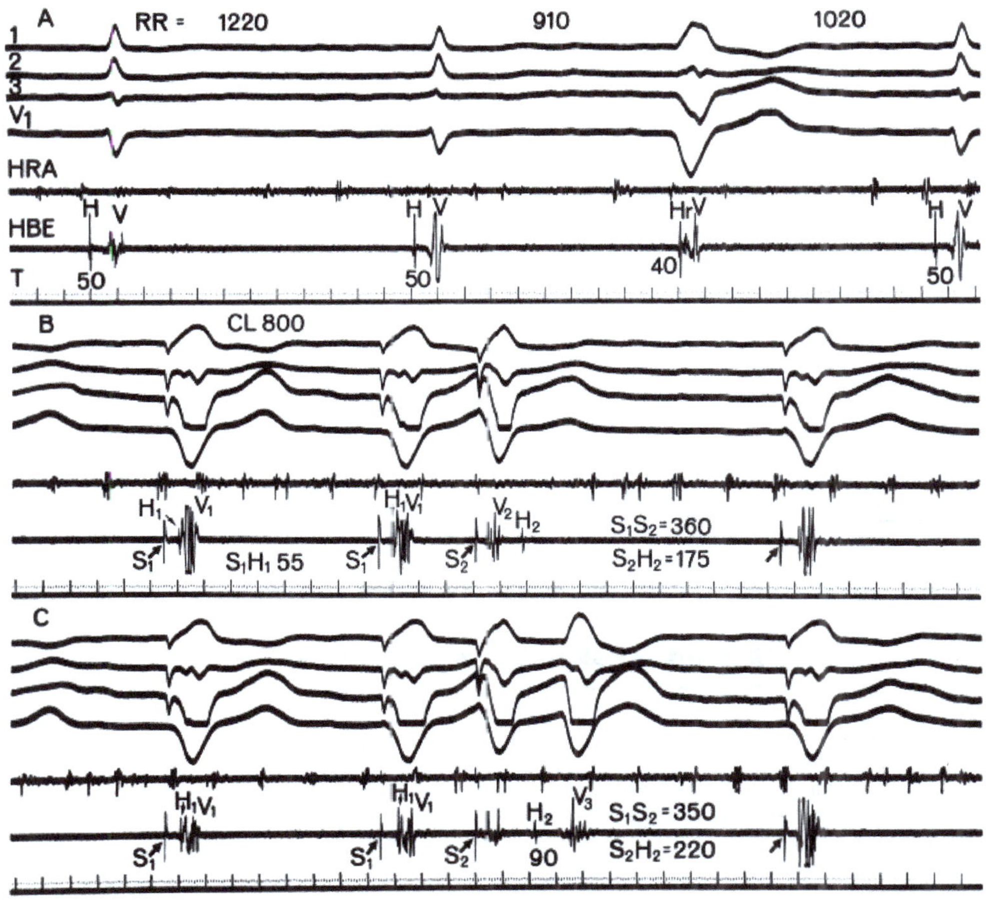

Figure 3.78. BBR in patients with AF. Spontaneous rhythm is AF (A). A spontaneous ventricular premature complex (HV minus 40 ms) is noted. Retrograde H_2 can be identified during V pacing (B and C). Occurrence of V_3 due to BBR with an H_2-V_3 of 90 ms can be appreciated. (Reproduced from Akhtar et al with permission from Wolters Kluwer Publishing and the American Heart Association.[13])

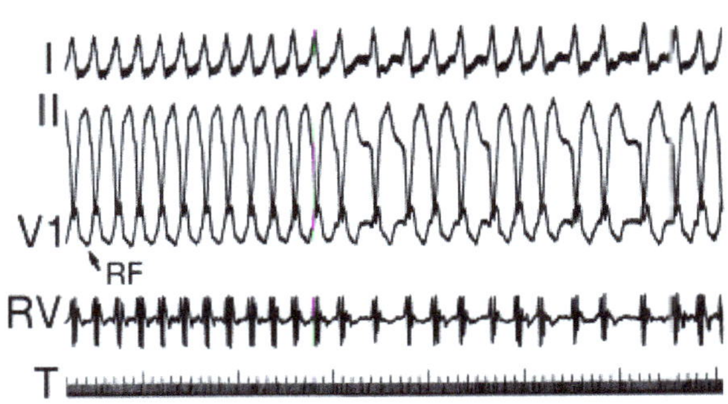

Figure 3.79. RBB ablation in BBR. The underlying atrial rhythm is AF. The initial part of the tracing shows an ongoing BBR VT with a RBB block and LAD artery. RF is initiated (arrow), and VT terminates. Note there is no change in QRS morphology but AF with irregular R-R can be observed. It is worth mentioning that, in this case and in the patient in Figure 3.80, there already existed an identical BBB pattern before ablation, which had to have been unidirectional.

12. With short-to-long CL change, BBR occurs at a longer V_1-V_2, either due to greater V_2-H_2 or more distal block in the RBB; and sometimes a nonsustained BBR may be noted, while only a single V_3 occurs at constant CL (Figure 3.81).[89] During RV pacing, the V_3 phenomenon is uncommon in patients with a preexisting RBBB pattern.[12]
13. A spontaneously prolonged HV interval, regardless of the underlying structural heart disease, can result in sustained BBR as a clinical tachycardia, which has additional characteristics as discussed in the next section.[90]

Sustained BB Reentry[91–99]

Since the initial description of BBR as a physiological phenomenon seen in nearly 50% of patients with baseline normal IVC, many reports have been

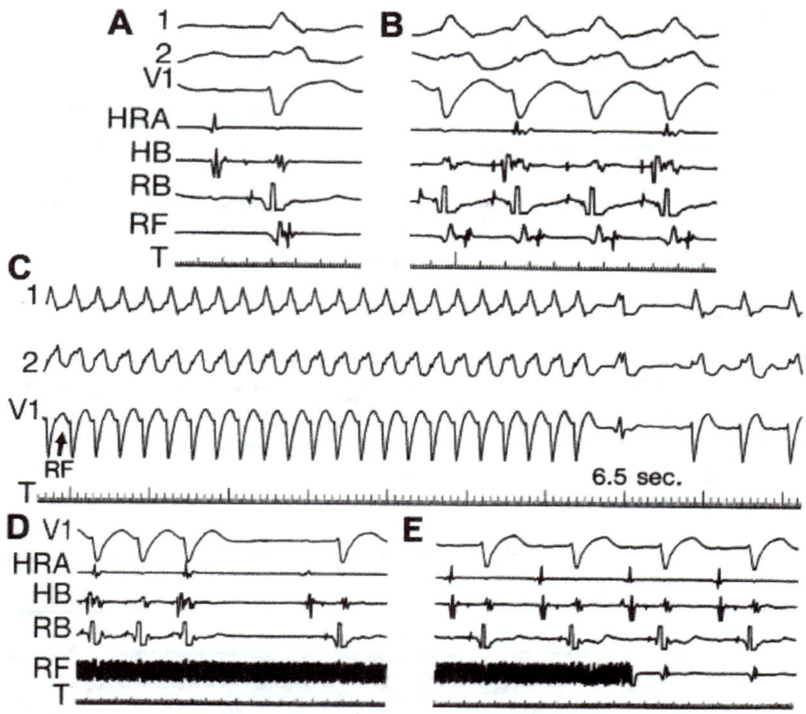

Figure 3.80. LBB ablation in BBR on baseline and tachycardia (VT). SR with H-RB sequence (**A**). The second deflection or RF tracing is LB potential. BBR-VT is shown in **Panel B**. The sequence is LB, followed by H-RB, activating almost simultaneously. **Panel C** is a surface ECG showing ongoing BBR-VT and terminated with RF. **Panels D** and **E** show ongoing and the end of LBB ablation. VT terminates in Panel E. LB can no longer be identified on the ablation catheter in Panel D. Note how identical the QRS is during VT versus baseline ECG with preexisting LBB block (Panels A–C). (Modified from Blanck et al with permission from John Wiley and Sons Publishing.[95])

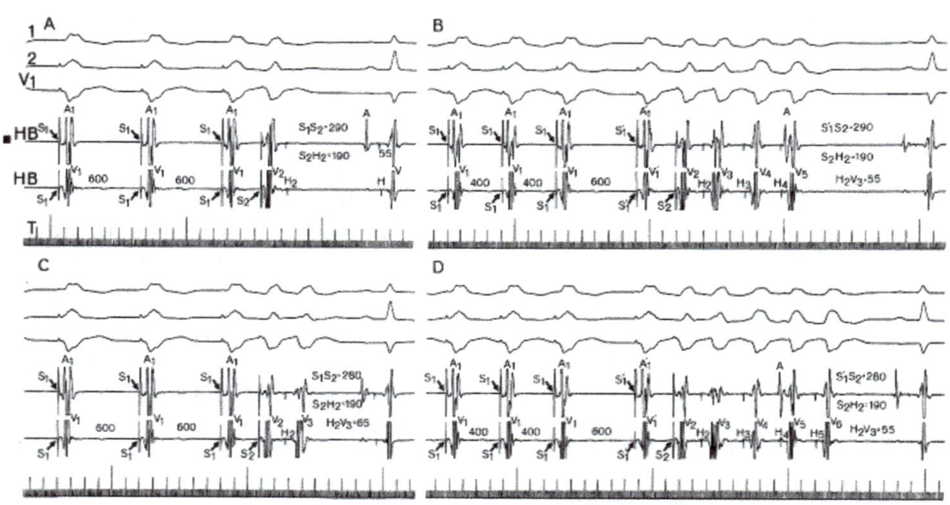

Figure 3.81. BBR with short-to-long cycle-length (CL) change. **Panels A** and **C** show a constant CL of 600 ms before V_2. The atria and ventricle are paced simultaneously in both panels during the basic drive. No V_3 occurs in Panel A and a single BBR-V_3 is seen in Panel C. However, a nonsustained BBR-VT occurs in **Panels B** and **D**. Note also that the H_2-V_3 is longer in Panel C because of greater delay distal to recording of H_2 following V_2; while in Panels B and D, an H_2-V_3 of 55 ms is equal to that during SR (labeled in Panel A), which probably reflects a more distal retrograde block in the RBB and faster recovery following V_2. The recovery of conduction in LBB following V_3 in Panel D is most likely due to a shorter V_2-V_3 CL compared to Panel C. Progressive shortening with each CL causes progressive decrease in the HH cycle, with eventual retrograde block in the LBB and termination of reentry. (Reproduced from Denker et al with permission from Wolters Kluwer Publishing and the American Heart Association.[89])

published regarding BBR as a clinical tachycardia. Over the years, we have diagnosed BBR in close to 100 patients. The summary of that experience is outlined below.[91,93–99]

1. Sustained VT due to BBR should be suspected in patients with dilated cardiomyopathy regardless of the etiology. Though most patients will have coronary heart disease due to its high prevalence, BBR is actually infrequent in patients with coronary heart disease. Overall incidence of BBR in our experience is about 6% of patients with induced sustained monomorphic VT.
2. The occurrence of BBR-VT is a much higher percentage of the sustained VT seen in several pathological states, that is, 30% incidence in valvular disease, 39% in idiopathic cardiomyopathy, and even higher in myotonic dystrophy.[95,98]
3. The underlying abnormality in all of the above conditions is prolonged HV interval or diseased HPS. The VM is not always abnormal. In fact, several of our patients had no detectable cardiovascular disease and many of the individuals with valvular disease had no left ventricular (LV) dysfunction.[91,98] Nonetheless, all of our patients had an abnormal QRS complex and a prolonged HV interval in the baseline state.
4. Since the RV is the pacing site for BBR-VT induction in most cases, LBBB pattern is much more common than RBBB pattern. However, this may not be the case if the LV is used as the pacing site.
5. Clinically it is not clear which morphology dominates in spontaneous sustained BBR since many of these patients are referred postcardiac arrest, syncope, post-DC cardioversion for rapid VT, and only a rhythm strip is available. However, from the available data, it seems both morphologies exist.[91,95]
6. The baseline ECG is almost always abnormal. AF is common, and the QRS complex shows a nonspecific IVC defect of LBBB type most of the time. Incomplete RBBB and LBBB are the other common findings on surface ECG. Surprisingly, some patients have a complete RBBB or LBBB pattern on baseline ECG (see Figures 3.78 and 3.79). Since these patients have documented BBR-VT on ECG, one has to assume that the BBB pattern reflects conduction delay rather than complete block. Ironically, a true complete BBB (as achieved by BB ablation) makes it impossible to have BBR, whereas prolonged conduction in the HPS is the ideal milieu for sustained BBR to occur (see Figures 3.78 and 3.79).
7. Sustained BBR is more reliably induced with a short-to-long CL change than constant CL and a short-to-long CL change should be part of pacing protocol to diagnose or exclude BBR in a given patient.[93,94]
8. The QRS morphology in these patients is quite variable and intriguing. The initial part of the QRS is rapid due to HPS-muscle spread rather than myocardial origin. When the VT CL is slower, that is, >300 ms, a classic BBB pattern is observed, but with rapid VT some degree of myocardial impulse spread could be uneven, producing unusual QRS morphology.
9. Post-RBB ablation eliminates any ventricular activation via the right bundle, which exposes the trifascicular nature of the LBB.[100] Invariably this leads to either or both of the following: pathological Q wave in V_1 (due to delay in the septal division of the LBB [Figure 3.82]); and right-axis or left-axis shift (Figures 3.83 and 3.84). This indicates that: LBB has 2 divisions, and conduction delay in these divisions is not equal (see Figures 3.82–3.84).
10. The above hypothesis is strengthened by the fact that even though the baseline HV is prolonged in these patients, it increases further after RBB ablation.[101] This means that the right bundle was contributing toward ventricular activation, masking the magnitude of conduction delay in the LBB system. This was revealed by complete interruption of conduction in the RBB. Furthermore, in patients with RBBB and normal HV interval, such as traumatic RBBB (ie, from catheter), no Q wave in V_1 nor axis shift is noted. In other words, the trifascicular nature of the LBB cannot be ascertained unless there is a

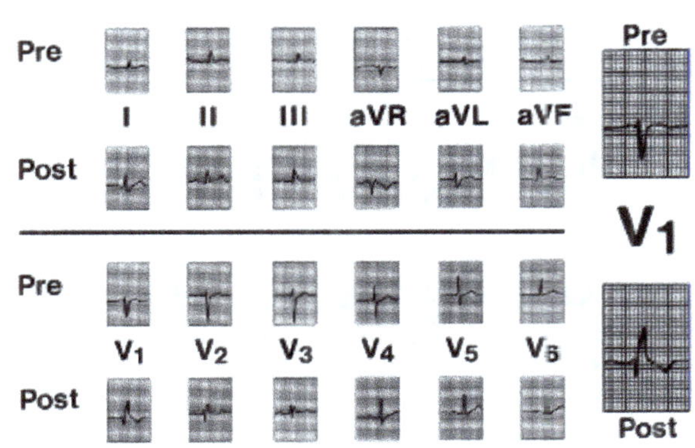

Figure 3.82. Twelve-lead ECG pre- and post-RBB ablation for BBR in a patient with no structural heart disease. Note there is no change in the axis but typical RBBB following ablation. (Reproduced from Dhala et al with permission from Elsevier Publishing.[100])

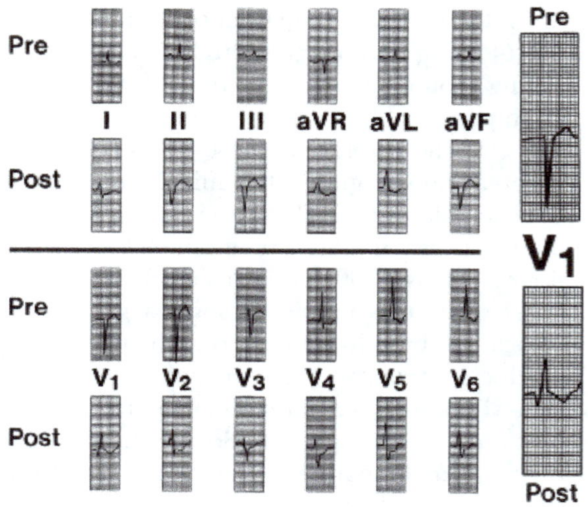

Figure 3.83. Twelve-lead ECG pre- and post-RBB ablation in a patient with BBR and dilated cardiomyopathy. Development of RBB block, shift in axis to the left (from +60 to –80), and pathological Q wave in lead V_1 can be observed, exposing septal and left anterior divisional conduction delay greater than left posterior division of the LBB. (Reproduced from Dhala et al with permission from Elsevier Publishing.[100])

variable degree of conduction delay in the LBB Purkinje system (see Figures 3.82–3.84). In fact, post-RBB ablation in patients with BBR, further HPS conduction delays are noted (long HV interval), which exposes the trifascicular nature of the LBBB and perhaps is the reason why it is not universally known or accepted, because it cannot be convincingly demonstrated in any other scenario. The anatomic description of LBB as a trifascicular system does not prove its functional involvement.

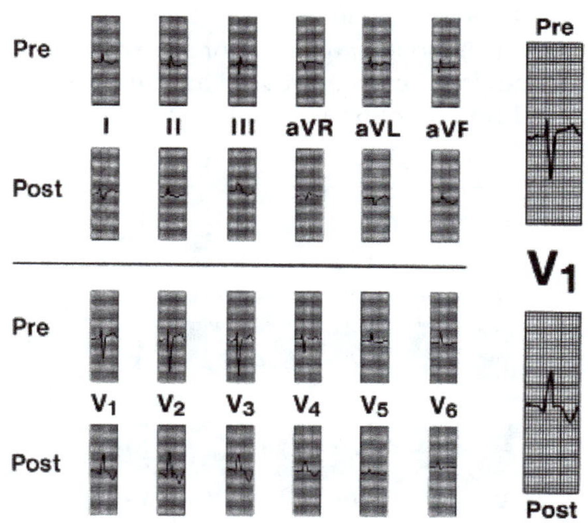

Figure 3.84. Twelve-lead ECG pre- and post-RBB ablation in a patient with BBR and dilated cardiomyopathy. These tracings show development of RBB block, right axis deviation (RAD), and new pathological Q waves in V_1-V_2, revealing greater conduction delay in the left posterior and septal divisions of the LBB than in the left anterior fascicle. (Reproduced from Dhala et al with permission from Elsevier Publishing.[100])

11. The prolonged and dissimilar degree of conduction delay in the left superior and inferior division of the LBB also explains interfascicular reentry.[97] Although BBR and interfascicular reentry often co-exist, it is after RBB ablation that the true nature of VT (interfascicular) becomes apparent, because the VT is not abolished and may even become incessant (Figure 3.85).[91] This is primarily the case with VT that has a RBBB pattern. Further mapping of the fascicles and ablation ultimately will abolish the reentry. Postfascicular ablation, the RBBB persists and an axis shift further prolonging PR (HV) is anticipated, depending upon the fascicle that is ablated (Figure 3.86).
12. Criteria for the diagnoses of BBR have been extensively published in detail and will not be repeated here.[91–100]

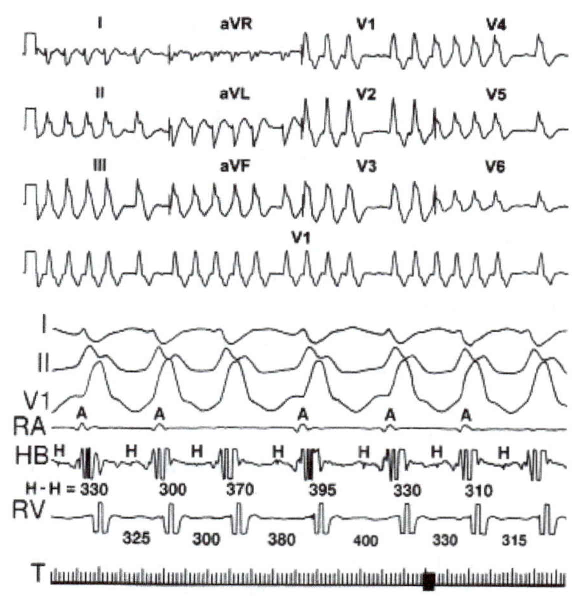

Figure 3.85. Interfascicular VT post-RBB ablation for BBR. This patient developed incessant VT (top panel) with RBB block plus RA pattern following RBB ablation. The intracardiac EGM (bottom panel) shows that HH CL drives the subsequent VV CL (diagnostic clue for BBR). Since the RBB was previously ablated, it was most likely interfascicular reentry with the circuit being antegrade conduction along the left anterior fascicle and retrograde conduction via the left posterior fascicle. (Reproduced from Blanck et al with permission from John Wiley and Sons Publishing.[96])

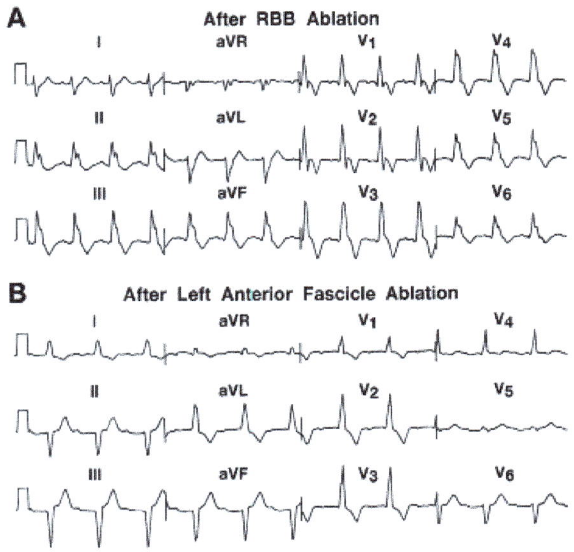

Figure 3.86. Twelve-lead ECG before and after LAF ablation. Same patient as Figure 3.85. The baseline ECG is post-RBB ablation and shows RBBB and left posterior fascicular (LPF) block (A). Interestingly, the LAF shows minimum delay, since the PR interval is relatively normal. Ablation of the left anterior fascicle terminated the interfascicular reentry (see Figure 3.84) and now the ECG shows RBBB, LAF block, and a long PR (B) due to HV delay along the LPF. (Reproduced from Blanck et al with permission from John Wiley and Sons Publishing.[96])

BBR complex (V_3) configuration is similar to paced V_2, hence at times it has been considered local reentry, with retrograde H_2 being incidental. While local reentry is possible, it is less likely in patients with no structural heart disease. However, in the circumstances outlined below, local reentry may be considered as follows:

1. There is no relationship with H_2 (ie, no H_2 preceding V_3), H_2-V_3 is shorter than SR, or there is a marked variation in H_2-V_3 interval.
2. The patient has local pathology, particularly near the pacing site.
3. The H_2 and RB_2 activation sequences are incompatible with QRS morphology of V_3.
4. V_3 is very dissimilar to V_2, unless there is significant (measurable) S_2-V_2 stimulus-to-response latency.
5. It occurs at longer V_1-V_2 but not at shorter V_1-V_2.
6. A nonsustained or sustained VT of V_3 morphology is triggered but no HH interval is identified before the QRS complexes.
7. Catheter-induced (mechanical) V_3 is excluded. Iatrogenic ventricular complexes tend to be random and are likely to occur at other times during the study.[19] Catheter repositioning or withdrawal will abolish traumatic ventricular complexes.
8. With decreasing V_1-V_2 and increasing S_2-V_2, the morphology of V_3 from local reentry may change sufficiently so that V_2-V_3 is considered of different origin. This pseudo-multifocal appearance of QRS originating from the paced site has been reported before and is likely to be from local reentry, unless it is preceded by an H-RB sequence that is appropriate for the QRS morphology.[101]
9. When recordings from the HB, right and/or LB are available, the origin of V_3 and the route of V_2 propagation are relatively easy to figure out.
10. With abrupt CL changes, short-to-long or vice versa HPS shows a similar electrophysiologic behavior as seen in the antegrade direction, that is, a short-to-long CL change before V_2 produces greater and earlier delays in the HPS (ie, at long V_1-V_2) than constant CL (Figure 3.87).[102,103] Similarly, a long–short CL change decreases V_2-H_2 delay, RRP, and ERP of HPS compared to a constant CL. V_3 due to BBR will be influenced by preceding HPS changes, that is, with short-to-long CL change, V_3 occurs at longer V_1-V_2, may extend to V_4, V_5, etc. (see Figure 3.81), and the reverse is noted with long-to-short sequence.[98] V_3 from local reentry will not be directly affected by prior HPS behavior.
11. Interestingly, VM shows a divergent behavior in response to V_2 with abrupt CL change (Figures 3.87 and 3.88) compared to HPS.[102,103] While the HPS shows exaggerated response in the direction of last CL, the VM reacts to the constancy of previously stable CL. In other words, VM shows retention of memory while HPS reacts in the direction of change. For example, with a 400- and 600-ms VM, ERP will occur as if the last CL (before S_2) is 500 ms, while with the same change, HPS and RRP-ERP would resemble the last CL at 700 to 800 ms. The exact electrophysiologic explanation for this divergent behavior is not clear, but the observations are consistent and reproducible.

Response of AP to ventricular extrastimulation: The reason for our earlier discussion regarding the response of AV node to V_2 is because AP is directly connected to the VM and HPS (unlike the AV node), and interplay between HPS and AP has a key role in tachycardia initiation in WPW syndrome. During the basic V-paced drive cycle, the retrograde atrial activation is often a fusion complex just as QRS is a fusion complex in manifest AP during SR. Due to the predictable response of NP to V_2, many phenomena are revealed (Figures 3.89 and 3.90) during scanning of the diastolic interval in patients with the rapidly

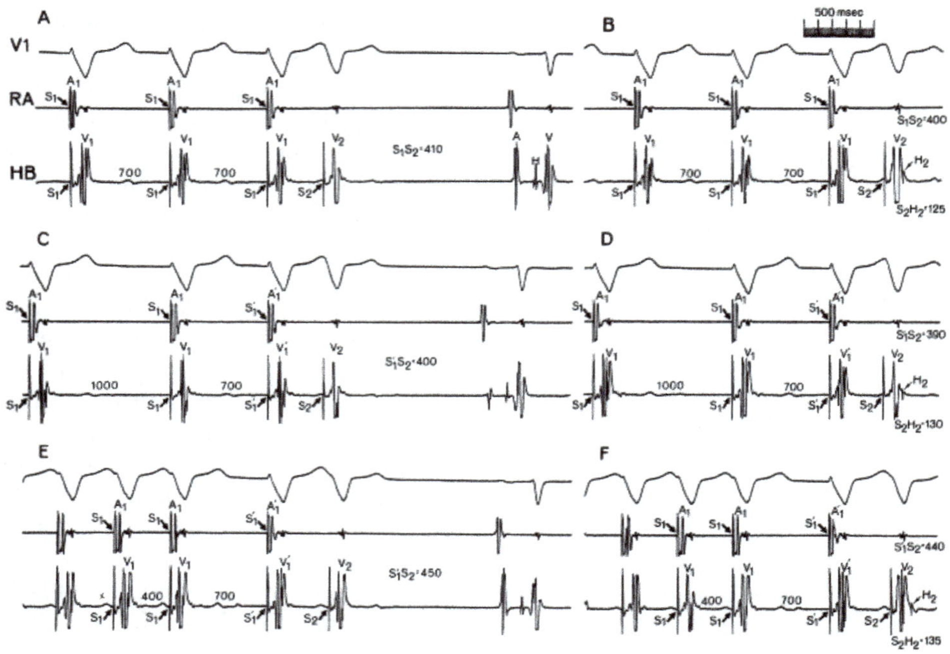

Figure 3.87. Retrograde HPS response to V_2 with sudden CL change. In all panels, the CL before V is the same and measures 700 ms. All basic CL are constant in **Panels A** and **B**, and long-to-short change in **Panels C** and **D**, whereas in **Panels E** and **F** the change is from short-to-long. Atria and ventricles (A_1, A_1, V_1, V_1) are paced simultaneously to maintain stability of basic CL. Note that compared to constant CL (Panel B), retrograde H_2 emerges at V_1-V_2 10 ms shorter in Panel D and 40 ms longer in Panel F, similar to the response in the antegrade direction. (Reproduced from Denker et al with permission from Wolters Kluwer Publishing and the American Heart Association.[102])

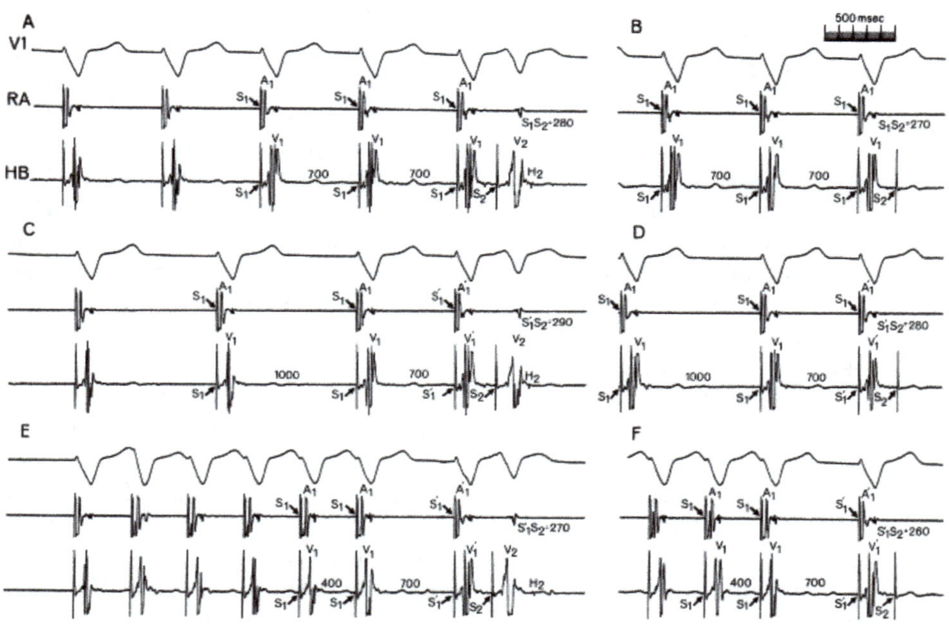

Figure 3.88. Response of VM to S_2 with sudden change in CL. Same patient and format as Figure 3.87. Note that the ERP of VM is now longer with a long-to-short change (**C** and **D**) compared to constant CL (**A** and **B**), and ERP-VM is shorter with short-to-long changes (**E** and **F**). In other words, there is a divergence in HPS and VM response to a single extrastimulus. (Reproduced from Denker et al with permission from Wolters Kluwer Publishing and the American Heart Association.[102])

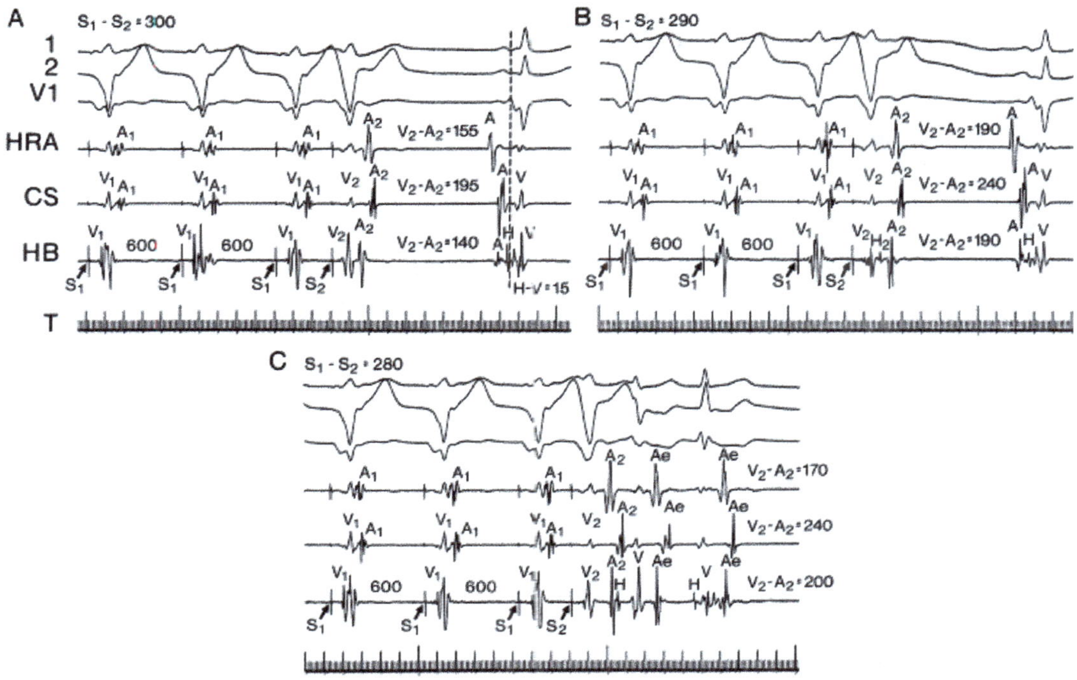

Figure 3.89. Response of right-sided AP and HPS to V_2. In a patient with manifest AP (last complex in **Panel A**), onset of retrograde A on HB is not recognizable and is obscured by the local V_1 EGM during the V_1-V_1 drive of 600 ms (in all panels). With V_2 at a V_1-V_2 of 300 ms (Panel A), the A_2 emerges from V_2 due to VH delay. Hence the retrograde atrial activation during A_1 was fusion complex because it changes following V_2. The H_2 emerges in **Panel B** and the V_2-A_2 prolongs further. Exclusive V_2-A_2 via the AP is seen in **Panel C** in which V_2 blocks in the HPS and the retrograde atrial activation sequence is different compared to Panels A and B. Orthodromic reentry is initiated with a RBBB and left-axis QRS. Short A_2-H is due to V_2 block in the HPS, no AVN penetration, and concealed conduction of V_2 in HPS, which prolongs the next HV, and the RBBB prolongs the next VA compared to the last complex in Panel C, which shows complete RBBB. All relevant values are labeled.

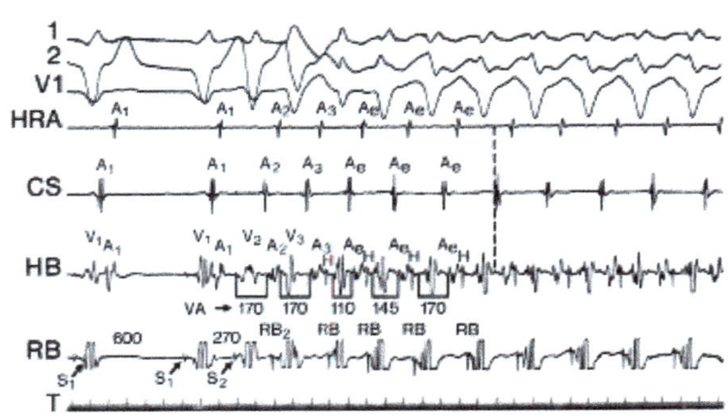

Figure 3.90. Response of left free wall AP and HPS to V_2. Basic format is the same in all intracardiac tracings. Note exclusive VA conduction via a left-sided AP during V_1 and V_2 (H_2 is not, but RB_2 is identifiable). At a V_1-V_2 of 270 ms, RB_2 emerges and BBR follows, which in turn blocks in the HPS, produces A_2 over AP and starts orthodromic reentry. Note the different VA interval with the narrow QRS and LBBB with and without LAD artery. It is also remarkable that VA during RV apical pacing, BBR, and spontaneous LBBB and LAD are the same. (Reproduced from Jazayeri et al with permission from John Wiley and Sons Publishing.[6])

conducting (Kent) type of AP. Some of these are discussed below:

1. In patients with no demonstrable VA or long VA conduction length through the NP, an eccentric atrial activation is noted both during V_1 and V_2 if there is retrogradely functioning AP (see Figure 3.90).[14] The only exception being anteroseptal AP with retrograde atrial breakthrough near the AV node, in which case atrial activation resembles retrograde NP conduction. While assessing retrograde conduction via AP, it is essential that left atrial recordings from the CS (proximal, middle, and distal) are tracked.
2. Distinction between the retrograde atrial breakthrough via normal versus anteroseptal or any other Kent bundle-type of AP is relatively easy with V_2 introduction. At V_1-V_2 interval ≤50% of the V_1-V_1, the VA interval through the NP will prolong V_2-A_2 > V_1-A_1. If this does not happen, there is an AP present (Figures 3.90 and 3.91).[104]

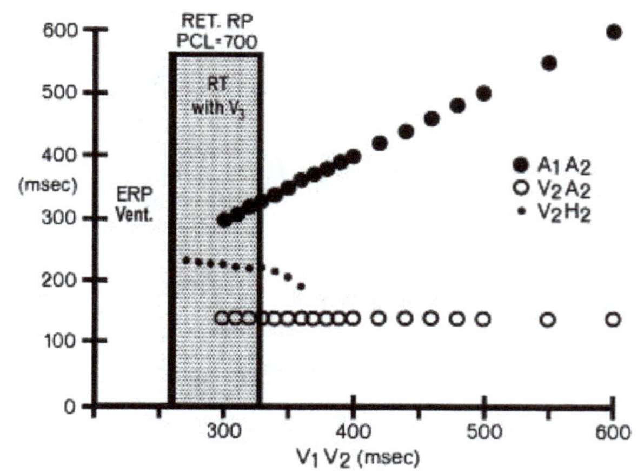

Figure 3.91. Retrograde HPS and AP response to V_2. The plot shows constant V_2-A_2 before and after emergence of the H_2. Orthodromic SVT could not be initiated as long as V_2-H_2 conduction is intact (due to concealed conduction in the AV node). Only when V_3 (BBR) blocks retrogradely in the HPS, is the SVT initiated. (Reproduced from Akhtar et al with permission from Elsevier Publishing.[51])

3. As the V_1-V_2 interval is shortened, there is a progressive dissociation between the VA conduction via the AP versus the NP due to VH and/or HA increase, while V_2-A_2 via the AP does not change. This can be diagnosed even if the atrial activation sequence is not available due to limited atrial recording sites. V_2-H_2 conduction measurement is helpful for this purpose and an identifiable retrograde His deflection is the key.
4. The presence of AP is further proven when the retrograde A_2 is registered before the retrograde H_2 (Figures 3.89–3.93).[59,104]
5. In individuals with longer retrograde RP of the NP (longer than AP), at some point, orthodromic SVT may be triggered (see Figures 3.89–3.91) due to retrograde block in the HPS.[59,104] When retrograde block occurs in the AV node (at V_1-V_2 > 70% of the basic CL), SVT is seldom initiated since retrograde AV nodal penetration is an impediment for atrial impulse (A_2) to propagate through the AV node (retrograde concealed conduction in the AV node). Continued block of V_2 in the AV node (with recognizable retrograde H_2) will continue to prevent SVT induction (see Figure 3.91).
6. In a small percentage of cases, particularly with long VA interval during RV pacing (ie, left free wall AP due to distance from pacing site), AV node may recover sufficiently for A_2 to navigate the AV node-HPS anterogradely and start an orthodromic SVT.
7. Due to penetration of the AV node by V_2 in the aforementioned scenario, the first orthodromic tachycardia complex will have a longer AH

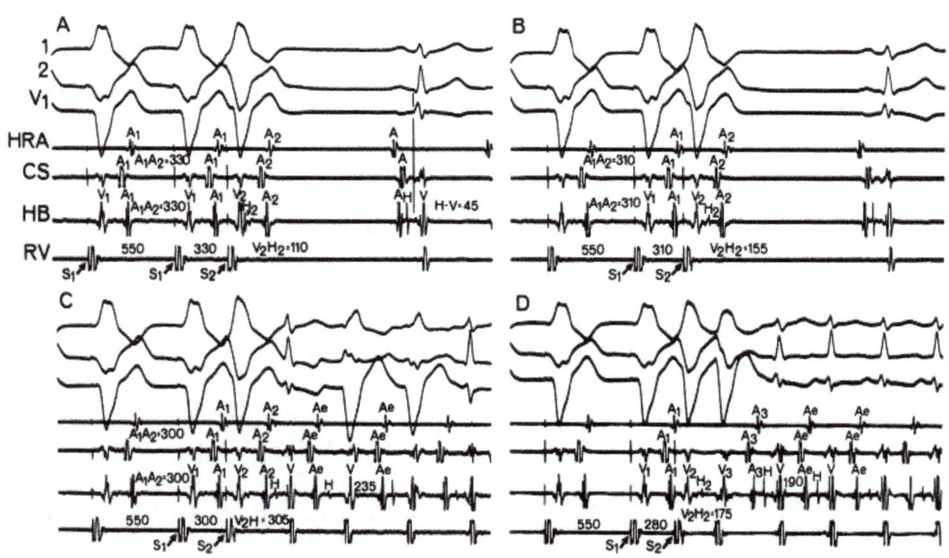

Figure 3.92. HPS block and orthodromic SVT. BCL is the same in all panels (550 ms). The patient has concealed left free wall AP as suggested by an atrial activation sequence which does not change (V_1-A_1 = V_2-A_2) despite H_2 emergence (**A** and **B**). V_2 blocks in the HPS in **Panel C**, conducts via the AP (V_2A_2-H sequence), and initiates orthodromic SVT. Note the V_2-A_2 prolongation with LBBB (235 ms) versus normal QRS (190 ms). **Panel D** shows that when V_2 reaches the node (ie, H_2) and also blocks in the AP, SVT is not triggered. However, HPS delay (V_2-H_2) results in V_3 from BBR, which in turn blocks in the HPS (no H_3), reaches A_3 via the AP and initiates orthodromic SVT. (Reproduced from Akhtar et al with permission from the American Society for Clinical Investigation.[104])

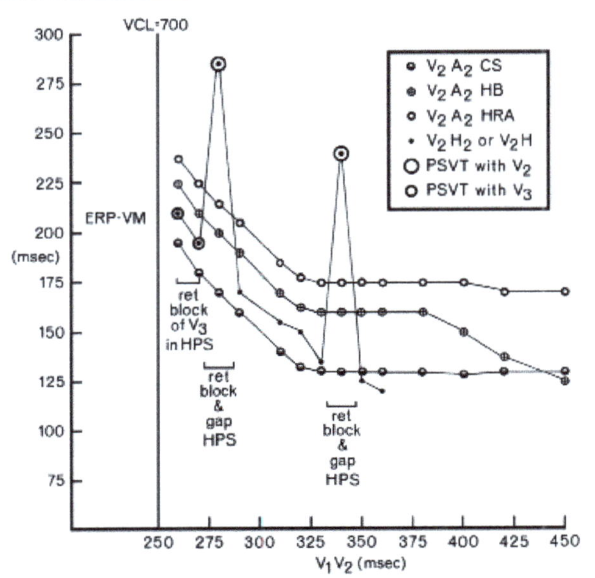

Figure 3.93. Ventricular CL (VCL) scanning with V_2 in WPW. This graph summarizes a typical interaction between HPS and left free wall AP. Note the shortest V_2-A_2 on CS and its parallel relationship with HRA. A_2 on the HB trace is initially via the NP but becomes parallel to the other AEGMs once the H_2 emerges or V_2 blocks in the HPS. Retrograde block of V_2 or V_3 in the HPS starts the orthodromic SVT. All values are in milliseconds. When V_2 blocks in the HPS, the jump in V_2-H reflects that the H follows A_2 and that H is anterograde. ERP-VM, effective refractory period-VM. (Reproduced from Akhtar et al with permission from the American Society for Clinical Investigation.[104])

interval compared to the rest of the tachycardia complexes (concealed conduction in the AV node) (see Figure 3.77). The sequence will be $V_2 - H_2 - A_2 - H_e - V_e$ (e = echo). In this type of induction, the first orthodromic QRS is usually not aberrant due to long H_2 to next H_e interval (in this situation, the HB is activated twice, once in the retrograde and then in the anterograde direction), allowing for longer HH cycle preceding the next QRS, which is then likely to exceed the RRP-ERP of the HPS.

8. If it occurs at all, aberrant conduction in a case like this is going to be in the second QRS complex, which could be either RBBB or LBBB.[105]
9. The most common mechanism of orthodromic SVT initiation with V_2 is a block in the HPS (V_2 no retrograde H_2), but with V_2-A_2 conduction via the AP and then unimpeded conduction through the AV node (see Figures 3.91–3.93).[104] The A_2-H_e interval of the first orthodromic cycle is relatively short, since V_2 never reached the AV node, retrograde A_2 is via AP, therefore, the first completed reentrant cycle of SVT may be aberrant (see Figures 3.90 and 3.92).
10. Aberrant conduction in this case will be LBBB because, while V_2 blocks in the HPS, it blocks later in the LBB than the RBB and, therefore, LBB has less time to recover and is more refractory to the unimpeded propagation of A_2 through the AV node.
11. If V_2 reaches the His, the SVT cannot be initiated; but when the H_2 is followed by V_3 from the BBR, it blocks in the HPS and the orthodromic SVT is initiated with somewhat the same set of circumstances as mentioned earlier, with V_3 blocking in the HPS, conducting through the AP (V_3-A_3) and starting the orthodromic SVT (see Figures 3.91–3.93).
12. In patients with a longer RP of the AP than the HPS, V_2 blocks in the AP and conducts to the His via the AV node (H_2), but may or may not reach the atria.
13. At this point orthodromic tachycardia cannot be initiated, but AV nodal reentry, intra-atrial reentry and antidromic SVT may be possible. For the latter 2 possibilities, V_2 must activate the atria (Figures 3.94 and 3.95).[14]
14. In patients with ventricular preexcitation, results of studies on A_1-A_2 during the antegrade RP parallel those on the V_1-V_2 interval at the point of AP insertion. Thus, HPS delays may be encountered retrogradely and hence the entire phenomenon described under application of the V_2 method from right or LV also applies. Therefore, when A_2 produces a dual ventricular response (V_2 and V_3), the HB potential (H_2) may be antegrade or retrograde and ATR after A_2 (A_2-H_2-V_3, with V_3 preceded by V_2 via AP) may in fact be V_1-V_2 – V_2-H_2 and V_3, that is, via the

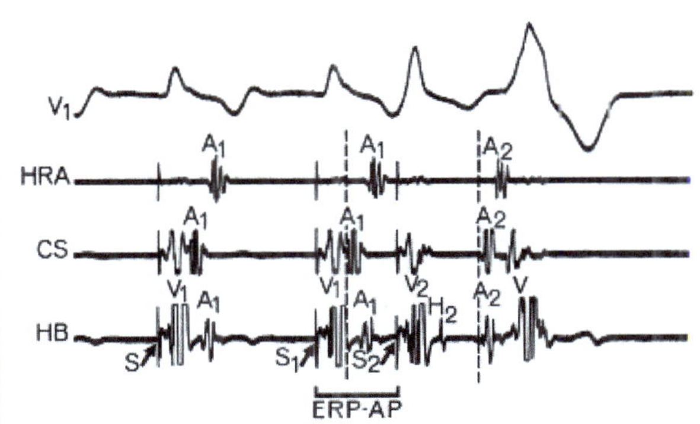

Figure 3.94. Antidromic echo in response to V_2. Retrograde A_1 is via a left free wall AP, which shows a block in response to V_2. The V_2 conducts via the AV node and activates the atria, resulting in an antidromic echo. ERP-AP, effective refractory period-AP.

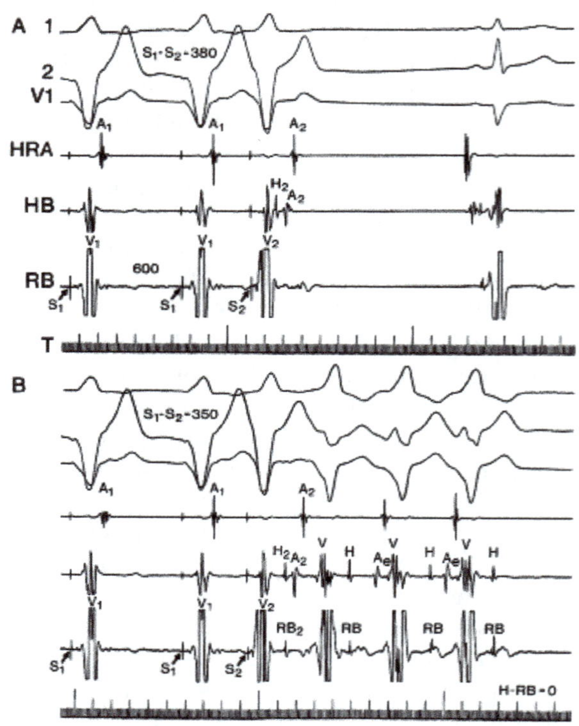

Figure 3.95. Antidromic reentry in response to V_2. **Panel A** shows emergence of H_2 with short H_2-A_2 values. The H-RB interval during sinus is 20 ms. At further V_2-H_2 delay, antidromic reentry is initiated. The H-RB interval of zero suggests the activation of H_2 via the LBB. The antidromic reentry is maintained as long as H_2 results in retrograde A (A_e). Retrograde block of impulse in the AV node terminates reentry (**B**). It should also be noted that HA interval variation leads to AA variation, which in turn drives the R-R intervals.

BBR rather than via the AP and then AV node antegradely (dual ventricular response) (Figures 3.51 and 3.96). This confusion can occur because the site of pacing is the atrium; and unless additional recordings from the BB are very helpful, it will be difficult to prove one way or the other.

15. Similarly, during AF and rapid ventricular response over the AP (ie, preexcited QRS complexes), some complexes may be due to BBR both with ipsilateral and contractual ventricular morphology and may be misinterpreted as evidence for additional AP. This can also be resolved with additional recordings from the BB.
16. Closer coupling of the V_1-V_2 interval may result in retrograde VA block (ie, no A_2) either in the AP and AV node or the AP and HPS (Figure 3.97).
17. A latency before S_2-V_2 could permit resumption of conduction through the HPS, AP, and perhaps even the AV node before ERP of the ventricular muscle (ie, S_2 but no V_2) is reached.
18. Patients with no VA conduction through the AP are not prone to orthodromic tachycardia but may experience other arrhythmias such as antidromic reentry, automatic junctional tachycardia, AV nodal reentry, AT, AFl, and AF with or without preexcited QRS complexes, as well as VT.
19. Other extranodal AP of clinical relevance include the so-called slow-conducting retrograde-only AP, which produces a long-RP tachycardia that may be paroxysmal or incessant.[7,106,107]
20. The incessant variety is often referred to as permanent junctional RT and can cause arrhythmic cardiomyopathy, which is generally reversible with control of the tachycardia.
21. This type of SVT is easily induced with the V_2 method, as is long-RP tachycardia due to fast–slow AV nodal reentry (uncommon AV nodal reentry).[8] The long-RP SVT due to the above 2 mechanisms can be distinguished by early or delayed excitation of the atria with one or 2 appropriately timed ventricular extrastimuli when the HB is refractory.
22. These slowly conducting AP (yet unnamed) have an earlier atrial breakthrough in the posteroseptal region and display decremental properties with the V_2 method.
23. The exact anatomic structure, the cause of decremented conduction, and the cellular orientation of these slowly conducting retrograde pathways are not fully understood at this time.

AV Nodal Response to V_2 Method

Unlike the HPS and AP, the AV node is relatively remote from all the V pacing sites. Similar to the HPS in the anterograde direction, where the response depends on the timing of the anterograde H_1-H_2 interval rather than A_1-A_2 CL; retrograde AV nodal response depends upon the timing of the retrograde H_1-H_2 duration and not the V_1-V_2 interval. Since the retrograde H_1 can seldom be measured directly (ie, V_1-H_1 interval) and assuming V_1-H_1 interval is stable at a constant pacing CL, V_1-H_2 interval can be taken instead and used as a substitute to assess timing of retrograde impulse input into the AV node and consequent H_2-A_2 response (Figure 3.98). Usual observations are as follows:

1. When there is no VA conduction during V_1 or V_2, the dissociated atrial signal (such as P wave, other atrial rhythms, etc.) shows retrograde impulse

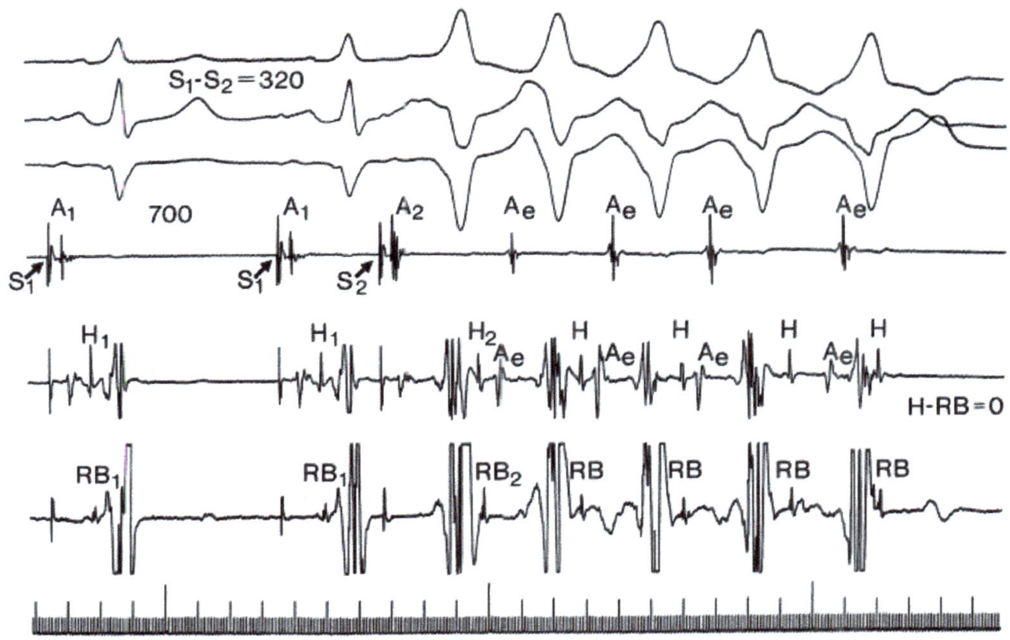

Figure 3.96. A_1-A_2 versus V_1-V_2 in WPW. Although the figure suggests that wide QRS tachycardia was initiated by A_2, the H_2-RB_2 interval of zero (H-RB at 15 ms during atrial pacing) suggests that H_2 was activated retrogradely via the LBB. The rest of the phenomena is quite similar to Figure 3.95. The HA variation causes AA variation, which controls VV interval. Retrograde block in the AV node stops the antidromic reentry.

concealment in the AV node as evidenced by block or conduction delay of dissociated atrial activity proximal to the HB deflection. This can be clearly observed by simultaneous A and V or sequential AV pacing during the basic drive; spontaneous or programed P that follows V_2 will show delay or block in the AV node.

2. Patients with a long VA time (V_1-A_1) will display a progressive V_2-A_2 increase at long V_1-V_2 interval (just shorter than V_1-V_1) even before the retrograde H_2 emerges from the V_2 EGM. If the retrograde H_2 has emerged from V_2, the V_1-H_2

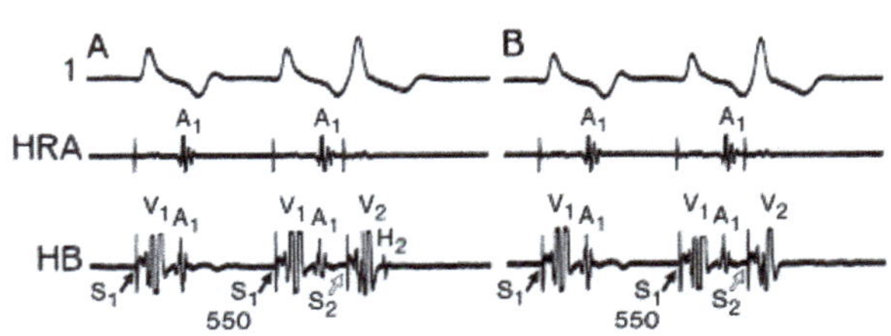

Figure 3.97. Block of V_2 in AP/AV node and HPS. Both panels display no A_2 in response to V_2 (same patient as Figure 3.94), but along the NP the block is in the AV node (**A**) and HPS (**B**).

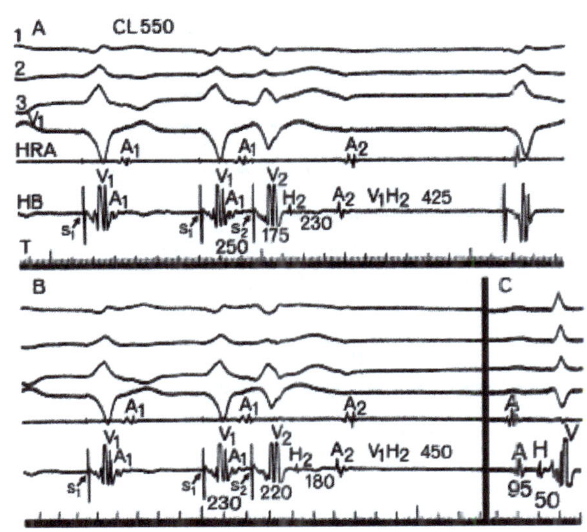

Figure 3.98. V_1-H_2 and retrograde AVN (H_2-A_2) response. During basic drive, retrograde H_1 is not identifiable but assuming V_1-H_1 is constant, one can substantiate V_1-H_2 rather than retrograde H_1-H_2 to assess H_2-A_2 conduction. It can be appreciated that a shorter V_1-H_2 of 425 ms (**A**) results in a longer H_2-A_2 compared to **Panel B**, despite shorter V_1-V_2. **Panel C** is a reference sinus beat. (Reproduced from Akhtar et al with permission from Wolters Kluwer Health and the American Heart Association.[64])

can be measured, and if the impulse blocks in the AV node, it can be demonstrated (H_2 but no A_2) (Figure 3.99). The V_2-H_2-A_2 conduction may resume at shorter V_1-V_2 if the V_1-H_2 intervals are sufficiently long for the AV node to recover (one form of retrograde gap phenomenon) (see Figure 3.99).

3. At certain degrees of H_2-A_2 delays, a ventricular echo or reciprocal complex may follow. This spontaneous QRS has the appearance of sinus complex and is preceded by HB activation twice, that is, once in the retrograde and then in the antegrade direction. Retrograde A_2 is sandwiched between the 2 His potentials and is less likely to encounter any delay in the HPS due to the sufficient time permitted by long intranodal conduction for the HPS to recover (see Figure 3.99).
4. However, if a V_3 from BBR also occurs following V_2 but preceding an AV nodal echo beat, the QRS complex during the AV nodal echo (V_e) may be aberrant (see Figure 3.99) and similar to the morphology of V_2, but could just be different depending on the relative recovery of the BB and HPS overall.[106]
5. The AV nodal ventricular echo phenomenon occurs quite often in patients with long VA (HA ≥100 ms) conduction, during retrograde WP as previously mentioned or following V_2, and does not require manifest retrograde dual AV nodal pathways. This should not be confused with the concept of reentry because all forms of reentry (manifest or concealed in any cardiac tissue) require a functional (not necessarily demonstrable) duality of pathways to occur.
6. In patients with slow–fast AV nodal reentry, V_2 may start the tachycardia (Figure 3.100) in a small percentage of cases with minimal VA or HA delay, unlike fast–slow AV nodal reentry or AP-mediated long-RP tachycardia (Figure 3.101), which can be easily triggered by V_2.[8]
7. When V_2 blocks above H_2 (ie, no A_2), AV nodal ventricular echo may still occur, although it is usually abolished. Atrial activation is not necessary, but the depth of H_2 penetration in the AV node is probably critical to engage the anterograde pathways.
8. When H_2-A_2 conduction resumes, ventricular echo may return (see Figure 3.99) if it occurred before the block. This will happen when there are greater V_2-H_2 delays so that V_1-H_2 is longer than the V_1-H_2 where H_2-A_2 block was observed. This is a form of retrograde gap phenomenon, in which the initial site of block is the AV node and return of conduction is due to delayed input of H_2 impulse into the AV node secondary to retrograde HPS delays (ie, V_2-H_2 increases and, consequently, V_1-H_2 increases) (see Figure 3.99).
9. Not unlike enhanced AV nodal conduction anterogradely, some patients have rapid conduction in the retrograde direction (HA ≤60 ms). The onset of the AEGM on the HB tracing cannot be detected and is obscured by the local ventricular EGM during the basic drive or long ventricular coupling (ie, V_1-V_2 > 70% of V_1-V_1).

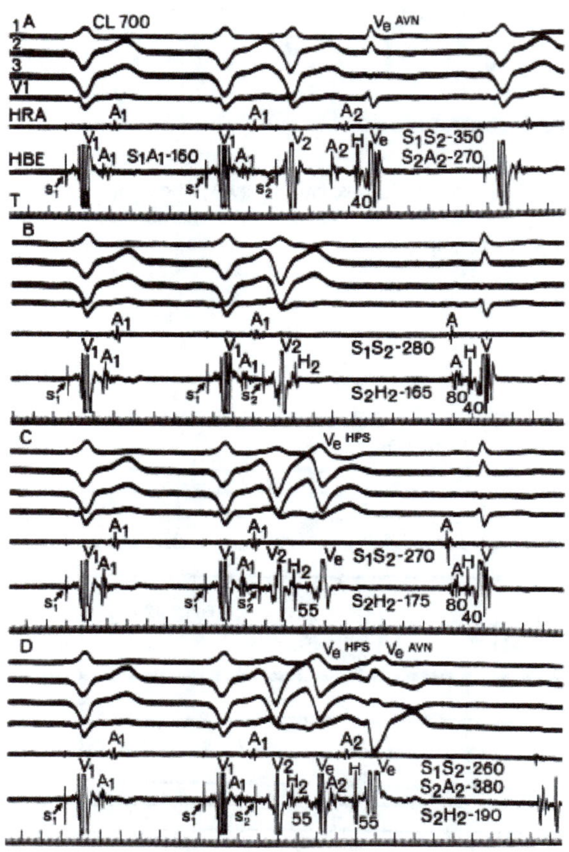

Figure 3.99. V_1-H_2 and H_2-A_2 block, gap and reentry in the AV node. **Panel A** displays long V_2-A_2, retrograde A is followed by a ventricular echo beat from AV nodal reentry. In **Panel B**, there is AV nodal block (H_2 no A_2) and no AV nodal echo. **Panel C** shows continued block in the node but V_2-H_2 delay causes BBR (V_e^{HPS}) with an appearance similar to V_2. Further shortening of V_1-V_2 (**D**) occurs in BBR again but now the V_1-H_2 is longer (450 ms compared to Panels B and C) and AV nodal reentry reappears; but since V_e^{AVN} is now preceded by V_{e3}^{HPS} (unlike in Panel A), V_e^{AVN} now shows a LBBB pattern. (Reproduced from Akhtar et al with permission from Elsevier Publishing.[106])

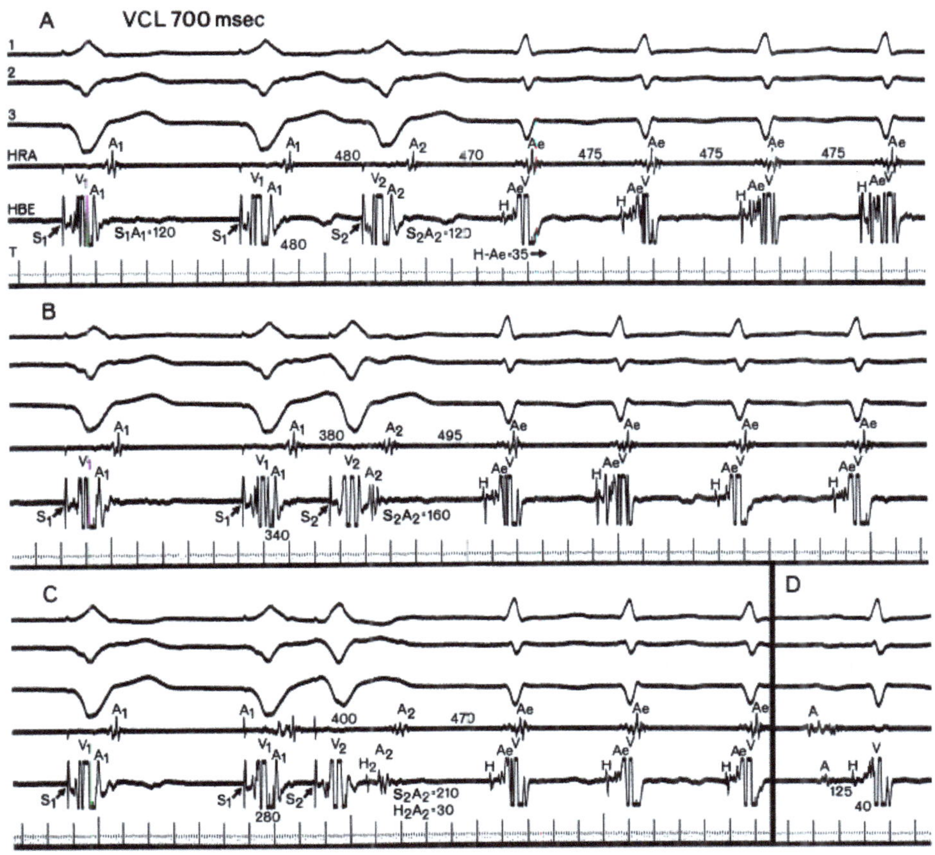

Figure 3.100. Slow–fast AVNRT with V_2 (A and B). The figure demonstrates the onset of common-type (short RP) AVNRT with V_2, without much delay in the H_2-A_2 interval (C). Panel D is a reference sinus beat. (Reproduced from Akhtar et al with permission from Elsevier Publishing.[8])

10. With shorter V_1-V_2, the A_2 and then the H_2 emerge from V_2 EGM (see Figure 3.100). At this point, H_2-A_2 can be measured and tends to remain constant as the V_1-V_2 is further decreased. This has led some investigators to believe this is an electrophysiologic behavior of the rapidly conducting AP, which partially or completely bypasses the AV node (see Figure 3.100). The arguments listed below counter this notion of the AP and favor an AV node-like response.

    1. During incremental V pacing, there is a small but measurable increase in VA time from the longest to the shortest V-PCL. If one assumes that V_1-H_1 remains constant during progressive shortening of the V_1-V_1 CL, the resulting increase in VA must be attributed to HA increase (not a Kent bundle-type of response).
    2. Similarly, when retrograde WP occurs in these patients, the longest and shortest VA intervals are with the last beat before and the first QRS after the block, respectively. Even if it is small, this is not the electrophysiologic response of normal HPS or rapidly conducting AP in either anterograde or retrograde direction.
    3. Since the V_1-H_2 curve is flat at a given basic CL, H_2-A_2 prolongation is unlikely during scanning of that CL. However, if the basic V_1-V_1 is shortened to V_1-H_2 (input into the AV node will be earlier) an increase in H_2-A_2 interval will be noted.
    4. Vagal stimulation, such as carotid massage, also results in HA prolongation or block, a typical AV nodal behavior (Figure 3.102).
    5. Medications traditionally known to suppress AV nodal conduction (such as beta-blockers, verapamil, etc.) have a

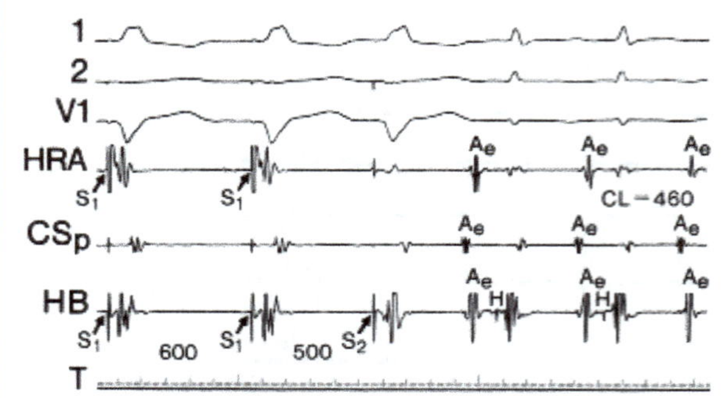

Figure 3.101. Long-RP SVT induction with V_2. The onset of long-RP SVT is depicted with a relatively late V_2. Note the earlier activation of the atrium in the proximal CS region. Long-RP SVT (fast–slow AVNRT or slow-conducting AP) induction with V_2 is relatively common.

similar effect on these retrogradely conducting fast pathways connected to the HB (Figures 3.103 and 3.104).

The HA interval in these cases is quite similar to that in patients with slow–fast (common) AV nodal reentry. The observation presented above, however, was noted in patients who did not have clinical tachycardia. Some of these individuals may develop SVT in the future unless they have an anatomically absent or physiologically ineffective common pathway, or lack of cross-connection, etc. In patients with existing slow–fast AV nodal reentry, the tachycardia can be initiated with no H_2-A_2 prolongation as pointed out earlier (see Figure 3.100). As to why such a short H_2-A_2 conduction can penetrate and navigate the slow pathway and start slow–fast AV nodal reentry is not clear. In this population with slow–fast AV node reentry, it can be postulated that the retrograde block in the slow pathway is quite distal, or perhaps these patients have no retrograde conduction through the slow pathway such that the AV nodal impulse can antegradely complete its transit to the HB to start the process.

Response of Atria to V_2

1. When V_2 reaches the atria via the AV node or AP, intra-atrial conduction delays block, intra-atrial reentry and intra-atrial gap phenomenon

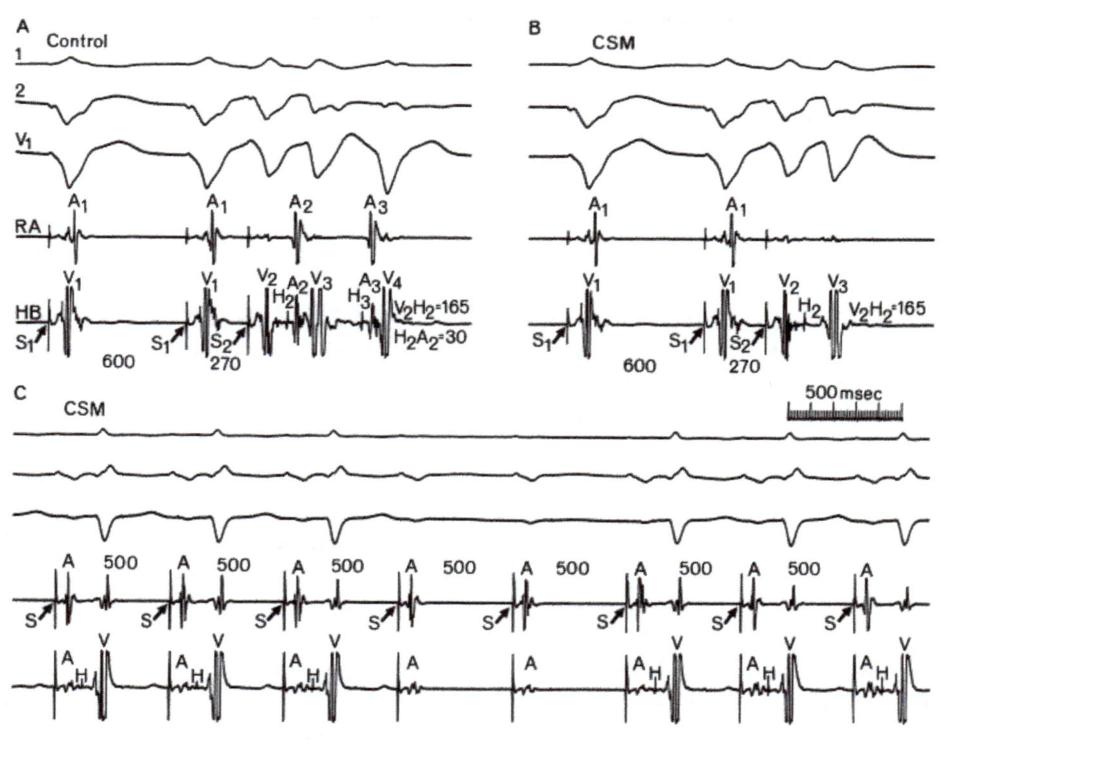

Figure 3.102. HA block with vagal stimulation. During the V_2 method in **Panel A**, the H_2 emerges and shows a relatively short H_2-A_2 interval of 30 ms. When the extrastimulus is delivered during CSM, H_2-A_2 block ensues, which is typical AVN response to vagal stimulation (B). **Panel C** shows AV nodal block during CS pacing in the anterograde direction with CSM. (Reproduced from Akhtar with permission from Wolters Kluwer Publishing.[44])

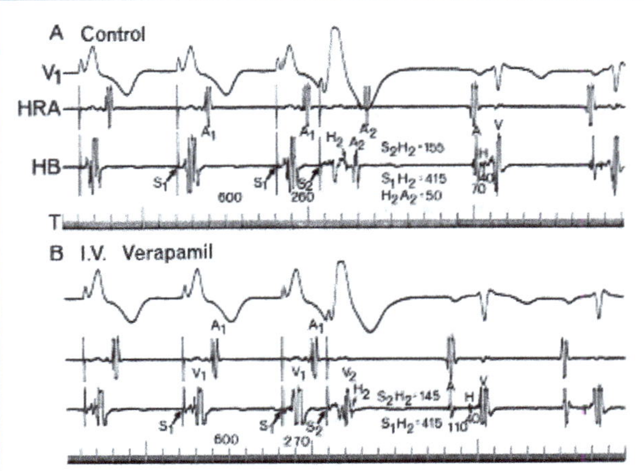

Figure 3.103. Verapamil effect on short HA conduction. **Panel A** is a control, and **Panel B** shows response to IV verapamil following V_2. Baseline H_2-A_2 interval is 50 ms (Panel A), and H_2 blocks in the AV node following verapamil administration (bottom panel). The timing of H_2 activation, that is, V_1-H_2, is comparable. It can also be noted that antegrade AV nodal conduction (AH) prolongs after verapamil, as would be expected.

(Figure 3.105) may occur on occasion and/or other atrial reentrant SVT may be triggered.[107]

2. In patients prone to AFl and/or AF, these arrhythmias may be initiated with V_2, but it is not common.

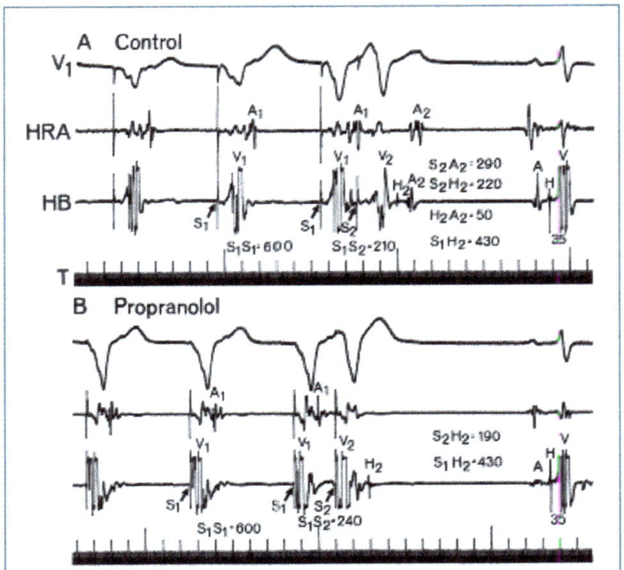

Figure 3.104. Response of HA conduction to IV propanolol. The format is the same as Figure 3.103. As compared to **panel A**, It can be appreciated in **Panel B** that H_2 is not followed by A_2 (AV nodal block). S_1-H_2 interval measures 430 ms in both panels.

3. Very short coupling intervals, that is, 250 ms or less at the atrial level, cannot be achieved with the V_2 method via the NP due to V_2, H_2 and H_2-A_2 delay.
4. Induction of AT with V_2 is less likely at longer atrial coupling (ie, V_1-$V_2 \rightarrow A_1$-$A_2 > 350$ ms) intervals but it may start VT or ventricular fibrillation (VF) in vulnerable patients. This should be kept in mind when attempting to induce an atrial process with V pacing.
5. Before the ERP of VM is reached, S_2-V_2 latency is encountered (RRP of VM), and further shortening of S_1-S_2 is not necessary unless determination of the ERP of VM is critical for EP evaluation in a given patient. Stimulus-to-ventricular-response latency with additional extrastimuli (ie, S_3, S_4, etc., to prevent inadvertent VF induction) should especially be avoided.

Iatrogenic Events During EP Studies[19]

Whenever a foreign object is introduced into the cardiac chamber, some mechanical stimulation will take place. In addition to the possible complications, many catheter-induced events can occur and, if not addressed, lead to limited study, misdiagnosis, and mismanagement. Examples that we have encountered are outlined below. It should be pointed out that this list consists of mechanically (not electrically) induced events.

1. AV and junctional premature (Figures 3.106–3.108) complexes are common and are not necessarily due to myocardial irritation from the tip of the catheter. The catheter could mechanically irritate along its entire intracardiac course in the cardiac chambers.
2. Traumatic RBBB and LBBB may be created and could last a few cycles to up to 24 hours. Also with traumatic BBB, the entire spectrum of phase-3 and phase-4 block (see Figure 3.9) may be encountered during the recovery process.
3. When patients are being investigated for specific symptoms, such as unexplained syncope, and have a preexisting LBBB or RBBB, intracardiac placement in close proximity to the HB and proximal BB via the contralateral ventricle should be done very carefully to avoid complete AV block (see Figure 3.16). One recommended alternative is to do incremental A pacing and extrastimulus study with placement of only the atrial catheter, determine the atrial CL and A_1-A_2 coupling interval where block in the anterograde direction is observed, and then place the HB and

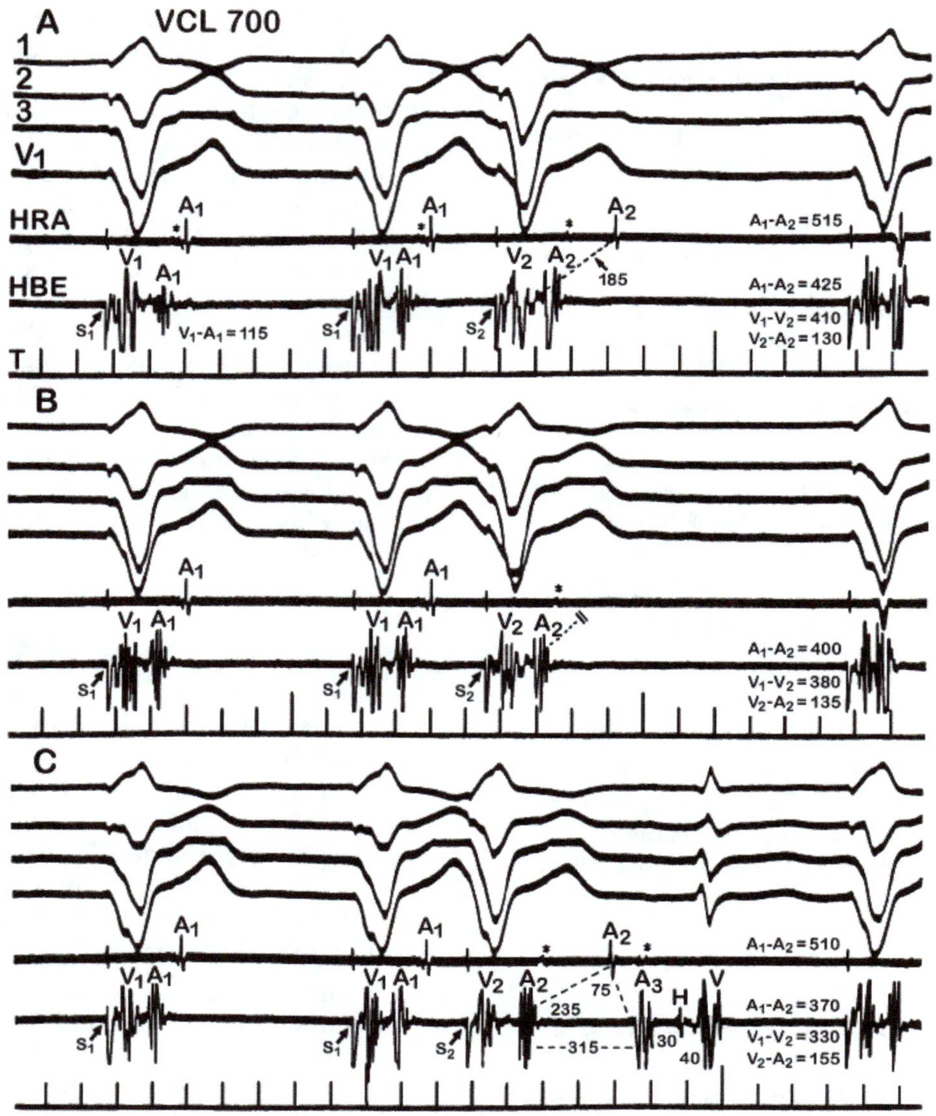

Figure 3.105. Intra-atrial delay, block, gap, and reentry in response to V_2. In all panels, the basic drive is 700 ms. V_1-V_2 is progressively shorter in **Panels A** to **C**. Note that there is a delay of conduction between the A_2 on HBE and HRA EGMs (185 ms). A far-field potential (noted by asterisk) maintains a constant relationship with A_2 in HBE. A block between A_2 on HBE and HRA is depicted in Panel B. Conduction to the HRA resumes at shorter V_1-V_2 with greater intra-atrial delay (235 ms), which re-excites the low atria. (Reproduced from Akhtar et al with permission from Wolters Kluwer Publishing and the American Heart Association.[107])

BB recording catheters. If the findings do not change after all catheters are in place, one can feel more confident that physical trauma was not responsible for infra-Hisian block, and HPS block may be the true cause of syncope.

4. In a similar way, subtle trauma to the HPS may prevent BBR from triggering; catheter-induced AP trauma damaging AP conduction in either direction could necessitate another procedure. A single-ventricle catheter study to initiate BBR-VT or SVT in the AP may be helpful to avoid such events if they do occur. At the very least, such precautions should be taken during the conduct of these studies and the electrophysiologist should be alert for these possibilities.
5. Another phenomenon that may occur is pseudonormalization of a preexisting BBB, in which a ventricular catheter is placed in the ipsilateral ventricle and ventricular irritation (or spontaneous ventricular impulse) coincides with arrival of impulse at the contralateral ventricular

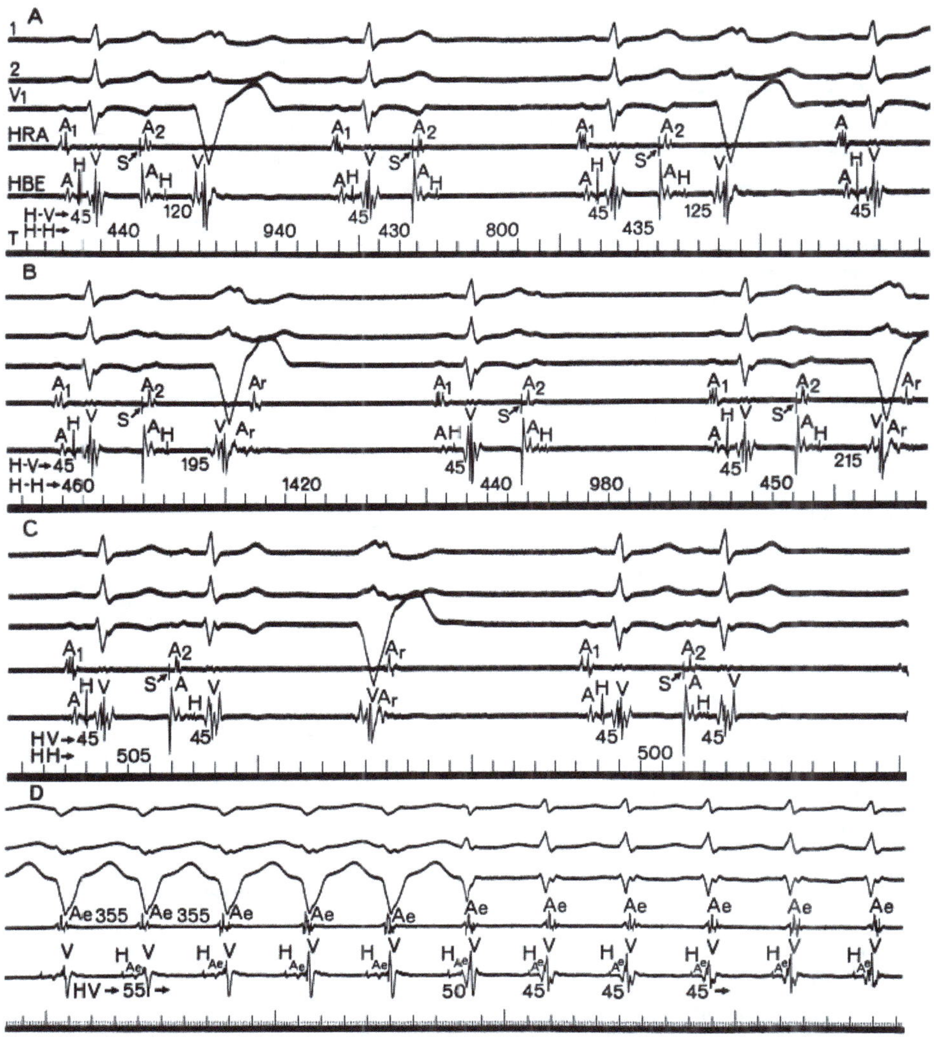

Figure 3.106. Iatrogenic events simulating EP phenomenon. **Panels A** to **C** show atrial bigeminy. In Panels A and B, block in HPS is seen following A_2 or conduction with long-HV interval and LBBB. Random recurrence of the extra ventricular complex in the middle of Panel C reveals its true nature, which is that all ventricular premature complexes are due to mechanical stimulation from the electrode catheter in the region of the HB. When the actual LBBB occurs, the QRS configuration (**D**) is quite different. (Reproduced from Akhtar et al with permission from Wolters Kluwer Publishing and the American Heart Association.[19])

breakthrough point, mimicking supernormal conduction (see Figure 3.107).

6. Ventricular irritation, when the catheter moves with atrial contraction, could produce ipsilateral preexcitation. It could mimic WPW and persist with the same degree of PR shortening if the catheter is not moved or withdrawn (Figure 3.109). In patients with SVT, it could lead to misdiagnosis; if the patient already has an AP, it may give the impression of 2 or more APs. This could lead to unnecessary and/or excessive ablative lesions with radiofrequency (RF) energy.

Complications Related to EP Studies[108,109]

Like other invasive intervention, EP studies carry risk of morbidity and possible mortality. These unfavorable outcomes can be characterized in 2 different categories: (1) complications related to the catheterization procedure; and (2) complications more specific to EP studies, *per se.*

1. Complications such as bleeding, hematoma, vessel dissection, vascular or cardiac chamber perforation, pericardial effusion, and tamponade, pneumothorax, hemothorax, venous or arterial thrombosis, early infection at the

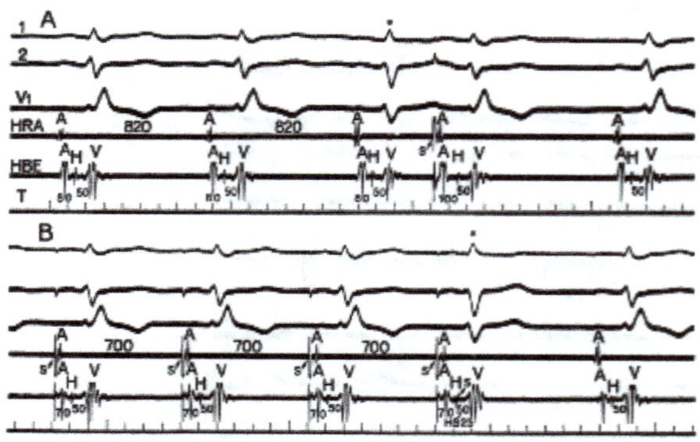

Figure 3.107. Catheter-induced events. **Panel A** shows SR with RBBB and LAD artery. The third complex shows loss of RBBB for no apparent reason, but the phenomenon can be reproduced by a properly timed electrically stimulated (S with HS interval of 25 ms) QRS (**B**), which reveals a possible explanation, that is, catheter-induced (mechanical) ventricular stimulation for loss of RBBB. (Reproduced from Akhtar et al with permission from Wolters Kluwer Publishing and the American Heart Association.[19])

puncture site, arrhythmias induced by mechanical irritation from catheter manipulation, and early or late thromboembolism (postdischarge) are generic and can happen with any catheter-related invasive procedure. Fortunately, these complications are relatively rare. Mortality during a routine study in a relatively stable patient is extremely uncommon but does occur; its incidence is <1/1000 studies. Since the introduction of resynchronization therapy as a part of the EP procedure for patients with systolic CHF, sicker patients are now seen by the EP staff. It is anticipated that, as a consequence, the mortality rate in EP laboratories will rise somewhat. It is more important than ever that patient selection is carried out very carefully and lengthy procedures in sicker patients are avoided. It is imperative that appropriate training and specialized skill are carefully assessed before privileges are granted to maintain a high standard of patient safety.

2. There are several other undesirable outcomes either specific to EP as a discipline or related to the length of the EP procedure, which make these patients more prone to complications. Catheter ablation using RF as the energy source, in the expert hand, will lead to correct diagnosis, precise identification of the arrhythmia source, and limited RF exposure, and will generally end with a good outcome. However, multiple lesions, high-wattage use of RF energy, and imprecise diagnosis of the arrhythmia source can lead to:

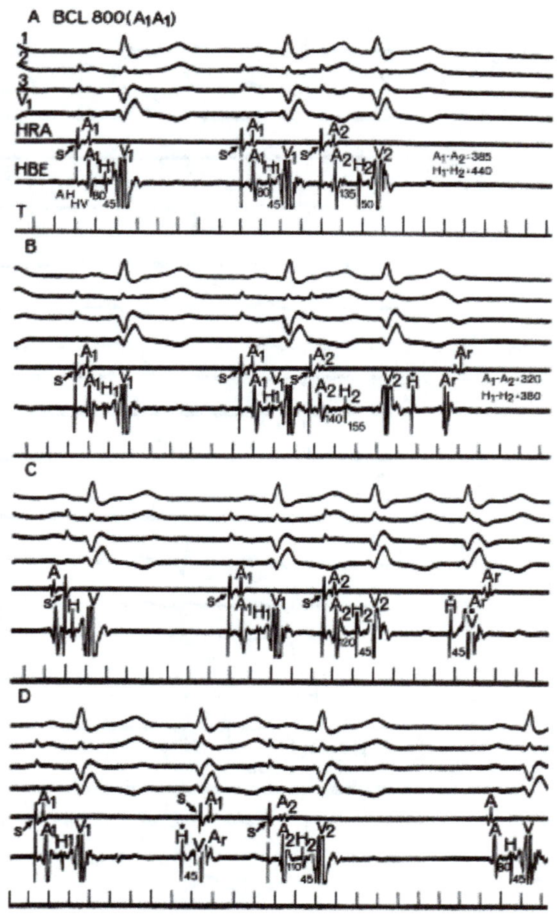

Figure 3.108. Simulation of EP phenomenon by catheter-induced junctional extra beats. **A**: BCL of 800 ms is seen with programed atrial extrastimuli. The QRS shows RBBB and normal axis. Some unexpected events are seen preceded by junctional extra beats. **Panel B** can also be interpreted as having long H_2-V_2 followed by the return of impulse to the His via transseptal conduction (the so-called HV echo), while **Panel C** simulates dual AV nodal physiology. However, a random complex (seen in **Panel D**) unmasks the actual event. The complexes were all resolved after repositioning of the catheter in the region of the AV junction.

(1) a high rate of failure; (2) excessive radiation exposure; (3) the possibility of repeat ablation; and (4) a higher chance of chamber perforation and creation of potential foci for new arrhythmia, such as creating a scar formation within healthy myocardium by unnecessary RF lesions.

Similarly, AF ablation brings its own set of complications: (1) pulmonary vein stenosis; (2) esophageal perforation; (3) thromboembolism secondary to gaps in anticoagulation by discontinued, reinitiated, or repeated cardioversions; and (4) length of the procedure. The latter predisposes patients to radiation-related

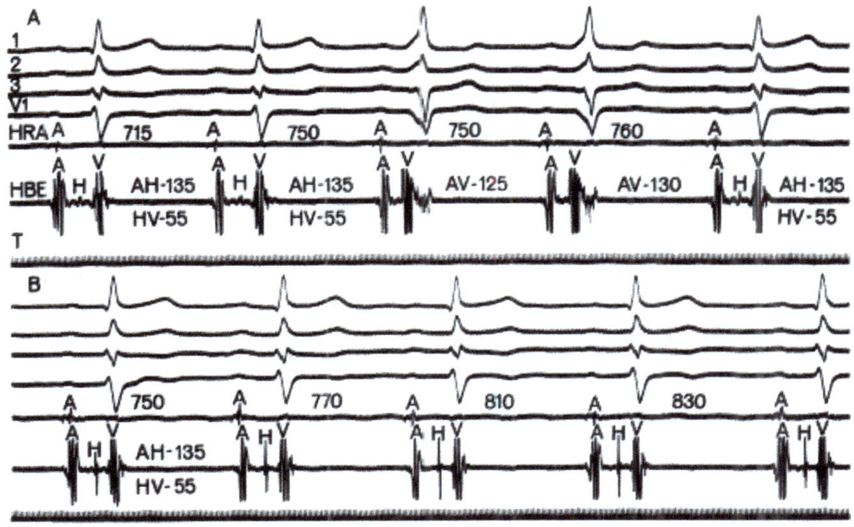

Figure 3.109. Simulation of preexcitation by mechanical stimulation. Two complexes in **Panel A** show ventricular preexcitation, which was abolished by slight change in catheter position in the AV junction (**B**).

complications in all invasive/interventional procedures, the ultimate cumulative risk of which is still unknown. Resynchronization therapy requires an additional lead in the CS and its tributaries and can lead to thrombosis, perforation, and vessel stenosis. The exact consequences of such occurrences are not completely clear. What is clear, however, is that repeat procedures are necessary in many patients with AF and CHF. Lead extraction and coronary venoplasty are being done more frequently and undoubtedly will bring their own set of issues.

Problems and complications related to implantable cardioverter-defibrillator/pacing therapy are covered elsewhere in this book and interested readers are referred to these sources.

SUMMARY

This chapter is a relatively truncated version of normal responses during the conduct of what may be considered a routine evaluation. Specific areas, some of which are covered elsewhere in the book, are beyond the scope of this writing, but serious students of EP, particularly those entering the field, will hopefully find this overview useful.

REFERENCES

1. Durrer D, Schoo L, Schuilenburg RM, Wellens HJ. The role of premature beats in the initiation and the termination of supraventricular tachycardia in the Wolff-Parkinson-White syndrome. *Circulation.* 1967;36:644–662.
2. Coumel P, Cabrol C, Fabiato A, Gourgon R, Slama R. Tachycardie permanente par rythme re'ciproque. *Arch Mal Coeur Vaiss.* 1967;60:1830–1864.
3. Scherlag BJ, Lau SH, Helfant RH, et al. Catheter technique for recording His bundle activity in man. *Circulation.* 1969;39:13–18.
4. Mandel WJ, Hayakawa H, Allen HN, Danzig R, Kermaier AI. Assessment of sinus node function in patients with the sick sinus syndrome. *Circulation.* 1972;46:761–769.
5. Narula OS, Shantha N, Vasquez M, Towne WD, Linhart JW. A new method for measurement of sinoatrial conduction time. *Circulation.* 1978;58:706–714.
6. Jazayeri MR, Tchou P, Caceres J, et al. Ventriculoatrial conduction time during bundle branch reentrant beat initiating orthodromic tachycardia: A simple and reliable method for localization of accessory pathways. *J Cardiovasc Electrophysiol.* 1990;1:121–131.
7. Coumel P. Junctional reciprocating tachycardias. The permanent and paroxysmal forms of A-V nodal reciprocating tachycardias. *J Electrocardiol.* 1975;8:79–90.
8. Akhtar M, Damato AN, Ruskin JN, et al. Antegrade and retrograde conduction characteristics in three patterns of paroxysmal atrioventricular junctional reentrant tachycardia. *Am Heart J.* 1978;95:22–42.
9. Wellens HJ, Lie KI, Durrer D. Further observations on ventricular tachycardia as studied by electrical stimulation of the heart. Chronic recurrent ventricular tachycardia and ventricular tachycardia during acute myocardial infarction. *Circulation.* 1974;49:647–653.

10. Gallagher JJ, Damato AN, Varghese PJ, et al. Alternative mechanisms of apparent supernormal atrioventricular conduction. *Am J Cardiol.* 1973;31:362–371.
11. Narula OS. Retrograde pre-excitation. Comparison of antegrade and retrograde conduction intervals in man. *Circulation.* 1974;50:1129–1143.
12. Akhtar M, Damato AN, Batsford WP, Ruskin JN, Ogunkelu JB. A comparative analysis of antegrade and retrograde conduction patterns in man. *Circulation.* 1975;52:766–778.
13. Akhtar M, Damato AN, Batsford WP, et al. Demonstration of re-entry within the His-Purkinje system in man. *Circulation.* 1974;50:1150–1162.
14. Akhtar M, Gilbert C, Al-Nouri M, Denker S. Site of conduction delay during functional block in the His-Purkinje system in man. *Circulation.* 1980;61:1239–1248.
15. Lehmann MH, Denker S, Mahmud R, Akhtar M. Patterns of human atrioventricular nodal accommodation to a sudden acceleration of atrial rate. *Am J Cardiol.* 1984;53:71–76.
16. Lehmann MH, Denker S, Mahmud R, Akhtar M. Functional His-Purkinje system behavior during sudden ventricular rate acceleration in man. *Circulation.* 1983;68:767–775.
17. Denker S, Shenasa M, Gilbert CJ, Akhtar M. Effects of abrupt changes in cycle length on refractoriness of the His-Purkinje system in man. *Circulation.* 1983;67:60–68.
18. Akhtar M, Gilbert CJ, Al-Nouri M, Schmidt DH. Electrophysiologic mechanisms for modification and abolition of atrioventricular junctional tachycardia with simultaneous and sequential atrial and ventricular pacing. *Circulation.* 1979;60:1443–1454.
19. Akhtar M, Damato AN, Gilbert-Leeds CJ, et al. Induction of iatrogenic electrocardiographic patterns during electrophysiologic studies. *Circulation.* 1977;56:60–65.
20. Akhtar M, Gilbert CJ, Wolf FG, Schmidt DH. Retrograde conduction in the His-Purkinje system. Analysis of the routes of impulse propagation using His and right bundle branch recordings. *Circulation.* 1979;59:1252–1265.
21. Akhtar M, Damato AN, Caracta AR, et al. Electrophysiologic effects of atropine on atrioventricular conduction studied by His bundle electrogram. *Am J Cardiol.* 1974;33:333–343.
22. Damato AN, Varghese PJ, Caracta AR, Akhtar M, Lau SH. Functional 2:1 A-V block within the His-Purkinje system. Simulation of type II second-degree A-V block. *Circulation.* 1973;47:534–542.
23. Reiffel JA, Gang E, Gliklich J, et al. The human sinus node electrogram: A transvenous catheter technique and a comparison of directly measured and indirectly estimated sinoatrial conduction time in adults. *Circulation.* 1980;62:1324–1334.
24. Gomes JA, Kang PS, El-Sherif N. The sinus node electrogram in patients with and without sick sinus syndrome: Techniques and correlation between directly measured and indirectly estimated sinoatrial conduction time. *Circulation.* 1982;66:864–873.
25. Narula OS, Samet P, Javier RP. Significance of the sinus-node recovery time. *Circulation.* 1972;45:140–158.
26. Strauss HC, Saroff AL, Bigger JT Jr, Giardina EG. Premature atrial stimulation as a key to the understanding of sinoatrial conduction in man. Presentation of data and critical review of the literature. *Circulation.* 1973;47:86–93.
27. Mandel W, Hayakawa H, Danzig R, Marcus HS. Evaluation of sinoatrial node function in man by overdrive suppression. *Circulation.* 1971;44:59–66.
28. Josephson ME. Sinus node function. In: Josephson ME, ed. *Clinical Cardiac Electrophysiology: Techniques and Interpretations.* 2nd ed. Philadelphia, PA: Lea & Febiger; 1993:71–95.
29. Breithardt G, Seipel L. Comparative study of two methods of estimating sinoatrial conduction time in man. *Am J Cardiol.* 1978;42:965–972.
30. Milstein S, Buetikofer J, Lesser J, et al. Cardiac asystole: A manifestation of neurally mediated hypotension-bradycardia. *J Am Coll Cardiol.* 1989;14:1626–1632.
31. Almquist A, Gornick C, Benson W Jr, Dunnigan A, Benditt DG. Carotid sinus hypersensitivity: Evaluation of the vasodepressor component. *Circulation.* 1985;71:927–936.
32. Sra JS, Anderson AJ, Sheikh SH, et al. Unexplained syncope evaluated by electrophysiologic studies and head-up tilt testing. *Ann Intern Med.* 1991;114:1013–1019.
33. Sra JS, Jazayeri MR, Avitall B, et al. Comparison of cardiac pacing with drug therapy in the treatment of neurocardiogenic (vasovagal) syncope with bradycardia or asystole. *N Engl J Med.* 1993;328:1085–1090.
34. Ogunkelu JB, Damato AN, Akhtar M, et al. Electrophysiologic effects of procainamide in subtherapeutic to therapeutic doses on human atrioventricular conduction system. *Am J Cardiol.* 1976;37:724–731.
35. Josephson ME, Seides SF, Batsford WP, et al. The electrophysiological effects of intramuscular quinidine on the atrioventricular conducting system in man. *Am Heart J.* 1974;87:55–64.
36. Ruskin, JN, Caracta AR, Akhtar M, Batsford WP, Damato AN. Electrophysiologic effects of tolamolol on atrioventricular conduction in man. *Am Heart J.* 1975;90:755–766.
37. Vargas G, Akhtar M, Damato AN. Electrophysiologic effects of isoproterenol on cardiac conduction system in man. *Am Heart J.* 1975;90:25–34.
38. Ruskin JN, Akhtar M, Damato AN, Foster JR. The effect of lidocaine on reentry within the His-Purkinje system in man. *Circulation.* 1980;62:388–400.
39. Dhatt MS, Akhtar M, Reddy P, et al. Modification and abolition of re-entry within the His-Purkinje system in man by diphenylhydantoin. *Circulation.* 1977;56:720–726.
40. Reddy CP, Damato AN, Akhtar M, et al. Time dependent changes in the functional properties of the atrioventricular conduction system in man. *Circulation.* 1975;52:1012–1022.

41. Denes P, Wu D, Dhingra RC, Chuquimia R, Rosen KM. Demonstration of dual A-V nodal pathways in patients with paroxysmal supraventricular tachycardia. *Circulation.* 1973;48:549–555.
42. Rosenbaum MB, Lazzari JO, Elizari MV. The role of phase 3 and phase 4 block in clinical electrocardiography. In: Wellens HJJ, Lie KI, Janse MJ, eds. *The Conduction System of the Heart: Structure, Function and Clinical Implications.* Philadelphia, PA: Lea & Febiger; 1976:126–142.
43. Narula O, Runge M. Accommodation of A-V nodal conduction and fatigue phenomenon in the His-Purkinje system, the conduction system of the heart. In: Wellens HJJ, Lie KI, Janse MJ, eds. *The Conduction of the System of the Heart: Structure, Function and Clinical Implications.* Philadelphia, PA: Lea & Febiger; 1976:529–544.
44. Akhtar M. Supraventricular tachycardias: Electrophysiologic mechanisms, diagnosis and pharmacologic therapy. In: Josephson ME, Wellens HJJ, eds. *Tachycardias: Mechanisms, Diagnosis, Treatment.* Philadelphia, PA: Lea & Febiger; 1984:137–169.
45. Akhtar M, Damato AN, Caracta AR. Clinical uses of His bundle electrocardiography. Part II. *Am Heart J.* 1976;91:660–664.
46. Akhtar M. Atrioventricular nodal reentry. *Circulation.* 1987;75:III26–III30.
47. Gang ES, Denton TA, Oseran DS, Mandel WJ, Peter T. Rate-dependent effects of procainamide on His-Purkinje conduction in man. *Am J Cardiol.* 1985;55:1525–1529.
48. Tonkin AM, Heddle WF, Tornos P. Intermittent atrioventricular block: Procainamide administration as a provocative test. *Aust N Z J Med.* 1978;8:594–602.
49. Akhtar M. Atrioventricular nodal reentrant tachycardia. *Med Clin North Am.* 1984;68:819–830.
50. Akhtar M. Techniques of electrophysiologic evaluation. In: Fuster V, O'Rourke RA, Walsh RA, Poole-Wilson P, eds. *Hurst's The Heart.* 12th ed. New York, NY: McGraw-Hill; 2008:1064–1076.
51. Akhtar M, Mahmud R, Tchou PJ, Denker ST, Gilbert CJ. Normal electrophysiologic responses of the human heart. *Cardiol Clin.* 1986;4:365–386.
52. Jazayeri MR, Sra JS, Akhtar M. Atrioventricular nodal reentrant tachycardia. Electrophysiologic characteristics, therapeutic interventions, and specific reference to anatomic boundary of the reentrant circuit. *Cardiol Clin.* 1993;11:151–181.
53. Strauss HC, Saroff AL, Bigger JT Jr, Giardina EGV. Premature atrial stimulation as a key to the understanding of sinoatrial conduction in man. Presentation of data and critical review of the literature. *Circulation.* 1973;47:86–93.
54. Narula OS. Sinus node re-entry: A mechanism for supraventricular tachycardia. *Circulation.* 1974;50:1114–1128.
55. Weisfogel GM, Batsford WP, Paulay KL, et al. Sinus node re-entrant tachycardia in man. *Am Heart J.* 1975;90:295–304.
56. Damato AN, Akhtar M, Ruskin J, Caracta A, Lau S. Gap phenomenon: Antegrade and retrograde. In: Wellens HJJ, Lie KI, Janse MJ, eds. *The Conduction of the System of the Heart: Structure, Function and Clinical Implications.* Philadelphia, PA: Lea & Febiger; 1976:504–528.
57. Moe GK, Mendez C, Abildskov JA. A complex manifestation of concealed A-V conduction in the dog heart. *Circ Res.* 1964;15:51–63.
58. Ticzon AR, Damato AN, Caracta AR, Lau SH, Bobb GA. The 1,2,3,4 phenomenon. Atrioventricular nodal gap in the dog. *Circ Res.* 1976;39:223–230.
59. Jazayeri MR, Sra J, Akhtar M. Wide QRS complexes: Electrophysiologic basis of a common electrocardiographic diagnosis. *J Cardiovasc Electrophysiol.* 1992;3:365–393.
60. Akhtar M, Jazayeri M, Sra J, et al. Electrophysiology of wide QRS tachycardia. In: Josephson ME, Wellens HJJ, eds. *Tachycardias: Mechanisms and Management.* Mount Kisco, NY: Futura Publishing Co.; 1993:215–235.
61. Denker S, Lehmann M, Mahmud R, Gilbert C, Akhtar M. Effects of alternating cycle lengths on refractoriness of the His-Purkinje system. *J Clin Invest.* 1984;74:559–570.
62. Miles WM, Prystowsky EN. Alteration of human right bundle branch refractoriness by changes in duration of the atrial drive train. *Circulation.* 1986;73:244–248.
63. Akhtar M, Damato AN, Batsford WP, et al. Unmasking and conversion of gap phenomenon in the human heart. *Circulation.* 1974;49:624–630.
64. Akhtar M, Damato AN, Caracta AR, Batsford WP, Lau SH. The gap phenomena during retrograde conduction in man. *Circulation.* 1974;49:811–817.
65. Gouaux JL, Ashman R. Auricular fibrillation with aberration simulating ventricular paroxysmal tachycardia. *Am Heart J.* 1947;34:366–373.
66. Akhtar M, Shenasa M, Jazayeri M, Caceres J, Tchou PJ. Wide QRS complex tachycardia. Reappraisal of a common clinical problem. *Ann Intern Med.* 1988;109:905–912.
67. Denker ST, Gilbert CJ, Shenasa M, Akhtar M. An electrocardiographic–electrophysiologic correlation of aberrant ventricular conduction in man. *J Electrocardiol.* 1983;16:269–277.
68. Akhtar M, Denker ST, Lehmann MH, Mahmud R. Effects of sudden cycle length alteration on refractoriness of human His-Purkinje system and ventricular myocardium. In: Zipes DP, Jalife J, eds. *Cardiac Electrophysiology and Arrhythmias.* Orlando, FL: Grune and Stratton; 1985:399–408.
69. Akhtar M, Lehmann MH, Denker ST, Mahmud R. Role of His-Purkinje system in the initiation of orthodromic tachycardia in Wolff-Parkinson-White syndrome. In: Benditt D, Woodrow BD, eds. *Cardiac Preexcitation Syndromes: Origins, Evaluation and Treatment.* Boston, MA: Martinus Nijhoff Publishing; 1986:289–304.
70. Rosenbaum MB, Sanchez RA, Goicoechea R. Aberrant ventricular conduction during atrial fibrillation. In: Touboul P, Waldo AL, eds. *Atrial Arrhythmias: Current*

Concepts and Management. St. Louis, MO: Mosby; 1990:163–168.
71. Narula OS. Longitudinal dissociation in the His bundle. Bundle branch block due to asynchronous conduction within the His bundle in man. *Circulation.* 1977;56:996–1006.
72. El-Sherif N, Scherlag BJ, Lazzara R. Conduction disorders in the canine proximal His-Purkinje system following acute myocardial ischemia. I. The pathophysiology of intra-His bundle block. *Circulation.* 1974;49:837–847.
73. Lehmann MH, Denker S, Mahmud R, Addas A, Akhtar M. Linking: A dynamic electrophysiologic phenomenon in macroreentry circuit. *Circulation.* 1985;71:254–265.
74. Svenson RH, Miller HC, Gallagher JJ, Wallace AG. Electrophysiological evaluation of the Wolff-Parkinson-White syndrome: Problems in assessing antegrade and retrograde conduction over the accessory pathway. *Circulation.* 1975;52:552–562.
75. Neuss H, Schlepper M, Thormann J. Analysis of re-entry mechanisms in the three patients with concealed Wolff-Parkinson-White syndrome. *Circulation.* 1975;51:75–81.
76. Tchou P, Lehman MH, Jazayeri M, Akhtar M. Atriofascicular connection or a nodoventricular Mahaim fiber? Electrophysiologic elucidation of the pathway and associated reentrant circuit. *Circulation.* 1988;77:837–848.
77. Klein GJ, Guiraudon GM, Kerr CR, et al. "Nodoventricular" accessory pathway: Evidence for a distinct accessory atrioventricular pathway with atrioventricular node-like properties. *J Am Coll Cardiol.* 1988;11:1035–1040.
78. Gallagher JJ, Sealy WC, Kasell J, Wallace AG. Multiple accessory pathways in patients with the pre-excitation syndrome. *Circulation.* 1976;54:571–591.
79. Lehmann MH, Tchou P, Mahmud R, Denker ST, Akhtar M. Electrophysiological determinants of antidromic reentry induced during atrial extrastimulation. Insights from a pacing model of Wolff-Parkinson-White syndrome. *Circ Res.* 1989;65:295–306.
80. Wellens HJ, Durrer D. Relation between refractory period of the accessory pathway and ventricular frequency during atrial fibrillation in patients with Wolff-Parkinson-White syndrome. *Am J Cardiol.* 1974;33:178.
81. Wellens HJ, Bär FW, Dassen WR, et al. Effect of drugs in the Wolff-Parkinson-White syndrome. Importance of initial length of effective refractory period of the accessory pathway. *Am J Cardiol.* 1980;46:665–669.
82. Akhtar M, Damato AN, Lau SH, Ticzon AR. Clinical uses of His bundle electrocardiography. Part III. *Am Heart J.* 1976;91:805–809.
83. Jazayeri MR, Caceres J, Tchou P, et al. Electrophysiologic characteristics of sudden QRS axis deviation during orthodromic tachycardia. Role of functional fascicular block in localization of accessory pathway. *J Clin Invest.* 1989;83:952–959.
84. Akhtar M, Lehmann MH, Denker ST, et al. Electrophysiologic mechanisms of orthodromic tachycardia initiation during ventricular pacing in the Wolff-Parkinson-White syndrome. *J Am Coll Cardiol.* 1987;9:89–100.
85. Mahmud R, Denker S, Lehmann MH, Akhtar M. Effect of atrioventricular sequential pacing in patients with no ventriculoatrial conduction. *J Am Coll Cardiol.* 1984;4:273–277.
86. Gallagher JJ, Damato AN, Varghese PJ, Lau SH. Manifest and concealed reentry: A mechanism of A-V nodal Wenckebach in man. *Circulation.* 1973;47:752–757.
87. Akhtar M, Gilbert C, Wolf FG, Schmidt DH. Reentry within the His-Purkinje system. Elucidation of reentrant circuit using right bundle branch and His bundle recordings. *Circulation.* 1978;58:295–304.
88. Reddy CP, Damato AN, Akhtar M, et al. Effect of procainamide on reentry within the His-Purkinje system in man. *Am J Cardiol.* 1977;40:957–964.
89. Denker S, Lehmann MH, Mahmud R, Gilbert C, Akhtar M. Facilitation of macroreentry within the His-Purkinje system with abrupt changes in cycle length. *Circulation.* 1984;69:26–32.
90. Blanck Z, Jazayeri M, Dhala A, et al. Bundle branch reentry: A mechanism of ventricular tachycardia in the absence of myocardial or valvular dysfunction. *J Am Coll Cardiol.* 1993;22:1718–1722.
91. Touboul P, Kirkorian G, Atallah G, et al. Bundle branch reentrant tachycardia treated by electrical ablation of the right bundle branch. *J Am Coll Cardiol.* 1986;7:1404–1409.
92. Tchou P, Jazayeri M, Denker S, et al. Transcatheter electrical ablation of right bundle branch. A method of treating macroreentrant ventricular tachycardia attributed to bundle branch reentry. *Circulation.* 1988;78:246–257.
93. Caceres J, Jazayeri M, McKinnie J, et al. Sustained bundle branch reentry as a mechanism of clinical tachycardia. *Circulation.* 1989;79:256–270.
94. Blanck Z, Dhala A, Deshpande S, et al. Bundle branch reentrant ventricular tachycardia: Cumulative experience in 48 patients. *J Cardiovasc Electrophysiol.* 1993;4:253–262.
95. Blanck Z, Deshpande S, Jazayeri MR, Akhtar M. Catheter ablation of the left bundle branch for the treatment of sustained bundle branch reentrant ventricular tachycardia. *J Cardiovasc Electrophysiol.* 1995;6:40–43.
96. Blanck Z, Sra J, Akhtar M. Incessant interfascicular reentrant ventricular tachycardia as a result of catheter ablation of the right bundle branch: Case report and review of the literature. *J Cardiovasc Electrophysiol.* 2009;20:1279–1283.
97. Narasimhan C, Jazayeri MR, Sra J, et al. Ventricular tachycardia in valvular heart disease: Facilitation of sustained bundle-branch reentry by valve surgery. *Circulation.* 1997;96:4307–4313.

98. Merino JL, Peinado R, Fernandez-Lozano I, et al. Bundle-branch reentry and the postpacing interval after entrainment by right ventricular apex stimulation: A new approach to elucidate the mechanism of wide-QRS-complex tachycardia with atrioventricular dissociation. *Circulation.* 2001;103:1102–1108.
99. Merino JL, Carmona JR, Fernández-Lozano I, et al. Mechanisms of sustained ventricular tachycardia in myotonic dystrophy: Implications for catheter ablation. *Circulation.* 1998;98:541–546.
100. Dhala A, Gonzalez-Zuelgaray J, Deshpande S, et al. Unmasking the trifascicular left intraventricular conduction system by ablation of the right bundle branch. *Am J Cardiol.* 1996;77:706–712.
101. Booth DC, Popio KA, Gettes LS. Multiformity if induced unifocal ventricular premature beats in human subjects: Electrocardiographic and angiographic correlations. *Am J Cardiol.* 1982;49:1643–1653.
102. Denker S, Lehmann MH, Mahmud R, Gilbert C, Akhtar M. Divergence between refractoriness of His-Purkinje system and ventricular muscle with abrupt changes in cycle length. *Circulation.* 1983;68:1212–1221.
103. Lehmann MH, Denker S, Mahmud R, Akhtar M. Postextrasystolic alterations in refractoriness of the His-Purkinje system and ventricular myocardium in man. *Circulation.* 1984;69:1096–1102.
104. Akhtar M, Shenasa M, Schmidt DH. Role of retrograde His Purkinje block in the initiation of supraventricular tachycardia by ventricular premature stimulation in the Wolff-Parkinson-White syndrome. *J Clin Invest.* 1981;67:1047–1055.
105. Gonzalez-Zuelgaray J, Sheikh S, Akhtar M, Jazayeri MR. Functional bundle branch block as a delayed manifestation of retrograde concealment in the His-Purkinje system. *J Cardiovasc Electrophysiol.* 1996;7:248–258.
106. Akhtar M, Damato AN, Ruskin JN, et al. Characteristics and coexistence of two forms of ventricular echo phenomena. *Am Heart J.* 1976;92:174–182.
107. Akhtar M, Caracta AR, Lau SH, Gilbert CJ, Damato AN. Demonstration of intra-atrial conduction delay, block, gap, and reentry: A report of two cases. *Circulation.* 1978;58:947–955.
108. Dimarco JP, Garan H, Ruskin JN. Complications in patients undergoing cardiac electrophysiologic procedures. *Ann Intern Med.* 1982;97:490–493.
109. Horowitz LN, Kay HR, Kutalek SP, et al. Risks and complications of clinical cardiac electrophysiologic studies: A prospective analysis of 1,000 consecutive patients. *J Am Coll Cardiol.* 1987;9:1261–1268.
110. Akhtar M. Electrophysiologic bases for wide QRS tachycardia. *Pacing Clin Electrophysiol.* 1983;6:81–98.
111. Akhtar M. Invasive cardiac electrophysiologic studies: An introduction. In: Parmley WW, Chatterjee K, eds. *Cardiology, vol. 1: Physiology, Pharmacology Diagnosis.* Philadelphia, PA: Lippincott; 1991:54–67.

CHAPTER 4

The Electrocardiogram in Ventricular Tachycardia

Hein J. J. Wellens, MD

INTRODUCTION

Although 90% of wide QRS tachycardias presenting in the emergency room are ventricular tachycardia (VT), errors are often made because physicians wrongly consider VT unlikely if the patient is hemodynamically stable.[1] In that setting, administering drugs under the assumption that the patient is suffering from a supraventricular tachycardia (SVT) may be deleterious.[2,3]

POSSIBLE CAUSES OF A WIDE (BROAD) QRS TACHYCARDIA (FIGURE 4.1)

1. SVT with preexisting or functional bundle branch block (BBB). This includes sinus tachycardia, atrial tachycardia, atrial flutter, atrial fibrillation, and atrioventricular (AV) nodal reentrant tachycardia. Functional BBB will occur once the SVT reaches the refractory period of the bundle branch, so-called tachycardia-dependent or phase 3 BBB, or because of retrograde invasion into the bundle branch.[4] (Figure 4.1A)
2. Circus movement tachycardia with AV conduction over the AV node and ventriculoatrial (VA) conduction over an accessory AV pathway with preexisting or functional BBB (Figure 4.1B).
3. SVTs as in Cause 1, but with AV conduction over an accessory AV pathway (Figure 4.1C).
4. Circus movement tachycardia with AV conduction over an accessory AV pathway and VA

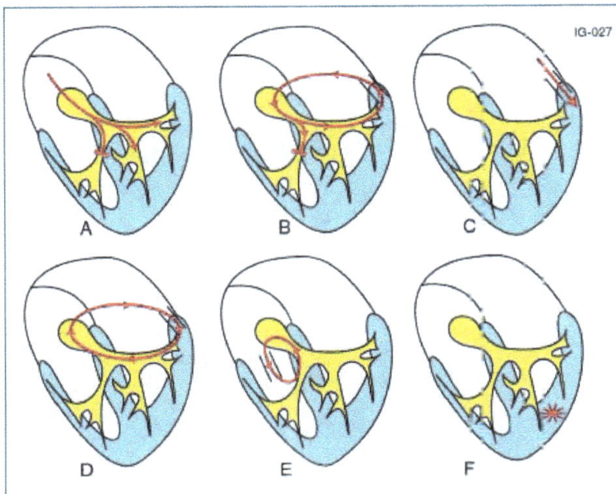

Figure 4.1. Possible causes of wide QRS tachycardia. See text for explanation.

conduction over the AV node or a second accessory AV pathway (Figure 4.1D).

5. A tachycardia using a nodoventricular fiber in anterograde direction and the bundle branch-His system or an accessory AV pathway in retrograde direction (Figure 4.1E).
6. Ventricular tachycardia (Figure 4.1F).

ELECTROCARDIOGRAPHIC DIAGNOSIS OF A WIDE QRS TACHYCARDIA

To come to a correct diagnosis, a 12-lead ECG is required. That ECG should be examined systematically, looking for AV dissociation and analyzing QRS characteristics such as width, axis, and configuration.[5]

ATRIOVENTRICULAR DISSOCIATION

Dissociation between atrial and ventricular activity during tachycardia is considered to be the hallmark of VT. On physical examination it leads to canon waves in the jugular pulse, varying loudness of the first heart sound, and beat-to-beat changes in the blood pressure. The challenge with the 12-lead ECG is to identify the P wave among the QRS complexes and to note that they are dissociated. Independent beating of atria and ventricles during a wide QRS tachycardia is the hallmark of VT.

Independent beating of atria and ventricles during a VT can result in "capture" and "fusion" beats (Figure 4.2). This occurs when the ventricular rate during VT is such that an appropriately timed atrial impulse, usually originating in the sinus node, is able to traverse the AV node to depolarize the ventricles completely ("capture") or partially ("fusion"). In the latter situation, fusion occurs because of concomitant ventricular activation from the VT focus. "Captures" and "fusion" beats are rare. They only occur when the VT rate is slow, excellent AV conduction is present, and no retrograde invasion into the intraventricular conduction system takes place during VT.

The occurrence of a more narrow QRS during VT is not always the result of a conducted supraventricular beat; it may also result from fusion with a ventricular depolarization arising in the ventricle contralateral to the ventricle in which the VT originates, or when fusion occurs with a ventricular echo beat when a retrogradely conducted impulse during VT travels to the AV node and reenters the ventricle[6] (Figure 4.3).

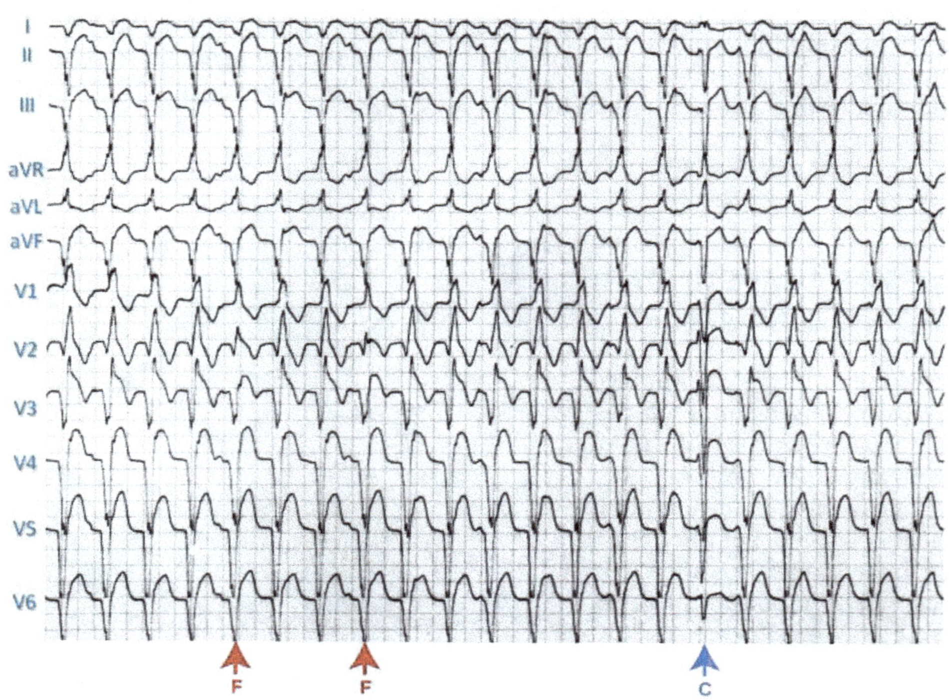

Figure 4.2. VT with "fusion" (F) and "capture" (C) beats.

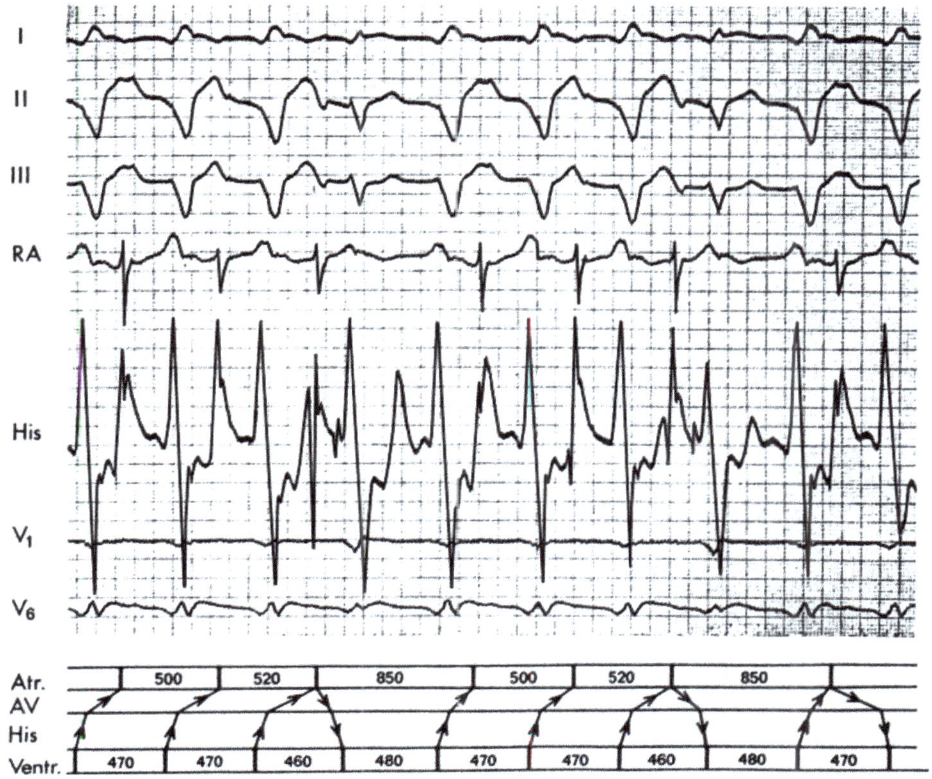

Figure 4.3. Example of a "fusion" beat during VT because of a ventricular echo beat. This occurs when a retrogradely conducted impulse during VT travels to the AV node and reenters the ventricle. As shown in the ladder diagram, VA conduction during VT shows progressive slowing followed by the echo beat.

Width of the QRS Complex

To determine the width of the QRS in a wide QRS tachycardia, it is essential to look for the widest one in the 12-lead ECG because cancellation by opposite activation forces may lead to a narrower QRS (Figure 4.4).

Bundle branch block will result in QRS widening, becoming maximal at 0.14 seconds for right BBB (RBBB) and 0.16 seconds for left BBB (LBBB).[5] A wider QRS complex should, therefore, lead to the consideration of a VT. However, there are three situations in which an SVT can have a wider QRS than the upper limits mentioned. The first one is preexistent BBB with additional intraventricular fibrosis or scarring. The second is the SVT with AV conduction over an accessory AV pathway. And the third is marked widening of the QRS during SVT because of the presence of an antiarrhythmic drug that prolongs (usually rate-related) intraventricular conduction, most commonly flecainide (Figure 4.5).

It is important to realize that width of the QRS during VT is related to its site of origin and the underlying disease. A VT originating in or close to the interventricular septum will have a smaller QRS than the VT arising in the lateral wall because of more synchronized activation of the two ventricles in the former. Also, the underlying disease plays a role.

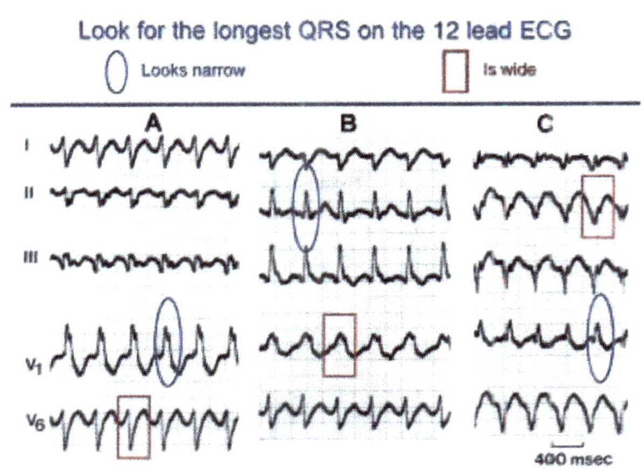

Figure 4.4. A depiction indicating the importance of scanning all electrocardiographic (ECG) leads for the widest QRS when determining the width of the QRS in a wide QRS tachycardia.

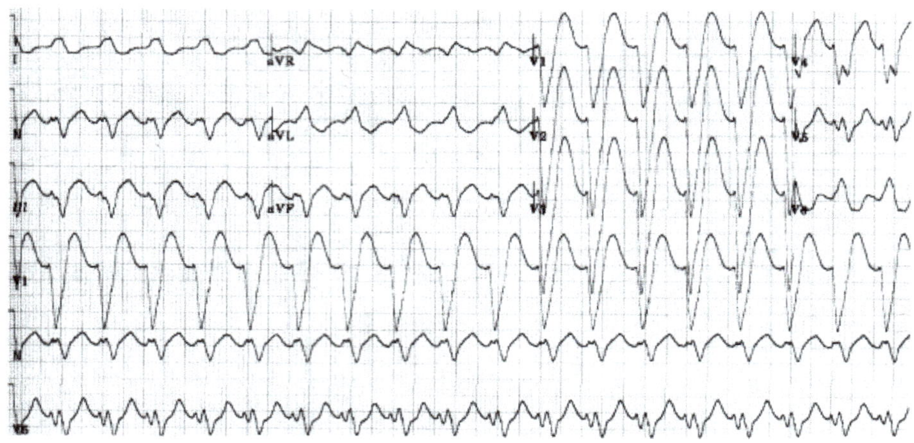

Figure 4.5. Marked widening of the QRS in a patient with an atrial tachycardia who is on flecainide medication.

A patient with a ventricular scar tends to have a wider QRS than the patient with an idiopathic VT. A VT originating in the interventricular septum may have a width of <0.12 seconds (Figure 4.6).

The QRS Axis

Patients with a VT frequently have an abnormal QRS axis in the frontal plane. In a VT with an RBBB-like configuration, the axis often points superiorly (marked left axis deviation), in contrast to SVT with RBBB in which the axis is more inferiorly directed.

In patients with an LBBB-shaped QRS tachycardia, right axis deviation suggests VT. Patients with an idiopathic VT can have a normal QRS axis in the frontal plane, for example, those with an idiopathic VT originating in the posterolateral outflow area of the right ventricle (Figure 4.7). However, idiopathic left and right ventricular VT often have an unusual frontal QRS axis depending on their site of origin (Figure 4.8).

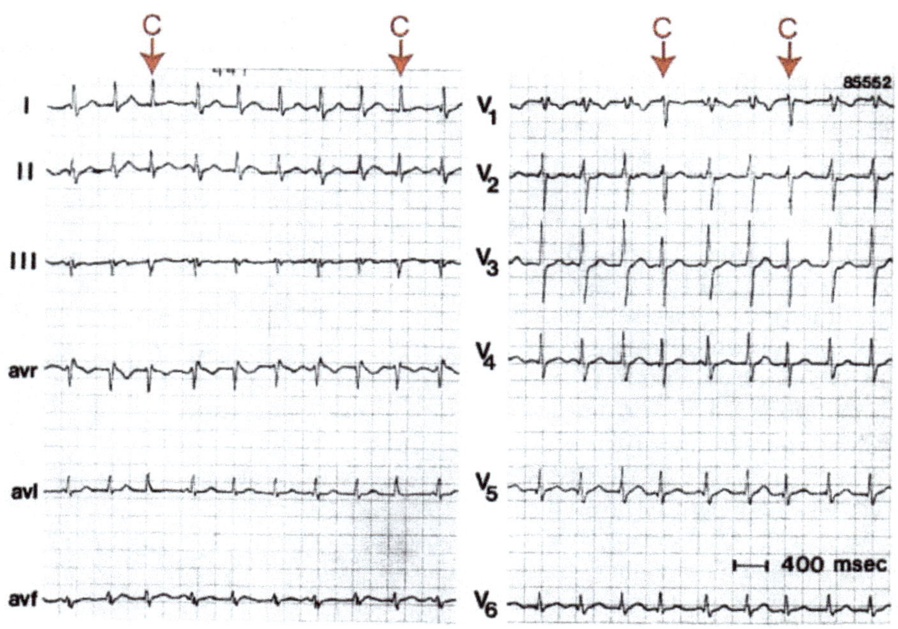

Figure 4.6. A VT with a QRS width of only 120 ms. The site of origin is on the left side of the middle part of the interventricular septum. Note also the "capture" beats (C).

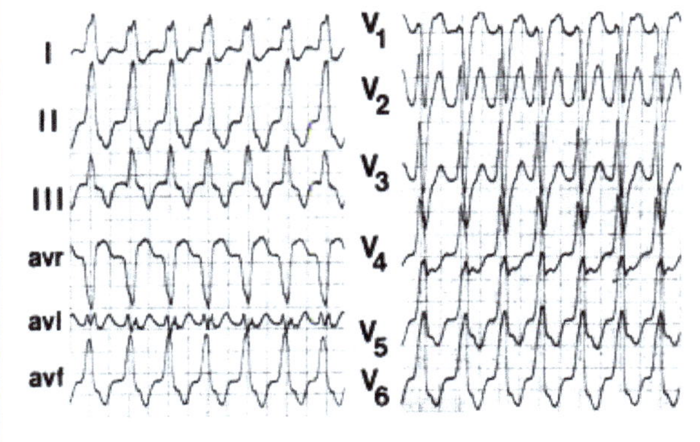

Figure 4.7. A: A VT with an origin in the inflow tract of the right ventricle. Note the intermediate frontal QRS axis. The characteristics of the initial part of the QRS in leads V_1 and V_2 reveal that the origin of the tachycardia is in the right ventricle. B: The (normal) ECG during sinus rhythm indicating that the VT is idiopathic.

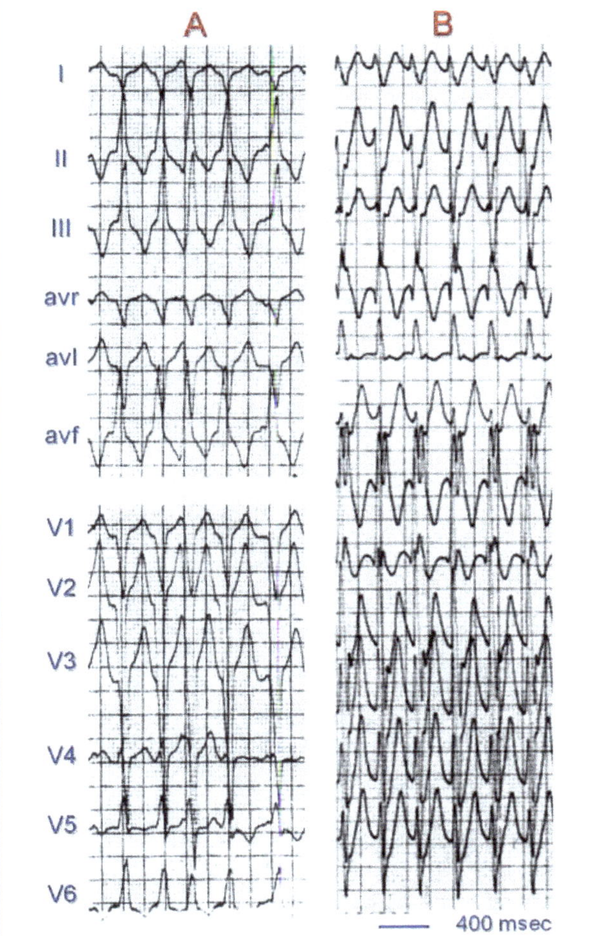

Figure 4.8. Two types of idiopathic VT. A: The origin is in the outflow tract of the right ventricle. The QRS shows an LBBB-shape and a vertical axis. B: The origin is in the inferoseptal portion of the left ventricle. The QRS shows an RBBB shape and a left axis deviation.

In the discussion of the importance of the frontal-plane QRS axis in the differential diagnosis of a wide QRS tachycardia, it is important to know that a markedly abnormal QRS axis can occur in patients with preexistent BBB during SVT (eg, the patient with preexistent RBBB and left anterior hemiblock), in patients with AV conduction over an accessory AV pathway (marked left axis deviation during AV conduction over a right-sided or posteroseptal accessory pathway, and marked right axis deviation in case of AV conduction over a left free wall accessory connection), and in patients with a tachycardia who are taking a drug that slows intraventricular conduction (eg, the class Ic antiarrhythmic drug flecainide).

As discussed below, the QRS axis is also important in determining the likely site of origin of a VT.

CONFIGURATIONAL CHARACTERISTICS OF THE QRS COMPLEX

RBBB-Shaped QRS Complexes

An SVT with RBBB aberration shows a triphasic rSR pattern in lead V_1 and a triphasic QRS pattern in lead V_6 (Figure 4.9). The initial R wave in lead V_1 reflects normal interventricular septal

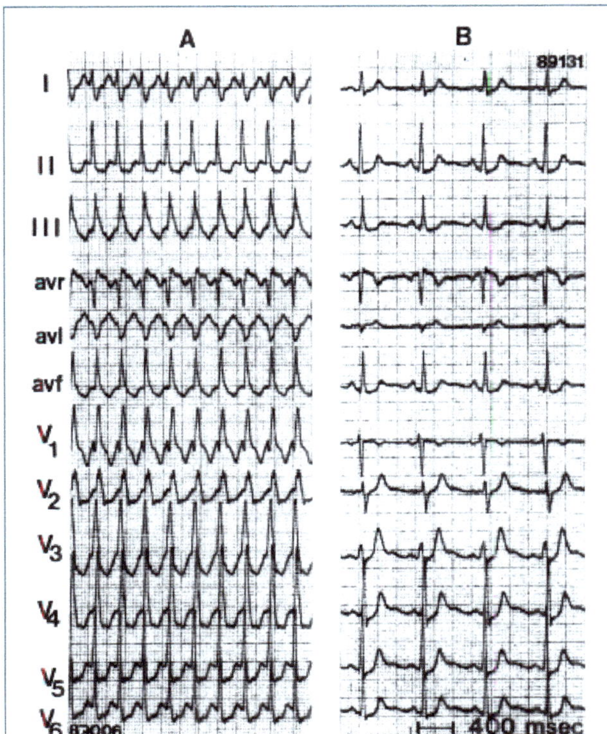

Figure 4.9. A: An SVT with RBBB. Note the triphasic QRS in lead V_1, and the QRS complex in V_6. B: Sinus rhythm.

activation, S-wave left ventricular activation and delayed R-wave activation of the right ventricle. Lead V_6 shows a narrow Q wave reflecting normal septal activation and an R:S ratio of >1.[5] In VT with an RBBB-like shape, lead V_1 usually shows a mono- or biphasic R wave. Figure 4.10 illustrates that the QRS configuration in lead V_6 is usually determined by the frontal-plane QRS axis. It shows a deep S wave (R:S ratio <1) in case of a superior axis and an R:S ratio >1 when an inferior axis is present.

LBBB-Shaped QRS Complexes

The leads to make the differential diagnosis between an LBBB-shaped VT and SVT with LBBB aberration are V_1, V_2, and V_6. LBBB aberrancy is characterized by a small and narrow (<0.04 seconds) R wave and a clean and swift (no slurs or notches) S wave, with a distance from the beginning of the QRS to the nadir of the S wave of <0.07 seconds.

In contrast, an R wave of >0.04 seconds, with a slurred downstroke and a delayed interval to the nadir of the S wave in V_1 and/or V_2 (>0.07 seconds) supports the diagnosis of VT.[7] A Q wave in lead V_6 confirms VT (Figure 4.11).

Presence of Q Waves during Tachycardia

Coumel et al[8] demonstrated the value of QR complexes during a wide QRS tachycardia as pointing to a ventricular origin for the arrhythmia. This is typically found in VT in patients with a localized ventricular scar because of a previous myocardial infarction. An example is given in Figure 4.12.

Algorithms to Diagnose VT

To facilitate diagnosing VT on the ECG, different algorithms have been published[9–12] (Table 4.1). In this author's experience, none of them is foolproof. For example, Brugada et al[9] postulated that the presence of an RS interval of more then 100 ms in one or more precordial leads is highly suggestive of VT. However, such a duration may occur in SVT with AV conduction over an accessory pathway, in SVT with preexistent BBB (especially LBBB), and in SVT during the administration of drugs that slow intraventricular conduction (see Figure 4.5). Vereckei et al[11] pointed to the value of the QRS configuration in lead aVR. However, as shown in Figure 4.13, a similar QRS configuration pointing to VT can usually be found in lead V_1.

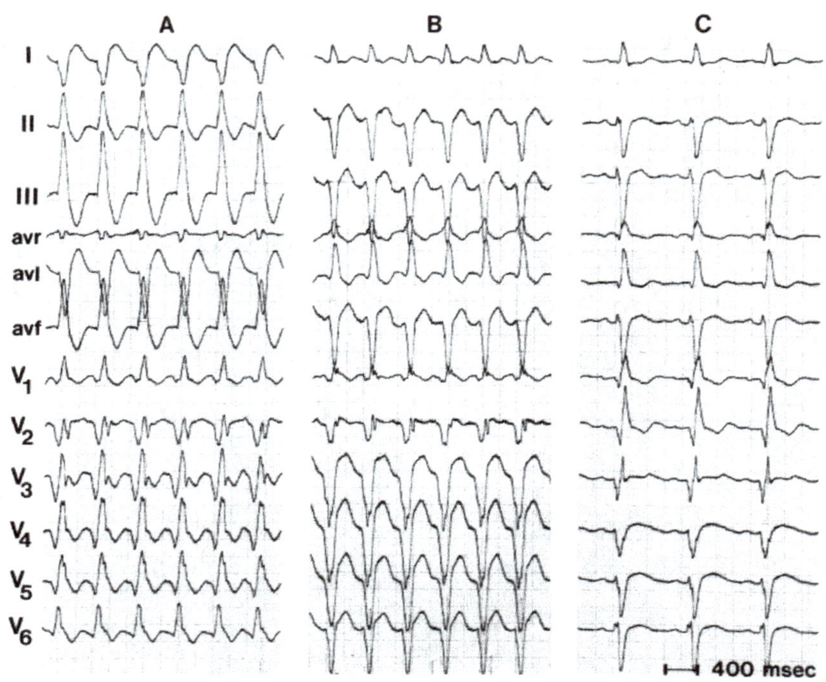

Figure 4.10. Two types of VT in a patient with an old anteroseptal myocardial infarction (**Panel C**). Both VTs (**Panels A** and **B**) have an RBBB shape with a QR in lead V_1. Note the effect of the frontal plane axis on the R:S ratio in lead V_6 in RBBB-shaped VT. An R:S < 1 is present in case of a superior axis (Panel B), but R:S > 1 with an inferior QRS axis (Panel A).

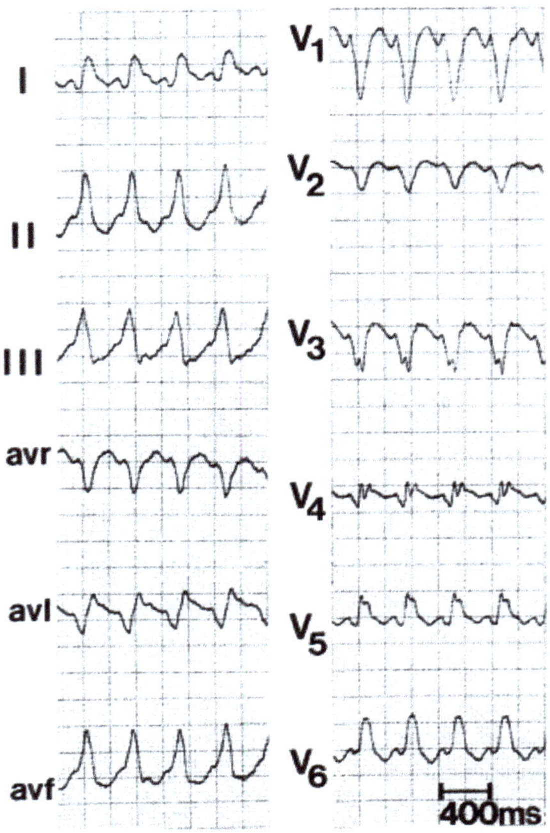

Figure 4.11. VT with a LBBB shape in a patient with an old anterior myocardial infarction showing initial positivity of the QRS in lead V_1 (>0.04 seconds), slurring of the S wave in lead V_1 and a distance from the beginning of the QRS to the nadir of the S wave in lead V_1 of 140 ms. Lead V_6 shows a QR complex.

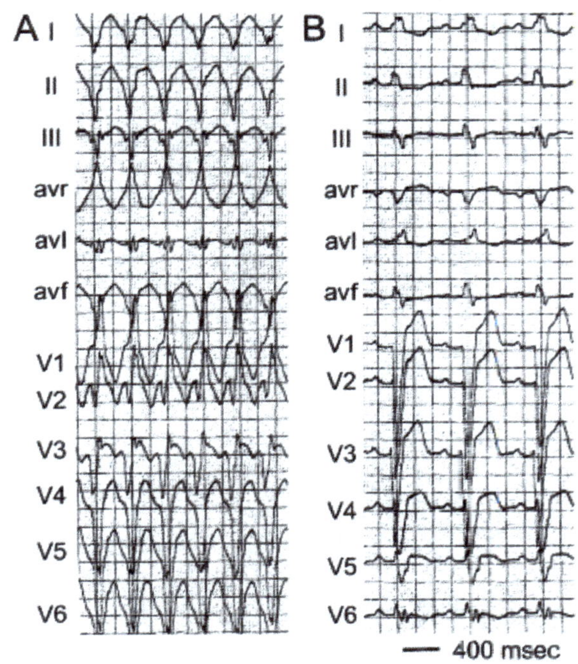

Figure 4.12. A: A VT with AV dissociation (best seen in lead aVL), a north-west frontal QRS axis and typical QRS configurational characteristics in V_1 and V_6. Note the QR complexes in leads V_2 and V_3, indicating a scar from a previous anterior wall myocardial infarction. B: Readings from the same patient showing that the signs of anterior wall myocardial infarction are completely masked during LBBB.

Localizing the Site of Origin of VT

The ability to treat VT by removing the site of origin or interrupting the reentrant pathway of the arrhythmia stresses the necessity to identify (when possible) the site of origin of VT.[13] The ECG can be helpful in doing so by looking at the configuration, the width and axis (in frontal and horizontal planes) of the QRS complex of the VT.

1. LBBB versus RBBB-like configuration: An LBBB-shaped VT indicates that the origin of the arrhythmia is in the right ventricle, resulting in delayed activation of the left ventricle. The reverse is true; when the site of origin of the VT is in the left ventricle, then the right ventricle is activated last, resulting in an RBBB-like QRS complex (Figure 4.14). The width of the QRS complex: As shown in Figure 4.15, when the origin of the VT is close to the interventricular septum, more simultaneous left and right ventricular activation occurs, leading to a narrower QRS complex. In contrast, a VT origin in the lateral wall of the origin results in sequential ventricular activation and a wider QRS complex. The QRS axis:
1. The frontal plane: As shown in Figure 4.16, a VT origin in the apical part of the ventricle has a superior axis (to the left of –30 degrees). An inferior axis is present when the VT has an origin in the basal area of the ventricle.
2. The horizontal plane: As shown in Figure 4.17, when a VT arises in the anteroapical area of the ventricle, all precordial QRS complexes will be negative (negative concordancy). When ventricular activation starts in the left posterior area, all precordial QRS complexes will be positive (positive concordancy). Positive concordancy can also be found when an SVT with AV conduction over a left posterior accessory pathway is present.

The ECG for localizing VT is most reliable when there is little or no cardiac damage. When scar tissue

Table 4.1. Limitations of electrocardiographic signs suggestive of a ventricular origin for a wide QRS tachycardia

Sign	Limitations
AV dissociation	• VA conduction may occur during VT • AV junctional rhythm with BBB and AV dissociation • Preexistent BBB (especially LBBB) • SVT with AV conduction over an accessory pathway • Use of drugs slowing intraventricular conduction (class Ia, class Ic, amiodarone)
Left axis deviation (to the left of −30 degrees)	• Not helpful in LBBB-shaped QRS • SVT with AV conduction over a right-sided or posteroseptal accessory pathway • SVT during use of class Ic drugs
Right axis deviation (to the right of +90 degrees)	• Not helpful in RBBB-shaped QRS
Presence of QR complexes in leads other than aVR	• Only in VT with localized myocardial scarring or infiltration (sarcoidosis or amyloidosis)
Concordant pattern in precordial leads	• Positive concordancy may occur during SVT with AV conduction over a left posterior accessory pathway
R-nadir S > 100 ms in one or more precordial leads	• SVT on drugs slowing intraventricular conduction • SVT with AV conduction over an accessory pathway • Preexistent BBB (especially LBBB)
Presence of a supraventricular impulse able to depolarize the ventricles completely ("capture beat") or partially (fusion complex)	• Occurs only at relatively slow VT rates • R/O fusion with a contralateral ventricular premature beat during VT • R/O fusion with a ventricular echo beat
Presence of premature beats during sinus rhythm with the same QRS configuration as during wide QRS tachycardia	• Requires availability of ECG during sinus rhythm • R/O atrial premature beat with aberrant conduction

R/O, rule out.

is present in the ventricle (as after myocardial infarction), localizing VT origin by ECG is often difficult and detailed intracardiac mapping will be necessary, especially when catheter ablation of VT is considered.

Etiology of VT

Most VTs have a previous myocardial infarction as their etiology and, as pointed out, a QRS complex during VT (see Figures 4.12 and 4.13) can be very helpful to make that diagnosis. However, characteristic ECG patterns can also be found in idiopathic VT[14] and VT in patients with arrhythmogenic right ventricular dysplasia (ARVD).[15] Figure 4.18 shows two patterns of idiopathic VT arising in or close to the outflow tract of the right ventricle. Both have an LBBB-like QRS complex indicating a right ventricular origin. In Figure 4.18A, the frontal QRS axis is +70 and lead I shows a positive QRS complex, indicating an origin of the tachycardia in the lateral part of the outflow tract of the right ventricle. In Figure 4.18B, the frontal QRS axis is inferior and the QRS is negative in lead I, pointing to an origin on the septal site in the right ventricular outflow tract (RVOT). Recently, it has become clear that VTs may originate in the root of the pulmonary artery and the aorta.[16–19] In those patients, epicardially located muscle fibers run to the base of the right or left ventricle. An example is given in Figure 4.19. These tachycardias require a different catheter ablative approach than the endocardially located idiopathic VTs.

Idiopathic left VTs all have an RBBB-like shape because of an origin in the left ventricle. Figure 4.20 shows three types. The most common one is shown in Figure 4.20A. The frontal QRS axis shows left axis deviation. The site of origin of the VT is in or close to the posterior fascicle of the left bundle branch (LBB). In Figure 4.20B, the frontal QRS axis is further leftward (a so-called north-west axis). This tachycardia arises more anteriorly, close to the interventricular septum. The least common

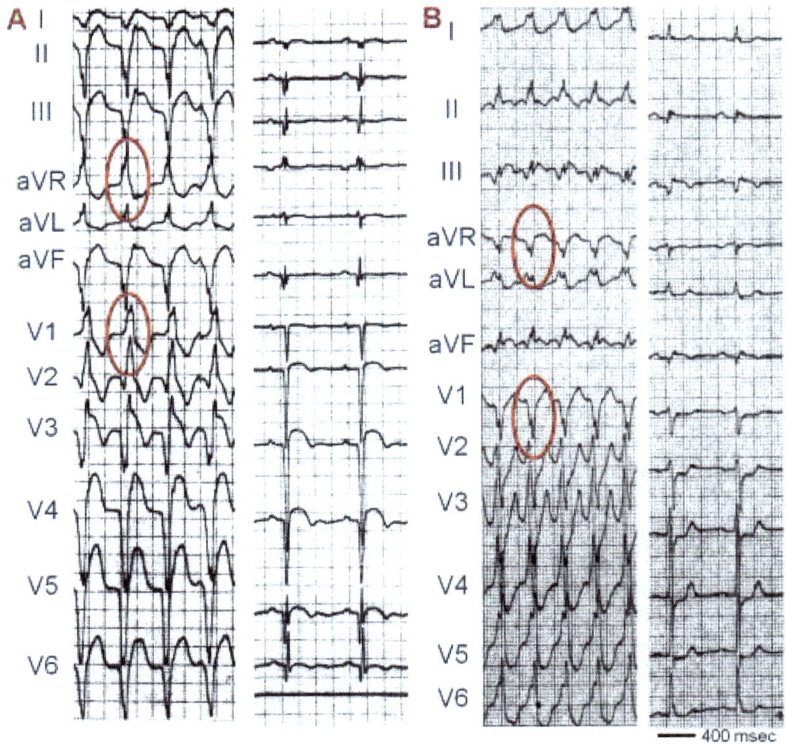

Figure 4.13. Two panels showing the similarity of the QRS in lead aVR and V_1 during VT. **Panel A** is from a patient with a scar from a large anterior wall myocardial infarction. The patient in **Panel B** has a scar from a previous inferior myocardial infarction. Note also that, during VT, the QR complexes are present in the leads corresponding to the location of the scar.

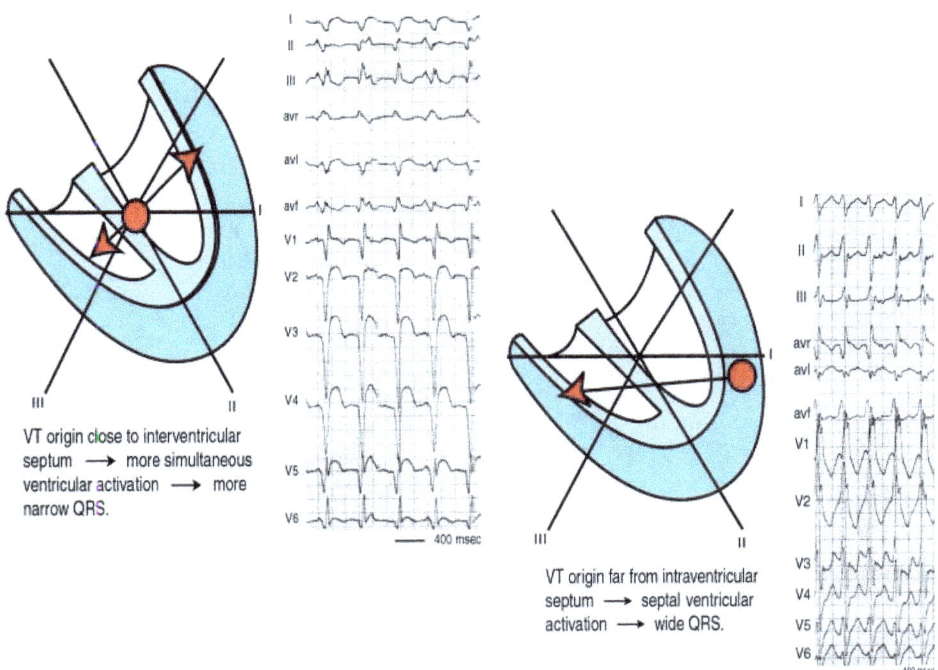

Figure 4.14. Ventricular tachycardia (VT) origin and QRS width. **Left panel:** An origin close to the interventricular septum results in more simultaneous right and left ventricular activation and, therefore, a narrower QRS complex. In contrast **(right panel)**, a VT origin in the lateral ventricular wall results in sequential ventricular activation and a wider QRS complex.

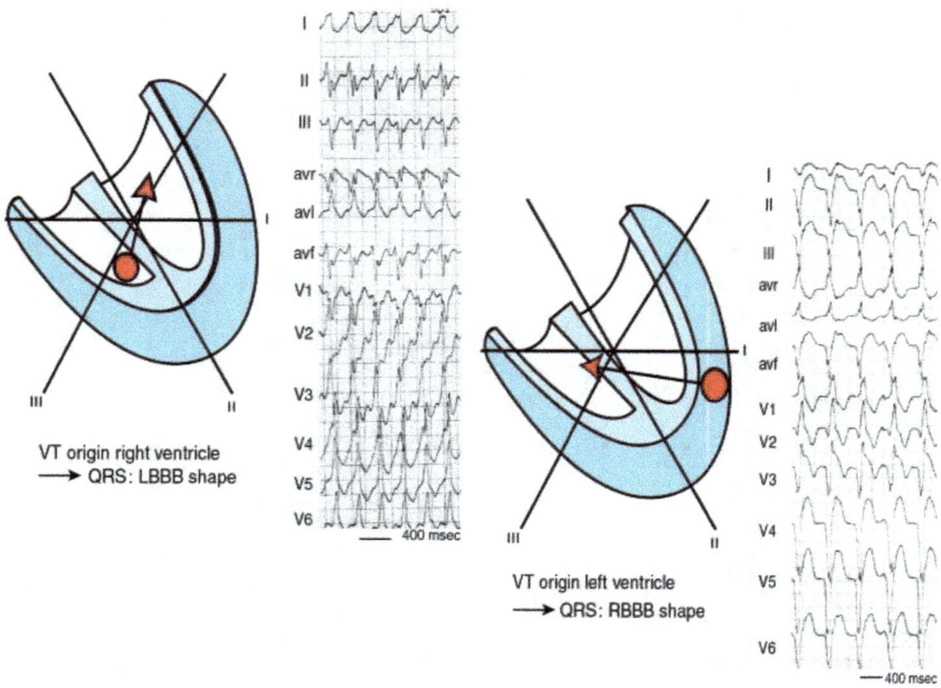

Figure 4.15. VT origin and shape of the QRS complex. **Left panel:** An origin in the right ventricle produces an LBBB-like shape. **Right panel:** A left ventricular origin gives an RBBB-like QRS configuration.

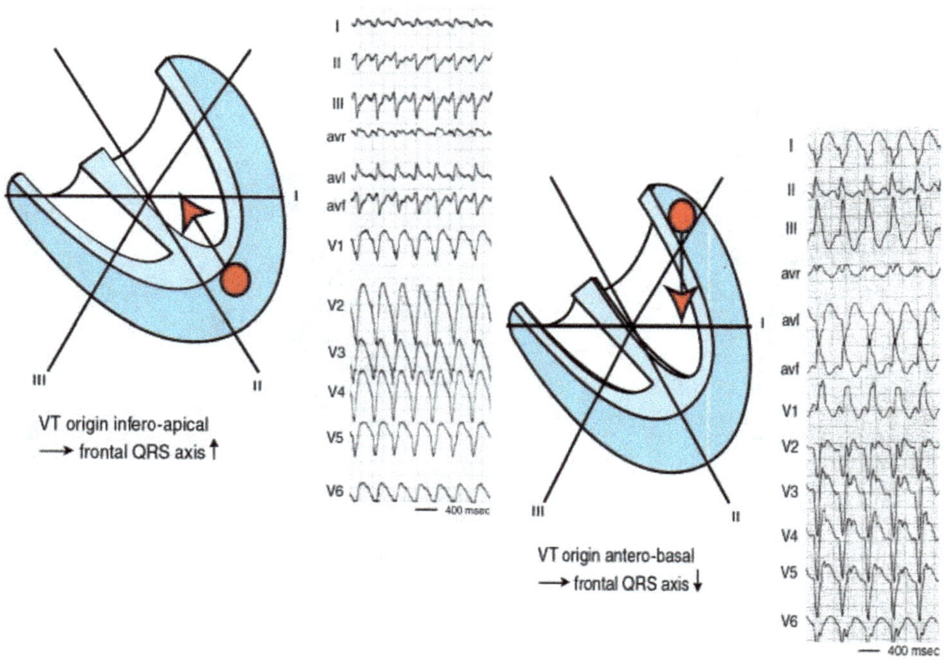

Figure 4.16. VT origin and QRS axis in the frontal plane. An apical origin results in a superiorly directed axis **(left panel)**. In contrast, a basal origin leads to an inferior QRS axis **(right panel)**.

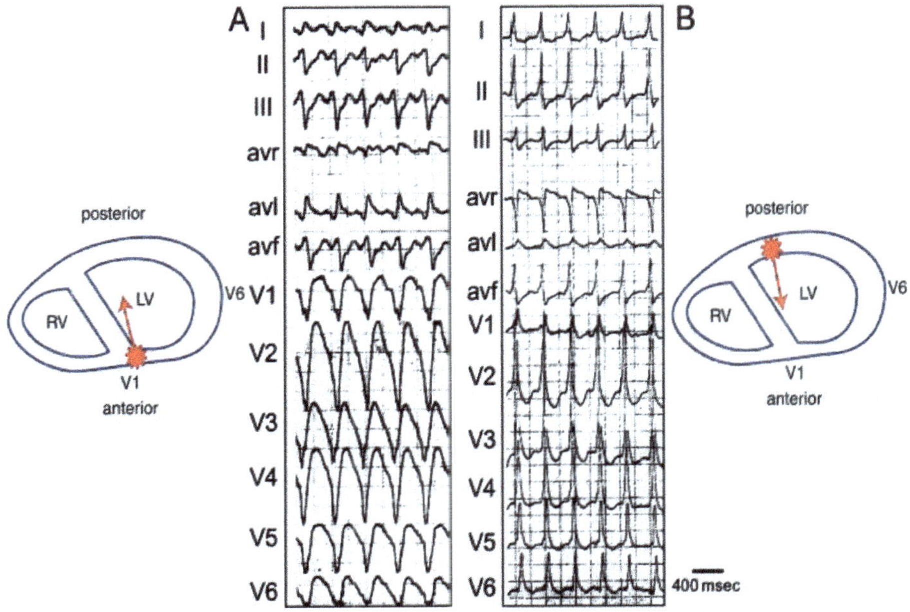

Figure 4.17. A concordant pattern in localizing the ventricular tachycardia (VT) origin in the horizontal plane. The **left panel** shows a VT arising in the apical area of the left ventricle (LV), resulting in a negative concordancy of all precordial leads. In the **right panel**, ventricular activation starts in the posterior area, resulting in positive concordancy of all precordial leads. The latter can be found in left posterior VT but also in SVT with AV conduction over a left posterior accessory AV pathway. RV, right ventricular.

idiopathic left VT is the one shown in Figure 4.20C. Now, the frontal QRS axis is inferiorly directed. This VT originates in or close to the anterior fascicle of the LBB.

In ARVD, there are three predilection sites in the right ventricle: the inflow and outflow tracts, and the apex. Whereas the first two sites have a QRS configuration during tachycardia that is difficult to differentiate from right ventricular idiopathic VT (Figure 4.21), left axis deviation in a young person with an LBBB-shaped VT should immediately lead to the suspicion of ARVD. In fact, there is an important rule in LBBB-shaped VT with left axis deviation that cardiac disease should be suspected and that idiopathic right ventricular VT is extremely unlikely. Figure 4.22 gives an example of an LBBB-shaped VT with left axis deviation in a patient with ARVD. The typical ECG features found in ARVD both during VT and during sinus rhythm are listed by Nasir et al.[15] Note the negative T waves in the precordial leads in both Figures 4.21 and 4.22.

When the broad QRS is identical during tachycardia and sinus rhythm, one has to differentiate SVT with preexistent BBB from bundle branch reentrant tachycardia.[20,21] In diseased hearts, especially when the bundle branches and the interventricular septum are

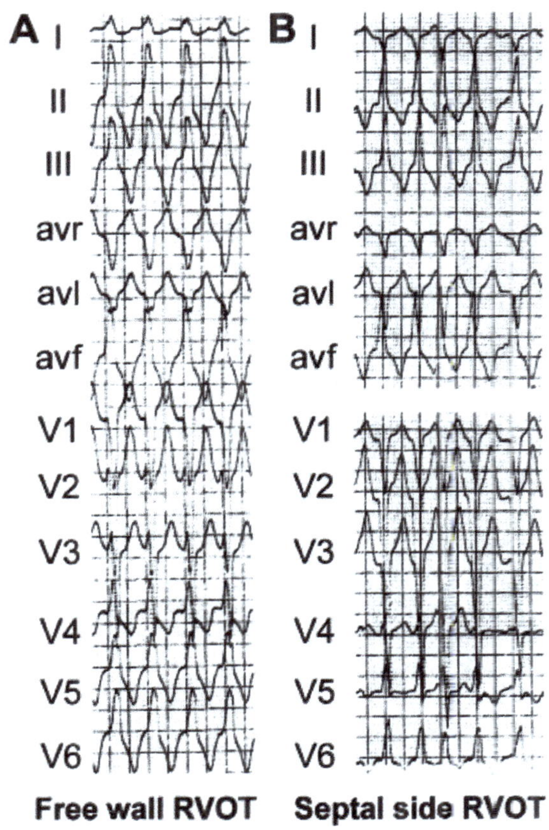

Figure 4.18. Two patterns of idiopathic VT arising in or close to the RVOT. See text for details.

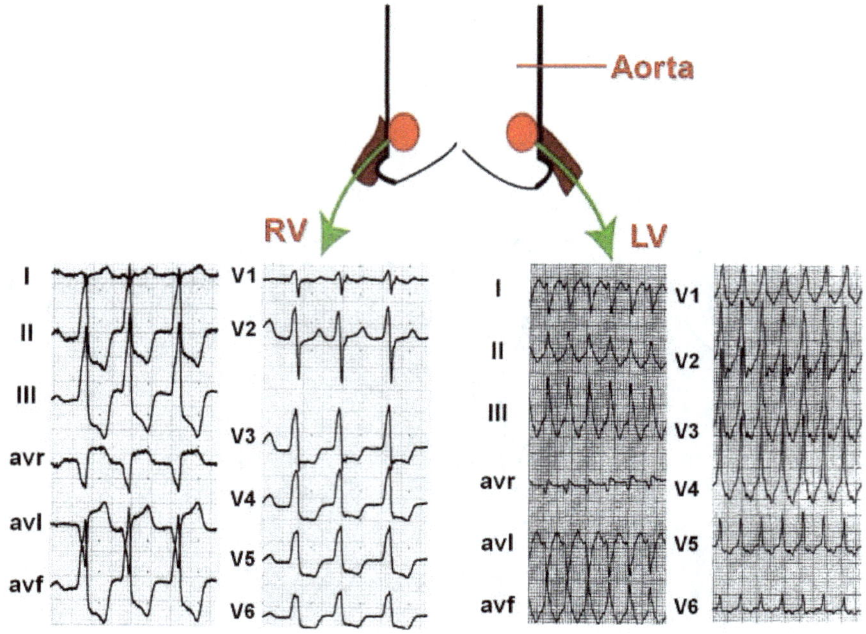

Figure 4.19. Two examples of idiopathic VT arising in the aortic root. The one on the left has its origin close to the right coronary cusp and is connected by a muscle band to the epicardium of the RVOT. The one on the right is originating close to the left coronary cusp and is connected to the posterior part of the left ventricular (LV) outflow tract.

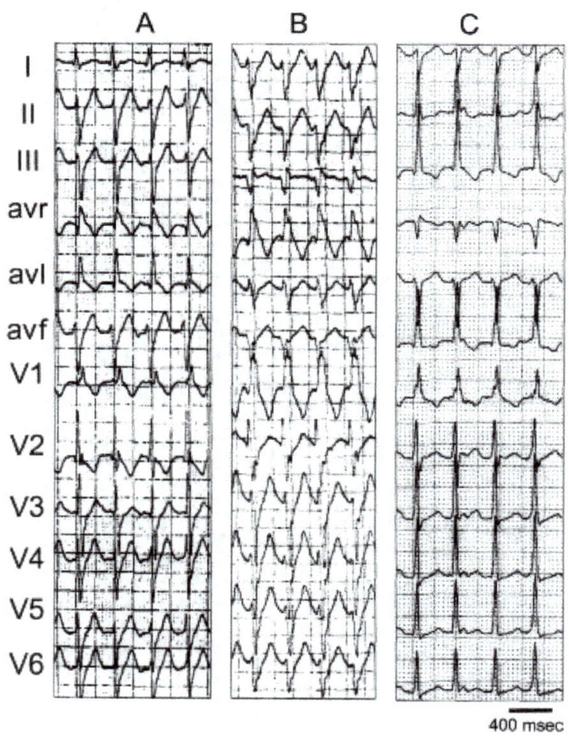

Figure 4.20. Three types of idiopathic left VT. See text for details.

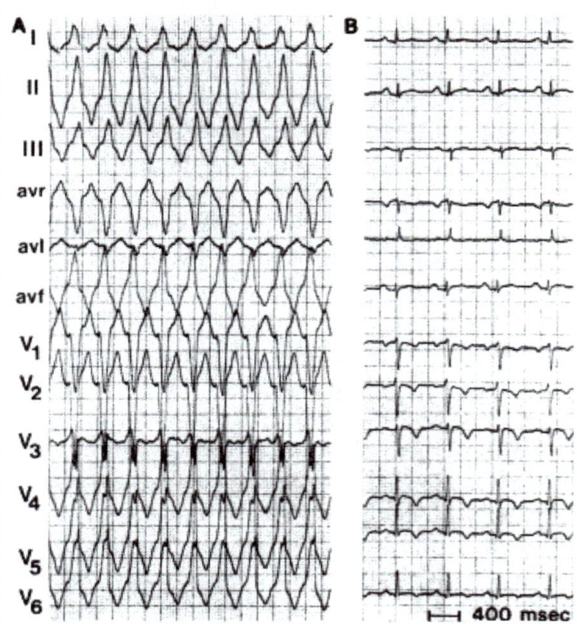

Figure 4.21. **A**: An LBBB-shaped VT in a patient with ARVD. The frontal QRS axis is intermediate, indicating that the site of origin is in the right ventricular inflow tract. Note the characteristic negative T waves in the precordial leads in **Panel B**.

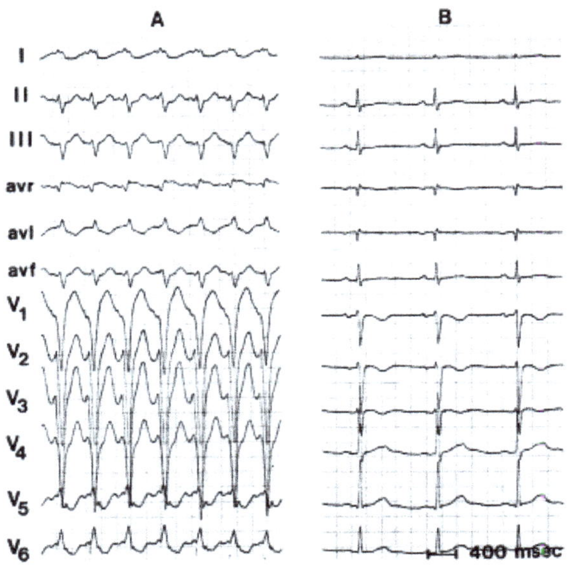

Figure 4.22. A: A VT with an LBBB-like shape and a frontal QRS axis of –60 degrees. The QRS width is 200 ms. B: The T-wave inversion in leads V_1 to V_3 and a different QRS width in V_1 and V_2 (120 ms) and V_6 (90 ms).

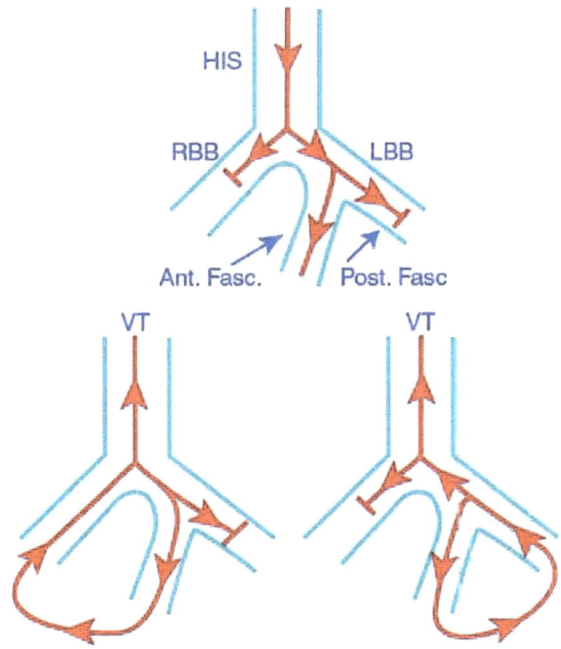

Figure 4.23. The tachycardia circuit in bundle branch or fascicular reentrant tachycardia. The top diagram shows the situation during sinus rhythm with anterograde block in the right bundle and the left posterior fascicle. During VT, the impulse uses either the right bundle branch (RBB) (bundle branch reentry) or the posterior fascicle (Post. Fasc) (fascicular reentry) for retrograde conduction. Ant. Fasc, anterior fascicle; HIS, His bundle; LBB, left bundle branch.

involved, a tachycardia may occur based on a circuit with anterograde conduction down one bundle branch or one of the left-sided fascicles and after septal activation, retrograde conduction over another branch of the bundle branch system (Figure 4.23). Figure 4.24 gives a clinical example in a patient with a previous anteroseptal myocardial infarction. Note initiation during sinus rhythm with progressive sub-AV nodal delay. Bundle branch reentry tachycardia may occur in anteroseptal myocardial infarction, idiopathic dilated cardiomyopathy, or myotonic dystrophy, and after aortic valve surgery or severe frontal chest trauma.

Value of the ECG during Sinus Rhythm

The ECG during sinus rhythm may show changes, such as preexistent BBB, ventricular preexcitation, or an old myocardial infarction, which are very helpful in correctly interpreting the ECG during broad QRS tachycardia. Also, the presence of AV conduction disturbances during sinus rhythm makes it very unlikely that a broad QRS tachycardia in that patient has a supraventricular origin. There are three other helpful signs pointing to VT: (1) as shown in Figure 4.12, a QRS showing a different BBB pattern in tachycardia and during sinus rhythm; (2) ventricular premature beats during sinus rhythm with the same QRS configuration as during tachycardia (Figure 4.25); and (3) a QRS width during tachycardia narrower than during sinus rhythm (Figure 4.26).

The Practical Approach

It is important not to panic when one is confronted with a wide QRS tachycardia. Statistically, VT is much more common than SVT. Also, when the tachycardia is hemodynamically well tolerated, the patient should be examined for clinical signs of AV dissociation and the 12-lead ECG should be systematically evaluated. When the tachycardia is hemodynamically not tolerated, cardioversion should be performed. If the tachycardia is tolerated but the diagnosis is in doubt, the patient should not be treated with adenosine or verapamil; procainamide should be given instead.

SUMMARY

Most wide QRS tachycardias have a ventricular origin. Correct diagnosis, which includes the identification of underlying heart disease, has

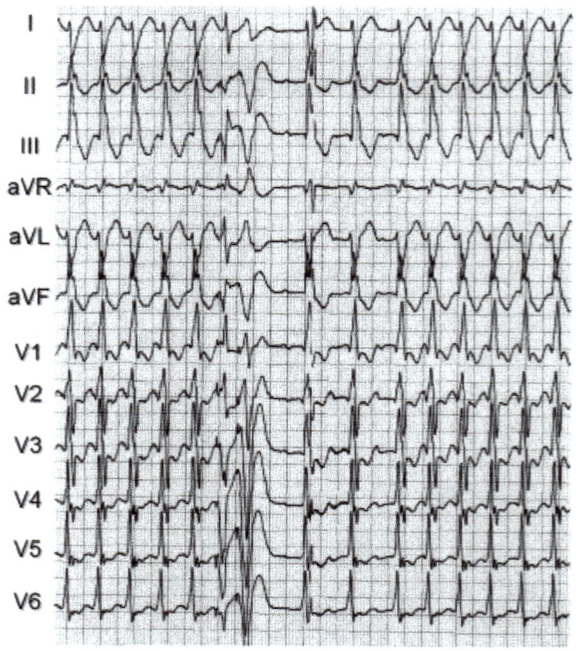

Figure 4.24. Example of a bundle branch reentrant VT in a patient with a previous anteroseptal myocardial infarction. During VT, the impulse goes down over the left anterior fascicle and back up over the RBB. The tachycardia is shortly interrupted by two paced ventricular beats, but resumes immediately after three conducted sinus beats. The sinus beats show that anterograde conduction to the ventricle went by way of the left anterior fascicle because of anterograde block in the left posterior fascicle and the RBB.

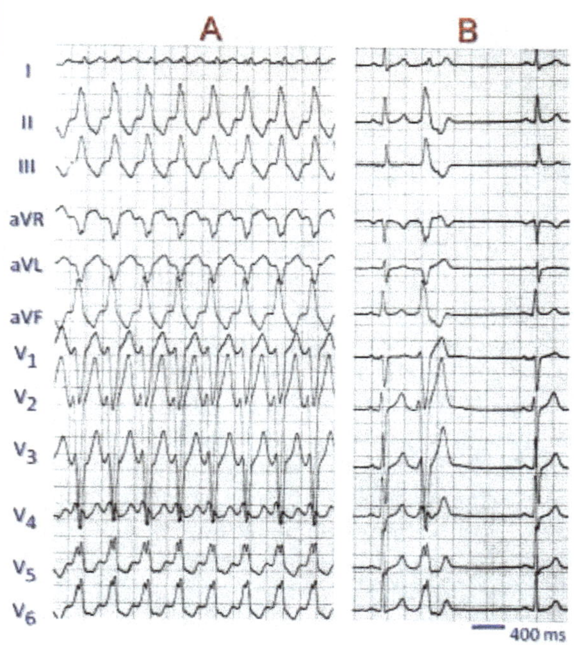

Figure 4.25. Example of how the diagnosis of VT **(Panel A)** can be helped by the presence of similarly shaped ventricular premature beats during sinus rhythm **(Panel B)**.

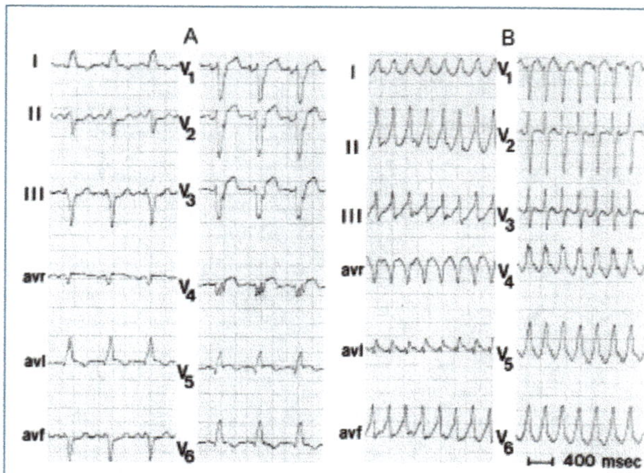

Figure 4.26. Shown here is a tachycardia QRS smaller than a QRS during sinus rhythm. On the left, sinus rhythm is present with a very wide QRS because of an anterolateral myocardial infarction and a marked delay in left ventricular activation. On the right, a VT arising on the right side of the interventricular septum results in more simultaneous activation of the right and left ventricles than during sinus rhythm and, therefore, a smaller QRS complex.

important prognostic and therapeutic implications. In VT, treatment during the arrhythmia depends upon the hemodynamic tolerance of the arrhythmia. When underlying heart disease is present, it should be corrected when possible. Currently, implantable cardioverter defibrillator implantation is often required. In some VT patients, cure can be obtained by catheter ablation.

REFERENCES

1. Stewart RB, Bardy GH, Greene HL. Wide complex tachycardia: Misdiagnosis and outcome after emergent therapy. *Ann Intern Med.* 1986;104:766–71.
2. Buxton AE, Marchlinski FE, Doherty JU, Flores B, Josephson ME. Hazard of intravenous verapamil for sustained ventricular tachycardia. *Am J Cardiol.* 1987;59:1107–1110.
3. Dancy M, Camm AJ, Ward D. Misdiagnosis of chronic recurrent ventricular tachycardia. *Lancet.* 1985;2:320–323.
4. Wellens HJJ, Ross DL, Farre J. Functional bundle branch block during supraventricular tachycardia in man: Observations on mechanisms and their incidence. In: Zipes DP, Jalife J, eds. *Cardiac Electrophysiology and Arrhythmias.* New York: Grune and Stratton; 1985: 435–441.
5. Wellens HJ, Bär FW, Lie KI. The value of the electrocardiogram in the differential diagnosis of a tachycardia with a widened QRS complex. *Am J Med.* 1978;64:27–33.
6. Vermeulen A, Wellens HJ. Paroxysmal ventricular tachycardia showing fusion with reciprocal ventricular beats. *Br Heart J.* 1971;33:320–322.

7. Kindwall E, Brown J, Josephson ME. Electrocardiographic criteria for ventricular tachycardia in wide complex left bundle branch block morphology tachycardias. *Am J Cardiol.* 1988;61:1279–1283.
8. Coumel P, Leclercq JF, Attuel P, Maisonblanche P. The QRS morphology in post-myocardial infarction ventricular tachycardia. A study of 100 tracings compared with 70 cases of idiopathic ventricular tachycardia. *Eur Heart J.* 1984;5:792–805.
9. Brugada P, Brugada J, Mont L, Smeets J, Andries EW. A new approach to the differential diagnosis of a regular tachycardia with a wide QRS complex. *Circulation.* 1991;83:1649–1659.
10. Lau EW, Pathamanathan RK, Ng GA, et al. The Bayesian approach improves the electrocardiographic diagnosis of broad complex tachycardia. *Pacing Clin Electrophysiol.* 2000;23:1519–1526.
11. Vereckei A, Duray G, Szénási G, Altemose GT, Miller JM. New algorithm using only lead aVR for differential diagnosis of wide QRS complex tachycardia. *Heart Rhythm.* 2008;5:89–98.
12. Pava LF, Perafán P, Badiel M, et al. R-wave peak time at DII: A new criterion for differentiating between wide complex QRS tachycardias. *Heart Rhythm.* 2010;7:922–926.
13. Wellens HJJ. The electrocardiographic diagnosis of arrhythmias. In: Topol EJ, ed. *Textbook of Cardiovascular Medicine.* 2nd ed. Philadelphia: Lippincott Williams & Wilkins; 2002:1365–1383.
14. Wellens HJJ, Rodriguez LM, Smeets JL. Ventricular tachycardia in structurally normal hearts. In: Zipes DP, Jalife J, eds. *Cardiac Electrophysiology: From Cell to Bedside.* 2nd ed. Philadelphia: WB Saunders; 1995:780–788.
15. Nasir K, Bomma C, Tandri H, et al. Electrocardiographic features of arrhythmogenic right ventricular dysplasia/cardiomyopathy according to disease severity: A need to broaden diagnostic criteria. *Circulation.* 2004;110:1527–1534.
16. Kanagaratnam L, Tomassoni G, Schweikert R, et al. Ventricular tachycardias arising from aortic sinus of valsalva: An under-recognized variant of left outflow tract ventricular tachycardia. *J Am Coll Cardiol.* 2001;37:1408–1414.
17. Ouyang F, Fotuhi P, Ho SY, et al. Repetitive monomorphic ventricular tachycardia originating from the aortic sinus cusp: Electrocardiographic characterization for guiding catheter ablation. *J Am Coll Cardiol.* 2002;39:500–508.
18. Timmermans C, Rodriquez LM, Crijns HJ, Moorman AF, Wellens HJ. Idiopathic left bundle-branch block-shaped ventricular tachycardia may originate above the pulmonary valve. *Circulation.* 2003;108:1960–1967.
19. Ito S, Tada H, Naito S, et al. Development and validation of an ECG algorithm for identifying the optimal ablation site for idiopathic ventricular outflow tract tachycardia. *J Cardiovasc Electrophysiol.* 2003;14:1280–1286.
20. Touboul P, Kirkorian G, Atallah G, et al. Bundle branch reentrant tachycardia treated by electrical ablation of the right bundle branch. *J Am Coll Cardiol.* 1986;7:1404–1409.
21. Blanck Z, Dhala A, Despande S, et al. Bundle branch reentrant ventricular tachycardia: Cumulative experience in 48 patients. *J Cardiovasc Electrophysiol.* 1993;4:253–262.

SECTION 1

ECGs, Syncope, Bradyarrhythmias, and Non-Invasive Testing

CASE STUDY 1

Interpretation of ECG

Jasbir S. Sra, MD and Masood Akhtar, MD

Question No. 1:

The electrocardiogram (ECG) in Figure 1.1 shows:

A. Normal sinus rhythm.
B. QRS axis at around +60 degrees.
C. Left axis deviation.
D. Left bundle branch block (LBBB).
E. Both A and B.

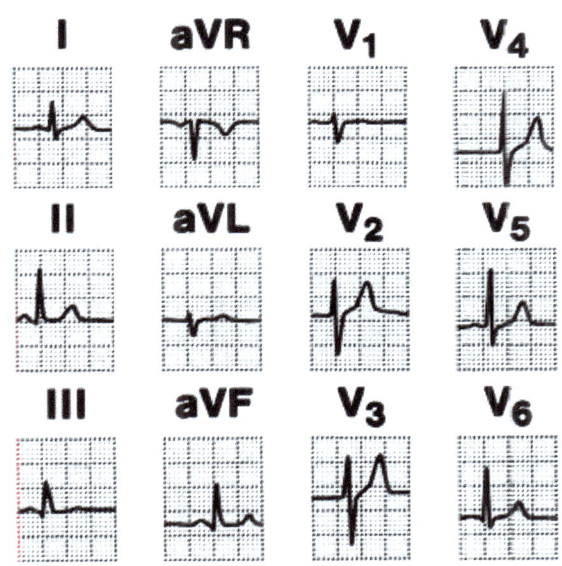

Figure 1.1. 12-lead ECG. (Reproduced with permission from the American Heart Association.[1])

Normal sinus rhythm ranges between 60 and 100 beats per minute (bpm). Sinus node depolarization is not recorded on the surface ECG (Figure 1.2). Activation of the atrial myocardium produces P waves on the surface ECG. Right atrial activation is represented on the initial part of the P wave, whereas the terminal part represents left atrial activation. The impulse then travels through the atrioventricular (AV) node, His bundle, bundle branches, Purkinje fibers, and the myocardium. The isoelectric part of the PR interval represents activation of the impulse through the AV node, His-Purkinje, and the Purkinje system, while the QRS complex represents ventricular depolarization. Ventricular repolarization is recorded as T wave.

Augmented unipolar leads related to Einthoven's triangle are depicted in Figure 1.3. Goldberger created modified unipolar leads by removing resistors from the central terminal and designating them aVR, aVL, and aVF. On a circular scale, the positive ends of aVL, aVR, and aVF can be projected as –30, –150, and +90 degrees, respectively. By producing the leads on negative halves, all 6 limb leads can be obtained on a hexaxial reference frame system. Figure 1.4 depicts an example of determination of axis on a hexaxial frame. All 6 limb leads and positive (solid lines, at right) and negative (dotted lines, at right)

Practical Electrophysiology © 2015 Jasbir S. Sra, Masood Akhtar, Andrea Natale, David J. Wilber
Cardiotext Publishing, ISBN: 978-1-935395-17-1

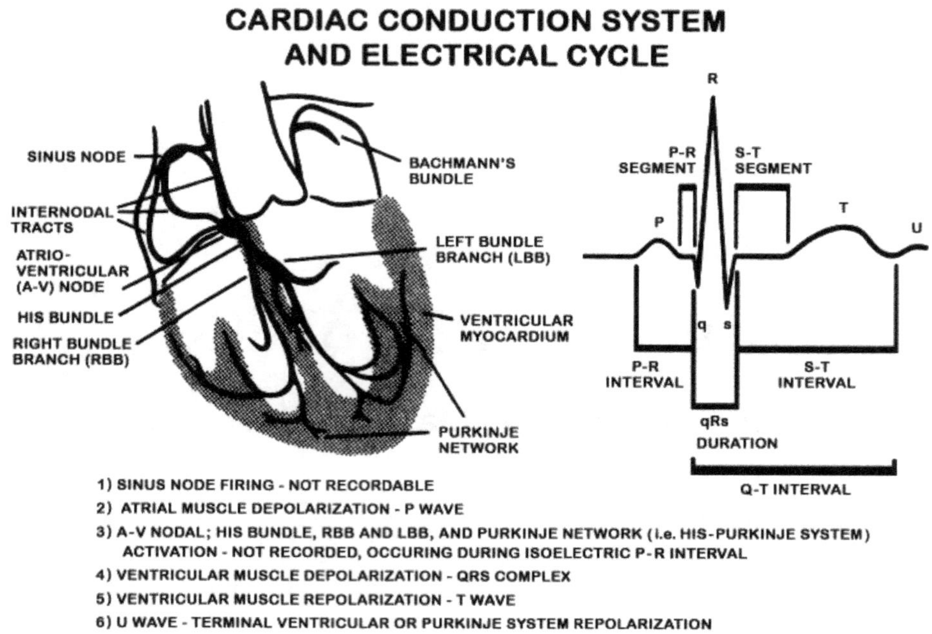

Figure 1.2. Cardiac conduction system and typical deflection of ECG. (Reproduced with permission from the American Heart Association.[1])

halves are shown. Each lead is divided into negative and positive hemicircles by another lead. For example, lead II is divided into positive and negative halves by aVL and vice versa. If an equiphasic lead is found, the QRS axis will be perpendicular to that lead. In a perpendicular lead, the positive or negative direction of the complexes can be easily identified. For example, if aVF is equiphasic, the axis is along lead I. If lead I is positive, the axis is 0 degrees; if negative, it will be ±180 degrees.

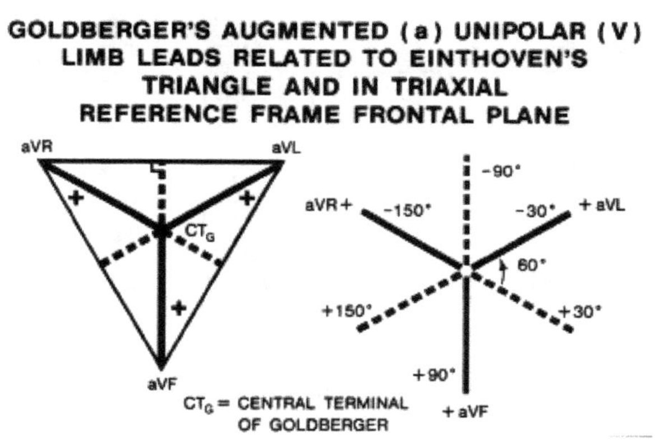

Figure 1.3. By connecting electrodes attached to the arms and the left leg, an indifferent central terminal at a zero potential can be obtained. An exploring electrode can then be applied to the extremities to obtain local electrical potentials. In clinical electrocardiography, such unipolar limb leads can be obtained by attaching pacing electrodes to the right arm (VR), left arm (VL), and the left leg (VF). On a triaxial frame system, connecting the apices of the Einthoven's triangle with the center of the equilateral triangle can project VR, VL, and VF. Unipolar leads generally register lower voltages than bipolar leads and thus need a correction factor. (Reproduced with permission from the American Heart Association.[1])

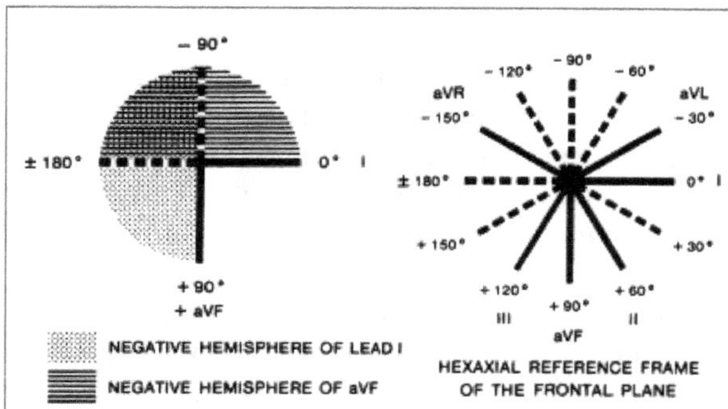

Figure 1.4. In the current example, the P wave is positive in leads II and aVF, biphasic in V_1, and negative in aVR. The P-axis is positive at +60 degrees, since it is isoelectric in aVL as this lead is perpendicular to II, which is positive. QRS complex shows normal transition from rS in V_1 to qRs pattern in V_6. There is a progressive increase in R wave from V_1 to V_6. T wave is upright in most leads and negative in aVR. QRS axis is approximately 100 degrees. (Reproduced with permission from the American Heart Association.[1])

Normal ventricular activation is usually completed in 60 to 80 ms due to simultaneous activation of both ventricles. In contrast, during bundle branch block, QRS duration is prolonged due to sequential activation of the ventricles.

Right Bundle Branch Block (RBBB): Presence of RBBB does not alter the initial part of the QRS as initial, normal left-to-right forces are intact (Figure 1.5). However, transseptal muscle-to-muscle conduction then generates significant electrical force, especially after 60 to 80 ms, when left ventricular activation is completed. The QRS vector is thus directed anteriorly and to the right, resulting in a positive terminal R pattern in V_1 and a terminal S wave in leads I and V_6. RBBB is designated complete when QRS duration equals or exceeds 120 ms and incomplete when it is less than 120 ms.

Left Bundle Branch Block: During LBBB, septal activation occurs through the right bundle, moving in the opposite direction of the activation occurring in the left bundle. This results in absence of Q waves in leads I and V_6. However, a small R wave may be present in V_1 due to an anterior component of the right-to-left activation. Ventricular activation during LBBB occurs posteriorly and to the left. In addition, secondary ST segment and T-wave abnormalities are universally seen (Figure 1.6).

Nonspecific Intraventricular Conduction Defect (IVCD): During IVCD, QRS duration is prolonged but there is no specific pattern of RBBB or LBBB (Figure 1.7). Class I drugs, ventricular hypertrophy, myocardial scarring, and hyperkalemia can also cause IVCD patterns.

Left Anterior (Superior) (LAFB) and Posterior (Inferior) (LPFB) Fascicular Blocks: During LAFB, septal activation generally occurs in a normal left-to-right fashion. Inferior orientation of the initial forces occurs, producing an initial R wave in leads II, III, and aVF, and Q waves in I and aVL (Figure 1.8). The mean QRS vector is directed superiorly and to the left, resulting in tall R waves in leads I and aVL as well as deep S waves in leads II, III, and aVF. As there is simultaneous activation of both the right and left ventricles, the QRS is not prolonged.

Isolated LPFB is unusual. The initial QRS vector is directed superiorly and to the left (Figure 1.9). This produces R waves in leads I and aVL and Q waves in leads, II, III, and aVF, with tall R waves in leads II, III, and aVF. RBBB is usually seen with LPFB.

Correct Answer:

Q1. A

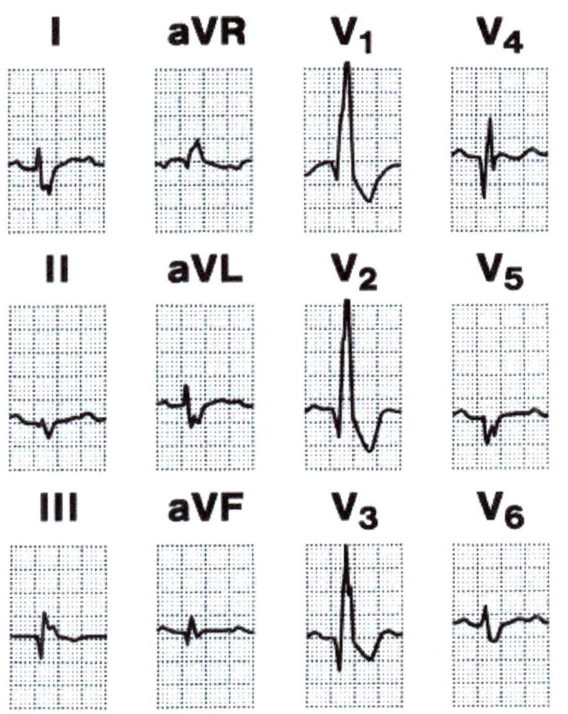

Figure 1.5. Typical 12-lead ECG in RBBB. Note normal q in leads I, aVL, and V_6, widening of QRS (160 ms), and rsR in V_1. The majority of the delay is in the latter part of the QRS complex. (Reproduced with permission from the American Heart Association.[1])

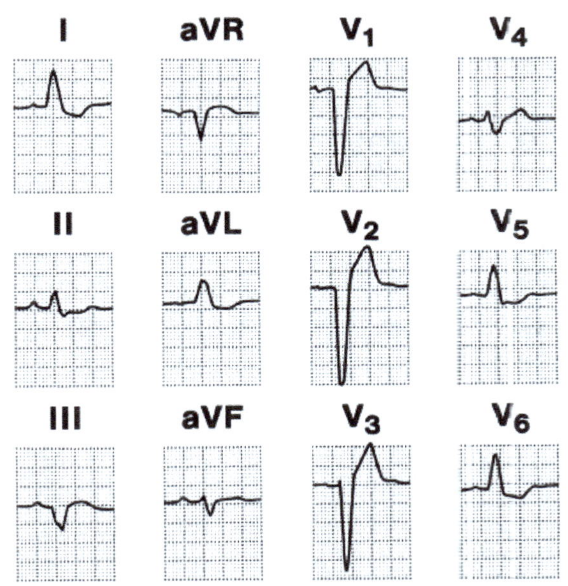

Figure 1.6. Typical 12-lead ECG in LBBB. Note absence of q in leads I and V_6 and wide QRS complex (170 ms) delay throughout inscription of QRS complex. Deep S in V_1 and V_2 and broad R in V_5 and V_6 are also typical. (Reproduced with permission from the American Heart Association.[1])

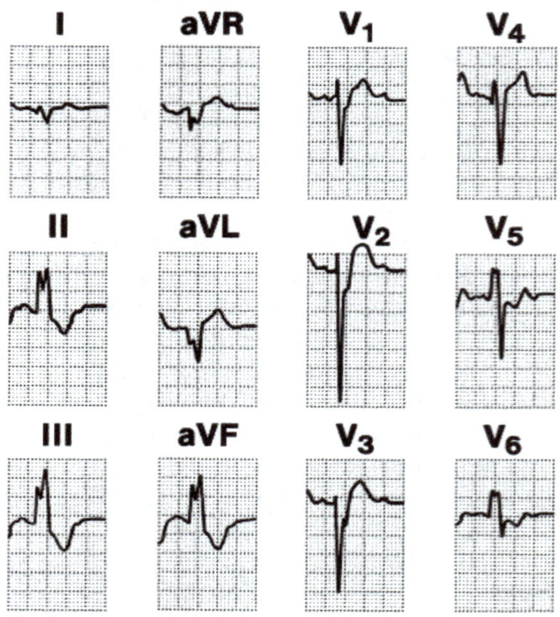

Figure 1.7. Nonspecific IVCD. Markedly slurred and broad QRS (190 ms) is seen, but septal forces, Q wave in leads I and V_6 are preserved. (Reproduced with permission from the American Heart Association.[1])

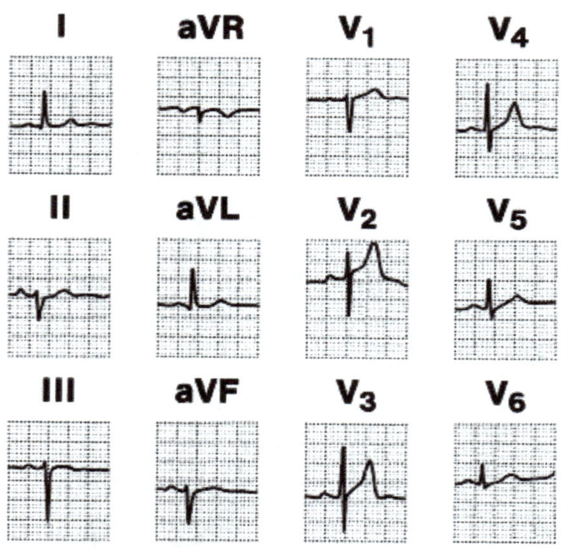

Figure 1.8. Left anterior fascicular block and hemiblock. Typical tracings are QRS axis of 0 degrees without significant QRS prolongation. (Reproduced with permission from American Heart Association.[1])

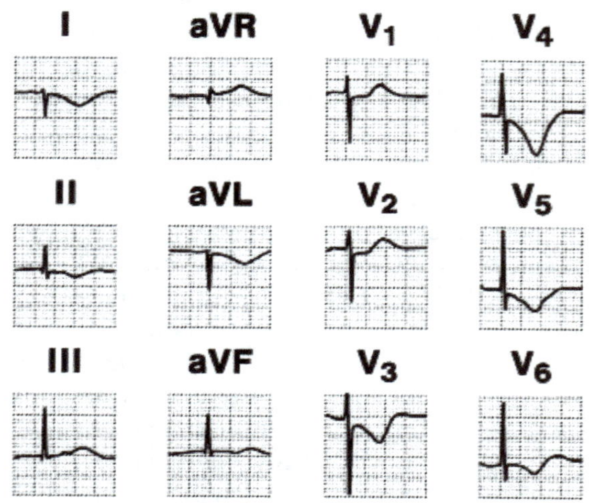

Figure 1.9. Left posterior fascicular block or hemiblock. QRS axis of +120 degrees can be appreciated. ECG also shows ST-T changes. (Reproduced with permission from the American Heart Association.[1])

QUESTIONS FROM THE EDITORS AND AUTHORS' REPLY

Question 1. What electrical force direction information do precordial leads provide?

Answer: Usual placement of precordial leads provides information about cardiac electrical forces along the horizontal plane, which is sensitive to forces directed anteriorly, posteriorly, right, and left but not as much to superior or inferior, seen primarily by limb leads of 12-lead ECG.

Question 2. How does QRS morphology vary at different electrode sites?

Answer: ECG configuration depends on location of electrode relative to direction of current flow. Positive signal is recorded when impulse approaches electrode and negative signal is registered at the electrode if the current is flowing away. Biphasic waveform is recorded if electrode is perpendicular to direction of impulse propagation, rather than parallel.

REFERENCES

1. Akhtar M. *Examination of the Heart: The Electrocardiogram, Part 5.* Dallas, TX: American Heart Association; 1990:16–21.

CASE STUDY

2

Syncope and Suspected Sinus Node Dysfunction

Wayne O. Adkisson, MD and David G. Benditt, MD

Mr. R is a 57-year-old gentleman seen in consultation regarding sinus bradycardia noted during evaluation for a minor surgical procedure.

Mr. R is employed as an airline pilot safety instructor. He reports one episode of loss of consciousness 3 years ago. After standing for a prolonged period of time he began to feel warm, diaphoretic, and nauseated. The next thing he remembers is waking up on the floor with co-workers asking him if he was okay. He was told he was "pale as a ghost." There was no incontinence and no injuries. After the event he felt fatigued but reports no other sequelae. His co-workers wanted to call an ambulance but he declined. He rested for several minutes, drank some water, and returned to work. He denies any other syncopal events. He reports no symptoms of near syncope.

Mr. R is physically fit. He runs on a daily basis and lifts weights 3 times a week. He reports no family history of cardiac disease or arrhythmia. He has no significant past medical or surgical history, other than obstructive sleep apnea. He has been fitted for, and wears, a CPAP mask. He takes no medications and uses no over-the-counter supplements. His physical examination, other than a resting heart rate of 50 bpm, is entirely normal. His blood pressure was 116/68 while at rest, seated.

Prior to electrophysiology (EP) consultation, an echocardiogram demonstrated a structurally normal heart. A 24-hour Holter monitor showed an average heart rate of 67 bpm. The maximum heart rate recorded was sinus at 152 bpm and was recorded while he was running on a treadmill. The minimum heart rate was 44 bpm recorded at 02:00. At that time, first-degree atrioventricular (AV) block was noted. No higher degree block was recorded. No ventricular ectopy was recorded. Rare atrial premature beats were seen with no reported associated symptoms. There were no pauses > 2 seconds.

Question No. 1:

As a consultant, you would recommend which of the following?

A. A tilt table study.

B. A 30-day event recorder (looping).

C. An implantable loop recorder (ILR).

D. An exercise study.

E. No further evaluation.

DISCUSSION

Review of the 12-lead ECG obtained during the patient's preoperative evaluation (Figure 2.1) shows a P-wave morphology and axis that are consistent

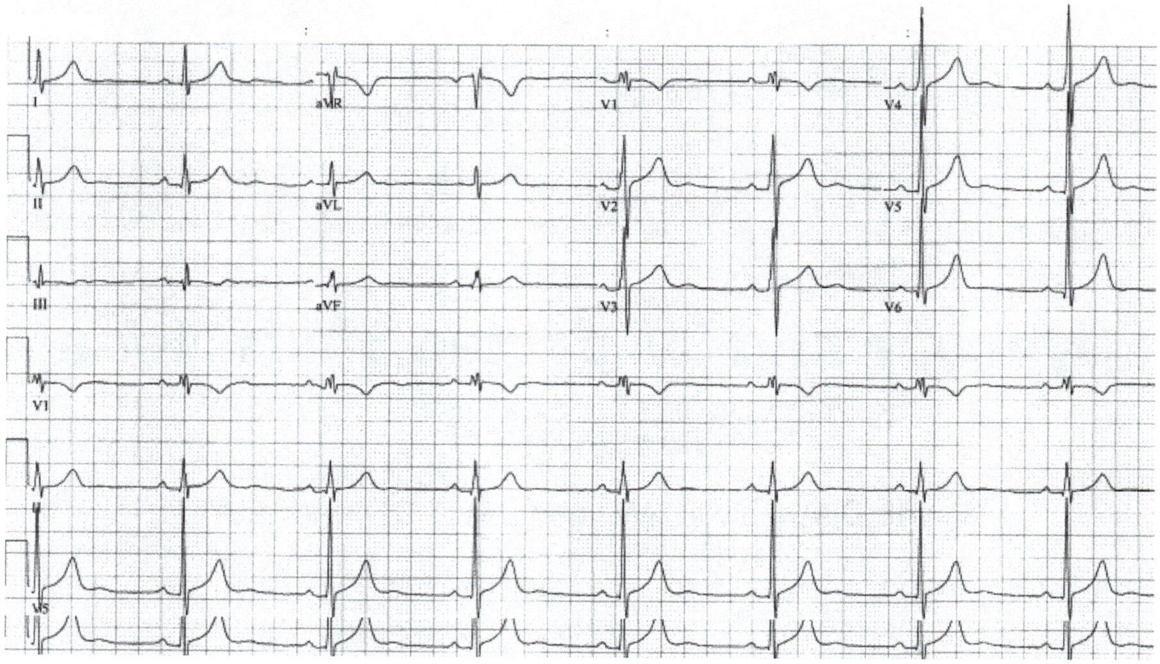

Figure 2.1. Baseline ECG obtained during preoperative evaluation.

with a sinus node origin at a rate of 47 bpm. The PR interval is normal at 180 ms; ST segment, T waves, and QT intervals are normal.

No pathologic Q waves are seen. The QRS duration and frontal axis are both normal. The QRS transition in the precordial leads is early. While this ECG finding raises the question of posterior infarction or right ventricular (RV) hypertrophy, an echocardiogram obtained prior to consultation demonstrated normal RV size with no wall motion abnormalities seen. In this case the early precordial transition noted is a normal variant.

Careful analysis of the QRS morphology demonstrates notching in the limb and a slurring in the upstroke of the QRS in the right precordial leads, suggesting mild nonspecific intraventricular conduction delay. This has not resulted in prolongation of the QRS. The slurring in V_1 and V_2 should not be confused with ventricular preexcitation. U waves are noted in the midprecordial leads. U waves can be seen in patients with hypokalemia, but Mr. R's chemistry evaluation was entirely normal.

It is difficult to formulate an efficient evaluation strategy without a clear understanding of the clinical question being asked. In this case we are being asked whether or not Mr. R's sinus bradycardia is pathologic, how does it relate to his earlier syncopal episode and does he need a pacemaker? In short, does Mr. R have sinus node dysfunction (SND)? SND can be broadly defined as a heart rate that is inappropriate for the physiological needs of the body. The incidence of SND increases with aging, but it has been described in all age groups. In the United States, SND is the indication for roughly half of all pacemaker implantations. In the face of an aging population, diagnosis and treatment of SND will consume an increasing proportion of health care resources. The need for an efficient and cost-effective approach to this syndrome is obvious.

Anatomy and Structure of the Sinus Node

Shortly after the turn of the last century Arthur Keith, then working on the question of whether the caval openings closed during contraction of the right atrium, noted in human hearts "a small condensed area of tissue, just where the cava sank into the auricle." A few years later, after learning of Tarawa's description of a "muscular" connection between the atria and the ventricles, Keith enlisted the aid of Martin Flack, a medical student at the time, in examining the hearts of a variety of animals for the structure described by Tarawa. When Flack described a structure he had identified in the right atrium of a mole Keith recognized this as being similar to the "condensed area" he had earlier identified in human hearts. They examined the remaining specimens and found a similar structure in each. In 1907, Keith and Flack published their paper describing what would later be referred to as the "sinus node."

The sinus node is located along the superior portion of the crista terminalis (CT) near the lateral aspect of the junction of the superior vena cava (SVC) and right atrium. However, the sinus node is not a small dot or sphere as often depicted in cartoons of the conduction system. In some mammals, the node extends along the entire intracaval region. In humans, histologically, the sinus node lies in the subepicardium and extends along the superior portion of the CT. The predominant pacemaking region is generally located in the superior portion of the CT. At lower heart rates the predominant pacemaking region shifts to more inferior locations along the CT. This shift in the location of the predominant pacemaking region may be large enough to result in a change in the P-wave axis on the surface ECG. Distinguishing whether a slow atrial rhythm, with a superiorly directed P-wave axis, is truly an ectopic atrial focus versus sinus bradycardia originating from the inferior aspect of the CT can be difficult on an isolated ECG. The presence of sinus node-like capability throughout the CT may explain the frequency with which focal atrial tachycardias arising in the right atrium are localized to the CT.

Histologically and functionally the sinus node is a complex structure, a structure of which basic EP is still debated despite a century of study. The mechanism(s) by which the pacemaking activity of the sinus node is regulated remains an area of controversy. Similarly, the question of how a small mass of cells is able to both electrically capture the much larger mass of the atrium and concomitantly protect itself from suppression by the larger and more hyperpolarized atrial myocardium remains incompletely understood. A great deal is known regarding details of ion channel expression, ionic currents, and electrical coupling within various regions of the sinus node but the larger question of how these components interact remains to be elucidated.

The primary pacemaking cells of the sinus node lie in the center of the sinus node. Cells in this region demonstrate typical Phase 4 depolarization that is characteristic of cells that demonstrate automaticity. As in other excitable tissues once a certain voltage threshold is reached an action potential is generated (Figure 2.2).

The upstroke of the action potential recorded from the center of the sinus node is slow due to the paucity of a Na current, I_{Na}. In these cells the primary current resulting in depolarization is the L-type Ca^{2+}, $I_{Ca,L}$. Other inward currents involved in the pacemaker action potential include, T-type Ca^{2+}, $I_{Ca,T}$ and the so-called "funny" current, I_f. A selective blocker of I_f, ivabradine is now available as a heart rate lowering agent. The Na^{2+}-Ca^{2+} exchanger responsible for Ca^{2+} extrusion from the cell is electrogenic, resulting in a net inward current, I_{NaCa}. This current is not insignificant, equaling roughly half of the amount of current carried by $I_{Ca,L}$ and $I_{Ca,T}$. Occurring during diastole I_{NaCa} may play a role in pacemaking. The role that Ca^{2+} release from the sarcoplasmic reticulum plays in the pacemaking activity of the sinus node remains uncertain.[3,4]

In myocardium the inward rectifying K^+ current, $I_{K,I}$, is primarily responsible for maintaining a stable diastolic resting potential, therefore it should not be surprising that this current is absent in pacemaking cells. The rapid and slow delayed rectifier K^+ currents, $I_{K,r}$ and $I_{K,s}$, are present helping to repolarize the cell and establishing the maximum diastolic potential.[3] The primary outward and inward currents and the ion channels carrying those currents in the central pacemaking cells of the sinus node are summarized in Table 2.1.

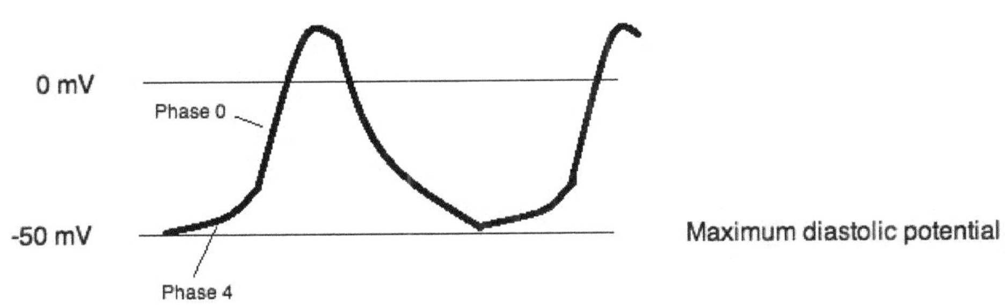

Figure 2.2. Sinus node action potential.

Table 2.1. **Principle Currents Involved in the Sinus Node Action Potential**

Outward currents (primary ion channel)	Inward currents (primary ion channel)
$I_{Ca,L}$ (Ca_v 1.3)	$I_{K,r}$ (ERG)
$I_{Ca,T}$ (Ca_v 3.1 and Ca_v 3.2)	$I_{K,s}$ (K,LQT1)
I_f (HCN_4, HCN_1, and HCN_2 also present)	
I_{NaCa}	

Once a pacemaker action potential has been generated it must then be conducted out of the sinus node if the atrium is to be activated. As discussed above, how the sinus node is able to accomplish this task yet also protect itself from suppression by the mass of atrial myocardium remains unresolved. There is a gradual change in cell type and cell morphology as one moves outward from the center of the sinus node. There is no clearly defined border between the node and the myocardium. Unlike working myocardium electrical coupling in the sinus node is poor and cell-to-cell conduction is slow. This is one of the likely mechanisms that protects the sinus node from suppression by the more polarized atrial tissue surrounding it. The poor electrical coupling found in the sinus node is related to a change in the connexin makeup of the gap junctions. This poor electrical coupling and slow conduction would make it difficult for the sinus node to capture the atrium suggesting that in the periphery electrical coupling should be stronger and more rapid. In the sinus node of the rabbit there is evidence that this is the case, with cells in the periphery expressing different connexins than cells in the central portion of the sinus node.[1]

A more detailed description of the structure and function of the sinoatrial (SA) node is beyond the scope of this chapter. Interested readers are referred to a recent review of this topic by Dobrzynski and colleagues.

SND can present in several ways—persistent inappropriate bradycardia, sinus pauses, atrial fibrillation (often with a slow ventricular response) or a combination of intermittent bradycardia and atrial tachyarrhythmias (the so-called bradycardia–tachycardia syndrome [BTS]). The tachyarrhythmia most commonly seen is atrial fibrillation, but atrial flutter and other reentrant or focal atrial tachycardias also occur. BTS is the most frequent manifestation of SND and is also the SND form with the highest incidence of syncope. Syncope most often results from the prolonged sinus pauses following the cessation of a paroxysmal supraventricular tachyarrhythmia. Sinus node pauses can also be seen without a preceding tachyarrhythmia. If prolonged, these pauses may also result in syncope or the feeling of impending syncope.

The monitor strips in Figure 2.3 are typical of BTS. The strips are continuous and demonstrate an episode of paroxysmal atrial fibrillation with a rapid ventricular rate. When the atrial fibrillation terminates, a 5.2-second pause follows. Although the episode occurs at 03:24 in the morning, this particular patient was awake, having been awakened by palpitations. The recorded pause was associated with this individual's feeling of impending collapse; permanent pacing was recommended.

The prolonged pauses that occur either spontaneously or after termination of a supraventricular tachycardia illustrate another facet of SND that is often overlooked, namely that this entity encompasses more than the sinus node alone. During such pauses one of the heart's subsidiary pacemakers should have become manifest. The absence of an "escape" rhythm suggests a more diffuse abnormality affecting the function of all cardiac pacemaker foci.

Persistent or inappropriate sinus bradycardia is rarely the cause of syncope. When SND presents in this fashion, symptoms are more often generalized fatigue or exercise intolerance. Patients frequently attribute such symptoms as a sign they are "getting old." Determining to what extent an older patient's declining exercise ability and fatigue are normal aging versus pathology is difficult.

In Mr. R's case, his ECG is borderline abnormal by definition; the heart rate is less than 60 bpm. The definition of what is and what is not a normal heart rate is arbitrary, a historical fait accompli. If SND were to be diagnosed on that basis alone many of the readers of this text, as well as most very fit citizens, would be considered to have SND. More than a heart rate is needed to determine the significance of sinus bradycardia in a particular individual. Is the sinus bradycardia associated with symptoms? Is the sinus rate inappropriate for the level of physical activity? Can the sinus bradycardia be attributed to other causes? Does the sinus bradycardia occur in the setting of other atrial tachyarrhythmias such as atrial fibrillation?

In the evaluation of bradycardia, the presence or absence of symptoms is of paramount importance. If a patient's symptoms can clearly be attributed to a bradyarrhythmia, no further diagnostic evaluation is needed. At that point the

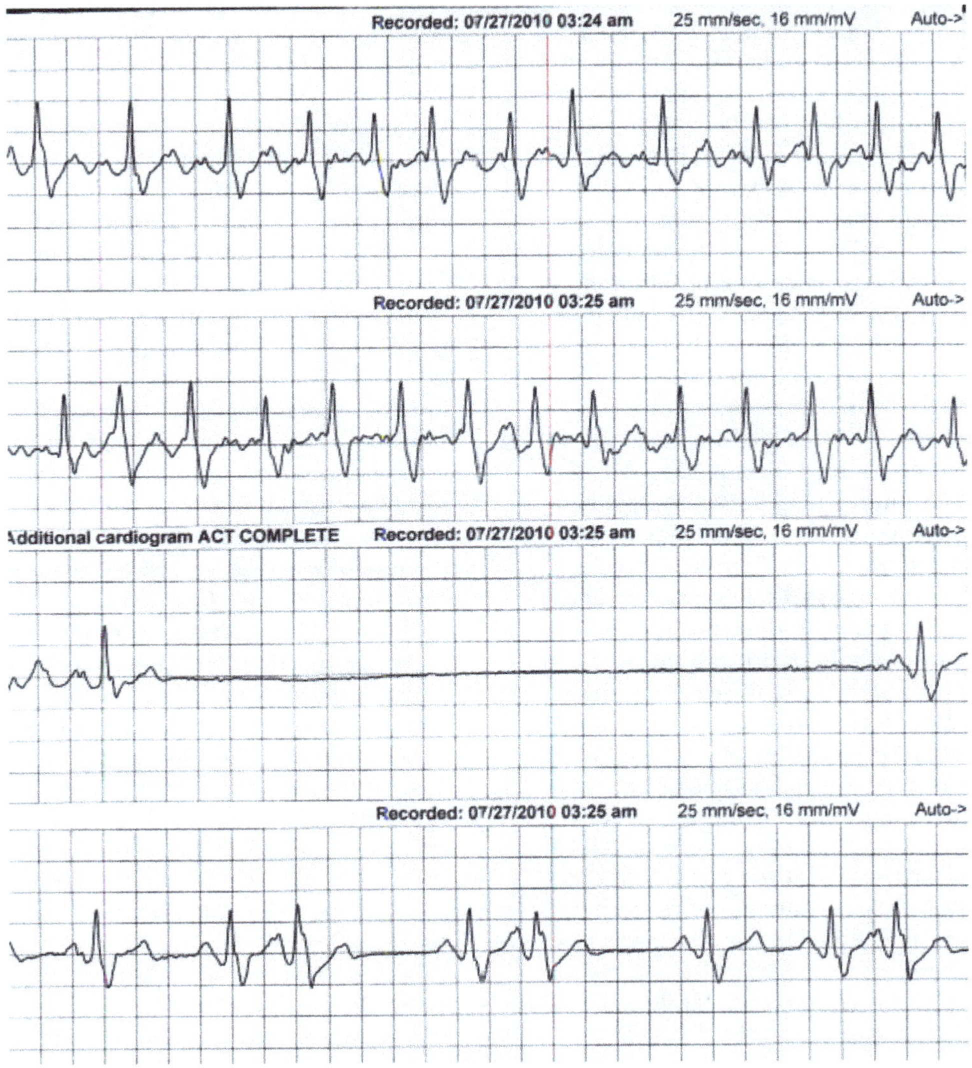

Figure 2.3. Bradycardia–tachycardia syndrome (BTS).

only pertinent question is what can be done to address the bradyarrhythmia.

Does Mr. R have symptoms? No. He reports feeling well. He has had no change in his exercise capacity. He is not fatigued. He does not have palpitations. He denies feeling unsteady, dizzy, or lightheaded. He states that he feels "perfect." What about his report of a syncopal episode? In fact, the description of that episode was classic for a vasovagal faint.

What evaluation, if any, is required in patients with asymptomatic sinus bradycardia? In general such patients, if truly asymptomatic, do not require evaluation. In Mr. R's case the need for surgery has raised concerns about the safety of surgery in the face of his sinus bradycardia. Granted he is asymptomatic but does he have SND? Is his heart rate in the 40s at rest because he is an aerobically fit man with high parasympathetic tone at rest or is his sinus node abnormal?

The sinus node is extensively innervated by both the parasympathetic and sympathetic nervous systems. Aerobically fit people have relative sinus bradycardia because at rest the parasympathetic nervous system input predominates. However, sinus bradycardia may be the result of intrinsic SND, or extrinsic factors (often termed "extrinsic" SND), particularly drugs. In any case, in health, the sinus node should respond normally to changes in autonomic tone. A withdrawal of parasympathetic tone or an increase in sympathetic tone will result in an increase in heart rate. How the sinus node responds to alteration in autonomic tone is critical in determining whether sinus bradycardia is a clinically important problem in a particular individual.

There are several methods of altering autonomic tone and studying its effects on sinus node function. The simplest test of this type is exercise. Physical activity results in withdrawal of parasympathetic tone and an increase in sympathetic outflow. The combined effect is an increase in the heart rate. Failure of the heart rate to increase appropriately in the face of increased metabolic demand is how we defined SND. If the inability to increase the heart rate appropriately is the primary manifestation of SND, the condition is referred to as "chronotropic incompetence." Ordering an exercise test was one possible answer in the question posed above. An exercise study allows a direct assessment of the ability of the sinus node to respond to physiological stress. In this case an exercise study is not needed. Mr. R underwent Holter monitoring that demonstrated a normal heart rate response to exercise making an exercise test unnecessary.

When a Holter monitor is used to evaluate heart rate response, patients must understand the importance of continuing to perform their usual activities and recording those activities in the diary provided. If the patient remains sedentary during the recording period, the Holter will tell us what we already know—the heart rate is low at rest. In sedentary patients, Holter monitoring often leads to a "the chicken or the egg" dilemma. Does the Holter demonstrate a low maximum heart rate because the patient is sedentary or is the patient sedentary because of the low heart rate? In these patients an exercise test is more useful.

When the goal of the exercise study is to evaluate heart rate response, low-level exercise is often sufficient. The exercise laboratory needs to be aware that the study is to assess heart rate response, not to assess for coronary artery disease. This is especially true in high-volume laboratories with a low threshold for converting exercise studies into pharmacological studies. If patients under evaluation require ongoing beta-adrenergic blocker therapy (coronary disease or congestive heart failure), the exercise study should be done while they remain on their baseline pharmacological regimen. The question is not what the patient's heart rate would be off beta-adrenergic blockers but what is the heart rate response to exercise in the presence of needed pharmacological therapy? In patients receiving needed beta-adrenergic blocker therapy, a blunted heart rate response to exercise in association with symptoms is an indication for permanent pacemaker placement.

Various definitions for chronotropic incompetence have been proposed. However, for practical purposes, the diagnosis can be made if there is failure of the exercise heart rate to exceed 65% of the predicted maximum heart rate (220-age) or failure to increase the heart rate to > 100 bpm. It is important to bear in mind that a blunted heart rate response to exercise is only meaningful if it occurs in a patient with symptoms of exercise intolerance, easy fatigability, or other symptoms suggestive of chronotropic incompetence. As a rule there is no reason to send a truly asymptomatic patient with resting sinus bradycardia for an exercise study.

Although not listed as an option in the question above, exercise is not the only means of altering autonomic tone. Autonomic tone can also be manipulated pharmacologically. Atropine given at a dose of 1 to 3 mg should result in a heart of > 90 bpm or, at minimum, a 20% increase in the heart rate above the baseline rate. A blunted response to atropine would suggest SND. Likewise, isoproterenol can be infused to increase sympathetic stimulation of the heart. In normal patients isoproterenol given at dosages of 1 to 3 mcg/min increases the heart rate by at least 25% over baseline.

One can also examine the sinus rate in the presence of "complete" pharmacological autonomic blockade by administering atropine (0.04 mg/kg) and propranolol (0.2 mg/kg). The heart rate after complete autonomic blockade is termed the "intrinsic heart rate" (IHR). IHR declines in an age-dependent fashion. This age dependency can be estimated with the following regression equation:

$$\text{IHR} = 117.2 - (0.57/\text{age})$$

In the presence of complete autonomic blockade, an IHR markedly lower than predicted would suggest that the sinus bradycardia seen clinically is related to an intrinsic disorder of the sinus node. Pharmacological assessment of sinus node autonomic responsiveness requires intravenous access and should be performed in a monitored setting with resuscitation equipment readily available.

The 2 options offered in the question above that have not yet been discussed are a 30-day event recorder or an ILR. External 30-day loop recorders are extremely useful in correlating patient symptoms with arrhythmias, but in an asymptomatic patient there are no symptoms to correlate. Likewise, ILRs are very useful in the evaluation of patients with *unexplained* syncope, but Mr. R does not have unexplained syncope. His solitary episode was a typical vasovagal syncope. Either type of loop recorder is likely to demonstrate that he has sinus bradycardia, a finding of which we are already aware.

To summarize the case to this point, Mr. R reports no exercise intolerance. He does report a syncopal event, but careful questioning identifies that event as a vasovagal faint. Given the typical description of vasovagal syncope and the fact it was a solitary event, a tilt table study was not felt necessary. His Holter monitor recording demonstrates a normal heart rate response to exercise, with only rare atrial ectopy and no sustained atrial arrhythmia. No sinus pauses were seen. Further, he is not taking any sinus node suppressing medications. Consequently, his sinus bradycardia is best attributed to elevated parasympathetic tone related to regular aerobic exercise. For these reasons option E, no further evaluation, was recommended.

Mr. R underwent his surgical procedure without complication. One year later an exercise study was ordered as part of his medical certification to maintain his pilot's license. He exercised on a Bruce protocol for 13:08. The maximum heart rate was 147 bpm, 91% of predicted. He reported no symptoms during exercise. The accompanying nuclear perfusion scan was normal.

Two years after his initial electrophysiologic consultation, he had an abrupt loss of consciousness. He was sitting on the side of his bed undressing after work. His wife reports he was unresponsive for several minutes. She had time to call 911 before he regained consciousness. She did not note any diaphoresis or pallor. When the paramedics arrived he was awake and alert. He sustained no injuries and was reluctant to be transported to the emergency department.

In the emergency department he reported no prodromal symptoms. He reported there were no associated symptoms with this syncopal event. Since he was last seen in the clinic he had remained active. He reports no change in his exercise capacity. He states that the current episode is in no way similar to the syncopal episode he had in 2001. No positive findings were reported on the review of systems. The physical examination was remarkable for a resting heart rate of 46 bpm. The oxygen saturation was 98% on room air. The 12-lead ECG demonstrated sinus bradycardia and is unchanged when compared with previous tracings. All intervals were within normal limits. Electrolytes, creatinine, and hemoglobin were normal. His initial troponin was normal. A computed tomography scan of the chest showed no evidence of pulmonary emboli.

The lack of prodromal symptoms does not exclude the possibility of vasovagal syncope, which it was thought that he had in the past. However, his report that this episode was unlike his prior episode of syncope is concerning. Although witnesses often overestimate the length of time a person is unconscious, the fact that Mr. R's wife had time to locate a phone, call 911 and talk with the emergency operator before he became responsive suggests a more prolonged episode than would be expected in vasovagal syncope. This is especially true given he was in a supine position immediately after the loss of consciousness.

The lack of a clear explanation for the current syncopal episode resulted in Mr. R being admitted to a monitored unit. He reported no symptoms overnight. The lowest recorded heart rate was 46 bpm. Figure 2.4 is a representative telemetry recording obtained while he was sleeping.

Patients with SND in whom pauses in the cardiac rhythm occur due to SA exit block may show evidence of this abnormality (SA block) on the surface ECG. Depolarization of the sinus node is not evident on the surface ECG, but inferences about SA block can be made through careful analysis of the surface ECG.

SA block can be described in terms analogous to AV conduction block. In the absence of an ECG marker of sinus node depolarization, one cannot observe SA conduction directly or measure SA conduction time. For this reason, there is no ECG pattern of SA block equivalent to first-degree AV block. Likewise, there is no ECG pattern of SA block comparable to third-degree AV block or complete AV block. On the surface ECG, "third-degree" SA block would result in prolonged atrial asystole. It would be impossible from the surface ECG to determine whether the resulting pauses were secondary to SA block or from a failure of sinus node automaticity (although, as discussed below, if the asystole is a precise whole number multiple of the previous sinus cycle length [SCL], one might infer "exit block" rather than depressed automaticity). The surface ECG permits only the 2 types of second-degree SA block to be described, Type I and Type II.

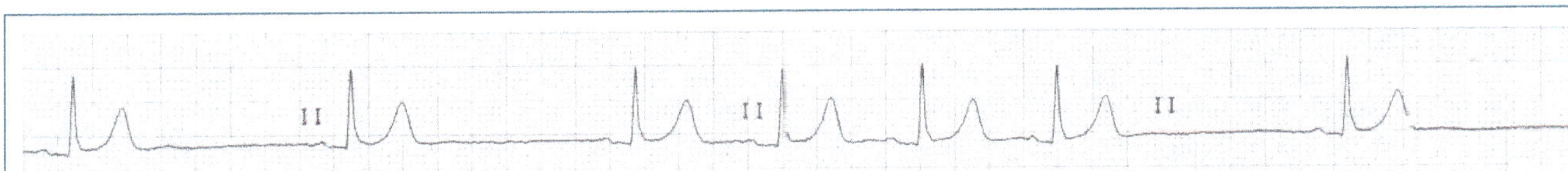

Figure 2.4. Rhythm strip with sinus pause.

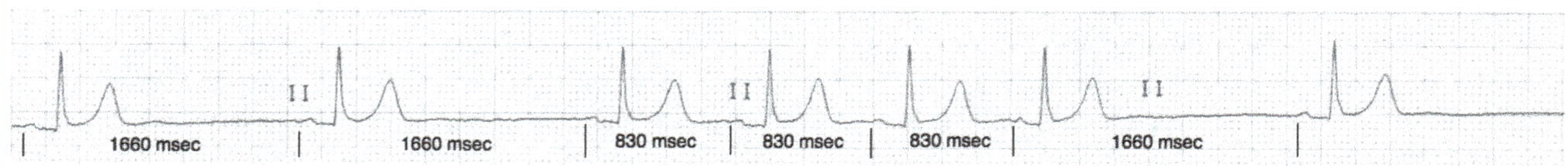

Figure 2.5. Type II SA block.

Type II SA block is the easiest to diagnosis from the surface ECG. If a sinus pause is noted and the length of the pause is a whole number multiple of the SCL, Type II SA block can be inferred as illustrated in Figure 2.4. In the example shown in Figure 2.5, the atrial pauses are essentially equal to 2 times the SCL. This suggests an intermittent and unanticipated failure of sinus node depolarization to be conducted to the atrium similar to the unanticipated failure of AV conduction in Type II second-degree AV block. Perfect whole number multiple differences are not expected as the sinus node tends to have some inherent variability. One can assume exit block if the relation is "close" to a whole number multiple.

Type I SA block can also be identified from the surface ECG. As with typical Type I second-degree AV block there is a pattern to the failure of conduction, "group beating" associated with a distinctive shortening of the conduction intervals of the tissue *below* the level of block. In classic Type I second-degree AV block, the conduction time, the PR interval, prolongs with each subsequent beat until block occurs. However, the incremental increase in the PR interval lessens with each subsequent beat resulting in what at first may seem a paradoxical decrease in the subsequent RR intervals.

Understanding the types of AV block is relatively straightforward. We have ECG markers of activation of both the tissue above the AV node, the P wave, and the tissue below the AV node, the QRS complex. In other words, we have ECG markers of both the input into the AV node and the output from the AV node. This is not the case with the sinus node. A surface ECG records only the output of the SA node, the P wave.

In AV block, as the PR intervals are prolonged, the subsequent *R* waves are delayed. In SA block, if the SA conduction time is prolonged, the subsequent *P* wave is delayed. Once again, as is seen in AV block, the incremental increase in SA conduction time is less with each beat, resulting in progressive shortening of the PP intervals. As in Type I second-degree AV block, Type I SA block is characterized by group beating with a progressive decrement of the PP intervals (analogous to the decrementing RR intervals in Type I AV block) followed by complete SA exit block, the atrial pause.

A review of all the telemetry data did not demonstrate any atrial pauses greater than the one shown in Figure 2.5. No atrial pauses were recorded while he was awake and active. Serial troponin measurements were normal. At this point, we are faced with a 59-year-old gentleman with a history of sinus bradycardia felt to be secondary to aerobic fitness, a remote vasovagal faint who now presents with a new syncopal event that clinically does not appear to be a vasovagal event.

Question No. 2:

You would recommend:

A. An electrophysiology study (EPS).
B. Discharge with an event recorder.
C. Dual-chamber pacemaker placement.
D. Discharge with mobile cardiac outpatient monitoring.
E. Implantable loop recorder.

Symptomatic bradycardia is a Class I indication for permanent pacing in the setting of SND.[2] Unfortunately, at this point Mr. R has only been documented to have *asymptomatic* bradycardia. A clear relationship between his symptom, syncope, and the bradycardia has not been established. Permanent pacemaker placement in patients with SND and heart rates less than 40 bpm is considered "reasonable" (Class IIa indication). Mr. R has not had documented sinus bradycardia of less than 40 bpm. The ACC/AHA/HRS guidelines include demonstration of "clinically significant abnormalities of SND" at EPS in patients with unexplained syncope as a Class IIa indication for permanent pacing.

Recall that SND may be the result of failure of sinus node automaticity or from an inability to conduct the sinus node impulse to the atrium. In the absence of cardioactive drugs, the health of sinus node automaticity can be estimated through determination of the sinus node recovery time (SNRT); this measurement is obtained by pacing the atrium at various cycle lengths for 30 to 60 seconds and measuring the time from the last paced beat to the first spontaneous beat to determine SNRTs.

Secondary pauses, pauses that occur after the first spontaneous beat, are also considered significant. A normal maximal SNRT is typically < 1500 ms. SNRT should be measured at a variety of cycle lengths in order to determine a maximum value. Apart from the duration of the SNRT alone, it is abnormal for the maximum value to be obtained at pacing rates < 100 bpm (ie, cycle lengths > 600 ms).

SNRTs are influenced by the baseline autonomic state, as is the resting heart rate. Subtracting the baseline SCL from the SNRT gives the corrected SNRT (CSNRT) and improves the sensitivity and specificity of the study. A normal CSNRT is < 550 ms. Even with correction, the SNRT, while reasonably specific, lacks sensitivity in identifying SND. Further, a normal SNRT or a normal CSNRT does not exclude the presence of clinically significant SND.

The following tracings are not from Mr. R. They are from an 89-year-old patient with unexplained episodes of decreased responsiveness and unexplained falls. Review of her medical records revealed that her typical resting pulse has been in the 40 to 50 bpm range. Holter monitoring demonstrated sinus bradycardia and occasional nocturnal pauses, the longest 2.8 seconds. After a long discussion with her family, the patient elected to proceed with an EPS to assess sinus node function. SNRT was measured as part of the EPS. The longest SNRT is shown in Figure 2.6.

Note that she does not have 1:1 AV conduction while being paced at 150 bpm. That is not an especially concerning observation in an elderly patient and does not affect determination of the SNRT. The uncorrected SNRT is given above and is quite prolonged at 2500 ms (see Figure 2.5). The patient's SCL was 1350 ms (44 bpm). Consequently, the CSNRT is also quite prolonged:

$$\text{CSNRT} = 2500 - 1350 = 1150 \text{ ms}$$

Determination of the SNRT allows for an estimation of the normalcy of sinus node automaticity. The health of the sinus node's ability to conduct impulses to the atrium is measured by determining the sinoatrial conduction time (SACT), a maneuver designed to evaluate the ability of impulses to conduct into and out of the sinus node. There are several different methods described for measuring SACT. The most widely known (so-called Strauss method) requires the introduction of a premature beat near the sinus node during sinus rhythm. Depending upon the timing of the premature impulse, several patterns of response have been described and are commonly referred to as "zones of collision." Two of the zones of collision are diagrammed in Figure 2.7. Premature impulses delivered late in diastole collide with the sinus node impulse and are unable to penetrate and reset the

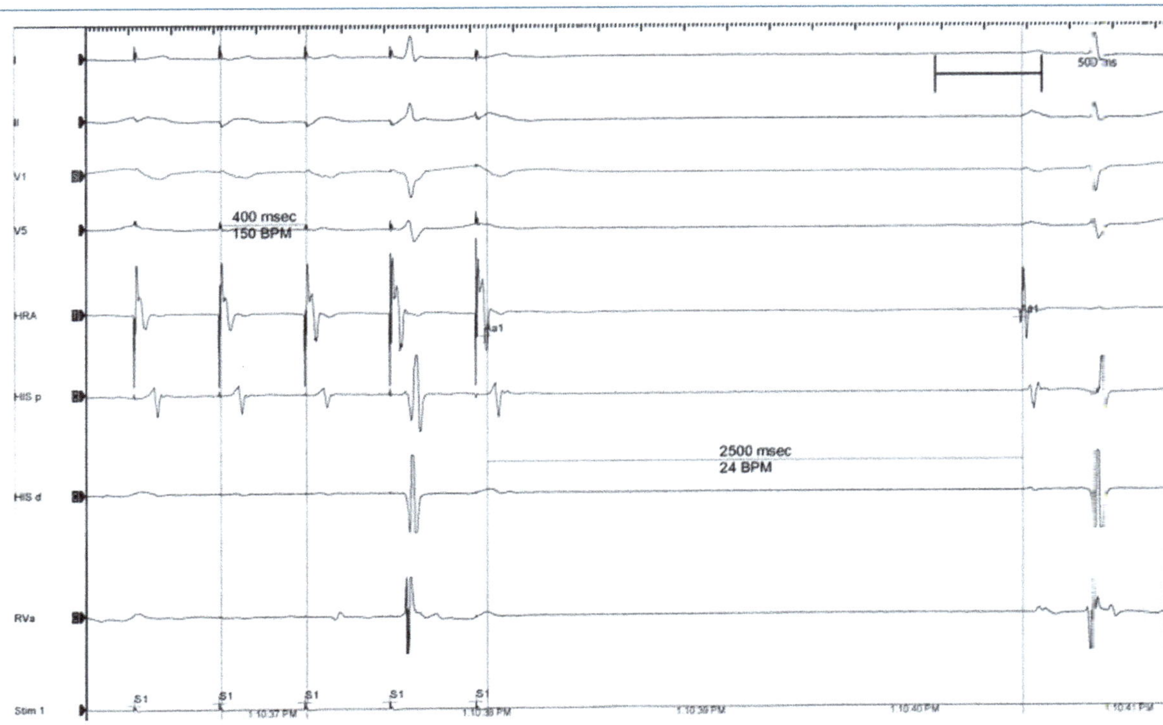

Figure 2.6. Evaluating SNRT.

sinus node. In that circumstance, referred to as Zone 1, the subsequent sinus impulse is unperturbed and the premature beat is fully interpolated. A premature beat delivered earlier in diastole may conduct into the sinus node and reset it, Zone 2. After resetting, the next sinus impulse will occur at the basic SCL *assuming* the premature beat did not result in suppression of the sinus node. This impulse will then need to be conducted through the perinodal region to excite the atrium. Thus, the time from the premature impulse to the next spontaneous sinus beat consists of the time it takes for the impulse to conduct into the node plus the basic SCL plus the time it takes for the sinus impulse to exit the node:

SACT = (return cycle length – sinus cycle length)/2

It is apparent that for the SACT to be valid 3 assumptions must be true. The first is that the basic SCL is relatively fixed at the time of the testing. The second is that the premature beat does not substantially overdrive or suppress sinus node automaticity, and the third is that the time it takes for an impulse to enter the sinus node is the same as the time it takes for an impulse to exit the sinus node (thus the division by "2" in the above equation).

SACT can only be determined from premature beats that are delivered in Zone 2 (reset zone). Other zones of collision can be seen with premature beats delivered even earlier in diastole but have little relevance to this discussion. Fortunately, premature beats delivered over a large segment of the diastolic interval demonstrate Zone 2 collisions, making determination of the SACT relatively straightforward.

SACT was determined during the EPS using the Strauss method (Figure 2.8). The baseline SCL is 1155 ms. The interval from the paced atrial

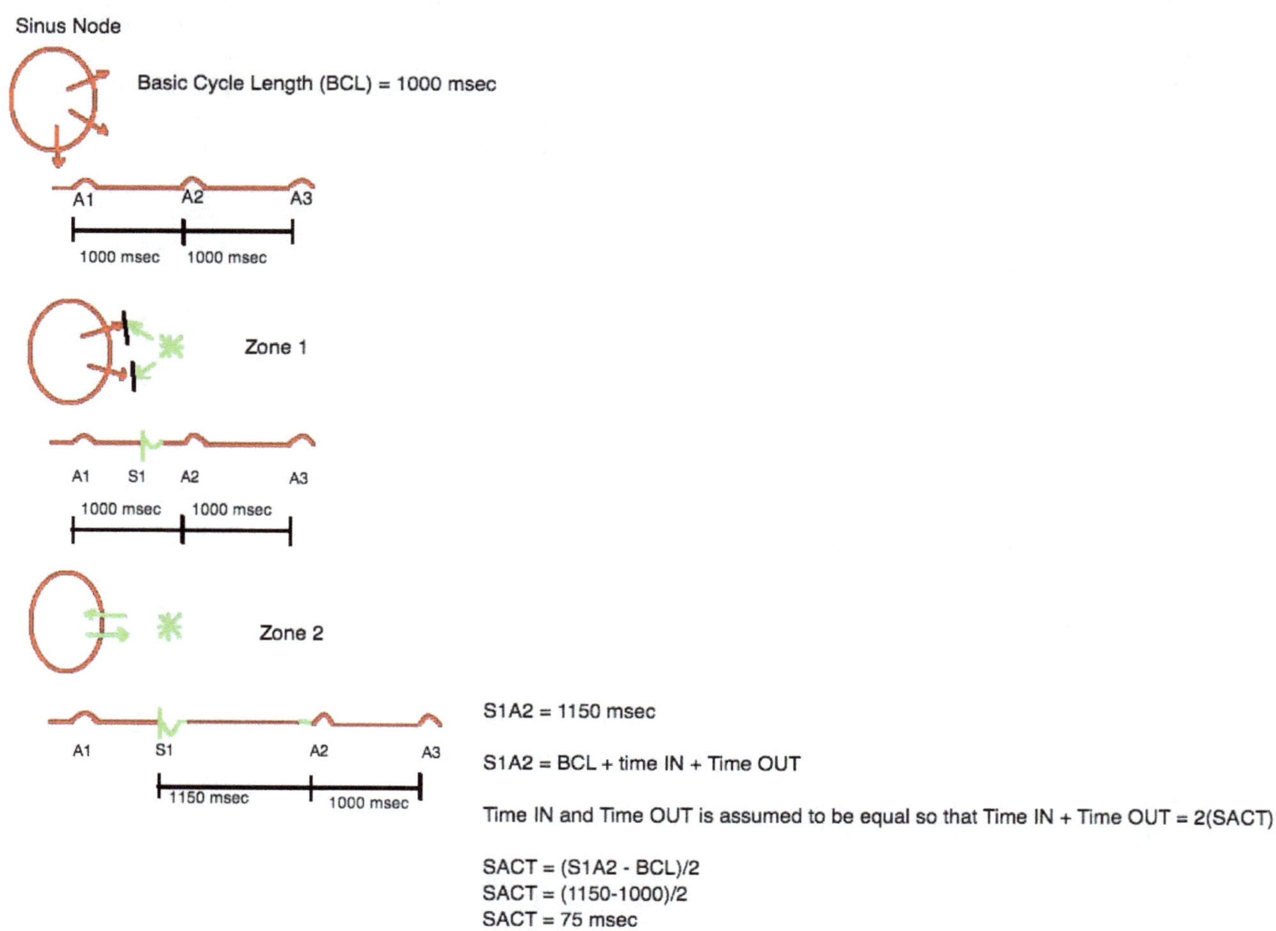

Figure 2.7. SACT diagram.

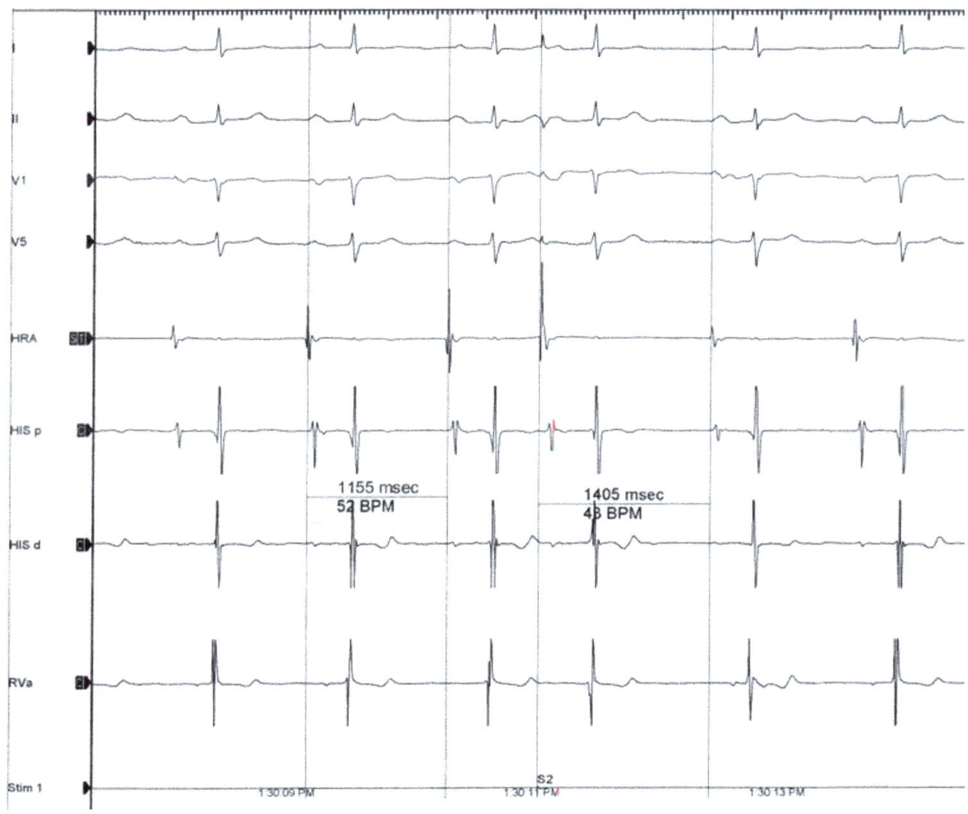

Figure 2.8. Example of SACT.

premature beat to the next spontaneous atrial beat, the return interval, is 1405 ms.

$$
\begin{aligned}
\text{SACT} &= (\text{Return interval} - \text{SCL})/2 \\
&= (1405 - 1155)/2 \\
&= 250\ \text{ms}/2 \\
&= 125\ \text{ms}
\end{aligned}
$$

Finally, SACT can be estimated indirectly by assessing the delay following a brief period of overdrive pacing at a rate just above the sinus rate (so-called Narula method), or directly by recording of the sinus node electrogram (EGM) using specialized recording techniques. In the latter case, the time from the onset of the EGM to the local atrial activation can be measured directly. Unfortunately, recording the sinus node EGM is a tedious procedure that is only very rarely undertaken outside of research laboratories.

A range of normal SACT values has been reported. In general, SACT measured by the methods most commonly used are considered abnormal if > 125 ms.

Overall, the principal weakness of invasive evaluation of sinus node function (ie, SNRT, CSNRT, and SACT) is poor sensitivity. In a review of these procedures, the calculated sensitivities for SNRT and CSNRT ranged from 18% to 69% with a combined sensitivity of 45%.[3] The combined sensitivity for SACT was 51%, and ranged from 29% to 75%. Several of the 8 studies reviewed had small number of patients and these values should be interpreted with caution. However, the specificity of SNRT and SACT determinations is high enough to be clinically useful. The total specificity for SNRT/CSNRT was 95%. The specificity for SACT was lower at 74%.[3]

Due to their lack of sensitivity and poor predictive value, EPS is not routinely done as a part of the evaluation of SND. The key to the clinical diagnosis of SND is a correlation of symptoms with the bradycardia. An EPS is not indicated in asymptomatic sinus bradycardia. Similarly, testing is not needed if an association has already been made between the patient's rhythm and the clinical symptoms. Despite these limitations, EPS can be helpful at times in symptomatic patients such as the patient studied above.

Question No. 3:

An EPS is reasonable in patients when there is:

A. Failure to obtain ECG correlation with spontaneous episodes.

B. Sufficient concern regarding potential injury during prolonged monitoring for ECG correlation during spontaneous episodes.

C. Both A and B.

The patient and family discussed above were very concerned about potential injury if her "falls" were in fact related to her bradycardia. Unfortunately, unlike the monitor recording shown in Figure 2.3, her Holter monitor did not provide a diagnosis. Prolonged monitoring with a 30-day event recorder or an ILR would be feasible but for the patient's concerns about injury. In patients such as this, EPS can be helpful. The patient needs to understand the limitations of the study. The poor sensitivity of the EPS for this disorder means a "normal" study does not exclude SND. We typically recommend proceeding with placement of an ILR if the EPS fails to provide a diagnosis. In this case the study was clearly abnormal. The SNRT and CSNRT were both markedly prolonged. The SACT was also prolonged. Based on these sinus node findings and evidence (not shown) of poor AV conduction, in the context of the patient's presentation, permanent dual-chamber pacing was recommended.

In patients in whom there is less concern of injury, we typically recommend placement of an ILR in order to try and obtain ECG-symptom correlation. Recent studies suggest that this approach may be the most cost-effective. In patients with syncope and ejection fractions > 35%, ILR placement has been shown to have a better diagnostic yield and be more cost-effective than conventional diagnostic approach that included invasive EPS.[4,5]

Recordings from ILRs need to be interpreted with some degree of caution. If Mr. R receives an ILR and has another episode of vasovagal syncope, the device will very likely demonstrate the episode to be associated with marked bradycardia, even asystole. Yet that type of neurally mediated bradycardia would not necessarily respond to pacing. Sinus bradycardia and asystolic pauses have to be interpreted in the overall context in which they occur. Reflex-mediated ("vagal") pauses are often first preceded by an increase in the sinus rate followed by rather abrupt slowing. A typical reflex-mediated atrial pause is shown in Figure 2.9. Asystolic pauses secondary to sinus node arrest or SA block are not typically preceded by a change in the sinus rate. Standardized guidelines for the interpretation of ILR tracings have recently been proposed.

In Mr. R's case, recommending placement of an ILR would be reasonable. The risk of sudden death related to SND is low. Mr. R would need to be cautioned regarding activities, especially driving. Driving restrictions in the setting of syncope vary from state-to-state. It is prudent to understand the law in the state in which one practices. Mr. R, while no longer an airline pilot, does fly privately. The Federal Aviation Administration (FAA) rules regarding syncope are quite strict. He certainly may not continue to fly as pilot-in-command until the etiology of his syncope has been identified and treated.

Proceeding with pacemaker placement assuming that his syncope "had" to be related to bradycardia is a less-acceptable option. Syncope in SND is relatively uncommon, other than in the setting of the BTS. In that syndrome the pause following cessation of the atrial tachyarrhythmia, usually atrial fibrillation, may be associated with presyncope or syncope. Other forms of SND more commonly present with nonspecific symptoms of fatigue and exercise intolerance. Mr. R has never had an atrial tachyarrhythmia identified. It is possible he has BTS, but that has not been established.

An event recorder or mobile cardiac outpatient monitoring might identify asymptomatic pauses or AV block, but the best indication for pacing is the identification of a correlation between the syncopal event and nonvagally mediated bradycardia. In that sense, outpatient monitoring is most helpful when patients have symptoms while being monitored. Mr. R has had 2 syncopal episodes over the course of 6 years. Even a 30-day monitor is unlikely to record a symptomatic episode.

Fortunately for his physicians, before a decision had been made regarding ILR placement versus EPS, he had a syncopal episode while on telemetry. A review of his telemetry data revealed a 12-second pause. Prior to the pause there had been no significant change in the sinus rate, suggesting that this was not a vagally mediated event but intrinsic SND. Fortunately, he was in bed at the time of the event and sustained no injuries. The following recording is not from the patient being discussed.

Contrast the tracing shown in Figure 2.9 with that shown in Figure 2.8, which demonstrated a vagally mediated sinus pause. In this case there is no evidence of sinus slowing prior to the pause. The pause in Figure 2.10 cannot be attributed to autonomic influence and is strongly suggestive of SND.

Mr. R conveniently demonstrated that he had symptomatic (syncope) bradycardia secondary to

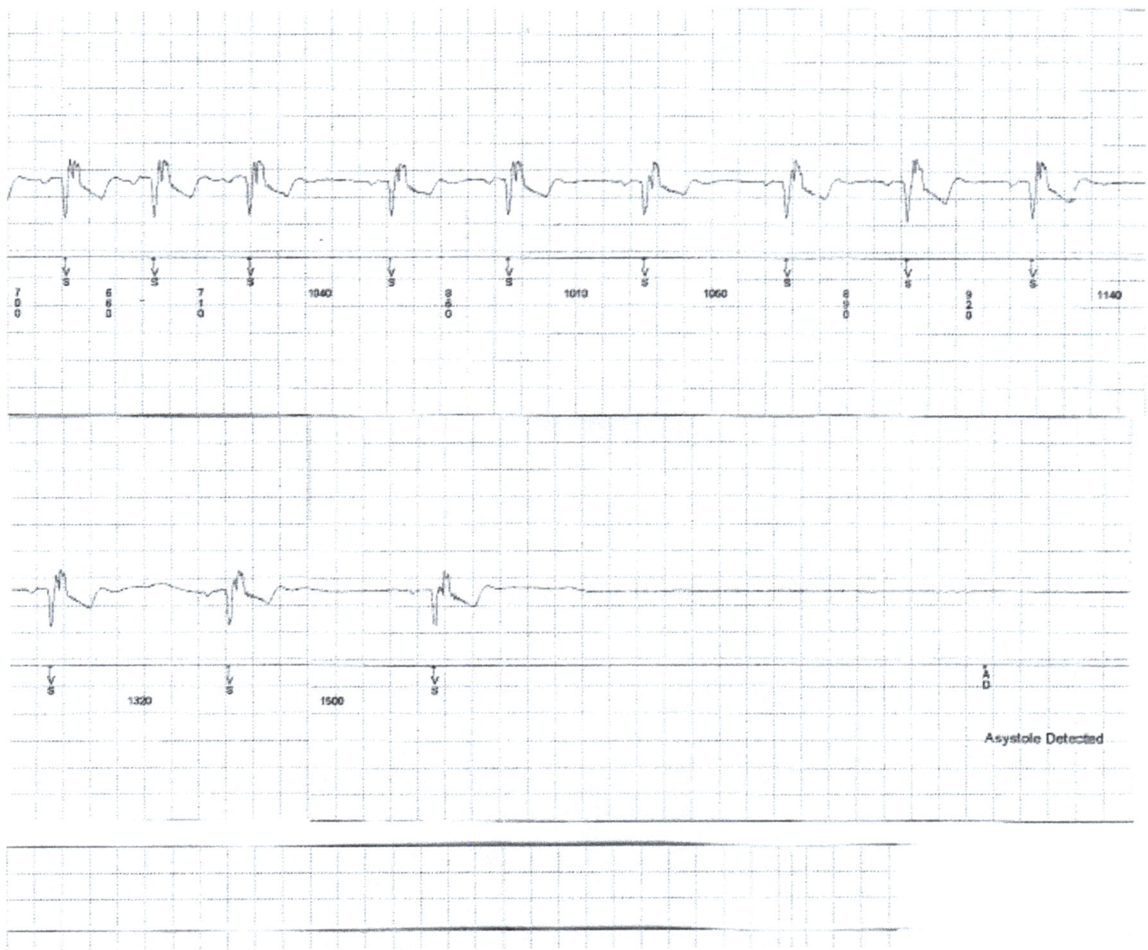

Figure 2.9. Implantable loop recording of vasovagal faint.

SND. That is a Class I indication for pacing, which is what was recommended to him. He received a dual-chamber permanent pacemaker. In 3 years of follow-up he has had no further syncopal episodes. Whether a single-chamber ventricular pacemaker would have been comparably effective cannot be stated with certainty in this case. However, several avenues of evidence indicate that "atrial-based" pacing (ie, atrial single-chamber or more commonly dual-chamber) are beneficial for long-term reduction of atrial fibrillation and stroke risk in SND patients.[7] Further discussion of this issue is beyond the scope of this chapter.

SUMMARY

SND is a common entity, especially in the elderly. Given the normal age-related changes in sinus node function, it can be challenging to distinguish SND from normal aging. Compounding this difficulty is the propensity for patients with SND to present with nonspecific complaints. Asymptomatic sinus bradycardia rarely requires additional evaluation. Demonstrating normal sinus node response to changes in autonomic tone is extremely useful when attempting to determine pathological versus physiological sinus bradycardia and is most easily accomplished with exercise. EPS evaluation is of limited utility in evaluating SND. The need for pacing is primarily based on a demonstration of a correlation between the patient's symptoms and the particular arrhythmic manifestation of SND (ie, tachycardia, bradycardia, or both). There is increasing evidence that the early placement of an ILR offers superior diagnostic yield and lower overall cost compared to conventional approaches.

Correct Answers:

Q1. E

Q2. E

Q3. C

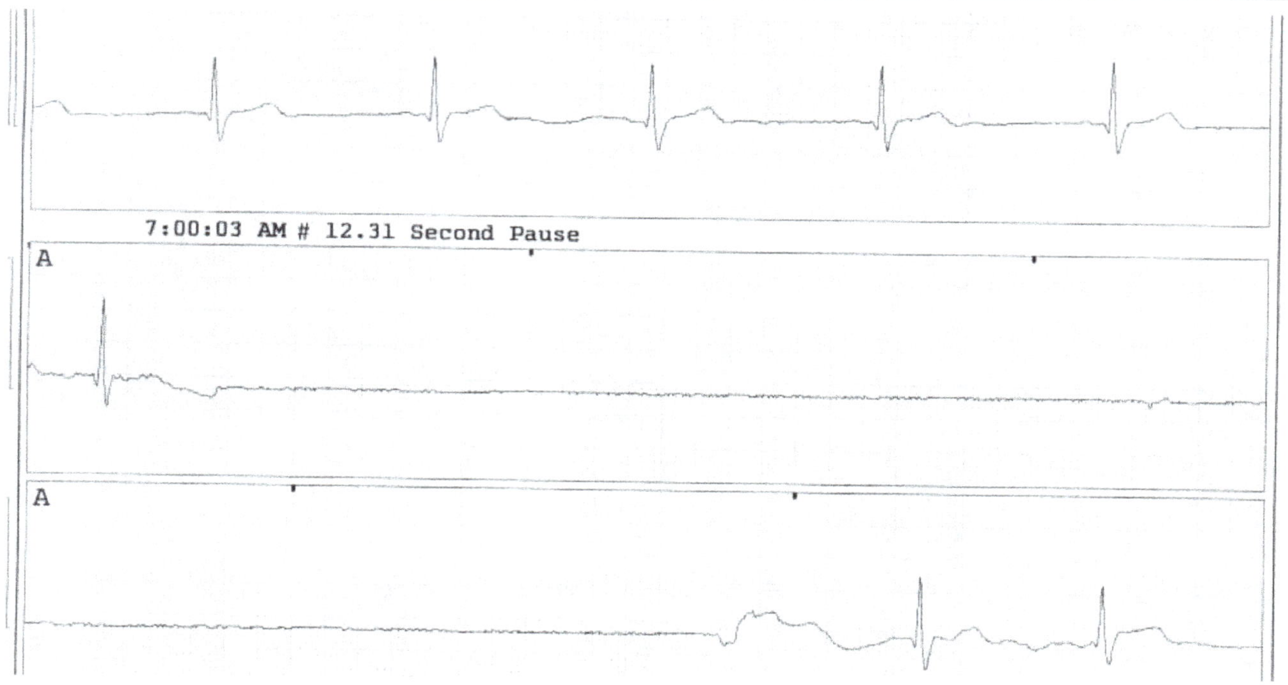

Figure 2.10. Sinus node pause.

QUESTION FROM THE EDITORS AND AUTHORS' REPLY:

Question 1. What are the other diseases of the sinus node?

Answer: Inappropriate sinus tachycardia (IST) and sinus node reentrant tachycardia (SNRe-T) are often not considered when discussing SND yet both clearly involve the sinus node. IST is an especially difficult arrhythmia to manage. It is a disorder that affects principally young women who present with resting sinus rates > 100 bpm and rapid increases in the sinus rate with minimal exertion. Before this condition can be diagnosed one must ensure that the sinus tachycardia is in fact *inappropriate.* Normal volume status, normal hematocrit, and normal thyroid status must be confirmed through appropriate physical exam and laboratory assessment.

Clinical Case

A 22-year-old female was seen in the Arrhythmia Center for consultation regarding persistent tachycardia. A TSH and CBC were normal. A 24-hour Holter demonstrated an average heart rate of 105 bpm and a maximum heart rate of 156 bpm. At the time of the maximum recorded heart rate the patient noted she had been climbing a flight of stairs. On physical exam she was determined to be euvolemic. Her cardiac exam was normal other than a mildly tachycardic rhythm. The striking finding on her examination was a body mass index (BMI) of 45. She reported being very sedentary. Climbing the stairs to the upper floor of her home was the most strenuous activity she described.

This clinical case highlights the high degree of overlap between IST and de-conditioning. One could argue that this young woman is sedentary because of symptoms related to IST but it is equally likely that the heart rate abnormalities noted on her Holter monitor are indicative of extreme de-conditioning.

Other entities that can be difficult to distinguish from IST unless care is taken are: (1) incessant ectopic atrial rhythms arising from the superior right atrium or right superior pulmonary vein; (2) sinus tachycardia related to postural tachycardia syndrome; and (3) SNRe-T. The latter 2 conditions are rarely persistent but can be impossible to exclude if one has only isolated ECG tracings available. Tachycardias originating from the right superior pulmonary vein can inscribe P waves that can be difficult to discern from sinus P waves, although again those tachycardias are not typically incessant.

The difficulty of caring for patients with IST does not end with the establishment of the diagnosis. Treatment for IST is difficult to say the least. Various approaches to ablation therapy have been attempted. Unfortunately, despite initial success there is an extremely high incidence of recurrence of symptoms within a few months of the procedure. Why this is the case is unclear but one possibility is that lower, previously subsidiary pacemaker sites, become predominant and display the same abnormality as the original pacemaker sites that were ablated. As we have discussed the sinus node extends for a variable distance along the CT. Ablation is focused on the site of earliest activation in this region. Frequently as the ablation procedure progresses the earliest activation site simply moves more inferiorly along the CT. Aggressive ablation can lead to significant sinus bradycardia and the need for pacemaker placement. Care must also be taken to avoid injury to the phrenic nerve. For all these reasons ablation therapy for IST is reserved for highly symptomatic patients who fail to respond adequately to medical therapy.

Medical therapy for IST is frequently vexing for the patient and physician. The standard approach has been beta-adrenergic blockade. This meets with limited success. Often very high doses are required and patients develop complaints attributable to the effects of the medication. In this regard, it is important to keep the goal of therapy firmly in the forefront of one's mind. The goal is to reduce the patient's symptoms not their heart rate. If on 100 mg of metoprolol succinate daily the patient reports a reduction in their palpitations and an improvement in their exercise abilities there is no particular need to increase the dose to 200 mg daily just because the resting heart rate in the clinic is still 96 bpm. On the other hand, beta-blocker side effects are often a major limiting factor in the quest for long-term heart rate control. The importance of hydration and adequate Na in the diet cannot be over emphasized to these patients. Another nonpharmacological approach that should be emphasized is exercise, preferably via a cardiac rehabilitation program if one is available and the patient qualifies for the service. It is important to stress to the patient that they are likely to continue to have heart rates that are "higher" than one would expect for a given level of exertion. It is easy for patients to become discouraged when they do not see the type of response they expect. Encouragement and management of patient expectations is a critical part of treatment.

The underlying mechanism(s) responsible for IST remain unclear which makes formulation of treatment strategies difficult. Recently, interest has focused on the use of ivabradine to treat IST. Experiential reports suggest that this agent may offer a valuable treatment option. Recall that ivabradine inhibits I_f and directly slows the heart rate. Ivabradine has been shown in IST patients to reduce the resting heart rate but more importantly to improve exercise time and quality of life measurements.[2] At this time ivabradine is not available in the United States.

Sinus node reentrant tachycardia is an arrhythmia in which reentry within the sinus node or sinus node and adjacent atrial tissue is felt to be the mechanism of the arrhythmia. Sinus node reentry should have a surface P wave that is identical to the sinus P wave, typically it is a long RP tachycardia with both abrupt initiation and termination. In the EP lab it is induced with an extra-stimulus, one of the reasons it is felt to be a reentrant arrhythmia. The reported incidence of SNRe-T varies greatly. In our experience it is an uncommon arrhythmia. Patients with reentrant arrhythmias such as AV nodal reentrant tachycardia or AV reentrant tachycardia do not have a higher incidence of underlying organic heart disease. This is not true of SNRe-T. Patients with SNRe-T often have underlying heart disease. As with IST it is important to exclude an ectopic atrial arrhythmia arising from the region of the sinus node. SNRe-T has been treated with ablation therapy. Response to therapy with verapamil, digoxin, beta-adrenergic blockade, and adenosine are inconsistent.

In summary, SND is a common entity especially in the elderly. Given the normal age-related changes in sinus node function it can be challenging to distinguish SND from normal aging. Compounding this difficulty is the propensity for patients with SND to present with nonspecific complaints. Asymptomatic sinus bradycardia rarely requires additional evaluation. Demonstrating normal sinus node response to changes in autonomic tone is extremely useful when attempting to determine pathological versus physiological sinus bradycardia and is most easily accomplished with exercise. EPS evaluation is of limited utility in evaluating SND. The need for pacing is primarily based on a demonstration of a correlation between the patient's symptoms and the particular arrhythmic manifestation of SND (ie, tachycardia, bradycardia, or both). There is increasing evidence that the early placement of an ILR offers superior diagnostic yield and lower overall cost compared to conventional approaches. Care must be taken to distinguish IST from other atrial arrhythmias with similar P-wave morphology and a physiological cause for sinus tachycardia must be excluded. Treatment for IST is only partial successful, time-consuming, and requires a great deal of encouragement from the physician and a motivated patient. SNRe-T is a rare arrhythmia that can be treated with ablation when identified.

REFERENCES

1. Keith A, Flack MW. The form and nature of the muscular connections between the primary divisions of the vertebrate heart. *J Anat Physiol.* 1907;41:172–189.
2. Dobrzynski H, Boyett MR, Anderson RH. New insights into pacemaker activity: promoting understanding of sick sinus syndrome. *Circulation.* 2007;115:1921–1932.
3. Epstein AE, DiMarco JP, Ellenbogen KA, et al. ACC/AHA/HRS 2008 Guidelines for Device-Based Therapy of Cardiac Rhythm Abnormalities: a report of the American College of Cardiology/American Heart Association Task Force on Practice Guidelines (Writing Committee to Revise the ACC/AHA/NASPE 2002 Guideline Update for Implantation of Cardiac Pacemakers and Antiarrhythmia Devices): developed in collaboration with the American Association for Thoracic Surgery and Society of Thoracic Surgeons. *Circulation.* 2008;117:e350–e408.
4. Benditt DG, Gornick CC, Dunbar D, Almquist A, Pool-Schneider S. Indications for electrophysiologic testing in the diagnosis and assessment of sinus node dysfunction. *Circulation.* 1987;75(4 Pt 2):III-93–III-102.
5. Krahn AD, Klein GJ, Yee R, Hoch JS, Skanes AC. Cost implications of testing strategy in patients with syncope: randomized assessment of syncope trial. *J Am Coll Cardiol.* 2003;42:495–501.
6. Farwell DJ, Freemantle N, Sulke AN. Use of implantable loop recorders in the diagnosis and management of syncope. *Eur Heart J.* 2004;25:1257–1263.
7. Brignole M, Moya A, Menozzi C, Garcia-Civera R, Sutton R. Proposed electrocardiographic classification of spontaneous syncope documented by an implantable loop recorder. *Europace.* 2005;7:14–18.
8. Nielsen JC. DANPACE: the Danish multicenter randomised trial on single-lead atrial versus dual-chamber pacing in sick sinus syndrome. *Presented at the European Society of Cardiology Congress,* August 31, 2010.
9. Kaplinsky E, Comes RP, Rrondo LSV, Ayma FP. Efficacy of ivabradine in four patients with inappropriate sinus tachycardia: a three month-long experience based on electrocardiographic, Holter monitoring, exercise tolerance and quality of life assessments. *Cardiol J.* 2010;17:166–171.
10. Gomes JA, Mehta D, Langan MN. Sinus node reentrant tachycardia. *PACE.* 1995;18:1045–1057.

CASE STUDY

3

Evaluating Transient-Loss of Consciousness and Suspected Neurally Mediated Reflex Syncope

David G. Benditt, MD and Wayne O. Adkisson, MD

A 37-year-old otherwise healthy woman presented to the syncope/arrhythmia clinic having been referred for consultation after a spell of transient-loss of consciousness (T-LOC). The most recent episode occurred at work. She is employed doing computer-based design and had been facing a major product deadline. The T-LOC occurred with no apparent warning; however in retrospect, she recalled she had been standing beside her desk looking at graphics and remembered feeling warm and noting nonspecific gastrointestinal upset. Witnesses found her collapsed on the floor beside her desk. They noted some "jerky movements" of her arms after her collapse, and estimated that she was unresponsive for < 1 minute. Thereafter, she responded when her name was called. Co-workers reported her to be sweaty, and it was evident that she had sustained a minor laceration on her forehead and had lost bladder control.

An emergency medical team arrived within a few minutes, and the patient was taken to a nearby emergency department (ED). Apart from documenting the episode history summarized above, the patient's past medical history revealed "one or two faints" as a child, but none in the last 25 years. The family history was unremarkable, and the patient was taking only nutritional supplements with no known cardioactive potential. In the ED, she was normotensive, alert, and fully oriented, with a normal physical examination. An ECG was obtained (Figure 3.1). Basic blood chemistry was normal. Due to the laceration, computed tomography of the head was undertaken and was normal.

Question No. 1:

With the information available to the ED physician, is it clear that the patient suffered a "syncope" event?

A. Yes

B. No

DID THE PATIENT SUFFER A "TRUE" SYNCOPE EPISODE?

The diagnostic evaluation of a patient who presents with an apparent T-LOC should begin by first *not assuming* that the episode was true "syncope." Syncope is defined as the temporary loss of consciousness due to transient diminution of cerebral nutrient flow, most often the result of transient hypotension. The assessment should be

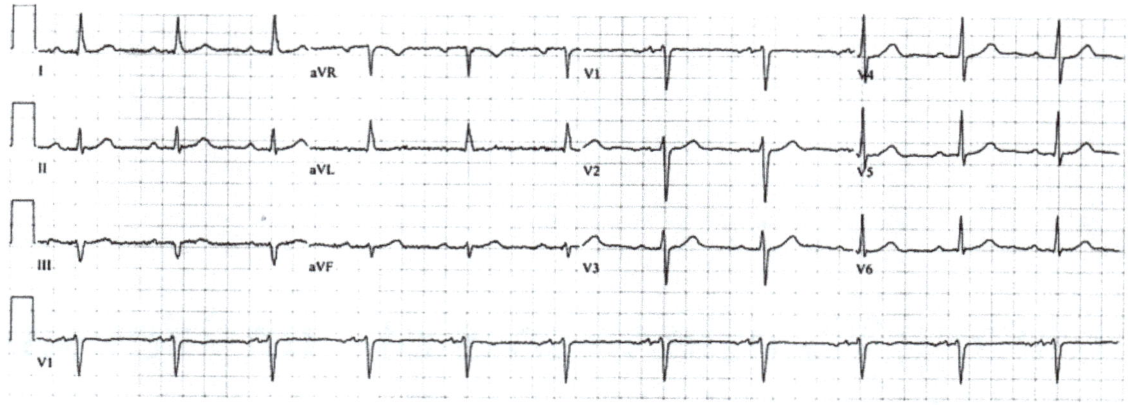

Figure 3.1. Baseline 12-lead ECG in the ED.

initiated with broad consideration of the range of possible causes for real (eg, seizures and concussion) or seemingly real (eg, narcolepsy and certain psychogenic disturbances) T-LOC[1] (Table 3.1). Syncope is only one of several possibilities. Consequently, the correct answer to Question #1 is: No.

It is evident that "syncope" cannot occur in the absence of true loss of consciousness, and the latter can only be determined from the medical history obtained from the patient and/or witnesses to the episode(s). In this context, loss of consciousness implies not only loss of awareness and appropriate responsiveness to external stimuli, but also loss of postural tone. As a result, if standing, the fainter falls down; if seated, he or she slumps over. Occasionally, however, syncope may occur when the patient is supine or prone (ie, loss of postural tone is not readily apparent); such cases suggest that a hemodynamically compromising tachy- or bradyarrhythmia has occurred.

If the history indicates that there has not been clear-cut loss of consciousness associated with the "spell," then "true" syncope is excluded. On the other hand, one can be misled. Due to lack of recall, cognitive impairment (especially in elderly "fallers"), or embarrassment, the patient may deny having lost consciousness. Reports from witnesses are often essential. Often, however, despite best efforts, it may not be possible to distinguish a faint from other conditions (eg, an accidental fall).

It is not unusual for patients to report having a sensation that "syncope" is imminent but that full T-LOC does not evolve at that time; this circumstance (which may be termed, "near-syncope") is often described as "dizziness" or

Table 3.1. Pathophysiology of Various Causes of Real or Seemingly Real T-LOC

Condition	Pathophysiology
Syncope	Transient self-terminating insufficiency of cerebral nutrient perfusion.
Concussion	Head trauma with consequent cerebral dysfunction of variable duration.
Epilepsy	Primary electrical disturbance of cerebral function.
Intoxication	Self-terminating depression of cerebral function by excess of specific agent (eg, alcohol, psychotropic drugs).
Metabolic	Self-terminating depression of cerebral function by metabolic disturbance (eg, hypoglycemia).
Psychogenic pseudosyncope, cataplexy	Not true T-LOC, mechanisms unknown.

T-LOC, transient-loss of consciousness.

"lightheadedness." However, these latter complaints (especially in the elderly) are nonspecific; while they may be due to an ill-defined functional cerebral disturbance triggered by a less severe transient hypotension than that needed to cause a complete faint, they may also be due to other less well-defined conditions (especially in elderly individuals). Distinguishing transient functional disturbances of cerebral function due to modest degrees of cerebral hypoperfusion, from lightheadedness or dizziness of nonspecific origin is often difficult and requires considerable attention to a detailed description of the events. One key differentiating point is that near-syncope is usually associated with a "dimming" or "graying out of vision," whereas that is not the case with less-specific complaints.

The timing of events surrounding an apparent syncope is notoriously unreliable; most fainters and witnesses are incapable of assessing the passage of time accurately. Nonetheless, true syncope tends to be characterized by a relatively abrupt onset, perhaps within 10 to 20 seconds of warning symptoms.

A spontaneous, complete, and prompt recovery from the faint excludes a number of nonsyncope conditions that may cause T-LOC, but which do not self-terminate promptly or require medical intervention in order to reverse. Examples include coma, states of intoxication, and stroke (which may cause loss of consciousness in some instances). On the other hand, in certain forms of syncope, particularly the vasovagal faint (VVS), recovery, while rapid, may be accompanied by fatigue and a general sense of diminished energy for a lengthy period of time (often hours in duration).

Cerebral hypoperfusion differentiates "true syncope" from T-LOC due to trauma (eg, concussion), seizures (epilepsy), intoxications, or metabolic disturbances. Both trauma and epilepsy may lead to loss of consciousness with complete and spontaneous recovery, but their origins are not inadequacy of cerebral perfusion (see Table 3.1). With regard to differentiating syncope from epilepsy, the aspect that causes most confusion is abnormal motor activity. "Jerky movements" of the arms and legs for a brief period of time is not unusual in syncope and may be incorrectly interpreted as epilepsy. However, unlike typical epilepsy, jerky movements during a faint are of shorter duration, tend to occur after collapse rather than before, and do not have tonic-clonic features. On the other hand, bladder and, less often, bowel incontinence is relatively common in syncope and should not be used to suggest that a seizure had occurred.

Question No. 2:

In regard to our patient in the ED, which of the following is the best next diagnostic step?

A. Echocardiogram.

B. Electrophysiologic study.

C. 48-hour Holter monitor recording.

D. Implantation of an insertable loop recorder (ILR).

E. Reassurance that this was a "benign faint," and no specific treatment is required.

INITIAL EVALUATION

Among patients in whom T-LOC is deemed to be due to "true" syncope, the next step is thorough evaluation of the underlying cause (Figure 3.2). The goal should be to determine the cause of syncope with sufficient confidence to provide a reliable assessment of prognosis, recurrence risk, and treatment options.[1]

A well-considered initial evaluation alone should permit an experienced physician to determine, with a high level of confidence, the cause of syncope in a large proportion of patients (perhaps 40% to 50%). As a rule, the initial evaluation begins with a comprehensive medical history including detailed description of symptomatic events (recognizing that often elderly patients and those with cognitive impairment may not remember or accurately relate

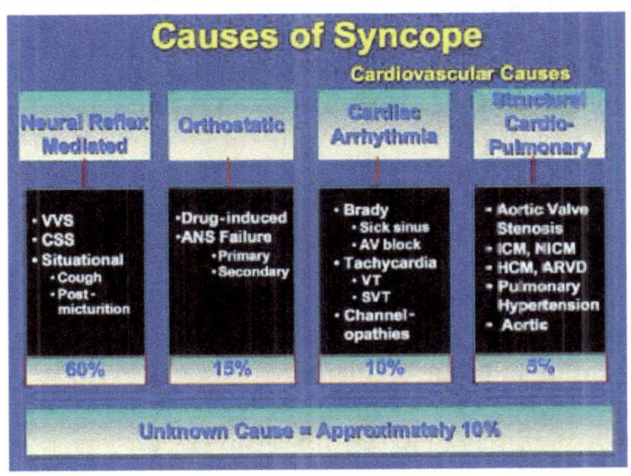

Figure 3.2. Causes of syncope. ANS, autonomic nervous system; ARVD, arrhythmogenic right ventricular dysplasia; AV, atrioventricular; CSS, carotid sinus syndrome; HCM, hypertrophic cardiomyopathy; ICM, ischemic cardiomyopathy; NICM, nonischemic cardiomyopathy; SVT, supraventricular tachycardia; VT, ventricular tachycardia; VVS, vasovagal syncope.

crucial details, while in other cases, patients may not volunteer information because of risk of losing jobs, avocations, or driving privileges) and physical examination. Additionally, several additional components are crucial: (1) assessment of known comorbidities (eg, diabetic neuropathy); (2) documentation of concomitant drug therapy (including recent dose changes); and (3) family medical history.

Among the most important factors to identify in the patient with suspected syncope is whether he/she has a history of or physical findings suggesting presence of underlying structural heart disease. In this regard, the inclusion of an echocardiogram as part of the initial evaluation of suspected syncope patients can be valuable. The presence of heart disease is an independent predictor of a "cardiac cause" for syncope (ie, a primary arrhythmic cause or a cause based on a structural cardiac abnormality leading to a transient hemodynamic disturbance), with a sensitivity of 95% and a specificity of 45%; by contrast, the absence of heart disease excludes a cardiac cause of syncope in 97% of the patients.

Syncope in conjunction with exertion raises special concerns. In particular, if the faint occurs in "full flight," one must consider the possibility of structural and/or dynamic heart lesions producing a relatively "fixed" cardiac output in the setting of vascular dilatation (eg, severe aortic or mitral valvular stenosis, hypertrophic cardiomyopathy) or arrhythmia-prone states such as myocardial ischemia or right ventricular cardiomyopathy. However, syncope during or shortly following exercise (even moderate exertion such as climbing stairs) can also occur in patients with severe autonomic dysfunction (eg, pure autonomic failure) in whom vascular control is unable to maintain adequate cerebral perfusion pressure. In addition, on rare occasions syncope occurs shortly after exertion as a consequence of a neurally mediated reflex faint (ie, a postexercise variant of VVS).

In regard to Question #2, given the circumstances, an echocardiogram is the best choice at this point of the evaluation in order to exclude structural heart disease; it proved to be entirely normal. In regard to the other choices, an electrophysiologic study (EPS) is usually reserved for patients with known structural heart disease or symptoms suggesting a paroxysmal tachycardia. Neither is the case. Holter monitoring is usually limited to 24 to 72 hours duration, and has almost no chance of capturing a spontaneous syncope recurrence as the events are so rare in this patient. An ILR might be considered; there is evidence suggesting that they may be cost-effective when used at an early stage. However, placement of such devices in the ED is not yet standard practice, although it may become more common in the future.

Question No. 3:

The next best step in this patient's assessment is:

A. Admit to a cardiology telemetry unit.

B. Admit to neurology.

C. Discharge with follow-up by internal medicine next week.

D. Discharge to a rapid-access syncope/falls clinic.

DOES MANAGEMENT REQUIRE ADMISSION TO HOSPITAL?

After the ED or urgent care physician has concluded that "syncope" was the most likely cause of T-LOC, a decision must be made as to whether the affected individual needs in-hospital evaluation or can be referred to an outpatient syncope evaluation clinic. In general, the issue that determines the answer to this question is an estimate of one or more of the following:

- the individual's immediate risk (usually measured as < 1 week) of a fatal event or potential for physical injury (eg, risk of falling);
- the patient's ability to care for him/herself; and
- whether appropriate treatment inherently requires in-hospital initiation (eg, antiarrhythmic drugs, device implantation, surgery, transfusions, etc.).

In instances when the etiology of syncope is known with certainty after initial clinical evaluation, the need for hospitalization depends in part on the immediate risk posed by the underlying problem, in addition to the treatment proposed.[2,3] Thus, for example, patients with syncope accompanying complete heart block, ventricular tachycardia, acute aortic dissection, or pulmonary embolism should be admitted to the hospital and preferably to an ECG-monitored unit. On the other hand, most vasovagal fainters can be sent home after careful discussion of the nature of the problem and simple preventive maneuvers (eg, hydration, avoidance of hot crowded environments, etc.); syncope evaluation clinic follow-up suffices in most of these cases.

The following breakdowns provide an overview of common circumstances for which hospitalization is recommended or, conversely, is usually not needed.

Patients at High Risk

Several markers identify syncope patients who should be considered for in-hospital evaluation. Syncope associated with symptoms suggestive of acute myocardial ischemia or acute aortic dissection, signs, or biochemical evidence (ie, elevated brain natriuretic peptide [BNP]) of congestive heart failure, and/or suspicion of hemodynamically concerning underlying structural heart disease (eg, valvular aortic stenosis, pulmonary hypertension) have the highest immediate mortality risk. At similar high risk are patients with certain ECG abnormalities, including high-grade atrioventricular (AV) block, cardiac pauses > 3 seconds, preexcitation syndromes (eg, Wolff-Parkinson-White syndrome), or suspected ventricular tachycardia including torsades de pointes (ie, marked QT interval prolongation). Patients with syncope during exercise, and syncope causing motor vehicle or major work accidents or severe injury should also be evaluated in the hospital.

Patients at Intermediate Risk

Syncope associated with age > 50 years, history of structural heart disease but without signs of active consequences of disease, certain ECG abnormalities (eg, bundle-branch block, Q waves), family history of sudden death, cardiac devices without evidence of dysfunction, symptoms not consistent with vasovagal or reflex-mediated syncope and physician's judgment that a cardiac syncope is possible constitute the patient group with intermediate risk for adverse outcome.

At present, guidelines for management of intermediate-risk cases are ill-defined. However, an ED-based observation unit may be useful initially; if there are no particularly worrisome physical examination, x-ray, ECG, or biochemical findings after 4 to 6 hours, the patient could be managed without hospital admission. These patients should be referred to and promptly seen (typically < 72 hours) in a syncope evaluation clinic.

Patients at Low Risk

Patients in this group typically have no evidence of structural heart disease and have a normal baseline ECG. The syncope is considered of a "relatively benign" nature in terms of mortality risk (although falls and injury risk remain a concern). As a rule, syncope in these cases is of neurally mediated reflex or orthostatic cause. However, rarer conditions that are more worrisome may present with a normal ECG; these include long-QT syndromes and preexcitation syndromes (in which the ECG abnormality is not always evident), Brugada syndrome and other less well-defined channelopathies.

Low-risk patients can generally be stabilized in the ED or clinic. Most can be reassured regarding immediate mortality risk, but early further assessment in a "syncope/falls" clinic is recommended.

In regard to our patient, there is no evidence of structural heart disease and her ECG is normal. Additionally, there is neither a worrisome past medical history nor a problematic family history. Finally, although perhaps not absolutely "classical" in terms of her presentation, the patient's history of her most current collapse is suggestive of a neurally mediated (vasovagal) reflex faint. Consequently, we can reasonably conclude that she is in the "low-risk" group; in-hospital cardiac telemetry is not needed, and she is certainly not going to be optimally served in a neurology unit. Discharge from the ED with an appointment in the very near future (preferably ≤ 72 hours) to a syncope clinic (sometimes referred to as a syncope management unit [SMU]) is the appropriate next step. This referral was arranged from the ED, and in the interim the patient was asked to avoid driving or other potentially hazardous activities.

SYNCOPE MANAGEMENT UNITS

SMUs have been advocated as a means of enhancing the management process for syncope patients while reducing cost.[1–3] In most institutions that have SMUs, it is in reality a "virtual unit" that operates as a diagnostic and treatment consortium comprising one or more of several key medical specialties (usually cardiac electrophysiology (EP), neurology, general medicine, and geriatrics). In some hospitals (principally in Italy), the SMU may be a physical space, incorporating an observation unit and possibly an outpatient facility.

One of the objectives of establishing an SMU is to help solve the problem of too many low- and intermediate-risk syncope patients being admitted to hospital where they often are submitted to unneeded

Table 3.2. Randomized Studies Assessing the Contribution of SMUs to Syncope Evaluation

Study	Location	Principal outcomes
SEEDS	USA, single center	Diagnosis of cause of syncope increased from 10% with usual care to 67% with SMU; hospital admission was reduced from 98% to 43% with SMU; length of patient-hospital days was reduced by > 50% for SMU.
EGSYS	Italy, multicenter	Compared outcomes of 745 patients managed with SMU to 929 patients managed with usual care. In SMU patients, hospitalizations were fewer (39% vs 47%), in-hospital stay was shorter (about 1 day), fewer tests were performed per patient (2.6 vs 3.4), and cost per patient, and mean cost per diagnosis were approximately 20% lower.

EGSYS, Evaluation of Guidelines in Syncope Study; SEEDS, Syncope Evaluation in the Emergency Department Study; SMU, syncope management units.

expensive diagnostic tests. In this regard, supportive evidence from two randomized trials is summarized in Table 3.2.

POSSIBLE CAUSES OF SYNCOPE

Syncope itself is not a sufficient diagnosis; in fact, syncope is better considered as a "syndrome" with many possible etiologies. Only after the correct cause is determined can one expect to make an accurate assessment of the patient's prognosis and initiate an effective therapy strategy. There are 3 relatively common variations of neurally mediated reflex syncope: vasovagal syncope (VVS, also often termed the "common faint"), carotid sinus syndrome (CSS), and the so-called "situational faints." The principal pathophysiological mechanism of these faints is the triggering of a neural reflex resulting in both hypotension due to vaso- and veno-dilation (including the splanchnic bed) and an inappropriate chronotropic response (occasionally resulting in severe bradycardia or asystole, but at other times causing a "relative" bradycardia in which the heart rate is less than expected for the degree of hypotension). Both vascular dilatation and/or bradycardia may diminish arterial pressure sufficiently (below cerebrovascular autoregulatory capability) to compromise perfusion to the brain, thereby causing syncope; however, the contribution of each of these two factors to systemic hypotension and cerebral hypoperfusion may vary considerably among patients and probably even within patients at different times.

The reflexes that cause VVS are thought to be present in all humans, but only about 20% to 30% of the population report VVS events in their lifetime; it seems, however, that certain individuals are particularly susceptible. In any case, VVS may be triggered by prolonged periods of upright posture, relative dehydration, excessively warm confining environments, extreme emotions, or acute pain. Common places for these events are religious services, restaurants, and long queues. Warning symptoms may occur, and include a variety of sensations, including the feeling of being: hot or cold, sweaty, tachycardic, short of air, hearing deficient, and nausea. Observations often reported by bystanders (if the physician actively solicits the information) include pallor ("white as a ghost," "death-like"), a cold "clammy" feel to the affected individual's skin, and confusion or marked fatigue on the part of the victim after return of consciousness. If "jerky" motor activity is reported (often mistaken as "seizures"), it occurs after loss of consciousness and is brief and incoordinate in character. After the collapse, if the patient is permitted to remain in a gravitationally neutral posture (eg, recumbent), recovery typically is prompt. However, post-event fatigue of variable duration is quite common.

Given the many possible causes for syncope, practitioners with experience in having cared for many syncope patients (such as those that would staff an SMU) have an advantage. Nevertheless, even after careful assessment, it may not be possible to assign a single cause for syncope; patients often have multiple comorbidities, and several equally probable causes may be identified (or several conditions may act together to cause the faint). For instance, individuals with severe heart disease may suffer from transient tachyarrhythmias, intermittent high-grade AV block or even the effects of being excessively medicated; on the other hand, they also remain susceptible to more benign causes of syncope such as VVSs. In essence, a prudent physician must not readily accept an observed abnormality during diagnostic testing as either the certain cause, or the sole cause, of fainting in a given individual.

Question No. 4:

Diagnostic evaluation of our patient was undertaken on an outpatient basis in a syncope/arrhythmia clinic (outpatient SMU). The following section briefly summarizes the principal diagnostic considerations. However, based on the information known so far, which of the following tests are now likely to be most helpful?

A. Carotid sinus massage.

B. Electrophysiologic study.

C. Head-up tilt test.

D. In-hospital combined electroencephalogram (EEG)-video monitoring.

Figure 3.2 provides a classification of the most important causes of syncope. In general, the neural-reflex causes and orthostatic causes are the most frequently encountered.[1]

Neurally Mediated Reflex Syncope

There are 3 relatively common variations of neurally mediated reflex syncope: VVS, also often termed the "common faint," CSS, and so-called "situational faints." The principal pathophysiological mechanism of these faints is the triggering of a neural reflex, resulting in both hypotension due to vaso- and venodilation (including the splanchnic bed) and an inappropriate chronotropic response (occasionally resulting in severe bradycardia or asystole, but at other times causing a "relative" bradycardia in which the heart rate is less than expected for the degree of hypotension). Both vascular dilatation and/or bradycardia may diminish arterial pressure sufficiently (below cerebrovascular autoregulatory capability) to compromise perfusion to the brain, thereby causing syncope; however, the contribution of each of these two factors to systemic hypotension and cerebral hypoperfusion may vary considerably among patients and probably even within patients at different times.

In terms of diagnostic testing, none may be needed other than a careful history review by an experienced practitioner. However, since warning symptoms may not always be present or, at least, not recollected after the fact, and if the medical history does not otherwise provide sufficient basis to make the diagnosis, head-up tilt-table testing (HUT) may prove helpful. Such testing, in the absence of pharmacological provocation, has a specificity of approximately 90%, but an uncertain sensitivity since there is no "gold standard" to rely upon. If the history is "typical" for VVS, HUT is not needed from a diagnostic perspective, but may nonetheless be used to teach the patient about warning symptoms so they become aware of future events.

CSS is a clinical condition in which syncope or near-syncope is due to bradycardia and/or vasodilatation secondary to carotid sinus baroreceptor hypersensitivity (CSH). Thus, CSS and CSH are related, but the mere presence of CSH (usually defined as carotid massage-induced asystole >3 to 5 seconds) does not establish the presence of CSS; the medical history must be compatible with the diagnosis (eg, syncope associated with accidental manipulation of the neck or neck movement that results in external pressure on the carotid sinus baroreceptors). As a rule, CSS is most often encountered in older males (> 60 years of age) and is highly correlated with atherosclerotic vascular disease and/or prior neck surgery or irradiation.

The third group of neurally mediated faints are the so-called "situational" faints; these are essentially identical to VVS except that the trigger is peripheral and identifiable. Thus, these faints include postmicturition syncope, defecation syncope, deglutition or swallowing syncope, etc. Cough may also trigger reflex hypotension, although in this case, both reflex (triggered by transient marked elevation of intravascular pressure) and nonreflex mechanisms participate. In any case, situational faints are diagnosed by their distinctive history; it is both difficult and usually unnecessary to evaluate these fainters in the clinical laboratory.

Orthostatic Syncope

As its name implies, orthostatic syncope occurs as a result of transient excessive cerebral hypotension that may occur when susceptible individuals move from a lying or sitting to a standing position (ie, orthostatic hypotension [OH]). There are two basic forms of OH. The first is the so-called "immediate hypotension." This occurs almost immediately upon "active" standing, and occurs in young healthy individuals as well as in older patients. Many healthy individuals experience a minor form of "immediate hypotension" as evidenced by the need to support themselves momentarily as they stand up. The second form of OH is the so-called "delayed form;" this is the classic situation in which symptoms occur several moments after standing up. The patient has already walked some distance, then collapses. The cause in both cases is deemed to be failure of autonomic nervous system to respond to a sudden upright posture, but the delayed form tends to be more severe (ie, the patient falls).

Lack of autonomic system response can be due to either extrinsic factors or primary autonomic failure. Extrinsic factors include dehydration from prolonged exposure to hot environments, inadequate fluid intake or excessive use of diuretics, antihypertensives or vasodilators. Additional extrinsic factors include chronic diseases such as diabetes, alcohol abuse, or other agents that induce a peripheral neuropathy and thereby predispose patients to orthostatic syncope. Less commonly, OH is the result of a primary autonomic failure with inadequate reflex adaptations to upright posture. These latter conditions include multisystem atrophy and Parkinson's disease.

Cardiovascular Causes of Syncope

The cardiovascular causes of syncope comprise both those that trigger syncope due to hypotension-triggering primary cardiac arrhythmias and those in which hypotension is due to structural disturbances of the heart or vascular system (see Figure 3.2).

Cardiac arrhythmias may be the primary cause of syncope if the heart rate becomes too fast (tachyarrhythmias) or too slow (bradyarrhythmias). Either or both may occur as a result of "intrinsic" disease of the cardiac conduction system, or as a consequence of structural cardiac or cardiopulmonary disease. Determining which, if any, of the cardiac arrhythmias are responsible for syncope in a given individual can be difficult, especially if symptomatic events are infrequent. Electrophysiologic testing may be helpful in some cases, but often the findings are nonspecific. Consequently, prolonged ambulatory ECG monitoring is usually safe and more effective, especially if life-threatening causes seem unlikely; external event recorders, mobile cardiac outpatient telemetry, and ILRs have become valuable diagnostic aids for this purpose.

The following subsections provide only a brief overview of the types of primary cardiac arrhythmias that may be responsible for syncope.

Sinus node dysfunction and AV conduction disorders: Syncope in patients with "sinus node dysfunction" may be due to either brady- or tachyarrhythmias. In either case, the heart rate may be inadequate to support adequate cerebral blood flow. In some instances both brady- and tachycardia may be contributory; for example, an abrupt spontaneous termination of an atrial tachycardia (especially atrial fibrillation) may be followed by long asystolic pause prior to recovery of the sinus mechanism. In other cases, sinus pauses or sinus arrest may occur sporadically resulting in transient hypotension and syncope.

Paroxysmal or persistent AV block can cause severe bradycardia and thereby lead to syncope. In practical terms, if a patient is found to have Mobitz type II second-degree AV block, third-degree AV block or alternating left and right bundle-branch block, a basis for syncope can be made with reasonable certainty and pacemaker implantation is justified. In the setting of nothing more than bifascicular block, or nonspecific intraventricular conduction disturbances, one should not assume that AV block was the cause; EPS may be helpful to evaluate the infranodal conduction system. However, in all of these cases, substantial underlying structural heart disease is often present, and it is important to consider concomitant susceptibility to tachyarrhythmias as a cause of symptoms.

Supraventricular (SVT) and ventricular tachyarrhythmias (VT): Paroxysmal SVT is not a very common cause of syncope. However, when syncope does occur in the setting of SVT, symptoms tend to occur at the onset of the tachycardia, before peripheral vessels have a chance to constrict and stabilize the circulation. Factors that increase risk of syncope during SVT include high heart rate (usually > 200 bpm), reduced circulating volume status (ie, relative dehydration), upright posture, presence of structural heart disease (especially with diminished ejection fraction, or significant valvular or other outflow obstruction), and cardioactive drugs that undermine vascular responsiveness (eg, beta-adrenergic blockers, vasodilators, and diuretics).

Ventricular tachyarrhythmias are more often triggers for syncope than are SVTs; this is primarily due to the fact that most VTs occur in the setting of more severe underlying heart disease (ie, a heart that is less capable of maintaining blood pressure when stressed). Thus, VT occurring in the setting of ischemic heart disease, valvular heart disease, dilated or obstructive cardiomyopathy, or arrhythmogenic right ventricular dysplasia (ARVD) may cause syncope. However, in addition, polymorphous VT (especially torsades de pointes) is known to cause syncope in some individuals with structurally relatively normal hearts, particularly in the setting of so-called "channelopathies" (eg, long-QT syndrome, Brugada syndrome, short-QT syndrome, and idiopathic ventricular fibrillation). In these latter cases, VT tends to be very fast and, as a result, abrupt hypotension ensues.

Structural cardiovascular disease as cause of syncope: Among all causes of syncope, structural cardiovascular disease etiologies are the least common, but are important as they have a worrisome prognosis. The most common scenarios in which structural disease is seemingly associated with triggering syncope in adult patients without congenital heart disease are acute coronary syndromes, acute myocardial infarction, acute aortic dissection, and severe pulmonary hypertension. In each of these cases, despite the potential for important hemodynamic consequences of the conditions, it is also believed that the syncope is primarily a product of neurally mediated reflexes secondary to pain, or excessive central mechanoreceptor activation, rather than solely a direct hemodynamic effect of the lesion. Syncope may also occur directly as a result of severe underlying cardiac or cardiopulmonary disease, although a concomitant neural reflex contribution is suspected of contributing in some cases. Perhaps the best examples are severe aortic stenosis and hypertrophic cardiomyopathy with outflow obstruction.

Despite obvious hemodynamic limitations imposed by structural heart and cardiopulmonary disease, the presence of such disease also predisposes to arrhythmia. It must always be borne in mind that syncope due to arrhythmia is far more common than syncope due to the structural cardiovascular disease itself. Further, arrhythmias (particularly VT) are often worrisome markers of sudden death risk in these patients.

Cerebrovascular disease and related conditions as cause of syncope: Cerebrovascular disease is a very rare cause of syncope, and evaluation of the cerebrovascular circulation (including the carotid vessels) should be reserved to those few instances in which there are evident neurological findings supporting the possibility. Otherwise, much energy will be wasted and needless expense incurred with low probability of success.

In order for syncope to occur as a result of cerebrovascular disease, large areas of both cerebral cortices, or at minimum critical zones within the reticular activating system, must be rendered hypoperfused. Thus, it is possible that vascular spasm associated with migraines may cause syncope. However, in most migraineurs, the extracranial vessels are principally involved, and if syncope occurs, it is vasovagal or orthostatic in nature. Similarly, carotid-distribution transient ischemic attacks (TIAs) do not cause syncope unless all other cerebral blood vessels are obstructed. Thus, carotid TIAs typically present with focal neurological deficit, but not global cerebral hypoperfusion. Further, vertebrobasilar TIAs, although more likely to present as apparent syncopal episodes than are carotid-distribution TIAs, are extremely rare and are associated with other symptoms (eg, vertigo) indicative of posterior circulation compromise.

Subclavian steal is both a rare condition and a very rare cause of syncope. Symptoms tend to occur when patients are exercising using the affected arm. Because of its rarity, it should only be considered when other causes have been eliminated.

Syncope mimics and pseudosyncope: There are 2 important groups of conditions that may present with real or apparent T-LOC but that should not be considered as "syncope." One group is composed of conditions that cause true T-LOC, but their mechanism differs from true syncope. We categorize these as "nonsyncope T-LOC," with epilepsy being the most important example, but trauma leading to concussion (such as might occur after an accidental fall in an older patient) being important as well. Second are conditions in which consciousness is never really lost. These we term "pseudosyncope," and the most important cause is psychiatric conversion disorders. Factitious disorders and malingering may also be encountered.

Pseudosyncope is a relatively common problem. As a rule, it is unassociated with any changes in heart rate or blood pressure. Most often the key suggestive finding on history-taking is a very high frequency of "symptomatic attacks." The frequency may be as high as many episodes per day or per week and is much greater than that of a true syncope, although these patients may in fact have occasional "true" faints as well. Pseudosyncope rarely results in serious injury (but cases with injury may occur).

As the name implies, pseudosyncope is not accompanied by true loss of consciousness, but the history from the patient and witnesses might suggest that it had. In this setting the history-taker must be alert in order to make the necessary distinction. In many cases the diagnosis is one of exclusion after other causes of presumed T-LOC have been eliminated.

"Drop attacks" also belong in the pseudosyncope group as true loss of consciousness does not occur. The basis for drop attacks is unclear. The drop attack is an uncommon condition in which postural tone is lost abruptly and patients (most often females) fall to the ground. Consciousness is never lost and the history-taker can usually determine this by careful questioning of the patient's recollection during the attack. Nevertheless, some patients claim they never lost consciousness when, in fact (based on observer report), they actually did. Consequently, it is sometimes difficult for syncope and drop attacks to be differentiated from one another.

SPECIFIC DIAGNOSTIC TESTING PROCEDURES

Table 3.3 provides an overview of certain potentially useful diagnostic tests for assessing the cause of syncope. It is recognized that the appropriate test selection must be based on careful assessment of clinical history, and that often none or, at most, only a few such tests are needed in the majority of syncope evaluations.[1]

The goal is to obtain a tentative diagnosis during the initial clinic evaluation and use selected tests only as necessary to confirm a suspected diagnosis. A more comprehensive discussion of the indications for and the risks and limitations of these various procedures are to be found in the Syncope Task Force Practice Guidelines document.[1]

Question No. 5 (see also Table 3.3):

Based on the information available to the SMU physician, which of the following is NOT useful at this stage?

- A. Head-up tilt test.
- B. Insertable loop recorder.
- C. Electrophysiologic study.
- D. No additional tests, with follow-up by her family physician as needed.

Table 3.3. Diagnostic Testing

Test	Indication/Use
Carotid sinus massage	Used as a means of confirming a suspected diagnosis of CSS. CSM response is classified as cardioinhibitory (ie, asystole), vasodepressive (fall in systolic blood pressure), or mixed. The mixed response is defined as asystole lasting ≥ 3 seconds and a decline of systolic blood pressure of ≥ 50 mmHg.[1] CSM is generally believed to be a safe test, complications are rare and transient in most instances. The main complications are neurological.
ECG monitoring	A 12-lead ECG is an essential part of the initial evaluation and is not discussed further here. An exercise ECG test has limited diagnostic utility, unless symptoms are clearly related to exercise, but may prove helpful in selected cases if findings reveal rate-dependent AV block, exercise-induced tachyarrhythmias, chronotropic incompetence, or hypotension in the setting of autonomic failure. If an arrhythmia is suspected to be the cause of syncope, extended-duration cardiac monitoring is essential to obtain ECG-symptom correlation (eg, ECG "event" recorders, mobile cardiac outpatient telemetry, and ILRs).
Echocardiography	Echocardiography rarely provides a definitive basis for syncope but is of value in confirming whether there is evidence of potentially contributing underlying SHD. SHD has an important impact on subsequent diagnostic strategy and, ultimately, on prognosis.
Head-up tilt table testing	HUT is useful in patients to help confirm a suspected diagnosis of VVS; if the medical history is clear, such testing adds no new diagnostic information. On the other hand, if HUT reproduces the faint, the experience may prove helpful to the patient to be more alert to warning symptoms. HUT should not be used to ascertain effectiveness of treatment interventions. HUT methodology is described in detail elsewhere in the text.
Electrophysiologic testing	EPS is appropriate when the initial evaluation points toward a primary cardiac rhythm abnormality (SVT, VT) or a conduction disturbance, and the diagnosis needs to be confirmed. EPS is most useful in the setting of SHD.
Neurological studies	Studies such as head CT and MRI and EEG are low yield in syncope patients and their use is discouraged. They are best reserved for cases in which there is concern that syncope may have resulted in head injury. Alternatively, they may be helpful in nonsyncope T-LOC patients.

CSM, carotid sinus massage; CT, computed tomography; ECG, electrocardiogram; EEG, electroencephalography; EPS, electrophysiologic testing; HUT, tilt table testing; MRI, magnetic resonance imaging; SHD, structural heart disease; SVT, supraventricular tachycardia; T-LOC, transient-loss of consciousness; VT, ventricular tachycardia.

At this stage of our patient's evaluation, the investigational strategy could reasonably take any one of several pathways, except for EPS. Given no reason to believe that she had a paroxysmal tachyarrhythmia (particularly paroxysmal SVT), EPS is highly unlikely to be helpful. On the other hand, since the presentation of the faint was not "classical" of VVS (although highly suggestive), one could try to reproduce the event by HUT in order to provide (albeit indirectly) support for the suspected diagnosis. Alternatively, the use of an early ILR strategy could be chosen. Such an approach has been advocated and shown to be cost-effective. Inasmuch as she did suffer injury and had jerky movements, one might opt for long-term ECG recording as a means of more definitively establishing the diagnosis when the next recurrence happens. Finally, since the presenting history suggests VVS with a very low frequency of recurrence and she is "low risk" by conventional risk stratification methods (see earlier), one could simply provide essential instructions regarding better hydration (salt and volume) and better protective physical maneuvers (eg, leg-crossing and muscle straining) with subsequent care by her family physician. In this case, the SMU physician chose to undertake HUT (Figure 3.3). In that test, after a baseline equilibration period during which the patient is allowed to rest, the table is tilted head-up to approximately 70 degrees with the patient gently restrained to prevent falling off.

There are a number of different published HUT protocols, but one of the most frequent is to maintain the upright posture for 20 minutes unless syncope occurs earlier. If syncope does not occur, then a single dose of sublingual nitroglycerine is administered and the test continued for a further 10 minutes. ECG and beat-to-beat blood pressure (usually noninvasive) are recorded throughout. A positive test is reproduction of the patient's symptoms of syncope with documentation of hypotension with or without bradycardia.

Figure 3.4 illustrates ECG and blood pressure recordings during a positive HUT and, although taken from a different patient, is representative of the findings recorded in our patient. The findings support susceptibility to VVS. Blood pressure recorded from a femoral arterial line is shown on the ordinate in mmHg. Syncope was associated with loss of postural tone; the latter caused the artifact on the pressure trace.

Question No. 6:

Inasmuch as the clinical suspicion of VVS was confirmed, the SMU physician would reasonably recommend which of the following courses of action?

A. Cardiac pacemaker with rate-drop response feature.

B. Beta-blocker and vasoconstrictor (eg, midodrine) therapy.

C. Education regarding increased salt and volume intake and leg-crossing maneuver.

D. Tilt-training (standing-training) exercise prescription.

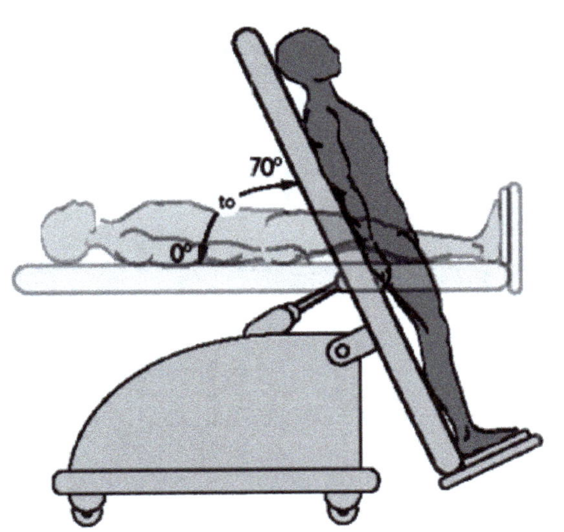

Figure 3.3. Head-up tilt test—schematic of procedure. See text for discussion.

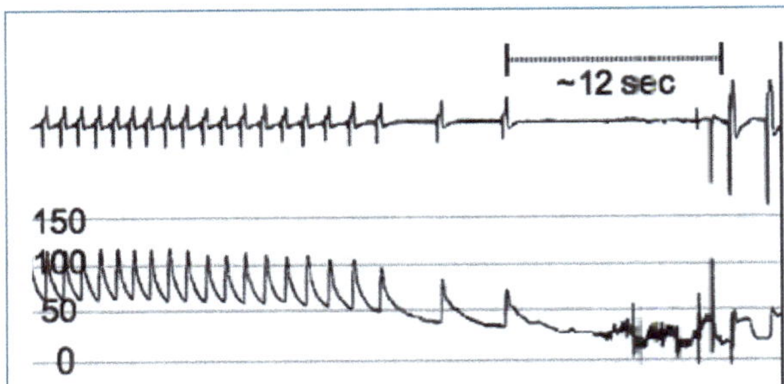

Figure 3.4. Positive head-up tilt test with hypotension and marked bradycardia.

Education regarding increased salt and volume intake, and instruction regarding physical maneuvers (eg, leg-crossing maneuver) designed to interrupt an impending VVS is the best answer at this stage. Since this patient's events are relatively rare, a simple treatment regimen is preferred. Increased salt intake has been shown to enhance orthostatic tolerance and should reduce VVS susceptibility. The risk of inducing hypertension is low in young patients, but blood pressure needs to be monitored during follow-up. Physical maneuvers, such as leg-crossing, have proven helpful to abort attacks when employed as soon as warning symptoms appear. Pacemakers, while occasionally helpful as suggested by the ISSUE-2 study, are not usually recommended for young patients and especially if VVS is infrequent. Furthermore, published randomized controlled trials (RCTs) (eg, VPS2 and Synpace) did not show a benefit for pacing in a broad population of VVS-susceptible individuals. Similarly, beta-adrenergic blockers were generally ineffective in the POST study, while exposure of a young person to vasoconstrictors would not seem prudent in this circumstance. Finally, while tilt-training or standing-training (ie, prescribed progressively longer periods of still upright posture to improve orthostatic tolerance) is helpful to some patients, it is unlikely that one could substantiate any benefit given the infrequent nature of this patient's syncope.

During the subsequent 2 years she had 3 or 4 episodes that she was able to terminate by leg-crossing. Then she experienced 3 more faints in the course of 6 months. In one instance she was at a business dinner in a crowded restaurant but was able to delay the episode by physical maneuvers and leave the room before collapsing. A second occurred during an argument with a family member, and the last seemed to evolve without any evident trigger but resulted in a sprained ankle and forearm contusion.

Question No. 7:

At this stage, what is likely to be the most effective next step?

- A. Reassurance and add serotonin re-uptake inhibitor (SSRI) to the treatment.
- B. Repeat HUT.
- C. Recommend placement of an ILR.
- D. Undertake EPS.
- E. Initiate 48-hour Holter monitor.

In response to Question #7, either repeating HUT or undertaking EPS is highly unlikely to help. Similarly, the odds of a 48-hour Holter capturing any useful information are low. Finally, providing reassurance and introducing an SSRI is at best a "shot in the dark." Consequently, an ILR was placed.

Approximately 7 months later the following ILR recording (Figure 3.5) was obtained after the patient collapsed at home.

Figure 3.5 illustrates the evolution of sinus bradycardia with subsequent development of an asystole of approximately 18 seconds duration. The baseline artifact may be the result of the physical effects of her collapsing. The overall picture is very compatible with a neurally mediated reflex faint and supports the diagnosis of VVS. However, it is evident that the initial treatment recommendations for our patient need to be revised.

TREATMENT OF VVS

In most cases, VVS or the closely related situational faints are a solitary or at most very infrequent event. Therefore, most patients need little more than reassurance and education about the "benign" nature of this condition from a mortality perspective. However, since certain drugs may increase VVS susceptibility (eg, diuretics, vasodilators), it may be reasonable to modify or, if possible, discontinue such treatment. Additionally, if it is apparent that certain triggers are playing a role (eg, prolonged upright posture, hot environments), then discussion of strategies to avoid these triggers is important.

In certain individuals vasovagal or situational faints become quite frequent and thereby adversely alter quality of life, as well as exposing the individual

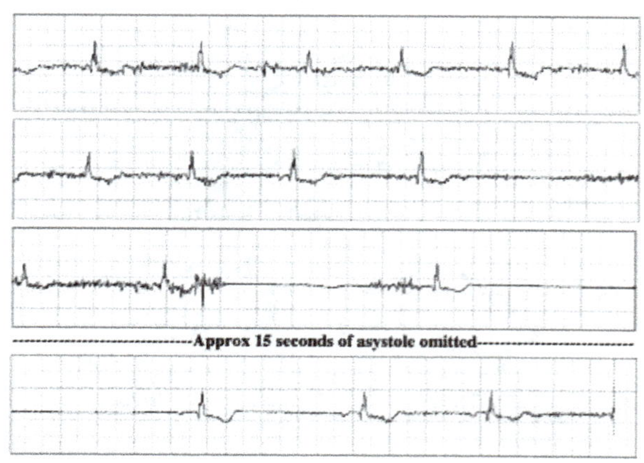

Figure 3.5. Insertable loop recording obtained during patient collapse. See text for discussion.

to risk of trauma. In these cases additional treatment steps may be necessary,[4] including:

1. Avoiding those environments that seem to provoke episodes.
2. Being alert to warning symptoms and taking action in order to prevent the episodes (eg, sitting or lying down with feet elevated).
3. Perform physical maneuvers to prevent the faint (eg, aggressive hydration, arm-tensing, leg crossing).
4. Incorporate fluids rich in salts and electrolytes into the daily diet. Examples would include sport drinks (although avoiding high-calorie versions) and salt tablets.

If these initial steps prove inadequate, as has apparently been the case for our patient, then the treatment strategy can be expanded to include, as needed, long-term physical training techniques to improve orthostatic tolerance, pharmacological interventions to prevent depletion of intravascular volume and/or enhance arterial and venous tone, and block the sympathetic neural excitation that often is the initial feature of the evolving faint. Very infrequently, cardiac pacing may be needed.

Long-term Physical Techniques to Improve Orthostatic Tolerance

The use of acute intervention physical maneuvers such as leg-crossing and muscle-tensing has already been discussed. Long-term physical maneuvers, specifically tilt-training (better termed standing-training), is a form of physical therapy intended to "re-train" abnormal neurovascular constriction responses. This technique has been advocated as a useful nonpharmacological treatment for VVS as well as other orthostatic intolerance syndromes, although the clinical evidence is inconsistent. In any case, the method we advocate entails home standing-training consisting of the patient standing with the upper back positioned lightly against a wall or a corner (preferably using a carpeted floor and no sharp-edged objects nearby) without moving arms or legs for progressively longer periods of stationary upright posture. Initially, the recommended standing duration is 3 to 5 minutes twice daily; the duration is then gradually lengthened every 3 to 4 days to as much as 30 to 40 minutes twice daily. Tilt-training has anecdotally proven helpful if undertaken twice daily for periods of 8 to 12 weeks, with subsequent maintenance 3 to 4 times weekly thereafter.

Sleeping with the head of the bed elevated provides nocturnal postural stress that has long been advocated for treatment of OH but may be similarly proposed in VVS (especially as adjunctive therapy in those who cannot take added salt due to comorbid conditions [eg, heart failure, hypertension] or who do not respond to other nonpharmacological interventions). The head of the patient's bed is raised to 5 to 20 degrees. However, there is no data showing sustained benefit in the long-term. Further, compliance may be a problem.

Pharmacological Therapy to Improve Orthostatic Intolerance

Volume expansion has long been, and remains, the foundation of medical therapy for VVS; conventional approaches entail increased dietary salt and electrolyte-rich sport drinks (although it is wise to avoid many high-caloric "sport" drinks). The primary safety concern is induction of hypertension. Fortunately, this is uncommon in younger patients but becomes a real concern in older patients.

A number of more conventional pharmacological agents have been proposed for preventing VVS; examples include fludrocortisone, beta-adrenergic blocking drugs, vasoconstrictor agents (eg, midodrine), and serotonin re-uptake inhibitors (Table 3.4). None are perfect.

Permanent Pacemakers

Cardiac pacing has received considerable study as a potential treatment option in patients experiencing very frequent VVS.[1,4,5] However, despite early favorable reports, placebo-controlled prospective trials in which all enrolled patients received pacemakers did not show a statistically significant benefit from pacing. A recent meta-analysis concluded that blinded trials do not show a benefit. Thus, pacing is not considered mainstream therapy for preventing VVS. However, the ISSUE 2 study suggests that pacing may be of value when other approaches fail in patients in whom ILR documentation of marked bradycardia or asystole is associated with spontaneous syncope. The ISSUE 3 study is assessing this latter ISSUE 2 finding in more detail. In any case, pacemaker therapy should not be embarked upon lightly in VVS fainters.

CAROTID SINUS SYNDROME

As discussed earlier, CSS is a particular form of neurally mediated reflex syncope that almost

Table 3.4. Principal Drugs Used for VVS Prevention

Drug	Action and typical dose (if applicable)
Midodrine	Midodrine is a pro-drug, the active metabolite of which is an alpha-agonist that stimulates both arterial and venous smooth muscle with minimal direct central nervous system effects. Midodrine has been most extensively studied in patients with OH, but has also been shown to be effective in VVS. Midodrine therapy can be started with 2.5 mg 3 times daily with the maximum dose being 10 to 15 mg 3 times daily, avoiding the evening. Midodrine is reasonably well-tolerated with dose-related and easily reversible side effects. The most potentially serious adverse reaction associated with midodrine is supine hypertension. The feelings of paresthesia, pruritus, piloerection, tingling of the scalp, and chills are associated with the action of midodrine on the alpha-adrenergic receptors of the hair follicles. Urinary urgency, retention, and frequency are associated with the action of midodrine on the alpha-receptors of the bladder neck.
Fludrocortisone	Fludrocortisone is a mineralocorticoid used to facilitate salt and water retention as an adjunct to the increase of dietary salt. As a rule the drug is well-tolerated at 0.1 to 0.2 mg/day. However, potassium depletion is a concern, and potassium concentration in the blood should be periodically assessed.
Beta-blockers	Beta-blockers were initially advocated based on the theoretical view that they would moderate the early phase sympathetic activation associated with the evolving VVS. However, evidence from the POST study suggests that this theoretical benefit may not translate into predictable clinical effectiveness; nevertheless they may be helpful in selected individuals.
SSRIs	SSRIs have been reported in small studies to be helpful in VVS, but the evidence is not well-substantiated. We only recommend their use in patients in whom a depressive overlay may be contributing to symptoms.

SSRI, serotonin re-uptake inhibitor; VVS, vasovagal syncope.

exclusively occurs in older individuals (mainly males). Susceptibility appears to be increased in patients with atherosclerotic vascular disease or previous neck dissection (usually for tumors) and/or neck irradiation. In such cases the value of lifestyle interventions as sole treatment is unknown. Therefore, while it may be prudent as a part of the treatment strategy to include avoidance of tight collars and neckties and abrupt neck movements, early initiation of cardiac pacing is usually an essential component of therapy as cardioinhibition is almost always a key component of the syndrome.[1] The vasodepressor aspect may need to be addressed as in the vasovagal fainter (see earlier discussion).

The patient described in this case, being a younger female with no known vascular disease and without previous neck surgery or irradiation, is highly unlikely to be experiencing CSS.

SITUATIONAL SYNCOPE

Avoiding or ameliorating the triggering event is the basic treatment for most forms of neurally mediated situational syncope. Sometimes, despite the fact that the trigger is readily recognized, suppressing it is not easily accomplished (eg, cough trigger in a patient with chronic obstructive pulmonary disease). As in the case of defecation or postmicturition syncope, it is impossible to avoid exposure to the triggering situation (ie, bowel movement and bladder emptying). In the latter situation, certain general treatment strategies may be advocated, including maintenance of central volume as discussed above, protected posture (sitting rather than standing), and slower changes of posture (eg, waiting after a bowel movement before arising). Certain additional advice may be helpful in specific conditions, such as use of stool softeners in patients with defecation syncope, avoidance of excessive fluid intake (especially alcohol) just prior to bedtime in postmicturition syncope, and elimination of extremely cold drinks in "swallow" syncope (the latter also merits assessment of the esophagus for structural abnormality).

The history in our patient does not support a diagnosis of any of the recognized situational faints. As such, these conditions will not be considered further.

ORTHOSTATIC SYNCOPE

Treatment of orthostatic syncope (and other orthostatic intolerance syndromes, such as postural orthostatic tachycardia syndrome, which are not relevant here), parallels the strategy discussed earlier for VVS. Treatment of orthostatic syncope will focus on the "delayed" form since it is clinically much more important than is the "immediate" form.

Typical delayed orthostatic syncope differ from VVS in several respects: (1) the duration of treatment is likely to be longer; (2) affected individuals are typically older and more frail, making physical maneuvers more difficult to employ; and (3) patients are more prone to supine hypertension thereby complicating the overall treatment strategy. In any event, treatment for prevention of orthostatic syncope should focus initially on maintenance of hydration and physical maneuvers as discussed earlier.

In many cases, syncope associated with postural change may be caused by low circulating plasma volume or inadequate vascular constriction upon moving to the upright posture (often drug-induced), or both. Consequently, one of the basic tenets of treatment is expanding central circulating volume (ie, encouraging increased salt and volume intake to the extent that this does not induce supine hypertension). Additionally, affected individuals should be instructed to try and avoid certain predisposing conditions (eg, prolonged exposure to hot environments) or medications that decrease volume status (eg, diuretics) or impair vasoconstriction (eg, vasodilators, beta-adrenergic blockers). In some severe cases, use of vasoconstrictors (like midodrine) or volume expanders (eg, fludrocortisone) may be necessary to maintain adequate cerebral perfusion. Finally, patients who exhibit poor autonomic function may benefit from standing-training (discussed earlier), counter-pressure clothing such as fitted stockings and/or "bikers' pants," and abdominal compression devices.

A special case is the patient with severe pure autonomic failure. In these situations, bolus water intake, especially before arising from bed in the morning, may result in a substantial and sustained increase in blood pressure. On the other hand, these same individuals tend to be very susceptible to supine hypertension. In such cases it may be necessary to tailor the treatment prescription to include salt, volume, and midodrine early in the day and antihypertensive at night. Bear in mind, however, the patient may be at risk if they get up in the middle of the night. A "walker" at the bedside, or use of a bedside commode, may be needed.

The clinical scenario presented by our patient does not fit with orthostatic syncope. In fact, the ILR documentation (see Figure 3.5) of progressive bradycardia followed by a prolonged asystole and subsequent spontaneous recovery is typical of VVS. Thus, the ILR has provided essential evidence with which to make a treatment recommendation.

CASE STUDY TREATMENT STRATEGY AND FOLLOW-UP

The ISSUE and ISSUE 2 studies have been instrumental in furthering insight into the nature of spontaneous syncope events in a variety of clinical circumstances (ie, normal hearts, underlying bundle-branch block, and underlying structural heart disease).[5] In particular, ISSUE 2 (albeit not an RCT) illustrated that not only were ILR findings highly reproducible within a given patient in whom more than one VVS episode was recorded, but that when marked bradycardia was the cause of symptoms, pacing was highly effective therapy (Figure 3.6).

Given the multiple recurrent faints and documented bradycardia in our patient, a dual-chamber pacemaker was recommended along with maintenance of baseline therapies as discussed above. The pacemaker was implanted using an anterior axillary line approach for aesthetic reasons. A device with rate-drop response capability was chosen. In essence, this programming feature is set to detect abrupt drops in cardiac rate suggesting onset of a VVS event, and react by dual-chamber pacing at a relatively high rate (approximately 110 bpm) for 1 to 2 minutes. The higher rate is chosen in order to not only prevent sustained

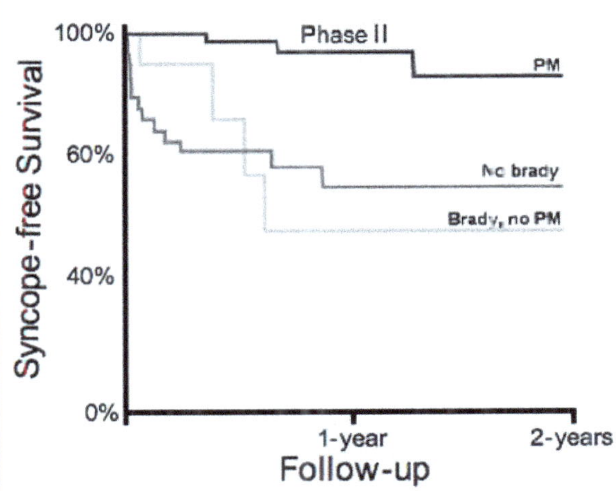

Figure 3.6. Impact of pacing on VVS with documented cardioinhibition versus no pacing. (Modified from Brignole et al[5] with permission from Oxford University Press.)

bradycardia but also to try to ameliorate the inevitable concomitant vasodepressor component of the VVS reflex.

During 2 years of follow-up the patient has had several lightheaded spells, but no syncope events. She did, however, note occasional periods of troublesome palpitations, usually when trying to go to sleep. Device interrogation confirmed that the rate-drop programming had activated on multiple occasions during daytime hours (probably effectively aborting an imminent VVS), but also from time-to-time was triggered at night; the latter was presumably due to natural heart rate slowing as she was falling asleep. Ultimately, it was possible to adjust the device programming to minimize this adverse effect.

CONCLUSION

This case illustrates the evaluation and treatment of a young woman who presented with recurrent syncope in the absence of any underlying structural heart disease or cardiac conduction system disease. Diagnostic findings, ultimately including ILR recordings, supported a diagnosis of VVS. Initial therapy was inadequate. Consequently, based on ILR documentation of marked bradycardia during a spontaneous faint, cardiac pacing was advised and appears to have been beneficial. It is understood that pacing therapy is only rarely advocated in VVS patients, and that it cannot completely eliminate episodic hypotension due to the vasodepressor component of the reflex. Consequently, it is essential to continue to focus on the basics of VVS treatment: education, reassurance, avoidance of triggers, salt/volume hydration, and physical maneuvers.

Correct Answers:

Q1. B
Q2. A
Q3. D
Q4. C
Q5. C
Q6. C
Q7. C

REFERENCES

1. Task Force for the Diagnosis and Management of Syncope, European Society of Cardiology (ESC), European Heart Rhythm Association (EHRA), et al. Guidelines for the diagnosis and management of syncope (version 2009). *Eur Heart J.* 2009;30:2631–2671.
2. Shen WK, Decker WW, Smars PA, et al. Syncope Evaluation in the Emergency Department study (SEEDS): a multidisciplinary approach to syncope management. *Circulation.* 2004;110:3636–3645.
3. Benditt DG, Can I. Initial evaluation of "syncope and collapse:" the need for a risk stratification consensus. *J Am Coll Cardiol.* 2010;55:722–724.
4. Benditt DG, Nguyen JT. Syncope: therapeutic approaches. *J Am Coll Cardiol.* 2009;53:1741–1751.
5. Brignole M, Sutton R, Menozzi C, et al. Early application of an implantable loop recorder allows effective specific therapy in patients with recurrent suspected neurally mediated syncope. *Eur Heart J.* 2006;27:1085–1092.

CASE STUDY

4

Noninvasive Testing for Sudden Cardiac Death

Ryan Cooley, MD

Mr. X was a 69-year-old Caucasian male with mitral valve disease and end-stage renal disease who later developed atrial fibrillation (AF) and a chronic nonischemic cardiomyopathy. His cardiac history started during 1999. He underwent mitral valve replacement for severe mitral regurgitation with a mechanical prosthesis. His left ventricular ejection fraction (LVEF) deteriorated over several years coincident with the development of persistent AF. In 2006, his LVEF measured 35% despite good ventricular rate control. He had New York Heart Association (NYHA) class II congestive heart failure (CHF). His AF was successfully treated by catheter ablation in 2006 after antiarrhythmic drugs failed. He was noninducible for sustained ventricular tachycardia (VT). Six months later, his EF was unchanged and he remained in class II CHF. A 24-hour Holter monitor revealed normal sinus rhythm with only occasional premature ventricular contractions.

Our patient was felt to be a candidate for implantable cardioverter-defibrillator (ICD) treatment based upon Sudden Cardiac Death in Heart Failure Trial (SCD-HeFT) indications, but he was undecided and reluctant. He therefore underwent microvolt T-wave alternans (MTWA) testing in 2007 for additional SCD risk stratification. The result was abnormal or positive (high risk for SCD). ICD treatment was recommended, however, other medical problems delayed implantation. He was having difficulty with his right-sided atrioventricular fistula and was being evaluated for a new left-sided fistula. He developed a large spontaneous left chest-wall hematoma that prevented implantation. He suffered SCD 6 months after his T-wave alternans test.

Question No. 1:

The absence of inducible VT in nonischemic cardiomyopathy patients is sufficiently predictive of low risk for SCD.

A. True.

B. False.

Question No. 2:

For which of the following are/is ambulatory electrocardiogram (ECG) monitoring not recommended?

A. Patients with unexplained recurrent palpitations.

B. Post-myocardial infarction (MI) patients with left ventricular dysfunction.

C. Patient with CHF.

D. Preoperative arrhythmia evaluation of patients for noncardiac surgery.

E. Patients with idiopathic hypertrophic cardiomyopathy.

Practical Electrophysiology © 2015 Jasbir S. Sra, Masood Akhtar, Andrea Natale, David J. Wilber
Cardiotext Publishing, ISBN: 978-1-935395-17-1

Question No. 3:

ICD therapy as primary prevention in nonischemic cardiomyopathy guided by SAECG has not been validated.

A. True.

B. False.

Question No. 4:

It is reasonable to use T-wave alternans for improving the diagnosis and risk stratification of patients with ventricular arrhythmias or who are at risk for developing life-threatening ventricular arrhythmias.

A. True.

B. False.

DISCUSSION

The unresolved public health challenge of predicting and preventing SCD remains a source of frustration. Accurate identification of individuals who are likely to die of cardiac arrhythmias would allow more selective use of ICDs to prevent SCD. Significant research energy has been spent in attempts to find noninvasive tests that can separate individuals into high- and low-risk groups.

Patients with low LVEF have consistently fallen into a high-risk category. Current guidelines reflect Multicenter Automatic Defibrillator Implantation Trial II (MADIT-II) findings, recommending ICD therapy for primary prevention of SCD in patients with ischemic heart disease and an LVEF of ≤ 30% without requiring other tests to guide therapy. As we know from MADIT-II,[1] post-MI patients with LVEF of ≤ 30% are at high risk and benefit from ICD therapy. Similarly, SCD-HeFT[2] showed that patients with reduced LVEF (≤ 35%) in the setting of CHF should be considered for ICD therapy to prevent SCD.

While LVEF has been the most consistent risk factor identified, there are well-recognized limitations to this approach. For example, there is often significant variation in LVEF determination based on the imaging modality utilized. In addition, the majority of SCD events occur in patients with ischemic heart disease but more preserved LVEF. Lastly, most patients who receive a primary prevention ICD based on low LVEF never require therapy from the device. For example, in SCD-HeFT,[2] 81% of patients in the ICD group did not use their device over 5 years of follow-up. The absolute risk reduction in the ICD group was a modest 7.2%. In MADIT-II,[3] the number-needed-to-treat to avoid one death over a 20-month period was 17.9. For these reasons, there has been keen interest to identify other means of identifying patients at risk for SCD that could potentially complement LVEF.

Three such noninvasive tests, beyond LVEF, evaluated to date measure:

1. Ventricular ectopy (long-term ambulatory ECG monitoring).
2. Slowed or fragmented conduction (signal-averaged ECG).
3. Variations in ventricular repolarization (T-wave alternans).

AMBULATORY ECG MONITORING

Ambulatory ECG monitoring is the continuous recording of ECG signals from patients while they are at home and engaged in routine daily activities. These noninvasive diagnostic tests are efficacious and cost-effective in the right clinical scenario. Many patients can have asymptomatic rhythm disturbances (such as nonsustained VT), and ambulatory monitoring may help define the existing burden of the arrhythmia. Ambulatory ECG monitoring has been used to estimate prognosis in patients with ischemic[4] or nonischemic cardiomyopathy.

The American College of Cardiology/American Heart Association (ACC/AHA), in collaboration with the North American Society of Pacing and Electrophysiology, has developed guidelines for the use of ambulatory ECG.[5] The guidelines include recommendations for the evaluation of symptoms of cardiac arrhythmias; for risk assessment in patients who have suffered a MI, have CHF or have hypertrophic cardiomyopathy; for the evaluation of antiarrhythmic therapy, pacemaker or ICD function; and for the evaluation of possible myocardial ischemia.

Three groups may benefit from ambulatory ECG to identify increased risk of future cardiac events (class IIb indication): patients with idiopathic hypertrophic cardiomyopathy, patients with CHF, and post-MI survivors with reduced ejection fraction.

SIGNAL-AVERAGED ELECTROCARDIOGRAPHY

Signal-averaged electrocardiography (SAECG) is a technique that averages multiple *QRS complexes* that are then digitalized, filtered, and further processed with spectral analysis to eliminate the background noise and detect ventricular late potentials (LPs). These LPs are low-amplitude, high-frequency signals representing areas of delayed myocardial activation

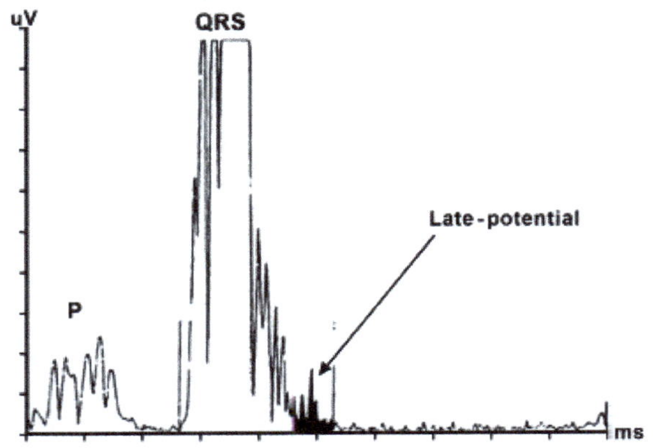

Figure 4.1. Example of a processed signal-averaged ECG revealing a high-frequency, low-amplitude LP.

due to slowed conduction from either regions of scar or fibrosis. Therefore, the LP signal represents the electrical substrate for reentry and VT (Figure 4.1).

Common criteria defining SAECG positivity include the following:

1. QRS duration: The filtered QRS complex is longer than 114 ms (in absence of bundle branch block).
2. Low-amplitude signal: The terminal filtered QRS complex remains below 40 microvolts for more than 38 ms.
3. Root mean square: There is less than 20 microvolts of signal in the last 40 ms of the filtered QRS complex.

It should be noted that SAECG has been validated only for those patients who are in sinus rhythm without ECG evidence of bundle branch block or intraventricular conduction delay. SAECG has been used to risk stratify the following populations for *sudden cardiac death:*

1. Post-MI patients.
2. Structural heart disease patients with depressed heart function.
3. Arrhythmogenic right ventricular dysplasia (ARVD).
4. Brugada syndrome.

There is insufficient reliable evidence to draw conclusions about the effectiveness of SAECG as a risk-stratification tool in patients with these cardiac conditions. Studies published to date fail to demonstrate clinical validity and utility.

For the purpose of post-MI risk stratification,[5] SAECG has a positive predictive value of 8% to 44% and a negative predictive value of 88% to 97%, but it has not been demonstrated to help impact clinical management of the patient. The Coronary Artery Bypass Graft (CABG)-Patch trial enrolled patients scheduled for CABG with an LVEF ≤ 35% and abnormalities on SAECG, and randomized them to an ICD group or control group.[7] There was no difference in overall mortality between groups. To date, no study has definitively reported a decrease in arrhythmic death as a direct result of using SAECG for risk stratification and subsequent treatment decisions.

The published ACC guidelines on implantable cardiac devices (1998) do not recommend SAECG for selecting patients for ICDs. A 2008 consensus document from the AHA, ACC Foundation, and Heart Rhythm Society indicates SAECG may identify patients with prior MI at risk for SCD, however, further studies are required to assess the utility of the test. The document states routine use of SAECG is not supported at this time.

SAECG testing is considered a standard part of the evaluation of patients with known or suspected ARVD. LPs by SAECG in 1 or more of 3 parameters in absence of QRS duration of ≥ 110 ms on ECG remains a minor criteria. The incidence of abnormal SAECG findings is directly proportional to the extent of anatomic right ventricular involvement. However, the presence of LPs does not give precise information as to electrical instability in these patients. Importantly, there have not been any studies directly comparing how ARVD patients do with and without ICDs.

LPs on SAECG have been shown to be useful (more than other arrhythmia markers) for identifying patients with Brugada syndrome at risk for ventricular fibrillation and SCD.

MICROVOLT T-WAVE ALTERNANS

MTWA testing is performed by placing high-resolution electrodes, designed to reduce electrical interference, on a patient's chest prior to a period of controlled exercise. The technology has been evaluated during pacing as well. These electrodes detect subtle beat-to-beat changes, on the order of one-millionth of a volt, in the ECG T-wave. Spectral analysis is used to calculate these minute voltage changes. Spectral analysis is a sensitive mathematical method of measuring and comparing time and the ECG signals. Software then analyzes these microvolt changes and produces a report to be read by a physician (Figure 4.2).

Only spectral-analytic-method MTWA tests for the evaluation of patients at risk for SCD are reimbursed by Medicare. The MTWA test is included in the Guidelines for Management of Patients with Ventricular Arrhythmias and the Prevention of SCD,

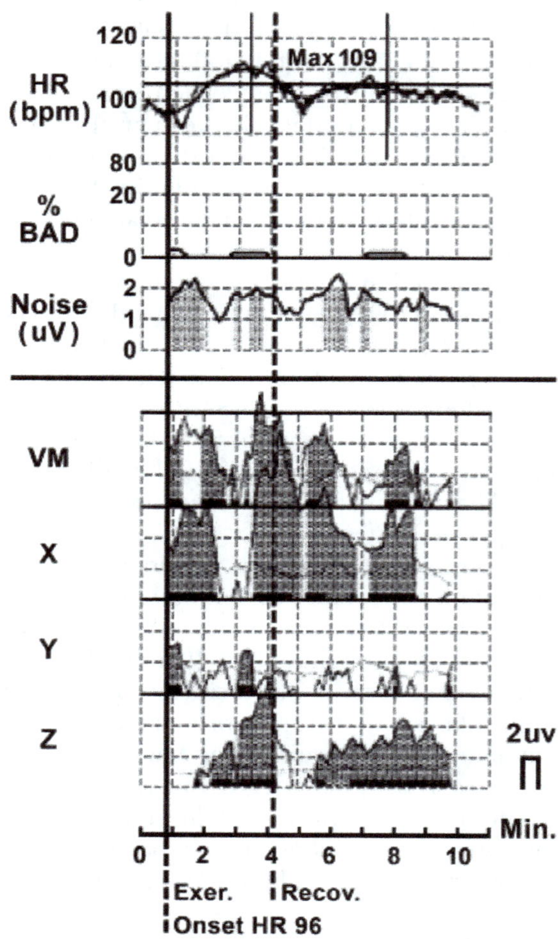

Figure 4.2. Example of a positive MTWA result. Top, Heart rate (HR) profile; ectopy (percentage of "bad" beats) and noise. Bottom, Alternans magnitude (microvolts, μV) in the vector magnitude (VM); and orthogonal leads (X,Y,Z) demonstrates sustained alternans at 96 bpm.

jointly developed by the ACC/AHA/European Society of Cardiology (ESC) as a class IIa indication.[6]

In short, the evidence for the use of MTWA in risk stratification after MI is not robust. The negative predictive value may be high in selected patients, but the results are conflicting in more representative cohorts. Recently published prospective studies raise big questions about the test's SCD risk-stratification value.

Results are now available from the MTWA Testing for Risk Stratification of Post-MI Patients (MASTER)-I study.[8] MASTER was conducted at 50 centers in the US and entered 575 patients who had MADIT-II indications for an ICD. All patients underwent MTWA testing before device implantation and were followed for an average of 2 years. An abnormal MTWA test did not predict the combined end point of arrhythmic death and ICD discharge. Interpretation of MASTER is complicated by its VT-event surrogate end point. The problem is we really don't know which patients would have gone on to die if they had not had an ICD and how many would have had arrhythmias that would have terminated on their own. In the MASTER-II trial,[9] the utility of MTWA was examined in 303 patients with LVEF between 31% and 40%, 48% of whom received an ICD. MASTER-II, which excluded patients with indeterminate test results, provided inconclusive results due to a lower-than-expected event rate.

The Alternans Before Cardioverter Defibrillator (ABCD) study[10] recruited 566 patients with ischemic cardiomyopathy and nonsustained VT and compared the ability of MTWA and electrophysiologic study to predict VT events. The investigators found that, at 1 year, non-negative MTWA was an important predictor of polymorphic VT, ventricular fibrillation, and SCD (2.7% vs 0%, $p = 0.04$) but not of monomorphic VT. At 2 years, non-negative MTWA was not associated with either outcome, suggesting time-dependent loss of value.

The ABCD study does have several limitations. This study demonstrated a very low event rate at 1 year of approximately 95% in both the negative MTWA and EPS patients. This contrasts with the relatively high event rate of 44% in the EPS-negative patients in the Multicenter UnSustained Tachycardia Trial (MUSTT) registry.[11] The difference in event rates between these 2 studies certainly requires further investigation, as the ABCD study suggests that MTWA alone, or in combination with EPS, may be used to avoid ICD in those not expected to benefit. One hypothesis is that the MUSTT study spanned a much longer time frame (5 years) and may have involved a dynamic SCD substrate. The value of serial MTWA testing would be intriguing but is unknown. It is also concerning that the ABCD study showed a loss of positive predictive value of MTWA after 12 months compared with EPS, which predicted events after 9 months for the duration of the study. Finally, this study did not evaluate patients with nonischemic cardiomyopathy at risk of SCD, in whom ICD implantation has also been shown to significantly reduce overall mortality.

Three early studies examined the prognostic value of MTWA in nonischemic cardiomyopathy. These studies were limited by small sample size, lack of adjustment for potential confounders, and no mortality end points. Results are now available from two larger studies of MTWA in patients with CHF. The MTWA in Patients with Heart Failure (ALPHA) study[12] recruited 446 patients with nonischemic cardiomyopathy (LVEF ≤ 40%) and stable NYHA class II/III CHF on optimal medical therapy.

A non-negative MTWA test was associated with an increased risk of cardiac death or VT events over 18 to 24 months. However, these patients were older, more symptomatic, and had a lower mean LVEF, imbalances that continue to limit the interpretation of these studies and highlight the need for careful multivariable adjustment.

In the SCD-HeFT T-wave alternans substudy,[13] MTWA testing was performed in 490 patients with LVEF ≤ 35% and NYHA class II/III CHF. The study had an unusually high indeterminacy rate of 41%. There was no difference in the rate of VT events between MTWA groups for patients who received either ICD or placebo over 35 months of follow-up. The substudy does, however, allow assessment of the MTWA marker in relation to the manifestation of the SCD phenotype.[14] In the main SCD-HeFT trial results, the mortality curves for the ICD and placebo groups separate at about 18 to 24 months. Since the ICD can only reduce total mortality by decreasing SCD rates, this represents the emergence of the SCD phenotype. The survival curves of MTWA-positive and MTWA-negative groups separate around the same time period, indicating the MTWA marker of risk closely tracks the SCD phenotype. The main conclusion of the study (MTWA did not predict outcomes) does not take into account the time-sensitive nature of risk evident in this population.

The confounding influence of surrogate end points for SCD cannot be overemphasized. For example, in the SCD-HeFT substudy, 38% of the end points were derived from "appropriate" ICD shocks. It is well recognized that the correlation between ICD-detected events and SCD is limited. ICD events have consistently overestimated actual risk for SCD. It is plausible that the lack of a statistical relationship between MTWA and VT events in the SCD-Heft substudy is due to this poor correlation. This issue was further examined by Hohnloser et al[15] in a meta-analysis of 14 clinical trials involving representative patient cohorts. The study population involved 6000 patients divided into low-ICD (≤ 15% of VT events were ICD-related) and high-ICD groups (majority of VT events were ICD-related). In the low-ICD group, total or SCD mortality was strongly correlated with MTWA positivity (average hazard ratio 16.6). Moreover, the annual event rate among MTWA negative patients was only 0.3%. In contrast, the trials that utilized ICD-related end points demonstrated very weak correlation between MTWA and events (average hazard ratio 1.6). These observations suggest that relevant clinical events, such as total or SCD mortality, can be predicted by the MTWA signal.

High MTWA indeterminacy rates remain a concern. Future refinements in data acquisition are needed to more broadly apply MTWA testing. An additional concern is that, to date, we still do not have a trial where ICD therapy is randomized on the basis of the MTWA signal and where clinically relevant end points are analyzed that do not include ICD-related events.

In summary, MTWA testing remains a scientifically valid tool for risk stratification, seems to hold the most promise, but is as yet incompletely tested. Clinical and experimental evidence suggest that it is a marker of electrophysiologic substrate for SCD. It helps physicians determine which patients are likely (or unlikely) to benefit from ICD therapy. It may be especially helpful for identifying high- or low-risk subsets of patients under circumstances where the ICD benefit is uncertain (advanced age, multiple comorbidities, etc.) or where patient reluctance to therapy exists. Lastly, it helps "stable" patients accept the need for an ICD and provides justification for exposing patients to surgical risks.

Correct Answers:

Q1. B

The predictive accuracy of electrophysiologic testing in nonischemic cardiomyopathy is limited. EP testing is less likely to induce VT in patients with nonischemic dilated cardiomyopathy and a presentation of aborted SCD or nonsustained VT than in patients with underlying coronary artery disease.

Q2. D

Preoperative arrhythmia evaluation of patients for noncardiac surgery. See ACC/AHA executive summary for more details.

Q3. A

No study definitively reported a decrease in fatal arrhythmias as a direct result of using SAECG for risk stratification and subsequent treatment decisions.

Q4. A

See ACC/AHA/ESC clinical guidelines for more details.

REFERENCES

1. Moss AJ, Zareba W, Hall WJ, et al. Prophylactic implantation of a defibrillator in patients with myocardial infarction and reduced ejection fraction. *N Engl J Med.* 2002;346:877–883.
2. Bardy GH, Lee KL, Mark DB, et al. Amiodarone or an implantable cardioverter-defibrillator for congestive heart failure. *N Engl J Med.* 2005;352:225–237.

3. Greenberg H, Case RB, Moss AJ, et al. Analysis of mortality events in the Multicenter Automatic Defibrillator Implantation Trial (MADIT-II). *J Am Coll Cardiol.* 2004;43:1459–1465.
4. Bailey JJ, Berson AS, Handelsman H, Hodges M. Utility of current risk stratification tests for predicting major arrhythmic events after myocardial infarction. *J Am Coll Cardiol.* 2001;38:1902–1911.
5. Crawford MH, Bernstein SJ, Deedwania PC, et al. ACC/AHA guidelines for ambulatory electrocardiography: executive summary and recommendations. A report of the American College of Cardiology/American Heart Association task force on practice guidelines (committee to revise the guidelines for ambulatory electrocardiography). *Circulation.* 1999;100:886–893.
6. Zipes DP, Camm AJ, Borggrefe M, et al. ACC/AHA/ESC 2006 Guidelines for Management of Patients With Ventricular Arrhythmias and the Prevention of Sudden Cardiac Death: a report of the American College of Cardiology/American Heart Association Task Force and the European Society of Cardiology Committee for Practice Guidelines (writing committee to develop Guidelines for Management of Patients With Ventricular Arrhythmias and the Prevention of Sudden Cardiac Death): developed in collaboration with the European Heart Rhythm Association and the Heart Rhythm Society. *Circulation.* 2006;114:e385–484.
7. Bigger JT Jr. Prophylactic use of implanted cardiac defibrillators in patients at high risk for ventricular arrhythmias after coronary-artery bypass graft surgery. Coronary Artery Bypass Graft (CABG) Patch Trial Investigators. *N Engl J Med.* 1997;337:1569–1575.
8. Chow T, Kereiakes DJ, Onufer J, et al. Does microvolt T-wave alternans testing predict ventricular tachyarrhythmias in patients with ischemic cardiomyopathy and prophylactic defibrillators? The MASTER (Microvolt T Wave Alternans Testing for Risk Stratification of Post-Myocardial Infarction Patients) trial. *J Am Coll Cardiol.* 2008;52:1607–1615.
9. Chow T, Kereiakes DJ, Onufer J, et al. Prognostic value of microvolt T-wave alternans in patients with moderate ischemic left ventricular dysfunction: results from the MASTER II trial (abstr). *J Am Coll Cardiol.* 2008; 51(10):A17.
10. Costantini O, Hohnloser SH, Kirk MM, et al. The ABCD (Alternans Before Cardioverter Defibrillator) Trial: strategies using T-wave alternans to improve efficiency of sudden cardiac death prevention. *J Am Coll Cardiol.* 2009;53:471–479.
11. Buxton AE, Lee KL, DiCarlo L, et al. Electrophysiologic testing to identify patients with coronary artery disease who are at risk for sudden death. Multicenter unsustained tachycardia trial investigators. *N Engl J Med.* 2000;342:1937–1945.
12. Salerno-Uriarte JA, De Ferrari GM, Klersy C, et al. Prognostic value of T-wave alternans in patients with heart failure due to nonischemic cardiomyopathy. *J Am Coll Cardiol.* 2007;50:1896–1904.
13. Gold MR, Ip JH, Costantini O, et al. Role of microvolt T-wave alternans in assessment of arrhythmia vulnerability among patients with heart failure and systolic dysfunction: primary results from the T-wave alternans sudden cardiac death in heart failure trial substudy. *Circulation.* 2008;118:2022–2028.
14. Rosenbaum DS. T-Wave alternans in the sudden cardiac death in heart failure trial population: signal or noise? *Circulation.* 2008;118:2015–2018.
15. Hohnloser SH, Ikeda T, Cohen RJ. Evidence regarding clinical use of microvolt T-wave alternans. *Heart Rhythm.* 2009;6:S36–S44.

SECTION 2

Supraventricular Tachycardias

CASE STUDY

5

Diagnostic Pacing Maneuvers During Paroxysmal Supraventricular Tachycardia

Aman Chugh, MD and Fred Morady, MD

A 25-year-old woman with recurrent palpitations and documented supraventricular tachycardia (SVT) presents to the electrophysiology (EP) laboratory. The electrocardiogram (ECG) shows no evidence of preexcitation. In the baseline state, there is no evidence of ventriculoatrial (VA) conduction or dual atrioventricular (AV) nodal pathways. There is no inducible tachycardia. During isoproterenol infusion (up to 10 mcg/min), there is an evidence of dual AV nodal physiology and a single AV nodal echo beat.

Question No. 1A:

Which of the following is the most appropriate next step?

A. Since there is no inducible tachycardia, terminate the procedure and prescribe low-dose propafenone and metoprolol.

B. Perform the induction protocol with atropine to increase the likelihood of sustained tachycardia.

C. Perform slow pathway ablation.

D. Perform fast pathway modification.

E. Target the earliest atrial activation during ventricular pacing and isoproterenol infusion since the clinical tachycardia is likely mediated by a decrementally conducting concealed accessory pathway.

Question No. 1B:

While performing radiofrequency ablation (RFA) of the slow pathway, you note junctional ectopy (cycle length 400 ms) with prolongation of the VA interval. What is the most appropriate next step?

A. Continue RF energy delivery as long as there is 1:1 VA conduction.

B. Increase the power to increase the likelihood of more rapid junctional ectopy.

C. Discontinue RF energy delivery and target a more superior site along the tricuspid annulus.

D. Discontinue RF energy delivery and target a more caudal site near the ostium of the coronary sinus.

E. Continue RF energy delivery since you have discussed the possibility of AV block and pacemaker implantation with the patient prior the procedure.

Question No. 2:

A 23-year-old woman is referred for an electrophysiologic study for a history of palpitations. The ECG reveals preexcitation compatible with an anteroseptal accessory pathway (QRS duration 90 ms). The baseline His-ventricular interval is 20 ms and remains unchanged during programmed atrial stimulation. With a single atrial extrastimulus, a tachycardia is induced with a cycle length of 320 ms. The HV interval remains unchanged during tachycardia. Diagnostic maneuvers during tachycardia are consistent with the diagnosis of typical atrioventricular nodal reentrant tachycardia (AVNRT). What is the next appropriate step?

A. First, ablate the accessory pathway, followed by the slow pathway, so that you can reliably document VA conduction during RFA of the slow pathway for AVNRT.
B. First, ablate the slow pathway, followed by the accessory pathway.
C. First, eliminate the accessory pathway using cryoablation, and then the slow pathway using conventional RF energy.
D. In order to minimize the risk of AV block, target the accessory pathway via the coronary cusp with conventional RF energy, and then target the slow pathway.
E. Target the slow pathway.

Question No. 3A:

A narrow complex tachycardia is induced in the electrophysiologic laboratory in a patient with a history of recurrent palpitations. The cycle length is 350 ms and the septal VA time is 140 ms. A His-synchronous premature ventricular complex (PVC) introduced during tachycardia advances the subsequent atrial electrogram (EGM) and the tachycardia continues at the same rate and with the same atrial activation pattern. Which of the following is true regarding this observation?

A. It is diagnostic of orthodromic reciprocating tachycardia (ORT) using a concealed AV accessory pathway.
B. It is diagnostic of the presence of an accessory pathway.
C. It is diagnostic of atypical (fast-slow) AVNRT with a bystander concealed accessory pathway.
D. It is diagnostic of ORT utilizing a concealed nodofascicular accessory pathway.

Question No. 3B:

Following ablation of a manifest midseptal accessory pathway, you perform high-output pacing from a catheter placed at the bundle of His. You note that the paced QRS complex suddenly widens as the pacing output is decreased. The stimulus-atrial (SA) interval with a narrow paced QRS complex is 100 ms compared to 130 ms with a wide paced QRS complex. Which of the following is true regarding these findings?

A. There is evidence of retrograde conduction over the accessory pathway requiring further ablation.
B. These findings are consistent with retrograde conduction over the AV node.
C. These findings are consistent with retrograde activation over the AV node and a partially ablated accessory pathway.
D. These findings are inconclusive; need more data.

DISCUSSION

Catheter ablation of SVT is a common indication for referral to an electrophysiologist. Patients with paroxysmal SVT tend to be relatively young and often are free of structural heart disease. Typical symptoms during a bout of SVT include abrupt onset and offset of rapid palpitations, chest discomfort, shortness of breath, lightheadedness or dizziness, and presyncope. Evaluation during an episode usually reveals a patient who is hemodynamically stable, with an elevated pulse typically in the range of 120 to 220 bpm. An ECG during a paroxysm of SVT usually shows a regular, narrow-QRS complex tachycardia, with P waves not readily apparent. In some cases, the ECG is diagnostic of the SVT mechanism. A pseudo R-prime (R′) or a pseudo S wave in the inferior leads during tachycardia that is not present during sinus rhythm is diagnostic of typical AVNRT. Vagal maneuvers may be helpful in terminating an episode of SVT. In the emergency room, intravenous adenosine is usually effective in terminating the tachycardia. Catheter ablation is first-line therapy for a patient with recurrent SVT given its favorable risk/benefit ratio.

The differential diagnosis of SVT during an EP study includes AVNRT, ORT, and atrial tachycardia (AT).

Less common diagnoses include junctional tachycardia and tachycardia using a concealed nodofascicular accessory pathway. Although pacing maneuvers are critical for arriving at the correct

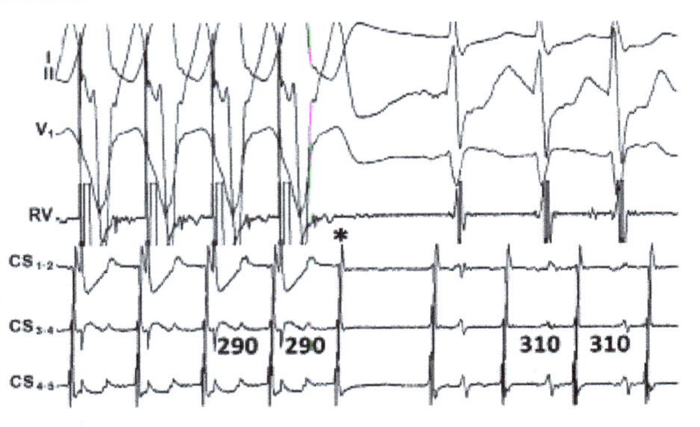

Figure 5.1. Response to entrainment from the ventricle in a patient with SVT. After cessation of ventricular pacing that entrains the tachycardia, there is an atrial-atrial-ventricular (A-A-V) response. This observation is diagnostic of AT as the mechanism. The asterisk identifies the last entrained atrial EGM. The AT was successfully ablated at the Para-Hisian area.

diagnosis, simply observing the tachycardia may provide some useful clues. For example, a variable VA interval makes AT more likely than AVNRT or ORT. A septal VA interval < 70 ms rules out ORT. AV block during tachycardia rules out ORT since ventricular myocardium is an integral part of the reentrant circuit. A tachycardia during which oscillations in the ventricular cycle length precede those in the atrial cycle length is likely to be typical AVNRT or ORT and unlikely to be AT.[1]

The mode of induction provides insight into the mechanism of the tachycardia. Induction of tachycardia with AV block rules out ORT and favors the diagnosis of AT. The observation that the initiation of tachycardia is dependent on a critical atrial-His (AH) interval is diagnostic of typical AVNRT (Figure 5.2).

Pacing from the ventricle is invaluable in establishing the mechanism in patients with paroxysmal SVT. A lack of VA conduction during the baseline state strongly argues against ORT. Even if the tachycardia is not entrained during ventricular overdrive pacing, the response may still be helpful. For example, if the tachycardia terminates without affecting the atrial EGM, AT is extremely unlikely. Insertion of a His-synchronous PVC is helpful in confirming the presence of an accessory pathway. If the PVC advances or delays the subsequent atrial EGM, the presence of an accessory pathway is confirmed (Figure 5.3). If the tachycardia terminates in the retrograde limb, the diagnosis of ORT is assured.

Ventricular overdrive pacing that entrains but does not terminate the tachycardia is a commonly performed maneuver in the EP laboratory for distinguishing between the various causes of SVT.[2,3] This is performed by pacing from a catheter in the right ventricular (RV) apex at a cycle length 20 ms shorter than that of the tachycardia. The P-wave morphology, which may be difficult to discern during tachycardia, may be unmasked after cessation of ventricular pacing. An upright P wave in the inferior ECG leads rules out AVNRT. An atrialventricular response upon cessation of ventricular pacing that entrains the tachycardia rules out AT (Figure 5.4). Entrainment from the ventricle during AVNRT or ORT occurs by VA conduction over the retrograde limb of the circuit. After cessation of pacing, the last ventricular paced complex results in an atrial EGM via conduction over the retrograde limb that is able to conduct down the anterograde limb, since it is not refractory, resulting in an AV response. Ventricular entrainment during AT advances the atrial EGM via retrograde conduction over the AV node. Upon

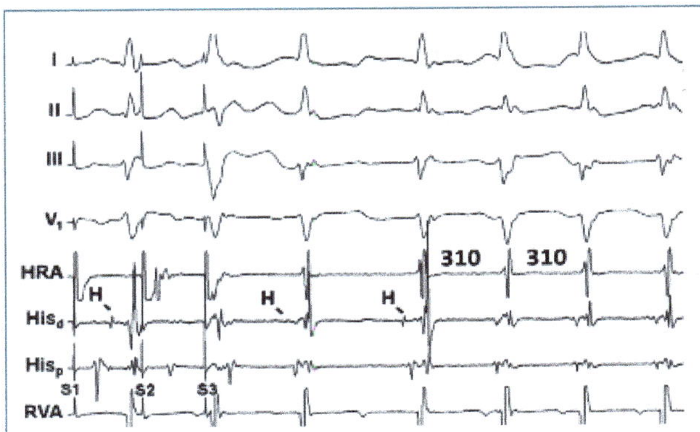

Figure 5.2. This tachycardia was only inducible with a critical delay in the AH interval, which is diagnostic of typical AVNRT. H, His potential.

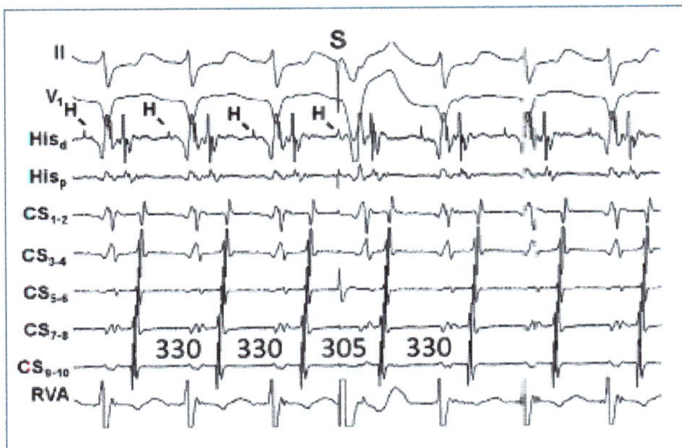

Figure 5.3. Response to His-synchronous PVC during SVT. The PVC advances the subsequent atrial EGM by 25 ms, confirming the presence of an accessory pathway. Other maneuvers confirmed the mechanism to be ORT utilizing a Para-Hisian accessory pathway. H, His potential; S, stimulus.

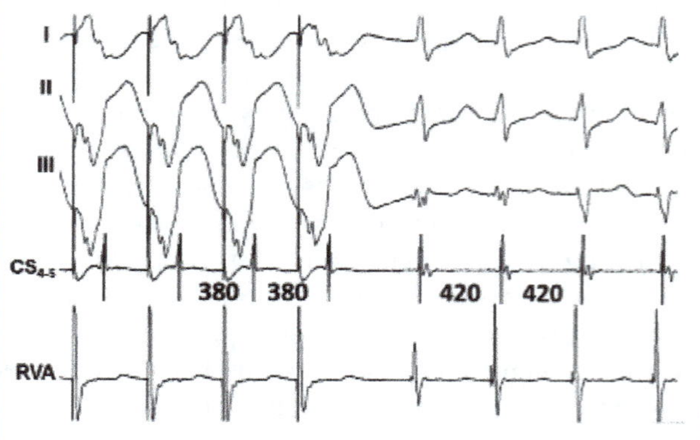

Figure 5.4. This tracing is from the same patient as in Figure 5.1. This tachycardia was induced after elimination of the Para-Hisian AT shown in Figure 5.1. Ventricular pacing advances the atrial EGMs to the pacing rate. Upon cessation of pacing, there is an AV response, ruling out AT. Furthermore, the simultaneous ventricular and atrial activation during tachycardia rules out ORT, leaving only typical atrioventricular nodal reentry tachycardia as the diagnosis.

cessation of ventricular pacing, the last entrained atrial EGM finds the AV node to be refractory and is unable to conduct anterogradely to the ventricle, resulting in an A-A-V response. This maneuver remains valid in a patient with AT and dual AV nodal pathway (Figures 5.1 and 5.4).

Entrainment of an SVT from the ventricle is also useful in discriminating between atypical (fast/slow) AVNRT and ORT using a septal accessory pathway. Since the ventricle is part of the macroreentrant circuit during ORT, the postpacing interval (PPI) minus the tachycardia cycle length (TCL) is shorter during the latter than the former. Specifically, a PPI-TCL < 110 ms is highly specific for ORT using a septal accessory pathway. This maneuver may not be helpful in distinguishing atypical AVNRT from ORT using a *very* slowly conducting accessory pathway.

The response to entrainment from the ventricle is only helpful if the tachycardia continues following cessation of pacing. However, ventricular pacing during SVT may still offer diagnostic clues despite tachycardia termination. At the onset of an entrainment attempt from the ventricle, the QRS complex may be identical to the QRS complex during tachycardia, may be fused, or may be identical to the fully paced QRS morphology. A recent study showed that in the case of ORT, the tachycardia is reset to the paced cycle length after 1 fully paced RV complex is observed on the ECG.[4] On the other hand, with AVNRT, an average of 3.7 fully paced QRS complexes occurred until the tachycardia was reset. The resetting response is quicker during ORT since there is only intervening ventricular tissue between the pacing site and the ventricular insertion of the pathway. During AVNRT, the paced wavefront must traverse the His-Purkinje system and AV nodal tissue before the atrium can be reset.

Para-Hisian pacing may be helpful in distinguishing between retrograde conduction over a septal accessory pathway and an AV nodal pathway. The His-bundle catheter is advanced into the ventricle to the point where the atrial EGM is not visible and the His potential still is recorded. The pacing output is gradually decreased from 15 to 20 mA while surveying the QRS morphology. Simultaneous capture of the His bundle and the RV myocardium at high output results in a relatively narrow QRS complex. As the pacing output is decreased, the QRS suddenly widens, consistent with capture of only the RV myocardium. In the presence of a septal accessory pathway, the SA EGM interval during the narrow and wide paced complex is the same (Figure 5.5). Since retrograde activation over the accessory pathway depends only on capture of ventricular tissue, the SA interval is the same. In the absence of an accessory pathway, the SA interval after the narrow paced ventricular complex is shorter than that after the wide complex. The SA interval is longer when the His bundle is not captured because atrial activation depends on a sequence of several steps: depolarization of the RV myocardium, followed by engagement of distal ramifications of the His-Purkinje system, the His bundle, and the AV node. But during His-bundle capture, atrial activation occurs earlier since the impulse only needs to conduct from the very

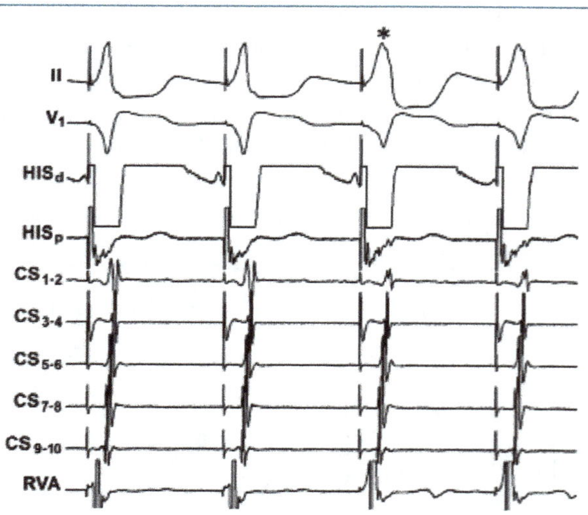

Figure 5.5. Para-Hisian pacing. As the pacing output is gradually decreased, the narrow QRS morphology gives way to a wide QRS morphology (*). Despite the change in the QRS morphology, the SA EGM interval and atrial activation remain the same, consistent with retrograde activation over a septal accessory pathway.

proximal aspect of the specialized conduction system. Para-Hisian pacing is also helpful in determining whether a septal accessory pathway has been successfully ablated.

Comparing the AH intervals during tachycardia and during atrial pacing at the TCL may be helpful in distinguishing between various causes of a long R-P tachycardia. In a prior study, the mean AH interval during atrial pacing was approximately 70 ms longer than that during tachycardia in patients with atypical AVNRT.[5] In contrast, in patients with ORT or AT, this difference never exceeded 20 ms and 10 ms, respectively.

UNUSUAL CAUSES OF SUPRAVENTRICULAR TACHYCARDIA

Besides AT, AVNRT and ORT, one may occasionally encounter more unusual causes of SVT in the EP laboratory. These include junctional tachycardia and ORT using a concealed nodofascicular/nodoventricular accessory pathway. With junctional tachycardia, one may observe periods of isorhythmic dissociation between sinus rhythm and junctional automaticity. On the ECG, one may be able to observe upright P waves in the inferior leads during isorhythmic dissociation, which rules out typical AVNRT, the major diagnosis to rule out in such a circumstance. Introduction of an atrial stimulus during tachycardia is also very helpful. If the atrial extrastimulus conducts to the ventricle with a relatively short AH interval and the tachycardia continues, the diagnosis is very likely junctional tachycardia. During typical AVNRT, the tachycardia would terminate after anterograde conduction occurs over the fast AV nodal pathway. A clue that one may be dealing with ORT using a concealed nodofascicular/nodoventricular pathway is the presence of VA block during tachycardia. A His-synchronous PVC that advances the subsequent His EGM supports the presence of such a pathway.

SUMMARY

In general, these observations and maneuvers are helpful in arriving at the correct diagnosis in the vast majority of cases of paroxysmal SVT. Occasionally, however, one may have to deal with conflicting information based on application of the various maneuvers. In such cases, one has to prioritize the results, based on the positive and negative predictive values of the observations. In difficult cases, there is no substitute for rigorously applying EP principles to elucidate the mechanism of arrhythmia.

Correct Answers:

Q1A. C

Q1B. D

Q2. E

Q3A. B

Q3B. B

REFERENCES

1. Crawford TC, Mukerji S, Good E, et al. Utility of atrial and ventricular cycle length variability in determining the mechanism of paroxysmal supraventricular tachycardia. *J Cardiovasc Electrophysiol.* 2007;18:698–703.
2. Knight BP, Zivin A, Souza J, et al. A technique for the rapid diagnosis of atrial tachycardia in the electrophysiology laboratory. *J Am Coll Cardiol.* 1999;33:775–781.
3. Knight BP, Ebinger M, Oral H, et al. Diagnostic value of tachycardia features and pacing maneuvers during paroxysmal supraventricular tachycardia. *J Am Coll Cardiol.* 2000;36:574–582.
4. Dandamudi G, Mokabberi R, Assal C, et al. A novel approach to differentiating orthodromic reciprocating tachycardia from atrioventricular nodal reentrant tachycardia. *Heart Rhythm.* 2010;7:1326–1329.
5. Man KC, Niebauer M, Daoud E, et al. Comparison of atrial-His intervals during tachycardia and atrial pacing in patients with long RP tachycardia. *J Cardiovasc Electrophysiol.* 1995;6:700–710.

CASE STUDY

6

Narrow-Complex Tachycardia with Features Suggesting Multiple Tachycardia Mechanisms–the Importance of Electrophysiology Fundamentals

Indrajit Choudhuri, MD and Jasbir S. Sra, MD

A 52-year-old woman without medical history presented with sustained palpitations provoking near-syncope. Diagnostic cardiac evaluation including electrocardiography and echocardiography were entirely normal, and event monitoring revealed narrow-complex tachycardia (Figure 6.1, top panel). Medical and interventional options were discussed and the patient opted for invasive evaluation. At electrophysiology (EP) study, the patient spontaneously developed tachycardia with similar cycle length (CL) and morphology (see Figure 6.1, bottom panel).

Question No. 1:

Which of the following supraventricular tachycardia (SVT) mechanisms is likely?

A. Atrioventricular node reentry tachycardia (AVNRT).

B. Atrial tachycardia (AT).

C. Orthodromic reentry tachycardia (ORT).

D. All of the above.

Baseline intervals included AH 150 ms and HV 55 ms at sinus CL 780 ms. During programmed atrial stimulation at a basic drive of 400 ms atrial

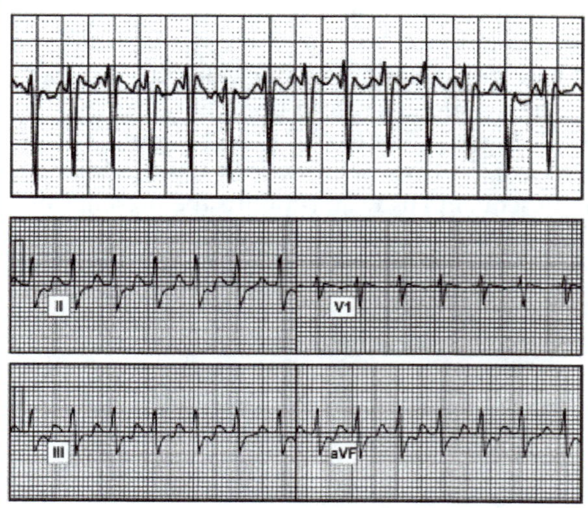

Figure 6.1. **Top panel:** Holter monitor recording showing narrow-complex tachycardia. **Bottom panel:** ECG recorded during EP study showing ECG leads II, III, aVF, and V_1.

Practical Electrophysiology © 2015 Jasbir S. Sra, Masood Akhtar, Andrea Natale, David J. Wilber
Cardiotext Publishing, ISBN: 978-1-935395-17-1

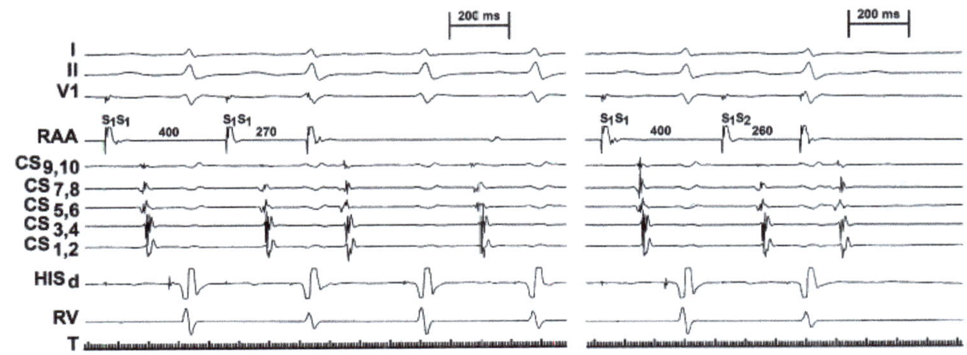

Figure 6.2. Recordings shown from top to bottom are electrocardiographic leads I, II, V_1 and right atrial appendage (RAA), coronary sinus (CS) proximal to distal, distal His Bundle (HIS_d), and right ventricle (RV).

extrastmulation at 270 ms reproducibly induced the tachycardia (first beat shown; Figure 6.2, left panel). A 10-ms decrement in the atrial extrastimulus coupling interval resulted in inability to induce the tachycardia (see Figure 6.2, right panel).

Question No. 2:

Which of the following tachycardia mechanisms is least likely?

A. Atrioventricular (AV) node reentry.

B. Atrial tachycardia.

C. Orthodromic reentry tachycardia.

Fixed CL ventricular pacing during EP study is depicted in Figure 6.3.

Question No. 3:

What can be inferred about the general tachycardia mechanism?

A. Retrograde conduction is critical to the tachycardia initiation.

B. Tachycardia is spontaneously induced during sinus rhythm and is independent of V pacing.

C. AT is eliminated as a mechanism of tachycardia.

During sinus rhythm, long ventricular pacing drives (pacing cycle length [PCL] 320 ms) were required to demonstrate 1:1 retrograde conduction (Figure 6.4, left panel); the tachycardia exhibited slight variability in CL, though with fixed mechanism–ventricular overdrive pacing at 330 ms

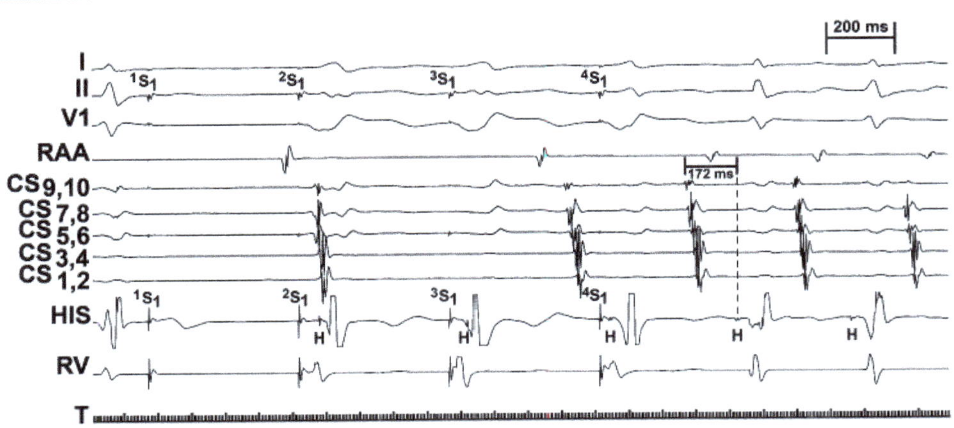

Figure 6.3. Fixed CL ventricular pacing. Tracings are similar to Figure 6.2.

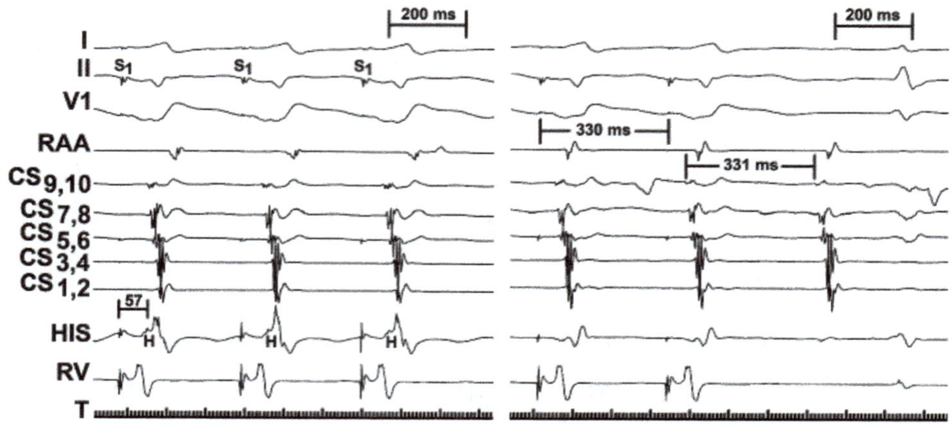

Figure 6.4. Ventricular pacing at cycle l (left panel) and ventricular overdrive pacing during tachycardia (right panel). Tracings are similar to Figure 6.2.

(tachycardia cycle length [TCL] 354 ms) is shown (see Figure 6.4, right panel).

Question No. 4:

What do the results of overdrive pacing demonstrate?

A. AT is the likely mechanism.

B. Entrainment response suggests AV node-dependent tachycardia as the mechanism.

C. Overdrive pacing fails to entrain the tachycardia.

During sustained tachycardia the phenomenon in Figure 6.5 was observed.

Question No. 5:

Which mechanism of tachycardia is ruled out?

A. Atrial tachycardia.

B. Orthodromic reentry tachycardia.

C. Atrioventricular node reentry.

DISCUSSION

The case presented illustrates several fundamental electrophysiologic properties that must be recognized in order to properly diagnose SVT.

One of the most critical pieces of information is a clinical recording of the arrhythmia (see Figure 6.1, upper panel). The Holter monitor demonstrates a narrow-complex tachycardia with unclear P-wave

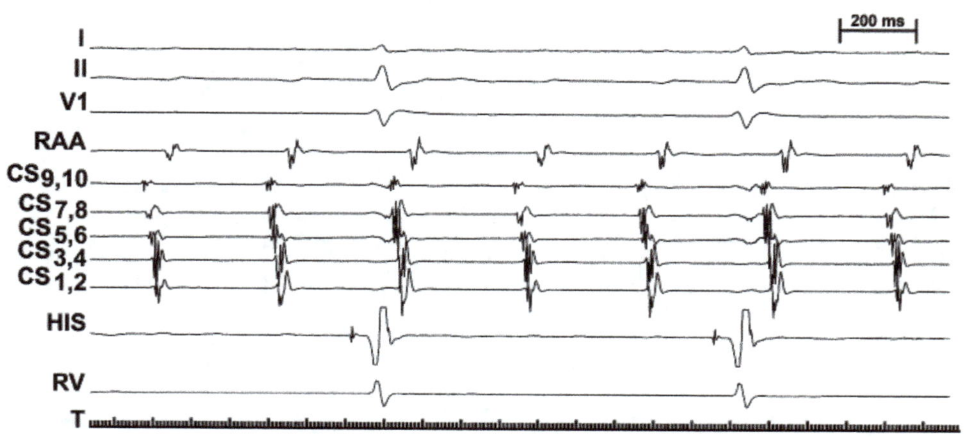

Figure 6.5. Advanced heart block during tachycardia. Tracings similar to Figure 6.2.

morphology, but the arrhythmia induced at EP study was of similar CL and the P waves were more easily discernable now with 12 leads available for scrutiny (see Figure 6.1, lower panel–ECG leads II, III, aV_F, and V_1 shown). The narrow QRS preceded by a His potential diagnoses supraventricular origin, and the P waves can be seen to have an inverted morphology in leads II, III, and aV_F: that is, the atria are activated with a superiorly directed axis, and occur closer to the *preceding* rather than *subsequent* QRS—a short to intermediate R-P tachycardia. The differential diagnosis includes AV node reentry, as well as ORT utilizing a somewhat slowly conducting accessory pathway, and AT. The normal HV interval in sinus rhythm and narrow QRS exclude antidromic tachycardia but do not eliminate the possibility of retrograde conduction over a concealed accessory pathway. Inappropriate sinus tachycardia, sinus node reentry or any other superiorly originating tachycardia such as superior pulmonary vein tachycardias can be eliminated because superiorly located sites depolarize the atria in an inferiorly directed manner.

With atrial extrastimuli the tachycardia was reproducibly induced. When a single extrastimulus at 270 ms was coupled to a 400-ms drive delivered from the RAA (see Figure 6.2, left panel), the tachycardia is induced with delay between the extrastimulus and the first tachycardia atrial activation. This may be seen with reentrant tachycardia because the coupling interval of the first spontaneous tachycardia beat varies inversely with the extrastimulus coupling interval due to engagement of a slow conduction zone that permits recovery and reexcitation at the stimulation site. In contradistinction, automatic and triggered rhythms initiate with rapidity of stimulation, which is felt to provoke intracellular calcium overload that dissipates with time from the last stimulus and, thus, the coupling interval of the first spontaneous beat varies *directly* with the coupling interval of the extrastimulus. By this feature alone, enhanced automaticity is least likely. AV node reentry and orthodromic reentry are certainly plausible, not only because as reentrant arrhythmias they would exhibit initiation pattern similar to that shown, but also because tachycardia initiation was accompanied by conduction through the AV node (see Figure 6.2, left panel). When the atrial coupling interval was decremented by 10 ms (see Figure 6.2, right panel), the AV node was refractory and the tachycardia could not be initiated. This tachycardia exhibits AV node dependence.

Two critical observations should be made from the 2 tracings in Figure 6.3, the first being whether retrograde conduction is present. Initially, it appears that there is no retrograde conduction, but retrograde conduction can be deduced. After the second ($^2S_1$) and third pacing stimuli ($^3S_1$), a retrograde His-bundle deflection is present at a coupling interval of roughly 55–60 ms, similar to the antegrade HV interval. The fourth pacing stimulus ($^4S_1$) is followed by a His with shorter coupling interval than the other retrograde His electrograms (EGMs), that is, this is an antegrade His produced by sinus activation of the second atrial beat. However, note the subtle changes in CS EGMs that are now similar to CS activation during tachycardia. This is because $^3S_1$ conducted retrogradely to the proximal CS.

The second observation is to determine the mechanism of tachycardia initiation. The tracing shows 5 atrial activations; the first 2 appear to be sinus and the remaining 3 are tachycardia beats. After the fourth pacing stimulus ($^4S_1$), which is followed by the antegrade His, the global atrial activation changes and the tachycardia appears to have been initiated, that is, the tachycardia was initiated when the His had already been activated antegradely and therefore could not have been available to accommodate retrograde conduction–the tracing appears to depict tachycardia induction with a ventricular stimulus during His refractoriness and therefore must represent retrograde activation over an accessory pathway. This is incorrect.

As stated, the second beat in fact is a fusion of sinus rhythm (RAA and His A's are as expected for sinus rhythm) and retrograde conduction from the previous pacing stimulus, $^3S_1$, which contributes the CS activation and is the true first beat of the tachycardia. Notice the particularly long VA interval from the $^3S_1$ that initiates the tachycardia. This implies that $^4S_1$ likely only captured local ventricular myocardium but did not propagate. The earliest atrial activation occurs at the proximal CS electrode, suggesting that this posteroseptal position is in the vicinity of the retrogradely conducting pathway that permitted early activation here—this does not definitively identify the tachycardia mechanism. Posteroseptal accessory pathways and slow AV nodal pathways are located in proximity to this region and would be the likely candidates to explain the tachycardia.

In Figure 6.4, ventricular pacing with 1:1 retrograde conduction during sinus rhythm (left panel) and overdrive pacing during tachycardia (TCL 354 ms, right panel) are shown. Observe first that the retrograde atrial activation from ventricular pacing during sinus rhythm is identical to the atria activation during tachycardia. This excludes AT except for the remote possibility of a focus coincidentally present at the proximal CS electrode activated by retrograde depolarization stimulated by ventricular pacing—a statistical improbability.

Second, it is important to recognize that ventricular overdrive pacing entrains the tachycardia—the tachycardia is accelerated to the PCL with preserved atrial activation sequence. With cessation of pacing, the pattern of return beats is 2 tachycardia atrial activations followed by conduction to the ventricle, an "A-A-V" response. This is suggestive of AT, which previously was felt to be unlikely and hence may serve as a source of confusion; AVNRT and ORT would be expected to exhibit an "AV" response to entrainment because both tachycardias exhibit 1:1 ventricular activation in the absence of AV block, which would also terminate ORT. Closer inspection reveals that not the first but rather the *second* atrial activation is the last atrial activation to occur at the PCL, suggesting that the second atrial activation after cessation of ventricular pacing is in fact the last entrained atrial beat and this is followed by ventricular activation, that is, an "AV" response to entrainment—the so-called "pseudo A-A-V" response—again suggesting AVNRT or ORT as possible mechanisms.

However, the postpacing interval (PPI) is much longer than tachycardia (PPI-TCL = 216 ms), and the difference between the stimulus to earliest retrograde atrial activation (stim-A) during overdrive pacing and the VA interval during tachycardia is also significant ([stim-A] – [VA] = 276 ms), intimating that the ventricular pacing site is remote from the circuit, a finding that does *not* suggest orthodromic reentry. With ORT, these values are typically < 115 ms and < 85 ms, respectively.

An easily overlooked observation is the dramatic retrograde conduction time. This is suggested in several places throughout the case—first in Figure 6.3 as alluded to above when ventriculoatrial (VA) conduction results from $^3S_1$, with the response to entrainment demonstrating a VA time ~385 ms (see Figure 6.4, right panel), and more subtly, by ventricular pacing (see Figure 6.4, left panel) in which the retrograde His can be observed to occur *after* the proximal CS, intimating that ventricular pacing activates the His retrogradely and then conducts to the *subsequent* atrial activation.

This dramatically prolonged VA time should suggest presence of a lower common pathway (LCP). In support of this, note that during tachycardia at 320 ms the HA time is 148 ms (320–172), whereas with ventricular pacing *at the same CL during sinus rhythm*, the HA time is ~290 ms. It is important to remember that the HA time during entrainment will be similar to the HA time during tachycardia resulting in "ΔH" < 40ms (= 0 ms in this case). In the presence of a LCP, the HA time during ventricular pacing is longer than the HA time during tachycardia, usually by at least 40–45 ms—in our case 142 ms—because during tachycardia the HA time is a "pseudo-interval" in which antegrade conduction from the lower turnaround point in the AV node through the LCP and then to the His is occurring simultaneously with retrograde activation, and hence, would be expected to be shorter than serial conduction from the His through the LCP, through the AV node and then to the earliest A.

Notice also that the VA time *exceeds* the TCL. This can only occur in AVNRT due to conduction delay *to* the tachycardia circuit because the right ventricular apex is remote from the AV node circuit. This cannot happen in AV reentry because ventricular pacing sites are close to or incorporated with the tachycardia circuits. In AT, overdrive ventricular pacing must conduct to the atrium with times less than the TCL to even entrain. The LCP contributes to conduction delay from the ventricular pacing site to the AV node circuit and exaggerates the degree the VA time exceeds the TCL.

Finally, AV block develops during tachycardia without change in atrial CL or activation sequence (see Figure 6.5). This immediately eliminates participation of an accessory pathway from the tachycardia mechanism, but AT and AVNRT can occur with AV block. Typically, when AVNRT occurs with AV block, there is a 2:1 conduction pattern with *infra*-Hisian block. However, the high-grade AV block occurred, which is itself unusual, *and* the site of block was *above* the His. This may cause the operator to lose faith in the diagnosis of AVNRT and reconsider AT as the final diagnosis, *despite the wealth of evidence to the contrary*. In fact, this *is* compatible with AVNRT in which block occurs between the lower turnaround point and the His, that is, in the LCP.

In view of these observations, the diagnosis was AVNRT. The specific subform was felt to be an atypical, fast-slow AVNRT in keeping with the current definition in which antegrade conduction occurs with AH < 200 ms and retrograde conduction is earliest at the proximal CS. Multiple antegrade and retrogradely conducting pathways may be identified, contributing to variability in TCL related to accommodation along the various AVN pathways. We performed extensive detailed catheter mapping during tachycardia to identify the retrograde slow pathway. The earliest retrograde atrial activation was found in the inferoposterior right atrium 77 ms earlier than the earliest activation at the proximal CS. Radiofrequency delivered there terminated the tachycardia and rendered it noninducible.

In summary, this case illustrates that a narrow-complex tachycardia with inverted P waves includes the differential of low AT, AVNRT, and ORT. The presence of retrograde conduction can be difficult to appreciate, and its absence implies that reentry cannot underlie the mechanism of a narrow-complex tachycardia and intimates AT, but the reader is advised to consider the presence of an LCP and its potential impact on the determination of VA conduction. A number of features make AT unlikely here, including retrograde atrial activation during ventricular pacing identical to tachycardia; tachycardia initiation dependent upon AV node conduction; and a "pseudo-A-A-V" response to ventricular entrainment making an AV node-dependent tachycardia, whether AVNRT or ORT, more likely. The seeming induction during His refractoriness was suggestive of accessory pathway conduction, but closer inspection revealed that this was not the case and there was no strong evidence to support presence of an accessory pathway. This was confirmed by the development of sustained AV block during tachycardia, conveying that the ventricle is not incorporated in the tachycardia circuit. Fast-slow AVNRT with presence of an LCP was felt most likely and reconciled with careful determination of the VA interval as well as further evaluation of the ΔHA interval during tachycardia compared to ventricular pacing and was also suggested by the inability to conduct retrogradely during ventricular pacing which was delivered at CLs far longer than tachycardia. The rigorous application of electrophysiologic principles permitted diagnosis of the tachycardia, and catheter mapping identified the site of earliest retrograde activation, leading to a successful outcome.

Correct Answers:

Q1. D
Q2. B
Q3. A
Q4. B
Q5. B

CASE STUDY 7

Slow–fast Atrioventricular Nodal Reentrant Tachycardia

Mohamed Djelmami-Hani, MD

A 55-year-old man with a history of recurrent palpitation and documented tachycardia underwent an exercise stress test. His clinical tachycardia was spontaneously induced during exercise (see ECG in Figure 7.1).

Question No. 1:

The ECG findings most likely support which of the following tachycardias:

A. Antidromic atrioventricular (AV) reentry tachycardia.
B. Sinus tachycardia.
C. Atrioventricular nodal reentrant tachycardia (AVNRT).
D. Typical atrial flutter.
E. Myocardial ventricular tachycardia (VT).

The test was stopped, but the tachycardia continued at rest. The patient became lightheaded. Blood pressure was recorded as 95/65 mmHg.

Question No. 2:

What is the best management of this recurrent tachycardia?

A. Hydrate and reassure the patient. No further treatment needed.
B. Start intravenous (IV) lidocaine and call for urgent coronary intervention.
C. Start IV diltiazem and warfarin, and plan for electrophysiology (EP) study.
D. Perform carotid sinus massage or start IV adenosine. Recommend EP study and ablation.
E. Urgent direct current cardioversion.

The tachycardia terminated with carotid sinus message. During EP study, slow–fast AVNRT was inducible with right ventricular (RV) pacing alone and not with atrial pacing.

Question No. 3:

These EP study results can be explained by which of the following?

A. Antegradely, the fast-pathway effective refractory period (ERP) is longer than slow-pathway ERP.
B. Retrogradely, the fast-pathway ERP is longer than the slow-pathway ERP.
C. Antegradely, the slow-pathway ERP is longer than the fast-pathway ERP.
D. Retrogradely, the slow-pathway ERP is longer than the fast-pathway ERP.

Practical Electrophysiology © 2015 Jasbir S. Sra, Masood Akhtar, Andrea Natale, David J. Wilber
Cardiotext Publishing, ISBN: 978-1-935395-17-1

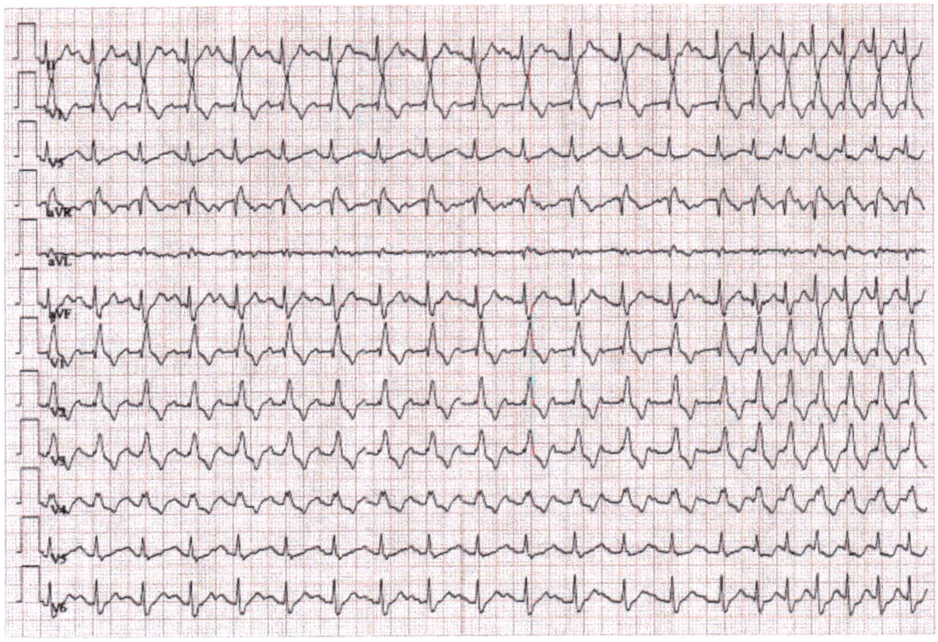

Figure 7.1. Continuous 12-lead electrocardiographic recording during exercise stress test.

The patient's ECG at rest is shown in Figure 7.2.

Question No. 4:

Based on Figure 7.2 and previous ECG findings, what is the best interventional approach?

A. Pacemaker insertion with beta-blocker therapy.

B. Careful slow-pathway ablation.

C. Ablation of the fast pathway.

D. AV nodal ablation with pacemaker insertion.

DISCUSSION

Slow–fast AVNRT, also known as typical or common AVNRT, is one of the most frequent regular

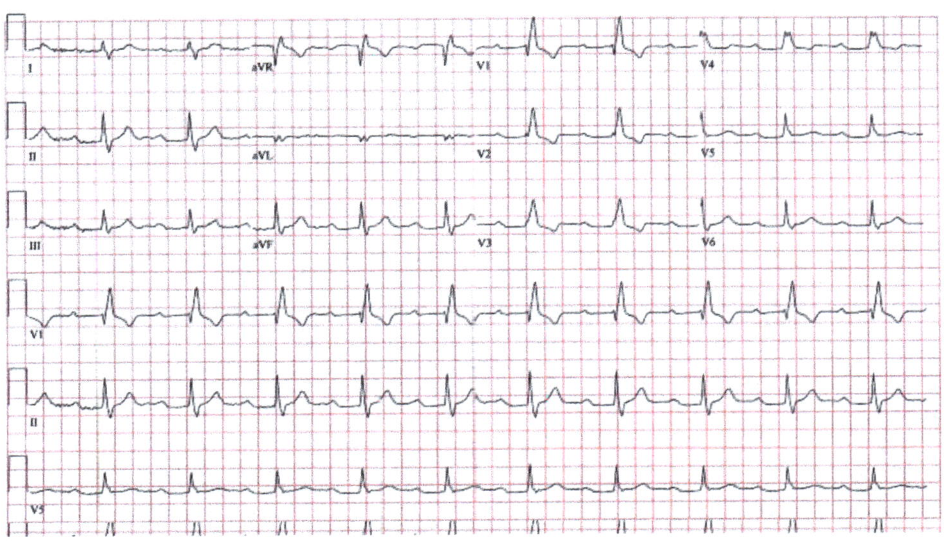

Figure 7.2. The ECG of the same patient at rest.

narrow-complex tachycardias in normal heart. It affects patients of different ages and tends to be more frequent in women. The clinical presentation in patients with normal heart usually includes palpitations, weakness, and dizziness.

ELECTROCARDIOGRAM

The ECG of typical AVNRT shows long-PR, narrow-complex regular tachycardia. Since the activation of both the ventricles and the atria occur simultaneously starting from the AV junction area, it is typically difficult to discern the P waves since they are obscured by the QRS complexes. The P waves can occur at the end of the QRS also and present as a terminal delay. To better observe the P waves at the end of QRS, it is helpful to carefully compare side by side the QRS complex in lead V1 during tachycardia and sinus rhythm.

ANATOMY, PHYSIOLOGY, AND INTRACARDIAC ELECTROPHYSIOLOGIC STUDIES

The anatomy of the AV node has been a subject of great debate in the EP arena and multiple models have been described. It is thought that the AV node has extensions into the atria with different conduction velocities. Typically, the slow-conducting pathway extends from the compact node in posterior direction toward the coronary sinus (CS) ostium. The fast pathway of the AV node takes a more anterior course before it connects to the compact node.

The physiology of AVNRT obeys the same electrophysiologic principles of reentry. An input into the AV node (typically premature compared to the previous inputs) blocks in the fast pathway and conducts down the slow pathway relatively slowly. Since the impulse reaches the distal end of the fast pathway as it connects to the compact node, it travels retrogradely in the fast pathway, causing a retrograde atrial activation before it turns around and reenters the slow pathway antegradely; the tachycardia is then initiated.

To allow the initiation of this reentry with a premature atrial complex (PAC), the antegrade ERP of the fast pathway should be higher than the ERP of the slow pathway. The block in the fast pathway and conduction down the slow pathway is seen and described as a "jump," which proves the existence of at least two different pathways with different ERPs. It is also known as "dual physiology" of the AV node. By definition, "jump" is when the A_2-H_2 interval prolongs by at least 50 ms with a 10-ms decrease in A_1-A_2. The presence of dual physiology of the AV node can also be observed with the sudden prolongation of the AH interval by at least 50 ms, when the pacing cycle is decreased by 10 ms during incremental atria pacing (Figure 7.3). Retrograde conduction with RV pacing is relatively fast and present down to at least the cycle length of the

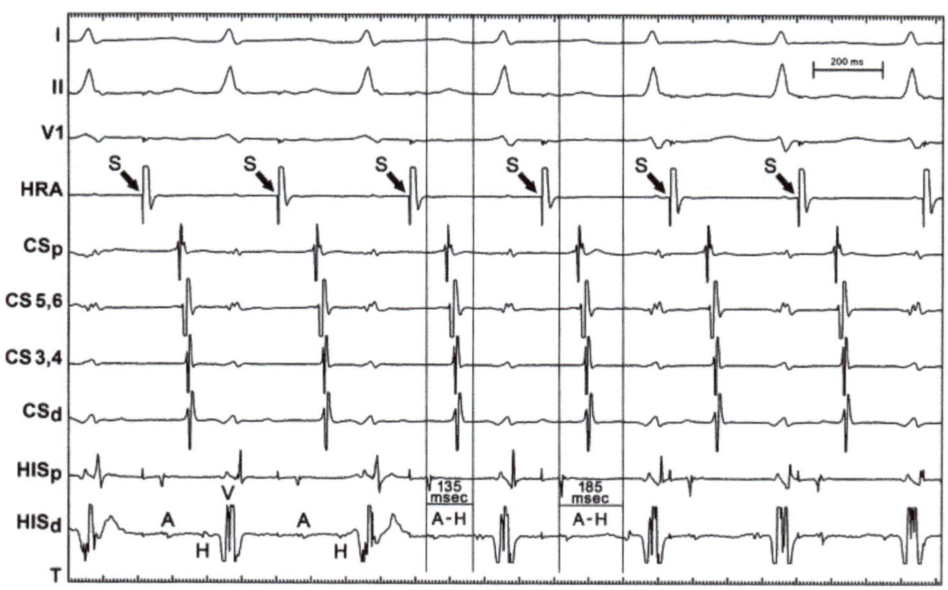

Figure 7.3. Sudden increase in AH interval during incremental atrial pacing from high right atrium.

tachycardia. However, the absence of this fast retrograde ventriculoatrial (VA) conduction at baseline does not necessarily exclude AVNRT. The tachycardia can also be initiated with RV pacing in some cases of fast retrograde VA conduction. This can be achieved with incremental RV pacing or, more likely, with premature RV stimulation.

Often initiation of the tachycardia requires IV isoproterenol. Typically a small dose (1 μ/min) is sufficient. At times, the tachycardia is initiated only with stimulation during the washout period after stopping isoproterenol. The tachycardia can be initiated with incremental atrial pacing. Pacing should be stopped immediately once the tachycardia starts. If incremental pacing is continued and the change in activation sequence and cycle length signaling the beginning of the tachycardia is missed, the latter can be terminated with faster pacing and overdrive.

Introducing PACs during different pacing drives is another way AVNRT can be induced (Figure 7.4). Sometimes an S_3 is necessary for the initiation of AVNRT. The slow–fast AVNRT has a typical activation with alignment of atrial (A) and ventricular (V) potentials when conduction is one-to-one.

The AVNRT usually conducts one-to-one to the atrium and ventricle. However, neither the ventricle nor the atrium is necessary for the tachycardia since neither is part of the circuit. The AV dissociation during tachycardia, spontaneous or with RV pacing, is very helpful in excluding accessory pathway and orthodromic AV reentry, but cannot exclude AVNRT. Dissociating the atrium from the tachycardia during AVNRT is more difficult albeit not impossible.

Spontaneous termination of the tachycardia usually occurs with a block in the antegrade limb (slow pathway) producing an A and no His (H) activation. Vagal maneuvers can slow down the tachycardia by affecting the AH interval, whereas the HA interval remains constant.

Introducing PAC during tachycardia can produce different responses depending on its timing, which can be helpful in differentiating AVNRT and atrial/junctional tachycardia. A PAC during AVNRT can capture the local myocardium but fail to penetrate the tachycardia circuit, thus continuing the AVNRT unaffected. A very early PAC can penetrate the circuit, collide with the wavefront in the fast pathway, and block in the refractory slow pathway. This results in termination of AVNRT. If the PAC is timed right, it can enter the circuit, collide with the wavefront in the fast pathway, but conduct down the recovered slow pathway. This results in resetting of the tachycardia. Another clue to help differentiate junctional tachycardia from AVNRT is that the atrial activation during RV pacing is exactly the same during AVNRT.

Dissociating the ventricles from the tachycardia by pacing slightly faster than the tachycardia cycle length without affecting the tachycardia rules out orthodromic AVRT via (anteroseptal) accessory pathway. Also during AVNRT, the atrial and ventricular activations are simultaneous, giving the typical alignment of As and Vs during intracardiac studies (Figure 7.5). In orthodromic AVRT, the

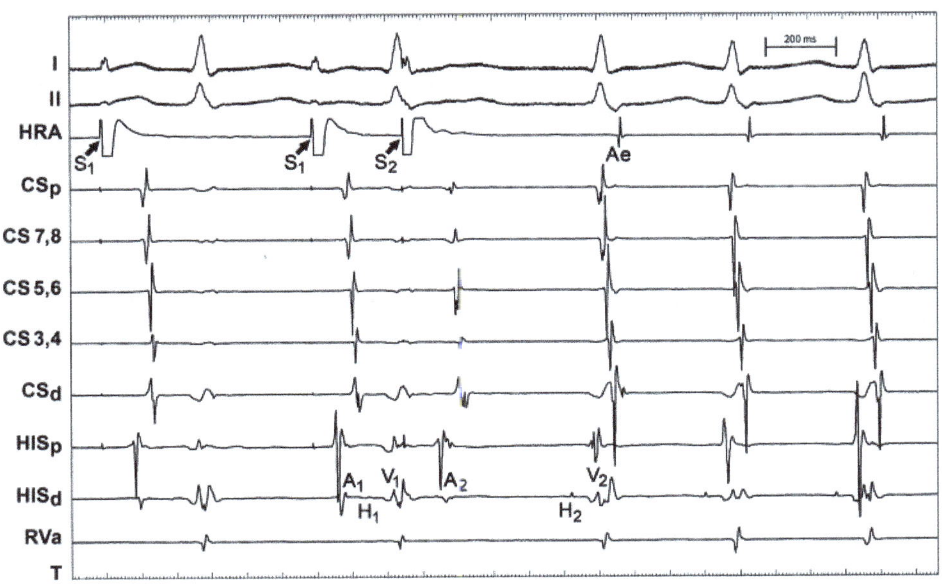

Figure 7.4. Initiation of slow–fast AVNRT with a premature atrial stimulation (S2) during base drive pacing (S1).

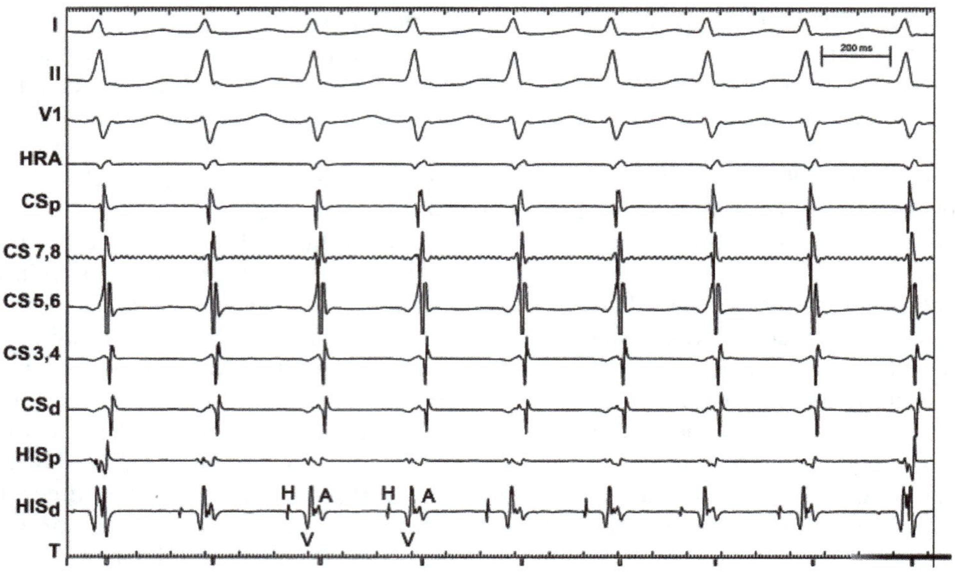

Figure 7.5. Intracardiac recording of slow–fast AVNRT with the typical alignment of As and Vs.

activation of the ventricles must precede the atrial activation. Hence, a VA interval during tachycardia shorter than 70 ms makes orthodromic AVRT very unlikely.

MANAGEMENT OF AVNRT

Acute Treatment

Vagal maneuvers such as Valsalva maneuver and carotid sinus massage are very helpful in terminating tachycardia. Patients are taught these maneuvers, and they frequently terminate AVNRT in a nonmedical setting. The IV adenosine push is frequently used in emergency rooms to terminate tachycardia. Termination of tachycardia with IV adenosine does not automatically diagnose AVNRT, since AVRT and some atrial tachycardias will terminate with adenosine. Beta-blockers as well as calcium channel blockers such as diltiazem and verapamil also can be used intravenously for acute termination of AVNRT. However, adenosine is preferred because of its very short half-life.

Chronic Therapy

Different antiarrhythmic medications can be used for this purpose, such as beta-blockers, calcium channel blockers, and class Ic agents. Because of the side effects of these medications, their long-term use is not recommended, especially in young patients. Patients with short and rare episodes of tachycardia with mild symptoms can be managed with the same medications on an as-needed basis if vagal maneuvers fail. However, with time, the majority of patients tend to experience longer and more frequent episodes despite being on medication.

Ablation

Successful ablation is a potential cure for AVNRT. Different ablation techniques and approaches have been described in literature and used in different labs. The goal is to ablate the slow pathway and render the tachycardia noninducible. Sra et al described a resetting technique where PACs are introduced from the ablation catheter in the inferior part of the triangle of Koch.[1] Ablation is applied when a PAC terminates the tachycardia. The study concluded that termination of tachycardia in the antegrade direction at the longest coupling interval by extrastimuli delivered from the ablating catheter can be helpful for identification of an optimal site for slow-pathway ablation in patients with common AVNRT.

Nakagawa and Jackman described an ablation technique that targets the slow-pathway potential.[2] However, the potential is not always present in patients with AVNRT, making this approach more difficult. A linear lesion from the tricuspid annulus to the CS ostium is used to interrupt the slow pathway as it curves superiorly toward the compact AV node and the His bundle.

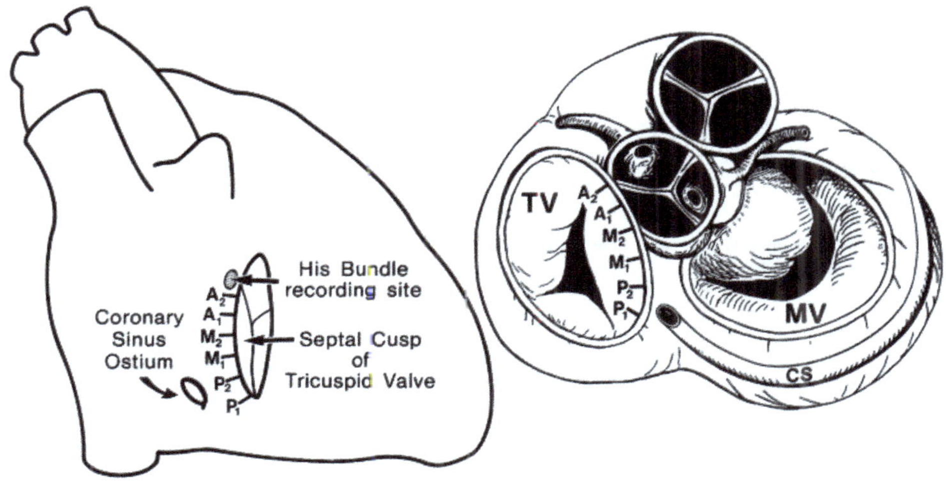

Figure 7.6. Diagrams showing anatomic orientation used for catheter modification of the AV node. The schemata depict right anterior oblique **(left image)** and left anterior oblique **(right image)** projections for anatomic localization of fast and slow pathways. Fast-pathway ablation is accomplished near the His bundle recording site. For the slow pathway, ablation is started at the most posterior (P) location (P_1), and additional exploration is done only if lesion at P_1 fails. P_2 is a more anterior location than P_1, as are M_1 and M_2 (medial). The areas labeled A_1 and A_2 (anterior) should be avoided because a more anterior location could carry a higher chance of AV block as one begins to approach the more compact portion of the AV node. TV, tricuspid valve; MV, mitral valve; CS, coronary sinus. (Modified with permission from Jazayeri et al.[5] and Akhtar et al.[3])

A more practical and successful anatomic approach has been described by Akhtar et al and successfully used in our laboratories.[3] Since the slow pathway is posterior in the CS ostium region, ablation is started first at the tricuspid annulus on the atrial side at the level of the CS (P_1 in Figure 7.6). If this fails, the catheter is moved gradually anteriorly with testing for inducibility after each ablation point (see Figure 7.6). Ablation anteriorly at A_1 and A_2 area should be avoided because of its proximity to the AV node and His bundle and the high risk of complete AV block. Placement of a CS catheter is extremely helpful in locating the CS ostium on fluoroscopy.

Induction of an accelerated junctional rhythm during radiofrequency application is a sign of a good site for successful ablation of a slow pathway. The most important task of the operator during ablation is to monitor AV conduction. Ablation should be stopped immediately if AV prolongation or block is observed. The retrograde conduction of the junctional rhythm does not necessarily mean that antegrade conduction is intact. To continuously monitor the antegrade AV conduction during ablation, pacing slightly faster from the high right atrium should be started immediately once the junctional rhythm is induced. It is prudent to start ablation at low power (25 watts), and then gradually increase the power if the AV conduction is conserved. If a fast junctional rhythm is induced, radiofrequency application should be stopped immediately since this is sign of a high-risk area with imminent permanent AV block.

Targeting the fast pathway for AVNRT ablation has been reported. However, the risk of AV block has been reported to be as high as 16%. Cryoablation can also be used with better safety since the recovery of AV conduction is more likely once freezing is stopped. However, the long-term success rate of tachycardia ablation is lower compared with radiofrequency ablation (86% vs 99%).

Successful ablation of a slow pathway can be verified by a higher AV nodal ERP as well as noninducibility of tachycardia. Single AV nodal echoes are usually accepted and are not considered a sign of ablation failure.

Long PR Interval and AVNRT

Some patients with slow–fast AVNRT have a prolonged PR interval at baseline (>200 ms). The ablation approach to these patients can present a challenge with the fear of causing complete AV block with slow-pathway ablation. However, studies and many cases have shown that ablation of the

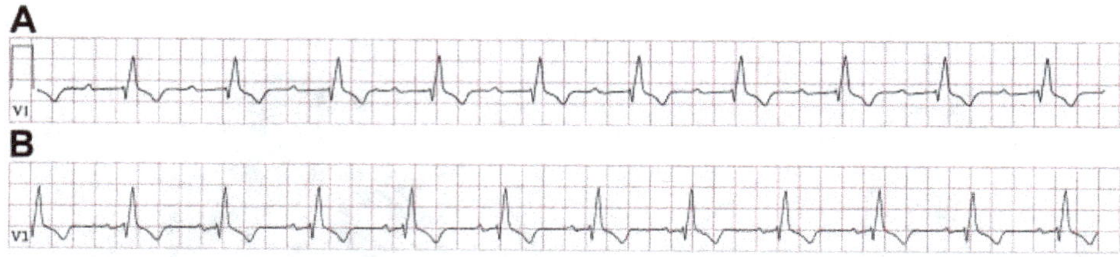

Figure 7.7. The ECG of a patient with AVNRT before **(Panel A)** and after **(Panel B)** slow-pathway ablation. Note the long PR interval in Panel A and its normalization in Panel B following ablation.

slow-pathway in these cases not only eliminates AVNRT, but can also result in shortening of PR interval. Natale et al described a series of 7 patients with long PR and long AH with AVNRT who were successfully ablated, resulting in shortening of PR interval in 5 of them.[4] The long PR interval can be explained by electrotonus. Electrotonus is a phenomenon in which there is passive current flow due to voltage gradients between cells with different conduction velocities and sequences of activation. Such electrotonic current, applied during repolarization, prolongs the action potential duration and myocardial refractoriness. Therefore, the fast-pathway conduction becomes prolonged due to the electrotonic interaction with the slow pathway. Ablation of the slow pathway eliminates such phenomena, resulting in faster conduction of the fast pathway (Figure 7.7).

CONCLUSION

Typical or slow–fast AVNRT is one of the most frequent regular supraventricular tachycardias (SVTs) encountered in normal heart. Although the anatomy of the AV node and its pathway is still debatable, the physiology of AVNRT has been well studied and understood. Medical management can help decrease the frequency and length of tachycardia episodes; however, ablation of the slow pathway is safe and effective in eliminating, and potentially curing, tachycardia.

Correct Answers:

Q1. C

The PR interval prolonged just before the tachycardia initiation without a significant increase in heart rate (Figure 7.8). This signals dual AV node physiology with conduction switching from the fast to the slow pathway, then starting tachycardia (AVNRT). Antidromic AVRT is unlikely given the absence of preexcitation. Tachycardia initiation is abrupt, which makes sinus tachycardia unlikely. The rate and the absence of typical flutter waves make typical atrial flutter also unlikely. The change in PR interval is not typical of myocardial VT initiation. Also note that the QRS morphology did not change (right bundle branch block [RBBB]). Given RBBB at baseline, bundle-branch VT cannot be ruled out; however, the PR changes leading to the tachycardia make AVNRT more likely.

Q2. D

The tachycardia observed is most likely AVNRT. Acute termination can be achieved with careful carotid massage if no contraindications are present. The IV adenosine is typically very effective in terminating AVNRT. Given the frequency of the episodes and the patient symptoms, EP study and ablation is recommended. Answer C would be correct except that anticoagulation is not indicated in AVNRT. Other choices are inappropriate in a stable patient with AVNRT.

Q3. D

To induce slow–fast AVNRT with RV pacing, the conduction will have to block retrogradely in the slow pathway and conduct up the fast pathway before it can reenter and initiate the tachycardia. For this to happen, the retrograde ERP of the fast pathway has to be shorter than that of the slow pathway. One explanation for the inability to induce slow–fast AVNRT with atrial pacing is that the ERP of the slow pathway is equal to or longer than the fast-pathway ERP.

Q4. B

The ECG at rest shows prolonged first-degree AV block. However, the PR interval shortened during stress test. The block is most likely in the AV node and not in the His

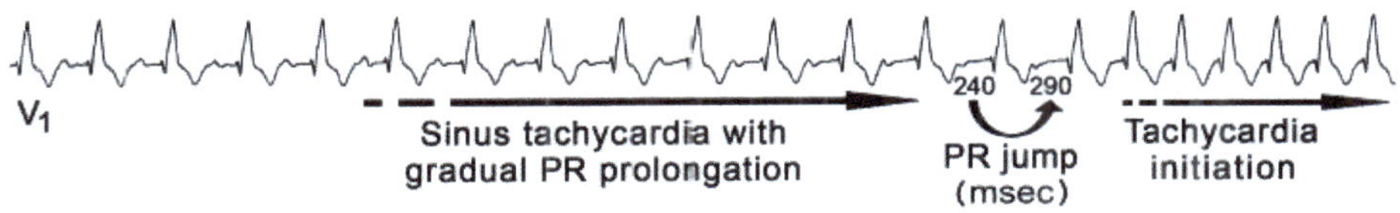

Figure 7.8. Gradual PR prolongation during sinus tachycardia followed by a sudden jump by about 50 ms. This is consistent with a probable dual AV node physiology with block in the fast pathway and conduction down the slow pathway. Based on these findings, the initiated tachycardia is most likely AVRNT.

bundle. Hence, pacemaker is not indicated. Ablation of fast pathway for AVNRT has been reported and carries a high risk of AV block. Medical management is not unreasonable as first step; however, the pacemaker is not indicated. A careful slow-pathway ablation is the preferred interventional approach in this patient. The PR-interval prolongation is not a contraindication. In fact, PR may shorten postablation due the electrotonus phenomenon as is the case in this patient (see Figure 7.7).

REFERENCES

1. Sra J, Jazayeri M, Natale A et al. Termination of atrioventricular nodal reentrant tachycardia by premature stimulation from ablating catheter: A reliable guide to identify site for slow-pathway ablation. *Circulation*. 1995;91:1095–1100.
2. Nakagawa H, Jackman WM. Catheter ablation of paroxysmal supraventricular tachycardia. *Circulation*. 2007;116:2465–2478.
3. Akhtar M, Jazayeri MR, Sra J, et al. Atrioventricular nodal reentry. Clinical, electrophysiological, and therapeutic considerations. *Circulation*. 1993;88;282–295.
4. Natale A, Greenfield RA, Geiger MJ et al. Safety of slow pathway ablation in patients with long PR interval: Further evidence of fast and slow pathway interaction. *Pacing Clin Electrophysiol*. 1997;20:1698–1703.
5. Jazayeri M, Hempe S, Sra J, et al. Selective transcatheter ablation of the fast and slow pathways using radiofrequency energy in patients with atrioventricular nodal reentrant tachycardia. *Circulation*. 1992;85:1318–1328.

CASE STUDY

8

Nonreentrant Atrioventricular Nodal Tachycardia

Atul Bhatia, MD

A 19-year-old female patient presented with complaints of recurrent palpitation at rest and on exertion for the past 2 years. These episodes were associated with dizziness and lightheadedness. Her medical history included surgery for atrial septal defect at age 4. Her clinical cardiovascular examination was normal.

A two-dimensional (2D) echocardiogram was normal with no evidence of any residual shunt across the atrial septum. A 12-lead ECG (Figure 8.1) is shown below.

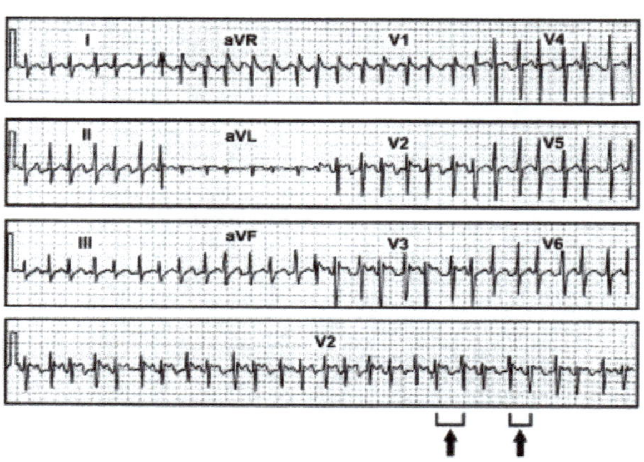

Figure 8.1. 12-lead ECG showing supraventricular tachycardia (SVT). Note the regularly irregular rhythm with 2 distant RR intervals (closed arrows).

Question No. 1:

What is the diagnosis?

A. Atrial fibrillation (AF).

B. Atrial flutter.

C. AV nodal reentrant tachycardia (AVNRT).

D. None of the above.

Subsequently, patient underwent electrophysiology (EP) study. Intracardiac electrogram (EGM) tracings from the study are shown in Figures 8.2 and 8.3.

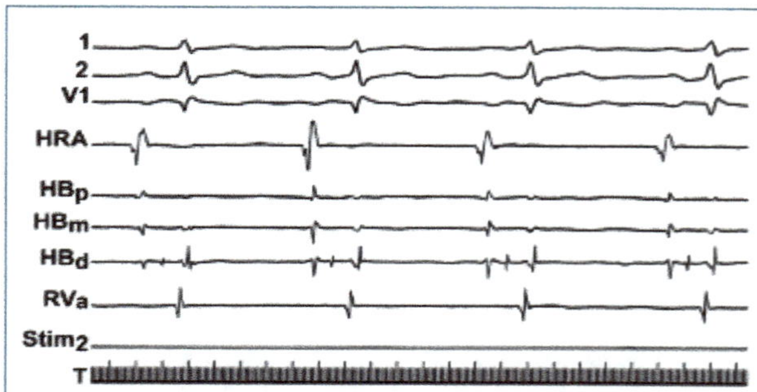

Figure 8.2. Shown from top to bottom are surface ECG leads I, II and V_1. HRA, high right atrial; HBp, His bundle proximal; HBm, His bundle mid; HBd, His bundle distal; RVa, right ventricular apex.

Practical Electrophysiology © 2015 Jasbir S. Sra, Masood Akhtar, Andrea Natale, David J. Wilber
Cardiotext Publishing, ISBN: 978-1-935395-17-1

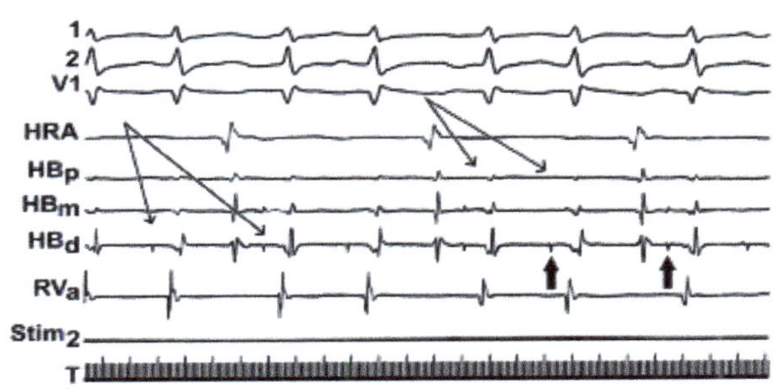

Figure 8.3. Intracardiac EGMs during SVT. Note 1:2 double ventricular response (solid arrows). Also, note that the HV intervals preceding the ventricular beats are the same throughout the tachycardia (thin arrows).

DISCUSSION

SVT-caused dual atrioventricular (AV) nodal physiology is typically reentrant in mechanism. However, tachycardia caused by simultaneous conduction over the slow and fast pathways can result in an incessant SVT, that is, nonreentrant by mechanism.

The following are salient clinical and electrophysiologic characteristics of nonreentrant AV nodal tachycardia (1:2 response/dual ventricular responses).

Clinical Features

1. Rare in clinical practice.
2. Present during sinus rhythm as well as during sinus tachycardia.
3. Incessant nature of the tachycardia can lead to tachycardia-induced cardiomyopathy.
4. Baseline ECG can mimic AF, also called "pseudo-AF."
5. Tachycardia is regularly irregular.
6. Recalcitrant to pharmacological therapy.
7. Radiofrequency ablation of slow pathway is curable and results in reversal of cardiomyopathy.

Electrophysiologic Features

1. Tachycardia present spontaneously or noted post isoprenaline infusion.
2. Each sinus beat is conducted simultaneously over the slow and fast pathway, resulting in double ventricular response.
3. Absent or poor retrograde ventriculoatrial (VA) conduction.
4. Usually do not have inducible AVNRT.
5. Crucial conduction delay in the slow pathway enabling the distal tissue to respond for a second time.
6. The effective refractory period (ERP) of the fast pathway must be shorter than that of the slow pathway.
7. The ERP of the His-Purkinje system is shorter than the difference of the ERP between the fast and slow pathways.
8. Retrograde unidirectional block in the slow pathway preventing the impulse in the fast pathway from entering and colliding with the oncoming slow pathway impulse.

Correct Answer:

Q1. D

REFERENCES

1. Csapo G. Paroxysmal nonreentrant tachycardias due to simultaneous conduction in dual atrioventricular nodal pathways. *Am J Cardiol.* 1979;43:1033–1045.
2. Neuss H, Buss J, Schlepper M, Mitrović V. Double ventricular response in dual AV nodal pathways mimicking supraventricular as well as ventricular tachycardia. *Eur Heart J.* 1982;3:146–154.
3. Li HG, Klein GJ, Natale A, Thakur RK, Yee R. Nonreentrant supraventricular tachycardia due to simultaneous conduction over fast and slow AV node pathways: Successful treatment with radiofrequency ablation. *Pacing Clin Electrophysiol.* 1994;17:1186–1193.
4. Mansour M, Marrouche N, Ruskin J, Natale A, Keane D. Incessant nonreentrant tachycardia due to simultaneous conduction over dual atrioventricular nodal pathways mimicking atrial fibrillation in patients referred for pulmonary veins isolation. *J Cardiovasc Electrophysiol.* 2003;14:752–755.
5. Fraticelli A, Saccomanno G, Pappone C, Oreto G. Paroxysmal supraventricular tachycardia caused by 1:2 atrioventricular conduction in the presence of dual atrioventricular nodal pathways. *J Electrocardiol.* 1999;32:347.
6. Clementy N, Cassel-Senon D, Giraudeau C, Cosnay P. Tachycardiomyopathy secondary to reentrant atrioventricular nodal tachycardia: Recovery after slow pathway ablation. *Pacing Clin Electrophysiol.* 2007;30:925–928.

CASE STUDY

9

Atrioventricular Nodal Reentrant Tachycardia with 2:1 AV Block

Atul Bhatia, MD

An 81-year-old female patient presents with recurrent episodes of palpitations with associated dizziness and presyncope. Episodes offset suddenly and usually occur at rest. Her medical history is significant for hypertension, currently under treatment with metoprolol 50 mg twice daily. Her physical examination is unremarkable. Echocardiography reveals normal left ventricular systolic function with no valvular heart disease.

A 12-lead ECG of the tachycardia is shown in Figures 9.1 and 9.2. Intracardiac electrograms (EMGs) of the tachycardia are shown in Figures 9.3 and 9.4.

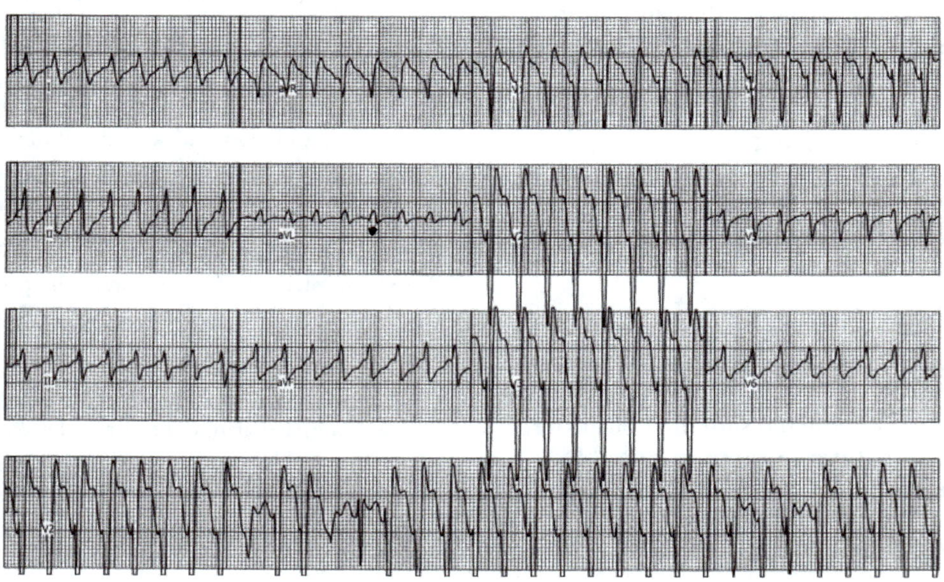

Figure 9.1. ECG of the patient's tachycardia at baseline.

Practical Electrophysiology © 2015 Jasbir S. Sra, Masood Akhtar, Andrea Natale, David J. Wilber
Cardiotext Publishing, ISBN: 978-1-935395-17-1

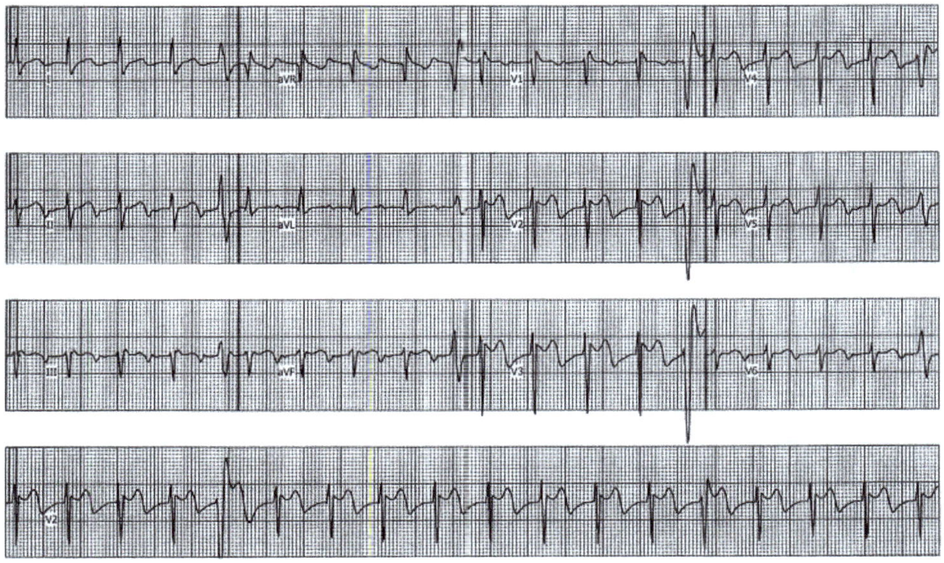

Figure 9.2. Continuation of the ECG at baseline.

Question No. 1:

1. What is the diagnosis?
2. Explain the mechanism of the tachycardia.
3. What diagnostic maneuvers should be used to confirm the diagnosis?

DISCUSSION

1. The 12-lead ECG during tachycardia (see Figure 9.1) is a regular wide-complex tachycardia. The differential diagnoses include supraventricular tachycardia (SVT) with aberrant conduction versus ventricular tachycardia (VT). A careful

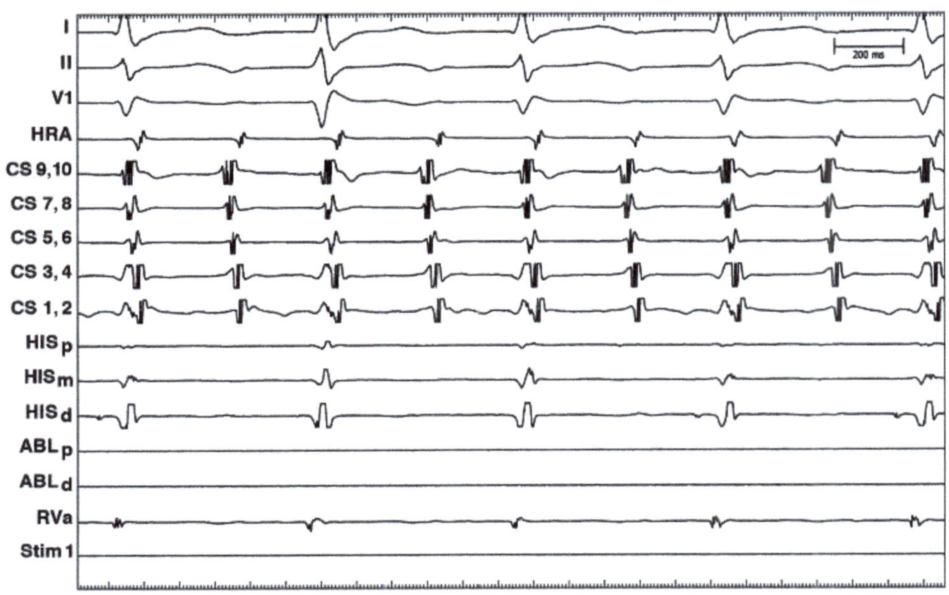

Figure 9.3. Intracardiac EMGs of the tachycardia.

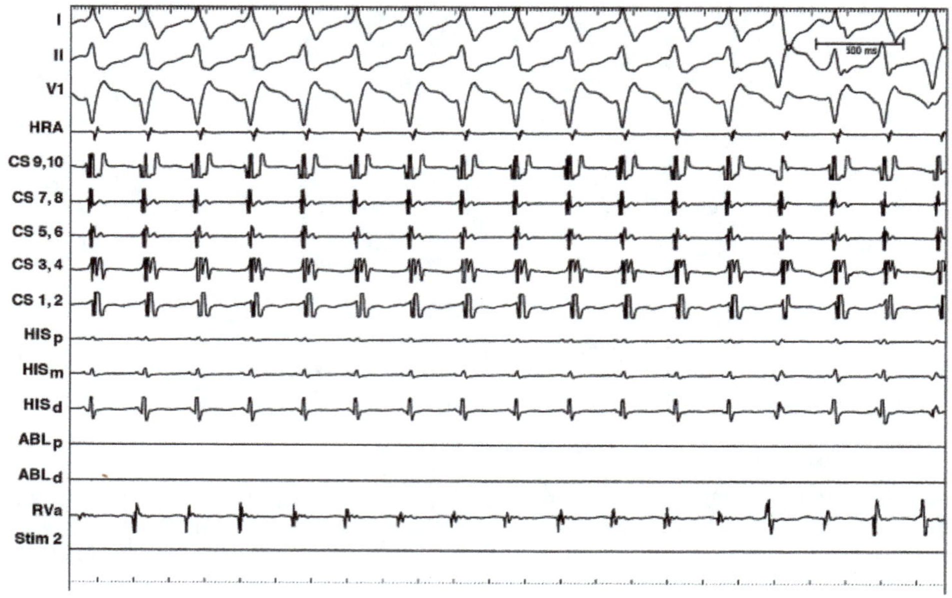

Figure 9.4. Continuation of intracardiac EMGs.

review of the ECG (see Figure 9.2) reveals narrow complex tachycardia with 2:1 conduction. This is consistent with atrioventricular nodal reentrant tachycardia (AVNRT) with 2:1 atrioventricular (AV) block.

2. Premature ventricular beats introduced during tachycardia resulted in the resolution of the block with resumption of 1:1 AV nodal conduction. The concentric retrograde sequence of activation during tachycardia with simultaneous atrial and ventricular activation is most consistent with common AV nodal reentry as the mechanism of the tachycardia.
3. Intracardiac EMGs during tachycardia confirm the diagnosis with the site of AV block appearing above the site of His-bundle recording (see Figures 9.3 and 9.4).

Salient Features

1. AV block is most commonly observed during common AVNRT, rarely in the uncommon variety.
2. Site of block is usually below the site of His-bundle recording, signifying a lower common pathway.
3. Less commonly (20%), the level of AV block is above the site of the His-bundle recording, as in the present case.
4. AV block typically occurs at the onset of tachycardia.
5. Differential diagnosis includes atrial tachycardia (AT) with 2:1 AV block.
6. Response to programmed ventricular stimulation during tachycardia with normalization of conduction is confirmatory of the diagnosis of AVNRT. In addition, initiation of the tachycardia with programmed ventricular stimulation and entrainment with ventricular pacing exclude AT.
7. Rarely, retrograde block to the atrium (ventriculoatrial [VA] block) can also be observed during AVNRT, signifying AVNRT lacks an atrial requirement.
8. Resumption of 1:1 AV or VA conduction can result in transient aberrancy with cycle length changes.

Correct Answers:

Q1. Answers appear in the Discussion section.

CASE STUDY

10

Slow/Fast AVNRT: A Group of 3 Tachycardias

Warren M. Jackman, MD, Hiroshi Nakagawa, MD, PhD, Sunny S. Po, MD, PhD, Benjamin J. Scherlag, PhD, Karen J. Beckman, MD, Deborah Lockwood, BM, BCh, Adrian Dusa, MD, Stavros Stavrakis, MD, PhD, and Ralph Lazzara, MD

A 27-year-old woman was undergoing an attempt at catheter ablation of slow/fast atrioventricular nodal reentrant tachycardia (AVNRT). The diagnosis of slow/fast AVNRT was based on: the response to a single ventricular extrastimulus delivered to the para-Hisian region (retrograde atrial activation was not advanced until His-bundle [HB] activation was advanced); and the retrograde atrial activation sequence (earliest activation was recorded *posterior* to the tendon of Todaro [ToT]), at a superior-inferior level approximately halfway between the level of the site recording the most proximal HB potential and the level of the roof of the coronary sinus (CS) ostium[1] (middle horizontal line in Figure 10.1A). Further, para-Hisian pacing (with intermittent His bundle-right bundle branch [HB-RB] capture), decremental ventricular pacing and ventricular extrastimulus testing demonstrated only retrograde fast atrioventricular (AV) nodal pathway conduction with the same atrial action sequence as the tachycardia (earliest activation was recorded *posterior* to the ToT). Not even a single beat of retrograde conduction over a slow AV nodal pathway was observed.

Linear radiofrequency (RF) ablation (continuous pullback, 40 W) was performed across the inferior triangle of Koch, between the tricuspid annulus (TA) and CS ostium at the level of the middle of the CS ostium (line #1, see Figure 10.1A), producing vigorous accelerated junctional rhythm (with retrograde fast pathway conduction) at 2 sites along the line. An A_{SP} potential (activation recorded between the TA and CS ostium in the inferior-to-superior direction, Figure 10.3A) continued to be recorded inferior to the ablation line, but not superior to the ablation line. Sustained slow/fast AVNRT continued to be induced, with no change in tachycardia cycle length (TCL), AH interval, or HA interval. There was also no change during decremental atrial pacing in the shortest atrial pacing cycle length, maintaining 1:1 antegrade conduction over the fast pathway or slow pathway.

A second RF ablation line was created between the TA and CS ostium at the level of the roof of the CS ostium (line #2, see Figure 10.1A). Accelerated junctional rhythm occurred at one of the sites during the linear RF application. Sustained slow/fast AVNRT was again induced with the same TCL, AH interval, and HA interval. Ablation was not attempted within the triangle of Koch further superior than the level of the roof of the CS ostium.

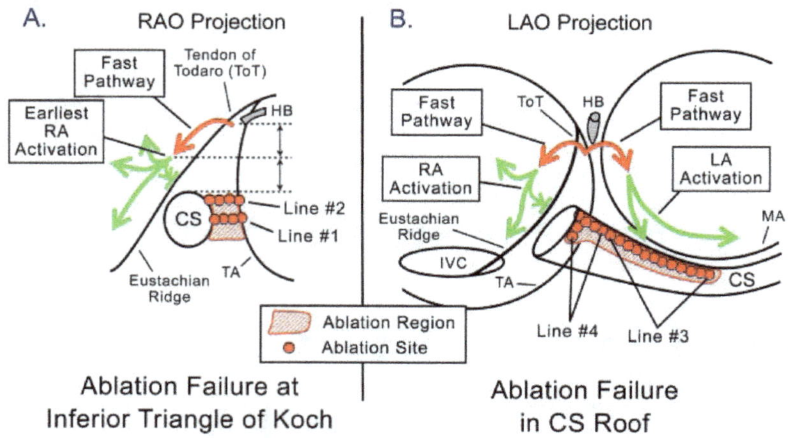

Figure 10.1. Schematic representation in the right anterior oblique (RAO) **(Panel A)** and left anterior oblique (LAO) **(Panel B)** projections of the initial pattern of right (RA) and left atrial (LA) activation during slow/fast AVNRT and the sites of ablation targeting the slow AV nodal pathway in this patient. *Red arrows* represent retrograde conduction over the fast AV nodal pathway. *Green arrows* represent the initial RA and LA activation. Earliest RA activation (Panel A) is recorded posterior to the ToT and eustachian ridge, at a height approximately halfway *(middle horizontal dashed line)* between the height of the site recording the most proximal HB potential *(upper horizontal dashed line)* and the height of the roof of the CS ostium *(lower horizontal dashed line)*. See text for description of the ablation sites. IVC, inferior vena cava; MA, mitral annulus; TA, tricuspid annulus.

Next attempted was a linear RF application along the roof of the CS as a pullback (RF power 15-20 W), beginning approximately 4 to 5 cm from the CS ostium and continuing until approximately 1 cm from the CS ostium (line #3, see Figure 10.1B. There was no accelerated junctional rhythm during that RF application. Sustained AVNRT remained inducible. The ablation line along the roof of the CS was extended to the CS ostium, and then along the superior half of the apical edge of the CS ostium (line #3, see Figure 10.1B). Sustained slow/fast AVNRT continued to be induced with the same TCL, AH interval, and HA interval. There was still no change in the shortest atrial pacing cycle length, maintaining 1:1 antegrade conduction over the fast pathway or slow pathway.

Question No. 1:

How could you now locate and ablate (with the least risk of AV block) the slow AV nodal pathway forming the antegrade limb of this reentrant circuit?

A. Return to the triangle of Koch, and ablate again between the posteroseptal TA and the CS ostium (region of lines #1 and #2 in Figure 10.1A) with improved electrode-tissue contact.

B. Return to the triangle of Koch, and ablate progressively further superior along the TA (superior to the lower dashed line in Figure 10.1A) until AVNRT is eliminated.

C. Terminate the procedure because there is no other safe place to ablate slow/fast AVNRT.

D. Use the resetting response to a single atrial extrastimulus at the locations of the 3 most common slow pathways (inferior triangle of Koch, roof of the CS, and left atrium [LA] close to the inferolateral mitral annulus [MA]) to identify the *atrial end* of the slow pathway forming the antegrade limb of this slow/fast reentrant circuit.

DISCUSSION

Slow/fast AVNRT is a group of 3 tachycardias, each using the fast AV nodal pathway for retrograde conduction but differing in the slow pathway used for the antegrade limb of the reentrant circuit.[1] The most prevalent form is "rightward inferior extension" (RIE) slow/fast AVNRT, using the RIE of the AV node for the antegrade slow pathway. The RIE has no direct connections with the right atrium. It connects to the LA through the CS myocardium. Importantly, the RIE connects to the CS *only through the floor of the CS ostium* (Figure 10.2).[1] As a result, the RIE can be interrupted (eliminating AVNRT) along a wide region between the TA and CS ostium. It is not necessary to ablate superior to the level of the roof of the CS ostium (see lower dashed line in Figure 10.1A). We prefer to deliver RF energy as a line at the level of the middle of the CS ostium (Figure 10.3), beginning at the ventricular edge of

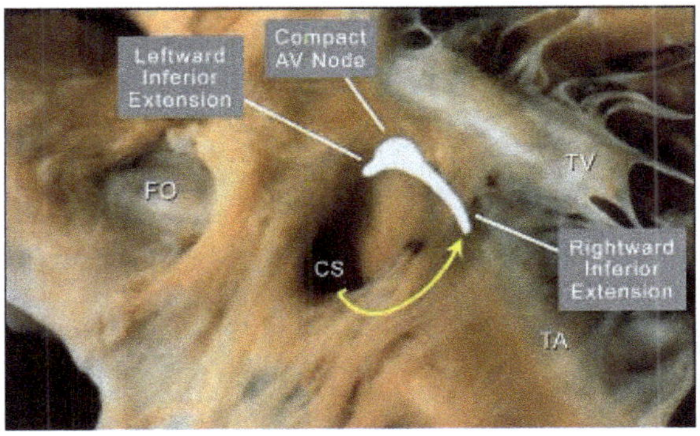

Figure 10.2. Photograph of the septum and AV junction in a human heart showing the locations of the RIE and LIE of the AV node,[2] which form the 2 slow AV nodal pathways most frequently found to participate in the circuit of the various forms of AVNRT.[1,2] The RIE has no direct connections with the right atrium. It connects to the LA through the CS myocardium and just the floor of the CS ostium (no connections with other regions of the CS ostium).[1] The atrial end of the LIE is located close to the roof of the CS, but it is unknown whether it connects to the CS myocardium or the LA myocardium close to the CS. FO, fossa ovalis; TA, tricuspid annulus; TV, tricuspid valve. (Modified from Inoue et al[2] with permission from Wolters Kluwer Health.)

the tricuspid valve (TV) (identified by the presence of the A_{SP} potential on the unipolar electrogram [EGM] recorded from the second electrode, but absent from the distal unipolar EGM) and extending to the anterior (apical) edge of the CS ostium. We then extend the line, by approximately 1 cm, inside the anterior margin of the CS. We avoid the floor of the CS (including the floor of the CS ostium) to avoid injuring the posterolateral branch of the right coronary artery (see Figure 16.6 in case 16). The occurrence of accelerated junctional rhythm usually indicates RIE injury and conduction block. Ablation further superiorly will continue to generate accelerated junctional rhythm, but this does not indicate that the RIE was not interrupted by the earlier ablation inferior to that site.

In this patient, the occurrence of accelerated junctional rhythm during ablation between the TA and CS ostium suggested that the RIE was interrupted, and that the leftward inferior extension (LIE) may form the antegrade limb of the reentrant circuit ("LIE" slow/fast AVNRT, Figure 10.4). Ventricular stimulation did not induce retrograde slow pathway conduction to help localize the atrial

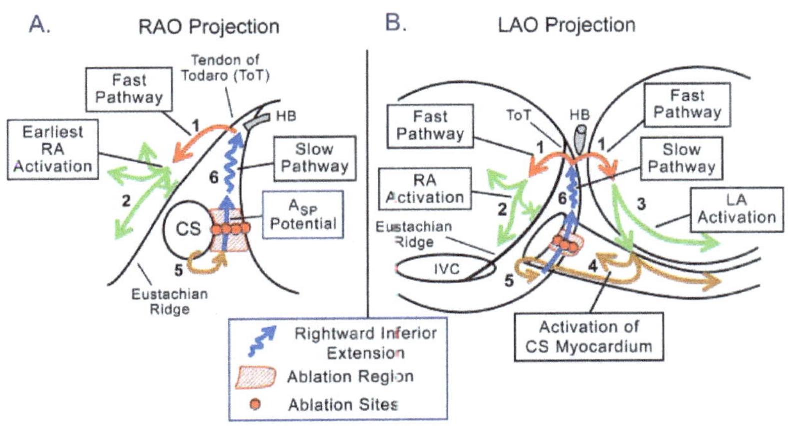

Figure 10.3. Schematic representation of the postulated circuit and recommended ablation sites for slow/fast AVNRT using the RIE of the AV node *(wavy blue arrow, #6—Slow Pathway)* for the antegrade limb of the reentrant circuit. Retrograde conduction over the fast AV nodal pathway *(red arrow, #1—Fast Pathway)* results in atrial activation posterior to the ToT (outside the triangle of Koch) at the interatrial septum. **A:** RA activation propagates away from the site of earliest RA activation in the posterior, superior, and inferior directions *(green arrows, #2)*. This results in superior-to-inferior RA activation along the eustachian ridge (outside the triangle of Koch), which is unable to cross the eustachian ridge and ToT to enter the triangle of Koch, presumably due to conduction block along the eustachian ridge *(short perpendicular green line)*. **B:** Retrograde conduction over the fast pathway also activates the LA septum *(red arrow, #1-Fast Pathway)*. LA activation propagating inferiorly and laterally along the MA *(green arrows, #3)* activates the CS myocardium, approximately 1 to 3 cm from the ostium *(brown arrows, #4)*. Propagation along the CS myocardium to the CS ostium *(brown arrows)* activates the inferior triangle of Koch in the inferior-to-superior direction *(nonwavy blue arrow)*, producing the A_{SP} potential. Only activation emerging from the floor of the CS ostium *(brown arrow, #5)* is able to activate the atrial end of the RIE to complete the circuit.[1] This allows interruption of the RIE over a wide area, accounting for the high success of ablation of all of the forms of AVNRT that use the RIE as one of the limbs of the reentrant circuit. A linear RF lesion (40 W) between the inferoseptal TA and the anterior (apical) margin of the CS ostium *(red dots in red hatched area in Panel A)* continuing at lower power (15-20 W) along a 1-cm length of the anterior (apical) aspect of the CS *(red dots in red hatched area in Panel B)* interrupts anterograde conduction over the RIE and eliminates "RIE" slow/fast AVNRT. Injury to the RIE is usually associated with periods of accelerated junctional rhythm, even if the RIE is not participating in the reentrant circuit of the AVNRT. HB, His bundle; IVC, inferior vena cava; LAO, left anterior oblique; RAO, right anterior oblique. (Reproduced from Lockwood et al[1] with permission from Elsevier.)

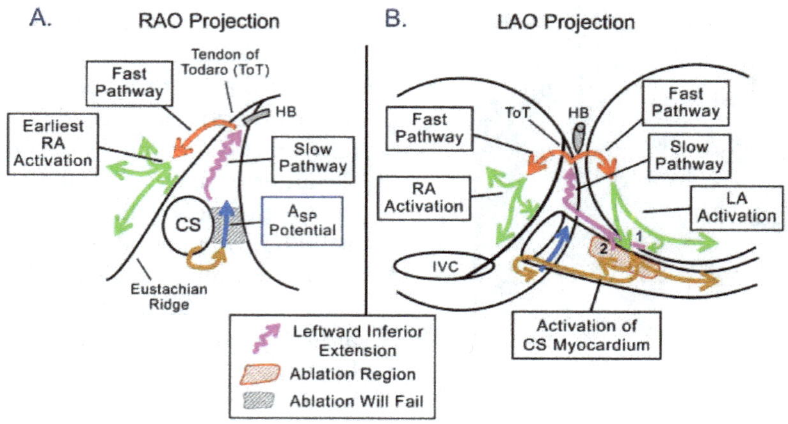

Figure 10.4. Schematic representation of the postulated circuit and recommended ablation sites for slow/fast AVNRT using the LIE of the AV node *(wavy purple arrow, Slow Pathway)* for the antegrade limb of the reentrant circuit. Retrograde conduction over the fast pathway *(red arrows)* activates the right and left sides of the interatrial septum, posterior to the ToT, similar to "RIE" slow/fast AVNRT (see Figure 10.3). Right atrial activation *(green arrows)* fails to reenter the triangle of Koch. Left atrial activation propagates around the MA *(green arrows in **Panel** B)* to activate the CS myocardium *(brown arrows)* and the atrial end of the LIE of the AV node *(purple arrow, Slow Pathway)*. It is not known whether the atrial end of the LIE is in the LA adjacent to the CS *(#1 in Panel B)* or the roof of the CS *(#2 in Panel B)*. Propagation along the LIE toward the compact AV node *(wavy purple arrow)* forms the antegrade slow pathway and completes the reentrant circuit. For ablation of the atrial end of the LIE, we empirically target the roof of the proximal CS, beginning the ablation 3 to 4 cm from the CS ostium and continue proximally toward the CS ostium *(red hatched area in Panel B)*. Accelerated junctional rhythm is usually not observed. Ablation of the RIE at the inferior triangle of Koch *(gray hatched area in **Panel** A)* at the site of the A_{SP} potential *(blue arrow)* will produce accelerated junctional rhythm, but fail to eliminate the "LIE" slow/fast AVNRT. HB, His bundle; IVC, inferior vena cava; LAO, left anterior oblique; RAO, right anterior oblique. (Reproduced from Lockwood et al[1] with permission from Elsevier.)

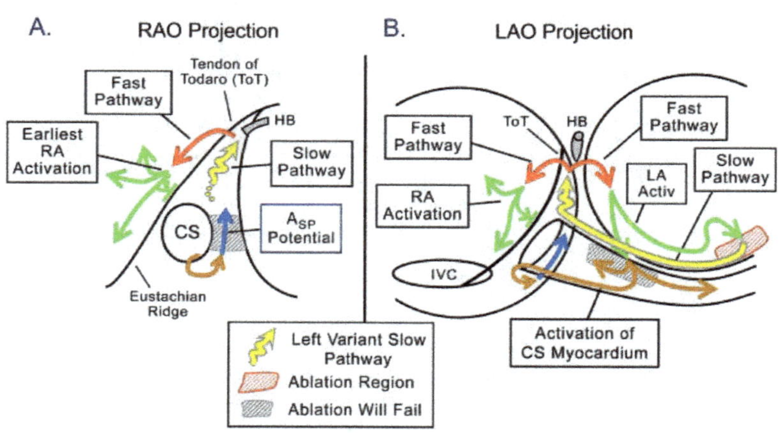

Figure 10.5. Schematic representation of the postulated circuit and recommended ablation sites for slow/fast AVNRT using the left variant slow pathway *(wavy yellow arrow, Slow Pathway)* for the antegrade limb of the reentrant circuit. The retrograde atrial activation sequence is same as other forms of slow/fast AVNRT (see Figures 10.3 and 10.4), including activation of the CS myocardium *(brown arrows)* that activates the inferior triangle of Koch generating an A_{SP} potential *(blue arrow)*. Propagation along the LA adjacent to the MA *(tip of long green arrow in **Panel** B)* activates the atrial end of an as yet anatomically unidentified slow pathway originating close to the inferolateral MA *(red hatched area)*, the left variant slow pathway *(yellow arrow)*. Propagation along the left variant toward the compact AV node *(wavy yellow arrow)* on the right and/or left side of the septum (unknown) forms the antegrade slow pathway and completes the reentrant circuit. For ablation, the atrial end of the left variant slow pathway is precisely located using the resetting response to a single LA extrastimulus (see Figure 10.6). One or two RF applications at the site, close to the inferolateral MA, to reset the tachycardia *(red hatched area)* eliminated AVNRT in 8 of 10 patients with "left variant" slow/fast AVNRT.[1] Ablation of the RIE at the inferior triangle of Koch *(gray hatched area in **Panel** A)* or the LIE at the CS roof *(gray hatched area in Panel B)* fails to eliminate "left variant" slow/fast AVNRT. HB, His bundle; IVC, inferior vena cava; LAO, left anterior oblique; RA, right atrial; RAO, right anterior oblique. (Modified from Lockwood et al[1] with permission from Elsevier.)

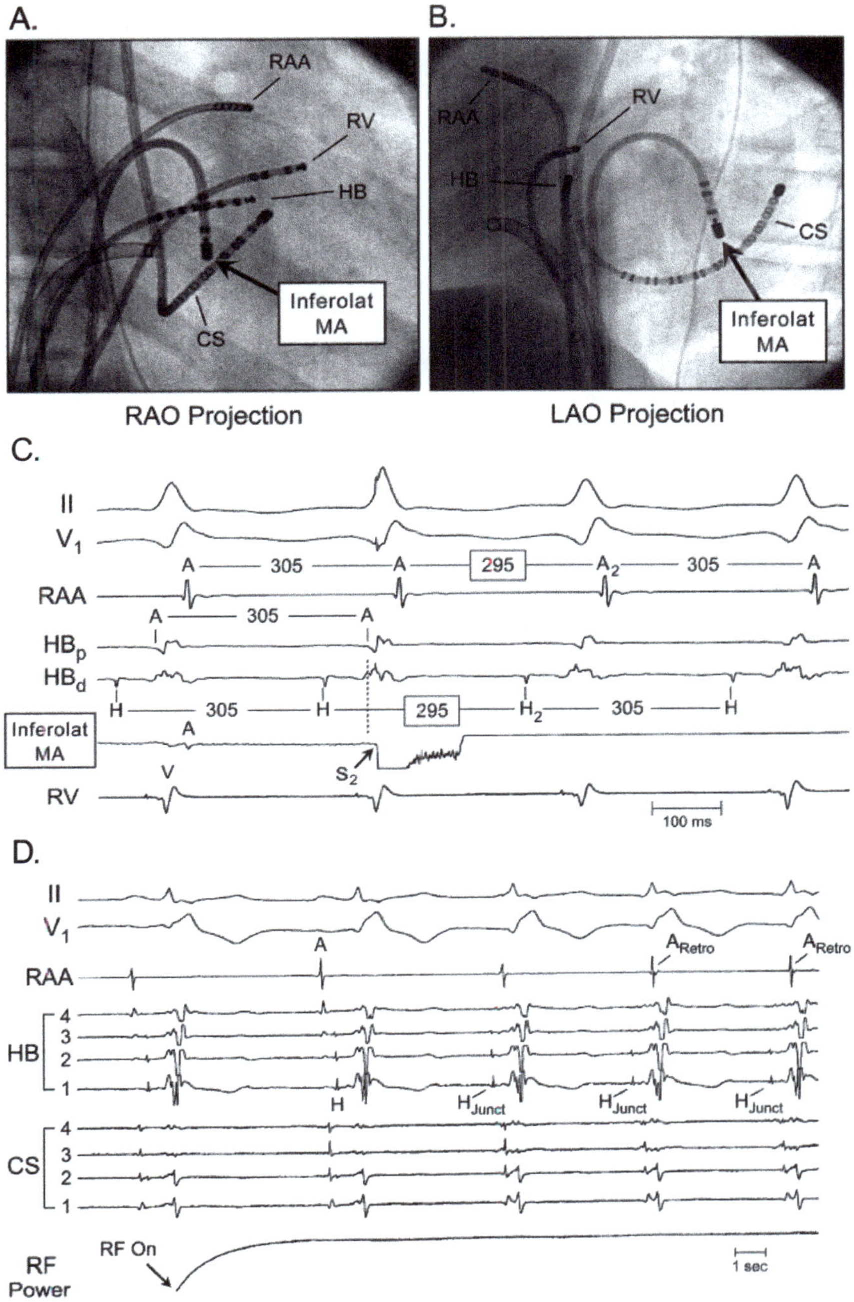

Figure 10.6. Localization of the atrial end of the left variant slow pathway close to the inferolateral MA using the resetting technique followed by catheter ablation of "left variant" slow/fast AVNRT in our patient. **A** and **B**: Radiographs in the RAO and LAO projections show catheter positions, including the mapping catheter in the LA (*Inferolat MA,* transseptal technique) positioned close to the inferolateral MA at the site of resetting and ablation. **C**: Resetting. During slow/fast AVNRT with constant cycle length 305 ms, an atrial extrastimulus was delivered to the mapping catheter in the LA close to the inferolateral MA *(arrow, S_2 in Inferolat MA EGM)* after retrograde atrial activation had occurred (A-A = 305 ms in the HB_p EGM). The atrial extrastimulus advanced the next HB potential by 10 ms (H-H_2 = 295 ms in the HB_d EGM). The advanced HB potential *(H_2)* was followed by a similar 10-ms advance in atrial activation (A-A_2 = 295 ms in the right atrial appendage [RAA] EGM) and a subsequent H_2-H interval equal to the TCL (305 ms in the HB_d EGM), showing the resetting response. The resetting of the tachycardia suggests the site is located close to an atrial end of the slow pathway forming the anterograde limb of the reentrant circuit. **D**: An application of RF current (onset at *arrow, RF On in RF Power Tracing)* to the resetting site in the LA produced an immediate accelerated junctional rhythm. Each HB potential of the accelerated junctional rhythm *(H_{Junct})* is followed by retrograde conduction over the fast AV nodal pathway *(A_{Retro})*, consistent with heating-induced automaticity of the slow pathway. CS, coronary sinus; RV, right ventricle. (Reproduced from Lockwood et al[1] with permission from Elsevier.)

end of the slow pathway. Testing the resetting response at the roof of the CS would identify whether the LIE was participating in the tachycardia and provide localization for ablation of the LIE. Instead, a linear RF application was placed along the roof of the CS, near the atrial end of the LIE (see Figure 10.1B). This failed to alter the tachycardia, suggesting the rare left variant slow pathway may form the antegrade limb of the reentrant circuit ("left variant" slow/fast AVNRT, Figure 10.5).

Transseptal puncture was performed, placing the ablation catheter in the LA. The resetting response was tested in the LA, adjacent to the inferolateral MA (Figure 10.6A and B). A single extrastimulus, delivered sufficiently late that retrograde fast pathway activation of the atria was not affected, advanced the next HB potential by 10 ms and reset the tachycardia (see Figure 10.6C), indicating the antegrade slow pathway was nearby. A single RF application at the resetting site (see Figure 10.6D) immediately induced accelerated junctional rhythm with retrograde fast pathway conduction (indicating heating of a slow pathway) and eliminated antegrade slow pathway conduction and AVNRT. Antegrade and retrograde fast pathway conduction were unaffected. The patient has remained free of tachycardia symptoms.

Correct Answer:

Q1. D

REFERENCES

1. Lockwood D, Nakagawa H, Jackman WM. Electrophysiologic characteristics of atrioventricular nodal reentrant tachycardia: Implications for reentrant circuits. In: Zipes DP, Jalife J, eds. *Cardiac Electrophysiology: From Cell to Bedside.* 5th ed. Philadelphia, PA: Saunders Elsevier; 2009: 615–646.
2. Inoue S and Becker AE. Posterior extensions of the human compact atrioventricular node: A neglected anatomic feature of potential clinical significance. *Circulation.* 1998;97:188–193.

CASE STUDY

11

Atrial Tachycardia: Differential Diagnosis

Melvin M. Scheinman, MD

A 59-year-old patient presents with an almost incessant tachycardia. He has no structural cardiac disease and the cardiac echocardiogram is normal. An example of tachycardia initiation is shown in Figure 11.1.

Question No. 1:

1. What are the possible mechanisms for this tachycardia?

A. Atrioventricular nodal reentry tachycardia (AVNRT).
B. Orthodromic reentrant tachycardia.
C. Atrial tachycardia (AT).

2. What is your next therapeutic maneuver?

A. Antiarrhythmic therapy.
B. Electrophysiology (EP) study and catheter ablation.

Question No. 2:

The mode of adenosine termination favors:

A. Atrial tachycardia.
B. Atrioventricular nodal reentry tachycardia.
C. Atrioventricular reentry tachycardia (AVRT).
D. Automatic junctional tachycardia.

Question No. 3:

Ventricular overdrive pacing is initiated during tachycardia. These observations prove:

A. AT is excluded.
B. AVNRT is a possible mechanism.
C. AVRT is a possible mechanism.
D. The data are not conclusive.

Question No. 4:

Pacing from the right ventricle close to the region of the CS:

A. Excludes permanent junctional reciprocating tachycardia (PJRT).
B. Proves AT.
C. Excludes AVNRT.
D. Excludes junctional tachycardia.

Question No. 5:

Atrial overdrive pacing from the CS:

A. Proves reentry is present.
B. Excludes a focal tachycardia.
C. Suggests that this is a good site for ablation.
D. Is compatible with a focal tachycardia.

Practical Electrophysiology © 2015 Jasbir S. Sra, Masood Akhtar, Andrea Natale, David J. Wilber
Cardiotext Publishing, ISBN: 978-1-935395-17-1

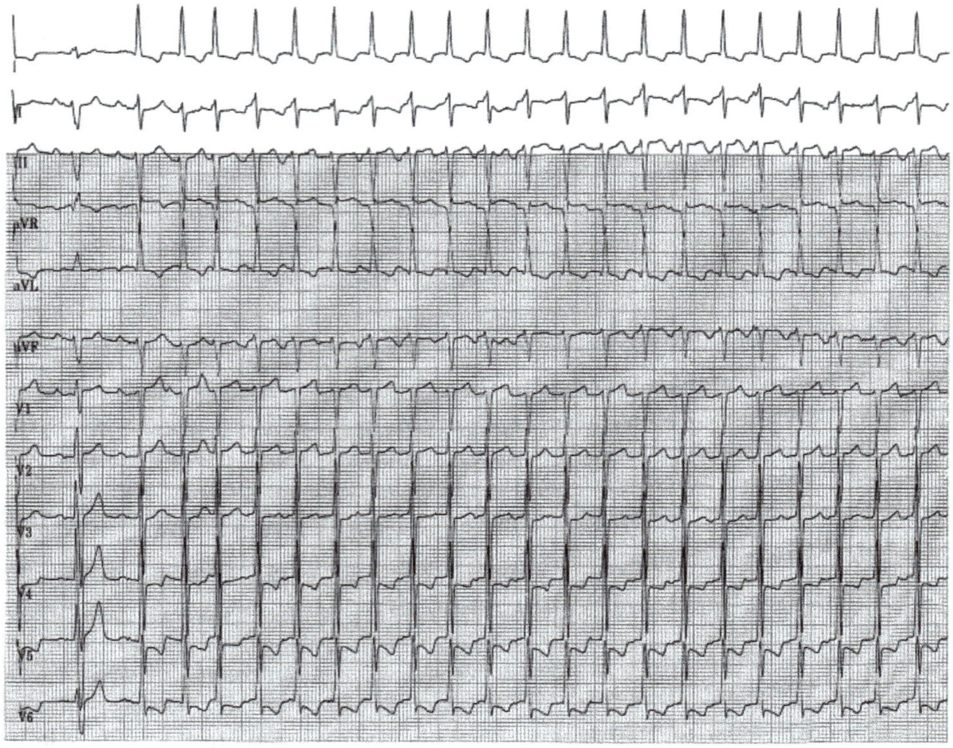

Figure 11.1. Tachycardia.

Question No. 6:

Mapping of tachycardia showed that the earliest atrial electrogram (EGM) was over at the 7 O'clock position at the tricuspid annulus. Entrainment at this site, shown in Figure 11.7:

A. Excludes a reentrant circuit.

B. Proves that the early site is not part of the circuit.

C. Proves presence of abnormal automaticity or triggered rhythm.

D. Is compatible with a focal tachycardia.

DISCUSSION

AT is the least common cause of paroxysmal AT. Note in Figure 11.1 that the tachycardia is initiated by several atrial ectopic complexes before settling into a long RP tachycardia. At first glance, the history of incessant tachycardia and the findings of a long RP configuration (Figures 11.1 and 11.2) with negative P waves in the inferior leads suggest the possibility of permanent junctional reentry tachycardia (PJRT). Other possibilities include AT or atypical (fast–slow) AVNRT.

Careful inspection of the 12-lead ECG gives additional and very valuable insight. Note in Figures 11.1 and 11.2 that the precordial P waves are negative in both the inferior as well as in the precordial leads. The data from Teh et al[1] using the P-wave contour to locate the atrial site of origin of tachycardia is very helpful. The negative P wave in the precordial leads suggests an anterior right atrial focus, whereas the negative P wave in the inferior leads suggests an inferior focus (Figure 11.3). Since the tricuspid annulus is the most anterior portion of the right atrium, we strongly suspect that the site of origin is over the inferior tricuspid annulus, probably in the annular portion of the cavotricuspid region. This P-wave configuration would be very unusual for atypical AVNRT since the atrial breakthrough site should be close to the ostium (os) of the CS and should show an upright P wave in V_1. We have not, however, excluded a decremental right-sided accessory pathway, and this is an important focus of our electrophysiologic study.

Figure 11.2 shows the termination of tachycardia with intravenous adenosine. Note that the tachycardia rate is slower and the negative P wave in V_1 is clearly evident. The tachycardia terminates with a QRS; this finding is most compatible with AT but

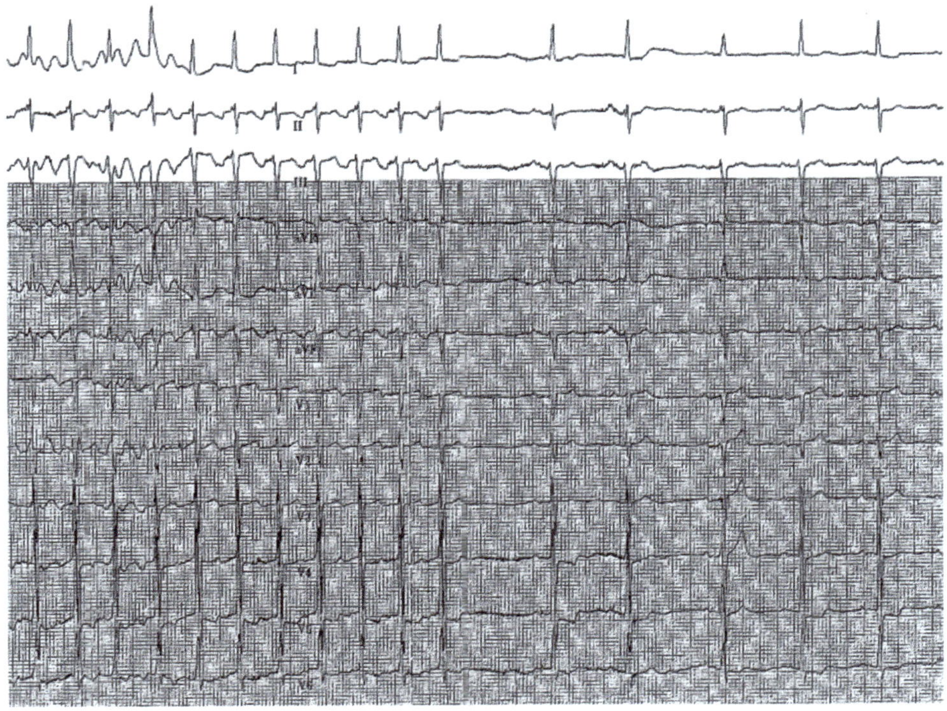

Figure 11.2. After adenosine.

certainly does not exclude either PJRT or atypical AVNRT as either of the latter may terminate with block in the retrograde pathway. Tachycardia termination with adenosine is the rule with junctional reentrant tachycardia, but some ATs are adenosine sensitive.

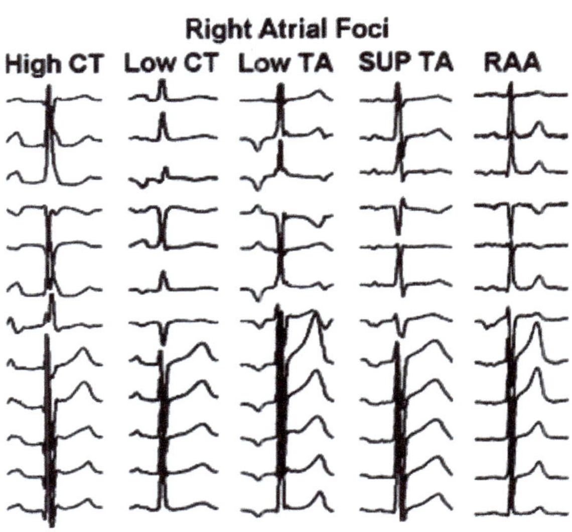

Figure 11.3. P-wave contour in determining the mechanism of the tachycardia. CT, crita terminalis; TA, tricuspid annulus; SUP, superior; RAA, right atrial appendage. (Reproduced with permission from Teh et al.[1])

Figure 11.4 shows the results of ventricular overdrive pacing during tachycardia. This maneuver is done first for the patient with supraventricular tachycardia (SVT), as emphasized by Morady's laboratory,[2] for the purposes of entrainment of the tachycardia. An AT will show a so-called "VAAV" response with termination of ventricular pacing, whereas those with AVNRT or AVRT will show a single A after pacing (VAV response).

Careful examination of Figure 11.4 shows that one cannot use the above criteria, since the atrium is not captured during ventricular pacing. The presence of ventriculoatrial (VA) dissociation excludes a right-sided accessory pathway and favors AT (AVNRT is not excluded by this maneuver).

An important maneuver especially useful to exclude PJRT is shown in Figure 11.5. In PJRT, the earliest atrial EGM is usually recorded near the CS. Pacing the ventricle from this region will always result in tachycardia entrainment. The presence of VA dissociation effectively excludes this diagnosis as well as any right-sided accessory pathway.

Atrial overdrive pacing from the region of the CS (or, in this patient, from the lateral right atrium) showed manifest entrainment with a very long postpacing interval (PPI; Figure 11.6).

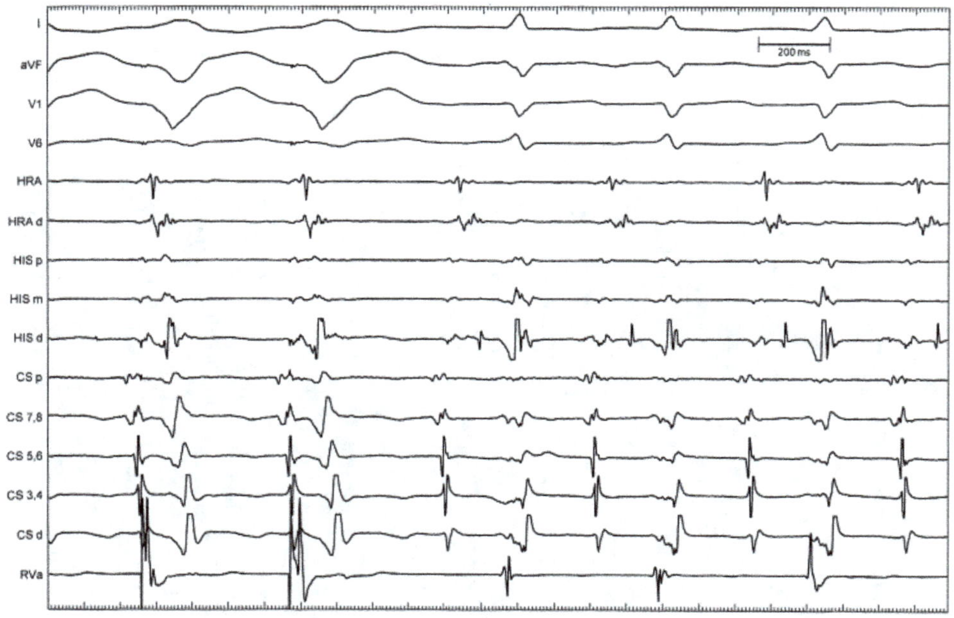

Figure 11.4. Ventricular pacing during tachycardia.

In Figure 11.7, pacing from the region of the earliest atrial EGM shows a much shorter PPI. Neither of these atrial pacing maneuvers proves reentry as proposed by Waldo et al[3] (ie, constant fusion and progressive fusion at different paced cycle lengths). More recently, Mohamed et al[4] showed that the shortest PPI is over the site of origin of focal tachycardia, which was true in our case. Note that the

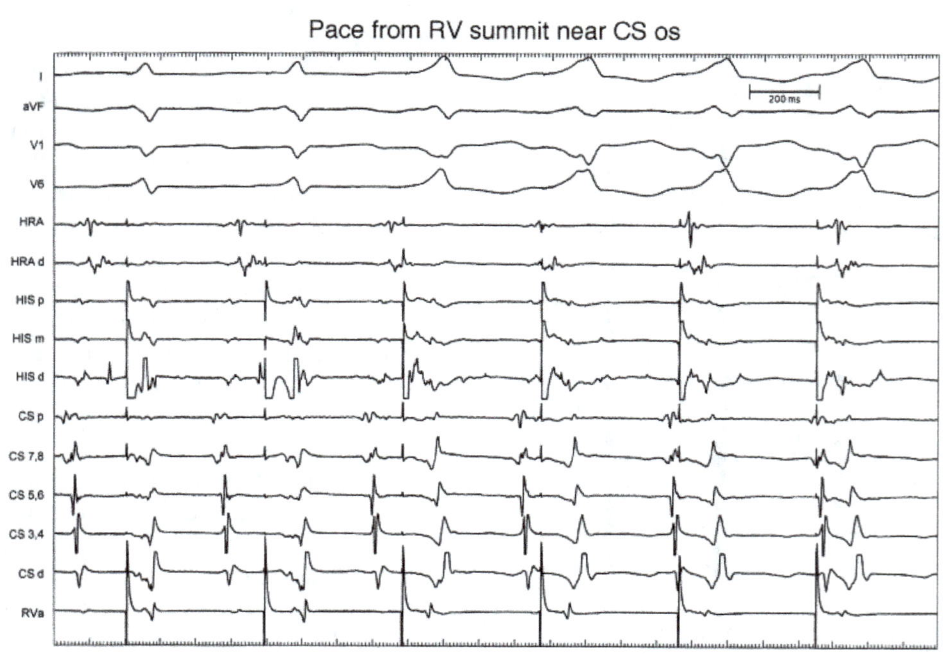

Figure 11.5. Pace from right ventricular (RV) summit near coronary sinus (CS) ostium (os).

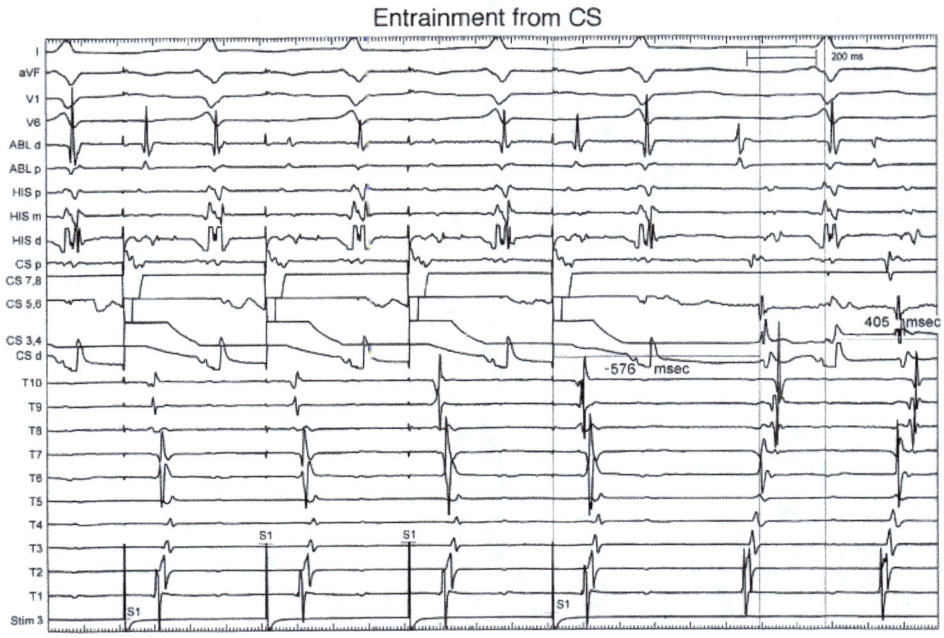

Figure 11.6. Entrainment from the CS.

earliest atrial endocardial potential is recorded over the right inferior annulus, thus excluding AVNRT.

Figure 11.8 shows the successful site of ablation for this AT. Note that the three-dimensional (3D) anatomic map shows a focal pattern with earliest activation over the inferior tricuspid annulus as predicted from analyses of the surface ECG recording (see Figure 11.3).

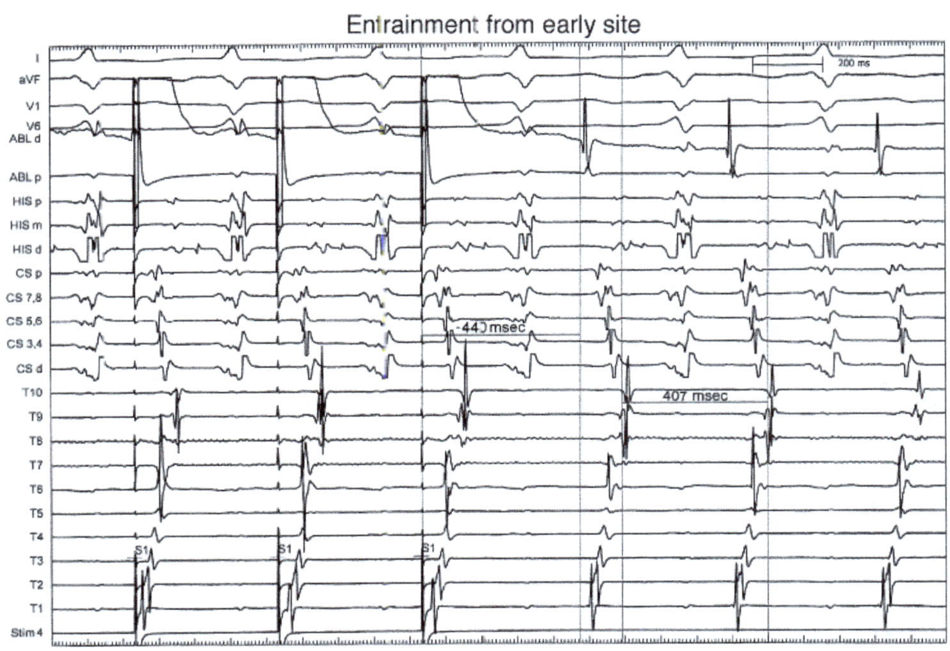

Figure 11.7. Entrainment from early site.

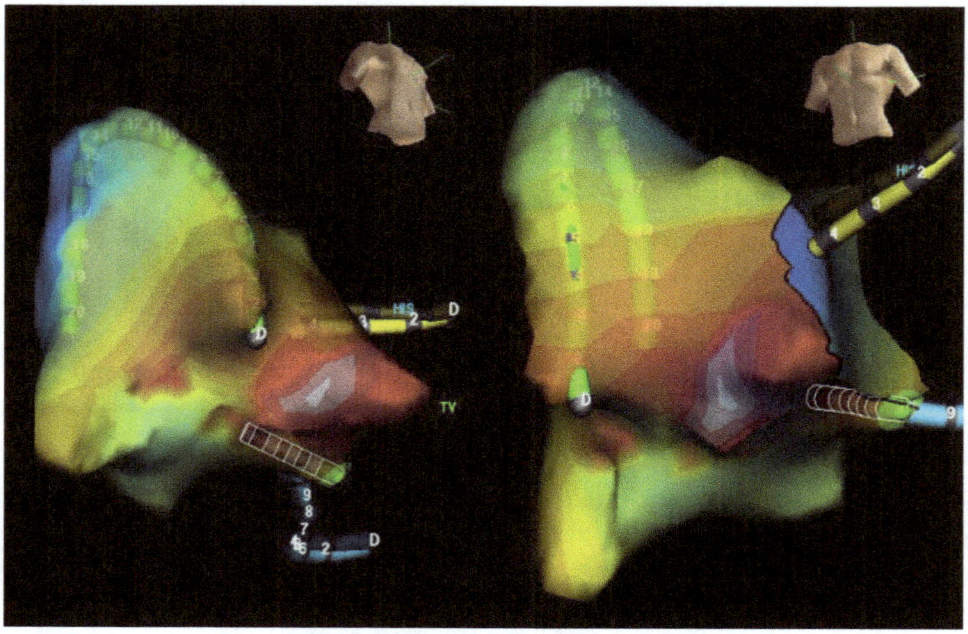

Figure 11.8. Site of successful ablation.

CONCLUSION

Although the history and several ECG features were suggestive of PJRT, careful analysis of the surface ECG nevertheless allowed for making the correct diagnosis. The electrophysiologic study was directed at excluding other causes of tachycardia like PJRT or atypical AVNRT. Once the latter were excluded, the focus then shifted to determining whether we are dealing with a focal versus macroreentrant mechanism. This is helped by several procedures, first, entrainment mapping as discussed above as well as 3D mapping. In this case, the latter showed a focal pattern with activation of <50% of the tachycardia cycle lengths. In the usual focal AT, the atrial EGM is found 20 to 40 ms prior to the surface P wave. In this case, the earliest endocardial atrial EGM recorded was 138 ms before the surface P wave. This finding and the shortest PPI tachycardia cycle length and concealed response (see Figure 11.6) found at the atrial site of origin suggested a microreentrant focal tachycardia. Against this mechanism is the response of the tachycardia to adenosine (see Figure 11.2). Adenosine termination of an AT is more reflective of a triggered mechanism.[5] This case highlights the often difficult problem of distinguishing microreentry versus triggered rhythm in the laboratory. Since the tachycardia was clearly focal, efforts were directed at searching for the site of origin. Use of 3D anatomic (see Figure 11.7) tools is extremely helpful for the diagnosis and ablation of this focus.

Correct Answers:

Q1. 1C, 2B
Q2. A
Q3. D
Q4. B
Q5. D
Q6. D

REFERENCES

1. Teh AW, Kistler PM, Kalman JM. Using the 12-Lead ECG to localize the origin of ventricular and atrial tachycardias: Part 1. Focal atrial tachycardia. *J Cardiovasc Electrophysiol.* 2009;20:706–709.
2. Knight BP, Ebinger M, Oral H, et al. Diagnostic value of tachycardia features and pacing maneuvers during paroxysmal supraventricular tachycardia. *J Am Coll Cardiol.* 2000;36:574–582.
3. Waldo AL, Henthorn RW, Plumb VJ, MacLean WA. Demonstration of the mechanism of transient entrainment and interruption of ventricular tachycardia with rapid atrial pacing. *J Am Coll Cardiol.* 1984;3:422–430.
4. Mohamed U, Skanes AC, Gula LJ, et al. A novel pacing maneuver to localize focal atrial tachycardia. *J Cardiovasc Electrophysiol.* 2007;18:1–6.
5. Markowitz SM, Nemirovsky D, Stein KM, et al. Adenosine-insensitive focal atrial tachycardia: Evidence for de novo micro-reentry in the human atrium. *J Am Coll Cardiol.* 2007;49:1324–1333.

CASE STUDY

12

Atrial Tachycardia from Above the Semilunar Valves

Ramprakash Balasundaram, MD and Calambur Narasimhan, MD, DM

Question No. 1:

In the patient with ectopic AT shown in Figure 12.1, the most likely focus is on:

A. Noncoronary sinus.
B. Left pulmonary vein (PV).
C. Superior mitral annulus (SMA).
D. Coronary sinus (CS) ostium.
E. Atrial septum.

DISCUSSION

When P waves during a tachycardia merge with the T wave, introducing premature ventricular complexes brings out the P waves better, as in this case. Morphology of P waves helps focus the mapping efforts to the area of interest.

Most foci responsible for focal AT tend to cluster at characteristic anatomical locations. In the right atrium (RA), these foci tend to occur along the crista terminalis (CT), tricuspid annulus (TA), ostium of the CS, and the para-Hisian region. In the left atrium (LA), foci occur predominantly at the ostia of the PVs and, less commonly, at the mitral annulus LA appendage and left atrial aspect of the interatrial septum (IAS).

The ECG in Figure 12.1 shows a tachycardia P wave narrower than the sinus rhythm P wave suggestive of a site of origin near the septum. The algorithm of Kistler et al[1] (Figure 12.2) may be applied here.

The tachycardia was mapped from the RA and showed diffuse area of septal breakthrough. The CARTO™ map (Biosense Webster Inc., Diamond Bar, CA, USA) of the RA (Figure 12.3) showed area of diffuse septal breakthrough of the tachycardia focus, and the unipolar electrogram (EGM) showed an initial positive deflection in the point of earliest activation from the right septal aspect.

This patient had a negative positive P wave in lead V_1 with a negative P wave in lead aVL. This corresponds to a site of origin from the SMA from the above algorithm. However, several anatomic sites that are in close proximity, such as the right superior pulmonary vein (RSPV), ostium of superior lip of the CS, SMA, non-CS of the aortic valve, and left atrial aspect of the septum, could not always be separated (Figure 12.4). Knowledge of these

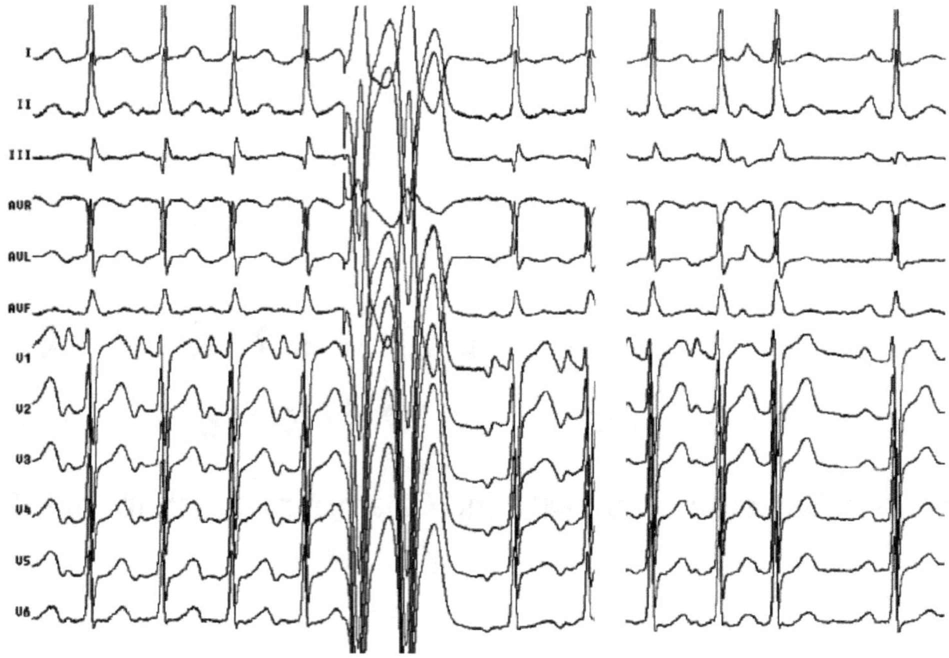

Figure 12.1. ECG of a patient with ectopic atrial tachycardia (AT).

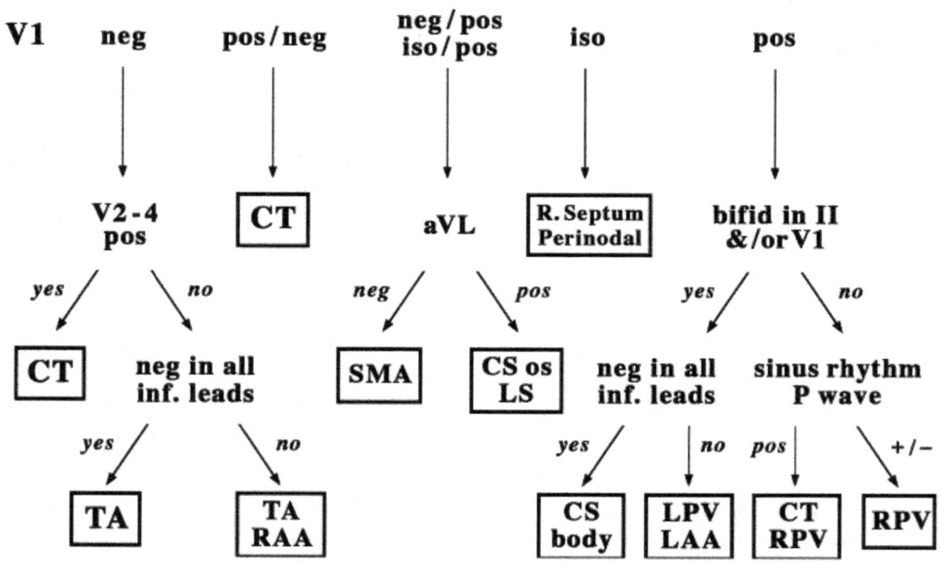

Figure 12.2. Algorithm for detecting location of tachycardia. See text for abbreviations and other details.

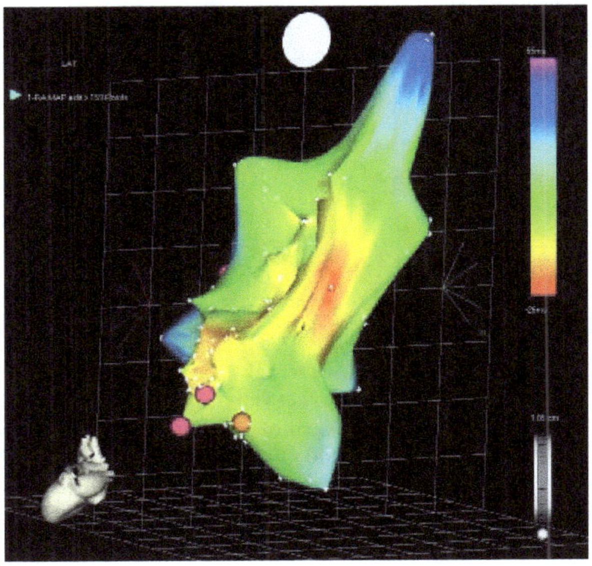

Figure 12.3. The CARTO map.

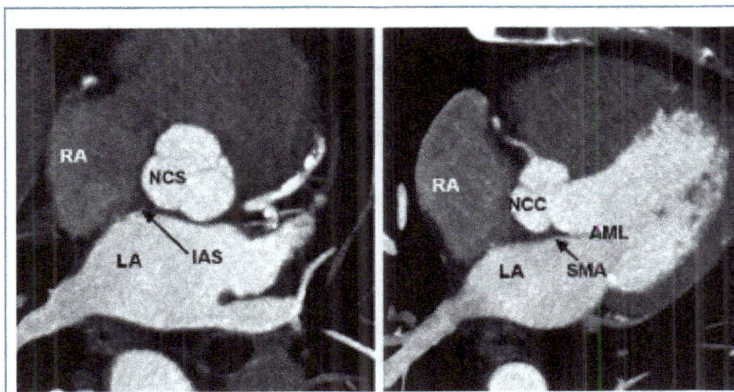

Figure 12.4. CT scan of the LA illustrating the relationship of noncoronary cusp (NCC) of the aortic valve to the septal aspect of RA, LA, and SMA/RA with IAS. See text for abbreviations and other details.

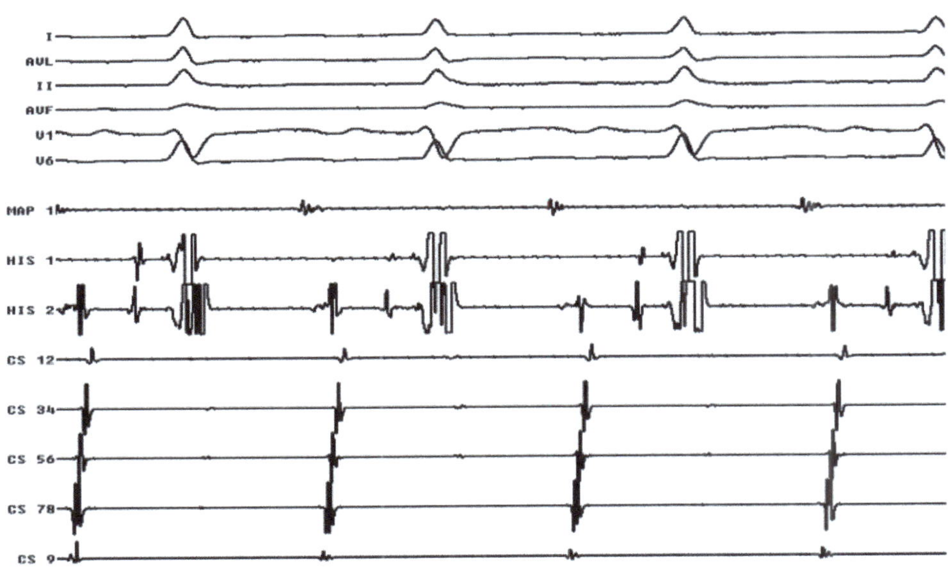

Figure 12.5. Mapping in the non-CS showed atrial signals that were 74 ms earlier than surface P and 42 ms earlier than atrial activation at the His-bundle area.

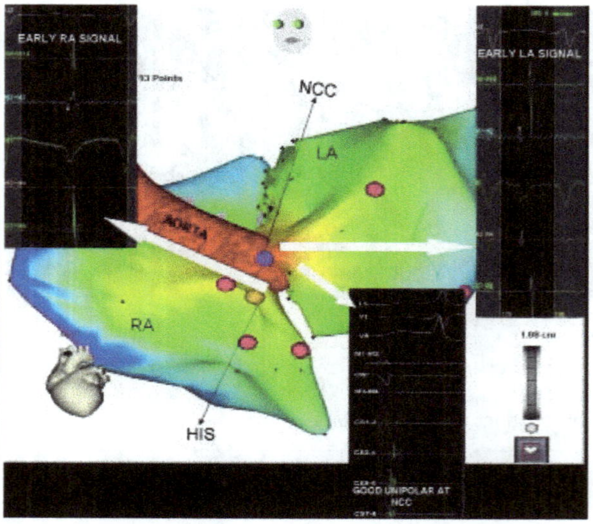

Figure 12.6. Superimposition of the mapping points from the NCC shows the close proximity of the NCC to the IAS. There is a clear unipolar negative deflection in the mapping catheter, and ablation here resulted in termination of the tachycardia.

overlapping sites is important to the clinical electrophysiologist.

Figure 12.6 depicts the relationship of NCC to other structures.

Correct Answer:

Q1. A

REFERENCES

1. Kistler PM, Sanders P, Fynn SP, et al. Electrophysiologic and electroanatomic changes in the human atrium associated with age. *J Am Coll Cardiol.* 2004;44:109–116.

CASE STUDY

13

ECG Localization of Accessory Pathways

Nitish Badhwar, MD and Melvin M. Scheinman, MD

A 49-year-old male presents with long-standing history of palpitations and a recent syncopal episode. Echocardiogram did not show any structural heart disease. The baseline ECG is shown in Figure 13.1.

Question No. 1:

Your diagnosis is ventricular preexcitation over a:

A. Right free wall accessory pathway (AP).
B. Right septal AP.
C. Left posterior AP.
D. Left lateral AP.

Question No. 2:

The most likely diagnosis is:

A. Atrial tachycardia (AT) arising from the superior crista terminalis.
B. Typical atrioventricular nodal reentrant tachycardia (AVNRT).
C. Atrioventricular reentrant tachycardia (AVRT) using left posterior AP.
D. AVRT using a right anterior AP.

Question No. 3:

The most likely diagnosis is:

A. AT arising from the coronary sinus (CS).
B. Typical AVNRT.
C. AVRT using left posterior AP.
D. AVRT using a right anterior AP.

Question No. 4:

This finding would rule out which one of the following:

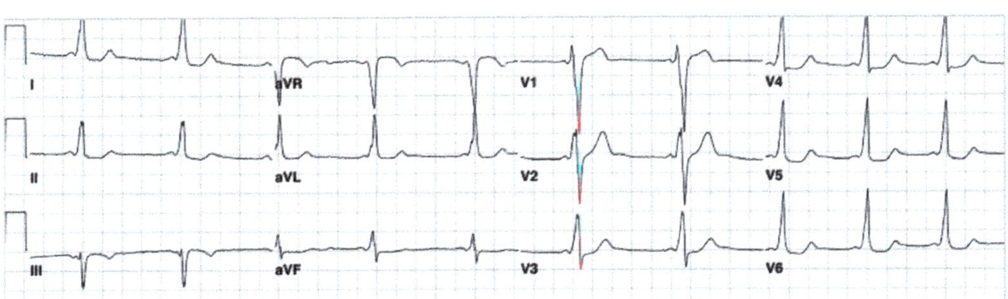

Figure 13.1 Baseline ECG.

Practical Electrophysiology © 2015 Jasbir S. Sra, Masood Akhtar, Andrea Natale, David J. Wilber
Cardiotext Publishing, ISBN: 978-1-935395-17-1

A. Atrial tachycardia.
B. Atypical AVNRT.
C. AVRT using a septal AP.

Question No. 5:

This observation proves:

A. AP pathway is present in this patient.
B. AP is integral to the tachycardia hence proving a diagnosis of AVRT.
C. AVNRT cannot be excluded.
D. AT cannot be excluded.

Question No. 6:

Findings shown in Figure 13.10 demonstrate:

A. Retrograde conduction over AV node only.
B. Retrograde conduction over septal AP only.
C. Retrograde conduction over AV node and AP.
D. Absence of retrograde conduction.

Question No. 7:

The ablation site shown in Figure 13.12:

A. Is an ideal site for ablation.
B. Can lead to AV block.
C. Will not affect conduction over septal AP.

DISCUSSION

The diagnosis of WPW syndrome is based on the presence of a short PR interval, delta wave, and a wide QRS complex that is associated with SVT. Possible AP locations are displayed in Figure 13.2. Most APs (65%) are located over the left free wall, 20% are located in the septum, and the remainder over the right free wall.[1] ECG localization of the site of ventricular insertions of the AP is listed in Table 13.1 and based on the study by Fitzpatrick et al.[2] Based on this algorithm, one would predict that the pathway in this patient is located on the right midseptal region (see Figure 13.1). The early precordial transitions (V_3) and intermediate frontal axis supports this diagnosis.

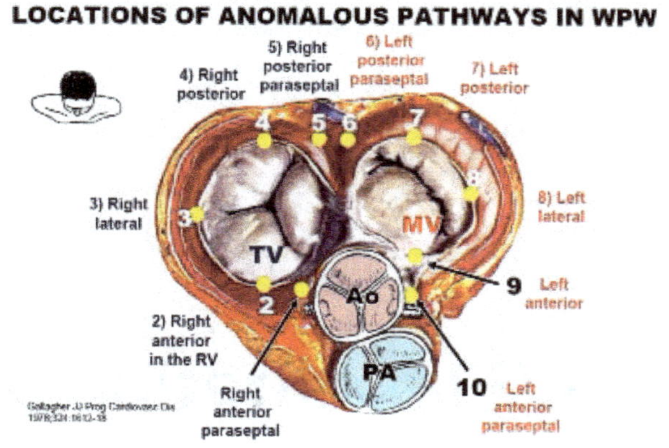

Figure 13.2. Location of various APs in patients with Wolff-Parkinson-White (WPW) syndrome. (Reproduced from Gallagher et al[1] with permission from Elsevier.)

Figure 13.3 shows an example of a right postero-septal AP. Again, note the early transition (V_3) with dominant negative delta waves in the interior leads. These pathways may be ablated over the right posterior free wall, over the posterior septum or in a branch of the CS. Arruda et al[3] have emphasized the importance of a negative delta wave in lead II pointing to pathway involving a CS venous tributary. Posterior pathways show initial superior forces.

Figure 13.4 is an example of a 12-lead ECG showing a classic right anterior AP. The precordial plane shows a later transition (V_4) and the initial

Table 13.1. ECG Localization of AP

Right free wall AP	Septal AP	Left free wall AP
Left bundle branch block (LBBB) pattern	Early transition V_1/V_2 or V_2/V_3	Transition in V_1
Late transition ($>V_3$)	Superior axis-posteroseptal	Q wave in lead I, aVL—lateral
Superior axis (inferior location)	Inferior axis—anteroseptal	Superior axis—inferolateral
Interior axis (superior location)	Intermediate axis—midseptal (para-Hisian)	Inferior axis-anterolateral
	Delta Q waves in inferior leads specific not sensitive for CS vein origin	
	Delta Q waves inferior leads + positive delta in aVR—CS vein origin	

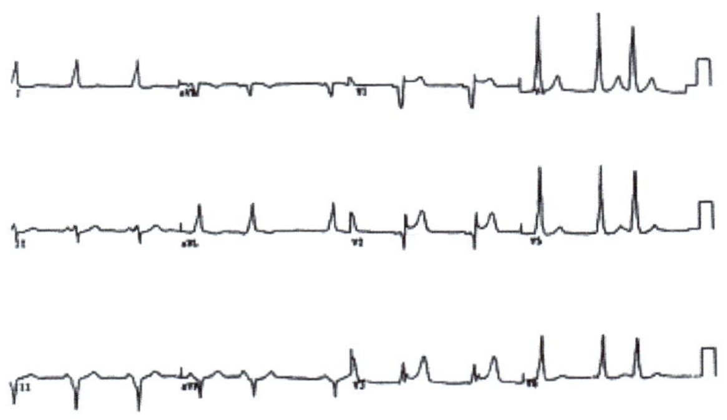

Figure 13.3. 12-lead ECGs of patients with WPW.

forces are directed inferiorly (positive delta in 2,3,F). More anterior pathways will project an inferior axis.

Figure 13.5 shows an ECG pattern that is classic for a left anterolateral AP. Note the dominant and positive initial forces (delta wave) in V_1 with an inferior axis. The negative delta waves in leads I and aVL suggest a left lateral AP.

The ECG localization of APs is both sensitive and specific compared with endocardial mapping, but the clinician should be aware of potential confounders. These include tracings showing minimal preexcitation (presence of delta wave after a fully inscribed P wave in right-sided APs suggests absence of full preexcitation). Pathway localization may be obfuscated by dominant conduction over the AV node—His axis. In addition, if an AP is located over the far left lateral region and the patient has rapid conduction over the AV node, then sinus impulses may result in minimal (or no) preexcitation. This may occur even with robust, life-threatening left-sided APs. Absence of septal Q waves with a short PR interval can be a subtle clue to the presence of preexcitation that normalizes after catheter ablation.[4] If the patient has dual AP, ventricular fusion from these sources may obscure pathway location. In patients with dual pathways, atrial fibrillation usually unmasks the presence of 2 pathways. Patients with congenital cardiac disease, thoracic cage abnormalities, or electrode misplacement may all obscure proper AP localization.

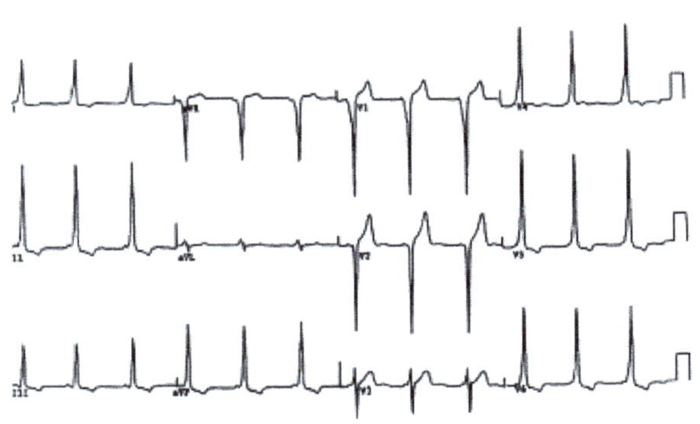

Figure 13.4. 12-lead ECGs of patients with WPW.

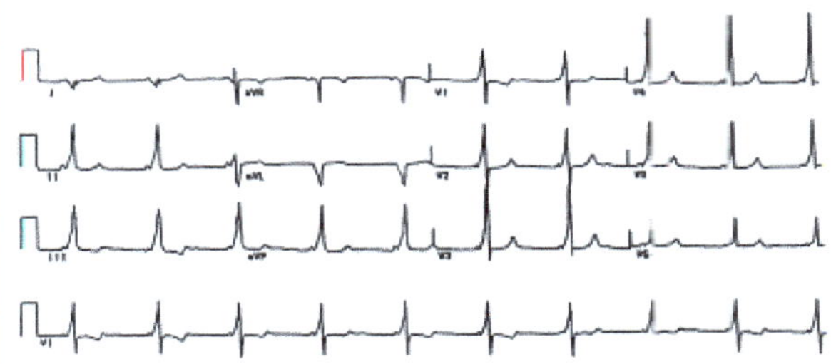

Figure 13.5. 12-lead ECGs of patients with WPW.

The 3 most common causes of SVT include AVNRT (50%-60%), followed by AVRT in patients with WPW syndrome (25%-30%), and AT (10%). The 12-lead ECG during SVT can be used to differentiate these based on the relationship of the P wave to the R wave. Typical AVNRT presents with A on V tachycardia, AVRT is associated with short RP tachycardia and AT usually presents as long RP tachycardia. Figure 13.6 shows a short RP tachycardia with negative P waves in the inferior and lateral leads. This would argue against typical AVNRT. The negative P waves in the inferior leads rule out AT arising from superior crista terminalis. The most likely diagnosis for a short RP tachycardia is AVRT. The P-wave configuration during tachycardia is also used as an aid for confirming AP location. Based on the P-wave morphology, the most likely diagnosis is AVRT using a left posterior AP. In this ECG, there is a disparity between the baseline ECG suggesting a right anteroseptal AP and the 12-lead ECG during tachycardia that indicates presence of posterior atrial insertion site. This finding should arouse suspicion for the presence of dual APs.

Figure 13.7 shows a 12-lead ECG of a short RP tachycardia recorded in our patient. Note that the P-wave polarity is now positive in the inferior and lateral leads. This is consistent with AVRT using an anteroseptal AP that correlates with the baseline ECG localization of the septal AP.

Figure 13.8 shows a narrow complex tachycardia that was regularly initiated by a single atrial premature beat that blocked in the AP and conducted back over the AV node-His axis (see Figure 13.11). The earliest site of atrial activation is over the septal region (His A). This tracing shows the result of ventricular overdrive pacing during SVT. AT is excluded by showing that the first ventricular premature beat (PVC) (*) of the drive

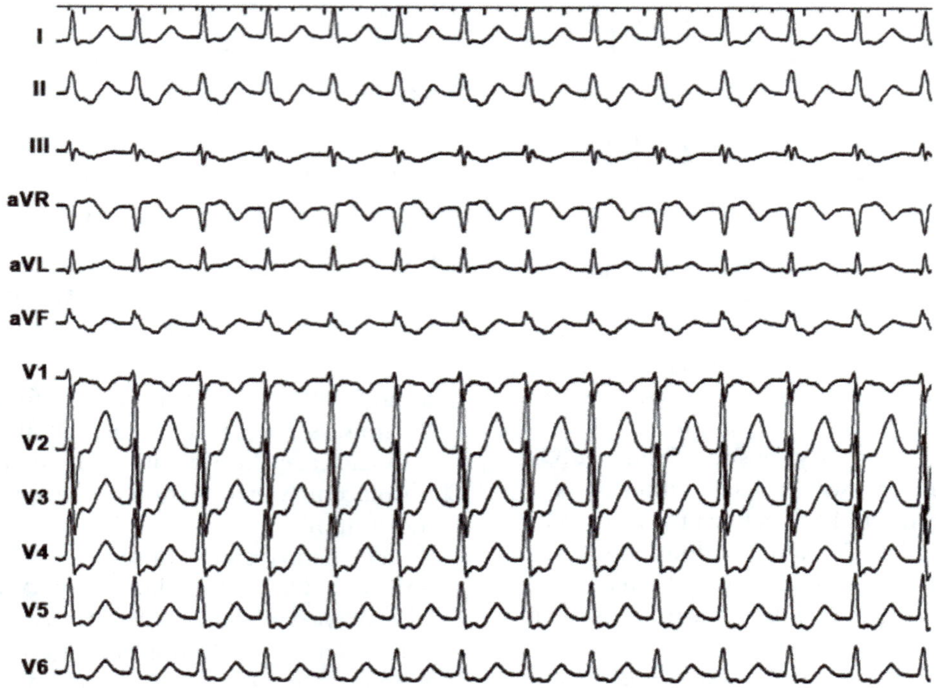

Figure 13.6. 12-lead ECG with narrow QRS complex supraventricular tachycardia (SVT) that was induced in the laboratory to the patient.

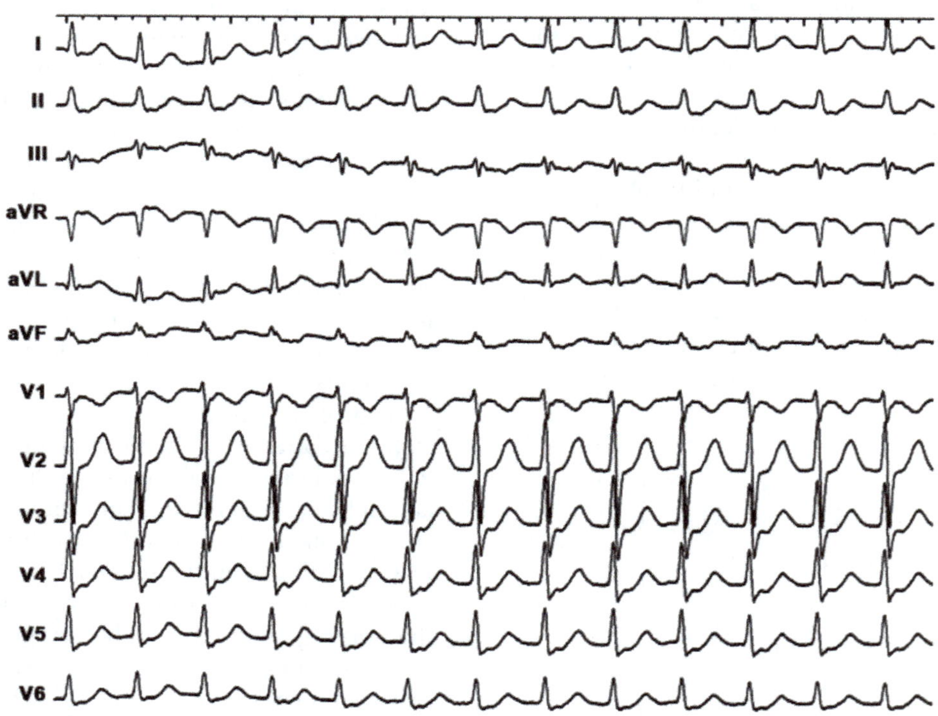

Figure 13.7. An example showing a 12-lead ECG of another SVT that was initiated in the laboratory in this patient.

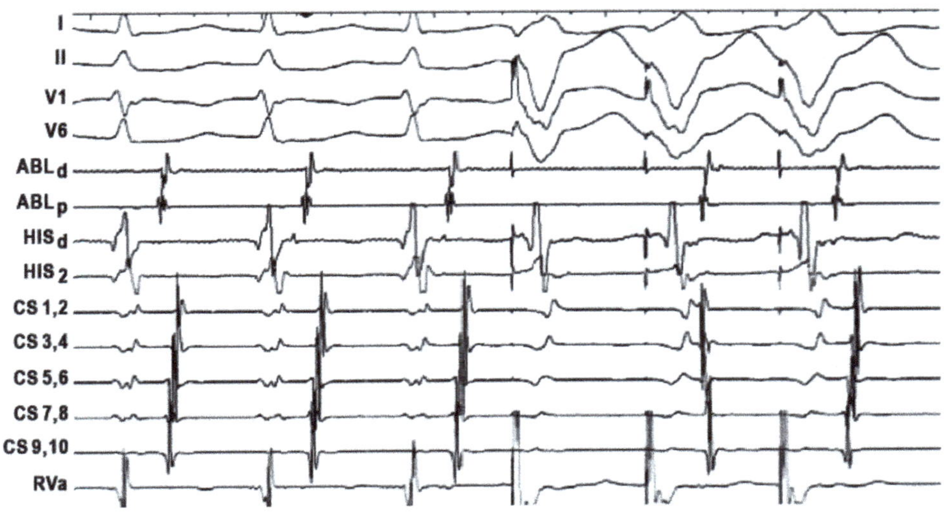

Figure 13.8. Surface ECGs together with recordings from the ablation catheter (Abl) in the high right atrium (HRA), the His bundle (His D, P, and 2), the proximal CS 9,10 to distal CS 1,2, and from the right ventricle (RVa).

train terminates the tachycardia without a change in the AA cycle length. An AT can be terminated with a single ventricular premature complex but only if the latter advances the atrial electrogram (EGM).

Figure 13.9 shows a single premature ventricular impulse (*) delivered while the His bundle is refractory (note the fusion complex), terminating the tachycardia. The latter conclusively proves the presence of AP and that it is part of the tachycardia circuit. It is important from an ablation standpoint to confirm the participation of an AP participation in AVRT. The presence of any of the 4 criteria can

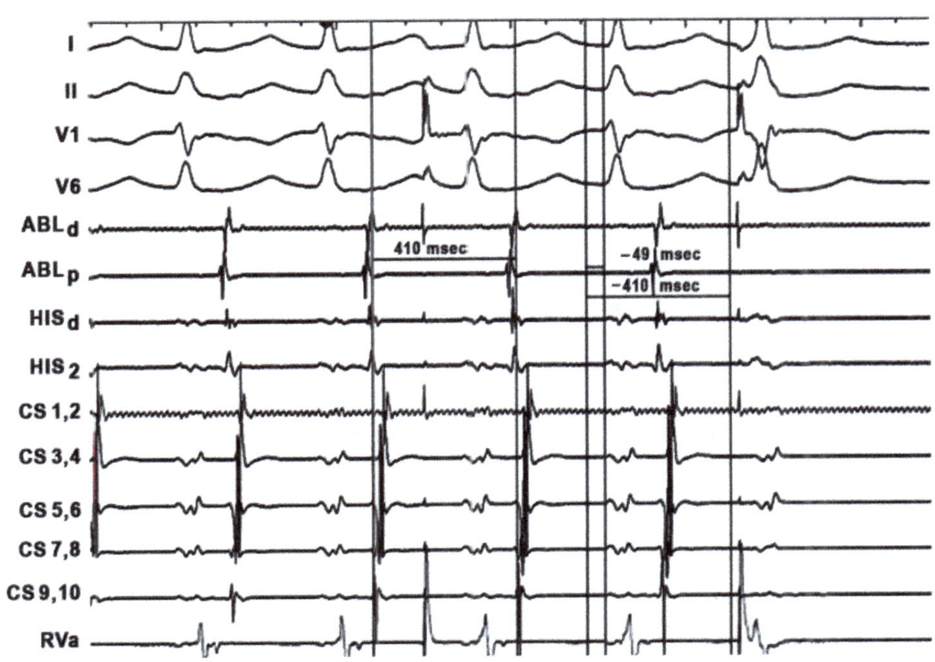

Figure 13.9. Delivery of single premature ventricular impulse during tachycardia.

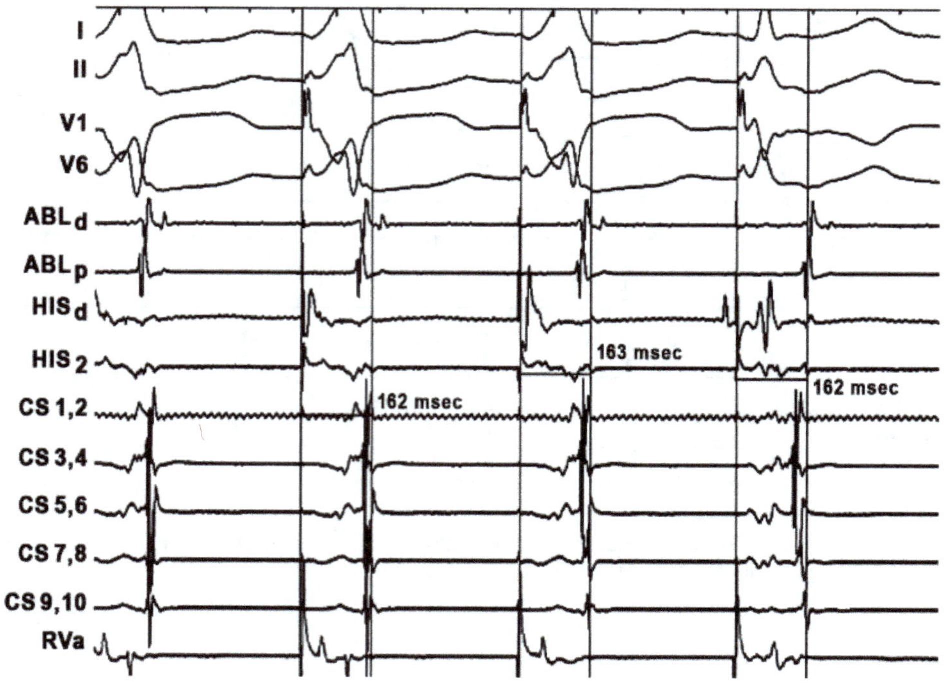

Figure 13.10. An example of para-Hisian pacing.

be used to establish this, and they include: (1) tachycardia termination obtained with a ventricular stimulus delivered during His refractoriness and without atrial depolarization; (2) delay in the next atrial depolarization when a ventricular stimulus is delivered during His refractoriness; (3) increase in VA interval with ipsilateral bundle branch block (BBB) during tachycardia;[2] and (4) an increase in HV interval results in an increase in HA interval during tachycardia. It is important to deliver the PVC in close vicinity to the AP so that it has the maximal chance of invading the retrograde limb and terminating tachycardia. In this case, a PVC delivered from the right ventricular (RV) apex did not perturb the tachycardia, whereas a PVC delivered from the RV septum (close to the AP insertion site) terminated tachycardia.

Figure 13.10 shows findings of Para-Hisian pacing.[5] This maneuver involves pacing from the RV summit at low output (RV capture) and high output (His capture) and is designed to differentiate retrograde conduction over a septal AP from conduction over the AV node. The first 3 complexes are wide and are indicative of sole excitation of the RV summit without His capture (note the retrograde His inscribed within the QRS complex). The fourth paced complex (*) is narrow denoting His capture. The stimulus to atrial deflection remains unchanged whether His capture is achieved or not. This finding is diagnostic of a septal AP since with normal retrograde nodal conduction the VA interval with summit pacing alone would be longer (stimulus must enter through the branches of the right bundle) compared to that of His capture with retrograde nodal and atrial activation.

Figure 13.11 shows the intracardiac EGMs at the site of successful ablation of the right septal AP. The first complex (*) is preexcited with a fused AV complex and possible Kent potential.[6] The latter disappears when an atrial premature complex initiates tachycardia and no His deflection is seen on the ablator catheter. Note that the VA complex is fused during tachycardia. Ablation at this site terminated the SVT mediated by the septal pathway and allowed for diagnosis of a separate concealed left posterior pathway.

Figure 13.12 shows fluoroscopic appearance (right anterior oblique (RAO) projection) of the site of ablation for the 2 pathways. The manifest right septal pathway was ablated over the anterior septum (A), while the concealed postero-septal AP was ablated via retrograde transaortic approach in the left ventricle (B).

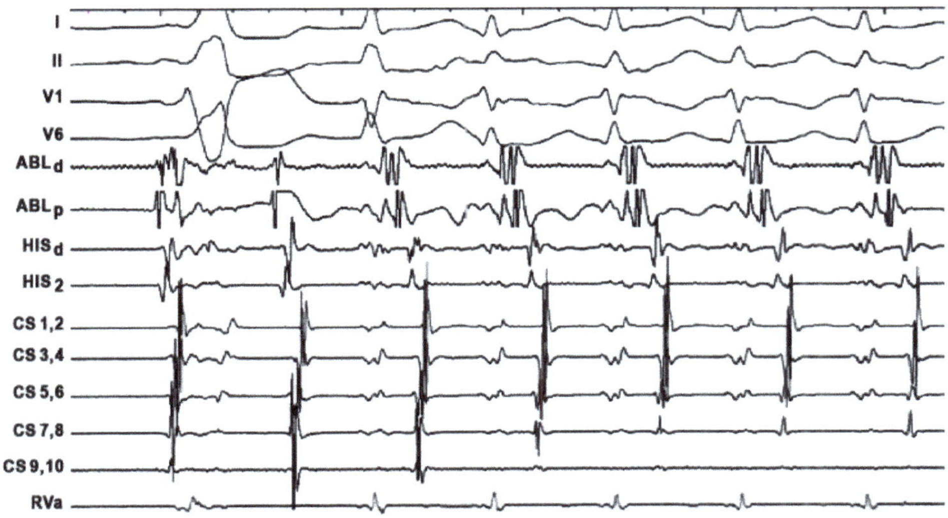

Figure 13.11. An example showing surface ECGs together with recordings from the Abl in the right septal region, the His bundle (His D and 2), the proximal CS 9,10 to distal CS 1,2 and from the RVa.

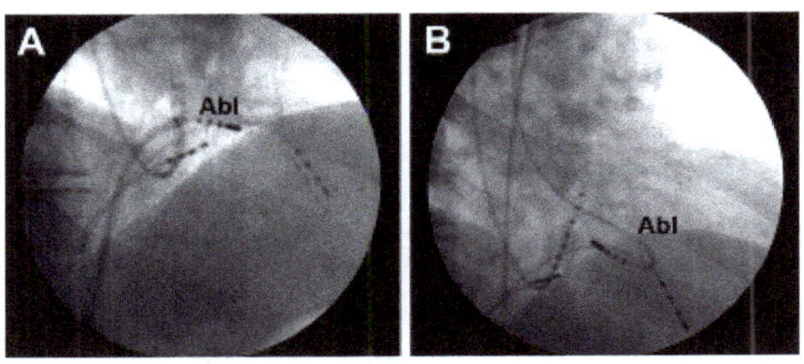

Figure 13.12. Ablation site shown in 2 views.

SUMMARY

This case demonstrates the surface ECG and intracardiac findings in a patient with WPW who had 2 APs. Multiple APs can be found in 10% of the patients with WPW, and dual AV node physiology is also common in these patients. The ECG is an excellent tool that allows for localization of manifest APs. Application of a few simple rules allows for a relatively confident diagnosis, providing attention is given to several confounding factors. The P wave contour during tachycardia may also aid in pathway localization as was nicely shown in the case presented. In the electrophysiology (EP) laboratory, detailed pacing and mapping maneuvers are required to exclude other tachycardia mechanisms and to allow for determining the successful ablation site.

Correct Answers:

Q1. B
Q2. C
Q3. D
Q4. A
Q5. B
Q6. A
Q7. A

REFERENCES

1. Gallagher JJ, Pritchett EL, Sealy WC, Kasell J, Wallace AG. The preexcitation syndromes. *Prog Cardiovasc Dis.* 1978;20:285–327.
2. Fitzpatrick AP, Gonzales RP, Lesh MD, et al. New algorithm for the localization of accessory atrioventricular connections using a baseline electrocardiogram. *J Am Coll Cardiol.* 1994;23:107–116.
3. Arruda MS, McClelland JH, Wang X, et al. Development and validation of an ECG algorithm for identifying accessory pathway ablation site in Wolff-Parkinson-White syndrome. *J Cardiovasc Electrophysiol.* 1998;9:2–12.
4. Liberman L, Pass RH, Starc TJ, Hordof AJ, Silver ES. Uncovering the septal Q wave and other electrocardiographic changes in pediatric patients with pre-excitation before and after ablation. *Am J Cardiol.* 2010;105:214–216.
5. Hirao K, Otomo K, Wang X, et al. Para-Hisian pacing. A new method for differentiating retrograde conduction over an accessory AV pathway from conduction over the AV node. *Circulation.* 1996;94:1027–1035.
6. Schwartz SM, Dick M, 2nd, Dorostkar PC, Serwer GA, LeRoy S. Electrogram patterns associated with successful radiofrequency ablation of accessory pathways in children. *Pediatr Cardiol.* 1996;17:137–142.

CASE STUDY

14

Permanent Junctional Reciprocating Tachycardia

George M. McDaniel, MD, MS and John P. DiMarco, MD, PhD

A 4-year-old male presents with a history of nausea and vomiting for a day. His parents note that he also has been complaining of his "heart hurting," and they have noticed a decrease in his activity level over the past couple of weeks. On examination, he is tachycardic with hepatomegaly and splenomegaly. A chest x-ray demonstrates an enlarged cardiac silhouette. An ECG and echocardiogram are performed. The echocardiogram demonstrates a dilated left ventricle (LV) with markedly reduced systolic function. The ECG is shown in Figure 14.1.

Question No. 1:

The rhythm shown in Figure 14.1 is consistent with the following diagnoses except:

A. Atrial tachycardia (AT).

B. Permanent junctional reciprocating tachycardia (PJRT).

C. Typical atrioventricular nodal reentry tachycardia (AVNRT).

In typical AVNRT, the antegrade limb of the circuit is the slow pathway of the atrioventricular

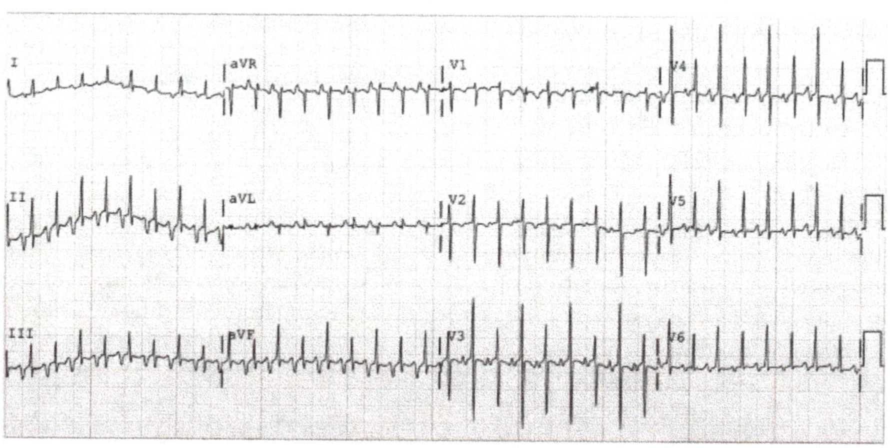

Figure 14.1. Standard 12-lead ECG of the patient.

Practical Electrophysiology © 2015 Jasbir S. Sra, Masood Akhtar, Andrea Natale, David J. Wilber
Cardiotext Publishing, ISBN: 978-1-935395-17-1

(AV) node and the retrograde limb is the fast pathway. Hence, atrial depolarization occurs almost simultaneously with ventricular depolarization. The larger voltage generated by the ventricles obscures the voltage generated by the atria, and P waves are not usually seen on a surface 12-lead ECG. In Figure 14.1, the T waves are not normal in their appearance. There is a P wave fused with the terminal portion of the T wave. This indicates that atrial depolarization is occurring closer in time to ventricular repolarization and not ventricular depolarization.[1] This is a long RP tachycardia, which may represent an AT or a pathway-mediated tachycardia, but it is not consistent with typical AVNRT.

Question No. 2:

Based upon the intracardiac tracing shown in Figure 14.2, the following conclusion can be made:

A. This is an AT.

B. There is an extranodal bypass tract present.

C. Retrograde conduction is through the AV node.

D. No conclusion can be made from this tracing.

In this tracing, the patient is in tachycardia. The atrial cycle length in tachycardia is ~375 ms as seen on the high right atrial catheter. A ventricular pacing stimulus, identified on the "Stim" channel, is delivered through the right ventricular catheter and captures the ventricle. Note the change in the morphology of the third QRS on the surface ECG; this confirms ventricular capture. The ventricle is paced at a time when the bundle of His is refractory. The subsequent atrial depolarization occurs 350 ms after the atrial depolarization prior to the delivery of the pacing spike, ~25 ms early relative to the tachycardia. The tachycardia is not interrupted and the atrial cycle length returns to the tachycardia

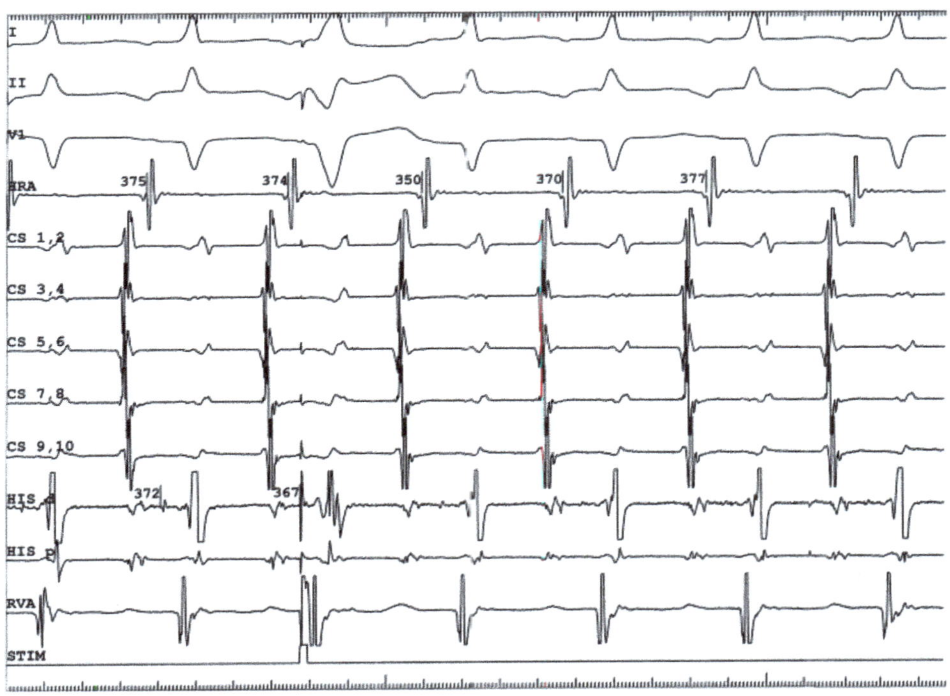

Figure 14.2. Intracardiac tracing of tachycardia. The sweep speed was 100 mm/s. Surface leads I, II, and V_1 are shown. Catheter in the high right atrium is labeled "HRA" and recorded the local atrial activation as a biphasic sharp spike. A quadrapolar catheter placed across the tricuspid valve captured the signal from the depolarization of the bundle of His. The signals captured by the proximal electrode pair are labeled "His p." The signals captured by the distal electrode pair are labeled "His d" and captured the local atrial depolarization followed by the His depolarization signal, and finally, the local ventricular depolarization signal. The right ventricular catheter was in the apex (RVA) of the right ventricle. The multiphasic signal represents the local depolarization signal. The decapolar coronary sinus (CS) catheter is a catheter with a series of 5 electrode pairs. It was placed in the coronary sinus with the 1,2 electrode pair in the distal coronary sinus and the 9,10 electrode pair near the ostium of the coronary sinus. It recorded the local atrial depolarization, represented by the larger signal on the tracing from each electrode pair. The smaller signal on each CS electrode pair is the local ventricular depolarization. The line labeled "Stim" indicates that a pacing stimulus is delivered.

cycle length. This is consistent with the ventricular impulse being conducted to the atria through an extranodal bypass tract, as the bundle of His is refractory.[2]

The tachycardia demonstrated in Questions 1 and 2 was near-incessant. When the tachycardia broke, the patient returned to normal sinus rhythm. However, the tachycardia would spontaneously restart. As noted in the case description, the patient presented with congestive heart failure. Following the ablation of the extranodal bypass tract, there was resolution of the tachycardia and rapid return of normal LV dimensions and systolic function.

Question No. 3:

The most likely diagnosis for these patients is:

A. Sinus tachycardia.

B. Atrial tachycardia.

C. Atypical AVNRT.

D. Permanent junctional reciprocating tachycardia.

Permanent junctional reciprocating tachycardia is an orthodromic reciprocating tachycardia mediated by an extranodal bypass tract, or tracts, with slow conduction properties.[3,4] The tachycardia circuit uses the AV node and the His-Purkinje network for antegrade conduction so that surface QRS complexes are normal in appearance. The retrograde limb is the slowly conducting bypass tract. As a result, a 12-lead ECG performed while the patient is in tachycardia often has discernable P waves. The interval from the preceding QRS to the P wave is longer than the interval from the P wave to the succeeding QRS (long RP tachycardia). As atrial depolarization begins at the AV valve annulus, P waves are often inverted in II, III, and aVF. The other hallmark of PJRT is that the rhythm is often incessant, and patients often present with decreased LV function, tachycardia-induced cardiomyopathy.[5]

Invasive electrophysiology (EP) study and ablation of the bypass tract allow for a definitive diagnosis and cure. These pathways are often located in the right posterior atrial septum.[3,6] However, they have been found on the left side of the septum or away from the septum across either AV valve ring.[7,8]

Correct Answers:

Q1. C

Q2. B

Q3. D

REFERENCES

1. Lockwood D, Otomo K, Wang Z, et al. Electrophysiologic characteristics of atrioventricular nodal reentrant tachycardia: Implications for the reentrant circuits. In: Zipes DP, Jalife J, eds. *Cardiac Electrophysiology: From Cell to Bedside.* 4th ed. Philadelphia: Saunders; 2004: 537–557.
2. Knight BP, Morady F. Atrioventricular reentry and variants. In: Zipes DP, Jalife J, eds. *Cardiac Electrophysiology: From Cell to Bedside.* 4th ed. Philadelphia: Saunders; 2004: 528–536.
3. Critelli G, Gallagher J, Monda V, et al. Anatomic and electrophysiologic substrate of the permanent form of junctional reciprocating tachycardia. *J Am Coll Cardiol.* 1984;4:601–610.
4. Coumel P. Junctional reciprocating tachycardias. The permanent and paroxysmal forms of A-V nodal reciprocating tachycardia. *J Electrocardiol.* 1975;8:79–90.
5. Lindinger A, Heisel A, von Bernuth G, et al. Permanent junction reentry tachycardia: A multicenter long-term follow-up study in infants, children and young adults. *Eur Heart J* 1998;19:936–942.
6. Critelli G, Gallagher JJ, Perticone F, et al. Transvenous catheter ablation of the accessory atrioventricular pathway in the permanent form of junctional reciprocating tachycardia. *Am J Cardiol* 1985;55:1639–1641.
7. Ticho BS, Saul JP, Hulse JE, et al. Variable location of accessory pathways associated with the permanent form of junctional reciprocating tachycardia and confirmation with radiofrequency ablation. *Am J Cardiol.* 1992;70:1559–1564.
8. Ali H, Vitali-Serdoz L, Ferrero P, et al. An unusual case of permanent junctional reciprocating tachycardia: Successful ablation at the mitral annulus-aorta junction. *J Interv Card Electrophysiol.* 2008;23:213–217.

CASE STUDY

15

Targeted Ablation of the Noncoronary Cusp of Aortic Valve for Incessant Orthodromic Atrioventricular Reentrant Tachycardia

Ramprakash Balasundaram, MD, Hygriv Rao, MD, DM, Samuel J. Asirvatham, MD, and Calambur Narasimhan, MD, DM

A 4-month-old infant was admitted with incessant tachycardia, failure to thrive, feeding difficulties, and episodic sweating. Clinical and echocardiographic examination showed a structurally normal heart. The patient's ECG during sinus rhythm and tachycardia is shown in Figure 15.1.

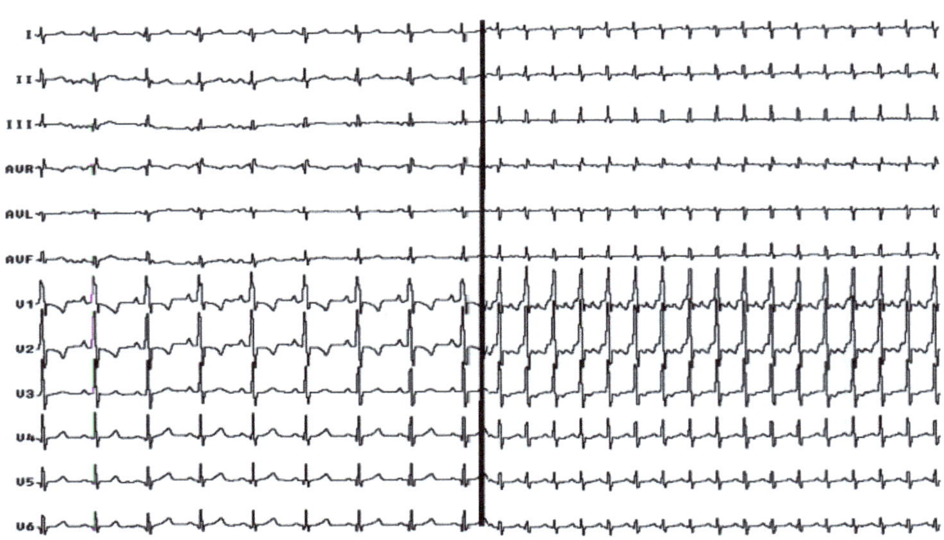

Figure 15.1. Patient's baseline ECG.

Practical Electrophysiology © 2015 Jasbir S. Sra, Masood Akhtar, Andrea Natale, David J. Wilber
Cardiotext Publishing, ISBN: 978-1-935395-17-1

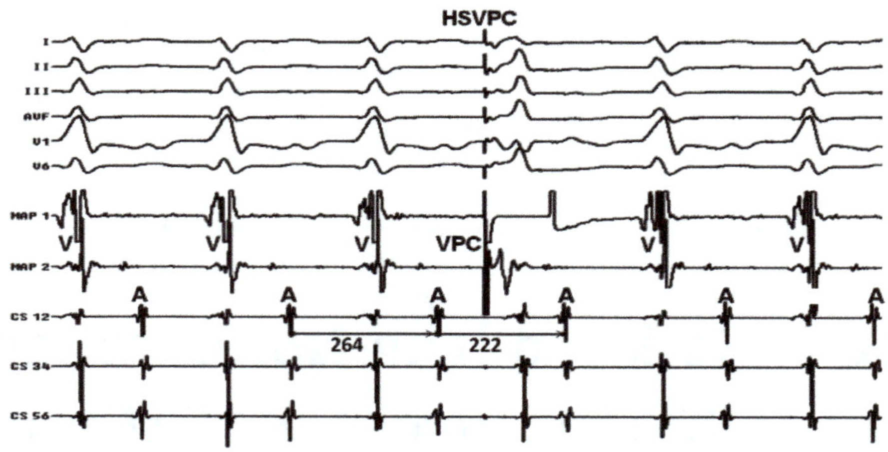

Figure 15.2. Intracardiac tracings during tachycardia.

Question No. 1:

The most likely mechanism of the tachycardia is:

A. Ectopic atrial tachycardia (AT).

B. Orthodromic atrioventricular reciprocating tachycardia.

C. Atypical atrioventricular nodal reentry tachycardia (AVNRT).

D. Atrial flutter.

Intracardiac tracings during the tachycardia are shown in Figures 15.2 and 15.3.

The tachycardia shows a concentric retrograde atrial activation pattern with a septal VA interval of 112 ms. A His-synchronous ventricular paced beat (VPC) advanced the subsequent atrial activation without changing the pattern of activation. An earlier VPC terminated the tachycardia without depolarizing the atrium. These features are typical of a tachycardia using an accessory pathway (AP) as the retrograde limb for reentry.

Question No. 2:

The next step would be to:

A. Map the coronary sinus ostium.

B. Map the anterior tricuspid annulus.

C. Map the anterior mitral annulus.

D. Deliver radiofrequency energy at the anteroseptal tricuspid valve (TV) annulus.

DISCUSSION

The most common tachyarrhythmias in infants are those that involve APs, accounting for 68% of all tachycardias in this age group. Mapping of the

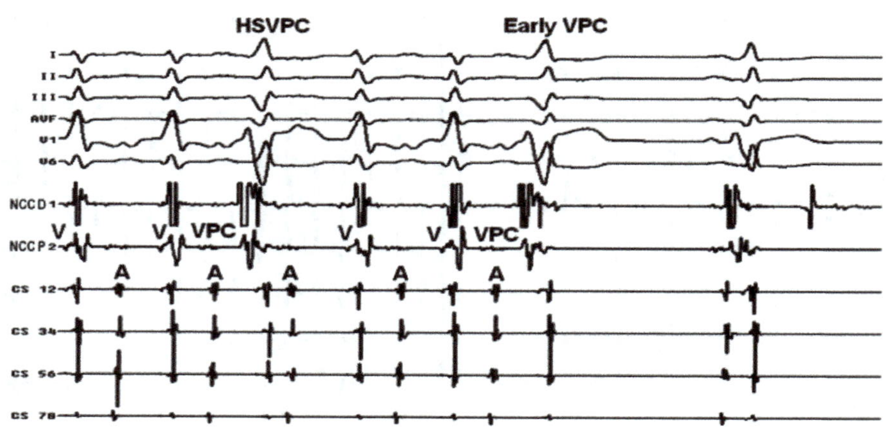

Figure 15.3. Continued intracardiac tracings during tachycardia.

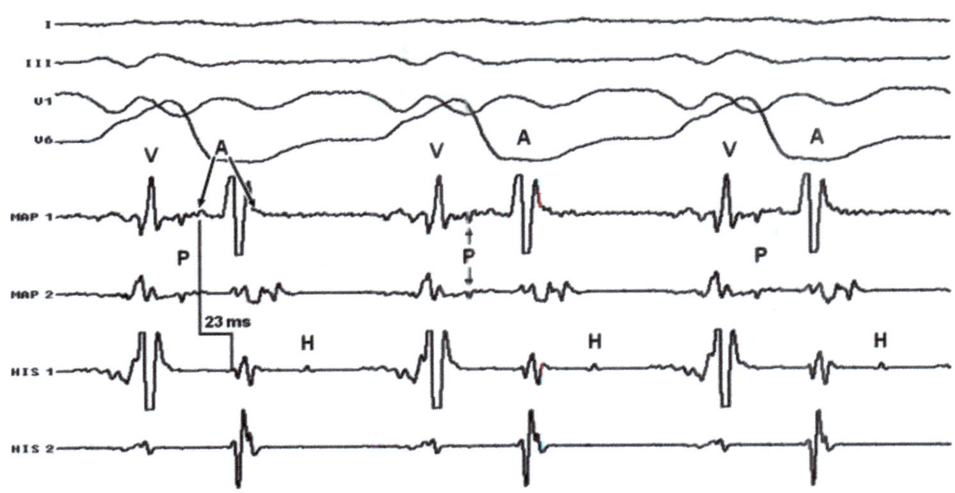

Figure 15.4. Intracardiac electrogram (EGM) during atrioventricular reentry tachycardia. Map 1 and Map 2 are distal and proximal EGMs, respectively, of mapping within the noncoronary cusp (NCC). Note that a near-field pathway potential recorded on both electrode pairs (P). This EGM could be dissociated from the ventricular and atrial EGMs. A small early atrial EGM possibly representing local atrial myocardial activation (A) that could not be dissociated from the atrium and earlier by 23 ms than atrial activation near the His bundle was also noted. Transient left bundle branch block during tachycardia was noted, and the VA interval increased slightly to 120 ms. The pathway potential occurs within the duration of the QRS and may have represented a separated ventricular EGM. However, the potential could be dissociated from the QRS, and in the normal heart an isoelectric period between 2 ventricular EGMs in this location would not be expected. H, HB EGM; HIS 1, distal His recording; HIS 2, proximal His recording; I/III/V./V_6, ECG leads).

tricuspid annulus during the tachycardia revealed earliest atrial activation at the His-bundle (HB) region, suggestive of an anteroseptal AP. Anteroseptal APs, defined as the APs situated above the HB with ventricular activation cephalad and anterior to the membranous septum, are relatively rare, constituting only 8% of all APs. The area of tricuspid annulus (anteroseptal) and anterolateral TV annulus were initially mapped. Later, the mitral annulus was mapped at the midseptal mitral annulus location and anterior mitral valve (MV) annulus. None of these areas revealed an activation earlier than His.

Nakagawa and Jackman reported a 5% incidence of unusual locations of APs in patients with failed

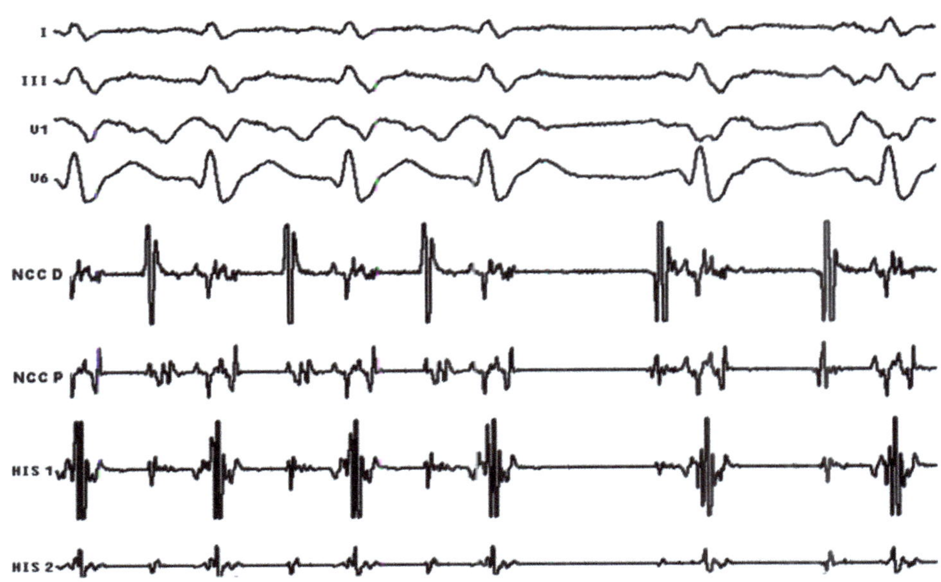

Figure 15.5. Termination of tachycardia with the block in AP conduction during radiofrequency delivery.

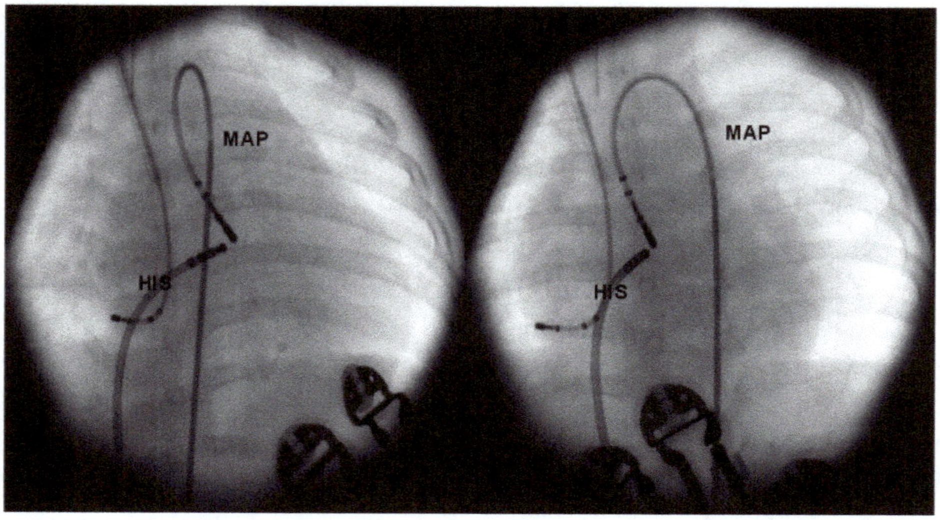

Figure 15.6. Right and left anterior oblique fluoroscopic views showing the position of the mapping catheter (MAP) at the site where the pathway potential was recorded and ablation successful. HIS, multipolar catheter in the HB region. The catheter placed via the jugular vein is also seen in the right atrium.

ablation. This includes epicardial anteroseptal pathways, epicardial pathways coursing anteriorly or posteriorly to the aortic root, epicardial connections between the right or left atrial appendage and epicardial surface of the ventricle, and along the anteroseptal and midseptal mitral annulus.

A 5-F quadripolar catheter was introduced from the femoral artery for mapping the aortic sinus and the left ventricle. Earliest retrograde atrial activation preceding the atrial activation in the HB region was observed while mapping the noncoronary cusp (NCS) region (Figure 15.4). At this location mechanical bumping of the AP resulted in termination of the tachycardia. Radiofrequency ablation (30 W at 55 C) was performed during tachycardia from noncoronary cusp of aortic valve, which terminated the tachycardia and the AP conduction (terminates with a V) (Figures 15.5 and 15.6).

Anatomically, the NCS of the aortic valve straddles the anteroseptal tricuspid annulus, and epicardial anteroseptal pathways coursing anteriorly or posteriorly to the aortic root can be approached from the NCS with lesser risk of heart block during ablation.

Correct Answers:

Q1. B

Q2. C

CASE STUDY 16

CS-LV Accessory Connection or Epicardial Posteroseptal Accessory Pathway

Warren M. Jackman, MD, Hiroshi Nakagawa, MD, PhD, Sunny S. Po, MD, PhD, Benjamin J. Scherlag, PhD, Karen J. Beckman, MD, Deborah Lockwood, BM, BCh, Adrian Dusa, MD, Stavros Stavrakis, MD, PhD, and Ralph Lazzara, MD

The electrograms (EGMs) in Figure 16.1 were recorded during supraventricular tachycardia (SVT) in an 18-year-old male who had undergone 3 prior failed catheter ablation procedures.

Question No. 1:

Are there any clues to suggest the reason(s) the prior ablation procedures were unsuccessful in eliminating the tachycardia?

A. Prolonged HV interval and right bundle branch block (RBBB).

B. Misleading atrial activation sequence in the proximal coronary sinus (CS) EGMs ($CS_5 - CS_7$).

C. The potential preceding atrial activation in EGMs $CS_5 - CS_7$.

D. Long VA interval at the site of earliest retrograde atrial activation.

E. No specific clue.

DISCUSSION

The differential diagnosis for a SVT with earliest atrial activation recorded from the proximal CS includes: (1) orthodromic atrioventricular reentrant tachycardia (AVRT) using a posteroseptal accessory atrioventricular (AV) pathway for retrograde conduction; (2) slow/slow AV nodal reentrant

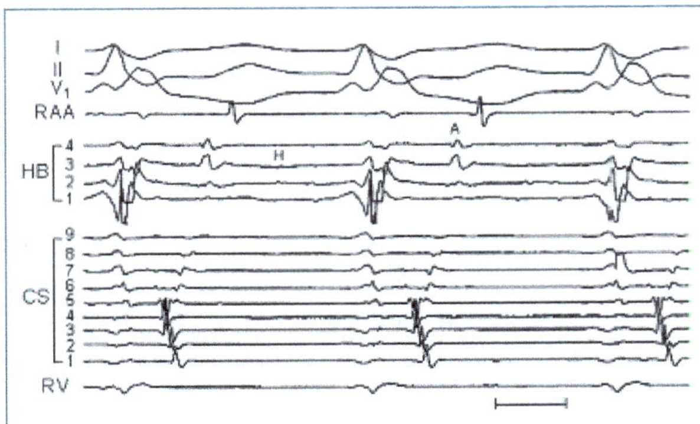

Figure 16.1. SVT with RBBB. Tracings from the top are ECG leads I, II, and V_1, and EGMs recorded from the right atrial appendage (RAA), His bundle (HB) region (HB_4 to HB_1, proximal to distal, respectively), coronary sinus (CS_9 to CS_1, proximal to distal, respectively, CS_8 is located at the CS ostium), and basal right ventricular (RV) septum. Retrograde atrial activation is recorded earlier in proximal CS EGMs than in the HB EGMs. Note the 2 potentials following the ventricular potential in CS_7 to CS_5 EGMs. A, atrial potential, H, HB potential. (Reproduced with permission from Nakagawa et al.[1])

Practical Electrophysiology © 2015 Jasbir S. Sra, Masood Akhtar, Andrea Natale, David J. Wilber
Cardiotext Publishing, ISBN: 978-1-935395-17-1

tachycardia (AVNRT); and (3) atrial tachycardia (AT) originating in the posterior basal left atrium (LA) or the myocardial coat of the CS. These 3 tachycardias can be differentiated by ventricular extrastimulus testing. Ideally, a single ventricular extrastimulus should be delivered to the base of the ventricle, close to the site of earliest atrial activation,[1] beginning approximately 20 ms before local ventricular activation (recorded at the pacing site) and then shortening the extrastimulus coupling interval (V_1-S_2) in 5–10 ms decrements until atrial activation is advanced by at least 20 ms, the tachycardia terminates or ventricular refractoriness is reached (whichever is first).

In AVRT, a ventricular extrastimulus—which does not propagate to the AV node (does not alter the timing or morphology of HB activation), but shortens the *local* ventricular coupling interval (V_1-V_2) by 20–60 ms *in the EGM recording earliest atrial activation*—will advance the timing of the next atrial activation by retrograde conduction over the accessory pathway (same retrograde atrial activation sequence) and reset the tachycardia. The failure to advance atrial activation by a ventricular extrastimulus which advances *local ventricular activation* (in the EGM recording earliest atrial activation) by ≥ 60 ms usually excludes the presence of retrograde conduction over an accessory pathway and orthodromic AVRT.

In slow/slow AVNRT, an earlier ventricular extrastimulus (or double ventricular extrastimuli), which results in retrograde activation of the HB (advancing the timing of HB activation by ≥*60 ms*), will advance the timing of the next atrial activation (without altering the retrograde atrial activation sequence) and reset the tachycardia. Importantly, in slow/slow AVNRT, advancing the timing of HB activation by < 60 ms may not advance atrial activation due to a long conduction time over the lower common pathway.[2] The failure of a ventricular extrastimulus to advance atrial activation and change the timing of the next tachycardia complex, despite retrogradely advancing HB activation by ≥ 60 ms, would exclude both AVRT and AVNRT and suggest the diagnosis of AT. Termination of the tachycardia by a ventricular extrastimulus associated with VA block excludes AT.

In this patient, a late ventricular extrastimulus delivered to the basal posterior RV septum (which did not change the timing or morphology of HB potential) advanced ventricular activation in the proximal CS EGMs by 20 ms. This ventricular extrastimulus advanced the timing of the next atrial activation (same atrial activation sequence) by 15 ms and reset the tachycardia, indicating: (1) the presence of retrograde conduction over an accessory pathway; (2) the accessory pathway was conducting in the retrograde direction during the tachycardia; and (3) the accessory pathway was part of the reentrant circuit of the tachycardia, that is, the tachycardia mechanism was orthodromic AVRT.

Another technique used to identify the presence of retrograde accessory pathway conduction is "para-Hisian pacing" (Figure 16.2).[3,4] In this technique, pacing is performed at the basal anterior RV septum, just anterior (superior) to the proximal right bundle branch (RB), such that slight movement of the pacing catheter during respiration results in a repetitive change between: (1) capture of both RV septum and the proximal right bundle branch (HB-RB); (2) capture of just RV; or (3) capture of just HB-RB. A change from capturing both RV and HB-RB to capturing just RV (loss of HB-RB capture) is associated with widening of the QRS complex (although the change in the QRS complex may be minimal as in Figure 16.2B) and the delay, and emergence of the retrograde RB and HB potentials (see Figure 16.2B). The loss of RV capture while continuing to capture HB-RB is associated with an abrupt delay in ventricular activation (stimulus-V interval is similar to the HV interval during sinus rhythm) and narrowing of the QRS with either a

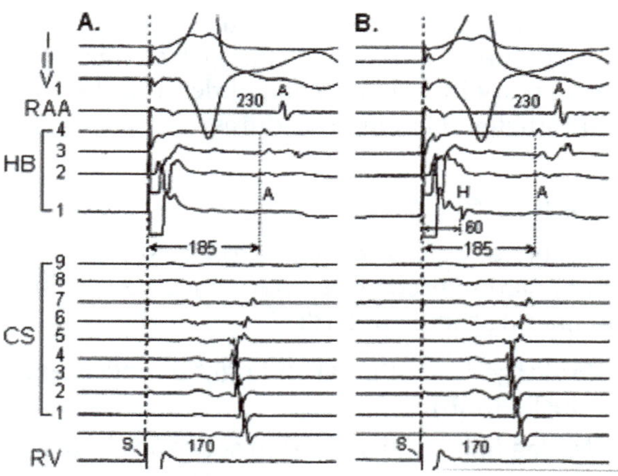

Figure 16.2. Para-Hisian pacing with intermittent HB-RB capture. Pacing was performed from the basal anterior RV septum, close to the very proximal right bundle branch (RB). Pacing captured both RV septum and HB-RB in **Panel A** and captured only RV septum in **Panel B**, as reflected by the appearance of the retrograde H potential *60 ms* after the pacing stimulus. The QRS is only slightly wider in Panel B despite the loss of HB-RB capture. Note that the retrograde atrial activation sequence and the stimulus-atrial intervals are identical in Panel B despite the loss of HB-RB capture, indicating that retrograde conduction is occurring only over an accessory AV pathway. EGM configuration is the same as Figure 16.1. The numbers 230, 185, and 170 indicate the stimulus-atrial intervals in the RAA, HB, and CS_1 EGMs, respectively.

normal (nonpreexcited) QRS complex (when capturing the very proximal RB) or a QRS complex simulating left bundle branch block (when capturing the RB farther distally).

Two variables should be measured, with each change in the structures captured. First examined is the retrograde atrial activation sequence. Second examined is the stimulus-atrial (SA) interval. The retrograde atrial activation sequence will be identical when retrograde conduction is occurring over a single route (AV node or accessory pathway). A change in the retrograde atrial activation sequence with loss of HB-RB or RV capture indicates the presence of retrograde conduction over both the AV node and an accessory AV pathway.

During para-Hisian pacing with loss of HB-RB capture (continued RV capture) and no change in the retrograde atrial activation sequence, an increase in the SA interval similar to the increase in S-HB interval indicates retrograde conduction is occurring over the AV node. The absence of a significant increase in SA interval, despite a clear increase in the SH interval, indicates retrograde conduction is occurring over an accessory pathway (see Figure 16.2). During para-Hisian pacing, the loss of RV capture with persistence of HB-RB capture and no change in the retrograde atrial activation sequence, the timing of atrial activation will remain constant during retrograde conduction over the AV node (SA interval is constant with shortening of the VA interval due to the delay in ventricular activation). With retrograde accessory pathway conduction, the delay in ventricular activation with loss of RV capture will be accompanied by a similar delay in atrial activation, such that the local VA interval near the site of earliest retrograde atrial activation remains similar (SA interval increases with constant local VA interval).

For best results in para-Hisian pacing, the HB EGM should be recorded using multiple closely spaced bipolar electrodes (1-mm spacing, edge-to-edge), and pacing should be performed using an independent catheter (rather than pacing from the distal electrodes on the HB catheter) with closely spaced bipolar electrodes (1-mm spacing).[3,4] The close bipolar electrode configuration used for recording significantly reduces the far-field ventricular signal, allowing identification of the retrograde RB-HB potentials when HB-RB capture is lost. The close bipolar configuration used for pacing allows capture with a relatively narrow current field, which facilitates intermittent loss of HB-RB capture and helps to reduce the stimulus artifact in the HB EGMs. Using deflectable catheters both for recording the HB EGMs and for pacing aids in determining optimal placement.[3,4]

Once a posteroseptal accessory pathway is identified, as in this patient, it is useful to determine whether the accessory pathway is: (1) a typical accessory AV connection (allowing ablation from the endocardium at the level of the tricuspid annulus or mitral annulus) as found in approximately 80% of patients; or (2) a connection between an extension of the CS myocardium running along the middle cardiac vein (MCV) (or another coronary vein or the neck of a CS diverticulum) and the epicardial surface of the left ventricle (CS-LV accessory connection or epicardial posteroseptal accessory pathway) found in approximately 20% of patients.[1,5] Ablation from the endocardium is generally ineffective for CS-LV connections. As a result, attempts at ablation are more frequently unsuccessful for CS-LV connections. In our experience, CS-LV connections were found in approximately 50% of patients who have had a failed attempt at ablation of a posteroseptal accessory pathway.[1,5] An oblique course, leading to incorrect localization of a typical accessory AV pathway, accounts for most of the remaining patients with failed ablation of a posteroseptal accessory pathway.[1,6]

CS-LV connections are differentiated from typical posteroseptal accessory AV pathways by identifying the order of the 2 potentials recorded along the floor of the proximal CS. During retrograde conduction over a typical posteroseptal accessory, left atrial activation precedes activation of the CS myocardium (first potential is LA myocardial activation and the second potential is CS myocardial activation). In contrast, retrograde conduction over a CS-LV connection activates the CS myocardium, which then activates the LA (first potential is CS myocardial activation and second potential is LA myocardial activation). The order of the potentials can be identified by delivering a late atrial extrastimulus and progressively shortening the extrastimulus coupling interval (V_1-S_2) in 5–10-ms decrements. The atrial extrastimulus selectively advances the atrial potential. Therefore, during retrograde conduction over a CS-LV connection, the extrastimulus will initially advance the second potential (identifying it as an atrial potential) without advancing the first potential (indicating the first potential does not represent atrial activation, Figures 16.3 and 16.4A). An earlier extrastimulus will advance the first potential without altering the timing or morphology of the ventricular potential (see Figures 16.4B and 16.4C), indicating the first potential does not represent ventricular activation. By elimination, the first potential recorded on the floor of the proximal CS EGMs is neither atrial nor ventricular activation, and must represent activation

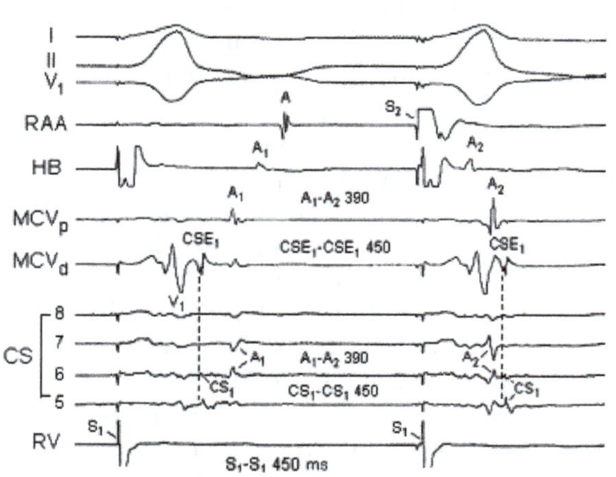

Figure 16.3. Use of a single extrastimulus (S_2) delivered to the RAA during RV pacing (para-Hisian region) at cycle length 450 ms (S_1-S_1 450 ms) to differentiate the 2 potentials recorded in the proximal CS EGMs (CS_1 and A_1) during retrograde conduction over a CS-LV accessory connection (epicardial posteroseptal accessory AV pathway). The late atrial extrastimulus (S_2) advances the timing of the second potential (A_2) in the proximal CS EGMs (CS_5-CS_8) by 60 ms (A_1-A_2 390), without altering the timing or morphology of the first potential (CS_1). This response indicates the first potential (A_1) represents atrial activation, and the second potential (CS_1) does not represent atrial activation. Note that the potential resulting from activation of the CS myocardial extension (CSE_1) recorded from the middle cardiac vein (MCV_d EGM) precedes activation of the CS myocardium (CS_1 in EGM CS_6) and that activation of the CS myocardium propagates laterally (CS_1 is earlier in EGM CS_6 than in CS_5) prior to activating the LA. Taken from a patient different than the presented case (see Figures 16.1 and 16.2), but who also had prior ablation at the posteroseptal mitral annulus and inside the CS, resulting in long CS_1-A_1 interval and fractionated potentials between CS_1 and A_1 (EGMs CS_5 and CS_6). CS-LV, CS myocardium-epicardial left ventricle connection; HB, His bundle.

of the CS myocardium. In the case of a typical accessory AV connection, a late atrial extrastimulus will initially advance the first potential before or simultaneously advancing the second potential, identifying the first potential as a left atrial potential and the second potential as CS myocardial activation.

In this patient, the CS myocardial and left atrial potentials recorded from the proximal CS could be differentiated during rapid ventricular pacing (Figure 16.5). The left atrial potentials delayed and separated from the CS potentials due to local conduction block at several different sites along the CS myocardial-left atrial junction. The origin of the left atrial potentials was identified by the similar delay in atrial activation recorded at other sites (HB and RAA EGMs). Since CS myocardial activation preceded left atrial activation during retrograde accessory pathway conduction, the accessory pathway was identified as a CS-LV connection. Therefore, transseptal puncture (for endocardial access to the mitral annulus) was not performed. Instead, CS angiography and mapping of the coronary veins was performed to identify the venous structure containing the CS myocardial extension, which connects to the LV, and coronary arteriography was performed prior to ablation to determine the presence or absence of a significant coronary artery within a few millimeters of the ideal ablation site.

CS-LV Accessory Connection or Epicardial Posteroseptal Accessory Pathway

The CS has a myocardial coat of multiple layers, encompassing the entire CS from the CS ostium in the triangle of Koch to a few millimeters lateral to the insertion of the great cardiac vein (Vieussen's valve, Figure 16.6).[7] The CS myocardial coat is capable of electrical conduction to (and from) the LA and the right atrium (within the triangle of Koch).[8] Von Ludinghausen and coworkers found that the CS myocardium extends a centimeter or more along the MCV (see Figure 16.6) or posterior coronary vein (PCV) in 3% and 2% of human hearts, respectively,[7] and is present along the neck of most or all CS diverticuli. This finding suggested that epicardial posteroseptal accessory pathways are formed by an anatomical and electrical connection between an extension of the CS myocardium (along a coronary vein) and the epicardial surface of the left ventricle (LV) (see labeled CSE-LV Connection in Figure 16.6).[5]

During antegrade accessory pathway conduction (Figure 16.7A), the LA activates the CS myocardium. Propagation along the CS myocardial extension (coronary vein or a CS diverticulum) activates the epicardial surface of the ventricle, 1–3 cm away from the mitral annulus (apical to the annulus). Endocardial mapping of the basal right and LVs using unipolar and bipolar EGMs finds only late local ventricular activation (downstroke of the unfiltered unipolar EGM), beginning at least 25 ms after the onset of far-field ventricular activation (initial positive component on the unipolar EGM). Earliest ventricular activation is recorded epicardially from the MCV, another coronary vein or a CS diverticulum, and is preceded by a potential resulting from antegrade activation of the CS myocardial extension (see large arrow, CSE in Figure 16.5B and CSE_1 in Figure 16.8A) running along the vein.[5] The role of the CSE potential in antegrade accessory pathway conduction can be verified by ventricular extrastimulus testing. During atrial pacing, a late ventricular extrastimulus is delivered

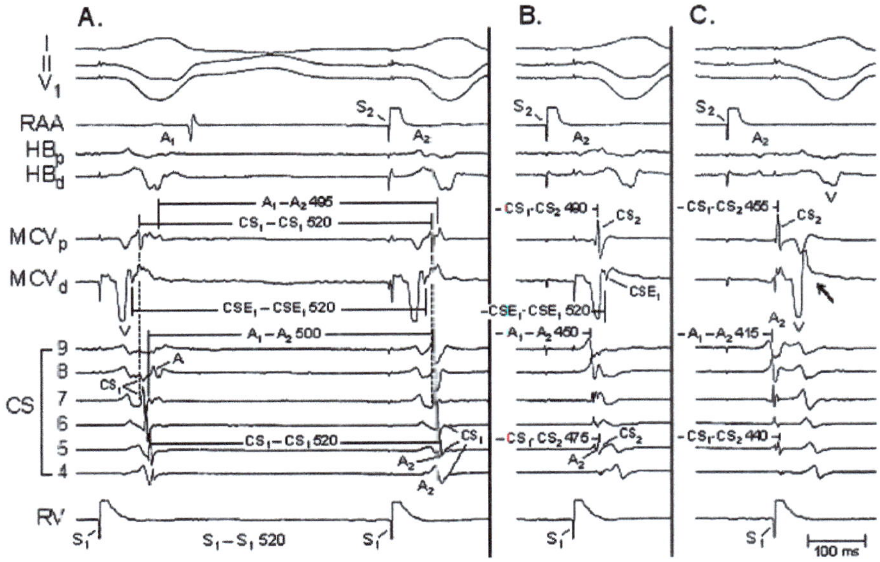

Figure 16.4. Single atrial extrastimulus (S_2) during RV pacing at cycle length 520 ms (S_1-S_1 520) to differentiate the 2 potentials (CS_1 and A_1) recorded in the proximal CS EGMs during retrograde conduction over a CS-left ventricle accessory connection (epicardial posteroseptal accessory AV pathway). **A:** Late atrial extrastimulus (S_2) advances the timing of the second potential (A_2) in the CS EGMs (A_1-A_2 500) without advancing the first potential (CS_1-CS_2 520), indicating the second potential represents atrial activation and the first potential does not represent atrial activation. Neither the CS myocardial extension potential (CSE_1 in MCV_d EGM) nor the CS myocardial potential (CS_1 in MCV_p EGM) recorded from the MCV were advanced by the atrial extrastimulus (CSE_1-CSE_1 520 and CS_1-CS_1 520). **B:** An earlier extrastimulus advances the first potential (CS_1-CS_2 475 in EGM CS_5 and CS_1-CS_2 490 in EGM MCV_p), without advancing CSE_1 in the MCV_d EGM, showing CSE_1 is separate from activation of the main body of the CS myocardium. **C:** Earliest extrastimulus antegradely activates CSE_1, advancing it into the ventricular potential (no longer visible, *arrow*), without altering the timing or morphology of the ventricular potential (V) in any EGM, indicating that CSE does not represent late ventricular activation. Taken from a patient different than the presented case (Figures 16.1 and 16.2), with no prior ablation procedure, demonstrating the typical, shorter conduction times between the retrograde CS myocardial and left atrial potentials (CS and A) in the CS EGMs. HB, His bundle; RAA, right atrial appendage. (Reproduced with permission from Sun et al.[5])

and the extrastimulus coupling interval (A_1–S_2) is shortened in 5–10-ms decrements. The ventricular extrastimulus initially advances the ventricular potential without altering the timing or morphology of the CSE potential (see Figure 16.8A), dissociating the CSE potential from ventricular activation. An earlier extrastimulus advances the CSE potential and begins advancing the CS myocardial potential recorded in the nearby CS EGMs (close to the insertion of the coronary vein of CS diverticulum), without altering the timing or morphology of the atrial potential (see Figure 16.8B), indicating that the CSE and CS potentials do not represent local atrial activation and that the CSE has a direct connection with the ventricle. An even earlier ventricular extrastimulus (not shown in Figure 16.8) further advances the CSE and CS myocardial potentials and begins to advance the atrial potential, starting at the lateral aspect of the LA-CS connections (near the insertion of the Marshall and PCVs).

The ECG during sinus rhythm in patients with CS-LV connections often exhibits a specific pattern of preexcitation, consisting of a V_1-V_2 transition, a negative delta wave in aVF, and a negative delta wave in lead II. A steep negative delta wave in lead II is highly specific for a CS-LV connection (few, if any, false positives) and is found in 70% of patients with an epicardial posteroseptal accessory pathway.[5] However, the absence of a negative delta wave in lead II does not exclude a CS-LV connection, as this sign is absent in approximately 30% of patients.

During retrograde accessory pathway conduction (see Figure 16.7B), the ventricle activates the CS myocardial extension located along the MCV, another coronary vein or the neck of a CS diverticulum. Activation of the CS myocardial extension propagates to the floor of the CS and activates the CS myocardial coat. Due to fiber orientation, the CS myocardium activates the LA at a location 2–4 cm leftward of the orifice of the MCV and results in rapid LA activation in the leftward direction. As a result, recordings from bipolar electrodes often suggest that earliest LA activation is located leftward of the site of LA activation (see "Pseudo Earliest LA Activ" in Figure 16.7B).

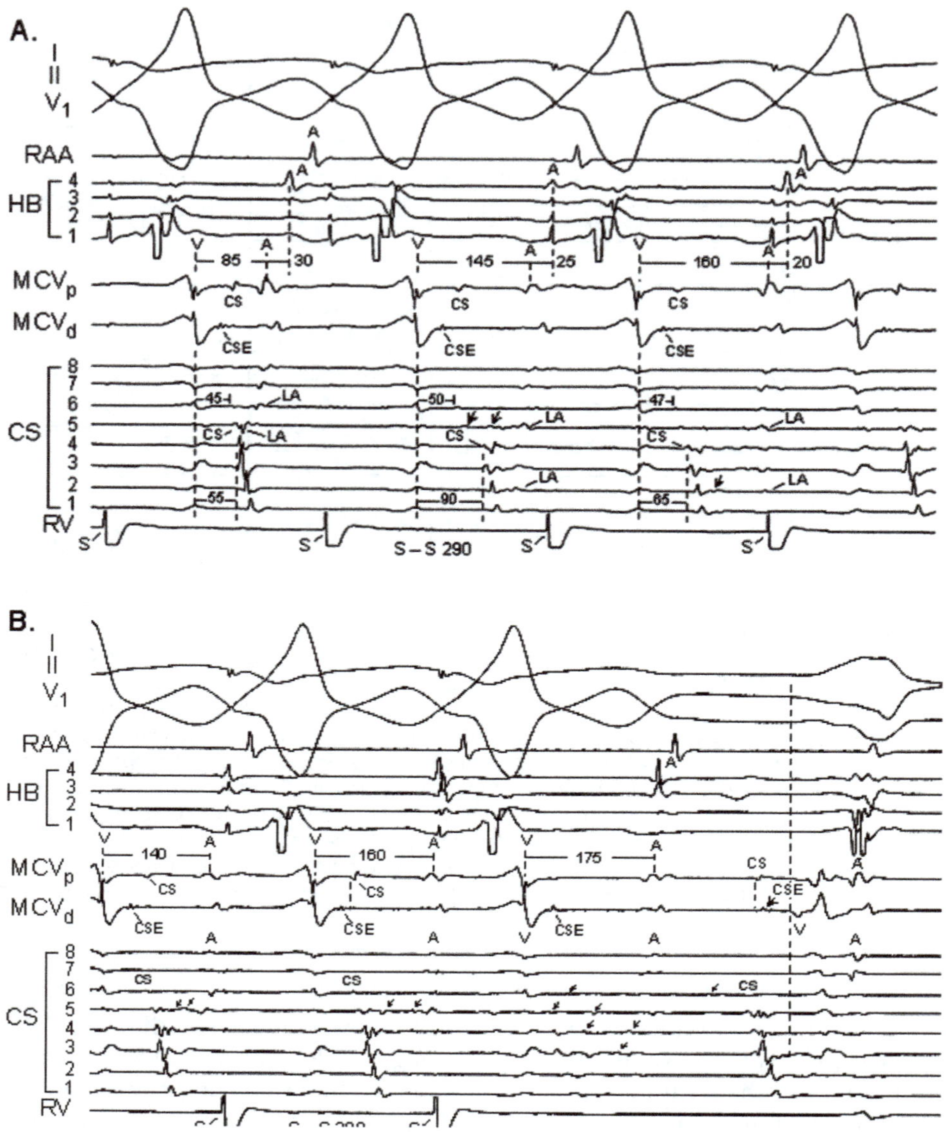

Figure 16.5. **A:** Separation of the retrograde CS and left atrial (LA) potentials during rapid ventricular pacing (cycle length 290 ms, S-S 290) due to progressive conduction block at different sites along the CS myocardial-left atrial (CS-LA) junction. For the first 3 ventricular complexes, the timing of retrograde activation of the CS myocardial extension (CSE potential in MCV_d EGM) and the earliest retrograde activation of the floor of the CS (CS potential in MCV_p EGM and the tiny first potential in EGM CS_6 beginning 45, 50, and 47 ms after ventricular activation in the MCV EGM, *long dashed vertical lines*) is relatively constant. For the first ventricular complex, earliest left atrial activation is recorded in EGM CS_5 (LA). For the second complex, local block within the CS myocardium near CS_5 (*left arrow*) delays CS activation leftward of CS_5 (local V-CS in EGM CS_4 increases from 55 ms to 90 ms). Local CS-LA conduction block also occurs near CS_5 (*right arrow*), delaying LA activation and shifting the site of earliest LA activation to a location near CS_2. These 2 local blocks result in a 60-ms delay in the timing of atrial activation in the MCV_p EGM (local VA increased from 85 to 145 ms). For the third ventricular complex, the local block within the CS myocardium near CS_5 has partially resolved (local V-CS in EGM CS_4 decreased from 90 to 65 ms), but local CS-LA conduction blocks near CS_5 persist and new local CS-LA conduction block occurring near CS_2 *(arrow)* greatly increases the delay in retrograde LA activation and further changes the retrograde atrial activation sequence (local VA in the MCV_p EGM increases from 145 to 160 ms). Atrial activation in the CS EGMs (LA) was differentiated from CS myocardial potentials (CS) by being linked to the atrial potentials in the HB and RAA EGMs. Delay in the LA potentials were associated with a similar delay in the atrial potential recorded in the MCV_p EGM (A), which was associated with a similar delay in the atrial potential in the HB and RAA EGMs (A in MCV_p to A in HB remained similar at 30, 25 and 20 ms). RV, right ventricle. **B:** Short burst of ventricular pacing at cycle length 280 ms induces further delay and fractionation in CS myocardial activation *(tiny arrows in CS EGMs)* and results in a ventricular echo complex with antegrade accessory pathway conduction, due either to reentry within the CS myocardial coat or reentry between the CS myocardium and LA myocardium. Note the reversal in CS and CSE activation from retrograde (CSE precedes CS in first 3 complexes) to antegrade (CS precedes CSE, *large arrow*) in producing the ventricular echo complex. Earliest ventricular activation, recorded in the MCV_d EGM *(vertical dashed line)*, precedes the onset of the delta wave in the 3 ECG leads and ventricular activation recorded in the CS EGMs. Note the negative delta wave in ECG lead II.

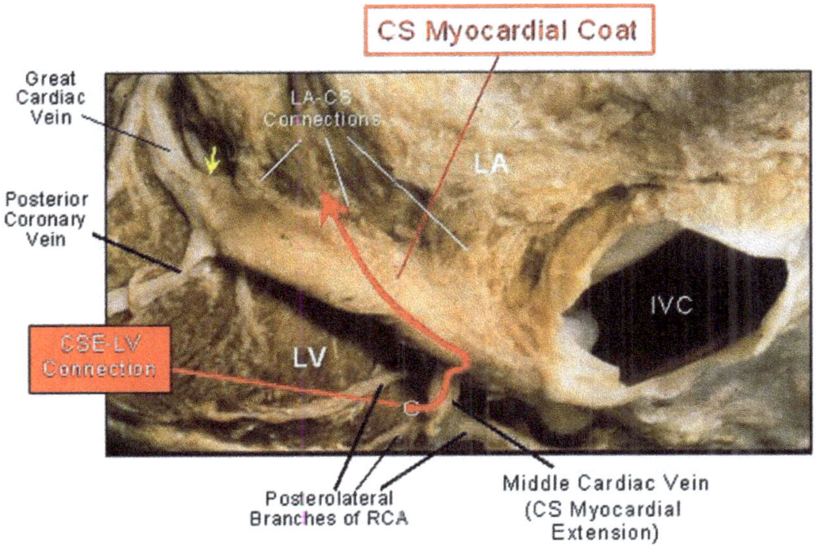

Figure 16.6. Photograph of a human heart (inferior view) with the epicardium removed showing the CS myocardial coat, extending laterally to the just beyond the insertion of the great cardiac vein *(yellow arrow).* The great cardiac vein and PCV are thin and translucent, with no myocardial coat. The reddish fibers of the CS myocardial coat connect extensively with the LA (LA-CS Connections). Note the reddish CS myocardial fibers extending along the MCV (CS Myocardial Extension). The long curved *red arrow* is a schematic representation of retrograde activation over a CS-LV connection (epicardial posteroseptal accessory pathway), originating at the connection *(red circle with white border)* between the CS myocardial extension (along the MCV) and the epicardial surface of the LV, and propagating superiorly along the MCV to the floor of the CS, leftward along the floor of the CS, and then toward the roof of the CS to activate the LA *(red arrowhead)*. The posterolateral branches of the right coronary artery (RCA) loop superiorly toward the CS at the level of the MCV, and are located between the veins and the epicardial surface of the LV. IVC, inferior vena cava. (Photograph courtesy of Anton E. Becker.)

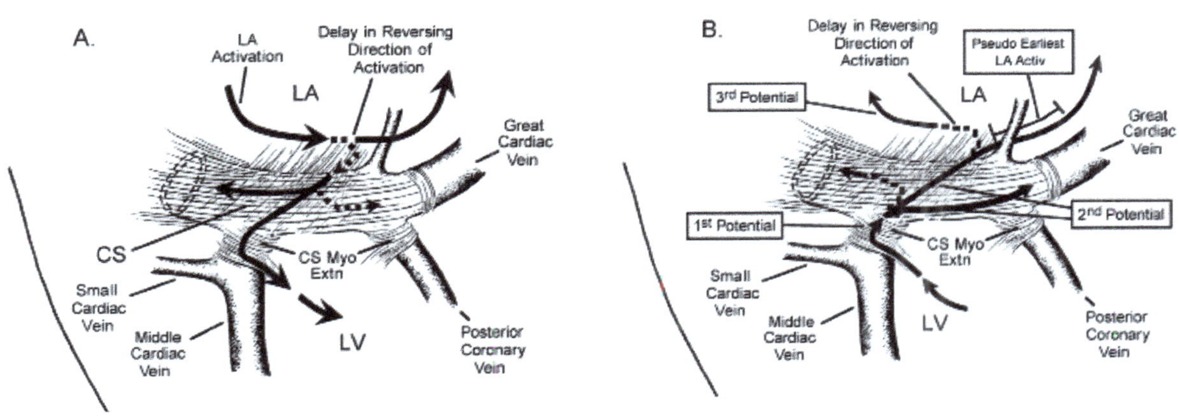

Figure 16.7. Schematic representation of the CS myocardial coat (as viewed in the left anterior oblique [LAO] projection) and conduction over a CS-LV connection (epicardial posteroseptal accessory pathway). **A:** Antegrade accessory pathway conduction. During sinus rhythm, the LA is activated via Bachmann's bundle (not shown). LA activation propagates counterclockwise around the septal and posterior mitral annulus (*arrows at top*, LA Activation). Due to CS fiber orientation at LA-CS connections, the direction of the wavefront is reversed as the CS myocardial coat is activated at its lateral aspect (*top dotted arrow* indicating conduction delay). Activation propagates along the CS myocardial coat in the inferior and septal directions (*solid arrows*), activating the CS myocardial extension (CS Myo Extn) running along the MCV (or other coronary vein or neck of a CS diverticulum). The CS Myo Extn activates the epicardial surface of the LV, completing conduction from the LA to the LV. **B:** Retrograde accessory pathway conduction. During AVRT or ventricular pacing, the epicardial LV wavefront *(bottom arrow)* activates the CS Myo Extn, which activates the floor of the CS myocardial coat. The CS myocardium is activated in all directions, but the dominant wavefront propagates leftward and superiorly. The CS myocardium activates the LA at its lateral aspect. Due to the parallel fiber orientation of the CS myocardium and LA in this area, left atrial activity propagates rapidly in the leftward direction and requires time to reverse direction to propagate in the septal direction (*top dotted arrow* indicating conduction delay). EGMs recorded from the MCV (or other vein containing the CS myocardial extension) during retrograde accessory pathway conduction will show 3 potentials: (1) retrograde activation of the CS Myo Extn *(first potential)*; (2) propagation along the CS myocardium *(second potential)*; and (3) left atrial activation *(third potential)*. (Reproduced with permission from Sun et al.[5])

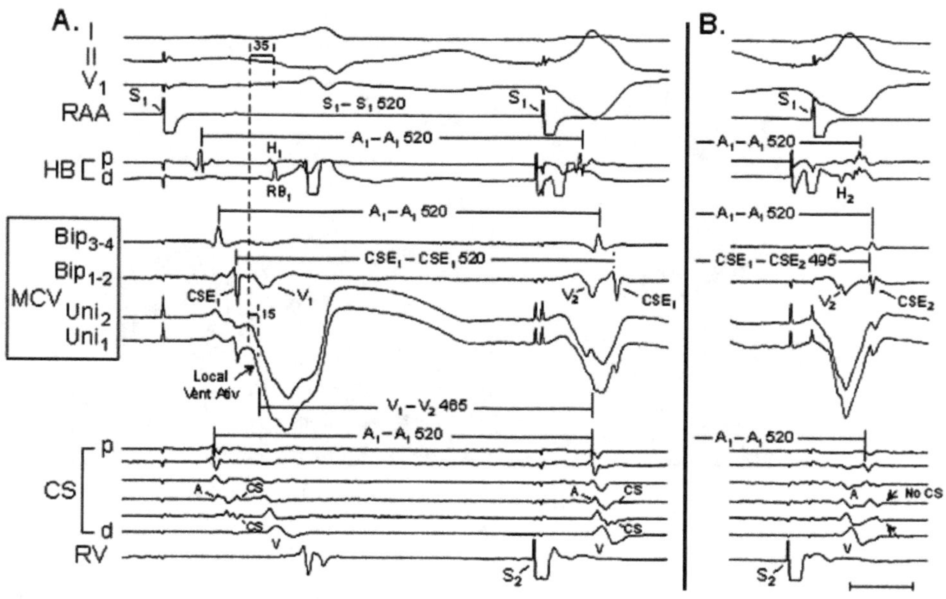

Figure 16.8. Single ventricular extrastimuli (S_2) during right atrial pacing at cycle length 520 ms (S_1-S_1 520) used to verify the origin of the antegrade CS myocardial extension potential (CSE) recorded from the MCV. **A:** During atrial pacing, earliest ventricular activation is recorded from the MCV. Far-field ventricular activation (*first vertical dashed line*) begins 35 ms before the onset of the delta wave (*upper short vertical dashed line*). The recording of local ventricular activation (*arrow, lower short vertical dashed line*) 15 ms after the onset of far-field activation, suggests the connection between the CSE and left ventricular epicardium is located a short distance from the MCV. A late ventricular extrastimulus (S_2) advances the timing of the ventricular potential (V_2) by 55 ms (V_1-V_2 465), without advancing the CSE_1 potential (CSE_1-CSE_1 520) or the CS myocardial potentials (CS), indicating the CSE_1 and CS potentials do not represent ventricular activation. **B:** An earlier extrastimulus (S_2) advances the CSE_2 potential by 25 ms (CSE_1-CSE_2 495) *and advances* the CS myocardial potentials (*arrows*, No CS) without changing the timing or morphology of the atrial potential in the His-bundle (HB), MCV, or CS EGMs (A_1-A_1 520), indicating that the CSE and CS potentials do not represent late or fractionated atrial activation. Taken from a patient different than the presented case (Figures 16.1, 16.2, and 16.5). Bip, bipolar EGM; RAA, right atrial appendage; RV, right ventricle; Uni, unipolar EGM. (Reproduced with permission from Sun et al.[5])

Rightward (septal) LA activation is delayed due to slowing of conduction during the reversal in direction of activation (see Figure 16.7B).

The sequence of retrograde activation produces a characteristic pattern of 3 distinct potentials recorded from the MCV (or other coronary vein) containing the CS myocardial extension (Figures 16.7B and 16.9). The first potential (see #1 CSE in Figure 16.9) is generated by retrograde activation of the CS myocardial extension. The second potential (#2 CS Myo) results from leftward activation of the CS myocardium. This potential is small and is recorded along the floor of the CS, beginning at the orifice of the MCV (or other coronary vein containing the CS myocardial extension) and extending laterally. The second potential, CS myocardial activation, is often recorded from the proximal bipolar electrode on the catheter in the MCV (see #2 CS Myo in MCV Bip_{3-4} EGM in Figure 16.9), as this bipolar electrode is often located close to the floor of the CS (see RAO and LAO radiographs in Figure 16.9). The third potential (#3 LA) is generated by late left atrial activation propagating toward the septum. The late timing results from slowing of conduction by the reversal in direction. Note this potential always precedes atrial activation in the HB EGMs.

CS angiography (balloon-occlusion technique) is a valuable tool to delineate the coronary venous anatomy associated with a CS-LV connection and as a guide for mapping within the coronary veins to locate the CS myocardial extension. For best results, the initial balloon-occlusion angiogram should be performed distal to the true CS, injecting contrast into the great cardiac vein. This allows *all* of the branches entering the CS to fill antegradely through collaterals, providing an image of the entire CS anatomy. CS angiography reveals a CS diverticulum in 20%, a distorted coronary vein (bulbous or fusiform dilatation) in 10%, and normal coronary venous anatomy in 70% of patients with a CS-LV connection.[5] The body of CS diverticula forming CS-LV connections are attached anatomically and electrically to the epicardial surface of the ventricle and contract with ventricular systole (Figure 16.10). However, the size, shape and location of CS

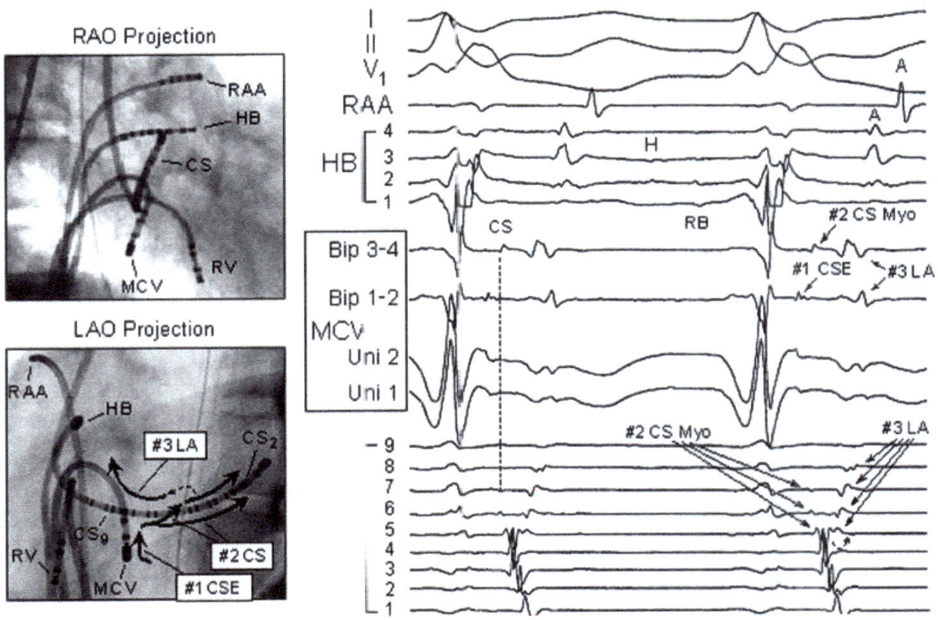

Figure 16.9. Orthodromic AVRT illustrating the pattern of retrograde conduction over the CS-LV connection (*arrows* on the radiographs in the right anterior oblique [RAO] and LAO projections) and the 3 characteristic potentials recorded from the MCV and CS EGMs (*#1 CSE, #2 CS Myo* and *#3 LA*). The *dashed arrow* in LAO radiograph represents conduction delay during reversal of direction of LA activation. (Reproduced with permission from Nakagawa et al.[1])

diverticula vary widely. CS diverticula may insert into the true CS at any location from the area of the CS ostium to the Vieussen's valve (Figure 16.11A and B), and approximately 25% of the diverticula insert into the MCV (see Figure 16.11C and D).[5] The neck of the diverticulum, the primary target for ablation, may be narrow or wide (see Figure 16.11E and F).

In the present case, CS angiography demonstrated no CS diverticulum. There was stenosis of both the MCV (at its insertion into the CS) and the proximal CS, close to the MCV

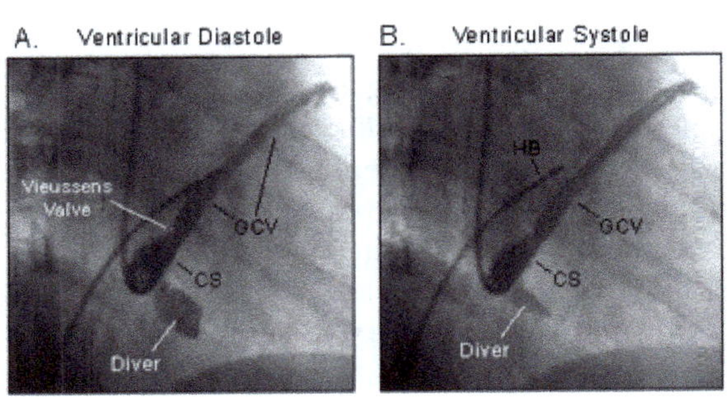

Figure 16.10. Radiographs in the RAO projection during CS angiography showing a CS diverticulum (Diver) filling with contrast media during ventricular diastole **(Panel A)** and contracting during ventricular systole **(Panel B)**. GCV, great cardiac vein. HB, His bundle.

(Figure 16.12) related to the prior ablation attempts. Mapping of the CS, MCV and PCV identified the CS myocardial extension located along the MCV (see #1 CSE in Figure 16.9).

Catheter Ablation of CS-LV Accessory Connections

Most of the failed attempts at ablation of CS-LV connections include delivering RF energy to the endocardial surface of the mitral annulus, 2–3 cm lateral to the insertion of the MCV, at the site of earliest retrograde left atrial activation (see Figure 16.7B). Those RF applications often eliminate the lateral CS-LA connections, resulting in a shift in the site of earliest retrograde atrial activation toward the septum, falsely suggesting there may have been 2 accessory pathways (left posterolateral and left posteroseptal accessory pathways). Further endocardial ablation often has little effect, due to the extensive connections between the CS myocardium and the LA (see LA-CS Connections in Figure 16.6). The optimal ablation site (narrowest point) is at the CS myocardial extension (see Figure 16.6), accessible from the MCV (or other coronary vein or neck of CS diverticulum). The target for ablation is the site, within the vein, recording the largest, sharpest *unipolar* potential generated by the CS myocardial extension. RF energy, using a

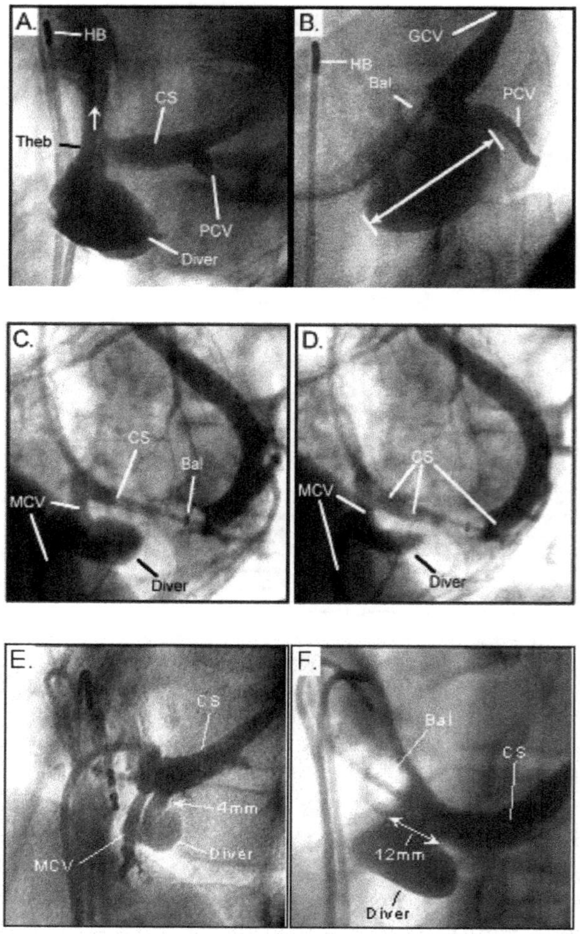

Figure 16.11. Radiographs in the LAO projection during CS angiography illustrating the marked variability in location, size, and shape of CS diverticula. **A:** Diverticulum (Diver) inserting into the CS at the CS ostium. Note the presence of a Thebesian valve (Theb) diverting the CS flow superiorly *(arrow)* and the absence of a MCV, both common findings with diverticula inserting into the CS at the CS ostium. HB, His bundle. **B:** Larger diverticulum (4 cm) inserting at the lateral aspect of the CS, close to the insertion of the PCV. **C and D:** Diverticulum (Diver) inserting into the MCV, recorded during ventricular diastole (Panel C) and systole (Panel D) to illustrate systolic contraction. The angiographic catheter balloon (Bal) is occluding the CS lateral to the MCV, resulting in antegrade filling of the MCV and diverticulum from collaterals. **E and F:** Diverticula inserting into the CS with narrow (4 mm) or wide (12 mm) neck. (Reproduced with permission from Sun et al.[5])

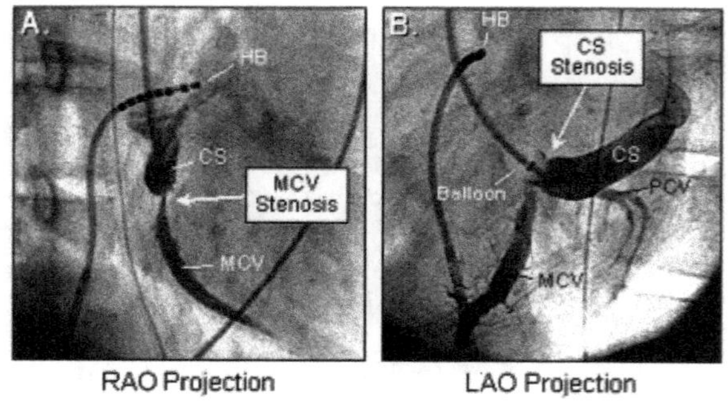

Figure 16.12. Radiographs in the RAO **(Panel A)** and LAO **(Panel B)** projections during balloon-occlusion CS angiography illustrating stenosis of the MCV *(arrow, MCV Stenosis in Panel A)* and the proximal CS *(arrow, CS Stenosis in Panel B)* as a result of prior radiofrequency (RF) ablation. HB, His bundle; PCV, posterior coronary vein.

saline-irrigated electrode, is preferred when coronary arteriography shows this site to be located at least 4 mm from the closest significant coronary artery. We usually begin RF delivery at 10–15 watts and increase power as required up to 25 watts. An impedance rise often occurs, despite saline irrigation. It is important to terminate the RF application as quickly as possible (preferably within 1 second) when an impedance rise occurs in order to prevent adherence of the electrode to the vein. An impedance rise has occurred when the impedance just begins to increase, that is, an increase of just 3–5 ohms above the lowest value (still well below the starting value for impedance). Accessory pathway conduction is usually eliminated by one or two RF applications, with a low long-term recurrence (less than 5%).

A limitation to the use of RF energy is the frequent close proximity of the posterolateral branch of the RCA (right dominant coronary arterial circulation) or posterior descending coronary artery (PDA) (left dominant circulation) to the MCV and floor of the CS near the MCV (Figure 16.13). Radiofrequency applications on the floor of the CS or inside a coronary vein within a few millimeters of a coronary artery have a high risk of acute arterial stenosis.[9] Therefore, coronary arteriography is recommended prior to ablation. Cryoablation is preferred to RF whenever the optimal ablation site in the coronary vein (of floor of the CS) is located within 3–4 mm of a significant coronary artery. In our early experience, using a 6-mm cryoelectrode, a large number of cryoapplications (median 9) were required, often with transient accessory pathway conduction block lasting up to 60 minutes. Long-term success was limited (approximately 80%), but there was no coronary artery stenosis.[10] Success has improved with larger cryoablation electrodes, but the recurrence of accessory pathway conduction remains higher with cryoablation than with RF ablation.

In the present case, coronary arteriography demonstrated a left dominant circulation with a 50% stenosis of the origin of the PDA (see Figure 16.13C) resulting from the prior ablation attempt with RF

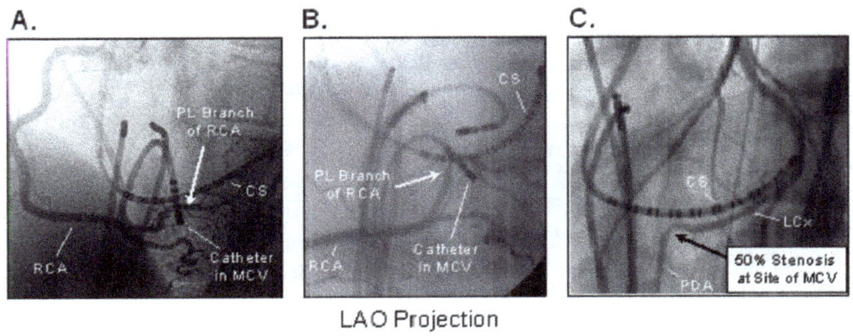

Figure 16.13. Radiographs in the LAO projection during coronary arteriography in 3 patients with CS-LV connections. **A** and **B**: Right coronary arteriography in 2 patients with right dominant coronary arterial circulation showing the typical curved deviation of the posterolateral branch (PL Branch) of the RCA toward the CS at the level of the MCV. Note in the second patient (Panel B) that RF ablation in the CS would have a greater risk of injuring the posterolateral branch of the RCA than ablation within the MCV. **C**: Left coronary arteriography in the present case showing a left dominant circulation with the typical curved deviation toward the CS at the origin of the PDA. Note the 50% stenosis of the origin of the PDA *(black arrow)*, near the insertion of the MCV into the CS, sites of venous stenosis due to prior RF applications (see Figure 16.12). LCx, left circumflex artery.

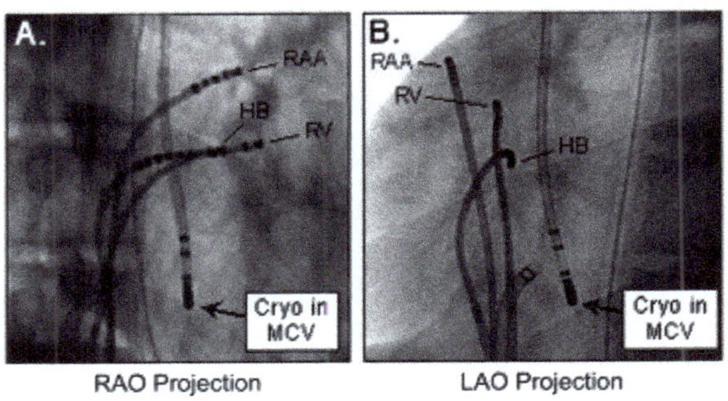

Figure 16.14. Radiographs in the RAO and LAO projections showing the position of the cryoablation catheter in the MCV *(arrows, Cryo in MCV)* during the third cryoapplication (see Figure 16.15). HB, His bundle; RAA, right atrial appendage; RV, right ventricle.

applications in the CS and MCV. Cryoablation was elected. The superior approach (right subclavian vein) was required to insert the cryoablation catheter into the MCV due to stenosis in both the proximal CS and orifice of the MCV (Figures 16.12 and 16.14). Accessory pathway conduction was eliminated by the third cryoapplication (Figure 16.15), and the patient has remained free of ventricular preexcitation and AV reentrant tachycardia.

Correct Answer:

Q1. A

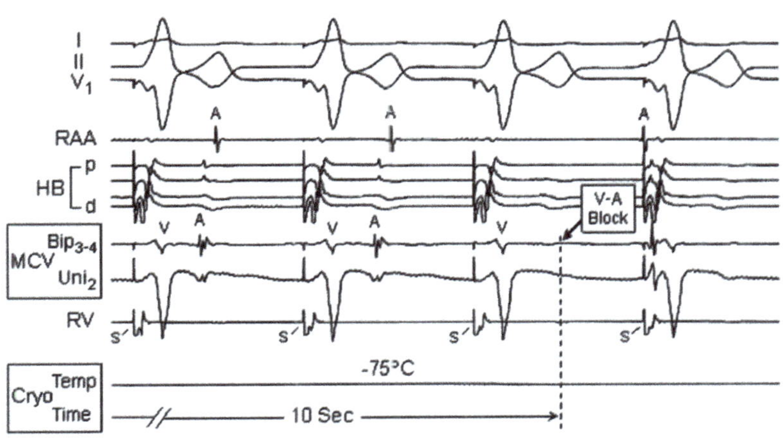

Figure 16.15. Intracardiac recordings during the third cryoapplication that eliminated accessory pathway conduction. Cryoablation was performed during RV pacing. Note that the loss of retrograde accessory pathway conduction (VA Block) at an electrode temperature (Temp) of −75° C, 10 seconds after the onset of the cryoapplication. The unipolar-tip (Uni) EGM and the bipolar-tip (Bip) EGM demonstrating the CSE potential are not displayed due to freezing artifact. HB, His bundle; MCV, middle cardiac vein; RAA, right atrial appendage.

REFERENCES

1. Nakagawa H, Jackman WM. Catheter ablation of paroxysmal supraventricular tachycardia. *Circulation.* 2007;116:2465–2478.
2. Lockwood D, Nakagawa H, Jackman WM. Electrophysiologic characteristics of atrioventricular nodal reentrant tachycardia: Implications for reentrant circuits. In: Zipes DP, Jalife J, eds. *Cardiac Electrophysiology: From Cell to Bedside,* 5th Ed. Philadelphia: Saunders Elsevier; 2009: 615–646.
3. Hirao K, Otomo K, Wang X, et al. Para-Hisian pacing. A new method for differentiating retrograde conduction over an accessory AV pathway from conduction over the AV node. *Circulation.* 1996;94:1027–1035.
4. Nakagawa H, Jackman WM. Para-Hisian pacing: Useful clinical technique to differentiate retrograde conduction between accessory atrioventricular pathways and atrioventricular nodal pathways. *Heart Rhythm.* 2005;2:667–672.
5. Sun Y, Arruda M, Otomo K, et al. Coronary sinus-ventricular accessory connections producing posteroseptal and left posterior accessory pathways: incidence and electrophysiological identification. *Circulation.* 2002;106:1362–1367.
6. Otomo K, Gonzalez MD, Beckman KJ, et al. Reversing the direction of paced ventricular and atrial wavefronts reveals an oblique course in accessory AV pathways and improves localization for catheter ablation. *Circulation.* 2001;104:550–556.
7. Von Lüdinghausen M, Ohmachi N, Boot C. Myocardial coverage of the coronary sinus and related veins. *Clin Anat.* 1992;5:1–15.
8. Antz M, Otomo K, Arruda M, et al. Electrical conduction between the right atrium and the left atrium via the musculature of the coronary sinus. *Circulation.* 1998;98:1790–1795.
9. Sun Y, Po S, Arruda M, et al. Risk of coronary artery stenosis with venous ablation for epicardial accessory pathways (abstr). *Pacing Clin Electrophysiol.* 2001;24:266.
10. Nakagawa H, Aoyama H, Beckman KJ, et al. Safety and efficacy of cryo-ablation of epicardial posteroseptal accessory pathways within the middle cardiac vein close (<2 mm) to a coronary artery (abstr). *Pacing Clin Electrophysiol.* 2003;26:978.

CASE STUDY

17

Mahaim Pathway-Mediated Tachycardia

Patrick Tchou, MD, Mohamed Kanj, MD, and Mohamed Bassiouny, MD

A 27-year-old woman was referred for evaluation of recurrent episodes of palpitation associated with mild shortness of breath. She had normal left ventricular ejection fraction and no evidence of valvular heart disease.

Question No. 1:

The intracardiac tracings in Figures 17.1 and 17.2 are most consistent with:

A. Bundle branch reentrant tachycardia.
B. Atrial tachycardia (AT) with left bundle branch block (LBBB).
C. Atrioventricular (AV) nodal reentry tachycardia with LBBB.
D. Antidromic AV reentrant tachycardia involving an atriofascicular pathway.
E. Ventricular tachycardia (VT).

DISCUSSION

Mahaim fibers were identified by Mahaim and Benatt in 1938[1] as fibers connecting the atrioventricular node (AVN) to the ventricular myocardium. The eponym Mahaim was used for generations and later expanded to include atypical pathways with AVN-like properties. However, with the discovery of bypass tracts, the true anatomic connections, it became more appropriate to describe them accordingly (Figure 17.3).

In general, atypical APs share common electrophysiologic properties:

1. Decremental antegrade conduction.
2. Variable degree of preexcitation.
3. His-bundle pacing or ectopic beats normalize the QRS except for fasciculoventricular pathways.
4. Predominantly located along the anterior and lateral free wall of the right ventricle (RV), giving rise to supraventricular tachycardia (SVT) with LBBB morphology.
5. Absence of retrograde conduction with rare exceptions.
6. May be associated with multiple AP including rapidly conducting AV pathways.

ATRIOFASCICULAR BYPASS TRACTS

Anatomy: Long insulated pathways with atrial insertion at the right atrium free wall and ventricular insertion at or near the RBB apically at the region of the moderator band (along the RV anterior free wall or at the septal insertion). Extensive arborization over a wide area of ventricular muscle is common.[2] A His-like structure usually can be identified near the annular region of the pathway, and these pathways are sensitive to adenosine and isoproterenol.[3,4]

ELECTROPHYSIOLOGIC CHARACTERISTICS

12-lead ECG: Minimal or no preexcitation at baseline (~60%) due to slow conduction down the

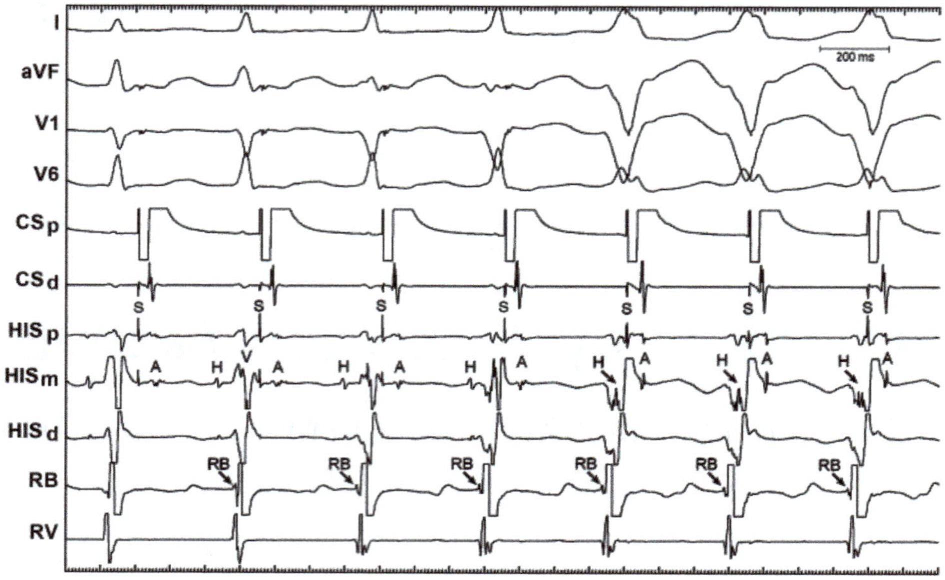

Figure 17.1. Tracings from the patient's ECG at presentation.

AP allowing preferential ventricular activation through the AVN. Subtle evidence of preexcitation includes absence of septal Q waves in leads I, aVF, and V_5–V_6, especially when associated with rS pattern in lead III.[5] Maximally preexcited QRS shows features similar to typical LBBB, with no or minimal R waves in V_1, precordial R-wave transition ≥ lead V_4, QRS duration ≤ 150 ms and axis between −30 and +60 (depending on location of ventricular insertion and retrograde RBB properties).

Intracardiac electrograms: At baseline the His-ventricular (HV) interval is normal or slightly short. A small discrete potential (accessory pathway potential [APP]) similar to His potential may be noted along the anterior/lateral tricuspid annulus. During sinus rhythm, the His potential precedes the

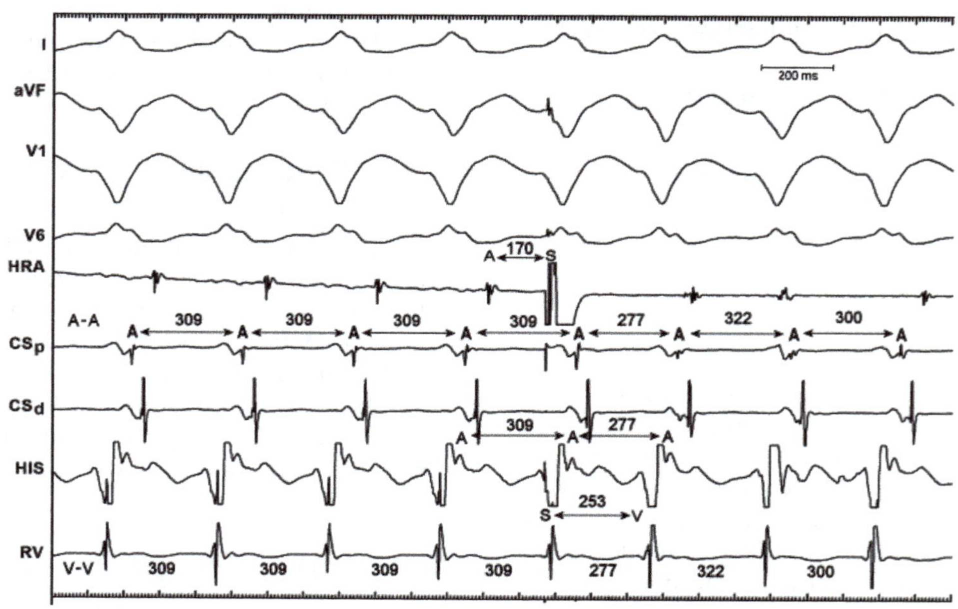

Figure 17.2. Tracings from the patient's ECG at presentation.

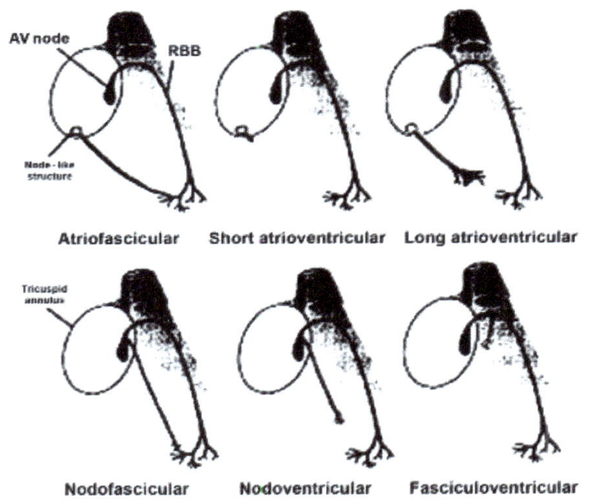

Figure 17.3. Different anatomical courses of atypical accessory pathways (APs). AV, atrioventricular; RBB, right bundle branch. (Reproduced with permission from Sternick et al.[5])

APP, and ventricular activation is predominantly over the normal conduction system (Figure 17.4). Preexcitation can be revealed with atrial pacing, leading to greater delay in conduction down the AVN compared to allowing a larger portion of the ventricles to be activated via the AP. With shorter coupling intervals, the atrial-His (AH) increases more than the AV interval (A-AP or A-Delta), leading to progressive shortening of the HV intervals and, eventually, retrograde RBB and His activation (see Figure 17.1).

Features associated with antidromic AV reentrant tachycardia utilizing an atriofascicular pathway include:

1. Initiation of tachycardia by atrial extra stimuli or burst pacing is facilitated by the antegrade conduction delay over the AP that allows recovery of retrograde conduction over the His-Purkinje system (HPS)-AVN. Initiation of tachycardia by ventricular pacing is also served by inherent retrograde conduction block in the AP.
2. The atrium and the ventricle are essential parts of the macro-reentrant circuit.
3. Long AV intervals more than 150 ms during preexcited tachycardia.
4. Short ventricular-His (VH) interval (5-25 ms). Atriofascicular pathways insert into the RBB, leading to rapid simultaneous activation of both His and ventricle with a short VH interval, except if there is a retrograde RBB block. Left bundle branch block morphology with axis −30 to +60. Rapid retrograde activation over the right bundle to the His-bundle can allow some degree of antegrade ventricular activation over the left anterior fascicle as well as the right bundle.
5. HA interval during tachycardia = HA interval with RV pacing at a similar cycle length.

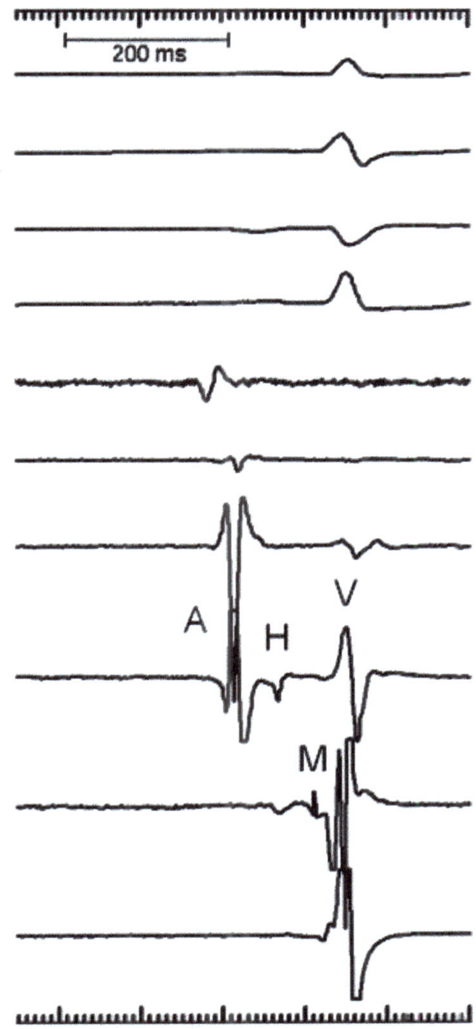

Figure 17.4. During sinus rhythm, the His potential precedes the APP, and ventricular activation is predominantly over the normal conduction system.

In this case, during proximal coronary sinus (CS) pacing at a cycle length of 360 ms, progressive AH prolongation was associated with gradual shortening of the HV interval and increase in QRS duration. In the first 4 beats of Figure 17.1, QRS fusion is seen with increasing contribution from AP conduction. On the fifth beat, maximal preexcitation is seen with an LBBB pattern and a left superior axis associated with a step-up of the AH interval. As maximal preexcitation is reached, HPS activation reverses. Right bundle precedes His, which becomes activated

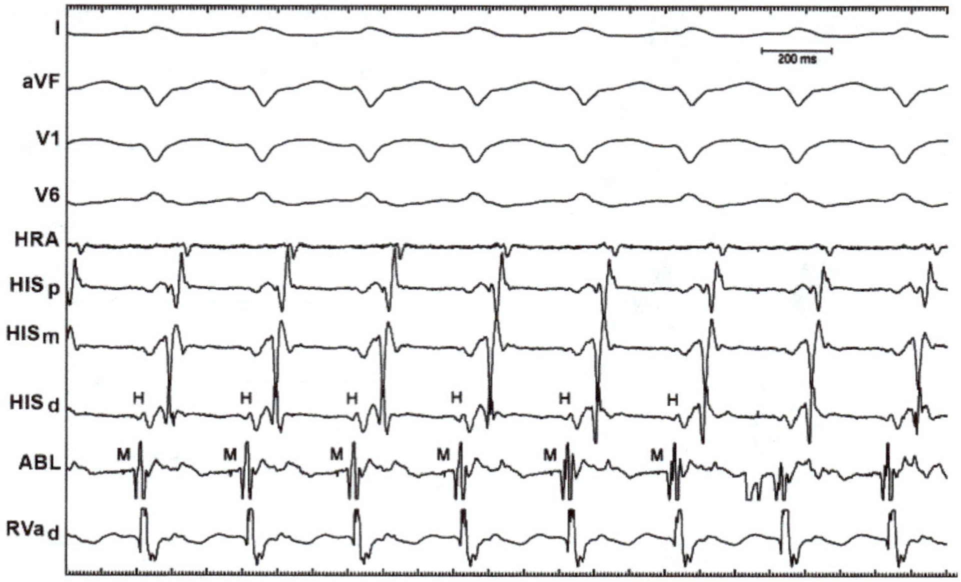

Figure 17.5. Successful ablation site at the 7 o'clock position along the tricuspid annulus. Note the sharp M potentials at that location.

distally to proximally. The VH interval remains short and fixed, indicating ventricular activation over an atriofascicular bypass tract.

Figure 17.2 illustrates the presence and active participation of the atriofascicular bypass tract in the preexcited tachycardia. An atrial premature contraction (APC) delivered from the high right atrium during preexcited tachycardia that advances the following ventricular activation by 32 ms without affecting atrial activation at the AVN region (His A and proximal CS A are not advanced) proves the presence of an atriofascicular bypass tract. Active participation of the bypass tract in the tachycardia is confirmed by the maintenance of the ventricular, HPS and atrial activation sequence and timing following the reset ventricular complex. Thus, the first A resulting from the advanced V is also advanced by a similar amount, resulting in a shortened AA interval of 277 ms. Decremental conduction of the AP is shown by the prolongation of AV interval, causing prolongation of the VV interval (322 ms) following the shortened AA interval (277 ms). Note that the subsequent AA interval is also prolonged (322 ms), resulting in a shortened AV interval and a shortened VV interval of 300 ms. These oscillating cycle lengths are characteristic of decremental pathway conduction when the tachycardia is perturbed by an APC.

Bypass tract was successfully ablated at the 7 o'clock position along the tricuspid annulus. Additional confirmatory lesions were delivered, guided by CARTO® navigation system (Biosense Webster Inc., Diamond Bar, CA) (Figure 17.5).

MAPPING AND ABLATION

The gold standard for ablation of atriofascicular bypass tracts is "M" potential mapping and ablation along the tricuspid annulus (see Figure 17.5).[6,7,8] Care must be taken to avoid mechanical trauma during mapping as it may lead to transient AP block that may last for hours. Mapping and ablation is preferably done during atrial pacing to allow better visualization of AP block with loss of preexcitation. Atrial pacing also minimizes catheter shift, which commonly occurs with termination of tachycardia if mapping and ablation are performed during tachycardia. "Mahaim" automatic tachycardia is commonly seen during atriofascicular pathway radiofrequency ablation. This is most likely similar to junctional rhythm during slow AVN pathway ablation.

Mapping and ablation of the ventricular insertion of atriofascicular and long AV pathways are difficult and less successful as extensive arborization is common. A failed attempt to ablate the ventricular insertion can be proarrhythmic if the right bundle is damaged and the pathway persists as the reentry circuit becomes longer involving the left bundle for retrograde conduction. Nevertheless, ablation of the ventricular insertion can be performed by mapping the site of earliest activation relative to surface QRS.

Risk of right bundle damage can be minimized by pace mapping and presence of a steep QS on unfiltered unipolar recording.

Correct Answer:

Q1. D

REFERENCES

1. Mahaim I, Benatt A. Nouvelles recherches sur les connexions superieures de la branche gauche du faisceau de His-Tarawa avec cloison interventriculaire. *Cardiologia.* 1938;1:61.
2. Haïssaguerre M, Cauchemez B, Marcus F, et al. Characteristics of the ventricular insertion sites of accessory pathways with anterograde decremental conduction properties. *Circulation.* 1995;91:1077–1085.
3. Tchou P, Lehmann MH, Jazayeri M, Akhtar M. Atriofascicular connection or a nodoventricular Mahaim fiber? Electrophysiologic elucidation of the pathway and associated reentrant circuit. *Circulation.* 1988;77:837–848.
4. Klein GJ, Guiraudon GM, Kerr CR, et al. "Nodoventricular" accessory pathway: Evidence for a distinct accessory atrioventricular pathway with atrioventricular node-like properties. *J Am Coll Cardiol.* 1988;11:1035–1040.
5. Sternick EB, Timmermans C, Sosa E, et al. The electrocardiogram during sinus rhythm and tachycardia in patients with Mahaim fibers: The importance of an "rS" pattern in lead III. *J Am Coll Cardiol.* 2004;44:1626–1635.
6. McClelland JH, Wang X, Beckman KJ, et al. Radiofrequency catheter ablation of right atriofascicular (Mahaim) accessory pathways guided by accessory pathway activation potentials. *Circulation.* 1994;89:2655–2666.
7. Klein LS, Hackett FK, Zipes DP, Miles WM. Radiofrequency catheter ablation of Mahaim fibers at the tricuspid annulus. *Circulation.* 1993;87:738–747.
8. Grogin HR, Lee RJ, Kwasman M, et al. Radiofrequency catheter ablation of atriofascicular and nodoventricular Mahaim tracts. *Circulation.* 1994;90:272–281.

CASE STUDY

18

Nodofascicular Pathway

Patrick Tchou, MD, Mohamed Kanj, MD, and Mohamed Bassiouny, MD

A 41-year-old gentleman presented with recurrent narrow-complex tachycardia, unresponsive to treatment with amiodarone or propafenone. The tachycardia was not terminable with adenosine, diltiazem, metoprolol, digoxin, and carotid sinus massage.

Question No. 1:

The tracings shown in Figure 18.1 are most consistent with:

A. Atrioventricular nodal reentry tachycardia (AVNRT).

B. Atrial tachycardia (AT).

C. Junctional tachycardia.

D. Orthodromic atrioventricular reentry tachycardia (AVRT).

E. Orthodromic tachycardia involving a nodofascicular or a nodoventricular bypass tract.

NODOVENTRICULAR AND NODOFASCICULAR BYPASS TRACTS

In 1971, Wellens et al[1] attributed the antegrade decremental conduction properties of atypical accessory pathways to their origin from the atrioventricular (AV) node. Atypical accessory pathways with decremental conduction were labeled as nodoventicular or nodofascicular pathways (Mahaim pathways). The tipping point was in 1988 when atriofascicular pathways were first described.[2,3] Over the next few years it became clear that the majority of patients with antidromic AVRT previously assumed to be secondary to nodoventricular pathways are, in fact, due to atriofascicular pathways, casting doubt over true participation of an anterogradely conducting nodoventricular/nodofascicular pathway in antidromic AVRT. Anterograde-conducting nodoventricular/nodofascicular pathways would not have a strong "lateralizing" manifestation. That is, pacing from the lateral right atrium near the tricuspid annulus should not generate more prexcitation than pacing from the septum or proximal coronary sinus. The presence of a preexcited tachycardia with ventriculoatrial (VA) dissociation and the same QRS morphology as elicited during accelerated atrial pacing would constitute proof of such a pathway. Alternatively, during AVRT with full preexcitation, a single premature ventricular complex (PVC) that generates conduction block to the atrium but leaves the tachycardia continuing unabated would be indicative of nonparticipation of the atria and, thus, the possibility that the tachycardia is AVRT utilizing an antegrade nodoventricular pathway. Because of the decremental nature of these pathways, the inability of a premature atrial complex (PAC) to advance or delay the next V during tachycardia is not adequate proof of a nodoventricular pathway,

Practical Electrophysiology © 2015 Jasbir S. Sra, Masood Akhtar, Andrea Natale, David J. Wilber
Cardiotext Publishing, ISBN: 978-1-935395-17-1

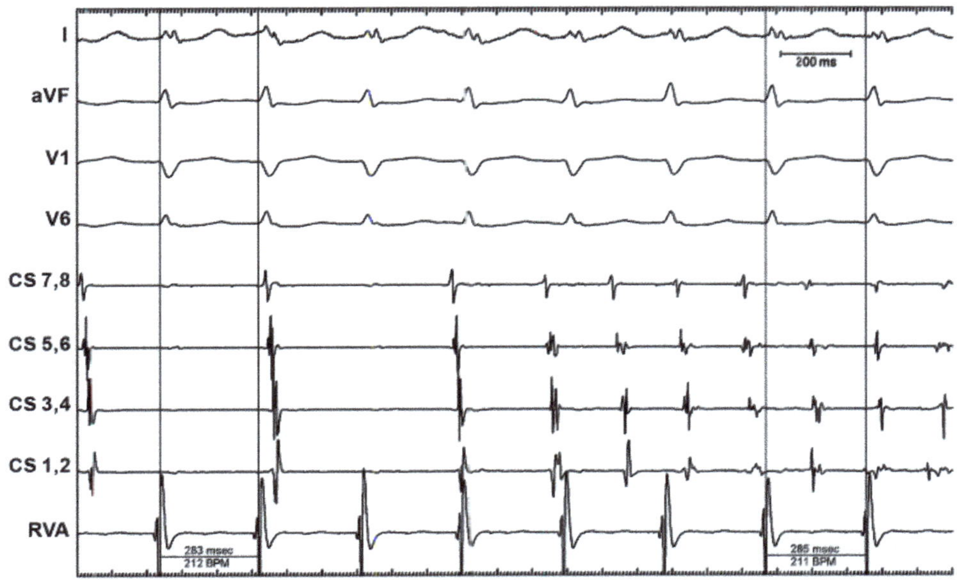

Figure 18.1. Tachycardia tracings.

even if one were quite confident that such a PAC did indeed capture the entire atria.

Electrophysiologic Characteristics

Nodofascicular and nodoventricular pathways that manifest only retrograde conduction have been associated with narrow-complex supraventricular tachycardia (SVT) with VA dissociation. This readily differentiates them from tachycardias using atriofascicular or other bypass tracts in which 1:1 AV association is a prerequisite. These bypass tracts are capable of retrograde but not anterograde conduction.

Our case presentation illustrates a narrow-complex tachycardia with VA dissociation. This tachycardia persisted during atrial fibrillation (AF) without any change in tachycardia cycle length (TCL), suggesting that the atrium is not a prerequisite for the maintenance of the tachycardia. This rules out AT and AVRT as a mechanism for the tachycardia (see Figure 18.1).

The mechanism of the SVT can be confirmed by terminating the tachycardia, or advancing/delaying the next His and QRS and resetting the tachycardia by a His refractory PVC, which excludes AVNRT as a mechanism of the tachycardia (Figure 18.2).[4,5] A ventricular extrastimulus delivered 28 ms prior to the His (during the His refractory period) advanced the next His and QRS by 26 ms.

Entrainment with right ventricular pacing with a postpacing interval (PPI) = TCL can also differentiate this tachycardia from AVNRT with VA block and from junctional tachycardia (Figure 18.3). If 1:1 AV association is present, the H and A relationship in the postentrainment beats should readily identify an accessory pathway-mediated tachycardia versus one utilizing a retrograde nodoventricular pathway. Tachycardias involving concealed nodofascicular or nodoventricular bypass tracts will often slow down with development of spontaneous catheter-induced proximal right bundle branch block (RBBB).

Ablation of nodofascicular and nodoventricular pathways can be challenging. Targeting the proximal insertion of the pathway carries a high risk of complete AV block. Successful ablation at the slow AV nodal pathway location has been reported.[6] We have also had successful ablation of these pathways from a subaortic position along the right coronary cusp. Presumably, the ventricular connection was to the distal node from the subaortic region rather than from the inferior basilar region of the ventricles connecting to the body of the AV node along the slow pathway.

Correct Answer:

Q1. E

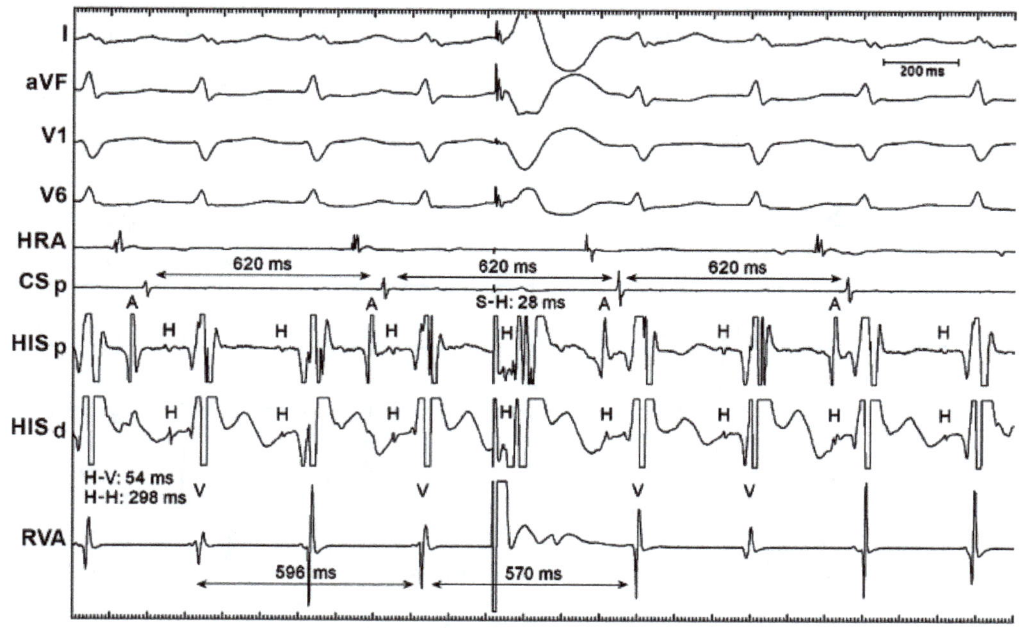

Figure 18.2. Narrow-complex SVT with VA dissociation. His-Purkinje system is utilized for antegrade conduction, leading to a narrow QRS complex with left inferior axis. Retrograde conduction occurs over a nodoventricular bypass tract. A His deflection precedes each ventricular potential by 54 ms. A ventricular extrastimulus delivered 28 ms prior to His advances the next His and V by 26 ms.

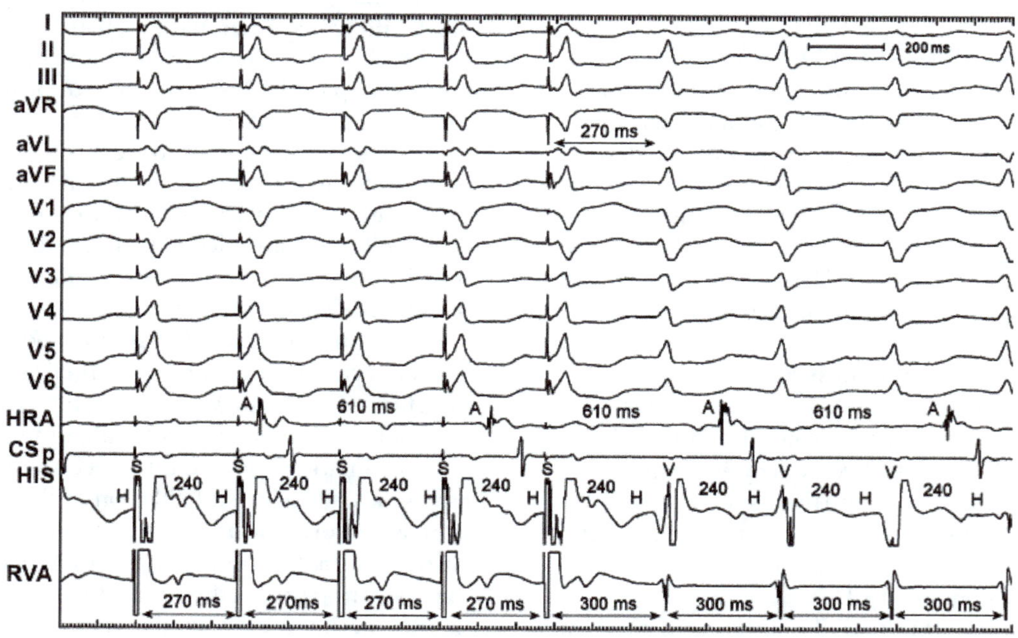

Figure 18.3. Narrow-complex SVT with VA dissociation utilizing His-Purkinje system for antegrade conduction and a nodoventricular bypass tract for retrograde conduction. The TCL is 300 ms. Pacing at the mid-right ventricular septum at a cycle length of 270 ms entrains the tachycardia. The paced QRS complexes are quite similar to QRS complexes following entrainment, indicating fusion from the paced and conducted wavefronts. Stimulus-to-His during entrainment is equal to V-to-His following cessation of pacing and the PPI is exactly equal to TCL, denoting that the site of pacing in the right ventricle is part of the tachycardia reentry circuit. This excludes AVNRT with VA block.

REFERENCES

1. Wellens HJJ. The preexcitation syndrome. In: Wellens HJJ, ed. *Electrical Stimulation of the Heart in the Study and Treatment of Tachycardias.* Baltimore, MD: University Park Press, 1971: 97–109.
2. Tchou P, Lehmann MH, Jazayeri M, Akhtar M. Atriofascicular connection or a nodoventricular Mahaim fiber? Electrophysiologic elucidation of the pathway and associated reentrant circuit. *Circulation.* 1988;77:837–848.
3. Klein GJ, Guiraudon GM, Kerr CR, et al. "Nodoventricular" accessory pathway: Evidence for a distinct accessory atrioventricular pathway with atrioventricular node-like properties. *J Am Coll Cardiol.* 1988;11:1035–1040.
4. Shimizu A, Ohe T, Takaki H, et al. Narrow QRS complex tachycardia with atrioventricular dissociation. *Pacing Clin Electrophysiol.* 1988;11:384–393.
5. Haïssaguerre M, Campos J, Marcus FI, Papouin G, Clémenty J. Involvement of a nodofascicular connection in supraventricular tachycardia with VA dissociation. *J Cardiovasc Electrophysiol.* 1994;5:854–862.
6. Grogin HR, Lee RJ, Kwasman M, et al. Radiofrequency catheter ablation of atriofascicular and nodoventricular Mahaim tracts. *Circulation.* 1994;90:272–281.

CASE STUDY

19

Preexcited Tachycardias: Differential Diagnosis, Mechanisms, and Diagnostic Clues

Masood Akhtar, MD and Jasbir S. Sra, MD

A 52-year-old patient presents with palpitations but no syncope; the 12-lead ECG shows wide QRS tachycardia.

Question No. 1:

In the presence of overt preexcitation, the mechanisms of wide QRS tachycardia include the following:

A. Aberrant conduction (Ab-C).

B. (1) Ventricular tachycardia (VT) (myocardial), and (2) VT due to bundle branch reentry (BBR).

C. Nonspecific intraventricular defect (IVCD).

D. Right or left ventricular pacing (via coronary sinus) and no detectable pacing artifact or VT in the presence of implantable cardioverter-defibrillator (ICD)/pacemaker leads and mechanical irritation.

E. Preexcited tachycardias.

F. All are true.

DISCUSSION

Wide QRS tachycardia challenges physicians in general and electrophysiologists in particular.[1–5] However, the existence of ventricular preexcitation makes a broad QRS tachycardia more complex. In some situations, it poses a real intellectual and clinical dilemma with difficulties regarding the exact mechanism and, hence, has therapeutic implications. This case report contains some (not all) examples of preexcited tachycardia, a relatively simple distinction from other wide QRS tachycardias that may or may not coexist with preexcitation using intracardiac electrocardiography.[5–15]

Since this communication is about preexcitation, the role of any accessory pathway (AP) functioning only in the retrograde direction with a narrow complex will not be discussed. In the presence of overt preexcitation, the mechanisms of wide QRS tachycardia include the following:

1. Aberrant conduction (Ab-C).[1–5]
2. a. VT (myocardial);[1–5] and b. VT due to BBR.[16–20]
3. Nonspecific IVCD.[4]
4. Right or left ventricular pacing (via coronary sinus) and no detectable pacing artifact[4] or VT in the presence of ICD/pacemaker leads and mechanical irritation.
5. Preexcited tachycardias.[6–15]

Aside from #1, #2b (BBR), and #3, the His-bundle potential will not precede the QRS with

Practical Electrophysiology © 2015 Jasbir S. Sra, Masood Akhtar, Andrea Natale, David J. Wilber
Cardiotext Publishing, ISBN: 978-1-935395-17-1

HV equal to or greater than the HV of the sinus rhythm. Brief descriptions of methods to eliminate #1–4 follow, and #5 will be discussed in detail.

1. Ab-C: This form of wide QRS can be distinguished from all other forms by a rapid and sharp initial part of the QRS regardless of the QRS morphology. There are only a limited number of configurations in patients with Ab-C. Physiologically speaking, if it does not fit one of the Ab-C patterns, it is not aberrant, which has been previously noted.[4] The presence of atrioventricular (AV) dissociation makes Ab-C unlikely. Realistically, IVCD and BBR with 1:1 AV relationships are the only entities that can be confused with Ab-C. The correct diagnosis of BBR can be made first with overdrive atrial pacing, which will not change the BBR-VT, second by recording from one or both of the bundle branches (BBs), and then studying the sequence of H-BB activation and interval in conjunction with QRS morphology. IVCD is discussed under #3.
2. Myocardial VT: Slurred wide QRS and AV dissociation, short HV to minus HV, no identifiable His potential before the QRS, axis between −90 and +180 are features fairly diagnostic of myocardial VT. Fibrosis-related VT may have no particular QRS configuration and axis combination because myocardial VT with variable origin does not have to fit into any particular configuration, unlike the Ab-C pattern or AP location. These observations, along with other published criteria regarding both clinical and surface ECG, make the diagnosis of VT relatively easy.[1–5] (Myocardial VTs in the absence of structural heart disease have certain surface ECG patterns and are detailed elsewhere in this book.)
3. Nonspecific IVCD:[4] This term is used when the QRS is wide but otherwise has a general orientation that is more or less normal. The usual causes are extensive scarring; antiarrhythmic drugs, particularly Class 1C (ie, flecainide, propafenone); and electrolyte abnormalities such as hyperkalemia. In these settings sinus tachycardia can cause confusion, particularly if the PR interval is long or the P wave is obscured. DC shock may be delivered not realizing that the underlying rhythm is sinus. Use of calcium chloride or bicarbonate will immediately change the QRS in case of hyperkalemia. It is important to recognize that the baseline ECG in these patients is also abnormal, may show slurring, and does not fit into any pattern of Ab-C or preexcitation.

    Ab-C-, BBR-VT-, and nonspecific IVCD-related tachycardias can be readily distinguished from others by the presence of His-bundle recording in which the HV interval equals or exceeds that in sinus rhythm. The atrial rate during Ab-C and IVCD either equals or exceeds VR, whereas in BBR the atrial rates (retrograde) either equals or is less than VR, or AV dissociation or ventriculoatrial (VA) block are noted. In a high percentage, the atria are in atrial fibrillation.[18–20]
4. Device-related wide QRS complex:[4] Prior history of device placement and palpation of the device, the location and course of the lead on x-ray and the QRS morphology will settle the issue. Further evidence of constancy of rate with the limits of what is programmed, identification of stimulus artifact, and the use of magnet would lead to accurate diagnosis. Fortunately these problems are relatively rare, but if not figured out correctly, they will not be resolved without lead manipulation and/or extraction.
5. Preexcited tachycardias (Table 19.1, Schemas I and II): Only those pathways producing regular clinical tachycardias will be discussed. Irregular tachycardias such as atrial fibrillation, which is easy to diagnose, will not be elaborated on here.

In all of these specified situations, if the A-to-V ratio is any higher than 1:1, the P wave may be identifiable, and hence the diagnosis might be relatively straightforward. With 1:1 ratio in atrial tachycardia and 1:1 or 2:1 with atrial flutter, the ventricular rate is likely to exceed 150 bpm, and the P-wave identification and its polarity may be hard to determine. A list of preexcited tachycardias likely to be encountered in clinical practice is outlined in Table 19.1 and shown schematically in Schema I (a, aa, b, c, d, dd, e, ee, eee, f) and Schema II in Figure 19.1 (a, aa, b, bb, c, cc, d). Although the surface ECG (preferably 12-lead) may give some clue to the mechanism and circuit and alert one to the various possibilities listed in Table 19.1, the gold standard for precise diagnosis is intracardiographic evaluation. Reasons for this include:

1. The complexity of the arrhythmia may not be fully appreciated until all necessary catheters are properly placed and appropriate electrophysiologic maneuvers are performed.
2. Incomplete assessment may lead to possible failure of ablative treatment, unnecessary recurrences, patient dissatisfaction, and prolonged procedure time and its related consequences.
3. Finally, accurate analysis leads to fewer ablative lesions, by itself a benefit, and better chances of a

Table 19.1. Preexcited Tachycardias

I. Preexcited tachycardia in which one or more APs play a bystander role
   - a. Sinus tachycardia (right-sided AP)
     - aa. Sinus tachycardia (left-sided AP)
   - b. Atrial tachycardia could be anywhere, and one or more AP may have different locations (left free wall shown here)
   - c. Common (counterclockwise) atrial flutter. Could be 1:1, 2:1, or 3:1 ratio, and the atrial flutter may originate in left atrium
   - d. Common (slow–fast) atrioventricular nodal reentrant tachycardia (AVNRT) with rapid-conduction AV-AP as bystander
   - e. Common (slow–fast) AVNRT and atriofascicular pathway (AFP) as the bystander (right side)
     - ee. Uncommon (fast–slow) with rapidly conducting AP
       - eee. Uncommon (fast–slow) with AFP as bystander (left side)
   - f. Atrial/sinus tachycardia/atrial flutter with conduction over fasciculoventricular pathway.

II. Preexcited tachycardia in which one or more AP is an essential part of the circuit
   - a. Antidromic reentry with anterograde conduction via fast-conducting AP (see Schema IIa) and retrograde conduction via the normal pathway (NP), for example, His-Purkinje system (HPS) and AV node (left free wall AP shown here but could be other locations)
     - aa. Is the same as (*a*), plus retrograde ipsilateral BB block
   - b. Antidromic reentry with anterograde conduction along AFP and retrograde via the NP
     - bb. The same as (*b*) with additional retrograde right bundle branch block (RBBB)
   - c. Preexcited circus movement (reentrant) tachycardia using 2 rapidly conducting (Kent) AP (can be used either anterogradely or retrogradely)
   - d. Circus movement tachycardia with anterograde conduction via an AFP and retrograde conduction via rapidly conducting AP

i. Sinus tachycardia (Schema Ia, Iaa); atrial tachycardia with 1:1 or 2:1 conduction (Schema Ib); atrial flutter with 1:1, 2:1, or 3:1 conductions (Schema Ic).

lasting cure not possible by any noninvasive means. The tachycardias are divided in 2 groups: Group 1, in which AP is a bystander, and Group 2, in which AP is an active participant. Further designation of a, aa, etc. show the various arrhythmia circuits. Labeling of schema and actual tracings (figures) in the same general pattern is to simplify the presentation.

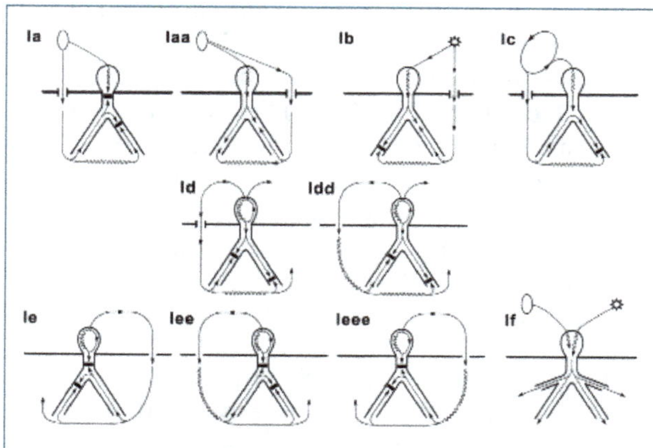

Schema I. All of the examples in the schema are of preexcited tachycardias that are regular and in which the AP functions as a bystander. Further designation is given to draw attention to the origin and mechanism of tachycardia. Ia, Iaa, Ib, and Ic are atrial. Id and Idd are slow–fast AVNRT with bystander AP. Ie, Iee, Ieee are examples of fast–slow reentry with slow-conducting AV, the usual depiction of AFP location and left-sided fast-conducting pathway, respectively. If is fasciculoventricular pathway. (See also Table 19.1.)

Important bullet points:

- Fast AV pathway locations depicted here are only for conceptual reasons. In terms of their prevalence, left free wall and posteroseptal locations are more common than right free wall and anteroseptal.
- AV nodal schema as shown here has no relationship to the existence of any particular AV nodal conduction, refractory period (RP) with any type of AP location or its conduction and RP properties.
- The site of HPS block is not intended to represent the actual site, which could vary, and depending upon many variables including pacing site, normal and abnormal HPS, AP location, cycle length (CL), sequence of pacing, etc.
- Wavy line represents slower conduction as opposed to fast conduction, but is not a reflection of the magnitude of slowing.
- Schema of the AV node, HB, BB, and AP is self-explanatory. The relative location of AFP versus rapid-conducting AP is not meant to be an accurate representation, but simply distinguishes the 2.
- All of these bullet points also apply to examples shown in Schema II.

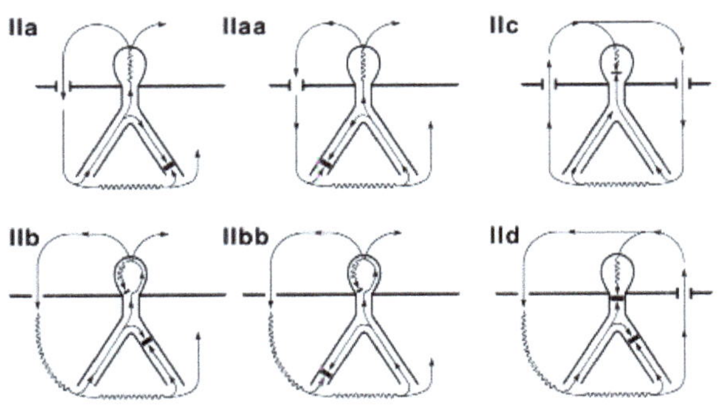

Schema II. This schema shows examples where the AP is an active participant in the reentry circuit. IIa, IIaa: Antidromic reentry using rapid AV pathway without (IIa) and with (IIaa) retrograde RBBB. IIb and IIbb: Antidromic reentry with AFP via RB (IIb) and through LBB (IIbb). Two APs involved in (rapidly conducting) circus reentrant tachycardia are displayed in IIc (anterograde AFP) and IId (rapid retrograde AV-AP). (See also Table 19.1.)

Preexcited Atrial Tachycardias (AP Bystander) (Table 19.1 and Schema Ia, Iaa, Ib, Ic)

With intracardiac electrocardiography, these tachycardias do not pose a diagnostic dilemma. Depending upon the origin of the atrial signal and location of the AP, AV interval and QRS configuration will vary. The maneuvers listed below, alone or in combination, will be helpful in arriving at a diagnosis.

1. When atrial rates exceed ventricular rates, that is, 2:1 or 3:1, a possible diagnosis may be myocardial VT with 1:2 VA response via dual retrograde AV nodal pathways, or some other retrograde functional AP with alternation of VA conduction between NP and AP. Another scenario may be 2 independent tachycardias with partial or complete interruption of AV conduction somewhere above the atrial end of AP. The foregoing situations are rare, but can be diagnosed by atrial overdrive pacing and observing no change in ventricular response.
2. A high–low atrial activation sequence makes it unlikely that the atrial response is secondary to ventricular response.
3. When the atria are driving the ventricles, overdrive pacing from the ventricle, and examining the atrial response showing ventricular rate acceleration, but no change in atrial rate (Schema Ic and Id) and/or atrial capture with a difference in activation sequence, will argue in favor of a preexcited tachycardia of atrial origin. This also can be accomplished with programmed single or double ventricular extrastimuli.
4. Another helpful maneuver is to simultaneously pace the atria and ventricle slightly faster than tachycardia, stop pacing and observe the return cycle for whether the A or V occurs first.
5. Single ventricular extrastimulus termination of preexcitation and continuation of tachycardia at the atrial level without CL change will also support bystander role of AP. This maneuver, *per se*, will not distinguish between the various atrial and/or junctional tachycardias with AP playing a bystander role.
6. The same maneuver from the atrial site may change the AV relationship, which also helps in the diagnosis. This could happen regardless of whether ventricular capture is achieved.
7. Entrainment of the atrial tachycardia not only will clarify the origin, the mechanism, but also, if there is concurrent ventricular response, that the atrium is driving the ventricle.

Preexcited Junctional Tachycardias (AP Bystander) (Table 19.1 and Schema Id, Idd, Ie, Iee, Ieee, If)

In this category are AV nodal reentry (slow–fast or fast–slow) with bystanding AP (which can present a significant problem in distinguishing from antidromic SVT (compare Ia and IIa). This is true with both a Kent bundle type, that is, rapidly conducting AP, as well as AFP. Since both of these tachycardias can cause true antidromic reentry as well, we later elaborate in a discussion of antidromic tachycardia. When an AP directly connects HPS to ventricular, the so-called fasciculoventricular Mahaim (Schema If) plays no role in clinical arrhythmias. Hence it is a cosmetic issue, defined by normal PR, QRS slurring, a short HV, and no change in HV or QRS during atrial pacing (Schema If).

Regular preexcited tachycardias in which AP is an active participant[6,8–11,14,21,22]

1. In antidromic tachycardia (see Table 19.1, Schema IIa, IIaa), sometimes referred to as true antidromic tachycardia, the QRS complex is formed by anterograde conduction over AP, and the retrograde limb is the NP. Although the QRS morphology is helpful in locating the ventricular insertion, the determination of retrograde pathway, that is, the NP (HPS/AV node vs AVNRT and bystander AP), could be difficult (Schema Id, Idd, Ie, Iee, Ieee vs IIa, IIaa, IIb, IIbb). In fact, among all of the preexcited

tachycardias listed in the Table 19.1, this differentiation is probably the most difficult. It is further complicated by the fact that around 15% of patients with fast-conducting AP, and almost all patients with AFP, have functional duality of pathways in the AV node. Antidromic limb (ie, AP) may utilize a fast-conducting Kent type of bundle or a slow-conducting pathway (AFP).

2. Antidromic reentry with fast-conducting pathway. When a fast-conducting pathway is the culprit, the following may be helpful to make the differentiation. If the His-bundle potential can be identified (see * on section vi and Figure 19.7 for details on identifying the His-bundle potential), there are several clues. These clues include:
   i. Retrograde His bundle-to-atrium conduction time (HA) with ventricular or His-bundle pacing longer than during the tachycardia, because in AVNRT, the HA is a spurious interval and not a reflection of true H-to-A conduction time. This has been pointed out by several investigations previously.[11, 23, 24] In the absence of a measurable His-bundle potential, this criterion is obviously not as helpful.
   ii. Conversion of a preexcited tachycardia to a narrow (or nonpreexcited) complex or aberrantly conducted complex with no change in CL is diagnostic of the inert role of AP. An example of such an occurrence is shown in Figure 19.1 (and Schema Id), in which the underlying arrhythmia is slow–fast (common) type of AVNRT.

      The initial part of the tracing shows a preexcited tachycardia (bystander AP), and a spontaneous VPC abolishes the linkage in the HPS, where the AV nodal anterograde impulse links to the impulse traveling over the AP. The site of the link is usually infranodal (HPS) but the distal part of the AV node cannot be excluded. The HPS link with AVNRT is necessary to activate the atria and then the AP so that the impulse can reach the ventricle. Long HV interval is the rule but is not representative of HPS conduction time in this situation. When the link breaks both normalization of the QRS, shortening of the HV interval (ie, H-A-V sequence will change to HV) will take place. In both of these instances, the local HV interval measurement may not be accurate because the HV with preexcited complex is related to the A and not to the H. On the other hand, when the QRS is normal the retrograde A may precede the local V electrogram, and therefore the HV in these circumstances should be measured from the surface ECG QRS.
   iii. Since the HPS is an essential part of true antidromic reentry and not in AVNRT, special attention should be paid to its role in the CL changes during preexcited tachycardia. Many times this happens at the onset of tachycardia, but it can be induced with programmed atrial and/or ventricular extrastimuli. With retrograde delay and/or block in the HPS (usually ipsilateral to AP location), the VA, VH, and the CL of tachycardia will temporarily prolong and all of the above will reverse when the retrograde HPS block is abolished. This may give one the spurious impression that the tachycardia is associated with different AP and not due to 2 sets of VH intervals. In most instances, 2 or more sets of VH intervals can simply be explained on the basis of HPS response to rapid changes in CL of tachycardia, particularly at the onset, but they may be sustained because of linking phenomena in the HPS[15] (Figures 19.2 and 19.3, Schema IIa, IIaa, IIb, IIbb). With VH changes and no CL changes, AVNRT with bystander AP participation is more likely. This is also true if the AFP constitutes the anterograde limb of the circuit.
   iv. Recording from one or both BBs (simultaneously, if feasible) with HB is particularly helpful because it provides the direction of impulse propagation through the

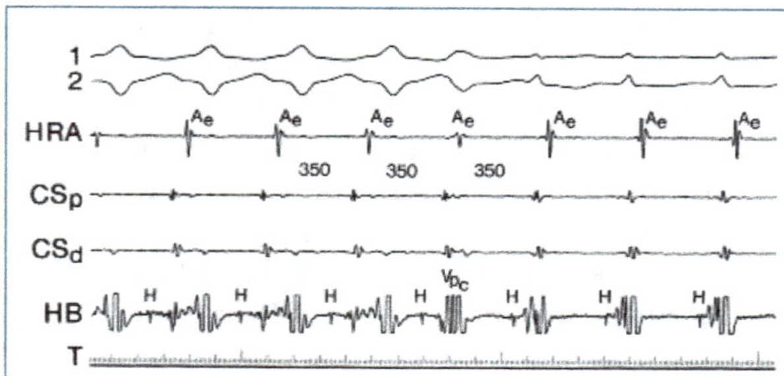

Figure 19.1. Bystander AP (also see Schema Id). Slow–fast AV nodal reentry is the underlying rhythm, that is, VA (RP) HA, atrial activation sequence (earliest on His-bundle electrogram (EGM)), and no change in tachycardia CL with disappearance of preexcitation (first 4 complexes). Note the HV internal shortening (sequence H-A-V change to HA. This reflects interruption of the link in the HPS between AV nodal impulse (anterogradely) and AP impulse (retrogradely) with a spontaneous right ventricular premature complex (VPC). Tracings from top to bottom are surface ECG leads I, II, high right atrial (HRA), coronary sinus proximal (CSp) and coronary sinus distal (CSd), His-bundle EGMs (HB), and timeline at 10 ms and 100 ms. Similar abbreviations are used for all forthcoming EGMs. (Reproduced from Jazayeri et al,[22] with permission from John Wiley and Sons.)

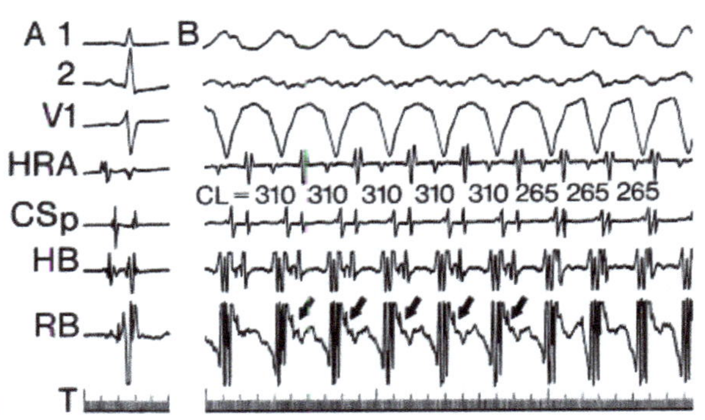

Figure 19.2. Retrograde RBBB during antidromic reentry (Schema IIaa with left free wall AP). **Panel A** is sinus rhythm with no visible preexcitation. Antidromic reentry using right-sided AP is shown in **Panel B**. Note the 5 bold arrows depicting the RB potential now following the local V electrogram due to retrograde RBBB. When the latter resolves, the VA shortens and now H and RB cannot be identified. The onset of atrial EGM on the HB merges with the local V electrogram and yet is identifiable.

HPS. The sinus H-RB/H-LB is the reference point. During any impulse conducting down from the atria or AV node, the H-RB/H-LB sequence and, more importantly, the interval would equal sinus rhythm (Figure 19.4, Schema IId). However, in antidromic tachycardia using a fast-conducting AV pathway, the ipsilateral BB deflection will precede the His potential. Furthermore, recording from the BB will make it easier to diagnose retrograde BB block as a cause of the CL change. In the case of antidromic tachycardia using AFP, RB recording is critical and shows not only the reversal of the H-RB sequence (ie, now RB-H), but RB potential precedes the QRS, whereas the HB deflection does not. Such tracings showing H-RB reversal have been noted before, but the entire circuit was not figured out. The recognition of AFP (rather than nodoventricular pathway) raises a serious question about the actual existence of nodoventricular pathways involved in clinical preexcited tachycardias. In fact, looking at many of the published examples (and our personal observations), we have not come across a single convincing example of nodoventricular connection used anterogradely in a preexcited tachycardia.

v. Properly timed atrial and/or ventricular premature stimuli near the ventricular and atrial insertion sites, respectively, can reveal the nature of reentrant circuits. For example, termination of preexcited tachycardia with a single atrial extrastimulus near the AP, which terminates the tachycardia without reaching the AV node, will be quite diagnostic of

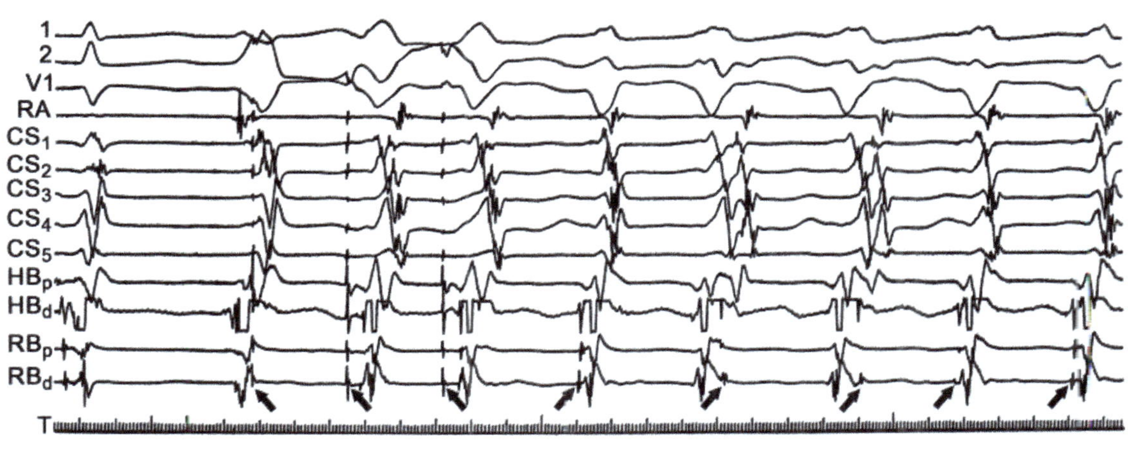

Figure 19.3. Antidromic reentry with AFP (see also Schema IIb and IIbb). The first complex is sinus with no preexcitation and the second one is spontaneous VPC; ventricular pacing is initiated (leftward directed bold arrows). The first stimulus is ineffective, and the next 2 capture the right ventricle and initiate the AFP-related antidromic reentry. Note the location of RB potential (rightward directed bold arrows). The next cycle of tachycardia (with left bundle branch block [LBBB] pattern) demonstrate no retrograde RBBB (Schema IIb). The next 2 depict net RBBB (Schema IIbb), which prolongs the VH and VA and tachycardia CL. Retrograde RBBB is then resolved, and the RB now precedes the local V with a change in axis toward normal. VH and VA shortens, and the tachycardia CL accelerates. Of special note is the fact that with right-sided AFP and normal RBB conduction, the onset of atrial EGM on the HB cannot be separated from local V's (the first and last 2 complexes). (Compare this with Figure 19.2.) In this example, the 2 sets of VH are simply a reflection of HPS conduction, which is an essential part of any true antidromic reentry (ie, AP anterogradely and NP retrogradely). Tracings are surface ECG CSp to distal (CS_1–CS_5), HB (proximal and distal) and RB right bundle potential (proximal and distal), and timelines.

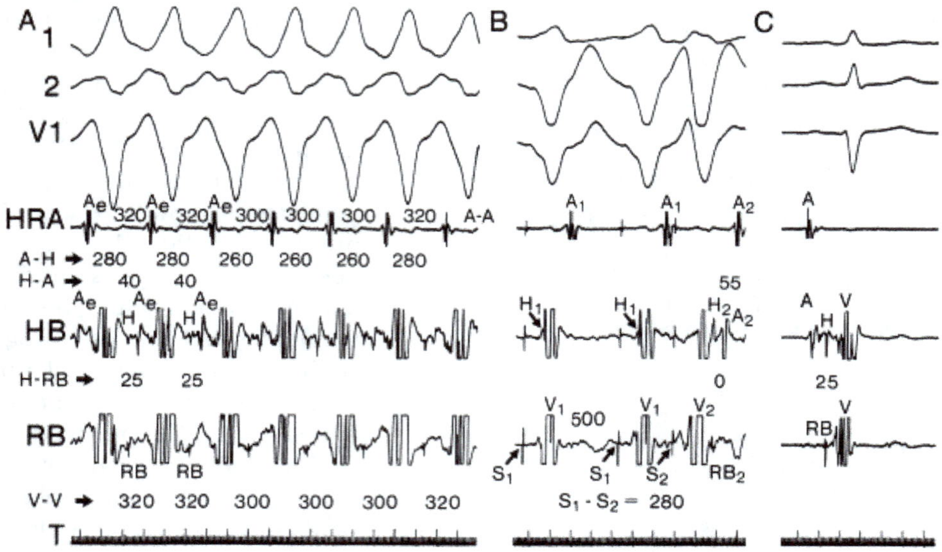

Figure 19.4. Preexcited tachycardia and H-RB recording (see also Schema Id). **Panel A** shows preexcited tachycardia using posteroseptal rapidly conducting AP. Ventricular pacing and sinus rhythm are displayed in **Panels B** and **C**, respectively. During sinus rhythm, the H-RB interval is 25 ms, the same as during preexcited tachycardia. As a reference, during RV pacing the retrograde H_2 emerges from V_2 (Panel B) because the V_2 impulse reaches the H_2 via LB; the H-RB measures zero. It is relatively easy to appreciate the value of RB potential recording with a visible HB potential which makes distinguishing between true antidromic reentry where RB or LB either precede the H or are activated at the same time. Hence the example in Figure 19.4 is that of slow–fast AVNRT with bystander AP. Appropriate values are labeled in milliseconds. (Reproduced from Jazayeri et al,[22] with permission from John Wiley and Sons.)

antidromic reentry versus AVNRT (Figure 19.5). The frequently used criterion of QRS, and subsequent atrial advancement with an atrial extrastimulus when it occurs, is helpful.[14] However, resultant shorter VV interval may lead to VH and/or HA prolongation with full compensation and thus not be diagnostic[25] (Figure 19.6). Response to late ventricular extrastimuli near the atrial insertion could be variable because an extrastimulus also could reach the atria via

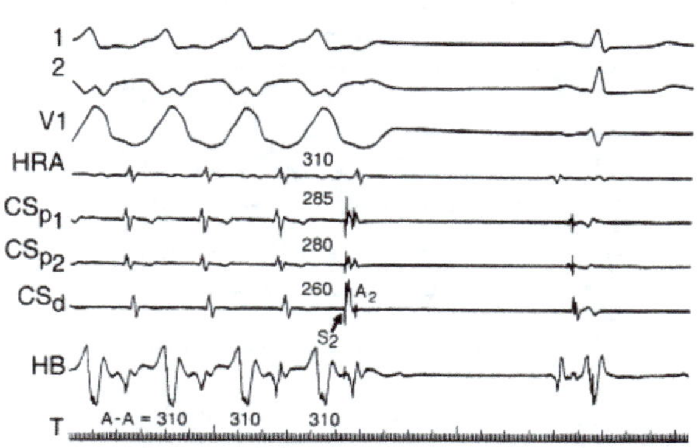

Figure 19.5. Timed atrial premature complex in preexcited tachycardia. A_2 delivered in CS at a coupling interval of 260 ms terminates a wide QRS (preexcited tachycardia). Note that the timing of HRA and atrial deflection on the HB are unchanged. This A_2 therefore could not have entered the AV node and yet it stops an antidromic reentry suggesting an active role of AP (see also Schema IIa).

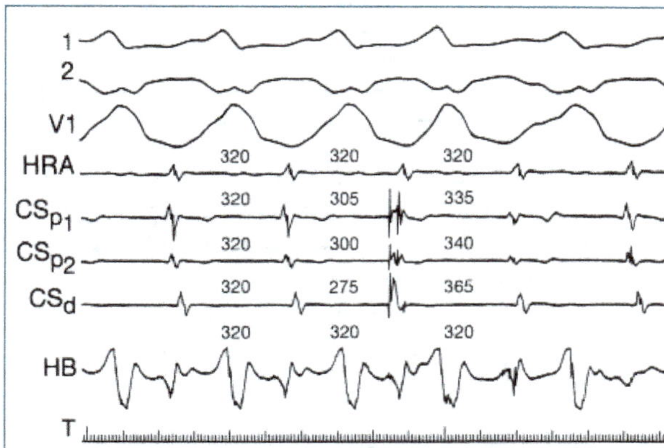

Figure 19.6. Resetting of true antidromic reentry with atrial extrastimulus. A_2 at 275 ms at CSd advances the QRS (ie, RR is shorter than previous RR). Despite this, however, the next atrial activation timing is not disturbed (complete compensatory interval). This is due to the fact that advancement of QRS produced additional delay in the NP. Since the HB cannot be clearly identified, the site of compensatory delay (ie, HPS vs AV node) cannot be stated. (Reproduced from Panatopoulos et al,[25] with permission from John Wiley and Sons.)

the fast pathway retrogradely in patients with functional duality of the AV node, that is, in AVNRT and bystander AP. On the other hand, early ventricular extrastimuli, by blocking retrogradely in the HPS or colliding with oncoming antidromic impulse via the AP, will provide a strong evidence for true antidromic reentry because the extrastimulus will not be able to penetrate the AV node and hence not affect AVNRT.

vi. AP ablation at the ventricular or atrial site after the atrial activity has been registered via the AV node and also will be diagnostic of true antidromic reentry (Figure 19.7). (*To improve the chances to record a recognizable His-bundle potential, an interelectrode distance of 1 ms to 2 ms makes it less likely that the His-bundle potential is obscured by the local V electrogram.)

Antidromic Tachycardia using AFP as the Anterograde Limb (Schema IIb/bb)[26, 27]

The exact nature of AFP has not been fully elucidated. What is clear, however, is that this pathway has been labeled in the past as nodoventricular and has decremental properties just

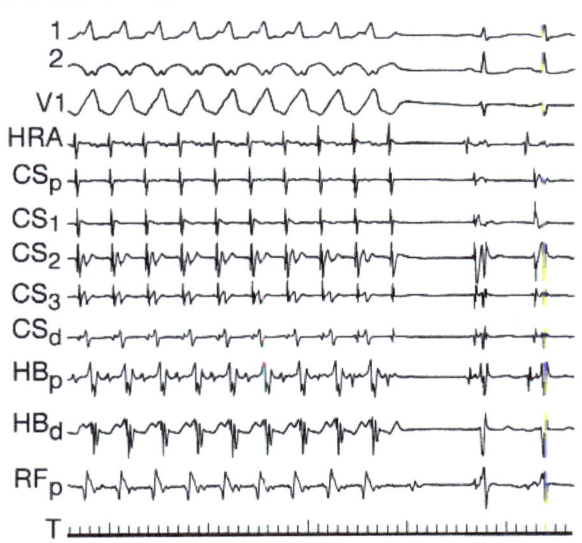

Figure 19.7. Radiofrequency (RF) ablation as a diagnostic tool. This figure demonstrates the termination of antidromic reentry (same patient as in Figure 19.6). The last-recorded complex is atrial followed by no QRS due to interruption of conduction over the rapidly conducting AP. RFp indicates RF delivery catheter recording from the proximal combination. Distal electrode is used for ablation. (Modified from Panatopoulos et al,[25] with permission from John Wiley and Sons.)

like the AV node and somewhat similar response to AV nodal suppressant drugs and atrial pacing. Other features include:

1. Most AFPs are located on the right side. The ventricular end either connects to the right bundle, right ventricle, or both.
2. The typical surface ECG at rest shows no preexcitation due to slow conduction in the AFP such that the sinus impulse completes the ventricular activation.
3. Response to atrial pacing is qualitatively similar but quantitatively different than the AV node. Although both AV node and AFP show progressive delay of conduction, it is generally greater in the AV node, which exposes the ventricular preexcitation.
4. At certain A_1A_2 or atrial pacing CL, block occurs in the AV node with exclusive conduction over the AFP, and frequently the onset of a wide QRS tachycardia shows a LBBB pattern.
5. The exact QRS morphology and axis depend upon the insertion of AFP. Either a normal or left-axis deviation may be observed. Insertion into the right bundle produces a typical LBBB pattern because the ventricular activation is via the HPS.
6. On the other hand, if the insertion is predominantly in the right ventricular myocardium, initial slurring of the QRS may be noted.
7. A retrograde block in the RB above the AFP insertion site will prolong VH > VA due to transseptal conduction and His-bundle activation through the LBB system (see Figure 19.3). The HA usually remains constant because most, if not all, of these patients have functional duality of the AV node with rapid conduction over the retrograde fast pathway.
8. During retrograde impulse conduction via the right side, that is, RB > His > atria, the VA internal is short and characteristically the onset of atrial EGM on His Bundle electrogram (HBE) cannot be clearly separated from the local ventricular EGM. This is in contrast to antridromic reentry using a fast-conducting AV (Kent bundle type), in which the onset of a low atrial EGM is usually identifiable and separated from the local ventricular EGM. This is due to the fact that antidromic (or orthodromic) reentry circuits must traverse the ventricular myocardium to reach the AV node or AP, respectively, and then the atria, which will prolong the VA interval. This observation is only true for atrial deflection on the HBE and applies to all fast-conducting AV nodal pathways,

regardless of the AP location (see Figure 19.3, the first and last 2 complexes of tachycardia).

9. As a general rule, under similar HPS delay or lack thereof, the antidromic reentry using AFP produces long PR/short RP as opposed to antidromic reentry with fast-conducting AV pathway where the PR-RP interval ratio is more balanced and the 2 intervals are closer to each other.
10. A unique feature of AFP-related antidromic reentry is not just the reversal of H-RB sequence, but the fact that RB deflection precedes the QRS in the absence of retrograde RBBB never seen in fast-conducting AV-related antidromic reentry.
11. Many of the distinguishing points discussed under antidromic reentry with fast-conducting AV pathway versus AVNRT (slow–fast) also apply to AFP with a particularly unique feature during the latter as follows: The RB-A (retrograde) interval with AVNRT will be exceedingly short compared with AFP-related antidromic reentry. Since AVNRT and bystander AFP distinction is a real possibility, this feature will be very useful for distinction between the 2, that is, AVNRT versus antidromic reentry using AFP (Schema Idd vs IIb), hence recording of RB potential is highly recommended.
12. Although the vast majority of AFPs are located on the right side, left-sided slow-conducting anterograde pathways do exist. An example is shown in Figures 19.8–19.14 (Schema Ieee, full antidromic circuit with left AFP not shown).

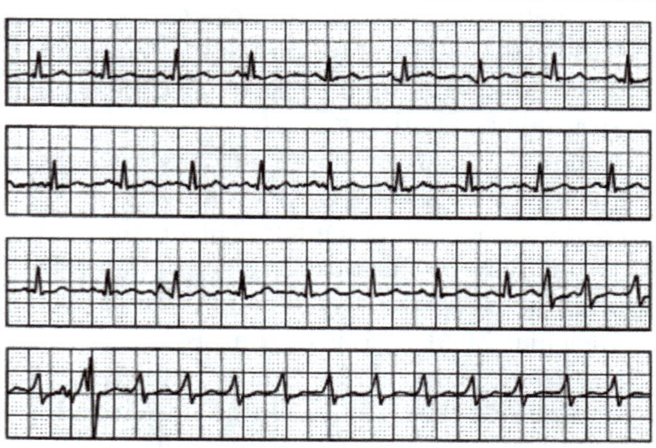

Figure 19.8. Wide QRS tachycardia on monitor (see also Figures 19.9–19.14, same patient). In a young patient with a history of palpitations but who was otherwise healthy, a monitor strip shows the onset of a wide QRS tachycardia (the last 3 complexes in the third panel and all of the fourth panel). Other than the fact that the QRS is slow during the R wave, there is no other clue to the true nature of this arrhythmia.

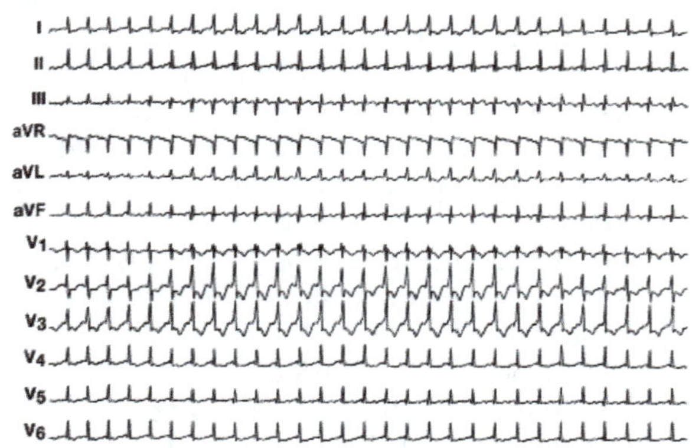

Figure 19.9. From top to bottom: The 12-lead ECG (simultaneous recording of all leads) leads I, II, III, aVR, aVL, aVF, and leads V_1 to V_6. Note that there is a change of QRS morphology, particularly obvious in lead V_2 and V_3. Initially the QRS seems normal, then there is preexcited QRS complex (perhaps fusion), and then near-normal QRS again.

Preexcited circus movement tachycardia using 2 rapidly conducting AP (Schema IIc): Although the exact prevalence of this tachycardia is not known, it probably is as common as the aforementioned 2 forms of antidromic tachycardias (see Figure 19.15, Schema IIc).[6,10,11,13] Some interesting, unique, and diagnostic features of this type of circus movement tachycardia are:

1. These patients harboring 2 or more pathways may have several different wide QRS configurations, such as preexcitation from one and the other pathway for anterograde conduction, Ab-C through NP, VT both from myocardial and BBR origin, and several sites of ventricular-paced complexes.
2. Just like with multiple QRS morphologies, there are several retrograde atrial activation combinations – namely, the 2 eccentric activation sequences with the 2 pathways (see Figure 19.15, Schema IIc) via the NP; fusion between NP and one or the other AP (depending on which is used anterogradely); and ventricular pacing (which could result in atrial fusion from 3 sources, ie, the 2 AP plus NP) – making 6 possible sequences of atrial activation. Accurate interpretation of these observations should be sufficient to diagnose the presence and participation of 2 pathways in this type of reentry.
3. During sinus rhythm, the proximity of one AP, such as anteroseptal AP, may obscure the presence of a coexisting left free wall AP or pathways at other locations. This could be readily

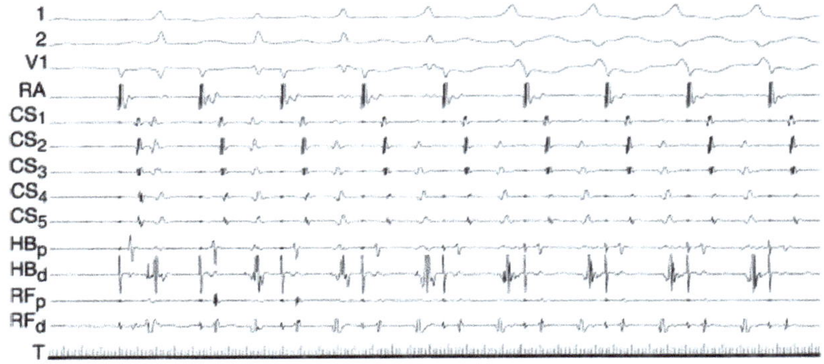

Figure 19.10. Accessory pathway (left-sided slow-conducting). Atrial pacing shows initial minimal to greater preexcitation, disappearance of the HB recording, and fairly long AV interval.

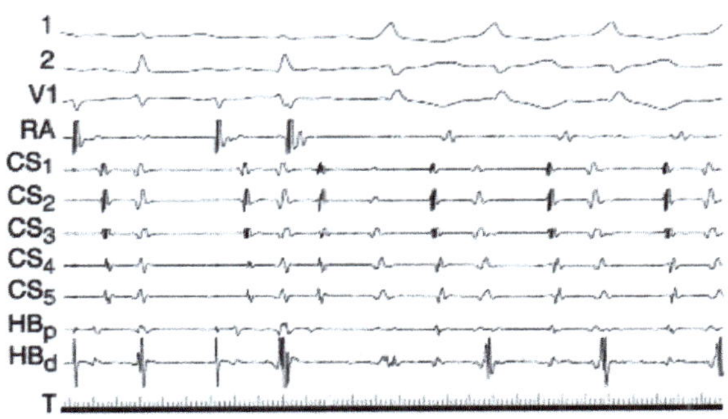

Figure 19.11. Preexcited tachycardia (antidromic reentry). This is an example of a slow-conducting left-sided pathway. No evidence of preexcitation on the basic drive and A_2 starts wide QRS tachycardia (RB and left axis), AV with no HB potential, and retrograde slow conduction with atrial activation sequence HB followed by proximal coronary sinus (NP conduction).

clarified by atrial pacing around the 2 valvular annuli, or simply by right atrial, proximal and distal CS pacing. This also facilitates the induction of orthodromic reentry because anterograde block in the AP closest to the pacing site is more often achieved. However, much of this also depends on the RP properties of the 2 AP.

4. There is a perception (though perhaps not widespread) that circus movement tachycardia using 2 AP (ie, Kent bundle type) seldom involves septal locations, that is, a free wall and septal AP combination is unusual. In this author's mind, this defies any logic, since a free wall and septal (including anteroseptal) combination has been observed many times, and true antidromic circuit with no HPS delay and fast retrograde AV nodal conduction (short HA) is not much different in terms of the proximity of the

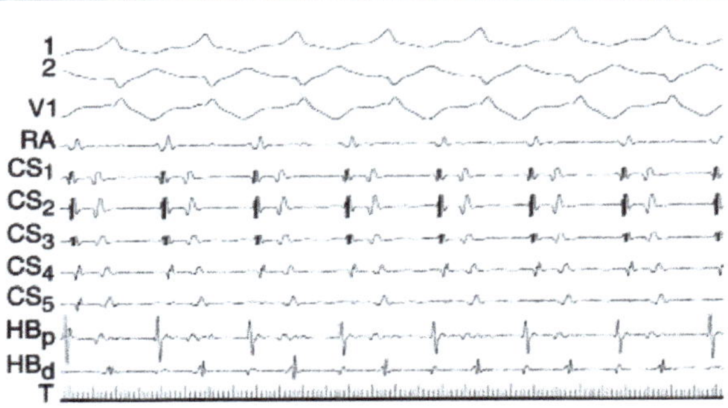

Figure 19.12. Sustained antidromic reentry. Example of the sustained tachycardia using slowly conducting anterograde left-side pathway (QRS configuration of RB and left-axis patterns), longer RP than PR, no HB potential preceding the QRS. Retrograde activation sequence via the NP. Compare this with Figure 19.3.

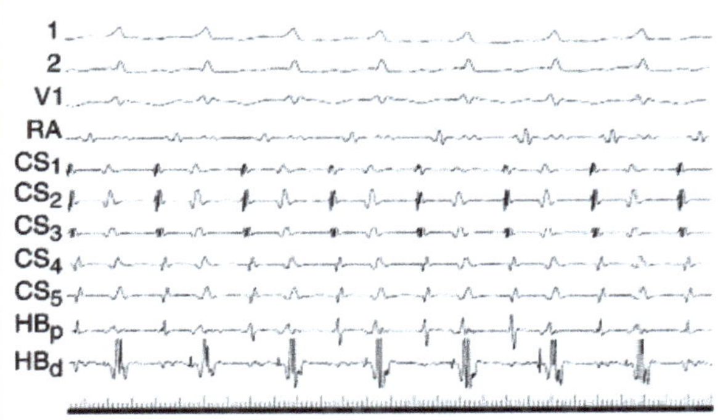

Figure 19.13. Fast–slow AVNRT with bystander slow-conducting AP. The QRS is less preexcited, the HB has reappeared with shorter HV, and the sequence of atrial activation shows proximal coronary sinus recording first (retrograde slow AVN pathway, long RP, or fast–slow AVNRT).

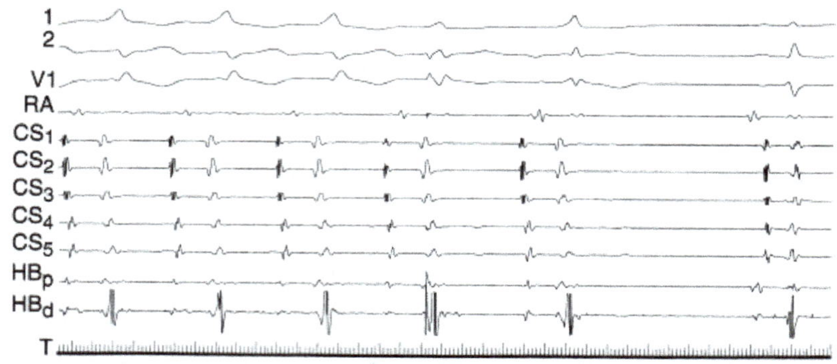

Figure 19.14. Termination of preexcited tachycardia. A single VPC terminates the initial tachycardia (preexcited) and engages retrogradely the slow pathway of the AV node with longer AA cycle; a fusion complex due to bystander AP activation follows for one cycle. This last complex is a reference sinus, with normal HV and no preexcitation. The behavior of this pathway is quite similar to AFP typically located on the right side. It is tempting to speculate on whether this represents an atriofascicular connection or AV connection on the left side with slow-conducting properties.

retrograde limb than a combination of free wall and septal AP. It seems, therefore, the primary determinant of which pathway will be involved in a given circuit is unknown, and one should remain open to the idea that any combination of pathways is possible.

5. Patients with 2 or more AP are not immune to true antidromic reentry – bystander AVNRT with fusion via one or both AP – and all of these may coexist. Hence, in all preexcited tachycardia, conducting delay and block in the HPS should be especially looked for and could be valuable to make an accurate assessment.
6. In addition to the several maneuvers listed earlier to confirm the involvement of the AP in a given circuit, including response to appropriate atrial and/or ventricular premature impulses will help to decipher the various components of the reentry circuit.
7. Finally, ablation of one AP should be followed up by testing for one or more additional AP until all possibilities have been accounted for.

Coexistence of AFP and Kent bundle type of AV pathways should not be ignored (Schema Id): The most likely combination would be AFP used anterogradely and a fast-conducting pathway retrogradely. This has been observed, but the opposite has not been reported.[28]

SUMMARY

This review is an attempt to cover some of the clinical preexcited tachycardias with illustrations and figures. By no means does it encompass all such arrhythmias, but perhaps more frequently seen tachycardias. For further details one must refer the original source. There is a strong possibility that new

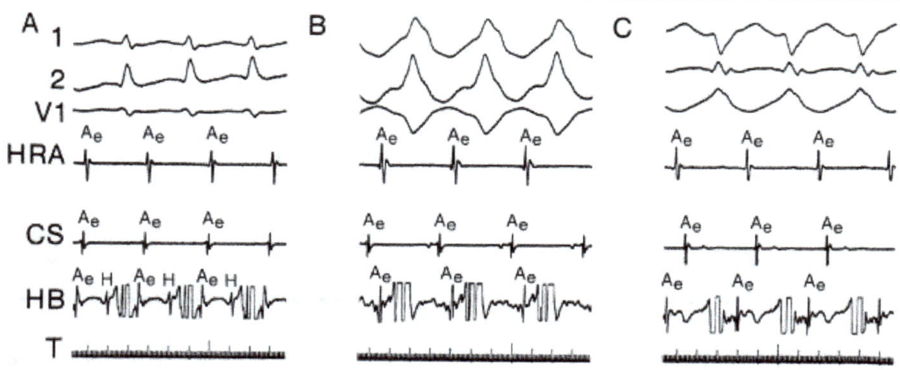

Figure 19.15. Circus movement (preexcited) (Schema IIc) using 2 APs. **Panel A** depicts orthodromic reentry with fusion atrial activation from rapidly conducting anteroseptal AV–AP (Panel B) and left free wall (Panel C). In **Panel B**, the retrograde atrial activation is via left free wall AP and, in **Panel C**, the direction reverses. It is unclear whether retrograde conduction via NP plays any role either during anterograde or retrograde conduction. (Reproduced from Jazayeri et al,[22] with permission from John Wiley and Sons.)

AP will be discovered, and the thinking about the understanding regarding the current role of a pathway could change. Misperceptions and old ideas could be corrected in due course, and our knowledge about APs will continue to evolve.

Correct Answer:

Q1. F (All are true.)

REFERENCES

1. Wellens HJJ, Bär FW, Lie KI. The value of the electrocardiogram in the differential diagnosis of a tachycardia with a widened QRS complex. *Am J Med.* 1978;64:27–33.
2. Akhtar M, Shenasa M, Jazayeri M, Caceres J, Tchou PJ. Wide QRS complex tachycardia. Reappraisal of a common clinical problem. *Ann Intern Med.* 1988;109:905–912.
3. Brugada P, Brugada J, Mont L, Smeets J, Andries EW. A new approach to the differential diagnosis of a regular tachycardia with a wide QRS complex. *Circulation.* 1991;83:1649–1659.
4. Miller JM, Das MK, Arora R, Alberte-Lista C, Wu J. Differential diagnosis of wide complex tachycardia. In: Zipes DP, Jalife J, Zorab R, eds. *Cardiac Electrophysiology: From Cell to Bedside.* 4th ed. Philadelphia: Saunders Elsevier; 2004: 747–757.
5. Akhtar M, Jazayeri M, Avitall B, McKinnie J, Tchou P. Clinical entities and syndromes: Electrophysiological spectrum of wide QRS complex tachycardia. In: Zipes DP, Jalife J, eds. *Cardiac Electrophysiology: From Cell to Bedside.* Philadelphia: Saunders; 1990: 635–646.
6. Gallagher JJ, Pritchett EL, Sealy WC, Kasell J, Wallace AG. The preexcitation syndromes. *Prog Cardiovasc Dis.* 1978;20:285–327.
7. Bardy GH, Packer DL, German LD, Gallagher JJ. Preexcited reciprocating tachycardia in patients with Wolff-Parkinson-White syndrome: Incidence and mechanisms. *Circulation.* 1984;70:377–391.
8. Gallagher JJ, Smith WM, Kasell JH, et al. Role of Mahaim fibers in cardiac arrhythmias in man. *Circulation.* 1981;64:176–189.
9. Josephson ME. Preexcitation syndromes. In: Josephson ME, ed. *Clinical Cardiac Electrophysiology: Techniques and Interpretations.* 2nd ed. Philadelphia: Lea & Febiger; 1993: 311–416.
10. Ward DE, Camm AJ, Spurrell RA. Ventricular preexcitation due to anomalous nodoventricular pathway: Report of 3 patients. *Eur J Cardiol.* 1979;9:111–127.
11. Smith WM, Broughton A, Reiter MJ, et al. Bystander accessory pathway during AV node re-entrant tachycardia. *Pacing Clin Electrophysiol.* 1983;6:537–547.
12. Touboul P, Vexler RM, Chatelain MT. Re-entry via Mahaim fibers as a possible basis for tachycardia. *Br Heart J.* 1978;40:806–811.
13. Trantham JL, Broughton A, Guarnieri T, et al. Two distinct V-H intervals in tachycardia associated with nodoventricular Mahaim fibers. *Circulation.* 1982;66:II-269.
14. Prystowsky EN, Packer DL. Preexcited tachycardia. In: Zipes DP, Jalife J, eds. *Cardiac Electrophysiology: From Cell to Bedside.* Philadelphia: Saunders; 1990: 472–479.
15. Prystowski EN, Klein GJ. *Cardiac Arrhythmias: An Integrated Approach for the Clinician.* New York: McGraw-Hill; 1994.
16. Touboul P, Kirkorian G, Atallah G, et al. Bundle branch reentry tachycardia treated by electrical ablation of the right bundle branch. *J Am Coll Cardiol.* 1986;7:1404–1409.
17. Blanck Z, Dhala A, Deshpande S, et al. Bundle branch reentrant ventricular tachycardia: Cumulative experience in 48 patients. *J Cardiovasc Electrophysiol.* 1993;4:253–262.
18. Blanck Z, Jazayeri M, Dhala A, et al. Bundle branch reentry: A mechanism of ventricular tachycardia in the absence of myocardial or valvular dysfunction. *J Am Coll Cardiol.* 1993;22:1718–1722.
19. Narasimhan C, Jazayeri MR, Sra J, et al. Ventricular tachycardia in valvular heart disease: Facilitation of sustained bundle-branch reentry by valve surgery. *Circulation.* 1997;96:4307–4313.
20. Merino JL, Carmona JR, Fernández-Lozano I, et al. Mechanisms of sustained ventricular tachycardia in myotonic dystrophy: Implications for catheter ablation. *Circulation.* 1998;98:541–546.
21. Akhtar M, Jazayeri M, Sra J, et al. Electrophysiology of wide QRS tachycardia. In: Josephson ME, Wellens HJJ, eds. *Tachycardias: Mechanisms and Management.* Mount Kisco, NY: Futura Publishing; 1993: 215–235.
22. Jazayeri MR, Sra J, Akhtar M. Wide QRS complexes: Electrophysiologic basis of a common electrocardiographic diagnosis. *J Cardiovasc Electrophysiol.* 1992;3:365–393.
23. Akhtar M, Damato AN, Ruskin JN, et al. Antegrade and retrograde conduction characteristics in three patterns of paroxysmal atrioventricular junctional reentrant tachycardia. *Am Heart J.* 1978;95:22–42.
24. Srivathsan K, Gami AS, Barrett R, et al. Differentiating atrioventricular nodal reentrant tachycardia from junctional tachycardia: Novel application of the delta H-A interval. *J Cardiovasc Electrophysiol.* 2008;19:1–6.
25. Panatopoulos P, Deshpande S, Akhtar M, Jazayeri MR. Wide QRS complex tachycardia in the presence of preexcitation: A diagnostic challenge. *Pacing Clin Electrophysiol.* 1997;20:1716–1720.
26. Tchou P, Lehmann MH, Jazayeri M, Akhtar M. Atriofascicular connection or a nodoventricular Mahaim fiber? Electrophysiologic elucidation of the pathway and associated reentrant circuit. *Circulation.* 1988;77:837–848.
27. Murdock CJ, Leitch JW, Klein GJ, et al. Epicardial mapping in patients with "nodoventricular" accessory pathways. *Am J Cardiol.* 1991;68:208–214.
28. Lerman BB, Waxman HL, Proclemer A, Josephson ME. Supraventricular tachycardia associated with nodoventricular and concealed atrioventricular bypass tracts. *Am Heart J.* 1982;104:1097–1102.

SECTION 3

Atrial Flutter, Atrial Fibrillation

CASE STUDY

20

Typical Cavotricuspid Isthmus-Dependent Flutter

Dipen Shah, MD

An 80-year-old male patient presented with fatigue and vertigo, and was found to have atrial flutter with a fast and irregular ventricular rate (130 bpm) and signs of heart failure. He had a history of a transient ischemic attack and 2 past episodes of atrial fibrillation (AF), which had been cardioverted electrically. His ECG is shown in Figures 20.1 and 20.2. He was anticoagulated and administered beta-blockers and diuretics, and hospitalized for catheter ablation of his atrial flutter. Intracardiac thrombi were excluded by a transesophageal echocardiogram and an electrophysiology (EP) study followed by radiofrequency (RF) catheter ablation was performed.

The ECG in Figure 20.1 shows atrial arrhythmia at a rate with variable atrioventricular (AV) conduction and complete right bundle branch and left axis deviation resulting with a rapid ventricular rate.

Question No. 1:

The atrial arrhythmia shown in this ECG is:

A. Clockwise typical atrial flutter.

B. Counterclockwise typical atrial flutter.

C. Atypical atrial flutter.

This intracardiac tracing (see Figure 20.2) at a sweep speed of 100 mm/s shows low right atrial free wall activation recorded from a vertically oriented quadripolar catheter (bipoles OD1 and OD2; OD1 is the lower of the 2), and a quadripolar RF ablation catheter (bipoles Abl dist and Abl prox, as well as Abl uni). Also shown at the top are surface ECG leads II, V_6 and aVF. Note that the atria are activated regularly at a cycle length of 250 ms with activation in OD2 preceding OD1. In view of the vertical position of the quadripolar catheter (Figure 20.3), this sequence indicates activation descending down the antero-lateral right atrial free wall, which is typical of counterclockwise typical atrial flutter. This arrhythmia is characterized by a counterclockwise-rotating macroreentrant circuit around the tricuspid valve annulus (when viewed from the left anterior oblique (LAO) perspective). The activation sequence has traditionally been documented with a multipolar (10–20 electrode) preformed loop-like catheter confirming to the tricuspid annulus. However, an appropriately placed (see Figure 20.3) quadripolar catheter can equally effectively document descending activation of the right atrial free wall. The correlation of low right atrial free wall activation with the beginning of the "plateau" segment of the surface ECG in inferior leads (II, III, and aVF) and that of the coronary sinus (CS) ostium (or the septal right atrium nearby the His bundle, see Figure 20.2) with the end of the "plateau" phase confirms the diagnosis of counterclockwise typical (cavotricuspid isthmus-dependent) atrial flutter. The ablation

Practical Electrophysiology © 2015 Jasbir S. Sra Masood Akhtar, Andrea Natale, David J. Wilber
Cardiotext Publishing, ISBN: 978-1-935395-17-1

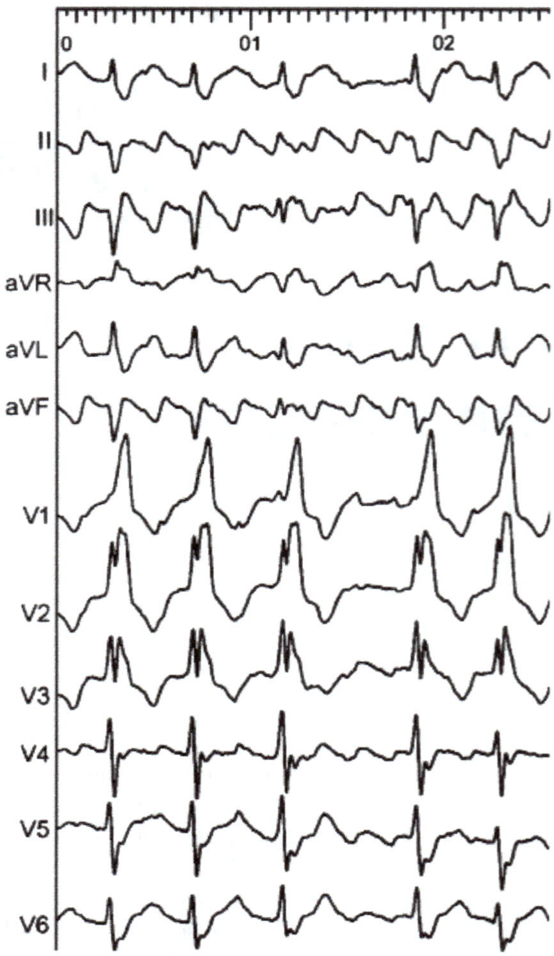

Figure 20.1. 12-lead ECG.

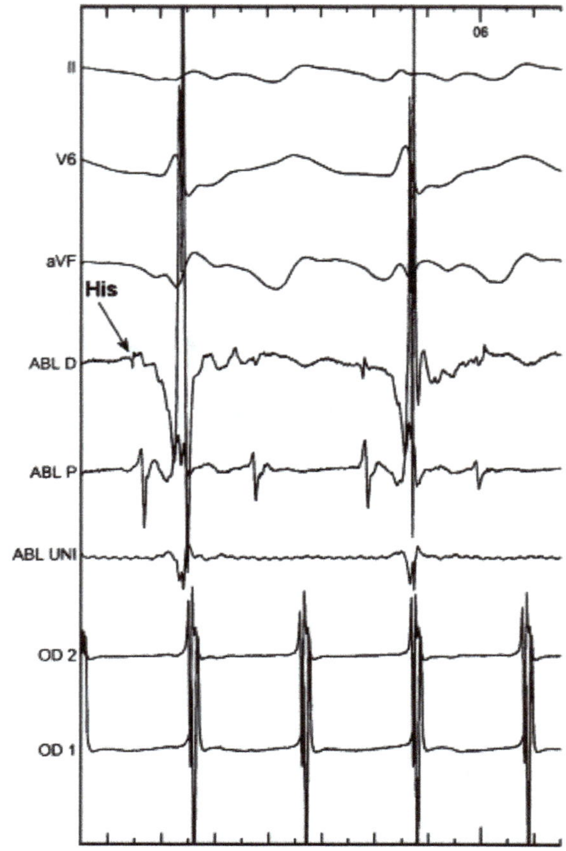

Figure 20.2. Intracardiac tracing during electrophysiology study.

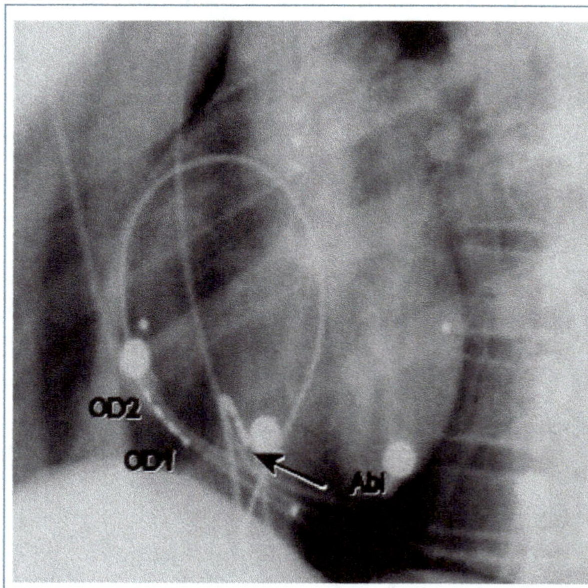

Figure 20.3. Fluoroscopic images of catheter positions during electrophysiologic evaluation and ablation of typical flutter. *(Reprinted from Shah et al[6] with permission from Wolters Kluwer Health.)*

catheter can be conveniently moved around to document this correlation.

Typical flutter is indicated in this ECG by:

- Negative deflections "saw-tooth" morphology in leads II, III, aVF, and V6;
- Narrow biphasic initial positive deflections in lead V1.

Question No. 2:

Clockwise typical atrial flutter is characterized by:

A. Negative deflections in V_1 and positive deflections in inferior leads II, III, and aVF.

B. A higher incidence compared to counterclockwise atrial flutter.

C. A different circuit compared to counterclockwise typical atrial flutter.

Typical atrial flutter is a macroreentrant circuit around the tricuspid valve annulus passing through the cavotricuspid isthmus and posteriorly anchored by a vertical zone of block extending from near the entrance of the superior vena cava (SVC) to the junction of the right atrium with the inferior vena.[1] The circuit of clockwise typical atrial flutter is the same as that for counterclockwise typical atrial flutter, therefore, successful catheter ablation of the cavotricuspid isthmus eliminates both forms of typical atrial flutter.

The surface ECG morphology, negative deflections in V_1 and positive deflections in inferior leads II, III, and aVF of typical counterclockwise flutter is the result of descending activation of the right atrial free wall and ascending activation of the septum and most of the left atrium. Conversely, the ECG of clockwise atrial flutter is the result of ascending activation of the right atrial free wall and descending activation of the septum and most of the left atrium. The ECG morphology of clockwise flutter is less distinctive and more variable than counterclockwise typical atrial flutter.

Clockwise typical atrial flutter is less commonly observed (10%) compared to counterclockwise typical atrial flutter (90%).

Question No. 3:

The diagnosis of typical atrial flutter can be established by: (Figures 20.4 and 20.5)

A. Entrainment mapping.

B. Activation mapping.

C. Successful catheter ablation (ie, flutter termination during RF energy delivery).

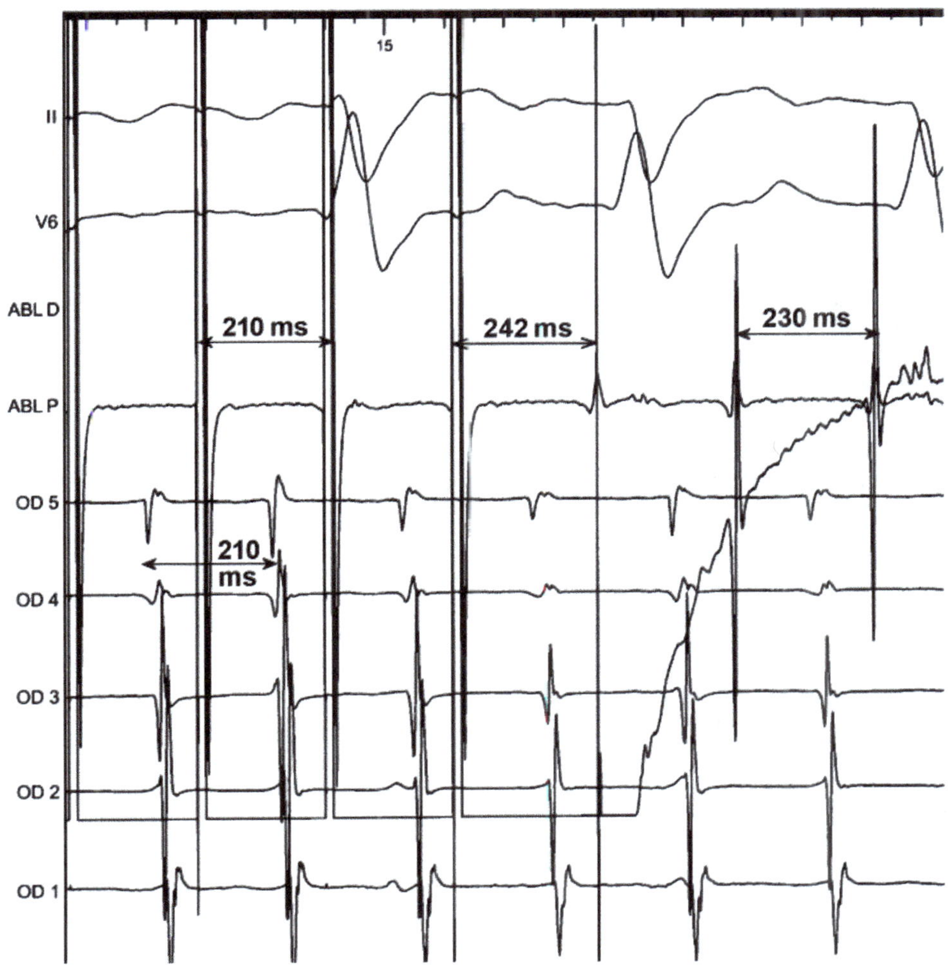

Figure 20.4. Entrainment (at a cycle length of 210 ms) from the distal bipole of the ablation catheter placed in the cavotricuspid isthmus. The post pacing interval is 242 ms for a baseline cycle length of 230 ms and during entrainment neither the intracardiac nor surface ECGs show any change apart from cycle length acceleration, indicating concealed entrainment from a critical part of the reentrant circuit.

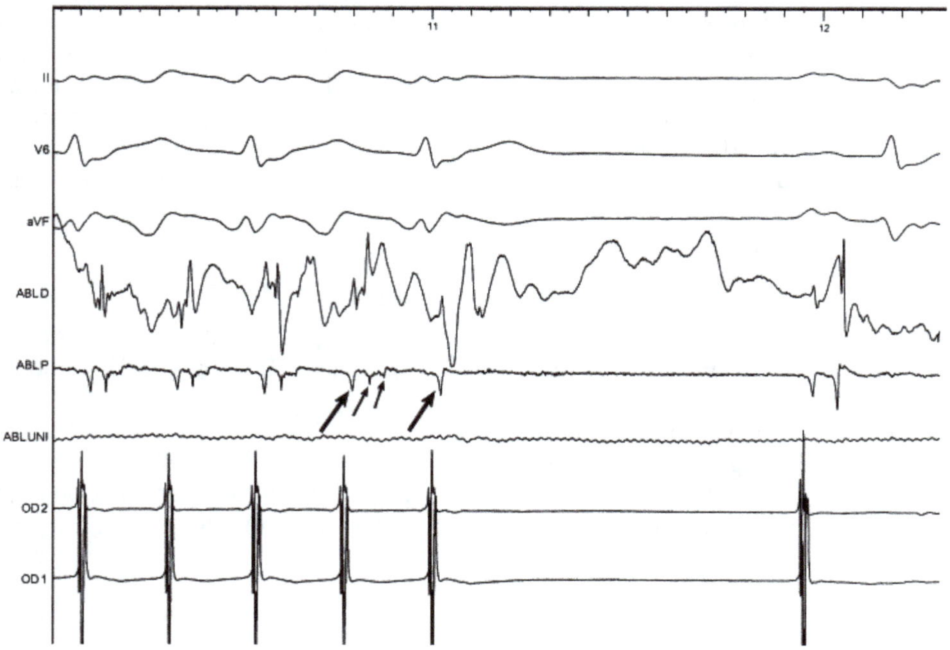

Figure 20.5. Termination of typical flutter during RF delivery in the cavotricuspid isthmus. Note the presence of a 3-component electrogram (EGM) from the proximal bipole of the ablation catheter and a complex EGM from the distal bipole during flutter. Termination of flutter coincides with block after the first components in both bipoles during RF delivery from the distal electrode. A double potential is observed in sinus rhythm from the ablation bipoles.

Question No. 4:

What is the most common mechanism of flutter termination during RF delivery at the cavotricuspid isthmus?

A. Stable cavotricuspid isthmus conduction block.

B. Transient conduction block or transient conduction slowing through the cavotricuspid isthmus.

C. Ectopic-induced termination of reentry

The most common mechanisms of termination of typical atrial flutter during RF delivery in the cavotricuspid isthmus are either transient conduction block or transient conduction slowing. Therefore, additional RF catheter ablation within the cavotricuspid isthmus is usually necessary even after flutter termination in order to produce stable conduction block.

Conduction block across the cavotricuspid isthmus is indicated by:

- Local double potentials;[2]
- Absence of an activation wave front penetrating the line of ablation.

During pacing from either side of the cavotricuspid isthmus, the presence of a continuous corridor of widely separated double potentials extending from the ventricular to the inferior vena caval edge of the isthmus indicates conduction block. The presence of a single or complex EGM bridging the interval between adjacent double potentials indicates a conducting gap through the ablation line.[3] The double potentials can be easily validated by differential pacing.

Because double potentials are recorded locally from the lesion, local EGM criteria based on double potentials and differential pacing are more sensitive in distinguishing slow conduction from conduction block than indirect criteria such as activation sequences of the right atrial free wall or of the septal right atrium.[4]

Figure 20.6 shows the achievement of conduction block during RF energy delivery while pacing from the low lateral right atrium. The local timing changes from 60 to 150 ms with the appearance of widely separated double potentials (arrows).

The double potentials are validated by differential pacing (Figures 20.7A and 20.7B). Comparison of the local activation timing of both components during pacing from the distal bipole of the right atrial catheter (OD1, see Figure 20.7A) followed by pacing from the proximal bipole (OD2, see Figure 20.7B) shows that moving the pacing site away from the ablation lesion (ie, changing from

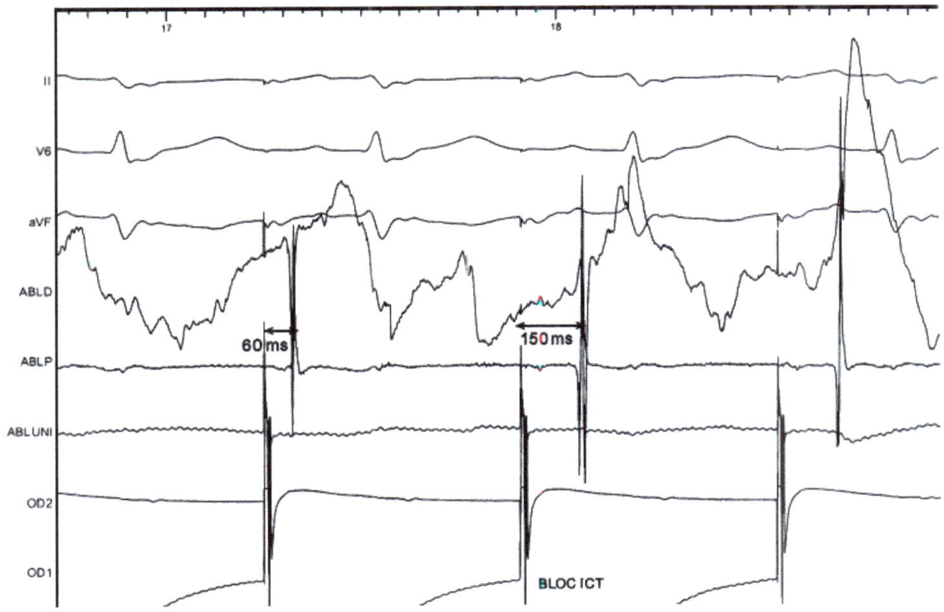

Figure 20.6. Conduction block during ablation.

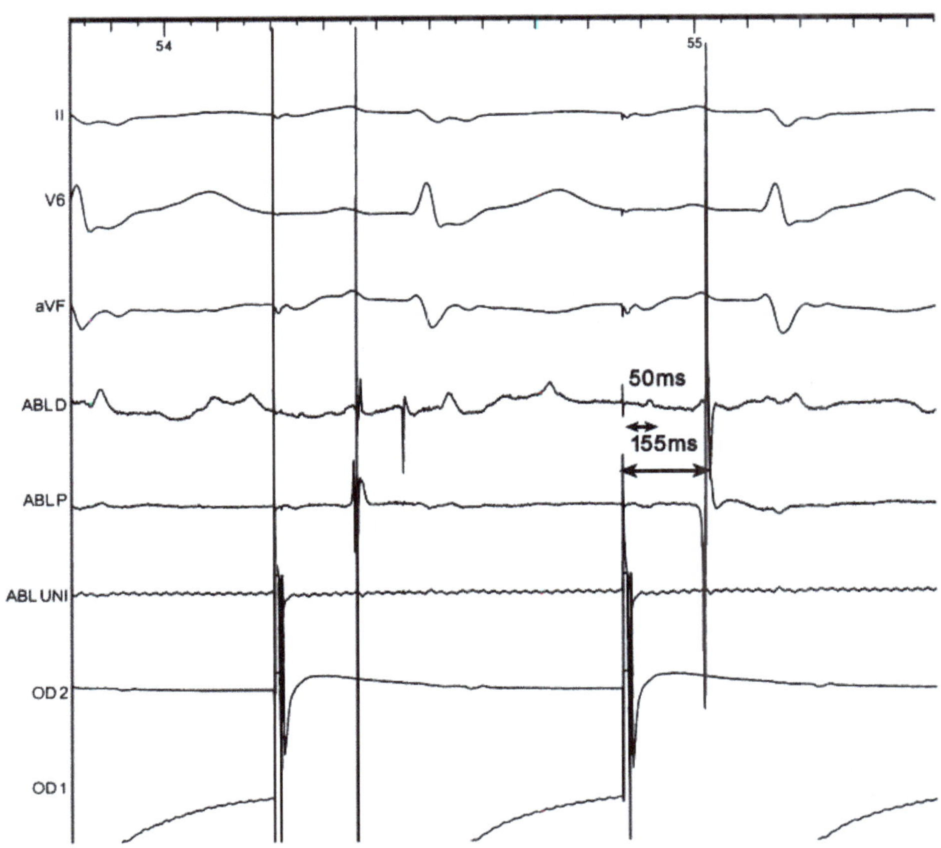

Figure 20.7A. Conduction block during pacing.

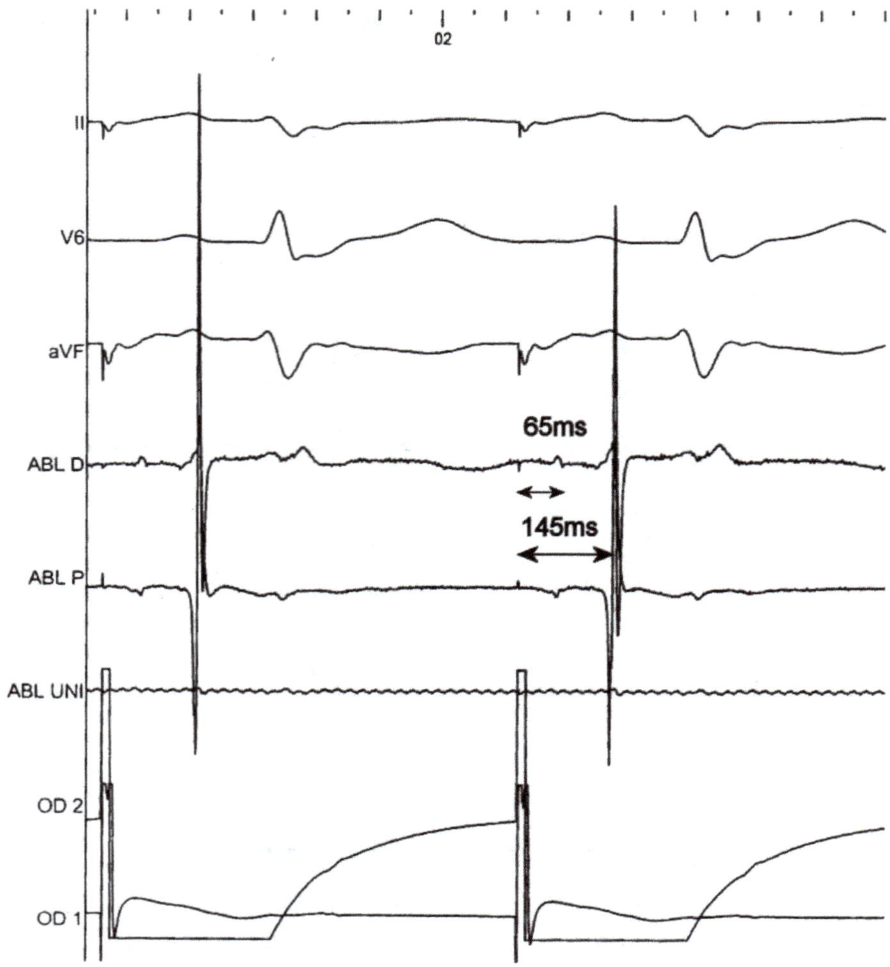

Figure 20.7B. Conduction block during pacing.

OD1 to OD2) increases the activation time to the first component (from 50 ms to 65 ms), but reduces the activation time to the second component (from 155 ms to 145 ms) because the second component is activated by the wave front going around the tricuspid annulus rather than through the cavotricuspid isthmus.[5]

Finally, bidirectional conduction block can be confirmed by pacing from the CS ostium from the distal bipole of the ablation catheter and recording a descending sequence on the low right atrium (OD1 succeeds OD2).[6]

Question No. 5:

The most important parameters for a sensitive and specific evaluation of conduction across the cavotricuspid isthmus in descending order are:

A. A large coverage of the tricuspid annulus with a 10 or more polar mapping catheter.

B. Assessing local activation on the lesion line.

C. Pacing from close to the lesion line.

Pacing from close to the lesion line minimizes the time to activate the opposite side of the lesion, if there is conduction through the lesion and therefore maximizes the sensitivity of detecting residual slow conduction through the lesion.

Correct Answers:

Q1. B

Q2. A

Q3. All 3 can be diagnostic, individually or separately. Clearly the greatest accuracy can be achieved by combining all 3 criteria.

Q4. B

Q5. C, B, A

QUESTION FROM THE EDITORS AND AUTHOR'S REPLY

Question. What are the causes of ablation failure and what precautions can be taken to prevent recurrence of atrial flutter after ablation?

Answer: The most frequent reason for procedural failure used to be (and still remains) unfavorable anatomy. A markedly concave cavotricuspid isthmus with a prominent Eustachian ridge makes it difficult to reach and target conducting myocardium at the bottom of the concave trough. Within a concave isthmus, the ablation electrode cannot be cooled efficiently by local blood flow thereby limiting the amount of RF energy that can be delivered by a nonirrigated 4-mm electrode tip catheter. Therefore, the availability of irrigated RF ablation catheters as well as 8-mm tip ablation catheters (which allow the delivery of higher power than a 4-mm conventional electrode) has reduced the procedural failure rates from 10–15% to 1%. Alternatively, the right ventricular or inferior vena caval edge of the cavotricuspid isthmus may be difficult to reach and a long sheath may be required. Unstable catheter contact may play an important role and can also be improved with a long sheath. Occasionally, achieving cavotricuspid isthmus conduction block may be difficult because of fibrosis in patients with a surgically repaired or operated tricuspid valve.

It is necessary to ensure persistent conduction block in order to avoid recurrence. Surveillance for 20 to 30 minutes with re-ablation if necessary *after* the achievement of isthmus conduction block is therefore an essential part of the procedure. Conduction recovery is frequently indicated by subtle changes and careful assessment of cavotricuspid isthmus conduction is often necessary. Adequate RF power delivery (35–40 W with an open irrigated tip catheter), good stable contact, and the documentation of prompt local EGM changes at ablation sites are all important for creating effective stable lesions.

REFERENCES

1. Shah DC, Jaïs P, Haïssaguerre M, et al. Three dimensional mapping of the common atrial flutter circuit in the right atrium. *Circulation.* 1997;96(11):3904–3912.
2. Shah DC, Takahashi A, Jaïs P, et al. Tracking dynamic conduction recovery across the cavotricuspid isthmus. *J Am Coll Cardiol.* 2000;35(6):1478–1484.
3. Shah DC, Takahashi A, Jaïs P, et al. Local electrogram-based criteria of cavotricuspid isthmus block. *J Cardiovasc Electrophysiol.* 1999;10(5):662–669.
4. Cauchemez B, Haissaguerre M, Fischer B, et al. Electrophysiological effects of catheter ablation of inferior vena cava-tricuspid annulus isthmus in common atrial flutter. *Circulation.* 1996;93(2):284–294.
5. Shah DC, Haïssaguerre M, Jaïs P, et al. High density mapping of activation through an incomplete isthmus ablation line. *Circulation.* 1999;99(2):211–215.
6. Shah DC, Haïssaguerre M, Takahashi A, et al. Differential pacing for distinguishing block from persistent conduction through an ablation line. *Circulation.* 2000;102(13):1517–1522.

CASE STUDY

21

Left Atrial Flutters After AF Ablation

Jasbir S. Sra, MD

A 54-year-old man with prior failed antiarrhythmic drug therapy and ablation for long-standing persistent atrial fibrillation (AF) presents with recurrent palpitations.

Question No. 1:

The ECG shown in Figure 21.1 suggests:

A. Typical atrial flutter (AFl).
B. Atypical left AFl.
C. Atrial fibrillation.
D. A and B are possible.

Intracardiac recordings of the tachycardia are shown in Figure 21.2.

Question No. 2:

The next best option would be:

A. Create a three-dimensional (3D) map of the tachycardia.
B. Entrain the flutter from right atrium (RA).
C. Entrain flutter from right and left atria.
D. Both A and C.
E. Perform cavotricuspid isthmus ablation.

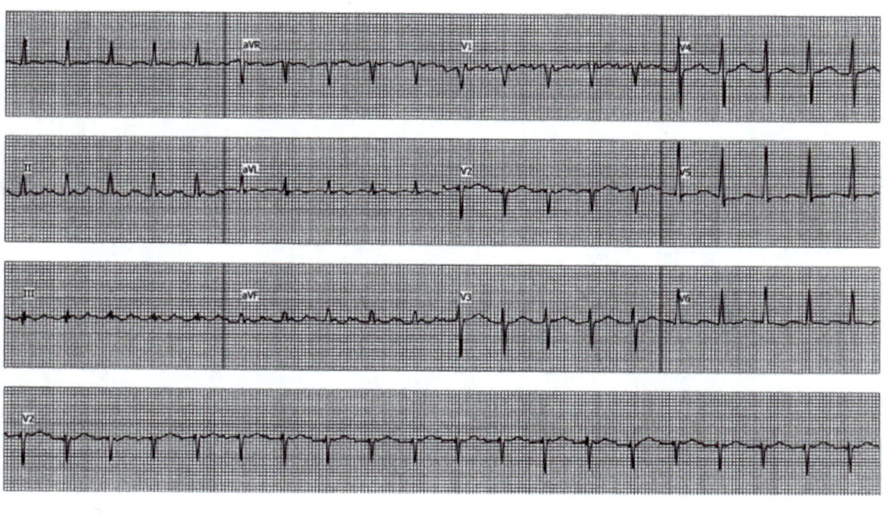

Figure 21.1. Rhythm strips recorded at presentation.

Practical Electrophysiology © 2015 Jasbir S. Sra, Masood Akhtar, Andrea Natale, David J. Wilber
Cardiotext Publishing, ISBN: 978-1-935395-17-1

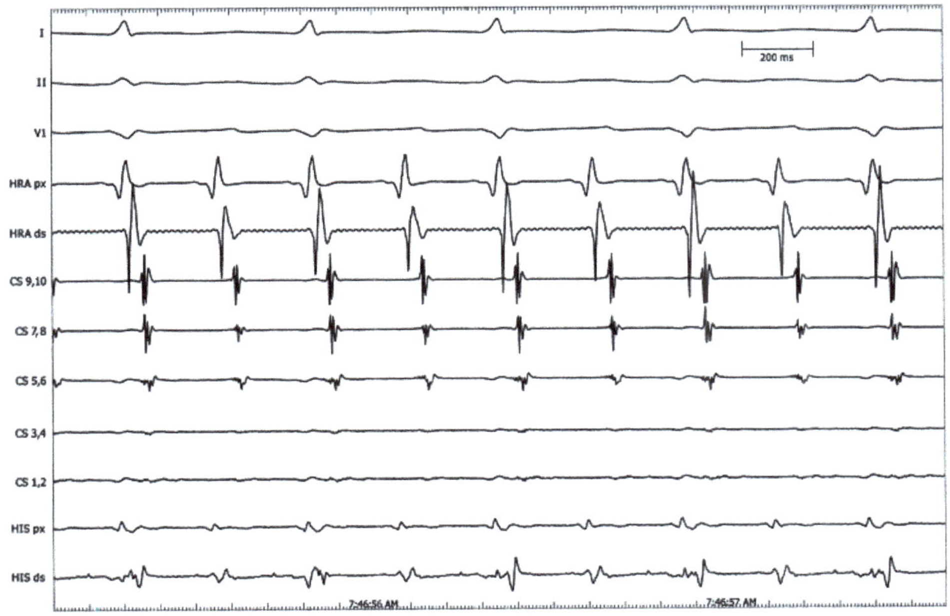

Figure 21.2. Intracardiac tracings of the tachycardia. Cycle length (CL) = 274 ms.

Question No. 3:

High right atrium (HRA) pacing suggests:

A. Tachycardia is right sided.

B. Tachycardia is left sided.

Question No. 4:

Based on Figures 21.4 and 21.5, ablation at which site is likely to be successful?

Question No. 5:

Figure 21.6A suggests that line of block across the mitral isthmus is:

A. Complete.

B. Incomplete; more ablation is needed in that area.

DISCUSSION

The patient in the current case report presents following AF ablation. Patients with AFl following extensive ablation for AF usually have uncommon types of ECGs; hence it is usually difficult to differentiate right versus left AFl at times. Intracardiac tracings in Figure 21.2 suggest that high right atrial and His bundle atrial deflections are the earliest, which goes against typical cavotricuspid isthmus ablation. Pacing from the HRA further corroborates this, as the postpacing interval (PPI) is significantly longer than the CL of the tachycardia (Figure 21.3).

Figures 21.4 and 21.5 depict pacing from different locations, including the mitral isthmus region. Entrainment is best seen at the mitral isthmus region. The best PPI is seen during pacing at mitral isthmus region suggesting that it is perimitral AFl. Ablation at the mitral isthmus region (see Figure 21.6) terminated the tachycardia.

As depicted in Figure 21.6A, ablation is performed from the mitral annulus to the left inferior pulmonary vein (PV). Pacing was then performed from the left atrial appendage to assess conduction across the mitral isthmus region. In the figure, both proximal and distal CS leads are activated simultaneously, suggesting conduction across the mitral isthmus. However, close observation on the CT-fluoro registered image suggests that distal poles of the CS are distal to the line of block. In Figure 21.6B, the CS is pulled back and proximal CS is activated earlier, suggesting complete line of block across the mitral isthmus. This is further confirmed in Figure 21.6C, where the ablation catheter is placed at the mitral annulus distal to the CS and pacing performed from the proximal and distal CS poles. Time from the distal CS pole to the ablation catheter recording (due to longer time the impulse has to travel) is longer than that during pacing from the proximal CS.

Both atrial tachycardias (AT) and AFls seen in the left atrium (LA) post-AF ablation have varying

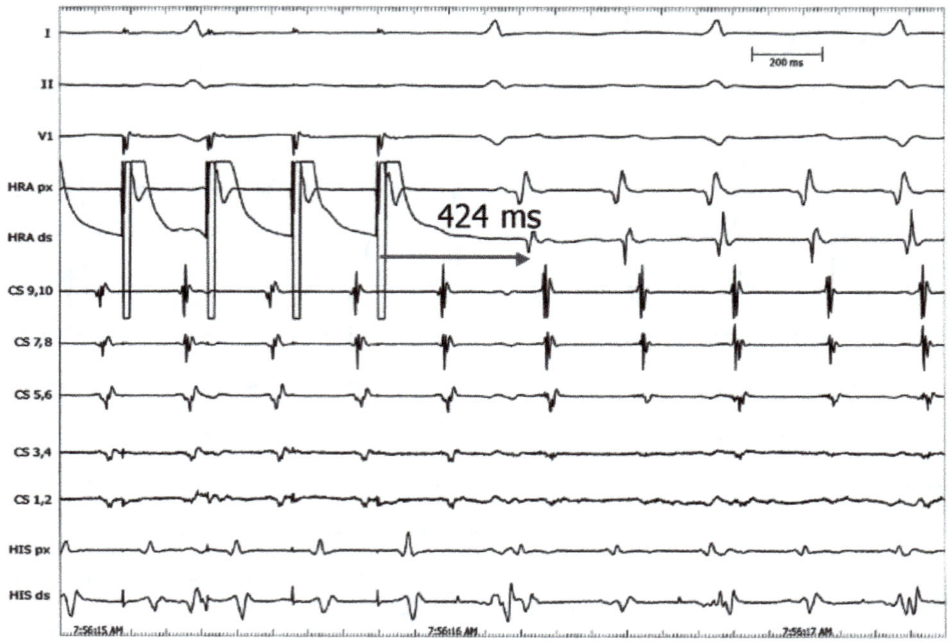

Figure 21.3. High right atrial pacing during tachycardia.

characteristics. Both macro- and microreentrant tachycardias are commonly seen, but at times focal tachycardias are not uncommon. 12-lead ECG for P-wave morphology, entrainment and activation mapping has been used to identify the mechanism and sites of these arrhythmias. Many of these tachycardias could originate from perimitral areas, could be roof-dependent, or could occur due to conduction through the previous ablation lines. A systematic approach is needed to identify the

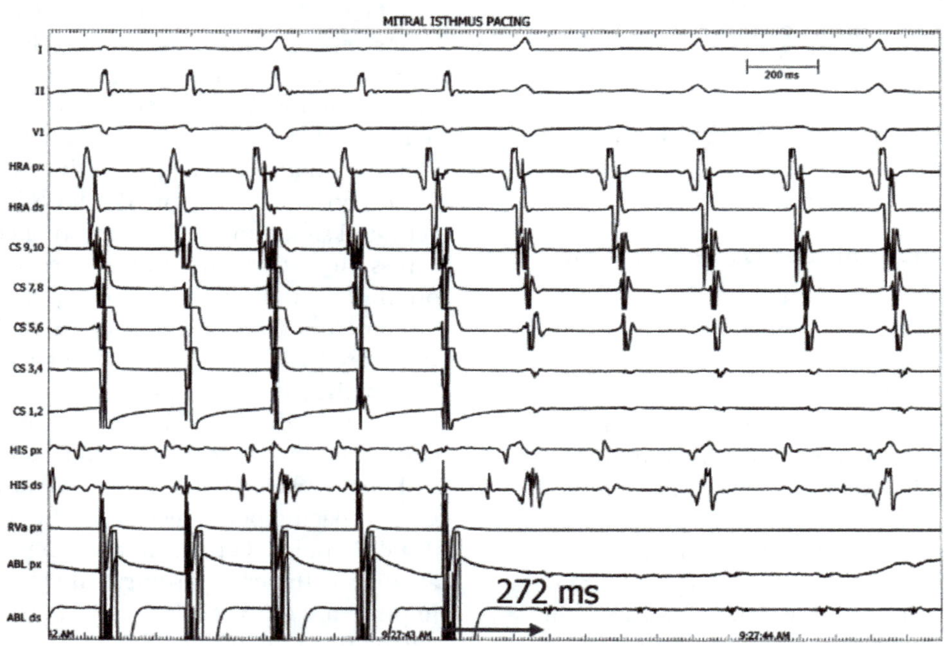

Figure 21.4. Mitral isthmus pacing.

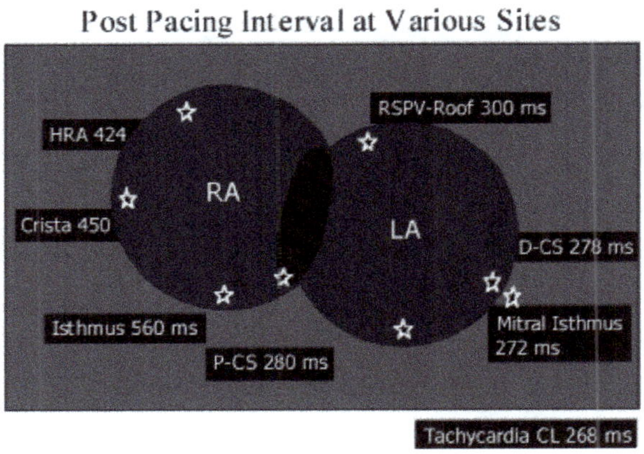

Figure 21.5. Pacing intervals at different locations.

mechanism and successfully ablate these complex arrhythmias.

Approach to Atrial Arrhythmias from the LA Following AF Ablation

Surface ECG Identification of Ectopic Foci Triggering AF: Surface ECG can be used to identify the location of the ectopic beats that initiate AF. Similar process can be used to try to identify the location of atrial arrhythmias post-ablation. Positive P waves in V_1 in the absence of typical cavotricuspid isthmus-dependent flutter (CTI-AFl) typically suggest LA-AFl. Bochoeyer et al compared the 12-lead ECG between those with LA-AFl and those with counterclockwise (CCW) CTI-AFl (Table 21.1). In all patients with the LA septal circuit pattern, inferior leads had almost flat-line recordings. The maximal amplitude in any of the 3 inferior leads (II, III, and aVF) was significantly smaller for those with mitral annulus or posterior wall scar circuits compared with those with CCW CTI-AFl (1.6 ± 1 vs 2.68 ± 0.7 mV, $p < 0.05$).

Definitions for Intracardiac Mapping

Several terminologies have been used to define atrial arrhythmias. The Heart Rhythm Society and European Society of Cardiology consensus document defined AT as macroreentrant and focal. These definitions become problematic in the context of prior ablation, especially extensive ablation in the context of prior long-standing persistent AF ablation.

Other Definitions: Other definitions used include:

1. **Electrically silent areas (ESA)**, defined as amplitude < 0.05 mV. These areas should be tagged as scar during mapping.
2. **Slow conduction**, defined as complex and fractionated electrical activity of > 50 ms duration.

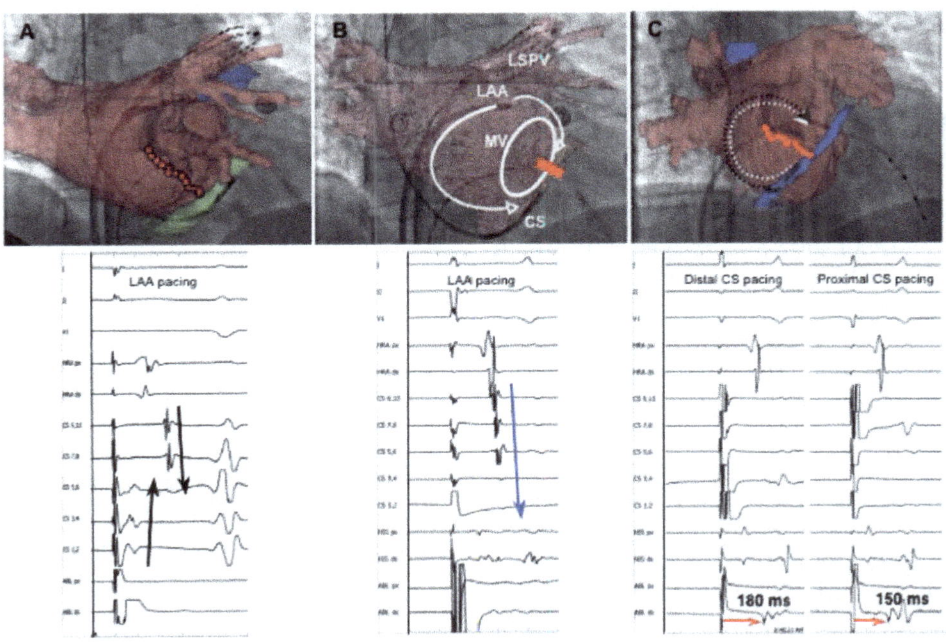

Figure 21.6. Pacing from left atrial appendage and coronary sinus (CS).

Table 21.1. Surface Electrocardiographic Features: Comparison of Mean Voltage (in mV)

	RA CCW (n=38)	LA Septal Circuit (n=11)	LA Mitral Annulus (n=3)	LA Posterior Wall Macroreentry (n=8)
Inferior leads*	2.62 ± 0.7	0.08 ± 0.1†	1.2 ± 0.35†	1.6 ± 1†
Lead I	0.005 ± 0.4	0	–0.3 ± 0.57‡	0.9 ± 1†
V_1	1.97 ± 0.35	2.3 ± 0.4§	1.6 ± 0.35§	1.6 ± 1.2§

*Mean ± SD of maximal amplitude in lead II, III, or aVF.
†$p < 0.05$ by *t* test compared with CCW.
‡Only one patient had negative flutter wave in lead I.
§p = NS compared with CCW.
CCW, counterclockwise; LA, left atrial; RA, right atrial.
(Adapted from Bochoeyer et al[6] with permission from Wolters Kluwer Health.)

3. **Zone of block**, defined by double potentials separated by ≥ 50 ms.
4. **Region of local block**: collision of 2 wavefronts traveling in different directions separated temporarily by 50 ms.
5. **Macro- and microreentrant circuit**: minimum diameter of the circuit being > 3 cm, and small reentrant circuit being defined as < 3 cm along with activation covering the entire CL.

STEP-BY-STEP APPROACH TO INTRACARDIAC MAPPING

Cycle Length Variability

Regularity of the tachycardia should be initially assessed. Recordings from the RA, LA and CS have been used for this purpose. CL variability greater than 100 ms in the RA with concomitant variation of < 20 ms in the LA suggests the tachycardia circuit is in the LA. In the case of CL variability > 15% in the LA, a focal tachycardia is deemed to be more likely.

In addition to the points described above, if the tachycardia is stable, further steps to exclude RA flutters include: < 50% of the CL during activation mapping, and PPI of ≥ 40 ms during pacing at 3 different sites in the RA excluding the septum (as depicted in this case). Once the RA tachycardias have been deemed less likely, the focus should be to differentiate macro- and macroreentry, as well as focal tachycardias.

Once the RA tachycardia has been eliminated as the cause, activation, entrainment and voltage maps are used to identify the focus and define the circuits. However, there are limitations to both entrainment and activation maps due to prior ablation, which include:

- High-output pacing leading to capture of adjacent myocardium;
- Change in the circuit or degeneration into AF during entrainment mapping; and
- Difficulty in interpreting the activation maps due to prior ablation/elimination of RA-AFl circuit.

Identifying Macro/Microreentry or Focal Tachycardia

In one study where mapping was performed in 155 patients, the mechanism was deemed to be reentry in 137 (88%) of the 155 tachycardias (Figure 21.7). Mitral isthmus was the most common site for ablation, accounting for 50 (42%) of the 120 LA tachycardias. Left atrium roof and septum accounted for 24 (20%) and 16 (13%), respectively. Twenty-one patients were deemed to have microrentry. Of these, the anterior periosteal aspect of the left superior PV was targeted in 6 (29%), the CS in 5 (24%), the anterior wall in 3 (14%), and the antrum of the right pulmonary veins (RPVs) in 2. Among the 18 focal tachycardias, 15 (83%) had their origins in one of the PVs. The study also suggested that more than 90% of these tachycardias were gap-related, traversing a prior ablation line. Multiple loop circuits were present in 22% of the tachycardias.

STUDY CASES

Macroreentry

To diagnose macroreentry, a reference channel on one of the mapping systems or recordings systems has been used for guidance. The number of potential macroreentrant circuits in the LA is usually limited to perimitral and roof-dependent tachycardias.

Differentiating Perimitral and Roof-Dependent Reentry: If there is a consistent activation seen on the CS recordings—proximal to distal or distal to proximal—demonstration of sequential

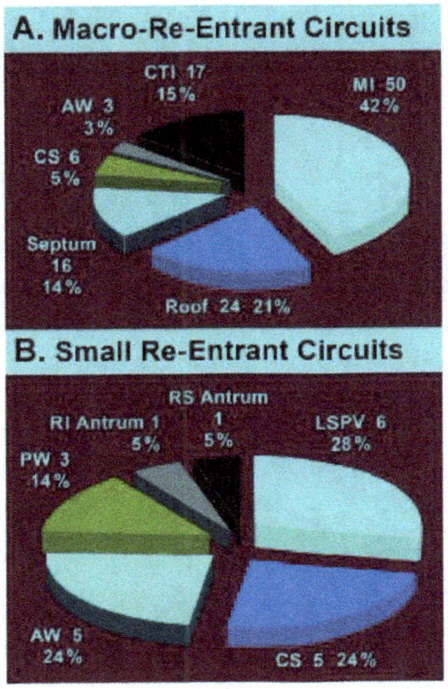

Figure 21.7. **A**: Pie chart showing the frequency and location of macroreentrant ATs. **B**: Pie chart showing the frequency and location of small reentrant ATs. AW, anterior wall; CS, coronary sinus; CTI, cavotricuspid isthmus; LSPV, around the left superior pulmonary vein; MI, mitral isthmus; PW, posterior wall; RI, right inferior; RS, right superior. (Reproduced from Chae et al[3] with permission from Elsevier.)

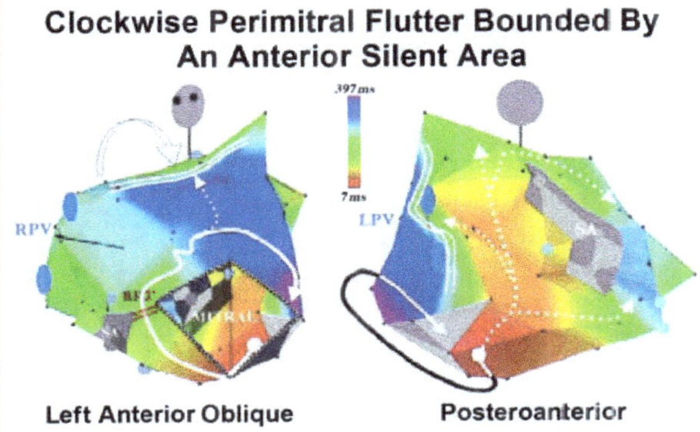

Figure 21.9. Two views of LA. Clockwise perimitral flutter (CL, 400 ms) was diagnosed by 3D mapping system. Circuit was bounded laterally by a silent area (SA), in gray. Posteroanterior view shows zone of block anchored in left pulmonary vein (LPV) and posterior SA with bystander activation proceeding superiorly and inferiorly, colliding on its right lateral aspect. Ablation was achieved with a 2-minute radiofrequency (RF) application to connect mitral annulus to anterior SA. Solid arrows indicate circuit loop(s); dotted arrows indicate passive activation; and double line indicates zone of block. RPV, right pulmonary vein. (Reproduced from Jaïs et al[1] with permission from Wolters Kluwer Health.)

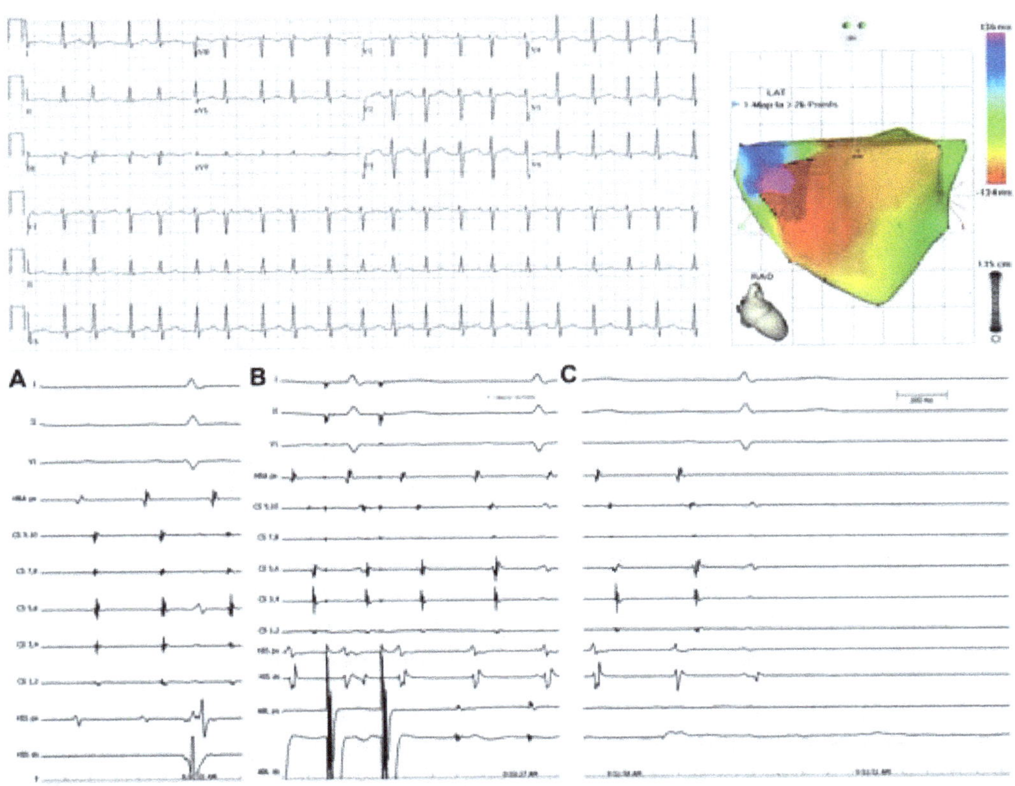

Figure 21.8. Macroreentry (figure-of-8) involving the antral area of right superior PV. Activation and entrainment mapping confirm the location. Ablation at this site terminated the tachycardia.

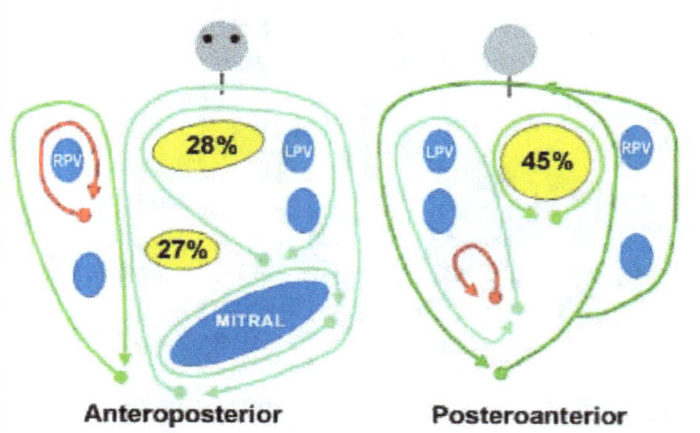

Figure 21.10. Schematic of locations of SAs and their percentages with circuits encountered in LA. Note that nearly half of 11 SAs were posteriorly located. LPV, left pulmonary vein; RPV, right pulmonary vein. (Reproduced from Jaïs et al[1] with permission from Wolters Kluwer Health.)

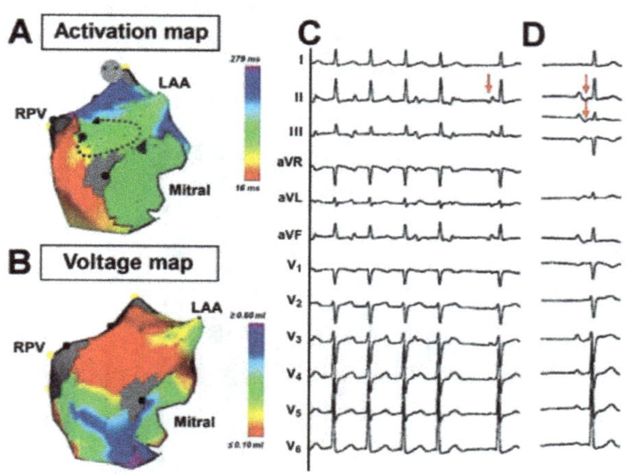

Figure 21.11. CARTO® (Biosense Webster Inc., Diamond Bar, CA) map **(Panel A)** showing a CCW small anterior circuit with a CL of 270 ms. Areas coded in gray are regions of scars with a bipolar voltage amplitude < 0.05 mV. This circuit is bounded by the ablation lesions isolating the RPV, the roof line, the mitral annulus, and an area of spontaneous conduction block at the anterior aspect of the left appendage (LAA). The voltage map for the same patient **(Panel B)** shows a large area of low voltage (<0.1 mV) displayed in red. Interestingly, the flutter circuit is entirely localized in the low-voltage area and the septal limit coincides with the transition in voltage. **Panel C** shows the typical surface electrocardiographic morphology of this tachycardia with a tiny negative followed by a larger positive component of the F wave in lead II (arrow). In **Panel D**, mechanical contact with the ablation catheter at the anterior LA terminated this flutter, demonstrating a P-wave morphology that is typically seen with delayed conduction at the anterior LA (arrows). The arrhythmia was reinduced and successfully ablated at the site where mechanical interruption was observed. (Reproduced from Jaïs et al[5] with permission from John Wiley & Sons.)

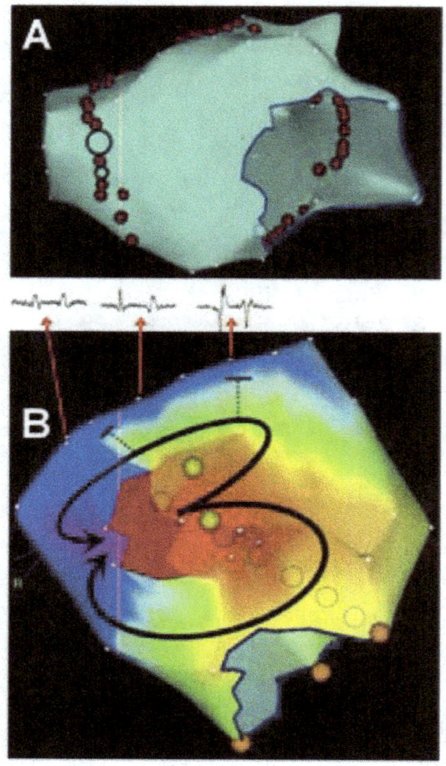

Figure 21.12. Gap-related tachycardia. **A:** Electroanatomical map of the LA from a patient with paroxysmal AF, in an anteroposterior (AP) projection. The red circles represent sites of RF energy delivery. The open circles indicate areas at which RF energy with an 8-mm-tip catheter could not be applied for more than a few seconds because of an abrupt decrease in the delivered power. **B:** An activation map performed several months later during a septal AT in the same patient, in an AP view with cranial angulation. There is a figure-of-8 activation pattern that is stabilized superiorly by a corridor of widely split (90 ms) double potentials (red arrows) related to prior ablation at the roof, by the mitral annulus, and by the right-sided PV. The tachycardia is likely utilizing gaps in the incomplete ablation line shown in Panel A. Gold circles represent sites at which the PPI matched the tachycardia CL. Linear ablation (dashed circles) from the mitral annulus to the right-sided PVs resulted in prolongation of the tachycardia CL but was unsuccessful in terminating the tachycardia. Entrainment mapping showed that the right atrial septum also was part of the circuit; however, ablation at this site also failed to terminate the tachycardia. L, left; R, right. (Reproduced from Chae et al[3] with permission from Elsevier.)

circumferential activation by recording from the anterior septal and lateral aspect of the mitral annulus has been used to quickly exclude or include perimitral reentry. If this is thought to be the case, then more detailed mapping can be performed. To identify roof-dependent reentry, ablation can be placed anteriorly and posteriorly to the LA wall. In the presence of similar sequence of activation, roof-dependent reentry can be ruled

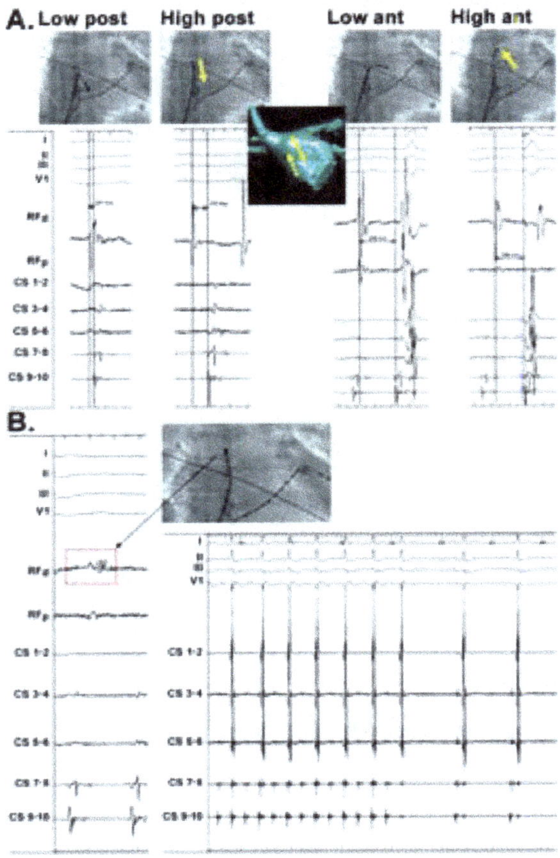

Figure 21.13. **A:** This figure demonstrates the technique for rapidly ruling in a roof-dependent circuit. The first interesting point to be noted is that the activation in the CS is complex but appears to be almost on time and does not appear to be compatible with a perimitral circuit. The other possible macroreentry in the LA being a roof-dependent circuit, this hypothesis is further investigated using CS 9-10 as a stable reference; the posterior left atrial activation (A) is determined to be cranial to caudal (yellow arrow), indicated by an earlier activation at the low posterior LA. The anterior left atrial activation is in the caudal to cranial direction (B). The activation fronts are of opposing directions in the anterior and posterior LA, which is compatible with a roof-dependent circuit. Entrainment performed at both the anterior and posterior LA showed a return cycle within 25 ms confirming a roof-dependent circuit. Note that the sum total of the signals recorded from the anterior and posterior LA cover almost the entire tachycardia CL, further supporting the suspicion that this is a roof-dependent macroreentrant AT. **B:** The successful site for ablation was located on the roof (AP fluoroscopic image) where a gap potential was identified (red rectangle). Ablation at this site successfully restored sinus rhythm (right panel). (Reproduced from Jaïs et al[4] with permission from John Wiley & Sons.)

out. Entrainment mapping should be limited to only a few sites in order to prevent degeneration of the flutter to AF or the circuit involving other channels.

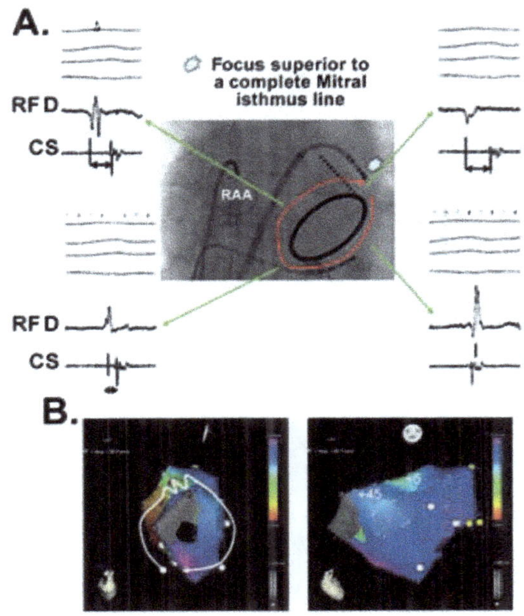

Figure 21.14. **A:** Complimentary roles of PPI and activation mapping. A focus adjacent to a complete ablation line may mimic macroreentry by activation mapping (red arrow): however, in this case, the PPI at the septum was long, thus excluding perimitral reentry. **B:** Complimentary roles of PPI and activation mapping. In this case, activation mapping demonstrates a discrete gap in the roof line. The PPI on the roof near the right superior PV was long (+45 ms), as the line is complete at this end, and this may lead to misinterpretation that the arrhythmia is not roof-dependent. The PPI immediately adjacent to the gap is short (+5 ms). CS, coronary sinus; RF D, radiofrequency delivery. (Reproduced from Jaïs et al[4] with permission from John Wiley & Sons.)

Focal Tachycardias

Once macroreentry is ruled out, localized reentry or focal tachycardia should be determined. Coronary sinus activation pattern is helpful in this regard. An ablation catheter can be used to identify areas exhibiting centrifugal activation pattern. Any fractionated areas described before are identified and entrainment mapping performed at CL 20 to 40 ms shorter than the tachycardia CL at these, and possibly other, sites to identify sites within the reentry circuit. Different examples of left sided tachycardias are depicted in figures 21.8 to 21.17.

SUMMARY

The most common left AFl seen are those arising after ablation of long-lasting persistent AF. Over 10% of patients with recurrence of atrial arrhythmias

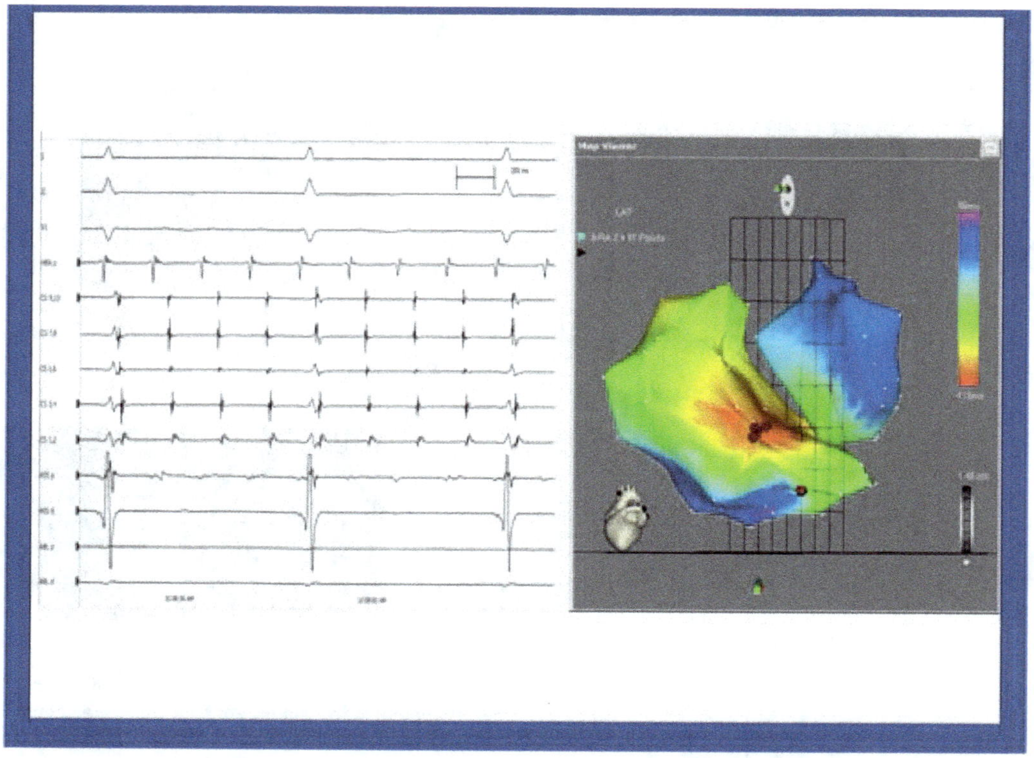

Figure 21.15. Focal tachycardia emanating from the septum.

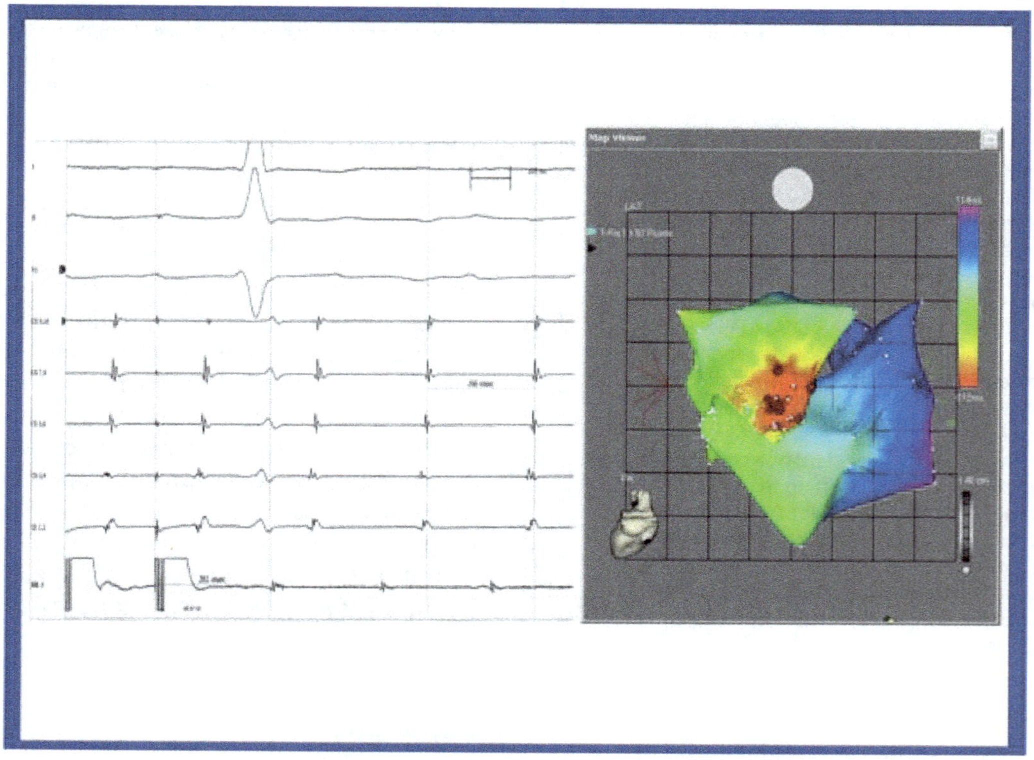

Figure 21.16. Activation and entrainment mapping. Broad front of activation suggests that tachycardia may be originating from the CS, where it was successfully terminated.

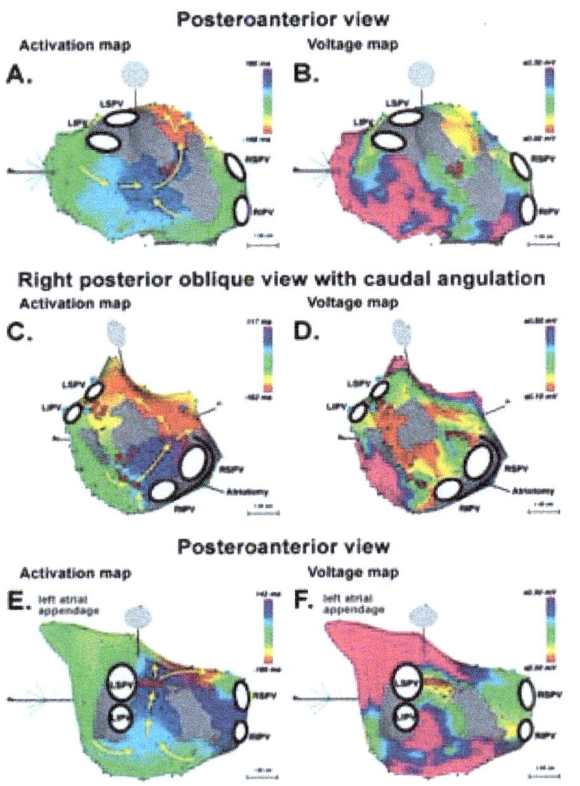

Figure 21.17. Illustration of dual-loop left atrial macroreentrant tachycardia (LAMRT) circuits in patients with mitral valve replacement. **A** and **B**: Demonstration of a protected isthmus between 2 posterior ESAs and a low amplitude within the isthmus. Radiofrequency pulses (brown tags) terminated the LAMRT and blocked the isthmus. **C** and **D**: Demonstration of a protected isthmus between a posterior ESA and the left atriotomy (heavy line) and a low amplitude within the isthmus. Radiofrequency pulses (brown tags) terminated the LAMRT and blocked the isthmus. Also, 3 linear lesions were applied to prevent different LAMRT by connecting 2 anatomic barriers. **E** and **F**: Demonstration of a protected isthmus between a posterior ESA and the left superior pulmonary venous ostium (LSPV), and a low amplitude within the isthmus. RF pulses (brown tags) terminated the LAMRT and blocked the isthmus. LIPV, left inferior pulmonary venous ostium; RIPV, right inferior pulmonary venous ostium; RSPV, right superior pulmonary venous ostium. (Reproduced from Ouyang et al[2] with permission from Wolters Kluwer Health.)

post-ablation have AT or AFl. Left AFl is usually characterized by a positive deflection in V_1 and low-voltage deflections in limb leads. These flutters and AT can occur in the mitral isthmus area, around the PVs, along the roof area connecting the superior PVs, anterior mitral annulus and LA septum. The usual cause of these flutters is incomplete lines without validation of conduction blocks.

Activation and entrainment mapping is used in an attempt to delineate the flutter circuit. However, in the presence of significant scarring from prior AF ablation, activation mapping may point to the wrong location of the reentrant circuit. High capture thresholds may also limit the accuracy of entrainment mapping over large areas previously ablated. Another challenging aspect is the presence of multiple simultaneous circuits. Recognition and ablation of the critical isthmus in these situations may be difficult. Despite these limitations, catheter ablation remains the treatment of choice in these patients.

Identification of reentrant circuits involving previously ablated but incomplete lines requires detailed mapping of these sites, which can be identified by low-amplitude fractionated signals. Identification of these gaps during the arrhythmia, however, is sometimes difficult and may require mapping during sinus rhythm. Localized reentry or focal AT, although difficult to recognize and map, are easy to ablate once the focus has been identified. Deductive strategy outlined by Jaïs et al can be extremely helpful in delineating these complex circuits and ablating them successfully.

Correct Answers:

Q1. D
Q2. D
Q3. B
Q4. Mitral isthmus
Q5. A

REFERENCES

1. Jaïs P, Shah DC, Haïssaguerre M, et al. Mapping and ablation of left atrial flutters. *Circulation.* 2000;101:2928–2934.
2. Ouyang F, Ernst S, Vogtmann T, et al. Characterization of reentrant circuits in left atrial macroreentrant tachycardia: critical isthmus block can prevent atrial tachycardia recurrence. *Circulation.* 2002;105:1934–1942.
3. Chae S, Oral H, Good E, et al. Atrial tachycardia after circumferential PV ablation of atrial fibrillation: mechanistic insights, results of catheter ablation, and risk factors for recurrence. *J Am Coll Cardiol.* 2007;50:1781–1787.
4. Jaïs P, Matsuo S, Knecht S, et al. A deductive mapping strategy for atrial tachycardia following atrial fibrillation ablation: importance of localized reentry. *J Cardiovasc Electrophysiol.* 2009;20:480–491.
5. Jaïs P, Sanders P, Hsu LF, et al. Flutter localized to the anterior left atrium after catheter ablation of atrial fibrillation. *J Cardiovasc Electrophysiol.* 2006;17:279–285.
6. Bochoeyer A, Yang Y, Cheng J, et al. Surface electrocardiographic characteristics of right and left atrial flutter. *Circulation.* 2003;108:60–66.

CASE STUDY

22

Targeting Complex Fractionated Atrial Electrograms to Eliminate Atrial Fibrillation

Hakan Oral, MD and José Jalife, MD

A 55-year-old man with persistent atrial fibrillation (AF) was first diagnosed 4 years prior to presentation. He has failed to maintain sinus rhythm despite treatment with propafenone and subsequently dofetilide and 3 cardioversions. Atrial fibrillation recurred within a week after each cardioversion and has been persistent over the last 18 months. He has fatigue, exertional dyspnea, and decreased exercise tolerance. His mean ventricular rate at rest is 78 bpm on a combination of metoprolol and diltiazem. His left atrial diameter is 48 mm on transthoracic echocardiography and ejection fraction is 0.50. He has a history of hypertension. After having discussed the treatment options, he expressed his interest to proceed with radiofrequency catheter ablation to eliminate AF.

During percutaneous radiofrequency catheter ablation, antral pulmonary vein (PV) isolation was performed and all PVs were electrically isolated. Linear ablation along the left atrial roof and mitral isthmus was also performed. Because AF still persisted, complex fractionated atrial electrograms (CFAEs) were targeted and resulted in termination of AF to sinus rhythm (Figure 22.1).

Question No. 1:

What are the potential mechanisms of AF that may contribute to CFAEs?

A. High-frequency sources with rapid activation rates (rotors).
B. Wavebreak and fibrillatory conduction.
C. Discordant activations from multiple layers of atrial myocardium and anisotropic conduction.
D. Shortened effective refractory period (ERP) at sites of autonomic innervation enabling rapid activation rates.
E. All of the above.

DISCUSSION

In patients with paroxysmal AF, the mechanisms of AF primarily depend on PV arrhythmogenicity.[1,2] Therefore, antral PV isolation is usually sufficient to maintain sinus rhythm in 75% to 85% of patients with paroxysmal AF.[3–6] The vast majority of recurrences appear to be due to recovery of conduction into the PVs or due to arrhythmogenic foci in the other thoracic veins. However, in patients

Practical Electrophysiology © 2015 Jasbir S. Sra, Masood Akhtar, Andrea Natale, David J. Wilber
Cardiotext Publishing, ISBN: 978-1-935395-17-1

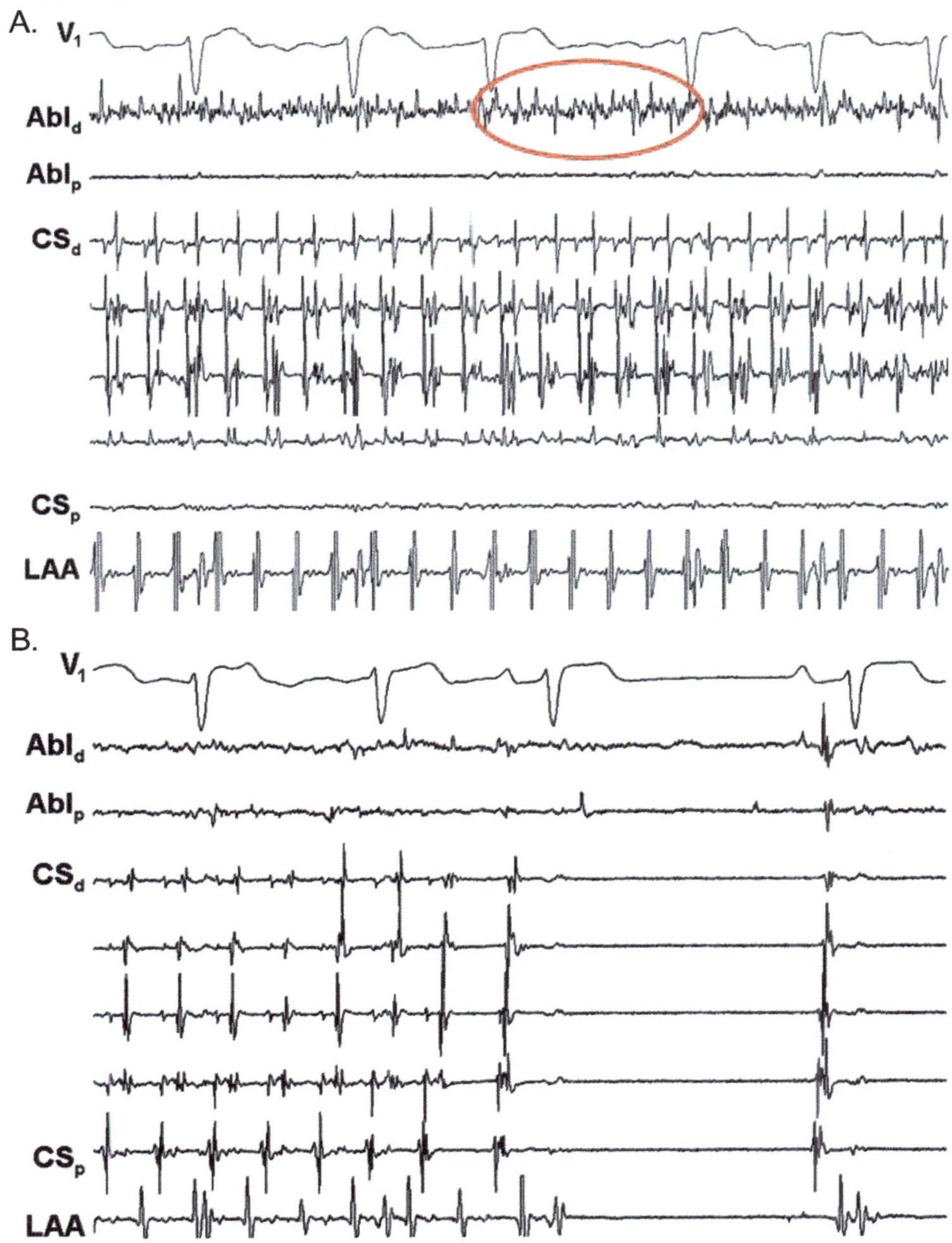

Figure 22.1. **A**: CFAEs at the base of the left atrial appendage (LAA). Shown are electrograms (EGMs) recorded from lead V_1 and mapping/ablation catheters positioned in the left atrium (Abl), coronary sinus (CS), and LAA. After antral PV isolation, AF persisted and application of radiofrequency energy at this site with CFAEs resulted in termination of AF (**Panel B**).

with persistent AF, extensive left or biatrial electroanatomical remodeling occurs, and the atrial tissue becomes capable of maintaining AF independent of the PVs. Therefore in these patients, antral PV isolation often has modest efficacy despite repeat ablation procedures that usually address PV reconnection.

To date, 4 major mechanisms have been suggested to play a role in the genesis of AF: (1) multiple, randomly wandering wavelets, ever changing in number and direction, which result in total disorganization of activity. This hypothesis was proposed to explain perpetuation of AF based on a computational model that was developed almost 50 years ago;[7] (2) high-frequency reentrant sources (rotors) generating spiral waves that interact with functional or structural heterogeneities in the atria, with resultant spatially distributed wavebreaks and fibrillatory conduction;[8–10] (3) thoracic vein hypothesis;[1,11] and (4) autonomic modulation with parasympathetic versus sympathetic imbalance that results in a decrease in atrial ERP, increased automaticity, and facilitated triggered activity that promotes premature atrial depolarizations and fast activation rates.[12]

To improve the efficacy of catheter ablation in patients with persistent AF, strategies to target and eliminate the residual mechanisms of AF beyond the PVs have been proposed. Linear ablation is often performed between well-defined anatomical landmarks such as along the left atrial roof between the left and right superior PVs, or along the mitral isthmus, or between the left inferior PV and lateral mitral annulus.[13–16] Lines at other sites such as the anterior left atrial wall, septum of the left atrium between the right superior or inferior PV, and mitral annulus have also been performed. Linear ablation may improve the efficacy of PV isolation by a variety of mechanisms: (1) linear ablation may partition the left atrium similar to a classic maze procedure so that the maximum wavelength that can be accommodated is shorter than the wavelength of the clinical reentrant circuit; (2) linear ablation includes sites where high-frequency sources may be more prevalent such as the left atrial roof; and (3) linear ablation may directly intercept the course of a macroreentrant circuit.

Few reports suggested a potentially beneficial role of linear ablation after antral PV isolation in patients with persistent AF. However, a previous report that utilized PV isolation, linear ablation, and ablation of CFAEs in a randomized order demonstrated that all 4 steps were necessary in >80% of the patients with long-standing persistent AF to terminate AF.[17] Linear ablation was often necessary to terminate AF after ablation of CFAEs.[14] During linear ablation, complete conduction block across the ablation line must be achieved and sought after as an end point.[13,18] Otherwise, linear ablation with incomplete conduction block increases the probability of proarrhythmia and has been associated with a higher prevalence of atrial flutter after the ablation procedure. Given the occasional challenge in achieving complete conduction block, particularly along the mitral isthmus, and potential risk of complications due to excess ablation (eg, coronary artery injury), linear ablation is often not the preferred first choice unless AF converts to a macroreentrant atrial flutter in contemporary practice. However, the precise role of linear ablation versus ablation of CFAEs as the next step after antral PV isolation remains to be determined.

A recent study that targeted CFAEs as the sole ablation strategy in patients with paroxysmal and persistent AF reported good clinical outcomes and rejuvenated the interest in CFAEs.[19] Based on analysis of epicardial unipolar EGMs recorded during AF, CFAEs have been suggested to indicate sites of slow conduction or conduction block, pivot points for reentrant circuits, and wavefront collision (Figure 22.2). Therefore, sites of CFAEs may indicate critical parts of reentrant circuits, and ablation of these sites may interfere with the reentrant mechanisms of AF. On the other hand, CFAEs may actually reflect fibrillatory conduction of broken spiraling wavefronts emanating from a nearby-high-frequency rotor.

The role of rotors as drivers of AF has been clearly established in experimental models (Figure 22.3). Furthermore, in a clinical study, left atrial sites where AF terminated during ablation was found to have a higher dominant frequency (DF) than the adjacent sites.

A major challenge in targeting CFAEs is how best to define and identify the CFAEs. CFAEs have been described in the time domain based on bipolar EGMs that show rapid activation rate (often <120 ms), fractionation and/or continuous electrical activity (Figure 22.4). Although all of these characteristics have been included in the definition, there has been large variability from one study to the next in describing CFAEs. Further, amplitude has been considered as a key criterion in some studies. Yet other objective criteria such as activation rate, extent of fractionation, voltage, or duration of continuous activation have not been well defined or validated.

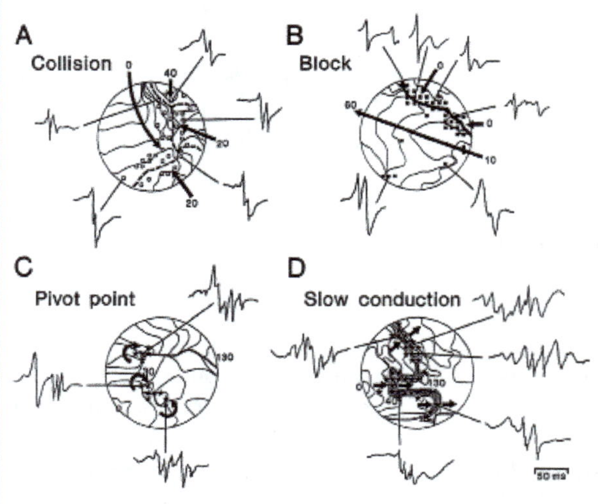

Figure 22.2. Complex EGMs. Isochronal maps of the unipolar EGMs acquired during epicardial mapping of the human atrium during AF demonstrated the role these EGMs may play in reentrant arrhythmias. See text for details. (Adapted from Konings et al[9] with permission from Wolters Kluwer Health.)

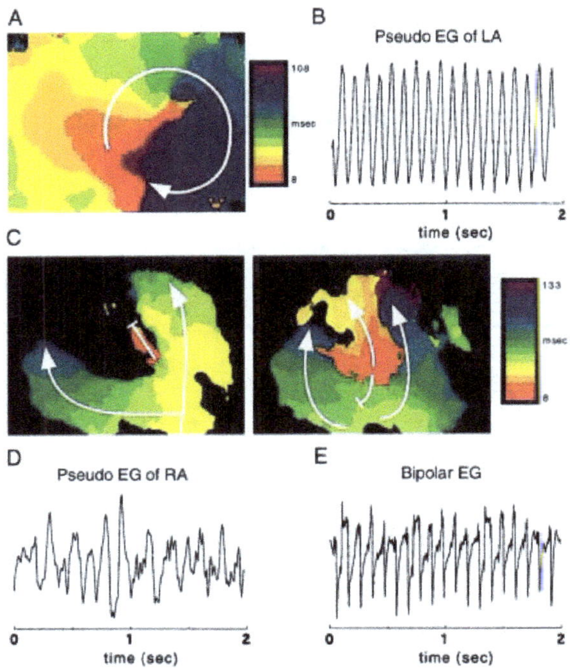

Figure 22.3. Spatiotemporal periodicity and rotors of AF. Shown are high-resolution optical maps of the left (LA) and right atrium (RA) with isochronal maps and pseudo electrograms (EG). A single rotation of stable rotor located in the LA is present in **Panel A** with its monomorphic pseudo EG in **Panel B**. Heterogeneous activation and fibrillatory conduction in the RA are shown in **Panels C–E**. (Adapted from Skanes et al[15] with permission Wolters Kluwer Health.)

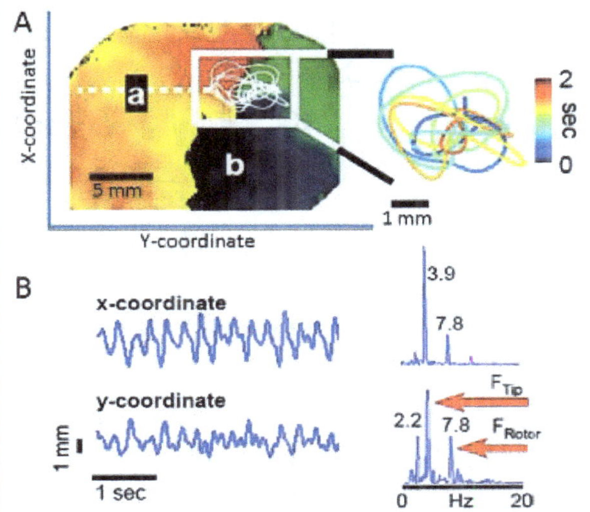

Figure 22.4. Rotor meandering and fractionation in optical mapping experiment during AF in isolated sheep heart. The left atrial (LA) phase snapshot demonstrates reentrant activity in the LA free wall. The time-space trajectory of the tip, the x- and y-coordinate signals, and their corresponding spectra are shown on the right. See text for more details. (Modified from Zlochiver et al[19] with permission from Elsevier.)

The EGMs often referred to as CFAEs based on the currently adopted time-domain criteria may in fact represent a rather heterogeneous group of electrophysiologic events. Examples follow:

1. EGMs recognized as CFAEs due to high activation rates may be the result of meandering of the rotor core.[10] This is illustrated by the example presented in Figure 22.4, which shows data from an optical mapping experiment in an isolated, Langendorff-perfused sheep heart. Figure 22.4A shows a phase map of a sustained rotor (left) of the left atrial free wall during an AF episode. The rotor had a frequency of ~7.8 Hz and was driving the overall atrial activity. Tracking in time the trajectory of the central tip of the rotor allowed demonstration of meandering over an area of ~3 to 4 mm, which resulted in significant complexity of the wavefront. The frequency spectra of the x- and y-axis coordinates (right), obtained by fast Fourier transform (FFT), showed a significant frequency component at 3.9 Hz, and additional, less prominent components at ~2.3 and 7.8 Hz. Clearly, the 7.8 Hz component relates to the rotor reverberating frequency, which is a harmonic of the tip meandering frequency of 3.9 Hz (Figures 22.3 and 22.5).[10] If the particular rotor is the sole driver of AF, appropriately targeted ablation at such a site would be expected to terminate AF. However, if there is a spectral hierarchy of multiple drivers (rotors), ablation of the rotor with the highest frequency may uncover the secondary drivers with lower frequencies. Sequential mapping and ablation of the residual drivers would terminate AF. A recent study also suggested that elimination of primary drivers of AF with the highest DF may uncover underlying reentrant circuits capable of perpetuating AF with fibrillatory conduction. It is also noteworthy that some rotors show little or no meandering, in which case the very center of the rotor core would be silent and the low-frequency component of the x- and y-coordinates (see Figure 22.4) would disappear. However, the relatively high-resolution multiple electrode mapping technology that is needed to precisely localize the rotating wavefront tip and the core, while potentially available, has not yet been used in the clinical electrophysiology laboratory.

Complex fractionated atrial EGMs may represent sites of wavebreak and fibrillatory conduction as the atrial tissue becomes unable to maintain high activation rates originating from a

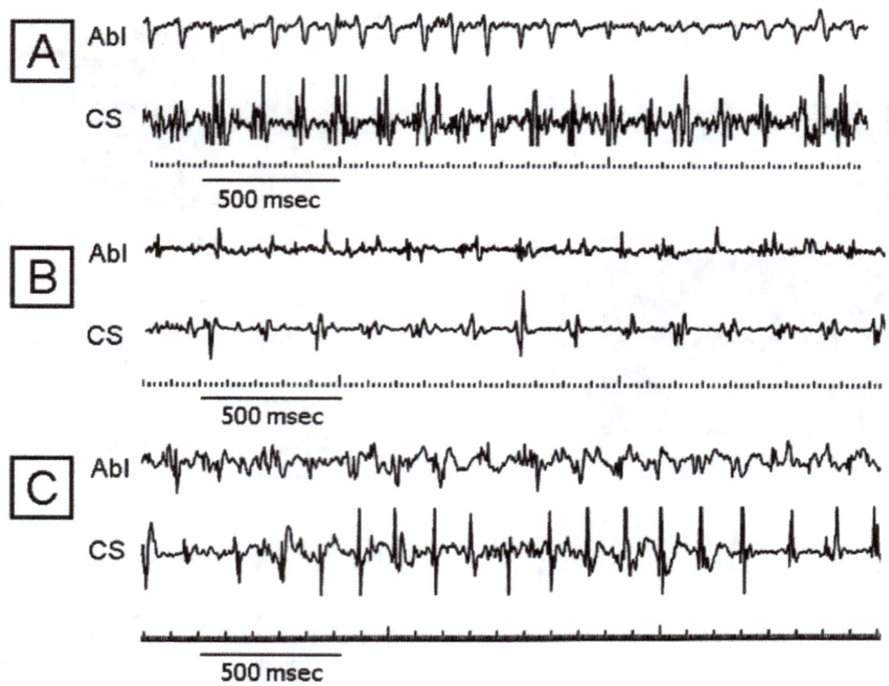

Figure 22.5. Examples of CFAEs. CFAEs may represent a heterogeneous group of EGMs with rapid activation rate (**Panel A**), fractionation (**Panel B**), and continuous electrical activity (**Panel C**). Shown are bipolar EGMs recorded from a mapping catheter (Abl) positioned in the left atrium and a reference catheter positioned in the CS. (Adapted from Oral H et al[14] with permission from Wolters Kluwer Health.)

rotor (see Figure 22.5). As illustrated by the numerical data presented in Figure 22.6, fibrillatory conduction and EGM fragmentation may ensue within a few millimeters of the core of a rotor.[10] Figure 22.6A shows a snapshot of a fast and stable rotor in the upper half of a 2D sheet of simulated atrial tissue. The upper half was programmed to have a shorter ERP by increasing

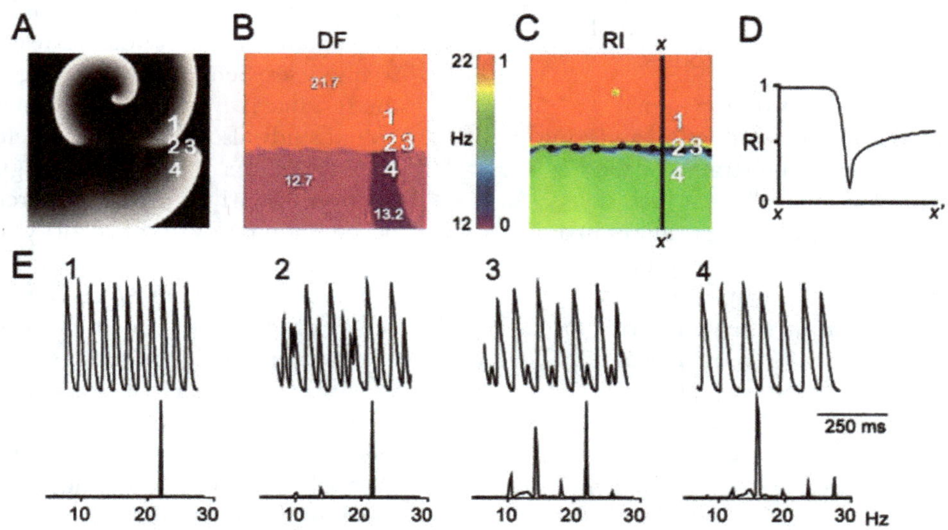

Figure 22.6. Numerical simulation of EGM fractionation at the periphery of a rotor. Shown is a fast rotor (**Panel A**) with waves intermittently blocking at the interface with the lower half of the panel. DF (**Panel B**) and RI (**Panel C**) maps are also shown. Region of the highest DF (excluding the core) corresponds to the region of highest RI. Circles indicate phase singularities. RI profile along x-x' line is shown in **Panel D**. Sample membrane potentials with corresponding power spectra from the DF_{max} (1), interface (2 and 3), and lowest-DF (4) areas are shown in **Panel E**. (Adapted from Kalifa et al[11] with permission from Wolters Kluwer Health.)

the density of the acetylcholine-activated potassium current ($I_{K,ACh}$). The stable rotor in the upper half generated high-frequency spiral waves that propagated rapidly toward the boundary with the bottom half. Not every wave made it through the boundary. Instead, intermittent, spatially distributed propagation was established as a result of the abrupt prolongation of the ERP. Consequently, 3 frequency domains became evident on the DF map in Figure 22.6B, corresponding to 4 seconds of simulation. In the upper half, a 21.7-Hz frequency domain (DF_{max}) matches the rotor frequency; in the lower half, domains with spectral peaks at 27.7, 13.2, and 12.7 Hz reflecting complex propagation patterns were identified; finally, the lower half showed a spectral peak at 16.3 Hz. Figure 22.6C shows a regularity index (RI) map, calculated as the ratio of the spectral power at the DF to the total power. Except for the very center of rotation (yellow dot), the highest-regularity region coincided with the DF_{max} domain. A profile of RI along the vertical line of Figure 22.6C is presented in Figure 22.6D. It reveals an area of very low regularity along a narrow band between the DF_{max} and the lower-DF domains. This correlates highly with the AP recordings in Figure 22.6E, obtained from locations 1 through 4 in the sheet. Clearly, activity in the DF_{max} domain is highly organized, whereas fragmentation (low RI) occurs at the boundary with the lower-frequency domains. Concomitantly, the vast majority of wavebreaks were clustered close to the limit within the boundary separating the DF_{max} from the low-DF domains (see Figure 22.6C). Note that fragmentation was again reduced (ie, increased RI) at the low-frequency domain, which was distal to the interface with the DF_{max} domain. Altogether, the above results suggest that, in theory, targeting the interface where wavebreak and most fragmentation occurs for ablation might result in sufficient collateral effect to eliminate the rotor. However, the ablation would need to include a significant portion of the highly regular DF_{max} site to obliterate the "elbow room" needed by the rotor to complete its rotations and to effectively terminate its activity. Clearly, in Figure 22.6, if the ablation is confined solely to the area of complex fractionation, the rotor might be insulated and waves would not propagate to the lower half. Yet, even if the entire boundary of fractionation is ablated, the rotor would not terminate but continue generating high-frequency spiral waves.

2. CFAEs may also represent sites of passive activation. The 3D atrial wall is highly heterogeneous and complex. In particular, the strikingly irregular topography of the endocardial surface of the RA is dominated by extensive pectinate muscles that form ridges and bridges, which when activated at the high frequencies that characterize AF result in structural discontinuities and destabilize propagation to result in multiple areas of intermittent block.[10] On the other hand, the left atrium, while apparently smooth, is a complex of overlapping anisotropic muscular bundles with varying orientations. In other areas there are abrupt changes in orientation, which when activated at high-excitation frequencies may result in discordance between subepicardial and subendocardial conduction, which could be reflected in the form of CFAEs. Similarly, complex EGMs may result from discordance of right atrial versus left atrial activation along the septum, or left atrial versus CS activation along the posterior mitral annulus.
3. Spectral analysis of atrial EGMs by FFT during sinus rhythm has led some investigators to distinguish areas of atrial muscle with apparently normal in-phase conduction from areas supposed to have fewer cellular connections and short refractoriness. The latter areas are inferred to undergo anisotropic out-of-phase conduction and a segmented right-FFT shift. Such fibrillar muscle (FM) areas are speculated to be associated with neural input or vein insertion, or even to be pathological tissue that has lost its intercellular connections. There, increased parasympathetic tone may facilitate shorter ERPs that enable faster activation rates and subsequently wavebreak and fibrillatory conduction. Hence, it is claimed that during AF, clusters of FM, dubbed "AF nests," represent CFAE sites that activate at higher frequencies than any surrounding tissue. This, however, remains highly speculative and requires confirmation.

Targeting CFAEs often involves extensive left and occasionally right atrial ablation to terminate AF.[17] As defined using widely adopted criteria based on time-domain parameters, CFAEs appear to lack sufficient specificity to indicate critical drivers of AF resulting in more than necessary ablation. Therefore, there remains a need to better identify underlying drivers of AF. A number of studies will be helpful to better determine EGM criteria:

1. Detailed *ex* and *in vivo* studies to determine specific EGM characteristics associated with drivers of AF.

2. Clinical studies to determine the role of real-time frequency-domain mapping for targeting high-frequency sources and its relationship to clinical outcome.
3. Clinical studies to determine the role of linear ablation alone or in conjunction with ablation of CFAEs.

Correct Answer:

Q1. E

REFERENCES

1. Haïssaguerre M, Jaïs P, Shah DC, et al. Spontaneous initiation of atrial fibrillation by ectopic beats originating in the pulmonary veins. *N Engl J Med.* 1998;339:659–666.
2. Oral H, Ozaydin M, Tada H, et al. Mechanistic significance of intermittent pulmonary vein tachycardia in patients with atrial fibrillation. *J Cardiovasc Electrophysiol.* 2002;13:645–650.
3. Douglas YL, Jongbloed MR, Gittenberger-de Groot AC, et al. Histology of vascular myocardial wall of left atrial body after pulmonary venous incorporation. *Am J Cardiol.* 2006;97:662–670.
4. Moe GK. A conceptual model of atrial fibrillation. *J Electrocardiol.* 1968;1:145–146.
5. Berenfeld O, Zaitsev AV, Mironov SF, Pertsov AM, Jalife J. Frequency-dependent breakdown of wave propagation into fibrillatory conduction across the pectinate muscle network in the isolated sheep right atrium. *Circ Res.* 2002;90:1173–1180.
6. Chen PS, Wu TJ, Hwang C, et al. Thoracic veins and the mechanisms of non-paroxysmal atrial fibrillation. *Cardiovasc Res.* 2002;54:295–301.
7. Hou Y, Scherlag BJ, Lin J, et al. Ganglionated plexi modulate extrinsic cardiac autonomic nerve input: Effects on sinus rate, atrioventricular conduction, refractoriness, and inducibility of atrial fibrillation. *J Am Coll Cardiol.* 2007;50:61–68.
8. Nademanee K, McKenzie J, Kosar E, et al. A new approach for catheter ablation of atrial fibrillation: Mapping of the electrophysiologic substrate. *J Am Coll Cardiol.* 2004;43:2044–2053.
9. Konings KT, Smeets JL, Penn OC, Wellens HJ, Allessie MA. Configuration of unipolar atrial electrograms during electrically induced atrial fibrillation in humans. *Circulation.* 1997;95:1231–1241.
10. Jalife J. Rotors and spiral waves in atrial fibrillation. *J Cardiovasc Electrophysiol.* 2003;14:776–780.
11. Kalifa J, Tanaka K, Zaitsev AV, et al. Mechanisms of wave fractionation at boundaries of high-frequency excitation in the posterior left atrium of the isolated sheep heart during atrial fibrillation. *Circulation.* 2006;113:626–633.
12. Mandapati R, Skanes A, Chen J, Berenfeld O, Jalife J. Stable microreentrant sources as a mechanism of atrial fibrillation in the isolated sheep heart. *Circulation.* 2000;101:194–199.
13. Sanders P, Berenfeld O, Hocini M, et al. Spectral analysis identifies sites of high-frequency activity maintaining atrial fibrillation in humans. *Circulation.* 2005;112:789–797.
14. Oral H, Chugh A, Good E, et al. Radiofrequency catheter ablation of chronic atrial fibrillation guided by complex electrograms. *Circulation.* 2007;115:2606–2612.
15. Skanes AC, Mandapati R, Berenfeld O, Davidenko JM, Jalife J. Spatiotemporal periodicity during atrial fibrillation in the isolated sheep heart. *Circulation.* 1998;98:1236–1248.
16. Ho SY, Anderson RH, Sánchez-Quintana D. Atrial structure and fibres: Morphologic bases of atrial conduction. *Cardiovasc Res.* 2002;54:325–336.
17. Pachon M JC, Pachon M El, Pachon M JC, et al. A new treatment for atrial fibrillation based on spectral analysis to guide the catheter RF-ablation. *Europace.* 2004;6:590–601.
18. Oral H, Chugh A, Good E, et al. Randomized evaluation of right atrial ablation after left atrial ablation of complex fractionated atrial electrograms for long-lasting persistent atrial fibrillation. *Circ Arrhythm Electrophysiol.* 2008;1:6–13.
19. Zlochiver A, Yamazaki M, Kalifa J, Beren O. Rotor meandering contributes to irregularity in electrograms during atrial fibrillation. *Heart Rhythm.* 2008;5:846–854.

CASE STUDY

23

Techniques of Segmental Pulmonary Vein Isolation and Circumferential Pulmonary Vein Ablation During Sinus Rhythm and Atrial Fibrillation

Aman Chugh, MD and Fred Morady, MD

A ring catheter is placed in the left superior pulmonary vein (LSPV) during sinus rhythm in a patient presenting for his first ablation procedure for idiopathic paroxysmal atrial fibrillation (AF). During extrastimulus testing while pacing from the distal coronary sinus (CS) (Figure 23.1), 2 distinct potentials emerge on the ring-catheter recording.

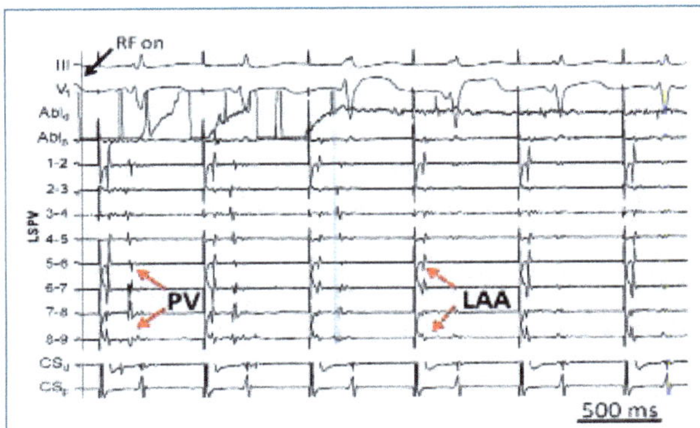

Figure 23.1. Coronary sinus pacing.

Question No. 1:

Which of the following is true?

A. The pulmonary vein (PV) potential is recorded prior to the left atrial appendage (LAA) potential.

B. The LAA potential precedes the PV potential.

C. The first potential represents near-field activation of the left superior PV, while the second represents far-field contribution from the left inferior PV.

D. The 2 potentials are generated by a conduction delay into the LAA.

Question No. 2:

Regarding the patient described in the previous question, what is the recommended ablation strategy?

A. PV isolation followed by linear ablation at the left atrial roof.

B. PV isolation followed by ablation of complex, fractionated electrograms (EGMs).

C. PV isolation followed by linear ablation at the mitral isthmus.

D. PV isolation only.

E. Circumferential PV ablation with the end point of voltage abatement along the ipsilateral circular lesions.

Question No. 3:

During antral ablation of the left-sided PVs, you note the esophagus is in close contact to the posterior aspect of the PVs. What is the most reasonable strategy that will facilitate PV isolation with minimal risk to the esophagus?

A. Since the likelihood of esophageal injury is so low, proceed with left antral ablation without regard to the esophagus, using 30 watts of irrigated radiofrequency (RF) energy.

B. Proceed with left antral ablation using 35 watts of irrigated RF energy.

C. Complete the anterior aspect of the circumferential lesion, followed by RF energy delivery at the crux between the upper and lower PVs before targeting the posterior aspect near the esophagus using 25 watts of irrigated RF energy.

D. Perform left antral ablation using 35 watts of irrigated RF energy, guided by esophageal temperature monitoring.

E. Complete the anterior aspect of the circumferential lesion using RF energy, followed by cryoballoon ablation of the left-sided PVs.

Question No. 4:

While ablating near the anterior aspect of the right superior PV in the patient described above, you note phrenic nerve capture with high-output pacing. What is a reasonable strategy to minimize the risk of phrenic nerve injury?

A. Perform right superior PV isolation using a 28-mm cryoballoon instead of conventional RF energy since the risk of phrenic nerve injury with the former is lower.

B. Since the risk of phrenic nerve injury is low, and reversible in many cases, proceed with ablation without further testing.

C. Withdraw the ablation catheter slightly more proximally, and repeat high-output pacing. If negative, then proceed with ablation.

D. Abort the procedure, and refer the patient for a surgical ablation procedure.

E. Forgo isolation of the right superior PV, and perform linear ablation at the left atrial roof instead to minimize the risk of AF recurrence.

Question No. 5:

A patient with idiopathic paroxysmal AF presents for a repeat procedure for recurrent AF. She did well for 6 months after the first procedure, which consisted of antral PV isolation. Mapping with a ring catheter shows PV reconnection in both upper PVs. Which of the following is the most reasonable strategy during the repeat procedure?

A. Perform complete circumferential ablation around both antra with the end point of PV isolation.

B. Since the patient has failed a PV-based procedure, perform linear ablation at the roof and mitral isthmus with demonstration of conduction block.

C. Map and ablate the discrete PV breakthroughs rather than recreating the entire circumferential lesion set.

D. Administer isoproterenol to determine which of the 2 reconnected PVs is arrhythmogenic, then isolate the culprit PV.

E. Induce AF with rapid atrial pacing, then target complex fractionated EGMs with the end point of AF termination.

DISCUSSION

Despite the promulgation of various ablation techniques for AF over the last decade, pulmonary vein remains the cornerstone for patients with both paroxysmal and persistent AF. The original description of elimination of PV triggers involved RF energy delivery within the culprit vein. Because ablation within a vein may be complicated by stenosis, ostial ablation of the PVs became the preferred method of PV isolation. Ostial or segmental PV isolation involves identifying PV breakthroughs on a circular mapping catheter placed at the PV ostium. Although PV potentials may be recorded on all bipoles of the circular catheter, identification of the earliest activation may obviate RF energy delivery around the entire circumference of the PV ostium (Figure 23.2). This may have important implications for procedure times and risk of complications, such as PV stenosis. However, in some patients the left atrial (LA)-PV connections may be quite broad, requiring ablation around much of the ostium.

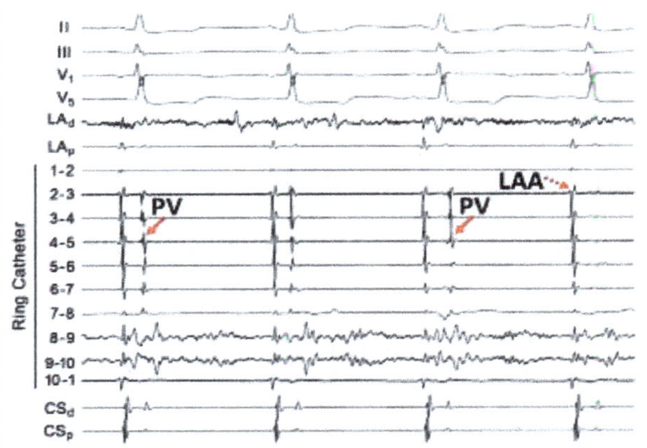

Figure 23.2. Electrical isolation of the LSPV during pacing from the CS. The ring-catheter recording shows 2 distinct potentials. The first potential is due to far-field activation of the LAA (dashed arrow). The second potential (red arrows) is generated by PV activation. Shortly after turning on RF current, the PV potential is eliminated, resulting in entrance block and electrical isolation of the PV. Also shown are electrocardiographic leads, I, III and V_1 and V_5.

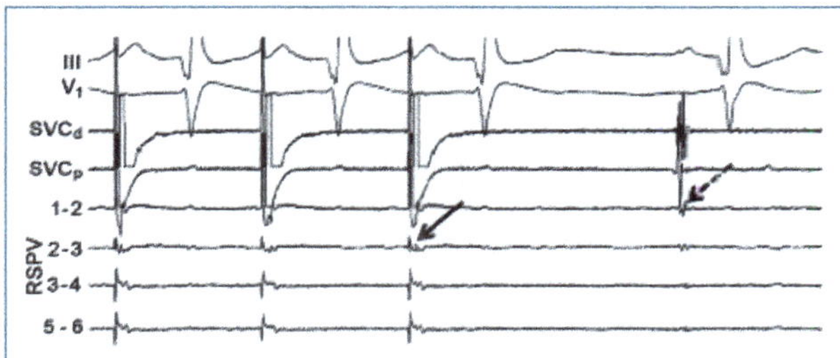

Figure 23.3. Far-field SVC potential recorded by the ring catheter in the right superior pulmonary vein (RSPV). The potential (solid arrow) is recorded 35 ms after the onset of the P wave. High-output pacing from the septal aspect of the SVC advances the potential in question (dashed arrow) to the pacing stimulus, confirming its origin from the SVC. This maneuver confirms that the PV is isolated and no additional ablation is required.

Discrete PV breakthroughs may be identified during sinus rhythm and atrial pacing. In a patient who has not previously undergone PV isolation, the atrial and venous potentials frequently appear fused on the circular mapping catheter. Atrial pacing, typically from the CS or the LA appendage, advances the atrial component, unmasking the PV potentials. The earliest activation is then identified and then targeted for ablation. After elimination of a PV breakthrough, the activation pattern frequently changes, suggesting the existence of multiple LA–PV connections. The end point is absence or dissociation of PV potentials at the ostium. Although some patients may only require isolation of only the "culprit" vein, that is, one that has been shown to be arrhythmogenic on provocative testing, it is reasonable to electrically isolate all PVs to prevent recurrent arrhythmias.

Pacing from the LA appendage or the CS is not as helpful in discriminating between atrial and venous potentials on tracings from the right-sided PVs. For the right superior PV, one can employ activation mapping with respect to the P wave on the ECG to determine if the residual potential is of atrial or venous origin. If the EGM in question occurs within 30 ms of the onset of the P wave, then it is very likely due to far-field activation of the superior vena cava (SVC).[1] If the EGM is advanced to the stimulus artifact during right atrial pacing, a far-field origin is confirmed (Figure 23.3). For the right inferior PV, posterior LA pacing may be helpful in discriminating between far-field LA potentials and near-field PV potentials.

PV Isolation during AF

Although it is possible to isolate the PVs during AF,[2] it is less efficient. During AF, it may be difficult to identify PV breakthroughs due to the constantly changing EGM morphology and activation pattern. However, there are some clues that may help the operator discern between LA and PV potentials. Placing the ablation catheter in the LA appendage and comparing the activation of the appendage with that of the circular mapping catheter in the left superior PV is helpful in identifying the appropriate PV target. The potentials on the circular mapping catheter that are synchronized to the LAA recording are obviously of appendage origin. The remaining PV potentials then may be targeted based on the earliest activation on the circular mapping catheter. Occasionally, the EGMs on the circular mapping catheter during AF may be nearly continuous and complex, which makes it difficult to identify PV activation pattern. In such cases, it is reasonable to start ablating by encircling the PV. During ablation, one begins to notice conduction block into the PVs, making it easier to distinguish the PV potentials from the far-field atrial potentials (Figure 23.4). The cycle length of the former will be slower than that of the latter.

Outcomes Following Ostial Isolation of the PVs

The midterm results of ostial PV isolation in patients with paroxysmal AF were modest, with an efficacy of about 65%. The modest efficacy could be due to lack of permanent PV isolation or to the presence of non-PV triggers. However, even if one targets non-PV foci in patients with paroxysmal AF, the AF is eliminated long-term in less than two-thirds of the patients after a single procedure.

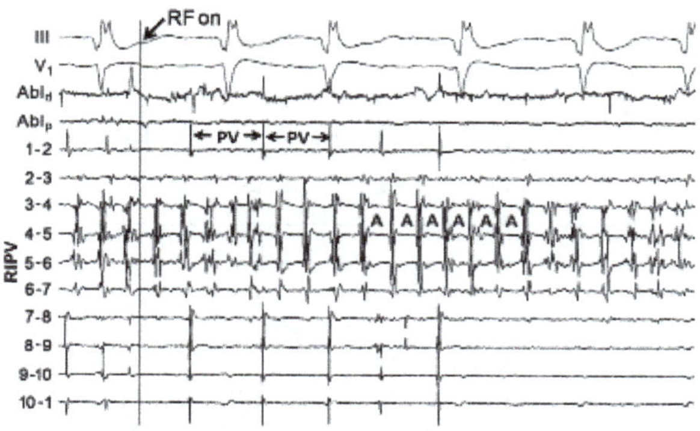

Figure 23.4. Distinguishing pulmonary venous from left atrial activity during AF. Prior to ablation around the right inferior pulmonary vein (RIPV), it was difficult to distinguish PV from far-field LA potentials. Since the PV breakthroughs could not be identified by activation mapping, the entire vein was encircled during RF energy delivery. As a result, conduction into the PV began to slow, making it easier to distinguish between atrial (A) and venous (PV) potentials. Thereafter, the last remaining connection could be identified and ablated (large arrow), resulting in complete isolation of the PV.

Circumferential Pulmonary Vein Ablation

The technique of circumferential pulmonary vein ablation (CPVA) was introduced in an effort to not only address the PV contribution but also that of the LA substrate. Each set of veins was encircled approximately 1 to 2 cm outside the PV ostia. Additional linear ablation was performed at the posterior LA connecting the circular lesions, and at the mitral isthmus (between the lateral mitral annulus and the left inferior PV ostium) (Figure 23.5A). The procedural end point consisted of only voltage abatement along the lesion set. There was no concerted effort to isolate the PVs or to demonstrate conduction block. A later study showed that the efficacy of CPVA was independent of PV isolation. Despite the simplicity and efficiency of the CPVA procedure, there was high prevalence of macro-reentrant atrial tachycardia (AT)[3] (Figure 23.5B) that required repeat ablation procedures in many of these patients. This is the major reason why the procedure is no longer performed in most laboratories.

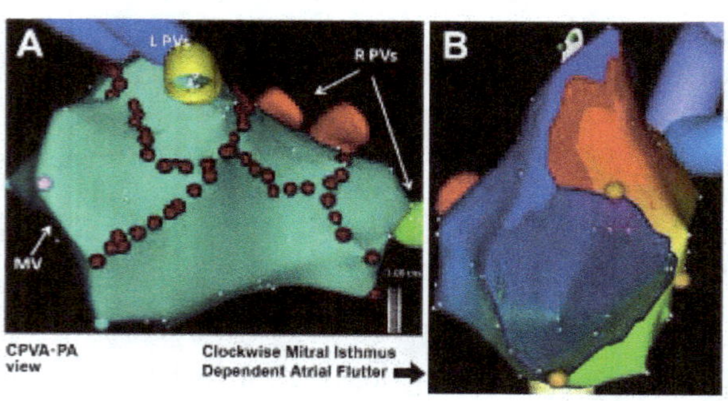

Figure 23.5. A: Posteroanterior (PA) view of the left atrium from a patient with paroxysmal AF who underwent CPVA. The circular lesion sets around each set of PVs are connected posteriorly. The left circular lesion is connected to the mitral valve (MV). B: An activation map (left anterior oblique view) from the same patient. The patient presented for a repeat procedure for atrial flutter 7 months after the initial procedure for AF. Activation mapping suggested presence of mitral isthmus-dependent atrial flutter.

Antral Isolation of the PVs

As alluded to above, the modest outcomes following ostial PV isolation in patients with paroxysmal AF may be attributable to a lack of durable PV disconnection. Several investigators showed that a more proximal approach to PV isolation is associated with improved efficacy. Instead of targeting the earliest PV breakthrough at the ostium, the lesion set was moved more proximally, about 1 cm away from the PV ostium (Figure 23.6). The improved efficacy of a proximal or antral approach to PV isolation is likely due to many factors. First, the myocardial sleeves that extend into the PVs originate from the LA myocardium and it is likely that an antral approach addresses antral drivers that are untreated by an ostial approach (Figure 23.7). Other possible mechanisms include modification of the intrinsic cardiac autonomic nervous system, elimination of small reentrant sources at the posterior LA, and a debulking effect of excluding LA myocardium. Another benefit of an antral approach is that it reduces the risk of PV stenosis.

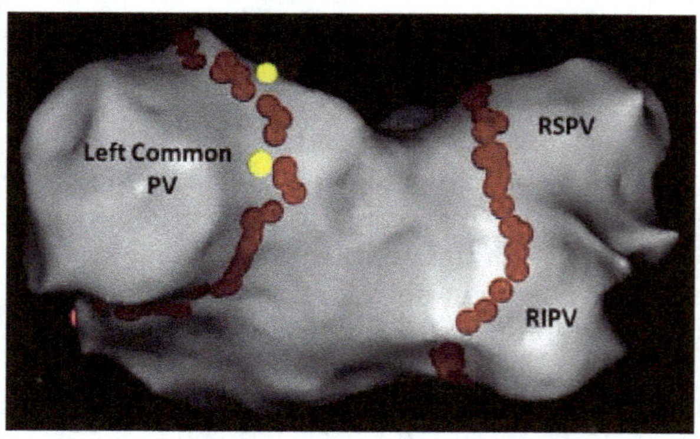

Figure 23.6A. An example of an antral lesion set in a patient with paroxysmal AF. The red tags correspond to sites where RF energy was delivered. The yellow tags correspond to sites where RF energy delivery resulted in a conduction delay or isolation of the PV. RIPV, right inferior pulmonary vein; RSPV, right superior pulmonary vein.

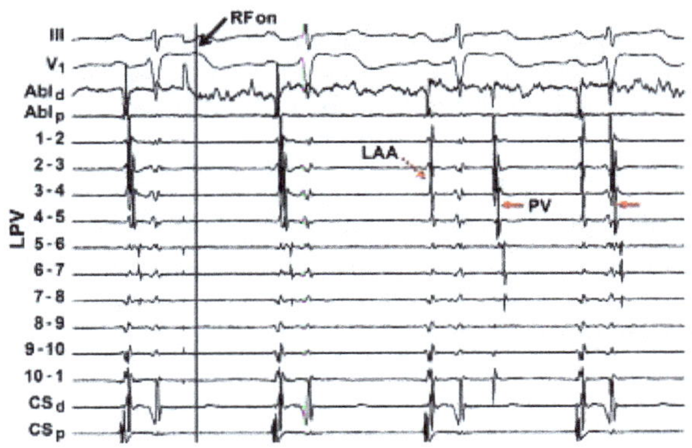

Figure 23.6B. Antral isolation of the left common PV in the same patient as in Figure 23.6A. RF energy was delivered at the superior aspect of the left antrum (yellow tag in Figure 23.6A), which readily resulted in antral isolation. Note that as compared to Figure 23.2, there is no artifact during RF energy delivery since ablation is performed proximal to the ring catheter as opposed to at the level of the ring catheter situated at the PV ostium. Note the dissociated PV potentials (solid red arrows).

Paroxysmal AF is eliminated in approximately 90% of patients with an antral isolation approach. Importantly, approximately one-third of these patients will require a repeat ablation procedure, usually for PV reconnection. The rate of serious complications such as tamponade, PV stenosis, and thromboembolism is approximately 1%.

Anatomic Considerations

Prior to ablation, it is critical to accurately delineate the extent of the PV ostium. This helps to expedite

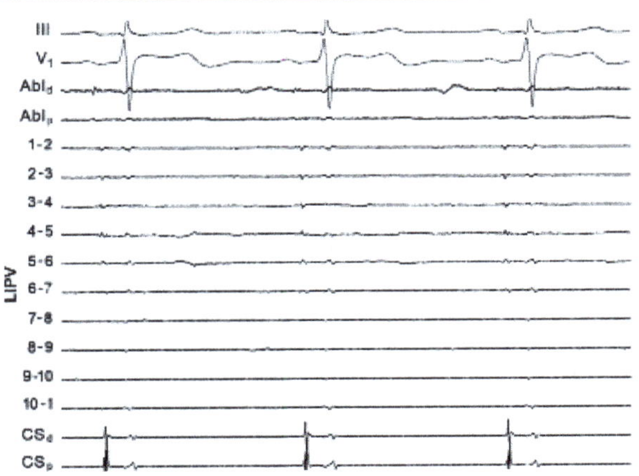

Figure 23.7A. Recording by the ring catheter placed in the left inferior pulmonary vein (LIPV). This patient had previously undergone PV isolation and presented for repeat procedure for recurrent AF. Mapping during sinus rhythm showed only a far-field atrial potential despite the catheter being withdrawn as proximally as possible.

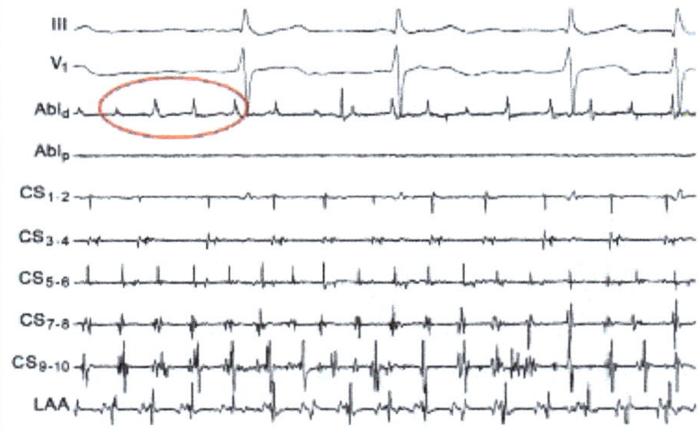

Figure 23.7B. (Continued from Figure 23.7A.) Mapping during AF revealed this EGM (oval) recorded from the inferoposterior left atrium, outside the LIPV despite documented isolation.

the isolation procedure and also helps to minimize the risk of PV stenosis. Pulmonary venograms can accurately identify the venous anatomy in most patients. This is best done after administration of intravenous adenosine to create transient atrioventricular conduction block. Other methods include analysis of the local EGM, real-time impedance monitoring or use of a three-dimensional (3D) mapping system with or without image integration. The current mapping systems allow for very detailed and reliable reconstruction of the PVs and their relationship to the LA. They also help identify the crux (or carina) between the ipsilateral veins. Ablation of this area is critical in antral isolation of the PVs. Use of a 3D mapping system has also been shown to reduce exposure to ionizing radiation. Intracardiac echocardiography is

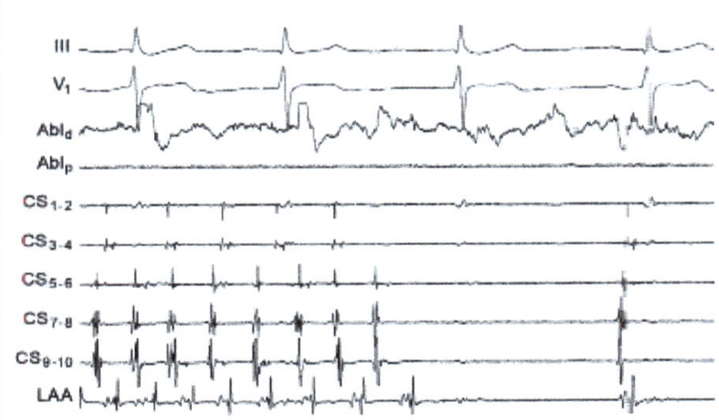

Figure 23.7C. (Continued from Figure 23.7B.) RF energy was delivered at the site shown in Figure 23.7B, which terminated AF within a few seconds. Even though PV was isolated at the level of the ostium, there was still evidence of a driver at the antrum of the LIPV. This example highlights the incremental value of antral versus ostial ablation.

used in some laboratories in identifying the PV ostia and also helps reduce fluoroscopy times.

Power Settings and Safety Considerations

Most laboratories use an externally irrigated ablation catheter during LA ablation. External irrigation prevents charring at the ablation electrode and may reduce the risk of thromboembolic complications. Power settings vary among operators. In most patients undergoing PV isolation, 25 watts is sufficient and probably helps reduce complications such as phrenic nerve or esophageal injury. When ablating near the anterior aspect of the right superior PV, high-output pacing is useful for identifying the course of the right phrenic nerve. If pacing results in phrenic nerve capture, the catheter is withdrawn slightly and the pacing maneuver is repeated. If one uses more than 25 watts, it is important to limit the duration to a few seconds prior to moving the catheter to the next site. Although lethal atrioesophageal fistula formation was reported more than 6 years ago, this complication continues to occur worldwide. This underscores the importance of esophageal monitoring during the procedure. The location of the esophagus may be ascertained in a variety of ways, including administration of oral barium in lightly sedated patients or insertion of a radio-opaque orogastric tube in deeply sedated patients. Areas along the posterior LA that are in close proximity to the esophagus can then be avoided during RF application. However, not infrequently, the course of the esophagus prevents completion of posterior aspect of the antral circular lesion. In such cases, it is helpful to complete the anterior aspect of the circular lesion, followed by ablation at the crux between the ipsilateral PVs. This strategy may result in PV isolation with minimal ablation along the posterior wall. The role of proton-pump inhibitors following the ablation procedure in preventing esophageal ulceration is not clear. Animal studies suggest they help prevent progression of thermal esophageal injury.

Extra-PV Approaches

Several investigators have examined the role of extra-PV targets in an effort to improve outcomes in patients with paroxysmal AF. These include the SVC, complex fractionated atrial electrograms (CFAE), and linear LA ablation. None of these approaches has been shown to be incremental to antral PV isolation. Therefore, for the vast majority of patients with paroxysmal AF, antral PV isolation suffices. SVC isolation is likely best left for the patient who demonstrates firing from this thoracic vein after PV isolation. Occasionally, a patient with paroxysmal AF may require atrial ablation for inducible or recurrent AF despite PV isolation. There may be a role for CFAE ablation in such patients. Although linear LA ablation has been shown to be incremental to ostial PV isolation in these patients, it is best avoided given the difficulty in achieving linear block and the risk of proarrhythmia even in the context of complete linear lesions.

Pulmonary Vein Isolation and CPVA in Persistent AF

While there is consensus regarding the ablation strategy in patients with paroxysmal AF, the optimal approach for patients with persistent AF remains elusive. Due to greater structural and electrical remodeling in patients with persistent AF compared to paroxysmal AF, these patients frequently require ablation of the atrial substrate. Which atrial targets and how they should be approached are matters of contention. CPVA was shown to be effective in more than 70% of patients with persistent AF.[4] However, lack of PV isolation and incomplete linear lesions remain significant limitations.

Several studies, but not all, have concluded that there is incremental benefit of CFAE ablation following PV isolation in patients with persistent AF. The arrhythmia may be eliminated in about 70% to 80% of patients, with many needing repeat procedures.

A number of years ago, a strategy of stepwise ablation in patients with persistent AF was described.[5] The end point of the ablation procedure is termination of AF without electrical or pharmacological cardioversion. The steps involve PV isolation, CFAE ablation, followed by linear ablation at the LA roof and mitral isthmus. If AF does not terminate after these steps, right atrial ablation is performed. Although the investigators usually were able to acutely terminate AF during RF ablation, it rarely terminated to sinus rhythm, and most often converted to an AT. The AT is then mapped and ablated. Linear block at the LA roof and the mitral isthmus and, preferably, the cavotricuspid isthmus should be confirmed at the end of the procedure. Haïssaguerre et al reported excellent results, exceeding 85% without the need for antiarrhythmic medications. Even with such an extensive ablation strategy, the LA transport function seems to be preserved in most patients.

The excellent results with the stepwise strategy notwithstanding, there are several limitations of the procedure. First, it may take 5 hours or longer to terminate long-standing, persistent AF. Second, linear LA ablation is required in most patients to terminate persistent AF during RF ablation. Although linear block can be achieved in most patients, it may be quite challenging in some patients and may be associated with potentially serious complications such as injury to the circumflex artery. Another limitation is that the technique may not be applicable to patients with continuous, uninterrupted persistent AF for more than a few years. These patients seem to fare poorly with any intervention, be it medical, catheter-based or surgical. Ultimately, randomized studies are needed to determine the ideal approach for patients with persistent AF.

Correct Answers:

Q1. B

Q2. D

Q3. C

Q4. C

Q5. C

REFERENCES

1. Shah D, Burri H, Sunthorn H, Gentil-Baron P. Identifying far-field superior vena cava potentials within the right superior pulmonary vein. *Heart Rhythm.* 2006;3:898–902.
2. Oral H, Knight BP, Ozaydin M, et al. Segmental ostial ablation to isolate the pulmonary veins during atrial fibrillation: Feasibility and mechanistic insights. *Circulation.* 2002;106: 1256–1262.
3. Chugh A, Oral H, Lemola K, et al. Prevalence, mechanisms, and clinical significance of macroreentrant atrial tachycardia during and following left atrial ablation for atrial fibrillation. *Heart Rhythm.* 2005;2:464–471.
4. Oral H, Pappone C, Chugh A, et al. Circumferential pulmonary-vein ablation for chronic atrial fibrillation. *N Engl J Med.* 2006;354:934–941.
5. Haïssaguerre M, Hocini M, Sanders P, et al. Catheter ablation of long-lasting persistent atrial fibrillation: Clinical outcome and mechanisms of subsequent arrhythmias. *J Cardiovasc Electrophysiol.* 2005;16:1138–1147.

CASE STUDY

24

Exertional Dyspnea After Atrial Fibrillation Ablation

Brian D. Powell, MD and Douglas L. Packer, MD

A 63-year-old male with a 10-year history of paroxysmal atrial fibrillation (AF) had frequent episodes with palpitations and fatigue in spite of antiarrhythmic drug therapy. He underwent a pulmonary vein isolation (PVI) ablation procedure without any immediate complications. His palpitations resolved; however, he returned 3 months later with dyspnea on exertion. Evaluation included a Holter monitor, which revealed normal sinus rhythm with a heart rate range of 66 to 110 bpm. Complete blood count and electrolytes were normal.

Question No. 1:

What is the most likely cause of His-dyspnea on exertion?

A. Pericardial effusion.

B. Pulmonary vein stenosis.

C. Pleural effusion.

D. Mitral regurgitation.

DISCUSSION

Patients can present with exertional dyspnea for several different reasons after PVI ablation for treatment of AF. When dyspnea occurs within the first week after an ablation, common causes include the following:

1. Fluid overload from an irrigated-tip ablation catheter.
2. Pericardial effusion with tamponade.
3. Recurrent atrial arrhythmias.

Patients who initially do well after a PVI, then develop dyspnea on exertion in the following weeks to months have the following differential diagnosis:

1. Pulmonary vein (PV) stenosis.
2. Delayed pericardial effusion.
3. Recurrent atrial arrhythmias.
4. Pneumonia, including aspiration pneumonia.
5. Exacerbation of previous comorbidities, such as congestive heart failure (CHF), coronary artery disease, or chronic obstructive pulmonary disease.

Symptoms can provide clues to the most likely etiology of the dyspnea that occurs in weeks to months following a PVI. Dyspnea secondary to PV stenosis may present with hemoptysis, cough, or chest pain. PV stenosis occurs in 1 to 2% of patients undergoing AF ablation.[1] Patients with a delayed pericardial effusion may develop edema and physical examination findings of tamponade (elevated jugular venous pressure, edema, and decreased heart sounds). Recurrent atrial arrhythmias may present

Practical Electrophysiology © 2015 Jasbir S. Sra, Masood Akhtar, Andrea Natale, David J. Wilber
Cardiotext Publishing, ISBN: 978-1-935395-17-1

with symptoms of palpitations and physical exam findings of an increased heart rate with an irregular rhythm. Pneumonia can occur in the setting of aspiration during conscious sedation or general anesthesia during an ablation. A productive cough or fever can be clues to this diagnosis.

Testing can be selected based on the presenting symptoms. An ECG, chest x-ray, and echocardiogram will often lead to the diagnosis. If PV stenosis is suspected, a computed tomography (CT) scan or magnetic resonance imaging (MRI) with contrast is the most useful test for diagnosis. When PV stenosis is detected, a ventilation perfusion scan can help determine the degree of ventilation perfusion mismatch, particularly if more than one stenosis is present.

When should an intervention be performed for PV stenosis? Intervention for PV stenosis is primarily driven by the severity of a patient's symptoms. Patients with single-vessel PV stenosis may present with few or no symptoms as long as the other ipsilateral PV is patent. Circumstances where early angioplasty or stent placement could be considered include symptomatic PV stenosis or a high-grade stenosis that may be at risk for complete occlusion. Patients should remain on anticoagulation with warfarin if moderate or severe PV stenosis is diagnosed. Thrombus formation in the area of stenosis can result in complete occlusion of the vessel. Once a complete occlusion occurs, a guide wire can sometimes be passed through a soft thrombus or sub-totally occluded vessel. Prior to intervention, a transesophageal echocardiogram is usually performed to exclude thrombus in the atria or PVs. Doppler signals from the stenosed PV usually reveal an increased velocity (> 1 m/s) with a loss of the normal phasic flow pattern on Doppler waveform (Figure 24.1).

Intervention for severe PV stenosis includes balloon angioplasty or stent placement. Almost 50% of patients develop restenosis with either approach.[2–5] For this reason, some centers first perform balloon angioplasty. If there is acute improvement in the degree of stenosis and the gradient, placement of a stent may not be required. If a stent is placed, it is ideal to use a stent with a diameter of 10 mm or greater, because smaller diameter stents have a higher risk of restenosis.[5] The diameters of the vessels proximal and distal to the stenosis are considered when selecting a stent size. A preprocedure CT (Figure 24.2) scan allows for measuring the vessel size in multiple planes. During the procedure, fluoroscopy and intracardiac echo can help guide angioplasty and stent placement.

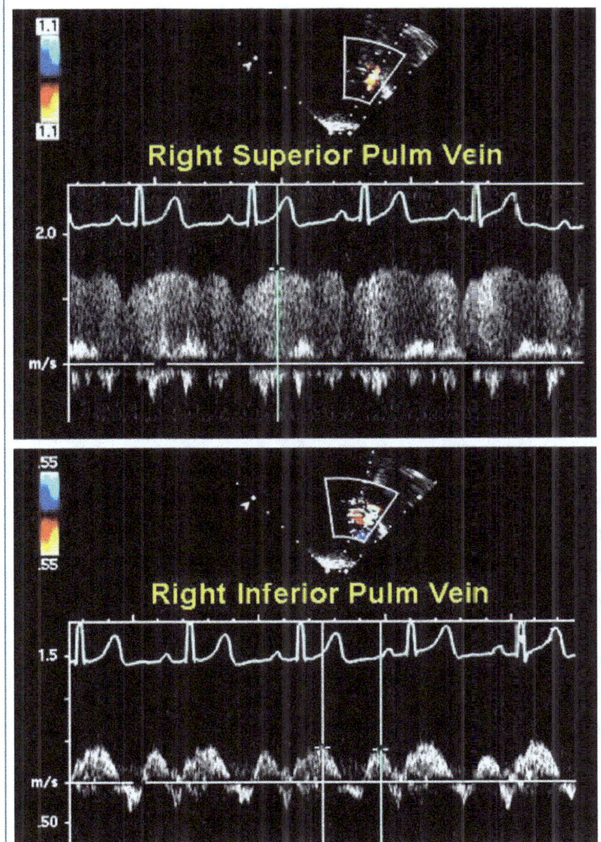

Figure 24.1. Doppler of the right superior pulmonary vein (RSPV) reveals an increased velocity of 1.5 m/s with continuous flow, consistent with PV stenosis. Normal PV velocities are seen in the right inferior PV (0.4 m/s).

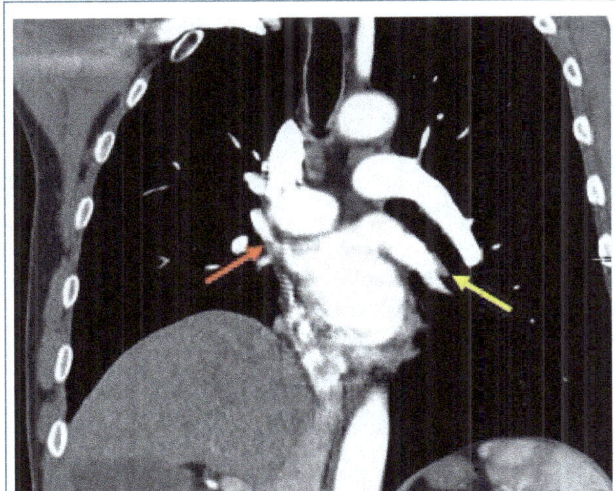

Figure 24.2. Cardiac CT with contrast shows RSPV stenosis measuring 3 mm in diameter (orange arrow). In comparison, the left inferior PV had a diameter of 15 mm (yellow arrow). There is also right hemidiaphragm elevation, which could be secondary to right phrenic nerve paralysis.

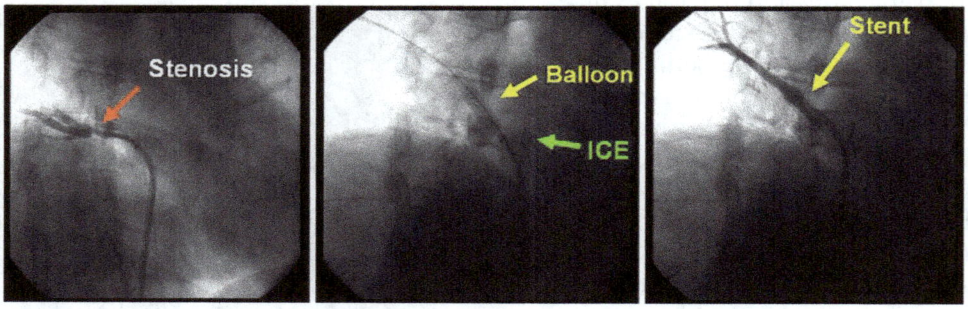

Figure 24.3. Right superior pulmonary vein stenosis followed by balloon dilatation, then stent placement. ICE, intracardiac echocardiography.

Anticoagulation and antiplatelet therapy after PV angioplasty or stenting is primarily based on extrapolation of data from studies of coronary artery stents and AF. The optimal duration of warfarin (target international normalized ratio 2:3) is unknown, but we typically continue it long-term for PV stenosis and risk of recurrent AF. Clopidogrel or aspirin can be used in combination with warfarin, particularly if a stent is placed.

The key to avoid PV stenosis is to avoid ablation inside the PVs. Rates of PV stenosis decreased over the past decade in part due to AF ablation transitioning from a focal ablation of AF triggers inside the PV to a wide-area circumferential ablation approach outside of the PV venoatrial junction. There also appear to be patient factors. Some patients are more prone to excessive inflammatory responses after ablation. This can lead to multiple PV stenoses. There are no good methods available at this time to accurately predict which patients are at increased risk of excessive inflammatory reactions.

Correct Answer:

Q1. B
PV stenosis.

The most likely etiology of progressive dyspnea 3 months after a PVI is PV stenosis. A delayed pericardial effusion presenting with symptoms 3 months after an ablation would be rare.

Treatment: After confirmation of PV stenosis by CT scan, the patient underwent pulmonary venography through a transseptal approach, which confirmed a 90% RSPV stenosis with a mean gradient of 10 mmHg (Figure 24.3). PV balloon dilatation was performed and a stent was placed in the RSPV with good results. Follow-up CT scan 3 months later revealed a patent RSPV. His-dyspnea significantly improved.

REFERENCES

1. Cappato R, Calkins H, Chen SA, et al. Worldwide survey on the methods, efficacy, and safety of catheter ablation for human atrial fibrillation. *Circulation.* 2005;111:1100–1105.
2. Saad EB, Marrouche NF, Saad CP, et al. Pulmonary vein stenosis after catheter ablation of atrial fibrillation: Emergence of a new clinical syndrome. *Ann Intern Med.* 2003;138:634–638.
3. Packer DL, Keelan P, Munger TM, et al. Clinical presentation, investigation and management of pulmonary vein stenosis complicating ablation for atrial fibrillation. *Circulation.* 2005;111:546–554.
4. Di Biase L, Fahmy TS, Wazni OM, et al. Pulmonary vein total occlusion following catheter ablation for atrial fibrillation: Clinical implications after long-term follow-up. *J Am Coll Cardiol.* 2006;48:2493–2499.
5. Prieto LR, Schoenhagen P, Arruda MJ, Natale A, Worley SE. Comparison of stent versus balloon angioplasty for pulmonary vein stenosis complicating pulmonary vein isolation. *J Cardiovasc Eletrophysiol.* 2008;19:673–678.

CASE STUDY

25

Paroxysmal Atrial Fibrillation with Pulmonary Vein Triggers

Pasquale Santangeli, MD, Luigi Di Biase, MD, PhD, and Andrea Natale, MD

A 62-year-old patient with weekly symptomatic episodes of paroxysmal atrial fibrillation (AF) after a previous failed ablation is referred for repeat catheter ablation. During the procedure, the intracardiac tracing shown in Figure 25.1 is recorded.

Question No. 1:

What does the intracardiac recording show?

A. Sinus rhythm.

B. Premature atrial contractions arising from the distal coronary sinus (CS).

C. Premature atrial contractions arising from the left superior pulmonary vein (LSPV).

D. Both A and B.

E. Both A and C.

DISCUSSION

Our approach at conventional intracardiac mapping during AF ablation involves placing a duo-decapolar catheter with the distal 10 poles in the CS and the proximal 10 poles along the high right atrium (HRA)/crista terminalis, and a decapolar circular mapping catheter within the PV antra.

During normal sinus rhythm, atrial activation begins from the HRA (more proximal poles of the 20-pole catheter), spreads toward the low-right atrium and then to the left atrium. The activation of the left atrium during normal sinus rhythm may occur through 3 different routes: (1) cranially through the Bachmann's bundle (up to 70% of patients); (2) at the level of the fossa ovalis; or (3) caudally at the region of the triangle of Koch. Predominant left atrial activation through the Bachmann's bundle can be demonstrated by distal CS activation (superior and lateral) preceding mid-CS activation (inferior and septal). On the other hand, predominant left atrial activation through the fossa ovalis/caudal right atrium is suggested by a proximal-to-distal activation of the CS.

In Figure 25.1, the first and third beats of the intracardiac tracing show normal sinus rhythm. The left atrial activation occurs predominantly at the level of the fossa ovalis/caudal right atrium. The contribution of the Bachmann's bundle to left atrial activation in this patient cannot be well-evaluated, due to the position of the CS catheter, which is not advanced to cover the more lateral aspect of the atrioventricular (AV) groove (Figure 25.2).

The second and fourth beats show an atrial activation sequence different from sinus rhythm. Indeed, the earliest endocardial activation is recorded at the level of the dipoles numbered 2,3 of the circular mapping catheter with a rapid spread toward the interatrial septum (Figure 25.3). Note that intracardiac recording precedes the onset of

Practical Electrophysiology © 2015 Jasbir S. Sra, Masood Akhtar, Andrea Natale, David J. Wilber
Cardiotext Publishing, ISBN: 978-1-935395-17-1

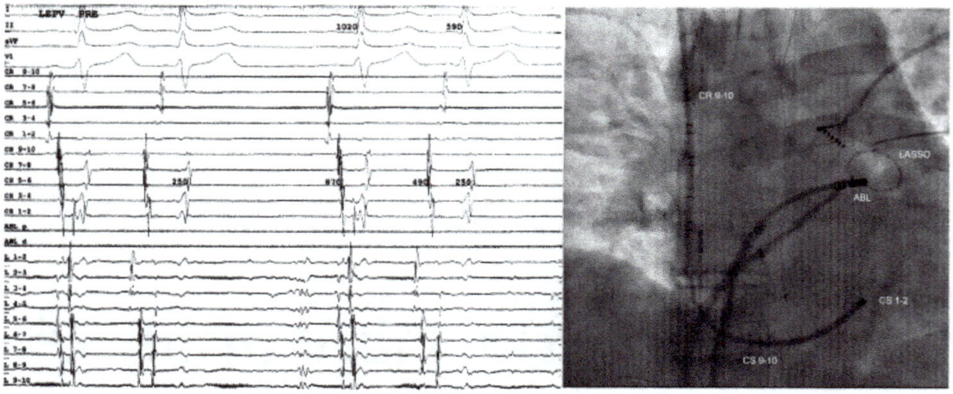

Figure 25.1. Intracardiac tracings during the procedure as well as the catheter location.

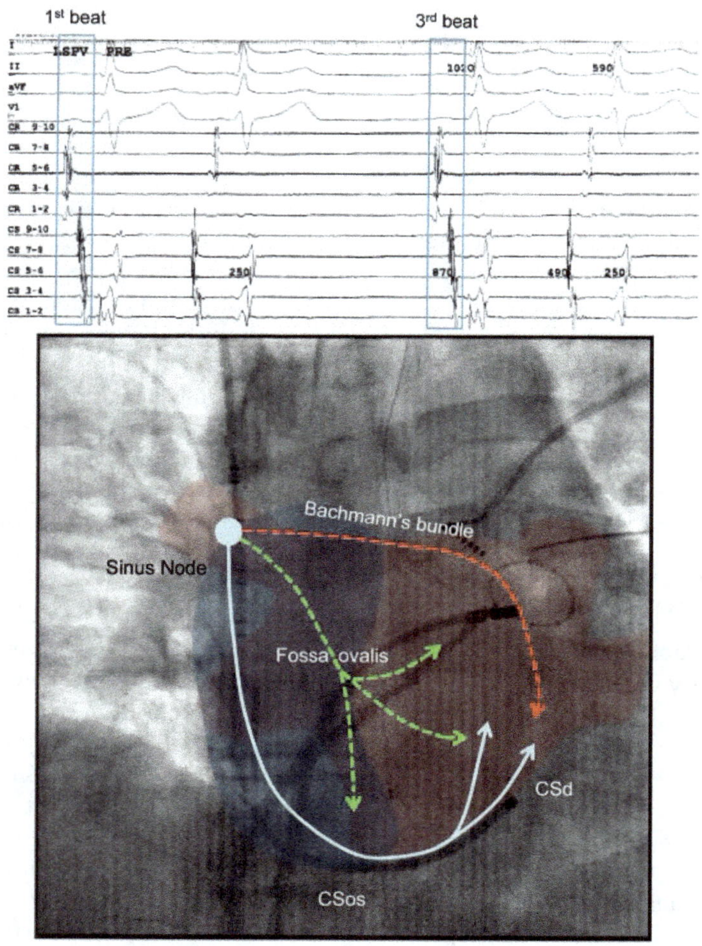

Figure 25.2. Different routes of atrial activation during sinus rhythm. The first and third beats of the intracardiac tracing (upper panel) are sinus beats, with right-to-left atrial activation most likely occurring through the caudal right atrium/fossa ovalis (blue and green lines) as suggested by proximal-to-distal CS activation. The contribution of the Bachmann's bundle to left atrial activation in this patient cannot be well-evaluated due to the position of the CS catheter (slightly withdrawn and not covering the lateral aspect of the AV groove). CSd, distal coronary sinus; CSos, coronary sinus ostium.

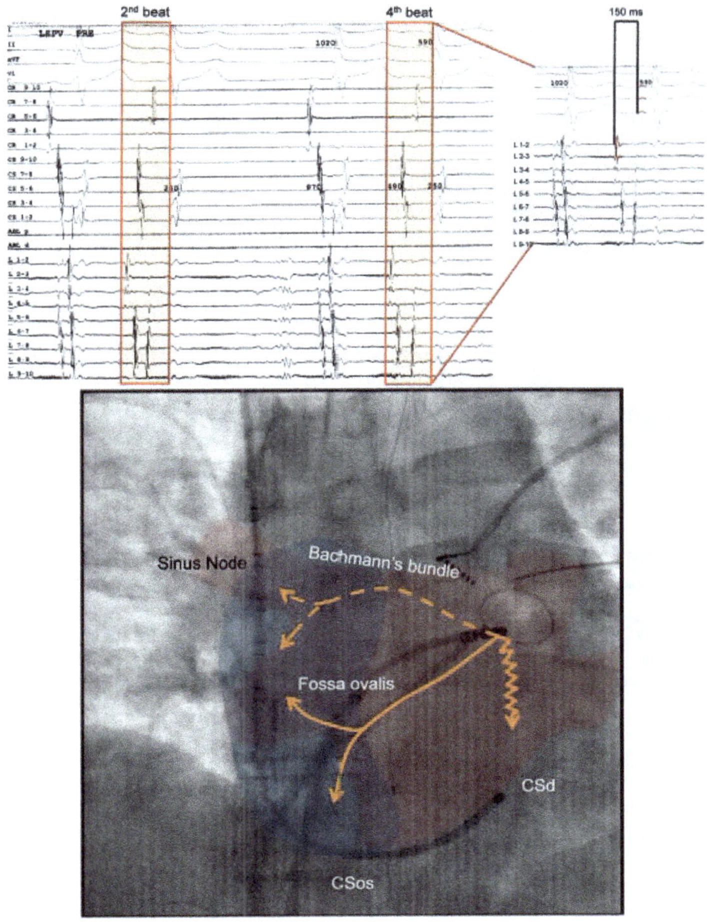

Figure 25.3. Evidence of LSPV origin of the atrial ectopic beats. The earliest potential at endocardial mapping is recorded at the dipoles 2,3 of the circular mapping catheter. Intra-atrial activation spreads rapidly toward the septum, and right atrium is activated most likely from the fossa ovalis. Bachmann's bundle does not contribute to right atrial activation in this case (dashed line). Delayed conduction in the posterolateral left atrial wall, most likely due to previous ablation, may explain the delayed activation of the distal CS (CSd). CSos, CS ostium.

ectopic P wave on the surface ECG by 150 ms (see Figure 25.3). Based on these findings, a diagnosis of premature atrial contractions originating from the LSPV is established.

Ectopic activity arising from the PVs is the most frequent AF trigger. The site of origin of atrial ectopic activity can be ascertained only by means of endocardial mapping, although several criteria based on surface ECG analysis have been proposed. Such criteria are based on the assumption that intra-atrial conduction is preserved and, therefore, are most useful in patients without significant left atrial scarring, such as those who already underwent left atrial ablation.

Leads aVL and V_1 are useful for distinguishing right from left atrial origin of ectopic activity. A left atrial origin is suggested by a negative (more lateral) or isoelectric (more septal) P wave in lead aVL and a positive wave in lead V_1. A positive P wave in leads V_2 to V_4 suggests a posterior origin. The width of the P wave is affected by the proximity to the septum. Left PVs are located in the posterolateral left atrium, whereas right PVs are posteroseptal left atrial structures. Therefore, ectopic activity from the left PVs may be suspected in the presence of a predominantly negative P wave in lead aVL and a positive P wave in leads V_1 to V_4; right PV ectopic activity may be suspected in the presence of a relatively narrow P wave—isoelectric or slightly positive in lead aVL and positive in leads V_1 to V_4. The distinction between superior and inferior PV origin by surface ECG is highly imprecise. As a general rule, however, beats originating from the superior PVs are characterized by a striking positive P wave in inferior leads.

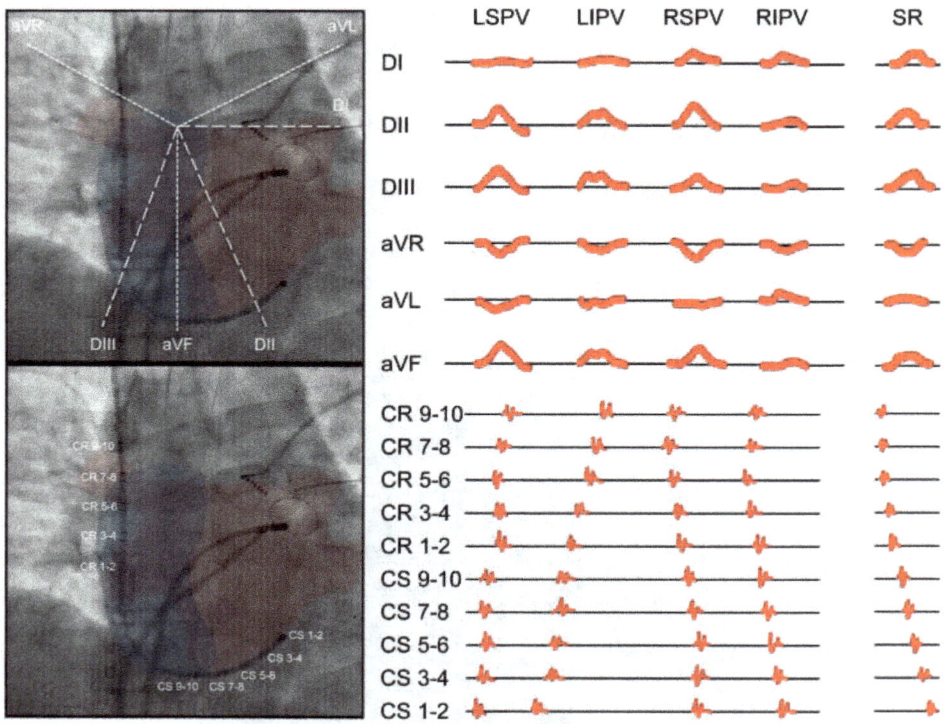

Figure 25.4. Example of surface ECG (frontal plane) and intracardiac activation sequence at the duo-decapolar catheter of ectopic beats arising from different PVs. The patterns showed in figure are nonspecific and do not obviate the need for more careful mapping with a circular mapping catheter. LSPV, left superior pulmonary vein LIPV, left inferior pulmonary vein; RSPV, right superior pulmonary vein; RIPV, right inferior pulmonary vein; SR, sinus rhythm.

As far as endocardial mapping is concerned, the pattern of atrial activation recorded at the duo-decapolar catheter is a useful tool to approximate the site of origin and to guide the positioning of the circular mapping catheter to definitely disclose a PV origin of ectopies. Typically, CS activation occurs from distal-to-proximal for left PVs, and from proximal-to-distal for right PVs (Figure 25.4). Variations in activation timing between the CS and crista terminalis poles can be used to further differentiate upper from lower PVs (see Figure 25.4). However, these activation patterns are not specific and do not obviate the need for a more careful mapping. For instance, the RSPV is in close proximity to the HRA, such that signals from both structures can display the same P-wave morphology at the surface ECG and the same activation pattern at the duo-decapolar catheter.

Therefore, a more detailed mapping at the level of PVs should be carried out to reach a definite diagnosis. To this aim, a circular mapping catheter is of invaluable aid. The circular mapping catheter should be positioned at the region of interest (ie, left or right PVs) to identify the earliest activation time relative to the onset of the ectopic P wave. Usually, when the activation time at the circular mapping catheter is > 30 to 40 ms before the onset of the ectopic P wave, a diagnosis of PV firing can be made with reasonable precision.

Spontaneous firing from the PVs has been consistently demonstrated to act as a trigger for AF, and radiofrequency catheter ablation is an established treatment to achieve cure in such patients. The initial experience with catheter ablation of AF relied on targeting solely the arrhythmogenic PVs responsible for precipitating AF. Induction of atrial ectopy or AF was required with this approach, which was a significant limitation, since up to 30% of patients actually have no sufficient ectopies to allow accurate mapping. Multiple periprocedural cardioversions were also required, as sustained episodes of AF were frequently induced. Even more importantly, the clinical outcome of such an approach was fairly disappointing and the risk of postablation PV stenosis was considerable.

Subsequent evolution of PV ablation involved empirical isolation of all PVs, since over 85% of patients actually have multiple arrhythmogenic PVs, and moving downstream in the atrium to minimize the risk of PV stenosis and triggers localized in the more proximal antral region of the PVs.

Accordingly, PV antrum isolation is thus far the most widespread approach to AF ablation. A circular mapping catheter is necessary to demonstrate effective electrical isolation of the PVs; this approach has been reported as an effective and reproducible ablation technique to achieve cure in a substantial proportion of patients.

Correct Answer:

Q1. E

REFERENCE

1. Natale A, Jalife J. *Atrial Fibrillation: From Bench to Bedside.* Totowa, NJ: Humana Press; 2008.

CASE STUDY

26

Identifying Extra Pulmonary Vein Triggers for AF Initiation

Luigi Di Biase, MD, PhD, Pasquale Santangeli, MD, and Andrea Natale, MD

A 51-year-old female patient with essential hypertension and preserved left ventricular ejection fraction presented for atrial fibrillation (AF) ablation after a previous failed ablation for paroxysmal AF. After left atrial instrumentation was achieved by double transseptal puncture, the circular mapping catheter was placed in all 4 pulmonary vein (PV) antra and along the posterior wall of the left atrium. All 4 PVs and the posterior wall did not regain conduction and were isolated. Challenge test with high dose of isoproterenol (up to 30 mcg/min for 15 minutes) showed no reconnection between the left atrium and the PVs. Firing from the left atrial appendage (LAA) was triggered by isoproterenol infusion at 30 mcg/min.

Question No. 1:

Based on history and the electrograms (EGMs) showed in Figure 26.1, what should be done in a patient with paroxysmal AF undergoing a redo procedure with all PVs isolated and evidence of extra PV triggers?

A. All demonstrated extra PV triggers should be ignored and the procedure aborted.

B. Demonstrated extra PV triggers that are initiating tachycardia or AF should be ablated.

C. Isolation of the superior vena cava should be performed.

D. All of the above.

E. None of the above.

DISCUSSION

Pulmonary vein antrum isolation (PVAI) is one of the most utilized techniques to achieve PV disconnection. Our approach is guided by a circular mapping catheter and by intracardiac echocardiography (ICE), along with the empirical isolation of the superior vena cava in either paroxysmal or nonparoxysmal patients.[1–4] Specifically, in patients with paroxysmal AF, the PV antra and the portion of the posterior wall contained within the PV area are targeted during the procedures with exit block as the end point.[1] After ablation, in all patients, a challenge with isoproterenol infusion up to 30 mcg/min for 15 minutes is performed to identify extra PV triggers.[4] In the case that atrial flutter or atrial tachycardias are induced, these are mapped and ablated. Radiofrequency energy is usually delivered with a 3.5-mm open-irrigated ablation catheter (Biosense Webster Inc., Diamond Bar, CA) with a maximum temperature of 42°C and power up to 45 watts and a flow rate of 30 cc/min. We recently reported the importance of the LAA as an underestimated

Practical Electrophysiology © 2015 Jasbir S. Sra, Masood Akhtar, Andrea Natale, David J. Wilber
Cardiotext Publishing, ISBN: 978-1-935395-17-1

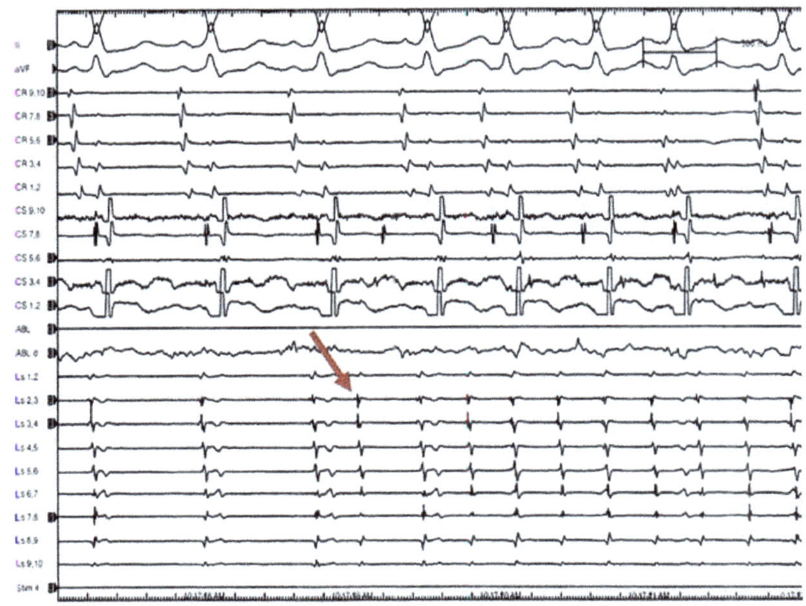

Figure 26.1. The circular mapping catheter is positioned at the base of the LAA. During isoproterenol infusion at 30 mcg/min, atrial tachycardia arising from the LAA is initiated after a premature atrial contraction (PAC, red arrow).

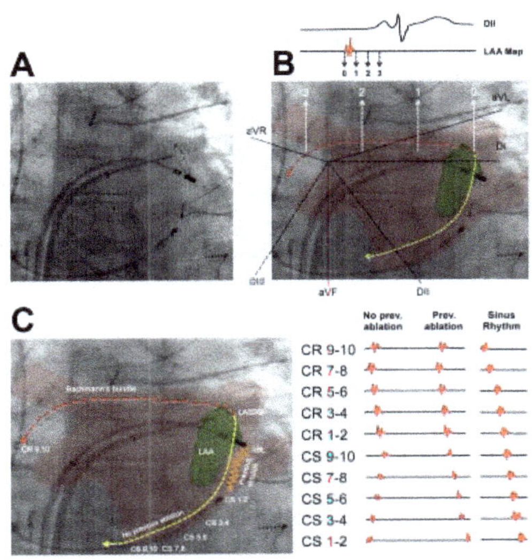

Figure 26.2. A: LAA isolation. The circular mapping catheter is positioned at the ostium of the LAA under ICE guidance. Ablation lesions are delivered proximally to the circular mapping catheter. B: Hypothetical activation sequence of a PAC originating from the LAA. In the frontal plane, the axis of the ectopic P wave is negative in D1-aVL and positive in the inferior leads. The interatrial activation occurs antero-superiorly through the Bachmann's bundle and infero-posteriorly through the coronary sinus musculature, which is in close proximity to the LAA at the level of the lateral great cardiac vein. This generates opposing activation vectors (antero-superior vs infero-posterior) causing a long isoelectric signal on the surface ECG (timings 1–2) despite the early local activation (timing 0) recorded at the circular mapping catheter. At the timing 3, the depolarization front of the PAC is similar to a septal high-right atrial PAC, but the AV node has been already engaged at the timing 2 by the depolarization front occurring through the coronary sinus musculature. Ultimately, this generates a P-wave morphology similar to a PAC originating from the septal high right atrium with an apparently short PR interval (but normal time between the earliest LAA EGM to the Q wave of the surface QRS). C: Intracardiac activation sequence at the level of a duodecapolar catheter (10 poles at crista terminalis and 10 poles in the coronary sinus) of a PAC originating from the LAA. In the absence of previous ablation at the level of the mitral isthmus the activation sequence proceeds through the coronary sinus musculature (distal-to-proximal) and the Bachmann's bundle. In the case of previous ablation at the level of the mitral isthmus, a PAC arising from the LAA cannot activate the distal coronary sinus due to block at the level of the mitral isthmus, and the activation proceeds only through the Bachmann's bundle producing a pattern similar to a high-right atrial PAC. In this latter case, the AV node can be engaged only after that the entire right atrium has been depolarized in a cranio-caudal fashion, and the PR interval on the surface ECG is expected to be normal.

trigger site for AF, even in patients with paroxysmal AF undergoing a repeat procedure.[5] In the case described in this chapter, firing from the LAA has been demonstrated. The need to ablate extra PV firings from this structure and from other structures is essential to improve long-term outcome in these patients.[5] In our recently published study, we reported the prevalence of AF triggers from the LAA in a series of 987 patients (29% paroxysmal AF) referred for redo catheter ablation. In 8.7% of patients, the LAA was found to be the only source of arrhythmia. We demonstrated that isolation of such structure was associated with a better outcome when compared to focal ablation. The technique employed for LAA isolation is similar to that utilized for PVAI. The circular catheter is positioned at the ostium of the LAA, guided by ICE, and the LAA is electrically isolated (Figure 26.2). Since ablation of the LAA could increase the risk of perforation during ablation, increased caution should be taken into account when isolating this structure.

In the present case, the first 3 beats of the intracardiac EGMs show normal sinus rhythm (see Figure 26.1). The fourth beat shows a PAC with an early activation at the level of the dipoles numbered 2,3 on the circular mapping catheter, which is positioned at the base of the LAA. The PAC initiates a left atrial tachycardia with a cycle length of 230 ms. Of interest, the activation pattern in the high right atrial catheter (CR 9,10 to CR 1–2; see Figures 26.1 and 26.2, respectively) and the proximal coronary sinus catheter are similar to the ones in sinus rhythm (first 3 beats). This feature, together with the short PR interval of the PAC, highlights the importance of recording from the LAA, since one could be misled by the PAC origin.

Correct Answer:

Q1. B

REFERENCES

1. Di Biase L, Elayi CS, Fahmy TS, et al. Atrial fibrillation ablation strategies for paroxysmal patients: Randomized comparison between different techniques. *Circ Arrhythm Electrophysiol.* 2009;2:113–119.
2. Elayi CS, Verma A, Di Biase L, et al. Ablation for longstanding permanent atrial fibrillation: Results from a randomized study comparing three different strategies. *Heart Rhythm.* 2008;5:1658–1664.
3. Bhargava M, Di Biase L, Mohanty P, et al. Impact of type of atrial fibrillation and repeat catheter ablation on long-term freedom from atrial fibrillation: Results from a multicenter study. *Heart Rhythm.* 2009;6:1403–1412.
4. Arruda M, Mlcochova H, Prasad SK, et al. Electrical isolation of the superior vena cava: An adjunctive strategy to pulmonary vein antrum isolation improving the outcome of AF ablation. *J Cardiovasc Electrophysiol.* 2007;18:1261–1266.
5. Di Biase L, Burkhardt JD, Mohanty P, et al. Left atrial appendage: An under recognized trigger site of atrial fibrillation. *Circulation.* 2010;122:109–118.

CASE STUDY

27

Catheter Ablation for Long-Standing Persistent AF

Méléze Hocini, MD, Ashok J. Shah, MD, Shinsuke Miyazaki, MD, Lena Rivard, MD, Amir S. Jadidi, MD, Daniel Scherr, MD, PD, Stephen B. Wilton, MD, MSc, Laurent Roten, MD, Patrizio Pascale, MD, Michala Pedersen, MD, DPhil, Nicolas Derval, MD, Sébastien Knecht, MD, PhD, Frederic Sacher, MD, Pierre Jaïs, MD, and Michel Haïssaguerre, MD

A 64-year-old hypertensive gentleman presented with symptomatic, 5-year-long permanent, drug-refractory atrial fibrillation (AF). Patient complained of dyspnea and effort intolerance—NYHA class II despite good rate control therapy. He was diagnosed with paroxysmal AF at age 30. Echocardiographic left atrial (LA) size was 49 mm and left ventricular ejection fraction was 58%. 12-lead ECG is shown in Figure 27.1.

Question No. 1:

What are the procedural end points of stepwise ablation approach?

Question No. 2:

What are the critical structures for termination of AF during stepwise ablation?

Question No. 3:

What are the short- and long-term clinical outcomes of stepwise ablation?

Question No. 4:

What are the characteristics of arrhythmia recurring after stepwise ablation?

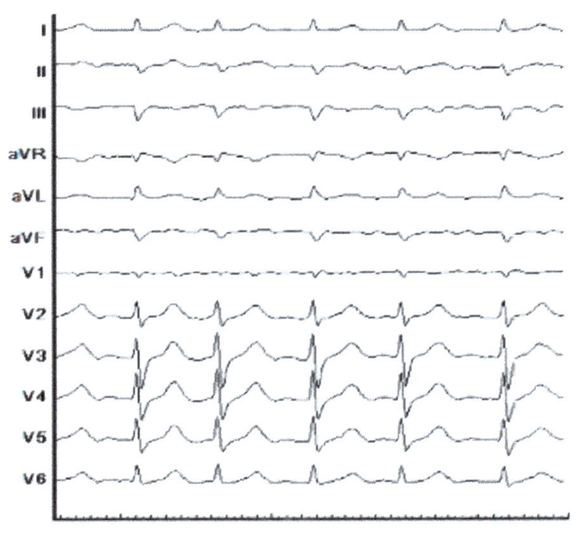

Figure 27.1 Baseline 12-lead ECG.

Practical Electrophysiology © 2015 Jasbir S. Sra, Masood Akhtar, Andrea Natale, David J. Wilber
Cardiotext Publishing, ISBN: 978-1-935395-17-1

DISCUSSION

Certain baseline electrocardiographic parameters can predict the procedural complexity and response to ablation in persistent AF. These parameters include: baseline average global fibrillatory cycle length (CL) in leads with best visible fibrillatory (F) waves' usually magnified lead V_1; and baseline average amplitude of the F wave measured in magnified leads II and V_1.[1] Ten discrete F waves were selected during a period of slow ventricular rate to avoid superimposition of QRST, and measurements were performed using electronic calipers.

After intracardiac introduction of the catheters, baseline atrial fibrillatory cycle length (AFCL) was measured simultaneously with the Lasso catheter in the left atrial appendage (LAA) and the mapping (ablation) catheter in the right atrial appendage (RAA), averaging a recording of at least 10 consecutive cycles to obtain local AFCL in the 2 chambers. Baseline LAA AFCL was 146 ms and RAA AFCL was 143 ms.

Using stepwise approach for persistent AF ablation, 4 pulmonary veins were isolated at first. Circumferential PVI was performed in a 2 × 2 manner, encircling 2 ipsilateral veins together under the guidance of a single Lasso catheter. Figure 27.2 shows left PVI. The procedural time for PVI was 50 minutes, with radiofrequency duration of 42 minutes and fluoroscopic time of 16 minutes.

Following PVI, AFCL was measured in the LAA/RAA regionally. LAA AFCL rose by 16 ms after successful PVI as shown in Figure 27.3. High gradient between local AFCLs recorded from the pulmonary vein and the LAA before ablation can substantially affect the regional and global AFCL after PVI.

Ablation of complex LA electrograms (EGMs) was the next step after PVI. Complex EGM targets included areas with continuous activity, long fractionated activity, activation gradient over small area, and frequency gradient between contiguous areas. The local organization in AF is the end point of ablation. The Lasso catheter was withdrawn into the right atrium and used for monitoring RAA AFCL continuously during this step.

Atrial complex EGM ablation was started endocardially, targeting the coronary sinus (CS) through ablation in the low inferior left atrium by looping the radiofrequency catheter using the sheath support as shown in Figure 27.4. This helps to avoid unnecessary ablation inside the CS. The ablation started on the septal side and was continued while dragging the catheter back toward the lateral inferoposterior LA. The

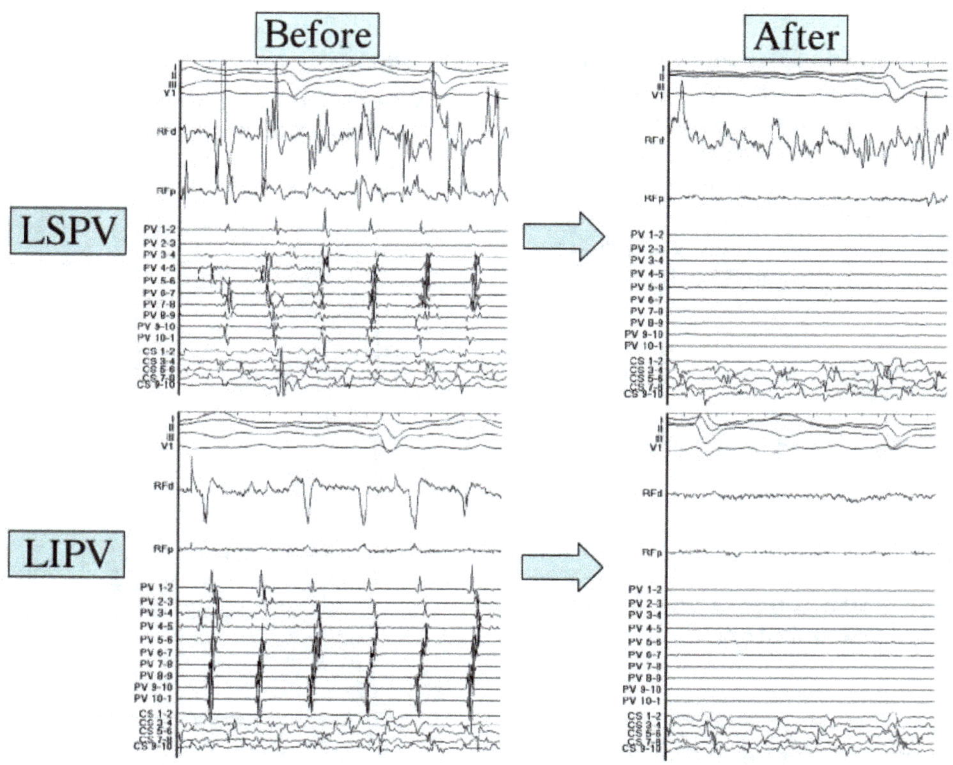

Figure 27.2. Example of left pulmonary vein isolation (PVI).

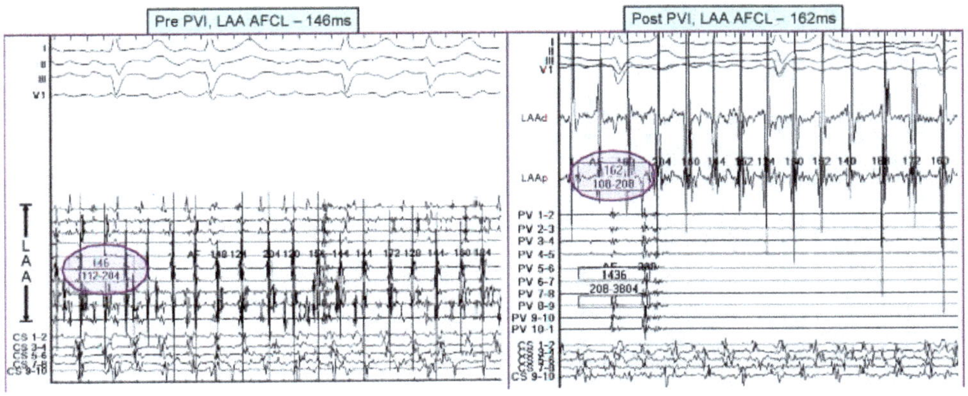

Figure 27.3. LAA AFCL both pre- and post-PVI.

organization of the EGMs and slowing down of the activity in the CS region was the end point of regional ablation. This ablation aimed to eliminate multiple LA-CS connections responsible for maintenance of AF. In cases of anatomical complexity and thick local tissue, it is necessary to ablate into the CS with power limited to 20 to 25 W.

Based on previous work, we targeted the left septum from posterior to superior to anterior and, finally, the inferior quadrant with the end point of local slowing and organization. The next target site

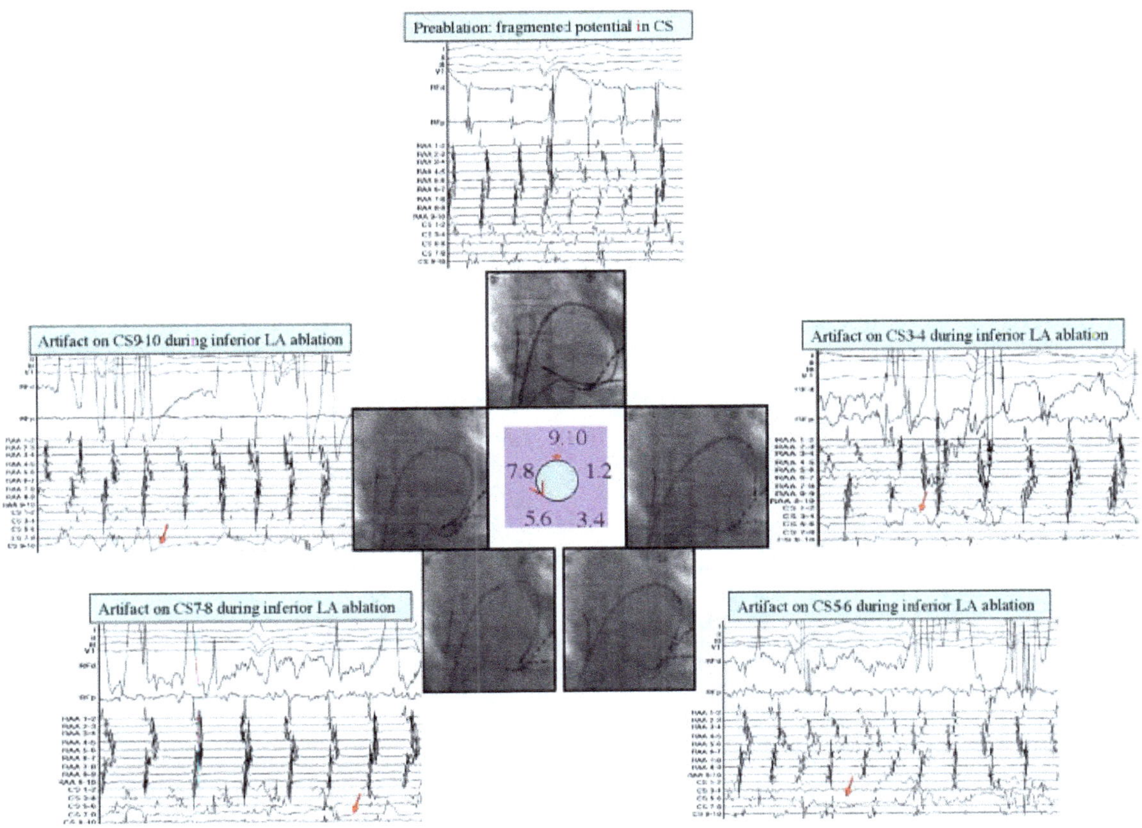

Figure 27.4. EGM-based ablation: Inferior LA ablation.

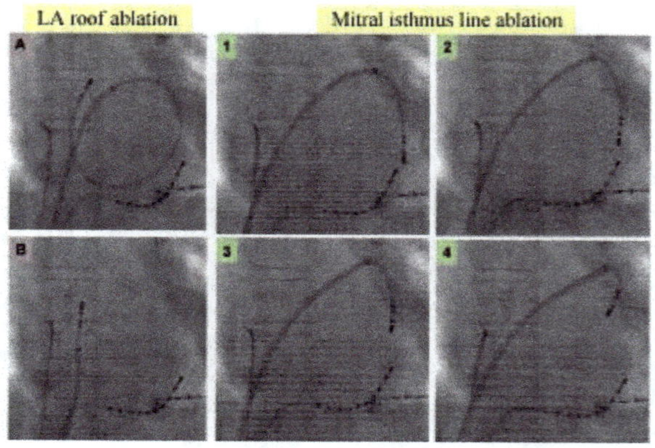

Figure 27.5. Catheter positions for linear ablation. Different positions for roof line (**Panels A** and **B**) and mitral isthmus line (1-4) are depicted in the figure.

was the mouth of LAA with the same end points and power delivery limited to 25 to 30 W. The other sites of complex atrial EGM-based ablation include posterior and anterior LA.[2]

Linear ablation was undertaken as the next step in LA ablation since AF continued after EGM-based ablation and PVI. LA roof linear ablation was undertaken in the superior part of the LA, followed by left posterior mitral isthmus linear ablation. The catheter positions for linear ablation are shown in Figure 27.5. These lines are aimed to create fixed anatomical obstacles in impulse propagation. The end point of linear ablation during AF was elimination of local atrial EGMs.

After complex atrial EGM and linear ablation in the left atrium, AFCL was rechecked in both appendages as shown in Figure 27.6. We had a frequency gradient of 20 ms from the right to the left atrium (LAA AFCL—178 ms and RAA AFCL—158 ms), indicating that the right atrium

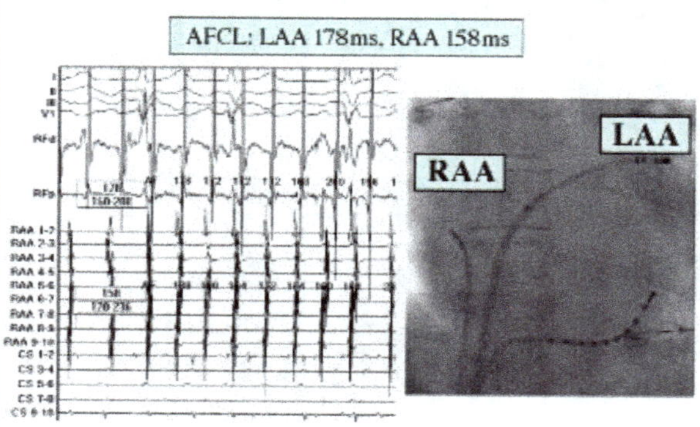

Figure 27.6. Atrial fibrillatory cycle length in the LAA and RAA.

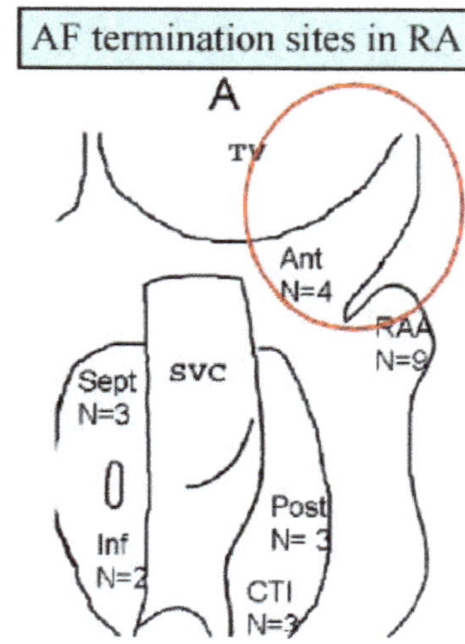

Figure 27.7. Atrial fibrillation termination sites in the right atrium. CTI, cavotricuspid isthmus; IVC, inferior vena cava; RAA, right atrial appendage; SVC, superior vena cava; TV, tricuspid valve. *(Reprinted from [3], with permission from Elsevier.)*

was driving the AF process, necessitating additional ablation in the RA.

Based on previously published work,[3] the right atrial sites displaying complex EGMs were targeted with similar end points as in the LA (Figure 27.7).

Ablation at the anterior/RAA resulted in conversion of AF into AT with CL of 208 ms (Figure 27.8). Mechanism and localization of AT was defined by activation and entrainment mapping. The AT was mapped and diagnosed to be focal from the CS ostium. Ablation at the CS ostium terminated AT, and sinus rhythm was restored. PVI was confirmed in sinus rhythm.

Left mitral isthmus ablation was assessed for complete bidirectional block using differential pacing techniques (Figure 27.9).

For the LA roof, activation detour with caudocranial pattern during LAA pacing is indicative of complete block (Figure 27.10). Cavotricuspid isthmus linear ablation is undertaken, finally, and complete block assessed during pacing. Induction of AF was not attempted.

Correct Answers:

Q1. Cessation of AF is the end point of persistent AF ablation using stepwise approach. Nonindűcibility is not the end point. When AF terminates into AT, all ATs are chased and ablated. PVI and linear blocks should be ascertained in sinus rhythm.

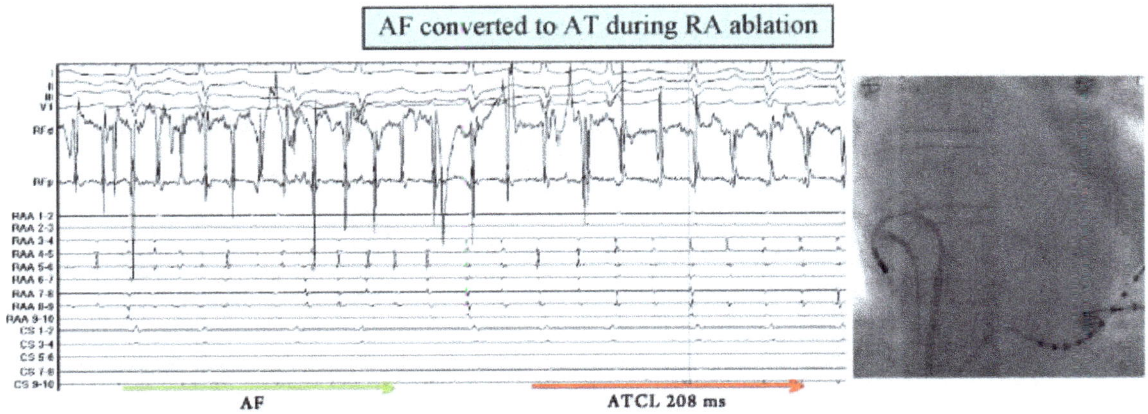

Figure 27.8. Atrial fibrillation converting to atrial tachycardia (AT) during right atrial ablation.

Q2. Using stepwise approach, the LAA, CS, pulmonary vein regions and the sites of linear ablation have the greatest impact in terms of slowing of the F rate before AF conversion to AT or sinus rhythm. Certain structures in the right atrium (see Figure 27.10) also contribute toward the maintenance of AF in 20% of patients, and ablation of these structures can substantially affect the AFCL and/or conversion to AT or SR. These critical locations sustain focal organized arrhythmias and have higher frequency and/or activation gradients, which make them important ablation targets to restore sinus rhythm.[2–4]

Q3. In our experience, acute termination could be achieved in more than 85% of patients having long-lasting persistent AF of ≤3 years duration and baseline AFCL ≥140 ms. Arrhythmia recurrence was observed in up to 40% of patients after the blanking period necessitating repeat ablation. One year after repeat procedure, up to 95% of patients maintained sinus rhythm.[5]

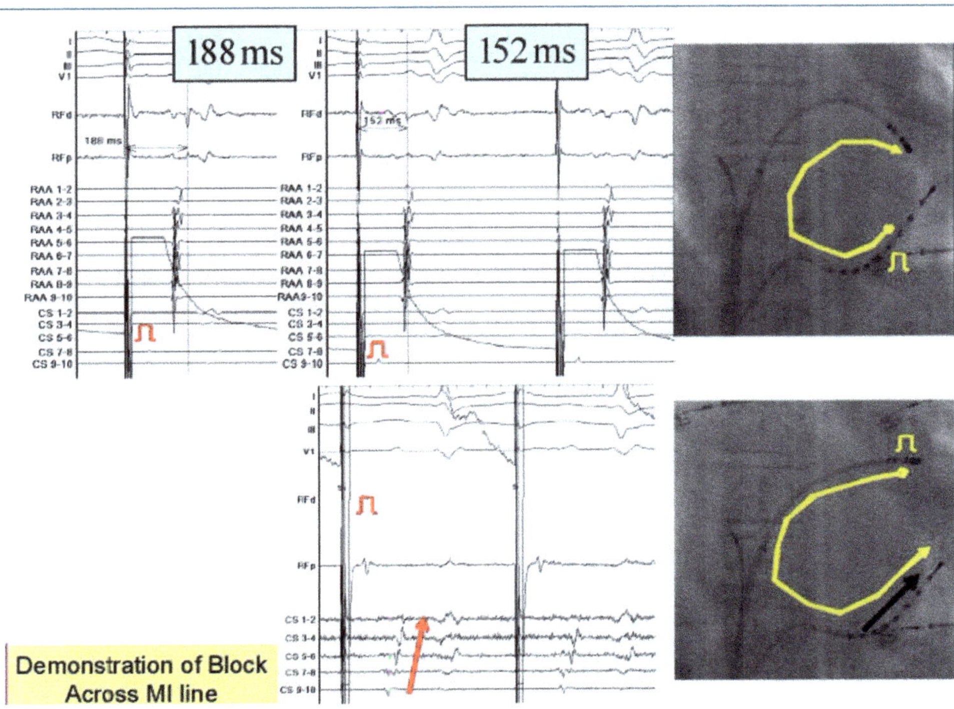

Figure 27.9. Demonstration of block across mitral isthmus line.

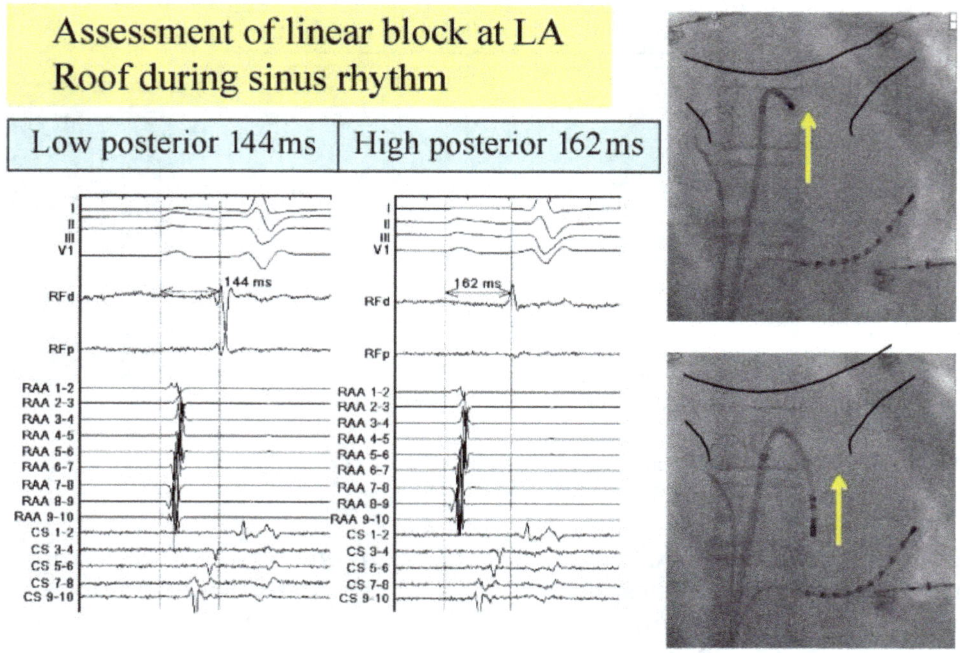

Figure 27.10. Assessment of linear block at LA roof during sinus rhythm.

Q4. More than 90% of the recurring arrhythmias after index stepwise ablation procedure are ATs. Mechanistically, they are focal or isthmus-dependent macroreentry. Notably, the successful AT ablation sites are similar to those where the greatest impact in terms of slowing or conversion of AF was observed during index stepwise procedure, *viz.* LAA, CS, pulmonary vein regions and sites of linear ablation.[3,5]

REFERENCES

1. Nault I, Lellouche N, Matsuo S, et al. Clinical value of fibrillatory wave amplitude on surface ECG in patients with persistent atrial fibrillation. *J Interv Card Electrophysiol.* 2009;26:11–19.
2. Haïssaguerre M, Sanders P, Hocini M, et al. Catheter ablation of long-lasting persistent atrial fibrillation: Critical structures for termination. *J Cardiovasc Electrophysiol.* 2005;16:1125–1137.
3. Hocini M, Nault I, Wright M, et al. Disparate evolution of right and left atrial rate during ablation of long-lasting persistent atrial fibrillation. *J Am Coll Cardiol.* 2010;55:1007–1016.
4. Takahashi Y, Hocini M, O'Neill MD, et al. Characterization of electrograms associated with termination of chronic atrial fibrillation by catheter ablation. *J Am Coll Cardiol.* 2008;47:2005–2012.
5. Haïssaguerre M, Hocini M, Sanders P, et al. Catheter ablation of long-lasting persistent atrial fibrillation: Clinical outcome and mechanisms of subsequent arrhythmias. *J Cardiovasc Electrophysiol.* 2005;16:1138–1147.

CASE STUDY

28

Hybrid Ablation for Recurrent Left Atrial Fibrillation Using Pericardioscopy and Transseptal Approach

David C. Kress, MD

A 68-year-old man presented with extreme symptoms of fatigue, palpitations, and dyspnea. He had a history of atrial tachyarrhythmias dating back to paroxysmal atrial fibrillation (AF) 11 years ago, and had been in persistent AF for the past 3 months. He failed 2 cardioversions 4 years ago, another cardioversion 7 months ago, and remained in AF despite sotalol.

The patient had a left atrial diameter of 5.7 cm. Left ventricular ejection fraction was normal, and there was no significant valvular disease, coronary artery disease, or underlying lung disease. Immediate preablation transesophageal echocardiogram (TEE) showed no left atrial appendage (LAA) thrombus.

Question No. 1:

What are the advantages of pericardioscopic surgical ablation to treat AF?

A. It avoids painful sternal or intercostal incisions.

B. It is the least invasive surgical approach into the pericardial space currently available.

C. It should enhance the efficiency, efficacy, and safety of percutaneous ablation.

D. All of the above.

DISCUSSION

Atrial fibrillation was present in approximately 3 million patients in 2005 and is responsible for more hospital admissions than any other heart rhythm disturbance in the United States.[1] It is a leading cause of stroke, and may require lifelong anticoagulation in patients with significant risk factors.

Patients who are symptomatic despite the use of one or more antiarrhythmic drugs may be candidates for ablative therapy. Procedural success with ablation is influenced by factors such as left atrial size, duration since onset and persistence despite cardioversion. A multidisciplinary approach in which patients are seen by both an electrophysiologist and a surgeon plays a critical role in the treatment of AF.

The patient described in this case underwent successful termination of persistent AF using a new hybrid ablation procedure[2] that adds epicardial ablation via subxiphoid pericardioscopy to conventional mapping-guided transseptal catheter ablation.

The procedure was carried out in an electrophysiology/surgical hybrid suite that provided simultaneous multichannel intracardiac (using high right atrial and decapolar coronary sinus [CS] catheters) and surface electrocardiographic (ECG) monitoring; high-resolution fluoroscopy with 3D computed tomography (CT) registration of the left atrium, esophagus, and CS; CARTO electroanatomical mapping (Biosense Webster Inc., Diamond Bar, CA) with both intra-atrial and epicardial data points; and full surgical capabilities including laparoscopic monitors, carts, and instruments.

A 2-cm subxiphoid midline incision was made and an 11-mm port placed using Hassan technique for a 10-mm laparoscope. Two 5-mm accessory ports were placed for instruments. Reverse Trendelenburg and CO_2 insufflation provided visualization of the central tendon of the diaphragm. It was opened 2 cm to the left of the falciform ligament using conventional laparoscopic techniques. The pericardioscope cannula was inserted in the midline incision and guided to the oblique sinus under direct visualization with the aid of a 7-mm laparoscope. Ablation of the anterior left and right pulmonary vein (PV) trunks was also performed through this access.

Epicardial radiofrequency ablation lesions were placed as shown in Figure 28.1, with epicardial and subsequent endocardial lesions highlighted. The 90-second lesions were placed using a flexible 3-cm saline-irrigated unipolar radiofrequency coil (Figure 28.2) that has the ability to attach to the atrium with vacuum. For curved lesions it can be placed on a guide wire rail that is built into the pericardioscope, or it can be used off the rail for more linear lesions. An esophageal temperature probe was placed with fluoroscopy guidance to monitor esophageal lumen temperature. Lesions are immediately terminated if the temperatures rise over 1°C.

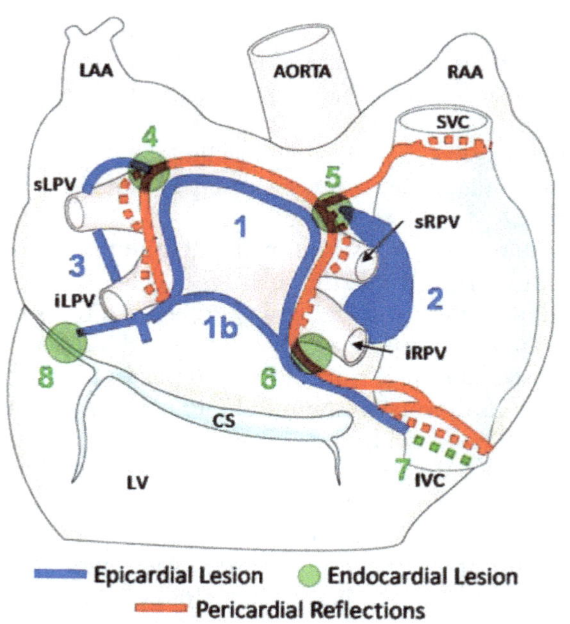

Figure 28.1. Hybrid lesion pattern with pericardial reflections (red), epicardial lesions (blue), and endocardial lesions (green) indicated. CS, coronary sinus; iLPV, inferior left PV; iRPV, inferior right PV; IVC, inferior vena cava; LAA, left atrial appendage; LV, left ventricle; RAA, right atrial appendage; sLPV, superior left PV; sRPV, superior right PV; SVC, superior vena cava. (Reproduced with permission from nContact, Inc.)

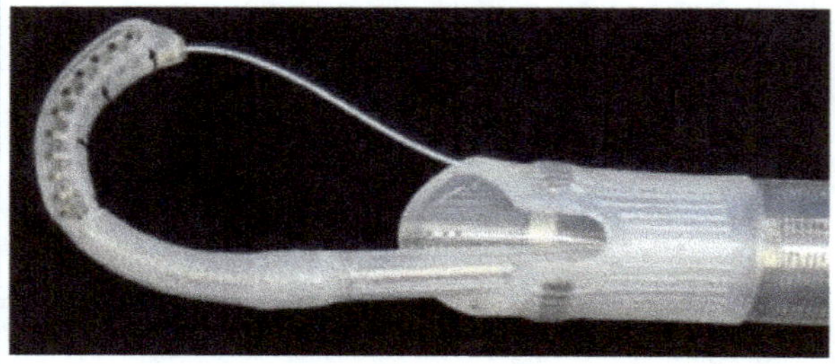

Figure 28.2. Pericardioscope (right) and radiofrequency ablation probe (left) used in epicardial ablation. (Reproduced with permission from nContact, Inc.)

Solu-Medrol (methylprednisolone) 125 mg was administered and a small drain was left in the pericardial space. It was placed next to a drainage system following radiofrequency catheter ablation and removed the following day.

Transseptal puncture was then carried out with TEE guidance and CARTO mapping performed in both the left and right atria. PV isolation was completed using a 4-mm deflectable saline-irrigated radiofrequency catheter. A roof line, mitral valve isthmus line, posterior mitral valve annular line, cavotricuspid isthmus line, and CS lesion were also performed. Programmed stimulation on increasing doses of Isuprel (isoproterenol) failed to induce sustained atrial dysrhythmias.

The patient was discharged home on postoperative day 2 with minimal discomfort. Discharge medications included Tikosyn (dofetilide), metoprolol, and Coumadin (warfarin). He had been free of atrial dysrhythmias and was otherwise asymptomatic at 3-month follow-up.

Pioneering work by Cox,[3] Haïssaguerre[4] and others has led to the recognition that triggers and common macro-reentry circuits can guide ablative therapy, whether percutaneous or surgical. Specific lesion patterns, mapping/intracardiac ECG-based lesions or a combination of both are applied. Unfortunately, some patients have an atrial substrate that is hard to completely normalize and attempts at ablation sometimes cause new atrial arrhythmias to arise from the left atrium.

Percutaneous ablation of the left atrium involves a TEE-guided transseptal puncture and placement of a multipolar mapping catheter and radiofrequency ablation catheter. Multiple contiguous lesions are usually placed with the goal of complete left and right PV trunk isolation as well as creation of connecting lesions from the PVs to the mitral annulus (isthmus lesion). Right atrial lesions are also sometimes placed.

Early surgical ablation for AF was actually "cut and sew" and involved long incisions that were made in the Maze 3 pattern. A new lesion pattern, developed in 1997, was a subset of the Maze 3[5] that could be applied with a variety of energy sources both endocardially on the arrested heart and epicardially on the beating heart. It continues to be used with modification for left atrial ablation today.

Epicardial ablation is performed at most centers using minimally invasive thoracic incisions. In the treatment of paroxysmal AF, the most reliable energy source is bipolar radiofrequency delivered via a clamp across the PV trunks. Used alone, however, this approach has an unacceptably high failure rate in persistent or long-standing persistent AF.[6]

Another disadvantage of the thoracic approach for epicardial ablation is that it breaches one or both pleural cavities, even when done in a thoracoscopic or even robotic manner. This leads to the need for chest tube drainage, creates the potential for deep incisional pain due to compression of or damage to intercostal nerves at the various port sites, and can result in pleural effusions, atelectasis or pneumonia.

The hybrid approach described here uses both surgical and percutaneous ablation techniques. In this way, both epicardial and endocardial lesions can be delivered. Endocardial lesions are guided by intracardiac mapping and fluoroscopy with overlay of registered CT images of the left atrium, CS, and esophagus. Epicardial lesions are placed using direct endoscopic monitor visualization of the posterior and lateral left atrium.

The surgical approach, pericardioscopy, allows access to the oblique sinus and anterior surface of the PV trunks via the central tendon of the diaphragm through a 2-cm midline incision. Its greatest advantages over other surgical approaches for the delivery of epicardial lesions are the simplicity of access and avoidance of entry into either pleural cavity. It is also able to access and deliver lesions to areas of the oblique sinus that are not visualizable through a purely thoracic approach. Areas of the left atrium covered by pericardial reflections are accessed using catheter ablation.

In some patients who have undergone prior catheter ablation, epicardial ablation alone may terminate an otherwise intractable left atrial tachyarrhythmia. A recent 57-year-old patient who had failed 3 previous catheter ablations presented for hybrid ablation with an intractable symptomatic rapid left atrial flutter. This rhythm had been previously terminated during the third catheter ablation with delivery of a lesion near the left anterior PV trunk.

This rhythm terminated during epicardial administration of a lesion in the same location, which presumably was due to either the transmural conversion of a previously nontransmural lesion or the closure of a gap from the earlier ablation(s). The patient had no inducible atrial arrhythmias during burst atrial pacing and programmed atrial stimulation at increasing Isuprel dosages of 2 to 20 mcg/min. Percutaneous ablation was not required in this patient. At 9-month follow-up, the patient, on Tikosyn, was asymptomatic with rare episodes of AF.

Figure 28.3 shows the surface and intracardiac ECG during epicardial ablation termination of left atrial flutter. Figure 28.4 demonstrates the utility of intraoperative fluoroscopy to allow registration of the epicardial ablation coil onto a CT image of the left atrium in this patient.

These patients illustrate the use of a new hybrid approach to the initial ablation of AF or postablation of atrial dysrhythmias. Epicardial surgical ablation is first performed using a minimally invasive approach to the pericardial space. Conventional percutaneous ablation is then carried out in the same setting in a procedure suite that combines complete electrophysiologic navigation, mapping, fluoroscopy,

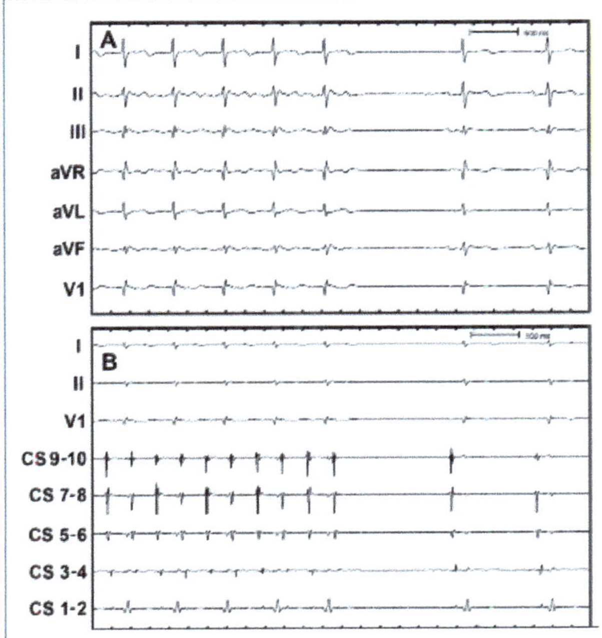

Figure 28.3. Surface (**Panel A**) and intracardiac (**Panel B**) ECGs during termination.

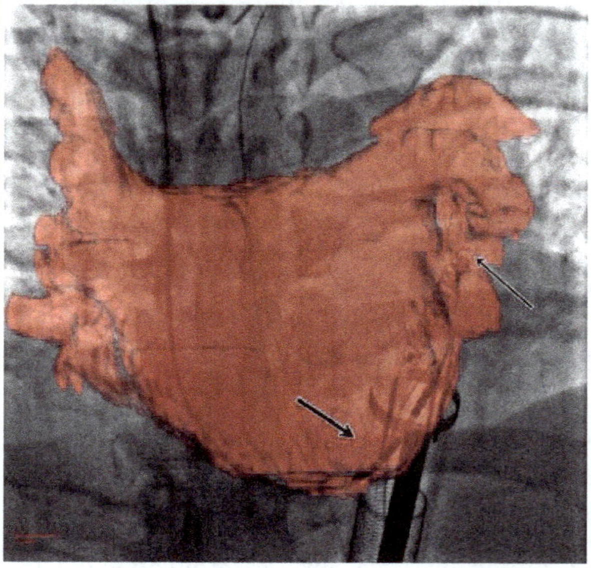

Figure 28.4. Registration image of previous left atrial CT-angiogram over real-time fluoroscopic image showing position of 3-cm epicardial radiofrequency ablation probe during termination of the lesion. Also shown are pericardioscope (large arrow) and ablation probe (small arrow).

and programmed stimulation resources with complete surgical endoscopic and ablative tools.

Pericardioscopic surgical ablation avoids painful sternal or intercostal incisions and is the least invasive surgical approach into the pericardial space currently available. It is becoming an increasingly frequent companion to conventional percutaneous ablation for drug refractory, symptomatic AF, or postablation tachycardias arising from the left atrium. A hybrid approach should enhance the efficiency, efficacy, and safety of percutaneous ablation.

Correct Answer:

Q1. D

REFERENCES

1. Naccarelli GV, Varker H, Lin J, Schulman KL. Increasing prevalence of atrial fibrillation and flutter in the United States. *Am J Cardiol.* 2009;104:1534–1539.
2. Kiser AC, Landers M, Horton R, et al. The convergent procedure: A multidisciplinary atrial fibrillation treatment. *Heart Surg Forum.* 2010;13:E317–E321.
3. Cox JL, Schuessler RB, D'Agostino HJ Jr, et al. The surgical treatment of atrial fibrillation. III. Development of a definitive surgical procedure. *J Thorac Cardiovasc Surg.* 1991;101:569–583.
4. Haïssaguerre M, Jaïs P, Shah DC, et al. Spontaneous initiation of atrial fibrillation by ectopic beats originating in the pulmonary veins. *N Engl J Med.* 1998;339:659–666.
5. Kress DC, Krum D, Chekanov V, et al. Validation of a left atrial lesion pattern for intraoperative ablation of atrial fibrillation. *Ann Thorac Surg.* 2002;73:1160–1168.
6. Castellá M, Pereda D, Mestres CA, et al. Thoracoscopic pulmonary vein isolation in patients with atrial fibrillation and failed percutaneous ablation. *J Thorac Cardiovasc Surg.* 2010;140:633–638.

SECTION 4

Ventricular Fibrillation, Ventricular Tachycardia

CASE STUDY

29

Mechanism of Ventricular Fibrillation

Rajesh Venkataraman, MD, MPH and Raymond Ideker, MD, PhD

A 73-year-old white gentleman is admitted to the intensive care unit post-coronary artery bypass grafting and is having an uneventful recovery. He reports no chest pain or shortness of breath and is doing well, until he suddenly develops the rhythm shown in Figure 29.1. He is now pulseless and a code-blue is activated.

Question No. 1:

What is the rhythm shown in Figure 29.1?

In Figure 29.1, the ECG tracing shows a sharp narrow premature ventricular complex (PVC), likely originating from the His-Purkinje system. The arrows show the similar PVCs that quickly degenerate into ventricular fibrillation (VF). VF is a ventricular tachyarrhythmia that has a constantly changing, disorganized-appearing morphology in the ECG.[1] A multitude of data exists to support the fact that VF is not entirely "chaotic or disorganized."[2] Although the 12-lead surface ECG appears disorganized during VF, extensive endocardial and epicardial

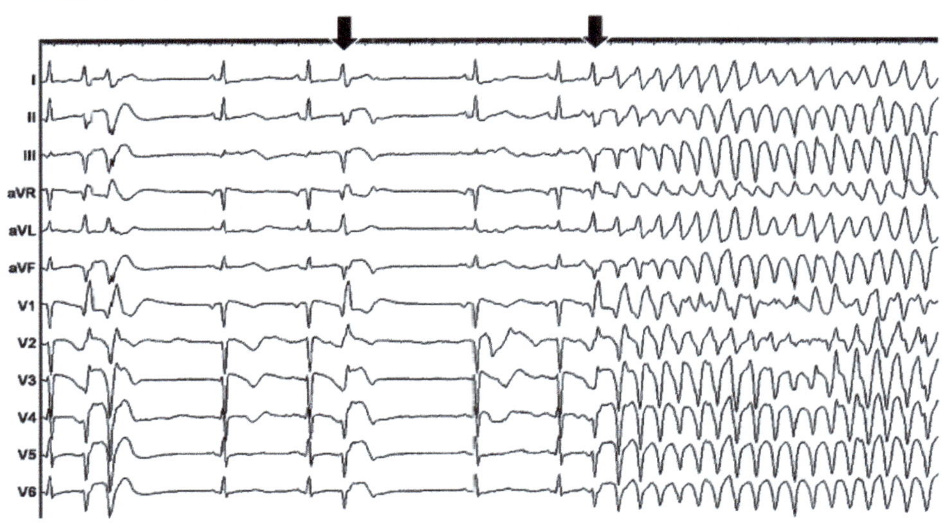

Figure 29.1. Clinical arrhythmia.

mapping have shown that dynamic patterns with variable degrees of organization maintain VF.[2]

There is no ventricular output during VF. Cardiopulmonary arrest ensues very rapidly in VF and immediate defibrillation with an electric shock is the only effective treatment. In a hospital setting, this can be accomplished very quickly. However, in the community, it takes several minutes for an emergency personnel to arrive and survival is low.

Question No. 2:

Is there more than one type of VF?

There are multiple classifications of VF, although for practical purposes, the final treatment of VF does not differ for different types of VF, since an electric shock is the only definitive treatment. One classification of VF is "fine" and another is "coarse." Coarse VF is frequently seen soon after the onset of VF (short-duration VF [SDVF]) and is associated with higher amplitude; hence it appears coarser on the surface ECG.[3] As VF persists (long-duration VF [LDVF]), the amplitude decreases and "fine VF" is seen. However, it has been reported that coarse VF can be seen in some ECG leads simultaneously with fine VF in other ECG leads, suggesting that this classification is not always directly related to the underlying degree of organization of cardiac electrical activity during VF.

Two other classifications of VF are based on how it changes with time (Figure 29.2). Wiggers first elucidated the progression of VF by high-speed cinematography in canine hearts.[4] He described 4 stages (see left side of Figure 29.2). Huang et al performed a cluster analysis of quantitative measurements of epicardial activation sequences obtained by recording simultaneously from over 500 electrodes in canine hearts (see right side of Figure 29.2).[5] In the Wiggers classification, VF organization decreases continuously with time, while in the Huang classification, a brief period of increased organization occurs after about 1 minute of VF.

Another classification of VF in Types 1 and 2 is discussed in the answer to question no. 4.

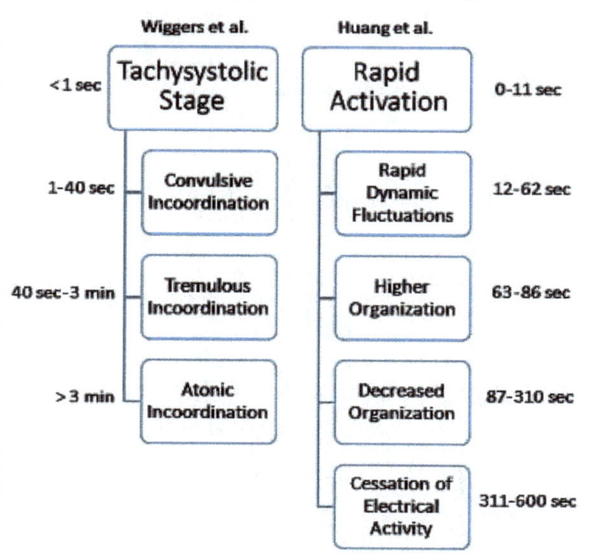

Figure 29.2. A schematic representation of the stages of VF, showing similarities in the stages as described by Wiggers et al[4] and Huang et al.[5]

Question No. 3:

What are the common causes of VF?

The most common etiologies of VF are ischemic heart disease followed closely by heart failure.[1] VF does occur in patients without structural heart disease as well, and idiopathic VF is now a well-known clinical entity.[1] In idiopathic VF, certain inherited ion channelopathies can cause electrical instability.

Question No. 4:

How is VF maintained?

There are 2 main theories on VF maintenance in humans: the multiple-wavelet hypothesis and the mother-rotor hypothesis.[2]

MULTIPLE-WAVELET HYPOTHESIS

This hypothesis was first proposed by Moe et al in 1964, to explain the activation patterns maintaining atrial fibrillation and was based on computer simulations.[6] In the simulations, inhomogeneity in the refractory periods of cardiac muscle caused conduction block, which led to constantly changing reentry and multiple wandering wavelets of activation fronts. Some of these wavelets would be extinguished when they collided with other wavefronts or totally blocked in refractory tissue. Other wavelets would partially block, with the surviving fragments of activation fronts forming daughter wavelets that would continue to sustain VF. If enough wavelets partially blocked and reentered, fibrillation would be sustained. Therefore, all parts of the tissue participated in the maintenance of fibrillation. Later simulations have shown that such multiple wavelets could also be formed in completely homogenous tissue.

MOTHER-ROTOR HYPOTHESIS

This hypothesis was first proposed by Lewis in 1925,[7] followed by Gurvich in 1975, and was more recently refined by Jalife and Pertsov.[8,9] This hypothesis proposes that a long-lived, stationary and stable reentrant circuit is present in one part of the heart during fibrillation, which is labeled as the "mother rotor." The mother rotor has a regular, repeating reentrant circuit in the portion of the tissue with the shortest refractory period. The mother rotor gives rise to daughter wavelets, which propagate away from the mother rotor as less organized, changing activation sequences that block in regions with a longer refractory period. These multiple daughter wavelets give rise to the disorganized pattern seen in the surface ECG. This hypothesis is supported by experiments in guinea pig hearts and isolated slabs of sheep right ventricle. According to this hypothesis, only the region containing the "mother rotor" is important for the maintenance of VF.

COMBINATION OF MULTIPLE WAVELETS AND A MOTHER ROTOR

At least 2 studies have shown that both "multiple wavelets" and "a mother rotor" can be present at the same time. Chen et al have reported 2 types of VF:[10] Type 1 VF is characterized by short-lived reentry of rapidly activating wandering wavelets; and Type 2 VF is characterized by large wavefronts with similar pathways and slower activation fronts. They have also shown that the type of VF can also be influenced by drugs, heart disease, and the duration of VF. Nash et al performed ventricular epicardial recordings for 20 to 49 seconds during VF in 10 patients undergoing cardiac surgery.[11] They found that human VF wavefronts were large and coherent with reentry from 1 or 2 distinct sources. Thus, both wandering wavelets and mother rotors appear to be present in human VF.

Question No. 5:

Is reentry the only mechanism for VF?

Recent studies indicate that during some portions of LDVF, VF may be maintained by activation fronts arising focally in Purkinje fibers (PF).[1,12] PF may remain electrically active during LDVF because they are more resistant to the global ischemia of VF due to their high glycogen content and confinement in humans to the sub-endocardium, which is less ischemic because of the oxygen content of the static blood in the ventricular cavities during VF.[12] Although the refractory period of PF is longer than that of the working myocardium during sinus rhythm, it is shorter than that of the working myocardium during the rapid activation rate of VF, and thus can give rise to activation fronts during VF.

Series of experiments have shown that the PF network is indeed active in VF. Newton et al used plunge needles to record LDVF in pigs and dogs and found that in dogs, in which the PF are confined to the sub-endocardium as in humans, the activation rate after 2 minutes of VF was highest in the endocardium.[13] In the pigs, in which PF extend almost to the epicardium, the activation rate was faster near the epicardium. Allison et al have shown similar findings with a rapid activation from the endocardium to the epicardium of dogs (Figure 29.3).[14] Taberaux et al have directly mapped retrograde activation of ventricular myocardium from the PF as well as antegrade propagation from PF to the ventricular myocardium during VF (Figure 29.4).[15] In a similar model, Dosdall et al applied lugol solution to the endocardium, which resulted in abolition of the rapid Purkinje deflections and also significantly decreased the time from onset of VF to spontaneous termination (4.9 minutes vs 9.2 minutes in controls).[16] Another recent study showed that after 3 to 7 minutes of VF, a highly synchronous, focal LV endocardial activation pattern called ventricular electrical synchrony frequently occurs, which is not consistent with reentry but probably represents triggered activity or abnormal automaticity in PF.[17]

Returning to our case in the first paragraph; the patient received prompt defibrillation, which converted him to normal sinus rhythm. However, he kept having repeated episodes of VF, which was refractory to standard antiarrhythmic medications. All episodes of VF were preceded by a sharp narrow complex beat suggesting Purkinje firing. He was taken to the electrophysiologic lab and ablation was performed in the midseptum, which prevented any VF recurrences.

Correct Answers:

Q1. VF

Q2. Yes

Q3. Ischemic heart disease is the most common cause of VF

Q4. There are 2 main theories on VF maintenance in humans: the multiple-wavelet hypothesis and the mother-rotor hypothesis

Q5. No; focal origin has been described

QUESTION FROM THE EDITORS AND AUTHORS' REPLY

Question. What is the mechanism for defibrillation of VF?

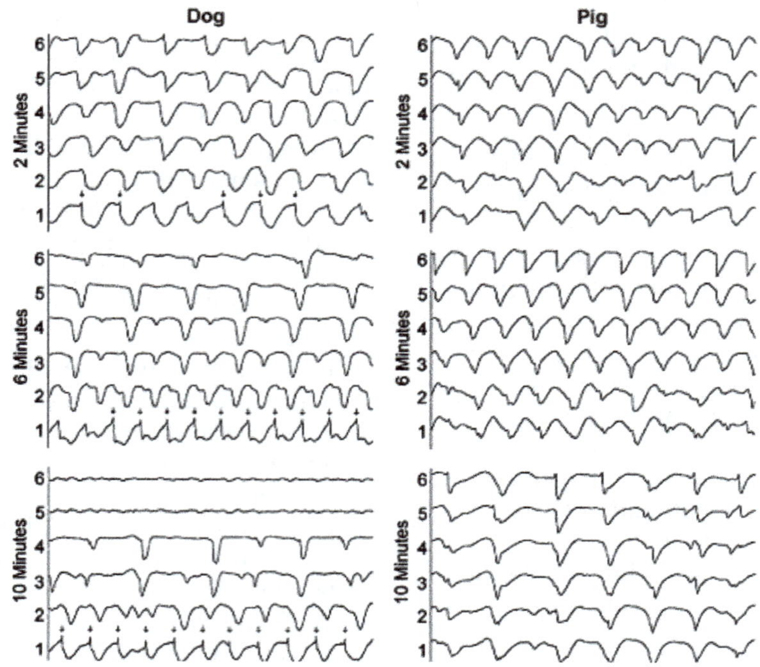

Figure 29.3. Example of transmural activation during long-duration ventricular fibrillation (LDVF) in a dog and a pig. Electrode 6 is near the epicardium and electrode 1 is near the endocardium. The transmural activation rate in the pig is uniform, while the transmural activation rate in the dog shows rapid Purkinje activations (arrows) preceding myocardial activation, and conduction block occurring progressively closer to the endocardium as LDVF progresses. Reprinted with permission.[14]

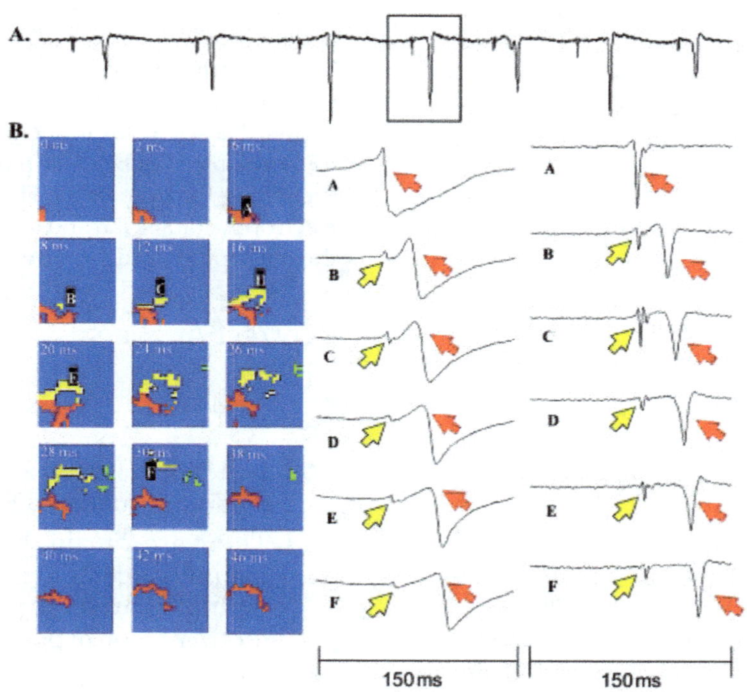

Figure 29.4. **A:** A 1.5-second example of dV/dt, with box highlighting the individual beat displayed in **panel B**. **B:** A wavefront in the working ventricular myocardium activating the PF. The Purkinje activation (yellow) appears to arise from the leading edge of the working ventricular myocardium (red). The selected panels represent frames from a dynamic display of the activation patterns. The frames are separated at various time intervals, with the time for each frame shown in white. Green represents an unrelated wavefront. In panel B, letters A to F indicate the locations of electrodes whose potentials (left) and temporal derivatives of the potentials (right) are shown during the wavefront propagation. Reprinted with permission.[15]

Answer: An electrical shock needs to alter the transmembrane potential such that it halts all VF wavefronts, yet it does not create new wavefronts which could reinduce VF. To achieve this, the electrical shock needs to have sufficient amplitude to create large virtual electrodes within the myocardium. These large virtual electrodes with a virtual anode and a virtual cathode in the myocardial fibers are created by changes in the intracellular resistivity. Multiple ionic currents interplay in the process of defibrillation and a detailed discussion is beyond the scope of this chapter. The alteration in the transmembrane potential is principally due to: a negative transmembrane potential when hyperpolarization produced by one shock polarity is greater than the depolarization caused by the other shock polarity or a nonmonotonic transmembrane potential in which the polarization first increases but then decreases while a square wave stimulus is being given.

Once the VF wavefront is halted, a focal activation front arises after a small "isoelectric" interval and spreads in all directions to activate the entire myocardium. If the shock failed to defibrillate, these foci may repeat for 4 cycles and then degenerate into VF again (Type A defibrillation success) or the focal activation pattern may only last 1 or 2 cycles before it stops spontaneously (Type B defibrillation success). Intramural mapping and electrical mapping has confirmed the existence of this isoelectric interval and this is not an artifact due to limitations of electrical recordings. These activation fronts, which arise after an electric shock, reinduce VF primarily by creating a rotor with reentry around a "critical point" where depolarization and hyperpolarization decrease such that no activation front arises at that point.

Returning to our case in the first paragraph, he received prompt defibrillation which converted him to normal sinus rhythm. However, he kept having repeated episodes of VF, which was refractory to standard antiarrhythmic medications. All episodes of VF were preceded by a sharp narrow complex beat suggesting Purkinje firing. He was taken to the EP lab and ablation was performed in the midseptum, which prevented any VF recurrences.

REFERENCES

1. Natale A, Raviele A, Al-Ahmad A, et al. Venice Chart International Consensus document on ventricular tachycardia/ventricular fibrillation ablation. *J Cardiovasc Electrophysiol.* 2011;21:339–379.
2. Ideker RE, Rogers JM. Human ventricular fibrillation: Wandering wavelets, mother rotors, or both? *Circulation.* 2006;114:530–532.
3. Jones DL, Klein GJ. Ventricular fibrillation: The importance of being coarse? *J Electrocardiol.* 1984;17:393–399.
4. Wiggers C. Studies of ventricular fibrillation caused by electric shock: Cinematographic and electrocardiographic observations of the natural process in the dog's heart: Its inhibition by potassium and the revival of coordinated beats by calcium. *Am Heart J.* 1930;5:351–365.
5. Huang J, Rogers JM, Killingsworth CR, et al. Evolution of activation patterns during long-duration ventricular fibrillation in dogs. *Am J Physiol Heart Circ Physiol.* 2004;286: H1193–H1200.
6. Moe GK, Rheinboldt WC, Abildskov JA. A computer model of atrial fibrillation. *Am Heart J.* 1964;67:200–220.
7. Lewis T. *The Mechanism and Graphic Registration of the Heart Beat.* London: Shaw and Sons, Ltd.; 1925.
8. Samie FH, Berenfeld O, Anumonwo J, et al. Rectification of the background potassium current: A determinant of rotor dynamics in ventricular fibrillation. *Circ Res.* 2001;89:1216–1223.
9. Zaitsev AV, Berenfeld O, Mironov SF, Jalife J, Pertsov AM. Distribution of excitation frequencies on the epicardial and endocardial surfaces of fibrillating ventricular wall of the sheep heart. *Circ Res.* 2000;86:408–417.
10. Chen PS, Wu TJ, Ting CT, et al. A tale of two fibrillations. *Circulation.* 2003;108:2298–2303.
11. Nash MP, Mourad A, Clayton RH, et al. Evidence for multiple mechanisms in human ventricular fibrillation. *Circulation.* 2006;114:536–542.
12. Tabereaux PB, Dosdall DJ, Ideker RE. Mechanisms of VF maintenance: Wandering wavelets, mother rotors, or foci. *Heart Rhythm.* 2009;6:405–415.
13. Newton JC, Smith WM, Ideker RE. Estimated global transmural distribution of activation rate and conduction block during porcine and canine ventricular fibrillation. *Circ Res.* 2004;94:836–842.
14. Allison JS, Qin H, Dosdall DJ, et al. The transmural activation sequence in porcine and canine left ventricle is markedly different during long-duration ventricular fibrillation. *J Cardiovasc Electrophysiol.* 2007;18:1306–1312.
15. Tabereaux PB, Walcott GP, Rogers JM, et al. Activation patterns of Purkinje fibers during long-duration ventricular fibrillation in an isolated canine heart model. *Circulation.* 2007;116:1113–1119.
16. Dosdall DJ, Tabereaux PB, Kim JJ, et al. Chemical ablation of the Purkinje system causes early termination and activation rate slowing of long-duration ventricular fibrillation in dogs. *Am J Physiol Heart Circ Physiol.* 2008;295:H883–H889.
17. Robichaux RP, Dosdall DJ, Osorio J, et al. Periods of highly synchronous, non-reentrant endocardial activation cycles occur during long-duration ventricular fibrillation. *J Cardiovasc Electrophysiol.* 2010;21(11):1266–1273.

CASE STUDY

30

Figure-of-8 Reentrant Circuits in VT

Andrew L. Wit, PhD

A 65-year-old man who had a transmural myocardial infarction 10 months ago experiences palpitations and syncope. He is brought to the hospital for an electrophysiologic study. The records in Figure 30.1 were taken during the study and show electrocardiographic (ECG) leads I, II and V_1 along with electrograms (EGMs) recorded with an electrode catheter from the right ventricular apex (RVA) and lateral wall of the left ventricle (LV lat), both at standard and high amplification (bottom

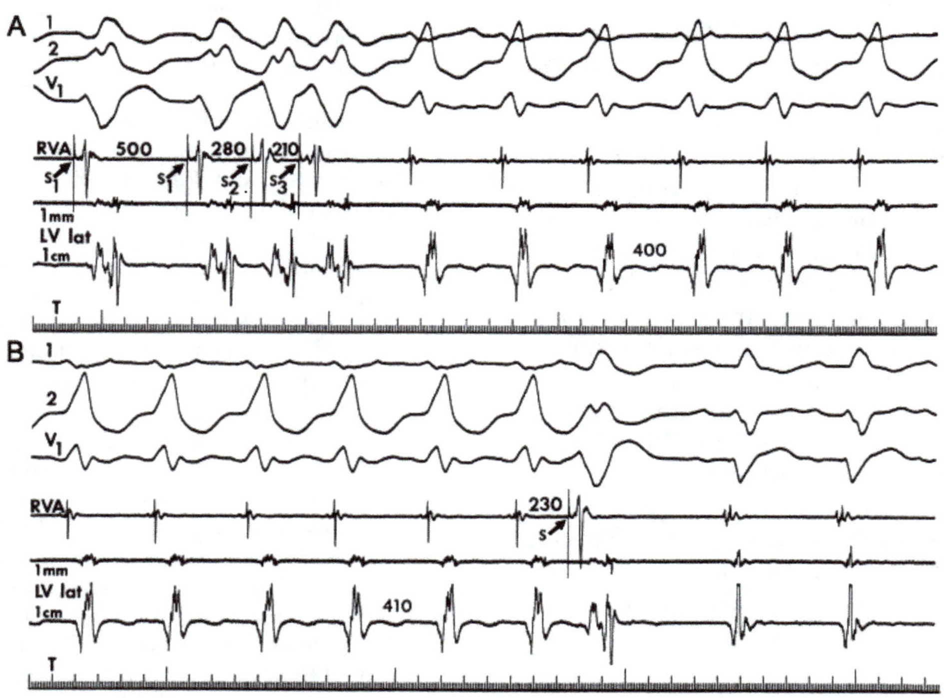

Figure 30.1. VT initiated and terminated by electrical stimulation during electrophysiologic study. (Reproduced from Josephson[1] with permission from Lippincott Williams and Wilkens.)

Practical Electrophysiology © 2015 Jasbir S. Sra, Masood Akhtar, Andrea Natale, David J. Wilber
Cardiotext Publishing, ISBN: 978-1-935395-17-1

trace). The LV was stimulated through an electrode catheter. As depicted in Figure 30.1, during a programmed stimulation protocol in which the ventricles were driven at a cycle length of 500 ms (Figure 30.1A, S_1), double premature stimuli were applied (Figure 30.1A, S_2, S_3) at varying coupling intervals. At coupling intervals of 280 and 210 ms, ventricular tachycardia (VT) was initiated that resembled the spontaneous VT that was causing the palpitations and captured on a Holter monitor. During tachycardia, single stimuli were applied to the LV at various coupling intervals to a preceding tachycardia impulse. At a coupling interval of 230 ms, tachycardia was terminated (Figure 30.1B).

Question No. 1:

Which of the following is/are possible electrophysiologic mechanisms causing VT that are initiated and terminated by programmed electrical stimulation during an electrophysiologic study?

A. Automaticity.

B. Triggered activity caused by early afterdepolarizations.

C. Triggered activity caused by delayed afterdepolarizations.

D. Reentrant excitation.

E. Both C and D.

Question No. 2:

VT associated with a healed myocardial infarction is most frequently caused by which of the following electrophysiologic mechanisms?

A. Automaticity.

B. Triggered activity caused by early afterdepolarizations.

C. Triggered activity caused by delayed afterdepolarizations.

D. Reentrant excitation.

Question No. 3:

The electrogram (EGM) in the left ventricle (LV lat) trace has a low amplitude and long duration and is fractionated compared to the EGM in right ventricular apex, which has a high amplitude and a short duration. Which of the following questions about the LV lat trace is incorrect?

A. It is always an artifact caused by poor contact of the catheter with the ventricular wall.

B. It indicates a region of sparse myocardial fibers with slow nonuniform conduction.

C. Low-amplitude, fractionated EGMs often are found at the site of tachycardia origin.

D. Low-amplitude, fractionated EGMs can be recorded in nonischemic heart disease.

DISCUSSION

When it was initially discovered in the 1970s by Wellens and by Josephson that programmed electrical stimulation could initiate and terminate VT in patients with healed myocardial infarction,[1] the mechanism causing the arrhythmia was speculated to be reentry, but additional proof was needed to support this hypothesis, in light of the recent discovery that triggered activity caused by delayed afterdepolarizations could also be initiated and terminated by stimulation. One way of acquiring additional proof was by making a laboratory animal model that would replicate this clinical arrhythmia.[2,3]

Transmural anteroseptal myocardial infarction was produced in the canine heart by surgical ligation of the left anterior descending coronary artery (LAD). Figure 30.2 shows the infarcted anterior surface of the LV 5 days after coronary occlusion; note the pale coloration in comparison to the

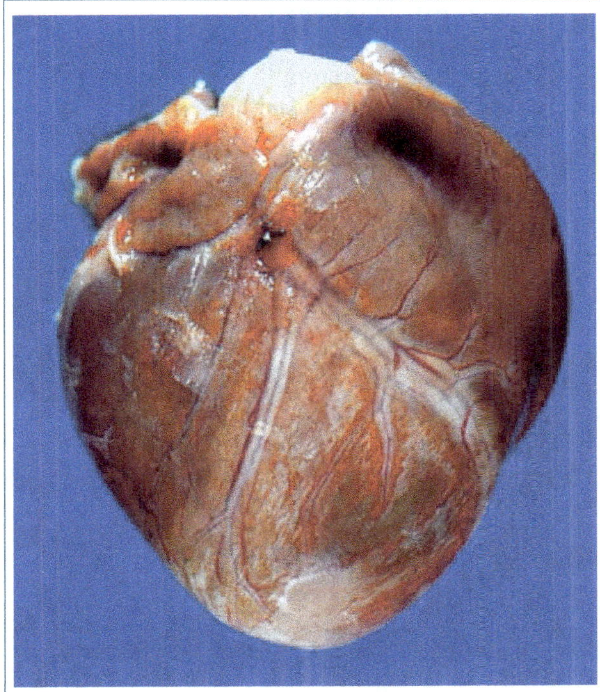

Figure 30.2. Anterior view of a canine heart 1 week after the left anterior coronary artery was occluded near its origin.

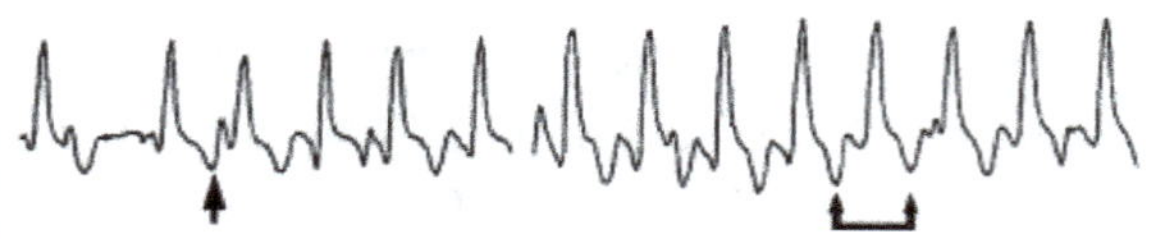

Figure 30.3. Induction of VT during programmed stimulation by a premature stimulus at the arrow. Time base indicated by arrows at right is 150 ms. (Reproduced from Dillon et al[2] with permission from Wolters Kluwer Health.)

normal right ventricle, which is at the upper right in the figure.

Programmed stimulation studies in canines with these myocardial infarcts showed that VT could be initiated by premature stimulation (arrow in Figure 30.3) in an identical manner as in the human clinical electrophysiologic studies, providing an animal model in which the mechanisms for inducible VT could be investigated.[2]

Electrophysiologic mapping during tachycardia showed the origin of tachycardia in the canine heart was in surviving myocardium on the surface of the transmural infarct, a region called the infarct epicardial border zone. This location enabled detailed electrophysiologic mapping of the pattern of electrical activation during tachycardia in the border zone, since the epicardial surface was readily accessible for electrical recording. Later mapping studies in humans showed tachycardia origin both in a subendocardial border zone of surviving muscle and in an epicardial border zone. Figure 30.4 at the left shows a diagram of the electrode array that was used for experimental mapping in the canine heart. It consists of a rubber sheet into which was implanted several hundred recording electrodes (small dots). Stimulating electrodes were positioned at the edges of the electrode array. The electrode array was sutured over the surviving muscle of the epicardial border zone. Histological sections taken from 2 different regions of the epicardial border zone are shown at the right, the upper from a region with widely separated myocardial fiber bundles and the lower from another region where the myocardial bundles were more compact.

Figure 30.5 shows an activation map of the epicardial border zone obtained with the electrode array shown in Figure 30.4 (top of the array is along the LAD) during one beat of sustained monomorphic tachycardia induced by premature stimulation. The same pattern was repeated for all tachycardia beats. Each small number on the map is an activation time relative to a reference "0" time as is standard procedure for activation mapping. Larger numbers are isochrones labels, isochrones are at 10-ms intervals. The pattern of activation can be discerned by following the sequence of isochrones from 0 to 240 ms as pointed out by the arrows. The activation pattern consists of 2 wavefronts, one moving toward the right and one toward the left (isochrones 30–180). Each circulates around a line of conduction block indicated by the parallel thick black lines. The 2 wavefronts merge at isochrone 180 and return to their origin after 240 ms. This pattern of reentry is referred to as a figure-of-8 pattern. There are 2 reentrant circuits, one around the right line of block and one around the left line of block. The 2 wavefronts merge in the region between the

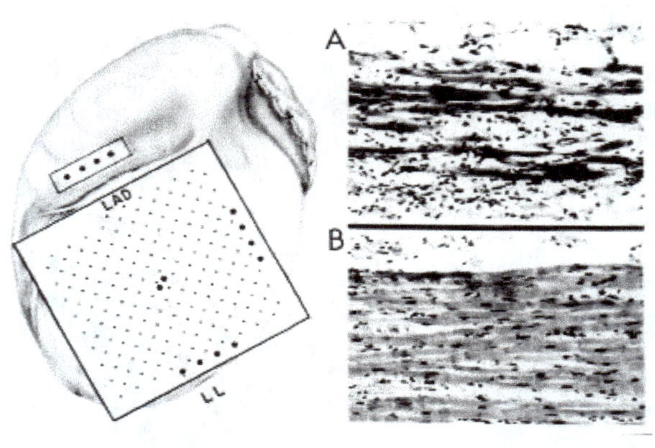

Figure 30.4. Drawing of anterior surface of the canine heart at the left with electrode array for mapping the epicardial border zone of an anterior infarct in place. **Panels A** and **B** at the right show histological sections of surviving border zone myocytes through 2 different regions of the epicardial border zone. LAD, left anterior descending artery.

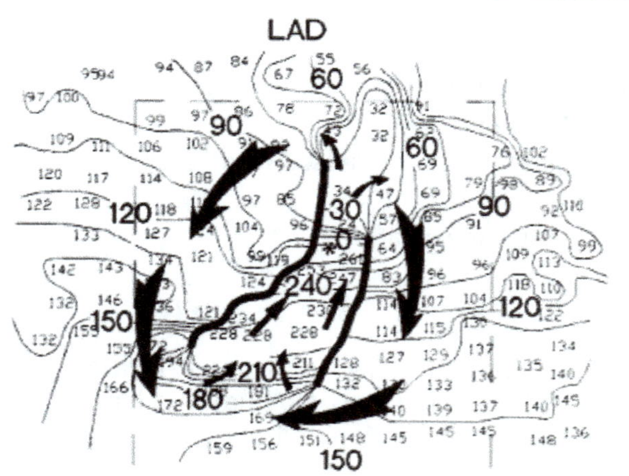

Figure 30.5. Activation map of figure-of-8 reentrant circuit in the epicardial border zone of the canine heart determined by mapping with the electrode array shown in Figure 30.4. (Reproduced from Peters et al[3] with permission from Wolters Kluwer Health.)

2 lines of block, a region called the central common pathway or isthmus.

Since the discovery of the "figure-of-8" pattern in the canine experimental studies, similar patterns have been found in human hearts with healed myocardial infarcts and sustained VT.[4]

Figure 30.6 shows schematic diagrams of 2 figure-of-8 reentry circuits based on clinical mapping studies. Gray stippled regions are not excitable during tachycardia. Arrows indicate propagation of excitation wavefronts through the circuit. In Figure 30.6A, there is an upper reentrant circuit called the outer loop (21–30) and a lower reentrant circuit called the inner loop (3–10) with a central common pathway (10–15). There is also a dead-end pathway, called a bystander pathway (C), which is not part of the reentrant circuit. The merged wavefront leaves the common pathway just before inscription of the QRS. In Figure 30.6B, another more complicated figure-of-8 circuit is shown with an outer loop (sites 21–30) and an inner loop (sites 1–10) with a central common pathway (sites 10–21). There are also several dead-end or bystander pathways (C and H) that are not part of the reentrant circuit.

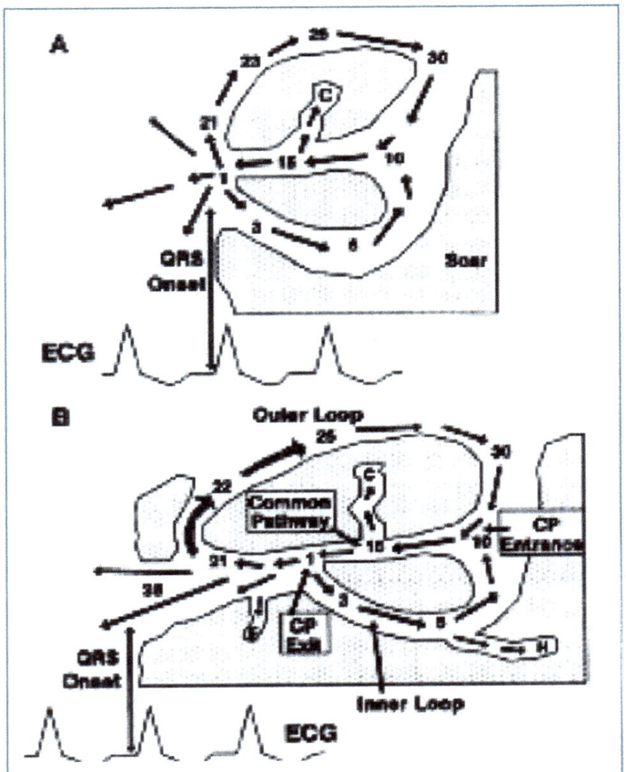

Figure 30.6. Diagrams of figure-of-8 reentrant circuits based on clinical mapping data. (Reproduced from Stevenson et al[4] with permission from Wolters Kluwer Health.)

CLINICAL SIGNIFICANCE OF THE FIGURE-OF-8 PATTERN OF REENTRANT EXCITATION

Question No. 4:

In Panel A of Figure 30.6, which is the numbered or lettered site on the diagram that is likely to be the most effective site for termination of tachycardia by ablation?

A. Site 25.

B. Site 8.

C. Site 15.

D. Site C.

The figure-of-8 reentrant pattern has a very important clinical significance in dictating the site for ablation of VT. It requires much more accurate criteria for locating the effective ablation site than circus movement reentry.

Figure 30.7 shows an example of circus movement (circular excitation) reentry in the subendocardial border zone of a patient with a healed myocardial infarction. Mapping was done with a probe electrode during open heart surgery with the endocardial surface of the LV exposed. Activation times relative to a reference 0 time (left dashed line on ECG above) during VT are indicated by the small numbers on a diagram of the endocardial surface (lower panel). Zero activation time is inferior (INF) and activation follows the sequence pointed out by the arrows, toward the septum and then anteriorly until it reaches its point of origin from which the sequence is repeated.

From the pattern of circus movement reentry shown in Figure 30.7, it appears that effective interruption of the reentrant circuit by ablation would be at any site in the circuit.[5] This would not be the case in the figure-of-8 circuits shown in Figures 30.5 and 30.6. The only effective ablation sites in those circuits would be in the central common pathway or isthmus, the region through which both reentrant wavefronts pass. Ablation of the outer or inner loop sites (see Figure 30.6) would still allow the opposite reentrant circuit to continue functioning. An important problem for the clinical electrophysiologist is how to locate the critical isthmus site.

There are a number of criteria for determining putative ablation sites in the isthmus of figure-of-8 reentrant circuits during electrophysiologic mapping. The characteristics of local EGMs are important with low-amplitude, fractionated EGMs often indicating sites within the reentrant circuit, although not specific for the isthmus. Other

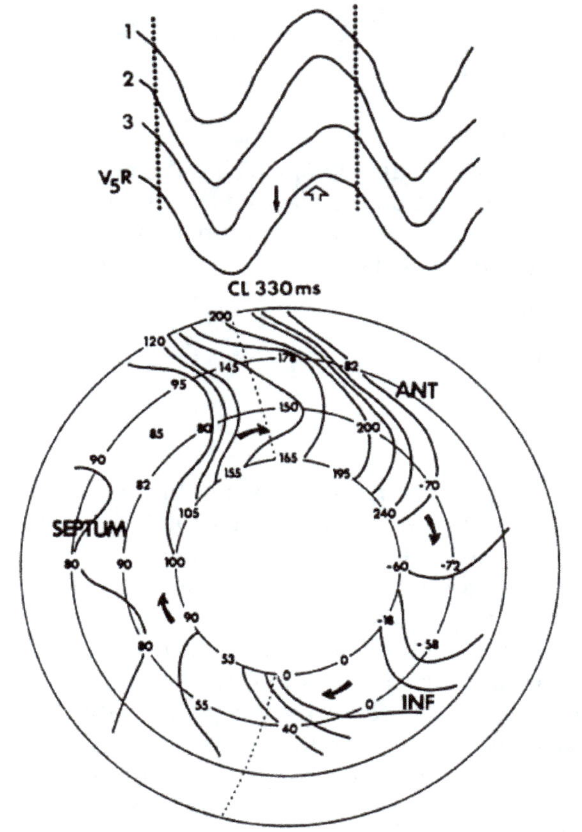

Figure 30.7. **Top** panel shows the ECG recording during sustained VT in the operating room. Vertical lines indicate the time period of the bottom panel activation map of circular reentrant circuit in aneurysm/subendocardial border zone. ANT, anterior; CL, cycle length; INF, inferior. (Reproduced from Miller et al[5] with permission from Elsevier.)

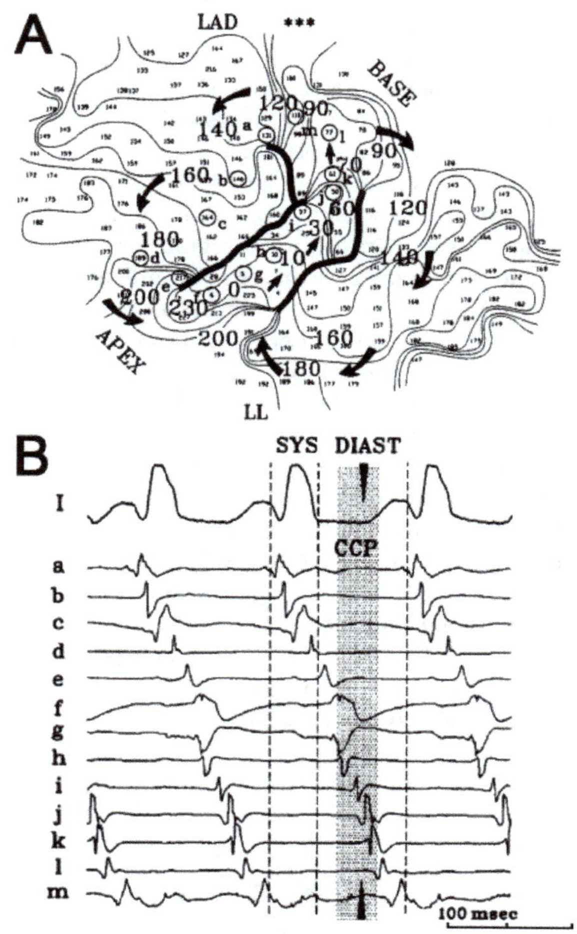

Figure 30.8. **Panel A** is an activation map of a figure-of-8 reentrant circuit with a similar format as in Figure 30.5. **Panel B** shows ECG and EGMs recorded from corresponding labeled sites on the activation map. (Reproduced from Cabo et al[6] with permission from Wolters Kluwer Health.)

predictors of effective ablation sites are sites where EGMs are recorded during the diastolic (DIAST) interval and sites that, when stimulated, result in exact entrainment of the tachycardia, ie, reproduce the QRS morphology of the tachycardia exactly without a significant delay between the stimulus artifact and the QRS complex.

Figure 30.8 shows the location of DIAST EGMs in an experimental mapping study of reentrant circuits during VT in the canine heart.[6] Figure 30.8A shows a figure-of-8 reentrant circuit. The reentrant pattern is indicated by black arrows. Thick black lines represent lines of functional block. Small numbers indicate activation times at recording sites; larger numbers are isochrones labels. EGMs recorded at circled electrodes (a through m) are shown in Figure 30.8B; surface ECG is at the top and EGMs in reentrant circuit (a through m) are beneath it. Systolic (SYS) and DIAST segments are indicated by vertical dashed lines. Figure 30.8 shows that EGMs that occur during the DIAST interval are located in the central common pathway of the reentrant circuits. DIAST EGMs are often effective ablation sites.

Correct Answers:

Q1. E

Q2. D

Q3. A

Q4. C

REFERENCES

1. Josephson ME. *Clinical Cardiac Electrophysiology: Techniques and Interpretations.* 3rd ed. Philadelphia, PA: Lippincott Williams and Wilkens; 2002.

2. Dillon SM, Allessie MA, Ursell PC, Wit AL. Influences of anisotropic tissue structure on reentrant circuits in the epicardial border zone of subacute canine infarcts. *Circ Res.* 1988;63:182–206.
3. Peters NS, Coromilas J, Severs NJ, Wit AL. Disturbed connexin43 gap junction distribution correlates with the location of reentrant circuits in the epicardial border zone of healing canine infarcts that cause ventricular tachycardia. *Circulation.* 1997;95:988–996.
4. Stevenson WG, Khan H, Sager P, et al. Identification of reentry circuit sites during catheter mapping and radiofrequency ablation of ventricular tachycardia late after myocardial infarction. *Circulation.* 1993;88:1647–1670.
5. Miller JM, Harken AH, Hargrove WC, Josephson ME. Pattern of endocardial activation during sustained ventricular tachycardia. *J Am Coll Cardiol.* 1985;6:1280–1287.
6. Cabo C, Schmitt H, Masters G, et al. Location of diastolic potentials in reentrant circuits causing sustained ventricular tachycardia in the infarcted canine heart: Relationship to predicted critical ablation sites. *Circulation.* 1998;98:2598–2607.

CASE STUDY

31

Anisotropic Conduction

Andrew L. Wit, PhD

A 55-year-old man who had a transmural anteroseptal myocardial infarction 10 months ago complains of palpitations. On 2 occasions, he has felt dizzy and almost fainted. Holter monitoring showed runs of nonsustained ventricular tachycardia (VT) associated with the palpitations and dizziness. He was brought into the hospital for an electrophysiologic study. The recordings in Figure 31.1 that were taken during this study show that premature stimulation either during sinus rhythm (see Figure 31.1A) or ventricular pacing (see Figure 31.1B) induced sustained monomorphic VT that could also be stopped by premature stimulation.

Question No. 1:

What is the mechanism for VT that is started and stopped by programmed stimulation in a patient with a 10-month-old anteroseptal myocardial infarction?

A. Automaticity.

B. Triggered activity caused by delayed afterdepolarizations.

C. Reentrant excitation.

D. Triggered activity caused by early afterdepolarizations.

Question No. 2:

Which of the following describe characteristics of EGMs recorded at the site of origin of the VT in a patient with a healed myocardial infarction?

A. High voltage amplitude.

B. Sharp (short duration).

C. Low voltage amplitude.

D. Fractionated (multiple components).

E. Both C and D.

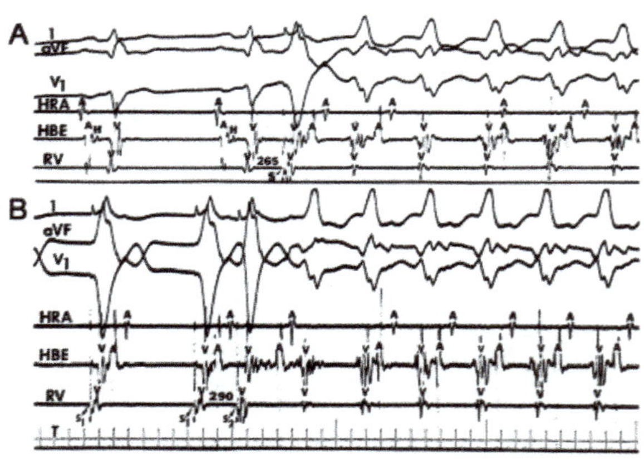

Figure 31.1. Electrocardiographic leads I, aVF and V_1, and electrograms (EGMs) high right atrium (HRA), His bundle region (HBE), and right ventricle (RV) recorded during electrophysiologic study in a patient with a healed myocardial infarction and sustained VT. In **Panel A**, tachycardia is induced by a single premature stimulated impulse (S and arrow in RV EGM) during sinus rhythm and, in **Panel B**, by a single premature stimulated impulse (S_2) during ventricular pacing (S_1) (RV EGM). (Reproduced from Josephson[1] with permission from Lippincott Williams and Wilkens.)

Practical Electrophysiology © 2015 Jasbir S. Sra, Masood Akhtar, Andrea Natale, David J. Wilber
Cardiotext Publishing, ISBN: 978-1-935395-17-1

Question No. 3:

Uniform anisotropy describes the relationship of conduction velocity to the orientation of the myocardial fiber bundles and is important for reentrant excitation. Which of the following accurately describes this relationship?

A. Generalized slow conduction in all directions, both along the long axis and transverse to the long axis of the bundles.

B. Conduction more rapid in the direction of the long axis and slower in the direction of the short axis of the bundles.

C. Conduction more rapid in the direction transverse to the long axis and slower in the direction of the long axis of the bundles.

D. Generalized rapid conduction in all directions, both along the long axis and transverse to the long axis of the bundles.

Question No. 4:

Which of the following is a mechanism for nonuniform anisotropic conduction that causes slow nonuniform conduction in structural heart disease?

A. Separation of myocardial fiber bundles by fibrosis.

B. Decreased gap junction coupling among myocardial fibers.

C. Neither of the above is correct.

D. Both of the above are correct.

MECHANISM OF TACHYCARDIA IN HEALED MYOCARDIAL INFARCTION

VT in patients with healed myocardial infarction, such as the patient described in the case, is caused by reentrant excitation.[1,2] A feature of reentry is that it can be started and stopped by premature electrical stimulation of the ventricles as shown in the records taken during the clinical study in Figure 31.1. Although triggered activity caused by delayed afterdepolarizations also can be started and stopped by electrical stimulation,[2] other features of tachycardia associated with healed myocardial infarction such as entrainment and failure to be stopped by adenosine or verapamil provide additional evidence for reentry.

Reentry that causes sustained VT in patients with healed myocardial infarction occurs in border zones of myocardium that are adjacent to the infarct scar and which have altered electrophysiologic properties that are caused by prolonged ischemia. Figure 31.2 shows an activation map of such reentrant activity.

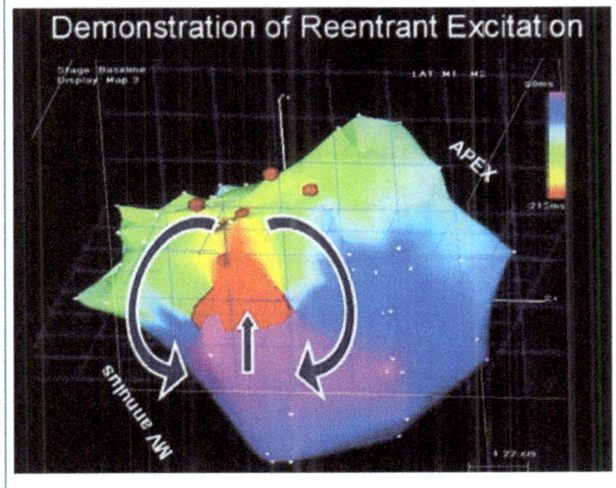

Figure 31.2. Figure-of-8 reentrant pattern in the subendocardial border zone during monomorphic sustained VT in a patient with a healed myocardial infarction. An anatomical representation of the surface of the left ventricle is shown with borders at mitral valve (MV) annulus and apex indicated. Activation pattern obtained from time relationships of multiple EGMs is indicated by colored isochrones from red being earliest to dark blue being latest. Arrows point out the sequence of figure-of-8 reentry characterized by 2 reentrant circuits, one clockwise and the other counterclockwise with a common central pathway. (Reproduced from Josephson[1] with permission from Lippincott Williams and Wilkens.)

Slow nonuniform conduction is an important cause of reentrant excitation. Its presence at sites of origin of VT is indicated by the characteristics of the EGMs recorded from this region (Figure 31.3).

These EGMs are characterized by low voltage amplitude and multiple components (fractionation) during sinus rhythm as shown in the left panel (NSR). During VT (right panel) the fractionated potentials precede the onset of the QRS complex, suggesting tachycardia origin at this site. This can be further confirmed by ablation guided by location of fractionated EGMs. However, fractionated EGMs can also be located at sites outside the reentrant circuit so other criteria must be used to verify that the fractionated EGMs are located in the reentrant circuit. Exact entrainment of the tachycardia from a site of fractionated EGMs confirms the importance of that site in the genesis of the reentrant tachycardia.

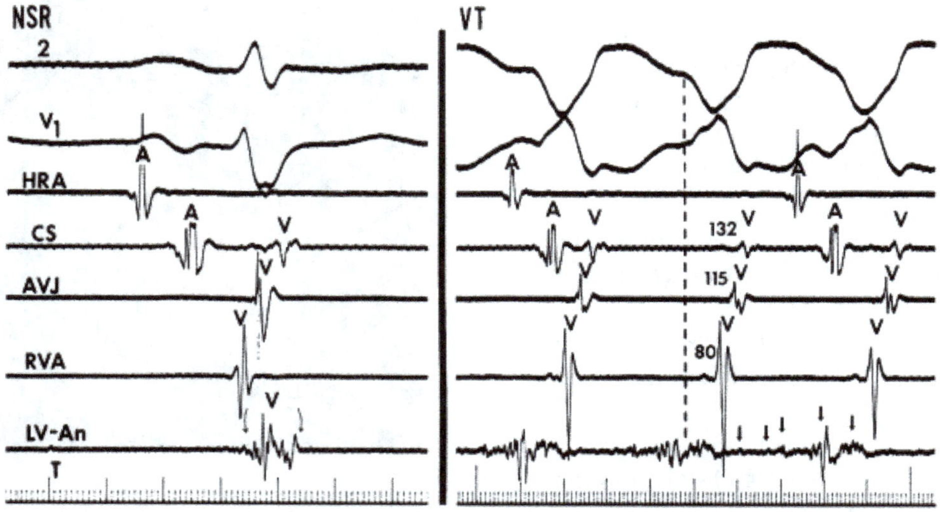

Figure 31.3. Fractionated EGMs recorded from the subendocardial border zone in a patient with a healed myocardial infarction during sinus rhythm (left, NSR) and ventricular tachycardia (right, VT). Traces are electrocardiographic leads II and V_1, intracardiac EGMs from HRA, coronary sinus (CS), atrioventricular junction (AVJ), right ventricular apex (RVA), and endocardial surface of the left ventricular aneurysm (LV-An). (Reproduced from Josephson[1] with permission from Lippincott Williams and Wilkens.)

RELATIONSHIP OF ANISOTROPY TO FRACTIONATED ELECTROGRAMS AND ARRHYTHMOGENESIS IN STRUCTURAL HEART DISEASE

Reentrant excitation is caused by slow conduction and unidirectional conduction block. One of the causes of slow conduction that facilitates the occurrence of reentrant tachycardia is the anisotropic properties of the infarct border zones. Current flow during propagation of the electrical excitatory wave occurs both in intracellular and extracellular space that form a complete electrical circuit. The resistance in this circuit governs the amount of current that can flow to cause propagation of the electrical impulse, with an inverse relationship between resistance (resistivity) and current flow. Intercellular current (axial current) flows from one myocardial cell (a myocyte) to the adjacent cell through the gap junctions of the intercalated disks. Gap junctions are channels connecting adjacent myocytes, through which the positive ions carry current flow. They are formed by proteins called connexins. The gap junctions form an intercellular resistance to current flow along a fiber bundle. Another source of resistance that resides within a myocyte (intracellular resistance) is the sarcoplasm of the myocyte. Gap junctional resistance is greater than sarcoplasmic resistance. The structure of the myocardium, which governs the extent and distribution of these gap junctions, has a profound influence on axial resistance and conduction.

Figure 31.4 is an electrical representation of ventricular myocardium that describes the distribution of gap junctions and how it affects conduction, the electrical basis for anisotropy.[3] It shows 3 myocardial cells (A, B, and C) coupled together by gap junctions represented by the bold resistor symbols (plicate and interplicate). The small resistor symbols represent the ion channels in the sarcolemma, which are involved in the generation of the action potential. Gap junctional connections are primarily at the ends of the long axis of the myocytes. Some of these junctions, called plicate

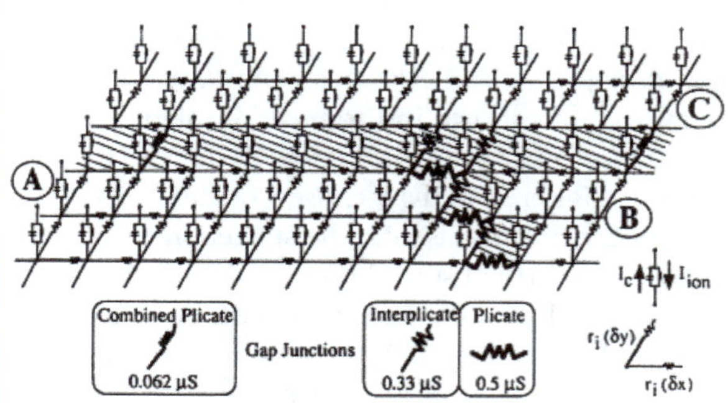

Figure 31.4. Diagram showing the electrical equivalent circuit of 3 interconnected myocytes. (Reproduced from Spach et al[3] with permission from Wolters Kluwer Health.)

junctions, connect myocytes in the end-to-end direction (cell A to cell B) (the long axis of conduction); other junctions at the ends of the myocytes connect them in the side-to-side direction (cell A to cell C) (transverse axis of conduction). There are only sparse gap junctions on lateral membranes connecting myocytes in the transverse direction (combined plicate junctions), so conduction in this direction is mostly supported by the junctions at the ends of the myocytes.

Conduction velocity in the longitudinal direction is more rapid than in the transverse direction, the definition of anisotropy in cardiac muscle, for the reasons shown in Figure 31.4. In the longitudinal direction, gap junctional connections occur over relatively long distances since they are at opposite ends of the myocyte, separated by a long segment of intracellular sarcoplasm. Therefore, axial resistance to longitudinal current flow, comprised of sarcoplasmic and gap junction resistance, has a relatively low per unit of distance. In the transverse direction, over an equivalent distance, there is less sarcoplasmic resistance and many more gap junction connections providing more gap junction resistance, because the diameter of myocytes is much less than their length, resulting in the impulse traveling over many more myocytes for the equivalent distance to longitudinal conduction.

This difference in conduction in longitudinal and transverse direction in ventricular muscle with uniform anisotropy is a normal property.[4] The left diagram in Figure 31.5 shows the distribution of isochrones during conduction of an impulse initiated at the electrical pulse symbol. The isochrones are elliptical because of the more rapid conduction longitudinally (Θ_L) than transversely (Θ_T). Also, shown in the figure are the EGMs recorded during impulse conduction in the different directions. Low-amplitude EGMs are characteristic of transverse conduction, high-amplitude EGMs are characteristic of longitudinal conduction.

In structural heart disease such as the border zones of the healed infarct, which is the pathology of this case, the uniform anisotropy is converted to nonuniform anisotropy as shown in Figure 31.6. The top 2 panels show longitudinal (left) and transverse (right) sections through normal ventricular muscle with uniform anisotropy described above.

The "squiggly" lines in the longitudinal section represent intercalated disks and the gap junctions in the disks are represented by the thickened

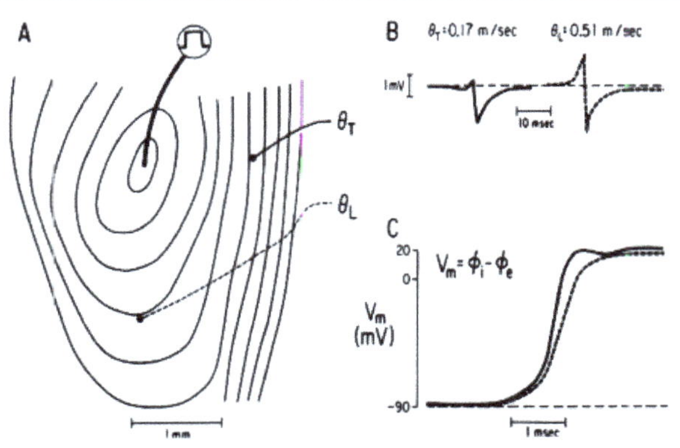

Figure 31.5. Characteristics of uniform anisotropic conduction in ventricular muscle. Excitation sequences in **Panel A** shows activation pattern characteristic of uniform anisotropy. Extracellular waveforms in **Panel B** were recorded at sites of transverse (low amplitude) and longitudinal (high amplitude) propagation indicated on activation map. **Panel C** shows effect of different directions of propagation on upstroke of action potential, longitudinal (slower upstroke) to transverse (more rapid upstroke). (Reproduced from Spach et al[4] with permission from Elsevier.)

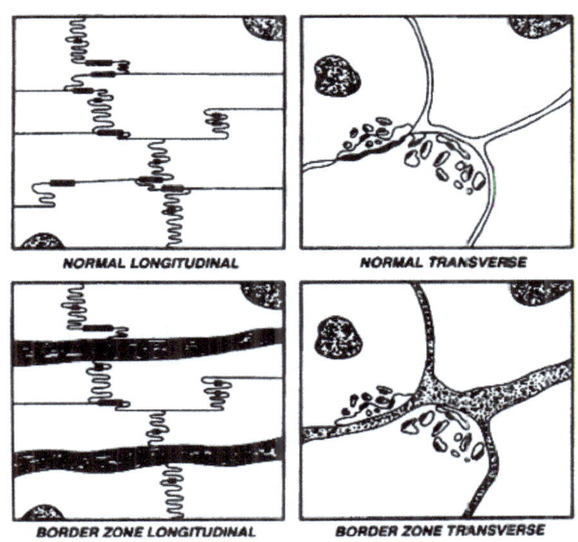

Figure 31.6. Diagram of the effects of interstitial collagen deposition on gap junction distribution and patterns of intercellular connection. Gap junctions are indicated by thickened segments of the junctional membrane connecting cells. A significant reduction in total gap junction membrane length occurs with selective disruption of long, ribbon-shaped interplicate gap junctions. Because the long gap junctions are seen in the short axis in longitudinal planes of section, their loss is observed as a modest reduction in gap junction profile number without a significant decrease in average profile length. In the transverse plane of section, however, their loss results in significant reduction in mean gap junction profile length, producing a more uniform distribution of short gap junction profiles. Reduced coupling is observed at the cellular level to occur disproportionately in cells connected in side-to-side orientation. End-to-end myocyte connections are relatively spared. (Reproduced from Luke et al[5] with permission from American Society for Clinical Investigation.)

segments connecting the myocytes in both the longitudinal and transverse directions as described in Figure 31.4. In the transverse section, a long gap junction is represented by the thickened area on the membrane. The bottom panels show a longitudinal (left) and transverse (right) section from the border zone of a healed infarct. The darkened regions in the left panel indicate fibrotic tissue that separates myocytes along their long axis, disrupting their transverse connections. This is also shown in the right panel by the darkly stippled region.

As a consequence of the disruption of mostly the transverse connections,[5] conduction becomes nonuniform as diagrammed in Figure 31.7. Three cardiac fibers are indicated by the cylindrical structures. These fibers are separated by connective tissue as shown by the darkened regions. The transverse connections have been disrupted in the region shown by the diagram. However, interconnections do occur at a distant site. The large circles represent the location of the bipoles and the dark arrows show impulse propagation. The wavefront moves through the recording region in a serpiginous pattern giving rise to multiple EGM deflections, a deflection that occurs each time the wavefront is within the recording field.[6]

In real border zone, there are many more myocardial cells within the recording field with much more complicated patterns of activity, resulting in more complex fractionated EGMs as shown in Figure 31.3. Figure 31.8 shows a longitudinal section through a subendocardial

Figure 31.8. Longitudinal histological section through a region of the subendocardial border zone from a patient with sustained VT. (Reproduced from Fenoglio et al[7] with permission from Wolters Kluwer Health.)

border zone in a subendocardial resection from a patient who had sustained VT. The resection was taken from the site of tachycardia origin.

Subendocardial muscle bundles in the resection contain large, pale-staining Purkinje fibers (arrows) located immediately beneath the thickened endocardium and separated from underlying ventricular myocardium by connective tissue.[7] The ventricular muscle cells are widely separated by connective tissue, as shown in the diagram in Figure 31.7, causing nonuniform anisotropy and fractionated EGMs. In addition to the fractionation, EGMs have low amplitude because of the reduced density of myocardial cells and increased connective tissue in the border zone.

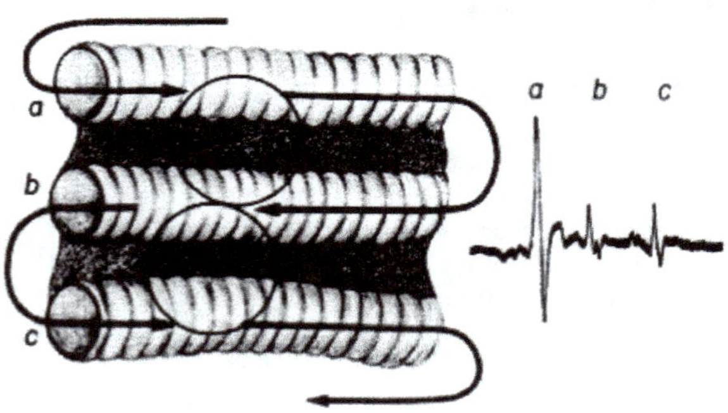

Figure 31.7. Three cardiac fibers or bundles are indicated by the cylindrical structures. These fibers are separated by connective tissue as shown by the results of the anatomic studies and are not interconnected in the region shown by the diagram. However, interconnections do occur at a distant site. The large circles represent the location of the bipoles and the dark arrows show impulse propagation. (Reproduced from Gardner et al[6] with permission from Wolters Kluwer Health.)

Correct Answers:

Q1. C

Q2. E

Q3. B

Q4. D

REFERENCES

1. Josephson ME. *Clinical Cardiac Electrophysiology: Techniques and Interpretations.* 4th ed. Philadelphia, PA: Lippincott, Williams & Wilkens; 2008.
2. Wit AL, Janse MJ. *The Ventricular Arrhythmias of Ischemia and Infarction: Electrophysiological Mechanisms.* Mount Kisco, NY: Futura Publishing Co.; 1993.
3. Spach MS, Heidlage JF. The stochastic nature of cardiac propagation at a microscopic level. Electrical description of myocardial architecture and its application to conduction. *Circ Res.* 1995;76:366–380.

4. Spach MS, Dolber PC. The relation between discontinuous propagation in anisotropic cardiac muscle and the "vulnerable period" of reentry. In: Zipes DP, Jalife J, eds. *Cardiac Electrophysiology and Arrhythmias.* Orlando, FL: Grune and Stratton; 1985:241–252.
5. Luke RA, Saffitz JE. Remodeling of ventricular conduction pathways in healed canine infarct border zones. *J Clin Invest.* 1991;87:1594–1602.
6. Gardner PI, Ursell PC, Fenoglio JJ Jr, Wit AL. Electrophysiologic and anatomic basis for fractionated electrograms recorded from healed myocardial infarcts. *Circulation.* 1985;72:596–611.
7. Fenoglio JJ Jr, Pham TD, Harken AH, et al. Recurrent sustained ventricular tachycardia: Structure and ultrastructure of subendocardial regions in which tachycardia originates. *Circulation.* 1983;68:518–533.

CASE STUDY

32

Electrophysiology of Epicardial, Endocardial, and M Cells: Implications for Arrhythmias

Serge Sicouri, MD and Charles Antzelevitch, PhD

Question No. 1:

What is the ionic basis for the longer action potential (AP) of M cells located in the deep subendocardial to midmyocardial regions of the ventricular wall?

A. A smaller I_{to} and $I_{K\text{-}ATP}$ than epicardium or endocardium.

B. A smaller I_{Ks} and larger late I_{Na} and $I_{Na\text{-}Ca}$ than epicardium or endocardium.

C. A smaller I_{K1} than epicardium or endocardium.

D. A larger I_{Kr} than epicardium or endocardium.

Question No. 2:

The Tpeak–Tend interval measured in the ECG of precordial leads provides an index of:

A. Conduction velocity within the ventricular myocardium.

B. Repolarization of epicardium.

C. Spatial or transmural dispersion of repolarization (TDR) within ventricular myocardium.

D. Repolarization of endocardium.

DISPERSION OF REPOLARIZATION WITHIN THE VENTRICULAR WALL: IMPLICATIONS FOR THE GENESIS OF THE T WAVE IN THE ECG

Epi, Endo, and M cells differ mainly with respect to repolarization characteristics. Epi and M cells display a prominent spike and dome morphology, largely absent in Endo cells. Moreover, M cells display more prolonged action potential duration (APD) than Epi or Endo cells, especially at slow rates of stimulation. Dispersion of repolarization across the ventricular wall has been suggested to underlie the inscription of the normal electrocardiographic T wave (Figure 32.1) and, when amplified, to contribute prominently to the development of cardiac arrhythmias. The magnitude of TDR is attributable to intrinsic differences in the APD of the 3 principal cell types that comprise the ventricular myocardium as well as the extent to which these repolarization differences are damped by electrotonic forces. Figure 32.1 illustrates the cellular basis for the inscription of the normal T wave[1] in the ECG. Shown are transmembrane potentials and a transmural ECG recorded from 2 arterially perfused canine LV

Practical Electrophysiology © 2015 Jasbir S. Sra, Masood Akhtar, Andrea Natale, David J. Wilber
Cardiotext Publishing, ISBN: 978-1-935395-17-1

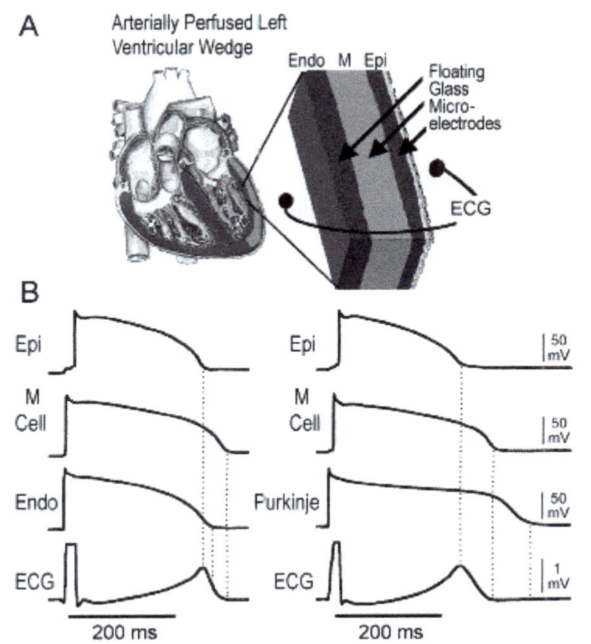

Figure 32.1. A: Schematic of left ventricular (LV) wedge preparation. Floating glass microelectrodes are inserted in the endocardial (Endo), M, and epicardial (Epi) regions together with a transmural ECG consisting of 2 Ag/AgCl half cells placed 1 cm from Endo and Epi surfaces. B: Cellular basis for the normal T wave APs and a transmural ECG recorded from 2 arterially perfused LV wedges. *Left*: APs from Epi, M, and Endo sites simultaneously recorded together with a transmural ECG. *Right*: APs from Epi, M, and a subendocardial Purkinje fiber simultaneously recorded together with a transmural ECG. In both cases, repolarization of Epi is coincident with the peak of the T wave of the ECG, whereas repolarization of the M cells is coincident with the end of the T wave. Endo AP duration is intermediate. Although repolarization of the Purkinje fiber occurs after that of the M cell, it does not register on the ECG. Basic cycle length (BCL) = 2000 ms. (Reproduced from Yan et al[1] with permission from Wolters Kluwer Health.)

wedges. In both, repolarization of Epi is coincident with the peak of the T wave of the ECG, whereas repolarization of the M cells is coincident with the end of the T wave. The Endo APD is intermediate. Although repolarization of the Purkinje fiber occurs after that of the M cell (see Figure 32.1B), it does not register on the ECG.

Thus, the degree of electrotonic coupling, together with intrinsic differences in APD between the distinct cell types, determines the extent to which TDR is expressed and its impact on arrhythmogenesis, as well as on the morphology of the T wave of the ECG. TDR is in large part attributable to the presence of M cells between the Endo and Epi layers of the heart. TDR represents the difference between the longest AP (longest M cell) and the shortest AP (generally Epi cell) of the wedge preparation and correlates with the Tpeak–Tend of the ECG (see Figure 32.1). The hallmark of the M cell, discovered in the early 1990s and named in memory of Gordon K. Moe, is its ability to prolong its AP more than that of Epi or Endo APs following a slowing of rate or exposure to agents that prolong APD.[2] The longer APD of the M cell results from the presence of a relatively weak I_{Ks} current density and the presence of relatively stronger I_{Na} and $I_{Na\text{-}Ca}$ current densities compared to Endo or Epi cells. These ionic distinctions sensitize the M cells to drugs that prolong the APD. Histologically, M cells are similar to Epi and Endo cells. Electrophysiologically and pharmacologically, they appear to be a hybrid between Purkinje and ventricular cells. Like Purkinje fibers, M cells show a prominent APD prolongation and develop early afterdepolarizations (EADs) in response to rapidly activating delayed rectifier potassium current (I_{Kr}) blockers, whereas Epi and Endo cells generally are less likely to do so. Like Purkinje fibers, M cells more readily develop delayed afterdepolarizations (DADs) in response to agents that calcium load or overload the cardiac cell. Thus, they may contribute to the development of both substrate and triggers associated with arrhythmogenesis.

Question No. 3:

The cellular basis for the development of arrhythmias associated with the LQT1 form of the long QT syndrome (LQTS) is due to:

A. An increase in TDR and development of EADs due to reduced I_{Ks}.

B. A decrease in TDR and development of DADs due to an increase in I_{Ks}.

C. A slowing of conduction due to a reduction in I_{Na}.

D. None of the above.

Question No. 4:

The electrocardiographic J wave is inscribed by a:

A. Transmural voltage gradient caused by differences in AP duration.

B. Transmural voltage gradient caused by differences in AP notch.

C. Transmural conduction delay.

D. None of the above.

AMPLIFICATION OF DISPERSION OF REPOLARIZATION: IMPLICATIONS FOR THE GENESIS OF CARDIAC ARRHYTHMIAS

Amplification of spatial dispersion of repolarization, particularly in the form of TDR, underlies the development of life-threatening ventricular arrhythmias associated with inherited ion channelopathies, such as the long QT, short QT, and Brugada syndromes. In LQTS, amplification of TDR is often secondary to preferential prolongation of the APD of M cells, whereas in Brugada syndrome, it is thought to be due to preferential abbreviation of the APD of right ventricular (RV) Epi. Preferential abbreviation of APD of either Endo or Epi is responsible for amplification of TDR in short QT syndrome (SQTS).

The Long QT Syndrome

Congenital LQTS are phenotypically and genotypically diverse, but have in common the appearance of long QT interval in the ECG, an atypical polymorphic ventricular tachycardia (VT) known as torsades de pointes (TdP) and, in many cases, a relatively high risk for sudden cardiac death. Acquired LQTS refers to a syndrome similar to the congenital form but caused by exposure to drugs that prolong the duration of the ventricular AP or QT prolongation secondary to cardiomyopathies such as dilated or hypertrophic cardiomyopathy, as well as to abnormal QT prolongation associated with bradycardia or electrolyte imbalance. The acquired form of the disease is far more prevalent than the congenital form and, in some cases, may have a genetic predisposition.

Models of the LQT1, LQT2, LQT3, LQT7, and LQT8 forms of congenital LQTS have been developed using the canine arterially perfused LV wedge preparation. These models suggest that preferential prolongation of the M-cell APD leads to an increase in the QT interval as well as an increase in TDR, which contributes to the development of spontaneous as well as stimulation-induced TdP. As shown in Figure 32.2A, full repolarization of the Epi AP coincides with the peak of the T wave, and repolarization of the M cells is coincident with the end of the T wave, under both normal and long QT conditions. The duration of the M-cell AP therefore determines the QT interval, whereas the duration of the Epi AP determines the QTpeak interval. Tpeak–Tend interval has been shown to provide an index of spatial dispersion of repolarization within the ventricular myocardium. Because precordial leads look across the ventricular wall, Tpeak–Tend interval in the precordial leads is thought to provide an index of TDR and thus may be prognostic of risk for development of arrhythmias under a variety of conditions.

Not all agents that prolong the QT interval increase TDR. Amiodarone, ranolazine, and pentobarbital are examples of drugs that prolong the QT but do not increase (or even decrease) TDR and are rarely associated with TdP. Hence, the spatial dispersion of repolarization is more important than the prolongation of the QT interval in determining the substrate for TdP. Chromanol 293B, a I_{Ks} blocker, is another example of an agent that increases QT without augmenting TDR.[3] As shown in Figure 32.2A, chromanol 293B prolongs APD of the 3 cell types homogeneously, neither increasing TDR nor widening the T wave. TdP is never observed under these conditions. However, in the presence of beta-adrenergic stimulation (see

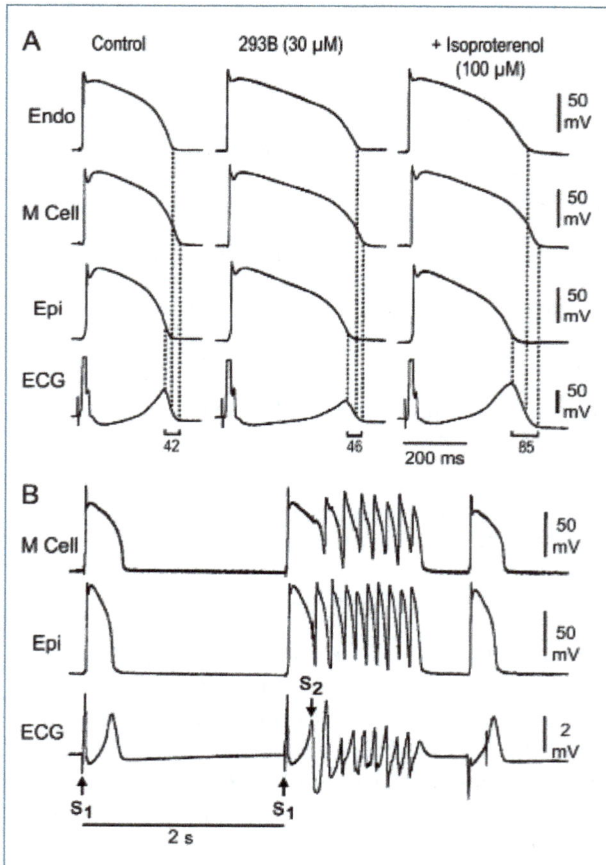

Figure 32.2. LV wedge preparation: LQT1 model of LQTS. **A:** Endo, M, and Epi APs together with a transmural ECG. From left to right: Control conditions, effect of chromanol 293B (30 μM) and chromanol 293B + isoproterenol (100 nM). BCL = 2000 ms. TDR and Tpeak–Tend increased from 42 (control) to 46 (chromanol 293B) to 85 ms (chromanol 293B + isoproterenol). **B:** A single extrastimulus (S_2) applied to the Epi surface (S_1–S_2 = 320 ms) initiated TdP. (Modified from Shimizu et al[3] with permission from Wolters Kluwer Health.)

Figure 32.2B), chromanol 293B abbreviates the APD of Epi and Endo cells but not that of the M cell, resulting in a marked accentuation of TDR and the development of TdP. Therefore, the arrhythmic risk associated with LQTS appears not to be due to the prolongation of the QT interval but rather to the increase in spatial dispersion of repolarization that usually, but not always, accompanies the prolongation of the QT interval. This creates a vulnerable window across the ventricular wall. The reduction in net repolarizing current also predisposes to the development of EAD-induced triggered activity, giving rise to extrasystoles that capture the vulnerable window and precipitate TdP arrhythmias.

Question No. 5:

The Brugada syndrome is characterized by:

A. An ST-segment elevation in the inferior and lateral ECG leads.

B. A reduced TDR.

C. Tall and peaked T waves.

D. None of the above.

Brugada Syndrome

Brugada syndrome is characterized by an ST-segment elevation in the right precordial ECG leads and a high incidence of sudden death in patients with structurally normal hearts. This syndrome has been linked to loss-of-function mutations in genes encoding sodium (mainly *SCN5A*, but also *SCN1B* and *SCN3B*) as well as calcium channels (*CACNA1C* and *CACNB2b*).

The arterially perfused canine RV wedge preparation has been used as a model of the ECG manifestations of the Brugada syndrome. The VT and ventricular fibrillation (VF) generated in these preparations are usually polymorphic, resembling a rapid form of TdP. Although sodium channel block is capable of recapitulating the ECG and arrhythmic manifestations of the Brugada syndrome, a combination of I_{Na} and I_{Ca} block appears more effective than I_{Na} inhibition alone in precipitating the Brugada syndrome phenotype in the arterially perfused RV wedge preparation (Figure 32.3).[4] High concentrations of terfenadine (5–10 μM) produce a potent block of I_{Na} and I_{Ca}, leading to accentuation of the Epi AP notch following acceleration. Accentuation of the notch is due to the effect of the drug to depress phase 0, augment the magnitude of phase 1, and delay the appearance of the second upstroke. Because of use-dependent inhibition of I_{Na} and I_{Ca}, with continued rapid pacing, phase 1

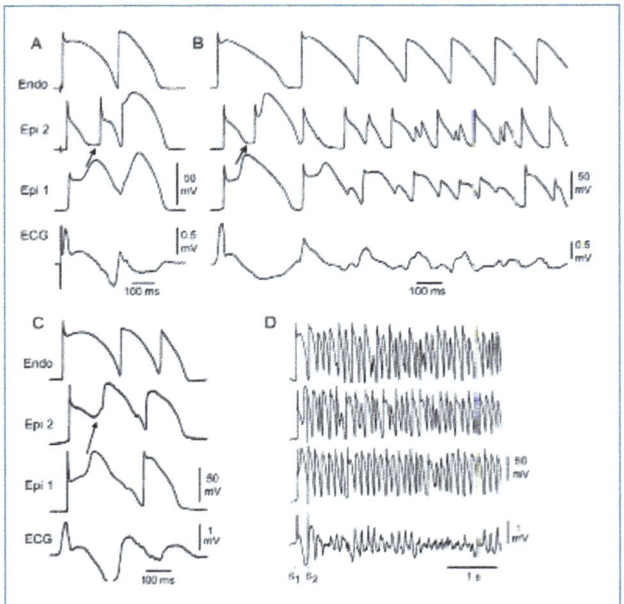

Figure 32.3. Model of Brugada syndrome. Spontaneous and programmed electrical stimulation induced polymorphic VT in RV wedge preparations pretreated with terfenadine (5 μM). **A:** Phase 2 reentry in Epi gives rise to a closely coupled extrasystole. **B:** Phase 2 reentrant extrasystole triggers a brief episode of polymorphic VT. **C:** Phase 2 reentrant extrasystole triggers brief reentry. **D:** Same impalements and pacing conditions as C; however, an extrastimulus (S_1–S_2 = 250 ms) applied to Epi triggers a polymorphic VT. (Modified from Fish et al[4] with permission from Elsevier.)

becomes more accentuated, until all-or-none repolarization occurs at the end of phase 1 at some Epi sites but not others, leading to the development of both Epi and TDR. Propagation of the dome from the region where it is maintained to the region at which it is lost then results in the development of local phase 2 reentry. Figure 32.3 shows the arrhythmogenic effects of terfenadine in 3 RV wedge preparations. In Figure 32.3A, phase 2 reentry in Epi gives rise to a closely coupled extrasystole. In Figure 32.3B and C, phase 2 reentrant extrasystole triggers a brief episode of polymorphic VT. In Figure 32.3D, in the same preparation as Figure 32.3C, an extrastimulus applied to Epi (S_1–S_2 = 250 ms) triggers a polymorphic VT.

Correct Answers:

Q1. B

Q2. C

Q3. A

Q4. B

Q5. D

REFERENCES

1. Yan GX, Antzelevitch C. Cellular basis for the normal T wave and the electrocardiographic manifestations of the long QT syndrome. *Circulation.* 1998;98:1928–1936.
2. Sicouri S, Antzelevitch C. A subpopulation of cells with unique electrophysiological properties in the deep subepicardium of the canine ventricle. The M cell. *Circ Res.* 1991;68:1729–1741.
3. Shimizu W, Antzelevitch C. Cellular basis for the ECG features of the LQT1 form of the long QT syndrome: Effects of b-adrenergic agonists and antagonists and sodium channel blockers on transmural dispersion of repolarization and torsade de pointes. *Circulation.* 1998;98:2314–2322.
4. Fish JM, Antzelevitch C. Role of sodium and calcium channel block in unmasking the Brugada syndrome. *Heart Rhythm.* 2004;1:210–217.

CASE STUDY

33 A

Right Ventricular Outflow Tract Ventricular Tachycardia

Rupa Bala, MD and Francis E. Marchlinski, MD

Question No. 1:

The ECG in Figure 33 A.1 shows:

A. Idiopathic right ventricular outflow tract (RVOT) tachycardia from an anterior, septal site.
B. Ventricular tachycardia (VT) in arrhythmogenic right ventricular cardiomyopathy/dysplasia (ARVC/D).
C. Idiopathic RVOT tachycardia from a free-wall site.
D. Idiopathic left ventricular (LV) outflow tract tachycardia.

CASE DISCUSSION

A 52-year-old male presented with palpitations and decreased exercise tolerance. His ECG revealed frequent left bundle, inferior-axis premature ventricular contractions (PVCs) as noted above. His cardiac workup included the following: (1) an echocardiogram that revealed an ejection fraction of 60% with a normal LV end-diastolic diameter; (2) a 24-hour Holter recording that revealed 30% ventricular ectopy; (3) magnetic resonance imaging (MRI) that revealed no evidence of delayed enhancement and mild global hypokinesis; and (4) a stress test that was negative for coronary artery disease, but did reveal an increase in PVCs with exercise. He was initially treated with beta-blockers, but experienced fatigue and the dose could not be titrated further. He was

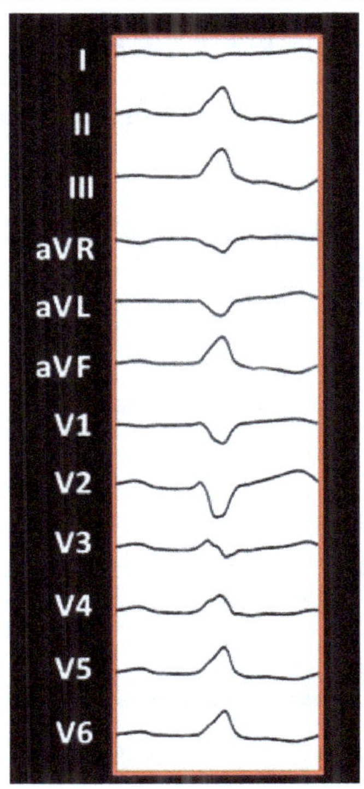

Figure 33 A.1. 12-lead ECG.

subsequently referred for electrophysiology (EP) study/ablation.

On arrival to the lab, the patient had infrequent, ventricular ectopy. The ECG morphology of the clinical ventricular premature depolarization was left bundle, inferior axis with a biphasic, QRS complex in lead I and a precordial transition by lead V_4. The first step was to perform activation mapping in the RVOT, given the PVC morphology. Due to the infrequent PVCs, limited activation mapping was performed and localized the PVC to the anterior, septal aspect of the RVOT. The site of earliest activation was 30 to 35 ms pre-QRS. Pace mapping at this site was a good match to the clinical PVC (see Figure 33 A.1). Radiofrequency ablation (RFA) at this site terminated the PVC (Figures 33 A.2 and 33 A.3). A detailed voltage map of the right ventricle (RV) revealed normal voltage. Post RFA, burst pacing, and isoproterenol failed to illicit any further PVCs.

The patient underwent telemetry monitoring postprocedure and no PVCs were noted. He was discharged off of beta-blockers. The patient was seen in follow-up in the outpatient EP clinic. A 24-hour Holter revealed < 1% PVC, and he reported no further palpitations as well as an improvement in his exercise tolerance.

DISCUSSION

VT arising from the RVOT is a common entity. RVOT tachycardias are a subgroup of idiopathic VT and occur in the absence of structural heart disease.

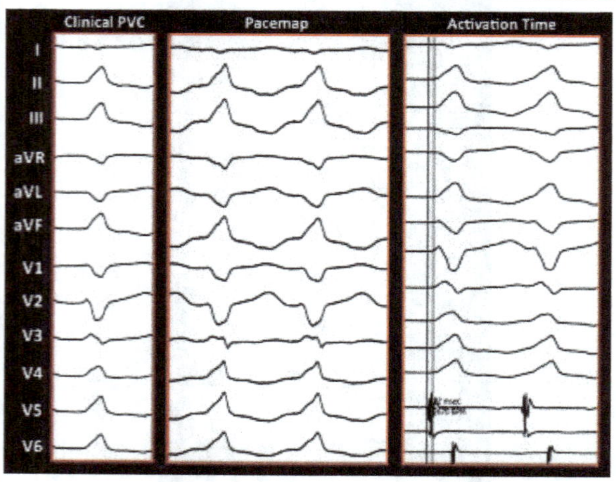

Figure 33 A.2. Clinical PVC is left bundle morphology with an inferior axis. The QRS complex in lead I is biphasic. The precordial transition is by lead V_4. Pace map obtained from RVOT site 2 to 3 (anterior, septal). The activation time is 30 to 35 ms pre-QRS.

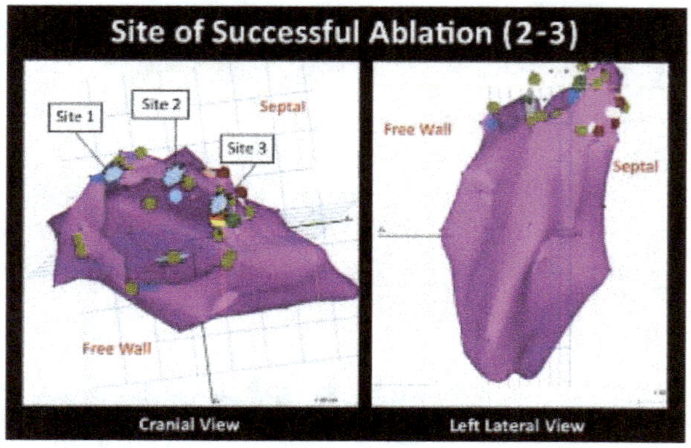

Figure 33 A.3. Electroanatomic map of the RV. The clinical PVC was successfully ablated from RVOT site 2 to 3 (anterior, septal). The site of ablation is marked by red tags on the electroanatomic map. RV voltage map is normal.

Episodes often occur in repetitive salvos and tend to increase in frequency and duration during exercise and emotional stress. The arrhythmia mechanism appears to be calcium-dependent triggered activity.[1,2]

The RVOT region is defined superiorly by the pulmonic valve and inferiorly by the RV inflow tract and the top of the tricuspid valve. The lateral aspect of the RVOT region is the RV free wall, and the medial aspect is formed by the interventricular septum at its base and RV musculature opposite the root of the aorta at the region just inferior to the pulmonic valve. From the coronal view above the pulmonic valve, the RVOT region is seen wrapping around the root of the aorta and extending leftward. The top of the RVOT may be convex or crescent-shaped with the posteroseptal region directed rightward and the anteroseptal region directed leftward. Occasionally, an extreme convexity of the superior, septal RVOT region can create a leftward direction for the most posterior and anterior aspects of the septal RVOT and result in a net negative QRS complex in lead I from either site. The anteroseptal aspect of the RVOT actually is located in close proximity to the LV epicardium, adjacent to the anterior interventricular vein and in proximity to the left anterior descending coronary artery. The aortic valve cusps sit squarely within the crescent-shaped septal region of the RVOT and are inferior to the pulmonic valve. The posteroseptal aspect of the RVOT is adjacent to the region of the right coronary cusp (RCC), and the anterior septal surface is adjacent to the anterior margin of the RCC or the medial aspect of the left coronary cusp.[3,4]

RVOT tachycardias have a left bundle branch block morphology with a precordial QRS transition that begins no earlier than V_3 and more typically occurs in V_4. The frontal plane axis, the precordial QRS transition, the QRS width, and the complexity of the QRS in the inferior leads can pinpoint the origin of VT in the RVOT. Tachycardias originating from the top of the tricuspid valve in the inflow area of the RVOT, which is lower and to the right in the outflow tract, will have a positive QRS deflection in aVL, a QRS amplitude in lead II > lead III and, typically, a very positive QRS complex in lead I.[4]

A study from our institution further divided the septal and free-wall areas into distinct anatomic sites and pace mapping was performed at each site. The 12-lead ECG QRS complexes from pace maps and from spontaneous VT that was successfully localized and ablated were analyzed.[5] The septal and free-wall sites just under the pulmonic valve were designated site 1 (posterior/right), site 2 (mid), and site 3 (anterior/left) to maintain simplicity with respect to the nomenclature used to describe the anatomic location. The QRS complex changed most notably in lead I when moving from site 3, anterior and leftward (negative QRS in lead I), to site 1, posterior and rightward (positive QRS in lead I). The QRS complex in lead I at site 2 (mid) demonstrated a biphasic or multiphasic QRS pattern. In contrast to the corresponding septal sites, the free-wall sites demonstrated a later QRS transition, a wider QRS duration and notching, and decreased amplitude in the inferior leads and the lateral precordial leads.[5]

The most common site of origin for RVOT tachycardias is from the left septal side of the outflow tract just underneath the pulmonic valve. These tachycardias produce a characteristic 12-lead ECG appearance with large positive QRS complexes in II, III, and aVF and large negative complexes in aVR and aVL. The QRS morphology in lead I will typically be multiphasic and have a net QRS vector that is equal to zero or is only modestly positive. Frequently, VT may originate at the leftward edge of the septal aspect of the outflow tract just under the pulmonic valve. This extreme "leftward" aspect of the RVOT actually produces a negative QRS complex in lead I, consistent with how far this aspect of the RVOT wraps around the aortic root and is located to the left of the apical septum[4] (Figure 33 A.4).

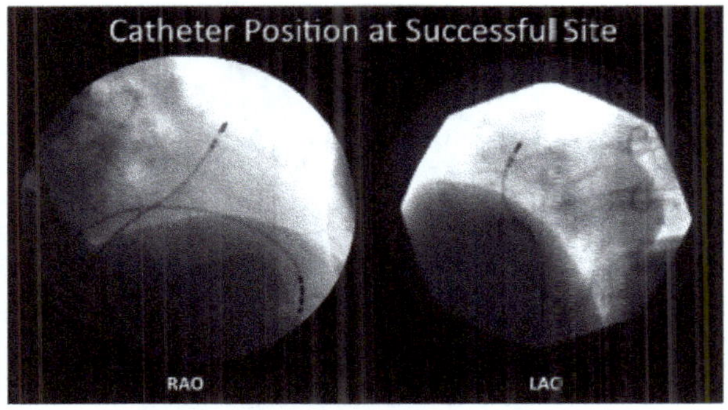

Figure 33 A.4. Fluoroscopic views of catheter position at successful site. RAO, right anterior oblique; LAO, left anterior oblique.

Ablation of RVOT tachycardias is traditionally reserved for patients who have had prolonged incapacitating symptoms and have failed pharmacological treatment. Careful analysis of 12-lead ECG during tachycardia is very useful and, in our experience, can guide catheter localization to within 0.5 to 1 cm of the site of successful ablation. Our technique consists of localization of the site of origin using earliest intracardiac activation and/or pace mapping. If the patient has frequent PVCs, then we construct an electroanatomic map during tachycardia to visualize the site of earliest activation. If PVCs are infrequent, then we construct an endocardial shell of the RVOT region and utilize both limited activation mapping and pace mapping to guide ablation. Pace mapping is performed at diastolic threshold and at a rate similar to tachycardia cycle length. The goal is to achieve an identical match between the clinical PVC and the pace map, including subtle features such as notches in the QRS complexes in various leads. During activation mapping, the site of successful ablation precedes the onset of the QRS by ~30 ms on a bipolar electrogram. In our experience, it is imperative to confirm site of earliest activation and the best pace-map site prior to ablation. In our laboratory, we use a 4-mm tip ablation catheter in a temperature-controlled mode with the typical settings: power of 40 to 50 watts, temperature not to exceed 55°C and duration of up to 60 seconds.[2]

In the case above, the ECG morphology revealed a left bundle, inferior-axis PVC with a precordial transition by lead V_4. There was no notching in the inferior leads, placing the PVC on the septal aspect of the RVOT. In lead I, the QRS complex was biphasic. Based on the results of our earlier study, we localized this PVC to septal site 2 to 3 from the ECG. Given that the PVCs were infrequent, we elected to construct an endocardial shell of the RVOT region and used limited activation mapping and pace mapping to localize the site of origin of the PVC. The site of earliest activation was 30 to 35 ms pre-QRS and the pace map at the site was a good match. RFA at site with a 4-mm tip catheter terminated the PVC, and the patient was noninducible post ablation.

In our experience, ECG analysis of RVOT tachycardias is helpful in planning the ablation procedure. Careful attention to the frontal plane axis, the precordial QRS transition, QRS

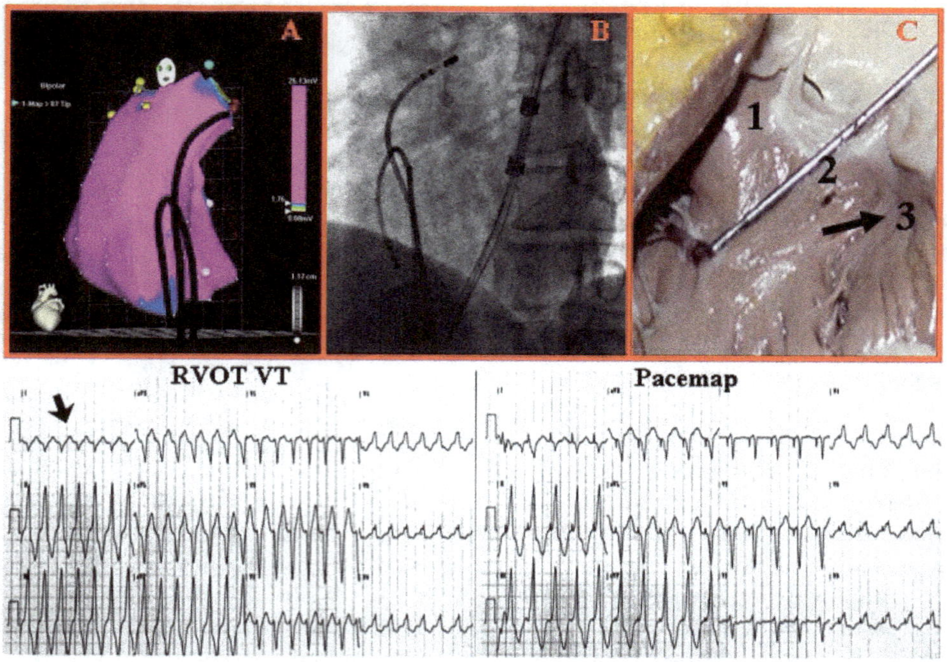

Figure 33 A.5. RVOT tachycardia from the left septal side of the outflow tract underneath the pulmonic valve. From the electroanatomic map (**A**) and fluoroscopic image (**B**), one can see how the RVOT wraps around the aortic root, is to the left of the apical septum and can generate a negative QRS complex in lead I during VT. The pace map from this leftward and anterior site is identical to the clinical tachycardia (bottom panel). The gross anatomic specimen (**C**) depicts the septal aspect of the RVOT underneath the pulmonic valve. In our laboratory, the septal and free wall sites just under the pulmonic valve are designated as site 1 (posterior/right), site 2 (mid), and site 3 (anterior/left) to simplify the nomenclature during mapping. (Adapted from Bala et al[4] with permission from Elsevier.)

morphology in the inferior leads, and QRS complex in lead I can further pinpoint the origin of VT. This analysis helps facilitate a near-uniform successful ablation strategy.

Correct Answer:

Q1. A

REFERENCES

1. Lerman BB, Stein KM, Markowitz SM. Mechanisms of idiopathic left ventricular tachycardia. *J Cardiovasc Electrophysiol.* 1997;8:571–583.
2. Dixit S, Lin D, Marchlinski FE. Ablation of ventricular outflow tract tachycardias. In: Huang SKS, Wood MA, eds. *Catheter Ablation of Cardiac Arrhythmias.* Philadelphia, PA: Saunders Elsevier; 2006: 473–490.
3. Marchlinski FE, Lin D, Dixit S, et al. Ventricular tachycardia from the aortic cusps: Localization and ablation. In: Raviele A, ed. *Cardiac Arrhythmias* 2003: *Proceedings of the 8th International Workshop on Cardiac Arrhythmias.* Milan: Springer-Verlag Italia; 2004: 357–370.
4. Bala R, Marchlinski FE. Electrocardiographic recognition and ablation of outflow tract ventricular tachycardia. *Heart Rhythm.* 2007;4:366–370.
5. Dixit S, Gerstenfeld EP, Callans DJ, Marchlinski FE. Electrocardiographic patterns of superior right ventricular outflow tract tachycardias: Distinguishing septal and free-wall sites of origin. *J Cardiovasc Electrophysiol.* 2003;14:1–7.

CASE STUDY

33 B

Idiopathic Left Ventricular Tachycardia

Indrajit Choudhuri, MD

A 51-year-old woman admitted for palpitations was found to have a wide-complex tachycardia (Figure 33 B.1). Sinus rhythm was restored medically and diagnostic cardiac workup included repeat ECG, serial cardiac enzymes, echocardiogram, and cardiac catheterization, all of which were normal.

Question No. 1:

The ECG in Figure 33 B.1 shows:

A. Supraventricular tachycardia (SVT) with right bundle branch block (RBBB) aberrant conduction.

B. Idiopathic left ventricular (LV) tachycardia.

C. Antidromic tachycardia involving a left-sided accessory pathway.

D. Bundle branch reentry.

Medical therapy effectively suppressed her tachycardia until a recurrence 1 year later that necessitated repeat hospitalization for severe palpitations and light-headedness. Invasive electrophysiology (EP) study and radiofrequency ablation were recommended. Baseline measurements in sinus rhythm included AH 89 ms and HV 55 ms, but tachycardia was not inducible.

Question No. 2:

Which of the following is correct about the expected physiology that can elucidate the tachycardia mechanism in sinus rhythm?

A. If dual atrioventricular (AV) node physiology and/or single AV nodal echoes are seen, AV node reentry is the tachycardia mechanism.

B. Ventricular pacing to identify the retrograde activation sequence will identify the tachycardia mechanism.

C. A multipolar catheter positioned along the posterior septum or catheter mapping in that region can suggest the diagnosis.

D. The tachycardia mechanism cannot be determined without induction.

Question No. 3:

Which of the following is correct about an ablation approach?

A. The tachycardia cannot be ablated without induction; the study should be aborted.

B. Empiric right bundle ablation is indicated.

C. Pacemapping will be associated with a 12/12-lead match at the successful ablation site.

D. A linear lesion delivered to transect sites along the posterior fascicle exhibiting Purkinje potentials can prevent tachycardia recurrence.

DISCUSSION

Q1: The patient's ECG demonstrates a wide-complex tachycardia with RBBB-like morphology and left-superior axis. In the absence of

Practical Electrophysiology © 2015 Jasbir S. Sra, Masood Akhtar, Andrea Natale, David J. Wilber
Cardiotext Publishing, ISBN: 978-1-935395-17-1

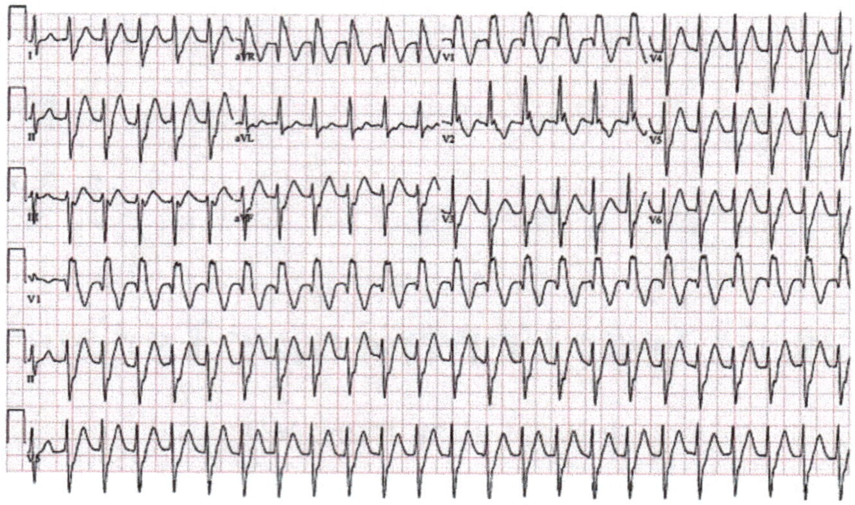

Figure 33 B.1. Patient's ECG at presentation.

structural heart disease, the presumptive diagnosis is idiopathic posterior fascicular left ventricular tachycardia (ILVT). AV dissociation is apparent, confirming the diagnosis of ventricular tachycardia (VT). Other relevant features include increased velocity of the intrinsicoid deflection, ie, sharpness of the initial QRS deflection, suggesting involvement of the His-Purkinje system. The posterior fascicle serves as a critical structure within the reentrant circuit, and earliest endocardial activation during VT is commonly seen in the ventricular septum,[1–3] leading to relatively narrow QRS. Other features include absence of characteristics associated with VT of myocardial origin, such as QRS duration <140 ms and time from onset of R to nadir of S <100 ms.

Aberrantly conducted beats conform to the morphology of true bundle branch blocks. The QRS morphology does not indicate the typical features of RBBB other than initial positive upstroke in V_1. In fact, the QRS is significantly different than typical RBBB with an initial Q wave in V_1 followed by an Rsr' seen best in V_2 and rS pattern in V_5 and V_6. Axis deviation from normal is not an expected feature of RBBB, though fascicular blocks can be seen with aberrancy. Reentrant tachycardia with antegrade conduction over an accessory pathway and retrograde conduction over the AV node (antidromic tachycardia) would be expected to exhibit 1:1 AV association. The identification of AV dissociation eliminates any form of SVT, including accessory pathway-mediated tachycardia.

Idiopathic LV tachycardia involves the Purkinje network of the left bundle branch system to support reentry. VT occurs in the *absence* of ostensible conduction disease and usually in the setting of structural normality, distinguishing idiopathic VT from bundle branch reentry and interfascicular reentry.

Q2: Dual AV node physiology would not necessarily be an unexpected finding as the majority of patients who present for EP study will have dual antegrade or retrograde AV nodal pathways, but this alone, and/or even if AV nodal echoes are seen, does not confirm a diagnosis of AV node reentry. Slow-pathway ablation is not indicated, particularly with the ECG features indicating VT. Similarly, ventricular pacing will likely demonstrate retrograde conduction over the fast AV nodal pathway if VA conduction is present, and will not provide insight to guide ablation because a retrogradely conducting accessory pathway does not participate in the tachycardia. The normal HV interval in sinus rhythm also makes antegrade accessory pathway conduction unlikely.

The tachycardia circuit of ILVT has been deduced through observations made during VT and sinus rhythm (Figure 33 B.2). In sinus, Purkinje potential (P2) recordings *preceding* ventricular activation demonstrate normal proximal-to-distal activation of the posterior fascicle, as in the example shown in Figure 33 B.3. The wavefront exits the Purkinje-network termini and retrogradely activates a specialized, decrementally conducting, verapamil-sensitive zone in the interventricular septum and propagates toward the basal septum, generating a high-frequency potential (P1) that may be identifiable in some patients *after* ventricular activation.[4,5] ILVT is typified by "Zipes' triad" of induction with atrial pacing, RBBB morphology during VT, and cardiac structural normality.[6] These

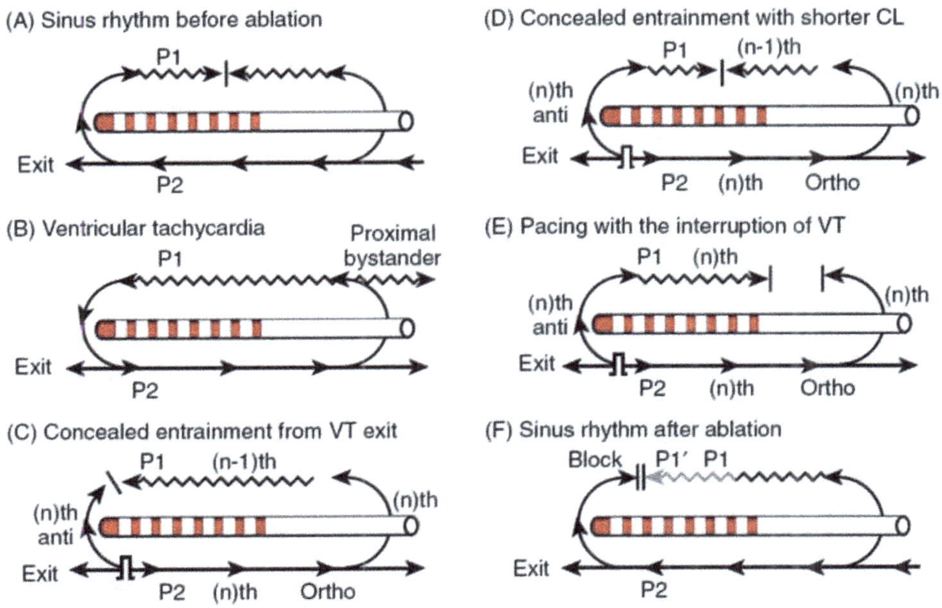

Figure 33 B.2. Schematic illustrating the mechanism of idiopathic verapamil-sensitive VT. **A:** During sinus rhythm, activation proceeds along the Purkinje network and depolarizes P2, and then retrograde activation along the septum results in P1 being fused and buried within the local ventricular activation. **B:** During VT, P1 precedes P2. **C:** During concealed entrainment from the VT exit, antidromic wavefront antiblock distal to P1 results in orthodromic activation (ortho) of P2 and P1. The orthodromic wavefront of the preceding (*n*−1)th beat blocks in the connection between P1 and P2, encountering refractoriness caused by antidromic penetration from the (*n*)th pacing impulse, while the orthodromic wavefront from the last pacing impulse continues to reset the tachycardia. **D:** Overdrive entrainment pacing with shorter cycle length (CL) advances the site of collision between the (*n*)th antidromic wavefront and the (*n*−1)th orthodromic wavefront to the middle portion of the area of slow conduction, with the interval from the last pacing stimulus to the orthodromically activated P1 prolonging because of rate-dependent conduction delay in the area of the slow conduction. **E:** Overdrive pacing with yet shorter CL results in the antidromic wavefront blocking at the proximal portion of the area of slow conduction and the orthodromic wavefront from the previous paced beat blocking as well, thereby interrupting VT. Radiofrequency delivery between P1 and P2 eliminates VT, and P1 activation then proceeds orthodromically around the circuit and subsequently from a proximal to distal direction during sinus rhythm; hence, after ablation **(Panel F)**, P1 appears in the mid-diastolic period with the same activation sequence as during VT. (Modified from Nogami et al[5] with permission from Elsevier.)

criteria were subsequently expanded by Belhassen to include demonstration of verapamil sensitivity, and these characteristics can be sought at EP study as well.[7]

During VT, the basal interventricular septum is captured antidromically and serves as the antegrade limb of the reentrant circuit,[4] allowing P1 to emerge as an isolated late diastolic potential *before* ventricular activation. The antegrade wavefront penetrates the posterior fascicle of the left bundle branch system in the distal third of the septum through the septal Purkinje-network termini at the so-called "lower turn-around point," creating diverging wavefronts that travel simultaneously *antegradely* to the myocardial exit site, with earliest activation seen at the posteroapical septum approximately 15 to 20 ms before QRS inscription,[8] and *retrogradely* activating the posterior fascicle, thereby *reversing* the direction of Purkinje activation compared to sinus rhythm (see Figure 33 B.2). This retrograde Purkinje activation may appear just before ventricular activation, resulting in the appearance of P1 preceding P2 during VT (Figure 33 B.4). The retrograde wavefront reaches the basal septum, where it exits the fascicle in close proximity to the main left bundle, the so-called "upper turn-around point," and then reenters the zone of slow conduction.

Q3: Bundle branch reentry with RBBB morphology is rare compared to the more common left bundle branch block (LBBB) variety. Hence, empiric right bundle branch ablation is not indicated without confirmation of the mechanism. Pacing from sites identifying the earliest Purkinje potential, as well as surrounding sites exhibiting later P2, all may produce QRS similar to VT morphology, suggesting activation of the Purkinje network with propagation to the exit site but without identifying the precise exit.[8] Pacemapping at sites recording P1 and/or P2 will usually produce, at best, 10/12 ECG lead matches due to difficulty in achieving isolated Purkinje capture without associated myocardial capture. Other reasons that pacemapping alone is

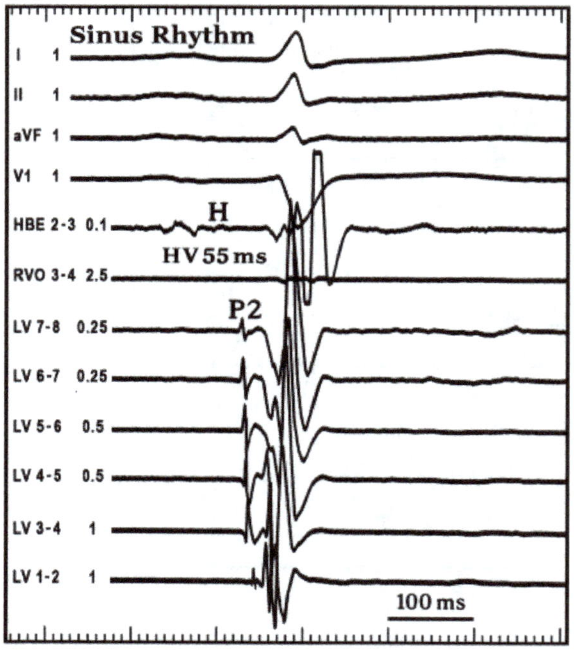

Figure 33 B.3. Intracardiac recordings from octapolar electrode catheter during sinus rhythm recording at the same site demonstrated the presystolic Purkinje potential (P2), which was recorded before the onset of QRS complex. (Modified from Nogami et al[5] with permission from Elsevier.)

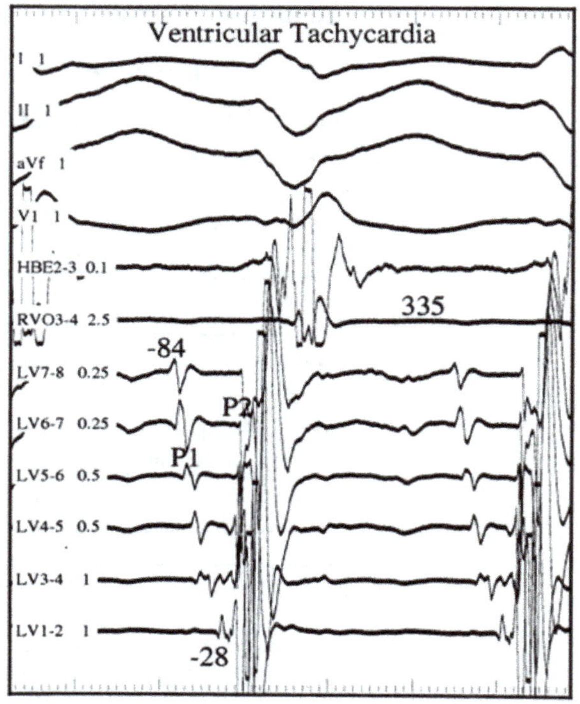

Figure 33 B.4. During VT, a diastolic potential (P1) and a presystolic Purkinje potential (P2) were recorded. While P1 was recorded earlier from the proximal than the distal electrodes, P2 was recorded earlier from the distal than the proximal electrodes. (Reproduced from Nogami et al[5] with permission from Elsevier.)

generally less effective in mapping ILVT and may be confusing include:

1. Differences between conduction characteristics during VT and pacing during sinus rhythm.
2. Disparity of locations, including bystander regions, from which pacing may engage the fascicular system and therefore produce excellent pacemap matches without identifying the true myocardial exit or critical segment of the reentrant circuit.
3. If the reentrant circuit is located proximal to the myocardial exit, then identification of the site of earliest activation may not terminate VT but only alter the exit site.
4. Selective capture of the fascicular system alone from sites recording P1 and/or P2 is difficult, manifesting poor pacemap matches due to capture of surrounding myocardium from sites demonstrated to be critical to the tachycardia circuit and successful at terminating VT.

Electroanatomic mapping studies during tachycardia in patients diagnosed with ILVT have confirmed the reentrant physiology of idiopathic posterior fascicular VT (Figure 33 B.5),[9] identifying retrograde activation of the septal region of slow conduction[4,5] as an approachable target for ablation that was not seen in controls. Apical-posterior septal sites exhibiting earliest activation and preceded by Purkinje potentials may be targeted, as per Nakagawa et al.[8] Tsuchiya et al[4] targeted basal septal and midseptal sites in which a late-diastolic potential analogous to P1, preceding P2, was identified. Nogami et al[5] and others[10] targeted sites where P1 and P2 may be recorded, or sites recording earliest P2 when P1 is not recorded, achieving tachycardia termination and noninducibility after ablation.[5] In patients with unmappable fascicular VT, electroanatomic mapping guided by Purkinje potentials and pacemapping has been employed to deliver a contiguous linear lesion across the inferior mid-to-apical LV septum to transect the posterior fascicle for VT control.[11,12]

Correct Answers:

Q1. B

Q2. C

Q3. D

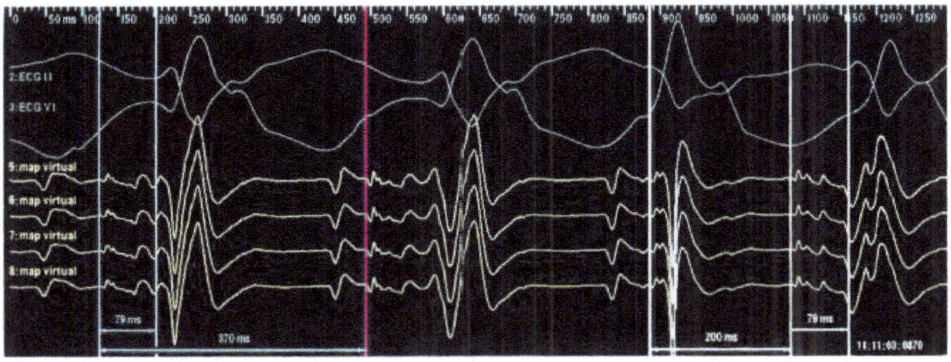

Figure 33 B.5. Mid-diastolic potential during VT. Shown from top to bottom are surface ECG leads II and V_1 and computed electrograms from the selected site. Time from mid-diastolic potential to onset of QRS is 79 ms. Tachycardia cycle length is 370 ms. Following a spontaneous premature ventricular beat, VT is reset, as evidenced by shortening of cycle length to 200 ms. However, potential to QRS duration of 79 ms is unaltered, suggesting this potential is part of the tachycardia circuit. (Reproduced from Sra et al[9] with permission from John Wiley and Sons.)

REFERENCES

1. Kottkamp H, Chen X, Hindricks G, et al. Idiopathic left ventricular tachycardia: New insights into electrophysiological characteristics and radiofrequency catheter ablation. *Pacing Clin Electrophysiol.* 1995;18:1285–1297.
2. Zardini M, Thakur RK, Klein GJ, Yee R. Catheter ablation of idiopathic left ventricular tachycardia. *Pacing Clin Electrophysiol.* 1995;18:1255–1265.
3. Bogun F, El-Atassi R, Daoud E, et al. Radiofrequency ablation of idiopathic left anterior fascicular tachycardia. *J Cardiovasc Electrophysiol.* 1995; 6:1113–1116.
4. Tsuchiya T, Okumura K, Honda T, et al. Significance of late diastolic potential preceding Purkinje potential in verapamil-sensitive idiopathic left ventricular tachycardia. *Circulation.* 1999;99:2408–2413.
5. Nogami A, Naito S, Tada H, et al. Demonstration of diastolic and presystolic Purkinje potentials as critical potentials in a macro-reentry circuit of verapamil-sensitive idiopathic left ventricular tachycardia. *J Am Coll Cardiol.* 2000;36:811–823.
6. Zipes DP, Foster PR, Troup PJ, Pedersen DH. Atrial induction of ventricular tachycardia: Reentry versus triggered automaticity. *Am J Cardiol.* 1979;44:1–8.
7. Belhassen B, Rotmensch HH, Laniado S. Response of recurrent sustained ventricular tachycardia to verapamil. *Br Heart J.* 1981;46:679–682.
8. Nakagawa H, Beckman KJ, McClelland JH, et al. Radiofrequency catheter ablation of idiopathic left ventricular tachycardia guided by a Purkinje potential. *Circulation.* 1993;88:2607–2617.
9. Sra J, Bhatia A, Krum D, Akhtar M. Endocardial noncontact activation mapping of idiopathic left ventricular tachycardia. *J Cardiovasc Electrophysiol.* 2000;11:1409–1412.
10. Aiba T, Suyama K, Aihara N, et al. The role of Purkinje and pre-Purkinje potentials in the reentrant circuit of verapamil-sensitive idiopathic LV tachycardia. *Pacing Clin Electrophysiol.* 2001;24:333–344.
11. Lin D, Hsia HH, Gerstenfeld EP, et al. Idiopathic fascicular left ventricular tachycardia: Linear ablation lesion strategy for noninducible or nonsustained tachycardia. *Heart Rhythm.* 2005;2:934–939.
12. Chen M, Yang B, Zou J, et al. Non-contact mapping and linear ablation of the left posterior fascicle during sinus rhythm in the treatment of idiopathic left ventricular tachycardia. *Europace.* 2005;7:138–144.

CASE STUDY

34

Mapping and Ablation of Stable Monomorphic Ventricular Tachycardia

Usha B. Tedrow, MD and William G. Stevenson, MD

A 65-year-old man is taken to the electrophysiology laboratory for catheter ablation of sustained monomorphic ventricular tachycardia (VT). He has a history of myocardial infarction 8 years ago. An implantable cardioverter-defibrillator (ICD) was placed 1 year ago after he presented with sustained VT requiring cardioversion. Amiodarone was added to his regimen after several episodes of VT required ICD shocks. ICD interrogation confirmed monomorphic VT that fails to terminate with antitachycardia pacing.

Programmed electrical stimulation induced the wide-complex tachycardia shown in Figure 34.1.

Question No. 1:

Based on the ECG, what is the likely location of the infarct scar?

A. Anterior lateral left ventricle (LV).
B. Septal LV.
C. Inferior LV.
D. Posterobasal LV.

DISCUSSION

The ECG shows a wide QRS tachycardia with a cycle length of 430 ms. The mechanism of inducible VT in patients with prior myocardial infarction is almost always scar-related reentry in the old infarct region, as illustrated in the top schematic of Figure 34.2.

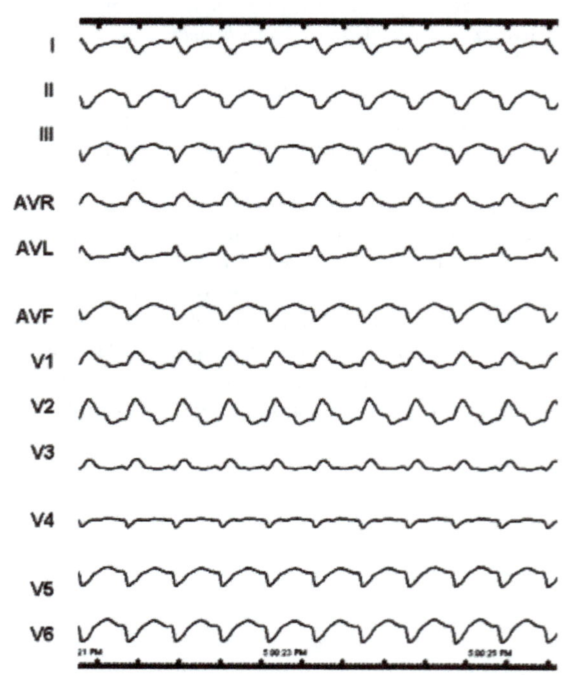

Figure 34.1. Wide-complex tachycardia induced by programmed electrical stimulation.

Practical Electrophysiology © 2015 Jasbir S. Sra, Masood Akhtar, Andrea Natale, David J. Wilber
Cardiotext Publishing, ISBN: 978-1-935395-17-1

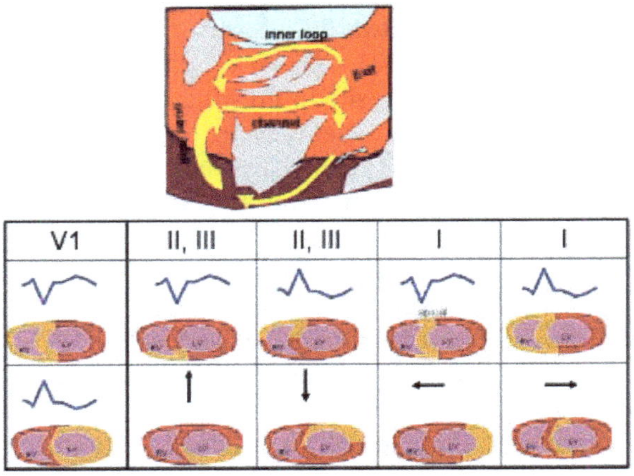

Figure 34.2. Top panel: Schematic of scar-related reentry in old infarct region. Bottom panel: An approach to estimating the location of the reentry circuit exit. LV, left ventricle; RV, right ventricle.

In these reentry circuits, propagation typically occurs through channels formed by surviving myocardium bordered by fibrosis and, in some cases, a valve annulus. Depolarization of the reentry circuit channel does not generate sufficient signal to be detected in the surface ECG. The QRS complex is inscribed as the wavefront emerges from the channel where it connects to the ventricular myocardium, a region that is identified as the exit from the channel. After emerging from the exit, the circulating wavefront propagates through loops, either along the border of the scar or within the scar, to return to the entrance of the channel.

The QRS morphology of scar-related reentrant VT reflects the location of the exit. An approach to estimate the location of the reentry circuit exit is shown in Figure 34.2. In lead V_1, a dominant S wave (left bundle branch block-like configuration) indicates the exit is in the septum or right ventricle (RV). A dominant R wave (right bundle branch block-like configuration), as in this case, indicates a left ventricular exit. The frontal plane axis identifies the position of the likely exit as viewed from the left anterior oblique projection of the ventricle.
A superior frontal plane axis, as shown here, is consistent with early activation of the inferior wall of the ventricle. For left ventricular VT, a rightward axis, as is present here, suggests the exit is farther from the septum, whereas an axis directed to the left places the exit toward the septum. However, very apical exits near the septum can have a rightward axis, as the LV sits progressively more leftward as one moves from base to apex. An exit near the apex generates S waves in the midprecordial leads (V_3–V_4), whereas basal exits near the mitral annulus produce dominant R waves in the midprecordial leads. In summary, this VT exit is likely at the inferolateral LV.

The approach to ablating this type of scar-related monomorphic VT proceeds in an organized stepwise manner. Catheters are inserted into the RV and, usually, right atrial locations. Programmed stimulation is used to induce VT and confirm the tachycardia is in fact ventricular and not due to bundle branch reentry. Atrioventricular (AV) dissociation indicates VT with rare exception. If a His deflection is present prior to the QRS, its relationship to ventricular activation is assessed to determine whether it is passively activated retrogradely, or whether it is part of the reentry circuit, in which case bundle branch reentry is likely. In this case, the morphology of the VT and known history of prior myocardial infarction favor a left ventricular origin of VT. The next step is to obtain access to the LV. This can be achieved via a retrograde aortic approach or, if peripheral vascular disease or aortic valve disease is present, an atrial transseptal approach, which allows the mapping catheter to reach the LV through the mitral valve. If VT is not incessant, initial mapping is performed during sinus or, due to poor AV conduction as in this case, paced rhythm.

The scar location is then identified from a voltage map of the LV (Figure 34.3). Electrograms

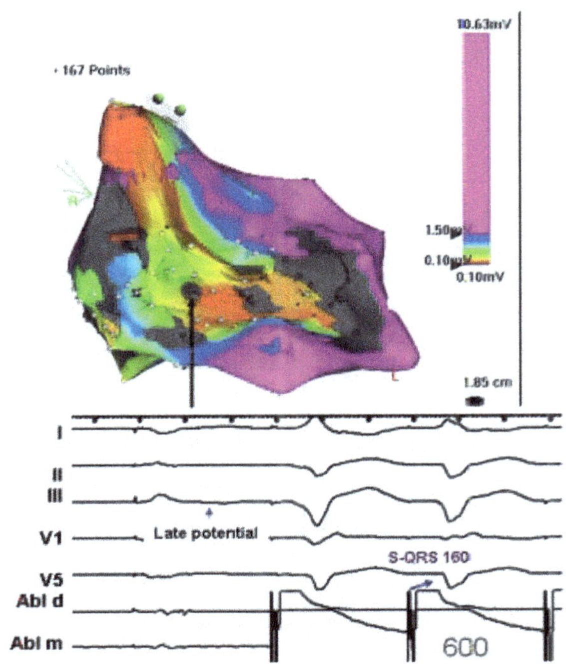

Figure 34.3. Top panel: Voltage map identifying scar location. Bottom panel: EGM findings at the site in the inferior wall.

(EGMs) are acquired from around the ventricle to create a voltage map. The color codes are set such that a peak-to-peak EGM amplitude that exceeds 1.5 mV is designated normal (see purple in Figure 34.3). EGM amplitude is progressively lower, moving from blue to green to yellow to red. A complete map of the entire ventricle is not required to define the anatomy, but rather focuses on completely marking the low-voltage region of infarct scar. The LV voltage map shows a low-voltage region in the inferior wall extending from the mitral annulus toward the apex. The location of this scar is consistent with that expected from the ECG morphology of the VT.

During creation of the voltage map several indicators of potential reentry circuit channels are sought and marked on the map. At each low-voltage site, pacing is performed to assess capture and the QRS morphology. Sites with pacing thresholds exceeding 10 mA (2-ms pulse width, and typically using unipolar [Uni] pacing in our laboratory) are designated unexcitable scar by gray regions. Figure 34.3 shows findings at an interesting site in the inferior wall. The EGM during paced rhythm (first beat of the tracing) is low-amplitude, fractionated and contains split potentials inscribed after the end of the QRS. These late potentials are indicative of delayed activation. Pacing at this site produces a QRS complex that differs from the VT (the last 2 beats of the tracing). The paced QRS does have a superiorly directed frontal plane axis with dominant S waves in the inferior leads, but the axis is directed leftward (dominant R wave in lead I). Thus, this site is probably not in the exit for VT. Furthermore, we can anticipate the VT exit is likely located more leftward from this site, where pacing would produce a rightward QRS axis (negative in lead I, as observed in VT). However, there is another feature that makes this site interesting. The delay from stimulus to QRS onset is relatively long at 160 ms, indicating the pacing site may be in a channel. When pacing in channels, the QRS will resemble that of VT, when the stimulated wavefront emerges from the channel at the same exit as the tachycardia. However, if there are other paths for the stimulated wavefront to follow to reach the border of the scar, the QRS morphology may differ from that of the VT. Therefore, we elected to leave the catheter at this site and induce VT to determine if this site is in the reentry circuit. If the site is in the circuit, ablation can be immediately performed. If the site is remote from the circuit, further mapping can continue in VT.

Figure 34.4 shows 5 surface ECG leads followed by intracardiac signal recordings. The recordings from the mapping catheter are labeled Abl (d = distal, m = middle electrode pair, p = proximal electrode pair). Bipolar and Uni signals are recorded. The first 2 Uni signals are high-pass filtered at 30 Hz to improve visualization of low-amplitude local signals. With a high-pass filter of 0.5 Hz (Uni-1 0.5), a substantial far-field signal obscures the low-amplitude local signals. Recordings from the His bundle region (His) and RV apex (RVA) are also shown, as well as the arterial blood pressure. Programmed stimulation induced a different VT than the previous VT (see Figure 34.4). The cycle length is 390 ms, and the frontal plane axis is approximately 90 degrees (nearly isoelectric in

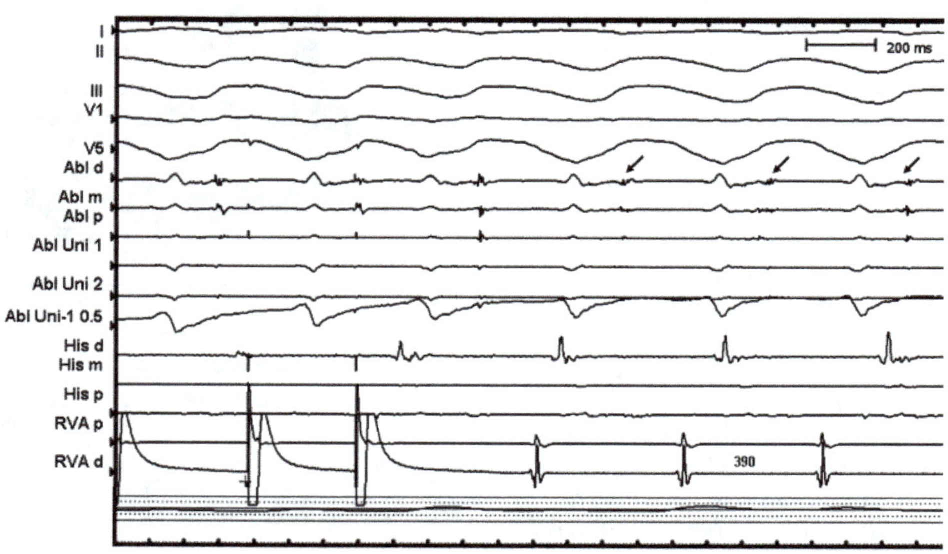

Figure 34.4. Programmed stimulation induced a different VT than the previous VT.

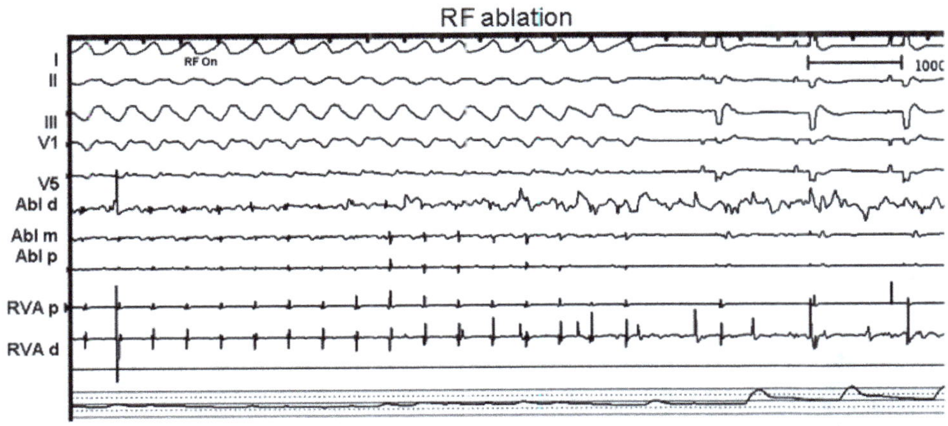

Figure 34.5. RF ablation terminated the VT.

lead I). There is diastolic electrical activation during VT. Diastolic potentials may indicate the site is in the reentry circuit channel but can also be seen at bystander regions. Ordinarily, we would pace to entrain the VT to assess whether this site is in the circuit. However, this VT is not hemodynamically tolerated, as indicated by the arterial blood pressure (bottom of tracing). If the blood pressure does not promptly improve, immediate termination will be needed. In this situation—the ablation catheter in a low-voltage scar, likely a reentry circuit channel based on the pace map, and diastolic activity during VT—immediate application of radiofrequency (RF) current was selected.

Application of RF current at this site terminated VT (Figure 34.5). We typically apply a 90-second RF application using an irrigated catheter and titrating power to achieve an impedance fall of approximately 10 ohms and a small increase in measured temperature. Power typically starts at 30 W and is titrated to a maximum of 50 W. Following the application, we pace at the site to confirm creation of a lesion, indicated by absence of capture with 10-mA, 2-ms pulse-width pacing stimuli. Multiple applications in the region may be required to render it electrically unexcitable.

Following ablation at this site, we do not know if other VTs, including the previously observed VT, have been abolished. Most patients have multiple VTs, reflecting multiple potential reentry circuits that can form in response to programmed stimulation. Furthermore, it is often uncertain which of these VTs have occurred spontaneously, as they have been promptly terminated by the ICD in most cases. Therefore, we target all sustained monomorphic VTs that are within 20 to 40 ms of the VT cycle length established from ICD interrogation of spontaneous VT and all slower VTs. During pacing at a site adjacent to the first ablation site in the inferior scar, we again induced the first VT (Figure 34.6, right panel).

The top left panel of Figure 34.6 shows a beat during RV pacing. The bottom left panel shows the LV voltage map with an arrow indicating the mapping site. The EGM is at low amplitude. Pacing at this LV site produces a QRS that resembles the first VT and has a stimulus to QRS delay of 140 ms, consistent with pacing in a channel for the first VT. Interestingly, pacing at this site induced the first VT (see Figure 34.6). This VT is sufficiently stable to attempt entrainment from the mapping site (Figure 34.7).

Figure 34.7A shows the findings at the same site as in Figure 34.6, now during the first VT. In the last 3 beats, a diastolic potential can be seen 130 ms prior to the QRS onset, consistent with activation of a site in the channel. Entrainment of VT from this site (as seen in the first 3 beats of the tracing) confirms the site is in the circuit and not a bystander. Pacing is performed at a cycle length of 420 ms. The QRS complexes and ventricular EGMs are accelerated to the pacing cycle length, and the QRS morphology is the same during pacing and during VT. Absence of QRS fusion was also confirmed in the 12-lead ECG (not shown). Thus, the stimulated wavefronts are exiting from the scar at the reentry circuit exit (Figure 34.7B and C). The stimulus-to-QRS (S-QRS) interval is 130 ms, which matches the EGM-to-QRS interval, a finding indicating the site is in the circuit. The postpacing interval (PPI), the time from the last stimulus to the next activation at the pacing site, is 470 ms (Figure 34.7C). This interval matches the first tachycardia cycle length after pacing (measured in the RVA distal-lead), but is longer than subsequent tachycardia cycle lengths of 435 ms. Variation of the

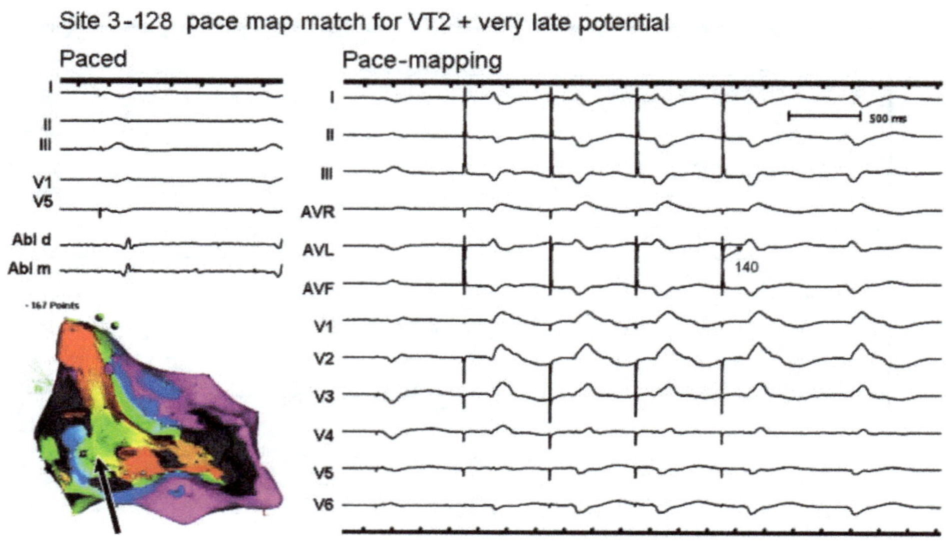

Figure 34.6. Top left panel: Paced beat. Bottom left panel: Left ventricular voltage map and arrow indicating the pacing site. Right panel: Induction of VT during pacing at a site adjacent to the first ablation site in the inferior scar.

PPI can be due to an increase in conduction time in the circuit during pacing, as commonly occurs in the presence of antiarrhythmic drugs. Cycle length oscillations may also occur when there are multiple potential reentry circuit loops that provide paths for the wavefront to return from the exit to the channel. In most cases, these 2 possibilities cannot be distinguished with single-catheter mapping.

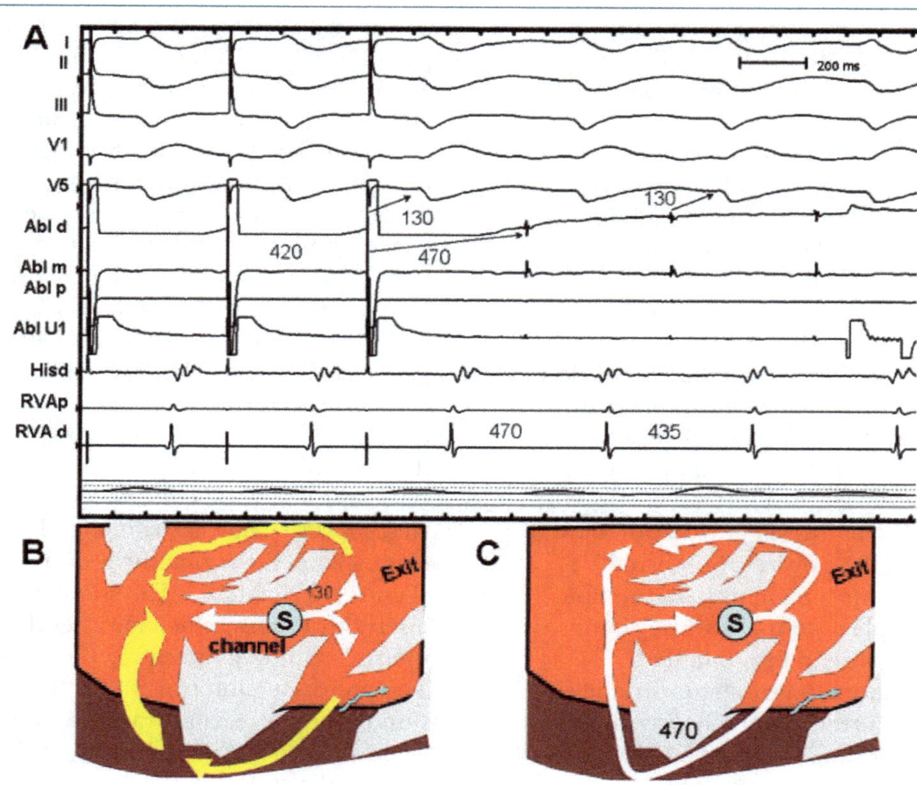

Figure 34.7. Entrainment of the VT. See text for details.

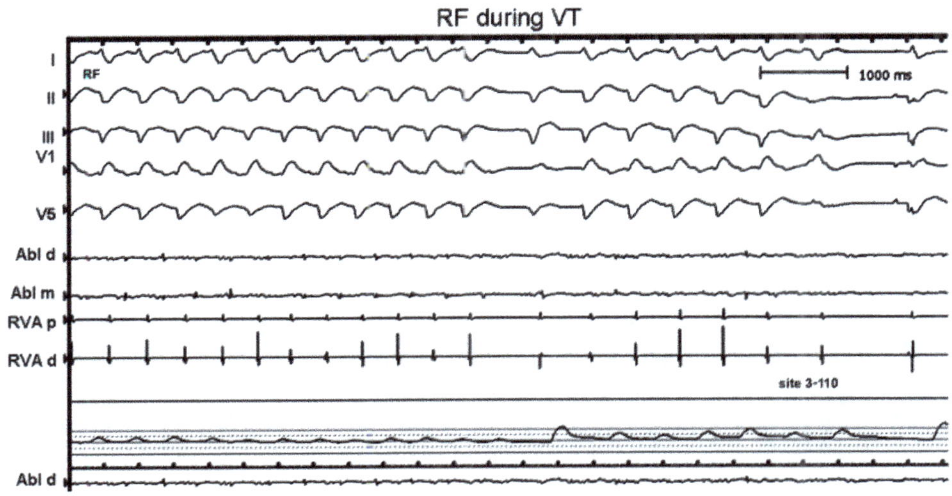

Figure 34.8. RF ablation of VT.

Because this site had several features consistent with reentry circuit site, RF ablation was performed during VT. Ablation terminated VT, as shown in Figure 34.8. Additional RF lesions were then placed in the region, rendering it electrically unexcitable. These sites were close to the mitral annulus; RF lesions were extended to the annulus to interrupt any potential reentry path running along the mitral annulus. Following ablation, no VT was inducible with programmed stimulation.

Correct Answer:

Q1. C

REFERENCES

1. Aliot EM, Stevenson WG, Almendral-Garrote JM, et al. EHRA/HRS Expert Consensus on Catheter Ablation of Ventricular Arrhythmias: Developed in a partnership with the EHRA, a Registered Branch of the ESC, and the HRS; in collaboration with the ACC and the AHA. *Heart Rhythm.* 2009;6:886–933.
2. Marchlinski FE, Callans DJ, Gottlieb CD, Zado E. Linear ablation lesions for control of unmappable ventricular tachycardia in patients with ischemic and nonischemic cardiomyopathy. *Circulation.* 2000;101:1288–1296.
3. Raymond JM, Sacher F, Winslow R, Tedrow U, Stevenson WG. Catheter ablation for scar-related ventricular tachycardias. *Curr Probl Cardiol.* 2009;34:225–270.
4. Soejima K, Stevenson WG, Maisel WH, Sapp JL, Epstein LM. Electrically unexcitable scar mapping based on pacing threshold for identification of the reentry circuit isthmus: Feasibility for guiding ventricular tachycardia ablation. *Circulation.* 2002;106:1678–1683.
5. Tedrow UB, Stevenson WG. Recording and interpreting unipolar electrograms to guide catheter ablation. *Heart Rhythm.* 2011;8:791–796.

CASE STUDY

35

Mapping, Imaging, Access, and Ablation: An Illustrative Case Presentation

Samuel J. Asirvatham, MD

A 56-year-old male presents with a history of documented ventricular tachycardia (VT) associated with presyncope and palpitations. A previously placed automated implantable cardioverter-defibrillator (ICD) has appropriately detected and delivered multiple therapies for VT. Continued ICD therapies have occurred despite attempts at medical therapy (dofetilide, mexiletine, amiodarone [toxicity], sotalol, and beta-blockers). A prior attempt at endocardial radiofrequency ablation was unsuccessful. The coronary arteries are normal. Computerized tomography (CT) (Figure 35.1) demonstrates dilatation of the right ventricle (RV) with microaneurysmal and fatty changes in the right ventricular free wall (hatched arrows).

Two Holter monitor tracings obtained during typical symptoms are shown in Figure 35.2. These episodes are very similar to multiple other documented VTs, including those that triggered ICD therapy.

Question No. 1:

After appropriate discussion of the risks and benefits involved, another attempt at radiofrequency ablation is planned. Which of the following criteria are indications for considering adjunctive epicardial mapping and ablation?

A. Failed prior ablation attempt.

B. Radiographic findings suggestive of arrhythmogenic right ventricular dysplasia (ARVD, arrhythmogenic right ventricular cardiomyopathy).

C. Electrocardiographic characteristics of the VT.

D. All of the above.

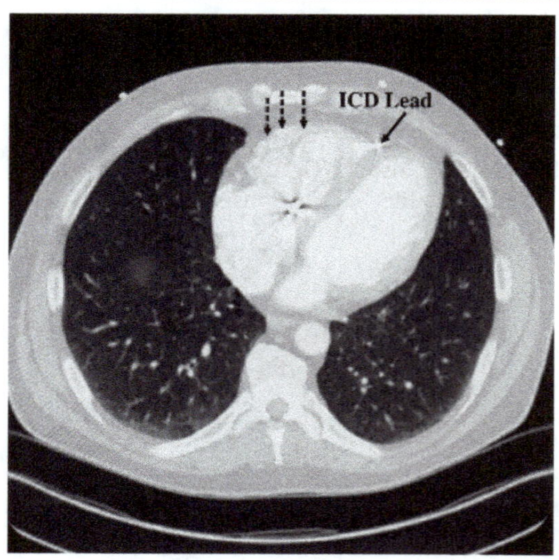

Figure 35.1. CT demonstrating dilatation of the RV with microaneurysmal and fatty changes in the right ventricular free wall (hatched arrows).

Practical Electrophysiology © 2015 Jasbir S. Sra, Masood Akhtar, Andrea Natale, David J. Wilber
Cardiotext Publishing, ISBN: 978-1-935395-17-1

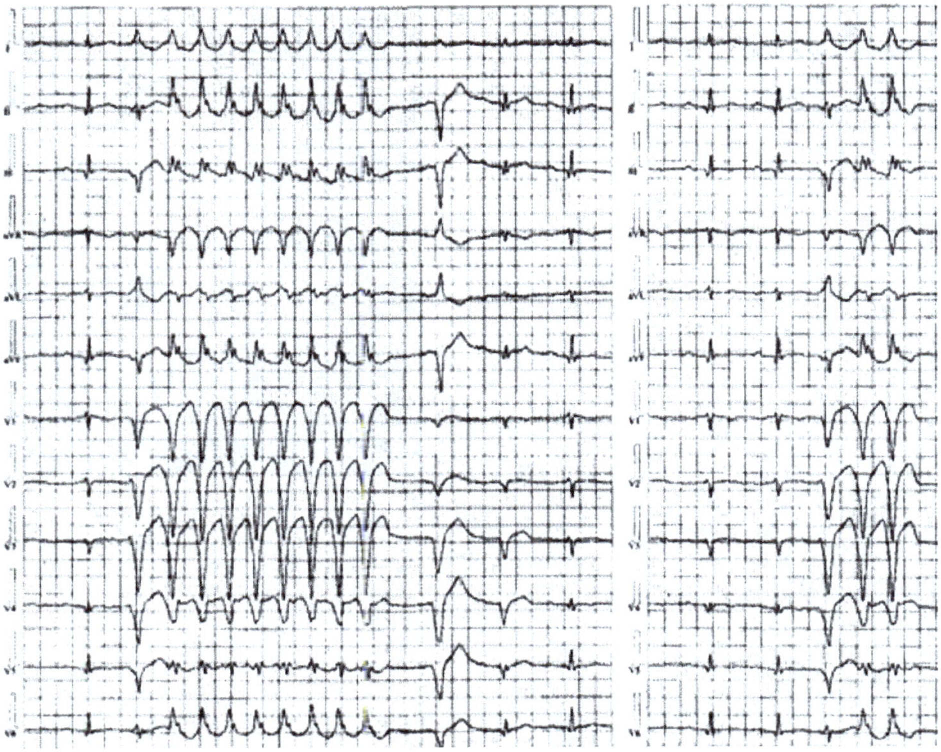

Figure 35.2. Holter monitor tracings obtained during typical symptoms.

DISCUSSION

- **Failed prior ablation attempt**: Success rates for VT ablation vary from about 50% to 70%,[1,2,3] and most patients in whom an attempted epicardial ablation is made have failed prior endocardial procedures. In these circumstances, approximately 75% will have an identifiable epicardial substrate that is targeted for ablation.[4] Reasons for failure of endocardial ablation include an epicardial substrate, epicardial focus, and limitations arising from endocardial energy delivery including coagulum formation and poor catheter contact. The latter specifically is an underappreciated cause of endocardial ablation failure, especially on the right ventricular free wall.

  The frequency of using an epicardial approach at tertiary centers varies with the underlying cardiac pathology (Figure 35.3). In coronary disease, epicardial ablation is more frequently required after inferior infarcts[5] as compared to anterior infarcts. Chagas disease—prevalent in South America, where the technique of subxiphoid pericardial access was first described[6]—typically requires epicardial mapping and ablation. In most of these instances, evidence of arrhythmogenic substrate enabling or constituting reentrant VT is epicardially targeted. In a few instances, automatic foci for premature ventricular contractions (PVCs) or VT in the setting of idiopathic outflow tract VT or other similar conditions may be required.[7,8]
- **Radiographic findings suggestive of ARVD (arrhythmogenic right ventricular cardiomyopathy)**: The cardiac CT findings in this

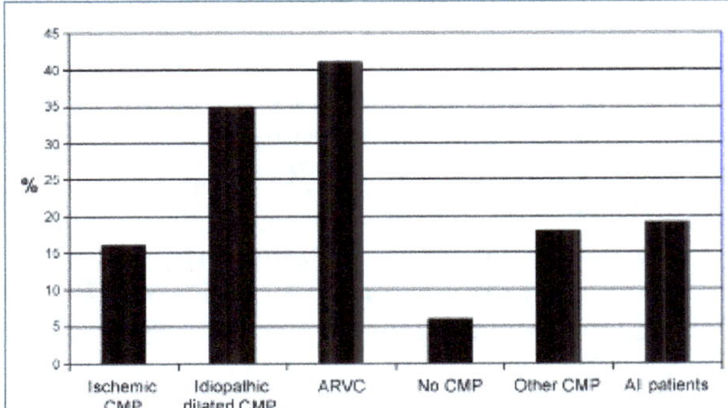

Figure 35.3. Frequency of epicardial ablation in cases of VT. Data from Sacher et al.[5] CMP, cardiomyopathy; ARVC, arrhythmogenic right ventricular cardiomyopathy.

patient are characteristic of ARVD. The fibrofatty displacement of the RV (occasionally left) begins in the subepicardium or midmyocardial layers and then progresses to the subendocardium.[9] Thus, a disproportionate number of patients with ARVD and VT require an epicardial approach to ablation.[5]

- **Electrocardiographic characteristics of the VT:** A critical and careful analysis of any obtained electrocardiographic (ECG) data can significantly facilitate the invasive approach for VT management. We note from the Holter data the following key points:
  1. The events are prototypical and suggest a higher likelihood of success with ablation.
  2. The inciting PVC is of different morphology from the induced, nonsustained VT, although each episode starts with the same PVC and produces the same VT, but the PVC and VT are different.
  3. Assuming that the PVC is automatic (focal source), analyzing the QRS axis and morphology can be very revealing in terms of identifying a suitable target for ablation (in this instance, strongly suggestive of RV apical free wall origin). On the other hand, with reentrant VT, ECG analysis may help identify the exit site (transition site to normal myocardium) but may be far removed (endocardial vs epicardial) from the critical substrate that requires targeting for successful ablation.

Can the ECG morphology of the focus/exit help identify epicardial origin of VT? Such criteria have been proposed; however, a clear understanding of the concept and limitations of these proposed criteria must be appreciated. The underlying principle is that the wavefront of a VT with an epicardial focus or exit site must initially travel slowly through the myocardium toward the endocardium, away from leads facing the focus, before it reaches the endocardial His-Purkinje system. This slow conduction results in a "pseudo delta wave" and a delay to the QRS peak (Figure 35.4). A number of parameters have been used to measure this delay, including the pseudo delta wave itself (≥34 ms; sensitivity 83%, specificity 95%), maximum intrinsicoid deflection (≥85 ms; sensitivity 87%, specificity 90%), and shortest RS complex time (>121 ms; sensitivity 76%, specificity 85%).[10] A Q wave may be present at the lead facing the tachycardia focus, suggesting an epicardial origin if absent during conduction via the Purkinje system.[11]

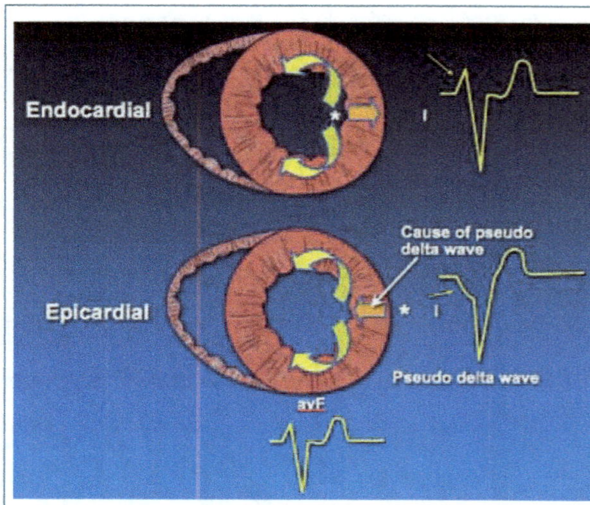

Figure 35.4. Origin of surface ECG characteristics of epicardial VT. Asterisk indicates the arrhythmogenic focus. (Reprinted from Swale et al[17] with permission from MediaSphere Medical.)

It is important to recognize the limitations of these surface ECG criteria. They will only be apparent in those leads aligned with the tachycardia exit site vector.[12] They are also dependent on a critical mass of myocardium to be activated to register on surface leads. A midmyocardial substrate may have an epicardial exit and be amenable to ablation from the endocardium. The presence of postinfarction Q waves on the baseline ECG may influence the arrhythmia morphology. Conduction delay in regions of scar, variations in cardiac rotation and antiarrhythmic drugs can lead to confounding activation sequences that have yet to be systematically evaluated. The ECG findings for right ventricular foci are also less well-defined because of the thinness of the myocardial wall and the lower density of the Purkinje network.

Question No. 2:

The patient proceeds to have epicardial ablation via a standard percutaneous subxiphoid approach. All of the following measures have been proposed to reduce procedural complications *except*:

A. Performing pericardial entry with placement of a sheath prior to endocardial transseptal puncture and systemic anticoagulation.

B. Maneuvering the liver and abdominal contents posteriorly with the guiding hand during needle entry.

C. Angling the needle toward the left shoulder at 45 degrees from the horizontal (posterior puncture), as opposed to toward the left midclavicle angled 15 to 30 degrees from the horizontal (anterior puncture).

D. Sequential aspiration and radiocontrast injection during pericardial entry with a blunt-tipped needle.

E. After needle entry, visualizing wire positioning in the pericardium using a left anterior oblique (LAO) projection of the heart.

DISCUSSION

Many of the potential mechanical complications of pericardial ablation occur at the time of pericardial entry, including right ventricular puncture, pericardial hemorrhage, coronary arterial injury, laceration of the liver, subdiaphragmatic vessels or abdominal contents, and pleural puncture.[4,5,13,14] An understanding of the relevant anatomy is critical to an optimal approach toward minimizing these risks. Anticoagulation is stopped in a timely fashion prior to all elective procedures, and subsequent anticoagulation, such as that required at the time of left-sided endocardial access, is delivered only after an epicardial sheath is in place. This allows easy drainage of any potential intrapericardial bleeding, which is usually seen initially and is self-limiting but may result in cardiac tamponade if not controlled. Placing standard right-sided catheters may aid in epicardial catheter guidance as the coronary sinus catheter clearly delineates the atrioventricular groove and position of the coronary arteries in the right anterior oblique (RAO) view, and the RV catheter marks the position of the septum and RV apex.

While obtaining pericardial access, the risk of RV puncture is minimized using the Seldinger-based technique described by Sosa,[6] where a blunt-tipped needle is slowly advanced via a subxiphoid entry, aspirating to identify inadvertent RV puncture while sequentially injecting radiocontrast material. Pericardial entry of the needle is identified by the contrast material forming a rim around the cardiac silhouette on fluoroscopy. Attaching a unipolar needle to the needle using a crocodile clip aids in identifying a current of injury on partial myocardial entry. Avoiding RV puncture and inadvertent mechanical trauma to other organs can be further facilitated by more recently described techniques using pressure frequency to assess pericardial entry or CT imaging to identify organ position, which are reviewed in further detail elsewhere.[17] An inferior puncture (obtained by angling the needle toward the left shoulder at 45 degrees from the horizontal) may at times be desirable based on the structures to be ablated or accessed, but may have a higher risk of passing through the left lobe of the liver or the diaphragm, and of coronary injury. An anterior puncture (obtained by angling the needle toward the left midclavicle angled 15 to 30 degrees from the horizontal) potentially avoids these risks. Using fluoroscopy, the needle tip is directed away from the septum and toward the lateral wall, within 1 to 2 cm of the apex, which minimizes the small risk of encountering a coronary artery.

After pericardial access is obtained, the guidewire should be advanced freely into the pericardial space. It is important to visualize the wire in the left anterior oblique fluoroscopic projection traversing the cardiac silhouette all the way to the lateral border. This is because inadvertent entry into the RV may allow free advancing of the wire, but the midleft lateral free wall will not be reached from this endocardial route. The RAO view can be inadequate, especially in ARVD, because of the thin right ventricular outflow tract free wall. It can be difficult to know in this projection whether the wire is epicardial or endocardial apposed to thin RV myocardium.

Question No. 3:

Which of the following findings during epicardial mapping is *most* supportive of an epicardial substrate?

A. Presence of low-voltage signals.

B. Presence of wide, fractionated or late near-field signals.

C. Earlier epicardial than endocardial signals.

D. All of the above.

E. None of the above.

DISCUSSION

- **Presence of low-voltage signals:** Electroanatomic mapping to identify abnormal substrate is now a well-established tool in the electrophysiologist's armamentarium, especially for treating patients with multiple and hemodynamically unstable VTs. Low voltage with endocardial mapping as long as catheter contact is assured generally denotes abnormal but not scarred myocardium—the frequent locations for critical zones maintaining reentrant VT.

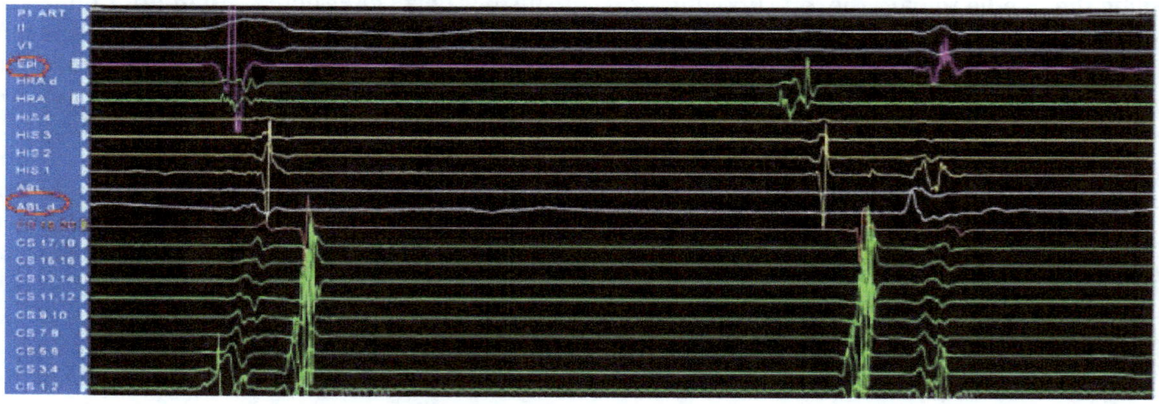

Figure 35.5. Early epicardial activation near the posterior right ventricular outflow tract (Epi) with a delay of 122 ms to the right ventricular endocardium (ABL) in the initial ectopic beat. The following sinus beat shows reversal of this activation pattern.

We see in Figure 35.5, for example, a PVC is mapped. Note there is clear evidence that the likely eventual site of ablation will be epicardial. First, we note the earliest ventricular activation is a far-field signal in the distal coronary sinus catheter (CS 1,2). The epicardial mapping catheter (Epi) is placed directly opposite to an ablation catheter on the endocardial surface (ABL d) and shows earlier activation. See the subsequent sinus beat with a reversal in this sequence.

In Figure 35.6, the ablation catheter is now moved to the epicardial surface. Note that a sharp near-field signal coincides with the previous earliest timing in the distal coronary sinus and was the site of successful ablation. An interesting phenomenon not uncommon with epicardial ablation can be seen in the next sinus beat in Figure 35.6. On the epicardial ablation catheter, note a sharp near-field signal is seen, even in sinus rhythm, that precedes the local ventricular activation. The etiology of such spikes has been presumed to represent local myocardial activation with significant exit delay to the rest of the myocardium. However, whether these represent epicardial, neural or conduction tissue is unknown.[15,16]

Question No. 4:

An epicardial focus over the anterior left ventricular wall was identified. Prior to or during delivering radiofrequency energy *at this site*, which of the following is *not* routinely indicated:

A. Coronary angiography.
B. Simultaneous pacing of the left phrenic nerve.
C. Monitoring of esophageal temperature.
D. None of the above.

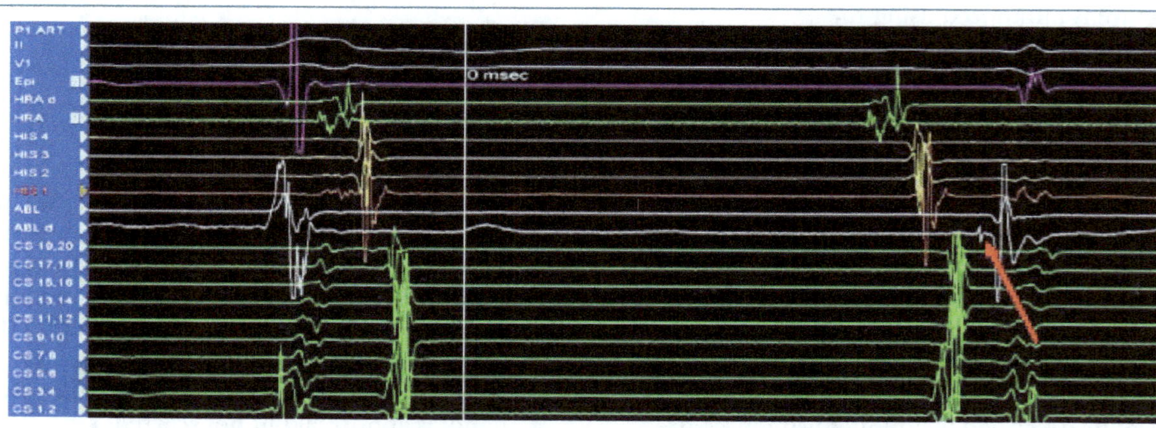

Figure 35.6. Successful ablation site. Pacemapping from this site showed good morphological characteristics.

DISCUSSION

In recent series of epicardial ablation, major complications were seen in up to 5% of patients.[4,5] In addition to the acute mechanical complications mentioned above, the other major instance when at risk of major complications is during ablating. As compared to endocardial ablation, epicardial ablation is limited by excessive overheating of tissue because of both the restricted space and lack of the cooling effect of blood flow. Therefore, irrigated catheters and cryoablation are increasingly used. Nevertheless, the structures that are particularly susceptible are the coronary arteries, the phrenic nerves, and the esophagus.[13] Coronary artery damage is unlikely to occur unless ablating directly over the artery; however, routine coronary angiography prior to ablating in the epicardium is advisable. The left phrenic nerve is susceptible as it courses along the lateral left ventricular wall. Pace capture of this nerve while ablating would identify involvement early during ablation, which can be withdrawn as soon as diaphragmatic stimulation becomes impaired. Separating the epicardial surface from the nerve with saline and air is one way of creating distance between catheter and nerve when proceeding with ablation. We routinely monitor esophageal temperature when ablating over the left atrium to detect esophageal involvement to prevent the late development of esophagoatrial fistula.

SUMMARY

When patients fail endocardial ablation, among the most significant challenges for the ablationist dealing with VT is presented. A careful understanding of mapping techniques, ECG localization and its limitations, minimizing risk with epicardial access, and appreciating the biophysicial differences inherent with epicardial locations (compared to endocardial) are essential for safety and success.

Correct Answers:

Q1. D
All of the above.

Q2. C
Angling the needle toward the left shoulder at 45 degrees from the horizontal (posterior puncture), as opposed to toward the left midclavicle angled 15 to 30 degrees from the horizontal (anterior puncture).

Q3. B
Presence of wide, fractionated or late near-field signals.

Q4. C
Monitoring of esophageal temperature.

REFERENCES

1. Calkins H, Epstein A, Packer D, et al. Catheter ablation of ventricular tachycardia in patients with structural heart disease using cooled radiofrequency energy: Results of a prospective multicenter study. Cooled RF Multi Center Investigators Group. *J Am Coll Cardiol.* 2000;35:1905–1914.
2. Sacher F, Tedrow UB, Field ME, et al. Ventricular tachycardia ablation: Evolution of patients and procedures over 8 years. *Circ Arrhythm Electrophysiol.* 2008;1:153–161.
3. Stevenson WG, Wilber DJ, Natale A, et al. Multicenter Thermocool VT Ablation Trial Investigators. Irrigated radiofrequency catheter ablation guided by electroanatomic mapping for recurrent ventricular tachycardia after myocardial infarction: The multicenter thermocool ventricular tachycardia ablation trial. *Circulation.* 2008;118:2773–2782.
4. Schmidt B, Chun KR, Baensch D, et al. Catheter ablation for ventricular tachycardia after failed endocardial ablation: Epicardial substrate or inappropriate endocardial ablation? *Heart Rhythm.* 2010;7:1746–1752.
5. Sacher F, Roberts-Thomson K, Maury P, et al. Epicardial ventricular tachycardia ablation a multicenter safety study. *J Am Coll Cardiol.* 2010;55:2366–2372.
6. Sosa E, Scanavacca M, d'Avila A, Pilleggi F. A new technique to perform epicardial mapping in the electrophysiology laboratory. *J Cardiovasc Electrophysiol.* 1996;7:531–536.
7. Aliot EM, Stevenson WG, Almendral-Garrote JM, et al. EHRA/HRS Expert Consensus on Catheter Ablation of Ventricular Arrhythmias: Developed in a partnership with the EHRA, a Registered Branch of the ESC, and the HRS; in collaboration with the ACC and the AHA. *Heart Rhythm.* 2009;6:886–933.
8. Baman TS, Ilg KJ, Gupta SK, et al. Mapping and ablation of epicardial idiopathic ventricular arrhythmias from within the coronary venous system. *Circ Arrhythm Electrophysiol.* 2010;3:274–279.
9. Kiès P, Bootsma M, Bax J, Schalij MJ, van der Wall EE. Arrhythmogenic right ventricular dysplasia/cardiomyopathy: Screening, diagnosis, and treatment. *Heart Rhythm.* 2006;3:225–234.
10. Berruezo A, Mont L, Nava S, Chueca E, Bartholomay E, Brugada J. Electrocardiographic recognition of the epicardial origin of ventricular tachycardias. *Circulation.* 2004;109:1842–1847.
11. Yamada T, McElderry HT, Okada T, et al. Idiopathic left ventricular arrhythmias originating adjacent to the

left aortic sinus of valsalva: Electrophysiological rationale for the surface electrocardiogram. *J Cardiovasc Electrophysiol.* 2010;21:170–176.
12. Bazan V, Bala R, Garcia FC, et al. Twelve-lead ECG features to identify ventricular tachycardia arising from the epicardial right ventricle. *Heart Rhythm.* 2006;3:1132–1139.
13. Syed F, Lachman N, Christensen K, et al. The pericardial space: Obtaining access and an approach to fluoroscopic anatomy. *Card Electrophysiol Clin.* 2010;2:9–23.
14. Tedrow U, Stevenson WG. Strategies for epicardial mapping and ablation of ventricular tachycardia. *J Cardiovasc Electrophysiol.* 2009;20:710–713.
15. Kapa S, McLeod CJ, Beyder A, et al. Identification of interstitial cells of Cajal and an ANO1-encoded chloride channel in the heart may offer novel targets for cardiac dysrhythmias (abstr). *Heart Rhythm.* 2010;7(5 Suppl.):S161.
16. Popescu LM, Gherghiceanu M, Hinescu ME, et al. Insights into the interstitium of ventricular myocardium: Interstitial Cajal-like cells (ICLC). *J Cell Mol Med.* 2006;10:429–458.
17. Swale M, Mikell S, Gard J, et al. Epicardial access: Patient selection, anatomy, and a stepwise approach. *J Innovat Cardiac Rhythm Manag.* 2011;2:239–249.

CASE STUDY

36

Epicardial Ventricular Tachycardia in ARVC/D

Rupa Bala, MD and Francis E. Marchlinski, MD

12-lead ECG of a patient with palpitations.

Question No. 1:

The ECG shows:

A. Idiopathic right ventricular tachycardia (VT).

B. Epicardial VT in arrhythmogenic right ventricular cardiomyopathy/dysplasia (ARVC/D).

C. Endocardial VT in ARVC/D.

CASE REPORT

A 39-year-old woman with a history of ARVC/D presented with recurrent VT and implantable cardioverter-defibrillator (ICD) shocks. She initially presented with sudden onset of palpitations during exercise. Her subsequent workup revealed the following: (1) right precordial T-wave inversions on electrocardiogram (ECG), (2) RV hypokinesia and dilatation on echocardiogram, (3) nonsustained VT on stress testing and event monitoring (left bundle morphology with both superior and inferior axis), (4) magnetic resonance imaging (MRI) consistent with ARVC/D, and (5) abnormal signal-averaged ECG. Two months prior to presentation, she had recurrent ICD shocks and was treated with beta-blocker therapy. Due to significant side effects, it was discontinued and she was referred for electrophysiology study (EPS)/ablation.

At the start of the procedure, intracardiac echocardiography (ICE) was used to assess the RV. It revealed moderate dilatation of the RV with moderate to severe hypokinesis. Left ventricular (LV) function was normal. The first step was to create a detailed voltage map of the RV. A 3D electroanatomical map (CARTO, Biosense Webster, Diamond Bar, CA) was constructed with a 7-Fr, 3.5-mm tip open-irrigated (Navistar Thermocool, Biosense Webster, Diamond Bar, CA) catheter. The map revealed a large area of low voltage on the RV basal free wall, extending inferiorly and superiorly to the right ventricular outflow tract (RVOT). This area of low voltage had late potentials, which were tagged on the endocardial map (Figure 36.1). Abnormal endocardial voltage was designated as <1.5 mV as previously described by our group. Programmed stimulation (PS) induced 4 different left bundle VTs that were not hemodynamically tolerated (Figure 36.2). VT_1 and VT_4 were similar, with a left bundle morphology (II > III) and a Q wave in lead V_2. Pacemapping and limited activation mapping localized these VTs to the basal superior segment of the RV and were used to guide substrate-based ablation on the basal RV free wall. A linear lesion set was performed from the basal RV free wall extending to the tricuspid valve annulus (TVA). VT_2 and VT_3 both had left bundle morphology with an inferior

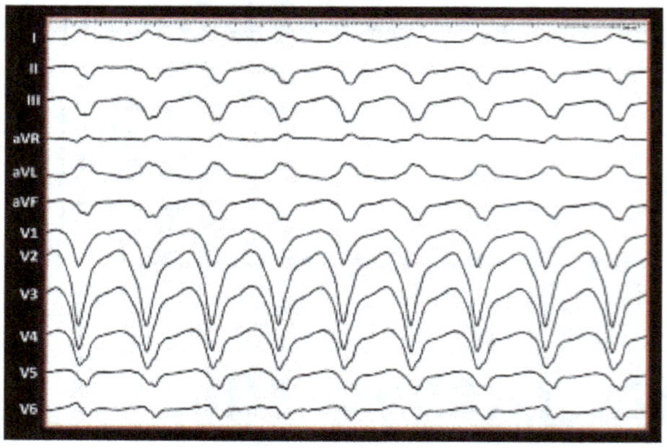

Figure 36.1. 12-lead ECG.

axis. Pacemapping and limited activation mapping localized these VTs to the free wall of the RVOT. A linear lesion set was performed from the anterior RVOT ($VT_{2/3}$ pacemaps) to the basal superior RV ($VT_{1/4}$ pacemaps). After the initial RV endocardial lesions, PS was performed and both VT_1 and VT_3 were still inducible. Thus, epicardial mapping was performed.

Access to the pericardial space was obtained by the technique described by Sosa et al. A Tuohy needle was introduced via a subxiphoid approach to gain access to the pericardial space. A long 8-Fr sheath was advanced over a wire after confirming intrapericardial access. Detailed electroanatomic mapping of the epicardium was performed during sinus rhythm with a 3.5-mm tip Thermocool™ catheter. The reference value for defining abnormal electrograms in the epicardium was previously established by our group, with normal epicardial electrograms defined as > 1.0 mV, which

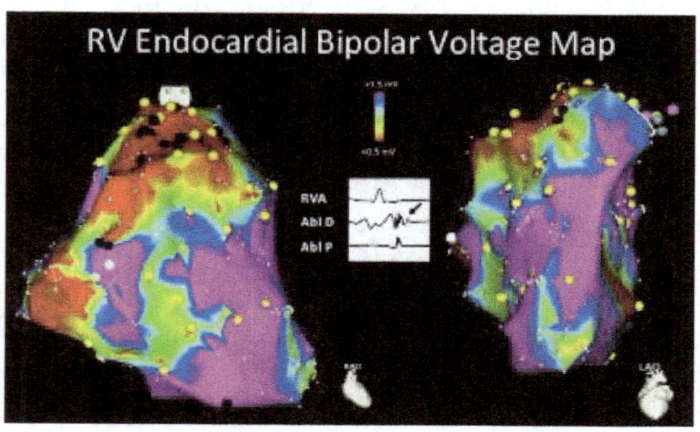

Figure 36.2. RV endocardial voltage map. Inset shows an example of a late potential (arrow) on the ablation catheter. Late potentials are denoted by black tags on the voltage map. RVA, right ventricular apical catheter; Abl D, ablator distal; Abl P, ablator proximal.

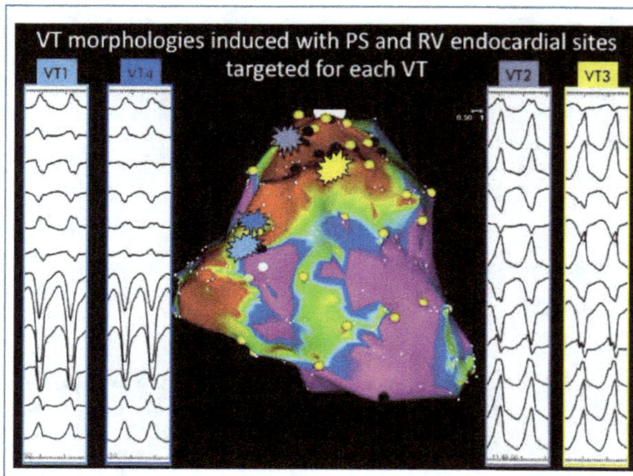

Figure 36.3. Four VTs were induced with PS. VT_1 and VT_4 were targeted on the basal, superior segment of the RV. VT_2 and VT_3 were targeted on the free wall of the (RV outflow tract).

corresponds to 95% of the signals from normal epicardium recorded at a distance of at least 1 cm from the defined large-vessel coronary vasculature.[1] The RV epicardial voltage map revealed a large area of low voltage from the free wall of the RVOT extending to the basal segment of the RV and inferiorly (Figure 36.3). This area of low voltage had late and fractionated potentials, which were annotated on the voltage map. The area of distribution of the RV endocardial and epicardial voltage abnormalities were similar. Pacemapping for VT_1 and VT_3 were performed on the epicardium. Both of these pacemaps had a long stimulus-to-QRS and were a good match to the clinical VT (Figure 36.4). A coronary angiogram was performed prior to radiofrequency ablation (RFA) on the epicardium. A linear lesion set was performed that extended from the free wall of the RVOT to the basal RV, targeting late potentials and guided by the pacemapping. PS was again performed and induced VT_5 (left bundle, superior axis). Pacemapping was performed on the basal inferior wall from the RV epicardium and endocardium for this VT (Figures 36.5 and 36.6). The epicardial pacemap had a longer stimulus-to-QRS and was an excellent match. The coronary angiogram revealed a moderate-sized marginal branch near the epicardial pacemap site for VT_5; thus limited epicardial RFA was performed and lesions were given on the RV endocardium opposite this site. Final PS was performed from the RV endocardium and epicardium and revealed no inducible VT.

After the procedure, the patient underwent noninvasive programmed stimulation (NIPS) and exercise stress testing, both of which were noninducible for VT. She was discharged on

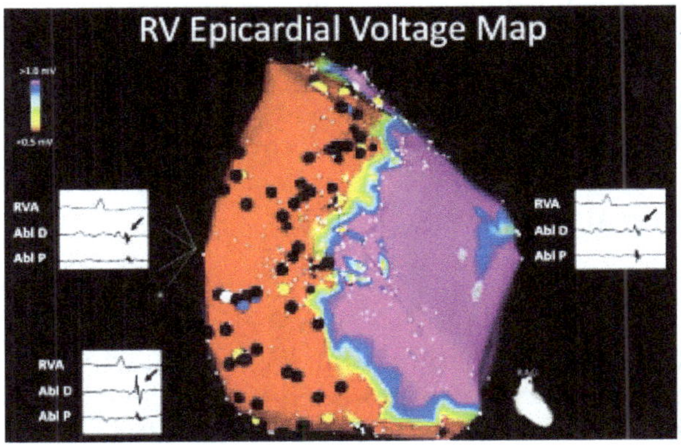

Figure 36.4. RV epicardial voltage map. Inset shows examples of late potentials (arrows) found on the epicardium, which are denoted by black tags on the voltage map. RVA, right ventricular apical catheter; Abl D, ablator distal; Abl P, ablator proximal.

low-dose beta-blocker. She has had no recurrent VT in 6 months of follow-up and is maintained on beta blocker therapy with no clinical arrhythmias noted on device interrogation.

DISCUSSION

ARVC/D is a genetically determined myocardial disease characterized by fibrofatty replacement of the RV, with progression to LV involvement in selected patients.[2] Myocardial involvement progresses from epicardium to endocardium with a predilection for the RV free wall extending from the perivalvular region. The triggers for progression are poorly understood, and many patients appear to demonstrate a stable substrate for years, supporting an aggressive approach to ablation. RV deterioration

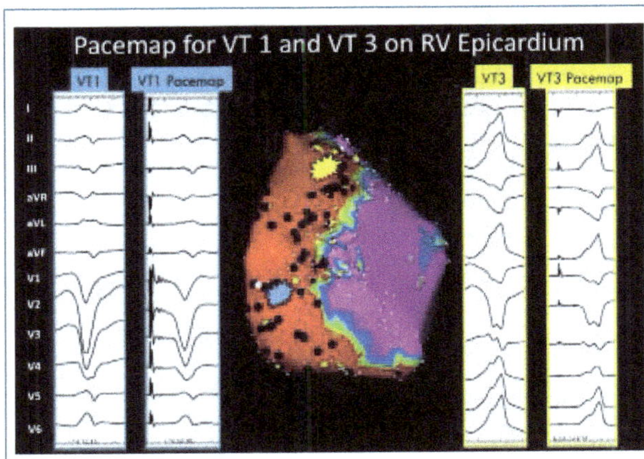

Figure 36.5. VT_1 and VT_3 are shown with corresponding epicardial pacemaps. Both pacemaps have a long stimulus-to-QRS and are a very close match to the clinical VTs.

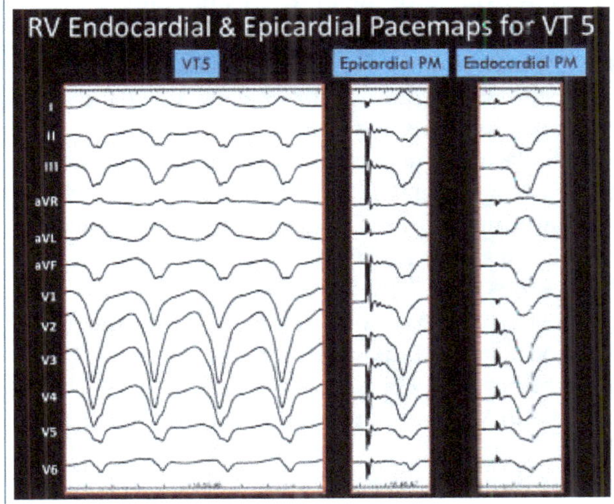

Figure 36.6. VT_5 is shown with the best epicardial and endocardial pacemap (PM). The epicardial pacemap has a long stimulus-to-QRS and is an excellent match to VT_5. Note the Q waves in the leads II, III, and aVF in both the clinical VT_5 and the epicardial pacemap.

leads to electrical instability, which predisposes patients to VT. Treatment of VT in the setting of ARVC/D includes antiarrhythmic medications and ablation. In the majority of cases, ablation is successful from an endocardial approach.[3] However, given the propensity for late recurrences and the lack of uniform success among published studies, a combined endocardial and epicardial approach has been developed and recently described.[4] There are distinct ECG features of epicardial VT in ARVC/D that may help guide the need for epicardial mapping and ablation.[5] This case illustrates a combined endocardial and epicardial approach for the ablation of VT in the setting of ARVC/D. The patient had multiple VTs arising from the endocardium and epicardium and we will discuss the ablation strategy in this case and the ECG features that distinguish endocardial and epicardial VT in ARVC/D.

Given the hemodynamic instability of the induced VTs, a substrate-based ablation strategy, guided by pacemapping and limited activation mapping, was performed. The patient had significant areas of low voltage on both the endocardium and epicardium, though the epicardial area of low voltage was larger in comparison. The ECG morphology of each induced VT was analyzed and this information was used to guide endocardial mapping and the need for epicardial mapping. VT_1 and VT_4 were pacemapped to the RV basal free wall (superior segment). Both VTs had a left bundle inferior axis (II > III) with a Q wave in lead V_2. Our prior work noted that pacemaps from both the endocardium and epicardium from the RV basal superior segment can have a Q wave in lead V_2, thus it was hard to define site-specific ECG findings for epicardial versus endocardial VT in this segment.[5] VT_1 was still inducible after endocardial ablation and the epicardial pacemap for VT_1 was a close match with a long stimulus-to-QRS. Thus, one can hypothesize that VT_1 may have involved an epicardial circuit. The degree of RV wall thickness may play a role in the pacemap features at this site, with a thinner free wall producing a similar pattern from both the endocardium and epicardium.

VT_2 and VT_3 were targeted in the RVOT region. VT_2 was successfully ablated from the RVOT endocardium and revealed a ratio >1 for II/III and aVL/aVR. VT_3 was still inducible after endocardial ablation and an excellent pacemap was found on the RVOT epicardium. The ECG morphology for VT_3 revealed a ratio <1 for II/III and aVL/aVR. Our prior work revealed that leads III and aVR tended to be more positive and less negative, respectively, when pacing from the epicardium compared to the endocardium. This study showed that there was a trend for II/III and aVL/aVR polarity amplitude ratios to be <1 for RVOT epicardial VTs.[5]

The ECG morphology for VT_5 was a left bundle morphology with a superior axis (LBLS). This VT had Q waves in the inferior leads, suggestive of an epicardial origin as described in our prior study.[5] Pacemapping from the epicardium was a close match for VT_5 (compared to the endocardium), and aggressive RFA was limited by coronary anatomy.

We recently published data on a series of 13 patients with ARVC/D undergoing combined endocardial and epicardial VT ablation after failed endocardial VT ablation. Ablation of both the RV endocardium and epicardium was necessary to achieve excellent long-term VT control in these patients.[4] This case report supports our earlier findings. This patient had a sizable area of RV epicardial and endocardial low voltage with multiple VTs displaying both endocardial and epicardial ECG features. Ablation targeting the endocardial and epicardial substrate was necessary to achieve success. A combined endocardial and epicardial mapping strategy in selected patients is necessary to achieve VT control and decrease VT recurrence.

Correct Answer:

Q1. B

This VT has a LBLS. It is not commonly associated with idiopathic RVOT VT, which typically demonstrates a left bundle, inferior-axis VT. Thus, one should always consider ARVC/D in the differential diagnosis for patients presenting with left bundle, superior-axis VT. The VT in this tracing has Q waves in the inferior leads, suggestive of an epicardial origin as described in our prior study.[5]

REFERENCES

1. Cano O, Hutchinson M, Lin D, et al. Electroanatomic substrate and ablation outcome for suspected epicardial ventricular tachycardia in left ventricular nonischemic cardiomyopathy. *J Am Coll Cardiol.* 2009;54:799–808.
2. Corrado D, Basso C, Thiene G, et al. Spectrum of clinicopathologic manifestations of arrhythmogenic right ventricular cardiomyopathy/dysplasia: A multicenter study. *J Am Coll Cardiol.* 1997; 30:1512–1520.
3. Marchlinski FE, Zado E, Dixit S, et al. Electroanatomic substrate and outcome of catheter ablative therapy for ventricular tachycardia in setting of right ventricular cardiomyopathy. *Circulation.* 2004;110:2293–2298.
4. Garcia FC, Bazan V, Zado ES, Ren JF, Marchlinski FE. Epicardial substrate and outcome with epicardial ablation of ventricular tachycardia in arrhythmogenic right ventricular cardiomyopathy/dysplasia. *Circulation.* 2009;120:366–375.
5. Bazan V, Bala R, Garcia FC, et al. Twelve-lead ECG features to identify ventricular tachycardia arising from the epicardial right ventricle. *Heart Rhythm.* 2006;3:1132–1139.

CASE STUDY

37

Magnetic Resonance Imaging in Ventricular Tachycardia

Takeshi Sasaki, MD, PhD and Saman Nazarian, MD, PhD

A 67-year-old man with history of coronary artery disease and normal ejection fraction (EF, 55%) post-coronary artery bypass graft surgery presented with recurrent episodes of sustained monomorphic ventricular tachycardia (VT) at a rate of 120 bpm. He complained of palpitations, but had no dizziness or syncope. After failure of multiple antiarrhythmic agents including amiodarone, the patient was referred for catheter ablation. Cardiac magnetic resonance imaging (CMR) was performed prior to VT ablation. Late gadolinium-enhanced (LGE) CMR demonstrated anterolateral scar in the left ventricle (LV) (Figure 37.1). During electrophysiology (EP) study, 4 distinct VT morphologies including the clinical VT (VT_1) were easily inducible with programmed ventricular stimulation from the right ventricular apex (RVA, Figure 37.2). All VT morphologies were terminable by overdrive pacing.

Question No. 1:

Reentry is generally considered to be the arrhythmogenic mechanism of VT in patients with remote myocardial infarction. Which of the following descriptions about the 4 induced VTs and their exit sites is most likely?

A. All induced VTs (VT_{1-4}) are scar-related and their exit sites are likely to be identified epicardially.

B. All induced VTs (VT_{1-4}) are scar-related and their exit sites are likely to be identified endocardially.

C. All induced VTs except VT_4 are scar-related and the exit sites of VT_1, VT_2, and VT_3 are likely to be identified endocardially.

D. None of the induced VTs are scar-related and the exit site locations of the induced VTs are not predictable without mapping.

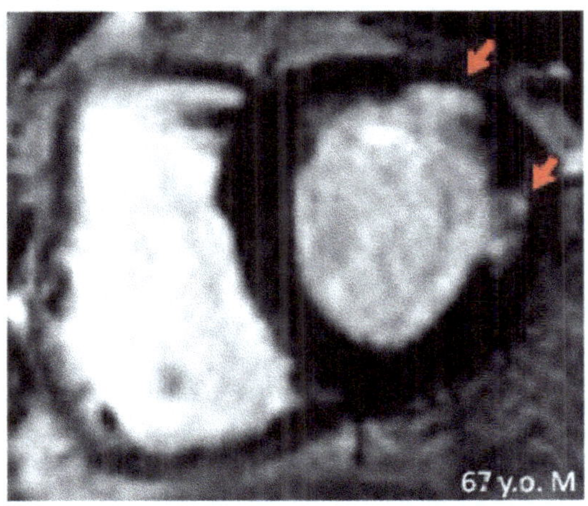

Figure 37.1 LGE-CMR imaging in a 67-year-old man (67 y.o. M) demonstrates anterolateral scar in the LV (red arrows).

Practical Electrophysiology © 2015 Jasbir S. Sra, Masood Akhtar, Andrea Natale, David J. Wilber
Cardiotext Publishing, ISBN: 978-1-935395-17-1

DISCUSSION

All VTs were reproducibly induced by single- or double-extra stimulations from the RVA, and all were terminable by overdrive pacing. Additionally, the patient has history of remote myocardial infarction. These findings strongly suggest that the mechanism underlying all of the induced VTs is reentry. 12-lead ECG demonstrated 4 different wide QRS complex tachycardias with right bundle branch block, negative R wave in leads I and aVL, and rS pattern in lead II, except VT_3, which demonstrated a positive R wave in the inferior leads. The QRS morphologies of these VTs are consistent with exit sites in the anterolateral myocardial region in the LV.[1] This is consistent with the scar location on LGE-CMR (see Figure 37.1). Berruezo et al[2] described ECG characteristics suggestive of VTs with epicardial exit sites. Epicardial VTs generally exhibit a pseudo-delta wave and long intrinsicoid deflection time. These ECG characteristics are likely due to the intramyocardial conduction delay from the epicardium to the endocardium. The ECG characteristics of the 4 induced VT morphologies in our patient do not meet the criteria that typically identify epicardial VTs. Endocardial scar location on LGE-CMR with ≤ 75% of scar transmurality (with respect to full wall thickness) also supports an endocardial exit site.

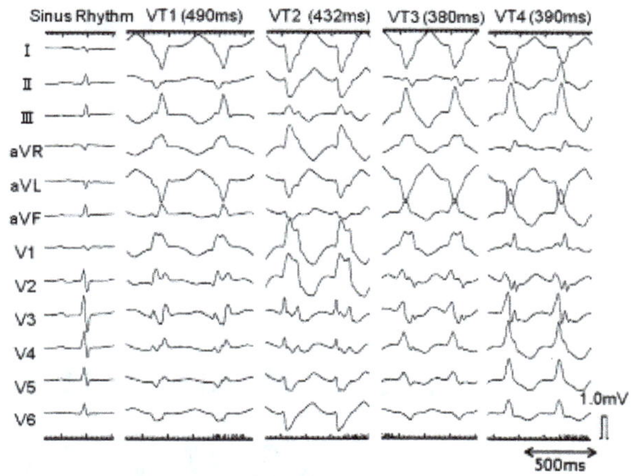

Figure 37.2. During EP study, 4 distinct VT morphologies with right bundle branch block morphology were inducible with extra stimulations from the RVA.

Question No. 2:

The fractionated potential or isolated delayed potential frequently observed in low-voltage areas on electroanatomic mapping may serve as ablation targets for scar-related VT. Critical sites of scar-related VT are identified by the following criteria except:

A. Concealed entrainment during tachycardia.

B. Postpacing interval (PPI) similar to VT cycle length (CL).

C. Perfect pace map or good pace map during sinus rhythm.

D. Scar size and location on LGE-CMR.

DISCUSSION

The significance of fractionated and isolated delayed potentials as characteristic electrograms (EGMs) often observed in critical sites for scar-related VT has been demonstrated.[3] The critical sites with fractionated or delayed potentials can be conventionally identified using entrainment mapping during tachycardia, comparison of PPI with VT CL and pace mapping during sinus rhythm. The endocardial bipolar voltage map in this case depicts the low-voltage area defined as <1.5 mV in the anterolateral region corresponding to the scar location on LGE-CMR (Figure 37.3). The critical site for VT_2 was identified by entrainment mapping at the scar border area on the voltage map. The local EGM at the critical site of VT_2 showed the delayed potential during sinus rhythm, which indicates the presence of slow conduction due to myocardial scar. VT_2 also changed into VT_3 by a premature ventricular contraction (PVC) delivered from the critical site of VT_2 with concealed entrainment. In the same mapping site, fractionated potentials during VT_2, mid-diastolic potential during VT_3, and delayed potentials during sinus rhythm were recorded (Figure 37.4). Recently, scar tissue heterogeneity on LGE-CMR has been reported to be significantly associated with all-cause mortality and inducibility of monomorphic VTs in patients with ischemic cardiomyopathy.[4] The fractionated or delayed potentials are likely to be associated with the scar tissue heterogeneity on LGE-CMR. This suggests that LGE-CMR may be capable of identifying critical sites in postinfarct scar-related VT prior to VT ablation. However, LGE-CMR often identifies many potential sites with scar and/or heterogeneous tissue, and is not yet capable of a priori identification of critical VT zones.

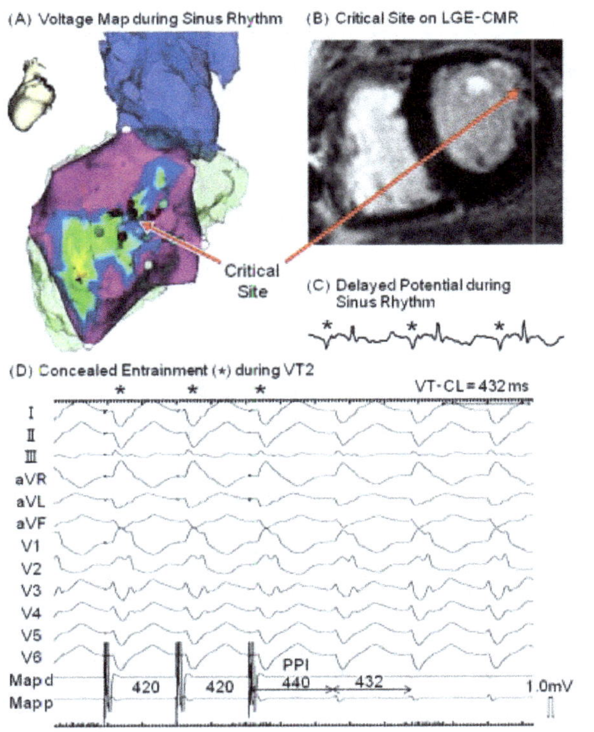

Figure 37.3. A: Endocardial bipolar voltage map demonstrates the low-voltage area defined at <1.5 mV in the anterolateral LV. B: The critical site of VT_2 on the voltage map is pointed out on corresponding LGE-CMR plane (red arrow). C: The local EGM of the critical site during sinus rhythm shows a delayed potential. D: Concealed entrainment is demonstrated by pacing at a site in the anterolateral region of the scar. The PPI approximates the VT CL.

Question No. 3:

Electroanatomic mapping demonstrates a low-voltage area in the anterolateral region corresponding to the scar on LGE-CMR. The scar transmurality defined by LGE-CMR images is significantly associated with the following local EGM characteristics during sinus rhythm:

A. Bipolar EGM amplitude.

B. Unipolar EGM amplitude.

C. EGM duration.

D. All of the above.

DISCUSSION

Previous reports have revealed a significant association between scar areas on LGE-CMR and electroanatomic mapping using a threshold of 1.3 to 1.5 mV for bipolar voltage and 5.5 to 6.5 mV for unipolar voltage.[5,6] The local EGM voltage is generated by viable myocardial tissue. Scar transmurality is defined as the ratio of scar thickness to LV wall thickness, which reflects the impact of the viable myocardial tissue and the scar tissue on local EGM attributes. Greater scar transmurality is significantly associated with lower bipolar and unipolar EGM amplitudes, and longer EGM duration (Figure 37.5).[7,8]

A CMR scan was performed 1 day after the radiofrequency catheter ablation (RFCA)

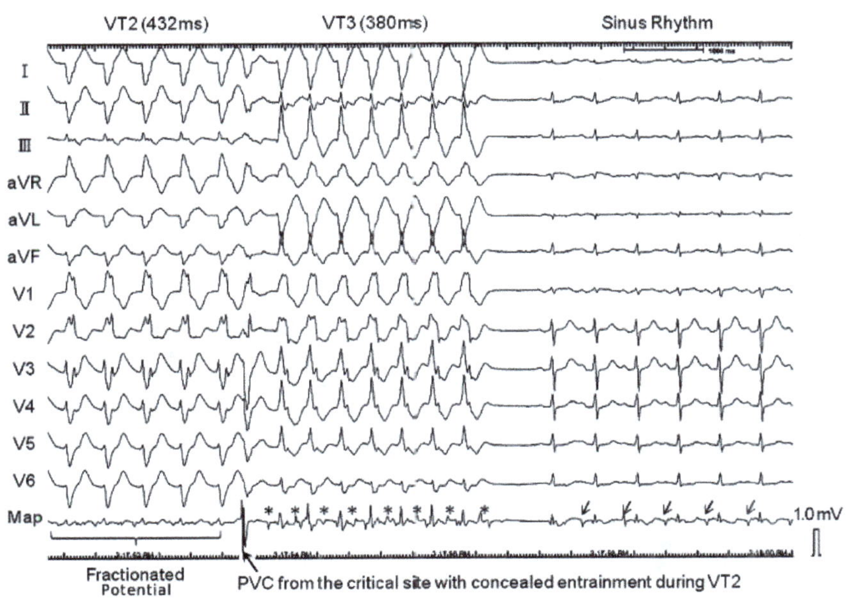

Figure 37.4. VT_2 changed into VT_3 by a PVC delivered from the critical site of VT_2. The fractionated potential during VT_2, mid-diastolic potentials during VT_3 (asterisks), and isolated delayed potentials during sinus rhythm (black arrows) are observed at the critical sites of VT_2.

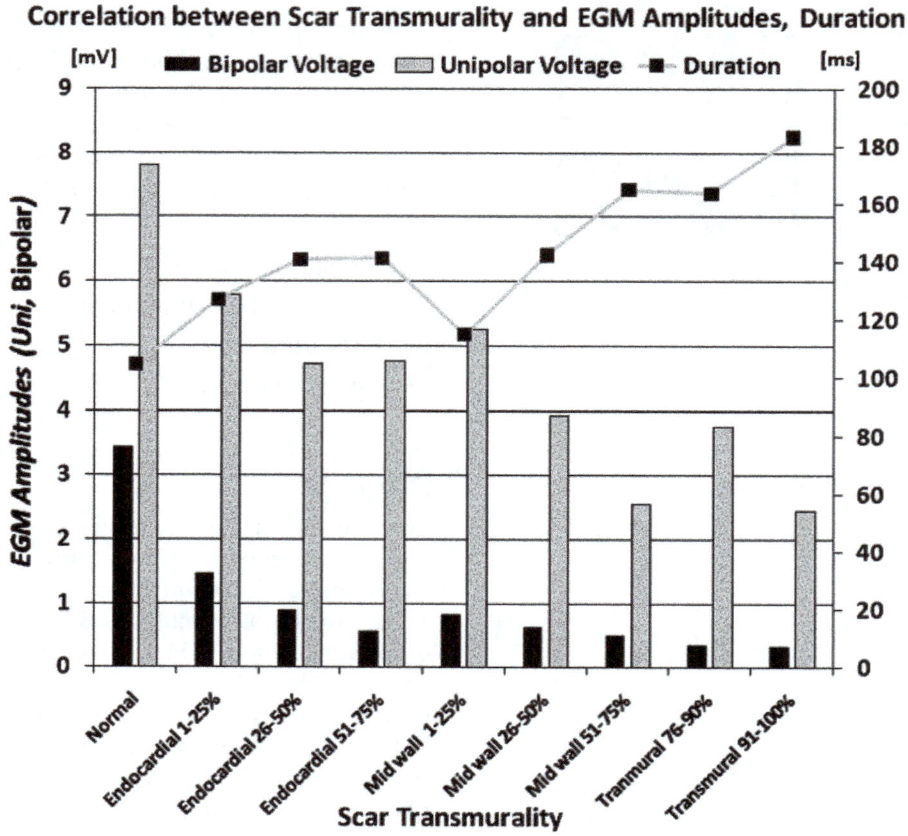

Figure 37.5. Scar transmurality is significantly associated with bipolar and unipolar EGM amplitudes, and EGM duration in patients with ischemic cardiomyopathy (*N* = 14).

procedure. Ablation regions are demonstrated as areas with low intensity on short-axis LGE-CMR and perfusion CMR planes (Figure 37.6). The patient underwent single-chamber implantable cardioverter-defibrillator (ICD) implantation for secondary prevention of VT after the procedure. A year later, the patient was referred for cardiac imaging for assessment of myocardial viability.

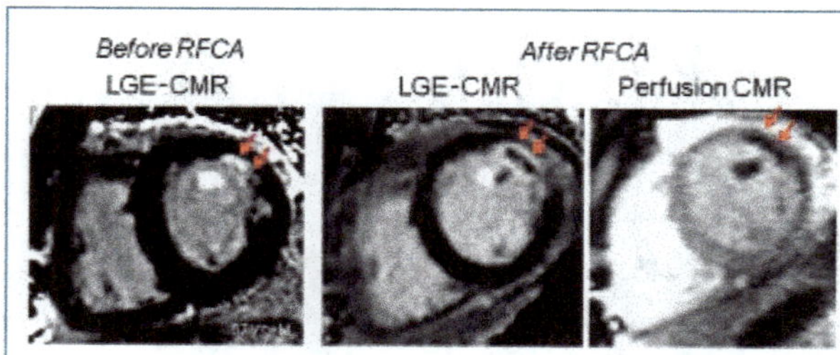

Figure 37.6. LGE-CMR before RFCA, and LGE and perfusion CMR 1 day after RFCA, demonstrate ablation regions (red arrows).

Question No. 4:

The safety and feasibility of CMR scans in patients with implantable cardiac devices have been investigated. However, the images may be associated with significant magnetic susceptibility artifacts by ferromagnetic components of device systems. Which of the descriptions about artifacts due to cardiac devices is appropriate?

A. The artifact effects on CMR images are different in each image sequence.

B. The anterior and apical myocardial regions are severely affected by artifact due to the ICD generator in patients with a left-sided ICD system.

C. Artifact effects due to ICD leads are much less compared with those due to ICD generators.

D. All of the above.

The safety and feasibility of cardiac and non-CMR in patients with implantable cardiac

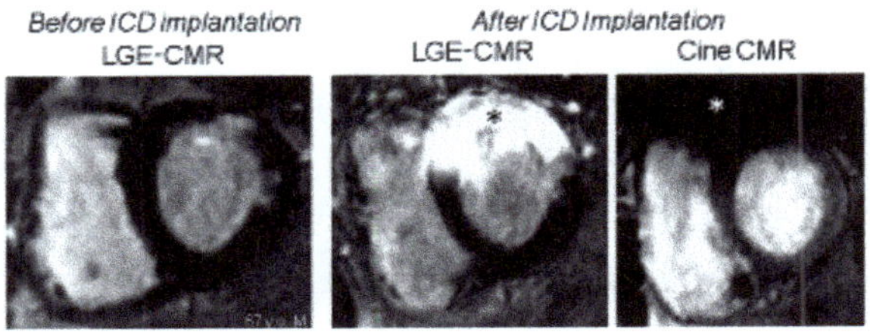

Figure 37.7. Artifacts due to the ICD generator on LGE-CMR and cine CMR sequences from a 67-year-old man (67 y.o. M). The anterior myocardial region on LGE-CMR is severely affected by artifact in a patient with a left-sided ICD system (asterisk). Artifacts due to ICD generator are also observed on cine CMR.

devices such as permanent pacemakers and ICDs have been demonstrated.[9] However, CMR images may be associated with significant metal susceptibility artifacts from implantable devices. Artifact effects due to the device generators are typically much greater than those due to device leads (Figure 37.7). Greater artifact effects are observed on LGE-CMR and steady-state free precession cine, compared with T_2-weighted, perfusion CMR and magnetic resonance angiography sequences. The anterior and apical myocardial regions on LGE-CMR are severely affected by artifact due to the device generator in patients with left-sided ICD systems.

Correct Answers:

Q1. C
All induced VTs except VT_4 are scar-related and the exit sites of VT_1, VT_2, and VT_3 are likely to be identified endocardially.

Q2. D
Scar size and location on LGE-CMR.

Q3. D
All of the above.

Q4. D
All of the above.

REFERENCES

1. Sippens Groenewegen A, Spekhorst H, van Hemel NM, et al. Localization of the site of origin of postinfarction ventricular tachycardia by endocardial pace mapping. Body surface mapping compared with the 12-lead electrocardiogram. *Circulation.* 1993;88:2290–2306.
2. Berruezo A, Mont L, Nava S, et al. Electrocardiographic recognition of the epicardial origin of ventricular tachycardias. *Circulation.* 2004;109:1842–1847.
3. Bogun F, Good E, Reich S, et al. Isolated potentials during sinus rhythm and pace-mapping within scars as guides for ablation of post-infarction ventricular tachycardia. *J Am Coll Cardiol.* 2006;47:2013–2019.
4. Schmidt A, Azevedo CF, Cheng A, et al. Infarct tissue heterogeneity by magnetic resonance imaging identifies enhanced cardiac arrhythmia susceptibility in patients with left ventricular dysfunction. *Circulation.* 2007;115:2006–2014.
5. Desjardins B, Crawford T, Good E, et al. Infarct architecture and characteristics on delayed enhanced magnetic resonance imaging and electroanatomic mapping in patients with postinfarction ventricular arrhythmia. *Heart Rhythm.* 2009;6:644–651.
6. Codreanu A, Odille F, Aliot E, et al. Electroanatomic characterization of post-infarct scars comparison with 3-dimensional myocardial scar reconstruction based on magnetic resonance imaging. *J Am Coll Cardiol.* 2008;52:839–842.
7. Wijnmaalen AP, van der Geest RJ, van Huls van Taxis CF, et al. Head-to-head comparison of contrast-enhanced magnetic resonance imaging and electroanatomical voltage mapping to assess post-infarct scar characteristics in patients with ventricular tachycardias: Real-time image integration and reversed registration. *Eur Heart J.* 2011;32:104–114.
8. Sasaki T, Hansford R, Miller CF, et al. Electrogram surrogates of the transmural extent of late gadolinium enhancement on cardiac magnetic resonance in patients with post-infarct ventricular tachycardia. *Circulation.* 2010;122:A10361.
9. Nazarian S, Roguin A, Zviman MM, et al. Clinical utility and safety of a protocol for noncardiac and cardiac magnetic resonance imaging of patients with permanent pacemakers and implantable-cardioverter defibrillators at 1.5 tesla. *Circulation.* 2006;114:1277–1284.

CASE STUDY

38

Bundle Branch Reentry Ventricular Tachycardia

Zalmen Blanck, MD

A 57-year-old man underwent mechanical aortic valve replacement for severe aortic insufficiency due to a bicuspid valve. Preoperative evaluation revealed a severely dilated left ventricle (LV) (6.9 cm at end diastole and 5.8 cm at end systole), left ventricular ejection fraction (LVEF) 54% and no coronary artery disease by coronary angiography. The ECG is shown in Figure 38.1. Three days after discharge, he had 2 syncopal episodes at home. Echocardiogram showed normal prosthetic valve function and LVEF of 49%; ECG is shown in Figure 38.2. Inpatient cardiac monitoring for 36 hours revealed sinus rhythm and no significant arrhythmias. On postoperative day 3, he had developed runs of self-terminating wide-complex tachycardia at 250 bpm (Figure 38.3). He was treated with amiodarone for the same.

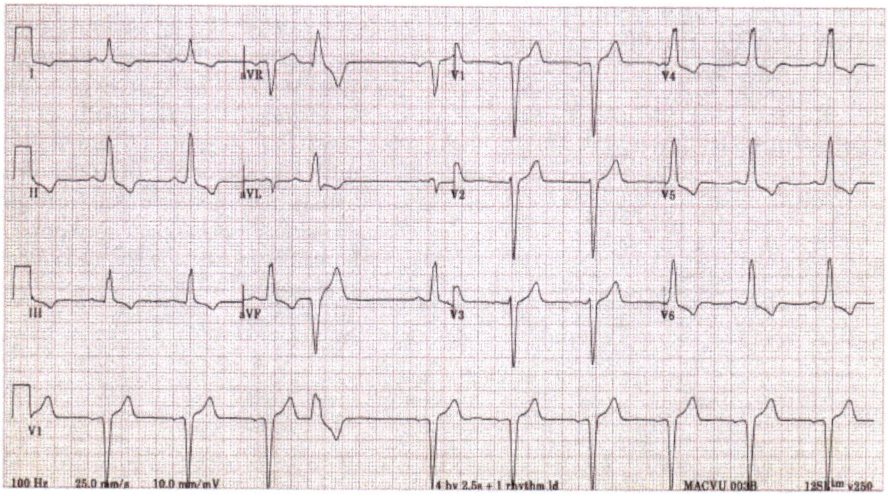

Figure 38.1. The 12-lead ECG before aortic valve surgery showed sinus rhythm with a PR interval of 170 ms, a QRS complex measuring 130 ms and a left bundle branch block (LBBB).

Practical Electrophysiology © 2015 Jasbir S. Sra, Masood Akhtar, Andrea Natale, David J. Wilber
Cardiotext Publishing, ISBN: 978-1-935395-17-1

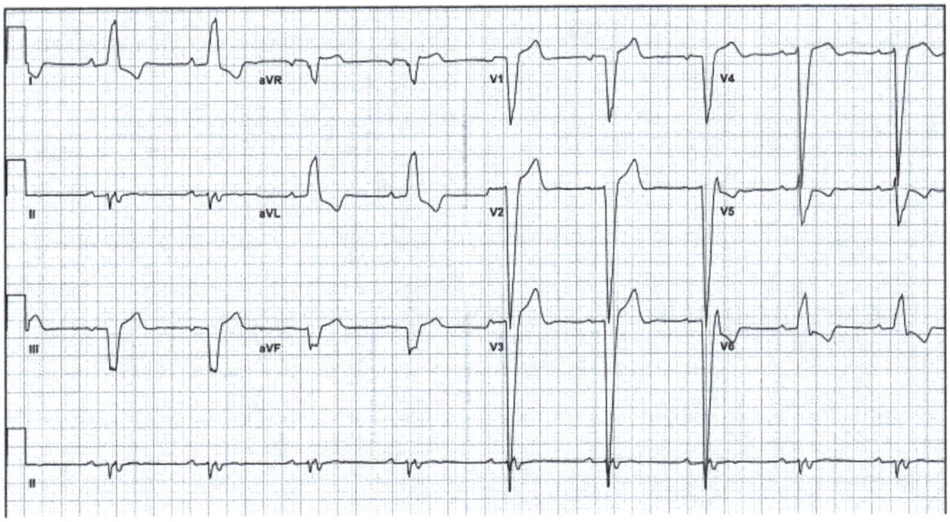

Figure 38.2. Three days after aortic valve surgery, the ECG showed a PR interval of 210 ms, widening of the QRS complex to 165 ms and leftward axis.

Question No. 1:

What is the likely cause of this patient's syncopal spells?

A. Vasovagal syncope.
B. Intermittent heart block.
C. Rapid atrial arrhythmia.
D. Monomorphic ventricular tachycardia (VT).
E. Bundle branch reentry (BBR) VT.

Question No. 2:

What would be the most appropriate management?

A. Increase amiodarone and metaprolol doses.
B. Outpatient long-term cardiac monitoring.
C. Transesophageal echocardiography to assess prosthetic valve function.
D. Electrophysiologic studies.
E. Dual-chamber implantable cardioverter-defibrillator (ICD) implantation.

DISCUSSION

Although BBR VT most commonly occurs in patients with idiopathic dilated cardiomyopathies,[1,2] it can also affect patients with valvular heart disease, particularly early in the postoperative period after aortic or mitral valve replacements.[3] Manipulation of the spectral area during valve surgery may alter the electrophysiologic properties of the conduction system and facilitate spontaneous BBR tachycardia; in some cases, significant damage to the conduction system results in postoperative heart block, making BBR VT less likely to develop. In our experience, patients who develop BBR VT after valvular replacement had more preserved LV function (ie, 43%)[3] compared with those who have idiopathic cardiomyopathies and BBR VT (ie, 23%).[2] Therefore, in the absence of myocardial scar from a prior myocardial infarction and with more preserved LV function, as seen in this patient, BBR VT should be considered instead of "scar-related" monomorphic VT.

During electrophysiologic studies, BBR VT can be induced by programmed ventricular stimulation confirming this mechanism of VT. Ablation of the right bundle branch (RBB) is the treatment of choice and can be easily performed in most patients with BBR VT.[1,2,3,4] In this patient, sustained BBR VT with a LBBB morphology was inducible using short-long pacing sequences (Figure 38.4),[1,2] and ablation of the RBB successfully eliminated the VT. After the ablation, the HV interval prolonged from 77 to 130 ms and a dual-chamber pacemaker was implanted.

Although it may be wondered how sustained BBR VT can occur when the 12-lead ECG shows a complete (ie, antegrade) LBBB, retrograde conduction over the LBB (a component of the reentrant circuit of BBR VT) can occur despite the absence of antegrade conduction. In these cases, successful ablation of the LBB, instead of the RBB, may also eliminate the VT, in which case complete heart block and the need for pacemaker

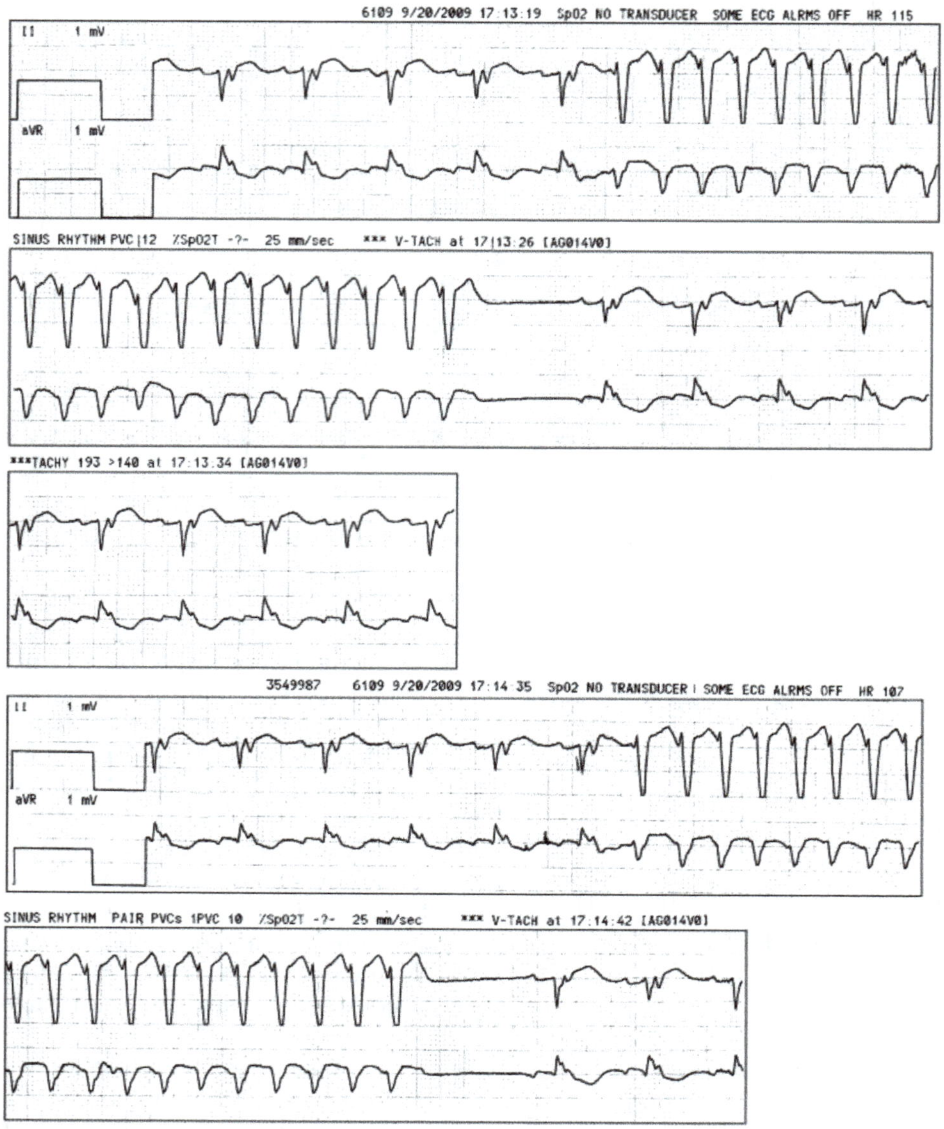

Figure 38.3. Telemetry strips 3 days after aortic valve replacement surgery showing runs of nonsustained VT.

implantation may be avoided.[5] However, in this patient, given the recent aortic valve replacement surgery and potentially higher complication rate targeting the LBB, ablation of the RBB was performed instead.

Diagnostic criteria for BBR VT include the following:

1. The His-bundle potential precedes the onset of the QRS during VT, and the HV interval is usually similar or slightly longer than the sinus rhythm HV interval.
2. Changes in the HH intervals during the V precede and dictate subsequent changes in VV intervals.
3. Postpacing intervals during entrainment from right ventricular apex are very similar (ie, difference <20 ms) to VT cycle length.
4. Sequence of activation of the His-Purkinje system (ie, His, right bundle and left bundle potentials, if available) are consistent with the QRS morphology during VT (ie, the His-bundle potential precedes the RBB potential during BBR VT with LBBB configuration).

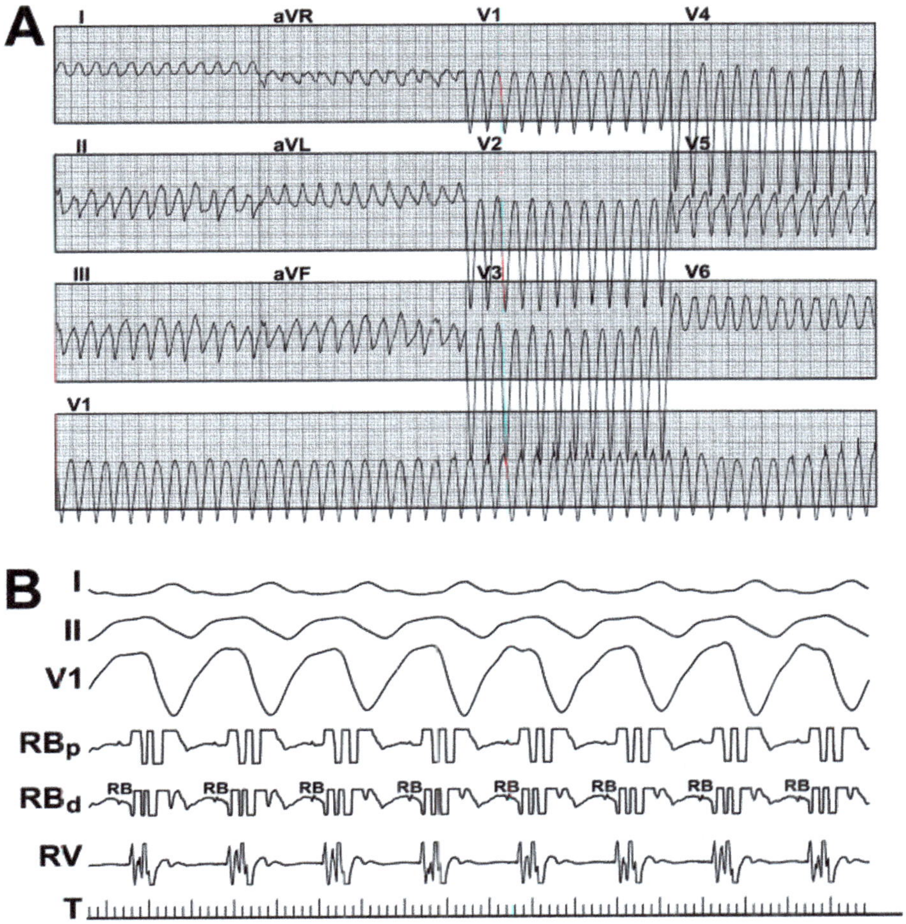

Figure 38.4. During electrophysiologic studies, sustained VT due to BBR with a cycle length of 230 ms was induced (**Panels A** and **B**). Note the RBB recording with an RB-V interval of 40 ms (Panel B, not labeled).

Correct Answers:

Q1. E

Q2. D

REFERENCES

1. Caceres J, Jazayeri M, McKinnie J, et al. Sustained bundle branch reentry as a mechanism of clinical tachycardia. *Circulation.* 1989;79:256–270.
2. Blanck Z, Dhala A, Deshpande S, et al. Bundle branch reentrant ventricular tachycardia: Cumulative experience in 48 patients. *J Cardiovasc Electrophysiol.* 1993;4:253–262.
3. Narasimhan C, Jazayeri MR, Sra J, et al. Ventricular tachycardia in valvular heart disease: Facilitation of sustained bundle-branch reentry by valve surgery. *Circulation.* 1997;96:4307–4313.
4. Tchou P, Jazayeri MR, Denker S, et al. Transcatheter electrical ablation of the right bundle branch: A method of treating macrorentrant ventricular tachycardia attributed to bundle branch reentry. *Circulation.* 1988; 78:246–257.
5. Blanck Z, Deshpande S, Jazayeri MR, Akhtar M. Catheter ablation of the left bundle branch for the treatment of sustained bundle branch reentrant ventricular tachycardia. *J Cardiovasc Electrophysiol.* 1995;6:40–43.

CASE STUDY

39

Interfascicular Reentrant Ventricular Tachycardia

Zalmen Blanck, MD

A 72-year-old woman with an ischemic cardiomyopathy (left ventricular ejection fraction (LVEF) 15%) received multiple implantable defibrillator shocks for sustained monomorphic ventricular tachycardia (VT) despite amiodarone therapy. Electrophysiologic studies revealed an HV interval of 70 ms during sinus rhythm and inducible bundle branch reentry (BBR) VT with a left bundle branch block (LBBB) pattern (cycle length 330 ms). Immediately after successful right bundle branch (RBB) ablation, multiple runs of nonsustained tachycardia occurred during sinus rhythm (ie, right bundle branch block [RBBB] with rightward axis, Figure 39.1). A His-bundle potential preceded each QRS complex, and the HV interval, which prolonged to 120 ms after the RBB ablation, shortened to 70 ms during the VT (Figures 39.1 and 39.2). Changes in HH intervals preceded and dictated subsequent VV intervals (see Figure 39.2).

Question No. 1:

What is the mechanism of this tachycardia?

A. Monomorphic "scar-related" VT.

B. Atrioventricular nodal reentrant tachycardia (AVNRT) with AV dissociation.

C. Atriofascicular reentrant tachycardia.

D. Interfascicular reentrant VT.

DISCUSSION

In the setting of atrioventricular (AV) dissociation, and when (irregular) HH intervals dictate subsequent VV intervals, one has to consider BBR VT or a VT originating in the His-Purkinje system as the mechanism of the VT. One would not expect this finding during a "scar-related" VT, where changes in VV cycle lengths would be more likely to dictate and precede changes in HH cycles.

Identical QRS morphology during sinus rhythm and during VT, as well as the His-bundle potential preceding the onset of the QRS complex during the tachycardia, would suggest a supraventricular tachycardia (SVT). However, AV nodal reentrant tachycardia with AV dissociation is extremely rare, and shortening of the HV interval during the tachycardia (from 120 ms to 70 ms) makes this or any other SVT unlikely.

True atriofascicular reentrant tachycardia is a type of antidromic SVT that cannot occur with AV dissociation; the QRS morphology is not the same during sinus rhythm and during preexcited tachycardia.

During interfascicular reentry, the irregular HH intervals dictate subsequent VV intervals, the HV interval shortens, and the same QRS morphology is seen during sinus rhythm and tachycardia.[1] During this type of VT, the (left-sided) reentry circuit incorporates the left anterior fascicle as the

Practical Electrophysiology © 2015 Jasbir S. Sra, Masood Akhtar, Andrea Natale, David J. Wilber
Cardiotext Publishing, ISBN: 978-1-935395-17-1

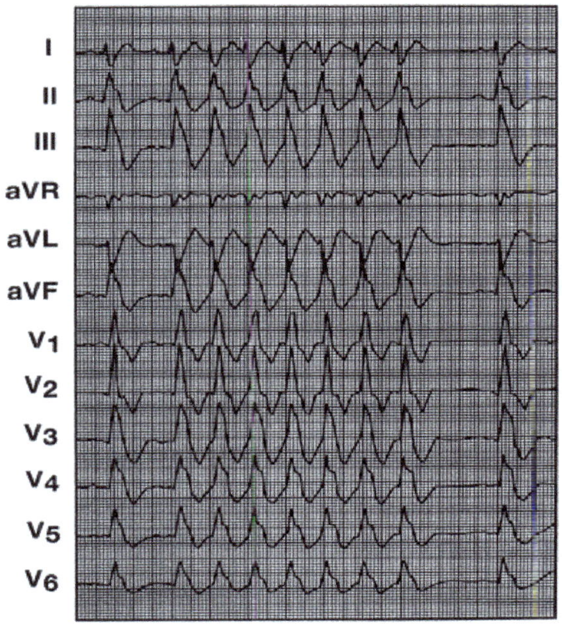

Figure 39.1. 12-lead ECG showing an identical QRS morphology during sinus rhythm (first and last beats) and during spontaneous, nonsustained interfascicular reentry after the RBB ablation. The QRS exhibits an RBBB and rightward axis, so the (left) ventricle is activated by way of the left anterior fascicle during sinus rhythm and during tachycardia. (Reproduced from Blanck et al[1] with permission from John Wiley and Sons.)

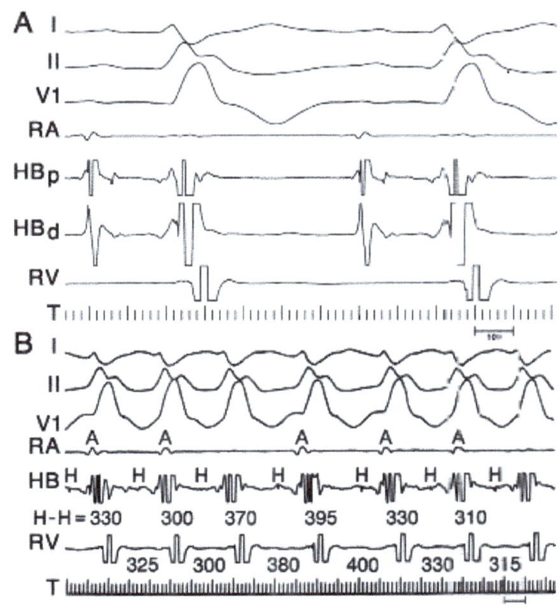

Figure 39.2. A: Intracardiac electrograms (EGMs) during sinus rhythm after the RBB ablation. Shown from top to bottom are surface ECG leads I, II, and V_1, and EGMs from the right atrium (RA), His bundle proximal (HBp), His bundle distal (HBd), and right ventricle (RV). Time lines (T) are also shown. Catheter ablation prolonged the HV interval (not labeled). B: Intracardiac EGMs during interfascicular reentrant tachycardia. All intervals are in milliseconds (ms). The HV interval shortens from 120 ms in sinus rhythm to 70 ms during tachycardia (not labeled). In addition, changes in HH intervals precede and dictate subsequent changes in VV intervals. (Reproduced with permission from Blanck et al.[1])

antegrade limb of the circuit and the left posterior fascicle as the retrograde limb of the circuit. Radiofrequency catheter ablation of the LBB was performed, which resulted in transient complete heart block and no further VT. Six months later, she again received defibrillator shocks for the same VT, and repeat catheter ablation of the left anterior fascicle was successfully accomplished. The ECG demonstrated prolongation of the PR interval from 180 ms to 300 ms and an RBBB with leftward axis (Figure 39.3). The HV interval prolonged to 250 ms. These findings reflected antegrade conduction over the posterior fascicle (after RBB and left anterior fascicular ablations). The VT was no longer inducible and the patient remained arrhythmia-free.

Interfascicular reentrant VT is a very rare mechanism of His-Purkinje VT. In contrast to patients with BBR VT, whose ECGs exhibit an LBBB pattern during sinus rhythm, most patients with interfascicular VT have an RBBB pattern, either after RBB ablation for BBR VT or a spontaneous RBBB. Although irregular HH intervals dictate and precede irregular VV changes as also seen during BBR VT, shortening of the HV interval is usually seen in interfascicular VT and not the former.

In our patient, antegrade conduction during sinus rhythm before the RBB ablation was by way of the fastest RBB (ie, the shortest HV interval among the 3 fascicles); the conduction over the left-sided fascicles was very slow (HV intervals of 120 and 25 ms),[1] which allowed the antegrade sinus impulse traveling down the RBB to cross the septum and penetrate the left-sided fascicles retrogradely, thereby preventing interfascicular reentry due to the collision of the antegrade and retrograde impulses. However, after RBB ablation, the transseptal retrograde left-sided penetration was abolished, and antegrade conduction proceeded through the fastest anterior fascicle, setting the stage for reentry through intermittent antegrade block and slow retrograde conduction via the posterior fascicle (Figure 39.4).[1] Although very rare, it is important to recognize that this type of reentry is left-sided, as ablation of the RBB will not eliminate this VT.

Correct Answer:

Q1. D

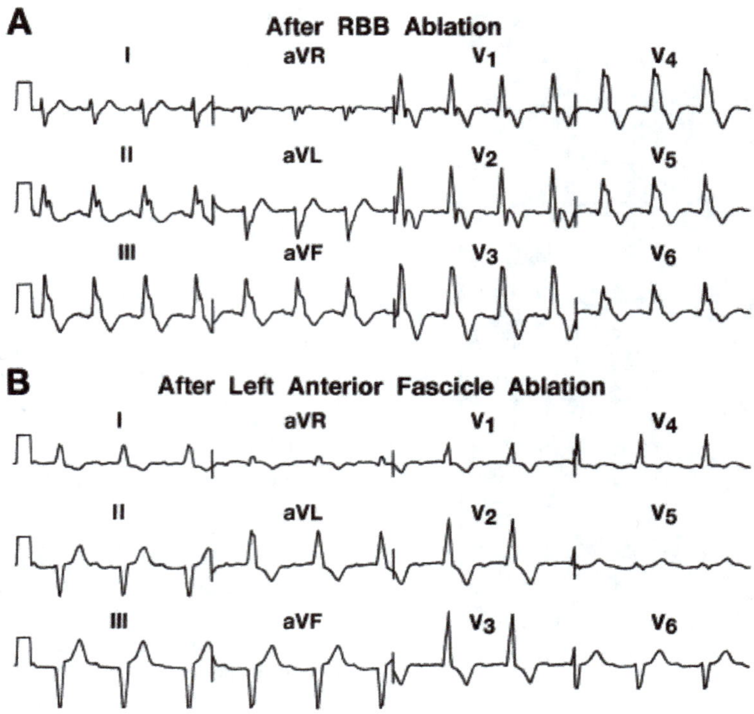

Figure 39.3. Surface ECG leads before and after ablation of the left anterior fascicle. **A:** ECG showing an RBBB (post-RBB ablation) with right posterior fascicular block (rightward axis) and a normal PR interval. **B:** After left anterior fascicular ablation, the PR interval significantly prolongs as a result of AV conduction now proceeding through the slower posterior fascicle (HV interval of 250 ms), which results in leftward axis. (Reproduced from Blanck et al[1] with permission from John Wiley and Sons.)

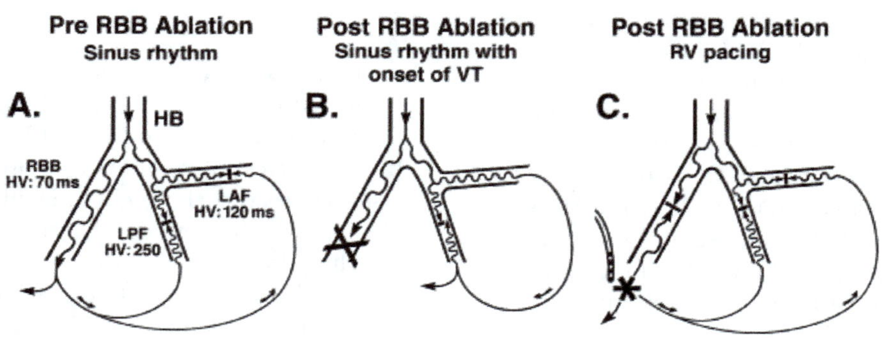

Figure 39.4. Pre-right bundle ablation pattern (**Panel A**). After RBB ablation, transseptal retrograde left-sided penetration was abolished, and antegrade conduction proceeded through the fastest anterior fascicle (**Panel B**), setting the stage for reentry through intermittent antegrade block and slow retrograde conduction via the posterior fascicle (**Panel C**). RV, right ventricular; VT, ventricular tachycardia. (Reproduced from Blanck et al[1] with permission from John Wiley and Sons.)

REFERENCE

1. Blanck Z, Sra J, Akhtar M. Incessant interfascicular reentrant ventricular tachycardia as a result of catheter ablation of the right bundle branch: Case report and review of the literature. *J Cardiovasc Electrophysiol.* 2009;20:1279–1283.

SECTION 5

Channelopathies, ECGs and other Markers for Cardiac Arrhythmias

CASE STUDY

40

LQTS

Ilan Goldenberg, MD and Arthur J. Moss, MD

M.T. is a sibling of a patient who has been recently diagnosed with long QT syndrome (LQTS) following an episode of syncope during swimming. M.T. is a 16-year-old male and has not experienced any episodes of syncope in the past. His corrected QT interval (QTc) is within the normal range (420–440 ms on 3 consecutive ECGs). However, genetic testing reveals that he is a carrier of his brother's LQTS-causing (*S349W*) mutation, which is located in the transmembrane region of the *KCNQ1* (LQT1) channel.

Question No. 1:

What therapy would you recommend to M.T. in light of the genetic diagnosis?

A. Avoid intense physical activity and QT-prolonging drugs and careful follow-up without medical therapy.

B. Beta-blocker therapy.

C. Left cardiac sympathetic innervations (LSCD).

D. Implantable cardioverter-defibrillator (ICD).

The hereditary LQTS is a familial disorder in which most affected family members have delayed ventricular repolarization manifest on the ECG as QT prolongation. This genetic channelopathy has variable penetrance with affected individuals having an increased propensity to syncope, polymorphous ventricular tachycardia (torsades de pointes), and sudden arrhythmic death. The estimated overt prevalence of this disorder is in the range of about 1:5,000 subjects. However, the prevalence of LQTS patients with overt or subclinical disorders is likely to be considerably greater than the currently estimated prevalence.[1]

Clinically, LQTS is identified by abnormal QT interval prolongation on the ECG. The QT prolongation may arise from either a decrease in repolarizing potassium currents or an inappropriate late entry of sodium into the myocyte. Most commonly, QT prolongation is produced by delayed repolarization due to mutations in the α-subunit of ion channels involving either the slowly (I_{Ks}; KCNQ1; LQT1) or rapidly (I_{Kr}; KCNH2; LQT2) acting repolarizing cardiac potassium currents.[1] Mutations of the sodium-channel protein are less frequent and are associated with prolonged depolarization due to a small persistent inward "leak" in cardiac sodium (Na^+) current I_{Na} (SCN5A; LQT3). To date, more than 600 mutations have been identified in 13 different LQTS genes, with the LQT1–3 making up approximately 95% of identified genotyped cases.[1]

Genotype–phenotype studies of LQTS have provided new insights into risk mechanisms, the electrocardiographic features, and the long-term clinical course associated with this inherited disorder. For example, each of the 3 major genotypes (LQT 1–3) has a distinctive T-wave repolarization pattern on the ECG. These include typical broad-based T waves in LQT1 patients, bifid T waves in LQT2 patients, and late-onset peaked T waves in LQT3 patients (Figure 40.1).

Practical Electrophysiology © 2015 Jasbir S. Sra, Masood Akhtar, Andrea Natale, David J. Wilber
Cardiotext Publishing, ISBN: 978-1-935395-17-1

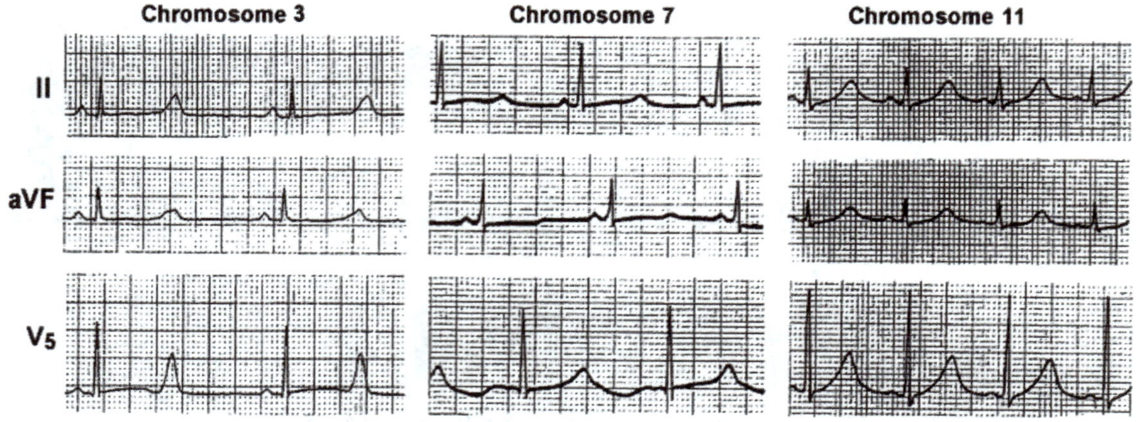

Figure 40.1. T-wave morphology by LQTS genotype: LQT1—right panel (QTc = 570 ms); LQT2—middle panel (QTc = 583 ms); and LQT3—left panel (QTc = 573 ms). (Reprinted with permission.[1])

The diagnosis of LQTS relies mainly on ECG findings and clinical history. An accurate measurement of the QT interval is essential for the diagnosis of LQTS. The QT interval should be determined as a mean value derived from at least 3 to 5 cardiac cycles (heart beats), and is measured from the beginning of the earliest onset of the QRS complex to the end of the T wave. The QT interval is usually corrected for heart rate using Bazett's formula ($QTc = QT/RR^{0.5}$ with all intervals in seconds), which remains the standard for clinical use. In general, **QTc is considered as prolonged if >450 ms and >470 ms in adult men and women**, respectively.[1] When marked QTc prolongation is present, the diagnosis is straightforward. For less clear cases, a scoring system has been introduced that takes into account ECG findings, prior symptoms, and family history.[1]

Genetic testing for LQTS has recently emerged as an additional important diagnostic and clinical tool. It should be noted that currently a genetic test can be expected to capture approximately three-fourth of phenotypically affected LQTS individuals, whereas a negative genetic test in a subject with clinical LQTS (ie, genotype negative/phenotype positive LQTS) provides no basis for removing the diagnosis.[1] However, a positive genetic test may influence treatment decisions and may identify relatives with concealed LQTS, since approximately 25% of individuals with genetically proven LQTS have a nondiagnostic QTc.[1]

The clinical course of patients with LQTS is variable due to incomplete penetrance. It is influenced by age, genotype, gender, environmental factors, therapy, and possibly other modifier genes. Based on published mortality rates, LQTS risk groups may be categorized as high risk (a history of aborted cardiac arrest and/or electrocardiographically documented episodes of torsades de pointes); intermediate risk (time-dependent syncopal history and/or QTc prolongation >500 ms); and low risk (affected subjects without a history of a prior syncope and with a QTc duration ≤0.50 second) (Figure 40.2).[1] In addition to this simplified approach to risk stratification, it should be noted that risk factors in LQTS are time-dependent and age-specific, warranting continuous risk assessment in this population.

Life-threatening arrhythmic events in LQTS patients have been shown to occur under specific circumstances in a gene-specific manner. Patients with the LQT1 genotype have a high frequency of cardiac events associated with vigorous physical activities, whereas patients with the LQT2 genotype

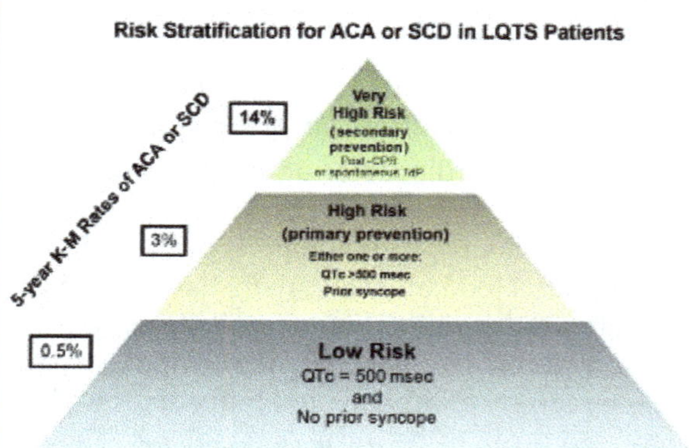

Figure 40.2. Risk stratification categories for LQTS patients based on published events rates. (Reprinted with permission.[1])

are at high risk of having arrhythmic events triggered by a sudden loud noise, such as the ringing of an alarm clock. LQT3 patients experience events without emotional arousal during sleep or at rest. Swimming has been described as an LQT1-specific arrhythmogenic trigger. It has also been shown that arrhythmic risk is related to the presence of specific LQTS genotypes. Thus, patients with LQT1 and LQT2 mutations have a higher risk for cardiac events as compared with those with LQT3 mutations (Figure 40.3). However, the lethality of cardiac events was found to be significantly higher in patients with LQT3 as compared with LQT2 and LQT1 patients.[2]

Recent genotype–phenotype studies from the International LQTS Registry have provided important information regarding the effect of the location of the LQTS-causing mutation in the LQTS channel on the phenotypic manifestations of LQTS (see Figure 40.3).[3,4] These studies have shown that mutations located in the transmembrane region of the *KCNQ1* channel and in the pore region of the *KCNH2* channel (which form the ion conduction pathway of the respective channels) are independently associated with a significantly higher risk for cardiac events as compared with mutations that are located in other regions of the LQTS channels (Figure 40.4).[3,4] Furthermore, recent data suggest that the presence of transmembrane mutations is the only independent risk factor for life-threatening arrhythmic events in LQTS patients who have a normal-range QTc. These findings further stress the importance of incorporating genotype–phenotype correlations in the risk assessment of LQTS patients.

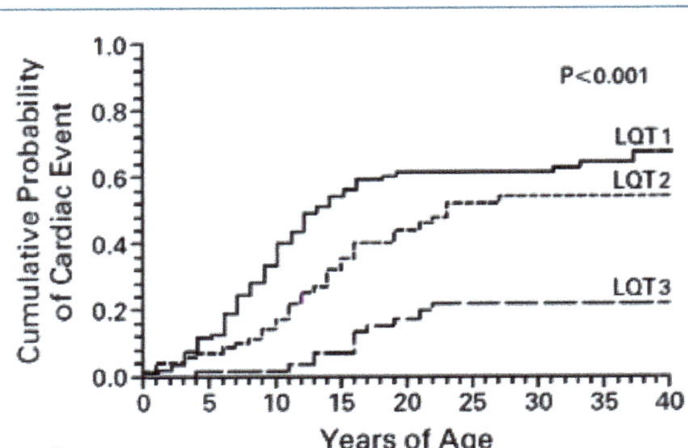

No. of Subjects					
LQT1 group	112	72	36	27	19
LQT2 group	72	56	29	16	11
LQT3 group	62	56	36	24	16

Figure 40.3. Kaplan–Meier estimates of the cumulative probability of cardiac events in LQT1, LQT2, and LQT3 patients. (Reprinted with permission.[2])

Guideline recommendations for the management of LQTS patients include a class I indication for: (1) lifestyle modifications in all LQTS patients; (2) treatment with beta-blocker in LQT1 and LQT2 patients with a clinical diagnosis of LQTS (ie, those with a prolonged QTc; the benefit of beta-blockers in LQT3 patients is less certain at this point in time); and (3) implantation of an ICD in those who experience cardiac arrest. Class IIa indications include: **beta-blockers in genetically diagnosed LQTS patients who have a normal QT interval; and implantation of an ICD in patients who remain symptomatic during beta-blocker therapy.**[5] Left cardiac sympathetic denervation was also shown to reduce the risk of cardiac events in LQTS patients and may be a therapeutic option in patients who experience frequent arrhythmias or ICD shocks despite medical therapy.[1]

Correct Answer:

Q1. B

QUESTIONS AND AUTHORS' REPLY:

Question 1. Does family history of sudden death predict risk for life-threatening arrhythmic events in LQTS patients?

Answer: The death of a sibling may be a marker of a more severe mutation. However, LQTS patients demonstrate variable penetrance within families, with a wide range of QT intervals and symptoms. Accordingly, a history of sudden cardiac death in a first-degree relative was not identified in previous studies as a significant risk factor.

Question 2. Does gender play a role in the clinical course of LQTS patients?

Answer: The phenotypic expression of LQTS displays major time-dependent gender differences in the risk of nonfatal and life-threatening cardiac events. During childhood, LQT1 males show a high risk for arrhythmic events, whereas after the onset of adolescence there is a gender-genotype risk-reversal showing a risk increase among LQT2 women. These effects may relate to hormonal factors since androgens were shown to be associated with QT shortening, whereas estrogens were shown to have a dose-dependent blocking effect on I_{Kr}.

Question 3. Which are the ideal leads to measure QT interval?

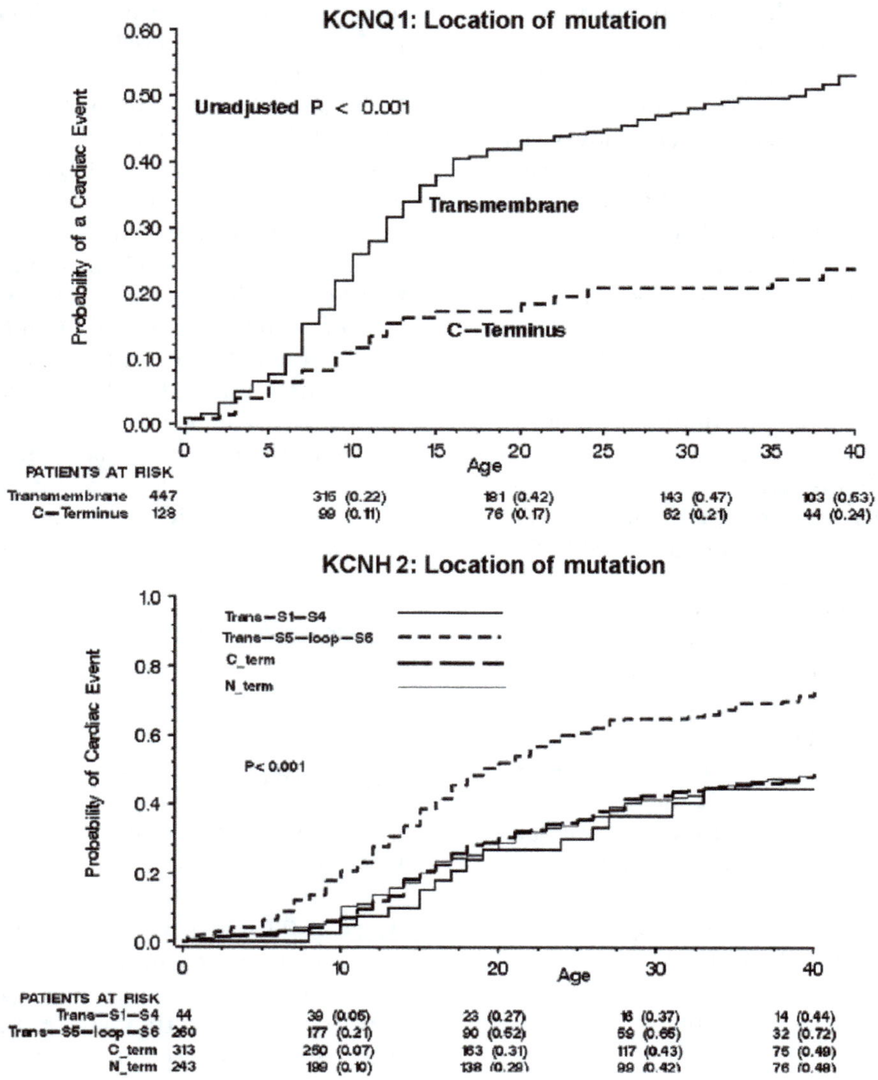

Figure 40.4. Kaplan–Meier estimate of the cumulative probability of a first cardiac event in *KCNQ*1 mutation carriers (LQT1 genotype: top) and in *KCNH2* mutation carriers (LQT2: bottom).

Answer: The QT measurement should preferably be made in leads II, and V_5 or V_6, with the longest value being used, from at least 3 to 5 cardiac cycles (ie, RR intervals). Measurement of the QT interval should be made from the earliest onset of the QRS complex to the end of the T wave when the descending limb returns to the TP baseline, if there is no U wave (Figure 40.5A) or if it is distinct from the following U wave (Figure 40.5B). In the case of a biphasic T wave, the QT interval is measured to the time of final return to baseline (Figure 40.5C). If a second low-amplitude repolarization wave interrupts the terminal portion of the T wave (Figure 40.5D), it is best to record both the QT interval (T-wave offset measured as the nadir between the T and U wave) and the QTU interval (repolarization offset measured at the end of the second wave). In general, biphasic T waves are frequently present in multiple leads, whereas discrete and separate low-amplitude U waves are best seen in the lateral precordial leads. The end of the U wave is defined as the intersection point of the descending limb of the U wave and the isoelectric baseline.

Question 4. What is the association of LQT syndromes with atrial fibrillation (AF) and congenital neuromuscular disorders?

Answer: LQTS patients were shown to have altered atrial electrical conduction and

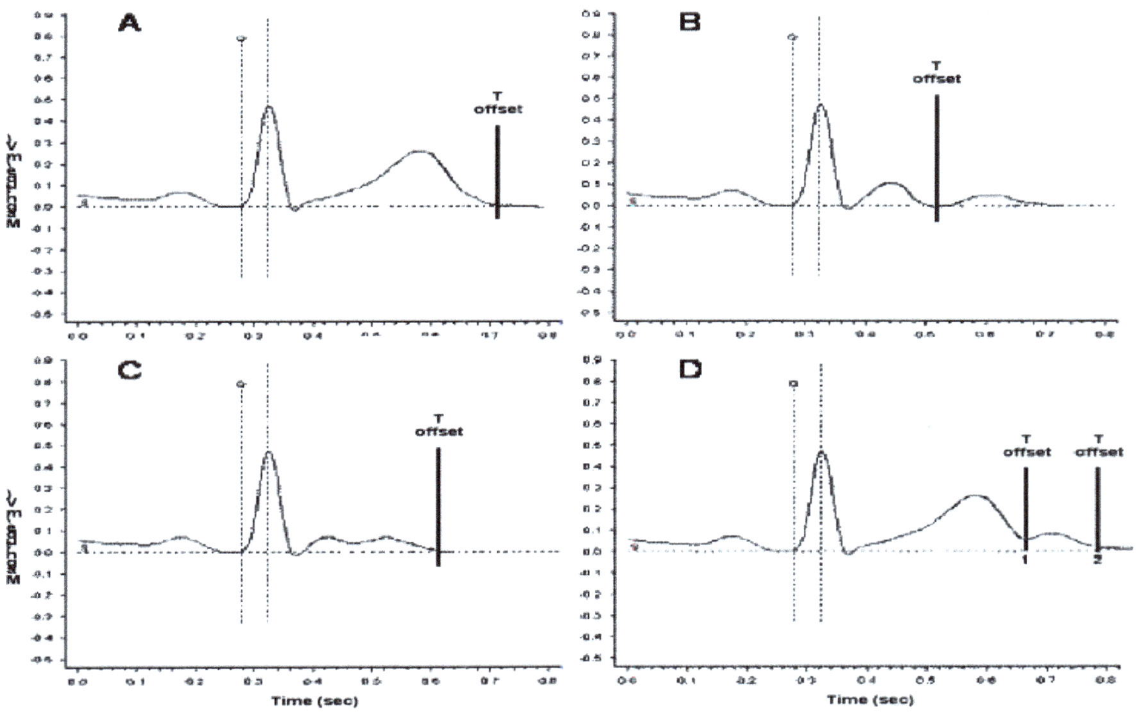

Figure 40.5. QT measurements by ECG presentation: A: normal T-wave morphology; B: T wave followed by a distinct U wave; C: biphasic T wave, with T1 and T2 waves of similar amplitude; and D: a second low-amplitude repolarization wave interrupts the terminal portion of the larger T wave. (Reprinted with permission.)

suggested to be at increased risk for polymorphic atrial tachyarrhythmias and AF. Notably, atrial torsade de pointes was first described nearly a decade ago in a canine model of Andersen–Tawil syndrome in which cesium-mediated block of Kir2.1 potassium channels induced atrial early afterdepolarizations that precipitated polymorphic atrial tachyarrhythmia and subsequent degeneration into AF. Consistent with these findings, P waves during episodes of polymorphic atrial tachyarrhythmia in LQTS patients were reported to have an undulating nature, not unlike those seen in ventricular fibrillation. Thus, it is possible that an atrial torsade mechanism mediates susceptibility for AF in LQTS.

There are also gene-specific mutations correlated with extracardiac phenotype: mutations in KCNJ2 gene with reduction in Kir2.1 current (Andersen–Tawil syndrome, LQT7) are associated with a phenotype dominated by skeletal abnormalities (typically affect the head, face, and limbs) and period paralysis; and mutations in CACNA1C gene with increase in Cav1.2 current (Timothy syndrome, LQT8) are associated with syndactyly in both hands and feet.

REFERENCES

1. Goldenberg I, Moss AJ. Long QT syndrome. *J Am Coll Cardiol.* 2008;51:2291–2300.
2. Zareba W, Moss AJ, Schwartz PJ, et al. Influence of genotype on the clinical course of the long-QT syndrome. International Long-QT Syndrome Registry Research Group. *N Engl J Med.* 1998;339:960–965.
3. Moss AJ, Shimizu W, Wilde AA, et al. Clinical aspects of type-1 long-QT syndrome by location, coding type, and biophysical function of mutations involving the KCNQ1 gene. *Circulation.* 2007;115:2481–2489.
4. Shimizu W, Moss AJ, Wilde AA, et al. Genotype-phenotype aspects of type 2 long QT syndrome. *J Am Coll Cardiol.* 2009;54:2052–2062
5. Zipes DP, Camm AJ, Borggrefe M, et al. ACC/AHA/ESC 2006 Guidelines for management of patients with ventricular arrhythmias and the prevention of sudden cardiac death. *Circulation.* 2006;114:e385–e484.

CASE STUDY

41

Short QT Syndrome

Chinmay P. Patel, MD and Charles Antzelevitch, PhD

A 35-year-old white man presented to the emergency department with an episode of syncope during exertion. He was previously healthy except for an episode of paroxysmal atrial fibrillation (AF) at the age of 25. He denied syncope prior to this episode but had frequent palpitations lasting for several minutes attributed to episodes of paroxysmal AF. He was not on any medications and his family history is significant for a brother who died suddenly at age of 3 months and father who died at age 39 due to sudden cardiac death (SCD). Autopsy of the father did not reveal any structural heart disease. His ECG is shown in Figure 41.1. He was admitted to the hospital for further workup and monitored on telemetry. He had frequent premature ventricular complexes (PVCs) on telemetry monitor and a couple of short-lived episodes of self-terminating polymorphic ventricular tachycardia (VT).

Question No. 1:

What is the diagnosis?

A. Catecholaminergic polymorphic VT.
B. Idiopathic ventricular fibrillation (VF)—J wave syndrome.
C. Hyperkalemia.
D. Short QT syndrome (SQTS).

Further workup including echocardiographic stress test did not reveal any evidence of structural heart disease. Results of signal-averaged ECG, heart rate (HR) variability and QT dispersion were within normal limits. Patient did not have any electrolyte abnormalities to explain persistently short QT interval. During stress testing, the QT interval abbreviated but failed to prolong to normal values upon recovery.

ECG obtained as a screening of family members revealed that the patient's 31-year-old sister and her 6-year-old son had similar ECG pattern showing extremely short QT intervals. The 6-year-old had an episode of aborted SCD at age 8 months.

On electrophysiologic study of the proband, both AF and VF were readily inducible using programmed atrial or ventricular stimulation. An implantable cardioverter-defibrillator (ICD) was implanted and the patient experienced 2 appropriate shocks during the first year of follow-up.

Question No. 2:

What QT interval on ECG will qualify as short QT interval?

A. 400 ms at HR of 60 bpm.
B. 360 ms at HR of 90 bpm.
C. 300 ms at HR of 120 bpm.
D. 280 ms at HR of 53 bpm.

Practical Electrophysiology © 2015 Jasbir S. Sra, Masood Akhtar, Andrea Natale, David J. Wilber
Cardiotext Publishing, ISBN: 978-1-935395-17-1

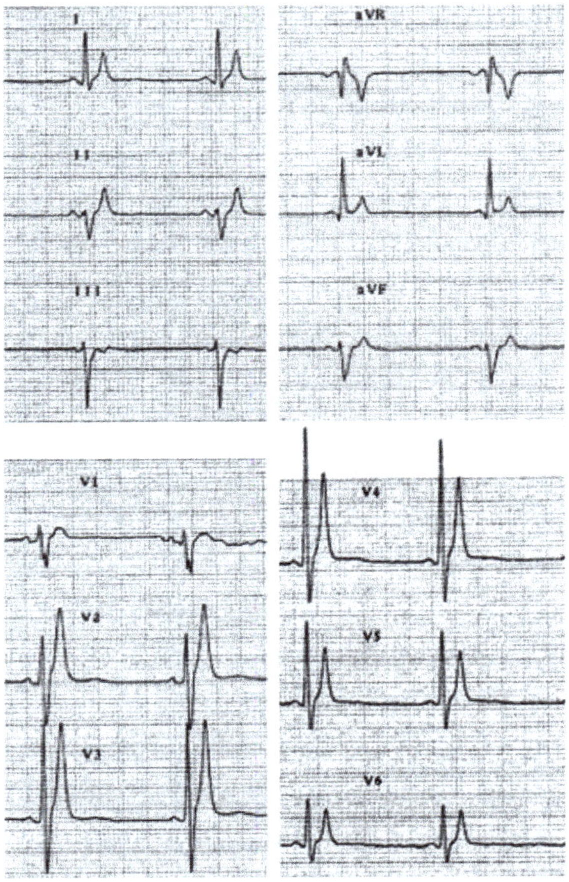

Figure 41.1. ECG of SQTS Proband. HR 52 bpm, QT 280 ms.

Question No. 3:

The SQTS is characterized by

A. Tall peaked T waves.

B. Gain of function mutations in potassium channel genes.

C. Loss of function mutations in calcium channel genes.

D. All of the above.

Question No. 4:

What is the cellular basis for development of arrhythmias in the SQT1 form of SQTS?

A. An increase in transmural dispersion of repolarization (TDR).

B. An increase in I_{Kr}.

C. A decrease in effective refractory period (ERP).

D. None of the above.

E. All of the above.

Question No. 5:

What drug is generally effective in prolonging the QT interval in SQT1 patients?

A. Quinidine.

B. Dofetilide.

C. Amiodarone.

D. Sotalol.

DISCUSSION

Short QT syndrome

The SQTS is an inherited syndrome characterized by a QTc ≤360 ms and high incidence of VT/VF in infants, children and young adults. It is a genetically heterogeneous disease, and thus far, mutations of 6 different genes encoding cardiac ion channels have been identified in familial or sporadic cases. Gain-of-function mutations in *KCNH2, KCNQ1,* and *KCNJ2* genes encoding for the rapidly and slowly activating potassium channel currents (I_{Kr} and I_{Ks}) and the inward rectifier potassium channel current (I_{K1}) give rise to the SQT1, SQT2, and SQT3 forms of SQTS. Loss of function mutations in *CACNA1c, CACNB2b,* and *CACNA2D1* genes encoding the α1, β, and α2δ subunits of cardiac L-type calcium channels is also responsible for SQTS. These clinical phenotypes, characterized by a clinical entity consisting of a combined Brugada-like ST-segment elevation and short QT intervals, have been designated SQT4, SQT5, and SQT6. In addition to an abbreviated QT interval, SQTS is characterized by the appearance of tall peaked and often symmetrical T waves in the ECG. The augmented T_{peak}-T_{end} interval associated with this electrocardiographic feature of the syndrome suggests that TDR is significantly increased.

Studies employing the left ventricular wedge model of the SQTS have provided evidence in support of the hypothesis that an increase in outward repolarizing current can preferentially abbreviate the epicardial action potential, thus increasing TDR and creating the substrate for reentry.[1] In addition to this augmented spatial dispersion of repolarization, arrhythmogenesis in SQTS is facilitated by the abbreviated ERP.

Diagnosis

SQTS is characterized by abnormally short QT intervals on the surface ECG (<360 ms) and an increased proclivity to develop atrial and/or ventricular tachyarrhythmias. SQTS patients generally have structurally normal hearts. Conditions that can cause physiological abbreviation of QT interval like hyperkalemia, acidosis, hypercalcemia, hyperthermia, effects of drugs like digitalis, effect of acetylcholine or catecholamine, and abbreviation of QT interval related to activation of ATP-sensitive potassium channel (K_{ATP}) current need to be ruled out before considering the diagnosis of SQTS. The dependence of the QT interval on HR is diminished in SQTS. The slope of the QT-HR relationship can be nearly half of that observed in healthy controls, making Bazett's correction of the QT interval inappropriate. Therefore, diagnosis is difficult to make at rapid HRs (>80 bpm; Figure 41.2).

Typical ECG in SQTS shows

1. Abnormally short QT intervals between 220 and 360 ms.
2. Tall, peaked, symmetrical or asymmetrical T waves in the precordial leads.
3. Prolonged T_{peak}-T_{end} interval and T_{peak}-T_{end}/QT ratio.
4. Diminished dependence of QT interval on HR.
5. In the case of calcium channel loss of function mutations, a Brugada-type ST-segment elevation in the right precordial leads V_1 and V_2 at baseline or after administration sodium channel blocker.
6. Higher prevalence of early repolarization.

Clinical presentation

A patient of SQTS may be completely asymptomatic and incidentally detected during routine physical examination or ECG done for screening purposes. The first clinical manifestation of the disease has been reported as early as first year of life and as late as 80 years of age. In the largest available case series of 29 SQTS patients,[2] approximately 62% of the patients were symptomatic. Cardiac arrest was the most frequently (34%) reported symptom and, in 28% of patients, it was the first clinical presentation. Cardiac arrest occurred in first year of life in 2 patients, suggesting that SQTS may be one of the causes of sudden infant death syndrome (SIDS). Palpitations were the second-most frequently reported symptom (31%) followed by syncope (24%). AF was the first presenting symptom in 17% of patients. Approximately 38% of patients were asymptomatic. Strong family history of arrhythmic symptoms, including SCD, is a common finding. Exact nature of factors that precipitate the arrhythmia is not clear and episodes of SCD have been reported following loud noise, at rest, during exercise and daily activities. Most patients with SQTS have QTc ≤340 ms with range of 210 to 320 ms. However, in patients with SQT4, SQT5, and SQT6, QTc intervals are generally a bit longer (330 to 360 ms). No correlation has been identified as yet between the extent of QT interval abbreviation and risk of arrhythmic events.

Treatment in SQTS

ICD: An ICD is strongly recommended for secondary prevention of SCD. There is scarcity of risk stratification data that can help identify patients at higher risk for SCD. The benefit of implanting an ICD for primary prevention of SCD must be weighed against a lifetime of risks associated with a lifetime of complications. The decision to implant an ICD should be based on clinical grounds (short QT interval on ECG in conjunction with arrhythmic symptoms and strong family history of SCD) rather than on genetic or electrophysiologic data, until

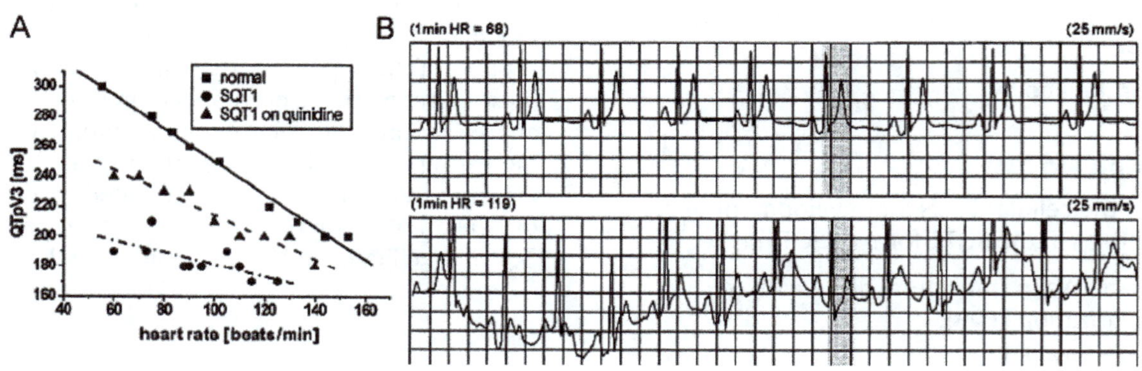

Figure 41.2. A: Poor rate-adaptation of QT interval. The QT-RR relationship slope is less steep in SQTS patient as compared to controls. Quinidine restores the relationship toward control values. QTpV3 denotes the interval from the beginning of QRS complex to peak of T wave measured in lead V_3. B: Holter monitoring showing impaired adjustment of QT interval with change in HR. (Modified from Wolpert et al[4] with permission from Wiley-Blackwell.)

further guidelines for risk stratification are available.

Oversensing of the T wave is a frequent clinical problem in patients with SQTS who receive ICD, as the tall, peaked, and closely coupled T waves are often mistakenly sensed as R waves, leading to inadvertent ICD shocks. Reprogramming the decay delay, sensitivity, or both generally prevents inappropriate discharges.

PHARMACOLOGICAL THERAPY

Data regarding pharmacological therapy in SQTS are emerging but limited. Pharmacological therapy may be useful as an adjunct to the ICD or may be used for primary prevention in cases in which the patient refuses an ICD or in young children in whom the implantation of an ICD may be problematic.

Information regarding pharmacological therapy for SQTS is fairly limited and much of the available data pertain to SQT1, the variant due to a gain of function of I_{Kr}. Traditional I_{Kr} blockers such as sotalol and dofetilide are often ineffective in SQT1. Gaita et al[3] examined the effectiveness of 4 different antiarrhythmic drugs including flecainide, sotalol, ibutilide, and hydroquinidine in 6 patients with SQT1. Only hydroquinidine prolonged the QT interval to normal levels, increased ventricular ERP and rendered VF noninducible. Class IC and III antiarrhythmic drugs failed to do so. Quinidine also restored QT-RR relationship toward the normal range (see Figure 41.2).[4] In a one-year follow-up, patients treated with hydroquinidine remained asymptomatic and no further episodes of ventricular arrhythmia were detected.

The efficacy of quinidine and the failure of Class III antiarrhythmic drugs in SQT1 is primarily related to a secondary effect of the N588K to reduce the availability of the inactivated state of the mutant channel, hence reducing the affinity of I_{Kr} channels to many drugs with Class III antiarrhythmic activity.[5] Electrophysiologic studies in heterologous expression systems have shown that N588K mutation reduces the affinity of the channel for E-4031 and d-sotalol much more than for quinidine.[5–7] Similar studies involving mutant channels have identified disopyramide as another effective agent.[8] Consistent with these bench-side observations, Schimpf et al reported clinical efficacy of disopyramide in 2 female patients with SQT1.[9]

Quinidine as well as the traditional Class III antiarrhythmic drugs are expected to be clinically useful in the other forms of SQTS, although clinical data are still very limited.

AF is another common clinical problem in SQTS. Some SQTS patients exhibit only AF.[10] Propafenone has been shown to be effective in preventing frequent paroxysms of AF with no recurrence of arrhythmia for more than 2 years without any effect on QT interval.[11] Quinidine is effective as well.

Correct Answers:

Q1. D

Q2. B

Q3. D

Q4. E

Q5. A

QUESTION AND AUTHORS' REPLY:

Question. What is the mechanistic and genetic correlation between the SQTS, AF, and Brugada syndrome (BrS)?

Answer: Loss of function mutations in the 3 subunits of the L-type calcium channel lead to the manifestation of BrS associated with AF as well as with a briefer than normal QT interval. Reduced calcium channel current (Ica) leads to a more prominent phase 1 of the action potential, particularly in the right ventricular epicardium where Ito is most prominent. This can lead to loss of the action potential dome at some epicardial sites, but not others creating epicardial dispersion of repolarization, thus setting the stage for phase 2 reentry, as well as TDR, thus setting the state for intramural reentry. The reduction in ICa also abbreviates action potential duration (APD) throughout the myocardium leading to an abbreviated QT interval.[1–3] These same electrophysiologic changes, resulting from ICa loss of function, are thought to underlie arrhythmogenesis in the atria, thus contributing to the development of AF.

REFERENCES

1. Patel C, Antzelevitch C. Cellular basis for arrhythmogenesis in an experimental model of the SQT1 form of the short QT syndrome. *Heart Rhythm.* 2008;5:585–590.
2. Giustetto C, Di Monte F, Wolpert C, et al. Short QT syndrome: clinical findings and diagnostic-therapeutic implications. *Eur Heart J.* 2006;27:2440–2447.
3. Gaita F, Giustetto C, Bianchi F, et al. Short QT syndrome: pharmacological treatment. *J Am Coll Cardiol.* 2004;43:1494–1499.

4. Wolpert C, Schimpf R, Giustetto C, et al. Further insights into the effect of quinidine in short QT syndrome caused by a mutation in HERG. *J Cardiovasc Electrophysiol.* 2005;16:54–58.
5. Brugada R, Hong K, Dumaine R, et al. Sudden death associated with short-QT syndrome linked to mutations in HERG. *Circulation.* 2004;109:30–35.
6. Cordeiro JM, Brugada R, Hong K, Antzelevitch C, Dumaine R. Short QT syndrome mutation in HERG abolishes inactivation (abstr). *Biophys J.* 2004;86:134a.
7. Cordeiro JM, Brugada R, Wu YS, Hong K, Dumaine R. Modulation of I(Kr) inactivation by mutation N588K in KCNH2: a link to arrhythmogenesis in short QT syndrome. *Cardiovasc Res.* 2005;67:498–509.
8. McPate MJ, Duncan RS, Witchel HJ, Hancock JC. Disopyramide is an effective inhibitor of mutant HERG K+ channels involved in variant 1 short QT syndrome. *J Mol Cell Cardiol.* 2006;41:563–566.
9. Schimpf R, Antzelevitch C, Hsu LF, et al. The QT-intervalin patients with a Brugada syndrome: is a shortening of the QT-time an existing and relevant ECG-pattern? (abstr). *Heart Rhythm.* 2007;4:S188.
10. Hong K, Bjerregaard P, Gussak I, Brugada R. Short QT syndrome and atrial fibrillation caused by mutation in KCNH2. *J Cardiovasc Electrophysiol.* 2005;16:394–396.
11. Bjerregaard P, Gussak I. Short QT syndrome. *Ann Noninvasive Electrocardiol.* 2005;10:436–440.

CASE STUDY

42

Brugada Syndrome: Different Presentations

Atul Bhatia, MD

FIRST CASE

A 68-year-old man who presented with palpitations and dizziness was noted to be in atrial fibrillation (AF) with fast ventricular rate, which spontaneously converted to sinus rhythm. An ECG obtained after conversion to sinus rhythm is shown in Figure 42.1. His past medical history included hypertension, which was well-controlled on cardiazem (240 mg daily). He was a nondiabetic smoker with no prior history of coronary artery disease, myocardial infarction (MI), congestive heart failure (CHF), syncope, transient ischemic attack, or cerebrovascular accident. His echocardiogram revealed normal left ventricular ejection fraction (LVEF) with no valve disease, and his cardiac catheterization revealed normal coronaries. The patient was on no other medications and his blood chemistries, cell count, and thyroid-stimulating hormone test were all normal. He was initiated on flecainide (100 mg orally twice daily) for suppression of paroxysmal AF and discharged home.

Question No. 1:

What is the most likely cause of the abnormality seen in the ECG shown in Figure 42.2?

A. Combined I_{Na} and I_{Ca} blockade leading to unmasking of underlying Brugada syndrome.

B. Flecainide toxicity.

C. Increased levels of calcium channel blocker due to drug interaction.

D. Acute pulmonary embolism from right atrial thrombus dislodgement.

Discussion

The typical ECG appearance (see Figure 42.2) is that of Brugada syndrome. Mutations in SCN5A associated with a loss of function of the sodium channel have been shown in up to 20% of patients with Brugada syndrome. This is consistent with the clinical finding that potent sodium channel blockers of class IC antiarrhythmic agents, such as flecainide, ajmaline or pilsicainide or class IA agents such as procainamide and disopyramide, can unmask the Brugada syndrome when the baseline ECG is normal. It has been shown, in vitro, that a combination of sodium and calcium channel block is much more effective in precipitating the ECG manifestation of Brugada syndrome experimentally.

Clinically, a variety of drugs have been used to unmask the so-called Type1 Brugada ECG pattern. Two of the commonly used drugs studied, flecainide and ajmaline, have shown variable sensitivity and specificity. Ajmaline (not available in United States) showed a sensitivity of 80%, specificity of 94%, and a negative predictive value of 83%. Flecainide showed a much lower negative predictive value of 36%, with

Practical Electrophysiology © 2015 Jasbir S. Sra, Masood Akhtar, Andrea Natale, David J. Wilber
Cardiotext Publishing, ISBN: 978-1-935395-17-1

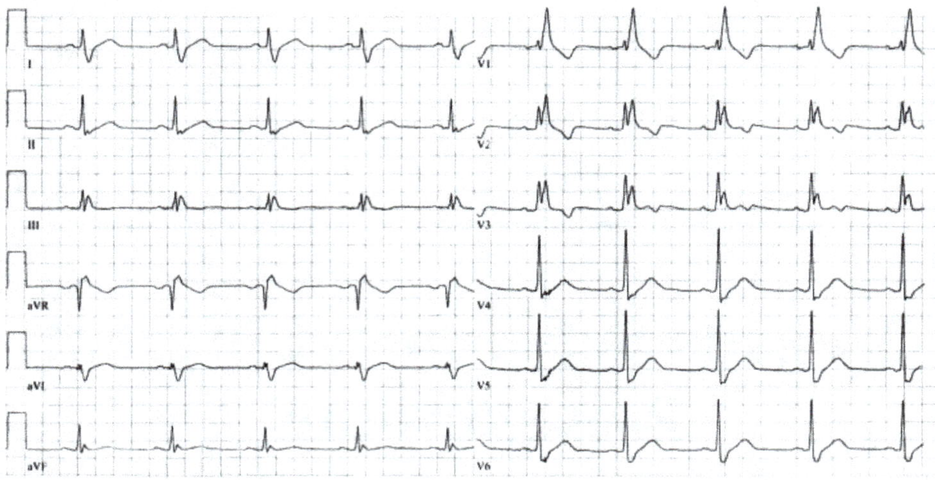

Figure 42.1. Baseline ECG with sinus rhythm, normal intervals, right bundle branch block (RBBB), and no evidence of ischemia.

sensitivity and specificity of about 80%. In light of these findings, ajmaline is more effective in unmasking Brugada syndrome.

A positive response to a sodium channel blocker is characterized by the following (Figure 42.3):

- Type 1-coving ST-segment elevation in leads V_1, V_2, or V_3
- Conversion of Type 2 or Type 3 to Type 1 pattern.

It should be noted that conversion of Type 3 to Type 2 pattern is not considered diagnostic of Brugada syndrome. The test should be performed under constant ECG monitoring in the presence of external defibrillator for termination of malignant ventricular arrhythmias. The test, ie, the drug infusion, should be stopped if premature ventricular beats or ventricular tachycardia (VT) (QRS widening >30% of baseline) is noted. Greater caution is needed in patients with baseline conduction abnormalities like first-degree atrioventricular (AV) block or wide QRS. In case of incessant VT, intravenous isoproterenol or sodium lactate infusion is effective in controlling the electrical ventricular storm.

Question No. 2:

In this patient, could the paroxysmal AF be:

A. A Brugada-associated phenomenon?

B. "Lone" AF?

C. Due to long-standing hypertension?

D. None of the above?

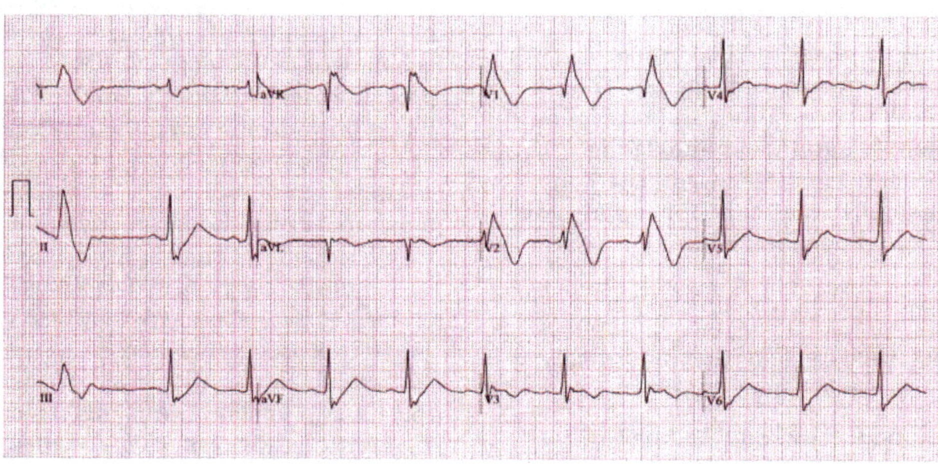

Figure 42.2. Follow-up ECG at 2-week posthospitalization visit.

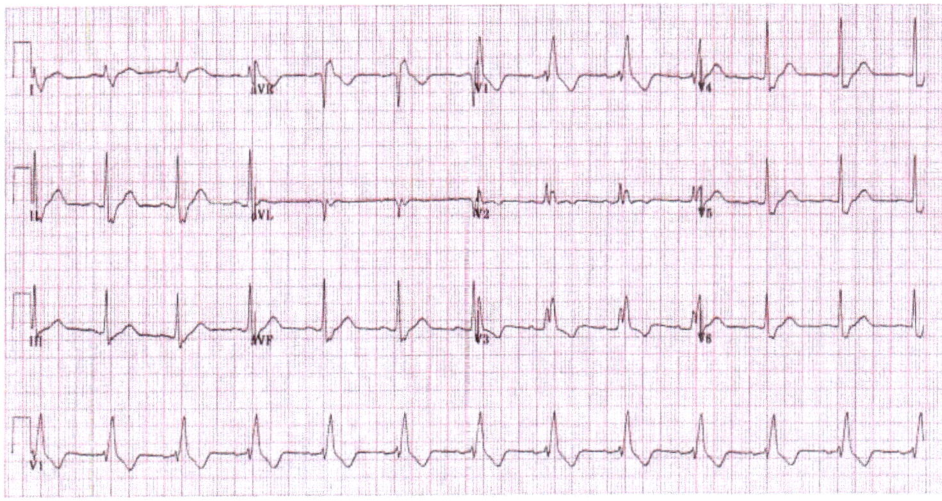

Figure 42.3. ECG obtained 2 weeks following the previous follow-up ECG (Figure 42.2).

Discussion

While the most common association with AF is hypertension, for this patient, we should keep in mind that a relationship between AF and Brugada syndrome has been described previously in 15 to 20% of patients and is associated with prolonged or biphasic P waves. Sodium current can be altered in both the atrial and ventricular tissues in a nonhomogenous manner in patients with Brugada syndrome.

Just as a Brugada pattern on the ECG does not itself equate with Brugada syndrome, AF in a patient with a Brugada pattern does not necessarily mean they are etiologically linked. Regardless, this possibility should be strongly considered. Lone AF would suggest the absence of any underlying heart disease, and clearly there is hypertension in this patient and probably some ventricular hypertrophy as well.

Question No. 3:

Which of the following is the most reasonable next step for this patient?

A. Electrophysiology (EP) study to further clarify patient's risk of sudden cardiac death.

B. Implantable cardioverter-defibrillator (ICD) implant.

C. Clinical observation.

D. None of the above.

Discussion

The role of EP study in risk stratification of Brugada syndrome is unclear. However, in this patient the unmasking of the typical ECG manifestation of Brugada syndrome with therapeutic doses of flecainide could be considered equivalent to a positive lab study, thus, ICD implantation would be the reasonable next step to take.

SECOND CASE

A 52-year-old man with no prior history of coronary artery disease, MI, and CHF presented for left shoulder surgery. Preoperative screening included echocardiogram and showed normal LVEF. Baseline ECG was noted to be abnormal as shown in Figure 42.4. Upon inquiry, the patient noted no prior history of syncope or palpitations. His only medication was ibuprofen, which he took for pain relief. His physical examination was notable for no significant cardiovascular abnormality.

Question No. 4:

The ECG is diagnostic of Brugada syndrome.

A. True.

B. False.

Discussion

Since the original description by Spanish cardiologists Pedro Brugada and Josep Brugada, 3 distinct ECG patterns of Brugada syndrome have emerged:

1. Type 1 pattern is characterized by J-point or coved ST-segment elevation >2 mm followed by a negative T wave.

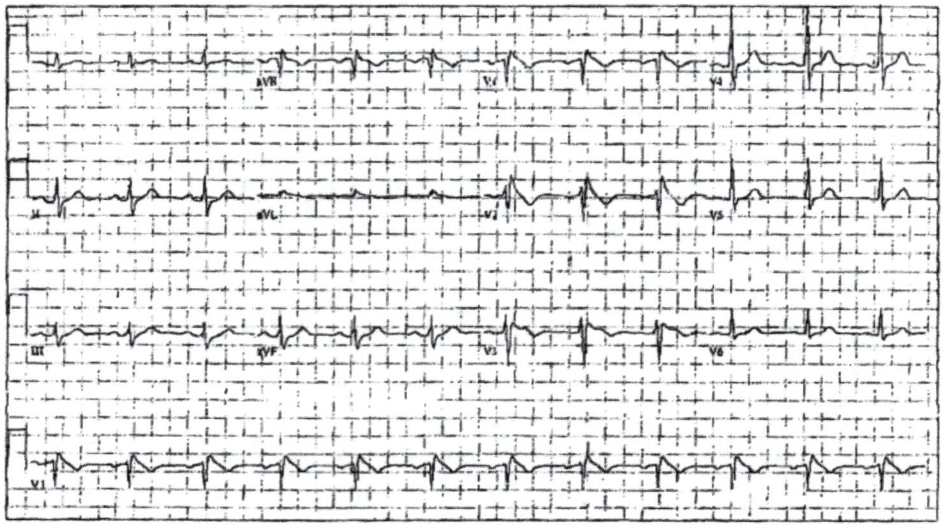

Figure 42.4. Baseline ECG.

2. Type 2 pattern is characterized by J-point or saddle-shaped ST-segment elevation >2 mm followed by positive or biphasic T wave.
3. Type 3 pattern is characterized by coved or saddle-shaped ST-segment elevation of <1 mm.

Type 1 ECG pattern is diagnostic of Brugada syndrome. Type 2 and Type 3 ECG patterns should not be considered diagnostic of Brugada syndrome. It cannot be overemphasized that Brugada syndrome is definitively diagnosed only if the ECG displays Type 1 pattern in more than one precordial lead (V_1–V_3), either spontaneously or provoked after administration of a sodium channel blocking agent, and is associated with one or more of the following clinical characteristics:

- Syncope
- Agonal nocturnal respiration
- Inducible VT/fibrillation (VT/VF) on electrical stimulation
- Family history of sudden cardiac death <45 years
- Type 1 ECG pattern in family member
- Spontaneous VT/VF

A recent study of 186 patients with Brugada syndrome noted that Type 1 ECG pattern in only one right precordial lead (V_1 or V_2) was associated with similar arrhythmic risk as compared with Type 1 ECG pattern in more than one right precordial lead. The authors also found that the V_3 lead in particular did not yield additional diagnostic information. It should be noted that the mere presence of RBBB pattern on an ECG may be associated with Brugada syndrome but not an integral part of the disease. Given the dynamic nature of the ECG observed in patients with Brugada syndrome, the aforementioned patterns (Types 1–3) can be found in the same patient in different settings.

In our patient, the ECG is Type 1 Brugada pattern. However, the absence of any associated clinical characteristics would disqualify it from being named Brugada syndrome. The patient should have clinical follow-up with ECG monitoring, and the aforementioned clinical characteristics should be actively sought. If not found, no further workup is indicated. While controversy does exist, the role of EP study is not proven to be useful in asymptomatic patients with Brugada pattern ECG alone.

THIRD CASE

A 32-year-old Burmese man presented with two episodes of syncope without warning. Clinical and echocardiographic evaluation did not reveal any structural heart disease, and the patient was on no medications.

Question No. 5:

Which of the following genetic mutations is most commonly associated with the ECG shown in Figure 42.5?

A. SCN5A.

B. Mutation of glycerol-3-phosphate dehydrogenase.

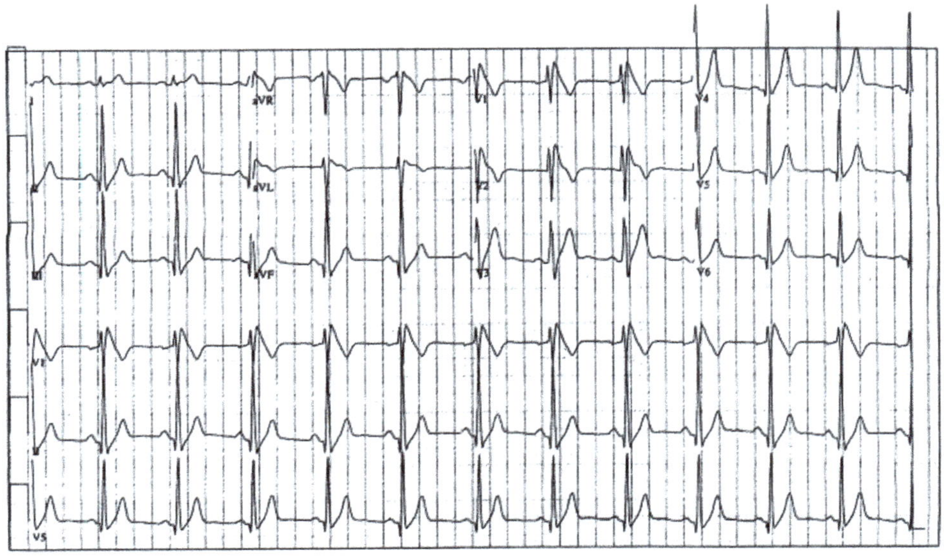

Figure 42.5. ECG at presentation.

C. SCN3B gene.
D. Any of the above.

Discussion

Currently, more than 300 mutations in the sodium channels have been identified in patients with Brugada syndrome. SCN5A mutation is the most common. However, it is to be noted that it is only present in 20% of patients with the syndrome.

Question No. 6:

What is the next step in management of this patient?

A. EP study.
B. ICD implant.
C. Clinical observation.
D. None of the above.

Discussion

In the presence of syncope and typical ECG pattern of Type 1 Brugada pattern, no further testing is warranted. ICD implant is a class I recommendation in this patient. Note that all 3 patients we have presented here were male. The incidence of Brugada syndrome carries a predilection for males (8:1 male to female ratio) and an autosomal dominant inheritance pattern.

CONCLUSIONS

In conclusion, the following clinical pearls regarding Brugada syndrome are important to remember:

- Potential cause of sudden cardiac death
- May present with syncope and sudden cardiac death
- Varied manifestations are observed
- Brugada Type I with syncope = ICD implant
- Brugada Type I with negative EP study = risk of death is small
- Provoked Brugada Type I with positive EP study = risk of death is intermediate
- Familial screening recommended
- 20% of Brugada syndrome associated with SCN5A mutation
- 15 to 20% of Brugada syndrome associated with AFEP study for prognosis controversial
- Asymptomatic Brugada = ECG with clinical follow-up
- Type 2 and Type 3 Brugada pattern ECGs are nondiagnostic; EP study may facilitate diagnosis

Correct Answers:

Q1. A
Q2. A
Q3. B
Q4. B
Q5. A
Q6. B

REFERENCES

1. Probst V, Veltmann C, Eckardt L, et al. Long-term prognosis of patients diagnosed with Brugada syndrome: Results from the FINGER Brugada Syndrome Registry. *Circulation.* 2010;121:635–643.
2. Brugada P, Brugada J. Right bundle branch block, persistent ST segment elevation and sudden cardiac death: a distinct clinical and electrocardiographic syndrome. A multicenter report. *J Am Coll Cardiol.* 1992;20:1392–1396.
3. Fish JM, Antzelevitch C. Role of sodium and calcium channel block in unmasking the Brugada syndrome. *Heart Rhythm.* 2004;1:210–217.
4. Antzelevitch C, Brugada P, Borggrefe M, et al. Brugada syndrome: report of the second consensus conference: endorsed by the Heart Rhythm Society and the European Heart Rhythm Association. *Circulation.* 2005;111:659–670.

CASE STUDY

43

Fragmented QRS

Saurabh Malhotra, MD, MPH and Mithilesh K. Das, MD, MRCP

A 55-year-old man with history of diabetes, hypertension, and hyperlipidemia presented with atypical chest discomfort. He denied any history of chest pain and did not receive any specific therapy for coronary artery disease (CAD) in the past. Examination revealed an obese gentleman with blood pressure of 142/88 mmHg and a heart rate of 75 beats per minute. Physical examination including lung and cardiac exam was unremarkable. He is on aspirin and lisinopril. An ECG was obtained in the emergency department and is shown in Figure 43.1.

Question No. 1:

The ECG in Figure 43.1 shows:

A. Acute myocardial infarction (AMI).

B. Interventricular conduction delay.

C. Fragmented QRS (fQRS).

D. Nonspecific ST-T wave abnormality.

Question No. 2:

Which of the following is associated with the abnormality in the above ECG (see Figure 43.1)?

A. Myocardial scar.

B. Arrhythmogenic right ventricular dysplasia (ARVD).

C. Brugada Syndrome.

D. All of the above.

Question No. 3:

The abnormality in the above ECG (see Figure 43.1) predicts which of the following in patients with ischemic cardiomyopathy?

A. All-cause mortality.

B. Cardiac mortality.

C. Arrhythmic events (ICD therapy).

D. All of the above.

The patient's chest pain was relieved by oral nitroglycerine. Three sets of cardiac markers were within normal limits. The patient underwent myocardial perfusion single photon emission computed tomography (SPECT) that revealed a fixed perfusion defect in the right coronary artery (RCA) territory (Figure 43.2).

A coronary angiogram was later performed that showed single-vessel CAD with an occluded RCA (Figures 43.3 and 43.4).

fQRS is defined as unexpected deviations in the QRS morphology on a 12-lead ECG. fQRS includes various RSR′ morphologies (see Figure 43.5) of the QRS complex with or without a Q wave and is defined by the presence of an additional R wave (R′ or R prime), or notching in the terminal portion of the S wave, notching of R wave or the presence of more than one R′ in 2 contiguous leads corresponding to a major coronary artery territory.[1] It is postulated that regional myocardial scar, which is associated with alterations in QRS morphology,

Practical Electrophysiology © 2015 Jasbir S. Sra, Masood Akhtar, Andrea Natale, David J. Wilber
Cardiotext Publishing, ISBN: 978-1-935395-17-1

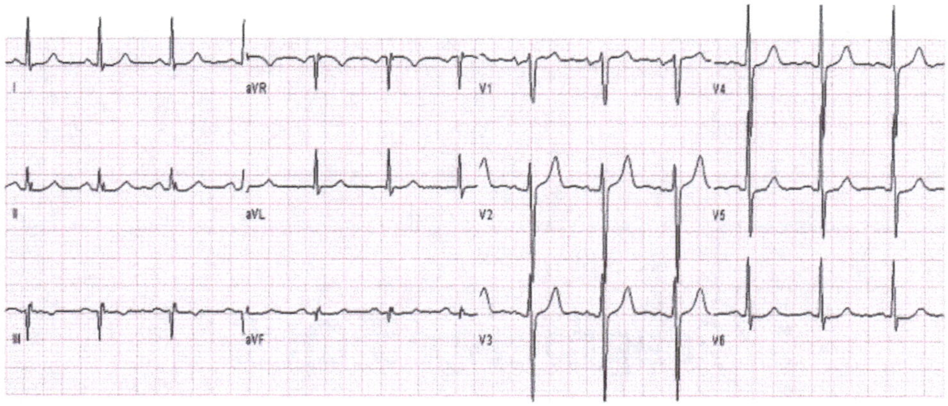

Figure 43.1. ECG on presentation.

causes terminal conduction delay, which is appreciated as fragmentation of QRS complexes.

The definition of fQRS has been extended to include fQRS in the presence of a wide QRS complex >120 ms (f-wQRS), as seen in bundle branch blocks (f-BBB), premature ventricular complexes (f-PVCs), and paced QRS complexes (pf-QRS). In the presence of a BBB, fQRS is defined as >2 notches in the R or S wave (Figure 43.6).

Similarly, pf-QRS complexes and f-PVCs are identified by the presence of >2 peaks or notches present in 2 contiguous leads (Figures 43.7 and 43.8). In addition, fragmented PVC also includes PVCs with only 2 notches in the R wave but that are >40 ms apart and present in 2 contiguous leads.

CLINICAL SIGNIFICANCE OF fQRS

Though the mechanisms of fQRS are speculative, it is thought to be the result of **inhomogeneous electrical activation** of the ventricles in the presence of a myocardial scar. Autopsy studies of individuals with myocardial infarction (MI) and left ventricular (LV) aneurysm have shown the presence of viable myocardium amidst significant myocardial necrosis and fibrosis. These areas of necrosis and fibrosis slow

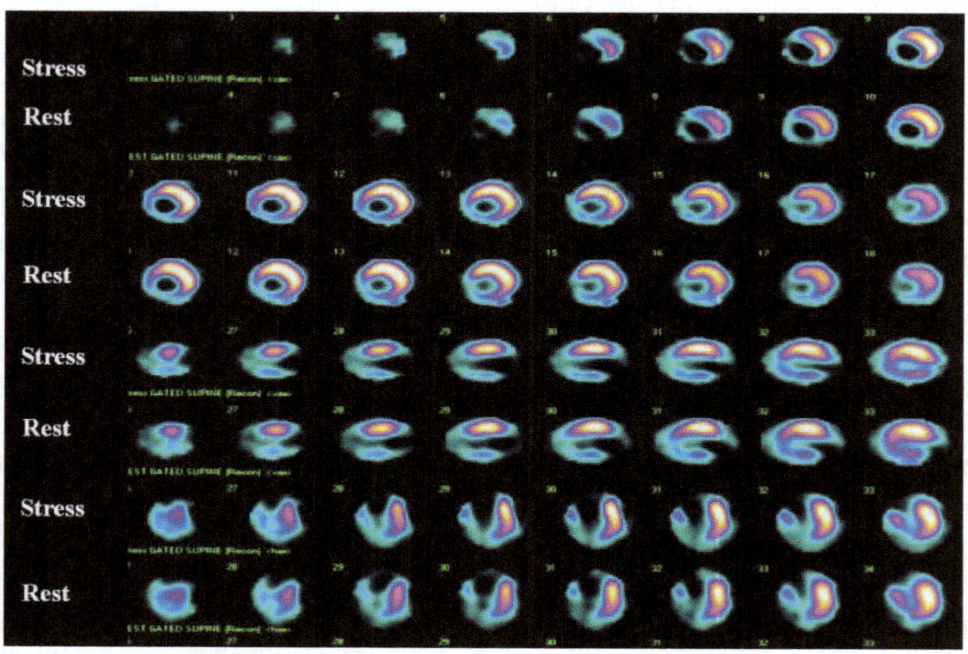

Figure 43.2. Myocardial perfusion SPECT depicting a fixed perfusion defect in the RCA territory.

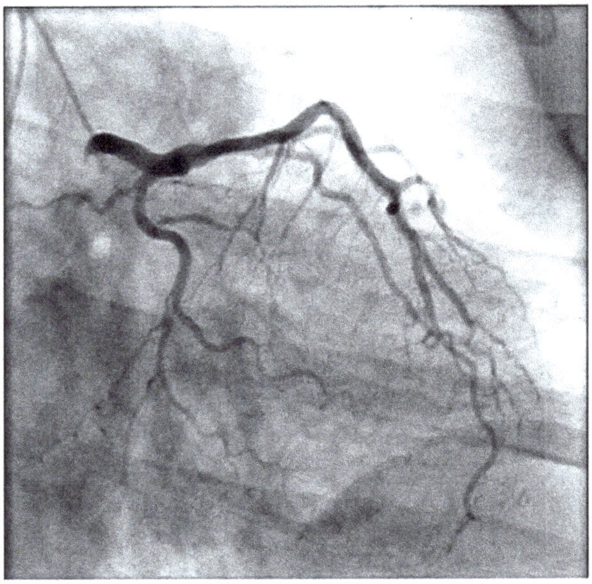

Figure 43.3. Coronary angiogram of the left coronary system showing patent coronary arteries.

activation due to partially depolarized and reduced action potential upstroke velocities resulting in fQRS. It has also been suggested that different RSR′ patterns of an fQRS are caused by differing QRS vectors that occur due to the terminal or midconduction delay of QRS forces arising from the scar area with conduction block or conduction slowing. Thus, fQRS signifies the presence of

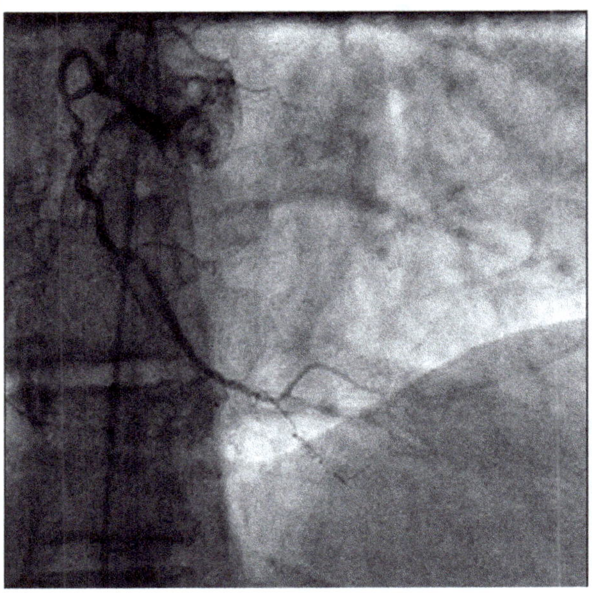

Figure 43.4. Coronary angiogram of the right coronary system showing an occluded left anterior descending (LAD) and left circumflex (LCX) coronary artery.

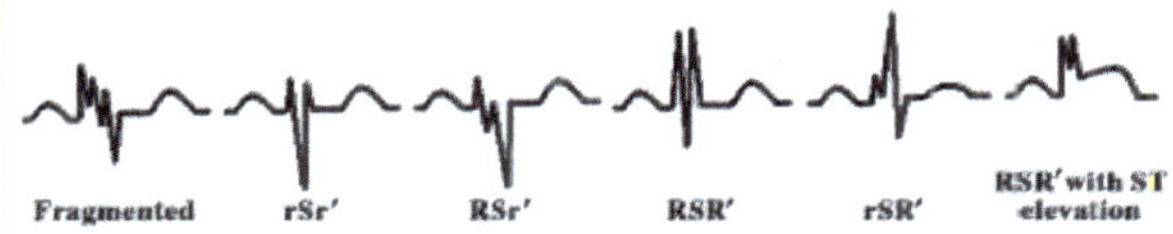

Figure 43.5. Types of fQRS.[2]

myocardial scar, which can be segmental with a subendocardial or transmural extent (as seen in patients with CAD) and follow the distribution of major epicardial coronary arteries, or can be patchy and predominantly occur in perivalvular areas of the ventricle in those with nonischemic dilated cardiomyopathy (NICMP). fQRS, which represents myocardial scarring, is not specific for CAD and is also encountered in patients with NICMP, ARVD, Brugada syndrome, and cardiac sarcoid.

In patients with CAD, fQRS has been shown to develop within 48 hours in over 50% of patients presenting with an AMI and is related to location of myocardial scar as determined by SPECT.[2] Presence of fQRS in ≥2 contiguous anterior leads (V_1–V_5), lateral leads (I, aVL, and V_6), and inferior leads (II, III, and aVF) predict scar in the LAD, LCX, and RCA territories, respectively. When compared to Q waves, fQRS has a greater prevalence among individuals referred for myocardial perfusion SPECT, with fQRS being associated with a larger perfusion defect and a lower left ventricular ejection fraction (LVEF). fQRS has also been shown to have a significantly greater sensitivity and negative predictive value (86% and 93%, respectively) in diagnosing myocardial scar on SPECT when compared to the Q wave (36% and 71%).[2] Similarly, f-wQRS complex (including f-BBB, pf-QRS, and f-PVC) was also found to be highly predictive of myocardial scar in patients with known or suspected CAD (sensitivity = 87% and specificity = 93%).[3] Thus, fQRS (both narrow and wide), which is a moderately sensitive but highly specific sign of myocardial scar in patients with CAD, can be used clinically as a marker of recent and remote MI. This is particularly important in those with no other evidence of recent (ST-T changes) and remote (Q waves) MI on ECG and in those who are unaware or unable to provide a history of a prior ischemic event.

Since fQRS is thought to be a result of myocardial scar tissue, it is reflective of significant myocardial disease (ischemic or nonischemic) and the presence of arrhythmic substrate, which can cause reentrant ventricular tachyarrhythmias resulting in sudden cardiac death (SCD). It has been shown that among individuals with known or suspected CAD, the all-cause mortality was significantly higher among those with a fQRS on

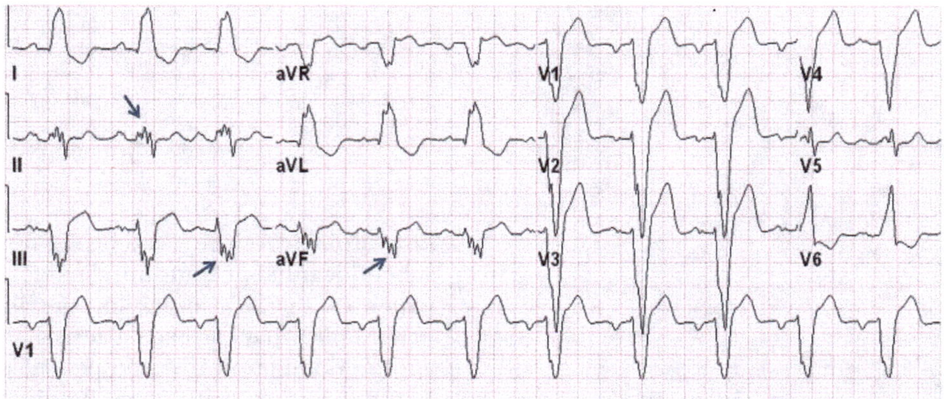

Figure 43.6. fQRS in the presence of a left bundle branch block (LBBB).

ECG (34%) compared to those without (26%).[1] Patients with fQRS also had a significantly greater cardiac event rate (MI, need for revascularization and cardiac death) when compared to those without (50% vs 28%).[1] fQRS was found to be a significant independent predictor of cardiac events but not of all-cause mortality. This is true for f-wQRS as well (when compared to wide non-fQRS), which, in addition, was found to be an independent predictor of mortality.[3] Few studies have assessed the role of fQRS in predicting only arrhythmic events. A study by Das et al revealed that among patients with ischemic (ICMP) or nonischemic cardiomyopathy (NICMP) who had an implantable cardioverter-defibrillator (ICD) implanted for primary and secondary prevention, fQRS was a significant independent predictor of arrhythmic events when compared to non f-QRS morphologies.[4] However, in a prospective study of patients with ICMP or NICMP, with or without ICD for primary prevention, fQRS was not found to be predictive of arrhythmic mortality.[5]

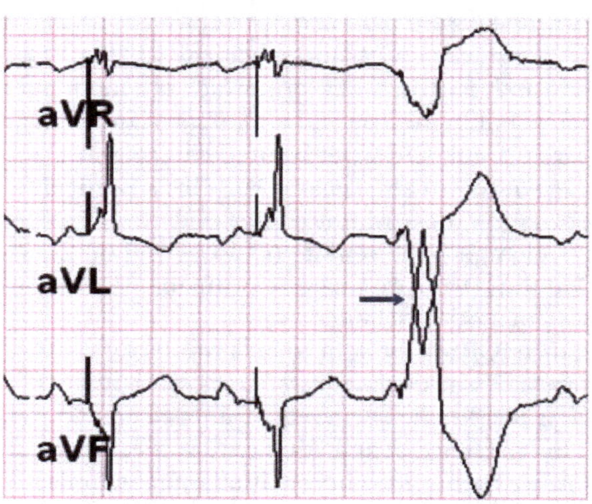

Figure 43.7. fQRS in a premature ventricular complex (PVC).

Thus, fQRS is a marker of poor prognosis among those with myocardial scar. The conflicting evidence regarding fQRS being a predictor of arrhythmic events and mortality in patients with CAD currently limits its utility in further risk stratification of patients who would benefit from an ICD for SCD prevention. **However, in patients with Brugada syndrome and ARVD, in whom fQRS is associated with a higher incidence of ventricular arrhythmic events, it might be prudent to consider ICD in the presence of fQRS on ECG.**

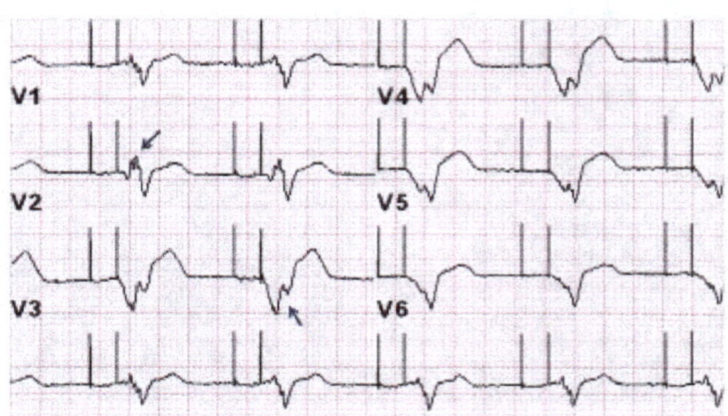

Figure 43.8. fQRS in a paced QRS complex.

Correct Answers:

Q1. C
(Arrows depict fragmented QRS complexes.)

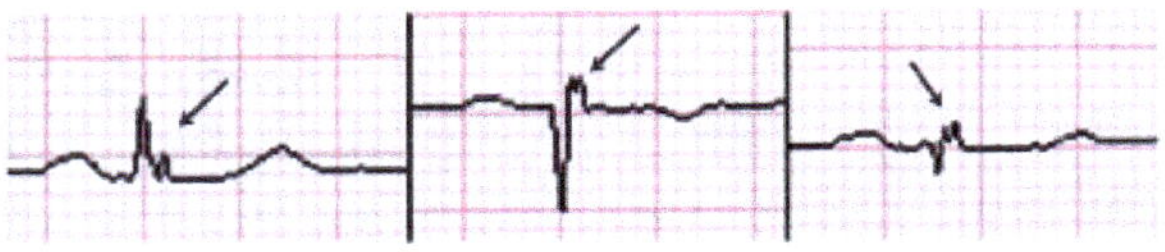

Q2. D

Q3. C

REFERENCES

1. Das MK, Saha C, El Masry H, et al. Fragmented QRS on a 12-lead ECG: a predictor of mortality and cardiac events in patients with coronary artery disease. *Heart Rhythm.* 2007;4:1385–1392.
2. Das MK, Khan B, Jacob S, Kumar A, Mahenthiran J. Significance of a fragmented QRS complex versus a Q wave in patients with coronary artery disease. *Circulation.* 2006;113:2495–2501.
3. Das MK, Suradi H, Maskoun W, et al. Fragmented wide QRS on a 12-lead ECG: a sign of myocardial scar and poor prognosis. *Circ Arrhythm Electrophysiol.* 2008;1:258–268.
4. Das MK, Maskoun W, Shen C, et al. Fragmented QRS on twelve-lead electrocardiogram predicts arrhythmic events in patients with ischemic and nonischemic cardiomyopathy. *Heart Rhythm.* 2010;7:74–80.
5. Cheema A, Khalid A, Wimmer A, et al. Fragmented QRS and mortality risk in patients with left ventricular dysfunction. *Circ Arrhythm Electrophysiol.* 2010;3:339-344.

CASE STUDY

44

An ECG of Early Repolarization

Ashok J. Shah, MD, Méléze Hocini, MD, Lena Rivard, MD, Shinsuke Miyazaki, MD, Amir S. Jadidi, MD, Daniel Scherr, MD, PD, Stephen B. Wilton, MD, MSc, Laurent Roten, MD, Patrizio Pascale, MD, Michala Pedersen, MD, DPhil, Nicolas Derval, MD, Sébastien Knecht, MD, PhD, Frederic Sacher, MD, Pierre Jaïs, MD, and Michel Haïssaguerre, MD

A 39-year-old gentleman presented with 3 noctural episodes of syncope and resuscitated cardiac arrest due to documented ventricular fibrillation (VF). He received multiple external defibrillations for recurrent VF unresponsive to antiarrhythmic drugs early during hospitalization. Coronary angiography and echocardiography were normal. Baseline PR interval and QTc were normal. No electrocardiographic (ECG) signs of Brugada syndrome were present at baseline and post flecainide challenge. Diagnostic electrophysiology (EP) study was normal for baseline intervals without inducible ventricular arrhythmias.

Question No. 1:

The ECG is suggestive of

A. Short QT syndrome.

B. Isoproterenol will be helpful in controlling VF episodes.

C. None of the above.

DISCUSSION

The baseline ECG showed >0.1mV J-point elevation with terminal QRS notch/slur in the inferior leads suggestive of abnormal early repolarization (ER) pattern (Figure 44.2).

Early in-hospital course was turbulent with recurrent uncontrollable VF. A continuous isoproterenol infusion was initiated and found to be effective in controlling recurrent VF. The abnormal ER pattern on the 12-lead ECG subsided.

Several drugs have been tested in VF storm. Isoprenaline has been the most effective in controlling VF storm (Figure 44.3). The infusion of isoproterenol can successfully manage electrical storms (>2 VF episodes per 24 hours), and can sometimes be used as a life-saving therapy. Sedation, parenteral amiodarone, and quinidine have also been effectively used. The arrhythmogenicity of ER syndrome is thought to be related to heterogeneity of action potentials (spike-and-dome morphology in epicardial tissue) across the ventricular wall at the end of phase 1. Any perturbation in the balance of currents may amplify the disparity in voltage gradient and precipitate local reentries and polymorphic ventricular arrhythmias, providing both the trigger and substrate of VF. Disappearance of abnormal ER and normalization of ECG following isoproterenol infusion is thought to be secondary to increase in $I_{Ca\text{-}L}$ current, thereby decreasing electrical gradient and

Practical Electrophysiology © 2015 Jasbir S. Sra, Masood Akhtar, Andrea Natale, David J. Wilber
Cardiotext Publishing, ISBN: 978-1-935395-17-1

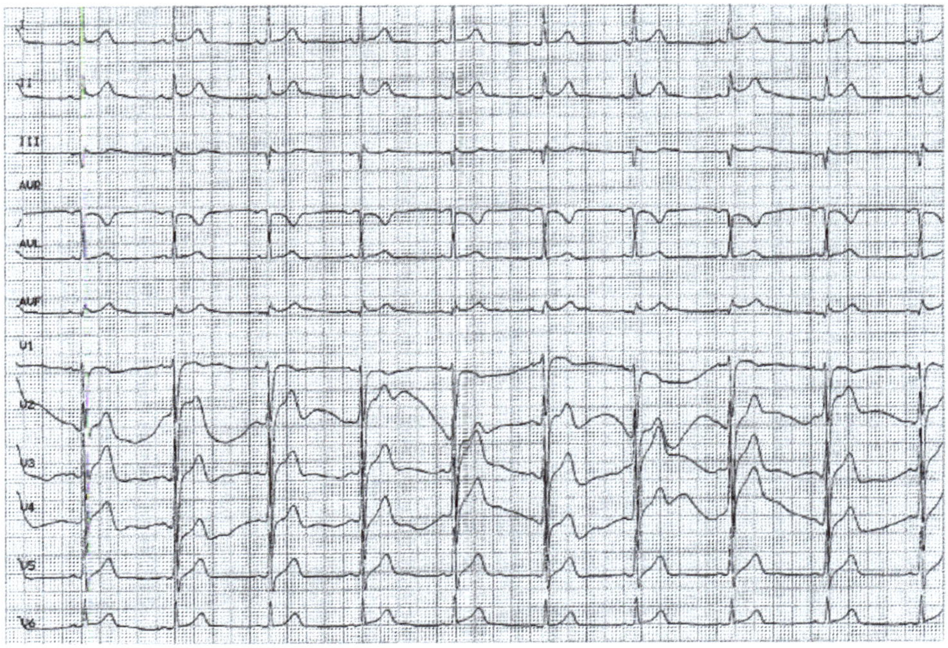

Figure 44.1. Baseline 12-lead ECG.

homogenizing action potential durations and, possibly, increasing the heart rate.[1]

The patient received an implantable defibrillator and was discharged home. Genetic tests were negative for known arrhythmogenic mutations. On follow-up, patient continued to receive appropriate internal defibrillations twice every month for the first year. Arrhythmic recurrences have been observed to be more frequent in patients with ER than in those without such repolarization (41% vs 23%), with a sex-adjusted hazard ratio of 2.1 (95% confidence interval [CI] 1.2–3.5; P=0.008) (Figure 44.4).[2]

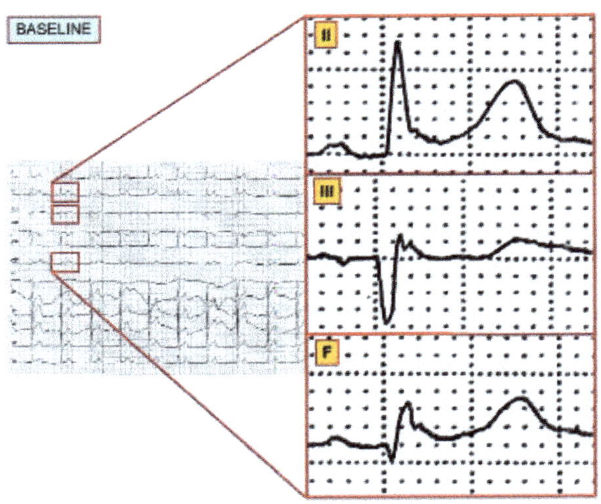

Figure 44.2. Baseline ECG highlighting the leads with slur and notch.

Question No. 2:

What should be done next to prevent frequent shocks?

A. **Start quinidine.**

B. **Ablate premature ventricular contractions (PVCs) originating from Purkinje fibers.**

C. **Start beta-blockers.**

Oral quinidine was started 14 months after the first episode. The abnormal ER pattern on ECG subsided after quinidine therapy was initiated. The patient has not had any episode of VF thereafter. The quinidine therapy in abnormal ER should be guided by the ECG ER pattern and serum levels. Of note, the disappearance or attenuation of ER on ECG is indicative of the therapeutic effect of quinidine. Quinidine inhibits several currents including an effect in blocking Ito current, which is pivotal in endoepicardial electrical gradient and the related arrhythmogenic risk. Therapeutic trough (drawn one hour before the next dose) serum levels are 2 to 5 mg/L when measured after 2 to 3 days of attainment of steady state (Source: 4th edition of

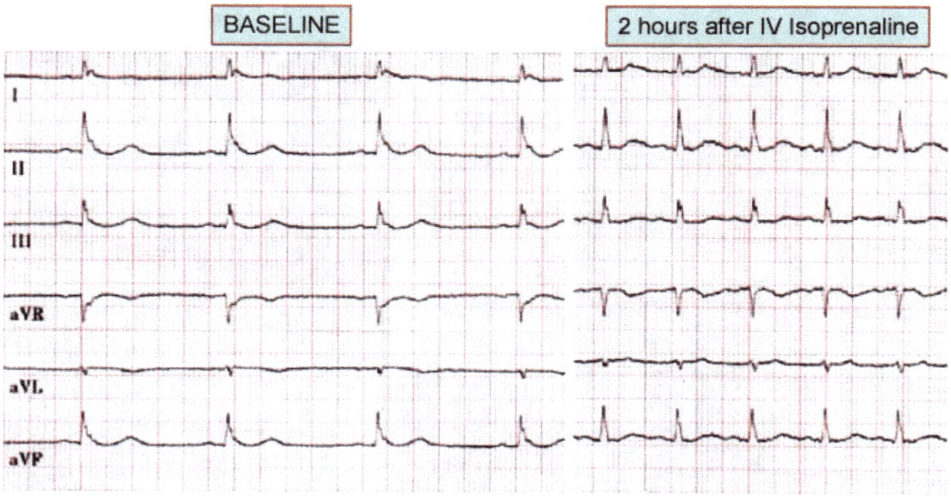

Figure 44.3. ECGs showing effect of isoprenaline in another patient with VF storm.

Basic Skills in Interpreting Laboratory Data by Mary Lee, 2009, Publisher: Special Publishing, Bethesda, MD, USA). While isoproterenol is effective in controlling acute VF storms, orally administered quinidine is effective for prevention of recurrence on a long-term basis (Figures 44.5 and 44.6).[1]

SUMMARY

1. How do you investigate a case of syncope?
2. What are the characteristics of ER associated with sudden unexplained cardiac death?
3. How do you counsel the blood relations of an ER-positive survivor of sudden cardiac arrest?

   A patient presenting with syncope should be attended with detailed clinical history, at the outset, focusing on the precipitating factors, suddenness of the episode, and how consciousness recovered. If the patient cannot narrate these details, a witness should be able to narrate the history or the rescue team members can provide information on the patient's condition on their arrival at the scene with circumstantial evidences. Detailed history should also include the history of episodes of syncope in the past and in family members. Neurocardiogenic syncope is the most common cause of syncope and diagnosed based solely on patient history. Tilt table test is not confirmatory but may help to reproduce the symptoms. Seizure is the most common differential diagnosis and should be ruled out, along with transient ischemic attack before proceeding further. No tests substitute for a detailed history-taking skill.

   Baseline ECG, echocardiography, cardiac enzymes, and serum biochemistry should be performed. Structural heart diseases and pulmonary vascular causes (coronary artery disease, valvular and myocardial diseases, and pulmonary embolism) associated with sudden cardiac death constitute more than 85% of the causes of profound arrhythmogenic syncope.[3] The remainder of cases include primary electrical disorders like, but not limited to, long-short QT and Brugada syndromes, catecholaminergic polymorphic ventricular tachycardia, ER syndrome, and idiopathic VF.
4. The prevalence of ER in the general population varies from 5 to 13%.[4] Currently, we have no definitive guidelines toward the risk stratification of a person with electrocardiographic ER. Based on current

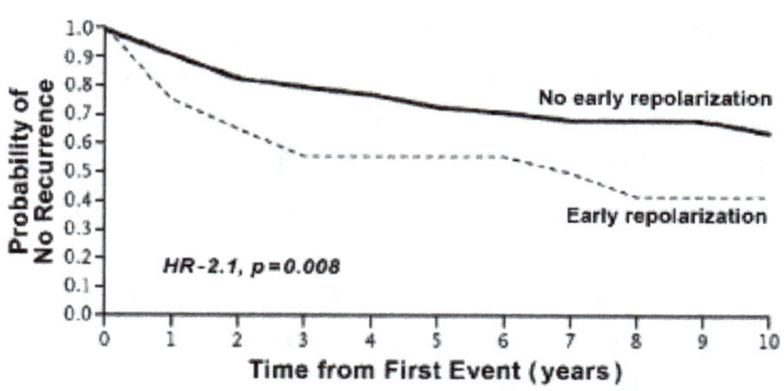

Figure 44.4. Comparison of probability of no arrhythmic recurrences after first event between patients who received ER and patients who did not.

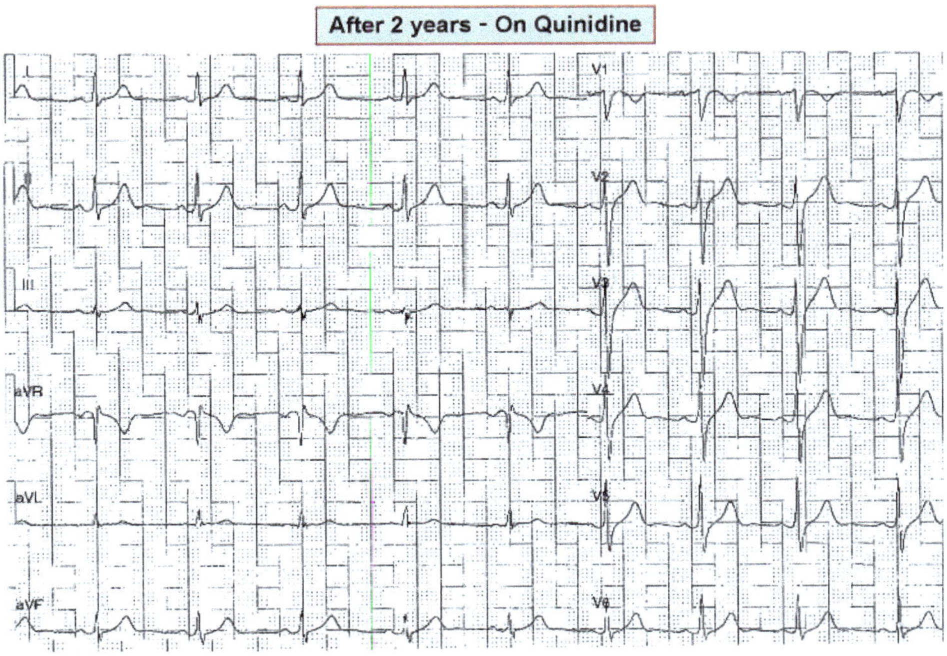

Figure 44.5. ECG of a patient on orally administered quinidine 2 years after acute occurrence of VF storms.

evidence from the studies involving survivors of sudden cardiac arrest and ER, ER appears to be a dynamic entity.[4] Absence of ER on single ECG does not rule out the possibility of abnormal ER syndrome. The amplitude and distribution of ER are often observed to dramatically increase around the period of VF storm. In a longitudinal study involving 10000+ persons followed for a mean 30 years, increased risk of arrhythmic death was observed in those with inferior ER >0.2 mV when compared with ER <0.2 mV.[4]

5. Counseling family of an ER patient with sudden cardiac death: In our cohort of 64 survivors of sudden cardiac arrest and ER, there were 16% with positive family history of

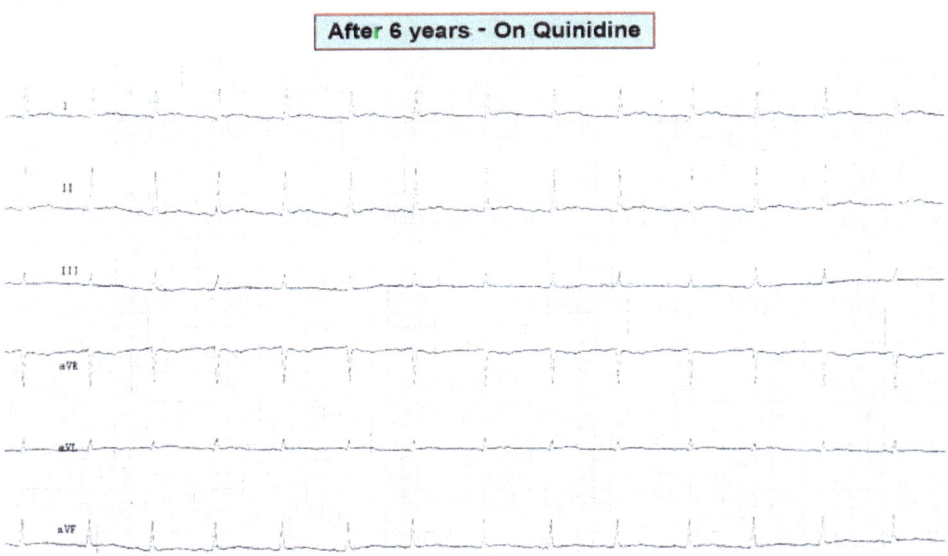

Figure 44.6. ECG of a patient on orally administered quinidine 6 years after acute occurrence of VF storms.

sudden cardiac arrest.[2] In one patient, genomic DNA sequencing of KATP channel genes showed missense variant in exon 3 (NC_000012) of the KCNJ8 gene, a subunit of the KATP channel, conferring predisposition to dramatic repolarization changes and ventricular vulnerability.[5] In our unpublished work, we found an association between ER and SCD at a familial level, demonstrating that abnormal ER could be considered as a malignant pattern in the blood relations of the ER-positive survivor of cardiac arrest. However limited, the evidence underlines the need for systematic familial screening for the identification of ER to limit the risk of sudden death within the family.

Correct Answers:

Q1. B

Q2. A

REFERENCES

1. Haïssaguerre M, Sacher F, Nogami A, et al. Characteristics of recurrent ventricular fibrillation associated with inferolateral early repolarization role of drug therapy. *J Am Coll Cardiol.* 2009;53:612–619.
2. Haïssaguerre M, Derval N, Sacher F, et al. Sudden cardiac arrest associated with early repolarization. *N Engl J Med.* 2008;358:2016–2023.
3. Zipes DP, Wellens HJJ. Sudden cardiac death. *Circulation.* 1998;98:2334–2351.
4. Tikkanen JT, Anttonen O, Junttila MJ, et al. Long-term outcome associated with early repolarization on electrocardiography. *N Engl J Med.* 2009;361:2529–2537.
5. Haïssaguerre M, Chatel S, Sacher F, et al. Ventricular fibrillation with prominent early repolarization associated with a rare variant of KCNJ8/KATP channel. *J Cardiovasc Electrophysiol.* 2009;20:93–98.

CASE STUDY

45

Diagnosing Catecholaminergic Polymorphic Ventricular Tachycardia

Anwer Dhala, MD

Tracings shown in Figure 45.1 (pre-exercise test) and Figure 45.2 (during exercise) were obtained in a 16-year-old female with recurrent syncope. Physical examination, baseline ECG and echocardiogram are within normal limits.

Question No. 1:

The most likely cause of this patient's symptoms is:

- **A.** Recordings show artifact and recurrent nature of symptoms suggests neurally mediated syncope.
- **B.** Preexcited atrial fibrillation (AF).
- **C.** Ventricular tachycardia (VT) secondary to hypertrophic cardiomyopathy (HCM).
- **D.** Ischemia-induced VT.
- **E.** Torsades de pointes secondary to congenital long QT syndrome.
- **F.** Catecholaminergic polymorphic ventricular tachycardia (CPVT).

DISCUSSION

Syncope and arrhythmias during exercise in young patients may be seen in a number of conditions, albeit all of them relatively uncommon.

Neurally mediated syncope is by far the most common cause of syncope in young individuals. Although many individuals claim to have syncope during exercise, with careful questioning it becomes clear that syncope occurs soon after a strenuous workout rather than during the exertion. Exercise-induced hypotension may occur but ventricular arrhythmias are not seen during exercise. Whenever a disorganized wide-complex rhythm is seen, spurious arrhythmia due to artifact should always be considered. This is particularly common when tracings are obtained during activity. In this patient, who is being monitored during a stress test, the hemodynamic consequences of polymorphic ventricular tachycardia (PMVT) will be apparent and, thus, exclude artifact.

Preexcited AF is characterized by an irregularly irregular rhythm and varying QRS morphology depending upon the degree of fusion between normal and accessory pathway conduction. The presence of multiple accessory pathways could result in multiple complex QRS morphologies. An important clue to the diagnosis in the patient presented is the absence of preexcitation in the baseline ECG. Careful review of the initiation of the tachycardia will also be helpful, as the onset of ventricular arrhythmias will often show atrioventricular (AV) dissociation, whereas preexcited tachycardia is likely to start with

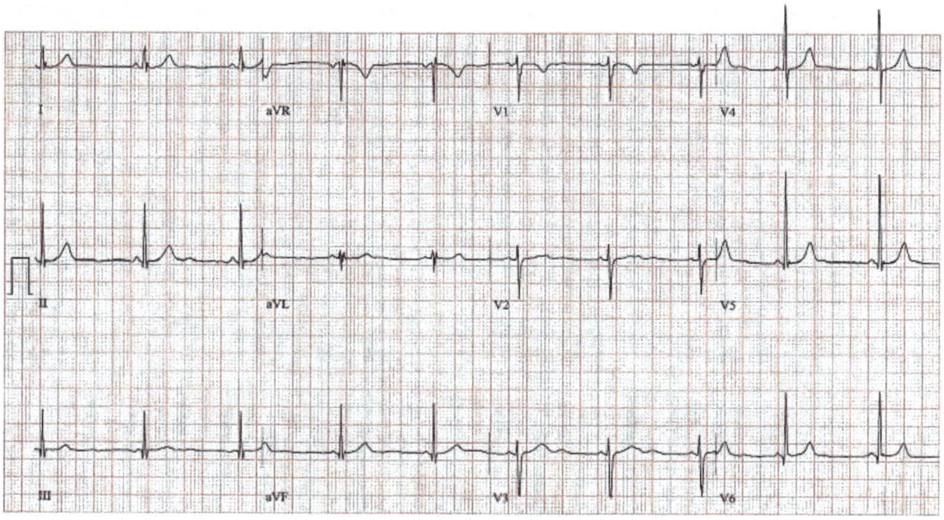

Figure 45.1. Patient's baseline ECG before exercise in supine position.

premature atrial contractions (PACs) or degenerate from an orthodromic tachycardia.

The commonest cause of exercise-induced syncope and sudden death in the young is probably HCM. With HCM, the baseline ECG is almost always abnormal (greater than 90%), although the findings may be nonspecific. The echocardiogram is usually diagnostic with the degree of hypertrophy correlating with risk level for arrhythmia in the majority of patients. Increasingly, magnetic resonance imaging (MRI) is used to quantify the extent of myocardial scarring and stratify risk. In this patient, normal ECG and echocardiogram results virtually rule out HCM.

Ischemia-induced VT is seen in children only rarely. Although coronary artery anomalies may be noted in up to 0.6% of patients undergoing angiography, ischemia as a result of the anomaly is less frequent. The most severe of these anomalies is an aberrant origin of the left main coronary artery from the pulmonary artery (ALCAPA). In this condition, blood flow follows the pressure gradient

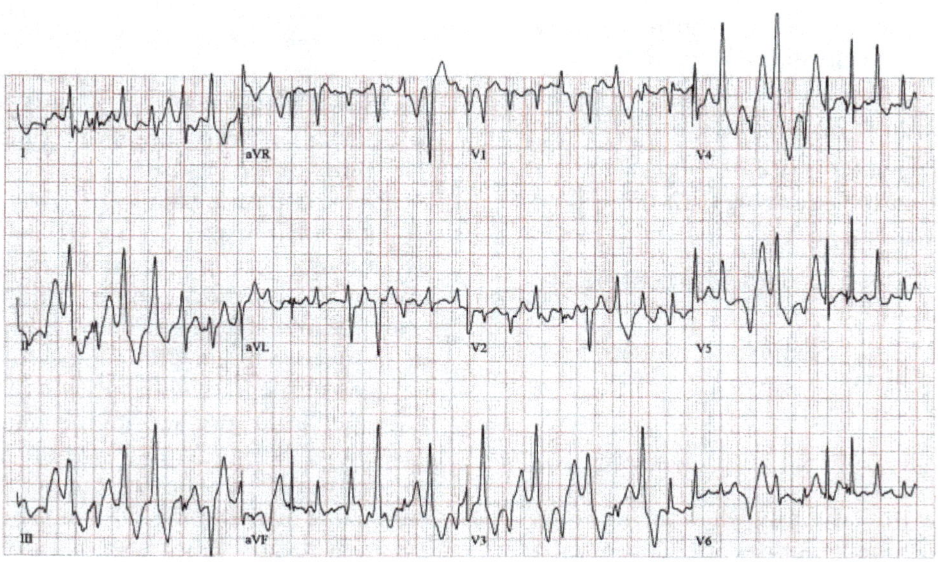

Figure 45.2. Patient's ECG during exercise test.

with the right coronary artery filling the left system via collaterals, which then drains into the low-pressure pulmonary artery. The majority of these patients are symptomatic and do not survive without correction in early childhood. Some patients with optimal collateralization survive to adulthood, albeit with left ventricular (LV) infarction and scarring, and are prone to ventricular arrhythmias.

When a coronary artery arises from the opposite sinus of Valsalva, ischemia may result from one of several mechanisms including narrowing of the ostium, intramural course of the proximal course of the artery, or compression of the artery when it courses between the aorta and the pulmonary outflow tract. Exercise-induced ischemic VT may therefore occur. A history of exertional chest pain may be present and ischemia may be demonstrable during stress testing. Echocardiography is an excellent tool for diagnosing these conditions, but the sonographer has to be made aware that this is clinically suspect, as generally little emphasis is placed on scanning the origins of the coronary arteries in institutions dealing primarily with adults. In Kawasaki's disease, the incidence of coronary artery involvement consisting of aneurysmal dilatation and focal stenosis was once reported as 15 to 30% but, due to current treatment with aspirin and gamma globulin, has decreased to less than 5%. Again, echocardiography is an excellent tool for diagnosis.

Congenital long QT syndrome is uncommon but not rare, with an incidence of 1:5000 of the population. Typical features include prolongation of the QT interval and runs of PMVT that may result in syncope or sudden death. Although the QT interval may be within the normal range in up to a third of individuals with genetically proven disease, the QT interval would be expected to be prolonged just prior to the onset of ventricular arrhythmias, which is not the case in the patient described here.

The most likely cause of PMVT in the absence of structural heart disease with a normal baseline ECG in young patients is CPVT. Patients with CPVT typically develop VT during exercise or emotional stress. Age at presentation is usually in the first or second decade of life. In the majority of patients, an autosomal dominant form of inheritance is noted due to a mutation in the ryanodine receptor (RyR2). A recessive variant due to mutation of the calsequestrin 2 gene (CASQ2) has also been described. In the presence of sympathetic stimulation, calcium overload occurs in the endoplasmic reticulum with store-induced calcium release. The baseline ECG in CPVT is normal as the ion channels involved are intracellular rather than on the cell membrane. However, sinus bradycardia and prominent U waves may occasionally be seen. During exercise, with increasing heart rates, the frequency and complexity of ventricular arrhythmias also increases. The most typical tachycardia is the so-called bidirectional VT with a 180-degree QRS morphology shift from beat to beat. Rapid supraventricular arrhythmias are also not uncommon. Confirmation of this diagnosis can currently be made by genetic testing, which can identify RyR2 mutations in about 70% and was identified in this patient.

Correct Answer:

Q1. F

CASE STUDY

46

A Case of Arrhythmogenic Right Ventricular Dysplasia

Hugh Calkins, MD

Patient J.D. is a 34-year-old man who presented with syncope. His ECG in the emergency department (ED) is shown in Figure 46.1. He reports that he had just finished running 5 miles and was in his cool-down phase when he abruptly lost consciousness. He scraped his knee and hit his head when he fell. He reports about 2 seconds of lightheadedness before passing out. When he awoke, he felt well other than his head and his knee. His physical examination in the ED was normal. He was not orthostatic. Rare premature ventricular contractions (PVCs) were observed on the monitor, which demonstrated a left bundle branch block (LBBB) superior axis. The patient takes no medications. His past medical history is unremarkable. He has always been an avid runner and was a 3-season varsity athlete in high school. He also ran in college on a track scholarship. Since

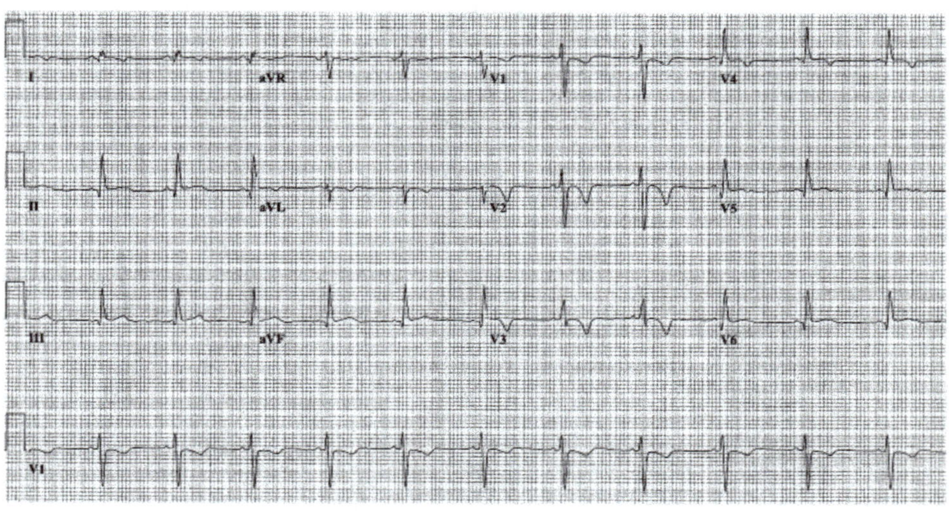

Figure 46.1. Patient's ECG at presentation.

Practical Electrophysiology © 2015 Jasbir S. Sra, Masood Akhtar, Andrea Natale, David J. Wilber
Cardiotext Publishing, ISBN: 978-1-935395-17-1

college, he has run 5 marathons. There is no family history of cardiac disease or sudden cardiac death (SCD). An echocardiogram obtained in the ED revealed a dilated right ventricle with reduced function, and normal left ventricular (LV) size and function.

Questions

1. Would you admit the patient or discharge him to home?
2. What is the most likely diagnosis?
3. What further testing would you advise?
4. Assuming your suspected diagnosis is confirmed, what treatment would you advise?

DISCUSSION

Arrhythmogenic right ventricular dysplasia (ARVD) is an inherited cardiomyopathy characterized by ventricular arrhythmias and structural abnormalities of the right ventricle. ARVD results from progressive replacement of right ventricular (RV) myocardium with fatty and fibrous tissue. The precise prevalence of ARVD in the United States has been estimated to be 1 in 5000 of the general population. Patients with ARVD typically present because of ventricular arrhythmias, syncope, or sudden death.

ARVD typically presents in young adults about 30 years of age. Diagnosis of ARVD prior to puberty is exceedingly uncommon. It is also uncommon for ARVD to initially present after the age of 50.

Diagnosis of ARVD

The diagnosis of ARVD is established based on the criteria developed by an expert writing group led by Dr. Frank Marcus. Specific cardiac tests that are recommended in all patients suspected of having ARVD include an ECG; a signal-averaged ECG (SAECG), a Holter monitor and an echocardiogram. Analysis of RV size and function can also be obtained by cardiac magnetic resonance (CMR) and/or computed tomography (CT). If the results of the noninvasive tests point toward a diagnosis of ARVD, invasive testing including RV angiography, endomyocardial biopsy, and electrophysiology testing are recommended to establish the diagnosis and help provide further information to guide treatment recommendations.

Electrocardiographic Evaluation

ECG abnormalities are detected in more than **90% of ARVD patients**. However, it is important to be aware that ARVD rarely can present in a patient with a totally normal ECG. The typical ECG features of ARVD are demonstrated by the ECG obtained from our patient. A diagnosing physician can appreciate the **T-wave inversion (TWI) in the right precordial leads from V_1 to V_4**. An epsilon wave is not present. This is not surprising as epsilon waves are observed in **less than 5%** of patients and usually in patients with advanced disease. **TWI in leads V_1 to V_3 is the most consistent ECG finding seen in ARVD patients.** The juvenile pattern of TWI in leads V_1 to V_5 or beyond is a normal variant in children less than 12 years of age and present in 1% to 3% of the healthy population 19 to 45 years of age, but it is present in 87% of patients with ARVD. TWI beyond V_1 in a patient in the above age group who has no apparent heart disease but does have ventricular arrhythmias of LBBB morphology should raise the suspicion of ARVD. Late potentials on SAECG recordings are the counterpart of the epsilon waves and are considered a minor criterion for the diagnosis of ARVD.

Imaging of the Right Ventricle

RV size and function can be evaluated using a variety of imaging modalities including echocardiography, angiography, CMR, and CT. RV angiography has historically been regarded as the best imaging test for the diagnosis of ARVD and has been shown to be highly specific (90%). Compared with RV angiography, echocardiography is noninvasive, and represents the first-line imaging approach in evaluating patients with suspected ARVD or in screening family members. Magnetic resonance (MR) is an attractive imaging method because it is noninvasive and has the ability to differentiate fat from muscle. Gadolinium-enhanced MR imaging can localize fibrosis in the RV myocardium. MR also allows for a highly accurate and quantitative evaluation of RV size and function. Regional wall motion abnormalities limited to the RV have been shown to have 75% sensitivity and 97% specificity for ARVD.

Although MR imaging is well suited for the evaluation of patients with suspected ARVD, it is also important not to "overinterpret" minor abnormalities detected on MR imaging. It is our experience that CMR is a common cause of over-diagnosis of ARVD. Physicians should be hesitant to diagnose ARVD when the only anatomic abnormalities are those seen on CMR. In our experience, an abnormality detected with MR imaging will virtually never be the sole abnormality detected in a patient with ARVD. It is also important to recognize that the presence of "fat" in the RV myocardium may be normal, and also that subtle wall motion abnormalities as well as thinning of the RV wall are common in normal populations, and that these findings should not be considered diagnostic of ARVD. Although CT imaging can also be used to image the right ventricle, there is limited experience with this imaging modality as compared with MR.

Myocardial Biopsy

Performance of an endomyocardial biopsy is recommended for patients in whom an alternative diagnosis, particularly when sarcoidosis is suspected. For typical cases of ARVD, such as this patient, biopsy is rarely performed. This reflects the low diagnostic yield of the test, as most biopsies are obtained from the septum and few cardiologists are comfortable biopsying the free wall of the RV.

Differential Diagnosis

Although it is not difficult to diagnose an overt case of ARVD, differentiation of ARVD at its early stages from idiopathic ventricular tachycardia (IVT), a usually benign and nonfamilial arrhythmic condition that also presents as VT with LBBB morphology, remains a clinical challenge. It is extremely important to determine if the patient has IVT or ARVD for 2 reasons. Firstly, IVT is a benign condition that is not associated with a risk of sudden death and, therefore, is not an indication for placement of an implantable cardioverter-defibrillator (ICD). In contrast, when the diagnosis of ARVD is established, there needs to be serious consideration for ICD implantation because of the risk of SCD. Secondly, ARVD is frequently hereditary, whereas IVT is not. As a result, **when a diagnosis of ARVD is made, screening of all first-degree family members is recommended.** As noted above, another condition commonly confused with ARVD is sarcoidosis. However, in our experience the results of MR imaging are fairly distinct. As such, we rely on MR imaging to help us determine when to suspect sarcoidosis and a cardiac biopsy.

Histology

The pathology of ARVD is characterized by **loss of myocytes and later replacement with fatty and/or fibrotic tissue,** mainly in the right ventricle. These changes are most prominent in the so-called "triangle of dysplasia:" the inflow, outflow, and apical areas. Patchy acute myocarditis with myocyte death and inflammatory infiltrates is also present in nearly two-thirds of the cases. As the disease progresses, myocytes are replaced by fatty and/or fibrous tissue, and changes extend to the posterolateral wall of the left ventricle and sometimes the interventricular septum.

Genetics

ARVD is a heritable condition, typically with an **autosomal dominant** pattern of inheritance and variable penetrance. There is a familial pattern of disease in approximately one-third of probands. **To date, 12 different genetic variants of ARVD have been identified.** Many of these variants involve a polymorphism in a gene encoding a desmosomal protein, which are part of the intercalated disc and are responsible for providing mechanical coupling and transmission of contractile force between cardiac myocytes. The desmosome consists of 3 parts: the transmembrane adhesion molecules, which provide the bridge between the 2 myocytes; the intracellular linker proteins, which link to the cytoskeleton; and, finally, the cytoskeletal proteins. Desmoglein and desmocollin, which are the genes responsible for ARVD 11 and ARVD 12, respectively, are the transmembrane adhesion molecules. Plakoglobin, responsible for ARVD 9, plakophilin, responsible for ARVD 10, and desmoplakin, linked to ARVD 8, are the intracellular linker proteins. Desmin, responsible for ARVD 7, is an intermediate filament that is part of the cytoskeleton. In our experience, a pathogenic mutation is identified in slightly more than half of patients who meet Task Force criteria for ARVD. The most common mutation is PKP2.

Implantable Defibrillators and Other Therapeutic Measures

Once the diagnosis of ARVD has been established, the main decision in regard to management concerns whether to implant an ICD for prevention of sudden death. ARVD patients who are at the highest risk for arrhythmic death include those patients with a history of having been resuscitated from SCD, patients with syncope and those with evidence of active disease, especially the presence of nonsustained VT or a high number of PVCs (>1000) on a Holter monitor. We also have found the results of electrophysiologic testing to be useful for risk stratification. In the past, the extent of RV disease was considered to be a marker for sudden death risk, but our most recent experience suggests that the risks of a malignant arrhythmia is high even in patients with mild RV disease, especially if they present with syncope, as our patient did, and/or have nonsustained VT or a high density of PVCs. Risk stratification based on any of the above characteristics singularly or in combination is statistically documented. However, the application of these risk factors to the decision to implant an ICD can only be used as a rough guide. The risk of SCD is low if these risk factors are not present; however, it is not zero.

Based on these findings, we concluded that patients who meet Task Force criteria for ARVD are at high risk for SCD and should undergo ICD placement for primary and secondary prevention, regardless of electrophysiologic testing results. However, this guideline applies mainly to probands. We know that family members are at reduced sudden death risk and are often diagnosed earlier in the course of the disease. For these reasons, we may not elect to implant an ICD in family members but rather follow them closely. In addition to placement of an ICD, we also recommend that patients with ARVD avoid competitive athletics. Additionally, we advise that patients with definite or probable ARVD avoid activities such as long-distance biking or running and long-distance swimming and/or weight training. We recommend that activity be limited to low-intensity activities such as walking or golf.

There is little data available concerning the use of pharmacological agents in the treatment of patients with ARVD for the prevention of SCD. Symptomatic ventricular arrhythmias are treated initially with beta-blocker therapy and/or sotalol. If this is inadequate to control a patient's symptoms or to prevent recurrent VT, membrane-active antiarrhythmic agents such as sotalol and, if necessary, amiodarone are considered. Further research is needed to determine if other commonly used treatments for heart failure such as angiotensin-converting enzyme (ACE) inhibitors are of value for treatment of patients with ARVD.

The progressive, widespread pathology of ARVD suggests that catheter ablation would not be a long-term curative procedure; most of the small clinical trials conducted thus far support this hypothesis. The use of substrate-based ablation and epicardial ablation has improved the outcomes of VT ablation during the past 5 years. We consider catheter ablation to be a valuable treatment approach for ARVD patients experiencing frequent recurrences of VT despite antiarrhythmic drug therapy. Many patients elect to proceed with catheter ablation to avoid taking amiodarone. However, it must be recognized that ablation is largely a palliative procedure.

Cardiac transplantation is considered in patients with progressive heart failure and intractable recurrent ventricular arrhythmias. In our experience, few patients with ARVD require cardiac transplantation.

CONCLUSION

ARVD is a genetic cardiomyopathy that leads to RV failure, arrhythmias, and sudden death. Thus far, 12 genes have been linked to ARVD. Most of these involve the cardiac desmosome. Diagnosis is based on finding a combination of characteristic abnormalities on cardiac imaging and electrocardiographic evaluation, as well as family history. Therapeutic options remain limited, though ICD implantation can prevent SCD.

Correct Answers:

Q1. This patient should be admitted for inpatient evaluation because he is at increased risk for sudden death as evidenced by an abnormal ECG, syncope with little warning and injury, and PVCs.

Q2. The most likely diagnosis is ARVD. Differential diagnosis would include sarcoidosis or other types of cardiomyopathy.

Q3. I would advise a SAECG, a Holter, CMR imaging, and an electrophysiologic study. In this patient each of these tests confirmed the diagnosis of ARVD. His EP study revealed easily inducible VT with multiple LBBB morphologies. Genetic testing was also ordered and confirmed a PKP2 mutation.

Q4. I would implant an ICD. I would also advise he stop running. I would start a beta-blocker and ACE inhibitor.

REFERENCES

1. Marcus FI, McKenna WJ, Sherrill D, et al. Diagnosis of arrhythmogenic right ventricular cardiomyopathy/dysplasia: proposed modification of the task force criteria. *Circulation.* 2010;121:1533–1541.
2. Dalal D, Nasir K, Bomma C, et al. Arrhythmogenic right ventricular dysplasia: a United States experience. *Circulation.* 2005;112:3823–3832.
3. Nasir K, Bomma C, Tandri H, et al. Electrocardiographic features of arrhythmogenic right ventricular dysplasia/cardiomyopathy according to disease severity: a need to broaden diagnostic criteria. *Circulation.* 2004;110:1527–1534.

4. Tandri H, Saranathan M, Rodriguez ER, et al. Noninvasive detection of myocardial fibrosis in arrhythmogenic right ventricular cardiomyopathy using delayed-enhancement magnetic resonance imaging. *J Am Coll Cardiol.* 2005;45:98–103.
5. Roguin A, Bomma CS, Nasir K, et al. Implantable cardioverter-defibrillators in patients with arrhythmogenic right ventricular dysplasia/cardiomyopathy. *J Am Coll Cardiol.* 2004;43:1843–1852.
6. Dalal D, Jain R, Tandri H, et al. Long-term efficacy of catheter ablation of ventricular tachycardia in patients with arrhythmogenic right ventricular dysplasia/cardiomyopathy. *J Am Coll Cardiol.* 2007;50:432–440.

CASE STUDY

47

Multimodality Approach to Diagnosing Arrhythmogenic Right Ventricular Dysplasia

Timothy E. Paterick, MD, JD, MBA and Matt M. Umland, RDCS

A 25-year-old male runs on the cross-country team for his college. He has had a long history of palpitation and rapid heart beating. He experienced a recent episode of complete syncope preceded by a rapid heart beat.

Question No. 1:

The ECG in Figure 47.1 reveals:

A. Inferior ischemia.
B. Intraventricular conduction delay.
C. Evidence of a genetic cardiomyopathy.
D. Wolff-Parkinson-White syndrome.
E. Long QT syndrome.

DISCUSSION/FOLLOW-UP QUESTIONS

Epsilon waves are caused by postexcitation of myocytes in the right ventricle (RV). Young patients with ventricular tachycardia (VT) or syncope and epsilon waves on electrocardiography (ECG) usually have arrhythmogenic right ventricular dysplasia (ARVD). In this condition myocytes are replaced by fat, producing islands of myocytes surrounded by variable fat. This causes the delay in excitation of some of the myocytes of the RV and causes the little "wiggles" seen during the ST segment of the ECG.

Epsilon waves have also been described in patients with posterior infarction, sickle cell disease with RV hypertrophy due to pulmonary hypertension, RV infarction, sarcoidosis, and infiltrative disease of the RV.

Question No. 2:

Which of the following is/are true?

A. The epsilon wave on the ECG is diagnostic of ARVD.
B. The epsilon wave occurs in 100% of cases of ARVD.
C. There are T-wave inversions in the V_1 to V_3 leads in 75% of cases of ARVD.
D. The ECG is frequently normal despite evidence of ARVD.
E. A and D are correct.
F. None are correct.

Diagnosis of ARVD involves the presence of various combinations of electrocardiographic (Table 47.1), structural and functional abnormalities, and a family

Practical Electrophysiology © 2015 Jasbir S. S-a, Masood Akhtar, Andrea Natale, David J. Wilber
Cardiotext Publishing, ISBN: 978-1-935395-17-1

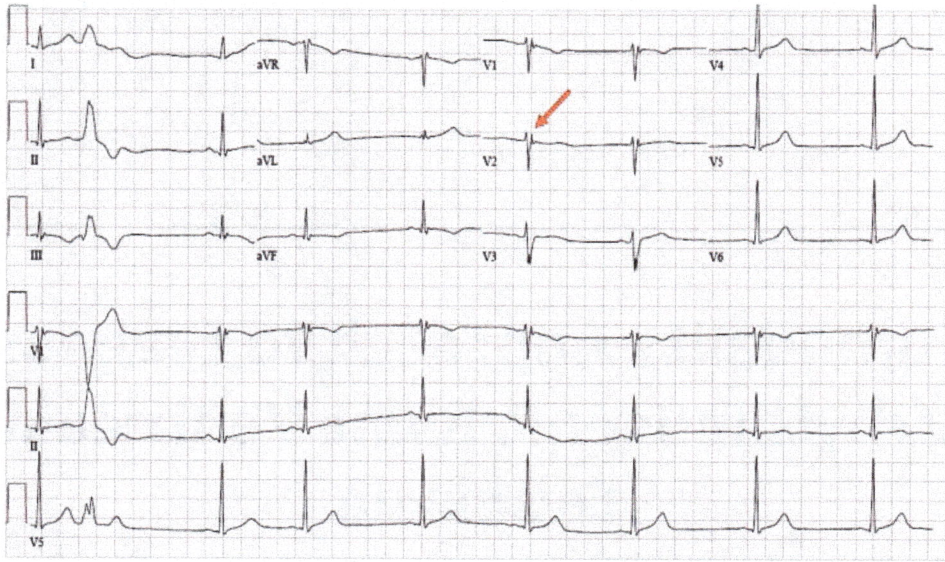

Figure 47.1a. Patient's ECG at baseline. See zoomed image for further detail.

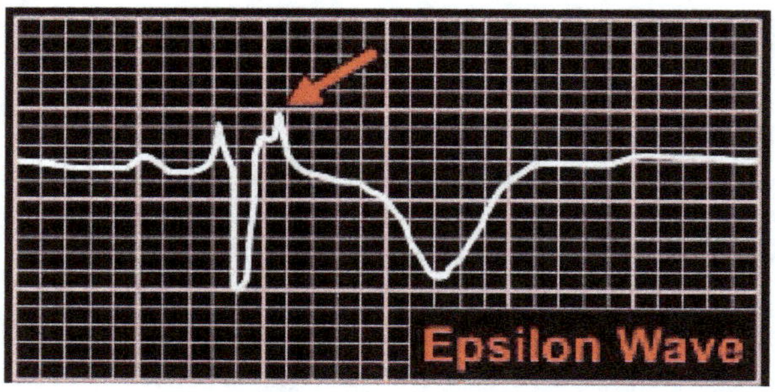

Figure 47.1b. **(Zoomed in).** Patient's ECG displaying an epsilon wave (red arrow).

history of the disease. Epsilon waves in leads V_1 to V_3 are a hallmark of the disease and occur in 30% of patients; 50% of patients have T-wave inversion in right precordial leads. (See also case 46.)

Question No. 3:

Which of the following is/are true?

A. Incidence of the disease is approximately 1:3000 to 1:10,000.

B. ARVD is a progressive disease.

C. ARVD is felt to be mostly an autosomal dominant disease.

D. Naxos disease is a unique form of ARVD and is autosomal recessive.

E. ARVD frequently manifests in young males.

F. All are true.

G. Three are true.

ARVD is characterized by fatty replacement and fibrosis of the heart. Most commonly, the right ventricle apex (RVA) and outflow tract are involved. However, the left ventricle (LV) can be affected too. As a result of the fatty replacement and fibrosis, ventricular arrhythmias are common in this disease and can lead to palpitations, syncope, and sudden death. At more advanced ages, right ventricular pump failure can occur.

ARVD is a progressive disease. The incidence is estimated to be 1:3000 to 1:10,000. Manifestations are usually seen in teenagers. Although the diagnosis is more often made in athletes, sports are not thought to have a causative relationship with the

Table 47.1. Diagnostic Criteria That Can be Diagnosed on Electrocardiography (ECG)

1. **(major)** Epsilon wave or localized prolongation (>110 ms) of the QRS complex in right precordial leads (V_1 to V_3)
2. **(minor)** Inverted T waves in right precordial leads (V_2 and V_3) (people aged more than 12 years); in absence of right bundle branch block (RBBB)
3. **(minor)** Late potentials (signal-averaged ECG)
4. **(minor)** Left bundle branch block (LBBB)-type VT (sustained and nonsustained) (ECG, Holter, and exercise testing)
5. **(minor)** Frequent ventricular extrasystoles (more than 1000/24 hours) (Holter)

disease. ARVD can occur in families; 12 different chromosomal defects have been described, most often with autosomal dominant inheritance. One unique form of ARVD called Naxos disease (after the Greek island where it was first diagnosed) has an autosomal recessive pattern of inheritance.

ARVD is a difficult diagnosis to make. ECG, echocardiography, and magnetic resonance imaging (MRI) can be used to validate the diagnosis.

Question No. 4:

Which of the following are diagnostic criteria for ARVD?

A. Dilatation of the RV with impaired diastolic and systolic performance.

B. Segmental dilation of the RV.

C. Fibrofatty replacement of the RV myocardium.

D. Diagnosis confirmed at surgery or necropsy.

E. All of the above.

See Table 47.2 for the major diagnostic criteria for ARVD. Figure 47.2 demonstrates evidence of ARVD on autopsy.

Question No. 5:

Which of the following are characteristic of echo findings in ARVD?

A. Thin RV wall.

B. RV dilatation with increase in diameter and length.

C. Depressed RV radial contractility.

D. Depressed RV annular motion (TAPSE).

E. Abnormal myocardial mechanics.

F. Abnormal RV systolic tissue velocity.

G. All of the above are possible findings.

Two-Dimensional Echocardiography (Figure 47.3)

Echocardiographic characteristics of ARVD include: severe dilatation and reduction of RV ejection fraction with no (or only mild) left ventricular impairment; localized RV aneurysms (akinetic or dyskinetic areas with diastolic bulging); and severe segmental dilatation of the RV. Another relevant abnormality to note is the combination of regional RV akinesia or dyskinesia, *and* one of the following (at end diastole):

- parasternal long-axis view at the level of RV outflow tract ≥29 to <32 mm (corrected for body size [PLAX/BSA] ≥16 to <19 mm/m^2);

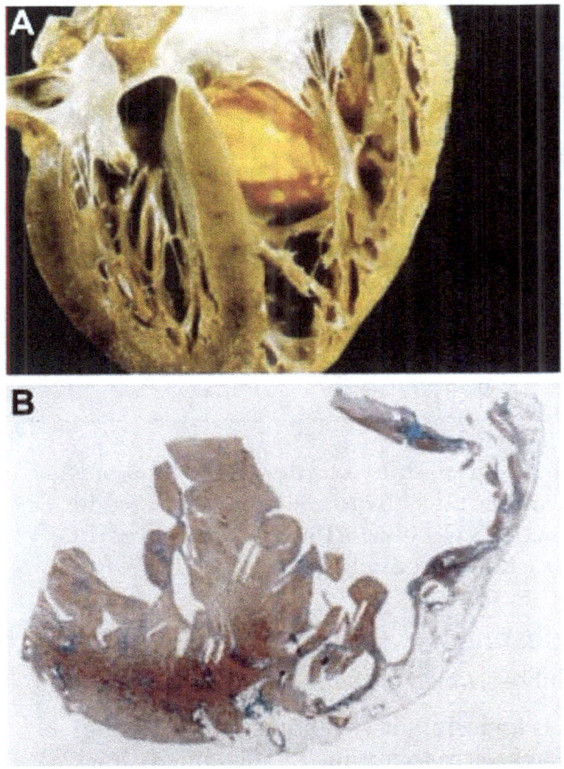

Figure 47.2. A: Autopsy specimen of the heart of an ARVD patient. Note the transmural fatty replacement of the right ventricular free wall. B: Myocardial atrophy is confined to the RV and substantially spares the interventricular septum as well as the left ventricular free wall. *(Images courtesy of William Edwards, Mayo Clinic Department of Pathology, Rochester, MN.)*

Table 47.2. Major Diagnostic Criteria for ARVD

1. Familial disease confirmed at necroscopy or surgery
2. Severe dilatation and reduction of RV ejection fraction with no (or only mild) left ventricular impairment
3. Localized RV aneurysms (akinetic or dyskinetic areas with diastolic bulging)
4. Severe segmental dilatation of the RV
5. Fibrofatty replacement of myocardium on endomyocardial biopsy

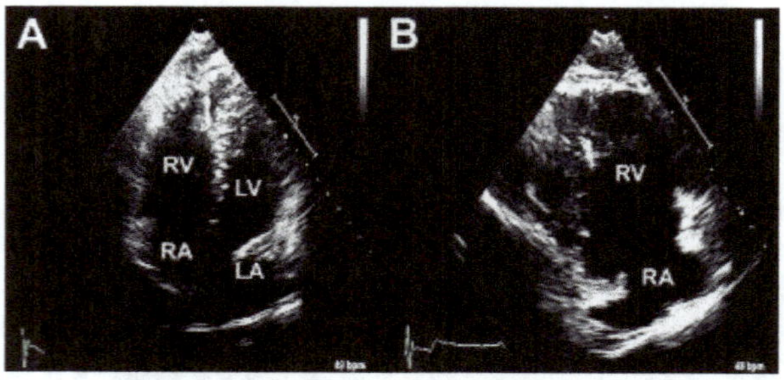

Figure 47.3. **A:** Apical four-chamber view demonstrating right ventricular (RV) dilation. **B:** RV inflow view demonstrating RV dilation and thin RV wall at the apex. LA, left atrium; LV, left ventricle; RA, right atrium.

- parasternal short-axis view at the level of RV outflow tract ≥32 to <36 mm (corrected for body size [PSAX/BSA] ≥18 to <21 mm/m^2); or
- fractional area change >33% to ≤40%.

Question No. 6:

MRI has which of the following roles in ARVD?

A. Is important in a multimodality imaging approach to making the diagnosis.

B. Is the diagnostic modality with the highest sensitivity and specificity.

C. Allows for assessment of fibrofatty infiltration of the RV myocardium.

D. Allows use of delayed gadolinium enhancement to assess for myocardial fibrosis.

E. Determines degree of myocardial fibrosis, which holds important prognostic information.

F. All of the above.

Magnetic Resonance Imaging (Figure 47.4)

Diagnosing ARVD using MRI requires the following: regional RV akinesia or dyskinesia or dyssynchronous RV contraction, *and* one of the following:

- Ratio of RV end-diastolic volume to body surface area (BSA) ≥100 to <110 mL/m^2 (male) or ≥90 to <100 mL/m^2 (female); or
- RV ejection fraction between 40% and 45%.

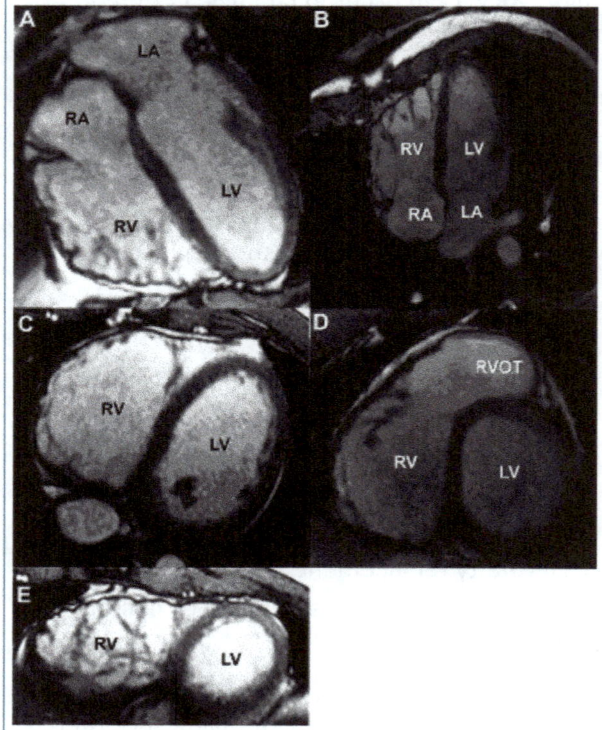

Figure 47.4. **A:** Cardiac MRI four-chamber view (apex down) demonstrating dilated RV with increased trabeculation and thinning of the RV free wall. **B:** Cardiac MRI four-chamber view (apex up) demonstrating RV dilation with thin RV wall and increased apical trabeculation. **C:** Cardiac MRI emphasizing RV dilation and paper-thin RV wall. **D:** Cardiac MRI demonstrating dilated RV and thin RV wall extending into a dilated RVOT with thin wall. **E:** Cardiac MRI coronal section emphasizing the thin RV wall and marked increase in RV trabeculations. LA, left atrium; LV, left ventricle; RA, right atrium; RVOT, right ventricular outflow tract.

DISCUSSION

Multimodality Approach to ARVD with ECG, Echocardiography, and MRI

ARVD is a genotypic disease that leads to a phenotypic expression of a cardiomyopathy with abnormal cardiac morphology and frequently abnormal diastolic function that exposes the affected individuals to risk for sudden cardiac death under conditions of extreme physical exertion. Progress has been made in the diagnosis, genetic basis, and pathophysiology of ARVD, but there persists a gap in the prevention or reversal of the disease. Endurance training can provoke or accelerate the development of ARVD.

ARVD, also called arrhythmogenic right ventricular cardiomyopathy, is characterized principally by fibro fatty replacement of the RV myocardium (the so-called triangle of dysplasia: apex, inflow tract, and outflow tract) producing myocardial thinning, RV dilation, and sometimes, tissue replacement involving areas of the LV with relative sparing of the septum.

The prevalence of ARVD in the general population is 1:1000 and is the leading cause (22%) of sudden cardiac death on the athletic field in Italy.

Presentation, physical examination, and genetics of ARVD are variable. Diagnosis of ARVD involves the presence of various combinations of electrocardiographic, structural, and functional abnormalities, and a family history of the disease. Epsilon waves in leads V_1 to V_3 are a hallmark of the disease and occur in 30% of patients; 50% of patients have T-wave inversion in right precordial leads.

Correct Answers:

Q1. C

Q2. E

Q3. F

Q4. E

Q5. G

Q6. F

REFERENCES

1. Peters S, Trümmel M. Diagnosis of arrhythmogenic right ventricular dysplasia-cardiomyopathy: value of standard ECG revisited. *Ann Noninvasive Electrocardiol.* 2003;8:238–245.
2. Marcus FI, McKenna WJ, Sherrill D, et al. Diagnosis of arrhythmogenic right ventricular cardiomyopathy/dysplasia: proposed modification of the task force criteria. *Circulation.* 2010;121:1533–1541.
3. Corrado D, Basso C, Pavei A, et al. Trends in sudden cardiovascular death in young competitive athletes after implementation of a preparticipation screening program. *JAMA.* 2006;296:1593–1601.
4. Fabritz L, Hoogendijk MG, Scicluna BP, et al. Load-reducing therapy prevents development of arrhythmogenic right ventricular cardiomyopathy in plakoglobin-deficient mice. *J Am Coll Cardiol.* 2011;57:740–750.
5. Marcus FI, Fontaine GH, Guiraudon G, et al. Right ventricular dysplasia: a report of 24 adult cases. *Circulation.* 1982;65:384–398.
6. Gemayel C, Pelliccia A, Thompson PD. Arrhythmogenic right ventricular cardiomyopathy. *J Am Coll Cardiol.* 2001;38:1773–1781.
7. Maron BJ, Shirani J, Poliac LC, et al. Sudden death in young competitive athletes. Clinical, demographic, and pathological profiles. *JAMA.* 1996;276:199–204.
8. Hermida JS, Minassian A, Jarry G, et al. Familial incidence of late ventricular potentials and electrocardiographic abnormalities in arrhythmogenic right ventricular dysplasia. *Am J Cardiol.* 1997;79:1375–1380.
9. Rampazzo A, Nava A, Malacrida S, et al. Mutation in human desmoplakin domain binding to plakoglobin causes a dominant form of arrhythmogenic right ventricular cardiomyopathy. *Am J Hum Genet.* 2002;71:1200–1206.
10. Yang Z, Bowles, NE, Scherer SE, et al. Desmosomal dysfunction due to mutations in desmoplakin causes arrhythmogenic right ventricular dysplasia/cardiomyopathy. *Circ Res.* 2006;99:646–655.
11. Yu CC, Yu CH, Hsueh CH, et al. Arrhythmogenic right ventricular dysplasia: clinical characteristics and identification of novel desmosome gene mutations. *J Formos Med Assoc.* 2008;107:548–558.
12. Douglas PS, O'Toole ML, Hiller WD, Reichek N. Different effects of prolonged exercise on the right and left ventricles. *J Am Coll Cardiol.* 1990;15:64–69.

CASE STUDY 48

Anomalous Coronary Artery and Ventricular Arrhythmias

Peter C. Frommelt, MD, Babak Haddadian, MD, and Tanvir Bajwa, MD

A 13-year-old male suddenly collapsed while running quarter-mile sprints in gym class at school and rapidly became pulseless and apneic. Cardiopulmonary resuscitation (CPR) was initiated at the scene and 911 was called. Emergency medical personnel arrived and identified ventricular fibrillation (VF) (Figure 48.1), which responded to epinephrine, atropine, and external defibrillation with return of spontaneous circulation and sinus rhythm. He was intubated in the field and transported to a local emergency department for stabilization.

Question No. 1:

Besides hypertrophic cardiomyopathy, what is the most common congenital cardiac anomaly that is associated with sudden cardiac death (SCD) in children and young adults?

- A. Aortic valve stenosis.
- B. Coarctation of the aorta.
- C. Congenital coronary artery abnormalities.
- D. Ventricular septal defect with pulmonary vascular disease.

The patient was started on dopamine for blood pressure support, and his first 12-lead ECG demonstrated probable sinus tachycardia with wide-complex QRS diagnosed as right bundle branch block (RBBB) (Figure 48.2).

He was transported to a children's medical center for further evaluation and, during transport, developed nonsustained ventricular tachycardia (VT) initially responsive to a lidocaine infusion. Transthoracic echocardiography showed significant global left ventricular (LV) hypokinesis (ejection fraction 30%) and an anomalous left coronary artery. From a short-axis view, two-dimensional imaging identified anomalous aortic origin of the left coronary artery from the right sinus of Valsalva (Figure 48.3) with an interarterial course between the aorta and the pulmonary artery (PA). The initial course of the anomalous coronary was intramural within the anterior aortic wall.

The intramural anomalous coronary course is frequently best appreciated by color Doppler and was clearly seen in this patient using this technique (Figure 48.4). A linear diastolic blue color signal identifies the intramural left coronary with flow coursing from the right sinus toward the left sinus of Valsalva (arrows) before exiting the aortic wall in the

Figure 48.1. ECG demonstrating patient's VF.

Cardiotext Publishing, ISBN: 978-1-935395-17-1

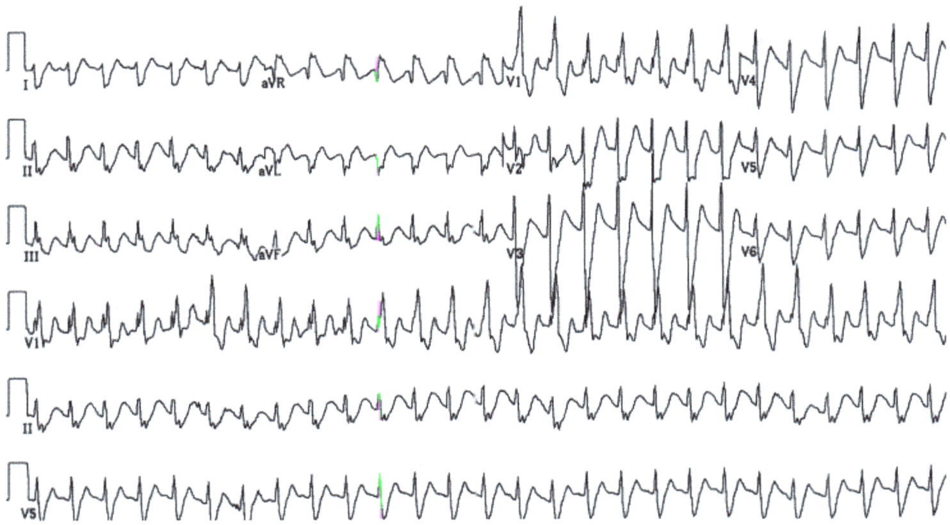

Figure 48.2. Probable sinus tachycardia with wide-complex QRS.

left sinus and giving rise to the circumflex and left anterior descending (LAD) branches.

The patient remained hemodynamically unstable over the initial 12 hours of hospitalization with persistent ventricular ectopy and nonsustained VT (Figure 48.5) that was poorly controlled with medical management and eventually deteriorated to sustained VT necessitating extracorporeal membrane oxygenation (ECMO) support.

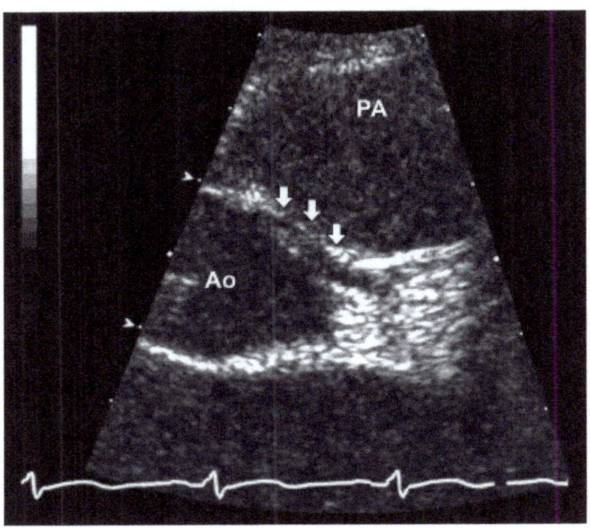

Figure 48.3. Short-axis view of anomalous aortic origin of the left coronary artery from the right sinus of Valsalva with an interarterial course between the aorta (Ao) and the PA. Initial course of the anomalous coronary is intramural within the anterior aortic wall (arrows).

After stabilization on ECMO support, he underwent successful surgical unroofing of the anomalous coronary to create a neo-orifice in the appropriate sinus, and was successfully weaned from ECMO at the completion of the cardiac surgery. His postoperative course was remarkable for resolution of the ventricular ectopy without the need for chronic antiarrhythmic therapy, progressive improvement in LV function with chronic LV regional dyskinesis, and persistent hypoxic-ischemic brain injury secondary to the aborted sudden death.

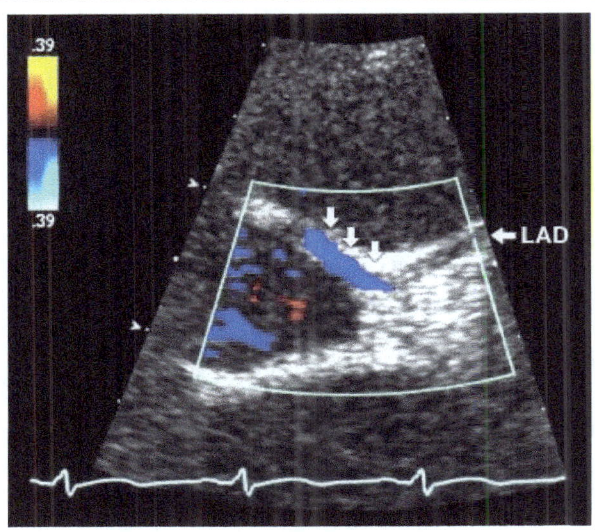

Figure 48.4. Color Doppler showing intramural anomalous coronary course. See text for details.

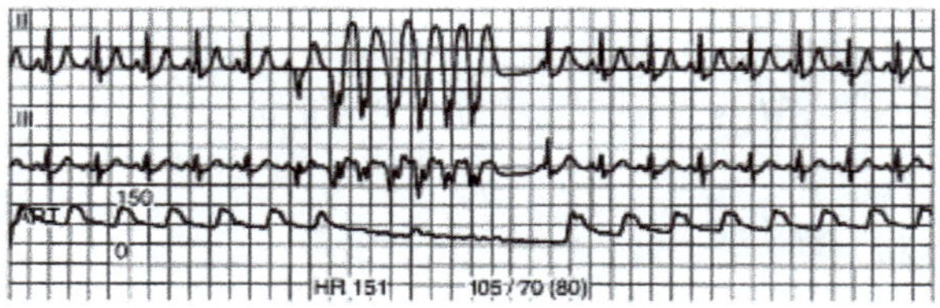

Figure 48.5. ECG showing nonsustained VT.

DISCUSSION

Congenital coronary artery anomalies (CCAA) as a cause of SCD has been well-documented, but the prevalence of CCAA in the general population and the risk associated with various anomalies remain unclear. Isolated CCAA have been described in approximately 1 to 5% of patients undergoing coronary angiography and approximately 0.3% of patients at autopsy. The prevalence of lesions felt to be most commonly associated with SCD, anomalous origin of the left coronary artery from the right sinus of Valsalva and anomalous origin of the right coronary artery from the left sinus of Valsalva with an interarterial course, were recently described in a prospective continuous angiographic series as 0.15% and 0.92%, respectively, suggesting a total prevalence of these potentially lethal anomalies of more than 1% in the general population.

Of all the CCAA, anomalous aortic origin of a coronary artery (AAOCA) from the opposite sinus has been most frequently associated with myocardial ischemia, ventricular arrhythmias, and sudden death, particularly when the anomalous coronary courses between the great arteries. AAOCA can occur from the opposite sinus of Valsalva (either the right coronary arising the left sinus or the left coronary arising the right sinus of Valsalva) but is not associated with myocardial ischemia unless the anomalous coronary courses in between the great arteries.

AAOCA of the left coronary artery from the right sinus of Valsalva with the anomalous coronary coursing between the great arteries is frequently associated with SCD and usually occurs in children and adolescents during vigorous exercise. In patients with this anomaly who had exercise-induced SCD, 50% are asymptomatic without previous complaints of chest pain, palpitations, syncope, or an identified arrhythmia. Sudden death in older patients is much less common, and cardiac death in the older group was generally associated with atherosclerotic disease. The lower risk of sudden death in older patients with this anomaly is likely related to the fact that they rarely participate in high-intensity competitive sports. AAOCA of the right coronary artery from the left sinus of Valsalva appears to be more common and has also been associated with SCD, although less commonly than AAOCA of the left coronary from the right sinus.

The mechanisms that lead to myocardial ischemia in the patient with AAOCA from the opposite sinus with an interarterial course between the pulmonary and aortic roots are unclear, but several theories have been proposed. The ostium of the anomalously arising coronary artery is frequently slit-like, likely compromising flow reserve. In addition, the anomalous coronary artery usually arises at an acute angle from the aorta, rather than perpendicularly, and this may alter flow patterns into that coronary artery bed. Finally, it has been hypothesized that the interarterial course places the anomalous coronary at risk of compression between the great arteries. This seems unlikely given the low pressure in the PA in normal individuals, even during exercise. The ischemia is more likely due to deformation of the anomalous coronary within the aortic wall during periods of arterial hypertension with exercise.

Echocardiography has been effectively utilized to identify coronary artery abnormalities, but this is challenging as a screening technique since CCAA identification usually requires sophisticated echo equipment and a laboratory examination with targeted views, higher-frequency transducers with harmonic capabilities, and high-quality color Doppler mapping. Other imaging modalities such as transesophageal echo, computed tomography (CT) imaging, coronary angiography, and magnetic resonance imaging (MRI) have been used as effective tools for delineation of coronary artery anatomy. The resting ECG is almost always normal in the face of CCAA except with anomalous origin of the left coronary artery from the PA (with resultant chronic ischemia), and so provides little benefit in coronary anomaly screening. Unfortunately, exercise treadmill testing has also been negative in the vast majority of patients identified prospectively as having a coronary artery abnormality, even in those who had symptoms of cardiac ischemia.

Multiple surgical techniques for CCAA have been developed, including patch coronary angioplasty of the anomalous origin, coronary bypass, and re-implantation of the anomalous coronary. All of these procedures carry substantial risk of later coronary compromise/bypass graft failure and have generally been reserved for individuals with documented myocardial ischemia. More recently, unroofing of the anomalous coronary has become the surgical intervention of choice in those with AAOCA from the opposite sinus with an interarterial course between the great arteries where the anomalous coronary runs intramural within the aortic wall. This procedure creates a large neo-orifice in the appropriate sinus and appears ideally suited for the patient with an intramural course of the anomalous coronary. Early results appear promising.

Correct Answer:

Q1. C

REFERENCE

1. Frommelt MA, Frommelt PC. Vascular abnormalities: Arterovenous fistula, pulmonary AV fistula, coronary artery abnormalities (anomalous origin of coronary artery, coronary fistula, congenital aneurysms, acquired coronary abnormalities – Kawasaki disease). In: Eidem BW, Cetta F, O'Leary PW, eds. *Echocardiography in Pediatric and Adult Congenital Heart Disease.* Philadelphia: Lippincott, Williams, and Wilkins, 2010.

CASE STUDY

49

Risk Stratification of Sudden Death in Hypertrophic Cardiomyopathy

M. Fuad Jan, MBBS, MD and A. Jamil Tajik, MD

Patient 1

A 59-year-old man with a known history of nonobstructive hypertrophic cardiomyopathy (HCM) (Figures 49.1, 49.2, and 49.3) (diagnosed about 26 years ago), maintained on verapamil 120 mg 3 times a day and quinidine 324 mg 3 times a day, sought attention after experiencing a syncopal episode while running. A 24-hour ambulatory ECG monitor revealed several runs of nonsustained ventricular tachycardia (NSVT) (Figure 49.4). Cardiac magnetic resonance (CMR) imaging revealed extensive areas of nodular delayed hyperenhancement (DHE) (Figure 49.5), and a sleep study showed evidence of severe obstructive sleep apnea (OSA)–hypopnea syndrome.

Question No. 1:

Which of the following is regarded as a high-risk marker for sudden death (SD) in a patient with HCM?

A. Age > 55 years.
B. Left ventricular outflow tract (LVOT) gradient >60 mmHg.
C. Atrial fibrillation (AF).
D. Syncope.

Question No. 2:

Based on your answer to the previous question, what would you recommend?

A. Implantation of an automatic implantable cardioverter-defibrillator (ICD).
B. Addition of a beta-blocker to therapy.
C. Discontinue verapamil and start disopyramide.
D. Exercise treadmill stress test.

Patient 2

A 36-year-old man was diagnosed with obstructive HCM after a transthoracic echocardiogram for evaluation of chest pain and dyspnea on exertion (Figures 49.6 and 49.7). Appropriate blood pressure response was noted on a treadmill exercise stress test and a 24-hour Holter monitor did not reveal any episodes of NSVT. He denied any family history of HCM or SD and any syncope. CMR obtained did not reveal any DHE (Figures 49.8 and 49.9).

Question No. 3:

Which of the following should be utilized in the risk stratification process for recommending prophylactic use of an ICD for primary prevention of SD in this patient?

Practical Electrophysiology © 2015 Jasbir S. Sra, Masood Akhtar, Andrea Natale, David J. Wilber
Cardiotext Publishing, ISBN: 978-1-935395-17-1

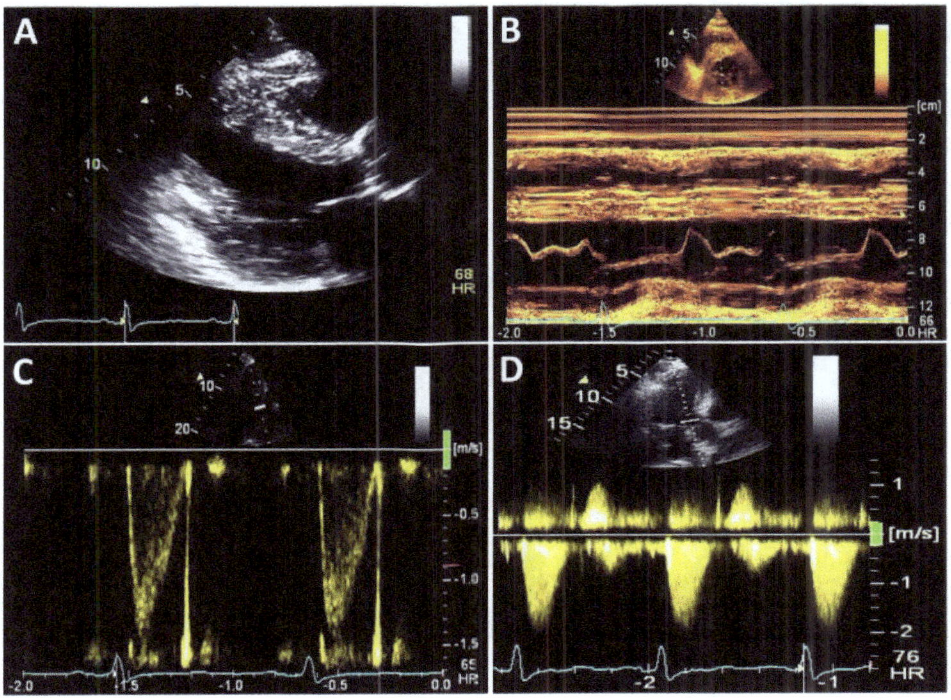

Figure 49.1. A: Parasternal long-axis view of the left ventricle (LV) in diastole in this patient with HCM. Note the asymmetric septal hypertrophy (=17 mm) and increased posterior wall thickness (=12 mm) with normal LV cavity size. B: M-mode echocardiogram recorded in this patient with HCM demonstrates the asymmetric septal hypertrophy; no systolic anterior motion (SAM) of the mitral valve is seen. C: Pulsed-wave Doppler tracing obtained with the cursor at the entrance of the LVOT and continuous-wave (CW) Doppler tracing through the LVOT entrance (D) after provocation with amyl nitrite in the same patient demonstrates the absence of LVOT gradient. This patient is classified as having nonobstructive HCM with no gradients either at rest or with provocation.

A. Electrophysiologic testing (programmed ventricular stimulation).

B. LV wall thickness.

C. Genetic testing for sarcomere gene mutations.

D. Coronary angiography.

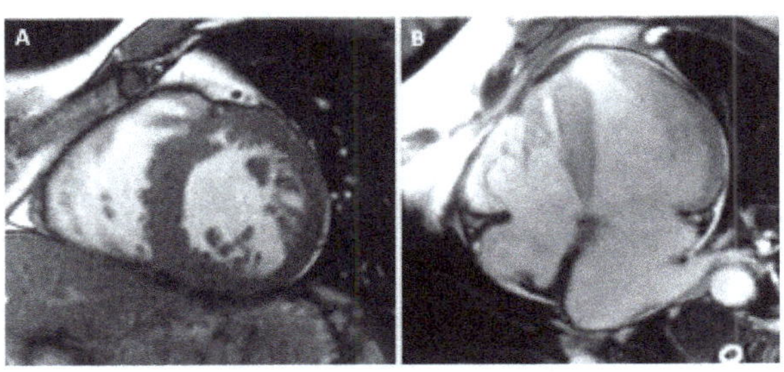

Figure 49.2. Steady-state free precession (bright blood) CMR images from a cine loop oriented in short axis (A) and long axis (B) in this patient with HCM. These images demonstrate the abnormal wall thickening of the septum at end diastole.

Question No. 4:

Based on the information provided and the answer to the previous question, what would you recommend?

A. Implantation of an automatic ICD.

B. Permanent pacemaker for alleviation of LVOT gradient.

C. Begin patient on a therapeutic trial of beta-blockade.

D. Invasive hemodynamic data assessment in the catheterization laboratory.

DISCUSSION

HCM is the most common inherited heart disease **(autosomal dominant with incomplete penetrance)** that primarily arises from mutations in the genes encoding the **cardiac sarcomere myofilament proteins** – "sarcomeric HCM" (Figure 49.10). However, additional subgroups (nonsarcomeric proteins), tentatively classified as "Z-disc HCM"

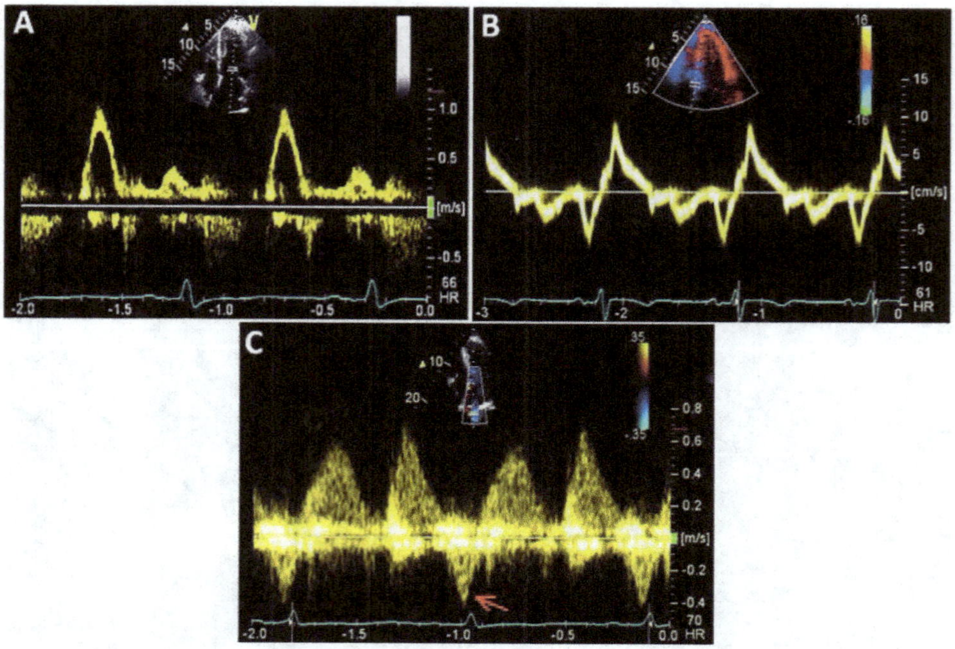

Figure 49.3. **A and B:** Mitral inflow and tissue Doppler recording of the medial mitral annular velocity in this patient with HCM. The early diastolic mitral annular velocity (E′) is markedly decreased, averaging 4.4 cm/s, indicating diastolic dysfunction. Also note the ratio of transmitral E velocity to E′ velocity, the E/E′ ratio, of around 17, indicating elevated LV filling pressures. The pulmonary vein Doppler tracing (C) also demonstrates increased "a" reversal velocity (42 cm/s) and prolonged duration, diastolic dysfunction with a_{dur}–A_{dur} > 20 ms, and $PV_S < PV_D$.

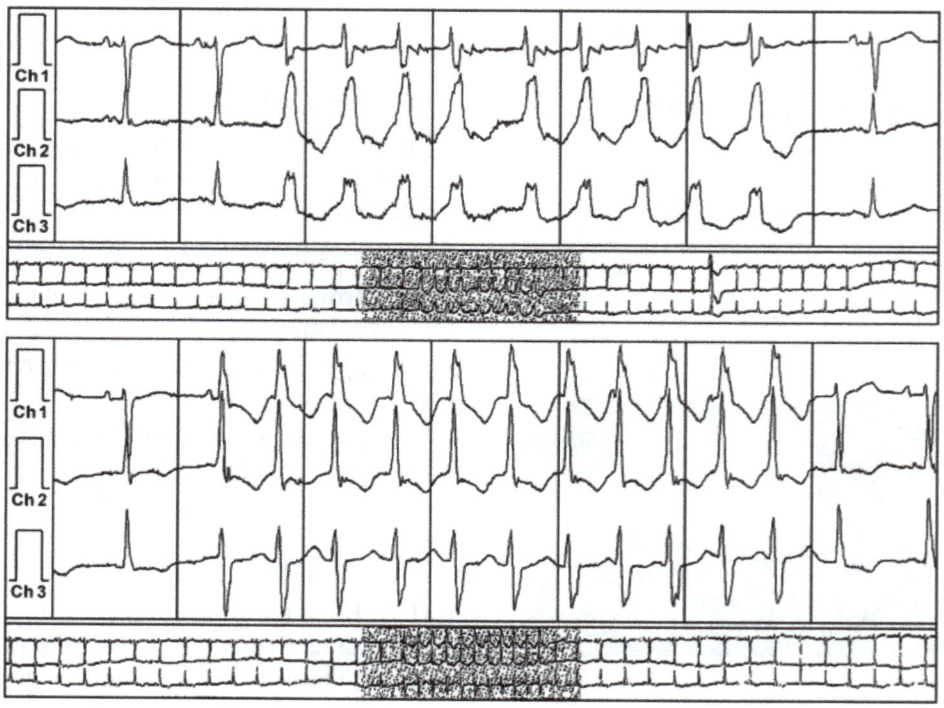

Figure 49.4. Recording from a 24-hour ambulatory ECG (Holter) monitor demonstrating several beats of NSVT in this patient with a diagnosis of nonobstructive HCM and a syncopal episode.

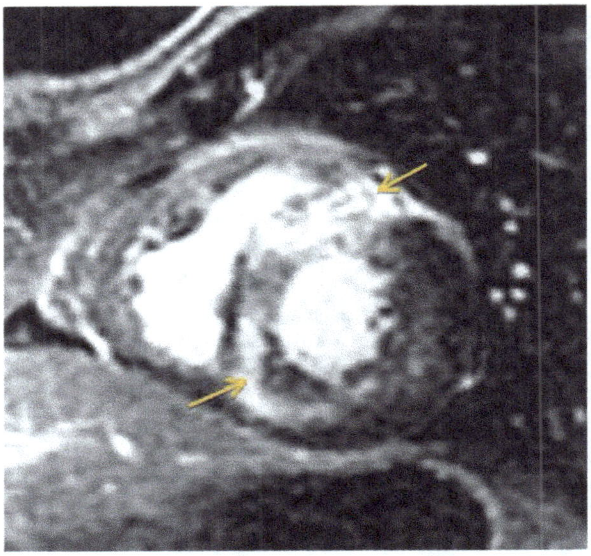

Figure 49.5. Short-axis inversion recovery gradient echo image (delayed CMR) obtained during systole, 15 minutes following the intravenous administration of gadolinium contrast (gadobenate dimeglumine). The normal myocardium has low signal. The patchy areas of bright signal, in the anterior and inferior septal walls (arrows) and at the insertion of the right ventricular free wall, correspond to areas of fibrosis (delayed gadolinium enhancement) seen in this patient with nonobstructive HCM.

and "calcium-binding HCM" (Table 49.1), also exist. These mutations alter the structure and function of the proteins resulting in myofibrillar disarray, marked ventricular hypertrophy (frequently asymmetric), diastolic dysfunction, and, in some patients, sudden cardiac death. The disease is characterized by extraordinary heterogeneity in its genetic substrate, its phenotypic expression, clinical presentation, and natural history, as well as management and specific therapy. HCM can present at any age and may be detected in infancy, childhood, adolescence, or as a coincidental finding at autopsy in the elderly.

Anatomically characterized by a thick (symmetrical or asymmetrical) hypertrophied (increased mass) LV, the classic form of HCM involves thickening of the basal anterior septum, which bulges beneath the aortic valve and causes narrowing of the LVOT. However, modern studies have demonstrated a spectrum of patterns of LV wall thickening that constitute the phenotypic expression of HCM (Figure 49.11). The morphological and functional changes associated with HCM result in complex and multiple interrelated changes in cardiac

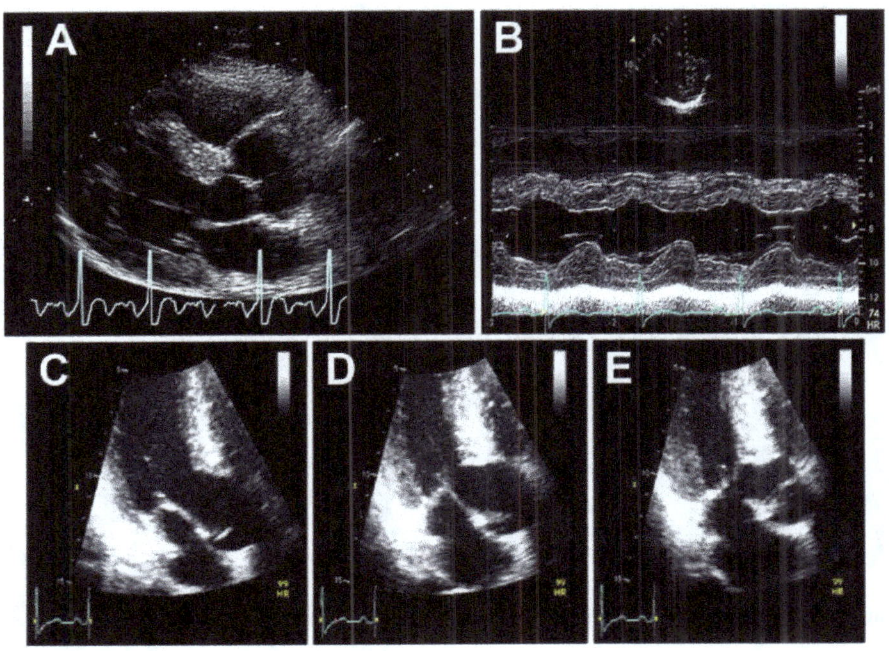

Figure 49.6. **A and B:** Parasternal long-axis view and M-mode echocardiogram of the LV in Patient 2. Note the moderate to severe increase in LV wall thickness and decreased LV cavity size. The septal wall is noted to be at least 18 mm in thickness, whereas the posterior wall is about 16-mm thick. **C, D, and E:** Expanded successive parasternal long-axis views (as systole progresses) in the same patient with HCM demonstrating the dramatic SAM of the mitral valve, indicative of significant LVOT obstruction. The greater the degree and duration of SAM, the larger the LVOT obstruction.

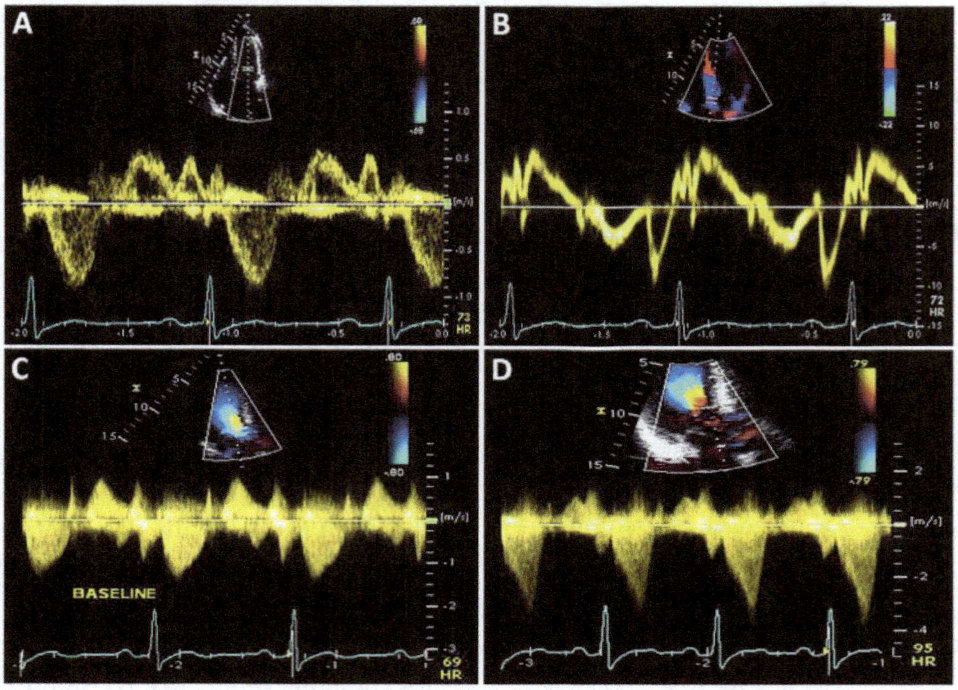

Figure 49.7. **A and B:** Mitral inflow and tissue Doppler recording of the medial mitral annular velocity in this patient with HCM. These Doppler tracings are suggestive of diastolic dysfunction (Grade I/IV – abnormal relaxation filling pattern with normal to mildly elevated filling pressures). **C and D:** CW Doppler recordings in this patient at rest (Panel C) and after a provocative maneuver of amyl nitrite inhalation (Panel D). Note that, at rest, there is no LVOT obstruction with a nonexistent resting gradient (Panel C), which increases significantly to 60 mmHg with amyl nitrite (Panel D). This corresponds to SAM where close approximation of the mitral valve with the septum occurs (see Figure 49.6E) in this patient. Note the typical shape on CW Doppler tracing – the "Dagger sign" – obtained through the orifice in the outflow tract.

physiology, including diastolic dysfunction, LVOT obstruction, mitral regurgitation, myocardial ischemia, arrhythmias, and, in a minority of patients over time, overt systolic dysfunction (Figure 49.12).

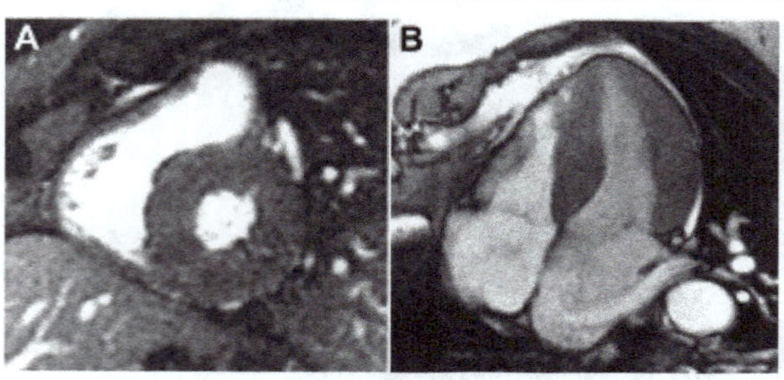

Figure 49.8. Steady-state free precession (bright blood) CMR images from a cine loop oriented in short axis (**A**) and long axis (**B**) in Patient 2. These images demonstrate the abnormally thick walls at end diastole. Note that the morphological subtype of HCM in this patient will be described as a reverse curve septal morphology with predominant midseptal convexity toward the LV cavity, with the cavity overall having a crescent shape.

SUDDEN DEATH

SD is the most devastating complication in patients with HCM. SD, however, occurs in only a small

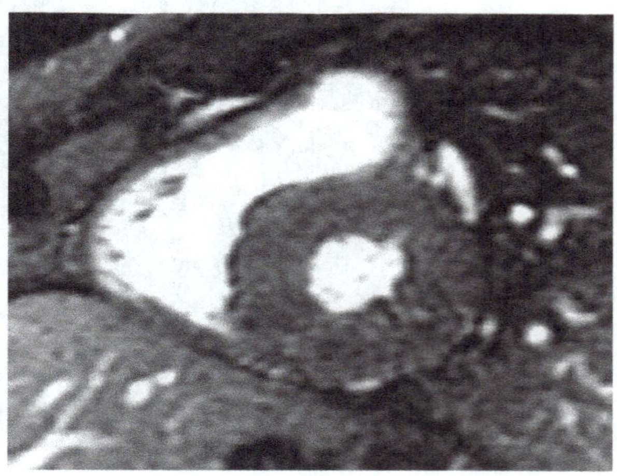

Figure 49.9. Short-axis delayed CMR image obtained during systole, 15 minutes after the intravenous administration of gadolinium contrast. Note the low signal of the myocardium without any fibrosis (compare with Patient 1).

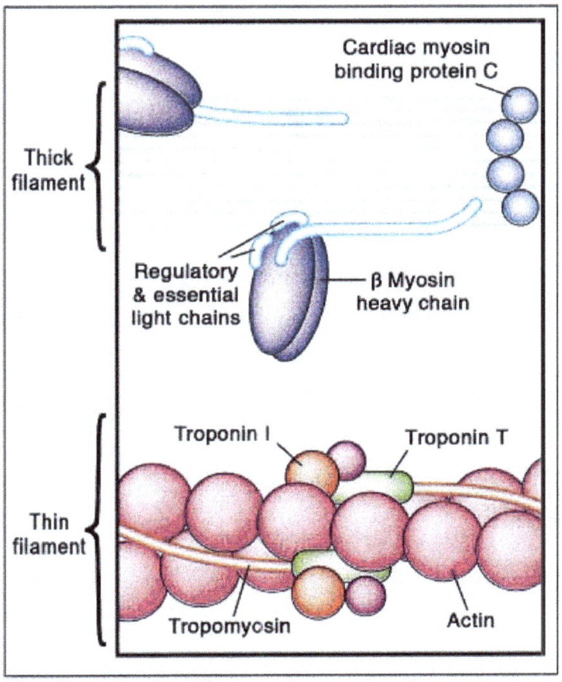

Figure 49.10. The cardiac sarcomere comprises the contractile proteins of the thick and thin filaments. Mutations in genes encoding each of the proteins labeled in this figure can cause HCM; thus, the disorder can result from abnormalities of proteins that have diverse structural, enzymatic, and regulatory functions. (Reproduced from Ashrafian et al[44] with permission from Elsevier.)

minority of patients and is less common than other adverse consequences, including AF and progressive heart failure. The spectrum of the natural history of HCM follows one of several profiles (Figure 49.13).

Epidemiology of SD in HCM

Although the early literature on HCM was dominated by reports of poor prognosis for patients with this disease – overall mortality rates of 3% to 6% per year[11-14] – this likely represented selection bias, since most clinical data emanated from select tertiary centers in North America and Europe. The described cohorts of these HCM patients disproportionately comprised patients referred because of their high-risk status or severe symptoms with the data underrepresenting clinically stable, asymptomatic, and elderly patients. On the other hand, more recent reports from unselected regional populations show lower mortality rates of ~1% per year, and HCM in the elderly has been shown to have mortality similar to control populations. Nevertheless, subgroups of patients within the broad HCM spectrum who are clearly at high risk for death (including SD) and in whom annual mortality exceeds 1% do exist, and it is this subgroup of HCM patients, particularly when diagnosed at a young age, which is the focus of intense debate and research.

Pathogenesis of SD – the Arrhythmogenic Substrate

Classically, HCM-related SD occurs due to sustained ventricular tachycardia (VT) and/or ventricular fibrillation (VF), with sinus tachycardia identified as an initiating rhythm in some cases, suggesting that in HCM patients, high sympathetic drive can be proarrhythmic. The architectural disorganization and scarring in HCM, in addition to microvascular dysfunction and ischemia, represents an unstable electrophysiologic substrate creating an arrhythmogenic milieu for tachy- and bradyarrhythmias (Table 49.2).

Interstitial fibrosis with increased cardiac collagen in HCM patients occurs prominently at sites of prior ischemic injury and myocyte disarray and is believed to play a prominent role in the pathogenesis of ventricular arrhythmias and SD. In clinical practice, the presence of interstitial fibrosis can be clinically utilized in risk stratification for SD as detected by delayed gadolinium (extracellular tracer) enhancement (DE) with CMR.

CLINICAL DOMAIN

In contemporary clinical practice, the classic profile of the patient with HCM who experiences SD continues to fit the traditional model where SD occurs without warning – "bolt out of the blue" – largely in an asymptomatic or mildly symptomatic young adult (predominantly <25 years of age). It continues to be the most common cause of SD in young adults including competitive athletes. SD occurs most commonly during mild exertion or sedentary activities (or during sleep), but may be triggered by vigorous physical exertion.[37-39] Currently, data on gender- and race-related differences in SD and HCM are unavailable.

Life-threatening ventricular tachyarrhythmias in HCM patients are unpredictable and can occur with varying periods (months to years) of dormancy. Documented long-term survival (up to 30 years) after VT/VF without recurrence of life-threatening arrhythmias has been reported. Thus, the arrhythmogenic substrate (Table 49.3) in HCM continues to elude predictable and proper interrogation.

Table 49.1. **Summary of HCM Susceptibility Genes**

Gene	Locus	Protein	Frequency (%)
Myofilament HCM			
TTN	2q24.3	Titin	<1
MYH7	14q11.2-q12	Beta-myosin heavy chain	15-25
MYH6	14q11.2-q12	Alpha-myosin heavy chain	<1
MYL2	12q23-q24.3	Ventricular regulatory myosin light chain	<2
MYL3	3p21.2-p21.3	Ventricular essential myosin light chain	<1
MYBPC3	11p11.2	Cardiac myosin-binding protein C	15-25
TNNT2	1q32	Cardiac troponin T	<5
TNNI3	19p13.4	Cardiac troponin I	<5
TPM1	15q22.1	Alpha-tropomyosin	<5
ACTC	15q14	Alpha-cardiac actin	<1
TNNC1	3p21.3-p14.3	Cardiac troponin C	<1
Z-disc HCM			
LBD3	10q22.2-q23.3	LIM binding domain 3 (alias: ZASP)	1-5
CSRP3	11p15.1	Muscle LIM protein	<1
TCAP	17q12-q21.1	Telethonin	<1
VCL	10q22.1-q23	Vinculin/metavinculin	<1
ACTN2	1q42-q43	Alpha-actinin 2	<1
MYOZ2	4q26-q27	Myozenin 2	<1
Calcium-handling HCM			
JPH2	20q12	Juntophilin-2	<1
PLN	6q22.1	Phospholamban	<1

Genes in **bold** are available as commercial genetic tests.
(Reprinted from Bos et al[45] with permission from Elsevier.)

RISK STRATIFICATION

The era of ICDs has ushered in an effective strategy for SCD prevention and enhanced focus on risk stratification of SD in HCM patients. Although universal agreement exists for implantation of an ICD in a HCM patient for secondary prevention after a sustained episode of VT or cardiac arrest, accurate selection of patients for ICD therapy for primary prevention of SD is not as precise and can be a contentious issue. The primary reason for the difficulty in precise risk stratification in HCM is due largely to its clinical heterogeneity of presentation and expression, its relatively low prevalence in general cardiology practice, and the complexity of potential pathophysiological mechanisms. Besides, a truly unselected HCM population for study is not available, and the precise proportion of HCM patients with a significant risk for SD remains elusive. Nevertheless, it is still possible to identify most high-risk patients by noninvasive clinical markers, and only a small minority of those HCM patients who die suddenly (about 3%) are without any of the currently acknowledged risk markers.

Risk Markers

In contemporary practice, several risk markers have emerged from observational studies and achieved

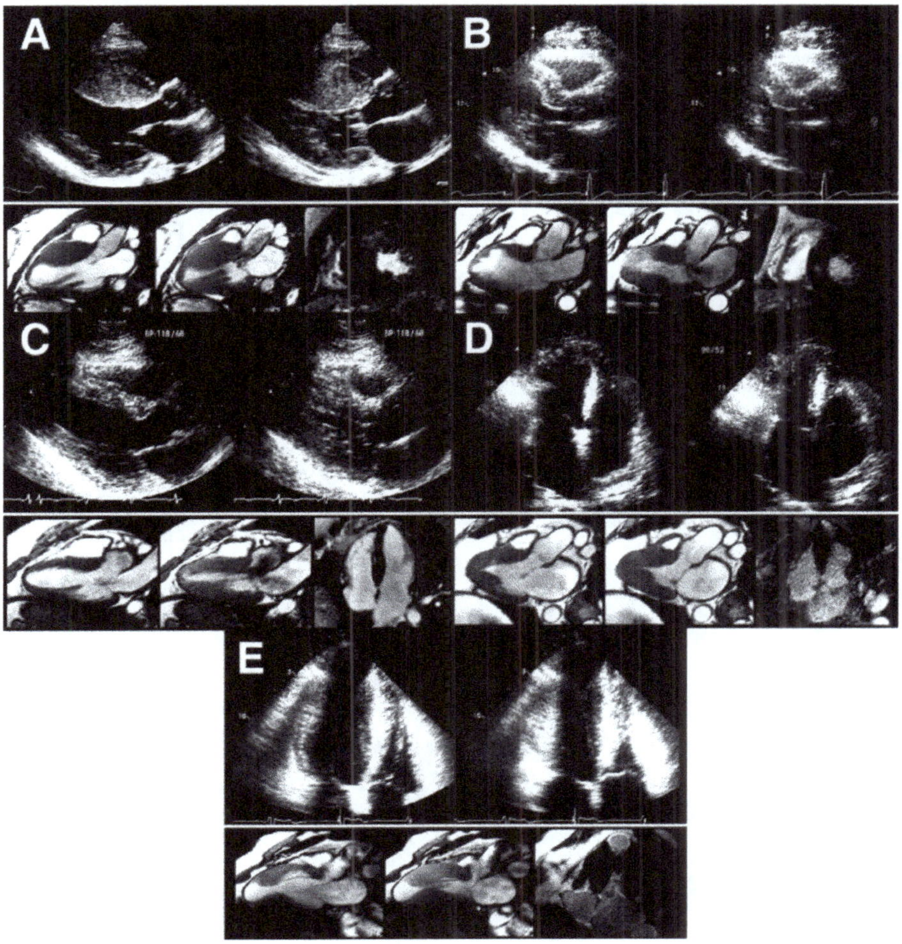

Figure 49.11. HCM morphological subtypes are demonstrated by echocardiography in end-diastolic (*left*) and end-systolic (*right*) images of the heart. The HCM subtypes are also demonstrated by magnetic resonance imaging in the same patients. Left and middle columns show steady-state free precession images of the heart in a three-chamber orientation in end diastole and end systole, respectively. Right column shows myocardial delayed enhancement (MDE) images using an inversion-recovery gradient-recalled echo technique. **A:** Reverse curvature septum HCM shows a predominant midseptal convexity toward the LV cavity with the cavity itself often having an overall crescent shape. Dynamic subaortic obstruction is present in this example with SAM of the mitral leaflets and turbulent flow in the outflow tract. Prominent foci of MDE that indicate myocardial fibrosis are present in the anteroseptum and inferoseptum. **B:** Sigmoid septum HCM shows a generally ovoid LV cavity with the septum being concave to the LV cavity and a prominent basal-septal bulge. Subaortic obstruction is present in this example with SAM of the mitral leaflets and a posteriorly directed jet of mitral regurgitation. A small amount of MDE is seen in the septum. **C:** Neutral septum HCM shows an overall straight septum that is neither predominantly convex nor concave toward the LV cavity. Subaortic obstruction is present in this example. A small focus of MDE is seen in the septum. **D:** Apical HCM shows a predominant apical distribution of hypertrophy. MDE is seen in the LV apex at the site of maximal hypertrophy in this example. **E:** Midventricular HCM shows predominant hypertrophy at the midventricular level. In this example, a thinned and dyskinetic apical pouch is also present. Obstruction is at the level of the papillary muscles, where turbulence was identified. No mitral SAM. MDE is seen in the dyskinetic apical pouch. (Reproduced from Syed et al[46] with permission from Elsevier.)

general acceptance in risk stratifying clinically stable HCM patients for SD. These can be classified into: (1) conventional risk factors; (2) potential risk factors; and (3) uncertain risk factors.

Conventional risk factors include:

1. Family history of one or more HCM-related SCD or history of resuscitated SCD (eg, sustained VT or VF).
2. One or more episodes of unexplained syncope.
3. LV wall thickness of ≥30 mm.
4. NSVT on 24-hour Holter monitor.
5. Hypotensive or attenuated blood pressure response on exercise stress testing.

Of the conventional risk markers, family history, syncope, wall thickness ≥30 mm, and bursts of NSVT identified over serial monitoring periods (as a matter

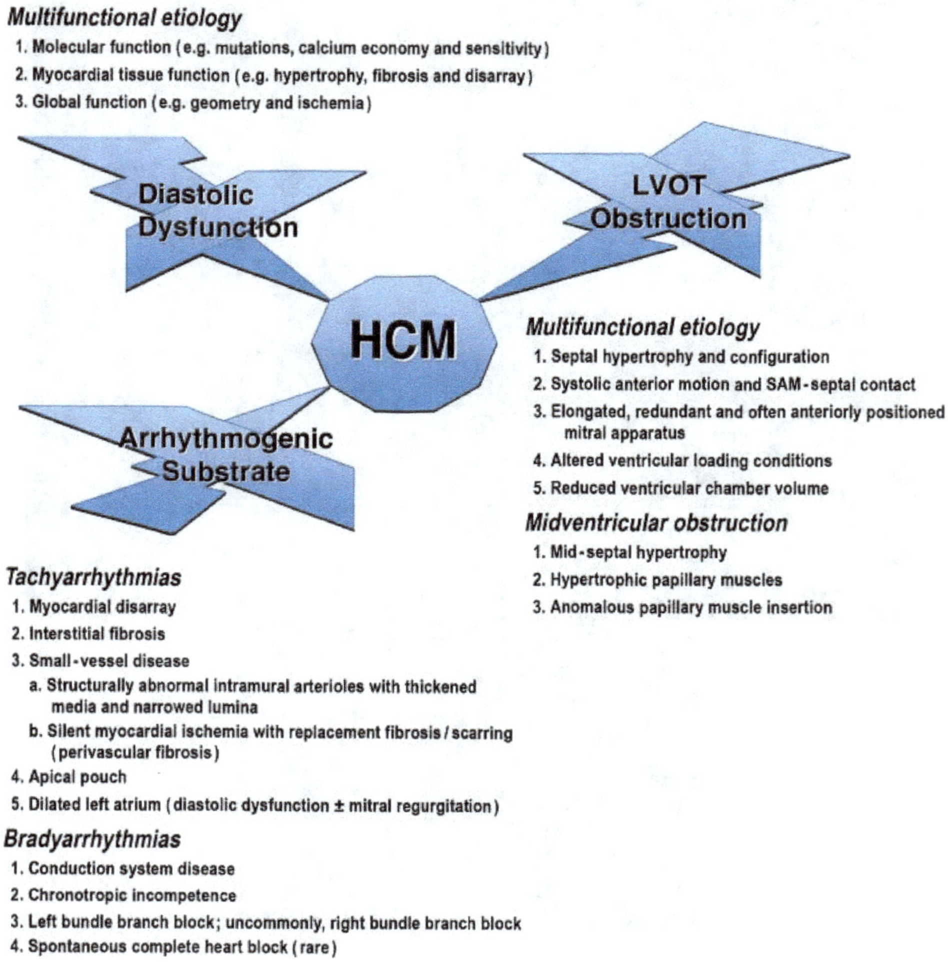

Figure 49.12. Key morphological and functional changes associated with HCM. LVOT, left ventricular outflow tract.

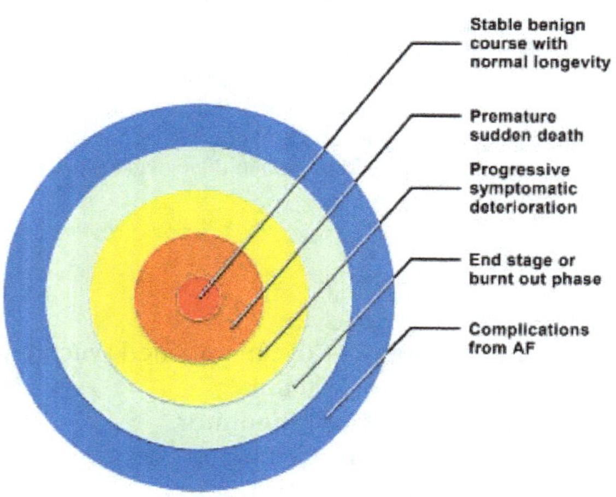

Figure 49.13. The spectrum of the natural history of HCM. AF, atrial fibrillation.

of clinical practice) are believed to carry greater weight as risk factors for SD. Of these, NSVT on ambulatory ECG monitoring is the only one that directly relates to the arrhythmogenic substrate in HCM. An attenuated or hypotensive blood pressure response to exercise can allude to significant LVOT obstruction but is rarely used as a sole indicator for primary implantation of an ICD in routine clinical practice.

Potential risk factors include:

1. Late gadolinium enhancement (LGE) by CMR. LV apical pouch/aneurysm.
2. End-stage phase with widespread LV scarring.
3. LVOT obstruction with resting gradient ≥30 mmHg.
4. Intense physical exertion.
5. Obstructive sleep apnea.
6. Associated epicardial coronary heart disease.

Table 49.2. Arrhythmogenic Substrate of HCM

Tachyarrhythmias
1. Myocardial disarray and interstitial fibrosis
2. Small-vessel disease
1. Structurally abnormal intramural arterioles with thickened media and narrowed lumina
2. Silent myocardial ischemic with replacement fibrosis/scarring (perivascular fibrosis)
3. Apical pouch with or without midventricular obstruction
4. Left atrial dilation (diastolic dysfunction ± mitral regurgitation)
Bradyarrhythmias
1. Conduction system disease
2. Chronotropic incompetence
3. Left bundle branch block; (rarely, right bundle branch block)
4. Complete heart block (rare)

Table 49.3. Characteristics of Clinical Unpredictability of the Arrhythmogenic Substrate in HCM

1. Unpredictable timing of life-threatening ventricular tachyarrhythmias
2. Varying periods of dormancy (months to years)
3. Substantial delay between ICD implant and device intervention
4. Lack of a circadian pattern of SD (however, VT during sleep is not uncommon)

ICD, implantable cardioverter-defibrillator; VT, ventricular tachycardia.

In clinical practice, potential risk markers are believed to be risk factors that by themselves do not carry enough weight to justify prophylactic insertion of an ICD, but are important disease features that can help resolve the issue when risk based on conventional markers borders on ambiguity. Of these, LGE by CMR has recently evolved as a noninvasive repeatable quantitative test that does not add to potential patient risk. LV pouch is associated with a higher event rate (10% annual) due to increased arrhythmogenicity, and the end-stage phase, which may morphologically resemble ischemic cardiomyopathy with irreversible systolic dysfunction, is usually considered an indication for ICD when it may serve as a bridge to heart transplant. Some practitioners also consider alcohol septal ablation as a risk arbitrator and implant ICDs when the commonly accepted risk markers are present before the ablation procedure. The rationale for this practice stems from the concern that alcohol-imposed infarcts could compound preexisting and underlying myocardial electric instability, increasing the chances for life-threatening sustained ventricular tachyarrhythmias over the short term.

Uncertain risk factors include:

1. Atrial fibrillation.
2. Malignant gene mutations (eg, nonsarcomere LAMP 2 or double sarcomere mutations).
3. Myocardial ischemia (± microvascular ischemia).
4. Myocardial bridging.
5. Troponin leak.

Uncertain risk markers include those that do not have compelling evidence as a specific predictor of SD. Thus, although AF is the most common sustained arrhythmia in HCM (20–25% of HCM patients) and is independently associated with HF-related death, occurrence of fatal and nonfatal stroke, long-term disease progression with heart failure symptoms and severe functional disability, strong evidence for its direct association in causing SD is lacking. Similarly, identification of malignant or benign mutations in genes encoding proteins of the cardiac sarcomere (genotypic risk assessment) has not been proven to be a reliable strategy for risk stratification in selecting patients for primary-prevention ICDs. This stems primarily from the heterogeneity of HCM, with more than 1000 mutations documented (in 11 genes), including many with unresolved pathogenicity. However, there are clinical circumstances where molecular diagnosis may predict prognosis, for example, nonsarcomeric LAMP2 cardiomyopathy (myofilament HCM; locus Xq24: lysosome-associated membrane protein), and sometimes multiple sarcomeric gene defects or "gene-dosage effect" (double, triple, or compound hetezygosity). Similarly, there is no direct evidence to substantiate use of either coronary bridging or microvascular ischemia as significant predictors or markers of SD in HCM patients, although the latter appears to be a determinant of progressive heart failure.

Of the above-mentioned risk markers, OSA and intense physical exertion may be regarded as somewhat modifiable risk factors. Sleep-disordered breathing (SDB), particularly OSA, has been strongly associated with increased incidence and prevalence of AF in patients with HCM, and severity of SDB may influence left atrial volume index and the prevalence of AF in this population. It is now known that the presence of OSA in patients with HCM could be an important contributor to drug-refractory symptoms and worsening LVOT obstruction as a result of heightened sympathetic activity. Treatment of OSA with continuous positive airway pressure (CPAP) is known to reduce blood pressure, sympathetic activity, and systemic inflammation as well as stimulate regression of LV hypertrophy. Thus, it is reasonable to screen HCM patients for OSA, and those at high risk may be considered for screening oximetry, polysomnography, or both. A formal sleep study should usually follow if the screening test(s) is positive.

Although the link between HCM and athletic-field SD has been established, intense physical exertion remains a modifiable factor. To reduce the risk of SD in athletes with HCM, the generally accepted recommendation of Bethesda Conference 36 is withdrawal from intense training and competition associated with most competitive sports in athletes with HCM, even in the absence of conventional markers. However, stringent lifestyle or employment modifications for other HCM patients (who are not participants in organized athletics) do not seem justified or practical, although intense physical activity involving burst exertion (eg, sprinting) or systematic isometric exercise (eg, heavy lifting) should always be discouraged. It needs to be mentioned, however, that ICD is not the preferred strategy if the sole purpose is continued participation in competitive athletics.

Although a single, strong, and unequivocal established marker of SD within the clinical profile of an individual HCM patient is sometimes sufficient to warrant ICD for primary prevention, it is important to remember that the independent weight of any given risk factor with respect to all others remains unknown, and interplay between markers in individual patients is complex. The one-risk factor model is complicated by the fact that not all patients with one risk marker are at the same magnitude of risk, and universal ICD recommendation can be a potentially flawed practice. On the other hand, a model of risk factor counting may appear to be more pragmatic since there is evidence that patients with multiple risk factors are at an increased risk of SD. However, risk factor counting can also lead to flawed practice when physicians rely solely on numeric summing, resulting in an artificial strategy. This is where the art of medicine meets the science and evidence in medicine, and individual and institutional experience dictate proper patient selection.

The following are examples of some of the inconsistencies relating to defining and weighting individual risk markers for SD in HCM, limitations that weaken the reliability of the risk stratification process.

1. *Imprecision of definition of a risk factor/marker:* **Family history of SD:** Multiple definitions exist in the literature for defining family history of HCM including: One first-degree relative, 2 or more relatives <40 years of age, one or more first-degree relatives <40 years of age, or one or more relatives <50 years of age.
Syncope: Inconsistency as to whether one should consider one or two syncopal events and whether time elapsed between these events affects significance makes use of this risk marker confusing.

2. *Low positive predictive values:* Although each conventional risk marker is associated with a high negative predictive value (≥90%), risk markers individually or collectively are limited by positive predictive value only in the range of 15% to 30%, largely resulting from the low event rate that is characteristic of HCM.

3. *Status changes over time:* Risk markers in HCM are not static disease components and can change over time (typically toward higher levels), underscoring the importance of ongoing clinical surveillance. For example, LV wall thickness may increase abruptly and substantially in young patients; syncope may occur for the first time; a family member may experience SD; or NSVT may appear on routine Holter monitors.

4. *LVOT obstruction with gradient of ≥30 mmHg:* Although a highly visible quantitative measure of elevated intraventricular pressures and wall stress, the major hurdle of this marker in risk stratification is its variability, dynamic nature, and frequency, with only 70% of patients capable of generating LVOT gradients at rest or with physiological exercise.

5. *LV apical pouch:* Defined as a discrete thin-walled dyskinetic or akinetic segment of the most distal portion of the LV chamber with a relatively wide communication to the LV cavity, LV apical pouch/aneurysm is associated with a 10% annual event rate. Apical pouches vary by size (large, >4 cm and small, <2 cm); and HCM-related complications are more common in patients with the apical pouch,

for example, SD, LV systolic dysfunction, progressive heart failure symptoms, and embolic stroke/LV apical thrombus. Contrast echocardiography must be performed in order to definitely diagnose an apical pouch. The low incidence of apical pouches reported in HCM patients (2%) is likely related to the under-use of contrast echocardiography in most clinical practices. In contemporary practice, CMR is very sensitive in identifying this pathology.

6. Delayed enhancement: Myocardial LGE in HCM has been demonstrated in both minimally symptomatic individuals and individuals at high risk of SD or myocardial thinning and dilation. Areas of LGE on CMR in HCM are attributable to foci of collagen deposition, and LGE has been shown to be present even in HCM carriers who have not yet developed overt hypertrophy. CMR can detect crypts in the inferoseptal LV wall, even when echocardiography and ECG are normal. In patients with HCM, presence of LGE on CMR has been shown to be common and more prevalent among gene-positive patients, is strongly associated with surrogates of arrhythmia, and is associated with subsequent SD and/or ICD discharge after controlling for other variables.

In contemporary clinical practice, HCM patients (particularly those aged <60 years old) should undergo comprehensive clinical assessments on an annual basis for risk stratification and evolution of symptoms, including: (1) careful personal and family history; (2) 2D/3D echocardiography (primarily for assessment of magnitude of LV hypertrophy, outflow obstruction, and severity of diastolic dysfunction); (3) a 24- or 48-hour ambulatory (Holter) ECG recording for VT; and (4) blood pressure response during maximal upright exercise (treadmill or bicycle) along with evaluation of exercise-induced severity of LVOT obstruction. Subsequent risk analysis **should be performed periodically** and when there is a perceived change in clinical status.

Currently, there is no convincing evidence that electrophysiologic testing (ie, programmed ventricular stimulation) has a routine role in identifying HCM patients at high risk for SD due to life-threatening arrhythmias. Both polymorphic VT and VF are generally regarded as nonspecific electrophysiologic testing responses to multiple ventricular extrastimuli. For example, stimulation with 3 ventricular premature depolarizations rarely triggers monomorphic VT in HCM (in contrast to coronary artery disease), but frequently induces polymorphic VT or VF, even in some patients at low risk for SD. Thus, risk stratification involving laboratory induction of ventricular arrhythmias is neither desirable in HCM patients on a routine basis nor, *per se*, to justify aggressive intervention. However, HCM patients with otherwise unexplained syncope may benefit from an electrophysiology study with or without programmed ventricular stimulation when a risk stratification dilemma exists.

PREVENTION OF SD IN HCM USING ICDs

Introduced more than 25 years ago for terminating lethal ventricular tachyarrhythmias, ICDs have created a paradigm shift in the effective strategy for SD prevention and, at the same time, an enhanced focus on risk stratification and need for reliable identification of high-risk patients. Although 2 of the 3 initial patients who were implanted with ICDs had HCM, ICDs were used sparingly in HCM patients until a decade ago. With their availability, prevention of SD in HCM by administration of drugs such as beta-adrenergic blocking agents, verapamil, and type 1A antiarrhythmic agents (ie, quinidine, procainamide) is only of historical interest now. However, worldwide, ICD implant rates differ considerably with regard to country and health care system (highest implant rate in the United States) because of a number of cultural, societal, and economic factors that unavoidably influence strategies for primary prevention of SD in HCM.

In contemporary clinical practice, ICD is the most effective and reliable treatment option available, harboring the potential for absolute protection and altering the natural history of this disease in some patients. Evidence over the last 15 years has clearly demonstrated that appropriate ICD interventions are not uncommon in HCM and that they are highly effective in terminating potentially lethal ventricular tachyarrhythmias despite the complex HCM phenotype, which may include extreme LV hypertrophy, LVOT obstruction, microvascular ischemia, and diastolic dysfunction.

The most robust evidence for ICDs comes from an international multicenter registry of 506 HCM patients from 42 centers with ICDs implanted on the clinical judgment of the managing cardiologist.

Data from this registry demonstrate that over an average follow-up of 3.7 years:

1. Twenty percent of patients experienced appropriate device therapy for VT/VF, equivalent to 5 ICDs implanted per intervention.
2. Discharge rates were 5.5% per year overall, 11% per year for secondary prevention (after cardiac arrest or sustained VT), and 4% per year for primary prevention (more than one risk factor).
3. ICD therapy was most common in young patients (average age 40 years) with the highest rates in children and adolescents (11% per year).

In this HCM registry, a significant proportion of appropriate ICD interventions for VT/VF occurred in patients implanted for only one risk factor (35%), and device therapy was as common in patients with one risk marker as in those with 2 or more markers, thus reiterating the dilemma of decision-making for primary prevention in HCM.

SUMMARY

SD remains the most dramatic and devastating complication of HCM and is caused primarily by sustained VT and/or VF. A comprehensive assessment should be made to risk stratify a patient with HCM with the currently available armamentarium of noninvasive diagnostic testing. An insightful decision-making process that involves patient preferences should be undertaken in selecting patients for an ICD implant for primary prevention. Because assignment of risk level in HCM is not straightforward but fraught with ambiguity, clinical judgment remains the most important tool available. Thus, in addition to the conventional risk markers, consideration should be given to both potential and uncertain risk markers, including the modifiable ones. For secondary prevention, the decision is usually unequivocal and ICDs are indicated in those HCM patients who have survived a cardiac arrest or sustained episodes of VT.

Correct Answers:

Q1. D

Q2. A

Q3. B

Q4. C

REFERENCES

1. Maron BJ, Roberts WC. Cardiomyopathies in the first two decades of life. *Cardiovasc Clin.* 1981;11:35–78.
2. Charron P, Carrier L, Dubourg O, et al. Penetrance of familial hypertrophic cardiomyopathy. *Genet Couns.* 1997;8:107–114.
3. Maron BJ, Spirito P, Wesley Y, Arce J. Development and progression of left ventricular hypertrophy in children with hypertrophic cardiomyopathy. *N Engl J Med.* 1986;315:610–614.
4. Maron BJ, McKenna WJ, Danielson GK, et al. American College of Cardiology/European Society of Cardiology clinical expert consensus document on hypertrophic cardiomyopathy. A report of the American College of Cardiology Foundation Task Force on Clinical Expert Consensus Documents and the European Society of Cardiology Committee for Practice Guidelines. *J Am Coll Cardiol.* 2003;42:1687–1713.
5. Maron BJ. Hypertrophic cardiomyopathy: A systematic review. *JAMA.* 2002;287:1308–1320.
6. Maron BJ, Estes NA 3rd, Maron MS, et al. Primary prevention of sudden death as a novel treatment strategy in hypertrophic cardiomyopathy. *Circulation.* 2003;107:2872–2875.
7. Maron BJ, Spirito P. Implantable defibrillators and prevention of sudden death in hypertrophic cardiomyopathy. *J Cardiovasc Electrophysiol.* 2008;19:1118–1126.
8. Olivotto I, Cecchi F, Casey SA, et al. Impact of atrial fibrillation on the clinical course of hypertrophic cardiomyopathy. *Circulation.* 2001;104:2517–2524.
9. McKenna W, Deanfield J, Faruqui A, et al. Prognosis in hypertrophic cardiomyopathy: Role of age and clinical, electrocardiographic and hemodynamic features. *Am J Cardiol.* 1981;47:532–538.
10. Frank MJ, Abdulla AM, Canedo MI, Saylors RE. Long-term medical management of hypertrophic obstructive cardiomyopathy. *Am J Cardiol.* 1978;42:993–1001.
11. Maron BJ, Casey SA, Poliac LC, et al. Clinical course of hypertrophic cardiomyopathy in a regional United States cohort. *JAMA.* 1999;281:650–655.
12. Cannan CR, Reeder GS, Bailey KR, Melton LJ 3rd, Gersh BJ. Natural history of hypertrophic cardiomyopathy. A population-based study, 1976 through 1990. *Circulation.* 1995;92:2488–2495.
13. Spirito P, Chiarella F, Carratino L, et al. Clinical course and prognosis of hypertrophic cardiomyopathy in an outpatient population. *N Engl J Med.* 1989;320:749–755.
14. Fay WP, Taliercio CP, Ilstrup DM, Tajik AJ, Gersch BJ. Natural history of hypertrophic cardiomyopathy in the elderly. *J Am Coll Cardiol.* 1990;16:821–826.
15. Cha YM, Gersh BJ, Maron BJ, et al. Electrophysiologic manifestations of ventricular tachyarrhythmias provoking appropriate defibrillator interventions in high-risk patients with hypertrophic cardiomyopathy. *J Cardiovasc Electrophysiol.* 2007;18:483–487.
16. Georgakopoulos D, Christe ME, Giewat M, et al. The pathogenesis of familial hypertrophic cardiomyopathy: Early and evolving effects from an alpha-cardiac myosin heavy chain missense mutation. *Nat Med.* 1999;5:327–330.
17. Moon JC, McKenna WJ, McCrohon JA, et al. Toward clinical risk assessment in hypertrophic cardiomyopathy with gadolinium cardiovascular magnetic resonance. *J Am Coll Cardiol.* 2003;41:1561–1567.
18. Nazarian S, Lima JA. Cardiovascular magnetic resonance for risk stratification of arrhythmia in hypertrophic cardiomyopathy. *J Am Coll Cardiol.* 2008;51:1375–1376.
19. Elliott PM, Gimeno Blanes JR, Mahon NG, Poloniecki JD, McKenna WJ. Relation between severity of left-ventricular hypertrophy and prognosis in patients with hypertrophic cardiomyopathy. *Lancet.* 2001;357:420–424.

20. Elliott PM, Poloniecki J, Dickie S, et al. Sudden death in hypertrophic cardiomyopathy: Identification of high risk patients. *J Am Coll Cardiol.* 2000;36:2212–2218.
21. Nishimura RA, Ommen SR. Hypertrophic cardiomyopathy, sudden death, and implantable cardiac defibrillators: How low the bar? *JAMA.* 2007;298:452–454.
22. Maron BJ. Sudden death in young athletes. *N Engl J Med.* 2003;349:1064–1075.
23. Maron BJ, Roberts WC, Epstein SE. Sudden death in hypertrophic cardiomyopathy: A profile of 78 patients. *Circulation.* 1982;65:1388–1394.
24. Maron BJ, Kogan J, Proschan MA, Hecht GM, Roberts WC. Circadian variability in the occurrence of sudden cardiac death in patients with hypertrophic cardiomyopathy. *J Am Coll Cardiol.* 1994;23:1405–1409.
25. Maron BJ, Spirito P, Shen W-K, et al. Implantable cardioverter-defibrillators and prevention of sudden cardiac death in hypertrophic cardiomyopathy. *JAMA.* 2007;298:405–412.
26. Epstein AE, DiMarco JP, Ellenbogen KA, et al. ACC/AHA/HRS 2008 guidelines for device-based therapy of cardiac rhythm abnormalities: A report of the American College of Cardiology/American Heart Association Task Force on Practice Guidelines (Writing Committee to Revise the ACC/AHA/NASPE 2002 Guideline Update for Implantation of Cardiac Pacemakers and Antiarrhythmia Devices) developed in collaboration with the American Association for Thoracic Surgery and Society of Thoracic Surgeons. *J Am Coll Cardiol.* 2008;51:e1–e62.
27. Maron MS, Finley JJ, Bos JM, et al. Prevalence, clinical significance, and natural history of left ventricular apical aneurysms in hypertrophic cardiomyopathy. *Circulation.* 2008;118:1541–1549.
28. Sorajja P, Valeti U, Nishimura R, et al. Outcome of alcohol septal ablation for obstructive hypertrophic cardiomyopathy. *Circulation.* 2008;118:131–139.
29. van der Lee C, ten Cate FJ, Geleijnse ML, et al. Percutaneous versus surgical treatment for patients with hypertrophic obstructive cardiomyopathy and enlarged anterior mitral valve leaflets. *Circulation.* 2005;112:482–488.
30. Robinson K, Frenneaux MP, Stockins B, et al. Atrial fibrillation in hypertrophic cardiomyopathy: A longitudinal study. *J Am Coll Cardiol.* 1990;15:1279–1285.
31. Ackerman MJ, VanDriest SL, Ommen SR, et al. Prevalence and age-dependence of malignant mutations in the beta-myosin heavy chain and troponin T genes in hypertrophic cardiomyopathy: A comprehensive outpatient perspective. *J Am Coll Cardiol.* 2002;39:2042–2048.
32. Alcalai R, Seidman JG, Seidman CE. Genetic basis of hypertrophic cardiomyopathy: From bench to the clinics. *J Cardiovasc Electrophysiol.* 2008;19:104–110.
33. Maron BJ, Zipes DP. Introduction: Eligibility recommendations for competitive athletes with cardiovascular abnormalities – general considerations. *J Am Coll Cardiol.* 2005;45:1318–1321.
34. Maron BJ. Contemporary insights and strategies for risk stratification and prevention of sudden death in hypertrophic cardiomyopathy. *Circulation.* 2010;121:445–456.
35. McKenna WJ, Behr ER. Hypertrophic cardiomyopathy: Management, risk stratification and prevention of sudden death. *Heart.* 2002;87:169–176.
36. Spirito P, Autore C, Rapezzi C, et al. Syncope and risk of sudden death in hypertrophic cardiomyopathy. *Circulation.* 2009;119:1703–1710.
37. McKenna WJ, England D, Doi YL, et al. Arrhythmia in hypertrophic cardiomyopathy. I. Influence on prognosis. *Br Heart J.* 1981;46:168–172.
38. Spirito P, Bellone P, Harris KM, et al. Magnitude of left ventricular hypertrophy and risk of sudden death in hypertrophic cardiomyopathy. *N Engl J Med.* 2000;342:1778–1785.
39. Monserrat L, Elliott PM, Gimeno JR, et al. Non-sustained ventricular tachycardia in hypertrophic cardiomyopathy: An independent marker of sudden death risk in young patients. *J Am Coll Cardiol.* 2003;42:873–879.
40. Germans T, Wilde AA, Dijkmans PA, et al. Structural abnormalities of the inferoseptal left ventricular wall detected by cardiac magnetic resonance imaging in carriers of hypertrophic cardiomyopathy mutations. *J Am Coll Cardiol.* 2006;48:2518–2523.
41. Rubinshtein R, Glockner JF, Ommen SR, et al. Characteristics and clinical significance of late gadolinium enhancement by contrast-enhanced magnetic resonance imaging in patients with hypertrophic cardiomyopathy. *Circ Heart Fail.* 2010;3:51–58.
42. Kuck KH, Kunze KP, Schlüter M, Nienaber CA, Costard A. Programmed electrical stimulation in hypertrophic cardiomyopathy. Results in patients with and without cardiac arrest or syncope. *Eur Heart J.* 1988;9:177–185.
43. Fananapazir L, Chang AC, Epstein SE, McAreavey D. Prognostic determinants in hypertrophic cardiomyopathy. Prospective evaluation of a therapeutic strategy based on clinical, Holter, hemodynamic, and electrophysiological findings. *Circulation.* 1992;86:730–740.
44. Ashrafian et al. Hypertrophic cardiomyopathy: A paradigm for myocardial energy depletion. *Trends Genet.* 2003;19:263–268.
45. Bos et al. Diagnostic, prognostic, and therapeutic implications of genetic testing for hypertrophic cardiomyopathy. *J Am Coll Cardiol.* 2009;54:201–211.
46. Syed et al. Hypertrophic cardiomyopathy: Identification of morphological subtypes by echocardiography and cardiac magnetic resonance imaging. *J Am Coll Cardiol Img.* 2008;1:377–379.

SECTION 6

Anticoagulation, Antiarrhythmic Drug Therapy, Catheter and Surgical Ablation

CASE STUDY

50

Risk-Based Anticoagulation Therapy in Patients with Atrial Fibrillation

Rohit Malhotra, MD and John P. DiMarco, MD, PhD

A 68-year-old woman with no prior medical history is referred because of atrial fibrillation (AF). She reports 3 prior episodes, which terminated on their own after 8 to 12 hours. The episodes are accompanied by shortness of breath and lightheadedness. During the third episode, she presented to the emergency department and was found to have AF. She presents to clinic following this emergency department visit to discuss her need for anticoagulation.

Question No. 1:

Should this woman be anticoagulated chronically?

Question No. 2:

Would testing this patient for warfarin point mutations be helpful when starting warfarin?

DISCUSSION

AF is a common medical problem worldwide, with an estimated prevalence of 0.95% in the United States and an estimated affected population of 12 million to 15 million patients in 2050.[1] Because hypertension, diabetes mellitus, obesity, and sleep apnea are increasingly prevalent in the US population and are risk factors for AF, controlling this rhythm and its associated morbidities is important.[2] AF is associated with a two- to fourfold increase in morbidity, mediated predominantly by an increase in stroke risk.[3] However, prevention of stroke, predominantly through anticoagulation, has its own risks. As newer antithrombotic methods become available, the risk–benefit ratio may change.

Risk of Stroke

AF confers at least a 1% annual stroke risk in patients younger than 60 years of age with lone AF. As risk factors increase, the risk of stroke increases, though not necessarily proportionately. As a result, numerous risk stratification scores have been developed to predict the annual chance of stroke based purely upon clinical factors. Whereas these algorithms are all effective predictors of risk, the $CHADS_2$ schema is most frequently used due to its relative simplicity.

In the $CHADS_2$ schema, one point is assigned for each risk factor as depicted in Table 50.1, expect for prior stroke or transient ischemic attack (TIA) of any cause, which gets 2 points. Annual stroke risk is estimated as roughly 1.9% for a score of 0, 3% for a score of 1, 4% for a score of 2, 6% for a score of 3, 8.5% for a score of 4, 12.5% for a score of 5, and 18.2% for a score of 6.[4] A point total of ≥ 2 suggests a benefit to anticoagulation with warfarin, whereas a score of 1 indicates a neutral risk–benefit ratio.

Table 50.1. Points per Risk Factor in the $CHADS_2$ Risk Score

Risk Factor	Point Score
Congestive heart failure	1
Hypertension	1
Age >75 years	1
Diabetes mellitus	1
History of stroke or TIA	2

A score of zero indicates that the stroke risk is low. However, aspirin is recommended for stroke prophylaxis in these patients.

Judging the risk–benefit ratio for patients with scores of 1 can be difficult, as patient preference regarding personal views on risk–benefit ratios can be difficult to ascertain. Recently, the CHA_2DS_2-VASc risk score was developed. This system recognizes that female gender, vascular disease, and age <75 years convey higher stroke risk than previously recognized. The $CHADS_2$ system is modified by incorporating 1 point each for ages 64 to 75 years, female gender, and vascular disease (including prior myocardial infarction, peripheral arterial disease, or aortic plaque), and 2 points for age >75 years. For patients with a score of zero, no stroke prophylaxis is recommended in either system. The CHA_2DS_2-VASc risk score moves many patients, especially women or those between the age of 55 and 74 years, up from a $CHADS_2$ score of 1 to a CHA_2DS_2-VASc score of ≥ 2, at which oral anticoagulation is required. However, uncertainty still remains for patients with a CHA_2DS_2-VASc score of 1 since oral anticoagulation is listed as preferred but aspirin is also an accepted option for these patients.

A retrospective evaluation of the CHA_2DS_2-VASc, $CHADS_2$, and Framingham risk scores applied to a group of patients followed for one year after hospital discharge for AF validated the stroke prediction efficiency of the CHA_2DS_2-VASc score. Assuming that patients enrolled were not taking aspirin, the annual risk of thromboembolic events for each CHA_2DS_2-VASc score is demonstrated in Table 50.2. The CHA_2DS_2-VASc scoring system was a better predictor than the $CHADS_2$ risk score though not as predictive as the Framingham scoring system. However, the Framingham scoring system is more difficult to use clinically than the $CHADS_2$ and CHA_2DS_2-VASc scoring systems. Overall, the CHA_2DS_2-VASc score dropped the stroke risk for a score 1 from 3% to 0.7%, allowing better stroke prediction. The recent European Society of Cardiology Guidelines for AF have switched to the CHA_2DS_2-VASc scoring systems.

Table 50.2. Annual Stroke Risk as Stratified by CHA_2DS_2-VASc Score

Total Score	Annual Stroke Risk (%)
0	0
1	0.7
2	1.9
3	4.7
4	2.3
5	3.9
6	4.5
7	10.1
8	14.2
9	100

Risk of Bleeding

These risk stratification schema analyze the stroke risk, but they do not address bleeding risks. As a result, bleeding risk scoring systems have been developed. Once again, several risk schema for bleeding have been proposed. Bleeding risk for warfarin therapy was first assessed using the Outpatient Bleeding Risk Index. This scoring system assessed the risk of bleeding in patients using warfarin, regardless of reason. Although this score was highly predictive, it is not specific for AF patients and has not come into common usage. Subsequently, other scoring systems have been developed.

Risk of bleeding in patients with AF has been assessed using primarily two scoring systems, $HEMORR_2HAGES$ and HAS-BLED. Three older bleeding risk algorithms were combined into one algorithm, $HEMORR_2HAGES$, which was then retrospectively applied to a population of 3791 Medicare patients and assessed for predictive value. The $HEMORR_2HAGES$ algorithm involves giving 1 point for each of the following risk factors – **H**epatic or kidney disease; **E**thanol abuse; **M**alignancy; **O**lder age (>75 years); **R**educed platelet count; **H**ypertension (not controlled); **A**nemia; **G**enetics (polymorphisms in CYP2C9); **E**xcessive fall risk; and **S**troke – and 2 points for **R**ebleeding risk (evidence of prior bleeding episode).

All of these risk factors were abstracted from implantable cardioverter-defibrillator (ICD) codes and, as a result, were decided upon by hospital coders, not physicians. Furthermore, the etiology of bleeding events, such as supratherapeutic international normalized ratios (INRs), could not be determined. The resulting scores correlate with a risk of bleeding per 100 patient years of: 1.9 for a score of zero; 2.5 for a score of 1; 5.3 for a score of 2; 8.4 for 3 points; 10.4 for 4 points; and 12.3 for >5 points. This scoring system appropriately predicted bleeding risk for patients with low versus high scores, with a c-statistic of 0.6.

However, perhaps due to the relative complexity of the algorithm, this scoring system has not come into common clinical use. As a result, a newer scoring system has been developed, based upon data derived from the Euro Heart Survey, a prospective trial on AF. The HAS-BLED scoring system is based upon a smaller set of laboratory studies, and excludes genetic testing. In this scoring system, **H**ypertension, **A**bnormal hepatic or renal function, **S**troke, **B**leeding tendency or history, **L**abile INR, **E**lderly patients (aged >65 years), and **D**rug/alcohol use each get 1 point, though 1 point is given for each liver and kidney function and 1 point each for drug and alcohol use for a total of 2 possible points for each of these categories, and a highest possible score of 9 points. Elevated blood pressure was defined as a systolic blood pressure of >160 mmHg. Abnormal kidney function was defined as on chronic dialysis, post-kidney transplant, or serum creatinine >200 μmol/L (convert mg/dL to μmol/L by multiplying by 88). Abnormal liver function was defined as cirrhosis or evidence of biochemical abnormalities in bilirubin or AST/ALT/alkaline phosphatase. Labile INR signifies wide variation in the INR or <60% of INR readings in the therapeutic range. Drugs identify patients using concomitant nonsteroidal anti-inflammatory drugs or antiplatelet agents.

Risk of bleeding per 100 patient years for each score was: 0 = 1.13 bleeds; 1 = 1.02 bleeds; 2 = 1.88 bleeds; 3 = 3.74 bleeds; 4 = 8.7 bleeds; and 5 = 12.5 bleeds. Above this score, as there were no bleeding events in the derivation cohort, the rate of bleeding events could not be predicted. The predictive accuracy of this scoring schema compares favorably with the HEMORR$_2$HAGES score, and it may be somewhat easier to calculate. Issues with assessment of INR in the therapeutic range prior to starting anticoagulation may be problematic, as acknowledged by the authors of the study.

Whereas some of these predictive algorithms are derived from prospective trial data, others are derived from real-world experience. Prospective trials have higher levels of patient compliance than "real-world" experience, thus, these predictive algorithms may not be truly predictive of the odds of bleeding or clotting. One factor that plays a role in the risk of bleeding or clotting is the time in therapeutic range. Unfortunately, time in therapeutic range averages between 40% and 55% depending upon participation in a clinical trial or "real-world" practice. In "real-world" practice, time above an INR of 3.0 was roughly 20%.[5] By improving the time in therapeutic range, the risk of bleeding or clotting should be reduced.

Unfortunately, some patients are not eligible for warfarin therapy, either due to risk of bleeding or physician preference. Since aspirin is known to reduce the risk of stroke in patients with AF and has a 70% lower risk of bleeding, the ACTIVE-A study tested the utility of combining aspirin and clopidogrel, an inhibitor of platelet aggregation, in patients with AF and at least one risk factor for stroke. Clopidogrel with aspirin was found to be better than aspirin alone. In the ACTIVE-W study, clopidogrel and aspirin (75–100 mg) were compared to warfarin therapy and found to have relative risk of 1.44 for stroke, myocardial infarction, systemic embolism, or vascular death, with the increase largely driven by stroke and non-central nervous system embolism. Overall, warfarin appears to be the most effective means of stroke reduction in the AF population.

Genetic Testing

One problematic issue with warfarin therapy is the initial dose. Although a number of different initiation algorithms have been developed, these algorithms do not adequately predict bleeding risk due to overshooting and stroke risk due to underdosing. Bleeding risk due to overdosing may be addressed by genetic testing for two mutations in enzymes involved in metabolism of vitamin K antagonists (VKAs). Vitamin K epoxide reductase (VKOR) is an enzyme, predominantly located in the liver, that oxides inactive vitamin K to active vitamin K. This enzyme is inhibited by warfarin. Mutations in subunit C1 (VKORC1) lead to reduced dosing requirements in Caucasians and Asians. In addition, vitamin K is metabolized by cytochrome p450 2C9 (CYP2CP) from the more active S-vitamin K into two less active R-vitamin K isoforms. Polymorphisms in this enzyme generate less metabolically active enzymes, leading to inadvertent overdosing. Although genetic testing has been available for some time, published studies are conflicting regarding the tests' utility. One study noted more stable dosing but did not affect the time in therapeutic range for the 3-month period of the study. A second study examining the utility of CYP2C9 testing only found that patients achieved a first therapeutic INR and stable INRs sooner than those patients who did not have pharmacogenetic dosing guidance. These patients had more time in therapeutic range and fewer minor bleeds. However, another study notes that genetic testing, when compared to historic controls, resulted in reduced hospitalization during warfarin initiation. A cost-effectiveness study suggests that genetic testing for mutations in enzymes involved in VKA metabolism will only be cost-effective if the time out of range is reduced by 5% to 9% or greater. Despite these cost-related data, U.S. Food and Drug Administration (FDA) package inserts suggest that genetic testing be considered when starting warfarin.

Newer anticoagulants, such as direct thrombin inhibitors, will require revisitation of these bleeding algorithms and will perhaps change the approach to anticoagulation in patients with AF. Furthermore, mechanical intervention is a promising intervention for stroke prevention that minimizes bleeding risk. The PROTECT-AF trial, utilizing the Watchman percutaneous appendage closure device, examined the noninferiority of a percutaneously placed appendage closure device. Although there was a higher risk of adverse events at the beginning of the study, largely due to pericardial effusion, the overall primary efficacy rate (evaluating for ischemic stroke, hemorrhagic stroke, embolic events, or cardiovascular death) was better in the appendage closure device group. Following implant, device implant patients were treated with warfarin for 45 days. This device is still under investigation in the United States, but is marketed in some European countries.

Overall, management of stroke and bleeding risks in the context of AF are in evolution, particularly in the face of both newer genetic insights into warfarin metabolism and newer anticoagulants. This dynamic area dramatically influences quality of life for patients, and given dramatic changes in the short term, close observation is necessary to monitor changes in management needs.

Correct Answers:

Q1. Under the $CHADS_2$ criteria, this patient would score a zero and anticoagulation with a VKA would not be recommended. However, based upon the newer European Society guidelines recommending the use of the CHA_2DS_2-VASc schema, she would score 2 points, suggesting an annual stroke risk of 1.9%. The HAS-BLED scoring system would estimate a bleeding risk of 1.02 bleeds per 100 patient years. Thus, the overall risk–benefit ratio would likely lie on the side of anticoagulating this patient with a VKA.

Q2. Whereas FDA guidelines suggest that genetic testing should be performed, as do the most recent AF guidelines from HRS/AHA/ACC, cost-effectiveness analyses have not demonstrated a cost benefit. Furthermore, Centers for Medicare & Medicaid Services (CMS) does not reimburse for testing, as it is not thought to have clinical benefit.

REFERENCES

1. Miyasaka Y, Barnes ME, Gersh BJ, et al. Secular trends in incidence of atrial fibrillation in Olmsted County, Minnesota, 1980 to 2000, and implications on the projections for future prevalence. *Circulation.* 2006;114:119–125.
2. Schoonderwoerd BA, Smit MD, Pen L, Van Gelder IC. New risk factors for atrial fibrillation: Causes of 'not-so-lone atrial fibrillation.' *Europace.* 2008;10:668–673.
3. Benjamin EJ, Wolf PA, D'Agostino RB, et al. Impact of atrial fibrillation on the risk of death: The Framingham Heart Study. *Circulation.* 1998;98:946–952.
4. Fuster V, Rydén LE, Cannom DS, et al., ACC/AHA TASK FORCE MEMBERS, Smith SC Jr, Jacobs AK, Adams CD, et al., ESC COMMITTEE FOR PRACTICE GUIDELINES, Priori SG, Blanc JJ, Budaj A, et al. ACC/AHA/ESC 2006 Guidelines for the Management of Patients With Atrial Fibrillation–Executive Summary: A Report of the American College of Cardiology/American Heart Association Task Force on Practice Guidelines and the European Society of Cardiology Committee for Practice Guidelines (Writing Committee to Revise the 2001 Guidelines for the Management of Patients With Atrial Fibrillation): Developed in Collaboration With the European Heart Rhythm Association and the Heart Rhythm Society. *Circulation.* 2006;114:700–752.
5. Samsa GP, Matchar DB, Goldstein LB, et al. Quality of anticoagulation management among patients with atrial fibrillation: Results of a review of medical records from 2 communities. *Arch Intern Med.* 2000;160:967–973.

CASE STUDY

51

Newer Oral Anticoagulants in Atrial Fibrillation

Andrew E. Darby, MD and John P. DiMarco MD, PhD

A 76-year-old woman is referred to discuss anticoagulation therapy for paroxysmal atrial fibrillation (AF). She had initially presented with AF 3 years earlier and indicates her episodes occur 4 to 5 times monthly and last 15 to 30 minutes. Her medical history also includes hypertension, dyslipidemia, former tobacco abuse, and peripheral vascular disease with a history of claudication symptoms in her left lower extremity. Her current medications include a daily aspirin (81 mg), losartan 100 mg daily, metoprolol 50 mg twice daily, and simvastatin 20 mg nightly. She has a structurally normal heart by transthoracic echocardiography. At the time of her office visit, she is in sinus rhythm with a heart rate of 65 bpm and her blood pressure is 130/85 mmHg. Her cardiac examination is normal with no murmurs or gallop and there are no signs of congestive heart failure. A bruit is heard over the left femoral artery.

You determine that the patient has a $CHADS_2$ risk score of 2 and, therefore, warrants systemic anticoagulation. The patient expresses reluctance to start warfarin since she wants to avoid the need for frequent blood tests.

Question No. 1:

Does patient need anticoagulation therapy?

A. Yes.

B. No.

Question No. 2:

What current and future alternatives to warfarin may be available for this patient?

DISCUSSION

AF confers an approximate fivefold increased risk of stroke, which is a major cause of disability and a leading cause of death among patients with AF. The vast majority of these strokes are cardioembolic. Systemic anticoagulation with warfarin has been the cornerstone of therapy for stroke prevention among patients who have a definite indication for anticoagulation (eg, patients with a $CHADS_2$ risk score of ≥ 2) and no contraindication. Treatment with warfarin provides a 60% to 70% reduction in stroke risk compared with no therapy and confers a 30% to 40% risk reduction when compared with aspirin. Despite its efficacy, warfarin has important limitations. The anticoagulant effects of the drug must be monitored regularly to ensure adequate stroke prevention, interactions with food and other drugs hamper its use, and the drug has a narrow therapeutic range. In randomized, controlled trials, patients treated with warfarin have a therapeutic international normalized ratio (INR) only two-thirds of the time. In routine clinical practice, an even lower percentage of patients are in range, leaving patients relatively unprotected from

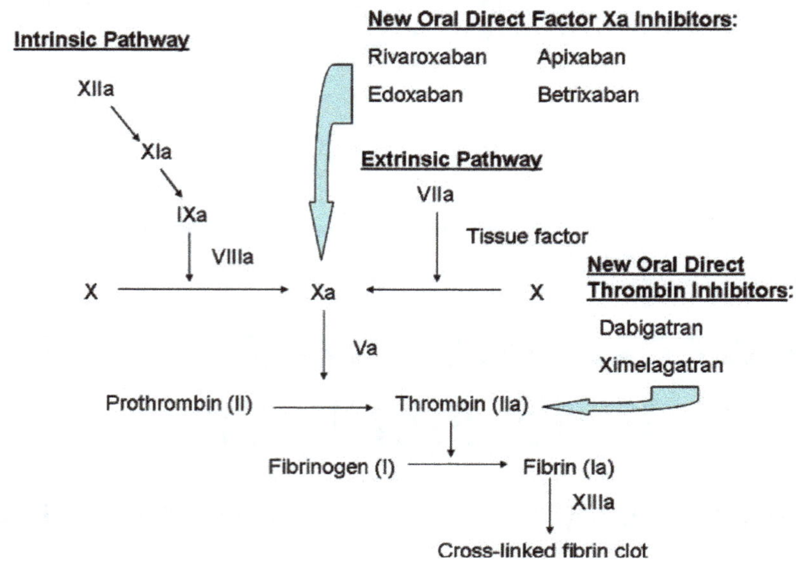

Figure 51.1. Site of action of newer anticoagulants.

thromboembolic complications.[2] An alternative to warfarin would clearly be desired.

New oral anticoagulants now undergoing trials for AF generally have different targets in the anticoagulation cascade than the vitamin K-dependent clotting factors inhibited by warfarin (Figure 51.1). The new oral anticoagulants closest to general release segregate into 2 classes: a direct thrombin (factor IIa) inhibitor (dabigatran) and direct factor Xa inhibitors (rivaroxaban, apixaban, betrixaban, and edoxaban).

Direct Thrombin Inhibitors

Direct thrombin inhibitors have a variety of effects on the coagulation cascade.[3] By inhibiting thrombin activation, the conversion of fibrinogen to fibrin is attenuated. Additional effects include the inhibition of thrombin-mediated activation of clotting factors V, VIII, XI, and XIII as well as the prevention of thrombin-mediated platelet activation. Ximelagatran was the first oral direct thrombin inhibitor studied for stroke prevention in patients with nonvalvular AF. Ximelagatran was shown to be noninferior to warfarin in both efficacy and safety end points.[4,5] However, a potentially higher risk for drug-induced liver injury with ximelagatran prompted its withdrawal from all markets and halted further development of the drug.

A second oral direct thrombin inhibitor, dabigatran, has recently completed clinical trials and has been approved for use in the United States by the Food and Drug Administration (FDA).

Dabigatran is a reversible direct inhibitor of thrombin with a rapid onset of action and 6.5% oral bioavailability not affected by food intake.[3] Maximum effects are achieved within 2 hours of administration, and the drug has a half-life of 14 to 17 hours. Approximately 80% of the drug is excreted renally. Importantly, it has shown no interaction with cytochrome p450 enzymes, although P-glycoprotein inhibitors (amiodarone, verapamil, and quinidine) may interfere with excretion and increase plasma concentrations of dabigatran. The effects of dabigatran on the prothrombin time (PT)/INR are variable, but it does prolong the activated partial thromboplastin time (aPTT), which may be used to assess for the anticoagulant effects of dabigatran. The largest study to date using dabigatran for stroke prevention in AF was the large, phase III study, RE-LY (Randomized Evaluation of Long-term Anticoagulation Therapy).[6] The study was a noninferiority trial comparing dabigatran with warfarin anticoagulation in 18,113 patients with AF and at least one traditional risk factor for stroke (mean $CHADS_2$ score of 2.1). Patients were randomized to receive either adjusted-dose warfarin to achieve an INR of 2.0 to 3.0 or a fixed dose of 110 mg or 150 mg dabigatran twice daily. The primary outcome was the rate of stroke or systemic embolism over a median follow-up of 2 years. Patients taking 110 mg dabigatran twice daily had similar rates of stroke or systemic embolism when compared with warfarin (1.5% vs 1.7%, $P = 0.34$), thereby meeting the criterion of noninferiority. Patients taking 150 mg dabigatran twice daily reached the primary end point less often when compared with warfarin (1.1% vs 1.7%, $P < 0.001$), resulting in this dose being not only noninferior but also superior to warfarin for stroke prevention. Patients taking dabigatran also had lower rates of hemorrhagic stroke (low dose 0.12% per year, high dose 0.10% per year) than did those taking warfarin (0.38% per year). The rate of major bleeding events was similar between high-dose dabigatran and warfarin (3.1% vs 3.4% per year), although more gastrointestinal bleeding events occurred with high-dose dabigatran (1.5% vs 1.0% per year). For unclear reasons, a higher rate of myocardial infarction was observed with both doses of dabigatran when compared with warfarin (0.72% and 0.74% vs 0.53% per year), and there was also a trend toward more pulmonary emboli in those receiving dabigatran. Further subanalysis and future trials will likely provide insight into these findings. Based upon the overall safety and efficacy of dabigatran, the FDA approved it for use in October 2010 for the anticoagulation of patients with nonvalvular AF. The FDA approved 150 mg dabigatran twice daily as studied in RE-LY as well as a lower dose, 75 mg twice daily, for patients with significant renal impairment (defined as a creatinine clearance of 15–30 mL/min). The drug is contraindicated in patients who have a creatinine clearance <15 mL/min.

Factor Xa Inhibitors

The second category of drugs under investigation for anticoagulation in AF are the factor Xa inhibitors. Inhibition of factor Xa prevents the conversion of prothrombin to thrombin (see Figure 51.1). Factor Xa inhibitors generally prolong the prothromin time in a dose-dependent fashion.[7] Several of these agents are currently being investigated in phase II or III trials and may be available for use in the near future. Drugs in this class include rivaroxaban, apixaban, betrixaban, and edoxaban.

Rivaroxaban is an oral direct factor Xa inhibitor with an oral bioavailability of 80%, a plasma half-life of 7 to 11 hours, and two-thirds of the drug undergoes metabolism in the liver while one-third is excreted renally.[8] The efficacy and safety of rivaroxaban in AF has recently been investigated in the phase III ROCKET AF study, a randomized, double-blind study comparing rivaroxaban with warfarin for stroke prevention in AF.[9] The results of the ROCKET AF study were presented at the 2010 American Heart Association meeting. The study randomized over 14,000 patients to 20 mg of rivaroxaban daily or dose-adjusted warfarin and determined that rivaroxaban was noninferior to warfarin with regard to all-cause stroke and systemic embolic events. This drug was recently approved by the FDA for clinical use.

Apixaban is another factor Xa inhibitor currently under investigation in patients with AF.[7] A phase III study, AVERROES, compared apixaban to aspirin for the prevention of stroke or embolic events in patients with AF deemed unsuitable for warfarin therapy. The study was halted early after enrolling 2809 patients, as apixaban was found to be superior with a 56% reduction in stroke or embolic events with no increase in bleeding complications. Another large phase III study, ARISTOTLE (Apixaban for the Prevention of Stroke in Subjects with Atrial Fibrillation), is comparing apixaban to warfarin and has enrolled over 18,000 patients.[10] Results of the study are expected to be reported in the near future. Two additional factor Xa inhibitors, betrixaban and edoxaban, are being studied in patients with AF. A phase II study of betrixaban in AF was presented at the 2010 American College of Cardiology meeting. The trial, EXPLORE-Xa, revealed the safety of betrixaban when compared with warfarin.[11] Of note, betrixaban is the only new oral anticoagulant that undergoes almost exclusive hepatic excretion, making it a potentially appealing option for patients with advanced kidney disease.[7] A phase III trial (ENGAGE-AF-TIMI 48) is currently comparing edoxaban with warfarin for the prevention of stroke in patients with AF.

Correct Answers:

Q1. A

Q2. Direct thrombin inhibitor (dabigatran) and factor Xa inhibitors (rivaroxaban, apixaban, betrixaban, and edoxaban).

QUESTIONS AND AUTHORS' REPLY

Question 1. How may one monitor the anticoagulant effects of dabigatran?

Answer: Dabigatran has variable effects on the INR. The test is therefore not reliable for monitoring the effects of the drug as is done routinely with warfarin. The ecarin clotting time (ECT) appears to increase linearly with increasing dabigatran concentrations, although the test is not performed in all laboratories and, therefore, not likely to be clinically useful. The aPTT increases in a nonlinear manner and is less precise than the ECT. It is useful, however, as a way to monitor for the presence of dabigatran in the patient's system, as values of 1.5 to 2 times control are indicative of the anticoagulant effects of dabigatran.

Question 2. For whom is the 75-mg twice daily dose of dabigatran indicated?

Answer: The lower dose of dabigatran is marketed for patients with advanced renal disease (specifically, a creatinine clearance of 15–30 mL/min). The drug is contraindicated in patients with a creatinine clearance < 15 mL/min. Patients with a creatinine clearance > 30 mL/min should receive the 150-mg twice daily dose.

Question 3. What are the most common side effects associated with dabigatran?

Answer: In the RE-LY study, 16% of patients receiving dabigatran 150 mg twice daily discontinued the drug at 1 year and 21% discontinued the drug at 2 years. In contrast, the discontinuation rates with warfarin were 10% and 17% at 1 and 2 years, respectively. The only side effect observed more frequently with dabigatran was dyspepsia, occurring in 11.3% of patients taking the higher dose of the drug. Dyspepsia is the most common side effect as dabigatran requires an acidic pH for absorption. Dabigatran capsules have a tartaric acid core, which is released in the stomach as the capsule dissolves, thereby lowering the gastric pH and potentially causing dyspepsia.

Question 4. Is there a reversal agent for patients having a hemorrhagic complication while taking dabigatran?

Answer: There is no intervention to specifically reverse the anticoagulant effects of dabigatran. Potential therapies include the transfusion of clotting factors through fresh frozen plasma (FFP), the administration of recombinant factor VII, and, potentially, hemodialysis as the drug primarily undergoes renal excretion.

REFERENCES

1. Hart RG, Pearce LA, Aguilar MI. Meta-analysis: Antithrombotic therapy to prevent stroke in patients who have nonvalvular atrial fibrillation. *Ann Int Med.* 2007;146:857–867.
2. Connolly SJ, Pogue J, Eikelboom J, et al.; ACTIVE W Investigators. Benefit of oral anticoagulant over antiplatelet therapy in atrial fibrillation depends on the quality of international normalized ratio control achieved by centers and countries as measured by time in therapeutic range. *Circulation.* 2008;118:2029–2037.
3. Siddiqui FM, Qureshi AI. Dabigatran etexilate, a new oral direct thrombin inhibitor, for stroke prevention in patients with atrial fibrillation. *Expert Opin Pharmacother.* 2010;11:1403–1411.
4. Olsson SB; Executive Steering Committee of the SPORTIF III Investigators. Stroke prevention with the oral direct thrombin inhibitor ximelagatran compared with warfarin in patients with nonvalvular atrial fibrillation (SPORTIF III): Randomised controlled trial. *Lancet.* 2003;362:1691–1698.
5. Albers GW, Diener HC, Frison L, et al.; SPORTIF Executive Steering Committee for the SPORTIF V Investigators. Ximelagatran vs warfarin for stroke prevention in patients with nonvalvular atrial fibrillation. *JAMA.* 2005;293:690–698.
6. Connolly SJ, Ezekowitz MD, Yusuf S, et al.; RE-LY Steering Committee and Investigators. Dabigatran versus warfarin in patients with atrial fibrillation. *N Engl J Med.* 2009;361:1139–1151.
7. Schirmer SH, Baumhäkel M, Neuberger HR, et al. Novel anticoagulants for stroke prevention in atrial fibrillation: Current clinical evidence and future developments. *J Am Coll Cardiol.* 2010;56:2067–2076.
8. Kubitza D, Becka M, Mueck W, Zuehlsdorf M. Safety, tolerability, pharmacodynamics, and pharmacokinetics of rivaroxaban – an oral, direct factor Xa inhibitor – are not affected by aspirin. *J Clin Pharmacol.* 2006;46:981–990.
9. ROCKET AF Study Investigators. Rivaroxaban – once daily, oral direct factor Xa inhibition compared with vitamin K antagonism for prevention of stroke and Embolism Trial in Atrial Fibrillation: Rationale and design of the ROCKET AF study. *Am Heart J.* 2010;159:340–347.
10. Lopes RD, Alexander JH, Al-Khatib SM, et al.; ARISTOTLE Investigators. Apixaban for reduction in stroke and other ThromboemboLic events in atrial fibrillation (ARISTOTLE) trial: Design and rationale. *Am Heart J.* 2010;159:331–339.
11. Jacobshagen C, Westermann D, Holubarsch C, Böhm M. Hotline update of clinical trials and registries presented at the American College of Cardiology Congress 2010: ACCORD, INVEST, NAVIGATOR, RACE II, SORT OUT III, CSP-474, DOSE, ASPIRE and more. *Clin Res Cardiol.* 2010;99:337–344.

CASE STUDY

52

Antiarrhythmic Drug Therapy

Jasbir S. Sra, MD

Question No. 1:

Which of the following is true regarding interaction between ion channels and antiarrhythmic drugs?

A. The Na channel shows transition between open and inactivated states.

B. At any point in time, the Na channel may be open or inactivated.

C. Tonic block refers to the drug-induced reduction of channel availability after a long rest period.

D. Use-dependent block implies progressive increase in Na^+ current during repetitive depolarization.

E. The association rate constant is more helpful in classifying antiarrhythmic drugs than the dissociation rate constant.

ANTIARRHYTHMIC DRUGS AND MOLECULAR AND CELLULAR TARGETS

Na^+ Channels

The modulated receptor hypothesis suggests that:

- Na channels exist in 3 primary states: rested (R, a closed state most prevalent at negative membrane potentials); inactivated (I, a closed state at depolarized membrane potentials); and open or activated (A). This means that the R state is more prevalent during diastole, the A state during the upstroke, and the I state more prevalent during the plateau or when the tissue is depolarized. Each Na channel shows a transition between the R, A, and I states. A substantial fraction of the Na channels undergo inactivation before they open. The implication of this is that both processes, that is, channel opening or inactivation, occur simultaneously. Tonic block refers to the drug-induced reduction of channel availability after a long rest period.
- Drugs can interact with resting, activated, and inactivated states of Na^+ channels. The voltage dependence of inactivation of drug-bound channels is shifted to more negative potentials.
- Affinity of the drugs is greater for activated and inactivated states than the resting state.
- Activated state blockers include class IA and class IC drugs, such as disopyramide, procainamide, quinidine, flecainide, and propafenone. Inactivated state blockers include amiodarone, lidocaine, and mexiletine.
- Presence of two or more drugs usually results in additive block. Mexiletine plus quinidine or disopyramide may have more effect than either of them alone. A drug with fast association and dissociation kinetics may displace a drug with slower kinetics from the receptor. As Na^+ ions cross the membrane by binding to multiple pores, Na^+ ion-containing solutions could be used to reverse the side effects of Na^+ channel-blocking drugs.

Practical Electrophysiology © 2015 Jasbir S. Sra, Masood Akhtar, Andrea Natale, David J. Wilber
Cardiotext Publishing, ISBN: 978-1-935395-17-1

- Conditions such as ischemia or hyperkalemia cause membrane depolarization, leading to an increased number of Na^+ channels in the inactivated state. This explains the efficacy of drugs such as lidocaine and amiodarone in these clinical settings.
- Drugs fall into three groups according to the rate at which they block. Class IB drugs, such as lidocaine, have fast kinetics. Class IA drugs, such as quinidine, have intermediate kinetics, and class IC drugs, such as flecainide, have slow kinetics. Drugs fall into three groups based upon the development of a steady state of frequencies over a wide range.
  1. *Group I:* Fast association and dissociation constants. Class IB drugs, such as lidocaine and mexiletine, fall into this group.
  2. *Group II:* Slow association and dissociation constants. This group includes class IA and class IC drugs, such as quinidine, procainamide, and flecainide.
  3. *Group III:* Slow association and fast dissociation constants. Disopyramide and propafenone are included in this group.
- Use-dependent block is seen in group II where, although there is a significant block at rest, the block is further increased during faster heart rates. There is a progressive decrease of Na^+ current during repetitive depolarization. Although Group III drugs have increased block at rest, similar to Group II, no additional block occurs at increased heart rates.
- Tonic block refers to continuous access to binding site in the absence of activation.

K^+ Channel Blockers

- K^+ channel blockers prolong action potential duration (APD) and increase refractoriness. Sotalol, amiodarone, and bretylium are relatively nonselective. Dofetilide selectively blocks the delayed rectifier I_{Kr}.
- As opposed to Na^+ channel blockers, pure K^+ channel blockers (eg, sotalol) produce maximal prolongation of APD at slower rates and have diminished effects at faster rates. This behavior is referred to as reverse use-dependent. Amiodarone, however, does not show reverse use-dependence and maintains its efficacy over a wide range of cycle lengths.

Ca^{++} Channel Blocking Agents

The L-type Ca^{++} channel has three different kinetic states depending upon the transmembrane voltage changes:

- As voltage descends below –40 mV, the channel is in resting state. It goes through the activation and inactivation stages during and after depolarization. Transition to inactivated state is voltage- and Ca^{++} dependent. Increased intracellular Ca^{++} accelerates the development of the inactivation state.
- Ca^{++} channels can be blocked at distinct binding sites by three major classes of drugs: dihydropyridines, phenylalkalamines, and benzothiazepines. Nifedipine belongs in the dihydropyridine class, verapamil in the phenylalkalamine class, and cardiazem in the benzothiazepine class.
- The receptor for dihydropyridines is located extracellularly. They have increased affinity for the channel in the inactivated state. They unbind rapidly from their receptors and do not show use-dependence.
- Phenylalkalamines bind to receptors located on the intracellular surface of Ca^{++} channels. They interact with the channels in the open (activated) state. They unbind slowly (slow dissociation constant) and show use-dependent block, that is, more channel blockade at faster rates.
- Benzodiazepines demonstrate kinetics intermediate between dihydropyridines and phenylalkalamines. They work in the inactivated state.

Question No. 2:

With regard to impulse propagation in the heart:

A. Enhanced normal automaticity results from Ca^{++} overload.

B. Accelerated ventricular rhythm postmyocardial infarction is an example of triggered activity and results from decreased refractoriness.

C. Early afterdepolarizations (EADs) occur during phase 2 of action potential.

D. Delayed afterdepolarizations (DADs) occur due to outward Na^+ current.

Mechanisms of arrhythmias include both enhanced normal automaticity and abnormal automaticity, triggered activity based upon EAD and DAD, and reentry. Some other mechanisms include reflection and parasystole.

CELLULAR MECHANISM OF NORMAL AND ABNORMAL AUTOMATICITY

Normal Automaticity

In the sinoatrial node, there is a slow diastolic (phase 4) depolarization from a maximum diastolic potential of about –60 mV. This is characteristic of automatic cells. There is closure of K^+ channels, and pacemaker-inward current I_f, carried by Na^+ and K^+ ions, is activated at about –45 mV. I_f is also a major current responsible for diastolic depolarization in Purkinje cells. Ca^{++} channels in the sinus node and Na^+ channels in the Purkinje cells cause rapid depolarization. Catecholamines increase the heart rate by moving the potential for I_f activation to more positive potentials and by increasing I_{Ca} and I_K. Parasympathetic stimulation causes bradycardia by: (1) activating K^+ current I_{KACH}; (2) reduction of I_{Ca}; and (3) reduction of I_f. Arrhythmias associated with enhanced normal automaticity include inappropriate sinus tachycardia. The vulnerable parameter is phase 4 depolarization.

Abnormal Automaticity

Abnormal automaticity is the mechanism by which spontaneous impulses are generated in tissues that are partially depolarized due to disease. During reduction of maximum diastolic potential from –75 mV to –50 mV as a result of ischemia, depolarization during phase 4 cannot occur due to I_f. Rather, it occurs due to decay of I_K and the action potential generated by $I_{Ca\text{-}L}$. Examples of arrhythmias caused by abnormal automaticity are: certain ectopic atrial tachycardias and accelerated idioventricular rhythms, and ventricular tachycardia (VT) occurring 24 to 72 hours after myocardial infarction.

Triggered Activity

Triggered activity refers to arrhythmias that arise as a consequence of depolarization. Afterdepolarizations can interrupt or occur after completion of repolarization and are termed EAD or DAD. EADs arise from the reduced level of membrane potential during phase 2 or from a more normal voltage potential during phase 3. The primary mechanisms include reduction of I_K and abnormal prolongation of inward currents carried by Ca^{++} or Na^+ or both. DAD is caused by the transient inward Na^+ current I_{TO} that is carried by the Na^+/Ca^{++} exchanger, and is small or absent under normal physiological conditions.

EADs are secondary depolarizations arising from the incomplete repolarized membrane. EADs are associated with APD prolongation, which is reflected as QT interval prolongation on the surface ECG. This may cause torsade de pointes. Blockade of outward K^+ current during repolarization with drugs, such as sotalol or dofetilide, or enhancement of inward current, such as Ca^{++} or Na^+, will cause QT prolongation. Rapid upstroke of EADs due to ionic currents may occur due to Ca^{++} channels in the early phase and Na^+ channels at more negative potentials.

DADs are secondary depolarizations that arise from a fully repolarized membrane. DADs may arise under conditions causing intracellular Ca^{++} overload such as ischemia or digitalis intoxication. DADs can also occur during catecholamines and positive inotropic agents. Acceleration of the heart rate can increase the amplitude of DADs, eventually making them reach threshold potential.

Reentry

Circus movement tachycardia causing reentry has an excitable gap. A premature extrastimulus can result in premature excitation of the reentrant loop. For reentry to occur, the length of the path must exceed the wavelength determined by effective refractoriness and there must be unidirectional block. Electrophysiologic parameters such as conduction are anisotropic. Anisotropy is determined by fiber orientation (see previous cases for more details).

Wavelength equals conduction velocity × refractory period: This relationship between wavelength and path length can be achieved by slowing conduction uniformly along the whole path length, slowing conduction locally, or shortening the refractory period. Cellular depression of the level of I_{Na} and I_{Ca} may cause the slowing of conduction. Depending upon the speed of impulse propagation and the effective refractory period, an excitable gap may be present in the reentrant circuit. The excitable gap may be present at some sites but not at others. Reentry can be suppressed by altering the two vulnerable periods, conduction and refractoriness. Examples of arrhythmias with a long excitable gap include typical atrial flutter and monomorphic VT. Depressing conduction will help alleviate these arrhythmias. Arrhythmias such as atrial fibrillation, polymorphic VT, and ventricular fibrillation (VF) have a short excitable gap. Prolongation of refractory period will be needed to suppress these arrhythmias.

Reflection

Reflection can be seen in a single bundle. Reflection is caused by electrotonic transmission across an inexcitable segment in a linear bundle. The proximal fibers recover excitability due to delay in impulse transmission. This allows these fibers to be made available for excitation by current flowing retrogradely across the inexcitable zone.

Parasystole

During parasystole, there is abnormal automatic impulse initiation, which is affected by exit block. For example, an automatic focus in Purkinje fibers may be surrounded by an area of inexcitable fibers, thus protecting this focus from sinus depolarizations. However, subthreshold depolarizations can delay or accelerate the automatic firing. Parasystole is usually benign but could conceivably provide the trigger for sustained arrhythmias.

Question No. 3:

During ischemia, the following is true:

A. There is increase in action potential amplitude and APD.

B. Channels, such as I_{KATP}, are open.

C. There is decrease in gap junction resistance.

ISCHEMIC CHANGES IN THE MYOCARDIUM

Several changes occur at the cellular level during ischemia, leading to ischemia-related electrophysiologic changes such as vulnerability to VT/VF.

- Initial ischemic changes occur in the subepicardium.
- There is a decrease in action potential amplitude and APD. The cells depolarize from –60 to –50 mV within 4 to 10 minutes.
- The K^+ channel I_{KATP} is normally closed by intracellular ATP but opens up when ATP levels are low, such as during ischemia or hypoxia. The channel is also present in pancreatic cells and is blocked by sulfonylureas, such as glibenclamide and glipizide. During myocardial ischemia, there is a rapid loss of intracellular K^+, causing membrane depolarization, slowing of conduction, and altered refractoriness. Increased K^+ conductance is due to low ATP levels reducing ATP-inhibition of I_{KATP} levels, leading to action potential shortening during acute myocardial ischemia.
- Metabolic acidification occurs due to hydrolysis of ATP and extracellular transfer of K^+.
- There is an increase in gap junction resistance. This causes electrical cell-to-cell uncoupling and conduction block. The initial phase of conduction block occurs due to cellular inexcitability.
- Sympathetic activity is enhanced—refractory period may increase due to inhomogeneous effects.
- Accelerated idioventricular rhythm is a form of reperfusion arrhythmia found after myocardial infarction. The mechanism is enhanced automaticity, and beta-adrenoceptor blockade can suppress this arrhythmia.

Question No. 4:

Which of the following is true for dofetilide?

A. Properties are concentration dependent.

B. It is predominantly excreted renally as a metabolite.

C. Verapamil use is contraindicated.

D. A and C.

DOFETILIDE

- Dofetilide has class III (APD prolongation) properties, which are concentration dependent. The drug acts by selectively blocking the delayed rectifier K^+ current I_{Kr}.
- It does not cause changes in PR interval or QRS duration. It does not increase defibrillation thresholds (DFTs). It has no negative ionotropic effect.
- Dofetilide has >90% bioavailability. About 80% of the substance is excreted by the kidneys in an unchanged state. Twenty percent is excreted as inactive or minimally active metabolites. A small amount of the drug is metabolized in the liver.
- Increased QT interval is directly related to dofetilide dose and plasma concentration.
- Dose-calculated creatinine clearance. In females, creatinine clearance is calculated by multiplying the formula result by 0.85.
- When creatinine clearance is <20 mL/min, dofetilide is contraindicated and recommended dose with 20 to 40 mL/min is 125 mcg twice daily; for creatinine clearance of 40 to 60 mL/min, it is 250 mcg twice daily; and for creatinine clearance > 60 mL/min, the dose can be 500 mcg twice daily.
- Dofetilide is contraindicated in patients with congenital or acquired long QT syndrome. Concomitant use of verapamil and cationic transport inhibitors such as cimetidine is also contraindicated.

Question No. 5a:

Which of the following is true?

A. Class I drugs depress phase 0.

B. Class I drugs can slow conduction.

C. Both are true.

CLASSIFICATION OF ANTIARRHYTHMIC DRUGS

The predominant approach to antiarrhythmic drug development and administration relies on the Singh–Vaughan-Williams classification. Despite several limitations, this classification has been useful in physician education and can be learned quickly. Based upon this classification, the available antiarrhythmic drugs are divided into those that block Na^+ channels, beta-blockers, K^+ channel blockers, and calcium channel blockers. Newer drugs such as dronaderone were not added initially in this classification.

Class I Drugs

IA (i) Depress phase 0
(ii) Slow conduction
(iii) Prolong repolarization

IB (i) Depress phase 0 (V_{max}) in diseased tissue only
(ii) Do not slow conduction
(iii) Shorten repolarization

IC (i) Markedly slow conduction
(ii) Markedly depress phase 0

Question No. 5b:

Which of the following is true with regard to onset/offset kinetics of antiarrhythmic drugs?

A. Mexiletine has fast-onset/slow-offset kinetics.
B. Flecainide has fast-onset/slow-offset kinetics.
C. Quinidine has slow-onset/fast-offset kinetics.
D. Disopyramide has fast-onset/slow-offset kinetics.

ONSET/OFFSET KINETICS

Free drug concentration may vary at different binding sites. However, dissociation of the drug is concentration independent and is also useful in classifying antiarrhythmic agents. The rate of dissociation of a drug from its receptor is a major determinant of the equilibrium block observed during repetitive stimulation. Dissociation may occur at depolarized potentials, for example, during the action potential plateau or during intervals between action potentials. Based upon the dissociation rate constants at plateau and normal resting potentials, antiarrhythmic drugs have been divided into three groups. Class IB drugs, such as mexiletine and lidocaine, have fast-onset/fast-offset kinetics (fast dissociation rates). Class IA and class IC drugs, such as quinidine, procainamide, and flecainide, have slow-onset/slow-offset kinetics. The third group consists of drugs, such as disopyramide and propafenone, which have fast-onset/slow-offset kinetics.

Question No. 6:

Which of the following is true with regard to calcium channel blockers and their effect on channel state and recovery kinetics?

A. Phenylalkalamines act primarily on the inactivated channels and have slow recovery kinetics.
B. Dihydropyridines act primarily on inactivated channels and have fast recovery kinetics.
C. Benzothiazepines act primarily on inactivated channels and have fast recovery kinetics.
D. All of the above are true.

CALCIUM CHANNEL BLOCKERS

The effects of phenylalkalamines (eg, verapamil), dihydropyridines (eg, nifedipine), and benzothiazepines (eg, cardiazem) on channels and recovery kinetics are shown in Table 52.1.

Question No. 7:

Which of the following drugs are most likely to result in maximum prolongation of action potential when cycle length changes from 500 ms to 1200 ms?

A. Flecainide.
B. Sotalol.
C. Propafenone.
D. A and C.

Table 52.1. Effects of Calcium Channel Blockers on Channel State and Recovery Kinetics

	Channel State		Recovery Kinetics
	Activated (Open)	Inactivated	
Phenylalkalamines (verapamil)	++	—	Slow
Dihydropyridines (nifedipine)	++	+	Fast
Benzothiazepines (cardiazem)	—	++	Intermediate

PROLONGATION OF APD

Class IA and class III drugs prolong APD and refractoriness by causing K^+ channel blockade. K^+ channel blocking drugs such as *N*-acetylprocainamide, quinidine, and sotalol cause less action potential prolongation at rapid heart rates than at slower heart rates. This is termed reverse use-dependence. Action potential prolongation and an increase in refractoriness are desirable at the rapid rates observed during tachyarrhythmias. There is a downside, however, as greater action potential prolongation at slow rates may precipitate torsade de pointes. This phenomenon is attributed to a reduction in K^+ channel blockade during the plateau or at positive potentials such that, during increased rates, the block decreases. In contrast, antiarrhythmic drugs such as flecainide demonstrate use-dependent block. Use-dependent block occurs whenever the affinity for open or inactivated channels exceeds the affinity for rested channels. Consequently, the more frequently the channels are used, the more block occurs, that is, block is frequency dependent.

Question No. 8:

Which of the following is true?

A. Most drugs are eliminated by zero-order kinetics.

B. Antiarrhythmic drugs have wide toxic-to-therapeutic ratio.

C. Loading dose determines the time to steady-state concentration of a drug.

D. None of the above.

PHARMACOKINETICS, PHARMACODYNAMICS, AND PHARMACOGENETICS

- Pharmacokinetics is the study of time-course of drug absorption, metabolism, distribution, and excretion. Pharmacodynamics is the relationship between the drug concentration at the active pharmacological site and the response. The term pharmacogenetics is used to describe the effect of drug-metabolizing enzymes, which may be genetically determined. Some adverse drug interactions may be related to this genetic alteration rather than to altered plasma concentrations alone. Most antiarrhythmic drugs are basic compounds, and hence they are ionized and poorly absorbed at normal gastric PH. Therefore, most drug absorption occurs in the intestinal tract.
- *Bioavailability:* Bioavailability is the amount of drug reaching the systemic circulation. Bioavailability is reduced if the drug undergoes extensive first-pass metabolism in the gut wall (ethanol) or liver (propranolol). One-compartment and two-compartment models are used to describe and predict the serum concentration of a drug at a given time. The central compartment is smaller and consists of blood volume and extracellular fluid volume. A larger, peripheral compartment consists of skin, muscle, and adipose tissue. The plasma drug concentration curve demonstrates two distinct phases: the alpha-phase, characterized by rapidly falling plasma drug concentration due to distribution between the central and peripheral compartments; and the beta-phase, the elimination phase, which is characterized by slower decline in concentration. For example, lidocaine, administered via intravenous bolus, has a biexponential drop in plasma concentration with a rapid (alpha) half-life of 8 minutes due to distribution and a slower (beta) half-life of 120 minutes due to elimination.

STEADY-STATE CLEARANCE AND ELIMINATION HALF-LIFE

- During drug administration, steady state is achieved when the amount of drug entering the plasma is equal to the drug leaving. Time to steady state of the drug depends entirely upon elimination half-life and is defined as dose-unit time divided by clearance. Thus, loading dose **does not reduce** the time to steady state. Most antiarrhythmic drugs are eliminated by **first-order** kinetics, which means that fraction of the drug eliminated per unit time remains constant, regardless of the amount of drug in the body. For example, 50% of the drug is eliminated after one half-life, 75% after two half-lives, and so on. Zero-order kinetics indicates that elimination occurs at a constant rate and cannot increase further despite increased drug concentrations. Clearance is expressed as volume of drug cleared per unit time. As opposed to elimination half-life, which determines the time to steady state, clearance and dose determine the actual steady state achieved.

VOLUME OF DISTRIBUTION

- Extent of extravascular distribution of a drug is obtained by measuring the apparent volume of distribution that is the hypothetical volume by which a drug would have to be diluted to give the observed plasma concentration. It is determined by dose administered divided by plasma concentration at time zero. Large volume of distribution indicates side distribution and extensive tissue uptake. During congestive heart failure, volume of distribution is decreased. Therefore, plasma concentration of drugs like lidocaine may increase.

PROTEIN BINDING AND EXCRETION

- Most drugs are bound to plasma proteins. Only the unbound drug is available for pharmacological effects. Some drugs such as lidocaine and quinidine are extensively bound, not only to albumin but also to acute-phase reactants, such as α-1-acid glycoprotein, which **rise** during stress and acute myocardial infarction. Thus, the drug concentrations may be lower during acute myocardial infarction. Most antiarrhythmic drugs have hepatic excretion except sotalol, digoxin, dofetilide, and *N*-acetylprocainamide, which predominantly have renal excretion.

DRUG DISPOSITION BASED UPON OXIDATION, CONJUGATION, AND P-GLYCOPROTEINS

- The process of hepatic metabolism involves oxidation and conjugation. Two enzymes, CYP2D6 and CYP3A4, are involved in oxidation. The process of conjugation involves acetylation, a good example being conversion of procainamide to *N*-acetylprocainamide. About 50% of Caucasians and African Americans lack *N*-acetyltransferase.
- *CYP2D6:* About 7% of Caucasians and African Americans lack the enzyme. Quinidine, fluoxetine, and some CYP2D6 substrates such as flecainide and propafenone are inhibitors of this enzyme. The substrates for P450 include: propafenone, debrisoquine, metoprolol, propranolol, phenytoin, and codeine.
- *CYP3A4:* CYP3A4 is the most abundant hepatic enzyme. Macrolide antibiotics such as erythromycin, the antifungal agent ketoconazole, grapefruit juice, and cimetidine inhibit it. CYP3A4 activity is increased by such agents as phenytoin, Phenobarbital, and rifampin. CYP3A4 biotransforms a large number of substrates including lidocaine, quinidine (which inhibits CYP2D6), Ca^{++} channel blockers, antihistamines, cisapride, cyclosporins, and HMG-CoA reductase P-glycoprotein. In addition to tumor cells, P-glycoprotein is seen at multiple sites, including renal tubular and intestinal cells, hepatocytes and blood vessels at the blood main barrier. P-glycoprotein helps to eliminate drug at these sites. A classic example involves increased digoxin concentration during concomitant quinidine administration, which inhibits P-glycoprotein in renal tubular cells. In addition to quinidine, the inhibitors of P-glycoprotein include verapamil and cyclosporin. In addition to digoxin, the substrates for P-glycoprotein are quinidine, verapamil, cyclosporin, and some anticancer drugs.

Question No. 9:

Principal blocking of sites occurs during the inactivated state in which of the following antiarrhythmic drugs?

A. Quinidine.
B. Disopyramide.
C. Lidocaine.
D. Propafenone.

Drugs such as propafenone, quinidine, and disopyramide produce an initial rapid phase of block during the first few milliseconds following depolarization with little increase in block occurring during further depolarization. These drugs are called open (activated) state blockers. Other drugs such as lidocaine and mexiletine produce a pattern of more slowly developing block and are termed inactivated state blockers. As action potential in the atrium is significantly shorter than in the ventricular myocardium and the His-Purkinje system, the inactivated state blockers have less time to access the Na channel. Therefore, agents such as lidocaine have less effect on the atrial arrhythmias. In contrast, agents that produce rapid block should have similar effects in the atrium and the His-Purkinje system. An agent that increases APD should, in principle, increase the time available for binding of agents that act during the inactivated state. A combination of quinidine and mexiletine are representative of this kind of favorable effect.

Question No. 10:

A 60-year-old gentleman with a history of recurrent monomorphic VT was treated with amiodarone. His condition remained stable for 2 years after which he began to experience episodes of breakthrough VT. Cardiac evaluation, including a cardiac catheterization, did not reveal progression of coronary artery disease or deterioration of left ventricular function. The patient states that he was compliant with use of his medication. Which of the following tests would be most helpful in directing further management of his arrhythmia?

A. Amiodarone levels.

B. Desethylamiodarone levels.

C. T4.

D. T3.

E. Reverse T3.

The effective plasma concentration of amiodarone is thought to be in the 1 to 3 µg/mL range. However, the therapeutic plasma concentration range of amiodarone, using a dose–response curve, is not known. Furthermore, the tissue concentration of amiodarone is significantly higher than the plasma concentration. Current ventricular arrhythmias do occur at levels <1 mg/mL during drug withdrawal. Thus, low levels could be useful in determining noncompliance or inadequate dosage. The relevance of desethylamiodarone as a clinically active metabolite and its plasma and tissue concentration levels are uncertain. In addition to its class I to class IV effects, amiodarone is known to inhibit the effect of thyroxine on the myocardium. Amiodarone can cause both hypothyroidism and hyperthyroidism. Sudden recurrence of VT, in a patient stable on amiodarone, may be suggestive of hyperthyroidism. Due to the drug's inhibition of peripheral conversion of T4 to T3, some biochemical abnormalities may occur initially then return to normal in a few months. Hypothyroidism is diagnosed if T3 has fallen, T4 has fallen or fails to rise, and/or thyroid-stimulating hormone has increased. Hyperthyroidism is diagnosed if serum T3 increases. Although reverse T3 also has been used to monitor amiodarone's effects, the electrophysiologic properties of the drug are not directly related to reverse T3. Due to the similarity of the molecular structure of amiodarone and desethylamiodarone to thyroid hormone, amiodarone mimics hypothyroidism by inhibiting the T3 nuclear receptor. This is a possible explanation for the delayed effect of amiodarone, that is, dynamic changes in thyroid function may be time dependent.

Question No. 11:

Which of the following is true with regard to procainamide-induced lupus?

A. It is seen in rapid acetylators.

B. It is seen in slow acetylators.

C. It occurs due to *N*-acetylprocainamide.

D. It affects predominantly the kidneys.

PROCAINAMIDE-INDUCED LUPUS

Two processes, oxidation and conjugation, are involved in hepatic metabolism of antiarrhythmic drugs. Conjugation involves acetylation of drugs like procainamide to *N*-acetylprocainamide, the result of which has different properties than the parent compound. Procainamide, not *N*-acetylprocainamide, causes drug-induced lupus prior to undergoing conjugation. Thus, individuals who lack *N*-acetyltransferase (about 50% of blacks and Caucasians) are more at risk for lupus since the presence of this enzyme affects rate of acetylation. Procainamide-induced lupus does not affect the kidneys or the brain.

Question No. 12:

A 65-year-old gentleman with new-onset atrial fibrillation was stable on propafenone for 6 months. He presented with history of dizziness and is found to have his heart rate ranging between 30 and 40 bpm. The patient's history was unremarkable except for the fact that he was started on fluoxetine for depression 6 weeks prior. The probable cause for the patient's bradycardia is:

A. Increased dealkylpropafenone levels.

B. Increased 5-hydroxypropafenone due to increased CYP2D6 activity caused by fluoxetine.

C. Increased levels of propafenone due to decreased metabolism of propafenone to *N*-dealkylpropafenone.

D. Increased propafenone levels.

PROPAFENONE

Propafenone is metabolized to *N*-deathylpropafenone and 5-hydroxypropafenone

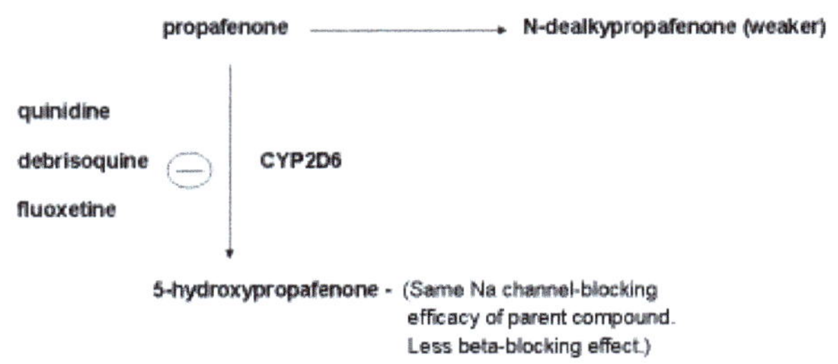

Figure 52.1. Propafenone metabolism.

(Figure 52.1). *N*-dealkylpropafenone has weaker antiarrhythmic properties than its parent compound. Metabolism of propafenone to 5-hydroxypropafenone involves the enzyme CYP2D6. 5-Hydroxypropafenone has the same Na^+ channel-blocking efficacy as the parent compound, but is a weak beta-adrenergic receptor blocker. Quinidine (low dose), debrisoquine, and fluoxetine can inhibit this enzyme. Patients on these medications, and patients who have genetic defects leading to impaired activity of the enzyme, often have increased levels of the parent compound and, thus, increased beta-adrenergic receptor-blocking activity.

Question No. 13:

Which one of the following drugs can inhibit the hepatic enzyme P4502D6, thus affecting the response to antiarrhythmic drugs metabolized by this enzyme?

A. Quinidine.
B. Metoprolol.
C. Erythromycin.
D. Ketoconazole.

Oxidation of drugs during hepatic metabolism involves two enzymes: CYP2D6 and CYP3A4. Deficiency of the CYP2D6 or drugs that inhibit this enzyme, such as quinidine, debrisoquine, and fluoxetine, can affect the response of antiarrhythmic drugs like propafenone. The substrates for CYP3A4 include terfenadine, astemizole, quinidine, lidocaine, dihydropyridines, cyclosporin, lovastatin, and midozolam.

Question No. 14:

Which of the following is false with regard to interaction of amiodarone with other drugs?

A. Increased digoxin concentration.
B. Increased flecainide concentration.
C. Decreased procainamide concentration.
D. Increased quinidine concentration.
E. Increased warfarin effect.

AMIODARONE INTERACTIONS

Amiodarone interacts with numerous drugs. The potential for interaction may persist for a long period due to the slow elimination half-life of amiodarone. The exact mechanism of increased digoxin concentration during concomitant amiodarone use is unclear, but is thought to be due to increased bioavailability or displacement of digoxin from storage sites. The digoxin dose should be halved and the plasma concentration of digoxin should be monitored during concomitant amiodarone therapy. The plasma concentration of other drugs such as quinidine, procainamide, flecainide, and phenytoin are also increased during amiodarone administration. Amiodarone may directly depress vitamin K-dependent clotting factors. Amiodarone is a noncompetitive beta-blocker that may lead to sinus bradycardia and hypotension. This is due to class II and class IV summation of effects of amiodarone rather than any pharmacokinetic interactions.

Question No. 15:

Which of the following drugs would increase quinidine plasma levels?

A. Phenytoin.
B. Cimetidine.
C. Phenobarbitone.
D. Rifampin.

QUINIDINE

Cimetidine increases quinidine levels by causing impaired hepatic metabolism. In contrast, drugs such as phenytoin, phenobarbitone, and rifampin decrease quinidine levels by causing increased hepatic metabolism. This is true for other drugs excreted through hepatic metabolism.

Question No. 16:

Which of the following drugs is not predominantly excreted by the kidneys?

A. Digoxin.
B. Sotalol.
C. Propafenone.
D. Dofetilide.

DRUG EXCRETION

Digoxin, sotalol, and dofetilide are predominantly excreted by the kidneys. Regardless of the route of administration, the major organ of excretion is the kidney and most of the digoxin is recovered as unchanged digoxin in the urine. The oral bioavailability of sotalol hydrochloride is 90% to 100%. Sotalol hydrochloride does not bind to plasma proteins and is not metabolized. Excretion of sotalol hydrochloride is predominantly via the kidneys in the unchanged form. The other drugs predominantly excreted via the kidneys include *N*-acetylprocainamide, the metabolite of procainamide, and the beta-adrenergic receptor-blocking agents acebutalol, atenolol, and nadolol. Other beta-blockers, such as esmolol, propranolol, and metoprolol, drugs like quinidine, procainamide, lidocaine, mexiletine, phenytoin, propafenone, amiodarone, and calcium channel blockers, such as verapamil, predominantly have hepatic excretion. Other drugs have mixed excretion patterns.

Question No. 17:

Which of the following drugs show nonlinear pharmacokinetics?

A. Mexiletine.

B. Disopyramide.

C. Sotalol.

Drugs exist in the plasma, both in free form and bound to plasma proteins. Only a free drug is capable of distributing into tissues and exerting a pharmacological action. With the exceptions of disopyramide, lidocaine, and propafenone, the fraction of drug bound to plasma proteins is fairly constant over a clinically relevant range of plasma concentrations. Thus, protein binding of disopyramide is concentration dependent. Disopyramide shows nonlinear pharmacokinetics with decreased protein binding as serum concentration increases. This is due to saturation of binding sites at a high concentration of the drug. Thus, a small increase in dose may lead to a significant increase in free drug available for pharmacological effects. Hence, side effects may occur even if therapeutic levels are shown on monitoring. Plasma monitoring is, therefore, of limited value. Quinidine, propafenone, verapamil, and digoxin, on the other hand, have the highest protein binding.

Question No. 18:

Which of the following drugs have active metabolites with pharmacological properties similar to the parent compound?

A. Propafenone.

B. Amiodarone.

C. Procainamide.

D. Flecainide.

E. Sotalol.

ACTIVE METABOLITES OF DRUGS

Propafenone has two active metabolites, *N*-desalkyl-propafenone and 5-hydroxypropafenone. *N*-desalkyl is a weaker metabolite, whereas 5-hydroxypropafenone has potency with the parent compound similar to Na^+ channel-blocking effects. Unlike propafenone, both metabolites have much less potent beta-adrenergic receptor-blocking properties. A major metabolite of amiodarone, desethylamiodarone, has an elimination half-life that is longer than that of amiodarone. It has properties similar to the parent compound. *N*-acetylprocainamide is a metabolite of procainamide. Unlike the parent compound, *N*-acetylprocainamide has less potent Na^+ channel-blocking effects but has predominant class III, that is, K^+, channel-blocking effects. Metabolites of quinidine, disopyramide, lidocaine, and verapamil have actions similar to the parent compound. Lidocaine has two active metabolites, MGx and Gx. MGx is 80% as effective as lidocaine, and Gx is 10% as effective as lidocaine. The Na^+ channel blocking effect of lidocaine can be reversed by glycine xylidide. Sotalol has no active metabolite and is excreted unchanged by the kidneys. Drugs such as mexiletine, flecainide, and ibutilide have clinically unimportant metabolites.

Question No. 19:

Which of the following does not increase digoxin levels?

A. Quinidine.

B. Amiodarone.

C. Verapamil.

D. Hypocalcemia.

E. Sotalol.

F. D and E.

G. C, D, and E.

DIGOXIN TOXICITY

Digoxin levels are increased by quinidine, verapamil, amiodarone, flecainide, propafenone, and diuretic agents. Quinidine reduces both renal and nonrenal elimination of digoxin and appears also to decrease the apparent volume of distribution. P-glycoprotein helps to eliminate drugs in the urine and the bile. Recent studies have shown that digoxin is a substrate for P-glycoprotein, and quinidine and verapamil are potent P-glycoprotein inhibitors. Diuretics can increase the incidence of digitalis toxicity by decreasing clearance and, through a variety of electrolyte disturbances, including hypokalemia, hypomagnesemia, and hypercalcemia. Verapamil can also increase the digoxin levels through inhibition of P-glycoprotein.

Question No. 20:

During which of the following conditions does the loading and maintenance dose of lidocaine need to be decreased?

A. Congestive heart failure.
B. Liver disease.
C. Myocardial infarction.
D. A and B.
E. A, B, and C.

LIDOCAINE

During congestive heart failure, both volume of distribution and clearance are reduced. This requires reduction of the loading and maintenance doses of lidocaine. Volume of distribution remains unchanged but clearance is reduced in liver disease and during concomitant use of drugs such as phenytoin, phenobarbitone, and rifampin; hence, only the maintenance dose needs to be reduced. α-1-Glycoprotein, an acute-phase reactant that binds drugs such as lidocaine and quinidine, increases during stressful conditions such as surgery or myocardial infarction. As only an unbound drug is available to tissues, the loading and maintenance doses of lidocaine will need to be increased during acute myocardial infarction.

Question No. 21:

A patient with an implantable cardioverter-defibrillator (ICD) is put on quinidine because of side effects on sotalol and amiodarone for sustained VT. He has been tolerating the drug well. He was started on phenytoin recently for seizures that occurred after a head injury. He presents with episodes of VT causing shocks after having remained stable for one year. Which is the best course of action to take next?

A. Start amiodarone.
B. VT ablation.
C. Increase quinidine dose.
D. Check ICD and program antitachycardia pacing.

Drugs such as cimetidine increase quinidine levels by causing impaired hepatic metabolism. In contrast, drugs such as phenytoin, phenobarbitone, and rifampin decrease quinidine levels by increasing hepatic metabolism. This is true for other drugs that are metabolized in the liver as well.

Question No. 22:

A 56-year-old female with atrial fibrillation, started on procainamide, presents with fever and sore throat. Which of the following tests is most likely to help diagnose her symptoms?

A. Antinuclear antibody (ANA).
B. Erythrocyte sedimentation rate (ESR).
C. Leukocyte count.
D. Lupus erythematosus (LE) cell preparation.

Procainamide-induced bone marrow depression, a rare side effect, can cause agranulocytosis. Decreased leukocyte count can lead to fever and sore throat.

Question No. 23:

A 63-year-old patient with an implantable defibrillator was started on mexiletine for nonsustained VT. The DFT one month later was 30 joules. The DFT at the time of implantation was <20 joules. What is the clinician's next best step?

A. Change lead position.
B. Change to a higher energy device.
C. Leave things as they are.
D. Discontinue mexiletine and retest in 2 weeks.

DFT AND DRUGS

Na^+ channel blockers such as quinidine, lidocaine, mexiletine, dilantin, flecainide, and propafenone can increase DFT. Beta-adrenergic receptor blockers and Ca^{++} channel blockers also increase DFT. Beta-agonists, such as isoproterenol, K^+ channel blockers, such as sotalol and *N*-acetylprocainamide, and other drugs, such as clofilium, cesium, and aminophylline, can also reduce DFT.

Question No. 24:

Shortening of APD by adenosine on the atrial tissue is mediated by:

A. AMP.
B. ATP.
C. IK.
D. GMP.

The adenosine action on the heart is mediated via receptor effector complex that includes the A_1 receptor and the guanine binding protein Gi. This interaction may produce either direct or indirect effects. The primary direct action of adenosine is activation of outward K current, IK Ach/Ado which is present in the atrium. Activation of this current causes shortening of the atrial action potential. Indirect actions of adenosine are mediated by inhibition of adenyl cyclase resulting in decreased intracellular cAMP concentrations.

Question No. 25:

Which drug should not be given in a patient with LVH, thickness ≥ 1.4 cm?

A. Sotalol.
B. Amiodarone.
C. Flecainide.
D. A and C.

ANTIARRHYTHMIC DRUG THERAPY AND GUIDELINES

See Figure 52.2.

Correct Answers:

Q1. C
Q2. C
Q3. B
Q4. D
Q5a. C
Q5b. D
Q6. B
Q7. B
Q8. D
Q9. C

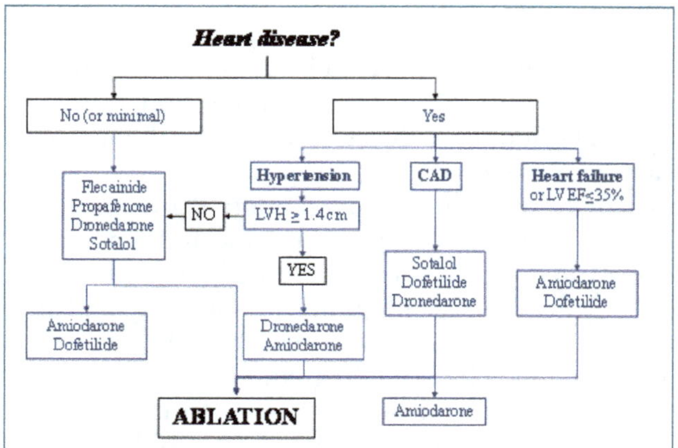

Figure 52.2. Current guidelines for antiarrhythmic drug therapy. Dronaderone is not indicated in patients with permanent AF.

Q10. D
Q11. B
Q12. D
Q13. A
Q14. C
Q15. B
Q16. C
Q17. B
Q18. B
Q19. F
Q20. A
Q21. C
Q22. C
Q23. D
Q24. C
Q25. D

REFERENCES

1. Fuster V, Ryden LE, Cannom DS, et al. ACC/AHA/ESC 2006 guidelines for the management of patients with atrial fibrillation—a report of the American College of Cardiology/American Heart Association Task Force on Practice Guidelines and the European Society of Cardiology Committee for Practice Guidelines (Writing Committee to Revise the 2001 Guidelines for the Management of Patients with Atrial Fibrillation). *J Am Coll Cardiol.* 2006;48:e149–e246.
2. Camm AJ, Kirchhof P, Lip GY, et al. Guidelines for the management of atrial fibrillation. The Task Force for the Management of Atrial Fibrillation of the European Society of Cardiology (ESC). *Eur Heart J.* 2010;31,2369–2429.

CASE STUDY

53

Lidocaine Use in a Patient with Old Myocardial Infarction

Dan M. Roden, MD

A 60-year-old man presents to the emergency department with a 3-hour history of palpitations. He has a history of 2 myocardial infarctions and is currently receiving appropriate drug therapy for symptoms of chronic congestive heart failure (CHF). His ECG shows ventricular tachycardia (VT) at 140/min, and lidocaine is started.

Question No. 1:

What is the optimal initial dosage scheme?

A. A large, rapid intravenous (IV) 100 mg bolus.

B. 100 mg bolus over 2 to 3 minutes.

C. Infusion of 3 mg/min (the usual maintenance dose).

D. Infusion of 20 mg/min.

The creation of coronary care units in the 1960s led to the observation that ventricular fibrillation (VF) was frequently preceded by ventricular ectopic beats. As a result, the electrophysiology community spent considerable effort testing various strategies to suppress ventricular ectopy, in the hope of reducing the incidence of VF. Among these strategies was the use of "prophylactic" lidocaine in the coronary care unit. Although this strategy has been abandoned because it does not reduce the incidence of VF (and may increase mortality due to bradyarrhythmias), the extensive use of the drug led to intensive characterization of its pharmacokinetics and an understanding of the way in which disease alters fundamental determinants of plasma concentration and effect. These principles apply to therapy with most drugs, whether administered oral or IV.

To achieve an immediate therapeutic effect, a loading dose is administered, but to maintain that effect, a chronic dose regimen (in this case a maintenance IV infusion) is required. Lidocaine pharmacokinetics can be described by parameters such as volume of distribution or clearance. Volumes of distribution include a central volume into which the drug distributes immediately upon bolus administration (often equivalent to plasma volume) and a total volume of distribution, which may be much larger than plasma and may even exceed total body volume. The usual loading dose of lidocaine in a normal adult subject is 3 to 4 mg/kg. Since administering a large dose such as this can lead to very high drug concentrations, the load is usually administered as a series of boluses over 10 to 20 minutes. Disease states can have major impacts on drug disposition. CHF leads to a decrease in both clearance and volumes of distribution of lidocaine. The decrease in volume of distribution mandates a decrease in load doses; otherwise, a usual loading dose will lead to excessive early plasma concentrations and toxicity.

Cardiotext Publishing, ISBN: 978-1-935395-17-1

Question No. 2:

Following an initial dose of lidocaine, the arrhythmia promptly stops. No further lidocaine is given and the arrhythmia recurs 5 minutes later. What can you infer about the elimination half-life of lidocaine?

A. It must be very short (5–10 minutes).

B. It is 4 to 5 times 5 minutes, or 20 to 25 minutes.

C. No inference about elimination can be made.

Lidocaine distribution and elimination are "first-order" processes; that is, they can be described by simple differential equations that state the rate of change of drug concentration which is determined by the drug concentration itself. The solution to such simple equations is in the form of an exponential function characterized by a rate constant or, more commonly (for clinical purposes), a half-life. Any first-order process, whether it is uranium decay or lidocaine elimination, can be characterized by a half-life: in one half-life, 50% of the process is complete; in 2 half-lives 75%; in 3 half-lives 87.5%, etc. For practical purposes, a first-order process can thus be viewed as "complete" in 4 to 5 half-lives. Steady-state plasma concentrations are attained 4 to 5 elimination half-lives after a constant rate infusion is started, and a drug can be viewed as absent from the circulation 4 to 5 elimination half-lives after a maintenance infusion is stopped.

When lidocaine is administered as a rapid IV bolus, plasma concentrations reach very high levels, and fall rapidly, in a biexponential fashion. The first time constant is very short, generally 5 to 10 minutes, and represents first-order distribution of lidocaine from sites in plasma to extravascular sites in tissue. The second time constant is longer, about 2 hours, and represents lidocaine elimination. Thus, in the case presented, no inference can be made about lidocaine elimination because the disappearance of drug effect is not due to elimination, but rather to rapid distribution.

Question No. 3:

After the patient has been given a total of 200 mg over 20 minutes, he starts to complain of a sensation of roaring in his ears; at that instant his rhythm reverts again to sinus. Based on this clinical scenario, it is decided that he developed transient side effects due to lidocaine but that the drug was probably effective in terminating his arrhythmia. Which of the following is an appropriate maintenance dose?

A. 3 mg/min.

B. 1.5 mg/min.

C. 6 mg/min.

A usual maintenance infusion is 3 mg/min. However, because heart failure decreases clearance, the maintenance infusion should be decreased.

Question No. 4:

Lidocaine elimination half-life is approximately 2 hours in normal individuals. Based on all the information available at this point, when do you predict lidocaine concentrations will be at steady state in this patient?

A. 8 to 10 hours.

B. 4 to 5 hours.

C. 20 to 25 minutes.

D. 16 to 20 hours.

E. Depends on the size of the loading dose.

Elimination half-life varies directly as volume of distribution and inversely as clearance. Since heart failure reduces both volume of distribution and clearance, the elimination half-life remains roughly the same as that in normal individuals, approximately 2 hours. Steady state occurs in 4 to 5 elimination half-lives, or 8 to 10 hours in this case.

It is a common misconception that the administration of loading doses abbreviates this time course. The administration of the "perfect" loading dose can, in fact, attain therapeutic concentrations much earlier than 4 to 5 elimination half-lives, but the perfect loading dose is not known *a priori* for an individual patient.

Liver disease also alters lidocaine pharmacokinetics but in a different fashion from CHF. In liver disease, clearance is profoundly decreased while volume of distribution remains the same. Therefore, elimination half-life is, in fact, much longer than in normal individuals leading to a much longer time course for drug accumulation to steady state. However, as with CHF, maintenance dosages must be decreased in liver disease to avoid high steady-state concentrations. Renal disease exerts no major effect on lidocaine kinetics.

Question No. 5:

Ten hours after the patient is admitted, the house staff is summoned to see him because he is disoriented, lethargic and dysarthric. The lidocaine plasma concentration is 12 μg/mL (usual therapeutic: 1.5–5 μg/mL), confirming the clinical suspicion of lidocaine toxicity. His maintenance infusion is 3 mg/min. Which one of the following is the most rational way of managing his lidocaine therapy?

A. Decrease the maintenance infusion by 50%.

B. Stop the lidocaine for 8 hours and then restart it.

C. Stop the lidocaine for 2 hours and then restart it at 2 mg/min.

D. Stop the lidocaine for 4 hours and then restart it at 1 mg/min.

Figure 53.1 shows the projected time course of lidocaine plasma concentrations with each of the suggested dose adjustments after a plasma concentration of 12 μg/mL is recognized. Decreasing the maintenance dose by 50% leads to a new steady-state plasma concentration – attained in 8 to 10 hours – of 6 μg/mL, which is still supertherapeutic. Stopping the drug for 8 hours (equal to 4 elimination half-lives) and then restarting it leads to a decrease in lidocaine concentrations to below the lower limit of the therapeutic range and then a return of concentrations back to the same steady state. Stopping the drug for one elimination half-life, 2 hours, leads to a fall in drug concentration from 12 μg/mL to 6 μg/mL. If the maintenance infusion is then started at two-thirds its original rate, the new steady state will be 8 μg/mL, which is still supertherapeutic. Stopping the drug for 2 elimination half-lives lets the plasma concentration fall from 12 μg/mL to 6 μg/mL (in one elimination half-life) and then to 3 μg/mL (in 2 elimination half-lives), within the therapeutic range. Restarting the maintenance infusion at 1 mg/min, one-third the original rate, will then lead to a stable plasma concentration of one-third the original steady-state level or 4 μg/mL, within the therapeutic range.

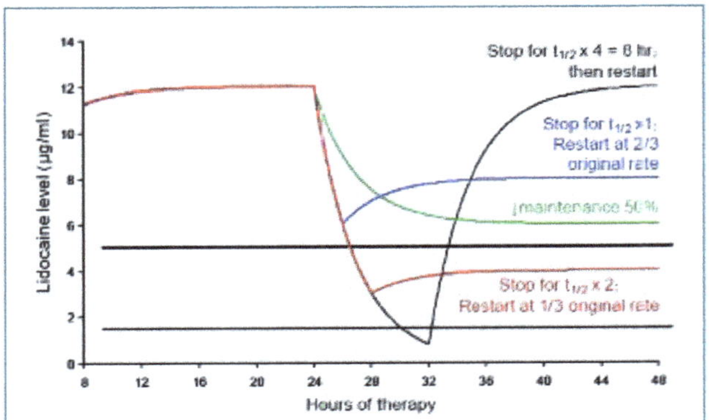

Figure 53.1. Adjusting lidocaine doses after toxicity is noted at steady state.

In Figure 53.1, the heavy bars delineate the target therapeutic range of 1.5 to 5 μg/mL. As can be seen here, the correct approach in this case is to stop for 2 elimination half-lives, allowing the concentration to decline from 12 μg/mL to 3 μg/mL. Resuming the maintenance infusion at 1 mg/min instead of 3 mg/min results in a new steady concentration of 4 μg/mL, within the therapeutic range.

SUMMARY

The principles outlined above apply not only to a drug like lidocaine, which is no longer widely used, but to virtually all drugs, since first-order elimination is the commonest pattern.

- Steady-state concentration is determined by drug clearance.
- The time course with which drug concentrations approach steady state is determined by the elimination half-life.
- Elimination half-life is a function of both clearance and volume of distribution.
- Disease-induced alterations in volume of distribution or clearance can lead to changes in maintenance dose requirement.
- Monitoring plasma drug concentrations, when available, is a useful adjunct to management of drugs with narrow therapeutic ranges, such as lidocaine.
- Loading doses do not accelerate the time to steady state and should only be used when a therapeutic drug effect is needed urgently, that is, before a maintenance dose of drug can accumulate the drug to steady state.
- Only a drug that is not bound to plasma proteins is available to exert a pharmacological effect. Changes in plasma protein binding may become important determinants of drug response, and this is especially likely for drugs that are highly protein-bound. Thus, for example, if a drug is 50% bound to plasma proteins and in disease this changes to 40%, the likely clinical consequences are small. However, if a drug is ordinarily 98% protein-bound and in disease this rises to 99%, the amount of drug

actually available to exert a pharmacological effect has been cut in half (from 2% to 1%). Lidocaine is bound to multiple plasma proteins, including acute-phase reactants. Thus, in acute disease, such as systemic infections or acute myocardial infarction, lidocaine binding may be increased and the net effect of drug may be reduced.

Correct Answers:

Q1. B

Q2. C

Q3. B

Q4. A

Q5. D

SECTION 7

Electrical Therapy—Pacing, Implantable Defibrillators, Cardiac Resynchronization Therapy

CASE STUDY

54

Cardiac Implantable Electronic Devices: Timing Cycles, Sensing, and Troubleshooting During Pacing

Carol J. Gilbert, RN

FIRST CASE

Cardiac implantable electronic devices (CIEDs) are named after their primary functions; they are pacemakers and defibrillators. The first permanent pacemakers (PPMs) in the 1960s were used only for extreme bradycardia and had only one function, to pace continuously, asynchronously, at a single rate, using a single lead placed in the ventricle. Implantable cardioverter-defibrillators (ICDs) initially had no pacing function, but later had backup emergency ventricular pacing at 30 bpm and now have fully functional single- or dual-chamber pacemakers.

Today, pacemakers and defibrillators are combined in devices with highly technical advances in the pacemakers' specifications with multiprogrammable and automated parameter features and multiple lead-placement capabilities. The more common automatic features of these physiological CIEDs are activity responses, capture management, and sensitivity adjustment. Because of the many automated features, the follow-up of patients with CIED requires diligent review of the patients' symptoms, rhythms, and device functioning as well as familiarity with the device-specific manufacturer's algorithms for these automated functions, keeping in mind that physiological functioning in a PPM or ICD is tied to these specific "robotic" controls whereas the patient is not. In the end, basics of CIEDs hold true for sensing and capture, battery longevity, and patient safety, and are usually evidenced by single beat-to-beat measures. Currently used abbreviations for pacemakers are depicted in Table 54.1.

CASE DESCRIPTION

A 68-year-old male with persistent, permanent atrial fibrillation presents with complaints of intermittent palpitations and lightheadedness for the past week. His VVI (see Table 54.1 for description of pacing mode abbreviations) pacemaker (PPM) was placed 5 years ago at the left pectoral region and was programmed to 72 ppm (pulses per minute). His updated history indicates that he was placed on beta-blocker and angiotensin-converting enzyme (ACE) inhibitor for congestive heart failure 1 month ago, and had arthroscopic surgery on his left shoulder 2 months ago, soon after his last pacemaker check. Current rhythm strip is shown in Figure 54.1.

Table 54.1. Abbreviations for Pacing Modes

Pacing Mode	Chamber Paced	Chamber Sensed	Mode of Response
VOO	Ventricle (V)	None	Asynchronous (continuous ventricular pacing).
AOO	Atrium (A)	None	Asynchronous (continuous atrial pacing).
VVT	V	V	Paces the ventricle on demand for pauses < low, escape rate. Triggered (paces in the QRS [R], marking R-wave detection).
VVI	V	V	Paces the ventricle on demand for pauses < low, escape rate. Inhibited on R-wave detection (holds V stimulus when R wave detected).
AAI	A	A	Paces the atrium on demand for pauses < low, escape rate. Inhibited (holds stimulus when P wave detected).
VAT	V	A	Paces the ventricle on demand for pauses < low, escape rate. Triggered (paces the ventricle following P-wave detection).
VDD	V	A and V	Paces the ventricle on demand for pauses < low, escape rate. Triggered (paces V following P-wave detection at a programmed sensed AV [SAV] interval); does not pace the atria. Inhibited (holds V stimulus when R wave detected).
VDI	V	A and V	Paces the ventricle on demand for pauses < low, escape rate. Inhibited (holds V stimulus when R wave detected). Inhibited (senses atrial activity only when P wave detected).
DDI	A and V	A and V	Paces the ventricle on demand for pauses < low, escape rate. Paces the atrium on demand for pauses < low, escape rate – the paced AV interval = VA interval. Inhibited (holds V stimulus when R wave detected). Inhibited (holds stimulus when P wave detected); does not initiate AV delay and AV sequential pacing on P-wave sensing.
DDD	A and V	A and V	Paces the ventricle on demand for pauses < low, escape rate; paces the atrium on demand for pauses < low, escape rate – the paced AV interval. Triggered (paces the ventricle following P-wave detection at a programmed SAV interval). Inhibited (holds stimulus when P wave detected). Inhibited (holds stimulus when R wave detected).
DDD-MS (Mode switch)	A and V	A and V	See DDD above. Atrial pacing is inhibited by rapid P-wave detection (eg, atrial fibrillation) and device automatically behaves in VDI mode.

Beats per minute (bpm) to cycle length (CL) conversion in milliseconds:
1 second = 1000 ms
60 seconds per minute × 1000 ms = 60,000 ms
60,000 ms/bpm = millisecond CL
60,000 ms/CL = bpm heart rate.

Question No. 1:

Your interpretation of the strip is:

A. Failure to capture the ventricle, appropriate VVI sensing of the escape beat. Possible fractured ventricular pacing wire.

B. Failure to capture the ventricle, appropriate VVI sensing of the escape beat. Possible loose set screw with intermittent contact and intermittent capture loss.

C. Failure to capture the ventricle, appropriate VVI sensing of the escape beat. Possible threshold rise from medication/device interaction.

D. Normal PPM behavior with isoelectric QRS resulting from fusion of intrinsic and paced complexes.

E. Normal PPM behavior with Recommended Replacement Time (RRT) Indices.

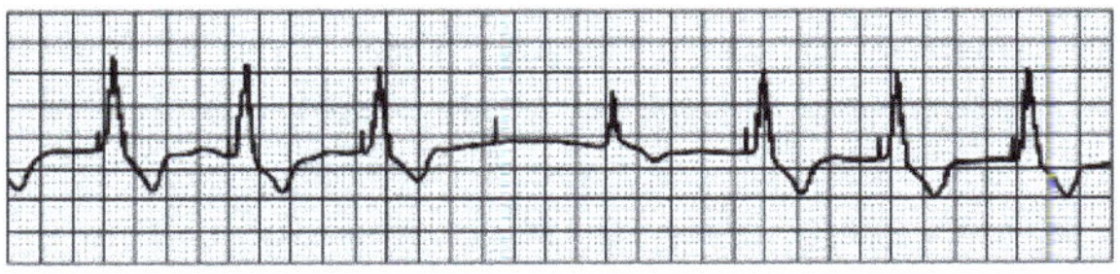

Figure 54.1. Patient's presenting rhythm-recorded strip.

DISCUSSION

The 4 repetitive 840-ms cycles of pacing stimuli can be measured as consistent. There is no ventricular response to the fourth stimulus (failure to capture), and 720 ms later, the patient has an escape beat (1560 ms from preceding QRS), which is sensed appropriately, resulting in resetting of the escape interval and a subsequent fifth pacemaker stimulus 880 ms after the beginning of the intrinsic QRS.

Late threshold rises are either clinical or mechanical lead failures. Clinically, medication changes (beta-blocker), metabolic or electrolyte imbalances, or a substrate change with myocardial fibrosis or infarction are differentials.[1] This patient's history did not reflect acute ischemic or infective process, but there was a progression to his cardiac disease and added medication therapy. In addition to beta-blockers, several medications can increase stimulation threshold with loss of capture; most antiarrhythmics will alter thresholds but usually only at very high serum levels. Type I antiarrhythmic drugs, flecainide, procainamide, and quinidine raise thresholds. Threshold rises can be seen in metabolic conditions such as hypercarbia, hyperglycemia, hyperkalemia, hypoxemia, and acidosis or alkalosis.[2] ACE inhibitor-induced hyperkalemia can occur in the setting of marginal renal function.[3] Generally these effects are seen in the older adult population, but serum electrolyte, serum urea nitrogen, and creatinine would be steps toward the differential diagnosis. In contrast, corticosteroids and catacholamines can reduce stimulation thresholds. Daily fluctuations of catecholamine levels can produce higher stimulation thresholds during sleep and after eating and decreased thresholds while awake, active, and exercising. This patient presents to the clinic, awake with intermittent capture. With the pacemaker device output close to the patient's capture threshold, more frequent episodes of failure to capture may be likely, especially at night.

Threshold rises that occur within weeks or a few months after implant are usually secondary to lead maturation, encapsulation, and polarization processes. This type of threshold rise due to the inflammatory process is reduced with the use of steroid-eluting electrodes and now rarely noted.

Points to Consider in Differential Diagnosis

A1: Possible fractured ventricular pacing wire: Mechanical failures do occur late after implant, often secondary to stress on the lead caused by friction, compression, or reactive failures. Insulation failures can cause extracardiac stimulation such as muscle twitching. Pacing stimulus artifacts can be attenuated with variation in amplitude; however, detection would require multichannel analog recordings. Digital 12-lead electrocardiographic sampling frequencies prohibit waveform analysis, as the sampling frequency is usually not rapid enough to trace a pacing spike accurately. Conductor coils come in contact with each other when the internal insulation fails to shunt energy from the myocardium, creating failure to capture. Electrical transients from the metallic connections, usually at very short coupling intervals of < 100 ms, can be detected by the CIED sensing circuit with corresponding oversensing or irregular inhibition and subsequent failure to pace in the absence of recorded ECG signal. Separation of fractured electrodes results in absence of stimuli and no visible pacing spikes. Neither is seen in this case. Of course, it is clinically prudent to run longer rhythm strips and manipulate the pacemaker pocket to see if any other manifestations of mechanical failure can be provoked. Using pacemaker programmers to interrogate lead impedances is important. Higher resistances may be indications of lead fractures or disconnections, whereas low impedances can indicate short circuits associated with insulation breaks. It is important to note that these problems can present sporadically. Long-term lead-trend monitoring can detect these early.

The history reports the patient as having had left shoulder arthroscopic surgery. The patient's symptoms are more recent and arthroscopic surgery is unlikely to reach the lead's intravascular course or pocket position. If the lead had been placed medially with subclavian puncture, a particular mechanical failure called subclavian crush syndrome could have occurred and, coincidentally, failed near the time of beta-blocker administration. Such lead injury would likely be accompanied by sensing abnormalities as well.

Another potentially late-occurring cause of failure to capture is twiddler's syndrome. This is lead dislodgment because the patient unconsciously turns the device in the pocket, resulting in the lead shortening as it winds upon itself. Similar lead dislodgments can occur when a pocket is too large, as an existing pocket can be when the pacemaker replacement is smaller, or if the pacemaker is loose, as in an obese patient. Chest x-ray is a good tool to find this diagnosis. The ECG presentation of lead dislodgment will depend on how far the lead has retracted. If the tip is no longer in contact, pacing capture is not possible; however, if the lead is still within the expected cardiac chamber, sensing may still occur. On the other hand, ventricular leads can retract to the atria and the CIED may behave like an AAI (atrial pacing, atrial sensing, and inhibited) for a time with intermittent capture of the atria, or it can be totally extracardiac, failing to both sense and capture. Chest x-rays can sometimes identify lead fractures and set screws poorly seated in the device header.

A2: Possible loose set screw with intermittent contact and intermittent capture loss: Knowing the lead has been planted for years essentially eliminates common lead dislodgment or loose set screws as a cause of the threshold rise. These threshold rises are generally noted within hours or days but may only have intermittently detectable impedance changes. Loose set screws will likely also have high-frequency, rapid, repetitive, "make/break" electrical signals or transients, similar to fracture electrodes. Irregular, rapid signals from broken atrial leads or loose set screws at the atrial lead can be interpreted as atrial fibrillation by the device rhythm discrimination software, resulting in a change in pacing behavior (mode) – switching to single-chamber (VVI) mode. Some newer devices will count these behaviors, identify the very short couplings of <100 ms and report them at the time of device interrogation for the clinician to evaluate. Remote monitoring with automated CIED testing and active online connections can deliver notice of these events to the clinician earlier, potentially in advance of any symptoms.

A3: Normal PPM behavior with isoelectric QRS resulting from fusion of paced and normal complexes: Isoelectric ventricular fusion beats can occur when the intrinsic beat and the paced beat have equal and opposite electrical axes in the monitored lead, but most electrically isoelectric beats would likely continue to have a T wave.[4] Using multichannel ECG recorders or intracardiac telemetry and markers can help differentiate this phenomenon. CIED sensing is a function of the difference in the signals from the two electrodes of the circuit. When a PPM is programmed to unipolar, the generator "can," or a portion of it, is used as the indifferent. It has a small signal exerting little effect. Unipolar electrograms are often large, easily detected by sense amplifiers. More often, bipolar sensing is used and the signal difference between the tip and ring electrodes is affected by timing and amplitudes. Depolarization forces parallel to the positioned bipolar electrodes have additive effects, whereas perpendicular forces arriving at the electrodes with little timing difference can have negative effect on the amplitude of the signal, resulting in no difference in amplitude and time.[5] In this way, even QRS complexes that can be appreciated on the surface ECG may go undetected by the CIED.

A4: RRT behavior: CIEDs are required to show an indication of battery depletion. The exact time the index is tripped depends upon the type of battery. Lithium batteries used most commonly today have longer decay characteristics and predictive behaviors, which allow sufficient windows of detection at routine CIED follow-up to allow for planned replacements. When the RRT (previously called: Elective Replacement Index or ERI) is reached, there are no usual changes in the device function. In the case illustrated above, we see loss of capture but no loss of other device functioning.

The RRT state is communicated with interrogation of the battery, either during clinic visits or with remote monitoring. The device will minimally report that the battery is good or has reached the RRT. It may additionally report the voltage and/or ampere hours available and may additionally estimate a range of minimum and maximum time before replacement can be expected to become an imminent need. For the PPM, magnet application may evoke the RRT indices when a CIED programmer or full remote communication is not available. Magnetic responses of PPMs differ for each manufacturer, and sometimes between models, but in most cases there will be a difference in the magnetic-influenced response at RRT compared to the response when the battery is at beginning of life (BOL). Device EOL (end of life) is often a second step of battery depletion with specific changes in programming. Such changes can include increased pulse durations and abandonment of automated functions including sensing. EOL behaviors may be performed in a stepwise fashion and are designed to provide output safety margins increasing capture reliability and reducing power drain. A dual-chamber, atrioventricular sequential PPM may, for example, switch mode first to VVI (ventricular pacing, ventricular sensing, inhibition response) at programmed rates, and subsequently change to VVI at 60 ppm, before abandoning all sensing to pace VOO (ventricular pacing only, no sensing functions).

The pacing function in most ICDs remains unaffected by magnetic application. Magnet use with the ICD is specifically designed to disable the tachycardia detection and treatment algorithms for the duration of application. Early devices were activated and deactivated by magnet applications, an operation now requiring programming. Battery measures for ERI must be interrogated but may also be part of the automated functions with audible patient alerts that advise the patient to seek attention.

Factors Influencing the Escape Interval

The increased escape interval is longer than the automatic pacing interval. This is not unexpected as the pacemaker electronic escape interval is initiated by the sensed endocardial signal. Figure 54.2 shows this increased escape interval. There is often a delay, referred to as a sensing latency, of as much as 40 ms or from the time the QRS is inscribed until the local endocardial activation at the sensing lead. Longest sensing latencies of as much as 120 ms are seen in the setting of right bundle branch block-pattern QRS, as sensing will not occur until after the impulse has crossed into the right ventricle (RV) from the left.

Sensing requires the amplitude of the depolarization across the sensing electrodes not only exceed the pacemaker's programmed sensitivity in mV amplitude but achieve that amplitude within a defined period of time. The slew rate or change in amplitude over time (dV/dt) must meet certain criteria. If this slew is <20 Hz or >80 Hz, the signal amplifier may not detect the intrinsic deflection.[6] The slew rate criterion is additive to amplitude detection criteria and designed to reduce oversensing of noncardiac signals, such as low-frequency T waves (usually at <5 Hz) and high-frequency myopotentials (10–200 Hz).[7] Since the frequency of myopotentials and intrinsic P and R waves can overlap, filtering may not always eliminate

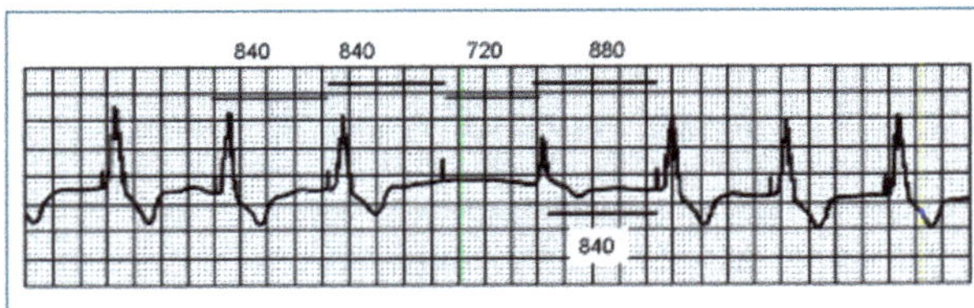

Figure 54.2. Solid bars mark the intervals. Low pacing rate cycles of 840 ms are seen for three pacing cycles. The escape beat (fifth QRS) ends the open sensing period of the pacemaker at about 720 ms from the fourth pacemaker spike that did not capture the ventricle. The next pacemaker spike is 880 ms from the onset of the escape beat QRS indicating that the intrinsic QRS was sensed by the endocardial lead at least 40 ms after it was inscribed on the surface ECG recording.

oversensing and amplitude sensitivity adjustments may help. The use of telemetry for intracardiac electrograms with markers can provide invaluable information about what and when device pacing and sensing occurs.

Patient Management

Restoring capture: Each programmer has a setting for STAT/Emergency pacing button (usually red) to manage acute episodes of asystole or extreme bradycardias. The device manual will list the precise specifications of emergency pacing, but as this patient is capturing most of the time, emergency pacing is unnecessary. Stat pacing usually changes the device to VVI at highest device output of 6 to 7 volts at 1.5-ms pulse durations, and at 60 or 70 ppm. It is a useful setting for asystole or extreme bradycardia.

Magnet application, placing the device in asynchronous mode, would be helpful in settings of electromagnetic interference (EMI) where the ambient noise is detected and interpreted by the CIED as physiological signals, thus inhibiting outputs or triggering rapid or irregular pacing. Magnet application restores pacing if the noise cannot be located and stopped. In the case of fractured electrodes, irregular firing due to oversensing the electronic noise of intermittent connections may be eliminated, but stimulus delivery is lost across high impedance breaks, resulting in prolonged failure to pace at multiples of the pacing CL. It is important to remember that this maneuver is ineffective in pacing ICDs.

Long term, the device needs to be reprogrammed to provide for a pacing threshold with adequate safety margin to cover the patient's daily needs. Basic concepts of electricity apply. Pacemaker output energy is a factor of voltage, pulse width, and pulse duration expressed by the equation: $E = V^2/Rt$, where V = volts, R = total pacing impedance (leads and myocardial interface, etc), and t = pulse duration. Clinically it is important to consider that the amplitude has an exponential relationship to the pulse duration. There is a steep rise in the curve at short pulse widths (Figure 54.3). Small changes in voltage at short pulse intervals can effect large changes in energy delivery. On the other end of the curve, the flattening of the line shows that increases in the pulse duration beyond a certain value (usually about 1.5 ms) do not increase the probability of capture but waste energy.

The output energy equation, therefore, defines voltage as having a direct and greater effect than pulse duration such that doubling the

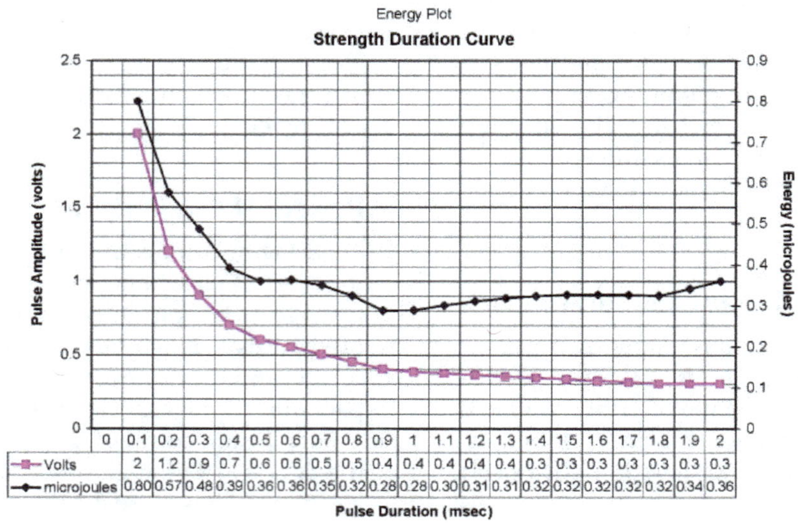

Figure 54.3. A strength duration curve plots capture threshold voltages against pulse duration (pink squares) and the calculated energy (black diamonds) in microjoules, assuming constant system impedance of 500 ohms. Notice the relative flattening of the strength duration curve after 1 ms and the associated increased energy expenditure.

stimulus voltage yields a fourfold increase in output. In contrast, increasing pulse width threefold may be necessary to achieve an adequate safety margin. Based upon characteristics of the curve, increasing pulse durations may be helpful in increasing output when the base pulse width is ≤0.15 ms, but it usually does not provide sufficient margins when programming from longer pulse width of ≥0.3 ms.

The patient's stimulation output was 1.25 volts at 0.4 ms on presentation to clinic. The voltage capture threshold was measured by decrementing the pacemaker voltage output every 6 beats by 4 volts, and consistent capture was noted until capture was lost. Capture was lost at 1.75 volts at 0.4-ms pulse width. Capture threshold was, therefore, determined by the last capture output of 2 volts at 0.4 ms. Capture safety margins are usually placed at twice the threshold value or at three times threshold if threshold rises are anticipated, as with medication changes, or the patient is pacemaker dependent. In this patient, underlying rhythm test showed the patient had a stable sinus bradycardia 45 to 50 bpm. The device was programmed to 4 volts at 0.4 ms to exceed twice the safety margin. Monitoring was increased to monthly for the short term to ensure that no further increases in threshold are observed.

Correct Answer to First Case:

Q1. C

REFERENCES

1. Love CJ. Pacemaker troubleshooting and follow-up. In: Ellenbogen KA, Kay GN, Lau CP, Wilkoff BL, eds. *Clinical Cardiac Pacing, Defibrillation and Resynchronization Therapy*. 3rd ed. Philadelphia: Saunders Elsevier; 2007:1026–1044.
2. Kay GN, Shepard RB. Cardiac electrical stimulation. In: Ellenbogen KA, Kay GN, Lau CP, Wilkoff BL, eds. *Clinical Cardiac Pacing, Defibrillation and Resynchronization Therapy*. 3rd ed. Philadelphia: Saunders Elsevier; 2007:3–58.
3. http://www.thomsonhc.com/micromedex2/librarian/ND_T/evidencexpert/ND_PR/evidencexpert/CS/0A04EA/ND_AppProduct/evidencexpert/DUPLICATIONSHIELDSYNC/3F798C/ND_PG/evidencexpert/ND_B/evidencexpert/ND_P/evidencexpert/PFActionId/evidencexpert.DisplayDrugdexDocument?docId=1341&contentSetId=31&title=BENAZEPRIL&servicesTitle=BENAZEPRIL&topicId=clinicalApplicationsSection&subtopicId=mechanismOfActionPharmacologySection.
4. Barold SS, Stroobandt RX, Sinnaeve AF. *Cardiac Pacemakers Step by Step: An Illustrated Guide*. Malden, MA: Blackwell Futura; 2004: 81.
5. Barold SS, Stroobandt RX, Sinnaeve AF. *Cardiac Pacemakers Step by Step: An Illustrated Guide*. Malden, MA: Blackwell Futura; 2004: 47–48.
6. Barold SS, Stroobandt RX, Sinnaeve AF. *Cardiac Pacemakers Step by Step: An Illustrated Guide*. Malden, MA: Blackwell Futura; 2004: 296.
7. Swerdlow CD, Gillberg JM, Olson WH. Sensing and detection. In: Ellenbogen KA, Kay GN, Lau CP, Wilkoff BL, eds. *Clinical Cardiac Pacing, Defibrillation and Resynchronization Therapy*. 3rd ed. Philadelphia: Saunders Elsevier; 2007: 75–160.

SECOND CASE

A 33-year-old man presents for follow-up of his DDD pacemaker (see Table 54.1 for description of pacing mode abbreviations), implanted 5 years ago for complete atrioventricular (AV) block. The programmed parameters are known:

- SAV pacing interval = 350 ms;
- Low (escape rate) 50 bpm; and
- Upper tracking rate limit of 160 ms.

Question No. 2:

Based upon the information at hand, your interpretation of the ECG in Figure 54.4 is:

A. Atrial output failure.

B. Normally functioning DDD pacemaker with appropriate atrial and ventricular sensing and appropriate ventricular pacing and capture.

C. Perforation of the RV by the pacing lead.

D. Possible abnormal atrial and/or ventricular sensing with prolonged escape pacing and prolonged SAV interval.

E. Unstable ventricular pacing lead position.

DISCUSSION

The pacemaker circuit senses each P wave at the point of programmed atrial sensitivity, which in this patient was found to be 0.5 mV. The site of the atrial lead implantation is the right atrial appendage, located away from the site of the sinus node, and the signal is sensed after the onset of the P inscribed by the surface ECG machine. The result is that the PV interval measures 380 ms, indicating that it took about 20 ms for intrinsic atrial conduction of the sinus P to reach the atria electrodes. The premature ventricular complex (PVC) (see second QRS in Figure 54.4) is sensed and inhibits the pacemaker. The next paced beat is expected to be an escape beat at the low rate of 50 bpm (1200 ms). An intrinsic sinus P occurs at about 880 ms after the PVC within the open sensing period. This sinus P is sensed and initiates the SAV delay of 350 ms (measured AV 380 ms), after which the ventricle is paced with a PVC-V interval totaling 1260 ms. It is not unusual to find 40- to 80-ms variations between visualized P and R waves compared to intracardiac time of detection (Figure 54.5).

Points to Consider in Differential Diagnosis

A1: Atrial output failure; and A4: Possible abnormal atrial and/or ventricular sensing with prolonged escape pacing and prolonged SAV interval: Pacemaker escape interval is determined by the pacemaker's AV and VA intervals.[1] At 50 bpm, the escape interval would be 1200 ms. The programmed AV interval of 350 ms allows the VA interval to be calculated as

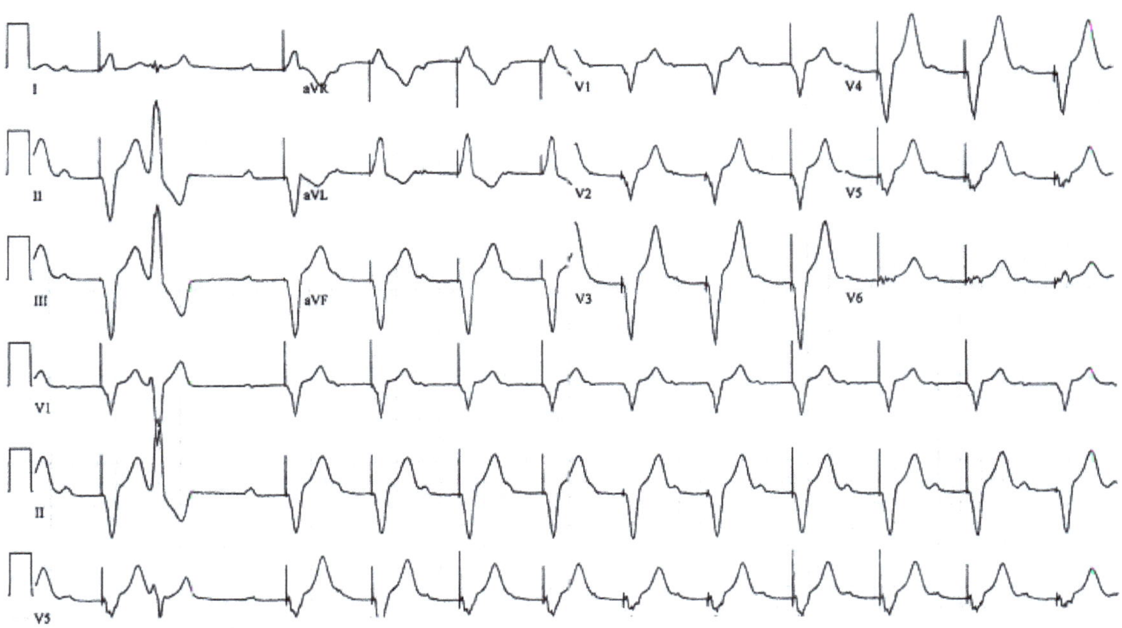

Figure 54.4. ECG at presentation.

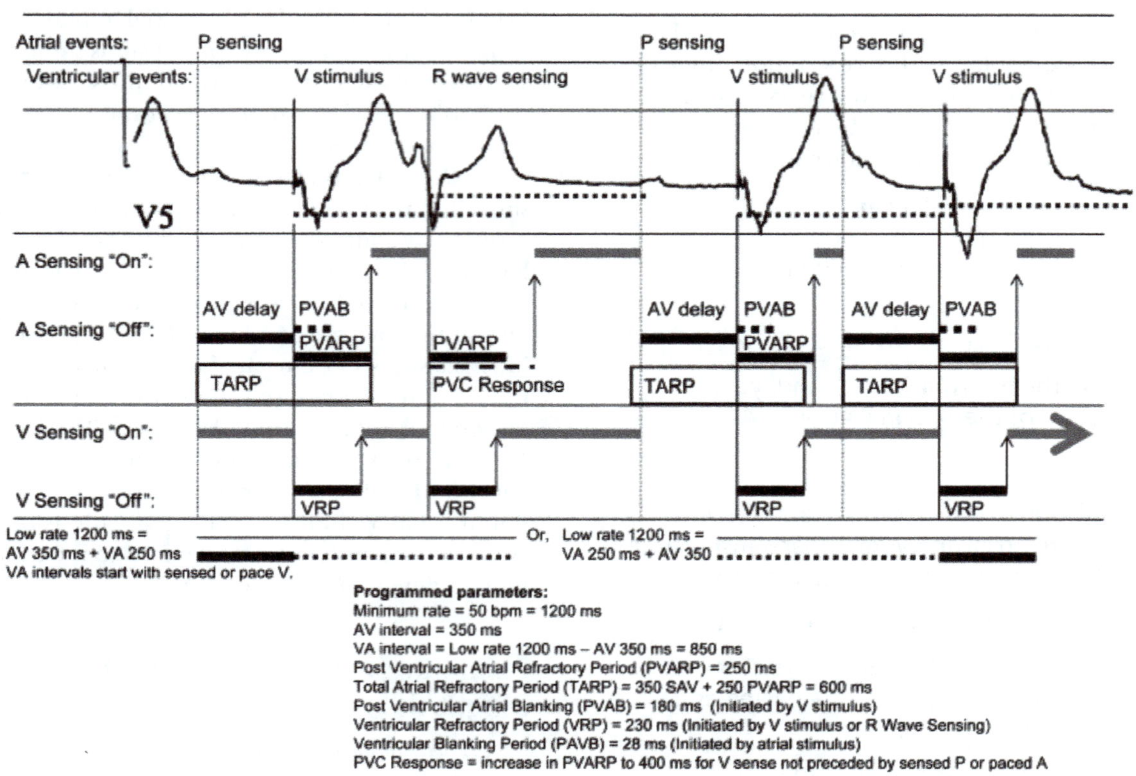

Figure 54.5. Interpretation of pacemaker parameters. See text for discussion.

1200 – 350 = 850 ms. The atria escape interval is equal to the VA interval and is measured from the onset of the ventricular pacemaker spike or sensed QRS to the next paced P; however, in this case an intrinsic P occurs and initiates an AV delay. Measuring 850 ms backward from the P to the QRS shows the intrinsic P to have occurred during the open atrial sensing within the VA interval following the PVC (beat 2), coincidentally close to the time a paced A was expected. Atrial output failure cannot be determined on the basis of this single interval. All other P waves clearly occur before the expiration of an 850-ms VA. All P waves are sensed as evidenced by the 350-ms PV interval (programmed SAV). The single intrinsic beat is sensed as evidenced by no paced P wave at 850 ms (VA) from the preceding paced QRS.

A3: Perforation of the RV by the pacing lead; and A5: Unstable ventricular pacing lead position: The ECG strip (see Figure 54.4) shows varying-sized pacemaker spikes inscribed during the short rhythm strip such that the stimulus artifacts on beats 6, 7, and 11 are very different from the others. The ventricular pacing stimulus at the time of this ECG was 5 volts at 0.5 ms. Digital ECG machine sampling is usually 500 Hz or once every 2 ms. Pacemaker stimuli of 0.5 ms may or may not be detected by digital sampling, and if sampled, detection could be at maximum output or at any point along the slope of the pacing output positive or negative slope. Visualization of the pacemaker artifacts is often compromised. Some equipment software will automatically enhance rapid deflection signals when a pacemaker detection algorithm is enabled. Such equipment may detect EMI and add multiple spikes creating false "pacemaker runaway" patterns.

Interrogation of the device parameters confirms:

- SAV pacing interval = 350 ms;
- Low (escape rate) 50 bpm; and
- Upper tracking rate limit of 160 ms,

and provides you with the following additional information:

- Postventricular atrial refractory period (PVARP) = 250 ms
- Postventricular atrial blanking period (PVAB) = 180 ms
- Atrial blanking period = 180 ms

- Ventricular refractory period = 230 ms
- Ventricular blanking period (after atrial pace) (PAVB) = 28 ms

Question No. 3:

Based upon this additional information, you determine the patient will have 2:1 block when the sinus rate reaches:

A. 100 bpm (600 ms).

B. 104 bpm (580 ms).

C. 113 bpm (540 ms).

D. 140 bpm (430 ms).

E. 160 bpm (375 ms).

DISCUSSION

A P wave sensed by the CIED initiates an AV delay (analogous to the PR interval) programmed to a value selected from among the range of available parameters in the device. In this case, a long AV delay of 350 ms was selected. The atrial sensing channel is closed to atrial activity (refractory) during AV delays and for a set period following a paced or intrinsic sensed QRS. The PVARP for this device was set at 250 ms. The AV delay (350 ms) and the PVARP (250 ms) come together to make up the total atrial refractory period (TARP) of 600 ms. P waves that occur within the TARP will not be tracked in the DDD mode. To calculate the bpm rate of 2:1 block for this patient, 60,000 milliseconds per minute is divided by the TARP of 600 ms to equal an upper rate limit of 100 bpm. The 160-bpm tracking rate cannot be achieved.

Patient Management

In a young person with goals to exercise, enabling the activity sensor and programming a suitable upper sensor rate will permit tracking or pacing up to 160 bpm. The CIED will automatically decrease the PVARP as the heart rate increases, therefore decreasing TARP. Further decreases in TARP can be achieved by programming shorter AV delays or enabling the sensor to initiate AV decreases from programming.

Correct Answers to Second Case:

Q2. B

Q3. A

REFERENCES

1. Wang PJ, Chen H, Okamura H, Al-Ahmad A, Hsia HH. Timing cycles of implantable devices. In: Ellenbogen KA, Kay GN, Lau CP, Wilkoff BL, eds. *Clinical Cardiac Pacing, Defibrillation and Resynchronization Therapy*. 3rd ed. Philadelphia: Saunders Elsevier; 2007:969–995.
2. Kligfield P, Gettes LS, Bailey JJ, et al. Recommendations for the standardization and interpretation of the electrocardiogram; Part I: The electrocardiogram and its technology: A scientific statement from the American Heart Association Electrocardiography and Arrhythmias Committee, Council on Clinical Cardiology; the American College of Cardiology Foundation; and the Heart Rhythm Society: endorsed by the International Society for Computerized Electrocardiography. *Circulation*. 2007;115:1306–1324.

CASE STUDY

55

Implantable Device Therapy for Prophylaxis of Sudden Cardiac Death and Management of Congestive Heart Failure

Sanjeev Saksena, MD and Kinjal Desai

A 49-year-old male was admitted to the hospital with chest pain and shortness of breath in August 2008. He had noted dyspnea on exertion and decreased exercise tolerance for several years. He was clinically noted to have cardiomegaly and rales at both lung bases. Cardiac enzymes and ECG documented a non-ST-segment elevation myocardial infarction (MI). Cardiac catheterization showed a patent coronary stent and medical treatment was recommended.

The patient had a long history of hypertension and juvenile onset diabetes mellitus of 20 years duration. In 2006, he had an acute anteroseptal MI and underwent a percutaneous coronary intervention (PCI) procedure for revascularization with placement of an intracoronary stent. He had a history of chronic heart failure and prerenal azotemia. An echocardiogram showed moderate left ventricular (LV) dysfunction with an estimated ejection fraction (EF) of 30%, while monitoring and electrocardiography showed ventricular ectopics and nonsustained ventricular tachycardia (VT). He was referred for prophylactic implantable cardioverter-defibrillator (ICD) insertion for potential sudden cardiac death (SCD).

Question No. 1:

The patient has evidence of moderate LV dysfunction and ventricular arrhythmias and is at risk for sudden death. The preferred therapy at this time is:

A. Prophylactic ICD insertion 40 days after discharge.
B. Prophylactic ICD insertion before hospital discharge.
C. Amiodarone.
D. None of the above.

CLINICAL COURSE

The patient was readmitted in October 2008. A preadmission ECG showed sinus rhythm with occasional premature ventricular contractions (PVC), QRS duration of 95 ms in sinus rhythm, healed anteroseptal MI QRS pattern, and a prolonged calculated QTc interval of 478 ms. Lateral T-wave inversions are consistent with ischemia (Figure 55.1). A gated blood pool scan showed LV dysfunction with a global EF of 40%. Electrophysiologic testing showed sinus rhythm with first-degree atrioventricular (AV) block,

Practical Electrophysiology © 2015 Jasbir S. Sra, Masood Akhtar, Andrea Natale, David J. Wilber
Cardiotext Publishing, ISBN: 978-1-935395-17-1

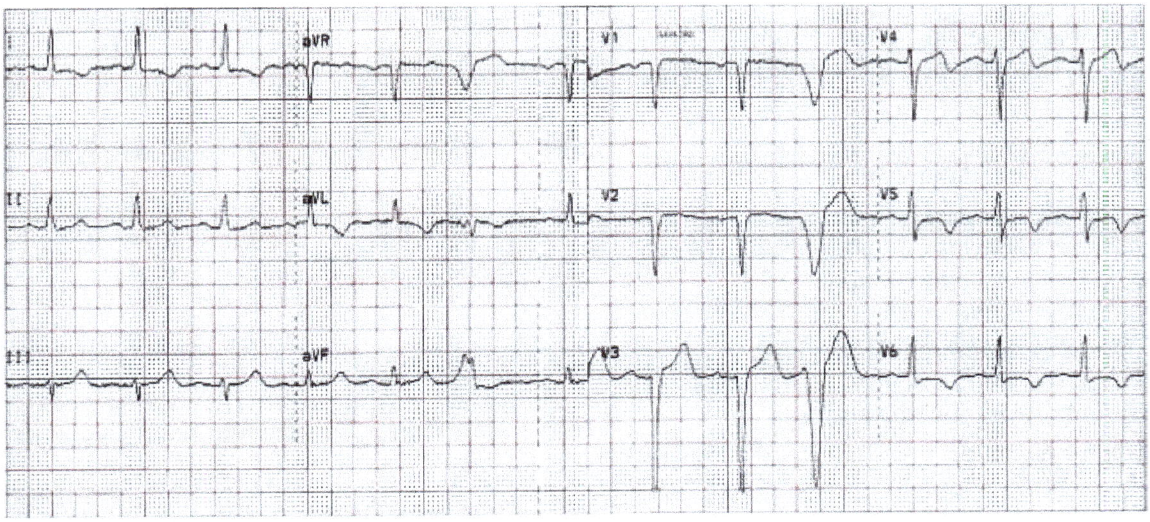

Figure 55.1. Sinus rhythm with evidence of a healed anteroseptal MI and lateral wall ischemia is present. Frequent PVC is noted.

QRS duration of 130 ms, AH interval of 102 ms, and HV interval of 80 ms. Atrial extrastimulation showed further HV prolongation and intermittent intra-His block. Programmed ventricular stimulation induced sustained monomorphic VT with a cycle length of 330 ms, right bundle branch block morphology, and was terminated by antitachycardia pacing (ATP) at 300 ms. A dual-chamber ICD was inserted. Five months later, sustained VT was terminated by ATP intervention from the ICD (Figure 55.2).

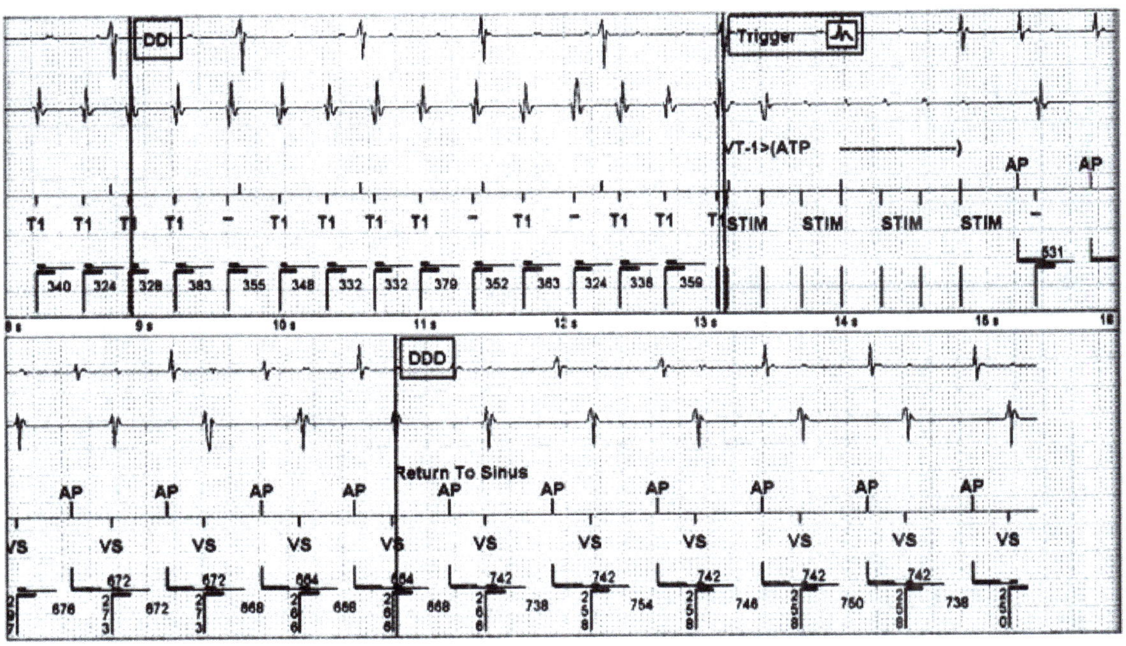

Figure 55.2. Spontaneous VT documented by intracardiac electrograms (EGMs) (upper panel) with a cycle length ranging from 324 to 379 ms is seen with AV dissociation. ATP commences (Trigger) and terminated the tachycardia. The patient returns to an atrial paced rhythm in the bottom channel. Upper panel: top EGM = right atrium; lower EGM = right ventricle (RV); top marker channel = interval and stimulation markers; bottom marker channel = ventricular events detection. Lower panel: top EGM = right atrium; lower EGM = RV; top marker channel = paced/sensed; bottom marker channel = A/V intervals.

Question No. 2:

Electrophysiologic testing is not routinely recommended before prophylactic ICD insertion because:

A. It has no prognostic value after acute myocardial infarction (AMI).

B. MADIT II showed that frequency of ICD therapy delivery was not different in patients who had VT induced with noninvasive programmed stimulation.

C. Moderate to severe reduction in left ventricular ejection fraction (LVEF) identifies all patients at risk for SCD after MI.

D. None of the above.

Question No. 3:

Which of the following statements accurately describe(s) risk stratification after AMI?

A. Depressed heart rate variability is a weak predictor of outcomes after AMI.

B. T-wave alternans during stress testing 6 weeks after MI is most predictive of the risk of sudden death when a patient can complete a maximal exercise test.

C. Induction of sustained VT > 10 days after AMI can predict recurrent VT, and ICD therapy can improve outcomes in these patients.

Question No. 4:

In patients manifesting nonsustained VT after AMI and depressed LV function:

A. Immediate ICD placement is appropriate.

B. Beta-blocker therapy should be maintained.

C. Wearable defibrillator is an alternative to immediate ICD implant.

D. Amiodarone is the best choice.

DISCUSSION

Patients with coronary artery disease make up the majority of ICD recipients in most studies. ICD implantation is accepted as reducing mortality and improving outcomes in patients with chronic ischemic heart disease. Patients with reduced LV function have increasing risk of sudden death with progressive impairment of ventricular function and have been shown to obtain greater benefit with ICD therapy than with drug therapy (Figure 55.3 and Table 55.1). A large number of clinical trials have demonstrated improvement in mortality with ICD therapy for secondary and primary prevention.

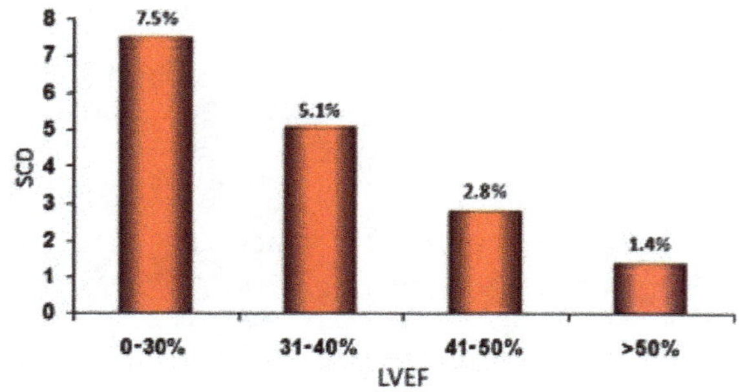

Figure 55.3. SCD incidences vary with degree of LV systolic dysfunction as assessed by EF. (Adapted from de Vreede-Swagemakers et al.[7])

Patients with AMI that complicates their coronary disease or preexisting LV dysfunction pose a special problem (see Table 55.1). Sudden death is an infrequent complication after AMI. Two large clinical trials have studied postinfarction populations with risk factors for benefits with ICD prophylaxis. In DINAMIT, patients had reduced LVEF (mean = 28%), depressed heart rate variability, or an elevated average 24-hour heart rate ≥ 80 bpm on Holter monitoring. In IRIS, patients enrolled had reduced LV function (mean LVEF = 35%) and a heart rate ≥ 90 bpm, or nonsustained VT at a heart rate ≥ 150 bpm on Holter monitoring with any LVEF. In these trials, ICD therapy was implemented in the immediate postinfarction period, defined as 6 to 40 days in DINAMIT and less than 30 days (mean 13 days) in IRIS. Neither trial demonstrated improved overall outcomes (see Table 55.1). In both trials, the ICD group had lower sudden death mortality. However, this result was offset by increased nonsudden mortality.

After AMI, delayed VT/ventricular fibrillation (VF) has been observed in patients with postinfarction ventricular arrhythmias, poor LV function, impaired heart rate variability, and other risk factors. In view of this long-standing observation, risk stratification may be appropriate to select patients for device insertion after the immediate postinfarction period (>40 days). Patients with relatively low risk of arrhythmic events can be identified for device insertion by an absence of:

- spontaneous ventricular arrhythmias;
- late potentials on signal-averaged ECG;
- T-wave alternans on exercise testing; and
- inducible VT/VF on programmed electrical stimulation >10 days, up to 45 days after MI.

Table 55.1. ICD Therapy Clinical Trials

Trial	Patients	Year	All-cause mortality	P value
AVID	1016	1997	39% relative risk reduction Hazard ratio (HR) 0.62	< 0.02
CIDS	659	2000	20% relative risk reduction HR 0.83	NS
CASH	191	2000	38% relative risk reduction HR 0.82	0.08
MADIT I	196	1996	54% relative risk reduction HR 0.46	0.009
MUSTT	704	1999	51% relative risk reduction HR 0.40	< 0.001
MADIT II	1232	2002	31% relative risk reduction HR 0.69	0.016
SCD-HeFT	1676	2005	23% relative risk reduction HR 0.77	0.007
DINAMIT	674	2004	HR 1.08	0.66
DEFINITE	485	2004	HR 1.04	0.78

Primary and secondary prevention trials. See text for details.

Unfortunately, the positive predictive accuracy of many risk stratification techniques is limited and most useful in identifying low-risk patients. Signal-averaged ECG has a strong negative predictive value but modest positive predictive accuracy. Gomes et al showed additive value with LV dysfunction in identifying higher-risk patients with chronic coronary artery disease and ventricular arrhythmia (Figure 55.4). Impaired heart rate variability has been shown to identify high-risk patients but may have a low sensitivity. T-wave alternans is limited by prior conduction defects, failure to maintain a stable heart rate in the target range during exercise, or drugs, such as beta-blockers or antiarrhythmics that can interfere with the test. Maximal exercise testing is not required for this procedure. In the ABCD trial, the positive predictive accuracy of the test was modest and inferior to electrophysiologic testing. The predictive accuracy of these 2 tests dissipates after 1 year for T-wave alternans and after 2 years for electrophysiologic testing. This approach has been used in different regions to combine clinical and laboratory observations to identify the highest-risk patients. Kearney and colleagues have used clinical, ECG, and radiographic criteria to achieve this outcome (Figure 55.5). An important element in assessing sudden death risk has emerged from recent clinical trials. Time dependence of risk based on the interval from MI as well as the severity of LV dysfunction play important roles in the absolute risk in a given patient (Figure 55.6). Models have included other noncardiac parameters such as creatinine clearance, occurrence of heart failure, and rehospitalization in projecting risk for an individual patient (Figure 55.7).

This patient had both a remote and a recent MI. While improvement in LV function after MI was demonstrated, suggesting lower risk, programmed stimulation induced monomorphic sustained VT. Iesaka et al demonstrated the prognostic

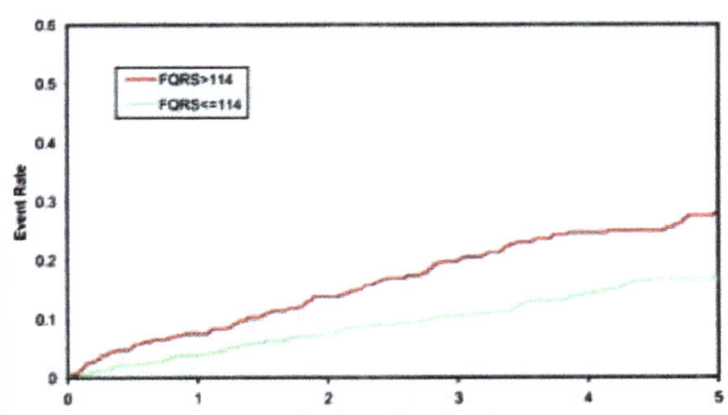

Figure 55.4. In patients with chronic coronary artery disease and ventricular arrhythmias, a LVEF of < 30% and the presence of late potentials identified the highest-risk groups. (Adapted from Gomes et al[8] with permission from Wolters Kluwer Health.)

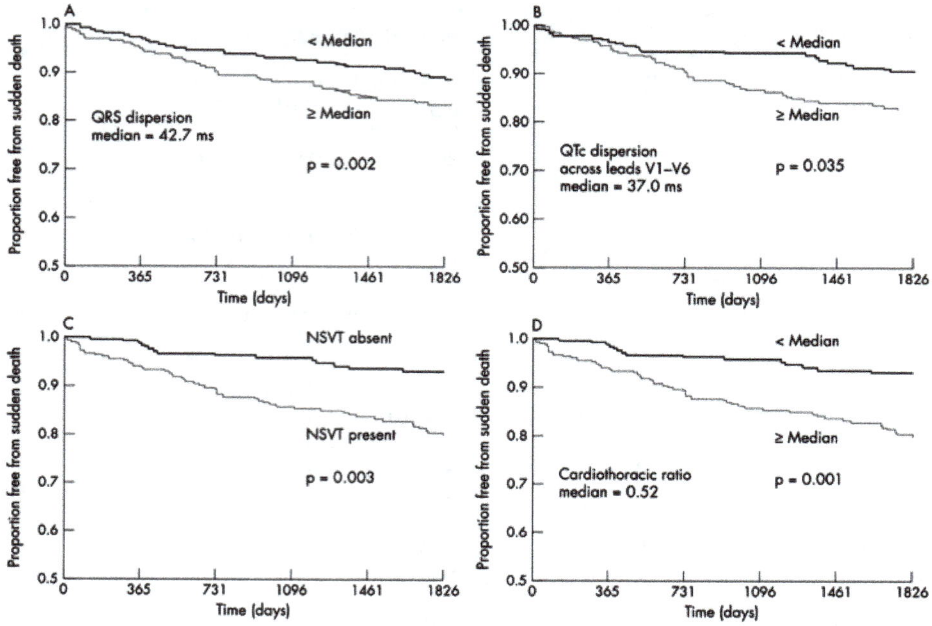

Figure 55.5. Electrocardiographic, clinical, and radiographic parameters that influence the risk of sudden death, used in a composite clinical risk assessment for sudden death. (Reproduced from Kearney et al[9] with permission from BMJ Publishing.)

implications of induction of monomorphic sustained VT (Figure 55.8); 19% of their patients in the pre-PCI era had this finding and a high risk of future VT/VF was noted. In the PCI era, a recent report from Zaman et al showed that this finding continues to have prognostic value and that implant of an ICD improved outcomes so comparable to patients who were noninducible at electrophysiologic study (Figure 55.9). In addition, this patient had a recent infarction that may have

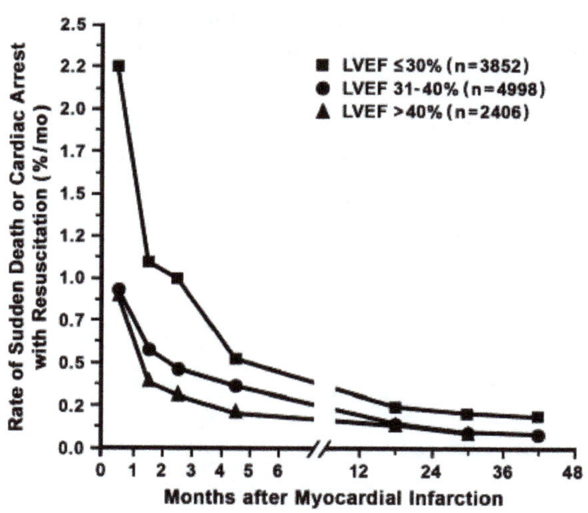

Figure 55.6. Time dependence of sudden death (SCD) risk after MI, stratified by LV dysfunction, and time interval from acute event. (Adapted from Solomon et al[10] with permission from Massachusetts Medical Society.)

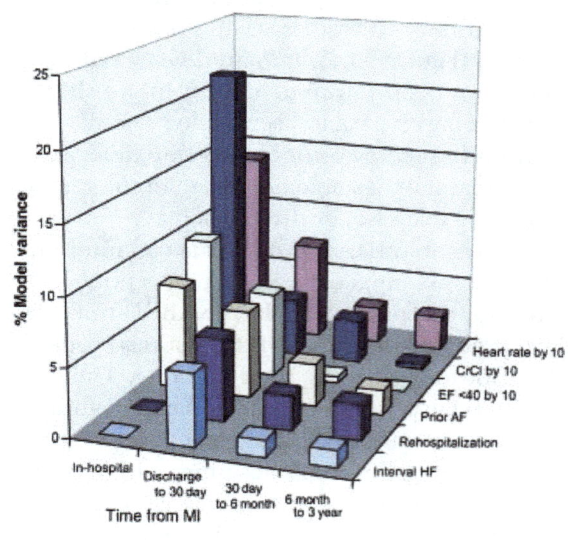

Figure 55.7. Development of a composite risk model for sudden death after MI with time dependence and impact of individual variables. AF, atrial fibrillation; CrCl, creatinine clearance; EF, ejection fraction; HF, heart failure. (Reproduced from Piccini et al[11] with permission from Oxford University Press.)

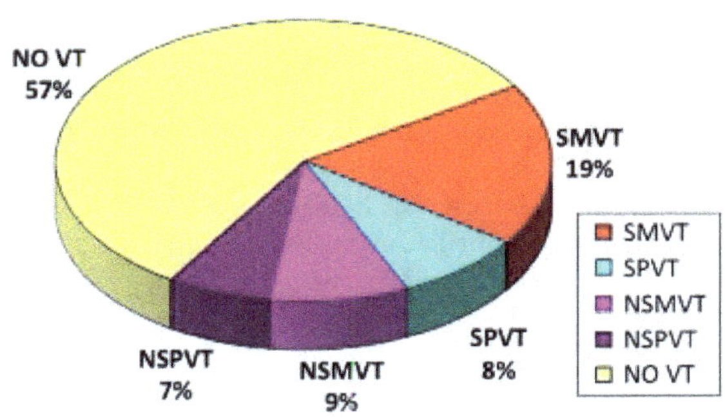

Figure 55.8. Programmed electrical stimulation performed 1.8 months after AMI. No antiarrhythmics or beta-blockers were administered. All VT or VF events occurred in the patients with SMVT, sustained monomorphic VT only. NSMVT, nonsustained monomorphic VT; NSPVT, nonsustained polymorphic VT; SMVT, sustained monomorphic VT; SPVT, sustained polymorphic VT. (Reproduced from Iesaka et al[12] with permission from Elsevier.)

increased risk. This was confirmed by the occurrence of spontaneous VT shortly after ICD implant.

MADIT II utilized a limited programmed stimulation protocol implemented after device insertion and did not show increased event rates in inducible patients. However, the single-site limited protocol had reduced sensitivity for VT induction compared to the above studies.

A continuing dilemma is that risk of delayed VT/VF is present in high-risk post-MI patients during the first 2 months after infarction. If risk stratification is delayed beyond 40 days, a wearable external defibrillator may permit monitoring for sustained arrhythmias and permit life-saving defibrillation in this period.

PRACTICE GUIDELINES

The relevant practice guidelines from the ACC/AHA indications for ICD therapy are as follows:

Class I Indication

1. ICD therapy is indicated in patients with LVEF less than 35% due to prior MI who are at least 40 days post-MI and are in New York Heart Association (NYHA) functional Class II or III (*Level of Evidence: A*).
2. ICD therapy is indicated in patients with LV dysfunction due to prior MI who are at least 40 days post-MI, have a LVEF less than 30%, and are in NYHA functional Class I (*Level of Evidence: A*).
3. ICD therapy is indicated in patients with nonsustained VT due to prior MI, LVEF less than 40%, and inducible VF or sustained VT at electrophysiological study (*Level of Evidence: B*).

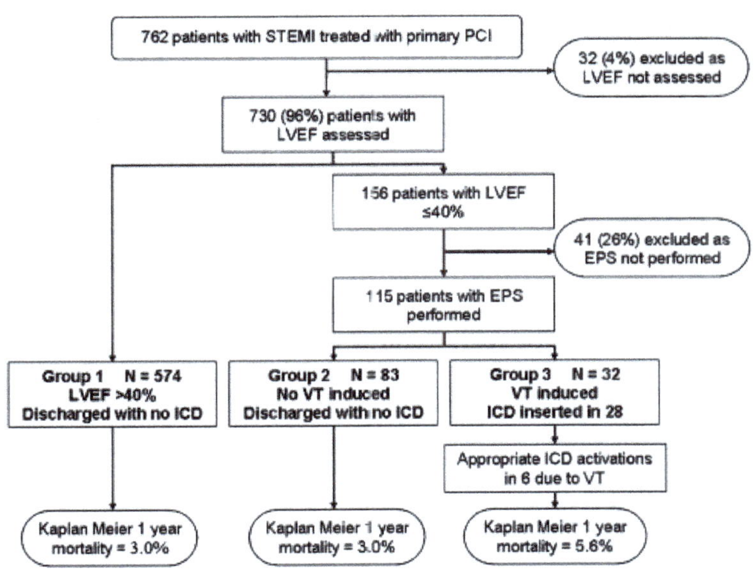

Figure 55.9. Flow diagram of outcomes of patients after AMI and PCI that underwent programmed electrical stimulation 9 days after MI and ICD implantation if VT was induced. Note that these patients after ICD implant had comparable outcomes to those who did not have this risk factor for sudden death. EPS, electrophysiology study; LVEF, left ventricular ejection fraction; STEMI, ST-segment elevation MI. (Reproduced from Zaman et al[13] with permission from Wolters Kluwer Health.)

Correct Answers:

Q1. D

Q2. D

Q3. C

Q4. C

REFERENCES

1. Bardy G, Lee K, Mark D, et al. Sudden Cardiac Death in Heart Failure Trial (SCD-HeFT) Investigators. Amiodarone or an implantable cardioverter-defibrillator for congestive heart failure. *N Engl J Med.* 2005;352:225–237.
2. Domanski MJ, Saksena S, Epstein AE, et al. Relative effectiveness of the implantable cardioverter-defibrillator and antiarrhythmic drugs in patients with varying degrees of left ventricular dysfunction who have survived malignant ventricular arrhythmias. AVID investigators. Antiarrhythmics versus implantable defibrillators. *J Am Coll Cardiol.* 1999;34:1090–1095.
3. Epstein AE, DiMarco JP, Ellenbogen KA, et al.; American College of Cardiology/American Heart Association Task Force on Practice Guidelines (Writing Committee to Revise the ACC/AHA/NASPE 2002 Guideline Update for Implantation of Cardiac Pacemakers and Antiarrhythmia Devices); American Association for Thoracic Surgery; Society of Thoracic Surgeons. ACC/AHA/HRS 2008 Guidelines for Device-Based Therapy of Cardiac Rhythm Abnormalities: A report of the American College of Cardiology/American Heart Association Task Force on Practice Guidelines developed in collaboration with the American Association for Thoracic Surgery and Society of Thoracic Surgeons. *J Am Coll Cardiol.* 2008;51:1–62.
4. Hohnloser SH, Kuck KH, Dorian P, et al.; DINAMIT Investigators. Prophylactic use of an implantable cardioverter-defibrillator after acute myocardial infarction. *N Engl J Med.* 2004;351:2481–2488.
5. Moss AJ, Zareba W, Hall WJ, et al.; Multicenter Automatic Defibrillator Implantation Trial II Investigators. Prophylactic implantation of a defibrillator in patients with myocardial infarction and reduced ejection fraction. *N Engl J Med.* 2002;346:877–883.
6. Steinbeck G, Andresen D, Seidl K, et al.; IRIS Investigators. Defibrillator implantation early after myocardial infarction. *N Engl J Med.* 2009;361:1427–1436.
7. de Vreede-Swagemakers, Gogeles AP, Dubois AW, et al. Out-of-hospital cardiac arrest in the 1990's: A population-based study in the Maastricht area on incidence, characteristics and survival. *J Am Coll Cardiol.* 1997;30:1500–1505.
8. Gomes et al. Prediction of long-term outcomes by signal-averaged electrocardiography in patients with unsustained ventricular tachycardia, coronary artery disease, and left ventricular dysfunction. *Circulation.* 2001;104:436–441.
9. Kearney et al. Predicting sudden death in patients with mild to moderate chronic heart failure. *Heart.* 2004;90:1137–1143.
10. Solomon et al. Sudden death in patients with myocardial infarction and left ventricular dysfunction, heart failure, or both. *New Engl J Med.* 2005;352:2581–2588.
11. Piccini et al. Predictors of sudden cardiac death change with time after myocardial infarction: Results from the VALIANT trial. *Eur Heart J.* 2010;31:211–221.
12. Iesaka et al. Prognostic significance of sustained monomorphic ventricular tachycardia induced by programmed ventricular stimulation using up to triple extrastimuli in survivors of acute myocardial infarction. *Am J Cardiol.* 1990;65:1057–1063.
13. Zaman et al. Outcomes of early risk stratification and targeted implantable cardioverter-defibrillator implantation after ST-elevation myocardial infarction treated with primary percutaneous coronary intervention. *Circulation.* 2009;120:194–200.

CASE STUDY

56

T-Wave Oversensing and Implantable Cardioverter-Defibrillator Shocks

Rajiv Mahajan, MD, DM, Manoj Kumar Rohit, MD, DM, and Kewal K. Talwar, MD, DM

A 65-year-old male was diagnosed to have dilated cardiomyopathy in 2000. He had presented with a history of syncope at the time of diagnosis. Coronary angiography had revealed normal coronaries. He had also undergone electrophysiology study in which ventricular tachycardia (VT) was induced; however, details of the procedure were not available. Subsequently, he had undergone single-chamber implantable cardioverter-defibrillator (ICD) implantation in 2000. The defibrillation test and R-wave details at the time of initial implant were not available. We saw him in 2008 when the generator had reached near-end of service. He was NYHA class II on an optimal dose of beta-blockers and angiotensin-converting enzyme (ACE) inhibitors. A repeat echocardiogram revealed left ventricular ejection fraction of 26%. He underwent ICD replacement (Entrust D154VRC, Medtronic Inc., Minneapolis, MN, USA). The previous lead (Medtronic Sprint 6942, a 65-cm, passive fixation, integrated, bipolar dual-coil lead) was retained. The R wave at the time of ICD replacement was 5 mV. Pacing and shock impedance were within normal range. Sensitivity was kept at the nominal 0.3 mV. Detection was programmed to 2 zones—VT cut-off was 150 bpm and ventricular fibrillation (VF) cut-off was 188 bpm. Antitachycardia pacing (ATP) during charging was programmed "on," thereby meaning one burst of ATP therapy would be delivered during charging for therapy in the VF zone (if cycle length is < 240 ms). Wavelet was programmed to "monitor," meaning the device will classify each beat in the tachycardia zone as supraventricular or ventricular but will not play a role in the delivery of therapy. The supraventricular tachycardia (SVT) limit was kept at 320 ms. In mid-2009, he received a cluster of shocks while bicycling furiously. He did not have presyncope/syncope symptoms at that time. Figure 56.1 shows the initial electrograms (EGM) when the device detected tachycardia and the first sequence of ATP was delivered; Figure 56.2 shows the EGMs after the first shock (Figure 56.2A), the second shock (Figure 56.2B), and the final shock (Figure 56.2C).

Question No. 1:

Was the therapy appropriate?

Question No. 2:

What happened after the second shock?

Question No. 3:

What are the measures that can be taken?

Practical Electrophysiology © 2015 Jasbir S. Sra, Masood Akhtar, Andrea Natale, David J. Wilber
Cardiotext Publishing, ISBN: 978-1-935395-17-1

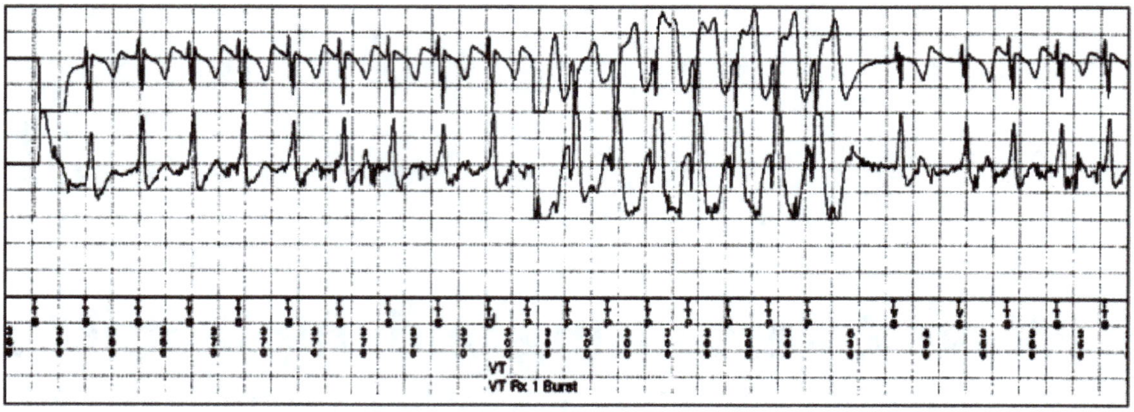

Figure 56.1. *From top to bottom:* Near-field EGM, far-field EGM, and marker channel. The initial part shows tachycardia at a cycle length of 370 ms. The wavelet suggested SVT. The far-field EGM was similar to the far-field EGM at the time of interrogation. Since the VT/SVT enhancements were not programmed "on," the device interpreted the rhythm as VT (wavelet was programmed to "monitor").

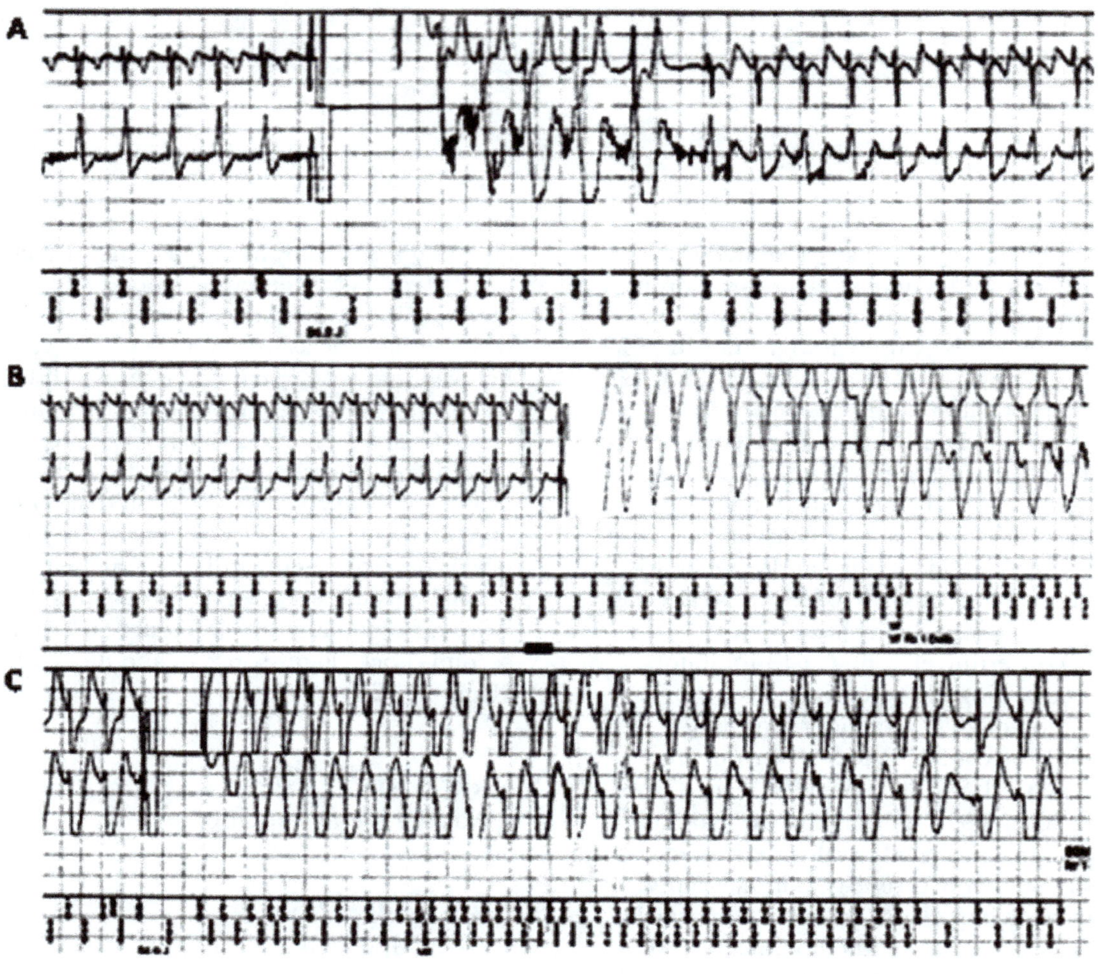

Figure 56.2. *From top to bottom in each panel:* Near-field EGM, far-field EGM, and marker channel. **A**: The initial EGM distortion quickly decreases with only mild broadening of EGM after the initial few beats. There is no oversensing of T wave. Sinus tachycardia, probably perpetuated by the shock, persists at the same cycle length of 370 ms. **B**: The distortion of the EGM is gross with T-wave oversensing. The distortion is so much that the EGM saturates the channel. Sinus tachycardia persists. **C**: The distortion of the EGM is even greater.

Question No. 4:

What type of internal and external noise can cause inappropriate shock?

T-wave oversensing can lead to inappropriate shocks. T waves are identified by the device as R waves and delivery of shock on T waves can induce VF. If this happens on the last shock, the consequences can be serious and can result in death.[12] Appropriate use of enhancements for better SVT/VT discrimination and ATP therapy can reduce the number of inappropriate and appropriate shocks and also reduce the risk for post-shock T-wave oversensing.

The size of the T wave relative to the R wave influences the choice of remedy. The presence of a good-sized R wave with a good R-to-T-wave ratio suggests that T-wave oversensing will be amenable to several measures like alteration of sensitivity, threshold start, and delay decay. However, a small R wave would indicate that insertion of a separate pace/sense lead would be required to solve the problem of T-wave oversensing. The lead and device implanted may also influence the remedy. In this case, the presence of an integrated bipolar lead was probably responsible for the distortion of EGMs post shock.

Correct Answers:

Q1. The patient was cycling at the time of receiving shocks. He did not have any symptoms of presyncope. He received a cluster of shocks. All these features suggest that the therapy may have been inappropriate. Moreover, the far-field EGM during the initial detection had a good wavelet match score and was similar on manual analysis to the far-field EGM recorded at the time of interrogation. The interval plot suggested that there was a gradual increase in heart rate prior to detection. These features suggest that the patient received the initial ATP therapies for sinus tachycardia.

Q2. After the first shock the sinus tachycardia persisted, probably perpetuated by the shock, and there was abnormal widening of the ventricular EGMs. However, after the second shock, the EGMs became grossly distorted and there was intermittent T-wave oversensing. The sinus tachycardia and T-wave oversensing persisted after each shock and multiple shocks were delivered. Lack of use of SVT discriminators and T-wave oversensing resulted in inappropriate shocks.

Q3. The wavelet had correctly classified the initial rhythm as SVT. Hence the wavelet was programmed "on" from "monitor." The SVT limit was changed from 320 to 300 ms. The patient was subjected to a treadmill test (TMT), however, he could achieve only a rate of 144/min. There was no rate-related bundle branch block (BBB). The patient was already on a maximally tolerated dose of beta-blockers. The VT cut-off was changed to 360 ms and the VF cut-off was changed to 300 ms. The ATP during charging was kept "on," effectively working like 3 zones. This programming would reduce the chance of inappropriate shock.

T-wave oversensing occurs broadly in 3 scenarios: post pacing leading to inhibition of pacing/ATP therapy; T-wave (spontaneous) oversensing with good R waves; and T-wave (spontaneous) oversensing with small R waves.[1] T-wave oversensing with good R waves may be amenable to reduction of sensitivity. Also, in St. Jude devices, threshold start and delay decay can be reprogrammed. These measures may not be as effective in T-wave oversensing with small R waves and, often, replacement of the lead or addition of a separate pace/sense lead is required. This situation was unique in the sense that T-wave oversensing occurred only in the post-shock scenario. The R and T waves, post shock, were both large and saturated the channel. Hence, alteration of sensitivity may not solve the post-shock oversensing as the T waves were as large as the R waves. Electroporation or appearance of BBB may result in broadening of the EGM post shock. Electroporation (creation of micro-holes in cardiac cell membranes by strong electric fields) has been proposed as a cause of EGM distortion post shock.[2] The bizarre and increasing grade of distortion following each subsequent shock and absence of rate-related BBB (on TMT) suggests that electroporation was probably the cause. Electroporation following ICD discharges, leading to broadening of the EGMs, is a problem with integrated bipolar leads and is less commonly seen with true bipolar defibrillation leads.[3] The Sprint 6942 (in this case) is an integrated bipolar lead. The patient was advised to undergo insertion of a separate pace/sense lead, since similar widening of EGMs could occur even after an appropriate shock. Another strategy in such a situation, which would also work after the first ICD therapy, is stability. However, the 2 intervals (RT and TR) were similar in our case and hence not a useful option.

Q4. Nonarrhythmic signals can be oversensed and detected as arrhythmias, resulting in inappropriate delivery of therapy. These signals can be extracardiac or intracardiac. Extracardiac signals may be physiological or nonphysiological. Implantable defibrillators use dynamic gain to appropriately sense R waves and avoid oversensing of T waves and still maintain adequate sensitivity to avoid undersensing of fine VF. Due to the dynamic nature, oversensing more often occurs after long diastolic intervals. Sensitivity is also rapidly maximized following a paced beat, increasing the chance of oversensing.

EXTRACARDIAC SIGNALS

The distinctive feature of extracardiac signals is the lack of relation to the cardiac cycle. Extracardiac signals may be nonphysiological or physiological.

Myopotentials

Myopotentials are physiological signals that typically produce high-frequency noise persisting for a variable fraction of the cardiac cycle. Pectoral myopotentials often occur during exercise and can be reproduced by pressing the hands forcibly against each other. The pectoral myopotentials are seen on the far-field EGM (Figure 56.3A). Hence, they do not usually cause ventricular oversensing in defibrillator devices. However, they can distort the far-field EGM and can interfere with the wavelet SVT/VT discrimination in Medtronic devices and

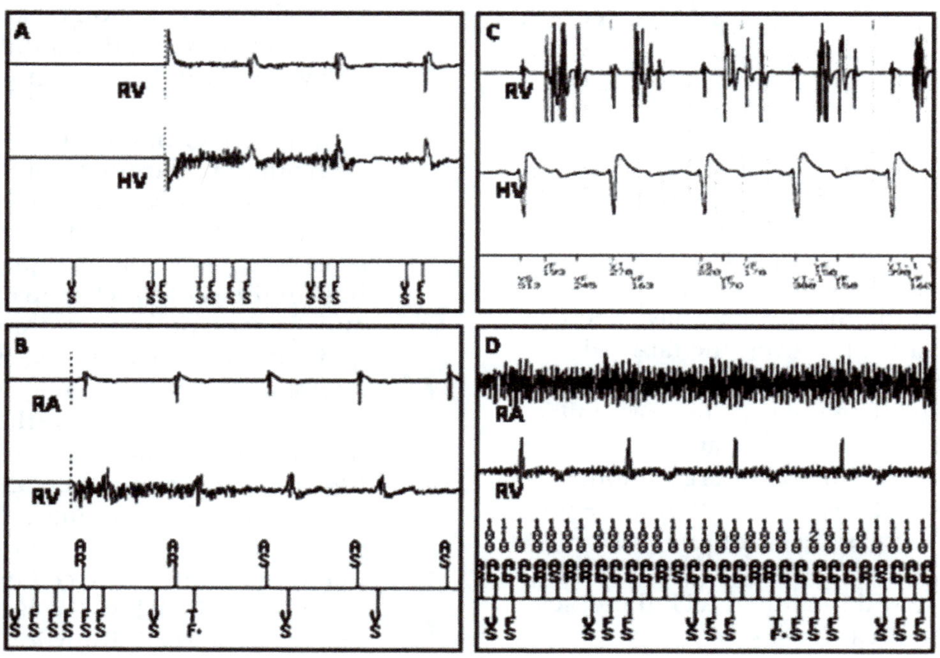

Figure 56.3. Types of oversensing leading to inappropriate detection of VT. **A:** *From top to bottom:* Near-field EGM, far-field EGM, and marker channel. Pectoral myopotentials recorded during exercise, seen more prominently on the far-field channel. Note the cluster of inappropriate sensing at end-diastole due to maximum sensitivity at end-diastole. **B:** *From top to bottom:* Atrial electrogram (AEGM), ventricular near-field EGM, and marker channel. Diaphragmatic myopotentials in a patient with an integrated bipolar lead implanted at the right ventricular apex (RVA). The noise is fine and seen only on the near-field EGM. The diaphragmatic myopotentials may be brought about by a deep breath. **C:** *From top to bottom:* Near-field EGM, far-field EGM, and marker channel. The noise caused by lead fracture is intermittent, often saturating the channels. It may be limited to the sensing channel and brought about by arm movement. **D:** *From top to bottom:* AEGM, ventricular near-field EGM, and marker channel. Electromagnetic interference produces a continuous noise, which is seen on all the channels.

may result in inappropriate shocks. St. Jude devices use the near-field EGM for morphology discrimination and are not affected by pectoral myopotentials. Pectoral myopotentials are of concern in unipolar pacemakers and can result in inappropriate inhibition of pacing.

Diaphragmatic myopotentials are low amplitude and typically seen on near-field EGMs. Since they are low amplitude, oversensing usually occurs after long diastolic intervals or after ventricular-paced events and tends to end with a sensed R, which abruptly reduces sensitivity. The diaphragmatic myopotentials are usually seen with integrated bipolar leads implanted at the RVA[3] (Figure 56.3C). Diaphragmatic myopotentials may cause persistent oversensing leading to inhibition of pacing, followed by inappropriate shocks. Altering sensitivity and, in some situations, replacing the lead with a true bipolar lead rectifies the issue.

Lead failure/Loose setscrew

Epicardial leads, abdominal implants, Sprint Fidelis and coaxial defibrillation leads, subclavian venous access, and dual-chamber (compared to single-chamber) ICD systems are predictive of lead failure.[4–8] The noise tends to occur for a fraction of the cardiac cycle and may saturate the channel (see Figure 56.3C). Occasionally, it may be detected after defibrillation. It may also be associated with rise in lead impedance, although the rise may be intermittent. The noise may be brought about by arm movement. Frequent sensing of nonphysiological VV intervals of less than 120 ms, increase in frequency of NSVT or impedance rise should alert to impending lead fracture. Loose setscrew may mimic lead fracture. However, it typically occurs post implant or battery change. Noise can also be produced due to interaction with retained leads.

Electromagnetic interference (EMI)

Radio waves, with frequencies between 0 and 109 Hz, including current electricity supplies and electrocautery, and microwaves with frequencies between 10^9 and 10^{11} Hz can affect implantable devices.[9] The characteristic of EMI is replacement of the isoelectric baseline with high-frequency noise, bearing no relation to the cardiac cycle (Figure 56.3D). Although the signal may be continuous, oversensing may be intermittent due to auto-adjusting gain. EMI signal amplitude is greater on the far-field EGM from widely spaced electrodes as compared to sensed EGM from closely spaced electrodes.

INTRACARDIAC SIGNALS

T-wave oversensing, R-wave double-counting, P-wave oversensing, and far-field R-wave (FFRW) oversensing are the intracardiac signals that may result in inappropriate shocks. These signals are most often seen with integrated bipolar defibrillation leads and have been nearly eliminated with true bipolar leads. T-wave oversensing is discussed elsewhere in the chapter.

R-wave double-counting occurs when the sensed ventricular EGM exceeds the ventricular blanking (VB) period. This may happen due to baseline slow ventricular depolarization and may be worsened with Class IC antiarrhythmic drugs or hyperkalemia, which further slow ventricular conduction. R-wave double-counting was also seen with the older-generation cardiac resynchronization therapy-defibrillation (CRT-D) devices using Y-adapted or extended bipolar sensing between right ventricular and left ventricular electrodes. R-wave double-counting results in a railroad track pattern of ventricular intervals.[10] R-wave double-counting can be overcome by increasing the VB period from a nominal of 120 ms to 150 to 160 ms. This feature is available in the St. Jude device and newer models of Medtronic.

P-wave oversensing results from sensing of P waves due to the distal coil of an integrated lead lying close to the tricuspid valve and the sensed PR interval exceeding the VB period. This is seen more often in children and when the lead is implanted in the tricuspid inflow or proximal septum or gets dislodged. If oversensing occurs in a 1:1 pattern, it also results in a railroad track pattern of ventricular intervals.

FFRW oversensing results from oversensing of R waves on the atrial channel and can confound SVT/VT discrimination algorithms. Intermittent FFRW is a common cause of inappropriate shocks during sinus tachycardia in the VT zone.[11]

REFERENCES

1. Swerdlow CD, Gillberg JM, Olson WH. Sensing and detection. In: Ellenbogen KA, Wilkoff BL, Kay GN, Lau CP, eds. *Clinical Cardiac Pacing, Defibrillation and Resynchronization Therapy.* Philadelphia, PA: Saunders; 2006:75–160.
2. Tung L, Tovar O, Neunlist M, Jain SK, O'Neill RJ. Effects of strong electrical shock on cardiac muscle tissue. *Ann N Y Acad Sci.* 1994;720:160–175.
3. Weretka S, Michaelsen J, Becker R, et al. Ventricular oversensing: A study of 101 patients implanted with dual chamber defibrillators and two different lead systems. *Pacing Clin Electrophysiol.* 2003;26:65–70.

4. Luria D, Glikson M, Brady PA, et al. Predictors and mode of detection of transvenous lead malfunction in implantable defibrillators. *Am J Cardiol.* 2001;87:901–904.
5. Glikson M, Luria D, Friedman PA, et al. Are routine arrhythmia inductions necessary in patients with pectoral implantable cardioverter defibrillators? *J Cardiovasc Electrophysiol.* 2000;11:127–135.
6. Kron J. Clinical significance of device-related complications in clinical trials and implications for future trials: Insights from the Antiarrhythmics Versus Implantable Defibrillators (AVID) trial. *Card Electrophysiol Rev.* 2003;7:473–478.
7. Gallik DM, Ben-Zur UM, Gross JN, Furman S. Lead fracture in cephalic versus subclavian approach with transvenous implantable cardioverter defibrillator systems. *Pacing Clin Electrophysiol.* 1996;19:1089–1094.
8. Ellenbogen KA, Wood MA, Swerdlow CD. The Sprint Fidelis lead fracture story: What do we really know and where do we go from here? *Heart Rhythm.* 2008;5:1380–1381.
9. Hayes DL, Strathmore NF. Electromagnetic interference with implantable devices. In: Ellenbogen KA, Kay GN, Wilkoff BL, eds. *Clinical Cardiac Pacing and Defibrillation,* 2nd ed. Philadelphia, PA: WB Saunders Co.; 2000: 939–952.
10. Garcia-Moran E, Mont L, Brugada J. Inappropriate tachycardia detection by a biventricular implantable cardioverter defibrillator. *Pacing Clin Electrophysiol.* 2002;25:123–124.
11. Wilkoff BL, Kühlkamp V, Volosin K, et al. Critical analysis of dual-chamber implantable cardioverter-defibrillator arrhythmia detection: Results and technical considerations. *Circulation.* 2001;103:381–386.
12. Dimitri H, John B, Young GD, Sanders P. Fatal outcome from inappropriate defibrillation. *Europace.* 2007;9:1059–1060.

CASE STUDY

57

Cardiac Resynchronization Therapy

José F. Huizar, MD, Karoly Kaszala, MD, PhD, Jayanthi N. Koneru, MBBS, and Kenneth A. Ellenbogen, MD

A 77-year-old African-American male with history of diabetes mellitus, gout, hypertension, and mild chronic renal insufficiency was diagnosed with dilated cardiomyopathy in March 2002. His echocardiogram revealed left ventricular (LV) ejection fraction (EF) of 10% to 15%, LV end-diastolic dimension (LVEDD) of 80 mm, and LV end-systolic dimension (LVESD) of 73 mm. He had several heart failure (HF) admissions in the ensuing months and remained in New York Heart Association (NYHA) class III despite optimal medical therapy (OPT). In July 2002, he received a biventricular pacemaker due to the presence of left bundle branch block (LBBB) with a QRS duration of 190 ms. Over the following weeks, he improved symptomatically to NYHA class II. A repeat echocardiogram the following year demonstrated improvement of his LVEF and dimensions (LVEF 20%, LVEDD 56 mm, LVESD 51 mm). His most recent echocardiograms (in 2005 and 2009) demonstrated gradual improvement of his LVEF to 35% and 55%, respectively (LVEDD 51 mm, LVESD 38 mm). He has been clinically stable over the past 8 years and has not had any HF admissions over the past 3 years.

Question No. 1:

Which of the following is true?

A. Cardiomyopathy and HF are associated with ventricular conduction delay.

B. Delayed electrical activation of left ventricle (LV) leads to abnormal pump efficiency.

C. Cardiac resynchronization therapy (CRT) can correct mechanical dyssynchrony in many patients.

D. All of the above are true.

DISCUSSION

Cardiomyopathy and HF have been commonly associated with ventricular conduction delays such as LBBB. LBBB typically results in regionally delayed electrical activation in the posterolateral LV wall. The early activation of the septum results in early systolic contraction that is unopposed due to the lack of simultaneous contraction of the LV free wall. Therefore, the lateral wall stretches in response to septal contraction, delaying the rise of LV pressure. Similarly, the lateral wall is activated (in late-systole) and reaches its peak force when the septum is relaxed (stretching this region), which could occur even after aortic valve closure (Figure 57.1A). Therefore, this delayed electrical activation of the LV results in abnormal myocardial shortening and pump efficiency,[1–3] which has led to the term mechanical dyssynchrony. LV dyssynchrony decreases cardiac output as this shifts the end-systolic pressure-volume relationship to the right, decreasing stroke volume and increasing LV end-systolic volume and wall stress (see Figure 57.1B).[1,3] Biventricular

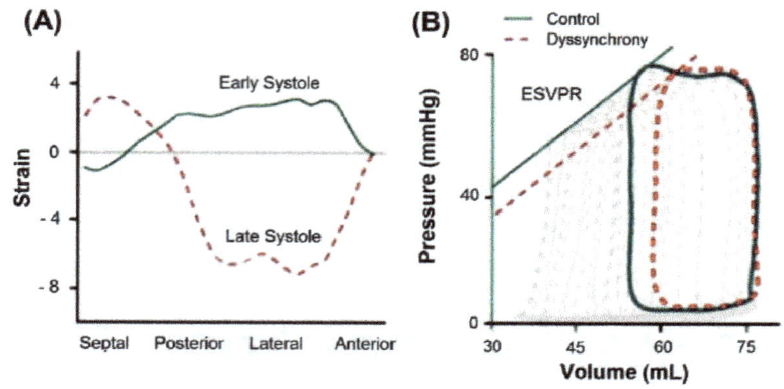

Figure 57.1. A: Plot of instantaneous circumferential strain at different regions across a short-axis section of the mid-LV in a dyssynchronous heart during early (solid line) and late (dashed line) systole.[3] This illustrates that septal wall strain increases while the posterior and lateral strain stretches in early systole. In contrast, the posterolateral wall strain increases, resulting in septal stretch in late-systole. (Reproduced from Park et al with permission.[1]) B: LV pressure-volume loops in dyssynchrony.[1] LV dyssynchrony decreases net cardiac output as the end-systolic pressure-volume relationship shifts to the right, with a reduction of stroke volume (loop width) and increase in end-systolic volume. (Reproduced from Byrne et al with permission.[3])

pacemakers or CRT were developed to correct LV mechanical dyssynchrony found in many patients with HF and ventricular conduction delays, principally LBBB. Over the past decade more than 9000 patients have been studied as part of randomized clinical trials, which have demonstrated the benefits of CRT (Table 57.1).[4]

The Multicenter InSync Randomized Clinical Evaluation (MIRACLE) trial[5] was one of the first landmark studies to demonstrate the clinical benefits of CRT in NYHA class III and IV HF. This trial demonstrated that CRT increased LVEF (+4.6% vs –0.2%, $p < 0.001$) and reduced QRS duration (–20 ms, $p < 0.001$), LVEDD (–3.5 mm, $p < 0.001$) and area of mitral regurgitation (–2.7 cm^2, $p < 0.001$) when compared to OPT. Consequently, functional capacity and quality of life (QOL) improved significantly in the majority of subjects. Furthermore, death or hospitalization for HF was significantly reduced with CRT alone (24.4% vs 15.7%, $p = 0.02$). Our patient received a CRT pacemaker (CRT-P) in 2002 based on the results of this trial and because the benefits of CRT with an implantable cardioverter-defibrillator (ICD) in NYHA class III and IV patients were unproven at that time.

The following year, the MIRACLE-ICD trial[6] compared for the first time ICD alone versus CRT plus ICD (CRT-D) in high-risk subjects for sudden cardiac death, LVEF < 35%, QRS > 130 ms and NYHA class III and IV HF. This trial demonstrated that CRT-D significantly improved QOL and functional class, peak oxygen consumption by 1.1 mL/kg/min ($p = 0.04$), and increased treadmill exercise duration by 56 seconds ($p < 0.001$) when compared to ICD alone. Interestingly, no difference was found in 6-minute walk distance, LV size or function, HF status, survival, or hospitalization. However, this trial had the smallest population size (see Table 57.1) and most likely was underpowered to find statistical difference in these variables.

In 2004, the Comparison of Medical Therapy, Pacing and Defibrillation in HF (COMPANION) study[7] compared the outcomes between medical therapy, CRT-P and CRT-D in more than 1500 patients with ischemic and nonischemic cardiomyopathy class III or IV HF, LVEF < 35% and ventricular conduction delay (QRS > 120 ms). This study demonstrated that both CRT-P and CRT-D reduced the composite of death or hospitalization for any cause by 20% (primary end point, hazard ratios (HRs) 0.81 and 0.80, $p = 0.014$ and $p = 0.01$, respectively), decreased the risk of death from or hospitalization for HF by 34% in CRT-P and 40% in CRT-D group (combined end point, $p < 0.002$ and $p = 0.001$, respectively) and reduced death from any cause (secondary end point) by 24% in CRT-P ($p = 0.059$) and by 36% in CRT-D group ($p = 0.003$) when compared to OPT (Figure 57.2). The trial was terminated prematurely due to the large clinical benefits from CRT-P and CRT-D. Subgroup analysis demonstrated that QRS > 148 ms predicted benefit for all end points. Similarly to MIRACLE,[5] the COMPANION trial confirmed improvement in NYHA class, QOL and 6-minute walk distance compared to OPT.

The Cardiac Resynchronization in Heart Failure (CARE-HF) study[8] compared CRT-P versus

Table 57.1. Summary of Landmark Trials in CRT (Modified from Kaszala et al)[4]

Trial	Patient no.	Inclusion criteria			Device	Follow-up (mo.)	Primary end point	Results	Summary
		NYHA class	EF (%)	QRS (ms)					
MIRACLE[5]	453	III–IV	35	130	CRT	6	NYHA class, 6MWT	Improvement in all primary end points	Large randomized trial to show symptomatic improvement with CRT
CONTAK CD	490	II–IV	35	120	CRT-D	6	Mortality or HF hospitalization or VT/VF	No difference in primary end point	Symptomatic improvement with CRT in NYHA class III and IV patients
MIRACLE ICD[6]	369	III–IV	35	130	CRT-D	6	NYHA class, 6MWT	Improved QOL and NYHA class	Symptomatic improvement with CRT in ICD candidates
COMPANION[7]	1520	III–IV	35	120	CRT, CRT-D	14.4	Mortality or hospitalization	Improved mortality and hospitalization	Symptomatic and mortality benefit with CRT-D therapy
CARE-HF[8]	813	III–IV	35	120*	CRT	29.4	Mortality or hospitalization	Improved mortality and hospitalization	Symptomatic and mortality benefit with CRT therapy
REVERSE[10]	610	I–II	40	120	CRT	12	HF progression or hospitalization	Reduction in admissions and progression of HF	CRT improves LV remodeling, decreases worsening, and admissions of HF in NYHA class I and II patients
MADIT-CRT[11]	1820	I–II	30	129	CRT-D, ICD	54	Mortality or HF	Reduction in HF events	Symptomatic benefit in NYHA class I and II without survival difference with ICD

*CARE-HF Trial included patients with QRS duration > 150 ms or patients with echocardiographic evidence of LV dyssynchrony and QRS duration between 120 and 149 ms. 6MWT, 6-minute walk test; CRT, cardiac resynchronization therapy; CRT-D, cardiac resynchronization therapy with additional defibrillation capability; EF, left ventricular ejection fraction; HF, heart failure; ICD, implantable cardioverter-defibrillator; LV, left ventricular; NYHA, New York Heart Association class; QOL, quality of life; VT/VF, ventricular tachycardia/ventricular fibrillation.

OPT in patients with HF class III and IV, LVEF < 35% and QRS > 120 ms. This was the first trial that required LV dyssynchrony as inclusion criteria for QRS duration between 120 and 149 ms. Death from any cause or hospitalization for a cardiovascular event (primary end point) significantly favored the CRT-P group (39% vs 55%, HR 0.63; $p < 0.001$) (Figure 57.3). Furthermore, CRT-P decreased pro-B-type natriuretic peptide by 1122 pg/ml (confidence interval [CI] –1815 to –429, $p < 0.002$), interventricular mechanical delay by 21 ms, LV end-systolic volume index by 26 ml/m^2 (CI –3.15 to –20.4, $p < 0.001$) and area of mitral regurgitation by 0.042 cm^2 (CI –0.070 to –0.014, $p = 0.003$), whereas it increased LVEF by 6.9% (CI 5.6 to 8.1, $p < 0.001$) and improved NYHA class HF (2.7 ± 0.9 vs 2.1 ± 1, $p < 0.001$) and QOL. Importantly, only 7% of the CRT-P group (29 out of 402 subjects) had sudden cardiac death, which could have been presumably prevented by the addition of a defibrillator.

A meta-analysis of CRT trials[9] with more than 9000 patients showed that biventricular pacing in appropriately selected patients results in an average 3% increase in LVEF and increase in NYHA

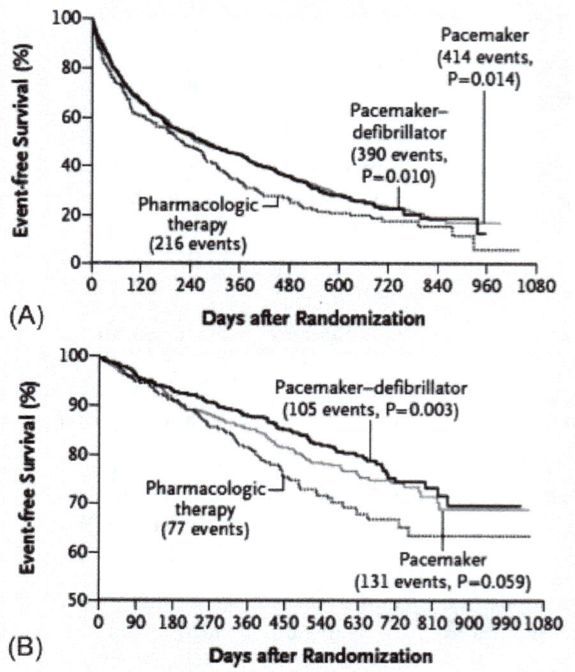

Figure 57.2. Kaplan–Meier estimates of primary **(A)** and secondary end points **(B)** at 3 years from COMPANION study.[7] Primary end point includes death from or hospitalization for any cause. Secondary end point includes death from any causes. (Reproduced from Bristow et al with permission.[7])

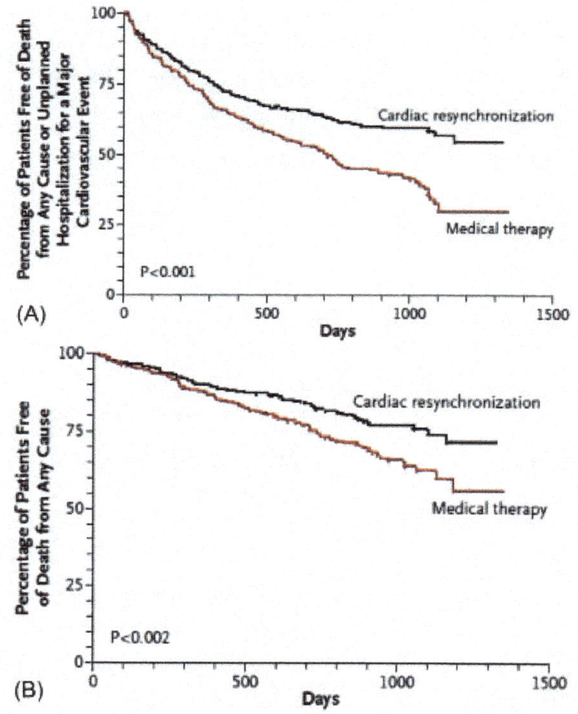

Figure 57.3. Kaplan–Meier estimates of the time to the primary end point **(A)** and secondary end point **(B)** in CRT from CARE-HF trial.[8] Primary end point includes death from any cause or unplanned hospitalization for a major cardiovascular event. Secondary end point consists of death from any cause. (Reproduced from Cleland et al with permission.[8])

functional class by 1, while it decreased hospitalization and all-cause mortality by 37% and 22%, respectively.

Most recently, the REVERSE[10] and MADIT-CRT[11] trials prove additional benefit of CRT in NYHA class I and II. The REVERSE trial assessed the benefit of CRT-P in subjects with LVEF < 40% and QRS > 120 ms. CRT-P decreased HF admissions, stabilized functional capacity such that patients remained in NYHA class I or II, and improved LV function after 12 months. MADIT-CRT[11] tested additional benefits of CRT-D in NYHA class I and II HF subjects with indication for ICD (LVEF < 30%) with QRS > 130 ms. Death from any cause or HF event (primary end point) was significantly decreased with the addition of CRT (17.2% vs 25.3%, HR 0.66; $p = 0.001$). The main benefit was driven by a 41% reduction in HF events, most evident in a prespecified subgroup of patients with a QRS > 150 ms. There was no significant difference in the risk of death between groups (3% annual mortality in both treatment arms). Benefit appeared equal in ischemic and nonischemic cardiomyopathy. Consistent with prior CRT trials, LV end-systolic and LV end-diastolic volumes were significantly reduced (LVESV –18 ml vs –57 ml, LVEDV –15 ml vs –52 ml; $p < 0.001$) and LVEF (3% vs 11%, $p < 0.001$) was significantly improved with the addition of CRT (Figure 57.4).

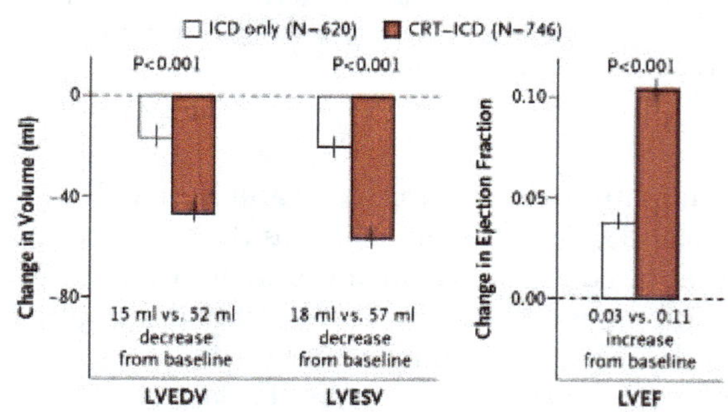

Figure 57.4. Changes in mean LV volumes and ejection fraction (LVEF) between baseline and 1-year follow-up from MADIT-CRT trial.[11] CRT, cardiac resynchronization therapy; ICD, implantable cardioverter-defibrillator; LVEDV, LV end-diastolic volume; LVESV, LV end-systolic volume. (Reproduced from Moss et al with permission.[11])

Several points need to be clarified regarding the results of these CRT trials:

1. At the present time, it remains unclear how to predict which patient will be a CRT nonresponder. It has been proposed that nonresponders may have extensive posterolateral LV scar and/or minimal LV dyssynchrony as noted in patients with isolated right bundle branch block (RBBB). In contrast to LBBB, RBBB has a different activation pattern that appears to cause less global dyssynchrony in the failing heart, with a subsequently reduced benefit of CRT.[3] This is consistent in subgroup analyses in CRT trials, which have demonstrated that the benefit of CRT in RBBB is less than in LBBB. Moreover, the clinical benefit of CRT[12] is also less pronounced or absent in subjects with narrow complex QRS and other non-LBBB interventricular conduction delays, even in the presence of echocardiographic LV dyssynchrony. Most recently, echocardiographic parameters of LV dyssynchrony used to predict CRT response have come into question due to the large variability, poor reproducibility, and limited predictive power.[13,14] Thus, echocardiographic LV dyssynchrony alone is not an indication for CRT implantation.
2. Subjects with specific comorbidities may not benefit from CRT devices, as most trials have excluded those patients such as: (1) severe HF requiring continuous intravenous inotropic therapy; (2) cardiovascular events or cardiac interventions (bypass surgery or percutaneous interventions) within 30 to 90 days prior to enrollment; (3) life expectancy <6 to 24 months; and (4) permanent atrial fibrillation (AF). In addition, some trials also excluded: (5) severe uncorrected valvular heart disease;[5,7] (6) chronic renal failure (blood urea nitrogen > 70 mg/dL or serum creatinine > 3 mg/dL); (7) liver disease (serum hepatic function > 3 times upper limit of normal); and (8) patients awaiting heart transplantation.[5,11] Furthermore, there is limited but impelling literature that ICDs do not impact survival in subjects with chronic renal failure on hemodialysis.[15,16] AF interferes with optimal biventricular pacing. Even patients with appropriate rate-controlled ventricular rates during AF have episodes of intrinsic rhythm that compete with biventricular pacing, which results in fusion beats and lack of "pure" biventricular pacing. It is a bigger problem than is apparent at first glance, because the device counters will report a high percentage of BiV pacing even when there is quite a significant fraction of fused and pseudo-fused beats.[17] Based on this premise, few studies[18,19] support ablation of the atrioventricular node in patients with CRT with permanent AF. Furthermore, all landmark CRT trials have excluded patients with AF, except for a small trial (MUSTIC-AF[20]) with 59 NYHA class III patients with permanent ventricular pacing (QRS > 200 ms). Even though this study was not powered to assess mortality, CRT increased 6-minute walk distance by 9.3% and peak oxygen uptake by 13%, whereas it decreased hospitalizations by 70%.
3. Clinical benefit of CRT in subjects with severe chronic obstructive pulmonary disease (COPD) and severe pulmonary hypertension may be difficult to assess due to their baseline dyspnea. Only limited data[21] support the benefit of ICD in patients with COPD. However, there is no literature that supports the clinical benefit of CRT in COPD patients, or benefit of ICD or CRT in patients with severe pulmonary hypertension. Most patients with severe pulmonary hypertension have been excluded due to their short life expectancy.
4. Finally, benefit from CRT-D over CRT alone in NYHA class IV remains controversial since patients die primarily of worsening HF. However, patients with class IV HF with prolonged QRS duration and optimal lead placement may return to class III status or better for both function and survival, at which point, prevention of sudden death again becomes a relevant goal. Application of CRT-D therapy in class IV HF patients is debated, since there is limited clinical data as only 10% of the almost 4000 patients in resynchronization trials had class IV symptoms.[5,6,8,22] A subgroup analysis of NYHA class IV HF in the COMPANION trial[22] demonstrated that CRT and CRT-D improved QOL and reduced hospitalizations and all-cause mortality when compared to OPT; however, no statistical difference was found between CRT and CRT-D arms. Furthermore, no significant difference in sudden cardiac death and HF death was noted after 2-year follow-up between CRT and CRT-D groups (Figure 57.5). Therefore, for a patient with class IV symptoms in whom resynchronization is inadequate to restore clinical stability, the presence of a defibrillator often complicates the impending transition to end-of-life care.

CRT-D should be considered, but not mandated, for all candidates for CRT. The benefit of ICD may be questionable in some subpopulations, particularly in the elderly or in

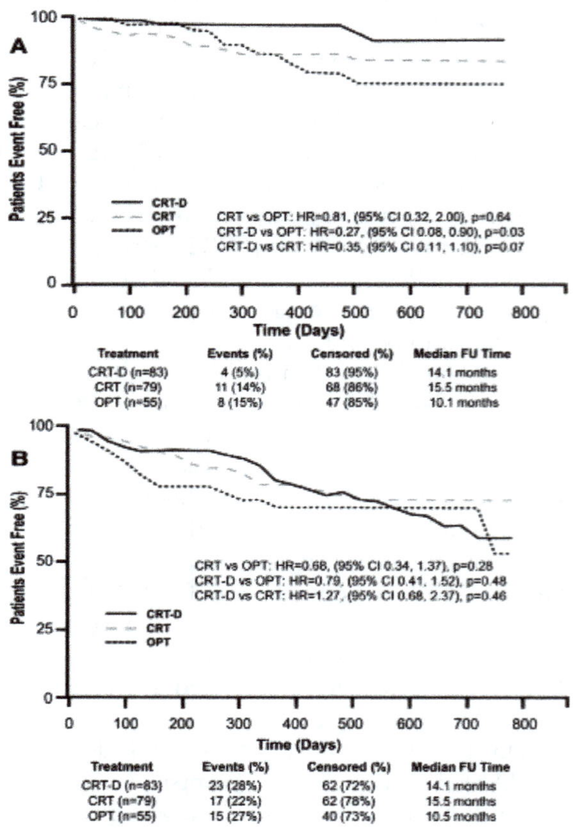

Figure 57.5. Subgroup analysis[22] of NYHA class IV patients and mode of death by treatment arm from COMPANION trial. **Panel A** illustrates the time to sudden cardiac death; **Panel B** depicts time to HF death. CI, confidence interval; CRT, cardiac resynchronization therapy; CRT-D, CRT-defibrillator; FU, follow-up; HR, hazard ratio; OPT, optimal medical therapy. (Reproduced from Lindenfeld et al with permission.[22])

patients with multiple complex concomitant medical comorbidities.[23] An ICD may not need to be added or replaced at the time of generator change in "super-responders," patients whose EFs normalize with CRT therapy because of a very low risk of sudden cardiac death. In addition, there is a significant collective burden that the ICD imposes, such as economic, psychological, and physical device-related issues. For instance, approximately one-third of all ICD shocks are inappropriate,[24] which affect QOL and cause significant anxiety, psychiatric disturbances, and even ventricular arrhythmias. Moreover, the presence of high-voltage coils and the tissue in-growth around them makes it easier for the ICD leads to fail and more difficult to remove if extraction were to be necessary.[25] Therefore, appropriate patient selection for CRT-D includes a careful assessment of various variables including the functional age of patients, comorbidities, and severity of HF.

In summary, both CRT-P and CRT-D benefit patients with QRS duration greater than 150 ms regardless of functional class as long as significant cardiomyopathy, either ischemic or nonischemic, is present. This benefit mostly consists of positive LV remodeling, decreased HF symptoms, and hospitalizations as well as decreased mortality from HF and ventricular arrhythmias.

Correct Answer:

Q1. D

REFERENCES

1. Park RC, Little WC, O'Rourke RA. Effect of alteration of left ventricular activation sequence on the left ventricular end-systolic pressure-volume relation in closed-chest dogs. *Circ Res.* 1985;57:706–717.
2. Cheng A, Helm RH, Abraham TP. Pathophysiological mechanisms underlying ventricular dyssynchrony. *Europace.* 2009;11 Suppl. 5:v10–v14.
3. Byrne MJ, Helm RH, Daya S, et al. Diminished left ventricular dyssynchrony and impact of resynchronization in failing hearts with right versus left bundle branch block. *J Am Coll Cardiol.* 2007;50:1484–1490.
4. Kaszala K, Huizar JF, Ellenbogen KA. Contemporary pacemakers: What the primary care physician needs to know. *Mayo Clin Proc.* 2008;83:1170–1186.
5. Abraham WT, Fisher WG, Smith AL, et al. MIRACLE Study Group. Multicenter InSync Randomized Clinical Evaluation. Cardiac resynchronization in chronic heart failure. *N Engl J Med.* 2002;346:1845–1853.
6. Young JB, Abraham WT, Smith AL, et al. Multicenter InSync ICD Randomized Clinical Evaluation (MIRACLE ICD) Trial Investigators. Combined cardiac resynchronization and implantable cardioversion defibrillation in advanced chronic heart failure: The MIRACLE ICD Trial. *JAMA.* 2003;289:2685–2694.
7. Bristow MR, Saxon LA, Boehmer J, et al. Comparison of Medical Therapy, Pacing, and Defibrillation in Heart Failure (COMPANION) Investigators. Cardiac-resynchronization therapy with or without an implantable defibrillator in advanced chronic heart failure. *N Engl J Med.* 2004;350:2140–2150.
8. Cleland JG, Daubert JC, Erdmann E, et al. Cardiac Resynchronization-Heart Failure (CARE-HF) Study Investigators. The effect of cardiac resynchronization on morbidity and mortality in heart failure. *N Engl J Med.* 2005;352:1539–1549.
9. McAlister FA, Ezekowitz J, Hooton N, et al. Cardiac resynchronization therapy for patients with left ventricular systolic dysfunction: A systematic review. *JAMA.* 2007;297:2502–2514.

10. Linde C, Abraham WT, Gold MR, et al.; REVERSE (REsynchroization reVErses Remodeling in Systolic left vEntricular dysfunction) Study Group Randomized trial of cardiac resynchronization in mildly symptomatic heart failure patients and in asymptomatic patients with left ventricular dysfunction and previous heart failure symptoms. *J Am Coll Cardiol.* 2008;52:1834–1843.
11. Moss AJ, Hall WJ, Cannom DS, et al. MADIT-CRT Trial Investigators. Cardiac-resynchronization therapy for the prevention of heart-failure events. *N Engl J Med.* 2009;361:1329–1338.
12. Beshai JF, Grimm RA, Nagueh SF, et al. RethinQ Study Investigators. Cardiac-resynchronization therapy in heart failure with narrow QRS complexes. *N Engl J Med.* 2007;357:2461–2471.
13. Chung ES, Leon AR, Tavazzi L, et al. Results of the Predictors of Response to CRT (PROSPECT) trial. *Circulation.* 2008;117:2608–2616.
14. Burri H, Müller H, Vieira I, Lerch R. Poor agreement of echographic measures of ventricular dyssynchrony. *Eur J Echocardiogr.* 2008;9:235–240.
15. Khan F, Adelstein E, Saba S. Implantable cardioverter defibrillators confer survival benefit in patients with renal insufficiency but not in dialysis-dependent patients. *J Interv Card Electrophysiol.* 2010;28:117–123.
16. Sakhuja R, Keebler M, Lai TS, et al. Meta-analysis of mortality in dialysis patients with an implantable cardioverter defibrillator. *Am J Cardiol.* 2009;103:735–741.
17. Kamath GS, Cotiga D, Koneru JN, et al. The utility of 12-lead Holter monitoring in patients with permanent atrial fibrillation for the identification of nonresponders after cardiac resynchronization therapy. *J Am Coll Cardiol.* 2009;53:1050–1055.
18. Doshi RN, Daoud EG, Fellows C, et al. PAVE Study Group. Left ventricular-based cardiac stimulation post AV nodal ablation evaluation (the PAVE study). *J Cardiovasc Electrophysiol.* 2005;16:1160–1165.
19. Dong K, Shen WK, Powell BD, et al. Atrioventricular nodal ablation predicts survival benefit in patients with atrial fibrillation receiving cardiac resynchronization therapy. *Heart Rhythm.* 2010;7:1240–1245.
20. Leclercq C, Walker S, Linde C, et al. Comparative effects of permanent biventricular and right-univentricular pacing in heart failure patients with chronic atrial fibrillation. *Eur Heart J.* 2002;23:1780–1787.
21. Razak E, Kamireddy S, Saba S. Implantable cardioverter-defibrillators confer survival benefit in patients with chronic obstructive pulmonary disease. *Pacing Clin Electrophysiol.* 2010;33:1125–1130.
22. Lindenfeld J, Feldman AM, Saxon L, et al. Effects of cardiac resynchronization therapy with or without a defibrillator on survival and hospitalizations in patients with New York Heart Association class IV heart failure. *Circulation.* 2007;115:204–212.
23. Theuns DA, Schaer BA, Soliman OI, et al. The prognosis of implantable defibrillator patients treated with cardiac resynchronization therapy: Comorbidity burden as predictor of mortality. *Europace.* 2010;13:62–69.
24. Daubert JP, Zareba W, Cannom DS, et al. MADIT II Investigators. Inappropriate implantable cardioverter-defibrillator shocks in MADIT II: Frequency, mechanisms, predictors, and survival impact. *J Am Coll Cardiol.* 2008;51:1357–1365.
25. Kowalski M, Huizar JF, Kaszala K, Wood MA. Problems with implantable cardiac device therapy. *Cardiol Clin.* 2008;26:441–458, vii.

CASE STUDY

58

Cardiac Resynchronization Therapy for the Treatment of Congestive Heart Failure

Sanjeev Saksena, MD and Kinjal Desai

CLINICAL PRESENTATION (PART 1)

A 74-year-old male complained of a notable increase in shortness of breath upon exertion in January 2007. He had previously documented dilated nonischemic cardiomyopathy diagnosed at a cardiac catheterization procedure in November 2006 with a markedly decreased left ventricular ejection fraction (LVEF) of 20%. The patient had a prior history of diabetes mellitus and chronic obstructive pulmonary disease. He had not been compliant with his medical regimen for the past year. An ECG showed sinus rhythm (SR) with an interatrial conduction defect, a PR interval of 206 ms and a prolonged QRS duration of 174 ms with a left bundle branch block (LBBB) pattern (Figure 58.1). Medical treatment for heart failure (HF) with triple drug therapy was initiated and symptoms of HF improved.

Question No. 1:

At this time, what should be the next steps?

A. The patient should be considered for single-chamber implantable cardioverter-defibrillator (ICD) for high risk of sudden cardiac death (SCD) or undergo risk stratification for SCD.

B. The patient should not be considered as a candidate for ICD due to comorbidities.

C. Cardiac resynchronization therapy (CRT) should be immediately instituted.

D. None of the above.

The patient subsequently underwent an electrophysiologic study for risk stratification for SCD. Programmed ventricular stimulation with rapid ventricular pacing at 230 ms in the outflow tract induced sustained monomorphic ventricular tachycardia (VT). The VT had a left axis, LBBB morphology, was rapid and hypotensive, and required a 150-J cardioversion shock for termination. Intracardiac echocardiography performed to assess left ventricular (LV) function showed an estimated LVEF of 35%, with no intracardiac thrombi or “smoke,” and regional wall motion abnormalities that resulted in septal and posterior wall dyssynchrony. Prophylactic ICD insertion was recommended. A single-chamber ICD was inserted.

Question No. 2:

CRT was not indicated in this patient because:

A. Patients with class III congestive heart failure (CHF) are not optimal candidates for CRT.

B. The LVEF had probably improved with HF resolution.

Practical Electrophysiology © 2015 Jasbir S. Sra, Masood Akhtar, Andrea Natale, David J. Wilber
Cardiotext Publishing, ISBN: 978-1-935395-17-1

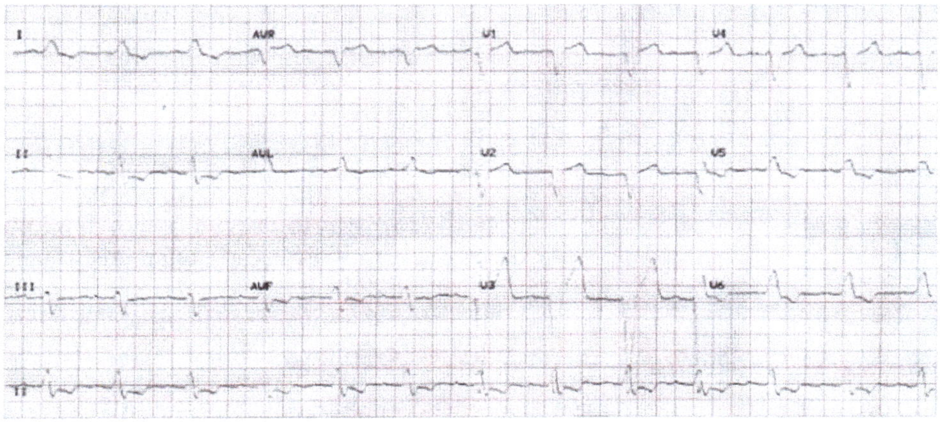

Figure 58.1. ECG showing SR with first-degree atrioventricular (AV) block, interatrial conduction defect, and a prolonged QRS duration with a LBBB pattern.

C. A stable medical regimen for HF consisting of beta-blockers, angiotensin-converting enzyme (ACE) inhibitors, and diuretics had not been implemented for 90 days.

D. Single-chamber ICD therapy is more cost-effective than CRT-ICD insertion.

DISCUSSION (PART 1)

The treatment of HF has markedly improved with advances in pharmacological management (Figure 58.2).[1–5]

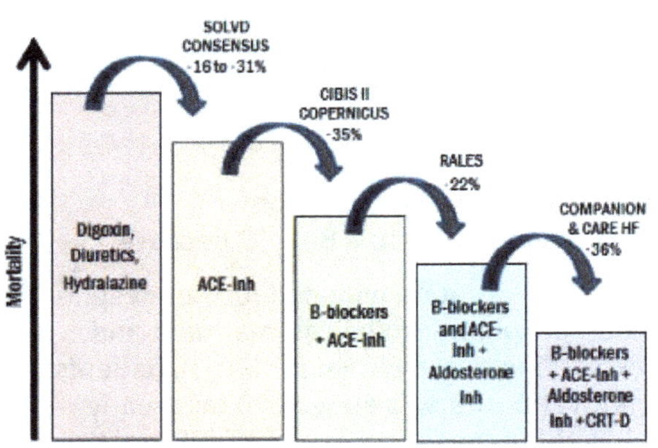

Figure 58.2. Add-on therapy in HF. Each added therapy incrementally decreases mortality when added on top of previous therapy. ACE-Inh, angiotensin-converting enzyme inhibitor; CRT-D, cardiac resynchronization therapy-defibrillator. (Adapted from Ellenbogen et al[25] with permission from Elsevier.)

The risk of SCD and pump failure is greatly increased in patients with HF. This risk has been stratified by LVEF (Figure 58.3).

The highest risk of SCD is seen in patients with increasing LV dysfunction. LBBB increases both sudden and all-cause mortality (Figure 58.4).

Mortality due to pump failure and sudden death caused by ventricular tachyarrhythmic events has also declined significantly with medical treatment.[5,6] For example, in the 2 trials with aldosterone antagonists, spironolactone (RALES), and eplerenone (EPHESUS), relative or risk of SCD was reduced by 29% and 21%, respectively (Figure 58.5).[7,8]

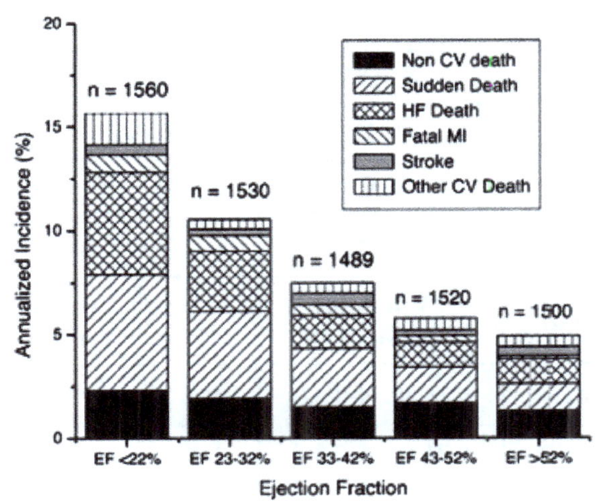

Figure 58.3. Annualized incidence of death, with components of death by LVEF. CV, cardiovascular; HF, heart failure; MI, myocardial infarction. (Adapted from Solomon et al[26] with permission from Wolters Kluwer Health.)

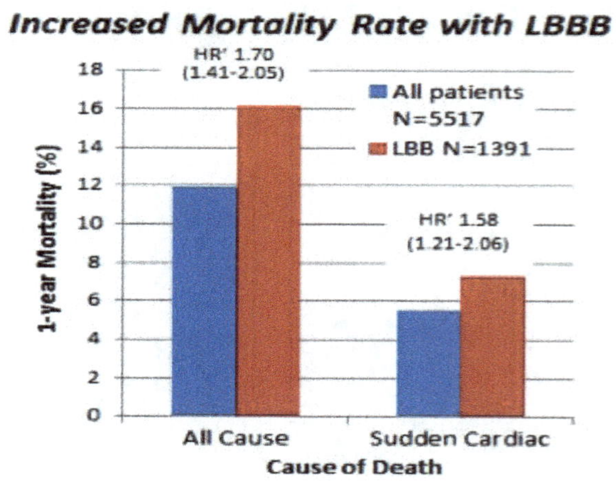

Figure 58.4. One-year mortality rate for all-cause death and sudden death in study population of 5517 patients with CHF (blue bars) and in a subgroup with complete LBBB (red bars). Unadjusted HR is also reported with 95% confidence interval. (Adapted from Baldasseroni et al[27] with permission from Elsevier.)

Hospitalizations for severe symptoms of HF have decreased after increasing the use of ACE inhibitors, beta-blockers, diuretics, and spironolactone. All existing studies on CRT have employed this therapy in symptomatic patients who are on stable background pharmacological therapy.[9] This has been defined as a stable regime of maximally tolerated triple drug therapy for 90 days.[10] In this patient, this medical approach had not been exhausted. In addition, the presence of AV block limited the use of optimal beta-blocker therapy. Insertion of a pacemaker-defibrillator device permitted prophylaxis for bradycardias and ventricular tachyarrhythmias as well as permitted optimizing medical therapy.

	RALES	***EPHESUS***
No	1663	6642
Path	CHF	MI + CHF
Drug	Aldactone	Eplerenone
F-U	24 months, stopped early	18 months, stopped early
ACM	0.70 (0.60-0.82)	0.85 (0.75-0.96)
SCD	0.71 (0.51-0.80)	0.79 (0.64-0.97)

Figure 58.5. Comparison of 2 aldosterone blocking agents and sudden death risk. ACM, all-cause mortality; CHF, congestive heart failure; F-U, follow-up; MI, myocardial infarction; SCD, sudden cardiac death. Reproduced from Pitt et al with permission.[7,8]

CLINICAL COURSE (PART 2)

In 2009, 2 years after the initial implant, the patient presented to the emergency department with severe shortness of breath, bilateral basilar rales, and marked jugular vein distension. An echocardiogram revealed further dilatation of the left ventricle with an ejection fraction of 17% along with severe and diffuse LV hypokinesis. Moderate mitral valve regurgitation and tricuspid regurgitation were also seen. The ECG now showed SR with premature ventricular and atrial complexes, a PR interval of 216 ms, a QRS duration of 180 ms with a LBBB pattern, a QT interval of 442 ms, and a QTc interval of 500 ms. The patient was recommended for an upgrade to a CRT-ICD, which was performed. At the implant, a venogram showed a thrombosed left cephalic vein with extension into the left subclavian, which was severely stenosed. A right cephalic approach was used to insert the LV lead and atrial lead. The coronary sinus was large with small branch veins, and placement of Attain® and Ability® leads (Medtronic Inc., Minneapolis, MN) were attempted without achieving a stable lead position in a lateral LV vein. A QuickFlex® lead (St. Jude Inc., St. Paul, MN) was placed and fixed in the target vein and a bipolar atrial pacing electrode was fixed in the right atrium. The leads were tunneled to the left pectoral pocket and a CRT-D (defibrillator) generator replaced the explanted single-chamber device (Figure 58.6).

He was anticoagulated with warfarin and a therapeutic regimen of carvedilol, digoxin, diuretic, aspirin, and enalapril therapy was continued.

Question No. 3:

The patient was upgraded from a single-chamber ICD to a BiV ICD because:

A. The QRS duration widened further despite optimal pharmacological treatment and according to the PATH-CHF-II trial, patients with QRS intervals above 150 ms greatly benefit immediately from CRT-ICD treatment.

B. His LVEF had dropped significantly despite the single-chamber ICD.

C. He fulfilled the minimum 2-year requirement with single-chamber ICD treatment before an upgrade to a BiV ICD can be considered.

D. None of the above.

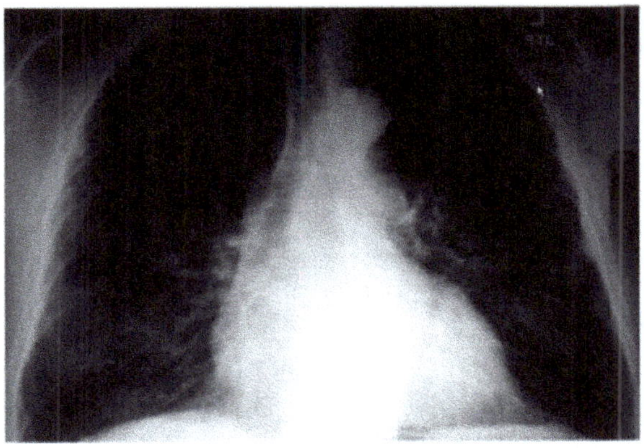

Panel 58.6A

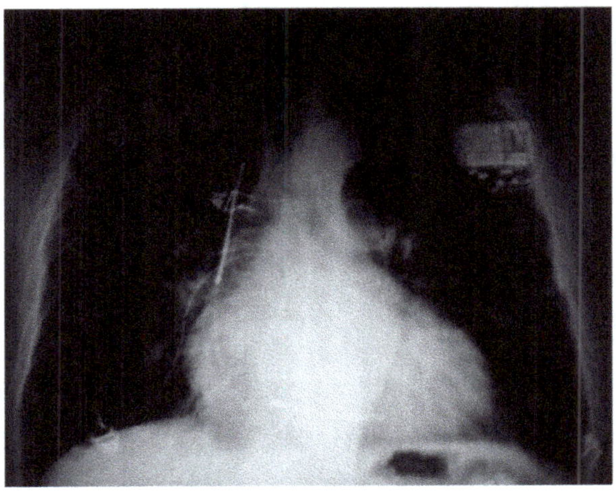

Panel 58.6B

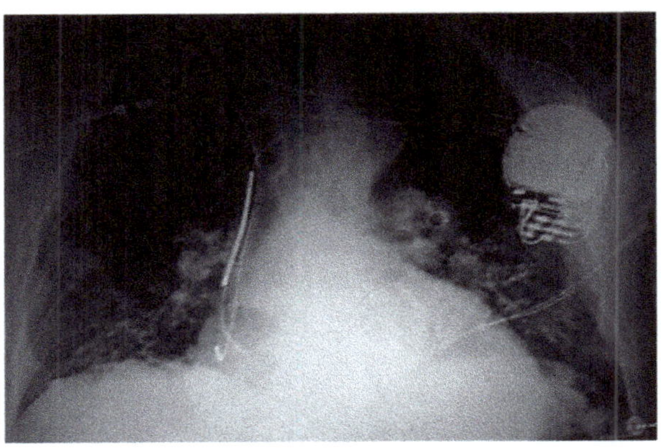

Panel 58.6C

Figure 58.6. **A**: Pre-implant x-ray. Baseline chest x-ray showing cardiomegady prior to initial ICD implant. **B**: Single-chamber ICD x-ray. Single-chamber ICD implant can be seen with the single right ventricular (RV) lead clearly visible. Cardiomegady is present with left atrial and pulmonary artery enlargement. **C**: CRT-D x-ray. x-ray showing biventricular (BiV) ICD implant with corresponding right atrial, RV and LV leads. Due to left subclavian vein thrombosis, the atrial and LV leads were placed from the cephalic vein and tunneled to the left pectoral device pocket. Note the increase in cardiac size and pulmonary congestion with the increasing HF.

Question No. 4:

In patients with progressive CHF, AV block, and intraventricular conduction delay:

A. A BiV ICD should be considered immediately for risk of SCD regardless of current medication regimens.

B. A BiV ICD prolongs survival and reduces hospitalizations even when the patient is on optimal pharmacological treatment, including beta-blockers, diuretics, and ACE inhibitors.

C. A BiV ICD should be considered only if the patient also has LBBB.

D. None of the above.

Question No. 5:

The MADIT-CRT and RAFT clinical trials found that CRT-ICD therapy reduced HF in patients with class I and II CHF. Would this patient have been considered for CRT therapy with a BiV ICD in 2007 if he was evaluated initially by the standards and trends found in MADIT-CRT and RAFT?

A. Yes, because his CHF was already stratified as class III, therefore he is at higher risk for SCD than the patients in the trial.

B. Yes, because posterior and septal wall dyssynchrony should be treated with BiV pacing as the trials have expanded the range of patients qualifying for CRT treatment.

C. No, because his LVEF still is not low enough to qualify for CRT treatment.

D. No, because he is not on an optimal pharmacological regimen.

DISCUSSION (PART 2)

Prospective randomized controlled studies have examined the effect of CRT-P (pacemaker) and CRT-D therapy on survival and morbidity (Table 58.1).[11] The COMPANION study randomized patients in New York Heart Association (NYHA) class III or IV with an LVEF less than 35% and a prolonged QRS duration (> 120 ms) and LV dilation (LV end-diastolic diameter > 60 mm). CRT reduced the composite end point of death or hospitalization for major CV event by 12%. However, significant mortality reduction was achieved only in the CRT-D arm. The impact of CRT in improving survival occurs along with ventricular reverse remodeling, which results in better LV systolic and diastolic function. This has now been also observed in patients with only mild HF (NYHA class II). In the REVERSE

Table 58.1. Randomized Clinical Trials of CRT with Defibrillation Backup

Description	NYHA class	Other criteria	End points	Results
Cardiac Resynchronization in Heart Failure (CARE-HF) trial				
Randomized, controlled trial of CRT with optimal medical therapy versus optimal medical therapy alone	III and IV	• LVEF ≤ 35% • QRS > 150 ms or QRS ≥ 120 ms and echocardiographic criteria of dyssynchrony • Stable optimal medical therapy	• All-cause mortality and hospitalization for HF • NYHA class • QOL • Echocardiographic LV function • Neurohormone levels • Economic impact	• Improvements in morbidity and mortality/CV hospitalization by CRT
Comparison of Medical Therapy Pacing and Defibrillation in Heart Failure (COMPANION) trial				
Randomized, three-arm study to compare optimal drug therapy and CRT or drug therapy and CRT-D	III and IV	• LVEF ≤ 35% • QRS ≥ 120 ms • PR interval > 150 ms • No indication for pacemaker or ICD • HF hospitalization in the past year	• All-cause mortality and hospitalization • QOL • Functional capacity • Peak exercise performance • Cardiac morbidity	• Early termination of the trial • Reduced all-cause mortality and hospitalization with CRT • Reduced all-cause mortality with CRT-D
CONTAK CD Trial				
Started as a 3-month crossover between BiV CRT and no CRT; modified to 6 months. Parallel, double-blind trial between CRT and no CRT, starting 1 month after implantation (6 months)	II, III and IV	• LVEF ≤ 35% • QRS ≥ 120 ms • ICD indication • Stable optimal medical therapy	• All-cause mortality • HF-related hospitalization • VT/VF resulting in device therapy • QOL/6MWD/peak VO_2 • NYHA class • Echocardiographic parameters • Neurohormone levels	• Improvements in peak VO_2, 6MWD/QOL/functional class in NYHA class III-IV patients with CRT
Multicenter InSync Randomized Clinical Evaluation (MIRACLE) trial				
Prospective, randomized, double-blinded, parallel, controlled trial	III and IV	• LVEF ≤ 35% • LVEDD ≥ 55 mm • QRS ≥ 130 ms • Pacing indication not admitted • Stable optimal medical therapy	• NYHA class • 6WMD/QOL/peak VO_2 • Echocardiography indices • Mortality • Hospitalization • QRS duration • Neurohormone levels	• Improvements in NYHA class/6MWD/QOL/peak VO_2 • LVEF • Ventricular volumes • Mitral regurgitation • Reduced hospitalizations
Multicenter InSync ICD Randomized Clinical Evaluation (MIRACLE-ICD) trial				
Prospective, randomized double-blind, parallel, controlled trial evaluating safety and efficacy of CRT in patients with HF and indication for ICD	III–IV	• LVEF ≤ 35% • LVEDD ≥ 55 mm • QRS ≥ 130 ms • ICD indication	• QOL/6MWD/peak VO_2 • NYHA class • Exercise duration • HF composite (death, HF, hospitalization, NYHA class, and patient global self-assessment) • Safety of CRT-D	• Improvements in QOL and NYHA class • Clinical composite end points • CRT-D safe to use

(*Continued*)

Table 58.1. (*Continued*)

Description	NYHA class	Other criteria	End points	Results
Multisite Stimulation in Cardiomyopathy Sinus Rhythm (MUSTIC SR) trial				
Prospective, randomized, single-blind crossover study	III	• LVEF < 35% • LVEDD > 60 mm • QRS ≥ 200 ms • 6MWD < 450 m	• 6MWD/peak VO_2/QOL • NYHA class • Hospitalization • Patient preference • All-cause mortality • Echocardiographic indices	• Improvements in 6MWD/peak VO_2/ QOL/NYHA class • Reduced hospitalizations • Patients preferred CRT
Multisite Stimulation in Cardiomyopathy Atrial Fibrillation (MUSTIC AF) trial				
Prospective, randomized, single-blind crossover VVIR-BiV study	> III	• LVEF < 35% • LVEDD > 60 mm • Paced QRS ≥ 200 ms in ventricular pacing • 6MWD < 450 m	• 6MWD/peak VO_2/QOL • NYHA class • Hospitalization • Patient preference • All-cause mortality • Echocardiographic indices	• Improvements in 6MWD/peak VO_2/ QOL/NYHA class • Reduced hospitalizations • Patients preferred CRT
Pacing Therapies in Congestive Heart Failure (PATH-CHF) trial				
Longitudinal study of CRT with 2nd placebo-control phase—1st and 3rd periods are crossovers between LV and BiV	III and IV	• QRS > 120ms • Sinus rate ≥ 55 bpm • PR interval ≥ 150ms	• 6MWD/QOL/peak VO_2 • NYHA class	• Improvements in exercise-capacity functional status and QOL
PATH-CHF II trial				
Randomized crossover of no CRT versus CRT in LV only—2 patient groups: QRS 120-150 ms and QRS > 150 ms	II, III and IV	• LVEF ≤ 30% • QRS ≥ 120 ms • Optimal therapy for HF • ICDs may be included	• AT/6MWD/QOL/peak VO_2 • NYHA class • Hospitalization	• QRS 120-150 ms: group: no improvement • QRS > 150 ms group: improvements in VO_2/AT/6MWD/ QOL
REVERSE trial				
Prospective, randomized, double-blind study with CRT-on versus CRT-off control	I and II	• QRS ≥ 120 ms • LVEF ≤ 40% • LVEDD ≥ 55 mm	• HF clinical composite response • LVESV • HF-related hospitalization	• CRT-on group improved LVESV and measures of LV reverse remodeling • Time to first HF hospitalization delayed in CRT-on group
Resynchronization/defibrillation for Ambulatory Heart Failure Trial (RAFT)				
Randomized, parallel double-blind study of ICD versus CRT-D therapy in ischemic and nonischemic cardiomyopathy patients	II and III	• QRS ≥ 120 ms • LVEF ≤ 30% • Optimal therapy for HF • ICD planned • SR or permanent AF/ AFl with controlled ventricular rate	• HF-related hospitalization • All-cause mortality • NYHA class	• CRT-D reduced relative rate of death by 25% and reduced hospitalization for HF • CRT-D had increased rates of short-term adverse effects

(*Continued*)

Table 58.1. (*Continued*)

Description	NYHA class	Other criteria	End points	Results
Multicenter Automatic Defibrillator Implantation Trial with CRT (MADIT-CRT)				
Randomized, parallel CRT-D versus ICD only	I and II	• QRS ≥ 130 ms • LVEF ≤ 30% • Optimal therapy for HF • NYHA class II with ischemic or nonischemic cardiomyopathy *or* class I with ischemic cardiomyopathy	• All-cause mortality • HF intervention	• CRT-D reduced relative risk of all-cause mortality or HF (with most success in LBBB patients) • Reduced LV volume and improved LVEF

6MWD, 6-minute walk distance; AF, atrial fibrillation; AFl, atrial flutter; AT, atrial tachycardia; BiV, biventricular; CRT, cardiac resynchronization therapy; CRT-D, CRT-defibrillator; HF, heart failure; ICD, implantable cardioverter-defibrillator; LBBB, left bundle branch block; LV, left ventricular; LVEDD, left ventricular end-diastolic diameter; LVEF, left ventricular ejection fraction; LVESV, left ventricular end-systolic volume; NYHA, New York Heart Association; QOL, quality of life; SR, sinus rhythm; VF, ventricular fibrillation; VO_2, volume of oxygen consumption; VT, ventricular tachycardia. (Adapted from Saksena et al[24] with permission from Elsevier.)

study, at 12 months, the composite end point was reduced to 16% in CRT-ON compared with 21% in CRT-OFF ($P = 0.10$).[12] Patients assigned to CRT-ON experienced a greater improvement in LV end-systolic volume index (–18.4 ± 29.5 ml/m^2 vs –1.3 ± 23.4 ml/m^2, $P < 0.0001$) and other LV remodeling indices. Time to first HF hospitalization was significantly delayed in CRT-ON (hazard ratio [HR]: 0.47, $P = 0.03$). MADIT-CRT study examined the effect of CRT-D in patients with no or only mild symptoms of HF (NYHA I/II), LVEF ≤ 30% and a QRS width of 130 ms compared with ICD therapy only. It showed a 34% reduction of HF or death after a mean F-U of 2.4 years. The benefit of CRT-D was mainly attributable to a 41% reduction of HF. Patients with LBBB and women had most benefit, while patients with ischemic and nonischemic cardiomyopathy had similar benefit. LV volumes decreased and a marked increase of LVEF was observed with CRT-D. Mortality was not significantly reduced, which may be due to F-U duration. The RAFT trial did show a reduction of overall mortality in a similar population. A summary of the major clinical trials involving CRT only or CRT-D is shown below.

HF can be associated with prolongation of the AV, intraventricular, and interventricular intervals, which can reduce diastolic ventricular filling and systolic LV performance. Mitral valve function is impaired by ventricular dyssynchrony as seen in LBBB ECG patterns. Increase in LV wall stress with increased cardiac work occurs with changes in myocardial cell architecture, regional hypertrophy, and even apoptosis. The mechanical contractile dysfunction caused by asynchrony or dyssynchrony is routinely assessed by standard echocardiographic indices. Greater precision is obtained with use of tissue Doppler imaging, magnetic resonance imaging, or intracardiac echocardiography.

CRT can correct many of these abnormalities. LV, and even RV, pacing-electrode placement can be critical to achieving effective resynchronization. Preexcitation of the left lateral ventricular wall with CRT in patients with LBBB bypasses ventricular conduction delay, improving synchrony, and performance. Optimizing AV synchrony is also necessary in order to achieve good hemodynamic outcomes in these patients.

Correct Answers:

Q1. A
Q2. C
Q3. A
Q4. B
Q5. D

REFERENCES

1. Hunt SA, Abraham WT, Chin MH, et al. 2009 focused update incorporated into the ACC/AHA 2005 Guidelines for the Diagnosis and Management of Heart Failure in Adults: A report of the American College of Cardiology Foundation/American Heart Association Task Force on Practice Guidelines: Developed in collaboration with the International Society for Heart and Lung Transplantation. *Circulation.* 2009;119:e391–e479.

2. Howlett JG, McKelvie RS, Costigan J, et al.; Canadian Cardiovascular Society. The 2010 Canadian Cardiovascular Society guidelines for the diagnosis and management of heart failure update: Heart failure in ethnic minority populations, heart failure and pregnancy, disease management, and quality improvement/assurance programs. *Can J Cardiol.* 2010;26:185–202.
3. Heart Failure Society of America; Lindenfeld J, Albert NM, Boehmer JP, et al. HFSA 2010 Comprehensive Heart Failure Practice Guideline. *J Card Fail.* 2010;16:e1–e194.
4. Cohn JN, Johnson G, Ziesche S, et al. A comparison of enalapril with hydralazine-isosorbide dinitrate in the treatment of chronic congestive heart failure. *N Engl J Med.* 1991;325:303–310.
5. Captopril Multicenter Research Group. A placebo-controlled trial of captopril in refractory chronic congestive heart failure. *J Am Coll Cardiol.* 1983;2:755–763.
6. Vardas PE, Auricchio A, Blanc JJ, et al.; European Society of Cardiology; European Heart Rhythm Association. Guidelines for cardiac pacing and cardiac resynchronization therapy: The Task Force for Cardiac Pacing and Cardiac Resynchronization Therapy of the European Society of Cardiology. Developed in collaboration with the European Heart Rhythm Association. *Eur Heart J.* 2007;28:2256–2295.
7. Pitt B, Remme W, Zannad F, et al.; Eplerenone Post-Acute Myocardial Infarction Heart Failure Efficacy and Survival Study Investigators. Eplerenone, a selective aldosterone blocker, in patients with left ventricular dysfunction after myocardial infarction. *N Engl J Med.* 2003;348:1309–1321.
8. Pitt B, Zannad F, Remme WJet al. The effect of spironolactone on morbidity and mortality in patients with severe heart failure. Randomized Aldactone Evaluation Study Investigators. *N Engl J Med.* 1999;341:709–717.
9. Epstein AE, DiMarco JP, Ellenbogen KA, et al. American College of Cardiology/American Heart Association Task Force on Practice Guidelines (Writing Committee to Revise the ACC/AHA/NASPE 2002 Guideline Update for Implantation of Cardiac Pacemakers and Antiarrhythmia Devices); American Association for Thoracic Surgery; Society of Thoracic Surgeons. ACC/AHA/HRS 2008 Guidelines for device-based therapy of cardiac rhythm abnormalities: A report of the American College of Cardiology/American Heart Association Task Force on Practice Guidelines (Writing Committee to Revise the ACC/AHA/NASPE 2002 Guideline Update for Implantation of Cardiac Pacemakers and Antiarrhythmia Devices): Developed in collaboration with the American Association for Thoracic Surgery and Society of Thoracic Surgeons. *Circulation.* 2008;117:e350–e408.
10. Authors/Task Force Members, Dickstein K, Vardas PE, Auricchio A, et al. 2010 Focused Update of ESC Guidelines on device therapy in heart failure: An update of the 2008 ESC Guidelines for the diagnosis and treatment of acute and chronic heart failure and the 2007 ESC guidelines for cardiac and resynchronization therapy. Developed with the special contribution of the Heart Failure Association and the European Heart Rhythm Association. *Eur Heart J.* 2010;31:2677–2687.
11. Bristow MR, Saxon LA, Boehmer J, et al. Comparison of Medical Therapy, Pacing, and Defibrillation in Heart Failure (COMPANION) Investigators. Cardiac-resynchronization therapy with or without an implantable defibrillator in advanced chronic heart failure. *N Engl J Med.* 2004;350:2140–2150.
12. Daubert C, Gold MR, Abraham WT, et al. REVERSE Study Group. Prevention of disease progression by cardiac resynchronization therapy in patients with asymptomatic or mildly symptomatic left ventricular dysfunction: Insights from the European cohort of the REVERSE (Resynchronization Reverses Remodeling in Systolic Left Ventricular Dysfunction) trial. *J Am Coll Cardiol.* 2009;54:1837–1846.
13. Moss AJ, Hall WJ, Cannom DS, et al.; MADIT-CRT Trial Investigators. Cardiac-resynchronization therapy for the prevention of heart-failure events. *N Engl J Med.* 2009;361:1329–1338.
14. Linde C, Abraham WT, Gold MR, et al.; REVERSE (REsynchronization reVErses Remodeling in Systolic left vEntricular dysfunction) Study Group. Randomized trial of cardiac resynchronization in asymptomatic and mildly symptomatic heart failure. *J Am Coll Cardiol.* 2008;52:1834–1843.
15. Tang AS, Wells GA, Arnold M, et al. Resynchronization/defibrillation for ambulatory heart failure trial: Rationale and trial design. *Curr Opin Cardiol.* 2009;24:1–8.
16. Gorcsan J 3rd, Abraham T, Agler DA, et al.; American Society of Echocardiography Dyssynchrony Writing Group. Echocardiography for cardiac resynchronization therapy: Recommendations for performance and reporting—a report from the American Society of Echocardiography Dyssynchrony Writing Group endorsed by the Heart Rhythm Society. *J Am Soc Echocardiogr.* 2008;21:191–213.
17. Abraham T, Kass D, Tonti G, et al. Imaging cardiac resynchronization therapy. *JACC Cardiovasc Imaging.* 2009;2:486–497.
18. Auricchio A, Salo RW. Acute hemodynamic improvements by pacing in patients with severe congestive heart failure. *Pacing Clin Electrophysiol.* 1997;20:313–324.
19. Xiao HB, Roy C, Gibson DG. Nature of ventricular activation in patients with dilated cardiomyopathy: Evidence for bilateral bundle branch block. *Br Heart J.* 1994;72:167–174.
20. Saxon LA, Kerwin WF, Cahalan MK, et al. Acute effects of intraoperative multisite ventricular pacing on left ventricular function and activation/contraction sequence in patients with depressed ventricular function. *J Cardiovasc Electrophysiol.* 1998;9:13–21.
21. Kass DA, Chen CH, Curry C, et al. Improved left ventricular mechanics form acute VDD pacing in

patients with dilated cardiomyopathy and ventricular conduction delay. *Circulation.* 1999;99:1567–1573.
22. Nelson GS, Berger RD, Fetics BJ, et al. Left ventricular or biventricular pacing improves cardiac function at diminished energy cost in patients with dilated cardiomyopathy and left bundle-branch block. *Circulation.* 2000;102:3053–3059.
23. Helm RH, Byrne M, Helm PA, et al. Three-dimensional mapping of optimal left ventricular pacing site for cardiac resynchronization. *Circulation.* 2007;115:953–961.
24. Saksena S, Simon AM, Mathew P, Nagarakanti R. Intracardiac echocardiography-guided cardiac resynchronization therapy: Technique and clinical application. *Pacing Clin Electrophysiol.* 2009;32:1030–1039.
25. Ellenbogen KA. Why should we care about CARE-HF? *J Am Coll Cardiol.* 2005;46:2199–2203.
26. Solomon SD, Foster E, Bourgoun M, et al. Effect of cardiac resynchronization therapy on reverse remodeling and relation to outcome: multicenter automatic defibrillator implantation trial: cardiac resynchronization therapy. *Circulation.* 2010;122:985–992.
27. Baldasseroni S, Di Biase L, Fresco C, et al. Cumulative effect of complete left bundle-branch block and chronic atrial fibrillation on 1-year mortality and hospitalization in patients with congestive heart failure: a report from the Italian network on congestive heart failure (in-CHF database). *European Heart Journal.* 2002;23(21):1692–1698.

CASE STUDY 59

Indications for Cardiac Resynchronization Therapy

Sitaramesh Emani, MD and William T. Abraham, MD

A 67-year-old female is referred to a heart failure clinic for further management. She was initially diagnosed with a nonischemic cardiomyopathy 2 years prior. Coronary angiography at that time confirmed the nonischemic etiology, showing only minimal plaque disease. She also has a history of an isolated episode of atrial fibrillation (AF). A recent echocardiogram demonstrated global hypokinesis with an ejection fraction (EF) of 15%. Left ventricular (LV) diameter at end-diastole measured 6.8 cm (normal 3.1–5.6 cm). Her cardiac medications at time of referral included carvedilol, losartan, spironolactone, and furosemide.

Clinically, the patient indicates that she has had a decline in her functional status over the past few months after an initial improvement with the initiation of therapy. Currently she describes significant, limiting dyspnea after 2 blocks of walking on level ground or after climbing one flight of stairs. She also has some worsening abdominal distention and lower extremity edema. Her ECG is shown in Figure 59.1.

Question No. 1:

Which of the following patient characteristics are included in the indications for cardiac resynchronization therapy (CRT)?

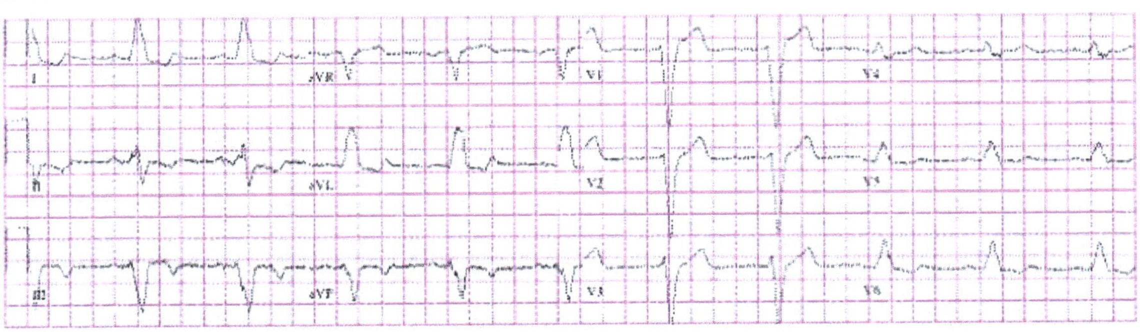

Figure 59 1. Patient's ECG at presentation.

Practical Electrophysiology © 2015 Jasbir S. Sra, Masood Akhtar, Andrea Natale, David J. Wilber
Cardiotext Publishing, ISBN: 978-1-935395-17-1

A. New York Heart Association (NYHA) functional class.
B. LV end-diastolic dimension.
C. History of medical therapy for heart failure.
D. Both A and C.

Question No. 2:

Which of the following ECG features predict response to CRT?

A. Absence of atrioventricular (AV) conduction delay.
B. A prolonged QRS duration.
C. Left-axis deviation.
D. Absence of Q waves.

Question No. 3:

Which of the following preimplant study findings would prompt an evaluation for CRT?

A. Electrocardiographic monitoring displaying chronic right ventricular (RV) pacing.
B. Findings consistent with a nonischemic etiology on cardiac magnetic resonance imaging.
C. Areas of active ischemia on stress imaging or coronary angiography.
D. Evidence of elevated filling pressures on echocardiography or cardiac catheterization.

Question No. 4:

Which of the following is a contraindication to CRT upgrade?

A. Existing dual-chamber pacing.
B. A preexisting implantable cardioverter-defibrillator (ICD).
C. Both A and B.
D. Neither A or B.

DISCUSSION

The use of biventricular pacing, or CRT, is considered as an integral part of the treatment of chronic systolic heart failure. Progressive remodeling of the LV along with dyssynchronous mechanical motion of the ventricular walls during systole can cause a decrease in cardiac output by impairing ventricular filling, reducing LV contractile forces, and worsening the degree of mitral regurgitation. Restoration of synchronous contraction between the right and left ventricles and within the LV can lead to clinical improvement for heart failure patients as well as promote reverse remodeling and improve morbidity and mortality. Central to the benefit derived from CRT is mechanical dyssynchrony, which is currently evaluated electrically by a prolonged QRS duration on surface ECGs (QRS duration > 0.12 ms).[1]

Current guidelines recommend consideration of CRT for patients with an EF < 35%, classified as NYHA functional class III/ambulatory IV despite optimal medical therapy, with the above-mentioned QRS duration. At present, sinus rhythm is a class I recommendation for CRT, as the major clinical trials showing the benefit of CRT carried this specification.[1] (Individual trials have been summarized in detail elsewhere.[2]) However, recent evidence has demonstrated an extension of benefit to those patients in AF and NYHA class II symptoms leading to a class IIa recommendation for CRT in this population.[3,4] In the presented case, the patient's functional status should be considered NYHA class III. With that classification, she meets recommendations based on her functional status, EF, current rhythm, underlying left bundle branch block, and appropriate medical therapy for her heart failure. Her LV end-diastolic dimension is irrelevant, as it is not considered a component of CRT evaluation. Similarly, the presence or absence of an AV conduction delay is not considered a relevant factor, although inappropriate AV intervals can lead to worse hemodynamic response to CRT. However, AV intervals can be readily adjusted by device programming to ensure optimal ventricular filling during mechanical diastole. Various parameters such as invasive hemodynamic monitoring, noninvasive hemodynamic monitoring, surface ECG tracings, and echocardiography can be used to find the patient's optimal AV delay.[5] The axis of depolarization is not considered a factor when evaluating a patient for CRT.

Patients presenting with NYHA class III/IV symptoms who have dual-chamber pacemakers can also benefit from an upgrade to CRT devices if they demonstrate a high burden of ventricular pacing (a class IIa recommendation). The iatrogenic intraventricular conduction delay created by chronic RV pacing can create the same detrimental mechanical dyssynchrony as a native conduction delay.[6] Consideration of a device upgrade should be given to these patients provided they meet the other criteria for CRT outlined in the cited guidelines. The underlying etiology for heart failure is not thought to be a limiting factor in patient selection as the major clinical trials demonstrating benefit of CRT included both ischemic and nonischemic heart failure patients. However, patients with clinical and objective evidence of active ischemia should be optimally treated for obstructive coronary disease as part of their heart failure management.[1] Current practice patterns generally stipulate that ischemic evaluations and therapies should be performed prior to the implant of a biventricular pacemaker as a part of optimal medical management. A reevaluation of the patient's clinical status may be warranted after an elapsed period of time since revascularization may improve LV function as well as heart failure classification.[1] Similarly, although no specific recommendations exist regarding a patient's volume status with regard to CRT, current practice patterns generally call for optimization of fluid management prior to implant. Finally, patients with an EF < 35% undergoing evaluation for CRT should simultaneously be evaluated for ICD therapy if not already present.

It should be noted that the current guideline recommendations are based on the enrollment criteria of the major randomized clinical trials that have shown benefit. Ongoing and future trials may expand or better define those patients with cardiomyopathies who will benefit from CRT. Recent data support an extension of the CRT indication to NYHA class II patients.[4]

Correct Answers:

Q1. D

Q2. B

Q3. A

Q4. D

REFERENCES

1. Hunt SA, Abraham WT, Chin MH, et al. 2009 focused update incorporated into the ACC/AHA 2005 Guidelines for the Diagnosis and Management of Heart Failure in Adults: A report of the American College of Cardiology Foundation/American Heart Association Task Force on Practice Guidelines: Developed in collaboration with the International Society for Heart and Lung Transplantation. *Circulation.* 2009;119:e391–e479.
2. Abraham WT. Devices for the treatment of heart failure. In: Abraham WT, Krum H, eds. *Heart Failure: A Practical Approach to Treatment.* New York, NY: McGraw Hill Medical; 2007:181–196.
3. Leclercq C, Walker S, Linde C, et al. Comparative effects of permanent biventricular and right-univentricular pacing in heart failure patients with chronic atrial fibrillation. *Eur Heart J.* 2002;23:1780–1787.
4. Tang AS, Wells GA, Talajic M, et al. Cardiac-resynchronization therapy for mild-to-moderate heart failure. *N Engl J Med.* 2010;363:2385–2395.
5. Jones RC, Svinarich T, Rubin A, et al. Optimal atrioventricular delay in CRT patients can be approximated using surface electrocardiography and device electrograms. *J Cardiovasc Electrophysiol.* 2010;21:1226–1232.
6. Fröhlich G, Steffel J, Hürlimann D, et al. Upgrading to resynchronization therapy after chronic right ventricular pacing improves left ventricular remodelling. *Eur Heart J.* 2010;31:1477–1485.

CASE STUDY

60

Left Ventricular Lead Placement in Cardiac Resynchronization Therapy

Sitaramesh Emani, MD and William T. Abraham, MD

A 63-year-old male presents for consideration of cardiac resynchronization therapy (CRT) upgrade. He has a history of ischemic heart disease and clinical heart failure with several admissions for acute decompensation in the past 2 years. At present, he is at his baseline volume status with ambulatory NYHA class IV symptoms. He denies any current chest pain or other signs of active ischemia, and his other chronic medical conditions (diabetes mellitus, hyperlipidemia, and sleep apnea) are controlled on good therapies. For his heart failure the patient is on carvedilol, lisinopril, spironolactone, and furosemide.

Reviewing objective data shows a recent echocardiogram with severe global hypokinesis of the left ventricle (LV) and an ejection fraction of 15% to 20%. His surface ECG is shown in Figure 60.1. His past surgical history is notable for two-vessel coronary bypass grafting 7 years ago.

Question No. 1:

Where is the current standard of optimal LV lead placement when implanting a CRT device?

A. Main coronary sinus (CS).

B. Middle cardiac vein.

C. Any position within the coronary venous system with low threshold values.

D. Anterior cardiac vein.

Question No. 2:

Which of the following characteristics are concerning for potential nonresponse to CRT?

A. QRS duration < 150 ms.

B. Evidence of lateral infarct.

C. Prior bypass surgery.

D. Severity of LV hypokinesis on echocardiography.

Question No. 3:

Which of the following tests may assist in the pre-implant process?

A. Computed tomographic (CT) angiogram.

B. Viability assessment.

C. CT venogram.

D. Cardiac magnetic resonance imaging (MRI).

Question No. 4:

An ischemic etiology of dysfunction, compared to a nonischemic etiology, is predictive of CRT response.

A. True.

B. False.

Practical Electrophysiology © 2015 Jasbir S. Sra, Masood Akhtar, Andrea Natale, David J. Wilber
Cardiotext Publishing, ISBN: 978-1-935395-17-1

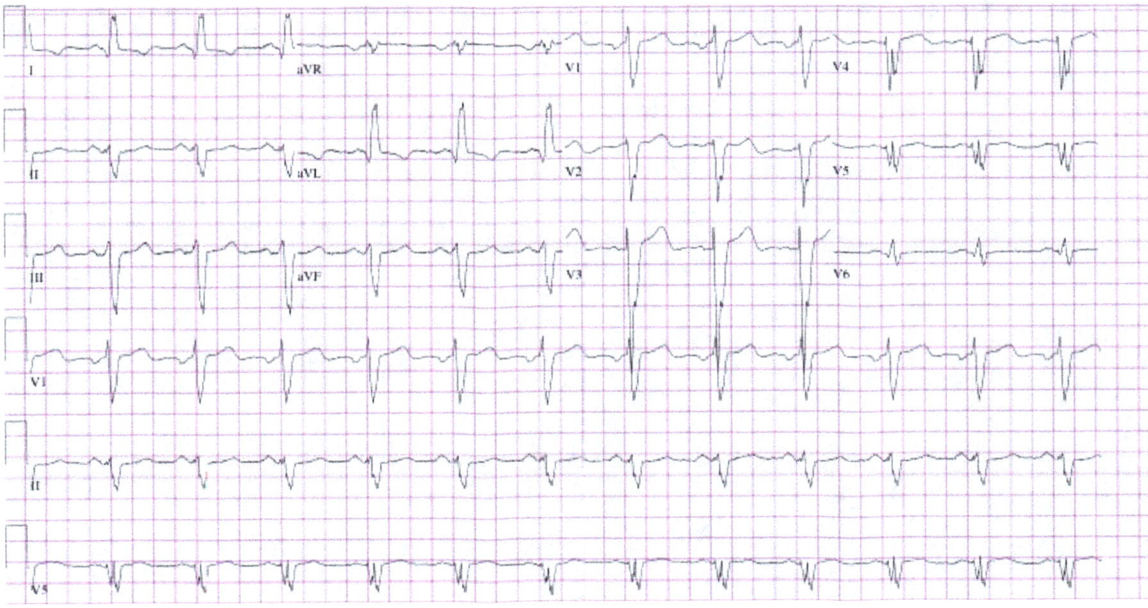

Figure 60.1. Patient's surface ECG at presentation.

DISCUSSION

Successful delivery of CRT is dependent on many factors. After selecting an appropriate patient (as outlined in the previous case), technical issues during implantation need to be considered. Typically, right atrial and right ventricular (RV) leads are placed in a fashion similar to standard dual-chamber device implants, via the venous system (generally accessed via the subclavian or axillary veins). The same access site is then used for the introduction of the LV lead. However, appropriately placing the LV lead can be challenging.

General standard is to implant the LV lead via the coronary venous system in the CS. A variety of guide catheters can be used to cannulate the CS from the right atrium, through which the lead can be advanced.[1] Because of the natural position of coronary venous anatomy, approaching the atrium from the left axillary or subclavian vein is preferred. Understanding the anatomical distribution of the venous anatomy is essential to place the LV lead. Anatomical variants are frequent and should be evaluated when possible. A particular challenge can be posed by the presence of a valve at the CS ostium, making cannulation of the vessel difficult. Pre-implant venous angiography or a CT venogram can often assist in evaluating and planning an anatomical approach to CRT implantation.[2]

At present, the preferred target for the LV lead is any vein that traverses the LV lateral or posterolateral wall.[2,3] However, other variables need to be considered when finding the optimal location for the LV lead. Recall that the mechanical goal of CRT is to restore mechanical interventricular and intraventricular synchrony as much as possible. Interventricular synchrony is created by timed electrical stimulation between the RV and LV leads. The mechanical contraction created by this timed stimulation can be improved by increasing the electrical distance between the RV and LV leads.[4] Therefore, LV anterior wall stimulation may not provide maximal benefit of CRT, particularly when the right-sided lead is placed in the RV apex along the interventricular septum. Additionally, electromechanical conversion is enhanced by the stimulation of viable myocardium. As a result, the placement of an LV lead directly into an area of scar is discouraged. Placement of an LV lead on the edge of scar tissue, though, may provide optimal benefit since scar borders can be the area of greatest mechanical delay.[4]

In the presented case, the patient has a significant history of ischemic heart disease as the etiology of his LV dysfunction, thus increasing the probability of a large scar burden in his myocardium. Evidence of scar can also be seen by the presence of Q waves in the lateral leads of his surface ECG. In this example, anatomically defining regions of viable myocardium as targets for an LV lead can help improve the potential for response. Viability assessment can be done with a number of imaging methods including nuclear perfusion scanning or

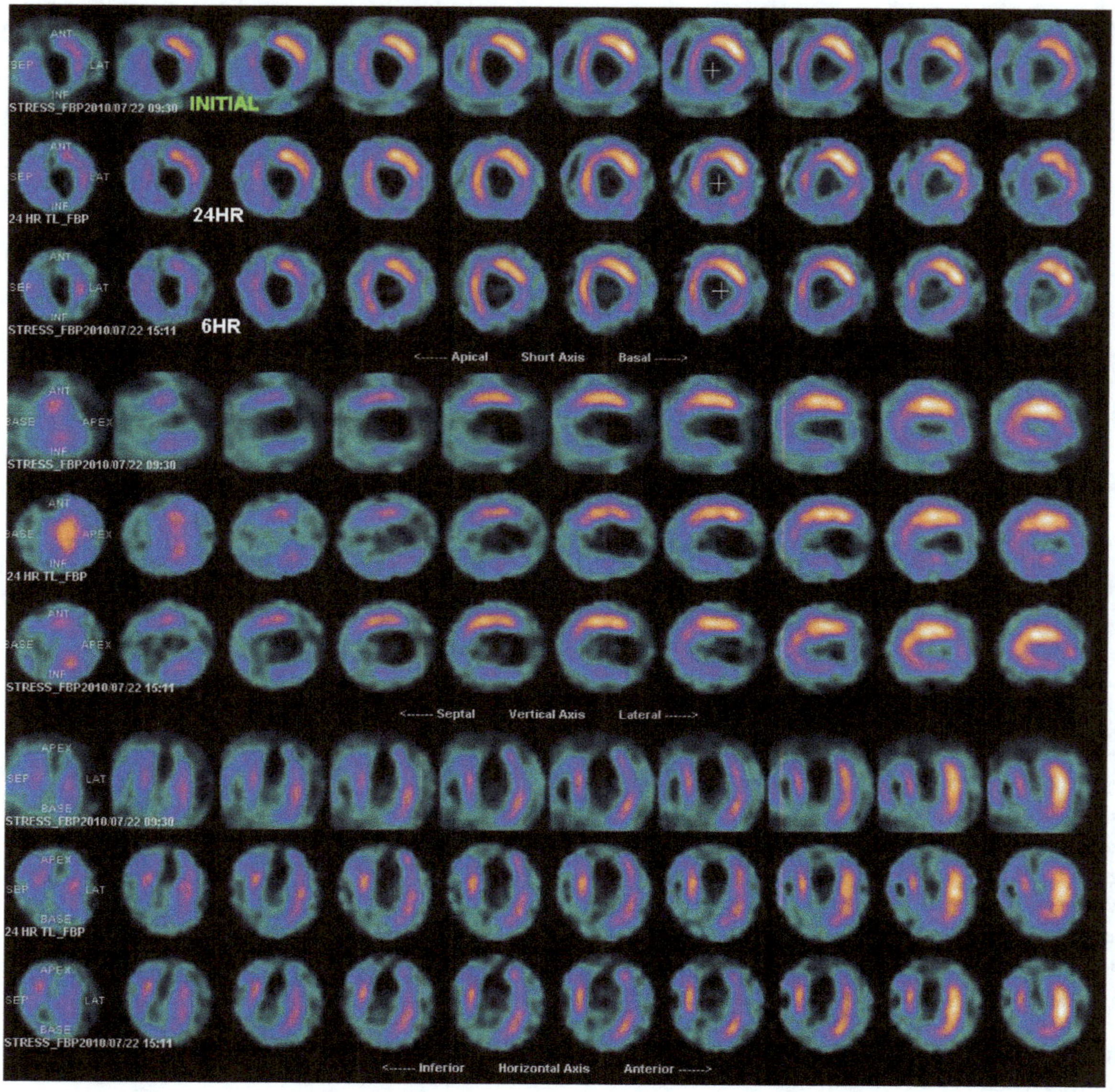

Figure 60.2. Note the large area of nonviable/scar tissue in the inferolateral walls.

cardiac MRI, when available (Figure 60.2). The large inferolateral infarct pattern seen on this patient's results would make that region a suboptimal target.

Emerging data suggest that other aspects of cardiac anatomy and physiology may play a role in optimal LV lead placement. Basal and midventricular segments along the lateral wall may provide better outcomes compared to apical segments, but this has not been consistently shown.[3] Similarly, lead placement into the area of most delayed mechanical activation, assessed during pre-implant imaging, has been suggested as a method to improve outcomes.[5] Individualized optimization of lead placement through segmental testing of multiple sites while undergoing hemodynamic monitoring may also improve function; but this approach has not yet been shown to improve long-term outcomes or clinical parameters.[6,7] As noted above, anatomical distance between the RV and LV leads may be an important factor as well.[4] Despite these various strategies to lead placement, no single method has proven to be the best approach based on current evidence. Further studies are needed to better evaluate optimal LV lead placement and will likely lead to improved recommendations for implant in the future.

In instances where successful LV lead placement cannot be placed transvenously via the CS, alternative approaches can be utilized including epicardial lead placement via thoracoscopy or limited thoracotomy as well as endocardial placement via a transseptal approach; however, these approaches are not frequently required.[1] Appropriate capture without diaphragmatic stimulation should be documented at the end of an implant case. As with standard pacemaker implants, inappropriate threshold values may indicate a loss of tissue contact or pacing within a nonviable segment.

Finally, current guidelines do not discriminate between ischemic and nonischemic etiologies of cardiomyopathy when recommending CRT as long as patients meet the other criteria stated.

Correct Answers:

Q1. B

Q2. A and B

Q3. B, C, D

Q4. B

REFERENCES

1. Dec GW, Pai RK, Shivkumar K, Wiener I. Cardiac resynchronization therapy. In: Shivkumar K, Weiss JN, Fonarow GC, Narula J, eds. *Atlas of Electrophysiology in Heart Failure.* Philadelphia, PA: Current Medicine LLC; 2005:106–125.
2. Singh JP, Houser S, Heist EK, Ruskin JN. The coronary venous anatomy: A segmental approach to aid cardiac resynchronization therapy. *J Am Coll Cardiol.* 2005;46(1):68–74.
3. Merchant FM, Heist EK, McCarty D, et al. Impact of segmental left ventricle lead position on cardiac resynchronization therapy outcomes. *Heart Rhythm.* 2010;7:639–644.
4. Singh JP, Heist EK, Ruskin JN, Harthorne JW. "Dialing-in" cardiac resynchronization therapy: Overcoming constraints of the coronary venous anatomy. *J Interv Card Electrophysiol.* 2006;17:51–58.
5. Ansalone G, Giannantoni P, Ricci R, et al. Doppler myocardial imaging to evaluate the effectiveness of pacing sites in patients receiving biventricular pacing. *J Am Coll Cardiol.* 2002;39:489–499.
6. Derval N, Steendijk P, Gula LJ, et al. Optimizing hemodynamics in heart failure patients by systematic screening of left ventricular pacing sites: The lateral left ventricular wall and the coronary sinus are rarely the best sites. *J Am Coll Cardiol.* 2010;55:566–575.
7. Singh JP, Abraham WT. Enhancing the response to cardiac resynchronization therapy: Is it time to individualize the left ventricular pacing site? *J Am Coll Cardiol.* 2010;55:576–578.

CASE STUDY

61

A Patient with NYHA Class II, Wide QRS, and Low EF

Arthur J. Moss, MD and Ilan Goldenberg, MD

A 67-year-old man with anterior myocardial infarction (MI) 3 years ago is currently in NYHA class II with normal sinus rhythm (NSR), left bundle branch block (LBBB; QRS duration 164 ms) on the ECG, and an ejection fraction (EF) of 28%. The patient is receiving a beta-blocker, an ACE-inhibitor, and a low-dose diuretic.

Question No. 1:

What would be the appropriate treatment to reduce the risk of the patient developing heart failure or death in the future?

A. Dual-chamber pacemaker.

B. Cardiac resynchronization therapy with defibrillator (CRT-D).

C. Implantable cardioverter-defibrillator (ICD).

D. Start an angiotensin II receptor blocker (ARB).

Following a significant MI, the heart undergoes adverse remodeling with cardiac enlargement and dilatation of the left ventricle with increase in the left ventricular (LV) end-diastolic pressure and reduction in the left ventricular ejection fraction (LVEF). This dysfunctional process is attenuated by beta-blocker and ACE-inhibitor therapy, but adverse remodeling may continue in some patients with the development of overt heart failure despite maximal medical management. Beginning in 2001, CRT was introduced for the treatment of patients with advanced NYHA class III/IV heart failure. With atrial synchronized biventricular pacing involving transvenous catheter electrodes in the right atrium, in the coronary vein for LV pacing, and in the right ventricular (RV) apex for nearly simultaneous RV pacing (Figure 61.1), approximately 70% of patients achieve significant improvement in LV function with increase in the LVEF and reduction in heart failure

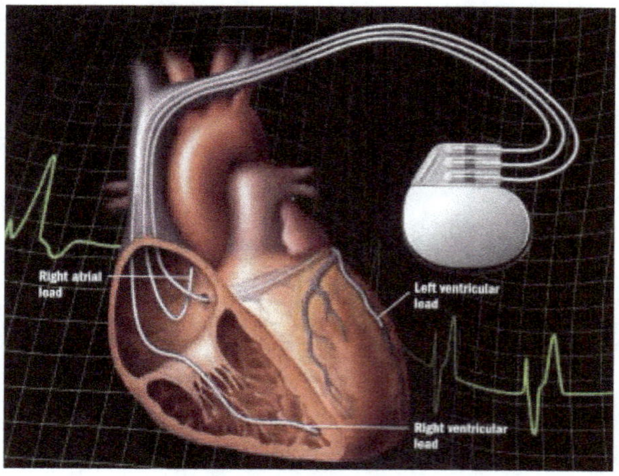

Figure 61.1. CRT with leads positioned in the right atrium, the right ventricle, and a coronary vein that overlies the left ventricle.

Practical Electrophysiology © 2015 Jasbir S. Sra, Masood Akhtar, Andrea Natale, David J. Wilber
Cardiotext Publishing, ISBN: 978-1-935395-17-1

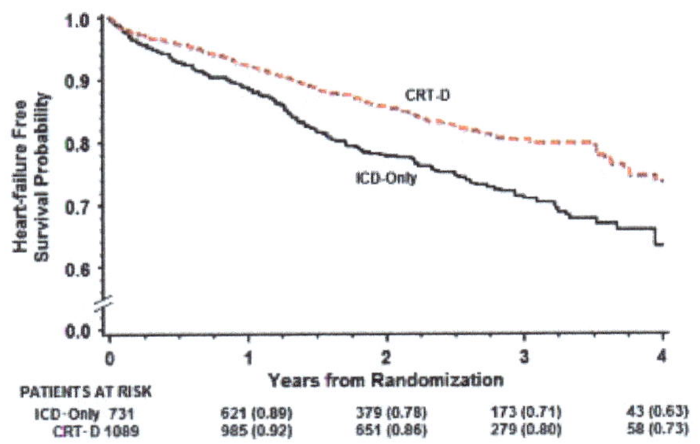

Figure 61.2. Kaplan–Meier estimates of the probability of survival free of heart failure. The hazard ratio of CRT-D to ICD was 0.66 with $P = 0.001$. CRT-D was associated with a 34% reduction in the primary end point of nonfatal heart failure. (Reprinted with permission.[3])

(reverse remodeling). Two major randomized trials substantiated the clinical improvement with CRT in patients with advanced heart failure, wide QRS complex, and low EF,[1,2] and AHA/ACC guidelines were established for this therapy in these patients with heart failure on optimal pharmacological management.

The question was then raised about the potential for CRT to reduce the probability of heart failure or death, whichever comes first, in cardiac patients who are essentially asymptomatic but have wide QRS complex on the ECG and low EF. The MADIT-CRT trial was conducted in 1820 at-risk cardiac patients (NYHA class I/II symptoms, QRS > 130 ms, and EF ≤ 30%) who were randomized to either a CRT with a defibrillator (CRT-D) or an ICD to prevent heart failure or death, whichever came first, with the benefit dominated by a 41% reduction in heart failure events (Figure 61.2).[3]

Additional analyses in the MADIT-CRT study revealed several important findings. Both women and men achieved benefit from CRT-D, but women achieved a significantly better benefit than men, with women having a remarkable 70% reduction in cardiac events (heart failure or death) during 2½-year follow-up. In addition, patients who had a QRS duration ≥ 150 ms or a LBBB conduction disturbance on the ECG (Figure 61.3) received the best benefit. In patients with LBBB, there is disordered dyssynchronous contraction of the left ventricle due to altered electrical activation of this chamber and increased myocardial strain. LV epicardial pacing through the electrode in the coronary vein creates a more synchronous contraction with restoration of the wringing and twisting motion that is an essential part of normal cardiac contractility.[4] The improvement in LV function in MADIT-CRT with CRT therapy was documented by echocardiographic studies obtained at baseline and again after 1 year of device implantation in the CRT-D and ICD-treated patients (Figure 61.4).

In September 2010, the FDA approved the MADIT-CRT indication for CRT-D therapy for cardiac patients who have ischemic (NYHA class I/II) or nonischemic (NYHA class II) heart disease with QRS ≥ 130 ms, LVEF ≤ 30%, and LBBB. Further documentation of the benefits of CRT-D

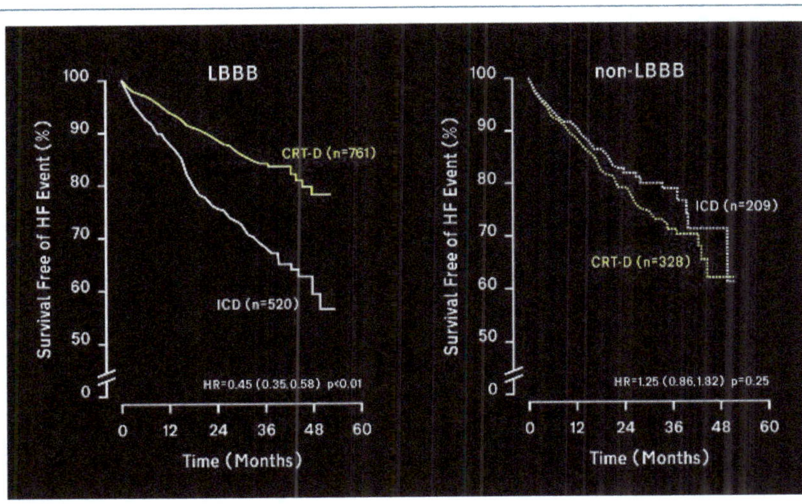

Figure 61.3. Kaplan–Meier estimates of the probability of survival free of heart failure by the presence of LBBB or non-LBBB conduction disturbance for patients randomized to CRT-D vs ICD. This graph was presented at the US FDA panel on March 23, 2010, and is available on the FDA government website.

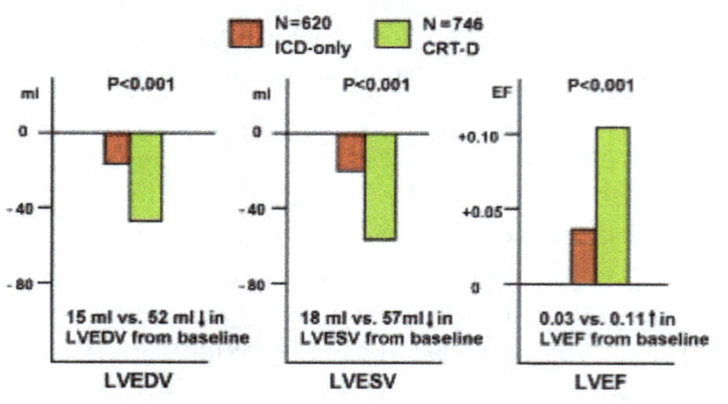

Figure 61.4. Changes in mean LV volumes and EF between baseline and 1-year follow-up in patients randomized to ICD and CRT-D therapy. (Reprinted with permission, with slight modification.[3])

therapy was shown in the RAFT trial that was published in December 2010,[5] with the population at risk and findings very similar to the MADIT-CRT trial. With FDA approval of CRT therapy based on the MADIT-CRT trial and 2 independent randomized trials showing essentially the same beneficial results in preventing heart failure with resynchronization therapy in NYHA class I/II patients, we are now able to recommend this therapy in these at-risk cardiac patients.

Correct Answer:

Q1. B

QUESTIONS AND AUTHORS' REPLY:

Question 1. Which subgroups of patients obtain the best benefit from CRT and why?

Answer: Women and patients with nonischemic cardiomyopathy obtain the best benefit with CRT. In both groups of patients, the LBBB conduction disturbance predominates, and CRT is particularly effective in improving the synchrony of LV contraction with this type of conduction abnormality.

Question 2. Does a reduction in heart failure precede a reduction in mortality with CRT therapy?

Answer: Yes! Several epidemiological studies have shown that the development of a heart failure event is associated with a 5-fold increase in mortality during the next 5 years. With longer follow-up, as occurred in the RAFT trial, the reduction in heart failure events was followed by a lower mortality rate over time.

REFERENCES

1. Bristow MR, Saxon LA, Boehmer J, et al. Cardiac-resynchronization therapy with or without an implantable defibrillator in advanced chronic heart failure. *N Engl J Med.* 2004;350:2140–2150.
2. Cleland JG, Daubert JC, Erdmann E, et al. The effect of cardiac resynchronization on morbidity and mortality in heart failure. *N Engl J Med.* 2005;352:1539–1549.
3. Moss AJ, Hall WJ, Cannom DS, et al. Cardiac-resynchronization therapy for the prevention of heart-failure events. *N Engl J Med.* 2009;361:1329–1338.
4. Bertini M, Marsan NA, Delgado V, et al. Effects of cardiac resynchronization therapy on left ventricular twist. *J Am Coll Cardiol.* 2009;54:1317–1325.
5. Tang AS, Wells GA, Talajic M, et al. Cardiac-resynchronization therapy for mild-to-moderate heart failure. *N Engl J Med.* 2010; 363:2385–2395.

CASE STUDY

62

Echocardiographic Parameters of Dyssynchrony

Khawaja Afzal Ammar, MD and Bijoy K. Khandheria, MD

A patient with wide QRS tachycardia and left ventricular ejection fraction (LVEF) of 26% prior to implantable cardioverter-defibrillation (ICD) implantation underwent three-dimensional speckle strain echocardiography (3DSTE). Time to peak systolic strain is shown in Figure 62.1.

Question No. 1:

This case shows presence of:

A. Intraventricular dyssynchrony.

B. Site of latest mechanical activation.

C. Interventricular dyssynchrony.

D. A and B.

E. A, B, and C.

As much as 30% of patients, selected based on the electrocardiographic (ECG), heart function (New York Heart Association class), and ejection fraction (EF) criteria, do not respond to cardiac resynchronization therapy (CRT). In addition, patients with left ventricular (LV) systolic dysfunction, narrow QRS complex and subtle signs of dyssynchrony are frequently encountered in clinical practice. Since CRT is a costly therapy, varying from $45,000 to $80,000, there is a need to identify appropriate responders, as well as accurately measure dyssynchrony.

While atrioventricular (AV) dyssynchrony and interventricular dyssynchrony are identified and optimized after biventricular (BiV) ICD implantation by selecting the AV interval and VV interval associated with the highest left ventricular outflow tract (LVOT) time velocity integral (TVI) and optimal mitral Doppler inflow pattern, the measurement of intraventricular dyssynchrony for optimal patient (responder) selection has gone

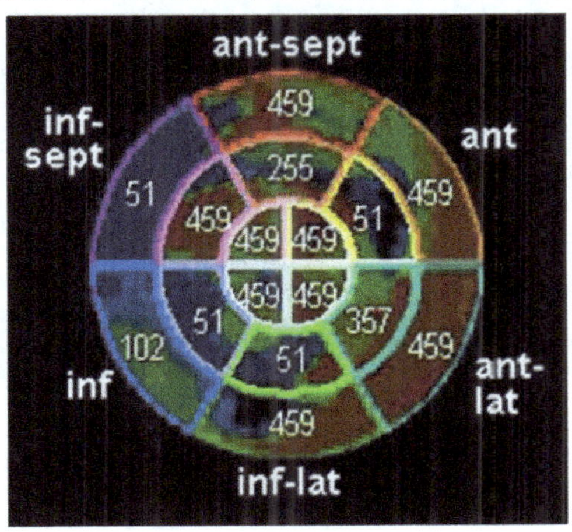

Figure 62.1. Polar plot showing time to peak systolic strain, measured in ms from onset of QRS.

through several developments as new technologies were introduced over time.

Common echocardiographic methods include:[1]

1. M-mode measurement of septal-to-posterior-wall delay of > 130 ms.
2. Tissue Doppler measurement of time to peak systolic velocity with septal-to-posterior-wall delay of > 60 ms (Figure 62.2).
3. Tissue Doppler measurement of time to peak systolic velocity in 12 segments and detection of standard deviation of 33 ms.
4. Speckle tracking measurement of septal-to-posterior-wall delay in time to peak radial strain of > 130 ms.
5. Speckle tracking measurement of time to peak systolic longitudinal strain in 12 segments and detection of standard deviation > 60 ms.

The interest in echocardiography to predict response to CRT markedly diminished after the PROSPECT and Mayo clinic studies in which none of the studied parameters could predict response to CRT in a reproducible manner.[2,3] These studies included not only tissue Doppler but, in the Mayo Clinic study, two-dimensional (2D) strain as well as 3D volume dyssynchrony indices. However, since the clinical need for measurement of dyssynchrony persists, there is continued interest in development of new methods.

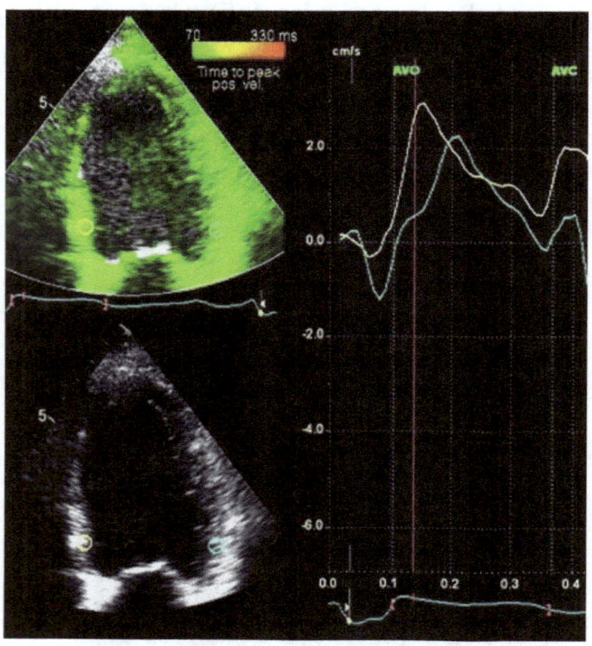

Figure 62.2. Septal-to-lateral-wall dyssynchrony as seen by tissue Doppler velocity imaging for measurement of time to peak systolic velocity. The yellow line, representing the region of interest on septum, peaks much earlier than the green line, which represents the velocity of region on interest of lateral wall. In this case it was 65 ms, indicating that the septum reaches peak systolic velocity 65 ms, hence it is dyssynchronous with the lateral wall by 65 ms.

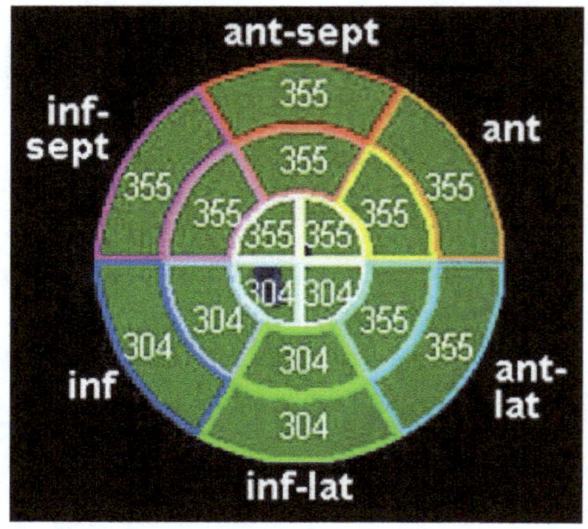

Figure 62.3. Time to peak systolic strain in a normal healthy volunteer.

Figure 62.1 represents the promise of 3DSTE for strain measurement, a new technique to assess dyssynchrony. 3DSTE was developed to conquer the limitations of 2DSTE, which was unable to track speckles in 3 dimensions. 3DSTE has a much larger scan area with the ability to track the fingerprints of myocardium, called speckles, in space and time. The time to peak contraction is measured by time to peak systolic strain. In case of dyssynchrony, the time to peak systolic strain is delayed in dyssynchronous segments. Time to peak systolic strain is similar in normal controls in all segments, between 300 and 400 ms, whether measured by 2D or 3D,[4] as shown in Figure 62.4.

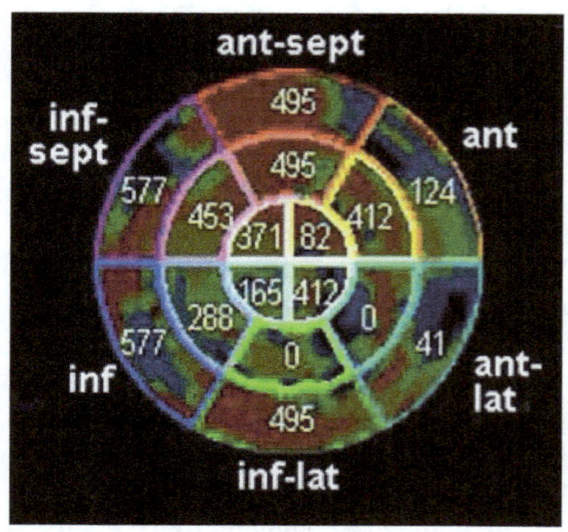

Figure 62.4. Polar plot of patient in Figure 62.1 after implantable cardioverter-defibrillator is turned on.

Compared to Figure 62.1, the time to peak systolic radial strain in Figure 62.3 is similar and between 300 to 355 ms. Apical and septal segments have the shortest time to peak systolic strain of 304 ms, as compared to the rest of the walls, which are about 350 ms. This is consistent with normal conduction pattern as the left bundle branch first depolarizes the septum, then the apex and free wall depolarize followed by the base, which is the last to depolarize.

In Figure 62.1 there is clear intraventricular dyssynchrony with different segments. In 2DSTE literature, a difference in time to peak radial systolic strain of opposing walls (midanteroseptal vs inferolateral) of > 130 ms is a predictor of response to CRT with a sensitivity of 89% and specificity of 83%.[5] In addition, the standard deviation of time to peak longitudinal strain of > 60 ms, in a 12-segment model, is considered in 2DSTE as an acceptable predictor of response to CRT.[1] In a recent study, Tanaka and colleagues showed the feasibility of quantification of dyssynchrony by 3DSTE (n = 64), as well as the close correlation of 3DSTE with 2DSTE parameters, including standard deviation in time to peak radial strain.[6]

In addition, the site of latest mechanical activation is provided in the same polar plot. This has the potential for guiding lead placement by placing the lead in the site of latest mechanical activation. In a recent study (n = 54), the 3D site of latest mechanical activation was successfully quantified in a manner not possible previously using 2D methods or by previous 3D methods that focused on passive endocardial motion alone.[6] Previous investigations have shown that the greatest immediate improvements from CRT result from coronary sinus (CS) lead placement at the most delayed segment.[6] By using 3DSTE, a color-coded polar map is generated within a few seconds, which allows rapid identification of the most delayed segment by several different methods, including all types of strain and area tracking. Another study (n = 57) from the same group demonstrated the site of latest activation to be similar from left bundle branch block (LBBB) or paced rhythm, supporting the argument for CRT in paced patients.[7]

In Figure 62.4, the ICD in the patient from Figure 62.1 is turned on. The lateral wall is first to reach peak systolic strain (indicating that the LV lead is closest to midinferolateral and midanterolateral walls where the time to peak systolic strain is zero). However, the septum is markedly delayed, 577 ms compared to 41 ms for the anterolateral wall. LVEF dropped to 15% as dyssynchrony is actually worse after BiV ICD placement. Lateral-wall delay has been exchanged for septal-wall delay by CRT. In Figure 62.5, the ICD has been optimized with LV lead trigger 20 ms behind the right ventricular (RV) lead in order to take care of the dyssynchrony introduced by the CS lead. Septal-to-lateral-wall delay is markedly reduced, as most walls reach peak systolic strain at the same time, increasing LVEF to 37%. This case shows the clinical application of 3DSTE in measuring dyssynchrony and guiding lead placement, as well as in device optimization with reduction of intraventricular dyssynchrony.

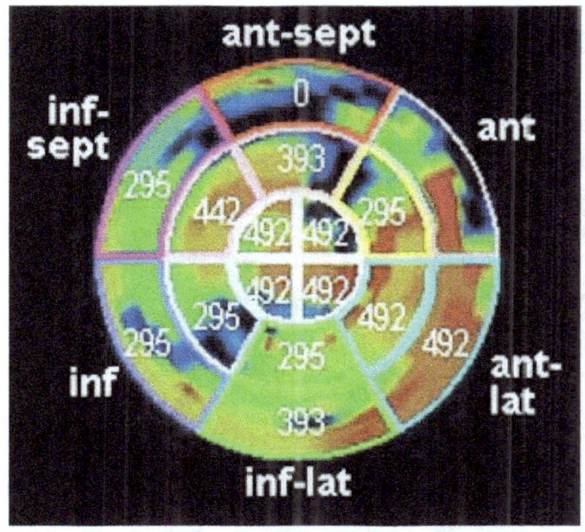

Figure 62.5. Polar plot of ICD optimization with left ventricular lead trigger 20 ms behind the RV lead in order to take care of the dyssynchrony introduced by the CS lead.

Question No. 2:

Figure 62.6 shows:

A. Base-to-apex rotational dyssynchrony.

B. Loss of apical twist.

STE has demonstrated the ability to measure LV twist mechanics. Effective contraction of the LV involves not only intraventricular septal-to-lateral-wall synchrony; it also needs base-to-apex rotational synchrony. In a normal left ventricle, the apex undergoes counterclockwise rotation, whereas the base undergoes a clockwise rotation in systole. This creates a towel-wringing motion of the heart, ensuring effective emptying of the LV. Figure 62.7 shows normal rotation of base (angles measured as negative values, therefore the red curve is below the base line) and apex (angles measured as positive values, therefore the white curve is above the base line). Both the curves peak at the same time near the end of systole.

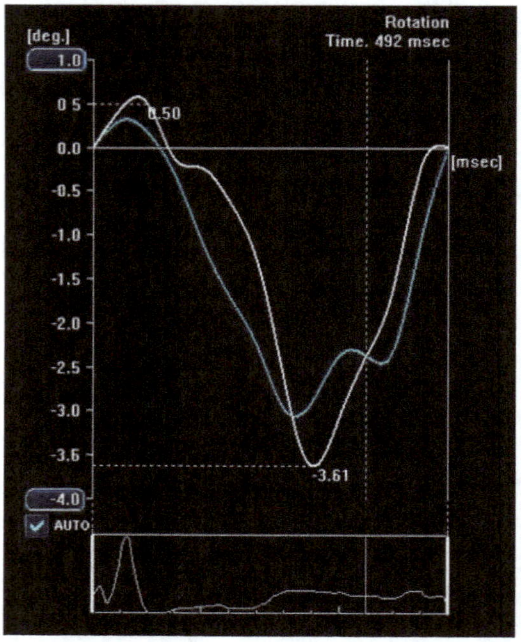

Figure 62.6. Prior to BiV activation, the rotation is markedly abnormal. The base is rotating clockwise and so is the apex, indicating that the towel-wringing motion of cardiac twist is lost.

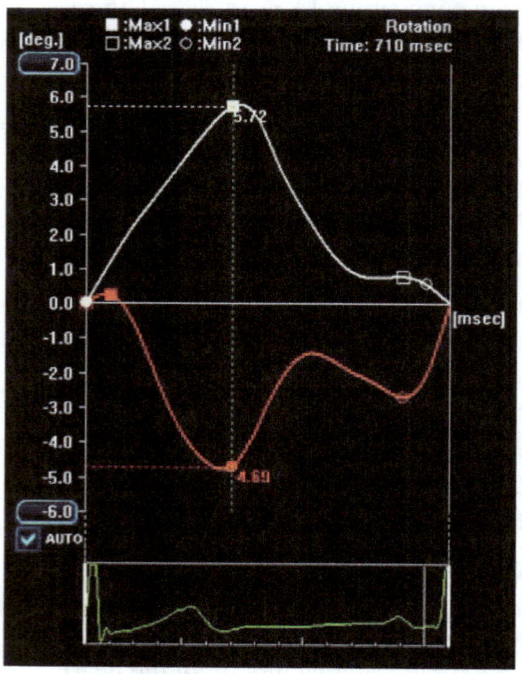

Figure 62.7. Normal rotation of base and apex in a normal heart. The Y axis depicts the rotation in degree and X axis depicts time in relation to the ECG below. The base of heart (red line) rotates clockwise, which is depicted below the baseline. The apex (white line) rotates clockwise, which is above the baseline. Both movements, opposite in direction, create the towel-wringing motion of heart.

In the case of base-to-apex dyssynchrony with LV systolic dysfunction, the peak rotation of both base and apex are not only reduced, but there is also a significant time delay in between peak basal and apical rotation. In patients with severe LV systolic dysfunction, the apical counterclockwise rotation is lost and the apex rotates in the same direction as the base. Using 2DSTE, Sade et al demonstrated that rotational mechanics outperformed radial and longitudinal strain in predicting CRT responders. This was based on the observation that in those patients with relatively preserved rotational mechanics, CRT did not lead to a clinical response. In addition, there was a meaningful improvement in rotational mechanics in responders. In some of the responders, not only were the rotational mechanics reduced, they were reversed (ie, apex was rotating with the base in a clockwise manner) prior to CRT. In normal controls, peak LV rotation (basal or apical) happened simultaneously at the end of systole, defined by aortic valve closure (AVC), at about 400 ms of cardiac cycle. In those with dyssynchrony, there was apex-to-base rotational dyssynchrony, with peak apical rotation occurring much later than peak basal rotation; peak rotation was significantly reduced at AVC. The best performing parameters were LV torsion (cutoff 0.1 degree/cm) and twist (cutoff 1 degree) at AVC with a sensitivity of 90% and specificity of 77% in predicting CRT responders. Nonresponders had an LV twist of > 1 degree at AVC and it decreased significantly with CRT, indicating that the rotational mechanics were relatively preserved at baseline, and were significantly reduced after CRT, providing proof that restoration of rotational mechanics may be central to identify responders.[8] This patient lost the apical counterclockwise rotation. This patient was known to have severe LV systolic dysfunction and a broad QRS complex. Following BiV ICD implantation, there was significant clinical improvement as well as EF response, and the rotational synchrony was restored, as shown in Figure 62.8.

3DSTE represents a promising new development in our ability to measure dyssynchrony. However, it awaits validation in further studies. Until such time, patient selection for dyssynchrony will continue to be based on standard EF and ECG criteria.

Correct Answers:

Q1. D

Q2. B

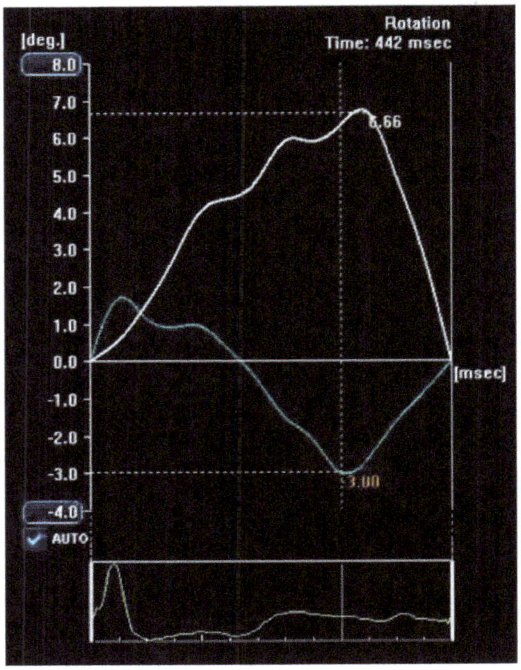

Figure 62.8. Rotational synchrony restored in the patient from Figure 62.6 after BiV implantable cardioverter-defibrillator implantation.

REFERENCES

1. Mor-Avi V, Lang RM, Badano LP, et al. Current and evolving echocardiographic techniques for the quantitative evaluation of cardiac mechanics: ASE/EAE consensus statement on methodology and indications endorsed by the Japanese Society of Echocardiography. *J Am Soc Echocardiogr.* 2011;24:277–313.
2. Chung E, Leon A, Tavazzi L, et al. Results of the Predictors of Response to CRT (PROSPECT) trial. *Circulation.* 2008;117:2608–2616.
3. Miyazaki C, Redfield MM, Powell BD, et al. Dyssynchrony indices to predict response to cardiac resynchronization therapy: a comprehensive prospective single-center study. *Circ Heart Fail.* 2010;3:565–573.
4. Saito K, Okura H, Watanabe N, et al. Comprehensive evaluation of left ventricular strain using speckle tracking echocardiography in normal adults: comparison of three-dimensional and two-dimensional approaches. *J Am Soc Echocardiogr.* 2009;22:1025–1030.
5. Suffoletto MS, Dohi K, Cannesson M, Saba S, Gorcsan J 3rd. Novel speckle-tracking radial strain from routine black-and-white echocardiographic images to quantify dyssynchrony and predict response to cardiac resynchronization therapy. *Circulation.* 2006;113:960–968.
6. Tanaka H, Hara H, Saba S, Gorcsan J 3rd. Usefulness of three-dimensional speckle tracking strain to quantify dyssynchrony and the site of latest mechanical activation. *Am J Cardiol.* 2010;105:235–242.
7. Tanaka H, Hara H, Adelstein EC, et al. Comparative mechanical activation mapping of RV pacing to LBBB by 2D and 3D speckle tracking and association with response to resynchronization therapy. *JACC Cardiovasc Imaging.* 2010;3:461–471.
8. Sade LE, Demir O, Atar I, Müderrisoglu H, Ozin B. Effect of mechanical dyssynchrony and cardiac resynchronization therapy on left ventricular rotational mechanics. *Am J Cardiol.* 2008;101:1163–1169.

CASE STUDY

63

Interventional Techniques to Successfully Add RV Pace-Sense and LV Leads Despite Multiple Pitfalls

Seth J. Worley, MD

The following 12 figures were taken from a single implant procedure in a patient with a Sprint Fidelis® lead (Medtronic Inc., Minneapolis, MN) who had a left bundle branch block (LBBB), class III congestive heart failure (CHF), and an ejection fraction of 27%. She had the addition of a left ventricular (LV) lead and a right ventricular (RV) pace-sense lead. To reduce the risk of nephropathy with the use of contrast, she was hydrated with normal saline 3 mL/kg for 1 hour before the procedure, 1 mL/kg per hour during the procedure, and 1 mL/kg for 6 hours after the procedure.[1] At the time of implant, a venogram was obtained via contrast injection through a peripheral vein in the left arm. The venogram (Figure 63.1A) demonstrates an occlusion of the axillary/subclavian vein with extensive collaterals.

Question No. 1:

Based on the venogram, which of the following is true?

- **A.** Access is not possible from the left; move to the right.
- **B.** Access is not possible; extract one of the leads to gain access on the left.
- **C.** Despite the apparent total occlusion, it may be possible to gain access on the left.

Figure 63.1B illustrates that the location of the occlusion was not where it appeared on the venogram. The peripheral venogram usually overestimates the severity and location of the occlusion associated with chronic leads.[2]

The venogram in Figure 63.2A was then obtained by injection of contrast through the dilator from a 5-Fr sheath advanced as far as possible over the J-wire.

Question No. 2:

Based on the local venogram, which of the following is true?

- **A.** Access is not possible from the left; move to the right.
- **B.** Access is not possible; extract one of the leads to gain access on the left.
- **C.** Despite the apparent total occlusion, it may be possible to gain access on the left.

Practical Electrophysiology © 2015 Jasbir S. Sra, Masood Akhtar, Andrea Natale, David J. Wilber
Cardiotext Publishing, ISBN: 978-1-935395-17-1

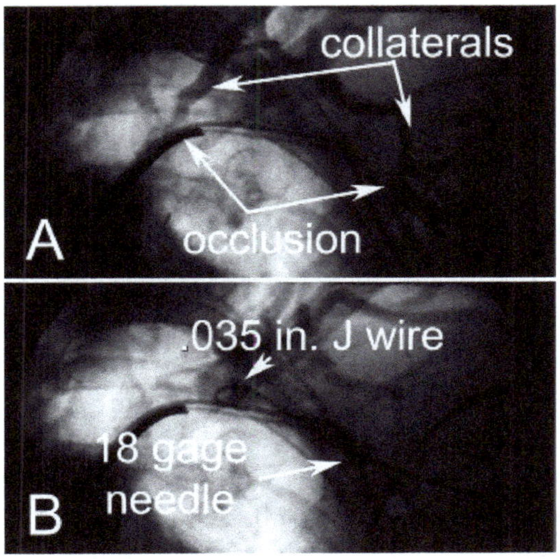

Figure 63.1. Insertion of a J-wire despite apparent long total occlusion on peripheral vein venogram. In **Panel A**, contrast injected through a peripheral vein demonstrates an apparent long total occlusion with extensive collaterals. In **Panel B**, the needle is inserted into the axillary vein at the site of apparent occlusion: The 0.035-inch J-wire advances through the site of apparent occlusion but will not advance beyond the clavicle into the subclavian vein. The peripheral vein venogram does not define the location or extent of occlusion.

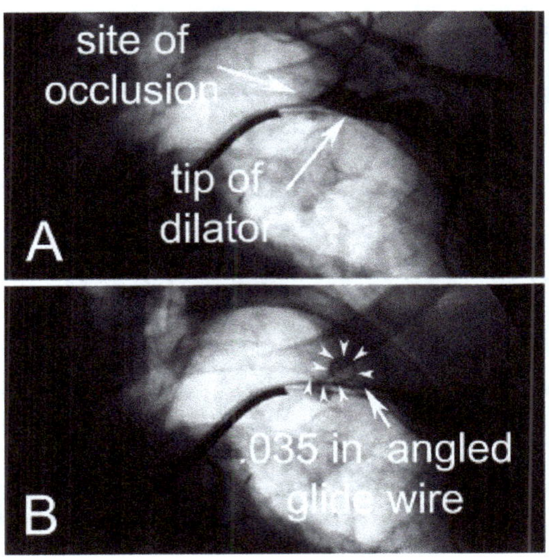

Figure 63.2. The angled 0.035-inch hydrophilic (glide) wire (Angled Glidewire [GR3506], Terumo Medical Corp., Somerset, NJ, or Angled Zipwire [46-151B], Boston Scientific Corp., Natick, MA) will not advance through the dilator from a 5-Fr sheath beyond the site of true obstruction. In **Panel A**, the dilator from a 5-Fr sheath advanced over the J-wire to the site of obstruction and contrast injected. There is no apparent opening in the occlusion when contrast is injected through the dilator. In **Panel B**, the glide wire (arrowheads) curls when advanced through the dilator and will not proceed beyond the obstruction.

Despite the apparent total occlusion, an attempt was made to advance a glide wire through the occlusion but was unsuccessful (see Figure 63.2B).

Question No. 3:

Based on the failure of the wire to advance, which of the following is indicated?

A. Access is not possible from the left; move to the right.

B. Access is not possible; extract one of the leads to gain access on the left.

C. Despite the failure of the glide wire to advance, it still may be possible to gain access on the left.

The venogram in Figure 63.3A was obtained by contract injection through a left vertebral diagnostic catheter and demonstrated an opening in the

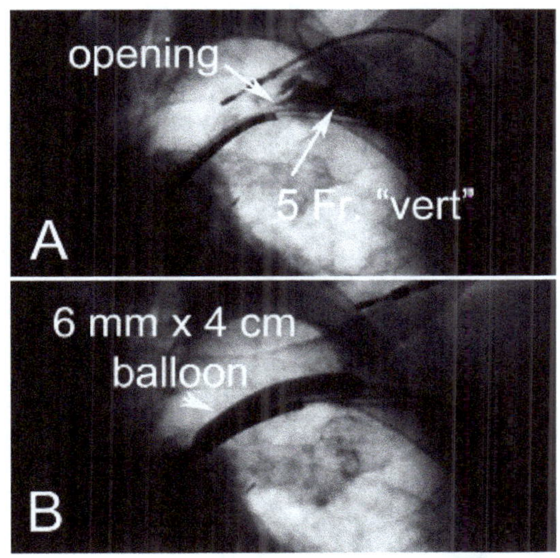

Figure 63.3. An opening is identified by injection of contrast through a 5-Fr left vertebral diagnostic catheter ("vert"), the glide wire advanced, and venoplasty performed. In **Panel A**, contrast injection through the "vert" catheter (Vert Slip-Cath Beacon® Tip [SCBR5.0-38-100-P-NS-VERT], Cook Medical Inc., Bloomington, IN, or Angled Taper Terumo Glidecath® [CG508], Terumo Medical Corp.) reveals an opening in the occlusion. Before inserting the "vert," the dilator is exchanged for a 5-Fr sheath. The tip of the "vert" is directed toward the leads where contrast injection defines the opening. The "vert" catheter provides direction and support to advance the glide wire through the opening. In **Panel B**, the obstruction is dilated with a 6-mm diameter by 4-cm length balloon inflated to rated burst pressure (14 atmospheres). With the glide wire in the pulmonary artery or the inferior vena cava, the "vert" catheter is advanced to the tip of the glide wire, and the glide wire exchanged for a 0.035-inch extra support wire (Amplatz Extra-Stiff Wire Guide, 0.035-inch, 145-cm, 3-mm tip curve [THSCF-35-145-3-AES], Cook Medical Inc.).

obstruction. The glide wire is advanced using the direction and support of the catheter, and venoplasty was performed to gain access.[3] Because the patient had a Sprint Fidelis defibrillator lead, an attempt was made to place an RV pace-sense lead in the setting of severe tricuspid regurgitation and massively enlarged right atrium. The lead repeatedly dislodged despite using stiff stylets and replacing the lead.

Question No. 4:

Based on the difficulty in placing the RV lead, which of the following is indicated?

- **A. Keep trying with stiff stylets; the lead will eventually stay put.**
- **B. Give up and continue to use the pace-sense portion of the Sprint Fidelis lead.**
- **C. Use a long anatomically shaped sheath to provide support for the lead as it is screwed in place.**

Figure 63.4 illustrates the use of a long anatomically shaped sheath to secure an RV lead despite severe tricuspid regurgitation and massively enlarged right atrium. Following placement of the RV lead, attempts were made to cannulate the coronary sinus (CS). The CS could not be located after 20 minutes, despite using an electrophysiology (EP) catheter and probing wire.

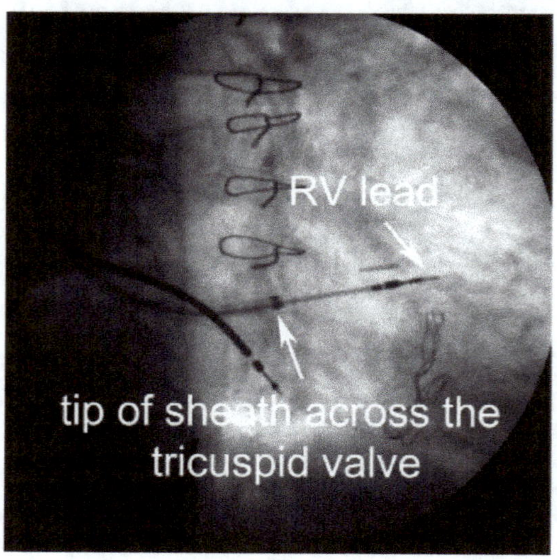

Figure 63.4. Long anatomically shaped sheath (SafeSheath® CSG® Worley Braided Core Series 40-cm long, 9-Fr I.D. [CSG/Worley/BCore-1-09], Pressure Products Inc., San Pedro, CA) providing support to secure RV lead despite severe tricuspid regurgitation.

Question No. 5:

Based on the failure to cannulate the CS, which of the following is indicated?

- **A. The patient does not have a CS; send the patient for an epicardial lead.**
- **B. Inject the coronary arteries and watch for venous return filling to outline the CS.**
- **C. Cannulate the CS from the leg to mark the location.**
- **D. Inject half-strength contrast to locate the CS.**
- **E. Inject full-strength contrast to locate the CS.**

Figure 63.5A illustrates the CS ostium (os) was easily located with a 3-mL puff of contrast. When half-strength contrast is used, it is more difficult to visualize the CS, thus, more total contrast is utilized. Despite easily locating the CS os with contrast, advancing beyond the os was difficult. Contrast divides the cannulation of the CS into 2 steps: locating the os, then advancing beyond the os. Without contrast, the CS is only

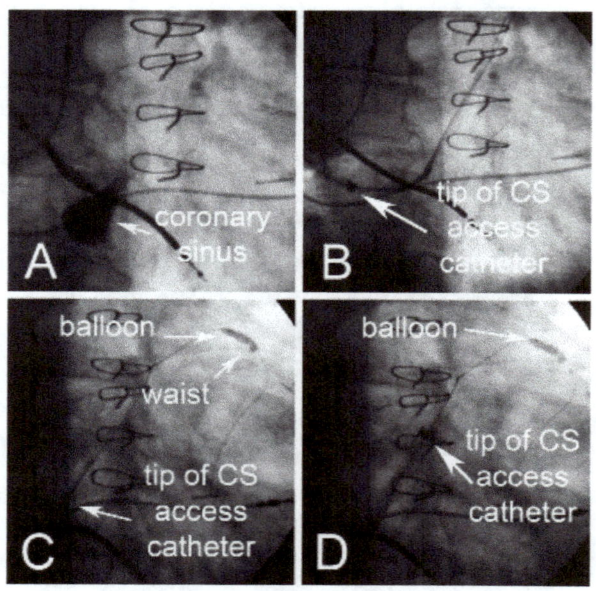

Figure 63.5. Anchor balloon techniques for cannulation of the CS. In **Panel A**, the CS os is visualized with a puff of contrast. In **Panel B**, the CS access catheter (SafeSheath® CSG® Worley Braided Core, Pressure Products Inc.) will not advance despite creating a variety of passive rails. The rail is augmented by placing tension on a venogram balloon inflated in the distal CS; however, the balloon dislodges as the CS access catheter is advanced. In **Panel C**, a 3-mm × 21-mm coronary balloon is advanced and inflated to 6 atmospheres in an anterior coronary vein. The waist indicates the coronary balloon is firmly anchored. In **Panel D**, the tension required to advance the CS access catheter does not dislodge the coronary balloon.

recognized when the catheter or wire advances beyond the os.[4-6]

The CS access catheter did not advance beyond the os despite multiple attempts to create a supportive rail.

Question No. 6:

Based on the inability to effectively cannulate the CS, which of the following is indicated?

A. Send the patient for an epicardial lead.

B. Advance a wire into the CS and place the lead blindly.

C. Use balloon anchoring techniques to cannulate the CS.

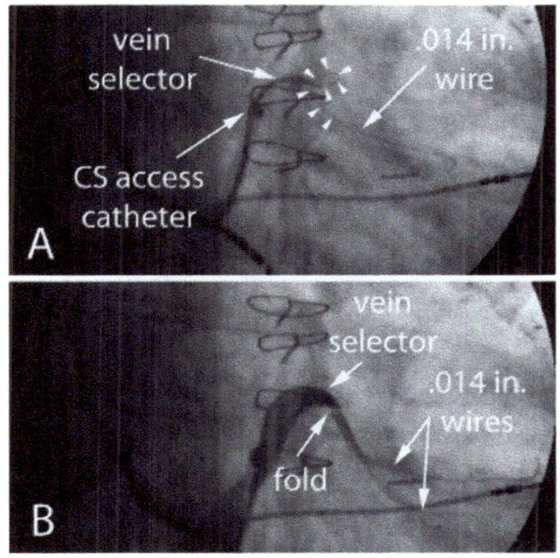

Figure 63.6. The target vein is identified and wires placed. In **Panel A**, the CS access catheter is at mid-CS level. The 5-Fr vein selector (Impress CS Standard [1628-Y8], Merit Medical Systems Inc., South Jordan, UT) in the target vein provides direction and support to advance the first wire. There is a loop (arrowheads) in the first wire advanced into the vein. In **Panel B**, the second wire advanced through the vein selector straightens the vein, eliminating the loop. A residual fold in the vein (arrow) is apparent when contrast is injected through the vein selector.

Figure 63.5C and D describe the balloon anchoring technique used for CS cannulation. Balloon anchoring techniques can also be used to facilitate entry into a target vein.[7]

Once the CS was cannulated, an occlusive CS venogram revealed a lateral wall target vein. A 5-Fr vein selector was telescoped inside a delivery guide, advanced into the CS, and turned laterally toward the target vein. When the tip of the vein selector was felt to have engaged, a puff of contrast confirmed its location in the target vein. Despite a fold in the vein that offers resistance, a 0.014-inch hydrophilic floppy angioplasty wire was advanced into the target vein. Resistance to advancing the wire was overcome by the support provided by the vein selector, however, a loop remains in the wire (Figure 63.6A). A 0.014-inch hydrophilic extra support wire was then advanced through the vein selector beside the first wire, straightening the vein. Contrast injection after the second wire was placed (see Figure 63.6B) revealed a lucent area, suggesting a fold in the vein. To provide optimum support for advancing the lead, a 5-Fr I.D. delivery guide was advanced over the wire-supported vein selector into the target vein (Figure 63.7A and B). The vein selector was removed from the delivery guide retaining the guide wires (Figure 63.8A). Despite support from the CS access catheter, the delivery guide and the wires, the LV lead will not advance (see Figure 63.8B).

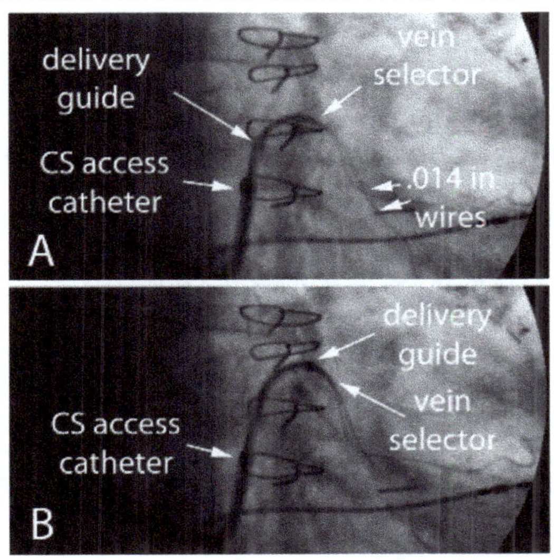

Figure 63.7. The delivery guide is advanced from the CS access catheter into the vein over the rail formed by the vein selector and 2 wires. In **Panel A**, the 5-Fr I.D. delivery guide (SafeSheath® Worley Series Telescopic Series Renal Shape lateral vein introducer [LVI/75-5-66-5.5-RE] Pressure Products Inc.) is advanced out of the CS access catheter over the rail formed by the vein selector and 2 wires. In **Panel B**, the delivery guide is in the target vein supported by the CS access catheter. The vein selector and wires remain in the target vein.

Question No. 7:

Based on the inability to advance the leads, which of the following is indicated?

A. Send the patient for an epicardial lead.

B. Advance a wire into the CS and place the lead blindly.

C. Use a snare to hold the distal end of the wire so the lead can be advanced.

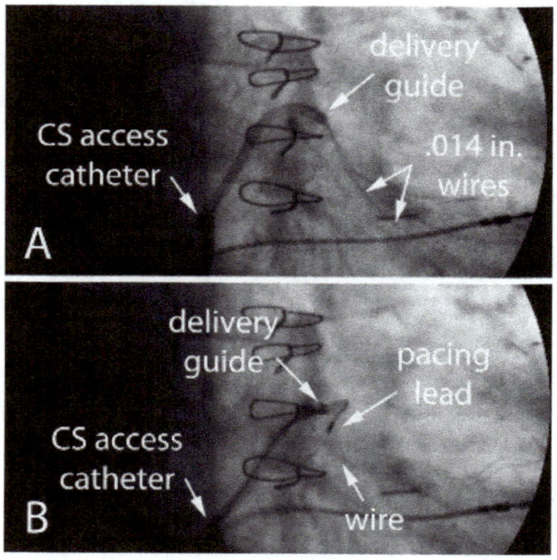

Figure 63.8. Despite support from the CS access catheter and the delivery guide, the pacing lead will not advance beyond the fold in the vein. In **Panel A**, the vein selector is removed, leaving the guide wires. The curve where the delivery guide exits the CS access catheter reflects the effect of the difficult-to-cannulate, tortuous CS. In **Panel B**, despite support from the delivery guide, the pacing lead will not advance beyond the vein fold seen in Figure 63.5B.

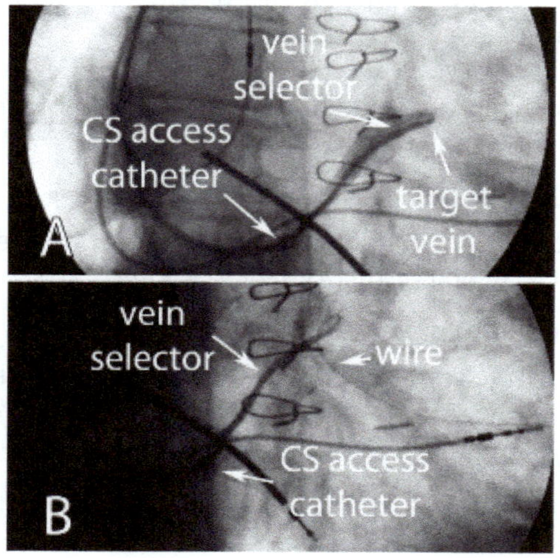

Figure 63.9. With the CS access catheter parked at the CS os, the vein selector is advanced into the CS to cannulate and wire the target vein. In **Panel A**, the CS access catheter is at the CS os. The vein selector is advanced into the CS, turned laterally, and contrast injected to identify and cannulate the target vein. In **Panel B**, using the direction and support provided by the vein selector, the 0.014-inch angioplasty wire is advanced. The acute bend in the vein is outlined by the loop in the wire at the tip of the vein selector.

Because it was very difficult to cannulate the CS initially and the LV lead did not advance despite optimal support, the snare technique was chosen. With the snare technique, the CS access catheter is parked at the CS os and the 5-Fr vein selector is advanced into the CS to cannulate the target vein and advance the wire (Figure 63.9A and B). See Figures 63.10 to 63.12 for a description of how to use the snare technique for LV lead placement by stabilizing the distal end of the wire. When the CS access catheter has a 9-Fr lumen, there is enough room for both the 3-Fr snare and the pacing lead. When the CS access catheter lumen is ≤7 Fr (all device companies' CS access catheters), the snare and distal end of the wire must be pulled out of the hub to provide room for the lead. The snare technique can also be used to advance leads over the distal end of the wire but requires a 300-cm wire.[8]

Correct Answers:

Q1. C

Q2. C

Q3. C

Q4. C

Q5. C

Q6. C

Q7. C

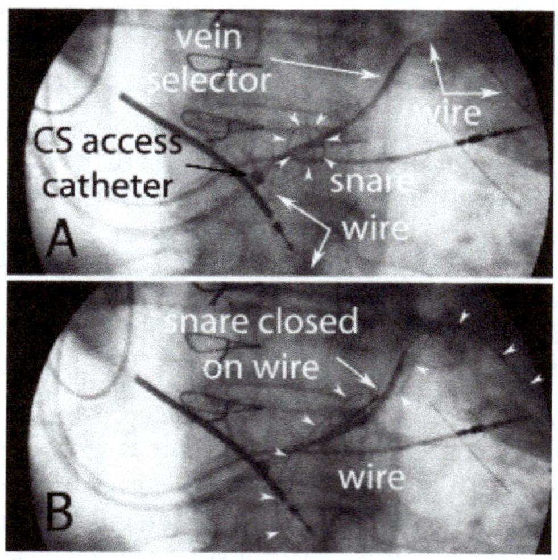

Figure 63.10. The distal end of the angioplasty wire reenters the CS via an adjacent vein and is snared. In **Panel A**, the CS access catheter is parked at the CS os. The vein selector is in the target vein. The angioplasty wire is advanced into the target vein through collaterals into an adjacent vein, then retrograde into the CS. The 7-mm loop snare and 3-Fr snare catheter (Amplatz GooseNeck® Microsnare Kit [SK 700], ev3 Inc., Plymouth, MN) are advanced beside the vein selector through the CS access catheter into the CS. The snare catheter is retracted to allow the loop snare to open (arrowheads). In **Panel B**, after the distal end of the angioplasty wire (arrowheads) is advanced through the open snare, the snare catheter is advanced to close the loop snare on the distal end of the angioplasty wire.

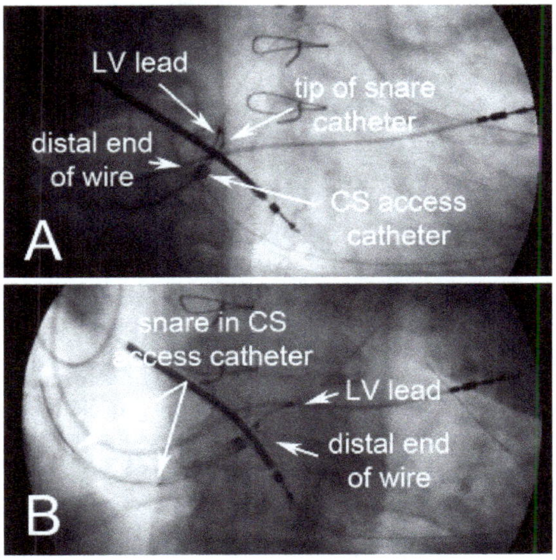

Figure 63.11. The snared distal end of the wire is pulled into the CS access catheter as the LV lead is advanced. In **Panel A**, distal end of the wire is in the tip of the snare catheter. The snare catheter is at the tip of the CS access catheter. The LV lead is entering the CS on the proximal end of the wire. In **Panel B**, the distal end of the wire and the snare system is within the CS access catheter. The LV lead is entering the CS on the proximal end of the wire.

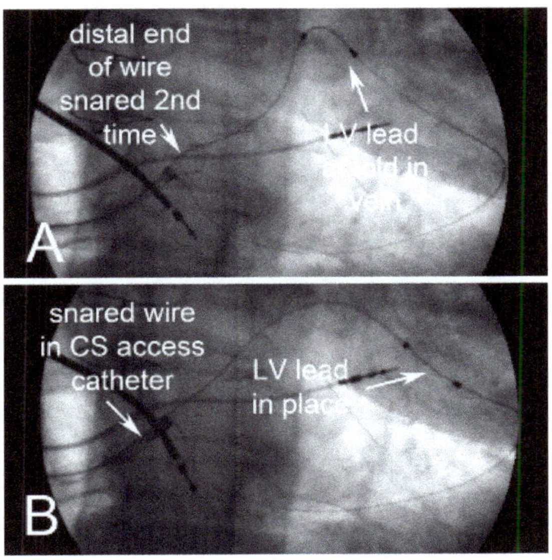

Figure 63.12. The wire is snared a second time and the lead advanced. In **Panel A**, the distal end of the wire is snared a second time. The tension applied during the first attempt to advance the lead was sufficient to pull the wire loose from the snare. A more proximal, stiffer segment of the distal end of the wire was snared the second time. In **Panel B**, the tension required to advance the lead did not dislodge the wire from the snare. The stiff section of the distal wire did not slip through the snare as easily as the soft-tip section.

QUESTION AND AUTHOR'S REPLY:

Question. What is the risk of angioplasty in the subclavian, superior vena cava, and CS and what precautions should be taken during these procedures?

Answer: The overall risk of doing angioplasty in the subclavian, superior vena cava, and CS to add a lead in the chronic setting is negligible, none in over 400 cases over 11 years in our experience. This applies to venoplasty to add a lead in the chronic setting not when venoplasty is used for patency. Venoplasty for patency is completely different. Much larger balloons are used (12–15 mm) that are used to add leads (6 mm). Also venoplasty for patency is often done in the sub-acute setting before the extensive fibrous tissue develops.

To avoid complications: There are 2 mechanisms by which a complication can be created and both are readily avoided. The first and most likely mechanism applies to progressively larger dilators as well as balloons and occurs when a wire is inadvertently advanced into a small branch vein or collateral but appears to be within the true lumen. Fortunately, when the wire enters a collateral or branch vein it will not advance into the pulmonary artery or inferior vena cava. Thus regardless of wire position seen on the AP projection, the operator must never perform venoplasty (or advance a dilator) until the wire is in the pulmonary artery or inferior vena cava, and orthogonal views confirm wire position adjacent to the leads. Second, inadvertent penetration of the fibrous tissue may occur when attempting to cross a wire-resistant total obstruction with microdissection ± needle directed reentry, excimer laser, or radiofrequency guide wire. As long as the perforation is recognized and the wire retracted, the consequence of simple perforation is similar to starting an IV or the vessel trauma associated with the initial axillary/cephalic vein access.

With regard to the CS and coronary veins the consequence of venous perforation or disruption in patients with OHS is negligible. We have never seen a complication of CS/coronary vein venoplasty in a patient with prior remote OHS because of the fibrous tissue. In patients without prior OHS, we have seen tamponade once in over 100 cases that was drained without adverse outcome to the patient. To avoid complications in patients without prior OHS, care must be taken when performing venoplasty near the takeoff of the vein from the main CS.

REFERENCES

1. Worley SJ. Chapter 23: Interventional device implantation. In: Ellenbogen KA, Kay GN, Lau CP, eds. *Clinical Cardiac Pacing, Defibrillation and Resynchronization Therapy*, 4th ed. Philadelphia: Saunders Elsevier, 2010.
2. Worley SJ, Charles Gohn DC, Pulliam RW, et al. Subclavian venoplasty by the implanting physicians in 373 patients over 11 years. *Heart Rhythm.* 2011;8:526–533.
3. Worley SJ. Implant venoplasty: dilation of subclavian and coronary veins to facilitate device implantation: Indications, frequency, methods and complications. *J Cardiovasc Electrophysiol.* 2008;19:1004–1007.
4. Worley SJ. Chapter 22: Left ventricular lead implantation. In: Ellenbogen KA, Kay GN, Lau CP, eds. *Clinical Cardiac Pacing, Defibrillation and Resynchronization Therapy*, 4th Ed. Philadelphia: Saunders Elsevier, 2010.
5. Worley SJ. Chapter 19: Left ventricular lead implantation. In: Ellenbogen KA, Kay GN, Lau CP, eds. *Clinical Cardiac Pacing, Defibrillation and Resynchronization Therapy*, 3rd Ed. Philadelphia: Saunders Elsevier, 2007.
6. Worley SJ, Leon A, Wilkoff BL. Chapter 6: Anatomy and implantation techniques for biventricular devices. In: Ellenbogen KA, Kay GN, Wilkoff BL, eds. *Device Therapy for Congestive Heart Failure.* Philadelphia: Saunders Elsevier, 2004.
7. Worley SJ. How to use balloons as anchors to facilitate cannulation of the coronary sinus left ventricular lead placement and to regain lost coronary sinus or target vein access? *Heart Rhythm.* 2009;6:1242–1246.
8. Worley SJ, Gohn DC, Pulliam RW. Goose neck snare for LV lead placement in difficult venous anatomy. *Pacing Clin Electrophysiol.* 2009;32:1577–1581.

CASE STUDY 64

Device Infection: Management Strategies

Khaldoun G. Tarakji, MD, MPH and Bruce L. Wilkoff, MD

A 67-year-old woman presented with fever and chills. Her past medical history was significant for diabetes mellitus type II requiring insulin, morbid obesity (body mass index = 55 kg/m^2), chronic obstructive pulmonary disease, and diverticulosis with recurrent lower gastrointestinal bleed requiring multiple transfusions. She also had history of sick sinus syndrome for which she underwent a permanent pacemaker implantation 8 years prior to her presentation. The device pocket was benign with no signs of inflammation. Two sets of blood cultures were sent and came back positive for enterococcus faecalis. Transesophageal echocardiography (TEE) showed evidence of vegetation on the pacing lead measuring 1.7 × 1.2 cm (Figure 64.1). It also showed evidence of small patent foramen ovale (PFO) with bidirectional flow, confirmed with bubble study (Figure 64.2). She was deemed not to be a surgical candidate due to her body habitus and comorbidities.

Question No. 1:

Which of the following statements is correct?

A. Cardiac implantable electronic device (CIED) infection is a common complication, affecting 25% of all implants.

B. CIED infections are usually benign and are rarely associated with any significant morbidity and mortality.

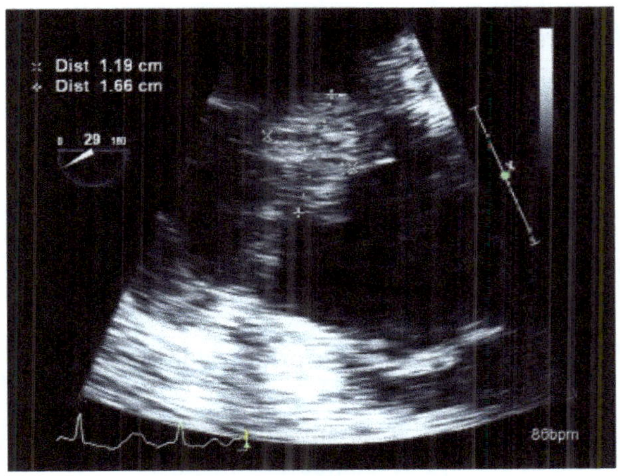

Figure 64.1. TEE view showing the largest vegetation on the pacemaker lead measuring 1.7 × 1.2 cm.

C. The lack of evidence of pocket inflammation rules out CIED infection.

D. CIED infection is rare, but the rate of increase in CIED infections each year has been out of proportion to the rate of increase in newly implanted devices.

Question No. 2:

Which of the following statements is correct?

A. New (first-time) CIED implant is associated with higher risk of infection than pulse generator change.

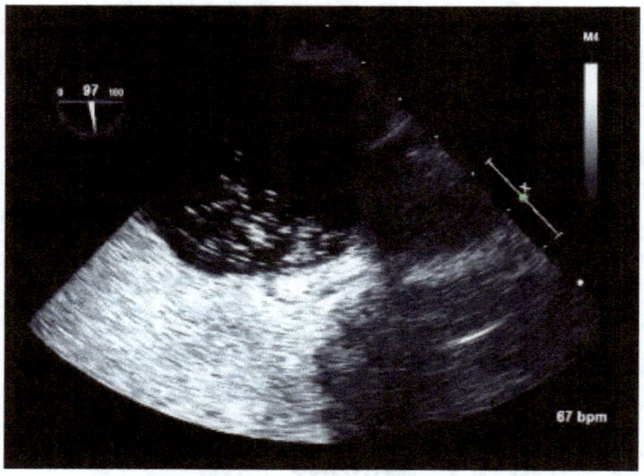

Figure 64.2. TEE view showing a PFO documented by bubble study.

B. Pulse generator change is associated with higher risk of infection than new CIED implant.

C. The infection risk is the same for new CIED implant and pulse generator change.

D. Pacemakers have a higher infection rate than implantable cardioverter-defibrillators (ICDs).

Question No. 3:

Which of the following tests is not recommended?

A. Obtain blood cultures prior to initiating antibiotic therapy.

B. If present, aspirate any fluid in the pocket and send it for gram stain and cultures.

C. Consider a TEE, especially in cases associated with bacteremia.

D. Obtain chest x-ray to identify the number and location of the leads.

Question No. 4:

In the management of CIED infections, which of the following statements is correct?

A. When the diagnosis of CIED infection is made, complete removal of all hardware is recommended regardless of whether it is a device pocket infection or endovascular infection.

B. CIED endocarditis can be managed conservatively with long-term antibiotic therapy.

C. CIED pocket infection can be managed by debriding the pocket along with antibiotic therapy without extracting the leads.

D. No need for antibiotic therapy as long as the CIED is removed.

DISCUSSION

Due to the growing evidence of the importance of CIEDs in improving both quality of life and survival among patients with heart disease, the number of newly implanted CIEDs has increased significantly over the last 2 decades. With this surge of CIED implants has come an increased recognition of associated complications, and infection is among the most important ones. Device-related infection is uncommon, but is associated with significant morbidity and mortality. The infection rate for patients with implanted CIEDs varies widely, but in general it ranges between 1 and 2%. The rate of increase in CIED infections each year has been out of proportion to the rate of increase in newly implanted devices. Multiple studies have identified several patient characteristics associated with increased risk for CIED infections. These include diabetes, renal dysfunction, heart failure, steroid use, and the presence of more than 2 leads. Additionally, several procedural characteristics could be associated with increased risk for CIED infections; these include pulse generator change (as opposed to new device implant), use of temporary pacing prior to the permanent device, and early reintervention. Pectoral transvenous device implant, perioperative antimicrobial prophylaxis, and physician experience are associated with lower risk for CIED infections.

The increasing number of CIED-related infections will continue to challenge both electrophysiologists and infectious disease specialists. The challenge begins with determining the diagnosis, as patients with CIED infections can present with a variety of manifestations. The diagnosis is clear when a patient presents with device pocket that is grossly infected (erythema, tenderness, drainage, erosion, etc.), which could be associated with systemic signs and symptoms (fever, chills, malaise, etc.) and occasionally positive blood cultures. On the other hand, the diagnosis might be delayed when a patient presents with only systemic signs and symptoms, with or without positive blood cultures, and a device pocket that appears grossly intact.

Despite these differences in presentations, it is important to recognize that the extent of infection could be underestimated in patients presenting with a localized pocket infection. Many of these patients (up to 20%) can have positive blood cultures and sometimes have evidence of vegetations on the leads. Therefore, it is not clinically useful to consider pocket infections separate from endovascular infection but better to consider an infected system in totality, with a component of the system in contact with endocardium and bloodstream. These findings also support the consensus that conservative management often is not optimal, and device removal is needed to eradicate the infection.

Etiology and Pathogenesis

Patients with CIED may become infected at the time of implantation or pocket revisions. Eroded devices are by default infected and can also be a secondary manifestation of an underlying infection. Pocket infection can track along the intravascular portion of the leads, leading to intravascular infection. On the other hand, the pocket or the intravascular portion of a CIED can become infected as a result of hematogenous seeding during an episode of bacteremia. Many patients with pocket infection have positive blood cultures and even evidence of vegetation on the leads; therefore, although the mechanism of infection could be different, the management of CIED infection is usually similar and involves considering it as infection of the whole system.

Microbiology

Most infections are monomicrobial, but some (about 10%) are polymicrobial. Few patients might have negative cultures despite clear presentation of CIED infection. This is often due to prior antibiotic therapy. Staphylococcus species represent the most common pathogen isolated in cases of CIED infections and account for almost 80% of cases; almost half of these are methicillin-resistant.

Diagnosis

CIED infections can have different presentations. Most patients present with inflammatory changes involving the device pocket itself. These changes include erythema, warmth, tenderness, drainage, and erosion of the device through the skin, and are sometimes accompanied by systemic signs and symptoms that include fever, chills, and malaise, etc. Other patients may present with a device pocket that looks intact but, with combination of signs and symptoms supported by blood culture findings and occasionally echo imaging data (especially TEE), suggests endovascular infection involving the CIED. Although the clinical presentation might be different, the management of both types, whether pocket infection or endovascular infection, is very similar and involves system removal and antibiotic therapy. The type of infection might affect the duration of the antibiotic therapy and the decision for the timing and type of reimplant.

All patients should have 2 sets of blood cultures drawn before the initiation of any antibiotic therapy. Percutaneous aspiration of the generator pocket should not be performed as part of the diagnostic evaluation. Transthoracic echocardiography is of great value in cases of CIED infections. However, one should be careful interpreting the findings A mass that is adherent to the lead could represent thrombus or vegetation. Making the distinction by morphology alone is not accurate and should be avoided. Rather, the interpretation should be based on the clinical situation. The accidental finding of an echo density adherent to the leads in the absence of any signs or symptoms of infection is likely to be thrombus and does not require lead removal or treatment. On the other hand, the lack of echo densities on the leads does not rule out the diagnosis of CIED infection. TEE represents a more sensitive tool to better visualize leads and valve to rule out endocarditis. TEE should be pursued in all patients with positive blood cultures and those suspected to have endovascular infection. Recent data suggest that even patients who present with pocket infection could have evidence of vegetations on the leads and valve. These findings lowered our threshold in obtaining TEE among all patients presenting with CIED infection.

Patients with bacteremia and without any evidence of pocket infection or lead or valvular involvement represent a clinical challenge. Relapsing infections after the completion of a full antibiotic course or persistent bacteremia despite antibiotic therapy support the diagnosis of CIED infection and therefore would require CIED removal. This is particularly true for gram-positive organisms, especially with the lack of other sources of bacteremia. On the other hand, gram-negative bacteremia, in the absence of any other evidence of infection, is usually treated with the appropriate antibiotic, and if the bacteremia relapses after the completion of the antibiotic course, the CIED removal should be performed.

Management

It is not rare for there to be some erythema at the incision site during the first week of healing after device implantation. Less frequently, but still common during the first 2 weeks after CIED implantation, there can be a small, superficial stitch abscess that does not involve the pocket. This usually responds to local measures and short course of oral antibiotics directed against staph species.

However, when the diagnosis of CIED infection is made, complete removal of all hardware is recommended regardless of whether it is a device pocket infection or endovascular infection. Conservative management with antibiotics does not work. Partial removal of the pulse generator and capping the lead should not be performed, even in cases of pocket infection, as the relapse rate due to the retained wire is high. Therefore, complete removal of all hardware is recommended.

The strategy of device removal should take into account different factors, and the decision-making process requires a team approach that involves the electrophysiologist, the infectious disease specialist, and oftentimes, the cardiothoracic surgeon. Percutaneous lead extraction has evolved over the years and became the method of choice for device removal. Despite a good safety record for percutaneous extraction, these procedures could be associated with significant risks that include perforation leading to tamponade, hemothorax, pulmonary embolism, and death. Therefore, lead extraction should be limited to centers that can provide the appropriate facilities, equipment, skills, and expertise. These centers should also provide surgical backup that would be ready for emergency surgery in the event of complications. Primary surgical extraction is usually reserved for specific cases—cases that already failed percutaneous extraction or retained leads remnants. Presence of epicardial leads requires surgical extraction and, oftentimes, combined or consecutive procedures (percutaneous approach for the endovascular leads and surgical removal of the epicardial leads). Surgical extraction also is often pursued for cases with large vegetation (>2 cm) on the leads. Although some data suggest that extraction of leads with large vegetations could be safely accomplished percutaneously, until more data are available, decisions regarding extraction of leads with vegetations > 2 cm should be individualized and should involve surgical option as an approach. Cases that have valve involvement may require surgical approach. However, a discussion between the surgeon and the electrophysiologist is of utmost importance. While the surgical approach can be successful in removing the leads, often this cannot be accomplished to extensive fibrosis. Cutting both sides of the lead and leaving a remnant of an infected lead should be avoided. These remnants could cause recurrent infection, and their extraction can represent a major challenge technically. Therefore, these cases require collaboration. If the leads are cut distally at time of surgery, it is important to maintain the proximal portion of the lead in order to enable the proceduralist to perform percutaneous removal of the remnant.

Antimicrobial therapy is important and plays an adjunctive role in patients with CIED infections. Since the majority of CIED infections are caused by Staphylococcal species and half of these are methicillin-resistant, vancomycin should be initiated empirically after the appropriate cultures are sent. Antibiotic therapy afterward should be tailored based on the culture and sensitivity data.

There are no clear data about the optimal duration of antibiotic therapy, but in general, patients with pocket infection are treated with antibiotics for 2 weeks, and patients with endovascular infection are treated for 4 to 6 weeks.

Reimplant

The decision to reimplant a new device after the removal of the infected one should be individualized. Careful assessment of new device implant should be performed even before the removal of the infected CIED as it might affect the management approach. A patient with CIED infection involving the lead and the tricuspid valve with vegetations on both might undergo percutaneous removal of the device, if the CIED is no longer needed and will not require reimplant. On the other hand, a pacemaker-dependent patient with vegetation on the tricuspid valve would require temporary pacemaker implant, which makes it very difficult to clear the infection that is involving the valve. Therefore, many of these patients end up with surgical extraction with debridement of the tricuspid valve and occasionally end up with epicardial lead implant to avoid implanting across the infected tricuspid valve.

A recent study showed that one-third of patients will not require new CIED placement due to different reasons, including improvement or reversal of the pathological processes that precipitated the initial CIED implantation (eg, improved left ventricular (LV) function with medical therapy in a patient with ICD that was implanted for primary prevention), emergence of new clinical conditions or patient preferences, and lack of absolute clinical indication initially. Wearable defibrillators represent a temporary solution that can provide enough time to complete the course of antibiotic before making a decision to reimplant a new ICD.

When reimplant is needed, it should be performed on the opposite side. Alternative approaches could be abdominal implant with transiliac leads or epicardial lead implant. There are no randomized trials to decide the timing of reimplant, but this usually depends on the type of infection, the presence of positive blood cultures, and the pathogen involved. Patients with bacteremia with no valve vegetation should have repeat blood cultures after CIED removal, and if the repeat cultures are negative for 72 hours, they can be reimplanted. On the other hand, patients with bacteremia and evidence of valve vegetation should not be reimplanted until at least 14 days from first negative blood cultures after device removal. For pacemaker-dependent patients, epicardial reimplant of pacing system is an alternative approach.

Case Presentation Continued

The management of this case was challenging. On one side, it is a clear case of pacemaker endocarditis with vegetations on the lead that would require removal of the whole system to achieve cure from the infection. On the other side, the sizable vegetation with the PFO carried a risk of embolization causing a stroke. The management required collaboration among an electrophysiologist, an interventional cardiologist, anesthesiologist, and an infectious disease specialist. Percutaneous extraction was performed while occluding the PFO temporarily using an Amplatzer® occluder device (AGA Medical Corp., Plymouth, MN) in order to avoid right-to-left embolization. The procedure was performed under general anesthesia. Right femoral venous sheath and right radial arterial line were placed. A transesophageal probe was inserted during the procedure to provide better visualization. The device pocket was opened and debrided. One bolus of heparin was given. A Goodale-Lubin catheter was used to cross the PFO and the Wholey wire (Covidien, Mansfield, MA) was placed into the left atrium. The catheter was advanced into the left atrium and the wire was exchanged for a stiff Amplatz wire (AGA Medical). The sheath was exchanged for a 9F Amplatzer delivery sheath (AGA Medical). Through this sheath a 25-mm Amplatzer occluder device was delivered into the PFO. The device was not deployed, and the sheath was in place until the pacer wires were extracted (Figure 64.3). The atrial and ventricular leads were prepped with a liberator, locking stylets, and sutures. The right atrial lead was removed with simple traction. The right ventricular (RV) lead required the use of a 14F laser sheath due to fibrosis at the level of superior vena cava. The PFO occluder was then removed and the device pocket was irrigated with antibiotic solution and closed. The patient tolerated the procedure well with no complications. Repeat blood cultures were negative and she was treated with intravenous ampicillin for 6 weeks. She did not require reimplant as she was not dependent and had no history of syncope, and she had good chronotropic response with ambulation. She was discharged home 2 days after the procedure.

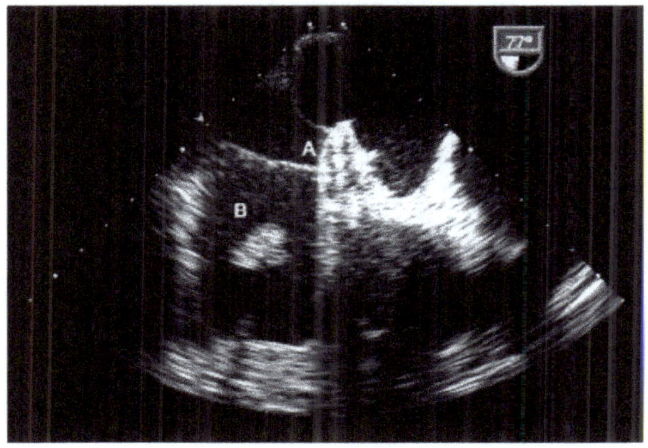

Figure 64.3. TEE view showing the deployment of the closure device (A) and the vegetation (B) on the pacemaker lead.

In summary, management of CIED infections can represent a challenge to electrophysiologists and often requires a team approach. While the general management of CIED infection is usually the same, each case should be studied carefully and managed individually, and the final strategy should be about patient management and not device management.

Correct Answers:

Q1. D
Q2. B
Q3. B
Q4. A

REFERENCES

1. Baddour LM, Epstein AE, Erickson CC, et al; American Heart Association Rheumatic Fever, Endocarditis, and Kawasaki Disease Committee; Council on Cardiovascular Disease in Young; Council on Cardiovascular Surgery and Anesthesia; Council on Cardiovascular Nursing; Council on Clinical Cardiology; Interdisciplinary Council on Quality of Care; American Heart Association. Update on cardiovascular implantable electronic device infections and their management. A scientific statement from the American Heart Association. *Circulation.* 2010;121:458–477.
2. Tarakji KG, Chan EJ, Cantillon DJ, et al. Cardiac implantable electronic device infections: presentation, management, and patient outcomes. *Heart Rhythm.* 2010;7:1043–1047.

3. Sohail MR, Uslan DZ, Khan AH, et al. Management and outcome of permanent pacemaker and implantable cardioverter-defibrillator infections. *J Am Coll Cardiol.* 2007;49:1851–1859.
4. Wilkoff BL. How to treat and identify device infections. *Heart Rhythm.* 2007;4:1467–1470.
5. Wilkoff BL, Love CJ, Byrd CL, et al; Heart Rhythm Society; American Heart Association. Transvenous lead extraction: Heart Rhythm Society expert consensus on facilities, training, indications, and patient management: this document was endorsed by the American Heart Association. *Heart Rhythm.* 2009;6:1085–1104.

CASE STUDY

65

Atrial Synchronous His-Bundle Pacing

Imran Niazi, MD

A 60-year-old patient with nonischemic cardiomyopathy (left ventricular ejection fraction [LVEF] 45%) required permanent pacemaker implantation for intermittent complete heart block. The following ECGs were recorded prior to and after pacemaker implant (Figure 65.1A and B, respectively).

Question No. 1:

The ECGs in Figure 65.1 demonstrate:

A. Atrial synchronous right ventricular (RV) pacing with ventricular fusion.
B. Atrial synchronous biventricular pacing.
C. Atrial synchronous direct His-bundle pacing.
D. Para-Hisian pacing.

DISCUSSION

Biventricular pacing results in a QRS complex > 120 ms, which is distinctly different from the intrinsic (unpaced) QRS complex. Biventricular pacing can, in some cases, result in a QRS complex narrower than during intrinsic rhythm, but only when the unpaced QRS is > 120 ms and almost always in the presence of left bundle branch block (LBBB). Typically, the paced QRS shows a positive deflection in V_1, reflecting capture of the left ventricle (LV), compared to a negative deflection seen in most cases of RV pacing.

Ventricular fusion occurs in the presence of intact conduction, when the programmed atrioventricular (AV) interval is similar to the intrinsic AV interval. In true ventricular fusion, the ventricular pacing stimulus lies at the beginning of the QRS complex,

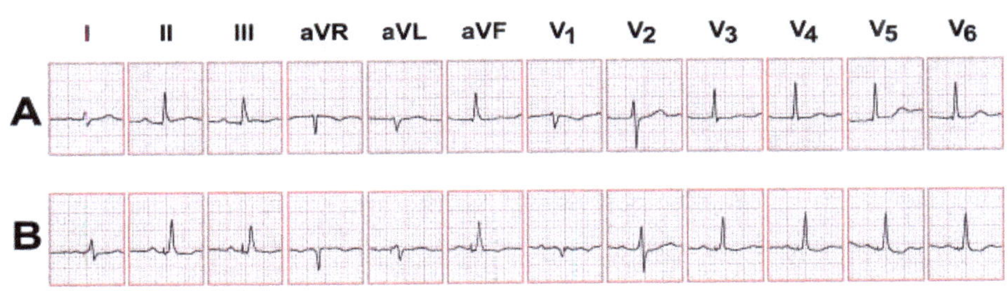

Figure 65.1. ECGs recorded during sinus rhythm (A) and after pacemaker implant (B).

Practical Electrophysiology © 2015 Jasbir S. Sra, Masood Akhtar, Andrea Natale, David J. Wilber
Cardiotext Publishing, ISBN: 978-1-935395-17-1

and the ventricles are excited simultaneously by intrinsic conduction through the His-Purkinje system as well as the ventricular pacing site. In such cases, the resulting QRS is never identical to the unpaced QRS.

On occasion, the ventricular pacing spike may be detected *after* the initiation of the QRS complex, as a result of it falling in the blanking period or due to sensing failure. The resultant QRS is identical to the unpaced QRS, because the ventricular stimulus has occurred when the ventricle is refractory and has not contributed to ventricular depolarization. This has been termed *pseudo*fusion.

In the above example, the paced AV interval is shorter than the intrinsic AV interval, yet the resultant QRS is identical to intrinsic rhythm; fusion is ruled out. Since the ventricular stimulus clearly lies at the beginning of the QRS, pseudofusion is also ruled out.

Direct or selective His-bundle pacing (SHBP) is arguably the most physiological form of ventricular pacing since both ventricles are activated via the His-Purkinje system, if the His-Purkinje system is intact. RV pacing has been shown to induce electrical and mechanical dyssynchrony and is associated with worse outcomes in patients with heart failure. It should be avoided in the presence of left ventricular dysfunction. SHBP does not induce electrical dyssynchrony and may be the preferred pacing mode in patients with an intact His-Purkinje system, although this has not been assessed in randomized trials.

It is technically difficult to position a permanent pacing lead to selectively pace the His bundle. However, the difficulties are not insuperable, and successful chronic His-bundle pacing has been reported by Deshmukh[1] and others. Specific lead and delivery systems have been designed for this application, and evolution of these designs may resolve the associated technical problems. A European registry comprising more than 200 patients has shown acceptable long-term pacing thresholds and lead retention.[2]

In general, pacing thresholds are higher for SHBP than for RV pacing. In our laboratory, permanent direct His-bundle pacing was successfully achieved in 16 of 22 patients. The mean His-bundle pacing threshold was 2.2 V at 0.5 ms pulse width. If an SHBP is not achieved with standard (cathodal) pacing, reversing polarity (making the tip the anode, ring the cathode) may allow His capture at adequate thresholds.

SHBP is characterized by: a paced QRS identical to the intrinsic QRS; and a stimulus-to-QRS (S-QRS) onset delay > 25 ms in patients with a normal QRS duration (see Figure 65.1B). The S-QRS delay is due to activation of the His bundle directly and represents conduction time through the His-Purkinje system.

The term "Para-Hisian pacing" is usually taken to mean nonselective His-bundle pacing (NSHBP), a more precise term describing pacing of the RV septum very close to the His bundle in such manner that there is capture of the His bundle but also capture of RV septal myocardium. It is characterized by a paced complex with an S-QRS delay > 25 ms, similar *but not identical* to the intrinsic QRS. This difference in the initial QRS is due to a slur on the upstroke of the QRS, caused by local capture of the septal myocardium as well as capture of the His bundle (Figure 65.2B). Frequently, NSHBP will be seen during pacing at lower amplitudes and SHBP at higher amplitudes in the same patient. In our series of 22 patients, NSHBP alone was possible in 3, while SHBP was achieved in 16.

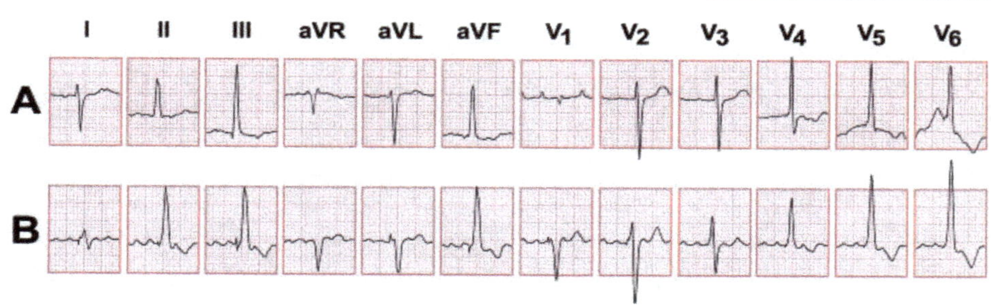

Figure 65.2. **A:** ECG during sinus rhythm. **B:** ECG during NSHBP. The paced QRS is similar, but not identical to that during sinus rhythm, due to a slur in the initial QRS deflection caused by activation of septal myocardium adjacent to the His bundle. There is a 25-ms delay between the pacing stimulus and QRS onset, a sine qua non for His-bundle pacing.

His-Bundle Pacing in LBBB

In 1978, El-Sherif et al reported that a narrow QRS complex could be achieved with SHBP in about two-thirds of patients with preexisting LBBB.[3] This phenomenon was ascribed to capture of the His bundle distal to the site of conduction block responsible for the LBBB seen during sinus rhythm. It implied that conduction delay usually occurred proximal to the bifurcation of the His bundle, in most cases, and was taken as further proof of the longitudinal dissociation of conduction fibers in the AV node and His-Purkinje system.[4]

Recently, this phenomenon has been described even in patients with LBBB associated with cardiomyopathy and severe LV dysfunction.[5] This may seem counterintuitive, since the site of conduction delay or block should presumably occur within the severely diseased and dilated LV in such patients. However, direct measurement of intraventricular conduction times in our laboratory has confirmed that direct His-bundle pacing, with a narrowing of the surface QRS complex and associated decrease in intraventricular conduction times, can be accomplished in about half of all patients with cardiomyopathy, LBBB, and severe heart failure. Further research is indicated to further clarify this apparent contradiction.

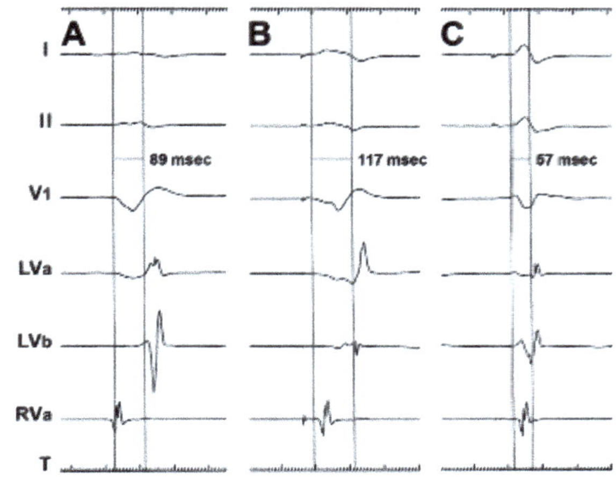

Figure 65.4. Intraventricular conduction during sinus rhythm (A), NSHBP (B), and His bundle pacing (C) in a patient with direct recordings from 2 epicardial LV sites and the RV apex. LVa = LV posterolateral wall, closer to apex than base; LVb = LV lateral wall, closer to base than apex; RVa = RV apex. SHBP produces a narrower QRS complex and shorter conduction times than sinus rhythm; NSHBP does not shorten conduction time in this case.

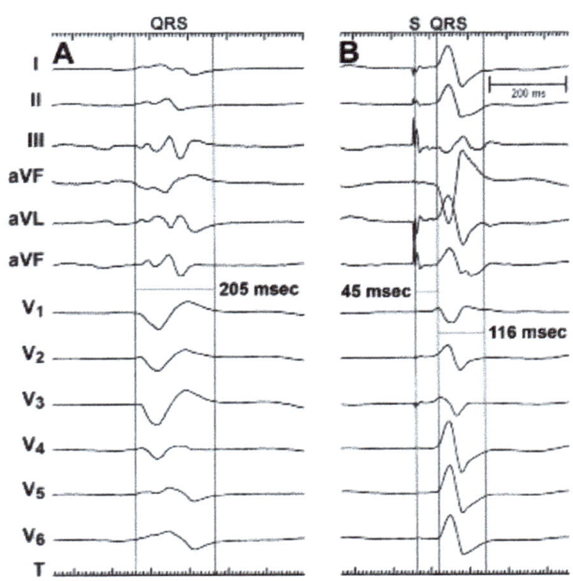

Figure 65.3. A: ECG during sinus rhythm demonstrating LBBB and a QRS duration of 205 ms. B: SHBP in the same patient, with narrowing of QRS to 116 ms. Note stimulus-QRS delay of 45 ms; in cases with LBBB, the resting HV interval is prolonged, and S-QRS interval is generally significantly longer than in cases without LBBB and slow His-Purkinje conduction. The above patient had an HV interval of 95 ms.

Figure 65.3 illustrates narrowing of the QRS complex with SHBP. Figure 65.3A represents sinus rhythm, and Figure 65.3B represents SHBP in the same patient. Note the S-QRS time of 45 ms seen with SHBP.

Figure 65.4 illustrates intraventricular conduction time during sinus rhythm, NSHBP, and SHBP. Direct recordings were obtained from 2 widely separated LV epicardial sites (LVa and LVb) and the right ventricular apex (RVA). LVa was closer to the apex, LVb was closer to the base of the LV. Conduction times are shorter during SHBP, but not NSHBP, than during sinus rhythm.

The clinical role of SHBP and NSHBP remains to be clarified; studies are ongoing in this regard.

Correct Answer:

Q1. C

REFERENCES

1. Deshmukh P, Casavant DA, Romanyshyn M, Anderson K. Permanent, direct His-bundle pacing: a novel approach to cardiac pacing in patients with normal His-Purkinje activation. *Circulation.* 2000;101:869–877.

2. Zanon F, Occhetta E, Cantù F, et al. South European Select Secure® Registry: acute and long term comparison of electrical performance in direct His bundle pacing vs. para-Hisian/inflow tract pacing. *Heart Rhythm.* 2007;4:S206.
3. El-Sherif N, Amay-Y-Leon F, Schonfield C, et al. Normalization of bundle branch block patterns by distal His bundle pacing: Clinical and experimental evidence of longitudinal dissociation in the pathologic his bundle. *Circulation.* 1978;57:473–483.
4. James TN, Sherf L. Fine structure of the His bundle. *Circulation.* 1971;44:9–28.
5. Lustgarten DL, Calame S, Crespo EM, et al. Electrical resynchronization induced by direct His-bundle pacing. *Heart Rhythm.* 2010;7:15–21.

CASE STUDY

66

Cardiac Resynchronization Therapy in Congenital Heart Disease

Imran Niazi, MD

A 51-year-old woman with asymptomatic congenital heart disease (left ventricular ejection fraction [LVEF] 64%) underwent chemotherapy with doxorubicin (Adriamycin) for breast cancer. Four weeks after completing chemotherapy, she presented with atrial fibrillation (AF), a rapid ventricular response, and heart failure. She was treated with amiodarone, digoxin, beta-blockers, and warfarin. One week later, she developed complete heart block with a junctional escape rhythm at 45 bpm, and worsening heart failure. Systemic LVEF was now 25%. Cardiac computed tomography (CT) was obtained (Figure 66.1).

Question No. 1:

The best course of action is:

A. Insert temporary pacemaker and stop amiodarone; await recovery of atrioventricular (AV) conduction.

B. Implant dual-chamber implantable cardioverter-defibrillator (ICD).

C. Implant transvenous cardiac resynchronization therapy-defibrillator (CRT-D).

D. Implant CRT-D with epicardial lead via thoracotomy.

DISCUSSION

Congenitally corrected transposition of the great vessels (CCTGV) accounts for less than 1% of congenital heart disease. It is usually associated with a ventricular septal defect or pulmonic stenosis, and survival depends upon the severity of the associated abnormalities. CCTGV *without* associated abnormalities is a very rare condition[1] and patients can survive up to middle or even old age.[2] Death usually occurs due to heart failure or complete heart block; the latter condition occurs in 10% of patients even without Adriamycin-induced cardiomyopathy.[3] Thus, implantation of a temporary pacemaker and discontinuing amiodarone is unlikely to resolve the heart block, and recurrence of AF is likely in the absence of antiarrhythmic therapy.

Since the patient is at risk of cardiac arrest (LVEF < 35%, heart failure) and has a reasonable expectation of surviving at least 1 year, a pacing ICD is indicated. The optimal pacing modality in CCTGV is not certain, although worsening heart failure has been documented with RV pacing in this population. The Dual-Chamber and VVI Implantable Defibrillator (DAVID) trial also documented worse outcomes in ICD patients with RV pacing and LVEF < 40%. In light of these findings, implantation of a dual-chamber ICD is not the best strategy. Biventricular pacing reverses the electromechanical

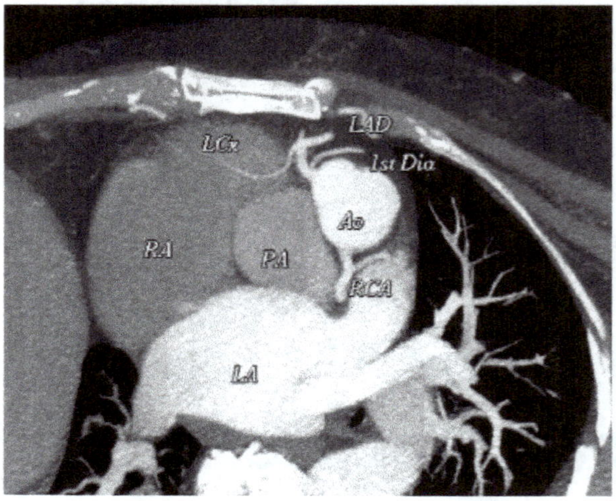

Panel A

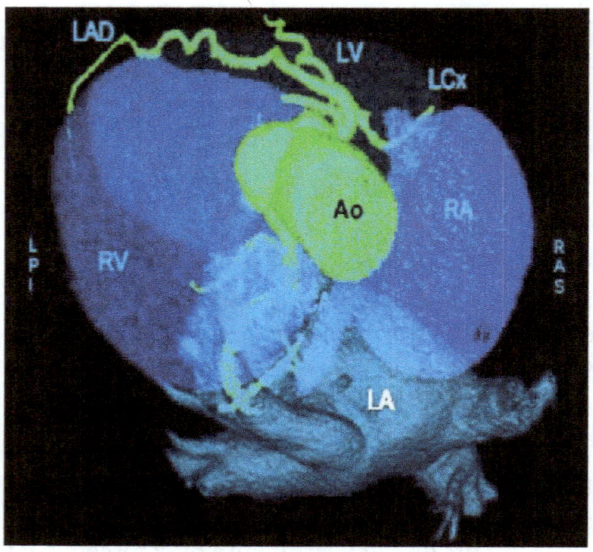

Panel B

Figure 66.1. Cardiac CT images showing location and course of the coronary sinus (CS), and coronary arteries (CA). **A:** The right (RV) and left ventricles (LV) were transposed as were the aorta (Ao) and pulmonary arteries (PA). The systemic LV was dilated (LVEF 30%). **B:** The CS was small and followed the course of the right coronary artery (RCA). LA, left atrium; LAD, left anterior descending artery; LCx, left circumflex artery; RA, right atrium.

dyssynchony caused by RV pacing and is the preferred pacing mode in this patient.[5]

The cardiac CT confirmed the diagnosis of CCTGV and showed the location and course of the CS and CAs (Figure 66.1). The RV and LVs were transposed as were the aorta and PAs. The aorta arose from the systemic LV and lay anteriorly and to the right; the PA arose from the systemic RV and lay posteriorly and to the left. The systemic LV was dilated with an ejection fraction of 30% (Figure 66.1A). The left main CA arose from the left coronary cusp. The LAD was large, coursing down the anterior AV groove; the circumflex was diminutive. The right coronary was very large, arose from the noncoronary cusp posterior, and supplied most of the systemic LV. The CS was small and followed the course of the RCA (Figure 66.1B).

Having determined the presence and location of the CS, it was appropriate to attempt transvenous implant of an epicardial lead before resorting to thoracotomy, which may have been poorly tolerated in this very sick patient.

At implant, CS cannulation was performed initially with a 0.035-inch glide wire directed through an Amplatz 3 coronary guide catheter (Merit Medical Systems Inc., South Jordan, UT). An 8-F guiding catheter could not be easily advanced distally over the guide wire, as the CS was a very small 3-mm vessel. Therefore, a 0.014-inch guide wire was advanced far into a CS branch and a 4-F glide catheter advanced over it. Selective CS branch angiography was performed, which visualized 2 small lateral branches (Figure 66.2A). A 0.014-inch Platinum Plus™ guide wire (Boston Scientific Corp., Natick, MA) was introduced into the smaller distal lateral branch and retained. Using this guide wire for support (to prevent decannulation), another 0.014-inch guide wire was introduced into the larger proximal lateral CS branch (Figure 66.2B). A 4F Acuity™ lead (Boston Scientific) was successfully placed over this with a threshold of 2.5 mA at 0.5 ms pulse width (Figure 66.2C). Right atrial and systemic RV positive fixation leads were also placed and connected to a CRT-D device (Medtronic Inc., St. Paul, MN).

When the guide catheter cannot be advanced far into the CS, it does not provide significant backup support; advancement of the LV lead can cause the guide catheter to prolapse into the RA. In such cases, a relatively stiff guide wire can be introduced distally into a CS branch and retained, thus offering support as the LV lead is advanced over another guide wire into the target branch. An 8-F guide catheter will accommodate a 0.014-inch guide wire alongside a 5-F LV lead. The Acuity lead was chosen as it had a diameter of < 5 F and a tapered tip, allowing it to be advanced far into the target branch.

Careful assessment of cardiac anatomy by CT scan and thoughtful choice of implant tools and techniques were key to this successful implant. At 6-month follow-up, the patient felt well with no symptoms of heart failure or AF. Systemic LVEF was maintained at 30%.

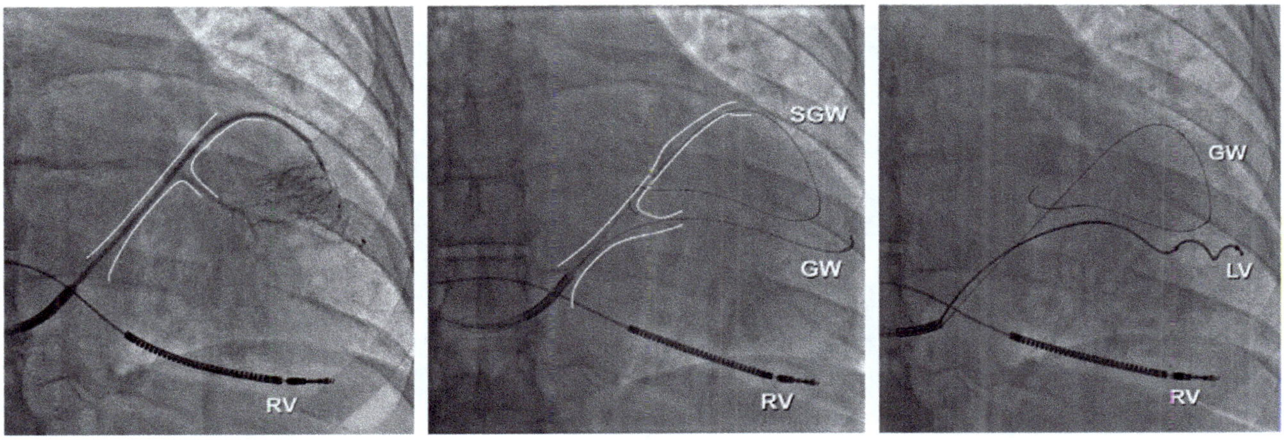

Figure 66.2. **A:** Selective CS branch angiography visualized 2 small lateral branches. **B:** Using a 0.014-inch guide wire (GW) for support (to prevent decannulation), another 0.014-inch guide wire (SGW) was introduced into the larger proximal lateral CS branch. **C:** A 4-F lead was successfully placed over this latter guide wire (GW) with a threshold of 2.5 mA at 0.5 ms pulse width. LV, left ventricle; RV, right ventricle.

Correct Answer:

Q1. C

REFERENCES

1. Connelly MS, Liu PP, Williams WG, et al. Congenitally corrected transposition of the great arteries in the adult: functional status and complications. *J Am Coll Cardiol.* 1996;27:1238–1243.
2. Kouchoukos NT, Blackstone EH, Doty DB, Hanley FL, Karp RB. Congenitally corrected transposition of the great arteries and other forms of atrioventricular discordant connection. In: *Kirklin/Barratt-Boyes Cardiac Surgery.* London, UK: Churchill Livingston; 2003:1549–1584.
3. Allwork SP, Bentall HH, Becker AE, et al. Congenitally corrected transposition of the great arteries: morphologic study of 32 cases. *Am J Cardiol.* 1976;38:910–923.
4. Ikeda U, Furuse M, Suzuki O, et al. Long-term survival in aged patients with corrected transposition of the great arteries. *Chest.* 1992;101:1382–1385.
5. Huhta JC, Maloney JD, Ritter DG, Ilstrup DM, Feldt RH. Complete atrioventricular block in patients with atrioventricular discordance. *Circulation.* 1983;67:1374–1377.
6. Karpawich PP. New directions in pacemaker therapy among children and adult patients with congenital heart disease. *Expert Rev Cardiovasc Ther.* 2007;5:611–613.
7. Van de Veire NR, Blom NA, Holman ER, Schalij MJ, Bax JJ. Triplane tissue Doppler imaging to evaluate mechanical dyssynchrony before and after cardiac resynchronization in a patient with congenitally corrected transposition of the great arteries. *J Cardiovasc Electrophysiol.* 2007;18:222–225.
8. Wilkoff BL, Cook JR, Epstein AE, et al; Dual Chamber and VVI Implantable Defibrillator Trial Investigators. Dual-chamber pacing or ventricular backup in patients with an implantable defibrillator: the Dual Chamber and VVI Implantable Defibrillator (DAVID) trial. *JAMA.* 2002;288:3115–3123.
9. Rodriguez-Cruz E, Karpawich PP, Lieberman RA, Tantengco VM. Biventricular pacing as alternative therapy for dilated cardiomyopathy associated with congenital heart disease. *Pacing Clin Electrophysiol.* 2001;24:235–237.

SECTION 8

Imaging—Fluoroscopy, Echocardiography, 3D Imaging, Remote Navigation, Image Integration

CASE STUDY 67

X-Ray Imaging System in the Electrophysiology Laboratory

Régis Vaillant, PhD

x-ray imaging equipment is a central piece in the electrophysiology (EP) laboratory. For many years, it has been the only way to get a view of the internal patient anatomy and the employed interventional tools. In the 1980s and 1990s, scientists developed new imaging equipment that is able to provide additional points of view of the theater of operation. This case explains how the x-ray system remains unique in its capability to deliver images of rich content for guidance during an electrophysiologic study.

Question No. 1:

x-ray images show the complete anatomy of the patient in the exposed field of view. How is the information combined?

Question No. 2:

x-ray images are the result of a conical projection. Are there any practical consequences that one should be aware of?

Question No. 3:

Is there any restriction to use x-ray images during clinical procedures?

Question No. 4:

x-ray images are conical projections of the patient anatomy. Are they limited and not useful to support accurate measurement or registration with the images provided by other modalities?

DESCRIPTION OF X-RAY IMAGING EQUIPMENT

The 2 main mechanical elements of x-ray imaging equipment are the table and the gantry.

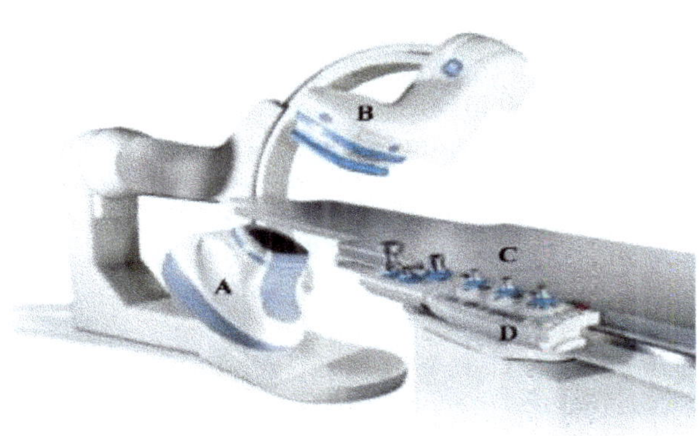

Figure 67.1. The main components of x-ray imaging equipment.

1. The table (Figure 67.1, Item A) is used to support the patient and key user interfaces (see Figure 67.1, Item D). These interfaces, which are used to operate the system, are hung on rails along the tableside. During the intervention, the tabletop (see Figure 67.1, Item C) is protected by sterile field, and the tools, such as the various catheters employed during the procedure, often lie on it.
2. The gantry supports the x-ray tube on one side and the x-ray detector on the other side (see Figure 67.1, Item B). It shall maintain these 2 elements in a perfect alignment to ensure image quality and safety. The x-ray tube is a complex piece of equipment that includes a cathode and an anode. These 2 elements are in a casing under reduced atmospheric pressure. A heated filament placed in front of the cathode produces electrons, which are accelerated by the strong electrical field created between the cathode and the anode. The electrons hit the anode and their interaction with the anode material produces x-ray photons. The geometry of the anode is defined such that the spatial extent interaction zone, as seen from the detector, is reduced. So the x-ray beam can be considered as a point source. Practically, the size of the x-ray emitting source is about 1 mm^2. The produced x-ray forms a fan beam. In the x-ray head, additional elements such as collimator blades are employed to control the opening of the beam to make sure that the irradiation is limited to the useful area of the anatomy.

UNIQUE PROPERTIES OF X-RAY IMAGES

Regarding the advantages of x-ray imaging, the main properties are as follows:

1. Capability to perform full-field imaging. The obtained image depicts the anatomy of a patient over a large volume, which is limited only by the size of the field of view employed. Typically the field of view employed during EP procedures is 20 cm × 20 cm at the detector level, which represents about 15 cm × 15 cm at the patient level. This size suffices to include the entire heart of a patient. Moreover, the image depicts completely the section of the patient, which is in the field of view. With respect to these specificities, the x-ray image is ideal to monitor the evolutions of a patient and potentially figure out any event that may happen, like a dislodged catheter for example.
2. Secondly, x-ray images can be produced in real time (<150 ms latency) at a high frame rate; 30 frames/second can be produced, which is higher than practically required to conduct an EP study. Each individual frame is obtained with a quite short pulse width, usually less than 10 ms. Thus, each image represents the state of the anatomy at a given instant. Other imaging modalities, in general, proceed by combining information taken at different instants.
3. The use of radio-opaque elements such as catheter and iodine considerably enlarges the potential of this imaging modality. Indeed, these "contrast agents" are used at different places and they highlight specific anatomical structures: vessel, chambers, etc.
4. The last advantage is more economical. Indeed, the most basic x-ray equipment is low-cost. Moreover, the cost of ownership is small since this type of equipment does not require complicated maintenance or supplies.

Principles of x-ray Imaging Physics

Some of the unique properties of x-ray imaging that we have listed above are a consequence of the physical properties of propagation of the x-ray photon through matter:

1. The x-ray photon propagates in a straight line. Consequently, the image formed on the detector at a precise location represents the matter encountered all along the ray that goes from the point of x-ray emission to this location.
2. Therefore, the intensity of the pixel is defined by the type and amount of materials (tissue, contrast agent, interventional tool) encountered along this path. Indeed, the attenuation of an x-ray beam is variable as a function of the atomic number and the density of the traversed tissue. This total attenuation, which varies from place to place, is, in the end, responsible for the contrast of the image. Figure 67.2 shows a typical x-ray image obtained in an EP lab.
3. A first side effect shall be considered: the attenuation of the x-ray beam by the matter produces secondary x-ray photons, which are produced at the precise place where the ray is attenuated. These secondary photons are emitted in all directions. These radiations are called scatter. As the source of this emission and the direction of this emission cannot be controlled, they can reach any point in the lab except if adequate protection measures are taken, such as lead aprons or radiation shields. This equipment contains elements with high atomic numbers that attenuate the x-ray beam and protect the catheterization lab team.

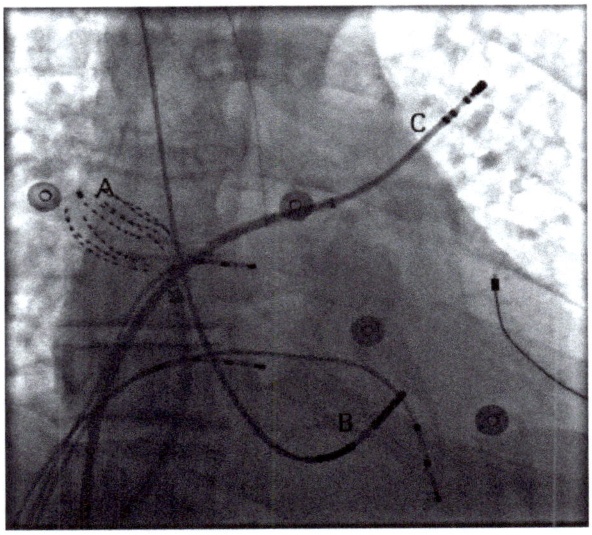

Figure 67.2. Typical x-ray image in an EP lab. Labels A, B, and C point out typical catheters used during the procedure. These catheters include radio-absorbing material and offer a good contrast versus the biological tissue. One can also see the contrast between the heart and the lung. The attenuation of the x-ray beam in the lung is inferior to the heart area because the lungs are filled with air, the density of which is inferior to regular anatomical tissue.

4. A second side effect is the irradiation delivered to the patient. This irradiation is necessary to ensure that a contrasted image is formed. Above some thresholds, these ionizing radiations may induce dermatological disease such as erythema and increase the stochastic risk of cancer for the population.

Geometrical Characteristics of x-ray Imaging

The x-ray image has a pretty good resolution. Current commercial standard is about 0.2 mm × 0.2 mm pixel size. This size is inferior to the diameter of the most widely used tool in the cath lab, the ablation catheter, the diameter of which is about 2 mm. Taking into account the magnification factor associated to the x-ray imaging geometry, its apparent diameter on the image is around 15 pixels. In principle, the resolution of the x-ray image could be increased. However, the formation of the image with an acceptable level of noise requires a minimum level of power in the x-ray beam at the level of the detector for each individual pixel. Consequently, decreasing the size of the pixels would require an increase in the power of the emitting tube. The irradiation received by the patient will also increase.

Another important geometrical characteristic of x-ray images is that the rays are issued from a single point, as explained above, and then form a diverging beam. The mathematical transform, which describes this geometry, is a conical projection. It is also called perspective projection. Conical projections are widely studied. They are characterized by the existence of a center of projection, which is defined as the point of intersection of the rays. Conical projections describe not only the geometry of the x-ray imaging system but other widely common imaging situations such as the human eyes or camera. In this case, the human eye or the focal point of the optical lens is the center of projection. So in an x-ray image, like in usual visual perception, the appearance of an object depends on its distance from the center of projection. Its apparent size decreases as the object is more distant from the center of projection. In the EP lab and considering an image done in frontal position, that is, with the line tube/detector being vertical and the tube placed below the patient, it means that anatomical structures on the back of a patient seem larger than the structures on the anterior side of the patient.

To illustrate this point, we have represented a picture like this one in Figure 67.3. We placed electrodes on the posterior side (labeled P) and anterior side (labeled A) of the patient. They are placed on his skin. The difference in the apparent size is about 25%. The relative variation inside the

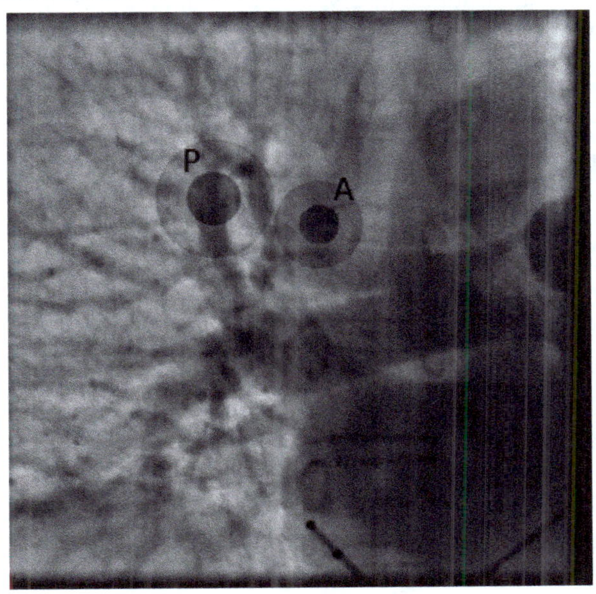

Figure 67.3. The difference in apparent size between identical electrodes placed either on the posterior side of the patient (P) or the anterior side (A) is shown. The x-ray image is done with a gantry occupying the anteroposterior position, that is, the tube is placed below the patient.

cardiac chamber is smaller since the thickness of the heart is much smaller.

This property has very practical consideration for the task of overlaying on the x-ray image some anatomical details that come from other imaging modalities such as magnetic resonance or computed tomography (CT). Usually, these images are rendered with highly optimized algorithms, which employ parallel projection to speed up the computation. So for the purpose of image fusion, dedicated algorithms implementing the computation of conical projection are required.

Three-Dimensional (3D) Imaging in the EP Lab

Regarding x-ray images, we have seen they are able to depict the entire field of view in a single shot. On the other side, they are limited because they do not give any hint on the relative position along the axis of projection of 2 objects. Looking to an x-ray image, it is then very hard to tell if an object is closer to the detector than another. The relative apparent size cannot be used unless one considers 2 identical objects and a significant difference of depth on the order of 10 cm. So the best way to overcome this limitation is to use 2 x-ray images taken at different viewing angles, which can be acquired simultaneously (or almost), as with biplane x-ray imaging equipment, or one after the other, as with a monoplane system. Looking at a specific point of interest, either an interventional tool or a precise anatomical landmark, it is possible to completely determine their spatial position. Mathematically, determining the point, which is at the intersection of the 2 following lines, stems from the following computation:

- Line formed by the source of the x-ray beam of the first image and the image of the point of interest in the first image;
- Line formed by the source of the x-ray beam of the second image and the image of the point of interest in the second image.

This geometrical construction is represented in Figure 67.4.

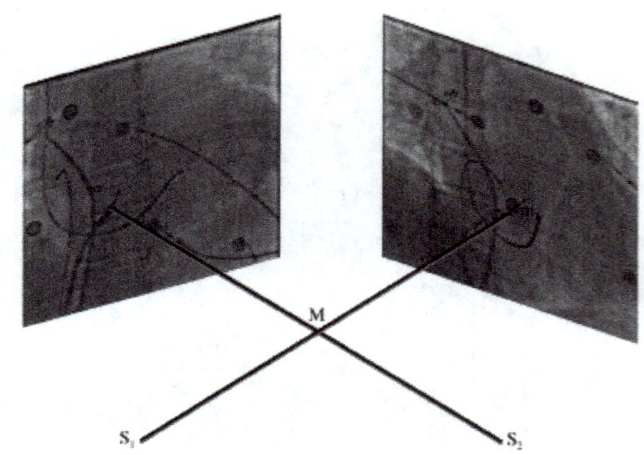

Figure 67.4. The point S_1 and S_2 represent the successive location of the focal source in 2 different angulations. Knowing this geometry, the spatial position of the point M is identified by taking the intersection of the 2 lines S_1–m_1 and S_2–m_2.

Another way to get 3D images in the cath lab is to perform a rotational acquisition. In this case, we collect a series of x-ray images and, during the time of acquisition, rotate the gantry around the patient. Each obtained individual image is a projection of the anatomy with a well-defined center of projection, the position occupied by the x-ray tube at the instant of acquisition. All these images can then be combined together using the principles of tomography as discovered in 1917 by the mathematician Radon, who studied the so-called Radon transform. The Radon transform is the integrals of an object along the different directions of a plane. This mathematical transformation corresponds to the data collected during a rotational acquisition. The formed images are the integrals of the attenuation properties of the traversed tissue. Scientists have also demonstrated that it is theoretically and also practically possible to inverse the Radon transform. This inversion gives the spatial distribution of the attenuation in a plane. This process is known as tomography and is widely used in CT imaging. In principle, it is also applicable to the x-ray image formed in an EP lab.

As seen in Figure 67.5, 3 images have been extracted from a set of 147 projections collected during a rotation of a gantry around a patient's chest right after the injection of iodine in the circulatory system and at the time of arrival of this contrast agent in the left atrium.

In Figure 67.6A, the determined distribution of the attenuation properties of the tissue is displayed. The distribution is shown in an axial slice from the middle of the left atrium.

In Figure 67.6B, the shape of the left atrium is displayed as a full object after segmentation of the other anatomical elements. The rendering of the object is done using a technique known as volume rendering.

Specificities Versus Other Cath Lab Imaging Equipment

After this review of x-ray imaging equipment, it may be interesting to summarize the capabilities of other

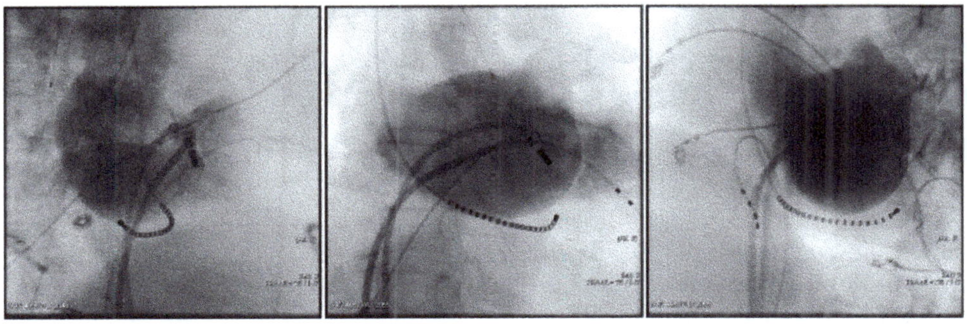

Figure 67.5. These images have been captured during a rotational acquisition after direct injection of iodine in the left atrium. The total series includes 147 images and can be used to perform 3D reconstruction of the chamber.

imaging equipments, which are widely used during EP intervention. We do not depict recording systems that are specifically designed to measure, record, and display electrical activity of cardiac tissue at the different locations reached by catheters.

Echocardiography systems can be employed during procedures for different purposes. They can be used either as a transesophageal probe (TEE) or an intracardiac echocardiography probe (ICE). TEE probes are relatively bulky and are introduced inside the esophagus of the patient. ICE probes are much smaller; their size is comparable to a catheter. They are introduced through the venous system like other interventional tools. These 2 types of probes use the physical principle of ultrasound imaging to form images of the anatomy. An ultrasound wave is emitted. Each time the waves encounter an interface between 2 tissues of different acoustic properties, a proportion of this wave is returned to its emitting source. The measure of the time difference between the emission and the reception of these echoes indicates the distance between the source of emission and the event. The returned energy is used to mark the spot in the formed image and, thus, responsible for the contrast in this image. The echo image, by construction, delivers the depth of the imaged tissue. On the other side, it is not equivalent to a freeze frame of the anatomy, since a sweep over the different imaging directions is performed in order to collect either a full plane or even a full volume.

Electromagnetic navigation equipment, by construction, is a true 3D imaging system. Indeed, it provides the location of a sensor, which can be placed, for example, at the tip of a catheter. It can also provide the spatial orientation of this sensor.

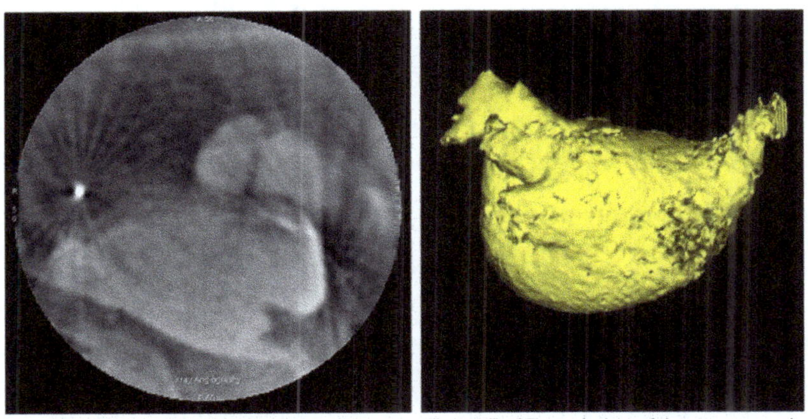

Panel A: One horizontal slice of the dataset reconstructed from the rotation acquisition. Both the shape of the left atrium and one of the employed catheters are visible.

Panel B: 3D rendering of the segmented left atrium.

Figure 67.6. Two representations of the 3D datasets obtained by applying a reconstruction algorithm to the images, which were obtained by rotational angiography.

This information is collected several times per second and can then be averaged to limit the impact of electronic noise or other sources of disturbance. The general principle of 3D location is to measure the intensity of an electromagnetic field with the sensor formed by solenoids and to produce a field in which the intensity and other properties vary in the volume occupied by the patient anatomy. In the end, one gets the distance between the source of the field and the catheter. By performing this measurement with 3 different emitting sources, the location of the sensor at the tip of the catheter is retrieved by taking the intersection of 3 spheres centered on each of the individual sources, the radius of which is determined by the 3 measurements. Different manufacturers implement this general principle with a number of variations. In comparison to x-ray imaging, this equipment delivers the position of the probe in the 3D space. This information is obtained without requiring the use of ionizing radiation. Its main limitation is the very narrow field of view. Only the location of the tip of the catheter is imaged. Additional technical elements have been developed to measure the position of additional points. In the end, only discrete sets of points, less than 100, are determined, and the only collected information is the presence of the probe at this place. Hence this tool is well suited to navigate the probes from place-to-place and to complement the x-ray imaging equipment.

CONCLUSION

This case describes the main properties of x-ray imaging equipment. A critical piece of the EP lab, it is able to deliver good-resolution images of the entire theater of the intervention. The contrast in the images is the result of the interaction between the x-ray beam and the traversed tissue. The contrast reflects the difference of attenuation properties of the traversed tissue. It is then possible to modify temporally the contrast at different locations and to visualize either anatomical structures, such as the cardiac chambers, or interventional tools. The x-ray image can be depicted as the result of a conical projection, the center of which is the source of the emitting beam. It has direct implication on the apparent size of the objects in the image. It also has direct implication for performing multimodal image integration. The precise geometry of the projection shall be taken into account to avoid mismatches.

By using multiple projections of the anatomy and by treating the data with special algorithms, the 3D shape of the anatomy is determined. This is very valuable information for complex procedures involving the left atrium, as the number, size, and shape of the pulmonary veins are variable from patient to patient.

Correct Answers:

Q1. The essence of the content of an x-ray image is very unique in comparison to other medical imaging modalities or the visual domain. Images are formed of pixels and usually the information carried by a single pixel is associated to a single spatial position. For example, the intensity of a pixel in a picture of a landscape corresponds to the point in space where the sunlight was reflected. In x-ray images, the intensity is the result of the interaction of the x-ray beam with the traversed tissue along the path. When the x-ray beam intersects an interventional tool in which the spatial extent is well defined and attenuation properties are high, the location is well defined. In other places, there may not be strong and localized interaction. As a consequence, the image has a fuzzy appearance and may be difficult to interpret. Specific training is required to work with those images.

Q2. Yes. It is important to always keep in mind this characteristic of x-ray images. This situation is very similar to the one encountered in the visual domain. Two objects, which appear close to each other in the image, may be far apart in the patient. It is just because they project to neighboring locations in the projected image that they look close. Spatial thinking shall always be used during the interpretation of x-ray images. Fortunately, this is something that the brain is used to doing with the human visual system.

Q3. x-ray images are obtained by using ionizing radiation. Stochastic and deterministic risks are associated with these radiations. So the physician should take care to keep the right balance between the risks for the patient and the benefit that he/she will get from the treatment. The clinical team involved in the procedure should also use all available protection means in order to minimize the potential impact on their own health.

Q4. Actually, over the years, scientists and clinicians have developed many tricks to overcome this fundamental limitation of x-ray images. Using images of known size, it is possible to determine the 3D position of anatomical structures. Actually, this is exactly what the brain of a driver does to figure out the distance between his car and the preceding vehicle.

In another approach, the use of multiple projections, from two to several hundred, makes it possible to determine the spatial location of different structures. In a number of cases, it is an important additional piece of information during the procedure.

During multimodal image integration, the geometry of the x-ray image formation shall be modeled accurately. Once it is done, it has many interesting advantages. For example, one can choose to display the integrated 3D model either from the x-ray tube side or the detector side. Two different views of the patient anatomy are available and can help the physician to understand the local anatomy. The physician will select the view depending on the orientation of the manipulated catheter.

CASE STUDY

68

Case Study Involving Radiation Exposure

Barry F. Belanger, PhD

A 75-year-old patient is undergoing a complex atrial fibrillation (AF) ablation procedure using a biplane x-ray imaging system oriented for anteroposterior (AP) and lateral (LAT) views.

Question No. 1:

Cumulative dose readings reported by the x-ray system are as follows:

	AP	LAT	TOTAL
Dose entrance skin air kerma (ESAK) in Gray	0.9	1.6	2.5
Dose-area product (DAP) in cGy-cm^2	5760	12018	17778

Given these cumulative dose readings mid-procedure, what decisions should be made concerning the rest of the procedure?

A. Terminate the procedure and reschedule the completion. Dangerous radiation levels have been reached.

B. Continue as planned. Radiation risk is minimal.

C. Proceed with caution, paying attention to cumulative dose and in making the following adjustments as appropriate or possible (select all that apply):

1. Change LAT view to oblique.
2. Verify whether AP and LAT detectors are as close to patient as possible.
3. Switch to a smaller field of view.
4. Verify that dose settings and frame rates are as low as possible without jeopardizing visibility required for procedure.
5. If adjustment is available, move x-ray tube ***away*** from patient.
6. If adjustment is available, move x-ray tube ***toward*** patient.
7. Store fluoroscopy (fluoro) sequences *in lieu* of digital cine recordings.

DISCUSSION

Types and Severities of Radiation Risks

Radiation risks are usually categorized as *Stochastic* versus *Deterministic.* Stochastic risk is an assessment of the increase in likelihood of the occurrence of cancers resulting from exposure to radiation. Lickfett et al[1] have reported that the additional lifetime risk for a fatal malignancy associated with the ablation of AF was 0.15% for female patients and 0.21% for male patients in their study. The average age of the patients in the study was 56 ± 11 (39–78). Special considerations may be made for individual patients based on their status and medical condition. For example, should a needed interventional procedure using x-ray be performed on a patient who is pregnant or can it be safely deferred until after delivery? Pediatric interventions are an important case of stochastic risk, since the likelihood of cancer expression increases with lifetime

Practical Electrophysiology © 2015 Jasbir S. Sra, Masood Akhtar, Andrea Natale, David J. Wilber
Cardiotext Publishing, ISBN: 978-1-935395-17-1

following exposure, and sensitivity to radiation increases with the rate of cell division. And stochastic risk is an important consideration for the physicians and staff working within the procedure room.

The main concern for patients in interventional procedures is usually *Deterministic* risk, which is the tissue damage that ***will*** occur if radiation exposure exceeds certain thresholds. The tissue at greatest risk is usually the skin at the entrance location of the incident x-ray beam, where the intensity is greatest.

Table 68.1 provides a list of the known injuries that can be caused by the type of radiation used in interventional procedures (x-rays in the normal diagnostic imaging energy range), and the thresholds at which they can be expected to occur. In the US, the Joint Commission (TJC), formerly the Joint Commission on Accreditation of Healthcare Organizations (JCAHO), has set a level of 15 Gray exposure to a single area of skin as the threshold of a sentinel event. A sentinel event is a serious medical error, such as operating on the wrong organ. To retain TJC accreditation, a medical facility must respond to the occurrence of a sentinel event with a thorough investigation and implementation of an effective remediation plan, including ongoing monitoring.

National regulatory agencies have also set upper limits on patient entrance dose ***rates*** in fluoroscopy, for example, 88 mGy/min in the US. However, with complex procedures often lasting hours, with significant dependence upon and use of fluoroscopy, skin injuries can still occur due to long fluoroscopy times and corresponding high cumulative radiation exposures.

Although the occurrence rate of *reported* deterministic radiation injuries is low, it is not zero,[4,5] and a significant fraction of these have resulted from cardiac radiofrequency (RF) catheter ablation procedures.[4] A very pertinent, well-conducted study on a small cohort of patients undergoing RF ablation procedures demonstrated that, even with careful attention to radiation exposure, including use of low-dose fluoroscopy techniques, the maximum measured skin exposures for AF ablation procedures were close to the threshold at which deterministic skin injuries occur.[1] The maximum skin doses accurately measured on a cohort of 15 patients was 1.48 ± 0.37 (0.8–2.05) Gray versus a threshold of 2.0 Gray for deterministic skin injuries. The weights and body mass indices (BMIs) of the patients were 82 ± 12 (64–108) kg and 26 ± 2 (23–32), respectively, which does not indicate a bias toward obesity. So it is not difficult to imagine how factors including case difficulty and patient obesity could conspire to drive cumulative skin doses into the range of deterministic injury during AF ablation procedures.

The fact that deterministic radiation injuries are a real risk in interventional procedures, combined with the increasing awareness of stochastic and deterministic radiation risks, have driven minimization efforts that are summarized in the As Low As Reasonably Achievable (ALARA) principle for use of radiation in medical procedures. A key premise, well expressed by Miller et al,[6] is that the physician is in charge of balancing all the risks to the patient in an interventional procedure, including radiation and the risk of not performing the procedure or aborting it. In this context, the use of radiation is analogous to application of a drug with known dose-dependent side effects.

In order to effectively balance the risks, the physician must be aware of radiation risks, how to control the radiation levels of the equipment, and have feedback from the x-ray system about dose rates and cumulative case doses while performing a procedure. The latter are particularly important, since interventional procedures are open-ended from a radiation standpoint, unlike a diagnostic imaging procedure with a specified set of imaging views.

Table 68.1. A List of the Known Injuries that Can Be Caused by x-Rays Used in Interventional Procedures, and the Thresholds at which They Can Be Expected to Occur

Entrance Skin Dose (ESD)	Injury	Time to appear
2 Gy (200 rad.)	Early transient erythema	224 hours
3 Gy (300 rad.)	Temporary epilation	~3 weeks
6 Gy (600 rad.)	Main erythema	~1.5 weeks
15–20 Gy (1500–2000 rad.)	Moist desquamation, dermal necrosis, and secondary ulceration	>52 weeks for necrosis

Sources: US FDA Public Health Advisory: Avoidance of Serious x-ray-induced Skin Injuries to Patients During Fluoroscopically-guided Procedures. May 2, 1996.[2] Interventional Fluoroscopy: Reducing Radiation Risks for Patients and Staff. Newsletter from NCI and SIR: 2005.[3]

The following is a list of recommendations to clinicians for minimizing the radiation exposure to patient and staff:

1. Maximize the distance between the patient and the x-ray tube.
2. Minimize the distance between the patient and the image detector/receptor.
3. Clinician and staff should stand as far away from the x-ray beam, particularly the x-ray tube side, as practical when radiation is on.
4. Remove the antiscatter grid when imaging infants.
5. Use the lowest-dose selection or operating mode that provides adequate image quality for the clinical task (ALARA principle). As per the above discussion, this includes automatic exposure program selections, including Low/Norm/High dose subselections.
6. Use the lowest fluoro and record image rates consistent with clinical needs. Many electrophysiologists find 7.5 frames/second to be acceptable.
7. Collimate to the area of interest.
8. After collimating, use the largest detector field of view consistent with clinical needs. (This is because automatic exposure control systems typically increase dose as field of view is decreased, under the assumption that smaller field of view selection by the operator signals a need to see smaller structures, which in turn would require higher dose, all other things being equal.)
9. Minimize time of radiation exposure in both fluoro and record modes.
10. Vary the x-ray viewing angle to the patient when possible to distribute the radiation exposure over the patient's skin.
11. When a choice is available in the direction of the beam passing through the patient, avoid directing the entrance beam at the patient's eyes and other sensitive organs, for example, female breast, ovaries.
12. Use stored fluoro sequences *in lieu* of higher-dose record sequences when practical.

Equipment manufacturers have responsibilities in the ALARA context as well. These include optimizing the dose efficiency of their imaging systems in all operating modes, providing a range of settings in imaging modes so dose can be adjusted to meet but not exceed clinical imaging needs, providing patient case dose data for the physician during procedures, and meeting regulatory requirements. These should be taken into consideration in equipment purchasing and upgrading decisions.

Radiation Measurements and Monitoring

In order to assess and manage radiation risks in the midst of complex interventional procedures, the clinician must have access to and understanding of the radiation values provided by the x-ray system.

Stochastic risk is usually assessed using measured or calculated DAP from the x-ray system. Figure 68.1

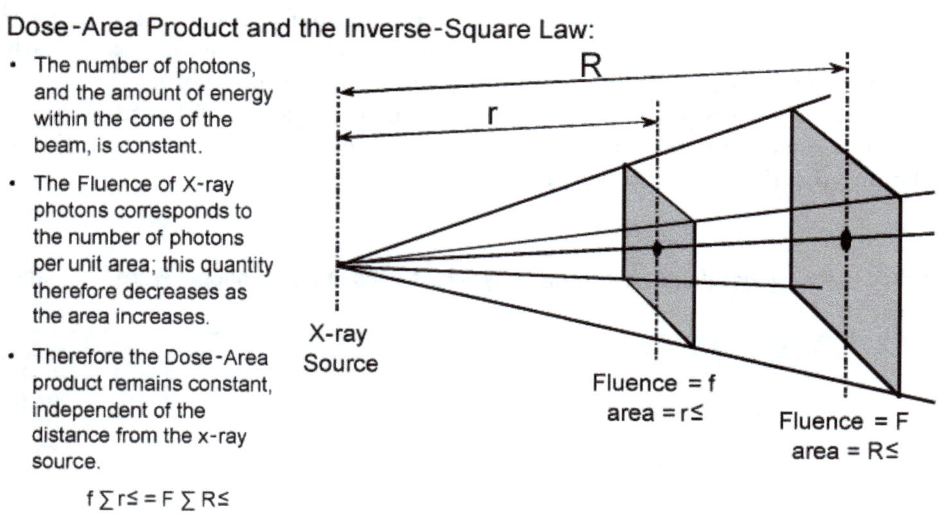

Figure 68.1. An illustration of how the measurement of DAP relies upon the inverse-square law to arrive at a simple measurement of radiation delivered to the patient for assessment of stochastic radiation risk. DAP is independent of the distance of the patient from the x-ray source.

illustrates the concept. DAP essentially estimates the total energy deposited in the patient by all of the x-ray photons that impinge upon the patient, without regard to the intensity of the radiation in terms of photons per unit area or energy per unit area.

To fully assess stochastic risk it is necessary to know what organs received radiation and to account for their specific sensitivities to radiation. A medical physicist can use procedure DAP values, along with viewing angles and anatomy irradiated to estimate stochastic risk.

Deterministic risk is usually assessed using measured or calculated skin dose, more precisely defined as ESAK. Unlike DAP, ESAK is an estimate of the *intensity* of the radiation, in terms of energy deposited per unit area exposed, which correlates to deterministic skin injuries as reported in Table 68.1. A very accurate measurement of ESAK would require a sensor to be placed at the entrance surface of the skin, which is not usually practical. But radiation intensity can be accurately estimated by placing a radiation sensor within the collimator and projecting the intensity to the skin entrance surface using the inverse-square law. Calibration tables and algorithms that take into account actual x-ray technique and spectral filtration can effectively substitute for having an actual x-ray sensor.

However, the distance of the skin entrance surface of the patient from the x-ray source is usually not known exactly. To address this problem, an international standard for the likely source-skin distance, the *interventional reference point* (IRP), has been defined for isocentric x-ray positioners (Figure 68.2). Most modern interventional x-ray systems report estimates of skin dose (ESAK) based on this standard. Depending on the actual positioning of the patient relative to the x-ray source, it is possible for ESAK estimates at the IRP to over- or under-estimate the actual skin dose.

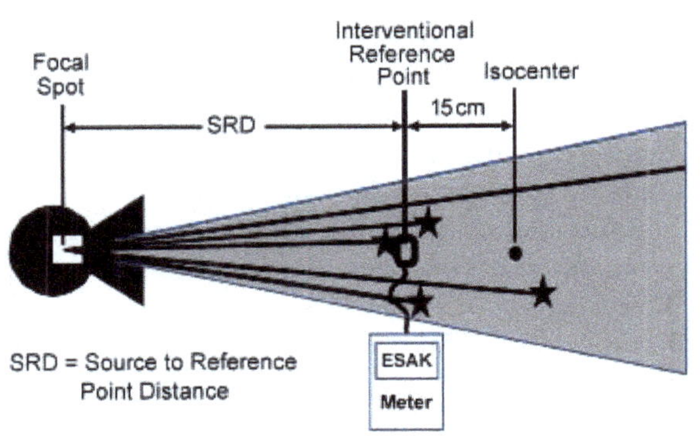

Figure 68.2. Imaging geometry for a modern isocentric positioning system, for example, C-arm, showing the location of the IRP at which skin dose (ESAK) is typically measured and reported. In this construct, the average patient anatomical "shell" is modeled as a 30-cm diameter sphere centered at system isocenter.

Furthermore, the maximum skin dose will be affected by changes in positioning of the x-ray beam around the patient, and the degree to which the various beam entrance areas overlap among views. If a single view is used in a case, the maximum skin dose will be the reported cumulative dose (by individual plane, if using a biplane system). If several views are used, the maximum skin dose will be less than the reported cumulative skin dose. Most systems provide cumulative case dose (ESAK), and in the case of biplane systems, the cumulative dose for each plane. Some systems keep track of maximum skin dose as well as cumulative dose by accounting for the total amount of radiation used as a function of viewing angles and projecting the related beam entrance areas onto a 3D patient model.

Table 68.2 summarizes the relative merits of dose (ESAK) and DAP in monitoring radiation exposure. For radiation risk monitoring during an invasive electrophysiology (EP) procedure, the clinician should usually focus on the cumulative dose (ESAK) levels—for each plane if biplane—relative to the deterministic skin injury thresholds given in Table 68.1. If cumulative dose approaches the injury threshold during a procedure, it is incumbent on the physician to consider ways to minimize radiation injury risk to the patient's skin, and to balance the radiation risk against the other risks related to the procedure and the patient's condition.

Analysis of the Hypothetical Case

The first observation should be the higher of the 2 individual plane cumulative exposures, which in this example is 1.6 Gray in the LAT plane. It should not be surprising that the exposure is higher in the LAT view obtained by the LAT plane, because the thickness of anatomy through which the x-ray beam must pass is greater than the AP view of the frontal plane. The demand for radiation is an exponential function of tissue thickness along the path of the beam. An additional 4 cm of tissue thickness typically requires a doubling of radiation to the patient to achieve similar imaging results.

It is not uncommon for a single set of views to be used with biplane systems in EP procedures, so it is likely that the 1.6 Gray cumulative exposures also represent the maximum skin dose. Comparing this level to the thresholds given in Table 68.1, it can be seen that the level where transient erythema could be induced is close, but has not yet been exceeded. Also, the risk of a transient injury may be acceptable

Table 68.2. **The Relative Merits of ESAK and DAP in Monitoring Radiation Exposure**

Quantity	Pros	Cons
Dose (ESAK)	• Estimates deterministic risk of skin injuries, particularly important for interventional procedures • Required reporting by TJC (a.k.a. JCAHO in some regions)	Subject to inaccuracies: • Source-skin distance uncertainty • Peak skin dose when using multiple gantry positions and FOVs
DAP	• Useful to estimate lifetime cancer stochastic risk • Easy to measure	• Doesn't estimate skin injury risk

DAP, dose-area product; ESAK, entrance skin air kerma; FOVs, fields of view; JCAHO, Joint Commission on Accreditation of Healthcare Organizations; TJC, the Joint Commission.

when compared with the patient's other medical risks. Stochastic risk is not a significant consideration, given the patient's age and need for the interventional procedure. So it is reasonable to continue the procedure, but it is also incumbent on the clinician to consider ways to minimize the additional radiation exposure to the area of the patient's skin, which has already likely received 1.6 Gray.

We can evaluate the following options, as posed at the beginning of this case study:

1. **Change LAT view to oblique**. This is a reasonable choice. Given that the LAT plane had been used to this point in a straight LAT projection, it might be possible to proceed using a 60-degree left anterior oblique (LAO) view *in lieu* of a 90-degree LAO, for example. This would move the entrance exposure area away from the area that has already received significant exposure. It might also be worthwhile to consider changing the frontal positioner to a 30-degree right anterior oblique, which would have the effect of reducing the maximum skin exposure delivered by the frontal plane, and keep the viewing angle between the 2 planes at 90 degrees.
2. **Verify AP and LAT detectors are as close to patient as possible.** This is basic good imaging practice and strongly recommended. Keeping the detector close to the patient reduces the amount of radiation demanded by the imaging system, without compromising the image quality.
3. **Switch to a smaller field of view**. This is **not** recommended. Changing to a smaller field of view usually causes an increase in the radiation *intensity* to the skin, even though the irradiated *area* may be smaller. (This is because most manufacturers associate a request for a smaller field of view with a need to visualize smaller objects better, which requires increased radiation intensity.) It would be preferable to simply collimate to a smaller imaged area, as possible within the constraints of the procedure, without selecting a smaller FOV.
4. **Verify dose settings and frame rates are as low as possible without jeopardizing visibility required for procedure**. This is always a good idea. It is natural to work with an image that is better than it needs to be, without testing whether a lower radiation level would produce adequate imaging. Most systems offer selection of several fluoroscopic dose levels and frame rates, often at tableside. These selections can have a substantial effect on the patient exposure rate.[5] Most electrophysiologists have found that a frame rate of 7.5 frames/second offers acceptable imaging, and substantial dose savings compared to higher rates often used in cardiology. It should also be noted that manufacturers often provide dose settings that are only accessible by service and other specialist personnel. It is worthwhile to investigate these options, perhaps with the aid of a medical physicist, to ensure that the lowest usable dose rates are readily available. This may be particularly important with multiuse systems, where other users may have higher image quality demands that will call for higher-dose imaging settings for their procedures.
5. **If adjustment is available, move x-ray tube *away* from patient**. This is always a good idea. Referring to the inverse-square law as illustrated in Figure 68.1, the dose to the skin will be reduced if the source-to-skin distance is increased while keeping the detector in the same position relative to the patient. This is because the irradiated entrance area becomes larger, spreading out the x-ray photons to produce a lower intensity (photons/unit area).
6. **If adjustment is available, move x-ray tube *toward* patient**. Not advised, in light of explanation of #5 above.

7. **Store fluoro sequences *in lieu* of digital cine recordings**. This is a very good idea if image quality is adequate medically. Digital cine record dose levels can be 5 to 20 times higher than fluoroscopic levels (this information can typically be found in manufacturers' operator manuals).

Correct Answer:

Q1. C (1, 2, 4, 5, 7)

REFERENCES

1. Lickfett L, Mahesh M, Vasamreddy C, et al. Radiation exposure during catheter ablation of atrial fibrillation. *Circulation.* 2004;110:3003–3010.
2. U.S. FDA Public Health Advisory: Avoidance of Serious X-ray-Induced Skin Injuries to Patients During Fluoroscopically-Guided Procedures. May 2, 1996. Accessed at: www.fda.gov/MedicalDevices/Safety/AlertsandNotices/PublicHealthNotifications/ucm063084.htm
3. Interventional Fluoroscopy: Reducing Radiation Risks for Patients and Staff. Newsletter from NCI and SIR: 2005. Accessed at: www.cancer.gov/cancertopics/causes/radiation/interventionalfluoroscopy/page1
4. Shope TB. Radiation-induced skin injuries from fluoroscopy. *Radiographics.* 1996;16:1195–1199. See also: www.fda.gov/Radiation-emittingProducts/RadiationEmittingProductsandProcedures/MedicalImaging/MedicalX-Rays/ucm116682.htm
5. Wagner LK, Archer BR, Cohen AM. Management of patient skin dose in fluoroscopically guided interventional procedures. *J Vasc Interv Radiol.* 2000;11:25–33.
6. Miller DL, Balter S, Noonan PT, Georgia JD. Minimizing radiation-induced skin injury in interventional radiology procedures. *Radiology.* 2002;225:329–336.

CASE STUDY

69

Electroanatomical Mapping: Combined Technologies for Advances in Electrophysiology

Shlomo Shpun, DSc and Meir Bar-Tal, MSc

Question No. 1:

The electroanatomical map in Figure 69.1 represents what type of arrhythmia?

A. Reentry confined to the right atrium.

B. Focal atrial tachycardia (AT) originating from the right atrium.

C. Focal AT originating from the left atrium.

D. Impossible to distinguish without the measured local activation time (LAT).

Based on the colors alone, red is the earliest site of activation, denoting a focal AT originating in the septal region. However, from the electroanatomical map shown, there is no way to distinguish what is the earliest since both the left and the right atrium presents red color.

Question No. 2:

Based on the answer to the previous question, what would you recommend?

A. To map only the right atrium.

B. To map extensively (high-points density) of both atria.

C. To map extensively both atria only in the area of interest (septum).

D. To create an anatomical map only without the electrical information.

In Figure 69.1, it can be noticed that the earliest site of activation (the origin of the focal tachycardia) is located in the septal area. In order to identify the source of the focal tachycardia and to minimize the number of RF applications, there is no need to map the whole chamber/s. It is recommended to highly dense map the area of interest only in the septal region from both atria in order to pinpoint the focus (in "EP language" this is called the "hot and cold approach").

Question No. 3:

Based on the LAT of points A and B delineated on the electroanatomical map:

A. Tachycardia originates on the right; the LAT is −80.

B. Tachycardia originates on the left; the LAT is −76.

C. Tachycardia originates on the right since the bipolar voltage is higher (3.16 mV).

D. There is more red color on the right.

Practical Electrophysiology © 2015 Jasbir S. Sra, Masood Akhtar, Andrea Natale, David J. Wilber
Cardiotext Publishing, ISBN: 978-1-935395-17-1

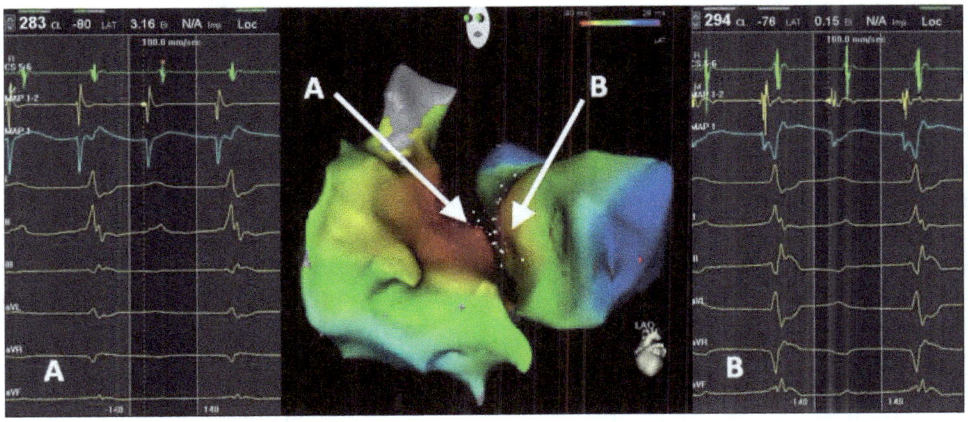

Figure 69.1. Electroanatomical map. (Image courtesy from Dr. Sabine Ernst, Royal Brampton Hospital, London, UK.)

INTRODUCTION

Previous CARTO® Systems consisted of only one technology, based on ultralow magnetic fields emitted by a location pad and an embedded passive sensor at the tip of a mapping catheter (NAVISTAR® Catheter, Biosense Webster, Inc., Diamond Bar, CA). The CARTO® 3 System technology combines these magnetic fields with electrical field technology. This hybrid technology not only provides accurate three-dimensional (3D) catheter location but also enables visualization of additional multielectrode catheters without a magnetic sensor. This paper provides a detailed description of the technology and its clinical applications.

MAGNETIC TECHNOLOGY

The CARTO® 3 System (hereinafter "System") magnetic technology consists of a miniature passive magnetic sensor located at the tip of an ablation catheter, an external ultralow magnetic field emitter, and a CARTO® System hardware unit. The System uses magnetic technology to accurately determine catheter location and orientation in 6 degrees of freedom (x, y, z, roll, pitch, and yaw), and simultaneously records the intracardiac local electrogram (EGM) from the catheter tip. The location pad placed beneath the patient table generates ultralow magnetic fields ($5 \times 10^{-6} - 5 \times 10^{-5}$ T), which define the mapping space around the patient's torso. The passive sensor detects the magnetic fields from the coils of the location pad and determines the distances from each coil.[1] These distances provide the 3D location and orientation in space. The catheter icon has 4 colors, red opposite to blue and yellow opposite to green (Figure 69.2). The colors provide a means of noting the rotation and orientation of the catheter tip. Moreover, since most catheters deflect to a single direction, determining the color toward which the catheter deflects may enhance catheter maneuverability without the use of fluoroscopy. To navigate to a specific location within the heart chamber, the operator first notes the color toward which the catheter deflects, and then turns that color toward the chamber wall and pushes until tissue contact is achieved. When using a bidirectional catheter, the operator must determine both colors toward which the catheter deflects in order to maximize navigation without fluoroscopy.

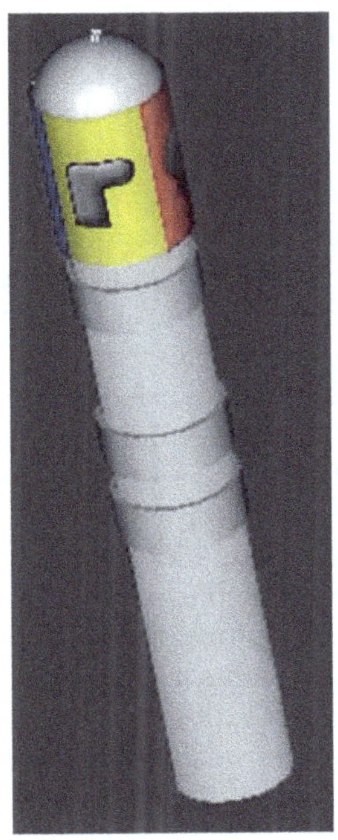

Figure 69.2. Catheter icon denoting colors around the tip.

Mapping

The mapping catheter is introduced with the aid of fluoroscopy and placed in the chamber to be mapped. As the catheter is moved within the heart chambers, the System continuously calculates the 3D location of the catheter and presents this data to the user. The visual location of the catheter can be gated to a fiducial point in the cardiac cycle (sequential mapping—or "point-by-point") or visualized in "stable" mode along the entire cardiac cycle, termed Fast Anatomical Mapping (FAM). These 2 options maximize operator treatment options in a clinical setting. FAM provides fast acquisition of the heart chamber boundaries for anatomically guided ablations. Sequential mapping provides point-by-point sequential maps with display of precise and detailed activation and voltage maps of focal and reentrant arrhythmias. Both mapping options result in 3D reconstructions of 3D heart chamber geometry with color-coded electrical information superimposed on the anatomical map.

Fast Anatomical Mapping (FAM)

Point-by-point acquisition allows for gating of a specific location to the cardiac cycle, whereas FAM is an average of locations throughout the entire cardiac cycle. Another difference between these mapping methods relates to how information is gathered for the reconstruction algorithm. With point-by-point mapping, locations determine the chamber surface reconstruction; however, with FAM, the catheter icon represents a "barrel" with a certain volume, determined by the catheter diameter and a length 10 mm from the tip. As the catheter is maneuvered in the heart chamber, the aggregate of these "barrels" and their volume determines the surface reconstruction. The user is presented with new "volume" information (colored orange) while older information is integrated into the reconstruction. Completeness of the acquired chamber can be readily determined by the absence of orange on the reconstruction. An additional feature that facilitates catheter navigation is the ability to apply transparency to the reconstruction such that formerly acquired "volumes" can be visualized. The System also presents users with several additional maps (activation, isochronal, propagation, and unipolar or bipolar voltage maps) resulting from information acquired at each point in the reconstruction. Each map displays unique information in relation to the mapped arrhythmia and chamber.

Ultrasound Maps

FAM is available with an ultrasound catheter equipped with a magnetic sensor. Ultrasound images provided by this catheter can be tracked in 3D space by the magnetic sensor in a manner similar to that described above. The ultrasound images are imported to the System; they can be then traced and define an anatomical structure or heart chamber. By tracing multiple frames of an anatomical structure in 3D space, the System can provide a 3D map of that structure. Moreover, an ablation catheter equipped with the same magnetic sensor can be navigated within the ultrasound map as if the reconstruction was provided by the other 2 mapping modes (point-by-point or FAM).

Activation Mapping

During sequential mapping the mapping catheter is dragged over the endocardium. While the catheter is in stable contact with the endocardium, tip location is acquired in 3D together with an EGM. The LAT at each site is determined from the intracardiac EGMs in relation to an electrical reference. The electrical reference serves as the fiducial EGM during the cardiac cycle to which all points are related. It is important to note, such an electrical reference must be selected by the user prior to collection of any sequential points. The signal derived from that reference electrode can be taken from any intracardiac electrode or any ECG lead. The reference EGM must also be derived from a stable location (usually the coronary sinus catheter) and needs to be acquired during a stable monomorphic rhythm. The System calculates the time difference between the reference EGM and the point EGM. The LAT is then displayed via color coding, red being the shortest interval and purple being the longest. The operator can review the quality of acquired data and either accept or reject this information. Electrophysiologic information is color-coded and superimposed on the 3D chamber geometry. The reconstruction is updated in real-time with the acquisition of each additional new point. If all the above criteria were followed closely, the result is a colored 3D map, which may present users with the arrhythmia indication. For focal arrhythmias, the map is displayed as one particular "red" (early) zone, surrounded by consecutive rainbow colors, with "purple" (late) at a remote distance. On the other hand, for a reentrant arrhythmia, red and purple meet and the rainbow of color spreads, indicating the direction of the reentrant circuit.[2] Acquisition of activation maps for FAM is somewhat different. A single click on the "Acquire" button acquires the electrical information, which is overlaid on the reconstruction with the same color coding.

One derivative of activation maps is *isochronal maps*—these maps represent the time zone within a determined range of 5, 10, 20 ms. Correct interpretation of these maps can help determine fast and slow conduction areas; this is visualized by the distances of the isochrone lines. Another means of visualizing the isochrones is via the *propagation map*, which is an active representation of the isochronal map. An activation wave is visualized propagating across the reconstruction, demonstrating regions of fast or slow conduction.

Voltage Mapping

Signal amplitude may provide an indication for substrate characteristics, especially in reentrant arrhythmias. The voltage map represents the amplitude of the acquired signal. The measurement, presented in millivolts, is from the highest to lowest peak of the signal; red indicates the lowest voltage and purple the highest. These maps can either be unipolar or bipolar and superimposed on the surface reconstruction. Voltage maps are mostly used for substrate-induced arrhythmias.[2]

Multielectrode Visualization: ACL Technology

Figure 69.3 shows a small sphere electrode in the center of a saline water tank (a volume conductor) (left panel). There are 2 planar electrodes on each side of the tank. Voltage is emitted from the center electrode with ground (zero voltage) on the planar electrodes. In this setup the current flows evenly from the center electrode to the plane electrodes. If the center electrode moves toward the left electrode, impedance to that electrode decreases and the current increases while the resistance increases and the current decreases on the right electrode.

Emitting current from the center electrode enforces the constant current flowing from the electrode, regardless of the surrounding resistance. By calculating the current ratio between the measured current at both sides to the sum of them, a linear relation exists between each of the current ratios to the location change of the center electrode (as illustrated in Figure 69.4). The current ratio serves as a good indicator of the center electrode position and can be used to determine the more complex 3D location, as will be explained next.

In fact, the current flowing out of the center electrode extends locally around the electrode in a radial fashion (depicted as a small gray sphere in the left panel of Figure 69.3). The current lines then divide into linear pathways until they reach the planar electrodes. This phenomenon is a result of the natural physical process to reduce the potential as fast as possible. For this saline tank, in order to reduce the surface area (ie, the resistance) the current flows outward toward the planar electrodes. This can be simplified as an "Exit Resistor" (indicated in gray in the left panel of Figure 69.3) accompanying the electrode, and is independent of the electrode location. Only the 2 resistors to the patches are a function of the electrode location. The "Exit Resistor" is much larger than the other 2 since current density near the electrode is quite large—usually ~ 90% of the resistance into the patches. The left and right resistors (R_{left}, R_{right}) are

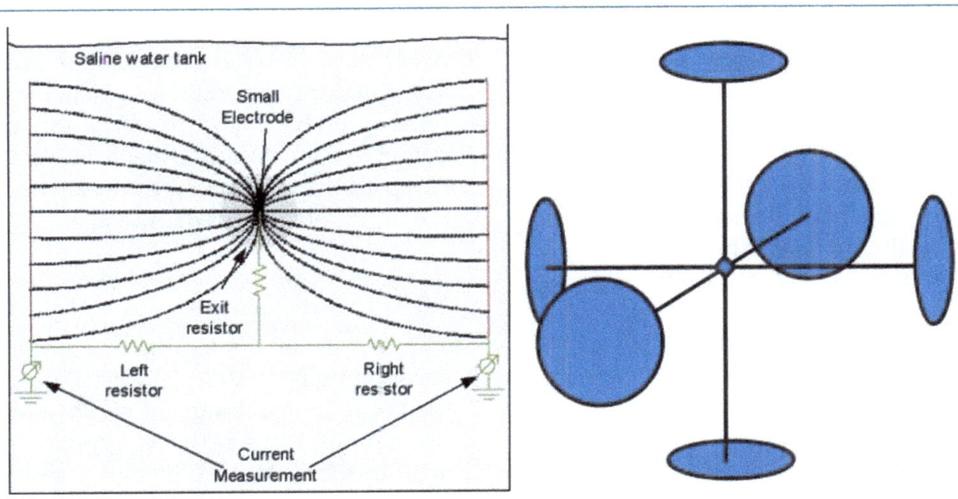

Figure 69.3. Electrode in a saline water tank (left) and 3D configuration model (right).

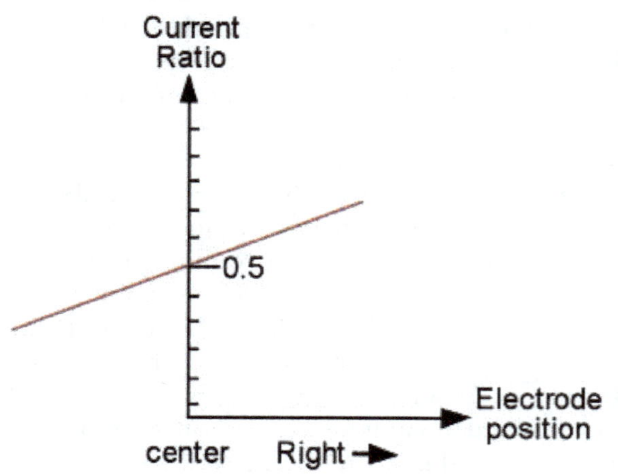

Figure 69.4. Current ratio versus electrode position in a saline tank.

proportional to the corresponding distances to the patches; thus the current ratio measurement is proportional to $\alpha 1/R_{left}$ and $\alpha 2/R_{right}$, where $\alpha 2/R_{right} = 1 - \alpha 1/R_{left}$. This can be calculated since the current measurements are normalized to the full amount of the current and actually only one of the currents is relevant as the other current is a direct outcome of the first.

SIMPLIFIED 3D MODEL

Additional patches are needed to apply this approach to the 3D location of the electrodes. Such a simplified model is illustrated in the right panel of Figure 69.3. Current from the center electrode is distributed evenly to 6 patches, assuming they are equidistant from the center electrode. Any displacement of the center electrode along the x, y, and z axes will only result in a change in the current distribution to the patches on both sides of that particular axis. However, the distance to the other axes patches will change slightly and their impedance from the center electrode will stay almost the same. In order to calculate the general resistance to the patches R1 to R6, the normalized current ratios must be measured [ie, $I1/(I1 + I2 + \ldots I6), \ldots I6/(I1 + I2 + \ldots I6)$], which are proportional to $\alpha 1/R1, \alpha 1/R2, \ldots, \alpha 1/R6$ with $\alpha 1/R6 = 1 - \alpha 1/R1 - \alpha 1/R2 \ldots - \alpha 1/R5$. As a result, only 5 relevant measurements need to be calculated. These equations provide enough information for the x, y, and z positions of the center electrode. By placing 6 patches equidistant around the center electrode, a good estimate of electrode movement in any direction is achieved. In clinical settings, given the position of the heart within the human body, 3 patches on the back and 3 on the chest provide similar information for detecting any electrode motion within the heart with a sensitivity of 0.1% per mm.[3] This sensitivity range is frequently measured for normalized current ratios in clinical settings. In order to maintain adequate sensitivity, note that patch sizes must be proportionally much larger than the center electrode, since the impedance input must be much smaller than the output impedance from the center electrode. Small patches (similar to the center electrode size) would encompass large impedance readings resulting in a dramatic drop in sensitivity. Patches must also be placed a fair distance from each other to avoid overlapping current readings.

APPLYING THE SIMPLIFIED MODEL TO REAL CLINICAL NEEDS—THE CARTO 3® SYSTEM

In the clinical environment the "volume conductor" is not homogeneous and the 6 patches are placed arbitrarily. This implies that impedance measurements and calculated locations are more complicated with too many unknown parameters. To provide accurate electrode location the relation between the impedance measurements and the electrode location must be calibrated. The System uses impedance measurements obtained from a catheter equipped with accurate position measurement (electromagnetic sensor) and normalized to the current ratios from the same catheter electrode. In other words, the magnetic-based catheter calibrates the impedance fields such that magnetic location is associated with an impedance measurement, termed Catheter Position Mapping. In brief, electromagnetic locations are associated to impedance readings at any given point in the heart. In return, when an electrode is associated with a certain impedance reading, it can be assigned backwards to a 3D position within the mapping zone. Multiple electrodes can be located in 3D space by preselecting the catheter structure, that is, quadripolar versus decapolar.

Respiration Gating

The System can track catheter motion, as described earlier, by gated mode (a fiducial point during the cardiac cycle) or nongated mode. For either mode, the catheter is subjected to respiratory movement. With former CARTO® Systems, a catheter placed in a stable position could be visually gated to respiratory motion.[2] With the System, using multiple patches permits introduction of a smart algorithm for respiratory gating. Respiration is tracked on the System using impedance readings derived from the interpatch measurements, called "respiratory indicators." The interpatch current (from one patch to the other) must pass through the lungs; thus, impedance changes are recorded based on air volume changes in the lungs. Heart motion is faster than respiratory movement, and its motion is filtered out to leave only the respiratory effect on the signals (both catheter location and respiratory indicators). For an algorithm to provide good gating performance, respiratory indicators need to correlate with catheter motion during respiration. The algorithm must be "trained" before the gating algorithm can start. Training is done when one catheter is placed in the heart, touching the chamber wall for recording heart movement during respiration. During training a correlation matrix is calculated that best correlates the respiratory indicators to catheter motion. Although training is sampled in one location, it is valid for the entire heart. A training graph is generated that selects the detection threshold. The preselected threshold provides the means to control the respiratory gating window and to acquire points or volumes in FAM mode only near end-expiration of the respiratory phase. A tighter threshold provides for significant respiratory gating, resulting in more accurate maps.

SUMMARY

In summary, the System offers clear advances toward improved mapping and treatment of heart arrhythmias. By combining the accurate 3D positioning of the mapping catheter by electromagnetic means with the additional real-time visualization of multiple electrodes via electrical fields, catheter visualization in the heart is improved, with minimal use of fluoroscopy.

Correct Answers:

Q1. D

Q2. C

Q3. A

QUESTIONS FROM THE EDITOR AND AUTHORS' REPLY

Question 1. Does using multiple catheters for FAM affects the accuracy of the map?

Answer: Multiple catheters do not affect the FAM accuracy since FAM is acquired by a single catheter equipped with a magnetic sensor (either a regular ablation catheter or a circular mapping catheter). Since the magnetic sensor is similar in all the catheters used with it, any such catheter used during FAM will be contributing equally to the FAM accuracy. The additional catheters (nonsensored) do not contribute to the FAM process.

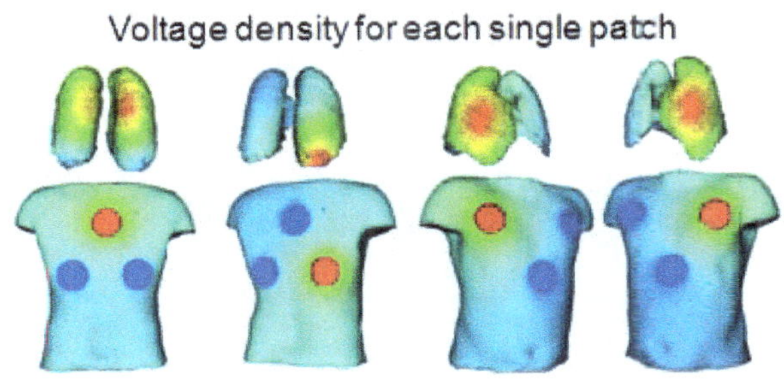

Figure 69.5. Back and front.

Question 2. How accurately does impedance translate into respiration gating?

Answer: The impedance measurements are based on correlation between multiple patches. Is it not calculated based on one single pair of electrodes? Current emitted from one patch is received by the other 5 patches, thus the sum of current is reflected in the total amount of emitted and received current. Each patch is emitting and receiving simultaneously. The respiration motion causes current to change according to the inspiration and expiration. Since the patches are placed on front and back of the patient, their position reflect local changes even before full inspiration is detected by all other patches (Figure 69.5). Those impedance changes generate a real-time graph denoting respiration cycle.

REFERENCES

1. Gepstein L, Hayam G, Ben-Haim SA. A novel method for nonfluoroscopic catheter-based electroanatomical mapping of the heart – *in vitro* and *in vivo* accuracy results. *Circulation*. 1997;95:1611–1622.
2. Soejima K. How to troubleshoot the electroanatomic map. *Heart Rhythm*. 2010;7:999–1003.
3. Jiang Y, Farina D, Bar-Tal M, Dosel O. Impedance based catheter positioning system for the cardiac mapping and navigation. *IEEE Trans. Biomed. Eng.* 2009;56:1963–1970.

CASE STUDY

70

Catheter Ablation of Persistent Atrial Fibrillation Using Ensite NavX™

Ross J. Hunter, BSc, MBBS and Richard J. Schilling, MD

Question No. 1:

Figure 70.1 shows the following lesion set:

A. Pulmonary vein (PV) isolation.

B. Complex fractionated electrogram (EGM) ablation.

C. Segmental ostial isolation.

D. Both A and B.

DISCUSSION

Catheter ablation of complex arrhythmia such as persistent atrial fibrillation (AF) usually requires prolonged procedures and fairly extensive ablation. The majority of centers now perform such procedures with the use of three-dimensional (3D) mapping systems, which allow manipulation of catheters without the constant use of fluoroscopy, facilitate accurate lesion placement and record lesion position, and can be used to map arrhythmia. The EnSite NavX™ system (St. Jude Medical Inc., St. Paul, MN) is one of the most popular 3D mapping systems in use.

The EnSite system passes a low-amplitude, 5.68-kHz "locator signal," which alternates between 3 different pairs of skin patches. These are placed on the patient in orthogonal planes, so that the voltage measured from each catheter electrode can be used to locate the 3 catheter planes (x, y, and z).

At the start of the case, while keeping all catheters still, a "respiratory compensation" is performed, where an electrode placed on the upper abdomen to track diaphragmatic movement allows the system to compensate for respiratory movement and keep catheters relatively still onscreen through

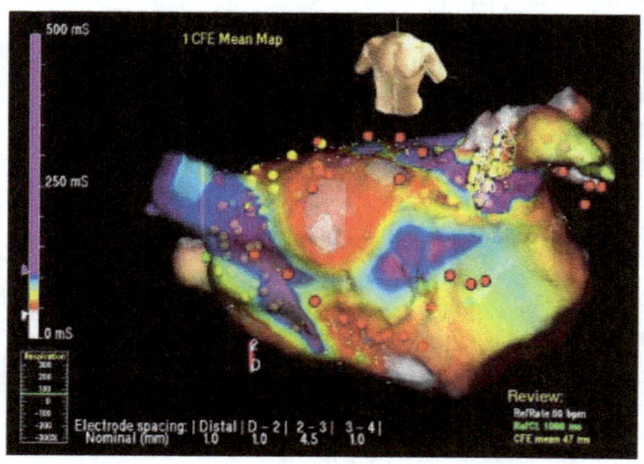

Figure 70.1. Left Atrial (LA) Geometry on the EnSite NavX™ System. Image shows a LA that has been fused with a computed tomographic (CT) reconstruction of the chamber. The torso at the top right indicates this is an anteroposterior view. Catheters are shown onscreen, including the PV mapping catheter (yellow), a quadripolar catheter in the right atrium (RA, red), and an ablation catheter (gray with a green tip). Lesions are marked in red and yellow.

Practical Electrophysiology © 2015 Jasbir S. Sra, Masood Akhtar, Andrea Natale, David J. Wilber
Cardiotext Publishing, ISBN: 978-1-935395-17-1

the respiratory cycle. A catheter is then selected, usually the PV mapping catheter, and is moved around the chamber of interest to collect points from the position of the electrodes to create a volume in 3D space. For catheter ablation of AF, most operators collect a LA geometry first, and only collect a RA geometry later if they intend to ablate in the RA. The acquired LA geometry is then "reassigned" in portions by clicking around parts of the geometry onscreen and using the mouse to designate parts as a PV or the LA appendage.

Using software on the mapping system called EnSite Verismo™, a CT or magnetic resonance imaging scan can be segmented to create an LA shell, which is then imported into the mapping system as shown in Figure 70.2. By placing fiducial points on corresponding locations of the LA geometry and the CT shell, the EnSite system will then "stretch" the LA geometry to fit the CT shell, and the operator can then visualize the catheters moving relative to the CT LA. This step is optional, however, since many operators find the initial geometry close enough to the anatomy to obviate the need for CT registration.

Most centers proceed to PV isolation by wide area circumferential ablation (WACA), where lesions are placed 1 to 2 cm from the PV ostia to isolate them in ipsilateral pairs (Figure 70.3). The PV mapping catheter is placed in the vein ostia and used to identify the points of earliest breakthrough, enabling the operator to target his ablation to achieve electrical isolation. Whatever other lesions are created, electrical isolation is an important procedural end point in catheter ablation of AF. Lesions causing a change in the activation sequence or that achieve electrical isolation can be marked using the mapping system, in this case by making the lesions a different color, such as those in green at the ostia of the right PVs in Figures 70.1 and 70.3.

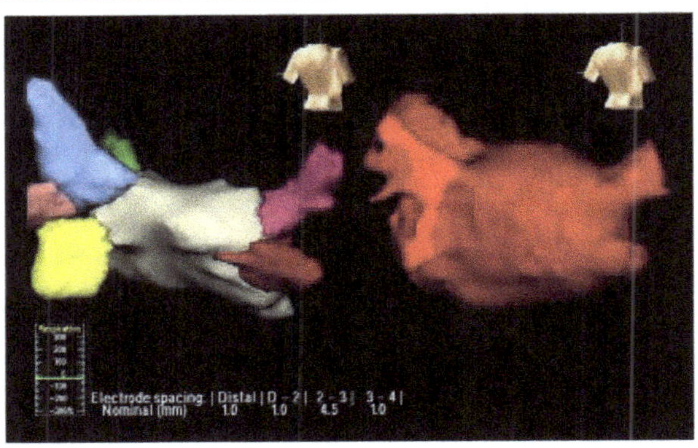

Figure 70.2. Fusion of the LA Geometry with CT Imaging. Imaging (in this case a CT scan) is segmented to give a reconstruction of the LA. This is imported alongside the acquired geometry of the LA. The operator places fiducial points at corresponding locations on the 2 surfaces. The system then "fuses" the 2 images by stretching the acquired geometry to best fit around the fiducial points.

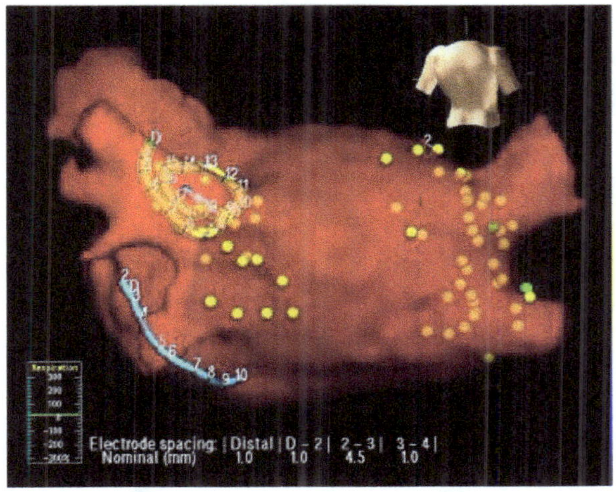

Figure 70.3. Wide Area Circumferential Ablation. EnSite NavX map showing LA geometry fused with CT. Yellow lesions show WACA. CT can be useful, particularly when anatomy is unusual such as with the left common trunk shown here. The PV mapping catheter is placed within the PV to identify the site of earliest breakthrough and to help in isolating the vein. Sites that cause a change in activation or isolation can be marked a different color, such as those in green.

Once the PVs are isolated, there is a debate as to the best way to proceed if AF persists. The most common options are ablation of complex fractionated atrial electrograms (CFAE) or the addition of linear lesions, usually between WACA lines at the roof or at the mitral isthmus between the left inferior PV and the mitral annulus. Other possibilities include further linear lesions, isolation of the posterior wall by placement of linear lesions at the upper and lower posterior wall between WACA lines, modification of the posterior wall (meaning extensive ablation to destroy all myocardium), or targeting of ganglionated plexi. At our center it is a common practice to target CFAE in the LA followed by the RA, and then add linear lesions at the mitral isthmus and roof. If at any point AF organizes into atrial tachycardia (AT), this is mapped and ablated. If sinus rhythm is not restored following these lesions, the patient is cardioverted with a direct current shock.

To guide CFAE ablation, a map of CFAE can be created using the EnSite NavX system (see Figure 70.1). The PV mapping catheter can be moved around the LA to acquire many points quickly, usually acquiring a full map in approximately 10 minutes. The PV mapping catheter can then be placed in the LA appendage to monitor AF cycle

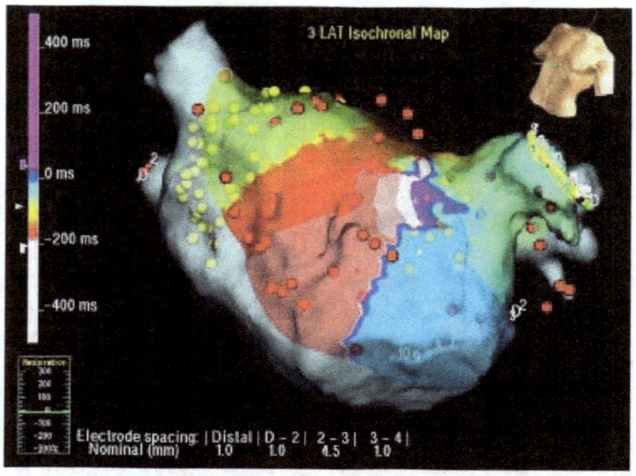

Figure 70.4. Isochronal Map of Left AT. Image showing LA geometry fused with CT. The torso shows the orientation is a left anterior oblique view. The color scale on the left shows the timing within the tachycardia cycle. The white arrows show the circular movement of the wavefront, which suggests a localized reentry circuit, possibly a rotor.

length. When mapping CFAE, the system calculates the mean interval between deflections or "CFAE mean," and a color-coded display then illustrates where CFAE are most prevalent. In Figure 70.1, the areas with CFAE mean < 120 ms are highlighted as CFAE. As in this case, much of the atrium can appear as CFAE, hence the operator may initially choose to concentrate ablation in areas with the lowest score, perhaps < 80 to 100 ms. If extensive CFAE ablation is required, the distribution of remaining CFAE may change. In this scenario, some operators will collect a repeat CFAE map to help target further ablation. However, it is noteworthy that many operators consider CFAE maps to be time-consuming and unnecessary, and instead prefer mapping with the ablation catheter and ablating any CFAE identified as the map is created.

In this case, we did not proceed to linear lesions as the rhythm slowed and EGMs became more organized, suggesting an AT. Although the PV mapping catheter can be used to acquire a map of the entire LA, we prefer to use the ablation catheter to create an activation map. The activation map shown in Figure 70.4 was produced, suggesting a localized macro-reentry circuit shown as arrows or possibly a rotor. Ablation in this region terminated the tachycardia to sinus rhythm.

Correct Answer:

Q1. D

REFERENCES

1. Calkins H, Brugada J, Packer DL, et al. HRS/EHRA/ECAS expert consensus statement on catheter and surgical ablation of atrial fibrillation: Recommendations for personnel, policy, procedures and follow-up. A report of the Heart Rhythm Society (HRS) Task Force on Catheter and Surgical Ablation of Atrial Fibrillation developed in partnership with the European Heart Rhythm Association (EHRA) and the European Cardiac Arrhythmia Society (ECAS); in collaboration with the American College of Cardiology (ACC), American Heart Association (AHA), and the Society of Thoracic Surgeons (STS). Endorsed and approved by the governing bodies of the American College of Cardiology, the American Heart Association, the European Cardiac Arrhythmia Society, the European Heart Rhythm Association, the Society of Thoracic Surgeons, and the Heart Rhythm Society. *Europace.* 2007;9:335–379.
2. Richmond L, Rajappan K, Voth E, et al. Validation of computed tomography image integration into the EnSite NavX mapping system to perform catheter ablation of atrial fibrillation. *J Cardiovasc Electrophysiol.* 2008;19:821–827.
3. Earley MJ, Showkathali R, Alzetani M, et al. Radiofrequency ablation of arrhythmias guided by non-fluoroscopic catheter location: A prospective randomized trial. *Eur Heart J.* 2006;27:1223–1229.

CASE STUDY

71

Mapping and Ablation of Right Atrial Tachyarrhythmia in a Patient with Complex Anatomy Using Noncontact Mapping

Ross J. Hunter, BSc, MBBS, Dominic J. Abrams, BSc, MD, and Richard J. Schilling, MD

Question No. 1:

Figure 71.1 shows the following tachycardia mechanism:

A. Focal tachycardia.
B. Atrial fibrillation (AF).
C. Typical cavotricuspid isthmus (CTI)-dependent atrial flutter.
D. Atypical atrial flutter.

CASE REPORT

A 17-year-old girl presented with recurrent tachycardia that was very poorly tolerated, requiring cardioversion on each occasion. Her background was of congenital tricuspid atresia and creation of a Fontan circulation at age 3 by anastomosing the right atrial appendage to the pulmonary artery (PA). She had since been symptomatically well until now.

Mapping in such a case is likely to be of a complex arrhythmia. The optimal site of ablation is usually customized to the patient's anatomy, with no standardized approach like in typical CTI-dependent atrial flutter. The arrhythmia is also poorly tolerated, and a lengthy mapping procedure would not be ideal.

For these reasons, noncontact mapping was performed using the EnSite Array™ system (St. Jude Medical Inc., St. Paul, MN). The multielectrode array is a mesh of 64 steel wires with a polyimide coating mounted on an 8-Fr catheter, and surrounds a 7.5-mL balloon (Figure 71.2). The 0.003-inch diameter wires have a 0.025-inch laser-etched defect in their insulation, forming unipolar electrodes. A low-amplitude, 5.68-Hz "locator signal" is passed through the proximal and distal ring electrodes (E1 and E2) on the array catheter and alternated between these electrodes and the ablation catheter (or any 4 electrodes on up to 2 catheters). This is used to locate catheters relative to the array. Assuming the array to be stationary, the ablation catheter can then be used to collect a geometry in three-dimensional(3D) space (in this case the right atrium).

The electrodes (see Figure 71.2c) make contact with the blood and sense far-field electrical signals generated by the myocardium. The electrodes are unipolar, measuring potential relative to an indifferent electrode (usually the proximal ring electrode on the array catheter situated in the inferior vena cava [IVC], otherwise, an external source). The electrograms (EGMs) are of lower voltage and amplitude compared to the

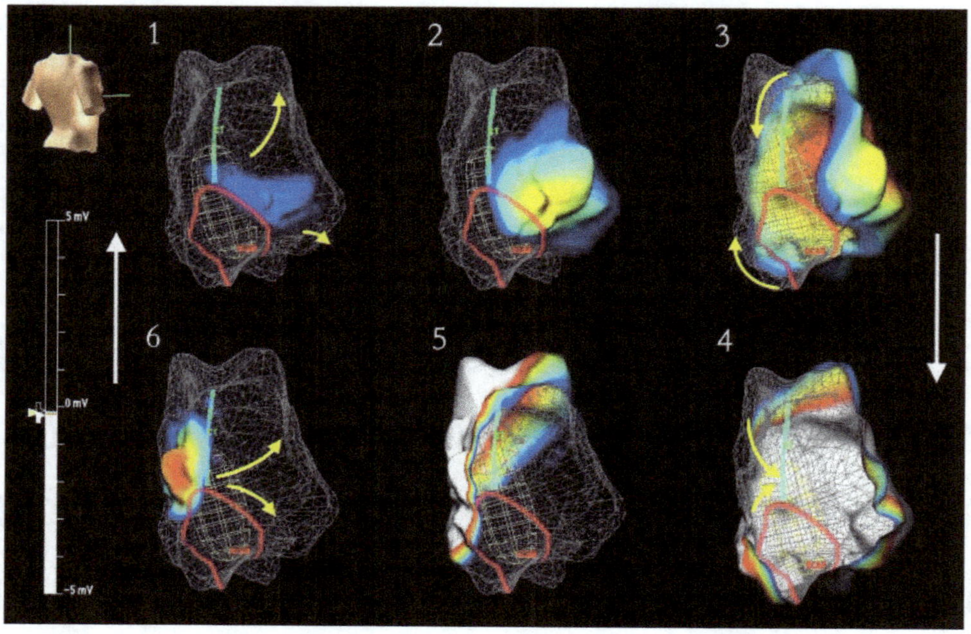

Figure 71.1. Noncontact Mapping Using the EnSite Array™ System. The wire mesh shows the right atrial geometry, with the oval mesh within this chamber representing the array. The torso in the top left shows this is a right lateral/posterior view. The green line is marked at the site of the crista terminalis, and the red line marks an area of scar (these were found on contact mapping and annotated onto the geometry, but can be quickly identified using a function called Dynamic Substrate Mapping). The scale on the left shows that depolarization will be represented by white, which will move around the chamber as the wavefront moves as shown from 1 to 6. Reproduced with permission.[1]

endocardial surface. A technique to reconstruct EGMs as they appear on the endocardial surface uses an inverse solution to the law of Laplace, with a boundary element method used to resolve a matrix of such signals. This is used to reconstruct 3360 "virtual" unipolar EGMs, which are superimposed onto the virtual endocardium.

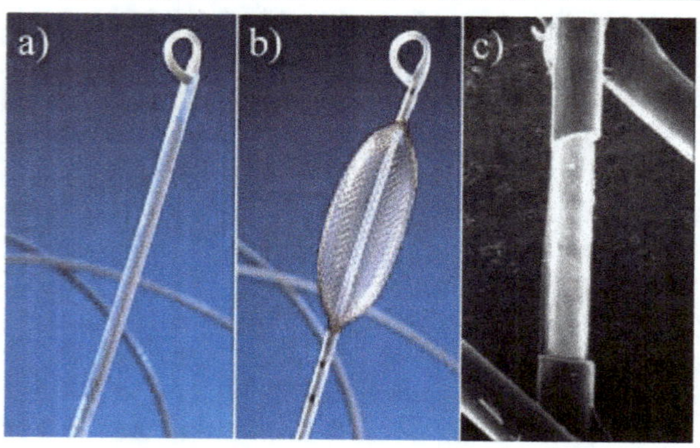

Figure 71.2. The EnSite Array System (The EnSite Array catheter). **Panel A** shows the catheter while the balloon is deflated. **Panel B** shows the wire mesh around the inflated balloon. **Panel C** shows a close-up of the wire mesh, with the defects in the polyimide wire coating in creating small electrodes.

Isopotential maps are produced in which a discrete front of endocardial depolarization is shown by the onset of a negative polarity. Recorded segments can be slowed to demonstrate the movement of wavefronts denoted by the onset of negative polarity (see Figure 71.1, which shows still images from 6 time points during the tachycardia cycle). Alternatively, it is possible to reconstruct an isochronal map or simply view EGMs.

With prior knowledge of the patient's anatomy and the isopotential maps obtained in Figure 71.1, it is possible to deduce the anatomy of the reentry circuit (Figure 71.3). In this case there is a "figure-of-8" tachycardia, which is divided by the anastamosis between the right atrial appendage and the PA. Entrainment from within either circuit would confirm which is supporting the tachycardia and which is "bystander," although it is clear in this case that the vulnerable point is between the crista terminalis and the area of scar near the IVC (a common site for scar in patients with a Fontan circulation caused by a surgical incision). Ablation here rendered the patient free of arrhythmia.

Correct Answer:

Q1. D

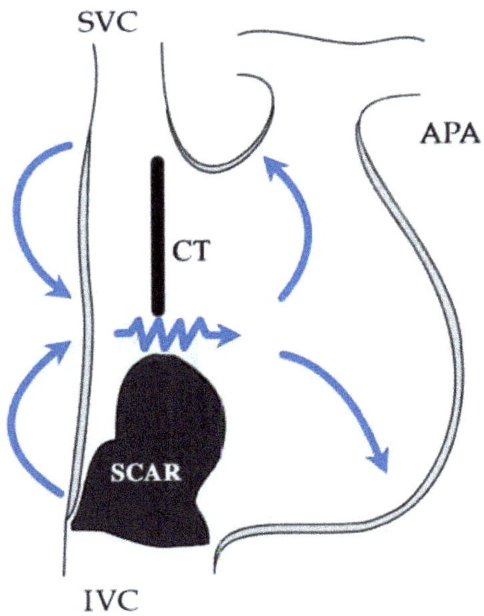

Figure 71.3. Anatomy of the reentry circuit. Right lateral/posterior view of the right atrium as in Figure 71.1. The blue arrow shows the tachycardia circuit. APA, atriopulmonary anastomosis; CT, crista terminalis; IVC, inferior vena cava; SVC, superior vena cava.

REFERENCES

1. Rajappan K, Schilling RJ. Principles of noncontact endocardial cardiac mapping. In: Shenasa M, Hindricks G, Borggrefe M, Breithardt G, eds. *Cardiac Mapping*, 3rd ed. West Sussex: Wiley-Blackwell; 2009: 37–48.
2. Earley MJ, Abrams DJ, Sporton SC, Schilling RJ. Validation of the noncontact mapping system in the left atrium during permanent atrial fibrillation and sinus rhythm. *J Am Coll Cardiol.* 2006;48:485–491.
3. Abrams DJ, Earley MJ, Sporton SC, et al. Comparison of noncontact and electroanatomic mapping to identify scar and arrhythmia late after the Fontan procedure. *Circulation.* 2007;115:1738–1746.

CASE STUDY

72

Three-Dimensional Echocardiography Cardiac Imaging

Khawaja Afzal Ammar, MD and Bijoy K. Khandheria, MD

The vast majority of emboli (90%) in patients with atrial fibrillation (AF) arise from the left atrial appendage (LAA) (Figure 72.1). LAA is a structural part of the left atrium, located anterior, lateral, and superior to rest of the left atrium. It is lined with pectinate muscles.

Question No. 1:

Figure 72.2 shows:

A. Normal LAA function.

B. Doppler evidence of organized atrial contraction.

C. Reduced LAA function with high propensity for thrombus formation.

Normal LAA function generates LAA emptying velocities of 50 to 80 cm/s. Two-dimensional echocardiography and Doppler imaging produce evidence of mechanical function during LAA contraction, which can be evaluated by transesophageal echocardiography (TEE). During TEE, LAA can be visualized in most patients midesophagus at an imaging plane of 90 degrees. Alternatively, it can be seen at zero degrees by slightly withdrawing the probe to a higher position followed by anteflexion. A normal LAA is small in size and visibly contracts. Color flow can be seen entering and leaving the LAA during the cardiac cycle, confirming lack of thrombus. The function is further evaluated by placing a Doppler pulsed-wave sample volume in the mouth of LAA. This reveals a Doppler spectral display with flow toward the probe, following each P wave. In case of AF, LAA velocity waveforms are irregular and of varying sizes due to fibrillation waves of different sizes and occur throughout the cardiac cycle, as displayed in Figure 72.2. Velocities below 40 cm/s are associated with LAA dysfunction, leading to increased risk of thrombus formation.

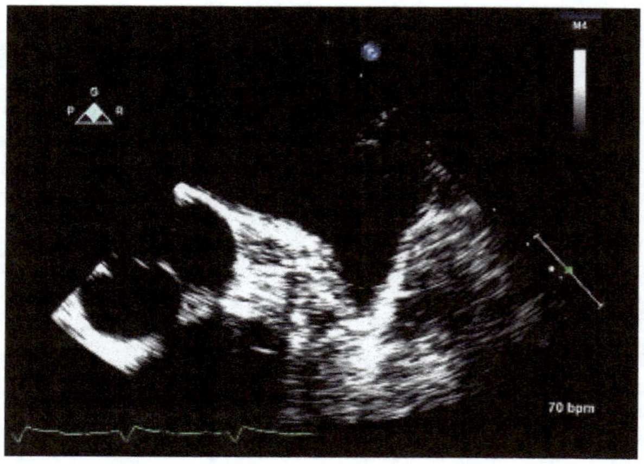

Figure 72.1. LAA with no clot.

Practical Electrophysiology © 2015 Jasbir S. Sra, Masood Akhtar, Andrea Natale, David J. Wilber
Cardiotext Publishing, ISBN: 978-1-935395-17-1

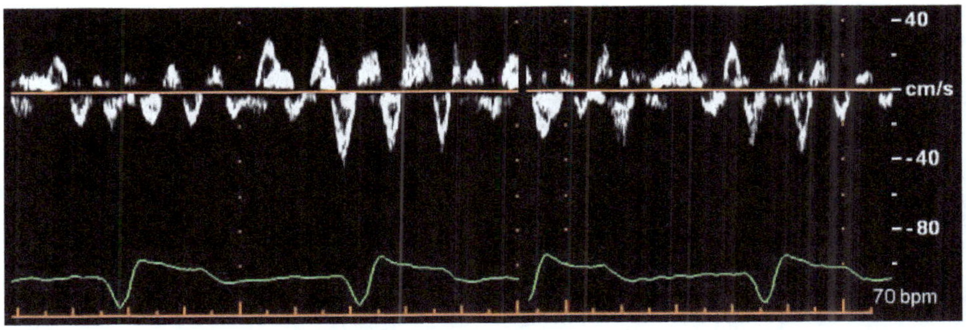

Figure 72.2. LAA emptying velocities.

Question No. 2:

An electrophysiologist often is faced with the need to cardiovert a patient in AF. Figure 72.3 shows:

A. LAA thrombus.

B. Side lobe artifact.

C. Reverberation artifact.

D. Pectinate muscles.

Apparent LAA thrombus is seen in Figure 72.3A. In Figure 72.3B, LAA is fully developed and the echocardiographic densities are the ends of pectinate muscles, which are recognized by their attachment to the LAA wall. A clot would not have such an attachment to the walls.

There is a relatively straight look into the LAA, and still an echocardiographic density is seen, suggestive of early thrombus (Figure 72.4). One of the best ways to avoid reverberation artifact, particularly those from the Coumadin ridge, is by looking straight down the barrel at the LAA.

Reverberation artifact from the Coumadin ridge is recognized by the acoustic shadow cast by a structure closer to the transducer (Coumadin ridge being the most common culprit) as compared to the LAA cavity (Figure 72.5A).

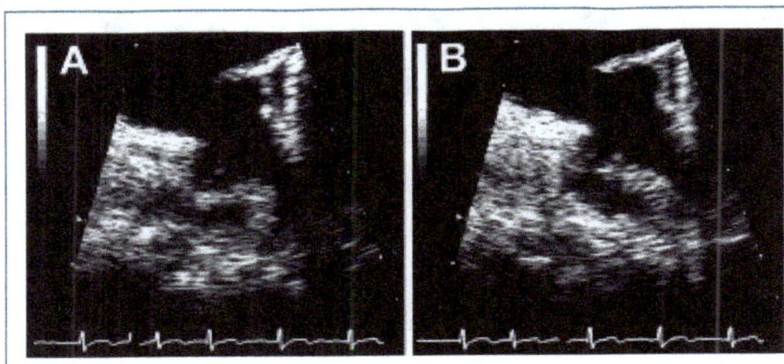

Figure 72.3. Echocardiogram of LAA.

A straight line shadow indicates its artifactual nature as nature does not create straight lines. The shadow is behind the Coumadin ridge and moves with it. Figure 72.5B shows a relatively straight look down the LAA cavity, which is empty of any shadows, confirming the earlier shadow to be a reverberation artifact.

The LAA is lined by multiple pectinate muscles, which appear as ridges on TEE, arising from the wall of LAA. These can be confused with a thrombus if only the tips are seen, but demonstration of these ridge-like structures arising from the wall of LAA, parallel to each other, confirms their true benign nature (Figure 72.6).

LAA may have one or multiple lobes. This necessitates its evaluation in several imaging planes, which is accomplished by dialing the angle between zero and 180 degrees, as well as looking at it from midesophageal and high midesophageal windows. This detailed evaluation is clinically important if the LAA emptying velocities are < 40 cm/s.

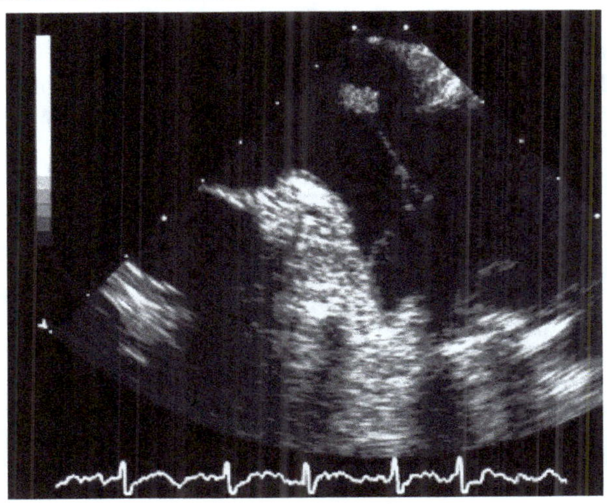

Figure 72.4. Early LAA thrombus.

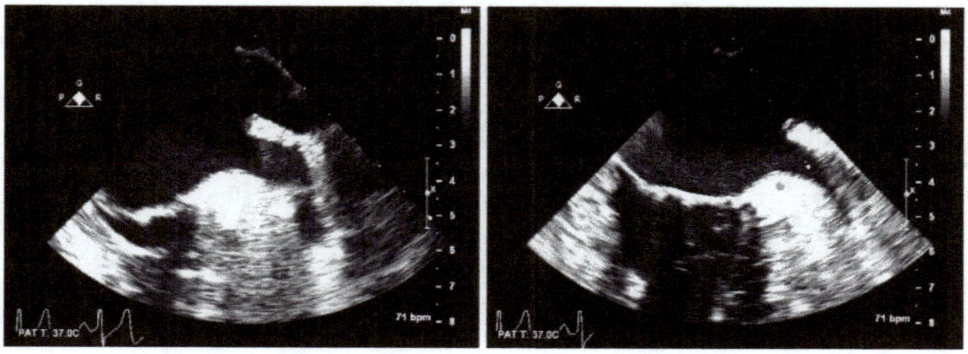

Figure 72.5. LAA with reverberation artifact from the Coumadin ridge.

Three-dimensional (3D) echocardiography provides the ability to visualize the entire LAA in one capture. A pyramidal volume 3D dataset is acquired via a special 3D TEE probe, commercially available only from Philips (iE33 machine, X72T probe; Andover, MA). This image can be sliced and post-processed to look through the entire LAA (Figure 72.7).

Question No. 3:

In patients with AF who are not candidates for long-term anticoagulation due to bleeding tendencies, the best treatment is:

- **A.** Radiofrequency ablation.
- **B.** Pharmacological therapy (with amiodarone, etc).
- **C.** Aspirin alone.
- **D.** LAA surgical isolation and amputation.
- **E.** LAA percutaneous isolation.

This is a controversial issue. Rhythm management with radiofrequency ablation or pharmacological therapy is effective for control of symptoms, including dyspnea and palpitations, but is not effective in preventing thromboembolic strokes. Aspirin is only half as effective in preventing strokes as compared to warfarin. Since 90% of clots arise in the LAA, isolation of LAA is attractive.[1] Studies have shown the value of this strategy.[1] LAA is technically easy to approach percutaneously using transfemoral vein approach; the transseptal wire invariably lands in the LAA. This has created the possibility of percutaneous LAA closure devices (Figure 72.8). 3D echocardiography provides the opportunity to visualize the LAA in three dimensions. Figure 72.9 shows the ability of 3D echocardiography to display multiple catheters in the LAA simultaneously, which is often not possible with 2D echocardiography due to thin imaging planes.

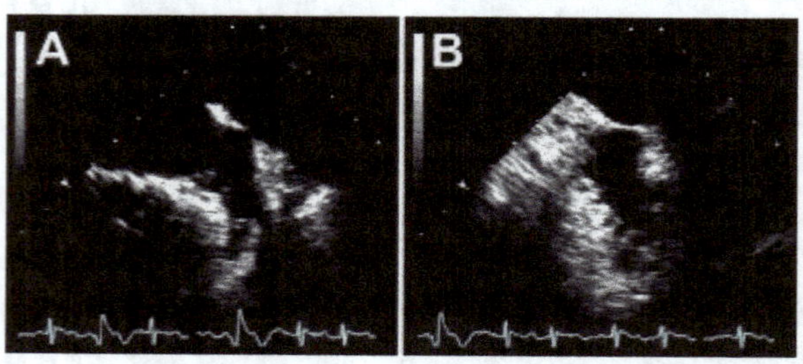

Figure 72.6. Another example of pectinate muscles masquerading as clot. Panel A suggests LAA clot. Panel B confirms pectinate muscles.

Correct Answers:

Q1. C

Q2. D

Q3. E; D also is acceptable if the patient is an acceptable surgical risk.

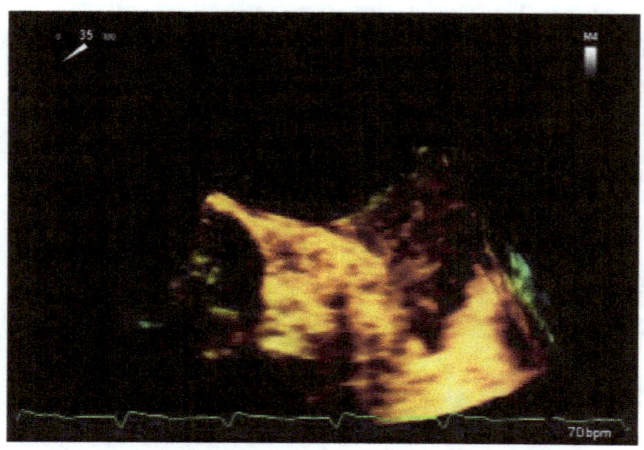

Figure 72.7. 3D transesophageal echocardiogram of LAA.

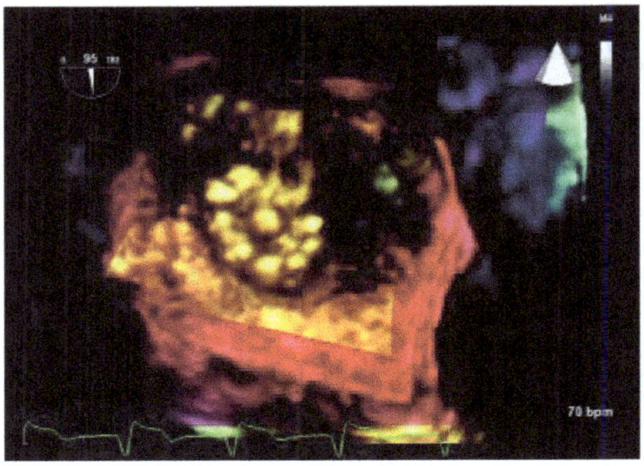

Figure 72.8. A Watchman device (Atritech Inc., Plymouth, MN) at the entrance of LAA.

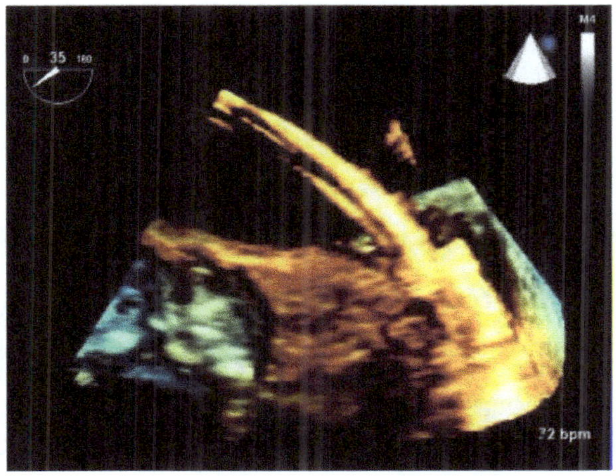

Figure 72.9. Two catheters entering the LAA show the strength of 3D imaging for simultaneously visualizing multiple catheters in multiple imaging planes.

REFERENCE

1. Meier B, Palacios I, Windecker S, et al. Transcatheter left atrial appendage occlusion with Amplatzer devices to obviate anticoagulation in patients with atrial fibrillation. *Catheter Cardiovasc Interv.* 2003;60:417–422.

CASE STUDY

73

CT Imaging of the Left Atrium

David Krum, MS and Jasbir S. Sra, MD

Question No. 1:

The images in Figure 73.1:

A. Show the left atrium (LA) of a patient with atrial fibrillation (AF) who has pulmonary vein (PV) narrowing.

B. Depict the anatomy from a patient with a very common two-right-by-two-left PV anatomy and are useful in preparing a road map for ablation.

C. Were obtained with the use of contrast.

D. B and C above.

DISCUSSION

Three-dimensional (3D) imaging of the heart in patients undergoing left atrial procedures is becoming popular because it accurately depicts the anatomy of the LA and PVs.[1] This information can assist in preprocedure planning, and the resultant segmented images may be integrated into a cardiac mapping system to aid in the ablation.[2,3] Cardiac CT has the advantage of better spatial resolution when compared to magnetic resonance imaging (MRI).

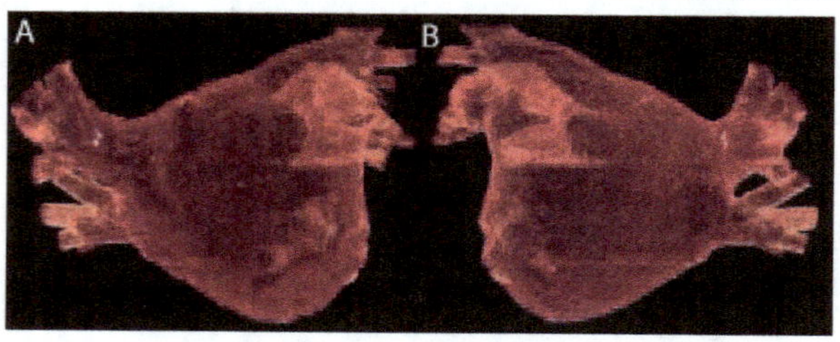

Figure 73.1. Anterior (**A**) and posterior (**B**) views of a segmented left atrial model from computed tomography (CT) scanning.

Compensation must be made for motion of the heart due to the respiratory and cardiac cycles to obtain good image quality. The CT scan is acquired quickly enough that it can be performed during a single breath hold for most patients. Cardiac cycle motion can be "frozen" through the use of ECG gating. During the scan, wherein image data are being acquired continuously, the patient's surface ECG is also being recorded. Then, a series of two-dimensional (2D) axial images (Figure 73.2) are retrospectively recreated from only the raw scan data that correspond to a particular point in the cardiac cycle. The images, which are acquired over several rotations of the gantry around the patient, will then correspond to one another, essentially eliminating motion of the heart from the series.

Each element, or pixel, of the 2D axial image has a grayscale intensity that corresponds to its density. More dense pixels will appear as brighter features on the image. Areas such as bone and blood infused with contrast are typically very bright, while low-density areas, such as the lungs, are very dark. A series of axial images, typically in the range of 0.5 to 1.0 mm apart, can then be assembled to form a 3D volume or space. Individual elements in the 3D

Practical Electrophysiology © 2015 Jasbir S. Sra, Masood Akhtar, Andrea Natale, David J. Wilber
Cardiotext Publishing, ISBN: 978-1-935395-17-1

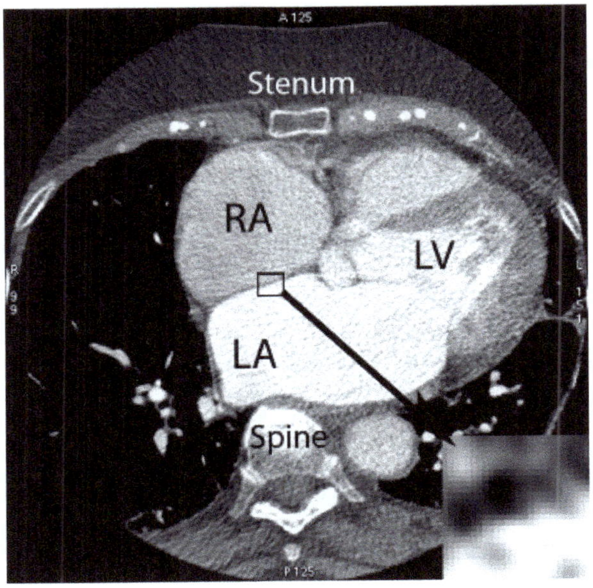

Figure 73.2. 2D axial slice from CT scanning showing the right atrium (RA), left atrium (LA) and left ventricle (LV), as well as the spine and sternum. The insert shows the individual pixels that make up the 2D image.

space, or volume, are called voxels, and have a 3D coordinate system.

Segmentation is performed by selecting or "growing" adjacent areas of comparable density. The endocardial surface, where contrast levels drop off precipitously from those of the blood pool, forms the limit of the 3D volume (Figure 73.3). Segmentation can also be thought of as the process of revealing the structure of interest, in this case the LA, by cutting away all of the surrounding anatomical structures in the scan volume. 2D images can also be created by cutting through the scan volume across the axial series. Sagital and coronal views (see Figure 73.3B and C) are commonly used, as are arbitrary oblique views.

The resultant 3D volume or object is very useful in cases involving left atrial ablation. Of primary interest is the pulmonary venous anatomy. In a series of more than 500 patients undergoing left atrial ablation, we saw that approximately 65% had 2 left and 2 right PVs.[4] The next most common variants had 2 left and 3 right PVs (17%) or 1 left common with 2 right PVs (12%). On a rare basis, we also observed some more unusual variants, such as 2 left with 4 right PVs, 1 common left with 3 right PVs, common inferior PVs, or anomalous PVs. Figure 73.4 shows examples of left atrial models from CT. Also of importance are the anatomy of the left atrial dome, the left inferior PV to mitral isthmus region, the left atrial size or volume, the presence of thrombus (which may be detected with CT scan), and the presence of PV stenosis in patients who have undergone prior left atrial procedures. Cut-away and endocardial or navigator views (Figure 73.5) are also helpful if they show the anatomy in a unique way to help plan the ablation.

The PVs tend to be elliptical rather than purely circular in shape. This consideration is important as designs for catheters to isolate the veins are implemented. The variability of the PV anatomy is also a significant factor. Common PVs, seen in approximately 30% of patients, tend to be significantly larger than either the superior or inferior PVs in patients with 2 or more veins on a side. Additionally, 80% of right PVs are larger than their corresponding left PVs, and 71% of superior PVs are larger than their corresponding inferior PVs.

After segmentation on an image processing workstation, the 3D models are saved in a standardized format (DICOM 3) that allows exchange and display of the images across different platforms or modalities. These images may then be registered with another system. Registration is the process of properly aligning, rotating, and scaling of the coordinate systems between 2 different

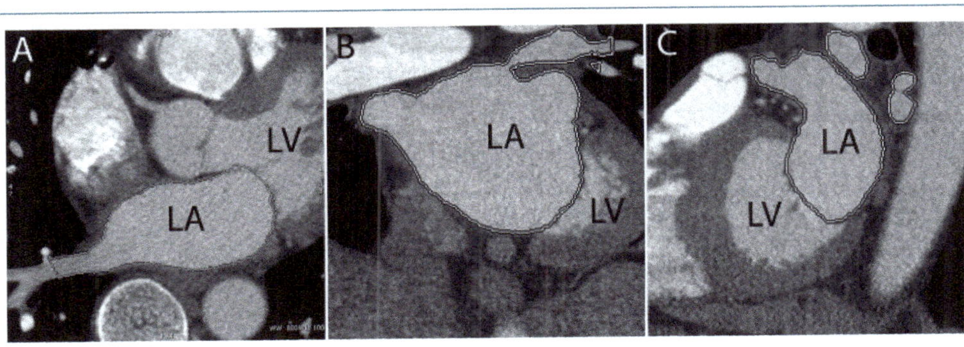

Figure 73.3. 2D images of the heart showing the limits of the segmented left atrium (LA). Axial (A), coronal (B) and sagital (C) images are shown. The left ventricle (LV) is also indicated.

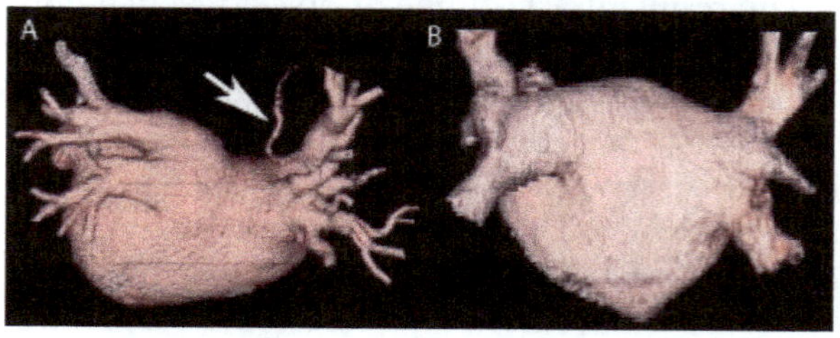

Figure 73.4. 3D volumes of the LA from CT scanning. Note the anomalous PV (arrow) and numerous proximal branches of the right inferior PV (**A**) and the left common PV (**B**).

modalities. This is typically done with a mapping system that permits navigation of catheters accurately over a correct representation of the anatomy (Figure 73.6). By using registered 3D anatomical images of the LA, electrophysiologists are better able to target the precise anatomical structures involved in the initiation and maintenance of AF and, thus, can improve the outcome of the ablation procedure.[5]

Correct Answer:

Q1. D

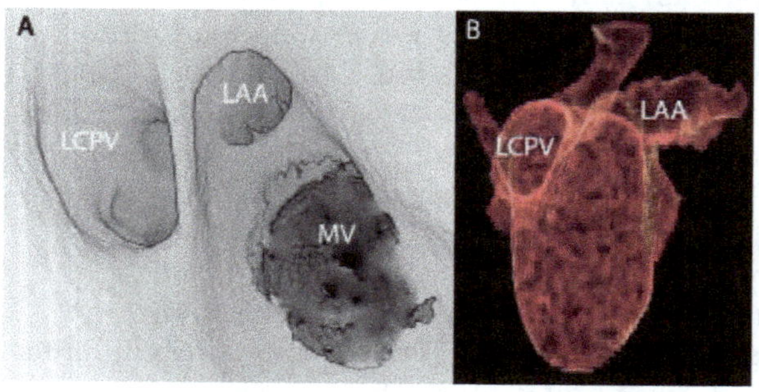

Figure 73.5. Endocardial or navigator view (**A**) and cut-away view (**B**) of the LA from a right lateral view, showing the left side of the LA. The left common pulmonary vein (LCPV), left atrial appendage (LAA) and mitral valve (MV) are shown clearly with these views.

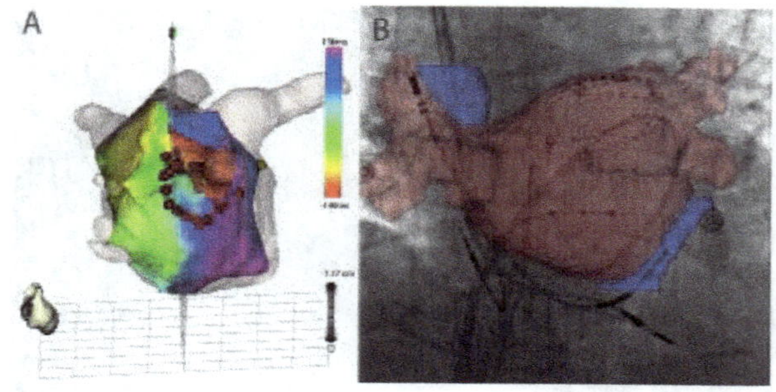

Figure 73.6. **A:** Merged CARTO (Biosense Webster Inc., Diamond Bar, CA) left atrial model and CT scan in the right lateral orientation showing an example of left atrial flutter. **B:** CT-fluoroscopy model showing navigation of an ablation catheter over a properly registered 3D volume rendering of the LA.

REFERENCES

1. Schwartzman D, Lacomis J, Wigginton WG. Characterization of left atrium and distal pulmonary vein morphology using multidimensional computed tomography. *J Am Coll Cardiol.* 2003;41:1349–1357.
2. Solomon SB, Dickfeld T, Calkins H. Real-time cardiac catheter navigation on three-dimensional CT images. *J Interv Card Electrophysiol.* 2003;8:27–36.
3. Sra J, Krum D, Malloy A, et al. Registration of three-dimensional left atrial computed tomographic images with projection images obtained using fluoroscopy. *Circulation.* 2005;112:3763–3768.
4. Hare J, Krum D, Mori N, et al. Characterization of left atrial anatomy in patients undergoing left atrial ablation (abstr). *Radiologic Soc North Am.* 2010:SSJ05–SSJ01.
5. Sra J, Narayan G, Krum D, et al. Computed tomography-fluoroscopy image integration-guided catheter ablation of atrial fibrillation. *J Cardiovasc Electrophysiol.* 2007;18:409–414.

CASE STUDY

74

Magnetic Resonance Imaging of the Cardiac Scar

Maggie Fung, MEng

Question No. 1:

In recent electrophysiology (EP) research studies, magnetic resonance (MR) imaging has been utilized for:

A. Contrast-enhanced MR angiography (CEMRA) for visualizing three-dimensional (3D) left atrial and pulmonary vein (PV) anatomy.

B. 3D delayed late gadolinium enhancement (LGE) for visualizing postablation scars in the left atrium.

C. Replacing electrical mapping of the left atrium.

D. A and B.

E. A and C.

DISCUSSION

CEMRA is a 3D spoiled gradient echocardiography-based technique that can be used with or without cardiac gating to visualize the vascular structures such as left atrium, PVs, or aorta. Cardiac gating is shown to improve cardiac motion suppression[1] and will result in sharper delineation of the PVs and left atrial border. Postprocessing software is often used to segment the left atrium from the rest of the nearby cardiac structures. The high-resolution left atrial 3D model can be reviewed preablation as a guiding roadmap, or integrated into an EP electroanatomic mapping system[2] during the ablation procedure for optimal accuracy in localization (Figure 74.1). A recent study showed that the integration of an EP electroanatomic mapping system with 3D CEMRA data could reduce fluoroscopy time, although it did not improve overall patient outcome.[3]

MR delayed LGE has been an established method to help assess myocardial scarring in various heart diseases such as myocardial infarction

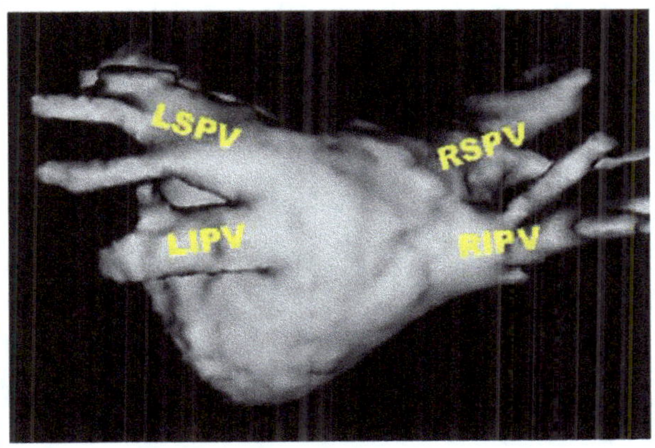

Figure 74.1. A left atrial 3D volume rendering of a MR angiogram. LIPV, left inferior pulmonary vein; LSPV, left superior pulmonary vein; RIPV, right inferior pulmonary vein; RSPV, right superior pulmonary vein.

and hypertrophic cardiomyopathies. The mechanism of LGE relies on the expanded extracellular space in fibrotic tissue and the slower perfusion of infarcted tissue. Since the infarcted/fibrotic tissue has a longer washout period post injection, the contrast media stays within these tissues longer than normal tissue. Therefore, the infarcted area will have shorter T1 (ie, elevated signal intensity in LGE) when imaged 10 to 20 min post injection. The LGE application employs an inversion recovery pulse to invert the Mz magnetization of all the tissue, and then images with a T1-weighted fast gradient echo sequence after a certain inversion (TI) time. The TI time is selected such that short T1 species (eg, scar) will recover faster and have positive signal, whereas longer T1 species (eg, normal tissue) will be at null point (ie, zero Mz magnetization).

Delayed Enhancement in Ventricle

LGE has been most commonly used to assess ventricular myocardial scarring in patients with ischemic and nonischemic cardiomyopathy. In cardiomyopathies, identifying the etiology is an important factor in determining the treatment and patient survival. In postinfarct patients (ischemic), the scar is usually located in the subendocardium at the beginning of an ischemia, and grows toward the epicardium into a transmural infarction if the ischemia is extensive or prolonged[4,5] (Figures 74.2A and B), whereas in nonischemic cardiomyopathy, the scar is usually located in the mid-myocardium or the epicardium[4] (see Figure 74.2C).

In regard to the role of ventricular delayed enhancement in EP, studies have also shown that ventricular arrhythmias are often associated with the presence of scar tissue in the ventricle. Therefore LGE, which can determine the extent and location of these scar tissues, might be useful in preablation planning for ventricular arrhythmia treatment.[6]

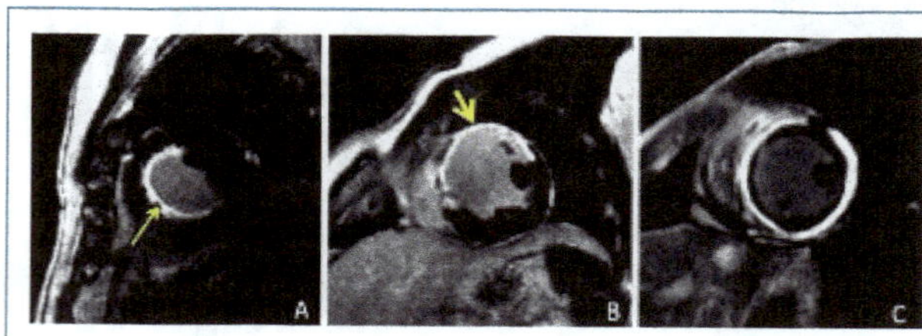

Figure 74.2. **A:** Subendocardial infarction. **B:** Transmural infarction. **C:** Epicardial enhancement in patient with myocarditis. *(Image courtesy of University of Wisconsin, Madison, USA and Center Cardiologique du Nord (CCN), France.)*

Delayed Enhancement in Atrium

Recent researches in the area of EP have explored the possibility of using LGE to assess left atrial scarring post ablation, aiming to better determine the effectiveness of ablation and identify ablation gaps that might cause recurrent atrial fibrillation (AF) post ablation.[7–9] Recent studies have shown that delayed-enhancement MR can help assess scar lesions after AF ablation and can potentially be used to target breaks in lesion sets during repeat ablation.[8,9]

Unlike left ventricular (LV) myocardial infarct imaging, imaging left atrial scar requires significantly higher spatial resolution, therefore, a 3D ECG-gated, respiratory-gated (navigator-echo technique), inversion-recovery fast gradient echo sequence is often used (Figure 74.3). It is usually acquired 15 to 20 minutes post injection, and the TI time is usually set to null the left atrial blood pool[8] or normal myocardium.[9] Typical scan time is 5 to 10 minutes depending on the patient's heart rate and respiratory cycle. Currently, manual or semiautomatic segmentation is used to segment out the atrial scar, which is then laid on top of the CEMRA left atrial 3D model (Figure 74.4).

Correct Answer:

Q1. D

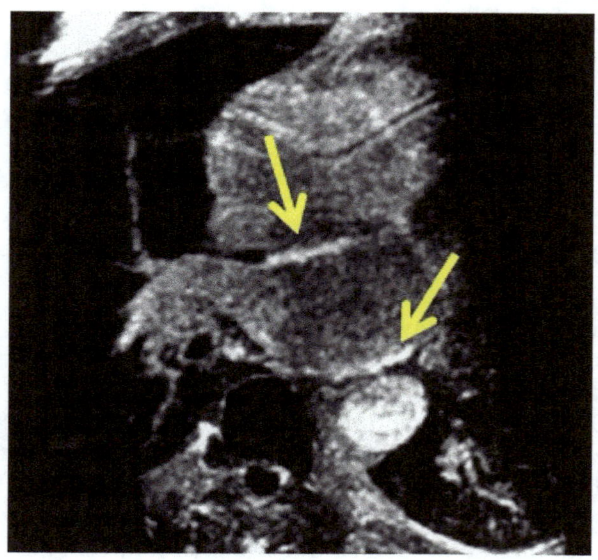

Figure 74.3. High-resolution 3D MR delayed-enhancement imaging showing scarring (arrows) in left atrium.

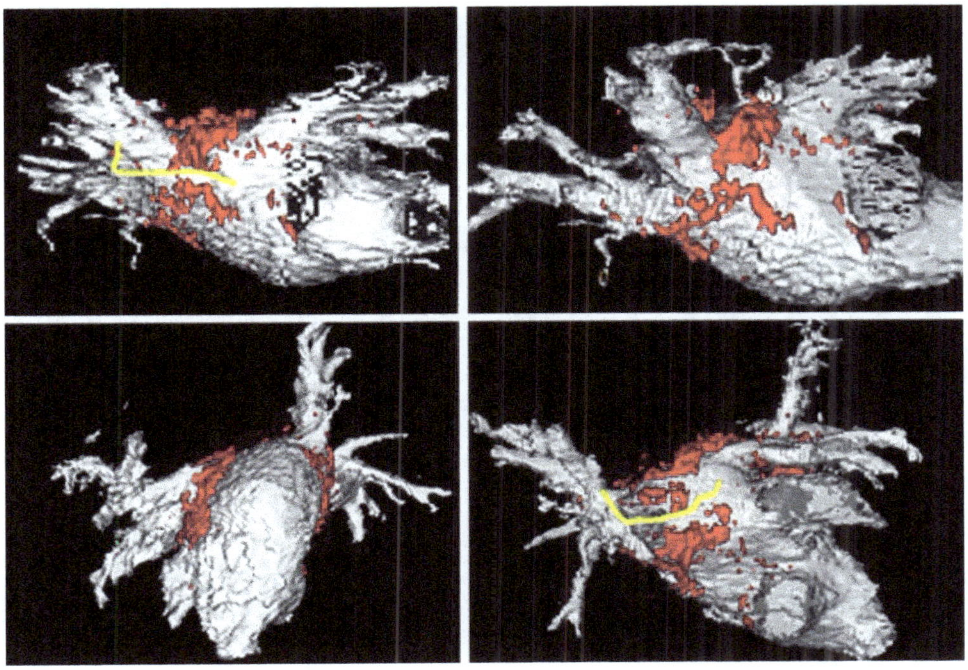

Figure 74.4. Patient with recurrent AF post ablation. CEMRA volume-rendered projections of the left atrium acquired in atrial diastole, overlaid with segmented ablation scars (red) acquired with high-resolution 3D delayed-enhancement. Yellow lines indicate gap in the ablation scar, which were found to be electrically conductive.[8] Repeat ablation procedure targets the gap area to terminate the conduction pathway.

REFERENCES

1. Groves EM, Bireley W, Dill K, Carroll TJ, Carr JC. Quantitative analysis of ECG-gated high-resolution contrast-enhanced MR angiography of the thoracic aorta. *AJR Am J Roentgenol.* 2007;188:522–528.
2. Ponti RD, Marazzi R, Lumia D, et al. Role of three-dimensional imaging integration in atrial fibrillation ablation. *World J Cardiol.* 2010;2:215–222.
3. Caponi D, Corleto A, Scaglione M, et al. Ablation of atrial fibrillation: does the addition of three-dimensional magnetic resonance imaging of the left atrium to electroanatomic mapping improve the clinical outcome?: A randomized comparison of Carto-Merge vs. Carto-XP three-dimensional mapping ablation in patients with paroxysmal and persistent atrial fibrillation. *Europace.* 2010;12:1098–1104.
4. Mahrholdt H, Wagner A, Judd RM, Sechtem U, Kim RJ. Delayed enhancement cardiovascular magnetic resonance assessment of non-ischaemic cardiomyopathies. *Eur Heart J.* 2005;26:1461–1474.
5. Kim RJ, Fieno DS, Parrish TB, et al. Relationship of MRI delayed contrast enhancement to irreversible injury, infarct age, and contractile function. *Circulation.* 1999;100:1992–2002.
6. Bogun FM, Desjardins B, Good E, et al. Delayed-enhanced magnetic resonance imaging in nonischemic cardiomyopathy: utility for identifying the ventricular arrhythmia substrate. *J Am Coll Cardiol.* 2009;53:1138–1145.
7. Peters DC, Wylie JV, Hauser TH, et al. Detection of PV and left atrial scar after catheter ablation with three-dimensional navigator-gated delayed enhancement MR imaging: initial experience. *Radiology.* 2007;243:690–695.
8. Reddy VY, Schmidt EJ, Holmvang G, Fung M. Arrhythmia recurrence after atrial fibrillation ablation: can magnetic resonance imaging identify gaps in atrial ablation lines? *J Cardiovasc Electrophysiol.* 2008;19:434–437.
9. Badger TJ, Daccarett M, Akoum NW, et al. Evaluation of left atrial lesions after initial and repeat atrial fibrillation ablation: lessons learned from delayed-enhancement MRI in repeat ablation procedures. *Circ Arrhythm Electrophysiol.* 2010;3:249–259.

CASE STUDY

75

3D Imaging to Guide Ablations

Brian D. Powell, MD and Douglas L. Packer, MD

A 57-year-old female presented with paroxysmal atrial fibrillation (AF). She reported palpitations, fatigue, and dyspnea occurring twice a week and lasting between 30 minutes and 12 hours. Symptoms continued in spite of treatment with propafenone. She has a past history of hypertension and mild left atrial enlargement. She has elected to undergo pulmonary vein (PV) isolation ablation.

Question No. 1:

Compared to performing an AF ablation guided by standard fluoroscopy alone, the addition of three-dimensional (3D) imaging has been shown to have which of the following benefits?

A. Increased achievement of PV entrance block.

B. Decreased recurrence of AF at 1-year follow-up.

C. Decreased radiation exposure to the patient.

D. Decreased procedure cost.

DISCUSSION

Various imaging modalities have been used to guide ablation procedures. Technology has rapidly progressed from two-dimensional (2D) fluoroscopic imaging to several different forms of 3D mapping systems using magnetic and impedance sensors, as well as automated imaging components derived from computed tomography (CT), magnetic resonance imaging (MRI), and ultrasound and rotational angiography. These have provided greater anatomic detail for accurate ablation. However, ablation outcomes rely not only on image-guided proper lesion location, but also adequate catheter contact, power, duration of energy delivery and overall ablation strategy.

CT and MRI provide detailed 3D geometries of the chambers of the heart. Carto® (Biosense Webster Inc., Diamond Bar, CA) and EnSite NavX™ (St. Jude Medical Inc., St. Paul, MN) mapping systems have the capability for integration of 3D CT and MRI images with electroanatomic maps. More recently, 3D rotational angiography can be used as an alternative to CT or MRI for image integration. Image registration is typically done by matching distinct landmarks or fiducial points on the electroanatomic map with the CT (or MRI) image. The registration process can be optimized by manual adjustment, or using a more global surface registration algorithm. The integrated image has an average accuracy of 1 to 3 mm. However, individual points can be inaccurate by more than 10 to 15 mm.[1]

Rotational angiography is typically performed on the day of the ablation, providing an up-to-date image of the left atrium and PVs. In contrast, CT and MRI are usually performed days to weeks prior to the ablation procedure. CT and MRI provide greater anatomic detail compared to rotational angiography, but each of the 3 imaging methods provides adequate images for guiding ablations. Individual studies comparing PV isolation guided by 3D image

Practical Electrophysiology © 2015 Jasbir S. Sra, Masood Akhtar, Andrea Natale, David J. Wilber
Cardiotext Publishing, ISBN: 978-1-935395-17-1

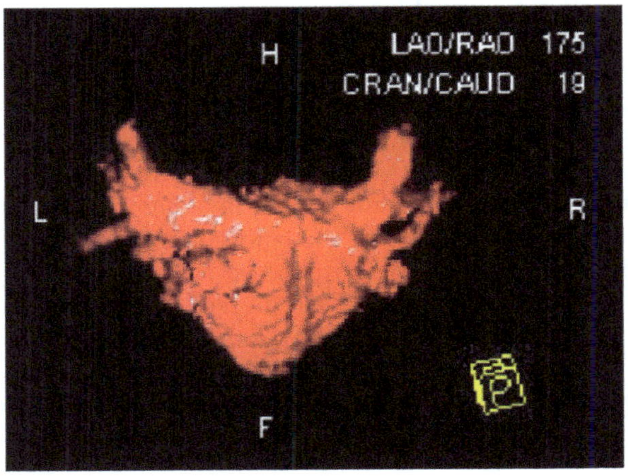

Figure 75.1. 3D rotational angiography of the left atrium and PVs.

integration to fluoroscopy alone have varied in regard to ablation outcomes. However, the majority of studies have shown no significant difference in outcomes whether the procedure is guided by 3D image integration or fluoroscopy alone.[2–5] More studies have found a decrease in fluoroscopy time with 3D electroanatomic map guidance, compared to fluoroscopy alone.

Two ultrasound-derived 3D imaging modalities have been developed in recent years. Intracardiac 3D echocardiography (CartoSound, Biosense Webster, Diamond Bar, CA) involves creation of a 3D geometry of the left atrium, PVs, or other cardiac structures of interest by compiling several 2D intracardiac echo (ICE) images to create a 3D static geometry. This is made possible by the 3D electromagnetic mapping system being able to track the position of the ICE probe in 3D space. A live 2D ICE image can be displayed within the 3D geometry. The advantage of this mapping system is that a 3D geometry can be created without requiring any fluoroscopy.

Another form of echocardiography that can be used to guide ablations is 3D transesophageal echo (TEE). This can provide a real-time image of a 3D sector of the left atrium and PVs. The advantage of 3D TEE is that it does not require creation of a 3D geometry. The disadvantages are the need for another physician to be present during the ablation to perform the TEE imaging; general anesthesia is generally needed; and there are some limitations in adequate image resolution of certain sectors of the left atrium and PVs. In a 20-patient study, we found that 3D TEE provided good to excellent image quality of the anterior aspect of the PVs, left atrial appendage, mitral valve, and mitral isthmus. Image quality of the posterior aspect of the PVs was fair, particularly in regions close to the esophagus. Care should be taken to pull the TEE probe back a few centimeters when ablation is performed on the posterior wall of the left atrium near the

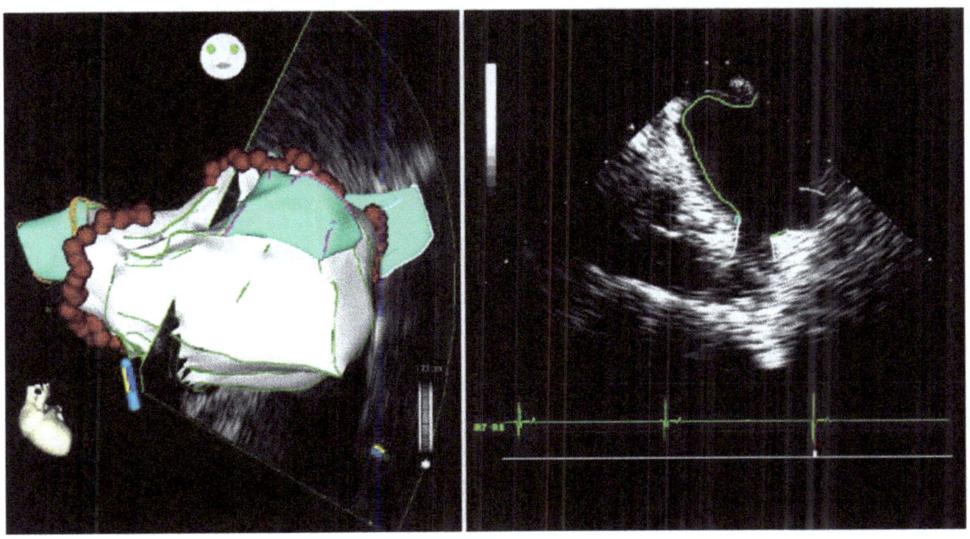

Figure 75.2. 3D ICE geometry (CartoSound) of the left atrium (gray color) and PVs (light blue) from a left anterior oblique view. A 2D image of the left-sided PVs is shown on the right, showing outlines of the structures used to create the 3D geometry. The brown dots on the 3D image are the sites of ablation.

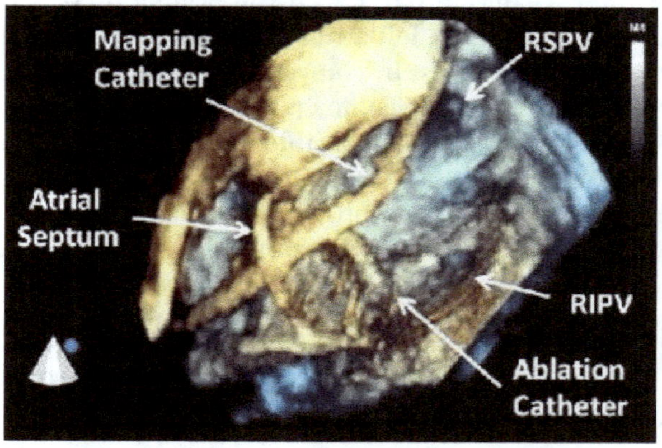

Figure 75.3. 3D TEE during PV isolation ablation. The view is similar to an anteroposterior view. Two catheters are crossing the atrial septum (ablation and circular mapping catheters). An irrigated-tip ablation catheter is in the left atrium on the septal aspect of the right inferior pulmonary vein (RIPV). RSPV, right superior pulmonary vein.

esophagus. This may decrease the risk of esophageal injury from displacement of the esophagus against the posterior wall of the left atrium.

SUMMARY

3D imaging has improved our understanding of location of ablation lesions relative to important structures such as the PV ostia and the esophagus. More anatomic detail may allow for more accurate ablation lesion application; however, the minimum amount of imaging detail needed to achieve excellent outcomes is still debated. Imaging serves as a guide, while the actual energy delivery to the myocardium is the treatment.

Correct Answer:

Q1. C
Decreased radiation exposure to the patient.

When all other factors were consistent between study groups, the addition of 3D imaging to guide ablations did not improve ablation outcomes. However, most studies have shown reduced fluoroscopy time with the addition of 3D imaging.

REFERENCES

1. Okumura Y, Johnson SB, Bunch TJ, et al. A systematical analysis of in vivo contact forces on virtual catheter tip/tissue surface contact during cardiac mapping and intervention. *J Cardiovasc Electrophysiol.* 2008;19:632–640.
2. Kistler PM, Rajappan K, Harris S, et al. The impact of image integration on catheter ablation of atrial fibrillation using electroanatomic mapping: a prospective randomized study. *Eur Heart J.* 2008;29:3029–3036.
3. Tang K, Ma J, Zhang S, et al. A randomized prospective comparison of CartoMerge and CartoXP to guide circumferential pulmonary vein isolation for the treatment of paroxysmal atrial fibrillation. *Chin Med J.* 2008;121:508–512.
4. Rotter M, Takahashi Y, Sanders P, et al. Reduction of fluoroscopy exposure and procedure duration during ablation of atrial fibrillation using a novel anatomic navigation system. *Eur Heart J.* 2005;26:1415–1421.
5. Knecht S, Wright M, Akrivakis S, et al. Prospective randomized comparison between the conventional electroanatomical system and three-dimensional rotational angiography during catheter ablation for atrial fibrillation. *Heart Rhythm.* 2010;7:459–465.

CASE STUDY

76

Cardiac Image Integration

Jasbir S. Sra, MD

Question No. 1:

Which of the following is true about integration of cardiac images?

A. Cardiac image integration (registration) is a rigid body registration as shown in Figure 76.1.

B. Cardiac motion and respiration can affect cardiac image integration.

C. Only 3D–3D image integration is possible.

DISCUSSION

Due to lack of contrast differentiation between the area of interest and surrounding structures in a moving structure like the heart, and as there is a lack of proper intraprocedure guidance using current imaging techniques for ablation, cardiac image registration using CT, magnetic resonance imaging (MRI), ultrasound, and fluoroscopy is currently being investigated and clinically used for atrial fibrillation (AF) ablation. Cardiac image registration, in the context of left atrium (LA), is intermodal, with the acquired image and the real-time reference image residing in different image spaces, and involves optimization in which one image space is transformed into the other. Geometry-based methods, which include using fiducial points and/or surface-based techniques, are usually used for cardiac image registration. During fiducial point registration, fiducial points are either external skin markers or manually determined by marking anatomical landmarks, using mapping catheters. Usually, a minimum of 3 noncollinear points are needed for optimal registration. Recently, a catheter placed inside the coronary sinus (CS) has also been used as a fiducial marker for the purpose of registration. During surface registration, the process

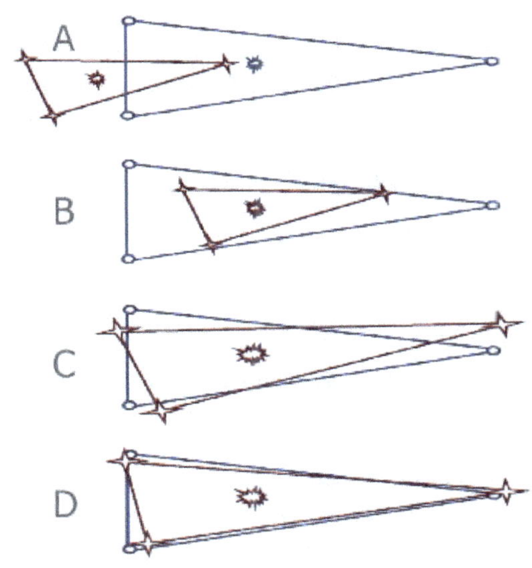

Figure 76.1. Simplified rigid body registration process. Panel A represents projection of computed tomography (CT) and reference image. Panel B depicts initial step, which involves translation of centroids. Panel C involves scaling to calibrate both images. Panel D shows rotation of the image to minimize the sum of the squared distances between fiducial points.

Practical Electrophysiology © 2015 Jasbir S. Sra, Masood Akhtar, Andrea Natale, David J. Wilber
Cardiotext Publishing, ISBN: 978-1-935395-17-1

involves characterizing surfaces in each of the images and deriving the best transformation between them. Unlike rigid body registration, such as has been extensively used in imaging the brain, cardiac image registration is unique and challenging. In addition to the errors inherent with intermodal registration, such as errors in pixel and voxel dimension and errors due to fiducial point selection, there are errors specific to cardiac image registration, that is, errors due to cardiac motion during the cardiac cycle and due to respiration.

Technical Considerations

Technical considerations regarding CT and MRI have been described in detail in previous cases. Cardiac registration, in the context of the LA, is intermodal, with the acquired images and real-time reference image residing in different imaged spaces. Registration algorithms involve the optimization of a cost function by the choice of a transformation T, which transforms one image space into the other. T can be either linear or nonlinear. Linear transformations are shape-preserving, and nonlinear transformations may deform both shape and size between images. A linear or nonlinear transformation decides the number of parameters involved in its definition. A linear transformation between three-dimensional (3D) spaces is defined by 6 parameters (or degrees of freedom), where 2 positions of a rigid body can always be related to one another in terms of 3 translations and 3 rotations. As the voxel sizes in each image may not be similar for calibration purposes, 3 extra degrees of freedom, equating to scaling in each direction, are needed. A nonlinear transformation, as it may deform both shape and size, will require more degrees of freedom.

Registration methods fall into 2 broad categories, those based on either geometric similarities between images or voxel-intensity similarities. Voxel-similarity methods are usually used to assess progression of disease. Geometry-based methods can be point-based or surface methods applied to both 3D–3D or 3D–2D registration methods.

Point-based registration methods depend on the appropriate definition of 2 sets of corresponding fiducial points in each set of images to be registered. Thus, if $\{x_i\}_1^N$ denotes the set of fiducial points in the acquired image, $\{y_i\}_1^N$ denotes the corresponding points on the real-time image, that is, the choice of T minimizes the error $\sum_{i=1}^{N} \|\tau(x)-y_1\|^2$. Fiducial point sets are either automatically determined based on external skin markers or manually determined based on anatomical landmarks. The latter method has been used in cardiac image registration. For rigid structures such as the brain, 3 noncollinear fiducial points may be sufficient to establish the transformation between the 3D image volumes. However, the larger the number of points used, the more any error occurring during collection of the points is averaged out. The algorithm for calculating the transformation between the 3D images involves first computing the average of the "centroids" of each set of points. The difference between the centroids in the 3D images tells us the translation that must be applied to one set of points. These points are then rotated about the new centroid until the sum of the squared distances between each corresponding point pair is minimized. The square root of the mean of this squared distance is referred to as fiducial registration error (FRE).

Several techniques are used to improve accuracy of the registration process. The process of registration becomes more complicated if one of the images represents "projection" images. One point in an x-ray image will correspond to some combination of the x-ray attenuation value along the line in the patient leading from the x-ray source to the imaging plane. This means that one point in the radiograph will correspond to a line of points through a CT or MRI. For computational efficiency, transformations are restricted to scaling translations and rigid rotations in the x-ray space, and are based on the optimized alignment of corresponding features or other image properties. The details may vary, but they are described mathematically as $\Re\,(af\,(x) + \tau) \approx y$, in which f is a choice function that selects an appropriate 2D slide of the CT volume. Based on the orientation of the x-ray C-arm, a scaling parameter that standardizes the sizes of the images, τ (or T) is a 2D translational parameter that aligns the centers of the images and $\Re$ is a rotational matrix that completes the alignment process. In the CT-x-ray registration, a, τ and $\Re$ are the usual parameters involved.

Clinical Applications

Registration using anatomical mapping systems: Anatomical mapping systems provide the 3D position of a navigational catheter within the cardiac chamber of interest and, in some instances, can also be used to construct 3D maps of the cardiac chamber. EnSite™ balloon (St. Jude Medical Inc., St. Paul, MN), CARTO® (Biosense Webster Inc. Diamond Bar, CA), ultrasound, and fluoroscopy have been used for registration and will be described in this section in the context of left atrial registration.

Point-based registration was also performed using CT images and the EnSite system. Fiducial points (≥ 3 noncollinear points) were manually inserted, in the form of pacing leads, into the LA of a canine model and CT images were recorded. Following segmentation of the LA, a wire-frame model of the LA was imported to the noncontact, multielectrode array-based EnSite mapping system. Goodness-of-fit was evaluated qualitatively by observing a cloud of points post-registration. When registration was tested by navigating the catheter to the buried electrodes and delivering radiofrequency pulses at these points, the mean FRE was 6.9 ± 2.2 mm and the target registration error was 2.0 ± 3.6 mm.

A clinical application of this technique is now available using CARTO with CT or MRI (Figure 76.2). A combination of landmark and surface registration is used to register CT or MRI with the CARTO system. Initially, several landmarks, usually 3, are manually chosen and annotated. Following this, the 3D left atrial reconstructed image from the CT or MRI is superimposed upon the electroanatomic map created by the CARTO mapping system.

Several studies have tested the accuracy of CT–MRI registration. Dong et al attached 2.3-mm CT fiducial markers to the epicardial surface of each cardiac chamber in 9 mongrel dogs. CT images were registered to the electroanatomic maps of each cardiac chamber. To assess accuracy, targeted ablations were performed at each of the fiducial markers, guided only by the reconstructed images. At autopsy, the position error was 1.92 ± 0.9 mm for the right atrium, and 1.8 ± 1.0 mm for the LA. Retrospective analysis revealed that a combination of landmark registration and the target chamber surface registration resulted in < 3 mm accuracy in all chambers.

However, in a study by Fahmy et al[6] involving 124 patients undergoing AF ablation guided by registration of images of the LA created using

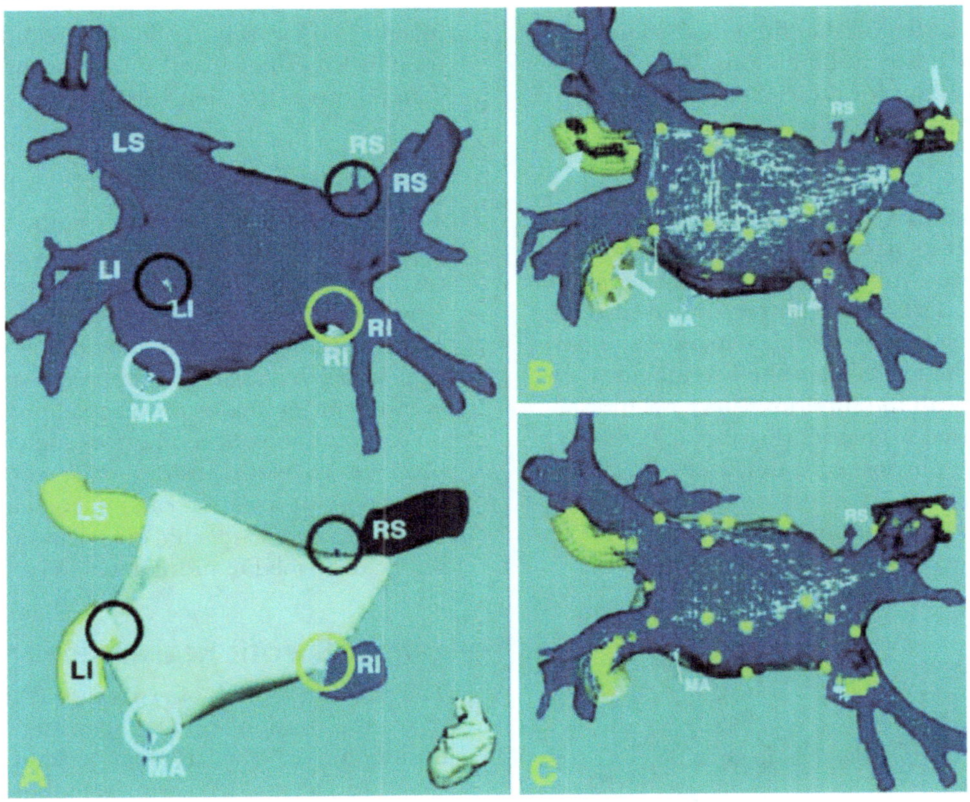

Figure 76.2. Registration using the CARTO system. **A:** Landmark pairs (highlighted with darker circles) at the 6 o'clock mitral annulus (MA) position and the junctions of the LA and right superior pulmonary vein (RS), right inferior pulmonary vein (RI), and left inferior pulmonary vein (LI) were annotated on the 3D CT LA surface reconstruction (upper image, shown as wire frame) and the LA electroanatomic map (lower image, shown as solid shell) with tubes representing pulmonary veins (PVs). **B:** After landmark registration, the 3D LA surface reconstruction was superimposed on the electroanatomic map (shown as mesh). Note the misalignment of the LS, LI, and RS between the 2 image datasets, indicated by the yellow or red color of the PV points (arrows) sampled in those PVs. **C:** After surface registration was executed, the PV alignment between the 2 image datasets was significantly improved, indicated by the PV points. The electroanatomic map points indicate their distance of <5 mm, 5 to 10 mm and >10 mm, respectively, from the registered CT reconstruction surface. (Reproduced from Dong et al[5] with permission from John Wiley and Sons Inc.)

CARTOMERGE™, the best landmark registration was achieved when the posterior points on the PVs were selected (5.6 ± 3.2 mm). Surface registration resulted in shifting of the initially registered landmarks, leading to a more significant level of error (from 5.6 ± 3.2 mm to 9.2 ± 2.1 mm; 95% confidence interval 4.2–3.05). Posteriorly located fiducial points reduced registration error in this study.

In another study on 16 patients undergoing AF ablation, CT images of the LA were merged with CARTO. Encircling of the left and right PV antra was performed, guided solely by intracardiac echocardiography, with point locations saved on a CARTOMERGE image to which the operator was blinded. The accuracy of the CARTOMERGE image was then assessed by measuring the distance from the location of each ablation point on the image to its actual anatomic location. A significant error of 1.6 ± 1.2 cm in the right antrum and 1.1 ± 0.7 cm from the planned path of ablation was observed. Accuracy was improved when an end-atrial contraction CT image was used.

These reports suggest that these registration techniques have inherent limitations. Use of appropriate locations for fiducial points, such as posteriorly, and cardiac cycle, and respiration gating at the time of imaging and registration could potentially help these approaches to registration.

Fluoroscopy-based registration: As fiducial selection (described in 3D-to-3D image registration) in each of the image spaces introduces several challenges, we have instead used a CS catheter positioned through the superior vena cava as a fiducial marker and have used a transformation to align the CS catheter seen on fluoroscopy with the superior vena cava and the CS in the exported 3D model (Figure 76.3). In addition, for the purpose of scaling, the CS catheter is placed at a location identified by the intersection of a horizontal plane and the line joining the focal point of the fluoroscopic image.

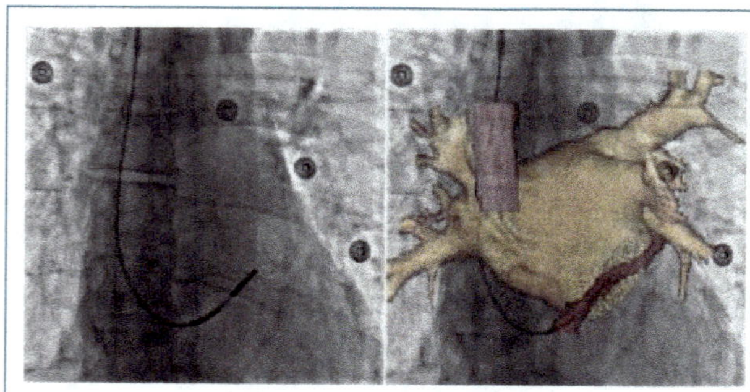

Figure 76.3. Computed tomography-fluoroscopy (CT-fluoro) registration. Left panel shows a fluoro image with CS catheter. Right panel depicts registered CT-fluoro image.

The horizontal plane is defined by the elevation above the anatomical reference system, the patient table. Once the location of the catheter and the actual and apparent dimensions of the elevation in the 3D model are known, the actual size of the same anatomical structure in the 3D and the fluoroscopic image can be determined. We recently described an implementation of a semiautomated 3D–2D CT-fluoroscopy registration strategy. The accuracy of this system was found to be within 1.4 mm in phantom studies. This strategy was also assessed in patients undergoing AF ablation. Accurate registration was confirmed by: PV angiography, and position and recordings on a 64-pole basket catheter located in the PVs. In a recent study, 50 patients with AF undergoing ablation were randomized to standard ablation techniques or ablation guided by the CT-fluoro registered images. Fluoroscopy time and total procedure time was significantly reduced in CT-fluoro group (p<0.01), and there was a trend toward better outcome. It is also possible to mark ablation points on these images and synchronize the images to the cardiac cycle. This technique has been used in 154 patients and should be available for general use in the next 18 months.

Sources of Error in Cardiac Registration

The registration techniques described above present a significant advantage over the less detailed surrogate geometry created by the mapping systems. However, in the context of cardiac registration, there are significant inherent limitations as registration algorithms that assume rigid transformations between image spaces have been applied in some of these techniques. Several errors such as reporting of pixel and voxel dimensions, scale computation, and fiducial point selection can effect image integration. However, cardiac-specific errors are important and will be described in more detail here.

Cardiac-Specific Registration Errors

The possible sources of error specific to cardiac image registration are: (1) inconsistency in patient positioning; (2) cardiac chamber movement due to cardiac cycle; and (3) cardiac chamber movement due to respiration.

Patient alignment: To help ensure same orientation of cardiac chambers during both imaging modalities, efforts should be made to align the patient identically on each imaging table. The orientation of the cardiac chamber changes with the orientation of the body. For this reason, this type of interscan movement can be described as a rigid body rotation of the heart with the spine. A change in the interscan angle of 5 degrees, for example, may give a rotational error of 0.44 cm in a 5-cm organ with a field-of-view of about 20 cm.

Errors due to cardiac and respiration cycle: The CT-gating technique involves electrocardiographic (ECG) gating of the acquired images and the patient holding breath for several heartbeats while 2D axial slices are sequentially obtained at fixed points of the cardiac cycle. These slices are stacked together and interpolated to create a smooth, static 3D CT image. Potentially, this technique could result in registration error. The assumption is that the generated image represents the actual cardiac chamber at a particular point of the cardiac cycle, corresponding exactly to a second image, for example, the x-ray image used for registration at the same point of the cardiac cycle. The likelihood that the heart returns to a fixed shape at fixed points of the cardiac cycle is questionable, due to the variability in cardiac deformation. As such, the CT image is more of an averaged view of several similar states rather than a "snapshot" of a particular state as assumed. Nevertheless, it is important that both the CT and the second image are ECG-gated to eliminate, as much as possible, errors due to cardiac motion.

Given the proximity of the heart to the diaphragm, it has been determined that translations in the craniocaudal direction are the main types of movement seen in the cardiac chamber due to respiration. Less significant movements include translation and rotation in other directions.

CONCLUSION

Cardiac image integration is increasingly used in patients undergoing AF ablations. The deformations caused by cardiac chamber motion present unique problems for cardiac image registration. The continued difficulty in defining the unpredictable and significant nature of cardiac movement suggests there is much room for development of improved strategies for cardiac image registration. ECG- and respiration-gating strategies attempting to achieve optimal registration are currently under investigation and should, along with the current improvements in navigational tools, help simplify therapy for complex arrhythmias like AF.

Correct Answer:

Q1. B

REFERENCES

1. Hajnal JV, Hill DLG, Hawkes DJ. *Medical Image Registration. The Biomedical Medical Engineering Series.* CRC Press: Boca Raton, FL; 2001: 1–70.
2. Sra J, Krum D, Hare J, et al. Feasibility and validation of registration of three-dimensional left atrial models derived from computed tomography with a noncontact cardiac mapping system. *Heart Rhythm.* 2005;2:55–63.
3. Sra J, Krum D, Malloy A, et al. Registration of three-dimensional left atrial computed tomographic images with projection images obtained using fluoroscopy. *Circulation.* 2005;112:3763–3768.
4. Ector J, De Buck S, Adams J, et al. Cardiac three-dimensional magnetic resonance imaging and fluoroscopy merging—a new approach for electroanatomic mapping to assist catheter ablation. *Circulation.* 2005;112:3769–3776.
5. Dong J, Calkins H, Solomon SB, et al. Integrated electroanatomic mapping with three-dimensional computed tomographic images for real-time guided ablations. *Circulation.* 2006;113:186–194.
6. Fahmy TS, Mlcochova H, Wazni OM, et al. Intracardiac echo-guided image integration: optimizing strategies for registration. *J Cardiovasc Electrophysiol.* 2007;18:276–282.
7. Zhong H, Lacomis JM, Schwartzman D. On the accuracy of CartoMerge for guiding posterior left atrial ablation in man. *Heart Rhythm.* 2007;4:595–602.

CASE STUDY

77

Mapping and Navigation Using Robotic Navigation System

Luigi Di Biase, MD, PhD, G. Joseph Gallinghouse, MD, and Andrea Natale, MD

A 41-year-old patient with essential hypertension and diabetes was admitted to our institution for catheter ablation of atrial fibrillation (AF) with the Sensei® robotic system (Hansen Medical Inc., Mountain View, CA). He had paroxysmal AF refractory to proprafenone and almarythm.

Question No. 1:

The operator is burning with an open-irrigated ablation catheter at a flow rate of 30 cc/min in power control mode with 40 watts at 42° C. The system shows an average contact force of 40 grams (Figure 77.1). Based on published data, the operator should:

A. Decrease the power to 35 W.

B. Decrease the power to 30 W and reduce the contact force to 20–30 grams.

C. Increase the contact force to 50 grams.

D. Increase the power to 45 W and the contact force to 45 grams.

E. None of the above.

DESCRIPTION OF THE HANSEN ROBOTIC CATHETER SYSTEM

The system is comprised of the Sensei Robotic Catheter System and the Artisan Control Catheter, a remotely controlled steerable sheath. The operator remotely directs the movement of the Artisan Guide Catheter through the Sensei Robotic Catheter System. The Sensei is comprised of a physician workstation, an electronics rack, and a patient-side Remote Catheter Manipulator (RCM). The system allows the clinician to direct the catheter tip to a desired location based on visual feedback from 3D electroanatomic maps, fluoroscopic images, and ICE images while seated at the workstation. The RCM electromechanically manipulates the steerable guide catheter in response to commands received from the physician through a special 3D joystick (Intuitive Motion Controller, or "IMC") at the physician workstation. The Artisan Control Catheter is a conventional pull-wire-actuated open-lumen guide catheter. The inner lumen is 8-F diameter, and the guide catheter fits through a standard 14-F introducer. The Artisan catheter attaches to the RCM through a sterile drape barrier and, in turn,

Practical Electrophysiology © 2015 Jasbir S. Sra, Masood Akhtar, Andrea Natale, David J. Wilber
Cardiotext Publishing, ISBN: 978-1-935395-17-1

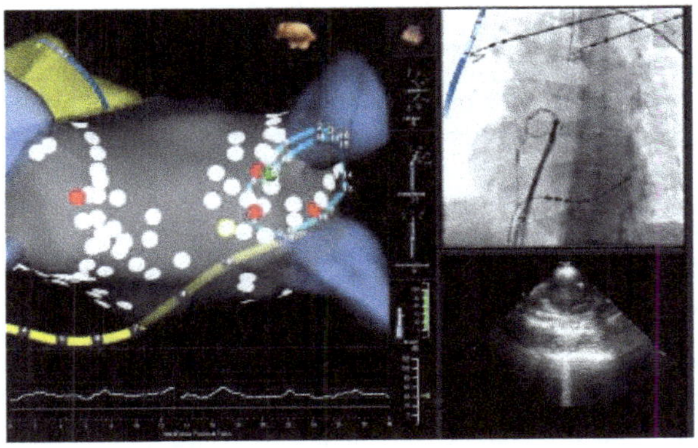

Figure 77.1. Left: NavX™ (St. Jude Medical Inc., St. Paul, MN) three-dimensional (3D) mapping posterior view showing the lesion set and mapping catheters (circular mapping catheter in blue; coronary sinus (CS) catheter in yellow). Below the map the "IntelliSense®" technology (Hansen Medical) shows an average contact force of 40 grams. Top right: Fluoroscopic left anterior oblique (LAO) view showing the ablation catheter and the circular mapping catheter at the level of the right pulmonary veins (PV). Bottom right: Intracardiac echocardiography (ICE) view on the console screen shot.

the RCM actuates the catheter pull-wires in response to commands from the physician. The Artisan is a guide catheter that can be used with any commercially available ablation catheter. The Artisan can articulate in any direction to 270 degrees, with a minimal working curve diameter of 30 mm.

DISCUSSION

Catheter ablation has proven to be safe and effective for the treatment of drug refractory AF.[1–4] Procedural results are often operator-dependent, so robotic navigation and ablation have been introduced into the market with the aim of eliminating the operator's dependency. New technologies in this field are generally designed to improve outcomes and reduce complications.[5] During manual ablation—although no information is present regarding the pressure applied by the ablation catheter against the targeted tissue—each operator develops a personal "tactile" feedback by which the operator tries to perform "effective" lesions while minimizing complications. This concept is widely applicable to manual ablation but more important for robotic navigation, because the operator loses his personal "feedback" and relies on a robotic navigation and ablation. This is why the Hansen robotic system has been integrated with a contact force sensor (IntelliSense) that is able to give a real-time measurement of the force applied by the ablation catheter against the tissue. This technology allows the operator to monitor contact force. From biophysics we know that the radiofrequency lesion formation results from a complex combination of tissue catheter contact, catheter stability, power settings, and duration of radiofrequency delivery. The platform provided by the inner guide of the Artisan sheath is pretty stable and prevents "rebound" of the ablation tip when placed against tissue, as typically happens with soft catheters during manual ablation. With this technology, the power required to achieve effective radiofrequency lesions is typically in the range of 15 to 30 watts, as opposed to 35 to 45 watts with manual ablation. A load sensor on the robotic arm calculates the contact force with the tissue and displays it graphically in real time for the operator. IntelliSense has been demonstrated to be highly accurate when the catheter is perpendicular to the tissue, otherwise the contact force reading might be inaccurate. Therefore, although not perfect, IntelliSense provides a safety mechanism where catheter forces on the myocardium are mainly perpendicular.[6] Taking into account this information, the Hansen Robotic Catheter System has shown satisfactory results that are at least comparable to manual ablation.[7,8] Until ablation catheters become available with contact sensors, operator awareness of contact pressure at the catheter tip-tissue interface will remain challenging. This is well-shown in preliminary findings of the TOCCATA study,[9] which showed that experienced operators underestimated contact in positions that they felt optimal for radiofrequency delivery when they were blinded to pressure data.

Correct Answer:

Q1. B

REFERENCES

1. European Heart Rhythm Association (EHRA); European Cardiac Arrhythmia Society (ECAS); American College of Cardiology (ACC); American Heart Association (AHA); Society of Thoracic Surgeons (STS), Calkins H, Brugada J, Packer DL, et al. HRS/EHRA/ECAS expert Consensus Statement on catheter and surgical ablation of atrial fibrillation: Recommendations for personnel, policy, procedures and follow-up. A report of the Heart Rhythm Society (HRS) Task Force on catheter and surgical ablation of atrial fibrillation. *Heart Rhythm.* 2007;4:816–861.
2. Natale A, Raviele A, Arentz T, et al. Venice Chart international consensus document on atrial fibrillation ablation. *J Cardiovasc Electrophysiol.* 2007;18:560–580.

3. Wilber DJ, Pappone C, Neuzil P, et al; ThermoCool AF Trial Investigators. Comparison of antiarrhythmic drug therapy and radiofrequency catheter ablation in patients with paroxysmal atrial fibrillation: A randomized controlled trial. *JAMA.* 2010;303:333–340.
4. Bhargava M, Di Biase L, Mohanty P, et al. Impact of type of atrial fibrillation and repeat catheter ablation on long-term freedom from atrial fibrillation: results from a multicenter study. *Heart Rhythm.* 2009;6:1403–1412.
5. Cappato R, Calkins H, Chen SA et al. Prevalence and causes of fatal outcome in catheter ablation of atrial fibrillation. *J Am Coll Cardiol.* 2009;53:1798–1803.
6. Di Biase L, Natale A, Barrett C, et al. Relationship between catheter forces, lesion characteristics, "popping," and char formation: Experience with robotic navigation system. *J Cardiovasc Electrophysiol.* 2009;20:436–440.
7. Di Biase L, Wang Y, Horton R, et al. Ablation of atrial fibrillation utilizing robotic catheter navigation in comparison to manual navigation and ablation: single-center experience. *J Cardiovasc Electrophysiol.* 2009;20:1328–1335.
8. Di Biase L, Davies W, Horton R, et al; on the behalf of Worldwide Hansen Robotic Investigators. Acute complication rate with the robotic navigation for catheter of atrial fibrilation: The worldwide experience. *Circulation.* 2008;118:S978–S979.
9. Kuck KH. TOCCATA European Clinical Study – First multiclinical study using irrigated ablation catheters with an integrated contact force sensor. Presented at Boston Atrial Fibrillation Symposium, 2010.

INDEX

www.ingramcontent.com/pod-product-compliance
Lightning Source LLC
LaVergne TN
LVHW060252011026
849161LV00011B/2696